Transmission and Population Genetics

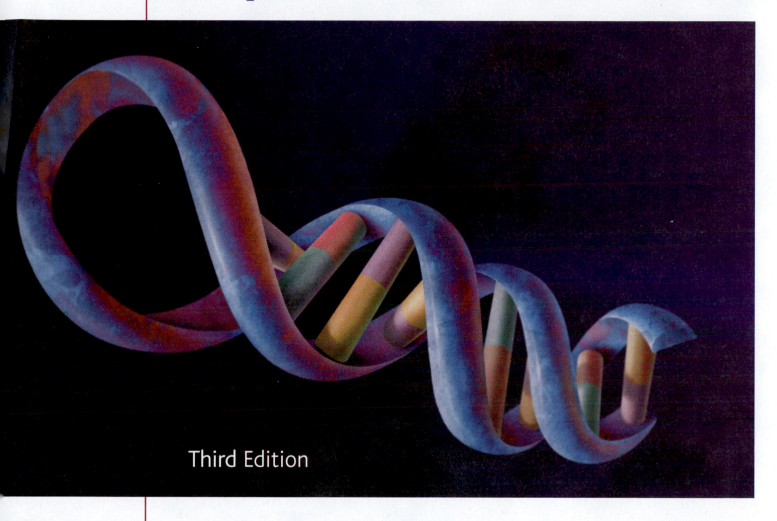

Third Edition

Benjamin A. Pierce

Southwestern University

W. H. Freeman and Company · New York

Acquisitions Editor: Jerry Correa

Development Editor: Lisa Samols

Media Editors: Patrick Shriner, Marc Mazzoni

Supplements Editors: Hannah Thonet, Beth McHenry

Project Editor: Leigh Renhard

Manuscript Editor: Patricia Zimmerman

Associate Director of Marketing: Debbie Clare

Publisher: Sara Tenney

Text Designer: Marsha Cohen/Parallelogram Graphics

Cover Illustration: Rosetta Genomics

Illustrations: Dragonfly Media Group

Senior Illustration Coordinator: Bill Page

Photo Editors: Ted Szczepanski, Christine Buese

Photo Researcher: Patty Catuera

Production Coordinator: Susan Wein

Composition: Black Dot Group

Printing and Binding: RR Donnelley

Library of Congress Control Number: 2007924869

ISBN-13: 978-1-4292-1118-5
ISBN-10: 1-4292-1118-0

Printed in the United States of America
First printing

W. H. Freeman and Company
41 Madison Avenue
New York, NY 10010
Houndmills, Basingstoke RG21 6XS, England

www.whfreeman.com

Student Solutions Manual

Multivariable Calculus

TENTH EDITION

Ron Larson
The Pennsylvania State University,
The Behrend College

Bruce Edwards
University of Florida

BROOKS/COLE
CENGAGE Learning·

Australia • Brazil • Japan • Korea • Mexico • Singapore • Spain • United Kingdom • United States

For product information and technology assistance, contact us at
**Cengage Learning Customer & Sales Support,
1-800-354-9706**

For permission to use material from this text or product, submit all requests online at **www.cengage.com/permissions**
Further permissions questions can be emailed to
permissionrequest@cengage.com

ISBN-13: 978-1-285-08575-3
ISBN-10: 1-285-08575-2

Brooks/Cole
20 Channel Center Street
Boston, MA 02210
USA

Cengage Learning is a leading provider of customized learning solutions with office locations around the globe, including Singapore, the United Kingdom, Australia, Mexico, Brazil, and Japan. Locate your local office at:
www.cengage.com/global

Cengage Learning products are represented in Canada by Nelson Education, Ltd.

To learn more about Brooks/Cole, visit
www.cengage.com/brookscole

Purchase any of our products at your local college store or at our preferred online store
www.cengagebrain.com

Printed in the United States of America
2 3 4 5 6 7 17 16 15

CONTENTS

CHAPTER 11
Vectors and the Geometry of Space

CHAPTER 11
Vectors and the Geometry of Space

Section 11.1   Vectors in the Plane

1. (a) $\mathbf{v} = \langle 5 - 1, 4 - 2 \rangle = \langle 4, 2 \rangle$

(b)

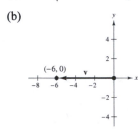

3. (a) $\mathbf{v} = \langle -4 - 2, -3 - (-3) \rangle = \langle -6, 0 \rangle$

(b)

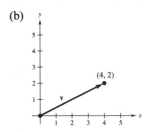

5. $\mathbf{u} = \langle 5 - 3, 6 - 2 \rangle = \langle 2, 4 \rangle$

$\mathbf{v} = \langle 3 - 1, 8 - 4 \rangle = \langle 2, 4 \rangle$

$\mathbf{u} = \mathbf{v}$

7. $\mathbf{u} = \langle 6 - 0, -2 - 3 \rangle = \langle 6, -5 \rangle$

$\mathbf{v} = \langle 9 - 3, 5 - 10 \rangle = \langle 6, -5 \rangle$

$\mathbf{u} = \mathbf{v}$

9. (b) $\mathbf{v} = \langle 5 - 2, 5 - 0 \rangle = \langle 3, 5 \rangle$

(c) $\mathbf{v} = 3\mathbf{i} + 5\mathbf{j}$

(a), (d)

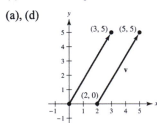

11. (b) $\mathbf{v} = \langle 6 - 8, -1 - 3 \rangle = \langle -2, -4 \rangle$

(c) $\mathbf{v} = -2\mathbf{i} - 4\mathbf{j}$

(a), (d)

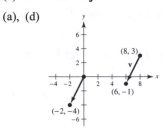

13. (b) $\mathbf{v} = \langle 6 - 6, 6 - 2 \rangle = \langle 0, 4 \rangle$

(c) $\mathbf{v} = 4\mathbf{j}$

(a) and (d).

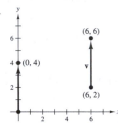

15. (b) $\mathbf{v} = \left\langle \frac{1}{2} - \frac{3}{2}, 3 - \frac{4}{3} \right\rangle = \left\langle -1, \frac{5}{3} \right\rangle$

(c) $\mathbf{v} = -\mathbf{i} + \frac{5}{3}\mathbf{j}$

(a) and (d)

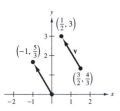

17. (a) $2\mathbf{v} = 2\langle 3, 5 \rangle = \langle 6, 10 \rangle$

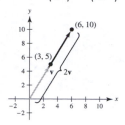

(b) $-3\mathbf{v} = \langle -9, -15 \rangle$

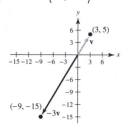

(c) $\frac{7}{2}\mathbf{v} = \left\langle \frac{21}{2}, \frac{35}{2} \right\rangle$

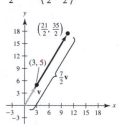

(d) $\frac{2}{3}\mathbf{v} = \left\langle 2, \frac{10}{3} \right\rangle$

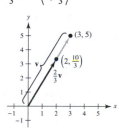

19. (a) $\frac{2}{3}\mathbf{u} = \frac{2}{3}\langle 4, 9 \rangle = \left\langle \frac{8}{3}, 6 \right\rangle$

(b) $3\mathbf{v} = 3\langle 2, -5 \rangle = \langle 6, -15 \rangle$

(c) $\mathbf{v} - \mathbf{u} = \langle 2, -5 \rangle - \langle 4, 9 \rangle = \langle -2, -14 \rangle$

(d) $2\mathbf{u} + 5\mathbf{v} = 2\langle 4, 9 \rangle + 5\langle 2, -5 \rangle = \langle 18, -7 \rangle$

21.

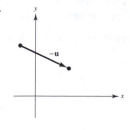

23.

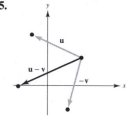

25.

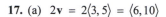

27. $u_1 - 4 = -1$ $u_1 = 3$

$u_2 - 2 = 3$ $u_2 = 5$

$Q = (3, 5)$

29. $\|\mathbf{v}\| = \sqrt{0 + 7^2} = 7$

31. $\|\mathbf{v}\| = \sqrt{4^2 + 3^2} = 5$

33. $\|\mathbf{v}\| = \sqrt{6^2 + (-5)^2} = \sqrt{61}$

35. $\mathbf{v} = \langle 3, 12 \rangle$

$\|\mathbf{v}\| = \sqrt{3^2 + 12^2} = \sqrt{153}$

$\mathbf{u} = \dfrac{\mathbf{v}}{\|\mathbf{v}\|} = \dfrac{\langle 3, 12 \rangle}{\sqrt{153}} = \left\langle \dfrac{3}{\sqrt{153}}, \dfrac{12}{\sqrt{153}} \right\rangle$

$\qquad = \left\langle \dfrac{\sqrt{17}}{17}, \dfrac{4\sqrt{17}}{17} \right\rangle$ unit vector

37. $\mathbf{v} = \left\langle \dfrac{3}{2}, \dfrac{5}{2} \right\rangle$

$\|\mathbf{v}\| = \sqrt{\left(\dfrac{3}{2}\right)^2 + \left(\dfrac{5}{2}\right)^2} = \dfrac{\sqrt{34}}{2}$

$\mathbf{u} = \dfrac{\mathbf{v}}{\|\mathbf{v}\|} = \dfrac{\left\langle \left(\dfrac{3}{2}\right), \left(\dfrac{5}{2}\right) \right\rangle}{\dfrac{\sqrt{34}}{2}} = \left\langle \dfrac{3}{\sqrt{34}}, \dfrac{5}{\sqrt{34}} \right\rangle$

$\qquad = \left\langle \dfrac{3\sqrt{34}}{34}, \dfrac{5\sqrt{34}}{34} \right\rangle$ unit vector

39. $\mathbf{u} = \langle 1, -1 \rangle$, $\mathbf{v} = \langle -1, 2 \rangle$

   (a) $\|\mathbf{u}\| = \sqrt{1+1} = \sqrt{2}$

   (b) $\|\mathbf{v}\| = \sqrt{1+4} = \sqrt{5}$

   (c) $\mathbf{u} + \mathbf{v} = \langle 0, 1 \rangle$

     $\|\mathbf{u} + \mathbf{v}\| = \sqrt{0+1} = 1$

   (d) $\dfrac{\mathbf{u}}{\|\mathbf{u}\|} = \dfrac{1}{\sqrt{2}} \langle 1, -1 \rangle$

     $\left\| \dfrac{\mathbf{u}}{\|\mathbf{u}\|} \right\| = 1$

   (e) $\dfrac{\mathbf{v}}{\|\mathbf{v}\|} = \dfrac{1}{\sqrt{5}} \langle -1, 2 \rangle$

     $\left\| \dfrac{\mathbf{v}}{\|\mathbf{v}\|} \right\| = 1$

   (f) $\dfrac{\mathbf{u}+\mathbf{v}}{\|\mathbf{u}+\mathbf{v}\|} = \langle 0, 1 \rangle$

     $\left\| \dfrac{\mathbf{u}+\mathbf{v}}{\|\mathbf{u}+\mathbf{v}\|} \right\| = 1$

41. $\mathbf{u} = \left\langle 1, \dfrac{1}{2} \right\rangle$, $\mathbf{v} = \langle 2, 3 \rangle$

   (a) $\|\mathbf{u}\| = \sqrt{1 + \dfrac{1}{4}} = \dfrac{\sqrt{5}}{2}$

   (b) $\|\mathbf{v}\| = \sqrt{4+9} = \sqrt{13}$

   (c) $\mathbf{u} + \mathbf{v} = \left\langle 3, \dfrac{7}{2} \right\rangle$

     $\|\mathbf{u}+\mathbf{v}\| = \sqrt{9 + \dfrac{49}{4}} = \dfrac{\sqrt{85}}{2}$

   (d) $\dfrac{\mathbf{u}}{\|\mathbf{u}\|} = \dfrac{2}{\sqrt{5}} \left\langle 1, \dfrac{1}{2} \right\rangle$

     $\left\| \dfrac{\mathbf{u}}{\|\mathbf{u}\|} \right\| = 1$

   (e) $\dfrac{\mathbf{v}}{\|\mathbf{v}\|} = \dfrac{1}{\sqrt{13}} \langle 2, 3 \rangle$

     $\left\| \dfrac{\mathbf{v}}{\|\mathbf{v}\|} \right\| = 1$

   (f) $\dfrac{\mathbf{u}+\mathbf{v}}{\|\mathbf{u}+\mathbf{v}\|} = \dfrac{2}{\sqrt{85}} \left\langle 3, \dfrac{7}{2} \right\rangle$

     $\left\| \dfrac{\mathbf{u}+\mathbf{v}}{\|\mathbf{u}+\mathbf{v}\|} \right\| = 1$

43. $\mathbf{u} = \langle 2, 1 \rangle$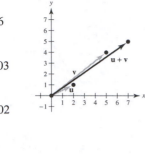

   $\|\mathbf{u}\| = \sqrt{5} \approx 2.236$

   $\mathbf{v} = \langle 5, 4 \rangle$

   $\|\mathbf{v}\| = \sqrt{41} \approx 6.403$

   $\mathbf{u} + \mathbf{v} = \langle 7, 5 \rangle$

   $\|\mathbf{u}+\mathbf{v}\| = \sqrt{74} \approx 8.602$

   $\|\mathbf{u}+\mathbf{v}\| \le \|\mathbf{u}\| + \|\mathbf{v}\|$

   $\sqrt{74} \le \sqrt{5} + \sqrt{41}$

45. $\dfrac{\mathbf{u}}{\|\mathbf{u}\|} = \dfrac{1}{3} \langle 0, 3 \rangle = \langle 0, 1 \rangle$

   $6 \left(\dfrac{\mathbf{u}}{\|\mathbf{u}\|} \right) = 6 \langle 0, 1 \rangle = \langle 0, 6 \rangle$

   $\mathbf{v} = \langle 0, 6 \rangle$

47. $\dfrac{\mathbf{u}}{\|\mathbf{u}\|} = \dfrac{1}{\sqrt{5}} \langle -1, 2 \rangle = \left\langle -\dfrac{1}{\sqrt{5}}, \dfrac{2}{\sqrt{5}} \right\rangle$

   $5 \left(\dfrac{\mathbf{u}}{\|\mathbf{u}\|} \right) = 5 \left\langle -\dfrac{1}{\sqrt{5}}, \dfrac{2}{\sqrt{5}} \right\rangle = \left\langle -\sqrt{5}, 2\sqrt{5} \right\rangle$

   $\mathbf{v} = \left\langle -\sqrt{5}, 2\sqrt{5} \right\rangle$

49. $\mathbf{v} = 3\left[(\cos 0°)\mathbf{i} + (\sin 0°)\mathbf{j} \right] = 3\mathbf{i} = \langle 3, 0 \rangle$

51. $\mathbf{v} = 2\left[(\cos 150°)\mathbf{i} + (\sin 150°)\mathbf{j} \right]$

   $= -\sqrt{3}\mathbf{i} + \mathbf{j} = \left\langle -\sqrt{3}, 1 \right\rangle$

53. $\mathbf{u} = (\cos 0°)\mathbf{i} + (\sin 0°)\mathbf{j} = \mathbf{i}$

   $\mathbf{v} = 3(\cos 45°)\mathbf{i} + 3(\sin 45°)\mathbf{j} = \dfrac{3\sqrt{2}}{2}\mathbf{i} + \dfrac{3\sqrt{2}}{2}\mathbf{j}$

   $\mathbf{u} + \mathbf{v} = \left(\dfrac{2+3\sqrt{2}}{2} \right)\mathbf{i} + \dfrac{3\sqrt{2}}{2}\mathbf{j} = \left\langle \dfrac{2+3\sqrt{2}}{2}, \dfrac{3\sqrt{2}}{2} \right\rangle$

55. $\mathbf{u} = 2(\cos 4)\mathbf{i} + 2(\sin 4)\mathbf{j}$

   $\mathbf{v} = (\cos 2)\mathbf{i} + (\sin 2)\mathbf{j}$

   $\mathbf{u} + \mathbf{v} = (2\cos 4 + \cos 2)\mathbf{i} + (2\sin 4 + \sin 2)\mathbf{j}$

   $= \langle 2\cos 4 + \cos 2, 2\sin 4 + \sin 2 \rangle$

57. Answers will vary. *Sample answer*: A scalar is a real number such as 2. A vector is represented by a directed line segment. A vector has both magnitude and direction. For example $\langle \sqrt{3}, 1 \rangle$ has direction $\dfrac{\pi}{6}$ and a magnitude of 2.

59. $(-4, -1), (6, 5), (10, 3)$

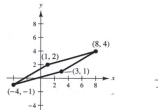 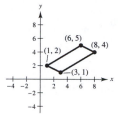 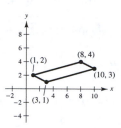

For Exercises 61–65,
$$\mathbf{au} + \mathbf{bw} = a(\mathbf{i} + 2\mathbf{j}) + b(\mathbf{i} - \mathbf{j}) = (a + b)\mathbf{i} + (2a - b)\mathbf{j}.$$

61. $\mathbf{v} = 2\mathbf{i} + \mathbf{j}$. So, $a + b = 2, 2a - b = 1$. Solving simultaneously, you have $a = 1, b = 1$.

63. $\mathbf{v} = 3\mathbf{i}$. So, $a + b = 3, 2a - b = 0$. Solving simultaneously, you have $a = 1, b = 2$.

65. $\mathbf{v} = \mathbf{i} + \mathbf{j}$. So, $a + b = 1, 2a - b = 1$. Solving simultaneously, you have $a = \frac{2}{3}, b = \frac{1}{3}$.

67. $f(x) = x^2, f'(x) = 2x, f'(3) = 6$

(a) $m = 6$. Let $\mathbf{w} = \langle 1, 6 \rangle, \|\mathbf{w}\| = \sqrt{37}$, then $\pm \dfrac{\mathbf{w}}{\|\mathbf{w}\|} = \pm \dfrac{1}{\sqrt{37}} \langle 1, 6 \rangle$.

(b) $m = -\frac{1}{6}$. Let $\mathbf{w} = \langle -6, 1 \rangle, \|\mathbf{w}\| = \sqrt{37}$, then $\pm \dfrac{\mathbf{w}}{\|\mathbf{w}\|} = \pm \dfrac{1}{\sqrt{37}} \langle -6, 1 \rangle$.

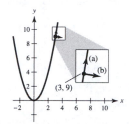

69. $f(x) = x^3, f'(x) = 3x^2 = 3$ at $x = 1$.

(a) $m = 3$. Let $\mathbf{w} = \langle 1, 3 \rangle, \|\mathbf{w}\| = \sqrt{10}$, then $\dfrac{\mathbf{w}}{\|\mathbf{w}\|} = \pm \dfrac{1}{\sqrt{10}} \langle 1, 3 \rangle$.

(b) $m = -\dfrac{1}{3}$. Let $\mathbf{w} = \langle 3, -1 \rangle, \|\mathbf{w}\| = \sqrt{10}$, then $\dfrac{\mathbf{w}}{\|\mathbf{w}\|} = \pm \dfrac{1}{\sqrt{10}} \langle 3, -1 \rangle$.

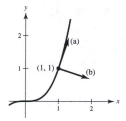

71. $f(x) = \sqrt{25 - x^2}$

$$f'(x) = \dfrac{-x}{\sqrt{25 - x^2}} = \dfrac{-3}{4} \text{ at } x = 3.$$

(a) $m = -\dfrac{3}{4}$. Let $\mathbf{w} = \langle -4, 3 \rangle, \|\mathbf{w}\| = 5$, then $\dfrac{\mathbf{w}}{\|\mathbf{w}\|} = \pm \dfrac{1}{5} \langle -4, 3 \rangle$.

(b) $m = \dfrac{4}{3}$. Let $\mathbf{w} = \langle 3, 4 \rangle, \|\mathbf{w}\| = 5$, then $\dfrac{\mathbf{w}}{\|\mathbf{w}\|} = \pm \dfrac{1}{5} \langle 3, 4 \rangle$.

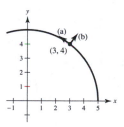

73. $\mathbf{u} = \dfrac{\sqrt{2}}{2}\mathbf{i} + \dfrac{\sqrt{2}}{2}\mathbf{j}$

$\mathbf{u} + \mathbf{v} = \sqrt{2}\mathbf{j}$

$$\mathbf{v} = (\mathbf{u} + \mathbf{v}) - \mathbf{u} = -\dfrac{\sqrt{2}}{2}\mathbf{i} + \dfrac{\sqrt{2}}{2}\mathbf{j} = \left\langle -\dfrac{\sqrt{2}}{2}, \dfrac{\sqrt{2}}{2} \right\rangle$$

75. $\mathbf{F}_1 + \mathbf{F}_2 = (500 \cos 30°\mathbf{i} + 500 \sin 30°\mathbf{j}) + (200 \cos(-45°)\mathbf{i} + 200 \sin(-45°)\mathbf{j}) = (250\sqrt{3} + 100\sqrt{2})\mathbf{i} + (250 - 100\sqrt{2})\mathbf{j}$

$\|\mathbf{F}_1 + \mathbf{F}_2\| = \sqrt{(250\sqrt{3} + 100\sqrt{2})^2 + (250 - 100\sqrt{2})^2} \approx 584.6$ lb

$\tan \theta = \dfrac{250 - 100\sqrt{2}}{250\sqrt{3} + 100\sqrt{2}} \Rightarrow \theta \approx 10.7°$

77. $\mathbf{F}_1 + \mathbf{F}_2 + \mathbf{F}_3 = (75 \cos 30°\mathbf{i} + 75 \sin 30°\mathbf{j}) + (100 \cos 45°\mathbf{i} + 100 \sin 45°\mathbf{j}) + (125 \cos 120°\mathbf{i} + 125 \sin 120°\mathbf{j})$

$= \left(\tfrac{75}{2}\sqrt{3} + 50\sqrt{2} - \tfrac{125}{2}\right)\mathbf{i} + \left(\tfrac{75}{2} + 50\sqrt{2} + \tfrac{125}{2}\sqrt{3}\right)\mathbf{j}$

$\|\mathbf{R}\| = \|\mathbf{F}_1 + \mathbf{F}_2 + \mathbf{F}_3\| \approx 228.5$ lb

$\theta_\mathbf{R} = \theta_{\mathbf{F}_1 + \mathbf{F}_2 + \mathbf{F}_3} \approx 71.3°$

79. (a) The forces act along the same direction. $\theta = 0°$.

    (b) The forces cancel out each other. $\theta = 180°$.

    (c) No, the magnitude of the resultant can not be greater than the sum.

81. Horizontal component $= \|\mathbf{v}\| \cos \theta$

$= 1200 \cos 6° \approx 1193.43$ ft/sec

Vertical component $= \|\mathbf{v}\| \sin \theta$

$= 1200 \sin 6° \approx 125.43$ ft/sec

83. $\mathbf{u} = 900(\cos 148°\mathbf{i} + \sin 148°\mathbf{j})$

$\mathbf{v} = 100(\cos 45°\mathbf{i} + \sin 45°\mathbf{j})$

$\mathbf{u} + \mathbf{v} = (900 \cos 148° + 100 \cos 45°)\mathbf{i} + (900 \sin 148° + 100 \sin 45°)\mathbf{j}$

$\approx -692.53\mathbf{i} + 547.64\mathbf{j}$

$\theta \approx \arctan\left(\dfrac{547.64}{-692.53}\right) \approx -38.34°; 38.34°$ North of West

$\|\mathbf{u} + \mathbf{v}\| \approx \sqrt{(-692.53)^2 + (547.64)^2} \approx 882.9$ km/h

85. True

87. True

89. False

$\|a\mathbf{i} + b\mathbf{j}\| = \sqrt{2}|a|$

91. $\|\mathbf{u}\| = \sqrt{\cos^2 \theta + \sin^2 \theta} = 1,$

$\|\mathbf{v}\| = \sqrt{\sin^2 \theta + \cos^2 \theta} = 1$

93. Let $\mathbf{u}$ and $\mathbf{v}$ be the vectors that determine the parallelogram, as indicated in the figure. The two diagonals are $\mathbf{u} + \mathbf{v}$ and $\mathbf{v} - \mathbf{u}$. So,

$\mathbf{r} = x(\mathbf{u} + \mathbf{v}), \mathbf{s} = 4(\mathbf{v} - \mathbf{u})$. But,

$\mathbf{u} = \mathbf{r} - \mathbf{s}$

$= x(\mathbf{u} + \mathbf{v}) - y(\mathbf{v} - \mathbf{u}) = (x + y)\mathbf{u} + (x - y)\mathbf{v}.$

So, $x + y = 1$ and $x - y = 0$. Solving you have

$x = y = \tfrac{1}{2}.$

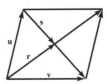

95. The set is a circle of radius 5, centered at the origin.

$\|\mathbf{u}\| = \|\langle x, y \rangle\| = \sqrt{x^2 + y^2} = 5 \Rightarrow x^2 + y^2 = 25$

Section 11.2   Space Coordinates and Vectors in Space

1.

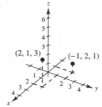

3.

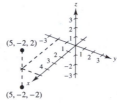

5. $x = -3,\ y = 4,\ z = 5$: $(-3, 4, 5)$

7. $y = z = 0,\ x = 12$: $(12, 0, 0)$

9. The z-coordinate is 0.

11. The point is 6 units above the xy-plane.

13. The point is on the plane parallel to the yz-plane that passes through $x = -3$.

15. The point is to the left of the xz-plane.

17. The point is on or between the planes $y = 3$ and $y = -3$.

19. The point (x, y, z) is 3 units below the xy-plane, and below either quadrant I or III.

21. The point could be above the xy-plane and so above quadrants II or IV, or below the xy-plane, and so below quadrants I or III.

23. $d = \sqrt{(-4 - 0)^2 + (2 - 0)^2 + (7 - 0)^2}$

$\quad = \sqrt{16 + 4 + 49} = \sqrt{69}$

25. $d = \sqrt{(6 - 1)^2 + (-2 - (-2))^2 + (-2 - 4)^2}$

$\quad = \sqrt{25 + 0 + 36} = \sqrt{61}$

27. $A(0, 0, 4),\ B(2, 6, 7),\ C(6, 4, -8)$

$|AB| = \sqrt{2^2 + 6^2 + 3^2} = \sqrt{49} = 7$

$|AC| = \sqrt{6^2 + 4^2 + (-12)^2} = \sqrt{196} = 14$

$|BC| = \sqrt{4^2 + (-2)^2 + (-15)^2} = \sqrt{245} = 7\sqrt{5}$

$|BC|^2 = 245 = 49 + 196 = |AB|^2 + |AC|^2$

Right triangle

29. $A(-1, 0, -2),\ B(-1, 5, 2),\ C(-3, -1, 1)$

$|AB| = \sqrt{0 + 25 + 16} = \sqrt{41}$

$|AC| = \sqrt{4 + 1 + 9} = \sqrt{14}$

$|BC| = \sqrt{4 + 36 + 1} = \sqrt{41}$

Because $|AB| = |BC|$, the triangle is isosceles.

31. The z-coordinate is changed by 5 units:

$(0, 0, 9),\ (2, 6, 12),\ (6, 4, -3)$

33. $\left(\dfrac{3 + 1}{2}, \dfrac{4 + 8}{2}, \dfrac{6 + 0}{2} \right) = (2, 6, 3)$

35. $\left(\dfrac{5 + (-2)}{2}, \dfrac{-9 + 3}{2}, \dfrac{7 + 3}{2} \right) = \left(\dfrac{3}{2}, -3, 5 \right)$

37. Center: $(0, 2, 5)$

Radius: 2

$(x - 0)^2 + (y - 2)^2 + (z - 5)^2 = 4$

39. Center: $\dfrac{(2, 0, 0) + (0, 6, 0)}{2} = (1, 3, 0)$

Radius: $\sqrt{10}$

$(x - 1)^2 + (y - 3)^2 + (z - 0)^2 = 10$

41. $\qquad x^2 + y^2 + z^2 - 2x + 6y + 8z + 1 = 0$

$(x^2 - 2x + 1) + (y^2 + 6y + 9) + (z^2 + 8z + 16) = -1 + 1 + 9 + 16$

$\qquad\qquad (x - 1)^2 + (y + 3)^2 + (z + 4)^2 = 25$

Center: $(1, -3, -4)$

Radius: 5

43. $9x^2 + 9y^2 + 9z^2 - 6x + 18y + 1 = 0$

$x^2 + y^2 + z^2 - \frac{2}{3}x + 2y + \frac{1}{9} = 0$

$\left(x^2 - \frac{2}{3}x + \frac{1}{9}\right) + \left(y^2 + 2y + 1\right) + z^2 = -\frac{1}{9} + \frac{1}{9} + 1$

$\left(x - \frac{1}{3}\right)^2 + (y + 1)^2 + (z - 0)^2 = 1$

Center: $\left(\frac{1}{3}, -1, 0\right)$

Radius: 1

45. (a) $\mathbf{v} = \langle 2 - 4, 4 - 2, 3 - 1 \rangle = \langle -2, 2, 2 \rangle$

(b) $\mathbf{v} = -2\mathbf{i} + 2\mathbf{j} + 2\mathbf{k}$

(c)

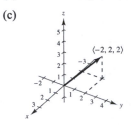

47. (a) $\mathbf{v} = \langle 0 - 3, 3 - 3, 3 - 0 \rangle = \langle -3, 0, 3 \rangle$

(b) $\mathbf{v} = -3\mathbf{i} + 3\mathbf{k}$

(c)

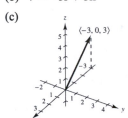

49. $\langle 4 - 3, 1 - 2, 6 - 0 \rangle = \langle 1, -1, 6 \rangle$

$\|\langle 1, -1, 6 \rangle\| = \sqrt{1 + 1 + 36} = \sqrt{38}$

Unit vector: $\dfrac{\langle 1, -1, 6 \rangle}{\sqrt{38}} = \left\langle \dfrac{1}{\sqrt{38}}, \dfrac{-1}{\sqrt{38}}, \dfrac{6}{\sqrt{38}} \right\rangle$

51. (b) $\mathbf{v} = \langle 3 - (-1), 3 - 2, 4 - 3 \rangle = \langle 4, 1, 1 \rangle$

(c) $\mathbf{v} = 4\mathbf{i} + \mathbf{j} + \mathbf{k}$

(a), (d)

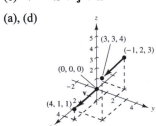

53. $(q_1, q_2, q_3) - (0, 6, 2) = (3, -5, 6)$

$Q = (3, 1, 8)$

55. (a) $2\mathbf{v} = \langle 2, 4, 4 \rangle$

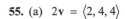
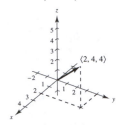

(b) $-\mathbf{v} = \langle -1, -2, -2 \rangle$

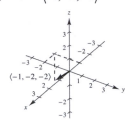

(c) $\frac{3}{2}\mathbf{v} = \left\langle \frac{3}{2}, 3, 3 \right\rangle$

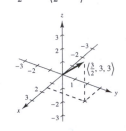

(d) $0\mathbf{v} = \langle 0, 0, 0 \rangle$

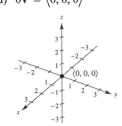

57. $\mathbf{z} = \mathbf{u} - \mathbf{v} + 2\mathbf{w}$

$= \langle 1, 2, 3 \rangle - \langle 2, 2, -1 \rangle + \langle 8, 0, -8 \rangle = \langle 7, 0, -4 \rangle$

59. $2\mathbf{z} - 3\mathbf{u} = 2\langle z_1, z_2, z_3 \rangle - 3\langle 1, 2, 3 \rangle = \langle 4, 0, -4 \rangle$

$2z_1 - 3 = 4 \Rightarrow z_1 = \frac{7}{2}$

$2z_2 - 6 = 0 \Rightarrow z_2 = 3$

$2z_3 - 9 = -4 \Rightarrow z_3 = \frac{5}{2}$

$\mathbf{z} = \left\langle \frac{7}{2}, 3, \frac{5}{2} \right\rangle$

61. (a) and (b) are parallel because

$\langle -6, -4, 10 \rangle = -2\langle 3, 2, -5 \rangle$ and

$\left\langle 2, \frac{4}{3}, -\frac{10}{3} \right\rangle = \frac{2}{3}\langle 3, 2, -5 \rangle$.

63. $\mathbf{z} = -3\mathbf{i} + 4\mathbf{j} + 2\mathbf{k}$

(a) is parallel because $-6\mathbf{i} + 8\mathbf{j} + 4\mathbf{k} = 2\mathbf{z}$.

65. $P(0, -2, -5), Q(3, 4, 4), R(2, 2, 1)$

$$\overrightarrow{PQ} = \langle 3, 6, 9 \rangle$$
$$\overrightarrow{PR} = \langle 2, 4, 6 \rangle$$
$$\langle 3, 6, 9 \rangle = \tfrac{3}{2}\langle 2, 4, 6 \rangle$$

So, $\overrightarrow{PQ}$ and $\overrightarrow{PR}$ are parallel, the points are collinear.

67. $P(1, 2, 4), Q(2, 5, 0), R(0, 1, 5)$

$$\overrightarrow{PQ} = \langle 1, 3, -4 \rangle$$
$$\overrightarrow{PR} = \langle -1, -1, 1 \rangle$$

Because $\overrightarrow{PQ}$ and $\overrightarrow{PR}$ are not parallel, the points are not collinear.

69. $A(2, 9, 1), B(3, 11, 4), C(0, 10, 2), D(1, 12, 5)$

$$\overrightarrow{AB} = \langle 1, 2, 3 \rangle$$
$$\overrightarrow{CD} = \langle 1, 2, 3 \rangle$$
$$\overrightarrow{AC} = \langle -2, 1, 1 \rangle$$
$$\overrightarrow{BD} = \langle -2, 1, 1 \rangle$$

Because $\overrightarrow{AB} = \overrightarrow{CD}$ and $\overrightarrow{AC} = \overrightarrow{BD}$, the given points form the vertices of a parallelogram.

71. $\mathbf{v} = \langle 0, 0, 0 \rangle$

$\|\mathbf{v}\| = 0$

73. $\mathbf{v} = 3\mathbf{j} - 5\mathbf{k} = \langle 0, 3, -5 \rangle$

$\|\mathbf{v}\| = \sqrt{0 + 9 + 25} = \sqrt{34}$

75. $\mathbf{v} = \mathbf{i} - 2\mathbf{j} - 3\mathbf{k} = \langle 1, -2, -3 \rangle$

$\|\mathbf{v}\| = \sqrt{1 + 4 + 9} = \sqrt{14}$

77. $\mathbf{v} = \langle 2, -1, 2 \rangle$

$\|\mathbf{v}\| = \sqrt{4 + 1 + 4} = 3$

(a) $\dfrac{\mathbf{v}}{\|\mathbf{v}\|} = \dfrac{1}{3}\langle 2, -1, 2 \rangle$

(b) $-\dfrac{\mathbf{v}}{\|\mathbf{v}\|} = -\dfrac{1}{3}\langle 2, -1, 2 \rangle$

79. $\mathbf{v} = 4\mathbf{i} - 5\mathbf{j} + 3\mathbf{k}$

$\|\mathbf{v}\| = \sqrt{16 + 25 + 9} = \sqrt{50} = 5\sqrt{2}$

(a) $\dfrac{\mathbf{v}}{\|\mathbf{v}\|} = \dfrac{1}{5\sqrt{2}}(4\mathbf{i} - 5\mathbf{j} + 3\mathbf{k})$

$\qquad = \dfrac{2\sqrt{2}}{5}\mathbf{i} - \dfrac{\sqrt{2}}{2}\mathbf{j} + \dfrac{3\sqrt{2}}{10}\mathbf{k}$

(b) $-\dfrac{\mathbf{v}}{\|\mathbf{v}\|} = -\dfrac{1}{5\sqrt{2}}(4\mathbf{i} - 5\mathbf{j} + 3\mathbf{k})$

$\qquad = -\dfrac{2\sqrt{2}}{5}\mathbf{i} + \dfrac{\sqrt{2}}{2}\mathbf{j} - \dfrac{3\sqrt{2}}{10}\mathbf{k}$

81. The terminal points of the vectors $t\mathbf{u}, \mathbf{u} + t\mathbf{v}$ and $s\mathbf{u} + t\mathbf{v}$ are collinear.

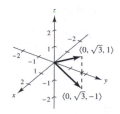

83. $\mathbf{v} = 10\dfrac{\mathbf{u}}{\|\mathbf{u}\|} = 10\dfrac{\langle 0, 3, 3 \rangle}{3\sqrt{2}}$

$\qquad = 10\left\langle 0, \dfrac{1}{\sqrt{2}}, \dfrac{1}{\sqrt{2}} \right\rangle = \left\langle 0, \dfrac{10}{\sqrt{2}}, \dfrac{10}{\sqrt{2}} \right\rangle$

85. $\mathbf{v} = \dfrac{3}{2}\dfrac{\mathbf{u}}{\|\mathbf{u}\|} = \dfrac{3}{2}\dfrac{(2, -2, 1)}{3} = \dfrac{3}{2}\left\langle \dfrac{2}{3}, \dfrac{-2}{3}, \dfrac{1}{3} \right\rangle = \left\langle 1, -1, \dfrac{1}{2} \right\rangle$

87. $\mathbf{v} = 2\left[\cos(\pm 30°)\mathbf{j} + \sin(\pm 30°)\mathbf{k} \right]$

$\qquad = \sqrt{3}\mathbf{j} \pm \mathbf{k} = \left\langle 0, \sqrt{3}, \pm 1 \right\rangle$

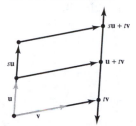

89. $\mathbf{v} = \langle -3, -6, 3 \rangle$

$\qquad\qquad \tfrac{2}{3}\mathbf{v} = \langle -2, -4, 2 \rangle$

$(4, 3, 0) + (-2, -4, 2) = (2, -1, 2)$

91. (a)

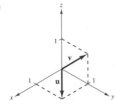

(b) $\mathbf{w} = a\mathbf{u} + b\mathbf{v} = a\mathbf{i} + (a + b)\mathbf{j} + b\mathbf{k} = 0$

$a = 0, a + b = 0, b = 0$

So, a and b are both zero.

(c) $a\mathbf{i} + (a + b)\mathbf{j} + b\mathbf{k} = \mathbf{i} + 2\mathbf{j} + \mathbf{k}$

$a = 1, a + b = 2, b = 1$

$\mathbf{w} = \mathbf{u} + \mathbf{v}$

(d) $a\mathbf{i} + (a + b)\mathbf{j} + b\mathbf{k} = \mathbf{i} + 2\mathbf{j} + 3\mathbf{k}$

$a = 1, a + b = 2, b = 3$

Not possible

93. x_0 is directed distance to yz-plane.

y_0 is directed distance to xz-plane.

z_0 is directed distance to xy-plane.

95. $(x - x_0)^2 + (y - y_0)^2 + (z - z_0)^2 = r^2$

97.

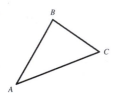

$\overrightarrow{AB} + \overrightarrow{BC} = \overrightarrow{AC}$

So, $\overrightarrow{AB} + \overrightarrow{BC} + \overrightarrow{CA} = \overrightarrow{AC} + \overrightarrow{CA} = \mathbf{0}$

99. Let α be the angle between $\mathbf{v}$ and the coordinate axes.

$\mathbf{v} = (\cos \alpha)\mathbf{i} + (\cos \alpha)\mathbf{j} + (\cos \alpha)\mathbf{k}$

$\|\mathbf{v}\| = \sqrt{3} \cos \alpha = 1$

$\cos \alpha = \dfrac{1}{\sqrt{3}} = \dfrac{\sqrt{3}}{3}$

$\mathbf{v} = \dfrac{\sqrt{3}}{3}(\mathbf{i} + \mathbf{j} + \mathbf{k}) = \dfrac{\sqrt{3}}{3}\langle 1, 1, 1 \rangle$

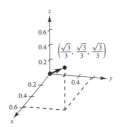

101. (a) The height of the right triangle is $h = \sqrt{L^2 - 18^2}$.

The vector $\overrightarrow{PQ}$ is given by

$\overrightarrow{PQ} = \langle 0, -18, h \rangle$.

The tension vector $\mathbf{T}$ in each wire is

$\mathbf{T} = c\langle 0, -18, h \rangle$ where $ch = \dfrac{24}{3} = 8$.

So, $\mathbf{T} = \dfrac{8}{h}\langle 0, -18, h \rangle$ and

$T = \|\mathbf{T}\| = \dfrac{8}{h}\sqrt{18^2 + h^2}$

$= \dfrac{8}{\sqrt{L^2 - 18^2}}\sqrt{18^2 + (L^2 - 18^2)}$

$= \dfrac{8L}{\sqrt{L^2 - 18^2}}, L > 18.$

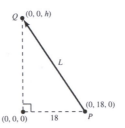

(b)

L	20	25	30	35	40	45	50
T	18.4	11.5	10	9.3	9.0	8.7	8.6

(c)

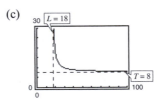

$x = 18$ is a vertical asymptote and $y = 8$ is a horizontal asymptote.

(d) $\lim\limits_{L \to 18^+} \dfrac{8L}{\sqrt{L^2 - 18^2}} = \infty$

$\lim\limits_{L \to \infty} \dfrac{8L}{\sqrt{L^2 - 18^2}} = \lim\limits_{L \to \infty} \dfrac{8}{\sqrt{1 - (18/L)^2}} = 8$

(e) From the table, $T = 10$ implies $L = 30$ inches.

103. $\overrightarrow{AB} = \langle 0, 70, 115 \rangle,\ \mathbf{F}_1 = C_1\langle 0, 70, 115 \rangle$

$\overrightarrow{AC} = \langle -60, 0, 115 \rangle,\ \mathbf{F}_2 = C_2\langle -60, 0, 115 \rangle$

$\overrightarrow{AD} = \langle 45, -65, 115 \rangle,\ \mathbf{F}_3 = C_3\langle 45, -65, 115 \rangle$

$\mathbf{F} = \mathbf{F}_1 + \mathbf{F}_2 + \mathbf{F}_3 = \langle 0, 0, 500 \rangle$

So:

$$-60C_2 + 45C_3 = 0$$
$$70C_1 \qquad\quad - 65C_3 = 0$$
$$115(C_1 + C_2 + C_3) = 500$$

Solving this system yields $C_1 = \frac{104}{69}, C_2 = \frac{28}{23}$, and $C_3 = \frac{112}{69}$. So:

$\|\mathbf{F}_1\| \approx 202.919\,\text{N}$

$\|\mathbf{F}_2\| \approx 157.909\,\text{N}$

$\|\mathbf{F}_3\| \approx 226.521\,\text{N}$

105. $d(AP) = 2d(BP)$

$$\sqrt{x^2 + (y+1)^2 + (z-1)^2} = 2\sqrt{(x-1)^2 + (y-2)^2 + z^2}$$

$$x^2 + y^2 + z^2 + 2y - 2z + 2 = 4(x^2 + y^2 + z^2 - 2x - 4y + 5)$$

$$0 = 3x^2 + 3y^2 + 3z^2 - 8x - 18y + 2z + 18$$

$$-6 + \frac{16}{9} + 9 + \frac{1}{9} = \left(x^2 - \frac{8}{3}x + \frac{16}{9}\right) + \left(y^2 - 6y + 9\right) + \left(z^2 + \frac{2}{3}z + \frac{1}{9}\right)$$

$$\frac{44}{9} = \left(x - \frac{4}{3}\right)^2 + (y - 3)^2 + \left(z + \frac{1}{3}\right)^2$$

Sphere; center: $\left(\frac{4}{3}, 3, -\frac{1}{3}\right)$, radius: $\dfrac{2\sqrt{11}}{3}$

Section 11.3   The Dot Product of Two Vectors

1. $\mathbf{u} = \langle 3, 4 \rangle,\ \mathbf{v} = \langle -1, 5 \rangle$

(a) $\mathbf{u} \cdot \mathbf{v} = 3(-1) + 4(5) = 17$

(b) $\mathbf{u} \cdot \mathbf{u} = 3(3) + 4(4) = 25$

(c) $\|\mathbf{u}\|^2 = 3^2 + 4^2 = 25$

(d) $(\mathbf{u} \cdot \mathbf{v})\mathbf{v} = 17\langle -1, 5 \rangle = \langle -17, 85 \rangle$

(e) $\mathbf{u} \cdot (2\mathbf{v}) = 2(\mathbf{u} \cdot \mathbf{v}) = 2(17) = 34$

3. $\mathbf{u} = \langle 6, -4 \rangle,\ \mathbf{v} = \langle -3, 2 \rangle$

(a) $\mathbf{u} \cdot \mathbf{v} = 6(-3) + (-4)(2) = -26$

(b) $\mathbf{u} \cdot \mathbf{u} = 6(6) + (-4)(-4) = 52$

(c) $\|\mathbf{u}\|^2 = 6^2 + (-4)^2 = 52$

(d) $(\mathbf{u} \cdot \mathbf{v})\mathbf{v} = -26\langle -3, 2 \rangle = \langle 78, -52 \rangle$

(e) $\mathbf{u} \cdot (2\mathbf{v}) = 2(\mathbf{u} \cdot \mathbf{v}) = 2(-26) = -52$

5. $\mathbf{u} = \langle 2, -3, 4 \rangle,\ \mathbf{v} = \langle 0, 6, 5 \rangle$

(a) $\mathbf{u} \cdot \mathbf{v} = 2(0) + (-3)(6) + (4)(5) = 2$

(b) $\mathbf{u} \cdot \mathbf{u} = 2(2) + (-3)(-3) + 4(4) = 29$

(c) $\|\mathbf{u}\|^2 = 2^2 + (-3)^2 + 4^2 = 29$

(d) $(\mathbf{u} \cdot \mathbf{v})\mathbf{v} = 2\langle 0, 6, 5 \rangle = \langle 0, 12, 10 \rangle$

(e) $\mathbf{u} \cdot (2\mathbf{v}) = 2(\mathbf{u} \cdot \mathbf{v}) = 2(2) = 4$

7. $\mathbf{u} = 2\mathbf{i} - \mathbf{j} + \mathbf{k},\ \mathbf{v} = \mathbf{i} - \mathbf{k}$

(a) $\mathbf{u} \cdot \mathbf{v} = 2(1) + (-1)(0) + 1(-1) = 1$

(b) $\mathbf{u} \cdot \mathbf{u} = 2(2) + (-1)(-1) + (1)(1) = 6$

(c) $\|\mathbf{u}\|^2 = 2^2 + (-1)^2 + 1^2 = 6$

(d) $(\mathbf{u} \cdot \mathbf{v})\mathbf{v} = \mathbf{v} = \mathbf{i} - \mathbf{k}$

(e) $\mathbf{u} \cdot (2\mathbf{v}) = 2(\mathbf{u} \cdot \mathbf{v}) = 2$

9. $\mathbf{u} = \langle 1, 1 \rangle$, $\mathbf{v} = \langle 2, -2 \rangle$

$$\cos\theta = \frac{\mathbf{u} \cdot \mathbf{v}}{\|\mathbf{u}\|\|\mathbf{v}\|} = \frac{0}{\sqrt{2}\sqrt{8}} = 0$$

(a) $\theta = \dfrac{\pi}{2}$ (b) $\theta = 90°$

11. $\mathbf{u} = 3\mathbf{i} + \mathbf{j}$, $\mathbf{v} = -2\mathbf{i} + 4\mathbf{j}$

$$\cos\theta = \frac{\mathbf{u} \cdot \mathbf{v}}{\|\mathbf{u}\|\|\mathbf{v}\|} = \frac{-2}{\sqrt{10}\sqrt{20}} = \frac{-1}{5\sqrt{2}}$$

(a) $\theta = \arccos\left(-\dfrac{1}{5\sqrt{2}}\right) \approx 1.713$

(b) $\theta \approx 98.1°$

13. $\mathbf{u} = \langle 1, 1, 1 \rangle$, $\mathbf{v} = \langle 2, 1, -1 \rangle$

$$\cos\theta = \frac{\mathbf{u} \cdot \mathbf{v}}{\|\mathbf{u}\|\|\mathbf{v}\|} = \frac{2}{\sqrt{3}\sqrt{6}} = \frac{\sqrt{2}}{3}$$

(a) $\theta = \arccos\dfrac{\sqrt{2}}{3} \approx 1.080$

(b) $\theta \approx 61.9°$

15. $\mathbf{u} = 3\mathbf{i} + 4\mathbf{j}$, $\mathbf{v} = -2\mathbf{j} + 3\mathbf{k}$

$$\cos\theta = \frac{\mathbf{u} \cdot \mathbf{v}}{\|\mathbf{u}\|\|\mathbf{v}\|} = \frac{-8}{5\sqrt{13}} = \frac{-8\sqrt{13}}{65}$$

(a) $\theta = \arccos\left(-\dfrac{8\sqrt{13}}{65}\right) \approx 2.031$

(b) $\theta \approx 116.3°$

17. $\dfrac{\mathbf{u} \cdot \mathbf{v}}{\|\mathbf{u}\|\|\mathbf{v}\|} = \cos\theta$

$$\mathbf{u} \cdot \mathbf{v} = (8)(5)\cos\frac{\pi}{3} = 20$$

19. $\mathbf{u} = \langle 4, 3 \rangle$, $\mathbf{v} = \left\langle \frac{1}{2}, -\frac{2}{3} \right\rangle$

$\mathbf{u} \neq c\mathbf{v} \Rightarrow$ not parallel

$\mathbf{u} \cdot \mathbf{v} = 0 \Rightarrow$ orthogonal

21. $\mathbf{u} = \mathbf{j} + 6\mathbf{k}$, $\mathbf{v} = \mathbf{i} - 2\mathbf{j} - \mathbf{k}$

$\mathbf{u} \neq c\mathbf{v} \Rightarrow$ not parallel

$\mathbf{u} \cdot \mathbf{v} = -8 \neq 0 \Rightarrow$ not orthogonal

Neither

23. $\mathbf{u} = \langle 2, -3, 1 \rangle$, $\mathbf{v} = \langle -1, -1, -1 \rangle$

$\mathbf{u} \neq c\mathbf{v} \Rightarrow$ not parallel

$\mathbf{u} \cdot \mathbf{v} = 0 \Rightarrow$ orthogonal

25. The vector $\langle 1, 2, 0 \rangle$ joining $(1, 2, 0)$ and $(0, 0, 0)$ is perpendicular to the vector $\langle -2, 1, 0 \rangle$ joining $(-2, 1, 0)$ and $(0, 0, 0)$: $\langle 1, 2, 0 \rangle \cdot \langle -2, 1, 0 \rangle = 0$

The triangle has a right angle, so it is a right triangle.

27. $A(2, 0, 1)$, $B(0, 1, 2)$, $C\left(-\frac{1}{2}, \frac{3}{2}, 0\right)$

$\overrightarrow{AB} = \langle -2, 1, 1 \rangle$ $\overrightarrow{BA} = \langle 2, -1, -1 \rangle$

$\overrightarrow{AC} = \left\langle -\frac{5}{2}, \frac{3}{2}, -1 \right\rangle$ $\overrightarrow{CA} = \left\langle \frac{5}{2}, -\frac{3}{2}, 1 \right\rangle$

$\overrightarrow{BC} = \left\langle -\frac{1}{2}, \frac{1}{2}, -2 \right\rangle$ $\overrightarrow{CB} = \left\langle \frac{1}{2}, -\frac{1}{2}, 2 \right\rangle$

$\overrightarrow{AB} \cdot \overrightarrow{AC} = 5 + \frac{3}{2} - 1 > 0$

$\overrightarrow{BA} \cdot \overrightarrow{BC} = -1 - \frac{1}{2} + 2 > 0$

$\overrightarrow{CA} \cdot \overrightarrow{CB} = \frac{5}{4} + \frac{3}{4} + 2 > 0$

The triangle has three acute angles, so it is an acute triangle.

29. $\mathbf{u} = \mathbf{i} + 2\mathbf{j} + 2\mathbf{k}$, $\|\mathbf{u}\| = \sqrt{1 + 4 + 4} = 3$

$\cos\alpha = \frac{1}{3} \Rightarrow \alpha \approx 1.2310$ or $70.5°$

$\cos\beta = \frac{2}{3} \Rightarrow \beta \approx 0.8411$ or $48.2°$

$\cos\gamma = \frac{2}{3} \Rightarrow \gamma \approx 0.8411$ or $48.2°$

$\cos^2\alpha + \cos^2\beta + \cos^2\gamma = \frac{1}{9} + \frac{4}{9} + \frac{4}{9} = 1$

31. $\mathbf{u} = 3\mathbf{i} + 2\mathbf{j} - 2\mathbf{k}$, $\|\mathbf{u}\| = \sqrt{9 + 4 + 4} = \sqrt{17}$

$\cos\alpha = \dfrac{3}{\sqrt{17}} \Rightarrow \alpha \approx 0.7560$ or $43.3°$

$\cos\beta = \dfrac{2}{\sqrt{17}} \Rightarrow \beta \approx 1.0644$ or $61.0°$

$\cos\gamma = \dfrac{-2}{\sqrt{17}} \Rightarrow y \approx 2.0772$ or $119.0°$

$\cos^2\alpha + \cos^2\beta + \cos^2\gamma = \dfrac{9}{17} + \dfrac{4}{17} + \dfrac{4}{17} = 1$

33. $\mathbf{u} = \langle 0, 6, -4 \rangle$, $\|\mathbf{u}\| = \sqrt{0 + 36 + 16} = \sqrt{52} = 2\sqrt{13}$

$\cos\alpha = 0 \Rightarrow \alpha = \dfrac{\pi}{2}$ or $90°$

$\cos\beta = \dfrac{3}{\sqrt{13}} \Rightarrow \beta \approx 0.5880$ or $33.7°$

$\cos\gamma = -\dfrac{2}{\sqrt{13}} \Rightarrow \gamma \approx 2.1588$ or $123.7°$

$\cos^2\alpha + \cos^2\beta + \cos^2\gamma = 0 + \dfrac{9}{13} + \dfrac{4}{13} = 1$

35. $\mathbf{u} = \langle 6, 7 \rangle$, $\mathbf{v} = \langle 1, 4 \rangle$

(a) $\mathbf{w}_1 = \text{proj}_\mathbf{v}\mathbf{u} = \left(\dfrac{\mathbf{u} \cdot \mathbf{v}}{\|\mathbf{v}\|^2} \right)\mathbf{v}$

$= \dfrac{6(1) + 7(4)}{1^2 + 4^2}\langle 1, 4 \rangle$

$= \dfrac{34}{17}\langle 1, 4 \rangle = \langle 2, 8 \rangle$

(b) $\mathbf{w}_2 = \mathbf{u} - \mathbf{w}_1 = \langle 6, 7 \rangle - \langle 2, 8 \rangle = \langle 4, -1 \rangle$

37. $\mathbf{u} = 2\mathbf{i} + 3\mathbf{j} = \langle 2, 3 \rangle$, $\mathbf{v} = 5\mathbf{i} + \mathbf{j} = \langle 5, 1 \rangle$

(a) $\mathbf{w}_1 = \text{proj}_\mathbf{v}\mathbf{u} = \left(\dfrac{\mathbf{u} \cdot \mathbf{v}}{\|\mathbf{v}\|^2} \right)\mathbf{v}$

$= \dfrac{2(5) + 3(1)}{5^2 + 1}\langle 5, 1 \rangle$

$= \dfrac{13}{26}\langle 5, 1 \rangle = \left\langle \dfrac{5}{2}, \dfrac{1}{2} \right\rangle$

(b) $\mathbf{w}_2 = \mathbf{u} - \mathbf{w}_1 = \langle 2, 3 \rangle - \left\langle \dfrac{5}{2}, \dfrac{1}{2} \right\rangle = \left\langle -\dfrac{1}{2}, \dfrac{5}{2} \right\rangle$

39. $\mathbf{u} = \langle 0, 3, 3 \rangle$, $\mathbf{v} = \langle -1, 1, 1 \rangle$

(a) $\mathbf{w}_1 = \text{proj}_\mathbf{v}\mathbf{u} = \left(\dfrac{\mathbf{u} \cdot \mathbf{v}}{\|\mathbf{v}\|^2} \right)\mathbf{v}$

$= \dfrac{0(-1) + 3(1) + 3(1)}{1 + 1 + 1}\langle -1, 1, 1 \rangle$

$= \dfrac{6}{3}\langle -1, 1, 1 \rangle = \langle -2, 2, 2 \rangle$

(b) $\mathbf{w}_2 = \mathbf{u} - \mathbf{w}_1 = \langle 0, 3, 3 \rangle - \langle -2, 2, 2 \rangle = \langle 2, 1, 1 \rangle$

41. $\mathbf{u} = 2\mathbf{i} + \mathbf{j} + 2\mathbf{k} = \langle 2, 1, 2 \rangle$

$\mathbf{v} = 3\mathbf{j} + 4\mathbf{k} = \langle 0, 3, 4 \rangle$

(a) $\mathbf{w}_1 = \text{proj}_\mathbf{v}\mathbf{u} = \left(\dfrac{\mathbf{u} \cdot \mathbf{v}}{\|\mathbf{v}\|^2} \right)\mathbf{v}$

$= \dfrac{2(0) + 1(3) + 2(4)}{3^2 + 4^2}\langle 0, 3, 4 \rangle$

$= \dfrac{11}{25}\langle 0, 3, 4 \rangle = \left\langle 0, \dfrac{33}{25}, \dfrac{44}{25} \right\rangle$

(b) $\mathbf{w}_2 = \mathbf{u} - \mathbf{w}_1 = \langle 2, 1, 2 \rangle - \left\langle 0, \dfrac{33}{25}, \dfrac{44}{25} \right\rangle = \left\langle 2, -\dfrac{8}{25}, \dfrac{6}{25} \right\rangle$

43. $\mathbf{u} \cdot \mathbf{v} = \langle u_1, u_2, u_3 \rangle \cdot \langle v_1, v_2, v_3 \rangle = u_1v_1 + u_2v_2 + u_3v_3$

45. (a) and (b) are defined. (c) and (d) are not defined because it is not possible to find the dot product of a scalar and a vector or to add a scalar to a vector.

47. See figure 11.29, page 770.

49. Yes, $\left\| \dfrac{\mathbf{u} \cdot \mathbf{v}}{\|\mathbf{v}\|^2}\mathbf{v} \right\| = \left\| \dfrac{\mathbf{v} \cdot \mathbf{u}}{\|\mathbf{u}\|^2}\mathbf{u} \right\|$

$|\mathbf{u} \cdot \mathbf{v}|\dfrac{\|\mathbf{v}\|}{\|\mathbf{v}\|^2} = |\mathbf{v} \cdot \mathbf{u}|\dfrac{\|\mathbf{u}\|}{\|\mathbf{u}\|^2}$

$\dfrac{1}{\|\mathbf{v}\|} = \dfrac{1}{\|\mathbf{u}\|}$

$\|\mathbf{u}\| = \|\mathbf{v}\|$

51. $\mathbf{u} = \langle 3240, 1450, 2235 \rangle$

$\mathbf{v} = \langle 2.25, 2.95, 2.65 \rangle$

$\mathbf{u} \cdot \mathbf{v} = 3240(2.25) + 1450(2.95) + 2235(2.65)$

$= \$17,490.25$

This represents the total revenue the restaurant earned on its three products.

53. Answers will vary. *Sample answer:*

$\mathbf{u} = -\dfrac{1}{4}\mathbf{i} + \dfrac{3}{2}\mathbf{j}$. Want $\mathbf{u} \cdot \mathbf{v} = 0$.

$\mathbf{v} = 12\mathbf{i} + 2\mathbf{j}$ and $-\mathbf{v} = -12\mathbf{i} - 2\mathbf{j}$ are orthogonal to $\mathbf{u}$.

55. Answers will vary. *Sample answer:*

$\mathbf{u} = \langle 3, 1, -2 \rangle$. Want $\mathbf{u} \cdot \mathbf{v} = 0$.

$\mathbf{v} = \langle 0, 2, 1 \rangle$ and $-\mathbf{v} = \langle 0, -2, -1 \rangle$ are orthogonal to $\mathbf{u}$.

57. Let $s = $ length of a side.

$\mathbf{v} = \langle s, s, s \rangle$

$\|\mathbf{v}\| = s\sqrt{3}$

$\cos \alpha = \cos \beta = \cos \gamma = \dfrac{s}{s\sqrt{3}} = \dfrac{1}{\sqrt{3}}$

$\alpha = \beta = \gamma = \arccos\left(\dfrac{1}{\sqrt{3}} \right) \approx 54.7°$

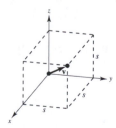

59. (a) Gravitational Force $\mathbf{F} = -48{,}000\mathbf{j}$

$\mathbf{v} = \cos 10°\mathbf{i} + \sin 10°\mathbf{j}$

$$\mathbf{w}_1 = \frac{\mathbf{F} \cdot \mathbf{v}}{\|\mathbf{v}\|^2}\mathbf{v} = (\mathbf{F} \cdot \mathbf{v})\mathbf{v}$$

$= (-48{,}000)(\sin 10°)\mathbf{v}$

$\approx -8335.1(\cos 10°\mathbf{i} + \sin 10°\mathbf{j})$

$\|\mathbf{w}_1\| \approx 8335.1 \text{ lb}$

(b) $\mathbf{w}_2 = \mathbf{F} - \mathbf{w}_1$

$= -48{,}000\mathbf{j} + 8335.1(\cos 10°\mathbf{i} + \sin 10°\mathbf{j})$

$= 8208.5\mathbf{i} - 46{,}552.6\mathbf{j}$

$\|\mathbf{w}_2\| \approx 47{,}270.8 \text{ lb}$

61. $\mathbf{F} = 85\left(\dfrac{1}{2}\mathbf{i} + \dfrac{\sqrt{3}}{2}\mathbf{j}\right)$

$\mathbf{v} = 10\mathbf{i}$

$W = \mathbf{F} \cdot \mathbf{v} = 425 \text{ ft-lb}$

63. $\mathbf{F} = 1600(\cos 25° \, \mathbf{i} + \sin 25° \, \mathbf{j})$

$\mathbf{v} = 2000\mathbf{i}$

$W = \mathbf{F} \cdot \mathbf{v} = 1600(2000)\cos 25°$

$\approx 2{,}900{,}184.9 \text{ Newton meters (Joules)}$

$\approx 2900.2 \quad \text{km-N}$

65. False.

For example, let $\mathbf{u} = \langle 1, 1 \rangle$, $\mathbf{v} = \langle 2, 3 \rangle$ and

$\mathbf{w} = \langle 1, 4 \rangle$. Then $\mathbf{u} \cdot \mathbf{v} = 2 + 3 = 5$ and

$\mathbf{u} \cdot \mathbf{w} = 1 + 4 = 5$.

67. (a) The graphs $y_1 = x^2$ and $y_2 = x^{1/3}$ intersect at $(0, 0)$ and $(1, 1)$.

(b) $y_1' = 2x$ and $y_2' = \dfrac{1}{3x^{2/3}}$.

At $(0, 0)$, $\pm\langle 1, 0 \rangle$ is tangent to y_1 and $\pm\langle 0, 1 \rangle$ is tangent to y_2.

At $(1, 1)$, $y_1' = 2$ and $y_2' = \dfrac{1}{3}$.

$\pm\dfrac{1}{\sqrt{5}}\langle 1, 2 \rangle$ is tangent to y_1, $\pm\dfrac{1}{\sqrt{10}}\langle 3, 1 \rangle$ is tangent to y_2.

(c) At $(0, 0)$, the vectors are perpendicular $(90°)$.

At $(1, 1)$,

$$\cos \theta = \frac{\dfrac{1}{\sqrt{5}}\langle 1, 2 \rangle \cdot \dfrac{1}{\sqrt{10}}\langle 3, 1 \rangle}{(1)(1)} = \frac{5}{\sqrt{50}} = \frac{1}{\sqrt{2}}$$

$\theta = 45°$

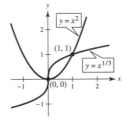

69. (a) The graphs of $y_1 = 1 - x^2$ and $y^2 = x^2 - 1$ intersect at $(1, 0)$ and $(-1, 0)$.

(b) $y_1' = -2x$ and $y_2' = 2x$.

At $(1, 0)$, $y_1' = -2$ and $y_2' = 2$. $\pm\dfrac{1}{\sqrt{5}}\langle 1, -2 \rangle$ is tangent to y_1, $\pm\dfrac{1}{\sqrt{5}}\langle 1, 2 \rangle$ is tangent to y_2.

At $(-1, 0)$, $y_1' = 2$ and $y_2' = -2$. $\pm\dfrac{1}{\sqrt{5}}\langle 1, 2 \rangle$ is tangent to y_1, $\pm\dfrac{1}{\sqrt{5}}\langle 1, -2 \rangle$ is tangent to y_2.

(c) At $(1, 0)$, $\cos \theta = \dfrac{1}{\sqrt{5}}\langle 1, -2 \rangle \cdot \dfrac{-1}{\sqrt{5}}\langle 1, -2 \rangle = \dfrac{3}{5}$.

$\theta \approx 0.9273$ or $53.13°$

By symmetry, the angle is the same at $(-1, 0)$.

71. In a rhombus, $\|\mathbf{u}\| = \|\mathbf{v}\|$. The diagonals are $\mathbf{u} + \mathbf{v}$ and $\mathbf{u} - \mathbf{v}$.

$$
\begin{aligned}
(\mathbf{u} + \mathbf{v}) \cdot (\mathbf{u} - \mathbf{v}) &= (\mathbf{u} + \mathbf{v}) \cdot \mathbf{u} - (\mathbf{u} + \mathbf{v}) \cdot \mathbf{v} \\
&= \mathbf{u} \cdot \mathbf{u} + \mathbf{v} \cdot \mathbf{u} - \mathbf{u} \cdot \mathbf{v} - \mathbf{v} \cdot \mathbf{v} \\
&= \|\mathbf{u}\|^2 - \|\mathbf{v}\|^2 = 0
\end{aligned}
$$

So, the diagonals are orthogonal.

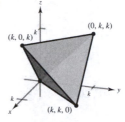

73. (a)

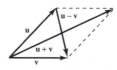

(b) Length of each edge: $\sqrt{k^2 + k^2 + 0^2} = k\sqrt{2}$

(c) $\cos\theta = \dfrac{k^2}{\left(k\sqrt{2}\right)\left(k\sqrt{2}\right)} = \dfrac{1}{2}$

$\theta = \arccos\left(\dfrac{1}{2}\right) = 60°$

(d) $\overline{r_1} = \langle k, k, 0 \rangle - \left\langle \dfrac{k}{2}, \dfrac{k}{2}, \dfrac{k}{2} \right\rangle = \left\langle \dfrac{k}{2}, \dfrac{k}{2}, -\dfrac{k}{2} \right\rangle$

$\overline{r_2} = \langle 0, 0, 0 \rangle - \left\langle \dfrac{k}{2}, \dfrac{k}{2}, \dfrac{k}{2} \right\rangle = \left\langle -\dfrac{k}{2}, -\dfrac{k}{2}, -\dfrac{k}{2} \right\rangle$

$\cos\theta = \dfrac{-\dfrac{k^2}{4}}{\left(\dfrac{k}{2}\right)^2 \cdot 3} = -\dfrac{1}{3}$

$\theta = 109.5°$

75. $\|\mathbf{u} - \mathbf{v}\|^2 = (\mathbf{u} - \mathbf{v}) \cdot (\mathbf{u} - \mathbf{v})$

$$
\begin{aligned}
&= (\mathbf{u} - \mathbf{v}) \cdot \mathbf{u} - (\mathbf{u} - \mathbf{v}) \cdot \mathbf{v} \\
&= \mathbf{u} \cdot \mathbf{u} - \mathbf{v} \cdot \mathbf{u} - \mathbf{u} \cdot \mathbf{v} + \mathbf{v} \cdot \mathbf{v} \\
&= \|\mathbf{u}\|^2 - \mathbf{u} \cdot \mathbf{v} - \mathbf{u} \cdot \mathbf{v} + \|\mathbf{v}\|^2 \\
&= \|\mathbf{u}\|^2 + \|\mathbf{v}\|^2 - 2\mathbf{u} \cdot \mathbf{v}
\end{aligned}
$$

77. $\|\mathbf{u} + \mathbf{v}\|^2 = (\mathbf{u} + \mathbf{v}) \cdot (\mathbf{u} + \mathbf{v})$

$$
\begin{aligned}
&= (\mathbf{u} + \mathbf{v}) \cdot \mathbf{u} + (\mathbf{u} + \mathbf{v}) \cdot \mathbf{v} \\
&= \mathbf{u} \cdot \mathbf{u} + \mathbf{v} \cdot \mathbf{u} + \mathbf{u} \cdot \mathbf{v} + \mathbf{v} \cdot \mathbf{v} \\
&= \|\mathbf{u}\|^2 + 2\mathbf{u} \cdot \mathbf{v} + \|\mathbf{v}\|^2 \\
&\leq \|\mathbf{u}\|^2 + 2\|\mathbf{u}\|\|\mathbf{v}\| + \|\mathbf{v}\|^2 \leq \left(\|\mathbf{u}\| + \|\mathbf{v}\|\right)^2
\end{aligned}
$$

So, $\|\mathbf{u} + \mathbf{v}\| \leq \|\mathbf{u}\| + \|\mathbf{v}\|$.

Section 11.4   The Cross Product of Two Vectors in Space

1. $\mathbf{j} \times \mathbf{i} = \begin{vmatrix} \mathbf{i} & \mathbf{j} & \mathbf{k} \\ 0 & 1 & 0 \\ 1 & 0 & 0 \end{vmatrix} = -\mathbf{k}$

3. $\mathbf{j} \times \mathbf{k} = \begin{vmatrix} \mathbf{i} & \mathbf{j} & \mathbf{k} \\ 0 & 1 & 0 \\ 0 & 0 & 1 \end{vmatrix} = \mathbf{i}$

5. $\mathbf{i} \times \mathbf{k} = \begin{vmatrix} \mathbf{i} & \mathbf{j} & \mathbf{k} \\ 1 & 0 & 0 \\ 0 & 0 & 1 \end{vmatrix} = -\mathbf{j}$

7. (a) $\mathbf{u} \times \mathbf{v} = \begin{vmatrix} \mathbf{i} & \mathbf{j} & \mathbf{k} \\ -2 & 4 & 0 \\ 3 & 2 & 5 \end{vmatrix} = 20\mathbf{i} + 10\mathbf{j} - 16\mathbf{k}$

  (b) $\mathbf{v} \times \mathbf{u} = -(\mathbf{u} \times \mathbf{v}) = -20\mathbf{i} - 10\mathbf{j} + 16\mathbf{k}$

  (c) $\mathbf{v} \times \mathbf{v} = \mathbf{0}$

9. (a) $\mathbf{u} \times \mathbf{v} = \begin{vmatrix} \mathbf{i} & \mathbf{j} & \mathbf{k} \\ 7 & 3 & 2 \\ 1 & -1 & 5 \end{vmatrix} = 17\mathbf{i} - 33\mathbf{j} - 10\mathbf{k}$

  (b) $\mathbf{v} \times \mathbf{u} = -(\mathbf{u} \times \mathbf{v}) = -17\mathbf{i} + 33\mathbf{j} + 10\mathbf{k}$

  (c) $\mathbf{v} \times \mathbf{v} = \mathbf{0}$

11. $\mathbf{u} = \langle 12, -3, 0 \rangle, \ \mathbf{v} = \langle -2, 5, 0 \rangle$

$\mathbf{u} \times \mathbf{v} = \begin{vmatrix} \mathbf{i} & \mathbf{j} & \mathbf{k} \\ 12 & -3 & 0 \\ -2 & 5 & 0 \end{vmatrix} = 54\mathbf{k} = \langle 0, 0, 54 \rangle$

$\mathbf{u} \cdot (\mathbf{u} \times \mathbf{v}) = 12(0) + (-3)(0) + 0(54)$

$\qquad\qquad = 0 \Rightarrow \mathbf{u} \perp \mathbf{u} \times \mathbf{v}$

$\mathbf{v} \cdot (\mathbf{u} \times \mathbf{v}) = -2(0) + 5(0) + 0(54)$

$\qquad\qquad = 0 \Rightarrow \mathbf{v} \perp \mathbf{u} \times \mathbf{v}$

13. $\mathbf{u} = \langle 2, -3, 1 \rangle, \ \mathbf{v} = \langle 1, -2, 1 \rangle$

$\mathbf{u} \times \mathbf{v} = \begin{vmatrix} \mathbf{i} & \mathbf{j} & \mathbf{k} \\ 2 & -3 & 1 \\ 1 & -2 & 1 \end{vmatrix} = -\mathbf{i} - \mathbf{j} - \mathbf{k} = \langle -1, -1, -1 \rangle$

$\mathbf{u} \cdot (\mathbf{u} \times \mathbf{v}) = 2(-1) + (-3)(-1) + (1)(-1)$

$\qquad\qquad = 0 \Rightarrow \mathbf{u} \perp \mathbf{u} \times \mathbf{v}$

$\mathbf{v} \cdot (\mathbf{u} \times \mathbf{v}) = 1(-1) + (-2)(-1) + (1)(-1)$

$\qquad\qquad = 0 \Rightarrow \mathbf{v} \perp \mathbf{u} \times \mathbf{v}$

15. $\mathbf{u} = \mathbf{i} + \mathbf{j} + \mathbf{k}, \ \mathbf{v} = 2\mathbf{i} + \mathbf{j} - \mathbf{k}$

$\mathbf{u} \times \mathbf{v} = \begin{vmatrix} \mathbf{i} & \mathbf{j} & \mathbf{k} \\ 1 & 1 & 1 \\ 2 & 1 & -1 \end{vmatrix} = -2\mathbf{i} + 3\mathbf{j} - \mathbf{k} = \langle -2, 3, -1 \rangle$

$\mathbf{u} \cdot (\mathbf{u} \times \mathbf{v}) = 1(-2) + 1(3) + 1(-1)$

$\qquad\qquad = 0 \Rightarrow \mathbf{u} \perp \mathbf{u} \times \mathbf{v}$

$\mathbf{v} \cdot (\mathbf{u} \times \mathbf{v}) = 2(-2) + 1(3) + (-1)(-1)$

$\qquad\qquad = 0 \Rightarrow \mathbf{v} \perp \mathbf{u} \times \mathbf{v}$

$(-\mathbf{v}) \times \mathbf{u} = -(\mathbf{v} \times \mathbf{u}) = \mathbf{u} \times \mathbf{v}$

17. $\qquad \mathbf{u} = \langle 4, -3, 1 \rangle$

$\qquad \mathbf{v} = \langle 2, 5, 3 \rangle$

$\mathbf{u} \times \mathbf{v} = \begin{vmatrix} \mathbf{i} & \mathbf{j} & \mathbf{k} \\ 4 & -3 & 1 \\ 2 & 5 & 3 \end{vmatrix} = -14\mathbf{i} - 10\mathbf{j} + 26\mathbf{k}$

$\dfrac{\mathbf{u} \times \mathbf{v}}{\|\mathbf{u} \times \mathbf{v}\|} = \dfrac{1}{\sqrt{972}} \langle -14, -10, 26 \rangle$

$\qquad\qquad = \dfrac{1}{18\sqrt{3}} \langle -14, -10, 26 \rangle$

$\qquad\qquad = \left\langle -\dfrac{7}{9\sqrt{3}}, -\dfrac{5}{9\sqrt{3}}, \dfrac{13}{9\sqrt{3}} \right\rangle$

19. $\qquad \mathbf{u} = -3\mathbf{i} + 2\mathbf{j} - 5\mathbf{k}$

$\qquad \mathbf{v} = \mathbf{i} - \mathbf{j} + 4\mathbf{k}$

$\mathbf{u} \times \mathbf{v} = \begin{vmatrix} \mathbf{i} & \mathbf{j} & \mathbf{k} \\ -3 & 2 & -5 \\ 1 & -1 & -4 \end{vmatrix} = 3\mathbf{i} + 7\mathbf{j} + \mathbf{k}$

$\dfrac{\mathbf{u} \times \mathbf{v}}{\|\mathbf{u} \times \mathbf{v}\|} = \dfrac{1}{\sqrt{59}} \langle 3, 7, 1 \rangle$

$\qquad\qquad = \left\langle \dfrac{3}{\sqrt{59}}, \dfrac{7}{\sqrt{59}}, \dfrac{1}{\sqrt{59}} \right\rangle$

21. $\qquad \mathbf{u} = \mathbf{j}$

$\qquad \mathbf{v} = \mathbf{j} + \mathbf{k}$

$\mathbf{u} \times \mathbf{v} = \begin{vmatrix} \mathbf{i} & \mathbf{j} & \mathbf{k} \\ 0 & 1 & 0 \\ 0 & 1 & 1 \end{vmatrix} = \mathbf{i}$

$A = \|\mathbf{u} \times \mathbf{v}\| = \|\mathbf{i}\| = 1$

23. $\qquad \mathbf{u} = \langle 3, 2, -1 \rangle$

$\qquad \mathbf{v} = \langle 1, 2, 3 \rangle$

$\mathbf{u} \times \mathbf{v} = \begin{vmatrix} \mathbf{i} & \mathbf{j} & \mathbf{k} \\ 3 & 2 & -1 \\ 1 & 2 & 3 \end{vmatrix} = \langle 8, -10, 4 \rangle$

$A = \|\mathbf{u} \times \mathbf{v}\| = \|\langle 8, -10, 4 \rangle\| = \sqrt{180} = 6\sqrt{5}$

25. $A(0, 3, 2), B(1, 5, 5), C(6, 9, 5), D(5, 7, 2)$

$\overline{AB} = \langle 1, 2, 3 \rangle$

$\overline{DC} = \langle 1, 2, 3 \rangle$

$\overline{BC} = \langle 5, 4, 0 \rangle$

$\overline{AD} = \langle 5, 4, 0 \rangle$

Because $\overline{AB} = \overline{DC}$ and $\overline{BC} = \overline{AD}$, the figure $ABCD$ is a parallelogram.

$\overline{AB}$ and $\overline{AD}$ are adjacent sides

$\overline{AB} \times \overline{AD} = \begin{vmatrix} \mathbf{i} & \mathbf{j} & \mathbf{k} \\ 1 & 2 & 3 \\ 5 & 4 & 0 \end{vmatrix} = \langle -12, 15, -6 \rangle$

$A = \left\| \overline{AB} \times \overline{AD} \right\| = \sqrt{144 + 225 + 36} = 9\sqrt{5}$

27. $A(0, 0, 0), B(1, 0, 3), C(-3, 2, 0)$

$\overline{AB} = \langle 1, 0, 3 \rangle, \overline{AC} = \langle -3, 2, 0 \rangle$

$\overline{AB} \times \overline{AC} = \begin{vmatrix} \mathbf{i} & \mathbf{j} & \mathbf{k} \\ 1 & 0 & 3 \\ -3 & 2 & 0 \end{vmatrix} = \langle -6, -9, 2 \rangle$

$A = \frac{1}{2} \left\| \overline{AB} \times \overline{AC} \right\| = \frac{1}{2}\sqrt{36 + 81 + 4} = \frac{11}{2}$

29. $\mathbf{F} = -20\mathbf{k}$

$\overline{PQ} = \frac{1}{2}(\cos 40°\mathbf{j} + \sin 40°\mathbf{k})$

$\overline{PQ} \times \mathbf{F} = \begin{vmatrix} \mathbf{i} & \mathbf{j} & \mathbf{k} \\ 0 & \cos 40°/2 & \sin 40°/2 \\ 0 & 0 & -20 \end{vmatrix} = -10 \cos 40°\mathbf{i}$

$\left\| \overline{PQ} \times \mathbf{F} \right\| = 10 \cos 40° \approx 7.66$ ft-lb

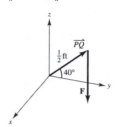

31. (a) $AC = 15$ inches $= \frac{5}{4}$ feet

$BC = 12$ inches $= 1$ foot

$\overline{AB} = -\frac{5}{4}\mathbf{j} + \mathbf{k}$

$\mathbf{F} = -180(\cos \theta\, \mathbf{j} + \sin \theta\, \mathbf{k})$

(b) $\overline{AB} \times \mathbf{F} = \begin{vmatrix} \mathbf{i} & \mathbf{j} & \mathbf{k} \\ 0 & -\frac{5}{4} & 1 \\ 0 & -180 \cos \theta & -180 \sin \theta \end{vmatrix}$

$= (225 \sin \theta + 180 \cos \theta)\mathbf{i}$

$\left\| \overline{AB} \times \mathbf{F} \right\| = \left| 225 \sin \theta + 180 \cos \theta \right|$

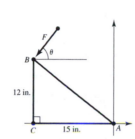

(c)   When $\theta = 30°, \left\| \overline{AB} \times \mathbf{F} \right\| = 225\left(\frac{1}{2}\right) + 180\left(\frac{\sqrt{3}}{2}\right) \approx 268.38$

(d)   If $T = \left| 225 \sin \theta + 180 \cos \theta \right|$, $T = 0$ for $225 \sin \theta = -180 \cos \theta \Rightarrow \tan \theta = -\frac{4}{5} \Rightarrow \theta \approx 141.34°$.

For $0 < \theta < 141.34$, $T'(\theta) = 225 \cos \theta - 180 \sin \theta = 0 \Rightarrow \tan \theta = \frac{5}{4} \Rightarrow \theta \approx 51.34°$. $\overline{AB}$ and $\mathbf{F}$ are perpendicular.

(e)

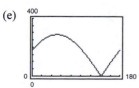

From part (d), the zero is $\theta \approx 141.34°$, when the vectors are parallel.

33. $\mathbf{u} \cdot (\mathbf{v} \times \mathbf{w}) = \begin{vmatrix} 1 & 0 & 0 \\ 0 & 1 & 0 \\ 0 & 0 & 1 \end{vmatrix} = 1$

35. $\mathbf{u} \cdot (\mathbf{v} \times \mathbf{w}) = \begin{vmatrix} 2 & 0 & 1 \\ 0 & 3 & 0 \\ 0 & 0 & 1 \end{vmatrix} = 6$

37. $\mathbf{u} \cdot (\mathbf{v} \times \mathbf{w}) = \begin{vmatrix} 1 & 1 & 0 \\ 0 & 1 & 1 \\ 1 & 0 & 1 \end{vmatrix} = 2$

$V = |\mathbf{u} \cdot (\mathbf{v} \times \mathbf{w})| = 2$

39. $\mathbf{u} = \langle 3, 0, 0 \rangle$

$\mathbf{v} = \langle 0, 5, 1 \rangle$

$\mathbf{w} = \langle 2, 0, 5 \rangle$

$\mathbf{u} \cdot (\mathbf{v} \times \mathbf{w}) = \begin{vmatrix} 3 & 0 & 0 \\ 0 & 5 & 1 \\ 2 & 0 & 5 \end{vmatrix} = 75$

$V = |\mathbf{u} \cdot (\mathbf{v} \times \mathbf{w})| = 75$

41. (a) $\mathbf{u} \cdot (\mathbf{v} \times \mathbf{w}) = (\mathbf{v} \times \mathbf{w}) \cdot \mathbf{u}$ (b)

$= \mathbf{w} \cdot (\mathbf{u} \times \mathbf{v}) = (\mathbf{u} \times \mathbf{v}) \cdot \mathbf{w}$ (c)

$= \mathbf{v} \cdot (\mathbf{w} \times \mathbf{u}) = (\mathbf{u}x - \mathbf{w}) \cdot \mathbf{v}$ (d)

$= \mathbf{v} \cdot (\mathbf{w} \times \mathbf{u}) = (\mathbf{w} \times \mathbf{u}) \cdot \mathbf{v}$ (h)

(e) $\mathbf{u} \cdot (\mathbf{w} \times \mathbf{v}) = \mathbf{w} \cdot (\mathbf{v} \times \mathbf{u})$ (f)

$= \mathbf{w} \cdot (\mathbf{v} \times \mathbf{u}) = (-\mathbf{u} \times \mathbf{v}) \cdot \mathbf{w}$ (g)

So, $a = b = c = d = h$ and $e = f = g$

43. $\mathbf{u} \times \mathbf{v} = \langle u_1, u_2, u_3 \rangle \cdot \langle v_1, v_2, v_3 \rangle$

$= (u_2 v_3 - u_3 v_2)\mathbf{i} - (u_1 v_3 - u_3 v_1)\mathbf{j} + (u_1 v_2 - u_2 v_1)\mathbf{k}$

45. The magnitude of the cross product will increase by a factor of 4.

47. False. If the vectors are ordered pairs, then the cross product does not exist.

49. False. Let $\mathbf{u} = \langle 1, 0, 0 \rangle$, $\mathbf{v} = \langle 1, 0, 0 \rangle$, $\mathbf{w} = \langle -1, 0, 0 \rangle$.

Then, $\mathbf{u} \times \mathbf{v} = \mathbf{u} \times \mathbf{w} = \mathbf{0}$, but $\mathbf{v} \neq \mathbf{w}$.

51. $\mathbf{u} = \langle u_1, u_2, u_3 \rangle$, $\mathbf{v} = \langle v_1, v_2, v_3 \rangle$, $\mathbf{w} = \langle w_1, w_2, w_3 \rangle$

$\mathbf{u} \times (\mathbf{v} + \mathbf{w}) = \begin{vmatrix} \mathbf{i} & \mathbf{j} & \mathbf{k} \\ u_1 & u_2 & u_3 \\ v_1 + w_1 & v_2 + w_2 & v_3 + w_3 \end{vmatrix}$

$= [u_2(v_3 + w_3) - u_3(v_2 + w_2)]\mathbf{i} - [u_1(v_3 + w_3) - u_3(v_1 + w_1)]\mathbf{j} + [u_1(v_2 + w_2) - u_2(v_1 + w_1)]\mathbf{k}$

$= (u_2 v_3 - u_3 v_2)\mathbf{i} - (u_1 v_3 - u_3 v_1)\mathbf{j} + (u_1 v_2 - u_2 v_1)\mathbf{k} + (u_2 w_3 - u_3 w_2)\mathbf{i} - (u_1 w_3 - u_3 w_1)\mathbf{j} + (u_1 w_2 - u_2 w_1)\mathbf{k}$

$= (\mathbf{u} \times \mathbf{v}) + (\mathbf{u} \times \mathbf{w})$

53. $\mathbf{u} = \langle u_1, u_2, u_3 \rangle$

$\mathbf{u} \times \mathbf{u} = \begin{vmatrix} \mathbf{i} & \mathbf{j} & \mathbf{k} \\ u_1 & u_2 & u_3 \\ u_1 & u_2 & u_3 \end{vmatrix} = (u_2 u_3 - u_3 u_2)\mathbf{i} - (u_1 u_3 - u_3 u_1)\mathbf{j} + (u_1 u_2 - u_2 u_1)\mathbf{k} = \mathbf{0}$

55. $\mathbf{u} \times \mathbf{v} = (u_2 v_3 - u_3 v_2)\mathbf{i} - (u_1 v_3 - u_3 v_1)\mathbf{j} + (u_1 v_2 - u_2 v_1)\mathbf{k}$

$(\mathbf{u} \times \mathbf{v}) \cdot \mathbf{u} = (u_2 v_3 - u_3 v_2)u_1 + (u_3 v_1 - u_1 v_3)u_2 + (u_1 v_2 - u_2 v_1)u_3 = 0$

$(\mathbf{u} \times \mathbf{v}) \cdot \mathbf{v} = (u_2 v_3 - u_3 v_2)v_1 + (u_3 v_1 - u_1 v_3)v_2 + (u_1 v_2 - u_2 v_1)v_3 = 0$

So, $\mathbf{u} \times \mathbf{v} \perp \mathbf{u}$ and $\mathbf{u} \times \mathbf{v} \perp \mathbf{v}$.

57. $\|\mathbf{u} \times \mathbf{v}\| = \|\mathbf{u}\|\|\mathbf{v}\| \sin \theta$

If $\mathbf{u}$ and $\mathbf{v}$ are orthogonal, $\theta = \pi/2$ and $\sin \theta = 1$. So, $\|\mathbf{u} \times \mathbf{v}\| = \|\mathbf{u}\|\|\mathbf{v}\|$.

59. $\mathbf{u} = u_1\mathbf{i} + u_2\mathbf{j} + u_3\mathbf{k}$, $\mathbf{v} = v_1\mathbf{i} + v_2\mathbf{j} + v_3\mathbf{k}$, $\mathbf{w} = w_1\mathbf{i} + w_2\mathbf{j} + w_3\mathbf{k}$

$$\mathbf{v} \times \mathbf{w} = \begin{vmatrix} \mathbf{i} & \mathbf{j} & \mathbf{k} \\ v_1 & v_2 & v_3 \\ w_1 & w_2 & w_3 \end{vmatrix} = \left\langle v_2w_3 - w_2v_3, -(v_1w_3 - w_1v_3), v_1w_2 - w_1v_2 \right\rangle$$

$$\mathbf{u} \times (\mathbf{v} \times \mathbf{w}) = \left\langle u_1, u_2, u_3 \right\rangle \cdot \left\langle v_2w_3 - w_2v_3, -(v_1w_3 - w_1v_3), v_1w_2 - w_1v_2 \right\rangle$$

$$= u_1v_2w_3 - u_1v_3w_2 - u_2v_1w_3 + u_2v_3w_1 + u_3v_1w_2 - u_3v_2w_1$$

$$= \begin{vmatrix} u_1 & u_2 & u_3 \\ v_1 & v_2 & v_3 \\ w_1 & w_2 & w_3 \end{vmatrix}$$

Section 11.5   Lines and Planes in Space

1. $x = -2 + t, y = 3t, z = 4 + t$

(a) $(0, 6, 6)$: For $x = 0 = -2 + t$, you have

$t = 2$. Then $y = 3(2) = 6$ and

$z = 4 + 2 = 6$. Yes, $(0, 6, 6)$ lies on the line.

(b) $(2, 3, 5)$: For $x = 2 = -2 + t$, you have

$t = 4$. Then $y = 3(4) = 12 \neq 3$. No, $(2, 3, 5)$ does

not lie on the line.

3. Point: $(0, 0, 0)$

Direction vector: $\langle 3, 1, 5 \rangle$

Direction numbers: 3, 1, 5

(a) Parametric: $x = 3t, y = t, z = 5t$

(b) Symmetric: $\dfrac{x}{3} = y = \dfrac{z}{5}$

5. Point: $(-2, 0, 3)$

Direction vector: $\mathbf{v} = \langle 2, 4, -2 \rangle$

Direction numbers: 2, 4, -2

(a) Parametric: $x = -2 + 2t, y = 4t, z = 3 - 2t$

(b) Symmetric: $\dfrac{x + 2}{2} = \dfrac{y}{4} = \dfrac{z - 3}{-2}$

7. Point: $(1, 0, 1)$

Direction vector: $\mathbf{v} = 3\mathbf{i} - 2\mathbf{j} + \mathbf{k}$

Direction numbers: 3, -2, 1

(a) Parametric: $x = 1 + 3t, y = -2t, z = 1 + t$

(b) Symmetric: $\dfrac{x - 1}{3} = \dfrac{y}{-2} = \dfrac{z - 1}{1}$

9. Points: $\left(5, -3, -2\right), \left(-\dfrac{2}{3}, \dfrac{2}{3}, 1\right)$

Direction vector: $\mathbf{v} = \dfrac{17}{3}\mathbf{i} - \dfrac{11}{3}\mathbf{j} - 3\mathbf{k}$

Direction numbers: 17, -11, -9

(a) Parametric:

$x = 5 + 17t, y = -3 - 11t, z = -2 - 9t$

(b) Symmetric: $\dfrac{x - 5}{17} = \dfrac{y + 3}{-11} = \dfrac{z + 2}{-9}$

11. Points: $(7, -2, 6), (-3, 0, 6)$

Direction vector: $\langle -10, 2, 0 \rangle$

Direction numbers: -10, 2, 0

(a) Parametric: $x = 7 - 10t, y = -2 + 2t, z = 6$

(b) Symmetric: Not possible because the direction
number for z is 0. But, you could describe the

line as $\dfrac{x - 7}{10} = \dfrac{y + 2}{-2}, z = 6$.

13. Point: $(2, 3, 4)$

Direction vector: $\mathbf{v} = \mathbf{k}$

Direction numbers: 0, 0, 1

Parametric: $x = 2, y = 3, z = 4 + t$

15. Point: $(2, 3, 4)$

Direction vector: $\mathbf{v} = 3\mathbf{i} + 2\mathbf{j} - \mathbf{k}$

Direction numbers: 3, 2, -1

Parametric: $x = 2 + 3t, y = 3 + 2t, z = 4 - t$

17. Point: $(5, -3, -4)$

Direction vector: $\mathbf{v} = \langle 2, -1, 3 \rangle$

Direction numbers: 2, -1, 3

Parametric: $x = 5 + 2t, y = -3 - t, z = -4 + 3t$

19. Point: $(2, 1, 2)$

Direction vector: $\langle -1, 1, 1 \rangle$

Direction numbers: $-1, 1, 1$

Parametric: $x = 2 - t, y = 1 + t, z = 2 + t$

21. Let $t = 0$: $P = (3, -1, -2)$ (other answers possible)

$\mathbf{v} = \langle -1, 2, 0 \rangle$ (any nonzero multiple of $\mathbf{v}$ is correct)

23. Let each quantity equal 0:

$P = (7, -6, -2)$ (other answers possible)

$\mathbf{v} = \langle 4, 2, 1 \rangle$ (any nonzero multiple of $\mathbf{v}$ is correct)

25. L_1: $\mathbf{v} = \langle -3, 2, 4 \rangle$ $(6, -2, 5)$ on line

L_2: $\mathbf{v} = \langle 6, -4, -8 \rangle$ $(6, -2, 5)$ on line

L_3: $\mathbf{v} = \langle -6, 4, 8 \rangle$ $(6, -2, 5)$ not online

L_4: $\mathbf{v} = \langle 6, 4, -6 \rangle$ not parallel to L_1, L_2, nor L_3

L_1 and L_2 are identical. $L_1 = L_2$ and is parallel to L_3.

27. L_1: $\mathbf{v} = \langle 4, -2, 3 \rangle$ $(8, -5, -9)$ on line

L_2: $\mathbf{v} = \langle 2, 1, 5 \rangle$

L_3: $\mathbf{v} = \langle -8, 4, -6 \rangle$ $(8, -5, -9)$ on line

L_4: $\mathbf{v} = \langle -2, 1, 1.5 \rangle$

L_1 and L_3 are identical.

29. At the point of intersection, the coordinates for one line equal the corresponding coordinates for the other line. So,

(i) $4t + 2 = 2s + 2$, (ii) $3 = 2s + 3$, and

(iii) $-t + 1 = s + 1$.

From (ii), you find that $s = 0$ and consequently, from (iii), $t = 0$. Letting $s = t = 0$, you see that equation (i) is satisfied and so the two lines intersect. Substituting zero for s or for t, you obtain the point $(2, 3, 1)$.

$\mathbf{u} = 4\mathbf{i} - \mathbf{k}$ (First line)

$\mathbf{v} = 2\mathbf{i} + 2\mathbf{j} + \mathbf{k}$ (Second line)

$$\cos \theta = \frac{|\mathbf{u} \cdot \mathbf{v}|}{\|\mathbf{u}\| \|\mathbf{v}\|} = \frac{8 - 1}{\sqrt{17}\sqrt{9}} = \frac{7}{3\sqrt{17}} = \frac{7\sqrt{17}}{51}$$

31. Writing the equations of the lines in parametric form you have

$x = 3t$ $y = 2 - t$ $z = -1 + t$

$x = 1 + 4s$ $y = -2 + s$ $z = -3 - 3s$.

For the coordinates to be equal, $3t = 1 + 4s$ and $2 - t = -2 + s$. Solving this system yields $t = \frac{17}{7}$ and $s = \frac{11}{7}$. When using these values for s and t, the z coordinates are not equal. The lines do not intersect.

33. $x + 2y - 4z - 1 = 0$

(a) $(-7, 2, -1)$: $(-7) + 2(2) - 4(-1) - 1 = 0$

Point is in plane

(b) $(5, 2, 2)$: $5 + 2(2) - 4(2) - 1 = 0$

Point is in plane

35. Point: $(1, 3, -7)$

Normal vector: $\mathbf{n} = \mathbf{j} = \langle 0, 1, 0 \rangle$

$$0(x - 1) + 1(y - 3) + 0(z - (-7)) = 0$$
$$y - 3 = 0$$

37. Point: $(3, 2, 2)$

Normal vector: $\mathbf{n} = 2\mathbf{i} + 3\mathbf{j} - \mathbf{k}$

$$2(x - 3) + 3(y - 2) - 1(z - 2) = 0$$
$$2x + 3y - z = 10$$

39. Point: $(-1, 4, 0)$

Normal vector: $\mathbf{v} = \langle 2, -1, -2 \rangle$

$$2(x + 1) - 1(y - 4) - 2(z - 0) = 0$$
$$2x - y - 2z + 6 = 0$$

41. Let $\mathbf{u}$ be the vector from $(0, 0, 0)$ to

$(2, 0, 3)$: $\mathbf{u} = \langle 2, 0, 3 \rangle$

Let $\mathbf{u}$ be the vector from $(0, 0, 0)$ to

$(-3, -1, 5)$: $\mathbf{v} = \langle -3, -1, 5 \rangle$

Normal vectors: $\mathbf{u} \times \mathbf{v} = \begin{vmatrix} \mathbf{i} & \mathbf{j} & \mathbf{k} \\ 2 & 0 & 3 \\ -3 & -1 & 5 \end{vmatrix} = \langle 3, -19, -2 \rangle$

$$3(x - 0) - 19(y - 0) - 2(z - 0) = 0$$
$$3x - 19y - 2z = 0$$

43. Let $\mathbf{u}$ be the vector from $(1, 2, 3)$ to

$(3, 2, 1)$: $\mathbf{u} = 2\mathbf{i} - 2\mathbf{k}$

Let $\mathbf{v}$ be the vector from $(1, 2, 3)$ to

$(-1, -2, 2)$: $\mathbf{v} = -2\mathbf{i} - 4\mathbf{j} - \mathbf{k}$

Normal vector:

$\left(\tfrac{1}{2}\mathbf{u}\right) \times (-\mathbf{v}) = \begin{vmatrix} \mathbf{i} & \mathbf{j} & \mathbf{k} \\ 1 & 0 & -1 \\ 2 & 4 & 1 \end{vmatrix} = 4\mathbf{i} - 3\mathbf{j} + 4\mathbf{k}$

$$4(x - 1) - 3(y - 2) + 4(z - 3) = 0$$
$$4x - 3y + 4z = 10$$

45. $(1, 2, 3)$, Normal vector: $\mathbf{v} = \mathbf{k}$, $1(z - 3) = 0$, $z = 3$

47. The direction vectors for the lines are
$$\mathbf{u} = -2\mathbf{i} + \mathbf{j} + \mathbf{k}, \quad \mathbf{v} = -3\mathbf{i} + 4\mathbf{j} - \mathbf{k}.$$

Normal vector: $\mathbf{u} \times \mathbf{v} = \begin{vmatrix} \mathbf{i} & \mathbf{j} & \mathbf{k} \\ -2 & 1 & 1 \\ -3 & 4 & -1 \end{vmatrix} = -5(\mathbf{i} + \mathbf{j} + \mathbf{k})$

Point of intersection of the lines: $(-1, 5, 1)$

$$(x + 1) + (y - 5) + (z - 1) = 0$$
$$x + y + z = 5$$

49. Let $\mathbf{v}$ be the vector from $(-1, 1, -1)$ to
$(2, 2, 1)$: $\mathbf{v} = 3\mathbf{i} + \mathbf{j} + 2\mathbf{k}$

Let $\mathbf{n}$ be a vector normal to the plane
$2x - 3y + z = 3$: $\mathbf{n} = 2\mathbf{i} - 3\mathbf{j} + \mathbf{k}$

Because $\mathbf{v}$ and $\mathbf{n}$ both lie in the plane P, the normal vector to P is

$\mathbf{v} \times \mathbf{n} = \begin{vmatrix} \mathbf{i} & \mathbf{j} & \mathbf{k} \\ 3 & 1 & 2 \\ 2 & -3 & 1 \end{vmatrix} = 7\mathbf{i} - \mathbf{j} - 11\mathbf{k}$

$$7(x - 2) + 1(y - 2) - 11(z - 1) = 0$$
$$7x + y - 11z = 5$$

51. Let $\mathbf{u} = \mathbf{i}$ and let $\mathbf{v}$ be the vector from $(1, -2, -1)$ to $(2, 5, 6)$: $\mathbf{v} = \mathbf{i} + 7\mathbf{j} + 7\mathbf{k}$

Because $\mathbf{u}$ and $\mathbf{v}$ both lie in the plane P, the normal vector to P is:

$\mathbf{u} \times \mathbf{v} = \begin{vmatrix} \mathbf{i} & \mathbf{j} & \mathbf{k} \\ 1 & 0 & 0 \\ 1 & 7 & 7 \end{vmatrix} = -7\mathbf{j} + 7\mathbf{k} = -7(\mathbf{j} - \mathbf{k})$

$$\left[y - (-2) \right] - \left[z - (-1) \right] = 0$$
$$y - z = -1$$

53. Let (x, y, z) be equidistant from $(2, 2, 0)$ and $(0, 2, 2)$.

$$\sqrt{(x - 2)^2 + (y - 2)^2 + (z - 0)^2} = \sqrt{(x - 0)^2 + (y - 2)^2 + (z - 2)^2}$$
$$x^2 - 4x + 4 + y^2 - 4y + 4 + z^2 = x^2 + y^2 - 4y + 4 + z^2 - 4z + 4$$
$$-4x + 8 = -4z + 8$$
$$x - z = 0 \quad \text{Plane}$$

55. Let (x, y, z) be equidistant from $(-3, 1, 2)$ and $(6, -2, 4)$.

$$\sqrt{(x + 3)^2 + (y - 1)^2 + (z - 2)^2} = \sqrt{(x - 6)^2 + (y + 2)^2 + (z - 4)^2}$$
$$x^2 + 6x + 9 + y^2 - 2y + 1 + z^2 - 4z + 4 = x^2 - 12x + 36 + y^2 + 4y + 4 + z^2 - 8z + 16$$
$$6x - 2y - 4z + 14 = -12x + 4y - 8z + 56$$
$$18x - 6y + 4z - 42 = 0$$
$$9x - 3y + 2z - 21 = 0 \quad \text{Plane}$$

57. The normal vectors to the planes are

$\mathbf{n}_1 = \langle 5, -3, 1 \rangle$, $\mathbf{n}_2 = \langle 1, 4, 7 \rangle$, $\cos \theta = \dfrac{|\mathbf{n}_1 \cdot \mathbf{n}_2|}{\|\mathbf{n}_1\| \|\mathbf{n}_2\|} = 0.$

So, $\theta = \pi/2$ and the planes are orthogonal.

59. The normal vectors to the planes are

$\mathbf{n}_1 = \mathbf{i} - 3\mathbf{j} + 6\mathbf{k}$, $\mathbf{n}_2 = 5\mathbf{i} + \mathbf{j} - \mathbf{k}$,

$\cos \theta = \dfrac{|\mathbf{n}_1 \cdot \mathbf{n}_2|}{\|\mathbf{n}_1\| \|\mathbf{n}_2\|} = \dfrac{|5 - 3 - 6|}{\sqrt{46}\sqrt{27}} = \dfrac{4\sqrt{138}}{414} = \dfrac{2\sqrt{138}}{207}.$

So, $\theta = \arccos\left(\dfrac{2\sqrt{138}}{207} \right) \approx 83.5°.$

61. The normal vectors to the planes are $\mathbf{n}_1 = \langle 1, -5, -1 \rangle$ and
$\mathbf{n}_2 = \langle 5, -25, -5 \rangle$. Because $\mathbf{n}_2 = 5\mathbf{n}_1$, the planes are parallel, but not equal.

63. $4x + 2y + 6z = 12$

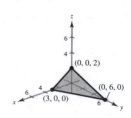

65. $2x - y + 3z = 4$

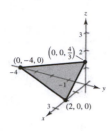

67. $x + z = 6$

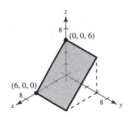

69. $x = 5$

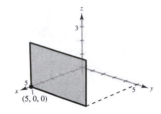

71. $P_1: \mathbf{n} = \langle -5, 2, -8 \rangle$ $(0, -1, -1)$ on plane

$P_2: \mathbf{n} = \langle 15, -6, 24 \rangle$ $(0, -1, -1)$ not on plane

$P_3: \mathbf{n} = \langle 6, -4, 4 \rangle$

$P_4: \mathbf{n} = \langle 3, -2, -2 \rangle$

Planes P_1 and P_2 are parallel.

73. $P_1: \mathbf{n} = \langle 3, -2, 5 \rangle$ $(1, -1, 1)$ on plane

$P_2: \mathbf{n} = \langle -6, 4, -10 \rangle$ $(1, -1, 1)$ not on plane

$P_3: \mathbf{n} = \langle -3, 2, 5 \rangle$

$P_4: \mathbf{n} = \langle 75, -50, 125 \rangle$ $(1, -1, 1)$ on plane

P_1 and P_4 are identical.

$P_1 = P_4$ and is parallel to P_2.

75. (a) $\mathbf{n}_1 = 3\mathbf{i} + 2\mathbf{j} - \mathbf{k}$ and $\mathbf{n}_2 = \mathbf{i} - 4\mathbf{j} + 2\mathbf{k}$

$$\cos \theta = \frac{|\mathbf{n}_1 \cdot \mathbf{n}_2|}{\|\mathbf{n}_1\| \|\mathbf{n}_2\|} = \frac{|-7|}{\sqrt{14}\sqrt{21}} = \frac{\sqrt{6}}{6}$$

$$\Rightarrow \theta \approx 1.1503 \approx 65.91°$$

(b) The direction vector for the line is

$$\mathbf{n}_2 \times \mathbf{n}_1 = \begin{vmatrix} \mathbf{i} & \mathbf{j} & \mathbf{k} \\ 1 & -4 & 2 \\ 3 & 2 & -1 \end{vmatrix} = 7(\mathbf{j} + 2\mathbf{k}).$$

Find a point of intersection of the planes.

$$6x + 4y - 2z = 14$$
$$x - 4y + 2z = 0$$
$$7x = 14$$
$$x = 2$$

Substituting 2 for x in the second equation, you have $-4y + 2z = -2$ or $z = 2y - 1$. Letting $y = 1$, a point of intersection is $(2, 1, 1)$.

$$x = 2, y = 1 + t, z = 1 + 2t$$

77. Writing the equation of the line in parametric form and substituting into the equation of the plane you have:

$$x = \frac{1}{2} + t, y = \frac{-3}{2} - t, z = -1 + 2t$$

$$2\left(\frac{1}{2} + t\right) - 2\left(\frac{-3}{2} - t\right) + (-1 + 2t) = 12, t = \frac{3}{2}$$

Substituting $t = 3/2$ into the parametric equations for the line you have the point of intersection $(2, -3, 2)$.

The line does not lie in the plane.

79. Writing the equation of the line in parametric form and substituting into the equation of the plane you have:

$$x = 1 + 3t, y = -1 - 2t, z = 3 + t$$

$$2(1 + 3t) + 3(-1 - 2t) = 10, -1 = 10, \text{ contradiction}$$

So, the line does not intersect the plane.

81. Point: $Q(0, 0, 0)$

Plane: $2x + 3y + z - 12 = 0$

Normal to plane: $\mathbf{n} = \langle 2, 3, 1 \rangle$

Point in plane: $P(6, 0, 0)$

Vector $\overrightarrow{PQ} = \langle -6, 0, 0 \rangle$

$$D = \frac{|\overrightarrow{PQ} \cdot \mathbf{n}|}{\|\mathbf{n}\|} = \frac{|-12|}{\sqrt{14}} = \frac{6\sqrt{14}}{7}$$

83. Point: $Q(2, 8, 4)$

Plane: $2x + y + z = 5$

Normal to plane: $\mathbf{n} = \langle 2, 1, 1 \rangle$

Point in plane: $P\langle 0, 0, 5 \rangle$

Vector: $\overline{PQ} = \langle 2, 8, -1 \rangle$

$$D = \frac{\left| \overline{PQ} \cdot \mathbf{n} \right|}{\| \mathbf{n} \|} = \frac{11}{\sqrt{6}} = \frac{11\sqrt{6}}{6}$$

85. The normal vectors to the planes are $\mathbf{n}_1 = \langle 1, -3, 4 \rangle$ and $\mathbf{n}_2 = \langle 1, -3, 4 \rangle$. Because $\mathbf{n}_1 = \mathbf{n}_2$, the planes are parallel. Choose a point in each plane.

$P(10, 0, 0)$ is a point in $x - 3y + 4z = 10$.

$Q(6, 0, 0)$ is a point in $x - 3y + 4z = 6$.

$\overline{PQ} = \langle -4, 0, 0 \rangle$, $D = \dfrac{\left| \overline{PQ} \cdot \mathbf{n}_1 \right|}{\| \mathbf{n}_1 \|} = \dfrac{4}{\sqrt{26}} = \dfrac{2\sqrt{26}}{13}$

87. The normal vectors to the planes are $\mathbf{n}_1 = \langle -3, 6, 7 \rangle$ and $\mathbf{n}_2 = \langle 6, -12, -14 \rangle$. Because $\mathbf{n}_2 = -2\mathbf{n}_1$, the planes are parallel. Choose a point in each plane.

$P(0, -1, 1)$ is a point in $-3x + 6y + 7z = 1$.

$Q\left(\dfrac{25}{6}, 0, 0 \right)$ is a point in $6x - 12y - 14z = 25$.

$\overline{PQ} = \left\langle \dfrac{25}{6}, 1, -1 \right\rangle$

$$D = \frac{\left| \overline{PQ} \cdot \mathbf{n}_1 \right|}{\| \mathbf{n}_1 \|} = \frac{\left| -27/2 \right|}{\sqrt{94}} = \frac{27}{2\sqrt{94}} = \frac{27\sqrt{94}}{188}$$

89. $\mathbf{u} = \langle 4, 0, -1 \rangle$ is the direction vector for the line.

$Q(1, 5, -2)$ is the given point, and $P(-2, 3, 1)$ is on the line.

$\overline{PQ} = \langle 3, 2, -3 \rangle$

$$\overline{PQ} \times \mathbf{u} = \begin{vmatrix} \mathbf{i} & \mathbf{j} & \mathbf{k} \\ 3 & 2 & -3 \\ 4 & 0 & -1 \end{vmatrix} = \langle -2, -9, -8 \rangle$$

$$D = \frac{\| \overline{PQ} \times \mathbf{u} \|}{\| \mathbf{u} \|} = \frac{\sqrt{149}}{\sqrt{17}} = \frac{\sqrt{2533}}{17}$$

91. $\mathbf{u} = \langle -1, 1, -2 \rangle$ is the direction vector for the line.

$Q(-2, 1, 3)$ is the given point, and $P(1, 2, 0)$ is on the line (let $t = 0$ in the parametric equations for the line).

$\overline{PQ} = \langle -3, -1, 3 \rangle$

$$\overline{PQ} \times \mathbf{u} = \begin{vmatrix} \mathbf{i} & \mathbf{j} & \mathbf{k} \\ -3 & -1 & 3 \\ -1 & 1 & -2 \end{vmatrix} = \langle -1, -9, -4 \rangle$$

$$D = \frac{\| \overline{PQ} \times \mathbf{u} \|}{\| \mathbf{u} \|} = \frac{\sqrt{1 + 81 + 16}}{\sqrt{1 + 1 + 4}} = \frac{\sqrt{98}}{6} = \frac{7}{\sqrt{3}} = \frac{7\sqrt{3}}{3}$$

93. The direction vector for L_1 is $\mathbf{v}_1 = \langle -1, 2, 1 \rangle$.

The direction vector for L_2 is $\mathbf{v}_2 = \langle 3, -6, -3 \rangle$.

Because $\mathbf{v}_2 = -3\mathbf{v}_1$, the lines are parallel.

Let $Q(2, 3, 4)$ to be a point on L_1 and $P(0, 1, 4)$ a point on L_2. $\overline{PQ} = \langle 2, 2, 0 \rangle$.

$\mathbf{u} = \mathbf{v}_2$ is the direction vector for L_2.

$$\overline{PQ} \times \mathbf{v}_2 = \begin{vmatrix} \mathbf{i} & \mathbf{j} & \mathbf{k} \\ 2 & 2 & 0 \\ 3 & -6 & -3 \end{vmatrix} = \langle -6, 6, -18 \rangle$$

$$D = \frac{\| \overline{PQ} \times \mathbf{v}_2 \|}{\| \mathbf{v}_2 \|}$$

$$= \frac{\sqrt{36 + 36 + 324}}{\sqrt{9 + 36 + 9}} = \sqrt{\frac{396}{54}} = \sqrt{\frac{22}{3}} = \frac{\sqrt{66}}{3}$$

95. The parametric equations of a line L parallel to $\mathbf{v} = \langle a, b, c, \rangle$ and passing through the point $P(x_1, y_1, z_1)$ are

$x = x_1 + at$, $y = y_1 + bt$, $z = z_1 + ct$.

The symmetric equations are

$\dfrac{x - x_1}{a} = \dfrac{y - y_1}{b} = \dfrac{z - z_1}{c}$.

97. Simultaneously solve the two linear equations representing the planes and substitute the values back into one of the original equations. Then choose a value for t and form the corresponding parametric equations for the line of intersection.

99. Yes. If $\mathbf{v}_1$ and $\mathbf{v}_2$ are the direction vectors for the lines L_1 and L_2, then $\mathbf{v} = \mathbf{v}_1 \times \mathbf{v}_2$ is perpendicular to both L_1 and L_2.

101. (a)

Year	2005	2006	2007	2008	2009	2010
x	36.4	39.0	42.4	44.7	43.0	45.2
y	15.3	16.6	17.4	17.5	17.0	17.3
z	16.4	18.1	20.0	20.5	20.1	21.4
Model z	16.39	17.98	19.78	20.87	19.94	21.04

The approximations are close to the actual values.

(b) According to the model, if x and y increase, then so does z.

103. L_1: $x_1 = 6 + t$; $y_1 = 8 - t$, $z_1 = 3 + t$

L_2: $x_2 = 1 + t$, $y_2 = 2 + t$, $z_2 = 2t$

(a) At $t = 0$, the first insect is at $P_1(6, 8, 3)$ and the second insect is at $P_2(1, 2, 0)$.

$$\text{Distance} = \sqrt{(6-1)^2 + (8-2)^2 + (3-0)^2} = \sqrt{70} \approx 8.37 \text{ inches}$$

(b) Distance $= \sqrt{(x_1 - x_2)^2 + (y_1 - y_2)^2 + (z_1 - z_2)^2} = \sqrt{5^2 + (6 - 2t)^2 + (3 - t)^2} = \sqrt{5t^2 - 30t + 70}$, $0 \le t \le 10$

(c) The distance is never zero.

(d) Using a graphing utility, the minimum distance is 5 inches when $t = 3$ minutes.

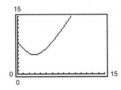

105. The direction vector $\mathbf{v}$ of the line is the normal to the plane, $\mathbf{v} = \langle 3, -1, 4 \rangle$.

The parametric equations of the line are $x = 5 + 3t$, $y = 4 - t$, $z = -3 + 4t$.

To find the point of intersection, solve for t in the following equation:

$$3(5 + 3t) - (4 - t) + 4(-3 + 4t) = 7$$
$$26t = 8$$
$$t = \frac{4}{13}$$

Point of intersection:

$$\left(5 + 3\left(\tfrac{4}{13}\right), 4 - \tfrac{4}{13}, -3 + 4\left(\tfrac{4}{13}\right)\right) = \left(\tfrac{77}{13}, \tfrac{48}{13}, -\tfrac{23}{13}\right)$$

107. The direction vector of the line L through $(1, -3, 1)$ and $(3, -4, 2)$ is $\mathbf{v} = \langle 2, -1, 1 \rangle$.

The parametric equations for L are
$x = 1 + 2t$, $y = -3 - t$, $z = 1 + t$.

Substituting these equations into the equation of the plane gives

$$(1 + 2t) - (-3 - t) + (1 + t) = 2$$
$$4t = -3$$
$$t = -\tfrac{3}{4}.$$

Point of intersection:

$$\left(1 + 2\left(-\tfrac{3}{4}\right), -3 + \tfrac{3}{4}, 1 - \tfrac{3}{4}\right) = \left(-\tfrac{1}{2}, -\tfrac{9}{4}, \tfrac{1}{4}\right)$$

109. True

111. True

113. False. Planes $7x + y - 11z = 5$ and $5x + 2y - 4z = 1$ are both perpendicular to plane $2x - 3y + z = 3$, but are not parallel.

Section 11.6   Surfaces in Space

1. Ellipsoid

Matches graph (c)

2. Hyperboloid of two sheets

Matches graph (e)

3. Hyperboloid of one sheet

Matches graph (f)

4. Elliptic cone

Matches graph (b)

5. Elliptic paraboloid

Matches graph (d)

6. Hyperbolic paraboloid

Matches graph (a)

7. $y = 5$

Plane is parallel to the xz-plane.

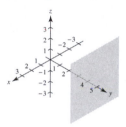

9. $y^2 + z^2 = 9$

The x-coordinate is missing so you have a right circular cylinder with rulings parallel to the x-axis. The generating curve is a circle.

11. $4x^2 + y^2 = 4$

$$\frac{x^2}{1} + \frac{y^2}{4} = 1$$

The z-coordinate is missing so you have an elliptic cylinder with rulings parallel to the z-axis. The generating curve is an ellipse.

13. $\dfrac{x^2}{1} + \dfrac{y^2}{4} + \dfrac{z^2}{1} = 1$

Ellipsoid

xy-trace: $\dfrac{x^2}{1} + \dfrac{y^2}{4} = 1$ ellipse

xz-trace: $x^2 + z^2 = 1$ circle

yz-trace: $\dfrac{y^2}{4} + \dfrac{z^2}{1} = 1$ ellipse

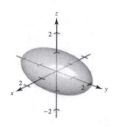

15. $16x^2 - y^2 + 16z^2 = 4$

$$4x^2 - \frac{y^2}{4} + 4z^2 = 1$$

Hyperboloid of one sheet

xy-trace: $4x^2 - \dfrac{y^2}{4} = 1$ hyperbola

xz-trace: $4\left(x^2 + z^2\right) = 1$ circle

yz-trace: $\dfrac{-y^2}{4} + 4z^2 = 1$ hyperbola

17. $4x^2 - y^2 - z^2 = 1$

Hyperboloid of two sheets

xy-trace: $4x^2 - y^2 = 1$ hyperbola

yz-trace: none

xz-trace: $4x^2 - z^2 = 1$ hyperbola

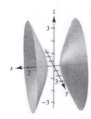

19. $x^2 - y + z^2 = 0$

Elliptic paraboloid

xy-trace: $y = x^2$

xz-trace: $x^2 + z^2 = 0$,
$\quad$ point $(0, 0, 0)$

yz-trace: $y = z^2$

$y = 1$: $x^2 + z^2 = 1$

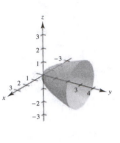

21. $x^2 - y^2 + z = 0$

Hyperbolic paraboloid

xy-trace: $y = \pm x$

xz-trace: $z = -x^2$

yz-trace: $z = y^2$

$y = \pm 1$: $z = 1 - x^2$

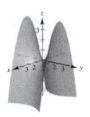

23. $z^2 = x^2 + \dfrac{y^2}{9}$

Elliptic cone

xy-trace: point $(0, 0, 0)$

xz-trace: $z = \pm x$

yz-trace: $z = \pm \dfrac{y}{3}$

When $z = \pm 1$, $x^2 + \dfrac{y^2}{9} = 1$ ellipse

25. Let C be a curve in a plane and let L be a line not in a parallel plane. The set of all lines parallel to L and intersecting C is called a cylinder. C is called the generating curve of the cylinder, and the parallel lines are called rulings.

27. See pages 796 and 797.

29. In the xy-plane, $4x^2 + 6y^2 - 3z^2 = 12$ is an ellipse.

In three-space, $4x^2 + 6y^2 - 3z^2 = 12$ is a hyperboloid of one sheet.

31. $x^2 + z^2 = \left[r(y)\right]^2$ and $z = r(y) = \pm 2\sqrt{y}$; so,

$x^2 + z^2 = 4y$.

33. $x^2 + y^2 = \left[r(z)\right]^2$ and $y = r(z) = \dfrac{z}{2}$; so,

$x^2 + y^2 = \dfrac{z^2}{4}$, $4x^2 + 4y^2 = z^2$.

35. $y^2 + z^2 = \left[r(x)\right]^2$ and $y = r(x) = \dfrac{2}{x}$; so,

$y^2 + z^2 = \left(\dfrac{2}{x}\right)^2$, $y^2 + z^2 = \dfrac{4}{x^2}$.

37. $x^2 + y^2 - 2z = 0$

$x^2 + y^2 = \left(\sqrt{2z}\right)^2$

Equation of generating curve: $y = \sqrt{2z}$ or $x = \sqrt{2z}$

39. $V = 2\pi \int_0^4 x\left(4x - x^2\right) dx = 2\pi \left[\dfrac{4x^3}{3} - \dfrac{x^4}{4}\right]_0^4 = \dfrac{218\pi}{3}$

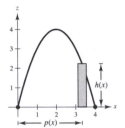

41. $z = \dfrac{x^2}{2} + \dfrac{y^2}{4}$

(a) When $z = 2$ we have $2 = \dfrac{x^2}{2} + \dfrac{y^2}{4}$, or

$1 = \dfrac{x^2}{4} + \dfrac{y^2}{8}$

Major axis: $2\sqrt{8} = 4\sqrt{2}$

Minor axis: $2\sqrt{4} = 4$

$c^2 = a^2 - b^2, c^2 = 4, c = 2$

Foci: $(0, \pm 2, 2)$

(b) When $z = 8$ we have $8 = \dfrac{x^2}{2} + \dfrac{y^2}{4}$, or

$1 = \dfrac{x^2}{16} + \dfrac{y^2}{32}$.

Major axis: $2\sqrt{32} = 8\sqrt{2}$

Minor axis: $2\sqrt{16} = 8$

$c^2 = 32 - 16 = 16, c = 4$

Foci: $(0, \pm 4, 8)$

43. If (x, y, z) is on the surface, then

$(y + 2)^2 = x^2 + (y - 2)^2 + z^2$

$y^2 + 4y + 4 = x^2 + y^2 - 4y + 4 + z^2$

$x^2 + z^2 = 8y$

Elliptic paraboloid

Traces parallel to xz-plane are circles.

45. $\dfrac{x^2}{3963^2} + \dfrac{y^2}{3963^2} + \dfrac{z^2}{3950^2} = 1$

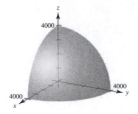

47. $z = \dfrac{y^2}{b^2} - \dfrac{x^2}{a^2}, \; z = bx + ay$

$$bx + ay = \dfrac{y^2}{b^2} - \dfrac{x^2}{a^2}$$

$$\dfrac{1}{a^2}\left(x^2 + a^2 bx + \dfrac{a^4 b^2}{4}\right) = \dfrac{1}{b^2}\left(y^2 - ab^2 y + \dfrac{a^2 b^4}{4}\right)$$

$$\dfrac{\left(x + \dfrac{a^2 b}{2}\right)^2}{a^2} = \dfrac{\left(y - \dfrac{ab^2}{2}\right)^2}{b^2}$$

$$y = \pm \dfrac{b}{a}\left(x + \dfrac{a^2 b}{2}\right) + \dfrac{ab^2}{2}$$

Letting $x = at$, you obtain the two intersecting lines

$x = at, \; y = -bt, \; z = 0$ and $x = at,$

$y = bt + ab^2, \; z = 2abt + a^2 b^2.$

49. True. A sphere is a special case of an ellipsoid (centered at origin, for example)

$$\dfrac{x^2}{a^2} + \dfrac{y^2}{b^2} + \dfrac{z^2}{c^2} = 1$$

having $a = b = c$.

51. False. The trace $x = 2$ of the ellipsoid

$$\dfrac{x^2}{4} + \dfrac{y^2}{9} + z^2 = 1 \text{ is the point } (2, 0, 0).$$

53. The Klein bottle *does not* have both an "inside" and an "outside." It is formed by inserting the small open end through the side of the bottle and making it contiguous with the top of the bottle.

Section 11.7   Cylindrical and Spherical Coordinates

1. $(-7, 0, 5)$, cylindrical

$x = r \cos \theta = -7 \cos 0 = -7$

$y = r \sin \theta = -7 \sin 0 = 0$

$z = 5$

$(-7, 0, 5)$, rectangular

3. $\left(3, \dfrac{\pi}{4}, 1\right)$, cylindrical

$x = 3 \cos \dfrac{\pi}{4} = \dfrac{3\sqrt{2}}{2}$

$y = 3 \sin \dfrac{\pi}{4} = \dfrac{3\sqrt{2}}{2}$

$z = 1$

$\left(\dfrac{3\sqrt{2}}{2}, \dfrac{3\sqrt{2}}{2}, 1\right)$, rectangular

5. $\left(4, \dfrac{7\pi}{6}, 3\right)$, cylindrical

$x = 4 \cos \dfrac{7\pi}{6} = -2\sqrt{3}$

$y = 4 \sin \dfrac{7\pi}{6} = -2$

$z = 3$

$\left(-2\sqrt{3}, -2, 3\right)$, rectangular

7. $(0, 5, 1)$, rectangular

$r = \sqrt{(0)^2 + (5)^2} = 5$

$\theta = \arctan \dfrac{5}{0} = \dfrac{\pi}{2}$

$z = 1$

$\left(5, \dfrac{\pi}{2}, 1\right)$, cylindrical

9. $(2, -2, -4)$, rectangular

$$r = \sqrt{2^2 + (-2)^2} = 2\sqrt{2}$$

$$\theta = \arctan(-1) = -\frac{\pi}{4}$$

$$z = -4$$

$\left(2\sqrt{2}, -\frac{\pi}{4}, -4\right)$, cylindrical

11. $\left(1, \sqrt{3}, 4\right)$, rectangular

$$r = \sqrt{1^2 + \left(\sqrt{3}\right)^2} = 2$$

$$\theta = \arctan\sqrt{3} = \frac{\pi}{3}$$

$$z = 4$$

$\left(2, \frac{\pi}{3}, 4\right)$, cylindrical

13. $z = 4$ is the equation in cylindrical coordinates.
(plane)

15. $x^2 + y^2 + z^2 = 17$, rectangular equation

$r^2 + z^2 = 17$, cylindrical equation

17. $y = x^2$, rectangular equation

$$r \sin \theta = \left(r \cos \theta\right)^2$$

$$\sin \theta = r \cos^2 \theta$$

$r = \sec \theta \cdot \tan \theta$, cylindrical equation

19. $y^2 = 10 - z^2$, rectangular equation

$$\left(r \sin \theta\right)^2 = 10 - z^2$$

$r^2 \sin^2 \theta + z^2 = 10$, cylindrical equation

21. $r = 3$

$$\sqrt{x^2 + y^2} = 3$$

$$x^2 + y^2 = 9$$

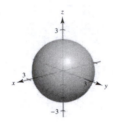

23. $\theta = \frac{\pi}{6}$

$$\tan \frac{\pi}{6} = \frac{y}{x}$$

$$\frac{1}{\sqrt{3}} = \frac{y}{x}$$

$$x = \sqrt{3}y$$

$$x - \sqrt{3}y = 0$$

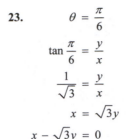

25. $r^2 + z^2 = 5$

$$x^2 + y^2 + z^2 = 5$$

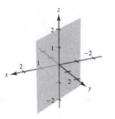

27. $r = 2 \sin \theta$

$$r^2 = 2r \sin \theta$$

$$x^2 + y^2 = 2y$$

$$x^2 + y^2 - 2y = 0$$

$$x^2 + \left(y - 1\right)^2 = 1$$

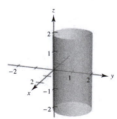

29. $(4, 0, 0)$, rectangular

$$\rho = \sqrt{4^2 + 0^2 + 0^2} = 4$$

$$\tan \theta = \frac{y}{x} = 0 \Rightarrow \theta = 0$$

$$\phi = \arccos 0 = \frac{\pi}{2}$$

$\left(4, 0, \frac{\pi}{2}\right)$, spherical

31. $\left(-2, 2\sqrt{3}, 4\right)$, rectangular

$$\rho = \sqrt{(-2)^2 + \left(2\sqrt{3}\right)^2 + 4^2} = 4\sqrt{2}$$

$$\tan \theta = \frac{y}{x} = \frac{2\sqrt{3}}{-2} = -\sqrt{3}$$

$$\theta = \frac{2\pi}{3}$$

$$\phi = \arccos \frac{1}{\sqrt{2}} = \frac{\pi}{4}$$

$$\left(4\sqrt{2}, \frac{2\pi}{3}, \frac{\pi}{4}\right), \text{ spherical}$$

33. $\left(\sqrt{3}, 1, 2\sqrt{3}\right)$, rectangular

$$\rho = \sqrt{3 + 1 + 12} = 4$$

$$\tan \theta = \frac{y}{x} = \frac{1}{\sqrt{3}}$$

$$\theta = \frac{\pi}{6}$$

$$\phi = \arccos \frac{\sqrt{3}}{2} = \frac{\pi}{6}$$

$$\left(4, \frac{\pi}{6}, \frac{\pi}{6}\right), \text{ spherical}$$

35. $\left(4, \frac{\pi}{6}, \frac{\pi}{4}\right)$, spherical

$$x = 4 \sin \frac{\pi}{4} \cos \frac{\pi}{6} = \sqrt{6}$$

$$y = 4 \sin \frac{\pi}{4} \sin \frac{\pi}{6} = \sqrt{2}$$

$$z = 4 \cos \frac{\pi}{4} = 2\sqrt{2}$$

$$\left(\sqrt{6}, \sqrt{2}, 2\sqrt{2}\right), \text{ rectangular}$$

37. $\left(12, -\frac{\pi}{4}, 0\right)$, spherical

$$x = 12 \sin 0 \cos\left(-\frac{\pi}{4}\right) = 0$$

$$y = 12 \sin 0 \sin\left(-\frac{\pi}{4}\right) = 0$$

$$z = 12 \cos 0 = 12$$

$$(0, 0, 12), \text{ rectangular}$$

39. $\left(5, \frac{\pi}{4}, \frac{3\pi}{4}\right)$, spherical

$$x = 5 \sin \frac{3\pi}{4} \cos \frac{\pi}{4} = \frac{5}{2}$$

$$y = 5 \sin \frac{3\pi}{4} \sin \frac{\pi}{4} = \frac{5}{2}$$

$$z = 5 \cos \frac{3\pi}{4} = -\frac{5\sqrt{2}}{2}$$

$$\left(\frac{5}{2}, \frac{5}{2}, -\frac{5\sqrt{2}}{2}\right), \text{ rectangular}$$

41. $y = 2$, rectangular equation

$$\rho \sin \phi \sin \theta = 2$$

$$\rho = 2 \csc \phi \csc \theta, \text{ spherical equation}$$

43. $x^2 + y^2 + z^2 = 49$, rectangular equation

$$\rho^2 = 49$$

$$\rho = 7, \quad \text{spherical equation}$$

45. $x^2 + y^2 = 16$, rectangular equation

$$\rho^2 \sin^2 \phi \sin^2 \theta + \rho^2 \sin^2 \phi \cos^2 \theta = 16$$

$$\rho^2 \sin^2 \phi \left(\sin^2 \theta + \cos^2 \theta\right) = 16$$

$$\rho^2 \sin^2 \phi = 16$$

$$\rho \sin \phi = 4$$

$$\rho = 4 \csc \phi, \text{ spherical equation}$$

47. $x^2 + y^2 = 2z^2$, rectangular equation

$$\rho^2 \sin^2 \phi \cos^2 \theta + \rho^2 \sin^2 \phi \sin^2 \theta = 2\rho^2 \cos^2 \phi$$

$$\rho^2 \sin^2 \phi \left[\cos^2 \theta + \sin^2 \theta \right] = 2\rho^2 \cos^2 \phi$$

$$\rho^2 \sin^2 \phi = 2\rho^2 \cos^2 \theta$$

$$\frac{\sin^2 \phi}{\cos^2 \phi} = 2$$

$$\tan^2 \phi = 2$$

$$\tan \phi = \pm\sqrt{2}, \text{ spherical equation}$$

49. $\rho = 5$

$x^2 + y^2 + z^2 = 25$

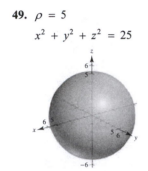

51. $\phi = \dfrac{\pi}{6}$

$$\cos \phi = \frac{z}{\sqrt{x^2 + y^2 + z^2}}$$

$$\frac{\sqrt{3}}{2} = \frac{z}{\sqrt{x^2 + y^2 + z^2}}$$

$$\frac{3}{4} = \frac{z^2}{x^2 + y^2 + z^2}$$

$$3x^2 + 3y^2 - z^2 = 0, z \geq 0$$

53. $\rho = 4 \cos \phi$

$$\sqrt{x^2 + y^2 + z^2} = \frac{4z}{\sqrt{x^2 + y^2 + z^2}}$$

$$x^2 + y^2 + z^2 - 4z = 0$$

$$x^2 + y^2 + (z - 2)^2 = 4, z \geq 0$$

55. $\rho = \csc \phi$

$$\rho \sin \phi = 1$$

$$\sqrt{x^2 + y^2} = 1$$

$$x^2 + y^2 = 1$$

57. $r = 5$

Cylinder

Matches graph (d)

58. $\theta = \dfrac{\pi}{4}$

Plane

Matches graph (e)

59. $\rho = 5$

Sphere

Matches graph (c)

60. $\phi = \dfrac{\pi}{4}$

Cone

Matches graph (a)

61. $r^2 = z, x^2 + y^2 = z$

Paraboloid

Matches graph (f)

62. $\rho = 4 \sec \phi, z = \rho \cos \phi = 4$

Plane

Matches graph (b)

63. $\left(4, \dfrac{\pi}{4}, 0\right)$, cylindrical

$\rho = \sqrt{4^2 + 0^2} = 4$

$\theta = \dfrac{\pi}{4}$

$\phi = \arccos 0 = \dfrac{\pi}{2}$

$\left(4, \dfrac{\pi}{4}, \dfrac{\pi}{2}\right)$, spherical

65. $\left(4, \dfrac{\pi}{2}, 4\right)$, cylindrical

$\rho = \sqrt{4^2 + 4^2} = 4\sqrt{2}$

$\theta = \dfrac{\pi}{2}$

$\phi = \arccos\left(\dfrac{4}{4\sqrt{2}}\right) = \dfrac{\pi}{4}$

$\left(4\sqrt{2}, \dfrac{\pi}{2}, \dfrac{\pi}{4}\right)$, spherical

67. $\left(4, -\dfrac{\pi}{6} - \dfrac{\pi}{6}, 6\right)$, cylindrical

$\rho = \sqrt{4^2 + 6^2} = 2\sqrt{13}$

$\theta = -\dfrac{\pi}{6}$

$\phi = \arccos\dfrac{3}{\sqrt{13}}$

$\left(2\sqrt{13}, -\dfrac{\pi}{6}, \arccos\dfrac{3}{\sqrt{13}}\right)$, spherical

69. $(12, \pi, 5)$, cylindrical

$\rho = \sqrt{12^2 + 5^2} = 13$

$\theta = \pi$

$\phi = \arccos\dfrac{5}{13}$

$\left(13, \pi, \arccos\dfrac{5}{13}\right)$, spherical

71. $\left(10, \dfrac{\pi}{6}, \dfrac{\pi}{2}\right)$, spherical

$r = 10\sin\dfrac{\pi}{2} = 10$

$\theta = \dfrac{\pi}{6}$

$z = 10\cos\dfrac{\pi}{2} = 0$

$\left(10, \dfrac{\pi}{6}, 0\right)$, cylindrical

73. $\left(36, \pi, \dfrac{\pi}{2}\right)$, spherical

$r = \rho\sin\phi = 36\sin\dfrac{\pi}{2} = 36$

$\theta = \pi$

$z = \rho\cos\phi = 36\cos\dfrac{\pi}{2} = 0$

$(36, \pi, 0)$, cylindrical

75. $\left(6, -\dfrac{\pi}{6}, \dfrac{\pi}{3}\right)$, spherical

$r = 6\sin\dfrac{\pi}{3} = 3\sqrt{3}$

$\theta = -\dfrac{\pi}{6}$

$z = 6\cos\dfrac{\pi}{3} = 3$

$\left(3\sqrt{3}, -\dfrac{\pi}{6}, 3\right)$, cylindrical

77. $\left(8, \dfrac{7\pi}{6}, \dfrac{\pi}{6}\right)$, spherical

$r = 8\sin\dfrac{\pi}{6} = 4$

$\theta = \dfrac{7\pi}{6}$

$z = 8\cos\dfrac{\pi}{6} = \dfrac{8\sqrt{3}}{2}$

$\left(4, \dfrac{7\pi}{6}, 4\sqrt{3}\right)$, cylindrical

79. Rectangular to cylindrical: $r^2 = x^2 + y^2$

$$\tan\theta = \dfrac{y}{x}$$

$$z = z$$

Cylindrical to rectangular: $x = r\cos\theta$

$$y = r\sin\theta$$

$$z = z$$

81. Rectangular to spherical: $\rho^2 = x^2 + y^2 + z^2$

$$\tan\theta = \dfrac{y}{x}$$

$$\phi = \arccos\left(\dfrac{z}{\sqrt{x^2 + y^2 + z^2}}\right)$$

Spherical to rectangular: $x = \rho\sin\phi\cos\theta$

$$y = \rho\sin\phi\sin\theta$$

$$z = \rho\cos\phi$$

83. $x^2 + y^2 + z^2 = 25$

    (a) $r^2 + z^2 = 25$

    (b) $\rho^2 = 25 \Rightarrow \rho = 5$

85. $x^2 + y^2 + z^2 - 2z = 0$

    (a) $r^2 + z^2 - 2z = 0 \Rightarrow r^2 + (z-1)^2 = 1$

    (b) $\rho^2 - 2\rho \cos\phi = 0$

        $\rho(\rho - 2\cos\phi) = 0$

            $\rho = 2\cos\phi$

87. $x^2 + y^2 = 4y$

    (a) $r^2 = 4r\sin\theta, r = 4\sin\theta$

    (b) $\rho^2 \sin^2\phi = 4\rho\sin\phi\sin\theta$

      $\rho\sin\phi(\rho\sin\phi - 4\sin\theta) = 0$

$$\rho = \frac{4\sin\theta}{\sin\phi}$$

$$\rho = 4\sin\theta\csc\phi$$

89. $x^2 - y^2 = 9$

    (a) $r^2\cos^2\theta - r^2\sin^2\theta = 9$

$$r^2 = \frac{9}{\cos^2\theta - \sin^2\theta}$$

    (b) $\rho^2\sin^2\phi\cos^2\theta - \rho^2\sin^2\phi\sin^2\theta = 9$

$$\rho^2\sin^2\phi = \frac{9}{\cos^2\theta - \sin^2\theta}$$

$$\rho^2 = \frac{9\csc^2\phi}{\cos^2\theta - \sin^2\theta}$$

91. $0 \le \theta \le \dfrac{\pi}{2}$

    $0 \le r \le 2$

    $0 \le z \le 4$

93. $0 \le \theta \le 2\pi$

    $0 \le r \le a$

    $r \le z \le a$

95. $0 \le \theta \le 2\pi$

    $0 \le \phi \le \dfrac{\pi}{6}$

    $0 \le \rho \le a\sec\phi$

97. $0 \le \theta \le \dfrac{\pi}{2}$

    $0 \le \phi \le \dfrac{\pi}{2}$

    $0 \le \rho \le 2$

99. Rectangular

    $0 \le x \le 10$

    $0 \le y \le 10$

    $0 \le z \le 10$

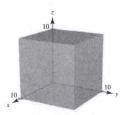

101. Spherical

    $4 \le \rho \le 6$

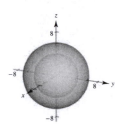

103. Cylindrical coordinates:

    $r^2 + z^2 \le 9,$

    $r \le 3\cos\theta, 0 \le \theta \le \pi$

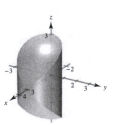

105. False. $r = z \Rightarrow x^2 + y^2 = z^2$ is a cone.

107. False. $(r, \theta, z) = (0, 0, 1)$ and $(r, \theta, z) = (0, \pi, 1)$

    represent the same point $(x, y, z) = (0, 0, 1)$.

109. $z = \sin\theta, r = 1$

$$z = \sin\theta = \frac{y}{r} = \frac{y}{1} = y$$

The curve of intersection is the ellipse formed by the intersection of the plane $z = y$ and the cylinder $r = 1$.

Review Exercises for Chapter 11

1. $P = (1, 2), Q = (4, 1), R = (5, 4)$

(a) $\mathbf{u} = \overline{PQ} = \langle 4 - 1, 1 - 2 \rangle = \langle 3, -1 \rangle$

$\mathbf{v} = \overline{PR} = \langle 5 - 1, 4 - 2 \rangle = \langle 4, 2 \rangle$

(b) $\mathbf{u} = 3\mathbf{i} - \mathbf{j}, \mathbf{v} = 4\mathbf{i} + 2\mathbf{j}$

(c) $\|\mathbf{u}\| = \sqrt{3^2 + (-1)^2} = \sqrt{10}$ $\|\mathbf{v}\| = \sqrt{4^2 + 2^2} = \sqrt{20} = 2\sqrt{5}$

(d) $2\mathbf{u} + \mathbf{v} = 2\langle 3, -1 \rangle + \langle 4, 2 \rangle = \langle 10, 0 \rangle$

3. $\mathbf{v} = \|\mathbf{v}\|(\cos\theta\,\mathbf{i} + \sin\theta\,\mathbf{j})$

$= 8(\cos 60°\,\mathbf{i} + \sin 60°\,\mathbf{j})$

$= 8\left(\dfrac{1}{2}\mathbf{i} + \dfrac{\sqrt{3}}{2}\mathbf{j}\right) = 4\mathbf{i} + 4\sqrt{3}\,\mathbf{j} = \langle 4, 4\sqrt{3} \rangle$

5. $z = 0, y = 4, x = -5: (-5, 4, 0)$

7. $d = \sqrt{(-2 - 1)^2 + (3 - 6)^2 + (5 - 3)^2}$

$= \sqrt{9 + 9 + 4} = \sqrt{22}$

9. $(x - 3)^2 + (y + 2)^2 + (z - 6)^2 = \left(\dfrac{15}{2}\right)^2$

11. $(x^2 - 4x + 4) + (y^2 - 6y + 9) + z^2 = -4 + 4 + 9$

$(x - 2)^2 + (y - 3)^3 + z^2 = 9$

Center: $(2, 3, 0)$

Radius: 3

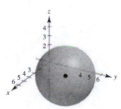

13. (a), (d)

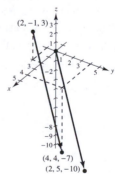

(b) $\mathbf{v} = \langle 4 - 2, 4 - (-1), -7 - 3 \rangle = \langle 2, 5, -10 \rangle$

(c) $\mathbf{v} = 2\mathbf{i} + 5\mathbf{j} - 10\mathbf{k}$

15. $\mathbf{v} = \langle -1 - 3, 6 - 4, 9 + 1 \rangle = \langle -4, 2, 10 \rangle$

$\mathbf{w} = \langle 5 - 3, 3 - 4, -6 + 1 \rangle = \langle 2, -1, -5 \rangle$

Because $-2\mathbf{w} = \mathbf{v}$, the points lie in a straight line.

17. Unit vector: $\dfrac{\mathbf{u}}{\|\mathbf{u}\|} = \left\langle \dfrac{2, 3, 5}{\sqrt{38}} \right\rangle = \left\langle \dfrac{2}{\sqrt{38}}, \dfrac{3}{\sqrt{38}}, \dfrac{5}{\sqrt{38}} \right\rangle$

19. $P = \langle 5, 0, 0 \rangle, Q = \langle 4, 4, 0 \rangle, R = \langle 2, 0, 6 \rangle$

(a) $\mathbf{u} = \overline{PQ} = \langle -1, 4, 0 \rangle$

$\mathbf{v} = \overline{PR} = \langle -3, 0, 6 \rangle$

(b) $\mathbf{u} \cdot \mathbf{v} = (-1)(-3) + 4(0) + 0(6) = 3$

(c) $\mathbf{v} \cdot \mathbf{v} = 9 + 36 = 45$

21. $\mathbf{u} = 5\left(\cos\dfrac{3\pi}{4}\mathbf{i} + \sin\dfrac{3\pi}{4}\mathbf{j}\right) = \dfrac{5\sqrt{2}}{2}[-\mathbf{i} + \mathbf{j}]$

$\mathbf{v} = 2\left(\cos\dfrac{2\pi}{3}\mathbf{i} + \sin\dfrac{2\pi}{3}\mathbf{j}\right) = -\mathbf{i} + \sqrt{3}\,\mathbf{j}$

$\mathbf{u} \cdot \mathbf{v} = \dfrac{5\sqrt{2}}{2}(1 + \sqrt{3})$

$\|\mathbf{u}\| = \sqrt{\dfrac{25}{2} + \dfrac{25}{2}} = 5$ $\|\mathbf{v}\| = \sqrt{1 + 3} = 2$

$\cos\theta = \dfrac{|\mathbf{u} \cdot \mathbf{v}|}{\|\mathbf{u}\|\|\mathbf{v}\|} = \dfrac{(5\sqrt{2}/2)(1 + \sqrt{3})}{5(2)} = \dfrac{\sqrt{2} + \sqrt{6}}{4}$

(a) $\theta = \arccos\dfrac{\sqrt{2} + \sqrt{6}}{4} = \dfrac{\pi}{12} \approx 0.262$

(b) $\theta \approx 15°$

23. $\mathbf{u} = \langle 10, -5, 15 \rangle, \mathbf{v} = \langle -2, 1, -3 \rangle$

$\mathbf{u} = -5\mathbf{v} \Rightarrow \mathbf{u}$ is parallel to $\mathbf{v}$ and in the opposite direction.

(a) $\theta = \pi$ (b) $\theta = 180°$

25. $\mathbf{u} = \langle 7, -2, 3 \rangle, \mathbf{v} = \langle -1, 4, 5 \rangle$

Because $\mathbf{u} \cdot \mathbf{v} = 0$, the vectors are orthogonal.

27. $\mathbf{u} = \langle 7, 9 \rangle$, $\mathbf{v} = \langle 1, 5 \rangle$

$$\text{proj}_\mathbf{v}\, \mathbf{u} = \frac{\mathbf{u} \cdot \mathbf{v}}{\|\mathbf{v}\|^2}\mathbf{v}$$

$$= \frac{7 + 45}{\left(\sqrt{26}\right)^2}\langle 1, 5 \rangle$$

$$= \frac{52}{26}\langle 1, 5 \rangle = \langle 2, 10 \rangle$$

29. $\mathbf{u} = \langle 1, -1, 1 \rangle$, $\mathbf{v} = \langle 2, 0, 2 \rangle$

$$\text{proj}_\mathbf{v}\, \mathbf{u} = \frac{\mathbf{u} \cdot \mathbf{v}}{\|\mathbf{v}\|^2}\mathbf{v}$$

$$= \frac{2 + 2}{\left(\sqrt{4 + 4}\right)^2}\langle 2, 0, 2 \rangle$$

$$= \frac{1}{2}\langle 2, 0, 2 \rangle = \langle 1, 0, 1 \rangle$$

31. There are many correct answers.

For example: $\mathbf{v} = \pm\langle 6, -5, 0 \rangle$.

33. (a) $\mathbf{u} \times \mathbf{v} = \begin{vmatrix} \mathbf{i} & \mathbf{j} & \mathbf{k} \\ 4 & 3 & 6 \\ 5 & 2 & 1 \end{vmatrix} = -9\mathbf{i} + 26\mathbf{j} - 7\mathbf{k}$

(b) $\mathbf{v} \times \mathbf{u} = -(\mathbf{u} \times \mathbf{v}) = 9\mathbf{i} - 26\mathbf{j} + 7\mathbf{k}$

(c) $\mathbf{v} \times \mathbf{v} = \mathbf{0}$

35. (a) $\mathbf{u} \times \mathbf{v} = \begin{vmatrix} \mathbf{i} & \mathbf{j} & \mathbf{k} \\ 2 & -4 & -4 \\ 1 & 1 & 3 \end{vmatrix} = -8\mathbf{i} - 10\mathbf{j} + 6\mathbf{k}$

(b) $\mathbf{v} \times \mathbf{u} = -(\mathbf{u} \times \mathbf{v}) = 8\mathbf{i} + 10\mathbf{j} - 6\mathbf{k}$

(c) $\mathbf{v} \times \mathbf{v} = \mathbf{0}$

37. $\mathbf{u} \times \mathbf{v} = \begin{vmatrix} \mathbf{i} & \mathbf{j} & \mathbf{k} \\ 2 & -10 & 8 \\ 4 & 6 & -8 \end{vmatrix} = 32\mathbf{i} + 48\mathbf{j} + 52\mathbf{k}$

$$\|\mathbf{u} \times \mathbf{v}\| = \sqrt{6032} = 4\sqrt{377}$$

Unit vector: $\dfrac{1}{\sqrt{377}}\langle 8, 12, 13 \rangle$

39. $\mathbf{F} = c(\cos 20°\mathbf{j} + \sin 20°\mathbf{k})$

$$\overrightarrow{PQ} = 2\mathbf{k}$$

$$\overrightarrow{PQ} \times \mathbf{F} = \begin{vmatrix} \mathbf{i} & \mathbf{j} & \mathbf{k} \\ 0 & 0 & 2 \\ 0 & c\cos 20° & c\sin 20° \end{vmatrix} = -2c\cos 20°\mathbf{i}$$

$$200 = \|\overrightarrow{PQ} \times \mathbf{F}\| = 2c\cos 20°$$

$$c = \frac{100}{\cos 20°}$$

$$\mathbf{F} = \frac{100}{\cos 20°}(\cos 20°\mathbf{j} + \sin 20°\mathbf{k}) = 100(\mathbf{j} + \tan 20°\mathbf{k})$$

$$\|\mathbf{F}\| = 100\sqrt{1 + \tan^2 20°} = 100\sec 20° \approx 106.4\ \text{lb}$$

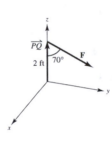

41. $\mathbf{v} = \langle 9 - 3, 11 - 0, 6 - 2 \rangle = \langle 6, 11, 4 \rangle$

(a) Parametric equations:
$x = 3 + 6t$, $y = 11t$, $z = 2 + 4t$

(b) Symmetric equations: $\dfrac{x - 3}{6} = \dfrac{y}{11} = \dfrac{z - 2}{4}$

43. $\mathbf{v} = \mathbf{j}$, $P(1, 2, 3)$

$x = 1$, $y = 2 + t$, $z = 3$

45. $3x - 3y - 7z = -4$, $x - y + 2z = 3$

Solving simultaneously, you have $z = 1$. Substituting $z = 1$ into the second equation, you have $y = x - 1$. Substituting for x in this equation you obtain two points on the line of intersection, $(0, -1, 1)$, $(1, 0, 1)$. The direction vector of the line of intersection is $\mathbf{v} = \mathbf{i} + \mathbf{j}$.

$x = t$, $y = -1 + t$, $z = 1$

47. $P = (-3, -4, 2), Q = (-3, 4, 1), R = (1, 1, -2)$

$\overrightarrow{PQ} = \langle 0, 8, -1 \rangle, \overrightarrow{PR} = [4, 5, -4]$

$$\mathbf{n} = \overrightarrow{PQ} \times \overrightarrow{PR} = \begin{vmatrix} \mathbf{i} & \mathbf{j} & \mathbf{k} \\ 0 & 8 & -1 \\ 4 & 5 & -4 \end{vmatrix} = -27\mathbf{i} - 4\mathbf{j} - 32\mathbf{k}$$

$$-27(x + 3) - 4(y + 4) - 32(z - 2) = 0$$
$$27x + 4y + 32z = -33$$

49. The two lines are parallel as they have the same direction numbers, $-2, 1, 1$. Therefore, a vector parallel to the plane is $\mathbf{v} = -2\mathbf{i} + \mathbf{j} + \mathbf{k}$. A point on the first line is $(1, 0, -1)$ and a point on the second line is $(-1, 1, 2)$. The vector $\mathbf{u} = 2\mathbf{i} - \mathbf{j} - 3\mathbf{k}$ connecting these two points is also parallel to the plane. Therefore, a normal to the plane is

$$\mathbf{v} \times \mathbf{u} = \begin{vmatrix} \mathbf{i} & \mathbf{j} & \mathbf{k} \\ -2 & 1 & 1 \\ 2 & -1 & -3 \end{vmatrix}$$
$$= -2\mathbf{i} - 4\mathbf{j} = -2(\mathbf{i} + 2\mathbf{j}).$$

Equation of the plane: $(x - 1) + 2y = 0$
$$x + 2y = 1$$

51. $Q(1, 0, 2)$ point

$2x - 3y + 6z = 6$

A point P on the plane is $(3, 0, 0)$.

$\overrightarrow{PQ} = \langle -2, 0, 2 \rangle$

$\mathbf{n} = \langle 2, -3, 6 \rangle$ normal to plane

$$D = \frac{|\overrightarrow{PQ} \cdot \mathbf{n}|}{\|\mathbf{n}\|} = \frac{8}{7}$$

53. The normal vectors to the planes are the same,
$\mathbf{n} = \langle 5, -3, 1 \rangle$.

Choose a point in the first plane $P(0, 0, 2)$. Choose a point in the second plane, $Q(0, 0, -3)$.

$\overrightarrow{PQ} = \langle 0, 0, -5 \rangle$

$$D = \frac{|\overrightarrow{PQ} \cdot \mathbf{n}|}{\|\mathbf{n}\|} = \frac{|-5|}{\sqrt{35}} = \frac{5}{\sqrt{35}} = \frac{\sqrt{35}}{7}$$

55. $x + 2y + 3z = 6$

Plane

Intercepts: $(6, 0, 0), (0, 3, 0), (0, 0, 2)$,

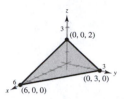

57. $y = \frac{1}{2}z$

Plane with rulings parallel to the x-axis.

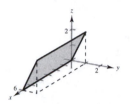

59. $\dfrac{x^2}{16} + \dfrac{y^2}{9} + z^2 = 1$

Ellipsoid

xy-trace: $\dfrac{x^2}{16} + \dfrac{y^2}{9} = 1$

xz-trace: $\dfrac{x^2}{16} + z^2 = 1$

yz-trace: $\dfrac{y^2}{9} + z^2 = 1$

61. $\dfrac{x^2}{16} - \dfrac{y^2}{9} + z^2 = -1$

$\dfrac{y^2}{9} - \dfrac{x^2}{16} - z^2 = 1$

Hyperboloid of two sheets

xy-trace: $\dfrac{y^2}{9} - \dfrac{x^2}{16} = 1$

xz-trace: None

yz-trace: $\dfrac{y^2}{9} - z^2 = 1$

63. $x^2 + z^2 = 4$.

Cylinder of radius 2 about y-axis

65. $z^2 = 2y$ revolved about y-axis

$z = \pm\sqrt{2y}$

$x^2 + z^2 = [r(y)]^2 = 2y$

$x^2 + z^2 = 2y$

67. $(-2\sqrt{2}, 2\sqrt{2}, 2)$, rectangular

(a) $r = \sqrt{(-2\sqrt{2})^2 + (2\sqrt{2})^2} = 4$,

$\theta = \arctan(-1) = \dfrac{3\pi}{4}, z = 2$,

$\left(4, \dfrac{3\pi}{4}, 2\right)$, cylindrical

(b) $\rho = \sqrt{(-2\sqrt{2})^2 + (2\sqrt{2})^2 + (2)^2} = 2\sqrt{5}$,

$\theta = \dfrac{3\pi}{4}, \phi = \arccos \dfrac{2}{2\sqrt{5}} = \arccos \dfrac{1}{\sqrt{5}}$,

$\left(2\sqrt{5}, \dfrac{3\pi}{4}, \arccos \dfrac{\sqrt{5}}{5}\right)$, spherical

69. $\left(100, -\dfrac{\pi}{6}, 50\right)$, cylindrical

$\rho = \sqrt{100^2 + 50^2} = 50\sqrt{5}$

$\theta = -\dfrac{\pi}{6}$

$\phi = \arccos\left(\dfrac{50}{50\sqrt{5}}\right) = \arccos\left(\dfrac{1}{\sqrt{5}}\right) \approx 63.4° \text{ or } 1.107$

$\left(50\sqrt{5}, -\dfrac{\pi}{6}, 63.4°\right)$, sperical or $\left(50\sqrt{5}, -\dfrac{\pi}{6}, 1.1071\right)$

71. $\left(25, -\dfrac{\pi}{4}, \dfrac{3\pi}{4}\right)$, spherical

$r^2 = \left(25 \sin\dfrac{3\pi}{4}\right)^2 \Rightarrow r = \dfrac{25\sqrt{2}}{2}$

$\theta = -\dfrac{\pi}{4}$

$z = \rho \cos \phi = 25 \cos \dfrac{3\pi}{4} = -\dfrac{25\sqrt{2}}{2}$

$\left(\dfrac{25\sqrt{2}}{2}, -\dfrac{\pi}{4}, -\dfrac{25\sqrt{2}}{2}\right)$, cylindrical

73. $x^2 - y^2 = 2z$

(a) Cylindrical:

$r^2 \cos^2 \theta - r^2 \sin^2 \theta = 2z \Rightarrow r^2 \cos 2\theta = 2z$

(b) Spherical:

$\rho^2 \sin^2 \phi \cos^2 \theta - \rho^2 \sin^2 \phi \sin^2 \theta = 2\rho \cos \phi$

$\rho \sin^2 \phi \cos 2\theta - 2 \cos \phi = 0$

$\rho = 2 \sec 2\theta \cos \phi \csc^2 \phi$

75. $r = 5 \cos \theta$, cylindrical equation

$r^2 = 5r \cos \theta$

$x^2 + y^2 = 5x$

$x^2 - 5x + \dfrac{25}{4} + y^2 = \dfrac{25}{4}$

$\left(x - \dfrac{5}{2}\right)^2 + y^2 = \left(\dfrac{5}{2}\right)^2$, rectangular equation

77. $\theta = \dfrac{\pi}{4}$, spherical coordinates

$\tan \theta = \tan \dfrac{\pi}{4} = 1$

$\dfrac{y}{x} = 1$

$y = x, x \geq 0$, rectangular coordinates, half-plane

Problem Solving for Chapter 11

1.
$$\mathbf{a} + \mathbf{b} + \mathbf{c} = \mathbf{0}$$
$$\mathbf{b} \times (\mathbf{a} + \mathbf{b} + \mathbf{c}) = \mathbf{0}$$
$$(\mathbf{b} \times \mathbf{a}) + (\mathbf{b} \times \mathbf{c}) = \mathbf{0}$$
$$\|\mathbf{a} \times \mathbf{b}\| = \|\mathbf{b} \times \mathbf{c}\|$$
$$\|\mathbf{b} \times \mathbf{c}\| = \|\mathbf{b}\|\|\mathbf{c}\| \sin A$$
$$\|\mathbf{a} \times \mathbf{b}\| = \|\mathbf{a}\|\|\mathbf{b}\| \sin C$$

Then,
$$\frac{\sin A}{\|\mathbf{a}\|} = \frac{\|\mathbf{b} \times \mathbf{c}\|}{\|\mathbf{a}\|\|\mathbf{b}\|\|\mathbf{c}\|}$$
$$= \frac{\|\mathbf{a} \times \mathbf{b}\|}{\|\mathbf{a}\|\|\mathbf{b}\|\|\mathbf{c}\|}$$
$$= \frac{\sin C}{\|\mathbf{c}\|}.$$

The other case, $\dfrac{\sin A}{\|\mathbf{a}\|} = \dfrac{\sin B}{\|\mathbf{b}\|}$ is similar.

3. Label the figure as indicated.
From the figure, you see that
$$\overline{SP} = \frac{1}{2}\mathbf{a} - \frac{1}{2}\mathbf{b} = \overline{RQ} \text{ and } \overline{SR} = \frac{1}{2}\mathbf{a} + \frac{1}{2}\mathbf{b} = \overline{PQ}.$$
Because $\overline{SP} = \overline{RQ}$ and $\overline{SR} = \overline{PQ}$,
$PSRQ$ is a parallelogram.

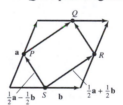

5. (a) $\mathbf{u} = \langle 0, 1, 1 \rangle$ is the direction vector of the line determined by P_1 and P_2.

$$D = \frac{\left\|\overline{P_1Q} \times \mathbf{u}\right\|}{\|\mathbf{u}\|}$$
$$= \frac{\|\langle 2, 0, -1 \rangle \times \langle 0, 1, 1 \rangle\|}{\sqrt{2}}$$
$$= \frac{\|\langle 1, -2, 2 \rangle\|}{\sqrt{2}} = \frac{3}{\sqrt{2}} = \frac{3\sqrt{2}}{2}$$

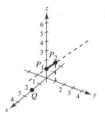

(b) The shortest distance to the line **segment**
is $\left\|\overline{P_1Q}\right\| = \|\langle 2, 0, -1 \rangle\| = \sqrt{5}$.

7. (a) $V = \pi \int_0^1 \left(\sqrt{z}\right)^2 dz = \left[\pi \dfrac{z^2}{2}\right]_0^1 = \dfrac{1}{2}\pi$

Note: $\dfrac{1}{2}$ (base)(altitude) $= \dfrac{1}{2}\pi(1) = \dfrac{1}{2}\pi$

(b) $\dfrac{x^2}{a^2} + \dfrac{y^2}{b^2} = z$: (slice at $z = c$)

$$\frac{x^2}{\left(\sqrt{c}a\right)^2} + \frac{y^2}{\left(\sqrt{c}b\right)^2} = 1$$

At $z = c$, figure is ellipse of area

$$\pi\left(\sqrt{c}a\right)\left(\sqrt{c}b\right) = \pi abc.$$

$$V = \int_0^k \pi abc \cdot dc = \left[\frac{\pi abc^2}{2}\right]_0^k = \frac{\pi abk^2}{2}$$

(c) $V = \dfrac{1}{2}(\pi abk)k = \dfrac{1}{2}$ (area of base)(height)

9. From Exercise 58, Section 11.4,

$$(\mathbf{u} \times \mathbf{v}) \times (\mathbf{w} \times \mathbf{z}) = \big[(\mathbf{u} \times \mathbf{v}) \cdot \mathbf{z}\big]\mathbf{w} - \big[(\mathbf{u} \times \mathbf{v}) \cdot \mathbf{w}\big]\mathbf{z}.$$

11. (a) $\rho = 2 \sin \phi$ **(b)** $\rho = 2 \cos \phi$

Torus   Sphere

13. (a) $\mathbf{u} = \|\mathbf{u}\|(\cos 0\,\mathbf{i} + \sin 0\,\mathbf{j}) = \|\mathbf{u}\|\mathbf{i}$

Downward force $\mathbf{w} = -\mathbf{j}$

$$\mathbf{T} = \|\mathbf{T}\|\big(\cos(90° + \theta)\mathbf{i} + \sin(90° + \theta)\mathbf{j}\big)$$
$$= \|\mathbf{T}\|(-\sin\theta\,\mathbf{i} + \cos\theta\,\mathbf{j})$$
$$\mathbf{0} = \mathbf{u} + \mathbf{w} + \mathbf{T} = \|\mathbf{u}\|\mathbf{i} - \mathbf{j} + \|\mathbf{T}\|(-\sin\theta\,\mathbf{i} + \cos\theta\,\mathbf{j})$$
$$\|\mathbf{u}\| = \sin\theta\|\mathbf{T}\|$$
$$1 = \cos\theta\|\mathbf{T}\|$$

If $\theta = 30°$, $\|\mathbf{u}\| = (1/2)\|\mathbf{T}\|$ and $1 = \big(\sqrt{3}/2\big)\|\mathbf{T}\| \Rightarrow \|\mathbf{T}\| = \dfrac{2}{\sqrt{3}} \approx 1.1547$ lb and $\|\mathbf{u}\| = \dfrac{1}{2}\left(\dfrac{2}{\sqrt{3}}\right) \approx 0.5774$ lb

(b) From part (a), $\|\mathbf{u}\| = \tan\theta$ and $\|\mathbf{T}\| = \sec\theta$.

Domain: $0 \le \theta \le 90°$

(c)

θ	0°	10°	20°	30°	40°	50°	60°
T	1	1.0154	1.0642	1.1547	1.3054	1.5557	2
$\|\mathbf{u}\|$	0	0.1763	0.3640	0.5774	0.8391	1.1918	1.7321

(d)

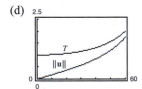

(e) Both are increasing functions.

(f) $\displaystyle \lim_{\theta \to \pi/2^-} T = \infty$ and $\displaystyle \lim_{\theta \to \pi/2^-} \|\mathbf{u}\| = \infty.$

Yes. As θ increases, both T and $\|\mathbf{u}\|$ increase.

15. Let $\theta = \alpha - \beta$, the angle between **u** and **v**. Then

$$\sin(\alpha - \beta) = \frac{\|\mathbf{u} \times \mathbf{v}\|}{\|\mathbf{u}\|\|\mathbf{v}\|} = \frac{\|\mathbf{v} \times \mathbf{u}\|}{\|\mathbf{u}\|\|\mathbf{v}\|}.$$

For $\mathbf{u} = \langle \cos\alpha, \sin\alpha, 0 \rangle$ and $\mathbf{v} = \langle \cos\beta, \sin\beta, 0 \rangle$, $\|\mathbf{u}\| = \|\mathbf{v}\| = 1$ and

$$\mathbf{v} \times \mathbf{u} = \begin{vmatrix} \mathbf{i} & \mathbf{j} & \mathbf{k} \\ \cos\beta & \sin\beta & 0 \\ \cos\alpha & \sin\alpha & 0 \end{vmatrix} = (\sin\alpha\cos\beta - \cos\alpha\sin\beta)\mathbf{k}.$$

So, $\sin(\alpha - \beta) = \|\mathbf{v} \times \mathbf{u}\| = \sin\alpha\cos\beta - \cos\alpha\sin\beta$.

17. From Theorem 11.13 and Theorem 11.7 (6) you have

$$D = \frac{\left|\overrightarrow{PQ} \cdot \mathbf{n}\right|}{\|\mathbf{n}\|}$$

$$= \frac{\left|\mathbf{w} \cdot (\mathbf{u} \times \mathbf{v})\right|}{\|\mathbf{u} \times \mathbf{v}\|} = \frac{\left|(\mathbf{u} \times \mathbf{v}) \cdot \mathbf{w}\right|}{\|\mathbf{u} \times \mathbf{v}\|} = \frac{\left|\mathbf{u} \cdot (\mathbf{v} \times \mathbf{w})\right|}{\|\mathbf{u} \times \mathbf{v}\|}.$$

19. $x^2 + y^2 = 1$ cylinder

$z = 2y$ plane

Introduce a coordinate system in the plane $z = 2y$.

The new u-axis is the original x-axis.

The new v-axis is the line $z = 2y, x = 0$.

Then the intersection of the cylinder and plane satisfies the equation of an ellipse:

$$x^2 + y^2 = 1$$

$$x^2 + \left(\frac{z}{2}\right)^2 = 1$$

$$x^2 + \frac{z^2}{4} = 1 \qquad \text{ellipse}$$

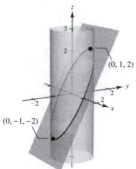

CHAPTER 12
Vector-Valued Functions

CHAPTER 12
Vector-Valued Functions

Section 12.1 Vector-Valued Functions

1. $\mathbf{r}(t) = \dfrac{1}{t+1}\mathbf{i} + \dfrac{t}{2}\mathbf{j} - 3t\mathbf{k}$

Component functions: $f(t) = \dfrac{1}{t+1}$

$g(t) = \dfrac{t}{2}$

$h(t) = -3t$

Domain: $(-\infty, -1) \cup (-1, \infty)$

3. $\mathbf{r}(t) = \ln t\mathbf{i} - e^t\mathbf{j} - t\mathbf{k}$

Component functions: $f(t) = \ln t$

$g(t) = -e^t$

$h(t) = -t$

Domain: $(0, \infty)$

5. $\mathbf{r}(t) = \mathbf{F}(t) + \mathbf{G}(t) = \left(\cos t\mathbf{i} - \sin t\mathbf{j} + \sqrt{t}\mathbf{k}\right) + \left(\cos t\mathbf{i} + \sin t\mathbf{j}\right) = 2\cos t\mathbf{i} + \sqrt{t}\mathbf{k}$

Domain: $[0, \infty)$

7. $\mathbf{r}(t) = \mathbf{F}(t) \times \mathbf{G}(t) = \begin{vmatrix} \mathbf{i} & \mathbf{j} & \mathbf{k} \\ \sin t & \cos t & 0 \\ 0 & \sin t & \cos t \end{vmatrix} = \cos^2 t\mathbf{i} - \sin t \cos t\mathbf{j} + \sin^2 t\mathbf{k}$

Domain: $(-\infty, \infty)$

9. $\mathbf{r}(t) = \frac{1}{2}t^2\mathbf{i} - (t-1)\mathbf{j}$

(a) $\mathbf{r}(1) = \frac{1}{2}\mathbf{i}$

(b) $\mathbf{r}(0) = \mathbf{j}$

(c) $\mathbf{r}(s+1) = \frac{1}{2}(s+1)^2\mathbf{i} - (s+1-1)\mathbf{j} = \frac{1}{2}(s+1)^2\mathbf{i} - s\mathbf{j}$

(d) $\mathbf{r}(2+\Delta t) - \mathbf{r}(2) = \frac{1}{2}(2+\Delta t)^2\mathbf{i} - (2+\Delta t - 1)\mathbf{j} - (2\mathbf{i} - \mathbf{j}) = \left(2 + 2\Delta t + \frac{1}{2}(\Delta t)^2\right)\mathbf{i} - (1+\Delta t)\mathbf{j} - 2\mathbf{i} + \mathbf{j}$

$= \left(2\Delta t + \frac{1}{2}(\Delta t)^2\right)\mathbf{i} - (\Delta t)\mathbf{j} = \frac{1}{2}\Delta t(\Delta t + 4)\mathbf{i} - \Delta t\mathbf{j}$

11. $\mathbf{r}(t) = \ln t\mathbf{i} + \dfrac{1}{t}\mathbf{j} + 3t\mathbf{k}$

(a) $\mathbf{r}(2) = \ln 2\mathbf{i} + \dfrac{1}{2}\mathbf{j} + 6\mathbf{k}$

(b) $\mathbf{r}(-3)$ is not defined. $\left(\ln(-3)\text{ does not exist.}\right)$

(c) $\mathbf{r}(t-4) = \ln(t-4)\mathbf{i} + \dfrac{1}{t-4}\mathbf{j} + 3(t-4)\mathbf{k}$

(d) $\mathbf{r}(1+\Delta t) - \mathbf{r}(1) = \ln(1+\Delta t)\mathbf{i} + \dfrac{1}{1+\Delta t}\mathbf{j} + 3(1+\Delta t)\mathbf{k} - (0\mathbf{i} + \mathbf{j} + 3\mathbf{k}) = \ln(1+\Delta t)\mathbf{i} + \left(\dfrac{-\Delta t}{1+\Delta t}\right)\mathbf{j} + (3\Delta t)\mathbf{k}$

13. $P(0,0,0), Q(3,1,2)$

$\mathbf{v} = \overline{PQ} = \langle 3,1,2 \rangle$

$\mathbf{r}(t) = 3t\mathbf{i} + t\mathbf{j} + 2t\mathbf{k}, \ 0 \le t \le 1$

$x = 3t, y = t, z = 2t, 0 \le t \le 1$, Parametric equation

(Answers may vary)

15. $P(-2,5,-3), Q(-1,4,9)$

$\mathbf{v} = \overline{PQ} = \langle 1,-1,12 \rangle$

$\mathbf{r}(t) = (-2+t)\mathbf{i} + (5-t)\mathbf{j} + (-3+12t)\mathbf{k}, 0 \le t \le 1$

$x = -2+t, y = 5-t, z = -3+12t,$

$0 \le t \le 1$, Parametric equation

(Answers may vary)

17. $\mathbf{r}(t) \cdot \mathbf{u}(t) = (3t - 1)(t^2) + \left(\frac{1}{4}t^3\right)(-8) + 4(t^3)$

$= 3t^3 - t^2 - 2t^3 + 4t^3 = 5t^3 - t^2$, a scalar.

No, the dot product is a scalar-valued function.

19. $\mathbf{r}(t) = t\mathbf{i} + 2t\mathbf{j} + t^2\mathbf{k}, -2 \le t \le 2$

$x = t, y = 2t, z = t^2$

So, $z = x^2$. Matches (b)

20. $\mathbf{r}(t) = \cos(\pi t)\mathbf{i} + \sin(\pi t)\mathbf{j} + t^2\mathbf{k}, -1 \le t \le 1$

$x = \cos(\pi t), y = \sin(\pi t), z = t^2$

So, $x^2 + y^2 = 1$. Matches (c)

21. $\mathbf{r}(t) = t\mathbf{i} + t^2\mathbf{j} + e^{0.75t}\mathbf{k}, -2 \le t \le 2$

$x = t, y = t^2, z = e^{0.75t}$

So, $y = x^2$. Matches (d)

22. $\mathbf{r}(t) = t\mathbf{i} + \ln t\mathbf{j} + \frac{2t}{3}\mathbf{k}, 0.1 \le t \le 5$

$x = t, y = \ln t, z = \frac{2t}{3}$

So, $z = \frac{2}{3}x$ and $y = \ln x$. Matches (a)

23. $x = \frac{t}{4} \Rightarrow t = 4x$

$y = t - 1$

$y = 4x - 1$

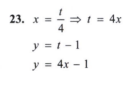

25. $x = t^3, y = t^2$

$y = x^{2/3}$

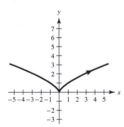

27. $x = \cos \theta, y = 3 \sin \theta$

$x^2 + \frac{y^2}{9} = 1$, Ellipse

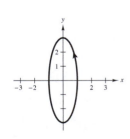

29. $x = 3 \sec \theta, y = 2 \tan \theta$

$\frac{x^2}{9} = \frac{y^2}{4} + 1$, Hyperbola

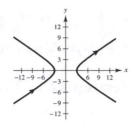

31. $x = -t + 1$

$y = 4t + 2$

$z = 2t + 3$

Line passing through
the points: $(0, 6, 5), (1, 2, 3)$

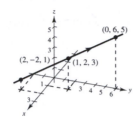

33. $x = 2 \cos t, y = 2 \sin t, z = t$

$\frac{x^2}{4} + \frac{y^2}{4} = 1$

$z = t$

Circular helix

35. $x = 2 \sin t, y = 2 \cos t, z = e^{-t}$

$x^2 + y^2 = 4$

$z = e^{-t}$

37. $x = t, y = t^2, z = \frac{2}{3}t^3$

$y = x^2, z = \frac{2}{3}x^3$

t	-2	-1	0	1	2
x	-2	-1	0	1	2
y	4	1	0	1	4
z	$-\frac{16}{3}$	$-\frac{2}{3}$	0	$\frac{2}{3}$	$\frac{16}{3}$

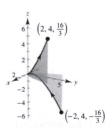

39. $\mathbf{r}(t) = -\dfrac{1}{2}t^2\mathbf{i} + t\mathbf{j} - \dfrac{\sqrt{3}}{2}t^2\mathbf{k}$

Parabola

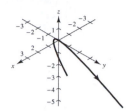

41. $\mathbf{r}(t) = \sin t\mathbf{i} + \left(\dfrac{\sqrt{3}}{2}\cos t - \dfrac{1}{2}t\right)\mathbf{j} + \left(\dfrac{1}{2}\cos t + \dfrac{\sqrt{3}}{2}\right)\mathbf{k}$

Helix

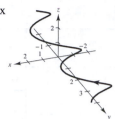

43.

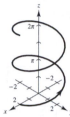

(a)

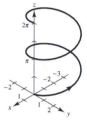

The helix is translated
2 units back on the *x*-axis.

(b)

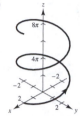

The height of the helix
increases at a faster rate.

(c)

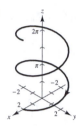

The orientation of the
helix is reversed.

(d)

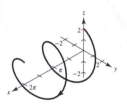

The axis of the helix is
the *x*-axis.

(e)

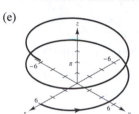

The radius of the helix is
increased from 2 to 6.

45. $y = x + 5$

Let $x = t$, then $y = t + 5$

$\mathbf{r}(t) = t\mathbf{i} + (t + 5)\mathbf{j}$

47. $y = (x - 2)^2$

Let $x = t$, then $y = (t - 2)^2$.

$\mathbf{r}(t) = t\mathbf{i} + (t - 2)^2\mathbf{j}$

49. $x^2 + y^2 = 25$

Let $x = 5\cos t$, then $y = 5\sin t$.

$\mathbf{r}(t) = 5\cos t\mathbf{i} + 5\sin t\mathbf{j}$

51. $\dfrac{x^2}{16} - \dfrac{y^2}{4} = 1$

Let $x = 4\sec t$, $y = 2\tan t$.

$\mathbf{r}(t) = 4\sec t\mathbf{i} + 2\tan t\mathbf{j}$

53. $z = x^2 + y^2,\ x + y = 0$

Let $x = t$, then $y = -x = -t$

and $z = x^2 + y^2 = 2t^2$.

So, $x = t,\ y = -t,\ z = 2t^2$.

$\mathbf{r}(t) = t\mathbf{i} - t\mathbf{j} + 2t^2\mathbf{k}$

55. $x^2 + y^2 = 4$, $z = x^2$

$x = 2 \sin t$, $y = 2 \cos t$

$z = x^2 = 4 \sin^2 t$

t	0	$\dfrac{\pi}{6}$	$\dfrac{\pi}{4}$	$\dfrac{\pi}{2}$	$\dfrac{3\pi}{4}$	π
x	0	1	$\sqrt{2}$	2	$\sqrt{2}$	0
y	2	$\sqrt{3}$	$\sqrt{2}$	0	$-\sqrt{2}$	-2
z	0	1	2	4	2	0

$\mathbf{r}(t) = 2 \sin t\,\mathbf{i} + 2 \cos t\,\mathbf{j} + 4 \sin^2 t\,\mathbf{k}$

57. $x^2 + y^2 + z^2 = 4$, $x + z = 2$

Let $x = 1 + \sin t$, then $z = 2 - x = 1 - \sin t$ and $x^2 + y^2 + z^2 = 4$.

$(1 + \sin t)^2 + y^2 + (1 - \sin t)^2 = 2 + 2 \sin^2 t + y^2 = 4$

$y^2 = 2 \cos^2 t$, $y = \pm\sqrt{2} \cos t$

$x = 1 + \sin t$, $y = \pm\sqrt{2} \cos t$

$z = 1 - \sin t$

$\mathbf{r}(t) = (1 + \sin t)\mathbf{i} + \sqrt{2} \cos t\,\mathbf{j} - (1 - \sin t)\mathbf{k}$ and

$\mathbf{r}(t) = (1 + \sin t)\mathbf{i} - \sqrt{2} \cos t\,\mathbf{j} + (1 - \sin t)\mathbf{k}$

t	$-\dfrac{\pi}{2}$	$-\dfrac{\pi}{6}$	0	$\dfrac{\pi}{6}$	$\dfrac{\pi}{2}$
x	0	$\dfrac{1}{2}$	1	$\dfrac{3}{2}$	2
y	0	$\pm\dfrac{\sqrt{6}}{2}$	$\pm\sqrt{2}$	$\pm\dfrac{\sqrt{6}}{2}$	0
z	2	$\dfrac{3}{2}$	1	$\dfrac{1}{2}$	0

59. $x^2 + z^2 = 4$, $y^2 + z^2 = 4$

Subtracting, you have $x^2 - y^2 = 0$ or $y = \pm x$.

So, in the first octant, if you let $x = t$, then $x = t$, $y = t$, $z = \sqrt{4 - t^2}$.

$\mathbf{r}(t) = t\mathbf{i} + t\mathbf{j} + \sqrt{4 - t^2}\,\mathbf{k}$

61. $y^2 + z^2 = (2t \cos t)^2 + (2t \sin t)^2 = 4t^2 = 4x^2$

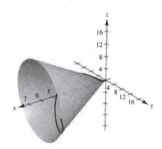

63. $\lim_{t \to \pi} (t\mathbf{i} + \cos t\,\mathbf{j} + \sin t\,\mathbf{k}) = \pi\mathbf{i} - \mathbf{j}$

65. $\lim_{t \to 0}\left[t^2\mathbf{i} + 3t\mathbf{j} + \dfrac{1 - \cos t}{t}\mathbf{k} \right] = 0$

because

$\lim_{t \to 0} \dfrac{1 - \cos t}{t} = \lim_{t \to 0} \dfrac{\sin t}{1} = 0.$ (L'Hôpital's Rule)

67. $\lim_{t \to 0}\left[e^t\mathbf{i} + \dfrac{\sin t}{t}\mathbf{j} + e^{-t}\mathbf{k} \right] = \mathbf{i} + \mathbf{j} + \mathbf{k}$

because

$\lim_{t \to 0} \dfrac{\sin t}{t} = \lim_{t \to 0} \dfrac{\cos t}{1} = 1$ (L'Hôpital's Rule)

69. $\mathbf{r}(t) = t\mathbf{i} + \dfrac{1}{t}\mathbf{j}$

Continuous on $(-\infty, 0), (0, \infty)$

71. $\mathbf{r}(t) = t\mathbf{i} + \arcsin t\,\mathbf{j} + (t - 1)\mathbf{k}$

Continuous on $[-1, 1]$

73. $\mathbf{r}(t) = \langle e^{-t}, t^2, \tan t \rangle$

Discontinuous at $t = \dfrac{\pi}{2} + n\pi$

Continuous on $\left(-\dfrac{\pi}{2} + n\pi, \dfrac{\pi}{2} + n\pi \right)$

75. $\mathbf{s}(t) = \mathbf{r}(t) + 3\mathbf{k} = t^2\mathbf{i} + (t - 3)\mathbf{j} + (t + 3)\mathbf{k}$

77. $\mathbf{s}(t) = \mathbf{r}(t) - 2\mathbf{i} = (t^2 - 2)\mathbf{i} + (t - 3)\mathbf{j} + t\mathbf{k}$

79. A vector-valued function $\mathbf{r}$ is continuous at $t = a$ if the limit of $\mathbf{r}(t)$ exists as $t \to a$ and $\lim\limits_{t \to a} \mathbf{r}(t) = \mathbf{r}(a)$.

The function $\mathbf{r}(t) = \begin{cases} \mathbf{i} + \mathbf{j} & t \geq 2 \\ -\mathbf{i} + \mathbf{j} & t < 2 \end{cases}$ is not continuous at $t = 0$.

81. One possible answer is

$$\mathbf{r}(t) = 1.5 \cos t\mathbf{i} + 1.5 \sin t\mathbf{j} + \frac{1}{\pi}t\mathbf{k}, 0 \leq t \leq 2\pi$$

Note that $\mathbf{r}(2\pi) = 1.5\mathbf{i} + 2\mathbf{k}$.

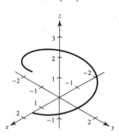

83. Let $\mathbf{r}(t) = x_1(t)\mathbf{i} + y_1(t)\mathbf{j} + z_1(t)\mathbf{k}$ and $\mathbf{u}(t) = x_2(t)\mathbf{i} + y_2(t)\mathbf{j} + z_2(t)\mathbf{k}$. Then:

$$\lim_{t \to c}\left[\mathbf{r}(t) \times \mathbf{u}(t)\right] = \lim_{t \to c}\left\{\left[y_1(t)z_2(t) - y_2(t)z_1(t)\right]\mathbf{i} - \left[x_1(t)z_2(t) - x_2(t)z_1(t)\right]\mathbf{j} + \left[x_1(t)y_2(t) - x_2(t)y_1(t)\right]\mathbf{k}\right\}$$

$$= \left[\lim_{t \to c}y_1(t)\lim_{t \to c}z_2(t) - \lim_{t \to c}y_2(t)\lim_{t \to c}z_1(t)\right]\mathbf{i} - \left[\lim_{t \to c}x_1(t)\lim_{t \to c}z_2(t) - \lim_{t \to c}x_2(t)\lim_{t \to c}z_1(t)\right]\mathbf{j}$$

$$+ \left[\lim_{t \to c}x_1(t)\lim_{t \to c}y_2(t) - \lim_{t \to c}x_2(t)\lim_{t \to c}y_1(t)\right]\mathbf{k}$$

$$= \left[\lim_{t \to c}x_1(t)\mathbf{i} + \lim_{t \to c}y_1(t)\mathbf{j} + \lim_{t \to c}z_1(t)\mathbf{k}\right] \times \left[\lim_{t \to c}x_2(t)\mathbf{i} + \lim_{t \to c}y_2(t)\mathbf{j} + \lim_{t \to c}z_2(t)\mathbf{k}\right]$$

$$= \lim_{t \to c}\mathbf{r}(t) \times \lim_{t \to c}\mathbf{u}(t)$$

85. Let $\mathbf{r}(t) = x(t)\mathbf{i} + y(t)\mathbf{j} + z(t)\mathbf{k}$. Because $\mathbf{r}$ is continuous at $t = c$, then $\lim\limits_{t \to c}\mathbf{r}(t) = \mathbf{r}(c)$.

$\mathbf{r}(c) = x(c)\mathbf{i} + y(c)\mathbf{j} + z(c)\mathbf{k} \Rightarrow x(c), y(c), z(c)$ are defined at c.

$$\|\mathbf{r}\| = \sqrt{(x(t))^2 + (y(t))^2 + (z(t))^2}$$

$$\lim_{t \to c}\|\mathbf{r}\| = \sqrt{(x(c))^2 + (y(c))^2 + (z(c))^2} = \|\mathbf{r}(c)\|$$

So, $\|\mathbf{r}\|$ is continuous at c.

87. $\mathbf{r}(t) = t^2\mathbf{i} + (9t - 20)\mathbf{j} + t^2\mathbf{k}$

$\mathbf{u}(s) = (3s + 4)\mathbf{i} + s^2\mathbf{j} + (5s - 4)\mathbf{k}$.

Equating components:

$$t^2 = 3s + 4$$
$$9t - 20 = s^2$$
$$t^2 = 5s - 4$$

So, $3s + 4 = 5s - 4 \Rightarrow s = 4$

$9t - 20 = s^2 = 16 \Rightarrow t = 4$.

The paths intersect at the same time $t = 4$ at the point $(16, 16, 16)$. The particles collide.

89. No, not necessarily. See Exercise 88.

91. True

93. True. See Exercises 87 and 88.

Section 12.2   Differentiation and Integration of Vector-Valued Functions

1. $\mathbf{r}(t) = t^2\mathbf{i} + t\mathbf{j}, t_0 = 2$

$x(t) = t^2, y(t) = t$

$x = y^2$

$\mathbf{r}(2) = 4\mathbf{i} + 2\mathbf{j}$

$\mathbf{r}'(t) = 2t\mathbf{i} + \mathbf{j}$

$\mathbf{r}'(2) = 4\mathbf{i} + \mathbf{j}$

$\mathbf{r}'(t_0)$ is tangent to the curve at t_0.

3. $\mathbf{r}(t) = \cos t\mathbf{i} + \sin t\mathbf{j}, \; t_0 = \dfrac{\pi}{2}$

$x(t) = \cos t, \; y(t) = \sin t$

$x^2 + y^2 = 1$

$\mathbf{r}\left(\dfrac{\pi}{2}\right) = \mathbf{j}$

$\mathbf{r}'(t) = -\sin t\mathbf{i} + \cos t\mathbf{j}$

$\mathbf{r}'\left(\dfrac{\pi}{2}\right) = -\mathbf{i}$

$\mathbf{r}'(t_0)$ is tangent to the curve at t_0.

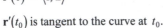
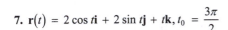

5. $\mathbf{r}(t) = \langle e^t, e^{2t} \rangle, \; t_0 = 0$

$x(t) = e^t, \; y(t) = e^{2t} = \left(e^t\right)^2$

$y = x^2, \; x > 0$

$\mathbf{r}(0) = \langle 1, 1 \rangle$

$\mathbf{r}'(t) = \langle e^t, 2e^{2t} \rangle$

$\mathbf{r}'(0) = \langle 1, 2 \rangle$

$\mathbf{r}'(t_0)$ is tangent to the curve at t_0.

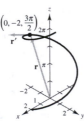

7. $\mathbf{r}(t) = 2\cos t\mathbf{i} + 2\sin t\mathbf{j} + t\mathbf{k}, \; t_0 = \dfrac{3\pi}{2}$

$x^2 + y^2 = 4, \; z = t$

$\mathbf{r}'(t) = -2\sin t\mathbf{i} + 2\cos t\mathbf{j} + \mathbf{k}$

$\mathbf{r}\left(\dfrac{3\pi}{2}\right) = -2\mathbf{j} + \dfrac{3\pi}{2}\mathbf{k}$

$\mathbf{r}'\left(\dfrac{3\pi}{2}\right) = 2\mathbf{i} + \mathbf{k}$

23. $\mathbf{r}(t) = 4\cos t\mathbf{i} + 4\sin t\mathbf{j}$

(a) $\mathbf{r}'(t) = -4\sin t\mathbf{i} + 4\cos t\mathbf{j}$

(b) $\mathbf{r}''(t) = -4\cos t\mathbf{i} - 4\sin t\mathbf{j}$

(c) $\mathbf{r}'(t) \cdot \mathbf{r}''(t) = (-4\sin t)(-4\cos t) + 4\cos t(-4\sin t) = 0$

25. $\mathbf{r}(t) = \dfrac{1}{2}t^2\mathbf{i} - t\mathbf{j} + \dfrac{1}{6}t^3\mathbf{k}$

(a) $\mathbf{r}'(t) = t\mathbf{i} - \mathbf{j} + \dfrac{1}{2}t^2\mathbf{k}$

(b) $\mathbf{r}''(t) = \mathbf{i} + t\mathbf{k}$

(c) $\mathbf{r}'(t) \cdot \mathbf{r}''(t) = t(1) + (-1)(0) + \dfrac{1}{2}t^2(t) = t + \dfrac{1}{2}t^3$

(d) $\mathbf{r}'(t) \times \mathbf{r}''(t) = \begin{vmatrix} \mathbf{i} & \mathbf{j} & \mathbf{k} \\ t & -1 & \dfrac{1}{2}t^2 \\ 1 & 0 & t \end{vmatrix} = (-t)\mathbf{i} - \left(t^2 - \dfrac{1}{2}t^2\right)\mathbf{j} + \mathbf{k} = -t\mathbf{i} - \dfrac{1}{2}t^2\mathbf{j} + \mathbf{k}$

9. $\mathbf{r}(t) = t^3\mathbf{i} - 3t\mathbf{j}$

$\mathbf{r}'(t) = 3t^2\mathbf{i} - 3\mathbf{j}$

11. $\mathbf{r}(t) = \langle 2\cos t, 5\sin t \rangle$

$\mathbf{r}'(t) = \langle -2\sin t, 5\cos t \rangle$

13. $\mathbf{r}(t) = 6t\mathbf{i} - 7t^2\mathbf{j} + t^3\mathbf{k}$

$\mathbf{r}'(t) = 6\mathbf{i} - 14t\mathbf{j} + 3t^2\mathbf{k}$

15. $\mathbf{r}(t) = a\cos^3 t\mathbf{i} + a\sin^3 t\mathbf{j} + \mathbf{k}$

$\mathbf{r}'(t) = -3a\cos^2 t\sin t\mathbf{i} + 3a\sin^2 t\cos t\mathbf{j}$

17. $\mathbf{r}(t) = e^{-t}\mathbf{i} + 4\mathbf{j} + 5te^t\mathbf{k}$

$\mathbf{r}'(t) = -e^{-t}\mathbf{i} + \left(5e^t + 5te^t\right)\mathbf{k}$

19. $\mathbf{r}(t) = \langle t\sin t, t\cos t, t \rangle$

$\mathbf{r}'(t) = \langle \sin t + t\cos t, \cos t - t\sin t, 1 \rangle$

21. $\mathbf{r}(t) = t^3\mathbf{i} + \dfrac{1}{2}t^2\mathbf{j}$

(a) $\mathbf{r}'(t) = 3t^2\mathbf{i} + t\mathbf{j}$

(b) $\mathbf{r}''(t) = 6t\mathbf{i} + \mathbf{j}$

(c) $\mathbf{r}'(t) \cdot \mathbf{r}''(t) = 3t^2(6t) + t = 18t^3 + t$

27. $\mathbf{r}(t) = \langle \cos t + t \sin t, \sin t - t \cos t, t \rangle$

    (a) $\mathbf{r}'(t) = \langle -\sin t + \sin t + t \cos t, \cos t - \cos t + t \sin t, 1 \rangle$

         $= \langle t \cos t, t \sin t, 1 \rangle$

    (b) $\mathbf{r}''(t) = \langle \cos t - t \sin t, \sin t + t \cos t, 0 \rangle$

    (c) $\mathbf{r}'(t) \cdot \mathbf{r}''(t) = (t \cos t)(\cos t - t \sin t) + t \sin t(\sin t + t \cos t) + 1(0)$

         $= t \cos^2 t - t^2 \cos t \sin t + t \sin^2 t + t^2 \sin t \cos t$

         $= t(\cos^2 t + \sin^2 t) = t$

    (d) $\mathbf{r}'(t) \times \mathbf{r}''(t) = \begin{bmatrix} \mathbf{i} & \mathbf{j} & \mathbf{k} \\ t \cos t & t \sin t & 1 \\ \cos t - t \sin t & \sin t + t \cos t & 0 \end{bmatrix}$

         $= (-\sin t - t \cos t)\mathbf{i} + (\cos t - t \sin t)\mathbf{j} + (t \cos t \sin t + t^2 \cos^2 t - t \sin t \cos t + t^2 \sin^2 t)\mathbf{k}$

         $= (-\sin t - t \cos t)\mathbf{i} + (\cos t - t \sin t)\mathbf{j} + t^2 \mathbf{k}$

         $= \langle -\sin t - t \cos t, \cos t - t \sin t, t^2 \rangle$

29. $\mathbf{r}(t) = t^2 \mathbf{i} + t^3 \mathbf{j}$

    $\mathbf{r}'(t) = 2t\mathbf{i} + 3t^2 \mathbf{j}$

    $\mathbf{r}'(0) = \mathbf{0}$

    Smooth on $(-\infty, 0), (0, \infty)$

31. $\mathbf{r}(\theta) = 2 \cos^3 \theta \mathbf{i} + 3 \sin^3 \theta \mathbf{j}$

    $\mathbf{r}'(\theta) = -6 \cos^2 \theta \sin \theta \mathbf{i} + 9 \sin^2 \theta \cos \theta \mathbf{j}$

    $\mathbf{r}'\left(\dfrac{n\pi}{2}\right) = \mathbf{0}$

    Smooth on $\left(\dfrac{n\pi}{2}, \dfrac{(n+1)\pi}{2}\right)$, n any integer.

33. $\mathbf{r}(\theta) = (\theta - 2 \sin \theta)\mathbf{i} + (1 - 2 \cos \theta)\mathbf{j}$

    $\mathbf{r}'(\theta) = (1 - 2 \cos \theta)\mathbf{i} + (2 \sin \theta)\mathbf{j}$

    $\mathbf{r}'(\theta) \neq \mathbf{0}$ for any value of θ

    Smooth on $(-\infty, \infty)$

35. $\mathbf{r}(t) = (t - 1)\mathbf{i} + \dfrac{1}{t}\mathbf{j} - t^2 \mathbf{k}$

    $\mathbf{r}'(t) = \mathbf{i} - \dfrac{1}{t^2}\mathbf{j} - 2t\mathbf{k} \neq \mathbf{0}$

    $\mathbf{r}$ is smooth or all $t \neq 0$: $(-\infty, 0), (0, \infty)$

37. $\mathbf{r}(t) = t\mathbf{i} - 3t\mathbf{j} + \tan t\mathbf{k}$

    $\mathbf{r}'(t) = \mathbf{i} - 3\mathbf{j} + \sec^2 t\mathbf{k} \neq \mathbf{0}$

    $\mathbf{r}$ is smooth for all $t \neq \dfrac{\pi}{2} + n\pi = \dfrac{2n+1}{2}\pi$.

    Smooth on intervals of form $\left(-\dfrac{\pi}{2} + n\pi, \dfrac{\pi}{2} + n\pi\right)$,

    n is an integer.

39. $\mathbf{r}(t) = t\mathbf{i} + 3t\mathbf{j} + t^2\mathbf{k}, \mathbf{u}(t) = 4t\mathbf{i} + t^2\mathbf{j} + t^3\mathbf{k}$

$\mathbf{r}'(t) = \mathbf{i} + 3\mathbf{j} + 2t\mathbf{k}, \mathbf{u}'(t) = 4\mathbf{i} + 2t\mathbf{j} + 3t^2\mathbf{k}$

(a) $\mathbf{r}'(t) = \mathbf{i} + 3\mathbf{j} + 2t\mathbf{k}$

(b) $\dfrac{d}{dt}\big[3\mathbf{r}(t) - \mathbf{u}(t)\big] = 3\mathbf{r}'(t) - \mathbf{u}'(t)$

$$= 3(\mathbf{i} + 3\mathbf{j} + 2t\mathbf{k}) - \left(4\mathbf{i} + 2t\mathbf{j} + 3t^2\mathbf{k}\right)$$

$$= (3 - 4)\mathbf{i} + (9 - 2t)\mathbf{j} + \left(6t - 3t^2\right)\mathbf{k}$$

$$= -\mathbf{i} + (9 - 2t)\mathbf{j} + \left(6t - 3t^2\right)\mathbf{k}$$

(c) $\dfrac{d}{dt}(5t)\mathbf{u}(t) = (5t)\mathbf{u}'(t) + 5\mathbf{u}(t)$

$$= 5t\left(4\mathbf{i} + 2t\mathbf{j} + 3t^2\mathbf{k}\right) + 5\left(4t\mathbf{i} + t^2\mathbf{j} + t^3\mathbf{k}\right)$$

$$= (20t + 20t)\mathbf{i} + \left(10t^2 + 5t^2\right)\mathbf{j} + \left(15t^3 + 5t^3\right)\mathbf{k}$$

$$= 40t\mathbf{i} + 15t^2\mathbf{j} + 20t^3\mathbf{k}$$

(d) $\dfrac{d}{dt}\big[\mathbf{r}(t) \cdot \mathbf{u}(t)\big] = \mathbf{r}(t) \cdot \mathbf{u}'(t) + \mathbf{r}'(t) \cdot \mathbf{u}(t)$

$$= \Big[(t)(4) + (3t)(2t) + \left(t^2\right)\left(3t^2\right)\Big] + \Big[(1)(4t) + (3)\left(t^2\right) + (2t)\left(t^3\right)\Big]$$

$$= \left(4t + 6t^2 + 3t^4\right) + \left(4t + 3t^2 + 2t^4\right)$$

$$= 8t + 9t^2 + 5t^4$$

(e) $\dfrac{d}{dt}\big[\mathbf{r}(t) \times \mathbf{u}(t)\big] = \mathbf{r}(t) \times \mathbf{u}'(t) + \mathbf{r}'(t) \times \mathbf{u}(t)$

$$= \Big[7t^3\mathbf{i} + \left(4t^2 - 3t^3\right)\mathbf{j} + \left(2t^2 - 12t\right)\mathbf{k}\Big] + \Big[t^3\mathbf{i} + \left(8t^2 - t^3\right)\mathbf{j} + \left(t^2 - 12t\right)\mathbf{k}\Big]$$

$$= 8t^3\mathbf{i} + \left(12t^2 - 4t^3\right)\mathbf{j} + \left(3t^2 - 24t\right)\mathbf{k}$$

(f) $\dfrac{d}{dt}\mathbf{r}(2t) = 2\mathbf{r}'(2t)$

$$= 2\big[\mathbf{i} + 3\mathbf{j} + 2(2t)\mathbf{k}\big]$$

$$= 2\mathbf{i} + 6\mathbf{j} + 8t\mathbf{k}$$

41. $\mathbf{r}(t) = t\mathbf{i} + 2t^2\mathbf{j} + t^3\mathbf{k}, \mathbf{u}(t) = t^4\mathbf{k}$

(a) $\mathbf{r}(t) \cdot \mathbf{u}(t) = t^7$

(i) $D_t\big[\mathbf{r}(t) \cdot \mathbf{u}(t)\big] = 7t^6$

(ii) Alternate Solution:

$$D_t\big[\mathbf{r}(t) \cdot \mathbf{u}(t)\big] = \mathbf{r}(t) \cdot \mathbf{u}'(t) + \mathbf{r}'(t) \cdot \mathbf{u}(t) = \left(t\mathbf{i} + 2t^2\mathbf{j} + t^3\mathbf{k}\right) \cdot \left(4t^3\mathbf{k}\right) + \left(\mathbf{i} + 4t\mathbf{j} + 3t^2\mathbf{k}\right) \cdot \left(t^4\mathbf{k}\right) = 4t^6 + 3t^6 = 7t^6$$

(b) $\mathbf{r}(t) \times \mathbf{u}(t) = \begin{vmatrix} \mathbf{i} & \mathbf{j} & \mathbf{k} \\ t & 2t^2 & t^3 \\ 0 & 0 & t^4 \end{vmatrix} = 2t^6\mathbf{i} - t^5\mathbf{j}$

(i) $D_t\big[\mathbf{r}(t) \times \mathbf{u}(t)\big] = 12t^5\mathbf{i} - 5t^4\mathbf{j}$

(ii) Alternate Solution: $D_t\big[\mathbf{r}(t) \times \mathbf{u}(t)\big] = \mathbf{r}(t) \times \mathbf{u}'(t) \times \mathbf{r}'(t) \times \mathbf{u}(t) = \begin{vmatrix} \mathbf{i} & \mathbf{j} & \mathbf{k} \\ t & 2t^2 & t^3 \\ 0 & 0 & 4t^3 \end{vmatrix} + \begin{vmatrix} \mathbf{i} & \mathbf{j} & \mathbf{k} \\ 1 & 4t & 3t^2 \\ 0 & 0 & t^4 \end{vmatrix} = 12t^5\mathbf{i} - 5t^4\mathbf{j}$

43. $\int (2t\mathbf{i} + \mathbf{j} + \mathbf{k})\,dt = t^2\mathbf{i} + t\mathbf{j} + t\mathbf{k} + \mathbf{C}$

45. $\int \left(\dfrac{1}{t}\mathbf{i} + \mathbf{j} - t^{3/2}\mathbf{k}\right) dt = \ln|t|\mathbf{i} + t\mathbf{j} - \dfrac{2}{5}t^{5/2}\mathbf{k} + \mathbf{C}$

47. $\int \left[(2t - 1)\mathbf{i} + 4t^3\mathbf{j} + 3\sqrt{t}\mathbf{k}\right] dt = \left(t^2 - t\right)\mathbf{i} + t^4\mathbf{j} + 2t^{3/2}\mathbf{k} + \mathbf{C}$

49. $\int \left[\sec^2 t\mathbf{i} + \dfrac{1}{1 + t^2}\mathbf{j}\right] dt = \tan t\mathbf{i} + \arctan t\mathbf{j} + \mathbf{C}$

51. $\int_0^1 (8t\mathbf{i} + t\mathbf{j} - \mathbf{k})\,dt = \left[4t^2\mathbf{i}\right]_0^1 + \left[\dfrac{t^2}{2}\mathbf{j}\right]_0^1 - \left[t\mathbf{k}\right]_0^1 = 4\mathbf{i} + \dfrac{1}{2}\mathbf{j} - \mathbf{k}$

53. $\int_0^{\pi/2} \left[(a\cos t)\mathbf{i} + (a\sin t)\mathbf{j} + \mathbf{k}\right] dt = \left[a\sin t\mathbf{i}\right]_0^{\pi/2} - \left[a\cos t\mathbf{j}\right]_0^{\pi/2} + \left[t\mathbf{k}\right]_0^{\pi/2} = a\mathbf{i} + a\mathbf{j} + \dfrac{\pi}{2}\mathbf{k}$

55. $\int_0^2 (t\mathbf{i} + e^t\mathbf{j} - te^t\mathbf{k})\,dt = \left[\dfrac{t^2}{2}\mathbf{i}\right]_0^2 + \left[e^t\mathbf{j}\right]_0^2 - \left[(t - 1)e^t\mathbf{k}\right]_0^2 = 2\mathbf{i} + \left(e^2 - 1\right)\mathbf{j} - \left(e^2 + 1\right)\mathbf{k}$

57. $\mathbf{r}(t) = \int\left(4e^{2t}\mathbf{i} + 3e^t\mathbf{j}\right) dt = 2e^{2t}\mathbf{i} + 3e^t\mathbf{j} + \mathbf{C}$

$\mathbf{r}(0) = 2\mathbf{i} + 3\mathbf{j} + \mathbf{C} = 2\mathbf{i} \Rightarrow \mathbf{C} = -3\mathbf{j}$

$\mathbf{r}(t) = 2e^{2t}\mathbf{i} + 3\left(e^t - 1\right)\mathbf{j}$

59. $\mathbf{r}'(t) = \int -32\mathbf{j}\,dt = -32t\mathbf{j} + \mathbf{C}_1$

$\mathbf{r}'(0) = \mathbf{C}_1 = 600\sqrt{3}\mathbf{i} + 600\mathbf{j}$

$\mathbf{r}'(t) = 600\sqrt{3}\mathbf{i} + (600 - 32t)\mathbf{j}$

$\mathbf{r}(t) = \int \left[600\sqrt{3}\mathbf{i} + (600 - 32t)\mathbf{j}\right] dt$

$\quad = 600\sqrt{3}t\mathbf{i} + \left(600t - 16t^2\right)\mathbf{j} + \mathbf{C}$

$\mathbf{r}(0) = \mathbf{C} = \mathbf{0}$

$\mathbf{r}(t) = 600\sqrt{3}t\mathbf{i} + \left(600t - 16t^2\right)\mathbf{j}$

61. $\mathbf{r}(t) = \int\left(te^{-t^2}\mathbf{i} - e^{-t}\mathbf{j} + \mathbf{k}\right) dt = -\dfrac{1}{2}e^{-t^2}\mathbf{i} + e^{-t}\mathbf{j} + t\mathbf{k} + \mathbf{C}$

$\mathbf{r}(0) = -\dfrac{1}{2}\mathbf{i} + \mathbf{j} + \mathbf{C} = \dfrac{1}{2}\mathbf{i} - \mathbf{j} + \mathbf{k} \Rightarrow \mathbf{C} = \mathbf{i} - 2\mathbf{j} + \mathbf{k}$

$\mathbf{r}(t) = \left(1 - \dfrac{1}{2}e^{-t^2}\right)\mathbf{i} + \left(e^{-t} - 2\right)\mathbf{j} + (t + 1)\mathbf{k}$

$\quad = \left(\dfrac{2 - e^{-t^2}}{2}\right)\mathbf{i} + \left(e^{-t} - 2\right)\mathbf{j} + (t + 1)\mathbf{k}$

63. See "Definition of the Derivative of a Vector-Valued Function" and Figure 12.8 on page 824.

65. At $t = t_0$, the graph of $\mathbf{u}(t)$ is increasing in the x, y, and z directions simultaneously.

67. Let $\mathbf{r}(t) = x(t)\mathbf{i} + y(t)\mathbf{j} + z(t)\mathbf{k}$. Then

$c\mathbf{r}(t) = cx(t)\mathbf{i} + cy(t)\mathbf{j} + cz(t)\mathbf{k}$ and

$\dfrac{d}{dt}\left[c\mathbf{r}(t)\right] = cx'(t)\mathbf{i} + cy'(t)\mathbf{j} + cz'(t)\mathbf{k}$

$\quad = c\left[x'(t)\mathbf{i} + y'(t)\mathbf{j} + z'(t)\mathbf{k}\right] = c\mathbf{r}'(t).$

69. Let $\mathbf{r}(t) = x(t)\mathbf{i} + y(t)\mathbf{j} + z(t)\mathbf{k}$, then $w(t)\mathbf{r}(t) = w(t)x(t)\mathbf{i} + w(t)y(t)\mathbf{j} + w(t)z(t)\mathbf{k}$.

$\dfrac{d}{dt}\left[w(t)\mathbf{r}(t)\right] = \left[w(t)x'(t) + w'(t)x(t)\right]\mathbf{i} + \left[w(t)y'(t) + w'(t)y(t)\right]\mathbf{j} + \left[w(t)z'(t) + w'(t)z(t)\right]\mathbf{k}$

$\quad = w(t)\left[x'(t)\mathbf{i} + y'(t)\mathbf{j} + z'(t)\mathbf{k}\right] + w'(t)\left[x(t)\mathbf{i} + y(t)\mathbf{j} + z(t)\mathbf{k}\right] = w(t)\mathbf{r}'(t) + w'(t)\mathbf{r}(t)$

71. Let $\mathbf{r}(t) = x(t)\mathbf{i} + y(t)\mathbf{j} + z(t)\mathbf{k}$. Then $\mathbf{r}(w(t)) = x(w(t))\mathbf{i} + y(w(t))\mathbf{j} + z(w(t))\mathbf{k}$ and

$$\frac{d}{dt}\big[\mathbf{r}(w(t))\big] = x'(w(t))w'(t)\mathbf{i} + y'(w(t))w'(t)\mathbf{j} + z'(w(t))w'(t)\mathbf{k} \quad \text{(Chain Rule)}$$

$$= w'(t)\big[x'(w(t))\mathbf{i} + y'(w(t))\mathbf{j} + z'(w(t))\mathbf{k}\big] = w'(t)\mathbf{r}'(w(t)).$$

73. Let $\mathbf{r}(t) = x_1(t)\mathbf{i} + y_1(t)\mathbf{j} + z_1(t)\mathbf{k}$, $\mathbf{u}(t) = x_2(t)\mathbf{i} + y_2(t)\mathbf{j} + z_2(t)\mathbf{k}$, and $\mathbf{v}(t) = x_3(t)\mathbf{i} + y_3(t)\mathbf{j} + z_3(t)\mathbf{k}$. Then:

$$\mathbf{r}(t) \cdot \big[\mathbf{u}(t) \times \mathbf{v}(t)\big] = x_1(t)\big[y_2(t)z_3(t) - z_2(t)y_3(t)\big] - y_1(t)\big[x_2(t)z_3(t) - z_2(t)x_3(t)\big] + z_1(t)\big[x_2(t)y_3(t) - y_2(t)x_3(t)\big]$$

$$\frac{d}{dt}\big[\mathbf{r}(t) \cdot (\mathbf{u}(t) \times \mathbf{v}(t))\big] = x_1(t)y_2(t)z_3'(t) + x_1(t)y_2'(t)z_3(t) + x_1'(t)y_2(t)z_3(t) - x_1(t)y_3(t)z_2'(t)$$

$$- x_1(t)y_3'(t)z_2(t) - x_1'(t)y_3(t)z_2(t) - y_1(t)x_2(t)z_3'(t) - y_1(t)x_2'(t)z_3(t) - y_1'(t)x_2(t)z_3(t)$$

$$+ y_1(t)z_2(t)x_3'(t) + y_1(t)z_2'(t)x_3(t) + y_1'(t)z_2(t)x_3(t) + z_1(t)x_2(t)y_3'(t) + z_1(t)x_2'(t)y_3(t)$$

$$+ z_1'(t)x_2(t)y_3(t) - z_1(t)y_2(t)x_3'(t) - z_1(t)y_2'(t)x_3(t) - z_1'(t)y_2(t)x_3(t)$$

$$= \Big\{x_1'(t)\big[y_2(t)z_3(t) - y_3(t)z_2(t)\big] + y_1'(t)\big[-x_2(t)z_3(t) + z_2(t)x_3(t)\big] + z_1'(t)\big[x_2(t)y_3(t) - y_2(t)x_3(t)\big]\Big\}$$

$$+ \Big\{x_1(t)\big[y_2'(t)z_3(t) - y_3(t)z_2'(t)\big] + y_1(t)\big[-x_2'(t)z_3(t) + z_2'(t)x_3(t)\big] + z_1(t)\big[x_2'(t)y_3(t) - y_2'(t)x_3(t)\big]\Big\}$$

$$+ \Big\{x_1(t)\big[y_2(t)z_3'(t) - y_3'(t)z_2(t)\big] + y_1(t)\big[-x_2(t)z_3'(t) + z_2(t)x_3'(t)\big] + z_1(t)\big[x_2(t)y_3'(t) - y_2(t)x_3'(t)\big]\Big\}$$

$$= \mathbf{r}'(t) \cdot \big[\mathbf{u}(t) \times \mathbf{v}(t)\big] + \mathbf{r}(t) \cdot \big[\mathbf{u}'(t) \times \mathbf{v}(t)\big] + \mathbf{r}(t) \cdot \big[\mathbf{u}(t) \times \mathbf{v}'(t)\big]$$

75. $\mathbf{r}(t) = (t - \sin t)\mathbf{i} + (1 - \cos t)\mathbf{j}$

(a)

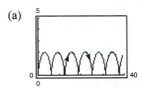

The curve is a cycloid.

(b) $\mathbf{r}'(t) = (1 - \cos t)\mathbf{i} + \sin t\,\mathbf{j}$

$\mathbf{r}''(t) = \sin t\,\mathbf{i} + \cos t\,\mathbf{j}$

$\|\mathbf{r}'(t)\| = \sqrt{1 - 2\cos t + \cos^2 t + \sin^2 t}$

$\qquad = \sqrt{2 - 2\cos t}$

Minimum of $\|\mathbf{r}'(t)\|$ is 0, $(t = 0)$.

Maximum of $\|\mathbf{r}'(t)\|$ is 2, $(t = \pi)$.

$\|\mathbf{r}''(t)\| = \sqrt{\sin^2 t + \cos^2 t} = 1$

Minimum and maximum of $\|\mathbf{r}'(t)\|$ is 1.

77. $\mathbf{r}(t) = e^t \sin t\,\mathbf{i} + e^t \cos t\,\mathbf{j}$

$\mathbf{r}'(t) = (e^t \cos t + e^t \sin t)\mathbf{i} + (e^t \cos t - e^t \sin t)\mathbf{j}$

$\mathbf{r}''(t) = (-e^t \sin t + e^t \cos t + e^t \sin t + e^t \cos t)\mathbf{i} + (e^t \cos t - e^t \sin t - e^t \sin t - e^t \cos t)\mathbf{j} = 2e^t \cos t\,\mathbf{i} - 2e^t \sin t\,\mathbf{j}$

$\mathbf{r}(t) \cdot \mathbf{r}''(t) = 2e^{2t} \sin t \cos t - 2e^{2t} \sin t \cos t = 0$

So, $\mathbf{r}(t)$ is always perpendicular to $\mathbf{r}''(t)$.

79. True

81. False. Let $\mathbf{r}(t) = \cos t\mathbf{i} + \sin t\mathbf{j} + \mathbf{k}$.

$$\|\mathbf{r}(t)\| = \sqrt{2}$$

$$\frac{d}{dt}\Big[\|\mathbf{r}(t)\|\Big] = 0$$

$$\mathbf{r}'(t) = -\sin t\mathbf{i} + \cos t\mathbf{j}$$

$$\|\mathbf{r}'(t)\| = 1$$

Section 12.3   Velocity and Acceleration

1. $\mathbf{r}(t) = 3t\mathbf{i} + (t - 1)\mathbf{j},\ (3, 0)$

  (a) $\mathbf{v}(t) = \mathbf{r}'(t) = 3\mathbf{i} + \mathbf{j}$

      Speed $= \|\mathbf{v}(t)\| = \sqrt{3^2 + 1^2} = \sqrt{10}$

      $\mathbf{a}(t) = \mathbf{r}''(t) = \mathbf{0}$

  (b) At $(3, 0), t = 1$.

      $\mathbf{v}(1) = 3\mathbf{i} + \mathbf{j}, \mathbf{a}(1) = \mathbf{0}$

  (c) $x = 3t,\ y = t - 1$

      $y = \dfrac{x}{3} - 1$, line

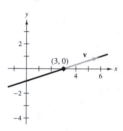

3. $\mathbf{r}(t) = t^2\mathbf{i} + t\mathbf{j},\ (4, 2)$

  (a) $\mathbf{v}(t) = \mathbf{r}'(t) = 2t\mathbf{i} + \mathbf{j}$

      Speed $= \|\mathbf{v}(t)\| = \sqrt{4t^2 + 1}$

      $\mathbf{a}(t) = \mathbf{r}''(t) = 2\mathbf{i}$

  (b) At $(4, 2), t = 2$.

      $\mathbf{v}(2) = 4\mathbf{i} + \mathbf{j}, \mathbf{a}(2) = 2\mathbf{j}$

  (c) $x = t^2,\ y = t$

      $x = y^2$, parabola

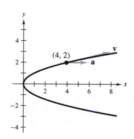

5. $\mathbf{r}(t) = 2\cos t\mathbf{i} + 2\sin t\mathbf{j},\ \left(\sqrt{2}, \sqrt{2}\right)$

  (a) $\mathbf{v}(t) = \mathbf{r}'(t) = -2\sin t\mathbf{i} + 2\cos t\mathbf{j}$

      Speed $= \|\mathbf{v}(t)\| = \sqrt{4\sin^2 t + 4\cos^2 t} = 2$

      $\mathbf{a}(t) = -2\cos t\mathbf{i} - 2\sin t\mathbf{j}$

  (b) At $\left(\sqrt{2}, \sqrt{2}\right), t = \dfrac{\pi}{4}$.

      $\mathbf{v}\left(\dfrac{\pi}{4}\right) = -\sqrt{2}\mathbf{i} + \sqrt{2}\mathbf{j}$

      $\mathbf{a}\left(\dfrac{\pi}{4}\right) = -\sqrt{2}\mathbf{i} - \sqrt{2}\mathbf{j}$

  (c) $x = 2\cos t,\ y = 2\sin t$

      $x^2 + y^2 = 4$, circle

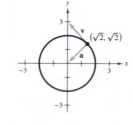

7. $\mathbf{r}(t) = \langle t - \sin t, 1 - \cos t \rangle, (\pi, 2)$

   (a) $\mathbf{v}(t) = \mathbf{v}'(t) = \langle 1 - \cos t, \sin t \rangle$

        Speed $= \|\mathbf{v}(t)\| = \sqrt{(1 - \cos t)^2 + \sin^2 t} = \sqrt{2 - 2 \cos t}$

        $\mathbf{a}(t) = \langle \sin t, \cos t \rangle$

   (b) At $(\pi, 2), t = \pi$.

        $\mathbf{v}(\pi) = \langle 2, 0 \rangle, \mathbf{a}(\pi) = \langle 0, -1 \rangle$

   (c) $x = t - \sin t, y = 1 - \cos t$

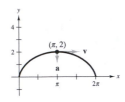

9. $\mathbf{r}(t) = t\mathbf{i} + 5t\mathbf{j} + 3t\mathbf{k}, t = 1$

   (a) $\mathbf{v}(t) = \mathbf{r}'(t) = \mathbf{i} + 5\mathbf{j} + 3\mathbf{k}$

        Speed $= \|\mathbf{v}(t)\| = \sqrt{1^2 + 5^2 + 3^2} = \sqrt{35}$

        $\mathbf{a}(t) = \mathbf{r}''(t) = \mathbf{0}$

   (b) $\mathbf{v}(1) = \mathbf{i} + 5\mathbf{j} + 3\mathbf{k}$

        $\mathbf{a}(1) = \mathbf{0}$

11. $\mathbf{r}(t) = t\mathbf{i} + t^2\mathbf{j} + \frac{1}{2}t^2\mathbf{k}, t = 4$

   (a) $\mathbf{v}(t) = \mathbf{r}'(t) = \mathbf{i} + 2t\mathbf{j} + t\mathbf{k}$

        Speed $= \|\mathbf{v}(t)\| = \sqrt{1 + 4t^2 + t^2} = \sqrt{1 + 5t^2}$

        $\mathbf{a}(t) = \mathbf{r}''(t) = 2\mathbf{j} + \mathbf{k}$

   (b) $\mathbf{v}(4) = \mathbf{i} + 8\mathbf{j} + 4\mathbf{k}$

        $\mathbf{a}(4) = 2\mathbf{j} + \mathbf{k}$

13. $\mathbf{r}(t) = t\mathbf{i} + t\mathbf{j} + \sqrt{9 - t^2}\,\mathbf{k}, t = 0$

   (a) $\mathbf{v}(t) = \mathbf{r}'(t) = \mathbf{i} + \mathbf{j} - \dfrac{t}{\sqrt{9 - t^2}}\mathbf{k}$

        Speed $= \|\mathbf{v}(t)\| = \sqrt{1 + 1 + \dfrac{t^2}{9 - t^2}} = \sqrt{\dfrac{18 - t^2}{9 - t^2}}$

        $\mathbf{a}(t) = \mathbf{r}''(t) = -\dfrac{9}{\left(9 - t^2\right)^{3/2}}\mathbf{k}$

   (b) $\mathbf{v}(0) = \mathbf{i} + \mathbf{j}$

        $\mathbf{a}(0) = -\dfrac{9}{9^{3/2}}\mathbf{k} = -\dfrac{1}{3}\mathbf{k}$

15. $\mathbf{r}(t) = \langle 4t, 3 \cos t, 3 \sin t \rangle, t = \pi$

   (a) $\mathbf{v}(t) = \mathbf{r}'(t) = \langle 4, -3 \sin t, 3 \cos t \rangle$

        Speed $= \|\mathbf{v}(t)\| = \sqrt{4^2 + (-3 \sin t)^2 + (3 \cos t)^2} = \sqrt{16 + 9} = 5$

        $\mathbf{a}(t) = \mathbf{r}''(t) = \langle 0, -3 \cos t, -3 \sin t \rangle$

   (b) $\mathbf{v}(\pi) = \langle 4, 0, -3 \rangle$

        $\mathbf{a}(\pi) = \langle 0, 3, 0 \rangle$

17. $\mathbf{r}(t) = \left\langle e^t \cos t, e^t \sin t, e^t \right\rangle, \ t = 0$

    (a) $\mathbf{v}(t) = \mathbf{r}'(t) = \left\langle e^t \cos t - e^t \sin t, e^t \sin t + e^t \cos t, e^t \right\rangle$

$$\text{Speed} = \|\mathbf{v}(t)\| = \sqrt{e^{2t}(\cos t - \sin t)^2 + e^{2t}(\cos t + \sin t)^2 + e^{2t}}$$
$$= e^t\sqrt{3}$$

$$\mathbf{a}(t) = \mathbf{r}''(t)$$
$$= \left\langle e^t \cos t - e^t \sin t - e^t \sin t - e^t \cos t, e^t \sin t + e^t \cos t + e^t \cos t - e^t \sin t, e^t \right\rangle$$
$$= \left\langle -2e^t \sin t, 2e^t \cos t, e^t \right\rangle$$

    (b) $\mathbf{v}(0) = \langle 1, 1, 1 \rangle$

$$\mathbf{a}(0) = \langle 0, 2, 1 \rangle$$

19. $\mathbf{a}(t) = \mathbf{i} + \mathbf{j} + \mathbf{k}, \mathbf{v}(0) = \mathbf{0}, \mathbf{r}(0) = \mathbf{0}$

$\mathbf{v}(t) = \int (\mathbf{i} + \mathbf{j} + \mathbf{k})\, dt = t\mathbf{i} + t\mathbf{j} + t\mathbf{k} + \mathbf{C}$

$\mathbf{v}(0) = \mathbf{C} = \mathbf{0}, \mathbf{v}(t) = t\mathbf{i} + t\mathbf{j} + t\mathbf{k}, \mathbf{v}(t) = t(\mathbf{i} + \mathbf{j} + \mathbf{k})$

$\mathbf{r}(t) = \int (t\mathbf{i} + t\mathbf{j} + t\mathbf{k})\, dt = \dfrac{t^2}{2}(\mathbf{i} + \mathbf{j} + \mathbf{k}) + \mathbf{C}$

$\mathbf{r}(0) = \mathbf{C} = \mathbf{0}, \mathbf{r}(t) = \dfrac{t^2}{2}(\mathbf{i} + \mathbf{j} + \mathbf{k}),$

$\mathbf{r}(2) = 2(\mathbf{i} + \mathbf{j} + \mathbf{k}) = 2\mathbf{i} + 2\mathbf{j} + 2\mathbf{k}$

21. $\mathbf{a}(t) = t\mathbf{j} + t\mathbf{k}, \mathbf{v}(1) = 5\mathbf{j}, \mathbf{r}(1) = \mathbf{0}$

$\mathbf{v}(t) = \int (t\mathbf{j} + t\mathbf{k})\, dt = \dfrac{t^2}{2}\mathbf{j} + \dfrac{t^2}{2}\mathbf{k} + \mathbf{C}$

$\mathbf{v}(1) = \dfrac{1}{2}\mathbf{j} + \dfrac{1}{2}\mathbf{k} + \mathbf{C} = 5\mathbf{j} \Rightarrow \mathbf{C} = \dfrac{9}{2}\mathbf{j} - \dfrac{1}{2}\mathbf{k}$

$\mathbf{v}(t) = \left(\dfrac{t^2}{2} + \dfrac{9}{2}\right)\mathbf{j} + \left(\dfrac{t^2}{2} - \dfrac{1}{2}\right)\mathbf{k}$

$\mathbf{r}(t) = \int\left[\left(\dfrac{t^2}{2} + \dfrac{9}{2}\right)\mathbf{j} + \left(\dfrac{t^2}{2} - \dfrac{1}{2}\right)\mathbf{k}\right] dt = \left(\dfrac{t^3}{6} + \dfrac{9}{2}t\right)\mathbf{j} + \left(\dfrac{t^3}{6} - \dfrac{1}{2}t\right)\mathbf{k} + \mathbf{C}$

$\mathbf{r}(1) = \dfrac{14}{3}\mathbf{j} - \dfrac{1}{3}\mathbf{k} + \mathbf{C} = \mathbf{0} \Rightarrow \mathbf{C} = -\dfrac{14}{3}\mathbf{j} + \dfrac{1}{3}\mathbf{k}$

$\mathbf{r}(t) = \left(\dfrac{t^3}{6} + \dfrac{9}{2}t - \dfrac{14}{3}\right)\mathbf{j} + \left(\dfrac{t^3}{6} - \dfrac{1}{2}t + \dfrac{1}{3}\right)\mathbf{k}$

$\mathbf{r}(2) = \dfrac{17}{3}\mathbf{j} + \dfrac{2}{3}\mathbf{k}$

23. $\mathbf{a}(t) = -\cos t\mathbf{i} - \sin t\mathbf{j}, \mathbf{v}(0) = \mathbf{j} + \mathbf{k}, \mathbf{r}(0) = \mathbf{i}$

$\mathbf{v}(t) = \int (-\cos t\mathbf{i} - \sin t\mathbf{j})\, dt = -\sin t\mathbf{i} + \cos t\mathbf{j} + \mathbf{C}$

$\mathbf{v}(0) = \mathbf{j} + \mathbf{C} = \mathbf{j} + \mathbf{k} \Rightarrow \mathbf{C} = \mathbf{k}$

$\mathbf{v}(t) = -\sin t\mathbf{i} + \cos t\mathbf{j} + \mathbf{k}$

$\mathbf{r}(t) = \int (-\sin t\mathbf{i} + \cos t\mathbf{j} + \mathbf{k})\, dt = \cos t\mathbf{i} + \sin t\mathbf{j} + t\mathbf{k} + \mathbf{C}$

$\mathbf{r}(0) = \mathbf{i} + \mathbf{C} = \mathbf{i} \Rightarrow \mathbf{C} = \mathbf{0}$

$\mathbf{r}(t) = \cos t\mathbf{i} + \sin t\mathbf{j} + t\mathbf{k}$

$\mathbf{r}(2) = (\cos 2)\mathbf{i} + (\sin 2)\mathbf{j} + 2\mathbf{k}$

25. $\mathbf{r}(t) = 140(\cos 22°)t\mathbf{i} + \left(2.5 + 140(\sin 22°)t - 16t^2\right)\mathbf{j}$

$\mathbf{v}(t) = \mathbf{r}'(t) = 140(\cos 22°)\mathbf{i} + \left(140(\sin 22°) - 32t\right)\mathbf{j}$

The maximum height occurs when

$$y'(t) = 140(\sin 22°) - 32t = 0 \Rightarrow t = \frac{140 \sin 22°}{32} = \frac{35}{8} \sin 22° \approx 1.639.$$

The maximum height is

$$y = 2.5 + 140(\sin 22°)\left(\frac{35}{8} \sin 22°\right) - 16\left(\frac{35}{8} \sin 22°\right)^2 \approx 45.5 \text{ feet.}$$

When $x = 375, t = \dfrac{375}{140 \cos 22°} \approx 2.889.$

For this value of t, $y \approx 20.47$ feet.

So the ball clears the 10-foot fence.

27. $\mathbf{r}(t) = (v_0 \cos \theta)t\mathbf{i} + \left[h + (v_0 \sin \theta)t - \frac{1}{2}gt^2\right]\mathbf{j} = \frac{v_0}{\sqrt{2}}t\mathbf{i} + \left(3 + \frac{v_0}{\sqrt{2}}t - 16t^2\right)\mathbf{j}$

$\dfrac{v_0}{\sqrt{2}}t = 300$ when $3 + \dfrac{v_0}{\sqrt{2}}t - 16t^2 = 3.$

$$t = \frac{300\sqrt{2}}{v_0}, \quad \frac{v_0}{\sqrt{2}}\left(\frac{300\sqrt{2}}{v_0}\right) - 16\left(\frac{300\sqrt{2}}{v_0}\right)^2 = 0, \quad 300 - \frac{300^2(32)}{v_0^2} = 0$$

$v_0^2 = 300(32), \ v_0 = \sqrt{9600} = 40\sqrt{6}, \ v_0 = 40\sqrt{6} \approx 97.98 \text{ ft/sec}$

The maximum height is reached when the derivative of the vertical component is zero.

$$y(t) = 3 + \frac{tv_0}{\sqrt{2}} - 16t^2 = 3 + \frac{40\sqrt{6}}{\sqrt{2}}t - 16t^2 = 3 + 40\sqrt{3}t - 16t^2$$

$$y'(t) = 40\sqrt{3} - 32t = 0$$

$$t = \frac{40\sqrt{3}}{32} = \frac{5\sqrt{3}}{4}$$

Maximum height: $y\left(\dfrac{5\sqrt{3}}{4}\right) = 3 + 40\sqrt{3}\left(\dfrac{5\sqrt{3}}{4}\right) - 16\left(\dfrac{5\sqrt{3}}{4}\right)^2 = 78 \text{ feet}$

29. $x(t) = t(v_0 \cos \theta)$ or $t = \dfrac{x}{v_0 \cos \theta}$

$y(t) = t(v_0 \sin \theta) - 16t^2 + h$

$y = \dfrac{x}{v_0 \cos \theta}(v_0 \sin \theta) - 16\left(\dfrac{x^2}{v_0^2 \cos^2 \theta}\right) + h = (\tan \theta)x - \left(\dfrac{16}{v_0^2}\sec^2 \theta\right)x^2 + h$

31. $100 \, \text{mi/h} = \left(100\dfrac{\text{miles}}{\text{hr}}\right)\left(5280\dfrac{\text{feet}}{\text{mile}}\right) \Big/ \left(3600\dfrac{\text{sec}}{\text{hour}}\right) = \dfrac{440}{3}\text{ft/sec}$

(a) $\mathbf{r}(t) = \left(\dfrac{440}{3}\cos\theta_0\right)t\mathbf{i} + \left[3 + \left(\dfrac{440}{3}\sin\theta_0\right)t - 16t^2\right]\mathbf{j}$

Graphing these curves together with $y = 10$ shows that $\theta_0 = 20°$.

(b)

(c) You want

$$x(t) = \left(\dfrac{440}{3}\cos\theta\right)t \geq 400 \text{ and } y(t) = 3 + \left(\dfrac{440}{3}\sin\theta\right)t - 16t^2 \geq 10.$$

From $x(t)$, the minimum angle occurs when $t = 30/(11\cos\theta)$. Substituting this for t in $y(t)$ yields:

$$3 + \left(\dfrac{440}{3}\sin\theta\right)\left(\dfrac{30}{11\cos\theta}\right) - 16\left(\dfrac{30}{11\cos\theta}\right)^2 = 10$$

$$400\tan\theta - \dfrac{14,400}{121}\sec^2\theta = 7$$

$$\dfrac{14,400}{121}\left(1 + \tan^2\theta\right) - 400\tan\theta + 7 = 0$$

$$14,400\tan^2\theta - 48,400\tan\theta + 15,247 = 0$$

$$\tan\theta = \dfrac{48,400 \pm \sqrt{48,400^2 - 4(14,400)(15,247)}}{2(14,400)}$$

$$\theta = \tan^{-1}\left(\dfrac{48,400 - \sqrt{1,464,332,800}}{28,800}\right) \approx 19.38°$$

33. $\mathbf{r}(t) = (v \cos \theta)t\mathbf{i} + \left[(v \sin \theta)t - 16t^2\right]\mathbf{j}$

(a) You want to find the minimum initial speed v as a function of the angle θ. Because the bale must be thrown to the position $(16, 8)$, you have

$$16 = (v \cos \theta)t$$
$$8 = (v \sin \theta)t - 16t^2.$$

$t = 16/(v \cos \theta)$ from the first equation. Substituting into the second equation and solving for v, you obtain:

$$8 = (v \sin \theta)\left(\frac{16}{v \cos \theta}\right) - 16\left(\frac{16}{v \cos \theta}\right)^2$$

$$1 = 2\left(\frac{\sin \theta}{\cos \theta}\right) - 512\left(\frac{1}{v^2 \cos^2 \theta}\right)$$

$$512\left(\frac{1}{v^2 \cos^2 \theta}\right) = 2\left(\frac{\sin \theta}{\cos \theta}\right) - 1$$

$$\frac{1}{v^2} = \left(2\frac{\sin \theta}{\cos \theta} - 1\right)\frac{\cos^2 \theta}{512} = \frac{2 \sin \theta \cos \theta - \cos^2 \theta}{512}$$

$$v^2 = \frac{512}{2 \sin \theta \cos \theta - \cos^2 \theta}$$

You minimize $f(\theta) = \dfrac{512}{2 \sin \theta \cos \theta - \cos^2 \theta}$.

$$f'(\theta) = -512\left[\frac{2\cos^2 \theta - 2 \sin^2 \theta + 2 \sin \theta \cos \theta}{\left(2 \sin \theta \cos \theta - \cos^2 \theta\right)^2}\right]$$

$$f'(\theta) = 0 \Rightarrow 2 \cos(2\theta) + \sin(2\theta) = 0$$

$$\tan(2\theta) = -2$$

$$\theta \approx 1.01722 \approx 58.28°$$

Substituting into the equation for v, $v \approx 28.78$ ft/sec.

(b) If $\theta = 45°$,

$$16 = (v \cos \theta)t = v\frac{\sqrt{2}}{2}t$$

$$8 = (v \sin \theta)t - 16t^2 = v\frac{\sqrt{2}}{2}t - 16t^2$$

From part (a), $v^2 = \dfrac{512}{2\left(\sqrt{2}/2\right)\left(\sqrt{2}/2\right) - \left(\sqrt{2}/2\right)^2} = \dfrac{512}{1/2} = 1024 \Rightarrow v = 32$ ft/sec.

35. $\mathbf{r}(t) = (v_0 \cos \theta)t\mathbf{i} + \left[(v_0 \sin \theta)t - 16t^2\right]\mathbf{j}$

$(v_0 \sin \theta)t - 16t^2 = 0$ when $t = 0$ and $t = \dfrac{v_0 \sin \theta}{16}$.

The range is

$$x = (v_0 \cos \theta)t = (v_0 \cos \theta)\frac{v_0 \sin \theta}{16} = \frac{v_0^2}{32} \sin 2\theta.$$

So,

$$x = \frac{1200^2}{32}\sin(2\theta) = 3000 \Rightarrow \sin 2\theta = \frac{1}{15} \Rightarrow \theta \approx 1.91°.$$

37. (a) $\theta = 10°$, $v_0 = 66$ ft/sec

$$\mathbf{r}(t) = (66 \cos 10°)t\mathbf{i} + \left[0 + (66 \sin 10°)t - 16t^2\right]\mathbf{j}$$

$$\mathbf{r}(t) \approx (65t)\mathbf{i} + \left(11.46t - 16t^2\right)\mathbf{j}$$

Maximum height: 2.052 feet

Range: 46.557 feet

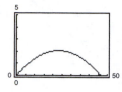

(b) $\theta = 10°$, $v_0 = 146$ ft/sec

$$\mathbf{r}(t) = (146 \cos 10°)t\mathbf{i} + \left[0 + (146 \sin 10°)t - 16t^2\right]\mathbf{j}$$

$$\mathbf{r}(t) \approx (143.78t)\mathbf{i} + \left(25.35t - 16t^2\right)\mathbf{j}$$

Maximum height: 10.043 feet

Range: 227.828 feet

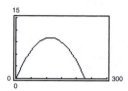

(c) $\theta = 45°$, $v_0 = 66$ ft/sec

$$\mathbf{r}(t) = (66 \cos 45°)t\mathbf{i} + \left[0 + (66 \sin 45°)t - 16t^2\right]\mathbf{j}$$

$$\mathbf{r}(t) \approx (46.67t)\mathbf{i} + \left(46.67t - 16t^2\right)\mathbf{j}$$

Maximum height: 34.031 feet

Range: 136.125 feet

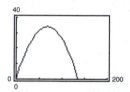

(d) $\theta = 45°$, $v_0 = 146$ ft/sec

$$\mathbf{r}(t) = (146 \cos 45°)t\mathbf{i} + \left[0 + (146 \sin 45°)t - 16t^2\right]\mathbf{j}$$

$$\mathbf{r}(t) \approx (103.24t)\mathbf{i} + \left(103.24t - 16t^2\right)\mathbf{j}$$

Maximum height: 166.531 feet

Range: 666.125 feet

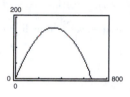

(e) $\theta = 60°$, $v_0 = 66$ ft/sec

$$\mathbf{r}(t) = (66 \cos 60°)t\mathbf{i} + \left[0 + (66 \sin 60°)t - 16t^2\right]\mathbf{j}$$

$$\mathbf{r}(t) \approx (33t)\mathbf{i} + \left(57.16t - 16t^2\right)\mathbf{j}$$

Maximum height: 51.047 feet

Range: 117.888 feet

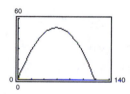

(f) $\theta = 60°$, $v_0 = 146$ ft/sec

$$\mathbf{r}(t) = (146 \cos 60°)t\mathbf{i} + \left[0 + (146 \sin 60°)t - 16t^2\right]\mathbf{j}$$

$$\mathbf{r}(t) \approx (73t)\mathbf{i} + \left(126.44t - 16t^2\right)\mathbf{j}$$

Maximum height: 249.797 feet

Range: 576.881 feet

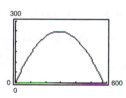

39. $\mathbf{r}(t) = (v_0 \cos \theta)t\mathbf{i} + \left[h + (v_0 \sin \theta)t - 4.9t^2\right]\mathbf{j}$

$$= (100 \cos 30°)t\mathbf{i} + \left[1.5 + (100 \sin 30°)t - 4.9t^2\right]\mathbf{j}$$

The projectile hits the ground when $-4.9t^2 + 100\left(\frac{1}{2}\right)t + 1.5 = 0 \Rightarrow t \approx 10.234$ seconds.

So the range is $(100 \cos 30°)(10.234) \approx 886.3$ meters.

The maximum height occurs when $dy/dt = 0$.

$100 \sin 30 = 9.8t \Rightarrow t \approx 5.102$ sec

The maximum height is $y = 1.5 + (100 \sin 30°)(5.102) - 4.9(5.102)^2$

$$\approx 129.1 \text{ meters.}$$

41. To find the range, set $y(t) = h + (v_0 \sin \theta)t - \frac{1}{2}gt^2 = 0$ then $0 = \left(\frac{1}{2}g\right)t^2 - (v_0 \sin \theta)t - h$. By the Quadratic Formula,

(discount the negative value)

$$t = \frac{v_0 \sin \theta + \sqrt{(-v_0 \sin \theta)^2 - 4[(1/2)g](-h)}}{2[(1/2)g]}$$

$$= \frac{v_0 \sin \theta + \sqrt{v_0^2 \sin^2 \theta + 2gh}}{g} \text{ second.}$$

At this time,

$$x(t) = v_0 \cos \theta \left(\frac{v_0 \sin \theta + \sqrt{v_0^2 \sin^2 \theta + 2gh}}{g} \right)$$

$$= \frac{v_0 \cos \theta}{g}\left(v_0 \sin \theta + \sqrt{v_0^2\left(\sin^2 \theta + \frac{2gh}{v_0^2}\right)} \right)$$

$$= \frac{v_0^2 \cos \theta}{g}\left(\sin \theta + \sqrt{\sin^2 \theta + \frac{2gh}{v_0^2}} \right) \text{ feet.}$$

43. $\mathbf{r}(t) = b(\omega t - \sin \omega t)\mathbf{i} + b(1 - \cos \omega t)\mathbf{j}$

$\quad \mathbf{v}(t) = b(\omega - \omega \cos \omega t)\mathbf{i} + b\omega \sin \omega t\mathbf{j} = b\omega(1 - \cos \omega t)\mathbf{i} + b\omega \sin \omega t\mathbf{j}$

$\quad \mathbf{a}(t) = (b\omega^2 \sin \omega t)\mathbf{i} + (b\omega^2 \cos \omega t)\mathbf{j} = b\omega^2[\sin(\omega t)\mathbf{i} + \cos(\omega t)\mathbf{j}]$

$\quad \|\mathbf{v}(t)\| = \sqrt{2}b\omega\sqrt{1 - \cos(\omega t)}$

$\quad \|\mathbf{a}(t)\| = b\omega^2$

(a) $\|\mathbf{v}(t)\| = 0$ when $\omega t = 0, 2\pi, 4\pi, \ldots$.

(b) $\|\mathbf{v}(t)\|$ is maximum when $\omega t = \pi, 3\pi, \ldots$, then $\|\mathbf{v}(t)\| = 2b\omega$.

45. $\qquad \mathbf{v}(t) = -b\omega \sin(\omega t)\mathbf{i} + b\omega \cos(\omega t)\mathbf{j}$

$\quad \mathbf{r}(t) \cdot \mathbf{v}(t) = -b^2\omega \sin(\omega t)\cos(\omega t) + b^2\omega \sin(vt)\cos(\omega t) = 0$

So, $\mathbf{r}(t)$ and $\mathbf{v}(t)$ are orthogonal.

47. $\mathbf{a}(t) = -b\omega^2 \cos(\omega t)\mathbf{i} - b\omega^2 \sin(\omega t)\mathbf{j} = -b\omega^2[\cos(\omega t)\mathbf{i} + \sin(\omega t)\mathbf{j}] = -\omega^2\mathbf{r}(t)$

$\quad \mathbf{a}(t)$ is a negative multiple of a unit vector from $(0, 0)$ to $(\cos \omega t, \sin \omega t)$ and so $\mathbf{a}(t)$ is directed toward the origin.

49. $\|\mathbf{a}(t)\| = \omega^2 b, b = 2$

$\quad 1 = m(32)$

$\quad \mathbf{F} = m(\omega^2 b) = \frac{1}{32}(2\omega^2) = 10$

$\quad \omega = 4\sqrt{10} \text{ rad/sec}$

$\quad \|\mathbf{v}(t)\| = b\omega = 8\sqrt{10} \text{ ft/sec}$

51. The velocity of an object involves both magnitude and direction of motion, whereas speed involves only magnitude.

53. $\mathbf{r}(t) = x(t)\mathbf{i} + y(t)\mathbf{j} + z(t)\mathbf{k}$ Position vector

$\quad \mathbf{v}(t) = x'(t)\mathbf{i} + y'(t)\mathbf{j} + z'(t)\mathbf{k}$ Velocity vector

$\quad \mathbf{a}(t) = x''(t)\mathbf{i} + y''(t)\mathbf{j} + z''(t)\mathbf{k}$ Acceleration vector

$\quad \text{Speed} = \|\mathbf{v}(t)\| = \sqrt{x'(t)^2 + y'(t)^2 + z'(t)^2}$

$\qquad\qquad = C, C \text{ is a constant.}$

$$\frac{d}{dt}\left[x'(t)^2 + y'(t)^2 + z'(t)^2\right] = 0$$

$$2x'(t)x''(t) + 2y'(t)y''(t) + 2z'(t)z''(t) = 0$$

$$2\left[x'(t)x''(t) + y'(t)y''(t) + z'(t)z''(t)\right] = 0$$

$$\mathbf{v}(t) \cdot \mathbf{a}(t) = 0$$

Orthogonal

55. $\mathbf{r}(t) = 6\cos t\mathbf{i} + 3\sin t\mathbf{j}$

(a) $\mathbf{v}(t) = \mathbf{r}'(t) = -6\sin t\mathbf{i}\ 3\cos t\mathbf{j}$

$\|\mathbf{v}(t)\| = \sqrt{36\sin^2 t + 9\cos^2 t}$

$\qquad = 3\sqrt{4\sin^2 t + \cos^2 t} = 3\sqrt{3\sin^2 t + 1}$

$\mathbf{a}(t) = \mathbf{v}'(t) = -6\cos t\mathbf{i} - 3\sin t\mathbf{j}$

(b)

t	0	$\dfrac{\pi}{4}$	$\dfrac{\pi}{2}$	$\dfrac{2\pi}{3}$	π
Speed	3	$\dfrac{3}{2}\sqrt{10}$	6	$\dfrac{3}{2}\sqrt{13}$	3

(c)

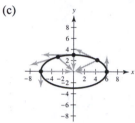

(d) The speed is increasing when the angle between **v** and **a** is in the interval

$$\left[0, \frac{\pi}{2}\right).$$

The speed is decreasing when the angle is in the interval

$$\left(\frac{\pi}{2}, \pi\right].$$

57. $\mathbf{a}(t) = \sin t\mathbf{i} - \cos t\mathbf{j}$

$\mathbf{v}(t) = \int \mathbf{a}(t)\,dt = -\cos t\mathbf{i} = \sin t\mathbf{j} + \mathbf{C}_1$

$\mathbf{v}(0) = -\mathbf{i} = -\mathbf{i} + \mathbf{C}_1 \Rightarrow \mathbf{C}_1 = \mathbf{0}$

$\mathbf{v}(t) = -\cos t\mathbf{i} - \sin t\mathbf{j}$

$\mathbf{r}(t) = \int \mathbf{v}(t)\,dt = -\sin t\mathbf{i} + \cos t\mathbf{j} + \mathbf{C}_2$

$\mathbf{r}(0) = \mathbf{j} = \mathbf{j} + \mathbf{C}_2 \Rightarrow \mathbf{C}_2 = \mathbf{0}$

$\mathbf{r}(t) = -\sin t\mathbf{i} + \cos t\mathbf{j}$

The path is a circle.

59. False. The acceleration is the derivative of the velocity.

61. True

Section 12.4 Tangent Vectors and Normal Vectors

1. $\mathbf{r}(t) = t^2\mathbf{i} + 2t\mathbf{j}, t = 1$

$\mathbf{r}'(t) = 2t\mathbf{i} + 2\mathbf{j}, \|\mathbf{r}'(t)\| = \sqrt{4t^2 + 4} = 2\sqrt{t^2 + 1}$

$\mathbf{T}(1) = \dfrac{\mathbf{r}'(1)}{\|\mathbf{r}'(1)\|} = \dfrac{1}{\sqrt{2}}(\mathbf{i} + \mathbf{j}) = \dfrac{\sqrt{2}}{2}\mathbf{i} + \dfrac{\sqrt{2}}{2}\mathbf{j}$

3. $\mathbf{r}(t) = 4\cos t\mathbf{i} + 4\sin t\mathbf{j}, t = \dfrac{\pi}{4}$

$\mathbf{r}'(t) = -4\sin t\mathbf{i} + 4\cos t\mathbf{j}$

$\|\mathbf{r}'(t)\| = \sqrt{16\sin^2 t + 16\cos^2 t} = 4$

$\mathbf{T}\left(\dfrac{\pi}{4}\right) = \dfrac{\mathbf{r}'\left(\dfrac{\pi}{4}\right)}{\left\|\mathbf{r}'\left(\dfrac{\pi}{4}\right)\right\|} = -\dfrac{\sqrt{2}}{2}\mathbf{i} + \dfrac{\sqrt{2}}{2}\mathbf{j}$

5. $\mathbf{r}(t) = 3t\mathbf{i} - \ln t\mathbf{j}, t = e$

$\mathbf{r}'(t) = 3\mathbf{i} - \dfrac{1}{t}\mathbf{j}$

$\mathbf{r}'(e) = 3\mathbf{i} - \dfrac{1}{e}\mathbf{j}$

$\mathbf{T}(e) = \dfrac{\mathbf{r}'(e)}{\|\mathbf{r}'(e)\|}$

$\qquad = \dfrac{3\mathbf{i} - \dfrac{1}{e}\mathbf{j}}{\sqrt{9 + \dfrac{1}{e^2}}} = \dfrac{3e\mathbf{i} - \mathbf{j}}{\sqrt{9e^2 + 1}} \approx 0.9926\mathbf{i} - 0.1217\mathbf{j}$

7. $\mathbf{r}(t) = t\mathbf{i} + t^2\mathbf{j} + t\mathbf{k}, P(0, 0, 0)$

$\mathbf{r}'(t) = \mathbf{i} + 2t\mathbf{j} + \mathbf{k}$

When $t = 0$, $\mathbf{r}'(0) = \mathbf{i} + \mathbf{k}$, $\left[t = 0 \text{ at } (0, 0, 0)\right]$.

$\mathbf{T}(0) = \dfrac{\mathbf{r}'(0)}{\|\mathbf{r}'(0)\|} = \dfrac{\sqrt{2}}{2}(\mathbf{i} + \mathbf{k})$

Direction numbers: $a = 1, b = 0, c = 1$

Parametric equations: $x = t, y = 0, z = t$

9. $\mathbf{r}(t) = 3\cos t\mathbf{i} + 3\sin t\mathbf{j} + t\mathbf{k}, \; P(3, 0, 0)$

$\mathbf{r}'(t) = -3\sin t\mathbf{i} + 3\cos t\mathbf{j} + \mathbf{k}$

$t = 0$ at $P(3, 0, 0)$

$\mathbf{r}'(0) = 3\mathbf{j} + \mathbf{k}$

$\mathbf{T}(0) = \dfrac{\mathbf{r}'(0)}{\|\mathbf{r}'(0)\|} = \dfrac{3\mathbf{j} + \mathbf{k}}{\sqrt{10}}$

Direction numbers: $a = 0, \; b = 3, \; c = 1$

Parametric equations: $x = 3, \; y = 3t, \; z = t$

11. $\mathbf{r}(t) = \langle 2\cos t, 2\sin t, 4\rangle, \; P\left(\sqrt{2}, \sqrt{2}, 4\right)$

$\mathbf{r}'(t) = \langle -2\sin t, 2\cos t, 0\rangle$

When $t = \dfrac{\pi}{4}, \; \mathbf{r}\left(\dfrac{\pi}{4}\right) = \langle -\sqrt{2}, \sqrt{2}, 0\rangle,$

$\left[t = \dfrac{\pi}{4} \text{ at } \left(\sqrt{2}, \sqrt{2}, 4\right) \right].$

$\mathbf{T}\left(\dfrac{\pi}{4}\right) = \dfrac{\mathbf{r}'(\pi/4)}{\|\mathbf{r}'(\pi/4)\|} = \dfrac{1}{2}\langle -\sqrt{2}, \sqrt{2}, 0\rangle$

Direction numbers: $a = -\sqrt{2}, b = \sqrt{2}, c = 0$

Parametric equations: $x = -\sqrt{2}t + \sqrt{2}, \; y = \sqrt{2}t + \sqrt{2},$
$z = 4$

13. $\mathbf{r}(t) = t\mathbf{i} + \dfrac{1}{2}t^2\mathbf{j}, \, t = 2$

$\mathbf{r}'(t) = \mathbf{i} + t\mathbf{j}$

$\mathbf{T}(t) = \dfrac{\mathbf{r}'(t)}{\|\mathbf{r}'(t)\|} = \dfrac{\mathbf{i} + t\mathbf{j}}{\sqrt{1 + t^2}}$

$\mathbf{T}'(t) = \dfrac{-t}{\left(t^2 + 1\right)^{3/2}}\mathbf{i} + \dfrac{1}{\left(t^2 + 1\right)^{3/2}}\mathbf{j}$

$\mathbf{T}'(2) = \dfrac{-2}{5^{3/2}}\mathbf{i} + \dfrac{1}{5^{3/2}}\mathbf{j}$

$\mathbf{N}(2) = \dfrac{\mathbf{T}'(2)}{\|\mathbf{T}'(2)\|} = \dfrac{1}{\sqrt{5}}(-2\mathbf{i} + \mathbf{j}) = \dfrac{-2\sqrt{5}}{5}\mathbf{i} + \dfrac{\sqrt{5}}{5}\mathbf{j}$

15. $\mathbf{r}(t) = \ln t\mathbf{i} + (t + 1)\mathbf{j}, \, t = 2$

$\mathbf{r}'(t) = \dfrac{1}{t}\mathbf{i} + \mathbf{j}$

$\mathbf{T}(t) = \dfrac{\mathbf{r}'(t)}{\|\mathbf{r}'(t)\|} = \dfrac{\dfrac{1}{t}\mathbf{i} + \mathbf{j}}{\sqrt{\dfrac{1}{t^2} + 1}} = \dfrac{\mathbf{i} + t\mathbf{j}}{\sqrt{1 + t^2}}$

$\mathbf{T}'(t) = \dfrac{-t}{\left(1 + t^2\right)^{3/2}}\mathbf{i} + \dfrac{1}{\left(1 + t^2\right)^{3/2}}\mathbf{j}$

$\mathbf{T}'(2) = \dfrac{-2}{5^{3/2}}\mathbf{i} + \dfrac{1}{5^{3/2}}\mathbf{j}$

$\mathbf{N}(2) = \dfrac{\mathbf{T}'(2)}{\|\mathbf{T}'(2)\|} = \dfrac{-2\sqrt{5}}{5}\mathbf{i} + \dfrac{\sqrt{5}}{5}\mathbf{j}$

17. $\mathbf{r}(t) = t\mathbf{i} + t^2\mathbf{j} + \ln t\mathbf{k}, t = 1$

$\mathbf{r}'(t) = \mathbf{i} + 2t\mathbf{j} + \dfrac{1}{t}\mathbf{k}$

$\mathbf{T}(t) = \dfrac{\mathbf{r}'(t)}{\|\mathbf{r}'(t)\|} = \dfrac{\mathbf{i} + 2t\mathbf{j} + \dfrac{1}{t}\mathbf{k}}{\sqrt{1 + 4t^2 + \dfrac{1}{t^2}}} = \dfrac{t\mathbf{i} + 2t^2\mathbf{j} + \mathbf{k}}{\sqrt{4t^4 + t^2 + 1}}$

$\mathbf{T}'(t) = \dfrac{1 - 4t^4}{\left(4t^4 + t^2 + 1\right)^{3/2}}\mathbf{i} + \dfrac{2t^3 + 4t}{\left(4t^4 + t^2 + 1\right)^{3/2}}\mathbf{j} + \dfrac{-8t^3 - t}{\left(4t^4 + t^2 + 1\right)^{3/2}}\mathbf{k}$

$\mathbf{T}'(1) = \dfrac{-3}{6^{3/2}}\mathbf{i} + \dfrac{6}{6^{3/2}}\mathbf{j} + \dfrac{-9}{6^{3/2}}\mathbf{k} = \dfrac{3}{6^{3/2}}[-\mathbf{i} + 2\mathbf{j} - 3\mathbf{k}]$

$\mathbf{N}(1) = \dfrac{-\mathbf{i} + 2\mathbf{j} - 3\mathbf{k}}{\sqrt{14}} = \dfrac{-\sqrt{14}}{14}\mathbf{i} + \dfrac{2\sqrt{14}}{14}\mathbf{j} - \dfrac{3\sqrt{14}}{14}\mathbf{k}$

19. $\mathbf{r}(t) = 6\cos t\mathbf{i} + 6\sin t\mathbf{j} + \mathbf{k}, t = \dfrac{3\pi}{4}$

$\mathbf{r}'(t) = -6\sin t\mathbf{i} + 6\cos t\mathbf{j}$

$\mathbf{T}(t) = \dfrac{\mathbf{r}'(t)}{\|\mathbf{r}'(t)\|} = -\sin t\mathbf{i} + \cos t\mathbf{j}$

$\mathbf{T}'(t) = -\cos t\mathbf{i} - \sin t\mathbf{j}, \|\mathbf{T}'(t)\| = 1$

$\mathbf{N}\left(\dfrac{3\pi}{4}\right) = \dfrac{\mathbf{T}'(3\pi/4)}{\|\mathbf{T}'(3\pi/4)\|} = \dfrac{\sqrt{2}}{2}\mathbf{i} - \dfrac{\sqrt{2}}{2}\mathbf{j}$

21. $\mathbf{r}(t) = t\mathbf{i} + \dfrac{1}{t}\mathbf{j}, \, t = 1$

$\mathbf{v}(t) = \mathbf{i} - \dfrac{1}{t^2}\mathbf{j}, \, \mathbf{v}(1) = \mathbf{i} - \mathbf{j},$

$\mathbf{a}(t) = \dfrac{2}{t^3}\mathbf{j}, \, \mathbf{a}(1) = 2\mathbf{j}$

$\mathbf{T}(t) = \dfrac{\mathbf{v}(t)}{\|\mathbf{v}(t)\|} = \dfrac{t^2}{\sqrt{t^4+1}}\left(\mathbf{i} - \dfrac{1}{t^2}\mathbf{j}\right) = \dfrac{1}{\sqrt{t^4+1}}(t^2\mathbf{i} - \mathbf{j})$

$\mathbf{T}(1) = \dfrac{1}{\sqrt{2}}(\mathbf{i}-\mathbf{j}) = \dfrac{\sqrt{2}}{2}(\mathbf{i}-\mathbf{j})$

$\mathbf{N}(t) = \dfrac{\mathbf{T}'(t)}{\|\mathbf{T}'(t)\|} = \dfrac{\dfrac{2t}{(t^4+1)^{3/2}}\mathbf{i} + \dfrac{2t^3}{(t^4+1)^{3/2}}\mathbf{j}}{\dfrac{2t}{(t^4+1)}}$

$= \dfrac{1}{\sqrt{t^4+1}}(\mathbf{i} + t^2\mathbf{j})$

$\mathbf{N}(1) = \dfrac{1}{\sqrt{2}}(\mathbf{i}+\mathbf{j}) = \dfrac{\sqrt{2}}{2}(\mathbf{i}+\mathbf{j})$

$a_{\mathbf{T}} = \mathbf{a} \cdot \mathbf{T} = -\sqrt{2}$

$a_{\mathbf{N}} = \mathbf{a} \cdot \mathbf{N} = \sqrt{2}$

23. $\mathbf{r}(t) = (t - t^3)\mathbf{i} + 2t^2\mathbf{j}, \, t = 1$

$\mathbf{v}(t) = (1 - 3t^2)\mathbf{i} + 4t\mathbf{j}, \, \mathbf{v}(1) = -2\mathbf{i} + 4\mathbf{j}$

$\mathbf{a}(t) = -6t\mathbf{i} + 4\mathbf{j}, \, \mathbf{a}(1) = -6\mathbf{i} + 4\mathbf{j}$

$\mathbf{T}(t) = \dfrac{\mathbf{v}(t)}{\|\mathbf{v}(t)\|} = \dfrac{(1-3t^2)\mathbf{i} + 4t\mathbf{j}}{\sqrt{9t^4 + 10t^2 + 1}}$

$\mathbf{T}(1) = \dfrac{-2\mathbf{i} + 4\mathbf{j}}{\sqrt{20}} = \dfrac{-\mathbf{i} + 2\mathbf{j}}{\sqrt{5}} = \dfrac{-\sqrt{5}}{5}(\mathbf{i} - 2\mathbf{j})$

$\mathbf{T}'(t) = \dfrac{-16t(3t^2 + 1)}{(9t^4 + 10t^2 + 1)^{3/2}}\mathbf{i} + \dfrac{4 - 36t^4}{(9t^4 + 10t^2 + 1)^{3/2}}\mathbf{j}$

$\mathbf{T}'(1) = \dfrac{-64}{20^{3/2}}\mathbf{i} + \dfrac{-32}{20^{3/2}}\mathbf{j}$

$\mathbf{N}(1) = \dfrac{-2\mathbf{i} - \mathbf{j}}{\sqrt{5}}$

$a_{\mathbf{T}} = \mathbf{a} \cdot \mathbf{T} = \dfrac{1}{\sqrt{5}}(6 + 8) = \dfrac{14\sqrt{5}}{5}$

$a_{\mathbf{N}} = \mathbf{a} \cdot \mathbf{N} = \dfrac{1}{\sqrt{5}}(12 - 4) = \dfrac{8\sqrt{5}}{5}$

25. $\mathbf{r}(t) = e^t\mathbf{i} + e^{-2t}\mathbf{j}, \, t = 0$

$\mathbf{v}(t) = e^t\mathbf{i} - 2e^{-2t}\mathbf{j}, \, \mathbf{v}(0) = \mathbf{i} - 2\mathbf{j}$

$\mathbf{a}(t) = e^t\mathbf{i} + 4e^{-2t}\mathbf{j}, \, \mathbf{a}(0) = \mathbf{i} + 4\mathbf{j}$

$\mathbf{T}(t) = \dfrac{\mathbf{v}(t)}{\|\mathbf{v}(t)\|} = \dfrac{e^t\mathbf{i} - 2e^{-2t}\mathbf{j}}{\sqrt{4e^{-4t} + e^{2t}}}$

$\mathbf{T}(0) = \dfrac{\mathbf{i} - 2\mathbf{j}}{\sqrt{5}}$

$\mathbf{N}(0) = \dfrac{2\mathbf{i} + \mathbf{j}}{\sqrt{5}}$

$a_{\mathbf{T}} = \mathbf{a} \cdot \mathbf{T} = \dfrac{1}{\sqrt{5}}(1 - 8) = \dfrac{-7\sqrt{5}}{5}$

$a_{\mathbf{N}} = \mathbf{a} \cdot \mathbf{N} = \dfrac{1}{\sqrt{5}}(2 + 4) = \dfrac{6\sqrt{5}}{5}$

27. $\mathbf{r}(t) = (e^t \cos t)\mathbf{i} + (e^t \sin t)\mathbf{j}, \, t = \dfrac{\pi}{2}$

$\mathbf{v}(t) = e^t(\cos t - \sin t)\mathbf{i} + e^t(\cos t + \sin t)\mathbf{j}$

$\mathbf{a}(t) = e^t(-2 \sin t)\mathbf{i} + e^t(2 \cos t)\mathbf{j}$

At $t = \dfrac{\pi}{2}$, $\mathbf{T} = \dfrac{\mathbf{v}}{\|\mathbf{v}\|} = \dfrac{1}{\sqrt{2}}(-\mathbf{i} + \mathbf{j}) = \dfrac{\sqrt{2}}{2}(-\mathbf{i} + \mathbf{j})$.

Motion along $\mathbf{r}$ is counterclockwise. So,

$\mathbf{N} = \dfrac{1}{\sqrt{2}}(-\mathbf{i} - \mathbf{j}) = -\dfrac{\sqrt{2}}{2}(\mathbf{i} + \mathbf{j})$.

$a_{\mathbf{T}} = \mathbf{a} \cdot \mathbf{T} = \sqrt{2}e^{\pi/2}$

$a_{\mathbf{N}} = \mathbf{a} \cdot \mathbf{N} = \sqrt{2}e^{\pi/2}$

29. $\mathbf{r}(t) = a \cos \omega t\mathbf{i} + a \sin \omega t\mathbf{j}$

$\mathbf{v}(t) = -a\omega \sin \omega t\mathbf{i} + a\omega \cos \omega t\mathbf{j}$

$\mathbf{a}(t) = -a\omega^2 \cos \omega t\mathbf{i} - a\omega^2 \sin \omega t\mathbf{j}$

$\mathbf{T}(t) = \dfrac{\mathbf{v}(t)}{\|\mathbf{v}(t)\|} = -\sin \omega t\mathbf{i} + \cos \omega t\mathbf{j}$

$\mathbf{N}(t) = \dfrac{\mathbf{T}'(t)}{\|\mathbf{T}'(t)\|} = -\cos \omega t\mathbf{i} - \sin \omega t\mathbf{j}$

$a_{\mathbf{T}} = \mathbf{a} \cdot \mathbf{T} = 0$

$a_{\mathbf{N}} = \mathbf{a} \cdot \mathbf{N} = a\omega^2$

31. Speed: $\|\mathbf{v}(t)\| = a\omega$

The speed is constant because $a_{\mathbf{T}} = 0$.

33. $\mathbf{r}(t) = t\mathbf{i} + \dfrac{1}{t}\mathbf{j}, t_0 = 2$

$x = t, y = \dfrac{1}{t} \Rightarrow xy = 1$

$\mathbf{r}'(t) = \mathbf{i} - \dfrac{1}{t^2}\mathbf{j}$

$\mathbf{T}(t) = \dfrac{t^2\mathbf{i} - \mathbf{j}}{\sqrt{t^4 + 1}}$

$\mathbf{N}(t) = \dfrac{\mathbf{i} + t^2\mathbf{j}}{\sqrt{t^4 + 1}}$

$\mathbf{r}(2) = 2\mathbf{i} + \dfrac{1}{2}\mathbf{j}$

$\mathbf{T}(2) = \dfrac{\sqrt{17}}{17}(4\mathbf{i} - \mathbf{j})$

$\mathbf{N}(2) = \dfrac{\sqrt{17}}{17}(\mathbf{i} + 4\mathbf{j})$

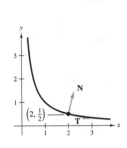

35. $\mathbf{r}(t) = (2t + 1)\mathbf{i} - t^2\mathbf{j}, t_0 = 2$

$x = 2t + 1,$

$y = -t^2 = -\left(\dfrac{x - 1}{2}\right)^2$

$\mathbf{r}(2) = 5\mathbf{i} - 4\mathbf{j}$

$\mathbf{r}'(t) = 2\mathbf{i} - 2t\mathbf{j}$

$\mathbf{T}(t) = \dfrac{2\mathbf{i} - 2t\mathbf{j}}{\sqrt{4 + 4t^2}} = \dfrac{\mathbf{i} - t\mathbf{j}}{\sqrt{1 + t^2}}$

$\mathbf{T}(2) = \dfrac{\mathbf{i} - 2\mathbf{j}}{\sqrt{5}}$

$\mathbf{N}(2) = \dfrac{-2\mathbf{i} - \mathbf{j}}{\sqrt{5}}$, perpendicular to $\mathbf{T}(2)$

37. $\mathbf{r}(t) = t\mathbf{i} + 2t\mathbf{j} - 3t\mathbf{k}, t = 1$

$\mathbf{v}(t) = \mathbf{i} + 2\mathbf{j} - 3\mathbf{k}$

$\mathbf{a}(t) = \mathbf{0}$

$\mathbf{T}(t) = \dfrac{\mathbf{v}}{\|\mathbf{v}\|}$

$\qquad = \dfrac{1}{\sqrt{14}}(\mathbf{i} + 2\mathbf{j} - 3\mathbf{k})$

$\qquad = \dfrac{\sqrt{14}}{14}(\mathbf{i} + 2\mathbf{j} - 3\mathbf{k}) = \mathbf{T}(1)$

$\mathbf{N}(t) = \dfrac{\mathbf{T}'}{\|\mathbf{T}'\|}$ is undefined.

$a_\mathbf{T}, a_\mathbf{N}$ are not defined.

39. $\mathbf{r}(t) = t\mathbf{i} + t^2\mathbf{j} + \dfrac{t^2}{2}\mathbf{k}, t = 1$

$\mathbf{v}(t) = \mathbf{i} + 2t\mathbf{j} + t\mathbf{k}$

$\mathbf{v}(1) = \mathbf{i} + 2\mathbf{j} + \mathbf{k}$

$\mathbf{a}(t) = 2\mathbf{j} + \mathbf{k}$

$\mathbf{T}(t) = \dfrac{\mathbf{v}}{\|\mathbf{v}\|} = \dfrac{1}{\sqrt{1 + 5t^2}}(\mathbf{i} + 2t\mathbf{j} + t\mathbf{k})$

$\mathbf{T}(1) = \dfrac{\sqrt{6}}{6}(\mathbf{i} + 2\mathbf{j} + \mathbf{k})$

$\mathbf{N}(t) = \dfrac{\mathbf{T}'}{\|\mathbf{T}'\|} = \dfrac{\dfrac{-5t\mathbf{i} + 2\mathbf{j} + \mathbf{k}}{(1 + 5t^2)^{3/2}}}{\dfrac{\sqrt{5}}{1 + 5t^2}} = \dfrac{-5t\mathbf{i} + 2\mathbf{j} + \mathbf{k}}{\sqrt{5}\sqrt{1 + 5t^2}}$

$\mathbf{N}(1) = \dfrac{\sqrt{30}}{30}(-5\mathbf{i} + 2\mathbf{j} + \mathbf{k})$

$a_\mathbf{T} = \mathbf{a} \cdot \mathbf{T} = \dfrac{5\sqrt{6}}{6}$

$a_\mathbf{N} = \mathbf{a} \cdot \mathbf{N} = \dfrac{\sqrt{30}}{6}$

41. $\mathbf{r}(t) = e^t \sin t\,\mathbf{i} + e^t \cos t\,\mathbf{j} + e^t\mathbf{k}, t = 0$

$\mathbf{v}(t) = (e^t \cos t + e^t \sin t)\mathbf{i} + (-e^t \sin t + e^t \cos t)\mathbf{j} + e^t\mathbf{k}$

$\mathbf{v}(0) = \mathbf{i} + \mathbf{j} + \mathbf{k}$

$\mathbf{a}(t) = 2e^t \cos t\,\mathbf{i} - 2e^t \sin t\,\mathbf{j} + e^t\mathbf{k}$

$\mathbf{a}(0) = 2\mathbf{i} + \mathbf{k}$

$\mathbf{T}(t) = \dfrac{\mathbf{v}}{\|\mathbf{v}\|}$

$\qquad = \dfrac{1}{\sqrt{3}}\left[(\cos t + \sin t)\mathbf{i} + (-\sin t + \cos t)\mathbf{j} + \mathbf{k}\right]$

$\mathbf{T}(0) = \dfrac{1}{\sqrt{3}}[\mathbf{i} + \mathbf{j} + \mathbf{k}]$

$\mathbf{N}(t) = \dfrac{1}{\sqrt{2}}\left[(-\sin t + \cos t)\mathbf{i} + (-\cos t - \sin t)\mathbf{j}\right]$

$\mathbf{N}(0) = \dfrac{\sqrt{2}}{2}\mathbf{i} - \dfrac{\sqrt{2}}{2}\mathbf{j}$

$a_\mathbf{T} = \mathbf{a} \cdot \mathbf{T} = \sqrt{3}$

$a_\mathbf{N} = \mathbf{a} \cdot \mathbf{N} = \sqrt{2}$

43. Let C be a smooth curve represented by $\mathbf{r}$ on an open interval I. The unit tangent vector $\mathbf{T}(t)$ at t is defined as

$$\mathbf{T}(t) = \frac{\mathbf{r}'(t)}{\|\mathbf{r}'(t)\|}, \ \mathbf{r}'(t) \neq 0.$$

The principal unit normal vector $\mathbf{N}(t)$ at t is defined as

$$\mathbf{N}(t) = \frac{\mathbf{T}'(t)}{\|\mathbf{T}'(t)\|}, \ \mathbf{T}'(t) \neq 0.$$

The tangential and normal components of acceleration are defined as $\mathbf{a}(t) = a_{\mathbf{T}}\mathbf{T}(t) + a_{\mathbf{N}}\mathbf{N}(t)$.

49. $\mathbf{r}(t) = \langle \pi t - \sin \pi t, 1 - \cos \pi t \rangle$

The graph is a cycloid.

(a) $\mathbf{r}(t) = \langle \pi t - \sin \pi t, 1 - \cos \pi t \rangle$

$\mathbf{v}(t) = \langle \pi - \pi \cos \pi t, \pi \sin \pi t \rangle$

$\mathbf{a}(t) = \langle \pi^2 \sin \pi t, \pi^2 \cos \pi t \rangle$

$\mathbf{T}(t) = \dfrac{\mathbf{v}(t)}{\|\mathbf{v}(t)\|} = \dfrac{1}{\sqrt{2(1 - \cos \pi t)}}\langle 1 - \cos \pi t, \sin \pi t \rangle$

$\mathbf{N}(t) = \dfrac{\mathbf{T}'(t)}{\|\mathbf{T}'(t)\|} = \dfrac{1}{\sqrt{2(1 - \cos \pi t)}}\langle \sin \pi t, -1 + \cos \pi t \rangle$

$a_{\mathbf{T}} = \mathbf{a} \cdot \mathbf{T} = \dfrac{1}{\sqrt{2(1 - \cos \pi t)}}\left[\pi^2 \sin \pi t (1 - \cos \pi t) + \pi^2 \cos \pi t \sin \pi t \right] = \dfrac{\pi^2 \sin \pi t}{\sqrt{2(1 - \cos \pi t)}}$

$a_{\mathbf{N}} = \mathbf{a} \cdot \mathbf{N} = \dfrac{1}{\sqrt{2(1 - \cos \pi t)}}\left[\pi^2 \sin^2 \pi t + \pi^2 \cos \pi t (-1 + \cos \pi t) \right] = \dfrac{\pi^2 (1 - \cos \pi t)}{\sqrt{2(1 - \cos \pi t)}} = \dfrac{\pi^2 \sqrt{2(1 - \cos \pi t)}}{2}$

When $t = \dfrac{1}{2}$: $a_{\mathbf{T}} = \dfrac{\pi^2}{\sqrt{2}} = \dfrac{\sqrt{2}\pi^2}{2}$, $a_{\mathbf{N}} = \dfrac{\sqrt{2}\pi^2}{2}$

When $t = 1$: $a_{\mathbf{T}} = 0$, $a_{\mathbf{N}} = \pi^2$

When $t = \dfrac{3}{2}$: $a_{\mathbf{T}} = -\dfrac{\sqrt{2}\pi^2}{2}$, $a_{\mathbf{N}} = \dfrac{\sqrt{2}\pi^2}{2}$

(b) Speed: $s = \|\mathbf{v}(t)\| = \pi\sqrt{2(1 - \cos \pi t)}$

$\dfrac{ds}{dt} = \dfrac{\pi^2 \sin \pi t}{\sqrt{2(1 - \cos \pi t)}} = a_{\mathbf{T}}$

When $t = \dfrac{1}{2}$: $a_{\mathbf{T}} = \dfrac{\sqrt{2}\pi^2}{2} > 0 \Rightarrow$ the speed in increasing.

When $t = 1$: $a_{\mathbf{T}} = 0 \Rightarrow$ the height is maximum.

When $t = \dfrac{3}{2}$: $a_{\mathbf{T}} = -\dfrac{\sqrt{2}\pi^2}{2} < 0 \Rightarrow$ the speed is decreasing.

45. If $a_{\mathbf{N}} = 0$, then the motion is in a straight line.

47. $\mathbf{r}(t) = 3t\mathbf{i} + 4t\mathbf{j}$

$\mathbf{v}(t) = \mathbf{r}'(t) = 3\mathbf{i} + 4\mathbf{j}, \|\mathbf{v}(t)\| = \sqrt{9 + 16} = 5$

$\mathbf{a}(t) = \mathbf{v}'(t) = \mathbf{0}$

$\mathbf{T}(t) = \dfrac{\mathbf{v}(t)}{\|\mathbf{v}(t)\|} = \dfrac{3}{5}\mathbf{i} + \dfrac{4}{5}\mathbf{j}$

$\mathbf{T}'(t) = 0 \Rightarrow \mathbf{N}(t)$ does not exist.

The path is a line. The speed is constant (5).

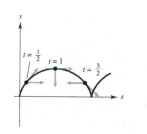

51. $\mathbf{r}(t) = 2\cos t\mathbf{i} + 2\sin t\mathbf{j} + \dfrac{t}{2}\mathbf{k}$, $t_0 = \dfrac{\pi}{2}$

$\mathbf{r}'(t) = -2\sin t\mathbf{i} + 2\cos t\mathbf{j} + \dfrac{1}{2}\mathbf{k}$

$\mathbf{T}(t) = \dfrac{2\sqrt{17}}{17}\left(-2\sin t\mathbf{i} + 2\cos t\mathbf{j} + \dfrac{1}{2}\mathbf{k}\right)$

$\mathbf{N}(t) = -\cos t\mathbf{i} - \sin t\mathbf{j}$

$\mathbf{r}\left(\dfrac{\pi}{2}\right) = 2\mathbf{j} + \dfrac{\pi}{4}\mathbf{k}$

$\mathbf{T}\left(\dfrac{\pi}{2}\right) = \dfrac{2\sqrt{17}}{17}\left(-2\mathbf{i} + \dfrac{1}{2}\mathbf{k}\right) = \dfrac{\sqrt{17}}{17}(-4\mathbf{i} + \mathbf{k})$

$\mathbf{N}\left(\dfrac{\pi}{2}\right) = -\mathbf{j}$

$\mathbf{B}\left(\dfrac{\pi}{2}\right) = \mathbf{T}\left(\dfrac{\pi}{2}\right) \times \mathbf{N}\left(\dfrac{\pi}{2}\right) = \begin{vmatrix} \mathbf{i} & \mathbf{j} & \mathbf{k} \\ -\dfrac{4\sqrt{17}}{17} & 0 & \dfrac{\sqrt{17}}{17} \\ 0 & -1 & 0 \end{vmatrix} = \dfrac{\sqrt{17}}{17}\mathbf{i} + \dfrac{4\sqrt{17}}{17}\mathbf{k} = \dfrac{\sqrt{17}}{17}(\mathbf{i} + 4\mathbf{k})$

53. $\mathbf{r}(t) = \mathbf{i} + \sin t\mathbf{j} + \cos t\mathbf{k}$, $t_0 = \dfrac{\pi}{4}$

$\mathbf{r}'(t) = \cos t\mathbf{j} - \sin t\mathbf{k}$,

$\|\mathbf{r}'(t)\| = 1$

$\mathbf{r}'\left(\dfrac{\pi}{4}\right) = \mathbf{T}\left(\dfrac{\pi}{4}\right) = \dfrac{\sqrt{2}}{2}\mathbf{j} - \dfrac{\sqrt{2}}{2}\mathbf{k}$

$\mathbf{T}'(t) = -\sin t\mathbf{j} - \cos t\mathbf{k}$,

$\mathbf{N}\left(\dfrac{\pi}{4}\right) = -\dfrac{\sqrt{2}}{2}\mathbf{j} - \dfrac{\sqrt{2}}{2}\mathbf{k}$

$\mathbf{B}\left(\dfrac{\pi}{4}\right) = \mathbf{T}\left(\dfrac{\pi}{4}\right) \times \mathbf{N}\left(\dfrac{\pi}{4}\right) = \begin{vmatrix} \mathbf{i} & \mathbf{j} & \mathbf{k} \\ 0 & \dfrac{\sqrt{2}}{2} & -\dfrac{\sqrt{2}}{2} \\ 0 & -\dfrac{\sqrt{2}}{2} & -\dfrac{\sqrt{2}}{2} \end{vmatrix} = -\mathbf{i}$

55. $\mathbf{r}(t) = 4\sin t\mathbf{i} + 4\cos t\mathbf{j} + 2t\mathbf{k}, \ t_0 = \dfrac{\pi}{3}$

$\mathbf{r}'(t) = 4\cos t\mathbf{i} - 4\sin t\mathbf{j} + 2\mathbf{k},$

$\|\mathbf{r}'(t)\| = \sqrt{16\cos^2 t + 16\sin^2 t + 4} = \sqrt{20} = 2\sqrt{5}$

$\mathbf{r}'\left(\dfrac{\pi}{3}\right) = 2\mathbf{i} - 2\sqrt{3}\mathbf{j} + 2\mathbf{k}$

$\mathbf{T}\left(\dfrac{\pi}{3}\right) = \dfrac{1}{2\sqrt{5}}\left(2\mathbf{i} - 2\sqrt{3}\mathbf{j} + 2\mathbf{k}\right) = \dfrac{\sqrt{5}}{5}\mathbf{i} - \dfrac{\sqrt{15}}{5}\mathbf{j} + \dfrac{\sqrt{5}}{5}\mathbf{k} = \dfrac{\sqrt{5}}{5}\left(\mathbf{i} - \sqrt{3}\mathbf{j} + \mathbf{k}\right)$

$\mathbf{T}'(t) = \dfrac{1}{2\sqrt{5}}\left(-4\sin t\mathbf{i} - 4\cos t\mathbf{j}\right)$

$\mathbf{N}\left(\dfrac{\pi}{3}\right) = -\dfrac{\sqrt{3}}{2}\mathbf{i} - \dfrac{1}{2}\mathbf{j}$

$\mathbf{B}\left(\dfrac{\pi}{3}\right) = \mathbf{T}\left(\dfrac{\pi}{3}\right) \times \mathbf{N}\left(\dfrac{\pi}{3}\right) = \begin{vmatrix} \mathbf{i} & \mathbf{j} & \mathbf{k} \\ \dfrac{\sqrt{5}}{5} & -\dfrac{\sqrt{15}}{5} & \dfrac{\sqrt{5}}{5} \\ -\dfrac{\sqrt{3}}{2} & -\dfrac{1}{2} & 0 \end{vmatrix} = \dfrac{\sqrt{5}}{10}\mathbf{i} - \dfrac{\sqrt{15}}{10}\mathbf{j} - \dfrac{4\sqrt{5}}{10}\mathbf{k} = \dfrac{\sqrt{5}}{10}\left(\mathbf{i} - \sqrt{3}\mathbf{j} - 4\mathbf{k}\right)$

57. $\mathbf{r}(t) = 3t\mathbf{i} + 2t^2\mathbf{j}$

$\mathbf{v}(t) = 3\mathbf{i} + 4t\mathbf{j}$

$\mathbf{a}(t) = 4\mathbf{j}$

$\mathbf{v} \cdot \mathbf{v} = 9 + 16t^2$

$\mathbf{v} \cdot \mathbf{a} = 16t$

$(\mathbf{v} \cdot \mathbf{v})\mathbf{a} - (\mathbf{v} \cdot \mathbf{a})\mathbf{v} = \left(9 + 16t^2\right)4\mathbf{j} - (16t)(3\mathbf{i} + 4t\mathbf{j})$

$\qquad\qquad\qquad\qquad = -48t\mathbf{i} + 36\mathbf{j}$

$\mathbf{N} = \dfrac{(\mathbf{v} \cdot \mathbf{v})\mathbf{a} - (\mathbf{v} \cdot \mathbf{a})\mathbf{v}}{\|(\mathbf{v} \cdot \mathbf{v})\mathbf{a} - (\mathbf{v} \cdot \mathbf{a})\mathbf{v}\|} = \dfrac{1}{\sqrt{9 + 16t^2}}(-4t\mathbf{i} + 3\mathbf{j})$

59. $\mathbf{r}(t) = 2t\mathbf{i} + 4t\mathbf{j} + t^2\mathbf{k}$

$\mathbf{v}(t) = 2\mathbf{i} + 4\mathbf{j} + 2t\mathbf{k}$

$\mathbf{a}(t) = 2\mathbf{k}$

$\mathbf{v} \cdot \mathbf{v} = 4 + 16 + 4t^2 = 20 + 4t^2$

$\mathbf{v} \cdot \mathbf{a} = 4t$

$(\mathbf{v} \cdot \mathbf{v})\mathbf{a} - (\mathbf{v} \cdot \mathbf{a})\mathbf{v} = \left(20 + 4t^2\right)2\mathbf{k} - 4t(2\mathbf{i} + 4\mathbf{j} + 2t\mathbf{k})$

$\qquad\qquad\qquad\qquad = -8t\mathbf{i} - 16t\mathbf{j} + 40\mathbf{k}$

$\mathbf{N} = \dfrac{(\mathbf{v} \cdot \mathbf{v})\mathbf{a} - (\mathbf{v} \cdot \mathbf{a})\mathbf{v}}{\|(\mathbf{v} \cdot \mathbf{v})\mathbf{a} - (\mathbf{v} \cdot \mathbf{a})\mathbf{v}\|} = \dfrac{1}{\sqrt{5t^2 + 25}}(-t\mathbf{i} - 2t\mathbf{j} + 5\mathbf{k})$

61. From Theorem 12.3 you have:

$\mathbf{r}(t) = \left(v_0 t \cos\theta\right)\mathbf{i} + \left(h + v_0 t \sin\theta - 16t^2\right)\mathbf{j}$

$\mathbf{v}(t) = v_0 \cos\theta\mathbf{i} + \left(v_0 \sin\theta - 32t\right)\mathbf{j}$

$\mathbf{a}(t) = -32\mathbf{j}$

$\mathbf{T}(t) = \dfrac{\left(v_0 \cos\theta\right)\mathbf{i} + \left(v_0 \sin\theta - 32t\right)\mathbf{j}}{\sqrt{v_0^2 \cos^2\theta + \left(v_0 \sin\theta - 32t\right)^2}}$

$\mathbf{N}(t) = \dfrac{\left(v_0 \sin\theta - 32t\right)\mathbf{i} - v_0 \cos\theta\mathbf{j}}{\sqrt{v_0^2 \cos^2\theta + \left(v_0 \sin\theta - 32t\right)^2}}$ (Motion is clockwise.)

$a_{\mathbf{T}} = \mathbf{a} \cdot \mathbf{T} = \dfrac{-32\left(v_0 \sin\theta - 32t\right)}{\sqrt{v_0^2 \cos^2\theta + \left(v_0 \sin\theta - 32t\right)^2}}$

$a_{\mathbf{N}} = \mathbf{a} \cdot \mathbf{N} = \dfrac{32 v_0 \cos\theta}{\sqrt{v_0^2 \cos^2\theta + \left(v_0 \sin\theta - 32t\right)^2}}$

Maximum height when $v_0 \sin\theta - 32t = 0$; (vertical component of velocity)

At maximum height, $a_{\mathbf{T}} = 0$ and $a_{\mathbf{N}} = 32$.

63. (a) $\mathbf{r}(t) = (v_0 \cos \theta)t\mathbf{i} + \left[h + (v_0 \sin \theta)t - \frac{1}{2}gt^2\right]\mathbf{j}$

$= (120 \cos 30°)t\mathbf{i} + \left[5 + (120 \sin 30°)t - 16t^2\right]\mathbf{j} = 60\sqrt{3}t\mathbf{i} + \left[5 + 60t - 16t^2\right]\mathbf{j}$

(b)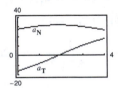

Maximum height $\approx$ 61.25 feet

range $\approx$ 398.2 feet

(c) $\mathbf{v}(t) = 60\sqrt{3}\mathbf{i} + (60 - 32t)\mathbf{j}$

Speed $= \|\mathbf{v}(t)\| = \sqrt{3600(3) + (60 - 32t)^2} = 8\sqrt{16t^2 - 60t + 225}$

$\mathbf{a}(t) = -32\mathbf{j}$

(d)

t	0.5	1.0	1.5	2.0	2.5	3.0
Speed	112.85	107.63	104.61	104.0	105.83	109.98

(e) From Exercise 61, using $v_0 = 120$ and $\theta = 30°$,

$$a_T = \frac{-32(60 - 32t)}{\sqrt{\left(60\sqrt{3}\right)^2 + (60 - 32t)^2}}$$

$$a_N = \frac{32\left(60\sqrt{3}\right)}{\sqrt{\left(60\sqrt{3}\right)^2 + (60 - 32t)^2}}$$

At $t = 1.875$, $a_T = 0$ and the projectile is at its maximum height. When a_T and a_N have opposite signs, the speed is decreasing.

65. $\mathbf{r}(t) = \langle 10 \cos 10\pi t, 10 \sin 10\pi t, 4 + 4t \rangle, 0 \leq t \leq \frac{1}{20}$

(a) $\mathbf{r}'(t) = \langle -100\pi \sin(10\pi t), 100\pi \cos(10\pi t), 4 \rangle$

$\|\mathbf{r}'(t)\| = \sqrt{(100\pi)^2 \sin^2(10\pi t) + (100\pi)^2 \cos^2(10\pi t) + 16}$

$= \sqrt{(100\pi)^2 + 16} = 4\sqrt{625\pi^2 + 1} \approx 314$ mi/h

(b) $a_T = 0$ and $a_N = 1000\pi^2$

$a_T = 0$ because the speed is constant.

67. $\mathbf{r}(t) = (a\cos\omega t)\mathbf{i} + (a\sin\omega t)\mathbf{j}$

From Exercise 29, we know $\mathbf{a}\cdot\mathbf{T} = 0$ and
$\mathbf{a}\cdot\mathbf{N} = a\omega^2$.

(a) Let $\omega_0 = 2\omega$. Then

$$\mathbf{a}\cdot\mathbf{N} = a\omega_0^2 = a(2\omega)^2 = 4a\omega^2$$

or the centripetal acceleration is increased by a factor of 4 when the velocity is doubled.

(b) Let $a_0 = a/2$. Then

$$\mathbf{a}\cdot\mathbf{N} = a_0\omega^2 = \left(\frac{a}{2}\right)\omega^2 = \left(\frac{1}{2}\right)a\omega^2$$

or the centripetal acceleration is halved when the radius is halved.

69. $v = \sqrt{\dfrac{GM}{r}} = \sqrt{\dfrac{9.56\times10^4}{4000+255}} \approx 4.74 \text{ mi/sec}$

71. $v = \sqrt{\dfrac{9.56\times10^4}{4000+385}} \approx 4.67 \text{ mi/sec}$

73. False. You could be turning.

75. (a) $\mathbf{r}(t) = \cosh(bt)\mathbf{i} + \sinh(bt)\mathbf{j}, b > 0$

$x = \cosh(bt), y = \sinh(bt)$

$x^2 - y^2 = \cosh^2(bt) - \sinh^2(bt) = 1$, hyperbola

(b) $\mathbf{v}(t) = b\sinh(bt)\mathbf{i} + b\cosh(bt)\mathbf{j}$

$\mathbf{a}(t) = b^2\cosh(bt)\mathbf{i} + b^2\sinh(bt)\mathbf{j} = b^2\mathbf{r}(t)$

77. $\mathbf{r}(t) = x(t)\mathbf{i} + y(t)\mathbf{j}$

$y(t) = m(x(t)) + b$, m and b are constants.

$\mathbf{r}(t) = x(t)\mathbf{i} + \big[m(x(t)) + b\big]\mathbf{j}$

$\mathbf{v}(t) = x'(t)\mathbf{i} + mx'(t)\mathbf{j}$

$\|\mathbf{v}(t)\| = \sqrt{[x'(t)]^2 + [mx'(t)]^2} = |x'(t)|\sqrt{1 + m^2}$

$\mathbf{T}(t) = \dfrac{\mathbf{v}(t)}{\|\mathbf{v}(t)\|} = \dfrac{\pm(\mathbf{i} + m\mathbf{j})}{\sqrt{1 + m^2}}$, constant

So, $\mathbf{T}'(t) = \mathbf{0}$.

79. $\|\mathbf{a}\|^2 = \mathbf{a}\cdot\mathbf{a}$

$\qquad = (a_\mathbf{T}\mathbf{T} + a_\mathbf{N}\mathbf{N})\cdot(a_\mathbf{T}\mathbf{T} + a_\mathbf{N}\mathbf{N})$

$\qquad = a_\mathbf{T}^2\|\mathbf{T}\|^2 + 2a_\mathbf{T}a_\mathbf{N}\mathbf{T}\cdot\mathbf{N} + a_\mathbf{N}^2\|\mathbf{N}\|^2$

$\qquad = a_\mathbf{T}^2 + a_\mathbf{N}^2$

$a_\mathbf{N}^2 = \|\mathbf{a}\|^2 - a_\mathbf{T}^2$

Because $a_\mathbf{N} > 0$, we have $a_\mathbf{N} = \sqrt{\|\mathbf{a}\|^2 - a_\mathbf{T}^2}$.

Section 12.5   Arc Length and Curvature

1. $\mathbf{r}(t) = 3t\mathbf{i} - t\mathbf{j}, [0, 3]$

$\dfrac{dx}{dt} = 3, \dfrac{dy}{dt} = -1, \dfrac{dz}{dt} = 0$

$s = \displaystyle\int_0^3 \sqrt{3^2 + (-1)^2}\, dt = \left[\sqrt{10}\,t\right]_0^3 = 3\sqrt{10}$

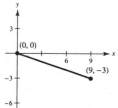

3. $\mathbf{r}(t) = t^3\mathbf{i} + t^2\mathbf{j}, [0, 1]$

$\dfrac{dx}{dt} = 3t^2, \dfrac{dy}{dt} = 2t, \dfrac{dz}{dt} = 0$

$s = \displaystyle\int_0^1 \sqrt{9t^4 + 4t^2}\, dt = \int_0^1 \sqrt{9t^2 + 4}\, t\, dt$

$\quad = \dfrac{1}{18}\displaystyle\int_0^1 (9t^2 + 4)^{1/2}(18t)\, dt = \dfrac{1}{27}\Big[(9t^2 + 4)^{3/2}\Big]_0^1 = \dfrac{1}{27}(13^{3/2} - 8) \approx 1.4397$

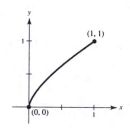

5. $\mathbf{r}(t) = a\cos^3 t\,\mathbf{i} + a\sin^3 t\,\mathbf{j},\ [0, 2\pi]$

$$\frac{dx}{dt} = -3a\cos^2 t\sin t,\ \frac{dy}{dt} = 3a\sin^2 t\cos t$$

$$s = 4\int_0^{\pi/2}\sqrt{\left[-3a\cos^2 t\sin t\right]^2 + \left[3a\sin^2 t\cos t\right]^2}\,dt$$

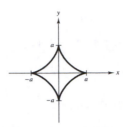

$$= 12a\int_0^{\pi/2}\sin t\cos t\,dt = 3a\int_0^{\pi/2}2\sin 2t\,dt = \left[-3a\cos 2t\right]_0^{\pi/2} = 6a$$

7. (a) $\mathbf{r}(t) = (v_0\cos\theta)t\,\mathbf{i} + \left[h + (v_0\sin\theta)t - \frac{1}{2}gt^2\right]\mathbf{j}$

$$= (100\cos 45°)t\,\mathbf{i} + \left[3 + (100\sin 45°)t - \frac{1}{2}(32)t^2\right]\mathbf{j} = 50\sqrt{2}t\,\mathbf{i} + \left[3 + 50\sqrt{2}t - 16t^2\right]\mathbf{j}$$

(b) $\mathbf{v}(t) = 50\sqrt{2}\,\mathbf{i} + \left(50\sqrt{2} - 32t\right)\mathbf{j}$

$$50\sqrt{2} - 32t = 0 \Rightarrow t = \frac{25\sqrt{2}}{16}$$

Maximum height: $3 + 50\sqrt{2}\left(\dfrac{25\sqrt{2}}{16}\right) - 16\left(\dfrac{15\sqrt{2}}{16}\right)^2 = 81.125$ feet

(c) $3 + 50\sqrt{2}t - 16t^2 = 0 \Rightarrow t \approx 4.4614$

Range: $50\sqrt{2}(4.4614) \approx 315.5$ feet

(d) $s = \int_0^{4.4614}\sqrt{\left(50\sqrt{2}\right)^2 + \left(50\sqrt{2} - 32t\right)^2}\,dt \approx 362.9$ feet

9. $\mathbf{r}(t) = -t\,\mathbf{i} + 4t\,\mathbf{j} + 3t\,\mathbf{k},\ [0, 1]$

$$\frac{dx}{dt} = -1,\ \frac{dy}{dt} = 4,\ \frac{dz}{dt} = 3$$

$$s = \int_0^1\sqrt{1 + 16 + 9}\,dt = \left[\sqrt{26}\,t\right]_0^1 = \sqrt{26}$$

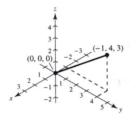

11. $\mathbf{r}(t) = \langle 4t, -\cos t, \sin t\rangle,\ \left[0, \dfrac{3\pi}{2}\right]$

$$\frac{dx}{dt} = 4,\ \frac{dy}{dt} = \sin t,\ \frac{dz}{dt} = \cos t$$

$$s = \int_0^{3\pi/2}\sqrt{16 + \sin^2 t + \cos^2 t}\,dt = \int_0^{3\pi/2}\sqrt{17}\,dt = \left[\sqrt{17}\,t\right]_0^{3\pi/2} = \frac{3\pi}{2}\sqrt{17}$$

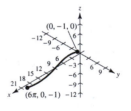

13. $\mathbf{r}(t) = a\cos t\,\mathbf{i} + a\sin t\,\mathbf{j} + bt\,\mathbf{k},\ [0, 2\pi]$

$$\frac{dx}{dt} = -a\sin t,\ \frac{dy}{dt} = a\cos t,\ \frac{dz}{dt} = b$$

$$s = \int_0^{2\pi}\sqrt{a^2\sin^2 t + a^2\cos^2 t + b^2}\,dt$$

$$= \int_0^{2\pi}\sqrt{a^2 + b^2}\,dt = \left[\sqrt{a^2 + b^2}\,t\right]_0^{2\pi} = 2\pi\sqrt{a^2 + b^2}$$

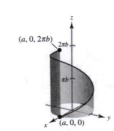

15. $\mathbf{r}(t) = t\mathbf{i} + \left(4 - t^2\right)\mathbf{j} + t^3\mathbf{k}$, $[0, 2]$

(a) $\mathbf{r}(0) = \langle 0, 4, 0 \rangle$, $\mathbf{r}(2) = \langle 2, 0, 8 \rangle$

distance $= \sqrt{2^2 + 4^2 + 8^2} = \sqrt{84}$

$= 2\sqrt{21} \approx 9.165$

(b) $\mathbf{r}(0) = \langle 0, 4, 0 \rangle$

$\mathbf{r}(0.5) = \langle 0.5, 3.75, 0.125 \rangle$

$\mathbf{r}(1) = \langle 1, 3, 1 \rangle$

$\mathbf{r}(1.5) = \langle 1.5, 1.75, 3.375 \rangle$

$\mathbf{r}(2) = \langle 2, 0, 8 \rangle$

distance $\approx \sqrt{(0.5)^2 + (0.25)^2 + (0.125)^2}$

$+ \sqrt{(0.5)^2 + (0.75)^2 + (0.875)^2}$

$+ \sqrt{(0.5)^2 + (1.25)^2 + (2.375)^2}$

$+ \sqrt{(0.5)^2 + (1.75)^2 + (4.625)^2}$

$\approx 0.5728 + 1.2562 + 2.7300 + 4.9702$

≈ 9.529

(c) Increase the number of line segments.

(d) Using a graphing utility, you obtain 9.57057.

17. $\mathbf{r}(t) = \langle 2 \cos t, 2 \sin t, t \rangle$

(a) $s = \int_0^t \sqrt{\left[x'(u)\right]^2 + \left[y'(u)\right]^2 + \left[z'(u)\right]^2} \, du = \int_0^t \sqrt{(-2 \sin u)^2 + (2 \cos u)^2 + (1)^2} \, du = \int_0^t \sqrt{5} \, du = \left[\sqrt{5}\,u\right]_0^t = \sqrt{5}\,t$

(b) $\dfrac{s}{\sqrt{5}} = t$

$x = 2 \cos\left(\dfrac{s}{\sqrt{5}}\right)$, $y = 2 \sin\left(\dfrac{s}{\sqrt{5}}\right)$, $z = \dfrac{s}{\sqrt{5}}$

$\mathbf{r}(s) = 2 \cos\left(\dfrac{s}{\sqrt{5}}\right)\mathbf{i} + 2 \sin\left(\dfrac{s}{\sqrt{5}}\right)\mathbf{j} + \dfrac{s}{\sqrt{5}}\mathbf{k}$

(c) When $s = \sqrt{5}$: $x = 2 \cos 1 \approx 1.081$

$\qquad\qquad\qquad y = 2 \sin 1 \approx 1.683$

$\qquad\qquad\qquad z = 1$

$\qquad\qquad (1.081, 1.683, 1.000)$

When $s = 4$: $x = 2 \cos \dfrac{4}{\sqrt{5}} \approx -0.433$

$\qquad\qquad y = 2 \sin \dfrac{4}{\sqrt{5}} \approx 1.953$

$\qquad\qquad z = \dfrac{4}{\sqrt{5}} \approx 1.789$

$\qquad\qquad (-0.433, 1.953, 1.789)$

(d) $\|\mathbf{r}'(s)\| = \sqrt{\left(-\dfrac{2}{\sqrt{5}} \sin\left(\dfrac{s}{\sqrt{5}}\right)\right)^2 + \left(\dfrac{2}{\sqrt{5}} \cos\left(\dfrac{\sqrt{s}}{\sqrt{5}}\right)\right)^2 + \left(\dfrac{1}{\sqrt{5}}\right)^2} = \sqrt{\dfrac{4}{5} + \dfrac{1}{5}} = 1$

19. $r(s) = \left(1 + \frac{\sqrt{2}}{2}s\right)i + \left(1 - \frac{\sqrt{2}}{2}s\right)j$

$r'(s) = \frac{\sqrt{2}}{2}i - \frac{\sqrt{2}}{2}j$ and $\|r'(s)\| = \sqrt{\frac{1}{2} + \frac{1}{2}} = 1$

$T(s) = \frac{r'(s)}{\|r'(s)\|} = r'(s)$

$T'(s) = 0 \Rightarrow K = \|T'(s)\| = 0$ (The curve is a line.)

21. $r(s) = 2\cos\left(\frac{s}{\sqrt{5}}\right)i + 2\sin\left(\frac{s}{\sqrt{5}}\right)j + \frac{s}{\sqrt{5}}k$

$T(s) = r'(s)$

$\quad = -\frac{2}{\sqrt{5}}\sin\left(\frac{s}{\sqrt{5}}\right)i + \frac{2}{\sqrt{5}}\cos\left(\frac{s}{\sqrt{5}}\right)j + \frac{1}{\sqrt{5}}k$

$T'(s) = -\frac{2}{5}\cos\left(\frac{s}{\sqrt{5}}\right)i - \frac{2}{5}\sin\left(\frac{s}{\sqrt{5}}\right)j$

$K = \|T'(s)\| = \frac{2}{5}$

23. $r(t) = 4ti - 2tj, \ t = 1$

$v(t) = 4i - 2j$

$T(t) = \frac{1}{\sqrt{5}}(2i - j)$

$T'(t) = 0$

$K = \frac{\|T'(t)\|}{\|r'(t)\|} = 0$

(The curve is a line.)

25. $r(t) = ti + \frac{1}{t}j, \ t = 1$

$v(t) = i - \frac{1}{t^2}j$

$v(1) = i - j$

$a(t) = \frac{2}{t^3}j$

$a(1) = 2j$

$T(t) = \frac{t^2 i - j}{\sqrt{t^4 + 1}}$

$N(t) = \frac{1}{(t^4 + 1)^{1/2}}(i + t^2 j)$

$N(1) = \frac{1}{\sqrt{2}}(i + j)$

$K = \frac{a \cdot N}{\|v\|^2} = \frac{\sqrt{2}}{2}$

27. $r(t) = \langle t, \sin t\rangle, \ t = \frac{\pi}{2}$

$r'(t) = \langle 1, \cos t\rangle, \|r'(t)\| = \sqrt{1 + \cos^2 t}$

$r'\left(\frac{\pi}{2}\right) = \langle 1, 0\rangle, \left\|r'\left(\frac{\pi}{2}\right)\right\| = 1$

$a(t) = \langle 0, -\sin t\rangle, \ a\left(\frac{\pi}{2}\right) = \langle 0, -1\rangle$

$T(t) = \frac{1}{\sqrt{1 + \cos^2 t}}\langle 1, \cos t\rangle$

$T\left(\frac{\pi}{2}\right) = \langle 1, 0\rangle$

$N\left(\frac{\pi}{2}\right) = \langle 0, -1\rangle$

$K = \frac{a \cdot N}{\|v\|^2} = \frac{1}{1} = 1$

29. $r(t) = 4\cos 2\pi t i + 4\sin 2\pi t j$

$r'(t) = -8\pi \sin 2\pi t i + 8\pi \cos 2\pi t j$

$T(t) = -\sin 2\pi t i + \cos 2\pi t j$

$T'(t) = -2\pi \cos 2\pi t i - 2\pi \sin 2\pi t j$

$K = \frac{\|T'(t)\|}{\|r'(t)\|} = \frac{2\pi}{8\pi} = \frac{1}{4}$

31. $r(t) = a\cos \omega t i + a\sin \omega t j$

$r'(t) = -a\omega \sin \omega t i + a\omega \cos \omega t j$

$T(t) = -\sin \omega t i + \cos \omega t j$

$T'(t) = -\omega \cos \omega t i - \omega \sin \omega t j$

$K = \frac{\|T'(t)\|}{\|r'(t)\|} = \frac{\omega}{a\omega} = \frac{1}{a}$

33. $r(t) = ti + t^2 j + \frac{t^2}{2}k$

$r'(t) = i + 2tj + tk$

$T(t) = \frac{i + 2tj + tk}{\sqrt{1 + 5t^2}}$

$T'(t) = \frac{-5ti + 2j + k}{(1 + 5t^2)^{3/2}}$

$K = \frac{\|T'(t)\|}{\|r'(t)\|} = \frac{\frac{\sqrt{5}}{(1 + 5t^2)}}{\sqrt{1 + 5t^2}} = \frac{\sqrt{5}}{(1 + 5t^2)^{3/2}}$

35. $\mathbf{r}(t) = 4t\mathbf{i} + 3\cos t\mathbf{j} + 3\sin t\mathbf{k}$

$\mathbf{r}'(t) = 4\mathbf{i} - 3\sin t\mathbf{j} + 3\cos t\mathbf{k}$

$\mathbf{T}(t) = \dfrac{1}{5}[4\mathbf{i} - 3\sin t\mathbf{j} + 3\cos t\mathbf{k}]$

$\mathbf{T}'(t) = \dfrac{1}{5}[-3\cos t\mathbf{j} - 3\sin t\mathbf{k}]$

$K = \dfrac{\|\mathbf{T}'(t)\|}{\mathbf{r}'(t)} = \dfrac{3/5}{5} = \dfrac{3}{25}$

37. $\mathbf{r}(t) = 3t\mathbf{i} + 2t^2\mathbf{j}, \ P(-3, 2) \Rightarrow t = -1$

$x = 3t, \ x' = 3, \ x'' = 0$

$y = 2t^2, \ y' = 4t, \ y'' = 4$

$K = \dfrac{|x'y'' - y'x''|}{\left[(x')^2 + (y')^2\right]^{3/2}} = \dfrac{|3(4) - 0|}{\left[9 + (4t)^2\right]^{3/2}}$

At $t = -1, \ K = \dfrac{12}{(9 + 16)^{3/2}} = \dfrac{12}{125}$

39. $\mathbf{r}(t) = t\mathbf{i} + t^2\mathbf{j} + \dfrac{t^3}{4}\mathbf{k}, \ P(2, 4, 2) \Rightarrow t = 2$

$\mathbf{r}'(t) = \mathbf{i} + 2t\mathbf{j} + \dfrac{3}{4}t^2\mathbf{k}$

$\mathbf{r}'(2) = \mathbf{i} + 4\mathbf{j} + 3\mathbf{k}, \|\mathbf{r}'(2)\| = \sqrt{26}$

$\mathbf{r}''(t) = 2\mathbf{j} + \dfrac{3}{2}t\mathbf{k}$

$\mathbf{r}''(2) = 2\mathbf{j} + 3\mathbf{k}$

$\mathbf{r}'(2) \times \mathbf{r}'(2) = \begin{vmatrix} \mathbf{i} & \mathbf{j} & \mathbf{k} \\ 1 & 4 & 3 \\ 0 & 2 & 3 \end{vmatrix} = 6\mathbf{i} - 3\mathbf{j} + 2\mathbf{k}$

$\|\mathbf{r}'(2) \times \mathbf{r}''(2)\| = \sqrt{49} = 7$

$K = \dfrac{\|\mathbf{r}' \times \mathbf{r}''\|}{\|\mathbf{r}'\|^3} = \dfrac{7}{26^{3/2}} = \dfrac{7\sqrt{26}}{676}$

41. $y = 3x - 2, \ x = a$

Because $y'' = 0, \ K = 0$, and the radius of curvature is undefined.

43. $y = 2x^2 + 3, \ x = -1$

$y' = 4x$

$y'' = 4$

$K = \dfrac{4}{\left[1 + (-4)^2\right]^{3/2}} = \dfrac{4}{17^{3/2}} \approx 0.057$

$\dfrac{1}{K} = \dfrac{17^{3/2}}{4} \approx 17.523 \ (\text{radius of curvature})$

45. $y = \cos 2x, \ x = 2\pi$

$y' = -2\sin 2x$

$y'' = -4\cos 2x$

At $x = 2\pi, \ y = 1, \ y' = 0, \ y'' = -4$

$K = \dfrac{|-4|}{\left[1 + 0^2\right]^{3/2}} = 4$

$\dfrac{1}{K} = \dfrac{1}{4}$

47. $y = x^3, \ x = 2$

$y' = 3x^2, \ y'' = 6x$

At $x = 2, \ y = 8, \ y' = 12, \ y'' = 12$

$K = \dfrac{12}{\left[1 + (12)^2\right]^{3/2}} = \dfrac{12}{(145)^{3/2}}$

$\dfrac{1}{K} = \dfrac{145\sqrt{145}}{12}$

49. $y = (x - 1)^2 + 3, \ y' = 2(x - 1), \ y'' = 2$

$K = \dfrac{2}{\left(1 + [2(x - 1)]^2\right)^{3/2}} = \dfrac{2}{\left[1 + 4(x - 1)^2\right]^{3/2}}$

(a) K is maximum when $x = 1$ or at the vertex $(1, 3)$.

(b) $\lim\limits_{x \to \infty} K = 0$

51. $y = x^{2/3}, \ y' = \dfrac{2}{3}x^{-1/3}, \ y'' = -\dfrac{2}{9}x^{-4/3}$

$K = \left|\dfrac{(-2/9)x^{-4/3}}{\left[1 + (4/9)x^{-2/3}\right]^{3/2}}\right| = \left|\dfrac{6}{x^{1/3}\left(9x^{2/3} + 4\right)^{3/2}}\right|$

(a) $K \to \infty$ as $x \to 0$. No maximum

(b) $\lim\limits_{x \to \infty} K = 0$

53. $y = \ln x, \ y' = \dfrac{1}{x}, \ y'' = -\dfrac{1}{x^2}$

$K = \left|\dfrac{-1/x^2}{\left[1 + (1/x)^2\right]^{3/2}}\right| = \dfrac{x}{\left(x^2 + 1\right)^{3/2}}$

$\dfrac{dK}{dx} = \dfrac{-2x^2 + 1}{\left(x^2 + 1\right)^{5/2}}$

(a) K has a maximum when $x = \dfrac{1}{\sqrt{2}}$.

(b) $\lim\limits_{x \to \infty} K = 0$

55. $y = 1 - x^3$, $y' = -3x^2$, $y'' = -6x$

$$K = \frac{|-6x|}{\left[1 + 9x^4\right]^{3/2}}$$

Curvature is 0 at $x = 0$: $(0, 1)$.

57. $y = \cos x$, $y' = -\sin x$, $y'' = -\cos x$

$$K = \frac{|y''|}{\left[1 + (y')^2\right]^{3/2}} = \frac{|-\cos x|}{\left(1 + \sin^2 x\right)^{3/2}} = 0 \text{ for}$$

$$x = \frac{\pi}{2} + K\pi.$$

Curvature is 0 at $\left(\dfrac{\pi}{2} + K\pi, 0\right)$.

59. $s = \displaystyle\int_a^b \sqrt{x'(t)^2 + y'(t)^2 + z'(t)^2}\; dt = \int_a^b \|r'(t)\|\; dt$

61. The curve is a line.

63. $f(x) = x^4 - x^2$

(a) $K = \dfrac{2\left|6x^2 - 1\right|}{\left|16x^6 - 16x^4 + 4x^2 + 1\right|^{3/2}}$

(b) For $x = 0$, $K = 2$. $f(0) = 0$. At $(0, 0)$, the circle of curvature has radius $\dfrac{1}{2}$. Using the symmetry of the graph of f, you

obtain $x^2 + \left(y + \dfrac{1}{2}\right)^2 = \dfrac{1}{4}$.

For $x = 1$, $K = \left(2\sqrt{5}\right)/5$. $f(1) = 0$. At $(1, 0)$, the circle of curvature has radius $\dfrac{\sqrt{5}}{2} = \dfrac{1}{K}$.

Using the graph of f, you see that the center of curvature is $\left(0, \dfrac{1}{2}\right)$. So,

$$x^2 + \left(y - \frac{1}{2}\right)^2 = \frac{5}{4}.$$

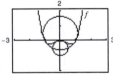

To graph these circles, use

$$y = -\frac{1}{2} \pm \sqrt{\frac{1}{4} - x^2} \text{ and } y = \frac{1}{2} \pm \sqrt{\frac{5}{4} - x^2}.$$

(c) The curvature tends to be greatest near the extrema of f, and K decreases as $x \to \pm\infty$. f and K, however, do not have the same critical numbers.

Critical numbers of f:

$$x = 0, \pm\frac{\sqrt{2}}{2} \approx \pm0.7071$$

Critical numbers of K:
$$x = 0, \pm0.7647, \pm0.4082$$

65. $y_1 = ax(b - x)$, $y_2 = \dfrac{x}{x + 2}$

You observe that $(0, 0)$ is a solution point to both equations. So, the point P is origin.

$y_1 = ax(b - x)$, $y_1' = a(b - 2x)$, $y_1'' = -2a$

$y_2 = \dfrac{x}{x + 2}$, $y_2' = \dfrac{2}{(x + 2)^2}$, $y_2'' = \dfrac{-4}{(x + 2)^3}$

At P, $y_1'(0) = ab$ and $y_2'(0) = \dfrac{2}{(0 + 2)^2} = \dfrac{1}{2}$.

Because the curves have a common tangent at P, $y_1'(0) = y_2'(0)$ or $ab = \dfrac{1}{2}$. So, $y_1'(0) = \dfrac{1}{2}$.

Because the curves have the same curvature at P, $K_1(0) = K_2(0)$.

$$K_1(0) = \left| \frac{y_1''(0)}{\left[1 + \left(y_1(0) \right)^2 \right]^{3/2}} \right| = \left| \frac{-2a}{\left[1 + (1/2)^2 \right]^{3/2}} \right|$$

$$K_2(0) = \left| \frac{y_2''(0)}{\left[1 + \left(y_2(0) \right)^2 \right]^{3/2}} \right| = \left| \frac{-1/2}{\left[1 + (1/2)^2 \right]^{3/2}} \right|$$

So, $2a = \pm\dfrac{1}{2}$ or $a = \pm\dfrac{1}{4}$. In order that the curves intersect at only one point, the parabola must be concave downward. So,

$a = \dfrac{1}{4}$ and $b = \dfrac{1}{2a} = 2$.

$y_1 = \dfrac{1}{4}x(2 - x)$ and $y_2 = \dfrac{x}{x + 2}$

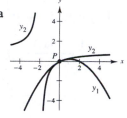

67. (a) Imagine dropping the circle $x^2 + (y - k)^2 = 16$

into the parabola $y = x^2$. The circle will drop to the point where the tangents to the circle and parabola are equal.

$y = x^2$ and $x^2 + (y - k)^2 = 16 \Rightarrow x^2 + (x^2 - k)^2 = 16$

Taking derivatives, $2x + 2(y - k)y' = 0$ and $y' = 2x$. So,

$(y - k)y' = -x \Rightarrow y' = \dfrac{-x}{y - k}$.

So, $\dfrac{-x}{y - k} = 2x \Rightarrow -x = 2x(y - k) \Rightarrow -1 = 2(x^2 - k) \Rightarrow x^2 - k = -\dfrac{1}{2}$.

So, $x^2 + (x^2 - k)^2 = x^2 + \left(-\dfrac{1}{2} \right)^2 = 16 \Rightarrow x^2 = 15.75$.

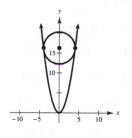

Finally, $k = x^2 + \dfrac{1}{2} = 16.25$, and the center of the circle is 16.25 units from the vertex of the parabola. Because the radius of the circle is 4, the circle is 12.25 units from the vertex.

(b) In 2-space, the parabola $z = y^2 \left(\text{or } z = x^2 \right)$ has a curvature of $K = 2$ at $(0, 0)$. The radius of the largest sphere that will touch the vertex has radius $= 1/K = \dfrac{1}{2}$.

69. $P(x_0, y_0)$ point on curve $y = f(x)$. Let (α, β) be the center of curvature. The radius of curvature is $\dfrac{1}{K}$.

$y' = f'(x)$. Slope of normal line at (x_0, y_0) is $\dfrac{-1}{f'(x_0)}$.

Equation of normal line: $y - y_0 = \dfrac{-1}{f'(x_0)}(x - x_0)$

(α, β) is on the normal line: $-f'(x_0)(\beta - y_0) = \alpha - x_0$ Equation 1

(x_0, y_0) lies on the circle: $(x_0 - \alpha)^2 + (y_0 - \beta)^2 = \left(\dfrac{1}{K}\right)^2 = \left[\dfrac{\left(1 + f'(x_0)^2\right)^{3/2}}{|f''(x_0)|}\right]^2$ Equation 2

Substituting Equation 1 into Equation 2:

$$\left[f'(x_0)(\beta - y_0)\right]^2 + (y_0 - \beta)^2 = \left(\dfrac{1}{K}\right)^2$$

$$(\beta - y_0)^2 + \left[1 + f'(x_0)^2\right] = \dfrac{\left(1 + f'(x_0)^2\right)^3}{\left(f''(x_0)\right)^2}$$

$$(\beta - y_0)^2 = \dfrac{\left[1 + f'(x_0)^2\right]^2}{f''(x_0)^2}$$

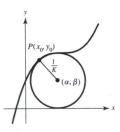

When $f''(x_0) > 0$, $\beta - y_0 > 0$, and if $f''(x_0) < 0$, then $\beta - y_0 < 0$.

So $\beta - y_0 = \dfrac{1 + f'(x_0)^2}{f''(x_0)}$

$$\beta = y_0 + \dfrac{1 + f'(x_0)^2}{f''(x_0)} = y_0 + z$$

Similarly, $\alpha = x_0 - f'(x_0)z$.

71. $r(\theta) = r\cos\theta\,\mathbf{i} + r\sin\theta\,\mathbf{j} = f(\theta)\cos\theta\,\mathbf{i} + f(\theta)\sin\theta\,\mathbf{j}$

$x(\theta) = f(\theta)\cos\theta$

$y(\theta) = f(\theta)\sin\theta$

$x'(\theta) = -f(\theta)\sin\theta + f'(\theta)\cos\theta$

$y'(\theta) = f(\theta)\cos\theta + f'(\theta)\sin\theta$

$x''(\theta) = -f(\theta)\cos\theta - f'(\theta)\sin\theta - f'(\theta)\sin\theta + f''(\theta)\cos\theta = -f(\theta)\cos\theta - 2f'(\theta)\sin\theta + f''(\theta)\cos\theta$

$y''(\theta) = -f(\theta)\sin\theta + f'(\theta)\cos\theta + f'(\theta)\cos\theta + f''(\theta)\sin\theta = -f(\theta)\sin\theta + 2f'(\theta)\cos\theta + f''(\theta)\sin\theta$

$$K = \dfrac{|x'y'' - y'x''|}{\left[(x')^2 + (y')^2\right]^{3/2}} = \dfrac{\left|f^2(\theta) - f(\theta)f''(\theta) + 2\left(f'(\theta)\right)^2\right|}{\left[f^2(\theta) + \left(f'(\theta)\right)^2\right]^{3/2}} = \dfrac{\left|r^2 - rr'' + 2(r')^2\right|}{\left[r^2 + (r')^2\right]^{3/2}}$$

73. $r = e^{a\theta}, a > 0$

$r' = ae^{a\theta}$

$r'' = a^2 e^{a\theta}$

$$K = \frac{\left|2(r')^2 - rr'' + r^2\right|}{\left[(r')^2 + r^2\right]^{3/2}} = \frac{\left|2a^2 e^{2a\theta} - a^2 e^{2a\theta} + e^{2a\theta}\right|}{\left[a^2 e^{2a\theta} + e^{2a\theta}\right]^{3/2}} = \frac{1}{e^{a\theta}\sqrt{a^2 + 1}}$$

(a) As $\theta \to \infty, K \to 0$.

(b) As $a \to \infty, K \to 0$.

75. $r = 4 \sin 2\theta$

$r' = 8 \cos 2\theta$

At the pole: $K = \dfrac{2}{|r'(0)|} = \dfrac{2}{8} = \dfrac{1}{4}$

77. $x = f(t), y = g(t)$

$$y' = \frac{dy}{dx} = \frac{\dfrac{dy}{dt}}{\dfrac{dx}{dt}} = \frac{g'(t)}{f'(t)}$$

$$y'' = \frac{\dfrac{d}{dt}\left[\dfrac{g'(t)}{f'(t)}\right]}{\dfrac{dx}{dt}}$$

$$= \frac{\dfrac{f'(t)g''(t) - g'(t)f''(t)}{\left[f'(t)\right]^2}}{f'(t)} = \frac{f'(t)g''(t) - g'(t)f''(t)}{\left[f'(t)\right]^3}$$

$$K = \frac{|y''|}{\left[1 + (y')^2\right]^{3/2}} = \frac{\left|\dfrac{f'(t)g''(t) - g'(t)f''(t)}{\left[f'(t)\right]^3}\right|}{\left[1 + \left(\dfrac{g'(t)}{f'(t)}\right)^2\right]^{3/2}}$$

$$= \frac{\left|\dfrac{f'(t)g''(t) - g'(t)f''(t)}{\left[f'(t)\right]^3}\right|}{\sqrt{\left\{\dfrac{\left[f'(t)\right]^2 + \left[g'(t)\right]^2}{\left[f'(t)\right]^2}\right\}^3}} = \frac{\left|f'(t)g''(t) - g'(t)f''(t)\right|}{\left(\left[f'(t)\right]^2 + \left[g'(t)\right]^2\right)^{3/2}}$$

79. $x(\theta) = a(\theta - \sin\theta)$ $y(\theta) = a(1 - \cos\theta)$

$x'(\theta) = a(1 - \cos\theta)$ $y'(\theta) = a\sin\theta$

$x''(\theta) = a\sin\theta$ $y''(\theta) = a\cos\theta$

$$K = \frac{\left|x'(\theta)y''(\theta) - y'(\theta)x''(\theta)\right|}{\left[x'(\theta)^2 + y'(\theta)^2\right]^{3/2}}$$

$$= \frac{\left|a^2(1 - \cos\theta)\cos\theta - a^2\sin^2\theta\right|}{\left[a^2(1 - \cos\theta)^2 + a^2\sin^2\theta\right]^{3/2}}$$

$$= \frac{1}{a}\frac{|\cos\theta - 1|}{\left[2 - 2\cos\theta\right]^{3/2}}$$

$$= \frac{1}{a}\frac{1 - \cos\theta}{2\sqrt{2}\left[1 - \cos\theta\right]^{3/2}} \quad (1 - \cos \geq 0)$$

$$= \frac{1}{2a\sqrt{2 - 2\cos\theta}} = \frac{1}{4a}\csc\left(\frac{\theta}{2}\right)$$

Minimum: $\dfrac{1}{4a}$ $(\theta = \pi)$

Maximum: none $(K \to \infty$ as $\theta \to 0)$

81. $F = ma_N = mK\left(\dfrac{ds}{dt}\right)^2$

$$= \left(\frac{5500\ \text{lb}}{32\ \text{ft/sec}^2}\right)\left(\frac{1}{100\ \text{ft}}\right)\left(\frac{30(5280)\ \text{ft}}{3600\ \text{sec}}\right)^2 = 3327.5\ \text{lb}$$

83. $y = \cosh x = \dfrac{e^x + e^{-x}}{2}$

$y' = \dfrac{e^x - e^{-x}}{2} = \sinh x$

$y'' = \dfrac{e^x + e^{-x}}{2} = \cosh x$

$$K = \frac{|\cosh x|}{\left[1 + (\sinh x)^2\right]^{3/2}} = \frac{\cosh x}{\left(\cosh^2 x\right)^{3/2}} = \frac{1}{\cosh^2 x} = \frac{1}{y^2}$$

85. False

87. True

89. Let $\mathbf{r} = x(t)\mathbf{i} + y(t)\mathbf{j} + z(t)\mathbf{k}$. Then $r = \|\mathbf{r}\| = \sqrt{\left[x(t)\right]^2 + \left[y(t)\right]^2 + \left[z(t)\right]^2}$ and $\mathbf{r}' = x'(t)\mathbf{i} + y'(t)\mathbf{j} + z'(t)\mathbf{k}$. Then,

$$r\left(\frac{dr}{dt}\right) = \sqrt{\left[x(t)\right]^2 + \left[y(t)\right]^2 + \left[z(t)\right]^2}\left[\frac{1}{2}\left\{\left[x(t)\right]^2 + \left[y(t)\right]^2 + \left[z(t)\right]^2\right\}^{-1/2} \cdot \left(2x(t)x'(t) + 2y(t)y'(t) + 2z(t)z'(t)\right)\right]$$

$$= x(t)x'(t) + y(t)y'(t) + z(t)z'(t) = \mathbf{r} \cdot \mathbf{r}'.$$

91. Let $\mathbf{r} = x\mathbf{i} + y\mathbf{j} + z\mathbf{k}$ where x, y, and z are function of t, and $r = \|\mathbf{r}\|$.

$$\frac{d}{dt}\left[\frac{\mathbf{r}}{r}\right] = \frac{r\mathbf{r}' - \mathbf{r}(dr/dt)}{r^2} = \frac{r\mathbf{r}' - \mathbf{r}\left[(\mathbf{r} \cdot \mathbf{r}')/r\right]}{r^2}$$

$$= \frac{r^2\mathbf{r}' - (\mathbf{r} \cdot \mathbf{r}')\mathbf{r}}{r^3} \text{ (using Exercise 105)}$$

$$= \frac{\left(x^2 + y^2 + z^2\right)\left(x'\mathbf{i} + y'\mathbf{j} + z'\mathbf{k}\right) - \left(xx' + yy' + zz'\right)\left(x\mathbf{i} + y\mathbf{j} + z\mathbf{k}\right)}{r^3}$$

$$= \frac{1}{r^3}\left[\left(x'y^2 + x'z^2 - xyy' - xzz'\right)\mathbf{i} + \left(x^2y' + z^2y' - xx'y - zz'y\right)\mathbf{j} + \left(x^2z' + y^2z' - xx'z - yy'z\right)\mathbf{k}\right]$$

$$= \frac{1}{r^3}\begin{vmatrix} \mathbf{i} & \mathbf{j} & \mathbf{k} \\ yz' - y'z & -(xz' - x'z) & xy' - x'y \\ x & y & z \end{vmatrix} = \frac{1}{r^3}\left\{\left[\mathbf{r} \times \mathbf{r}'\right] \times \mathbf{r}\right\}$$

93. From Exercise 90, you have concluded that planetary motion is planar. Assume that the planet moves in the xy-plane with the sum at the origin. From Exercise 92, you have

$$\mathbf{r}' \times \mathbf{L} = GM\left(\frac{\mathbf{r}}{r} + \mathbf{e}\right).$$

Because $\mathbf{r}' \times \mathbf{L}$ and $\mathbf{r}$ are both perpendicular to $\mathbf{L}$, so is $\mathbf{e}$.

So, $\mathbf{e}$ lies in the xy-plane. Situate the coordinate system so that $\mathbf{e}$ lies along the positive x-axis and θ is the angle between $\mathbf{e}$ and $\mathbf{r}$. Let $e = \|\mathbf{e}\|$. Then $\mathbf{r} \cdot \mathbf{e} = \|\mathbf{r}\|\|\mathbf{e}\|\cos\theta = re\cos\theta$. Also,

$$\|\mathbf{L}\|^2 = \mathbf{L} \cdot \mathbf{L}$$

$$= (\mathbf{r} \times \mathbf{r}') \cdot \mathbf{L}$$

$$= \mathbf{r} \cdot (\mathbf{r}' \times \mathbf{L})$$

$$= \mathbf{r} \cdot \left[GM\left(\mathbf{e} + \frac{\mathbf{r}}{r}\right)\right]$$

$$= GM\left[\mathbf{r} \cdot \mathbf{e} + \frac{\mathbf{r} \cdot \mathbf{r}}{r}\right]$$

$$= GM\left[re\cos\theta + r\right].$$

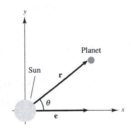

So, $\dfrac{\|\mathbf{L}\|^2/GM}{1 + e\cos\theta} = r$

and the planetary motion is a conic section. Because the planet returns to its initial position periodically, the conic is an ellipse.

95. $A = \dfrac{1}{2}\displaystyle\int_\alpha^\beta r^2 \, d\theta$

So,

$$\frac{dA}{dt} = \frac{dA}{d\theta}\frac{d\theta}{dt} = \frac{1}{2}r^2\frac{d\theta}{dt} = \frac{1}{2}\|\mathbf{L}\|$$

and $\mathbf{r}$ sweeps out area at a constant rate.

Review Exercises for Chapter 12

1. $\mathbf{r}(t) = \tan t\,\mathbf{i} + \mathbf{j} + t\mathbf{k}$

(a) Domain: $t \neq \dfrac{\pi}{2} + n\pi$, n an integer

(b) Continuous for all $t \neq \dfrac{\pi}{2} + n\pi$, n an integer

3. $\mathbf{r}(t) = \ln t\,\mathbf{i} + t\,\mathbf{j} + t\mathbf{k}$

(a) Domain: $(0, \infty)$

(b) Continuous for all $t > 0$

5. $\mathbf{r}(t) = (2t + 1)\mathbf{i} + t^2\,\mathbf{j} - \sqrt{t + 2}\,\mathbf{k}$

(a) $\mathbf{r}(0) = \mathbf{i} - \sqrt{2}\,\mathbf{k}$

(b) $\mathbf{r}(-2) = -3\mathbf{i} + 4\mathbf{j}$

(c) $\mathbf{r}(c - 1) = (2c - 1)\mathbf{i} + (c - 1)^2\,\mathbf{j} - \sqrt{c + 1}\,\mathbf{k}$

(d) $\mathbf{r}(1 + \Delta t) - \mathbf{r}(1) = \left[2(1 + \Delta t) + 1\right]\mathbf{i} + (1 + \Delta t)^2\,\mathbf{j} - \sqrt{3 + \Delta t}\,\mathbf{k} - \left(3\mathbf{i} + \mathbf{j} - \sqrt{3}\,\mathbf{k}\right)$

$\qquad\qquad\qquad\quad = 2\Delta t\,\mathbf{i} + \left(\Delta t^2 + 2\Delta t\right)\mathbf{j} - \left(\sqrt{3 + \Delta t} - \sqrt{3}\right)\mathbf{k}$

7. $P(3, 0, 5)$, $Q(2, -2, 3)$

$\qquad \mathbf{v} = \overline{PQ} = \langle -1, -2, -2 \rangle$

$\qquad \mathbf{r}(t) = (3 - t)\mathbf{i} - 2t\mathbf{j} + (5 - 2t)\mathbf{k}$, $0 \leq t \leq 1$

$\qquad x = 3 - t$, $y = -2t$, $z = 5 - 2t$, $0 \leq t \leq 1$

(Answers may vary)

9. $\mathbf{r}(t) = \langle \pi \cos t, \pi \sin t \rangle$

$\qquad x = \pi \cos t$, $y = \pi \sin t$

$\qquad x^2 + y^2 = \pi^2$, circle

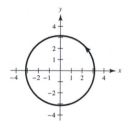

11. $\mathbf{r}(t) = (t + 1)\mathbf{i} + (3t - 1)\mathbf{j} + 2t\mathbf{k}$

$\qquad x = t + 1$, $y = 3t - 1$, $z = 2t$

This is a line passing through the points $(1, -1, 0)$ and $(2, 2, 2)$.

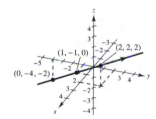

13. $3x + 4y - 12 = 0$

$\qquad$ Let $x = t$, then $y = \dfrac{12 - 3t}{4}$.

$\qquad r(t) = t\mathbf{i} + \dfrac{12 - 3t}{4}\mathbf{j}$

$\qquad$ Alternate solution: $x = 4t$, $y = 3 - 3t$

$\qquad r(t) = 4t\mathbf{i} + (3 - 3t)\mathbf{j}$

15. $z = x^2 + y^2, x + y = 0, t = x$

$x = t, y = -t, z = 2t^2$

$\mathbf{r}(t) = t\mathbf{i} - t\mathbf{j} + 2t^2\mathbf{k}$

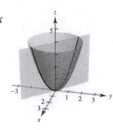

17. $\lim\limits_{t \to 4^-} \left(t\mathbf{i} + \sqrt{4 - t}\,\mathbf{j} + \mathbf{k}\right) = 4\mathbf{i} + \mathbf{k}$

19. $\mathbf{r}(t) = \left(t^2 + 4t\right)\mathbf{i} - 3t^2\mathbf{j}$

(a) $\mathbf{r}'(t) = (2t + 4)\mathbf{i} - 6t\mathbf{j}$

(b) $\mathbf{r}''(t) = 2\mathbf{i} - 6\mathbf{j}$

(c) $\mathbf{r}'(t) \cdot \mathbf{r}''(t) = (2t + 4)(2) + (-6t)(-6)$

$\qquad = 40t + 8$

21. $\mathbf{r}(t) = 2t^3\mathbf{i} + 4t\mathbf{j} - t^2\mathbf{k}$

(a) $\mathbf{r}'(t) = 6t^2\mathbf{i} + 4\mathbf{j} - 2t\mathbf{k}$

(b) $\mathbf{r}''(t) = 12t\mathbf{i} - 2\mathbf{k}$

(c) $\mathbf{r}'(t) \cdot \mathbf{r}''(t) = 6t^2(12t) + (-2t)(-2)$

$\qquad = 72t^3 + 4t$

(d) $\mathbf{r}'(t) \times \mathbf{r}''(t) = \begin{vmatrix} \mathbf{i} & \mathbf{j} & \mathbf{k} \\ 6t^2 & 4 & -2t \\ 12t & 0 & -2 \end{vmatrix}$

$\qquad = -8\mathbf{i} - \left(-12t^2 + 24t^2\right)\mathbf{j} + (-48t)\mathbf{k}$

$\qquad = -8\mathbf{i} - 12t^2\mathbf{j} - 48t\mathbf{k}$

23. $\mathbf{r}(t) = 3t\mathbf{i} + (t - 1)\mathbf{j}, \mathbf{u}(t) = t\mathbf{i} + t^2\mathbf{j} + \frac{2}{3}t^3\mathbf{k}$

$\mathbf{r}'(t) = 3\mathbf{i} + \mathbf{j}, \mathbf{u}'(t) = \mathbf{i} + 2t\mathbf{j} + 2t^2\mathbf{k}$

(a) $\mathbf{r}'(t) = 3\mathbf{i} + \mathbf{j}$

(b) $\dfrac{d}{dt}\big[\mathbf{u}(t) - 2\mathbf{r}(t)\big] = \mathbf{u}'(t) - 2\mathbf{r}'(t)$

$\qquad = \left(\mathbf{i} + 2t\mathbf{j} + 2t^2\mathbf{k}\right) - 2(3\mathbf{i} + \mathbf{j})$

$\qquad = (1 - 6)\mathbf{i} + (2t - 2)\mathbf{j} + 2t^2\mathbf{k}$

$\qquad = -5\mathbf{i} + (2t - 2)\mathbf{j} + 2t^2\mathbf{k}$

(c) $\dfrac{d}{dt}(3t)\mathbf{r}(t) = (3t)\mathbf{r}'(t) + 3\mathbf{r}(t)$

$\qquad = (3t)(3\mathbf{i} + \mathbf{j}) + 3\big[3t\mathbf{i} + (t - 1)\mathbf{j}\big]$

$\qquad = 9t\mathbf{i} + 3t\mathbf{j} + 9t\mathbf{i} + (3t - 3)\mathbf{j}$

$\qquad = 18t\mathbf{i} + (6t - 3)\mathbf{j}$

(d) $\dfrac{d}{dt}\big[\mathbf{r}(t) \cdot \mathbf{u}(t)\big] = \mathbf{r}(t) \cdot \mathbf{u}'(t) + \mathbf{r}'(t) \cdot \mathbf{u}(t)$

$\qquad = \left[(3t)(1) + (t - 1)(2t) + (0)(2t^2)\right] + \left[(3)(t) + (1)(t^2) + (0)(\tfrac{2}{3}t^3)\right]$

$\qquad = \left(3t + 2t^2 - 2t\right) + \left(3t + t^2\right)$

$\qquad = 4t + 3t^2$

(e) $\dfrac{d}{dt}\big[\mathbf{r}(t) \times \mathbf{u}(t)\big] = \mathbf{r}(t) \times \mathbf{u}'(t) + \mathbf{r}'(t) \times \mathbf{u}(t)$

$\qquad = \left[\left(2t^3 - 2t^2\right)\mathbf{i} - 6t^3\mathbf{j} + \left(6t^2 - t + 1\right)\mathbf{k}\right] + \left[\tfrac{2}{3}t^3\mathbf{i} - 2t^3\mathbf{j} + \left(3t^2 - t\right)\mathbf{k}\right]$

$\qquad = \left(\tfrac{8}{3}t^3 - 2t^2\right)\mathbf{i} - 8t^3\mathbf{j} + \left(9t^2 - 2t + 1\right)\mathbf{k}$

(f) $\dfrac{d}{dt}\big[\mathbf{u}(2t)\big] = 2\mathbf{u}'(2t)$

$\qquad = 2\left[\mathbf{i} + 2(2t)\mathbf{j} + 2(2t)^2\mathbf{k}\right]$

$\qquad = 2\mathbf{i} + 8t\mathbf{j} + 16t^2\mathbf{k}$

25. $\int (\mathbf{i} + 3\mathbf{j} + 4t\mathbf{k})\, dt = t\mathbf{i} + 3t\mathbf{j} + 2t^2\mathbf{k} + \mathbf{C}$

27. $\int \left(3\sqrt{t}\,\mathbf{i} + \dfrac{2}{t}\mathbf{j} + \mathbf{k} \right) dt = 2t^{3/2}\mathbf{i} + 2\ln|t|\mathbf{j} + t\mathbf{k} + \mathbf{C}$

29. $\int_{-2}^{2} \left(3t\mathbf{i} + 2t^2\,\mathbf{j} - t^3\mathbf{k} \right) dt = \left[\dfrac{3t^2}{2}\mathbf{i} + \dfrac{2t^3}{3}\mathbf{j} - \dfrac{t^4}{4}\mathbf{k} \right]_{-2}^{2} = \dfrac{32}{3}\mathbf{j}$

31. $\int_{0}^{2} \left(e^{t/2}\mathbf{i} - 3t^2\,\mathbf{j} - \mathbf{k} \right) dt = \left[2e^{t/2}\mathbf{i} - t^3\mathbf{j} - t\mathbf{k} \right]_{0}^{2} = (2e - 2)\mathbf{i} - 8\mathbf{j} - 2\mathbf{k}$

33. $\mathbf{r}(t) = \int \left(2t\mathbf{i} + e^t\mathbf{j} + e^{-t}\mathbf{k} \right) dt = t^2\mathbf{i} + e^t\mathbf{j} - e^{-t}\mathbf{k} + \mathbf{C}$

$\mathbf{r}(0) = \mathbf{j} - \mathbf{k} + \mathbf{C} = \mathbf{i} + 3\mathbf{j} - 5\mathbf{k} \Rightarrow \mathbf{C} = \mathbf{i} + 2\mathbf{j} - 4\mathbf{k}$

$\mathbf{r}(t) = (t^2 + 1)\mathbf{i} + (e^t + 2)\mathbf{j} - (e^{-t} + 4)\mathbf{k}$

35. $\mathbf{r}(t) = 4t\mathbf{i} + t^3\mathbf{j} - t\mathbf{k},\ t = 1$

(a) $\mathbf{v}(t) = \mathbf{r}'(t) = 4\mathbf{i} + 3t^2\mathbf{j} - \mathbf{k}$

     Speed $= \|\mathbf{v}(t)\| = \sqrt{16 + 9t^4 + 1} = \sqrt{17 + 9t^4}$

     $\mathbf{a}(t) = \mathbf{r}''(t) = 6t\mathbf{j}$

(b) $\mathbf{v}(1) = 4\mathbf{i} + 3\mathbf{j} - \mathbf{k}$

     $\mathbf{a}(1) = 6\mathbf{j}$

37. $\mathbf{r}(t) = \langle \cos^3 t,\ \sin^3 t,\ 3t \rangle,\ t = \pi$

(a) $\mathbf{v}(t) = \mathbf{r}'(t) = \langle -3\cos^2 t \sin t,\ 3\sin^2 t \cos t,\ 3 \rangle$

     Speed $= \|\mathbf{v}(t)\| = \sqrt{9\cos^4 t \sin^2 t + 9\sin^4 t \cos^2 t + 9}$

                  $= 3\sqrt{\cos^2 t \sin^2 t (\cos^2 t + \sin^2 t) + 1}$

                  $= 3\sqrt{\cos^2 t \sin^2 t + 1}$

     $\mathbf{a}(t) = \langle -6\cos t(-\sin^2 t) + (-3\cos^2 t)\cos t,\ 6\sin t \cos^2 t + 3\sin^2 t(-\sin t),\ 0 \rangle$

           $= \langle 3\cos t(2\sin^2 t - \cos^2 t),\ 3\sin t(2\cos^2 t - \sin^2 t),\ 0 \rangle$

(b) $\mathbf{v}(\pi) = \langle 0, 0, 3 \rangle$

     $\mathbf{a}(\pi) = \langle 3, 0, 0 \rangle$

39. $\mathbf{r}(t) = \left\langle (v_0 \cos \theta)t,\ (v_0 \sin \theta)t - \dfrac{1}{2}gt^2 \right\rangle$

     $= \langle 42\sqrt{3}\,t,\ 42t - 16t^2 \rangle$

     $42t = 16t^2 \Rightarrow t = 0, \dfrac{21}{8}$

     Range $= 42\sqrt{3}\left(\dfrac{21}{8} \right) = \dfrac{441\sqrt{3}}{4} \approx 190.96$ ft

41. Range $= x = \dfrac{v_0^2}{9.8}\sin 2\theta = 95$

     $v_0^2 = \dfrac{9.8(95)}{\sin(40°)}$

     $v_0 \approx 38.06$ m/sec

43. $\mathbf{r}(t) = 3t\mathbf{i} + 3t^3\mathbf{j},\ t = 1$

     $\mathbf{r}'(t) = 3\mathbf{i} + 9t^2\mathbf{j}$

     $\mathbf{r}'(1) = 3\mathbf{i} + 9\mathbf{j},\ \|\mathbf{r}'(1)\| = \sqrt{9 + 81} = \sqrt{90} = 3\sqrt{10}$

     $\mathbf{T}(1) = \dfrac{\mathbf{r}'(1)}{\|\mathbf{r}'(1)\|} = \dfrac{1}{3\sqrt{10}}(3\mathbf{i} + 9\mathbf{j}) = \dfrac{1}{\sqrt{10}}\mathbf{i} + \dfrac{3}{\sqrt{10}}\mathbf{j}$

45. $\mathbf{r}(t) = 2\cos t\mathbf{i} + 2\sin t\mathbf{j} + t\mathbf{k}, \; P\left(1, \sqrt{3}, \frac{\pi}{3}\right)$

$\mathbf{r}'(t) = -2\sin t\mathbf{i} + 2\cos t\mathbf{j} + \mathbf{k}$

$t = \frac{\pi}{3}$ at $P\left(1, \sqrt{3}, \frac{\pi}{3}\right)$

$\mathbf{r}'\left(\frac{\pi}{3}\right) = -\sqrt{3}\mathbf{i} + \mathbf{j} + \mathbf{k}$

$\mathbf{T}\left(\frac{\pi}{3}\right) = \dfrac{\mathbf{r}'\left(\frac{\pi}{3}\right)}{\left\|\mathbf{r}'\left(\frac{\pi}{3}\right)\right\|} = \dfrac{-\sqrt{3}\mathbf{i} + \mathbf{j} + \mathbf{k}}{\sqrt{5}} = \dfrac{\sqrt{15}}{5}\mathbf{i} + \dfrac{\sqrt{5}}{5}\mathbf{j} + \dfrac{\sqrt{5}}{5}\mathbf{k}$

Direction numbers: $-\sqrt{3}, 1, 1$

$x = 1 - \sqrt{3}t, \; y = \sqrt{3} + t, \; z = \frac{\pi}{3} + t$

47. $\mathbf{r}(t) = 2t\mathbf{i} + 3t^2\mathbf{j}, \; t = 1$

$\mathbf{r}'(t) = 2\mathbf{i} + 6t\mathbf{j}, \; \mathbf{r}'(1) = 2\mathbf{i} + 6\mathbf{j}$

$\|\mathbf{r}'(t)\| = \sqrt{4 + 36t^2}, \|\mathbf{r}'(1)\| = \sqrt{40} = 2\sqrt{10}$

$\mathbf{T}(1) = \dfrac{\mathbf{r}'(1)}{\|\mathbf{r}'(1)\|} = \dfrac{2\mathbf{i} + 6\mathbf{j}}{2\sqrt{10}} = \dfrac{1}{\sqrt{10}}(\mathbf{i} + 3\mathbf{j})$

$\mathbf{N}(1)$ is orthogonal to $\mathbf{T}(1)$ and points towards the concave side. Hence,

$\mathbf{N}(1) = \dfrac{1}{\sqrt{10}}(-3\mathbf{i} + \mathbf{j}).$

49. $\mathbf{r}(t) = 3\cos 2t\mathbf{i} + 3\sin 2t\mathbf{j} + 3\mathbf{k}, \; t = \frac{\pi}{4}$

$\mathbf{r}'(t) = -6\sin 2t\mathbf{i} + 6\cos 2t\mathbf{j}, \; \|\mathbf{r}'(t)\| = 6$

$\mathbf{T}(t) = -\sin 2t\mathbf{i} + \cos 2t\mathbf{j}, \; \mathbf{T}(\pi/4) = -\mathbf{i}$

$\mathbf{T}'(t) = -2\cos 2t\mathbf{i} - 2\sin 2t\mathbf{j}, \; \|\mathbf{T}'(t)\| = 2$

$\mathbf{N}(t) = -\cos 2t\mathbf{i} - \sin 2t\mathbf{j}$

$\mathbf{N}(\pi/4) = -\mathbf{j}$

51. $\mathbf{r}(t) = \dfrac{3}{t}\mathbf{i} - 6t\mathbf{j}, \; t = 3$

$\mathbf{v}(t) = -\dfrac{3}{t^2}\mathbf{i} - 6\mathbf{j}, \; \mathbf{v}(3) = -\dfrac{1}{3}\mathbf{i} - 6\mathbf{j}$

$\mathbf{a}(t) = \dfrac{6}{t^3}\mathbf{i}, \; \mathbf{a}(3) = \dfrac{2}{9}\mathbf{i}$

$\mathbf{T}(t) = \dfrac{\mathbf{v}(t)}{\|\mathbf{v}(t)\|} = \dfrac{\left(-\frac{3}{t^2}\right)\mathbf{i} - 6\mathbf{j}}{\sqrt{\frac{9}{t^4} + 36}} = \dfrac{-3\mathbf{i} - 6t^2\mathbf{j}}{3\sqrt{1 + 4t^4}}$

$\mathbf{T}(3) = \dfrac{-3\mathbf{i} - 54\mathbf{j}}{3\sqrt{1 + 324}} = \dfrac{-\mathbf{i} - 18\mathbf{j}}{\sqrt{325}} = -\dfrac{\sqrt{13}}{65}\mathbf{i} - \dfrac{18\sqrt{13}}{65}\mathbf{j}$

$\mathbf{N}(3)$ is orthogonal to $\mathbf{T}(3)$, and points in the direction the curve is bending. Hence,

$\mathbf{N}(3) = \dfrac{18\mathbf{i} - \mathbf{j}}{\sqrt{325}} = \dfrac{18\sqrt{13}}{65}\mathbf{i} - \dfrac{\sqrt{13}}{65}\mathbf{j}.$

$a_\mathbf{T} = \mathbf{a} \cdot \mathbf{T} = -\dfrac{2}{9\sqrt{325}} = -\dfrac{2\sqrt{13}}{585}$

$a_\mathbf{N} = \mathbf{a} \cdot \mathbf{N} = \dfrac{4}{\sqrt{325}} = \dfrac{4\sqrt{13}}{65}$

53. $\mathbf{r}(t) = 2t\mathbf{i} - 3t\mathbf{j}, [0, 5]$

$\mathbf{r}'(t) = 2\mathbf{i} - 3\mathbf{j}$

$s = \int_a^b \|\mathbf{r}'(t)\| \, dt = \int_0^5 \sqrt{4 + 9} \, dt = \left[\sqrt{13}\, t\right]_0^5 = 5\sqrt{13}$

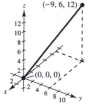

55. $\mathbf{r}(t) = 10\cos^3 t\,\mathbf{i} + 10\sin^3 t\,\mathbf{j}, [0, 2\pi]$

$\mathbf{r}'(t) = -30\cos^2 t \sin t\,\mathbf{i} + 30\sin^2 t \cos t\,\mathbf{j}$

$\|\mathbf{r}'(t)\| = 30\sqrt{\cos^4 t \sin^2 t + \sin^4 t \cos^2 t}$

$\qquad = 30|\cos t \sin t|$

$s = 4\int_0^{\pi/2} 30\cos t \cdot \sin t \, dt = \left[120\frac{\sin^2 t}{2}\right]_0^{\pi/2} = 60$

57. $\mathbf{r}(t) = -3t\mathbf{i} + 2t\mathbf{j} + 4t\mathbf{k}, [0, 3]$

$\mathbf{r}'(t) = -3\mathbf{i} + 2\mathbf{j} + 4\mathbf{k}$

$s = \int_a^b \|\mathbf{r}'(t)\| \, dt = \int_0^3 \sqrt{9 + 4 + 16} \, dt$

$\qquad = \int_0^3 \sqrt{29} \, dt = 3\sqrt{29}$

59. $\mathbf{r}(t) = \langle 8\cos t, 8\sin t, t \rangle, \left[0, \dfrac{\pi}{2}\right]$

$\mathbf{r}'(t) = \langle -8\sin t, 8\cos t, 1 \rangle, \|\mathbf{r}'(t)\| = \sqrt{65}$

$s = \int_a^b \|\mathbf{r}'(t)\| \, dt = \int_0^{\pi/2} \sqrt{65} \, dt = \dfrac{\pi\sqrt{65}}{2}$

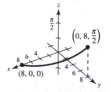

61. $\mathbf{r}(t) = 3t\mathbf{i} + 2t\mathbf{j}$

Line

$K = 0$

63. $\mathbf{r}(t) = 2t\mathbf{i} + \dfrac{1}{2}t^2\mathbf{j} + t^2\mathbf{k}$

$\mathbf{r}'(t) = 2\mathbf{i} + t\mathbf{j} + 2t\mathbf{k}, \|\mathbf{r}'\| = \sqrt{5t^2 + 4}$

$\mathbf{r}''(t) = \mathbf{j} + 2\mathbf{k}$

$\mathbf{r}' \times \mathbf{r}'' = \begin{vmatrix} \mathbf{i} & \mathbf{j} & \mathbf{k} \\ 2 & t & 2t \\ 0 & 1 & 2 \end{vmatrix} = -4\mathbf{j} + 2\mathbf{k}, \|\mathbf{r}' \times \mathbf{r}''\| = \sqrt{20}$

$K = \dfrac{\|\mathbf{r}' \times \mathbf{r}''\|}{\|\mathbf{r}'\|^3} = \dfrac{\sqrt{20}}{(5t^2 + 4)^{3/2}} = \dfrac{2\sqrt{5}}{(4 + 5t^2)^{3/2}}$

65. $\mathbf{r}(t) = \dfrac{1}{2}t^2\mathbf{i} + t\mathbf{j} + \dfrac{1}{3}t^3\mathbf{k}, P\left(\dfrac{1}{2}, 1, \dfrac{1}{3}\right) \Rightarrow t = 1$

$\mathbf{r}'(t) = t\mathbf{i} + \mathbf{j} + t^2\mathbf{k}, \mathbf{r}'(1) = \mathbf{i} + \mathbf{j} + \mathbf{k}$

$\mathbf{r}''(t) = \mathbf{i} + 2t\mathbf{k}, \mathbf{r}''(1) = \mathbf{i} + 2\mathbf{k}$

$\mathbf{r}' \times \mathbf{r}'' = \begin{vmatrix} \mathbf{i} & \mathbf{j} & \mathbf{k} \\ 1 & 1 & 1 \\ 1 & 0 & 2 \end{vmatrix} = 2\mathbf{i} - \mathbf{j} - \mathbf{k}$

$K = \dfrac{\|\mathbf{r}' \times \mathbf{r}''\|}{\|\mathbf{r}'\|^3} = \dfrac{\sqrt{4 + 1 + 1}}{(\sqrt{3})^3} = \dfrac{\sqrt{6}}{3\sqrt{3}} = \dfrac{\sqrt{2}}{3}$

67. $y = \dfrac{1}{2}x^2 + 2, x = 4$

$y' = x$

$y'' = 1$

$K = \dfrac{|y''|}{\left[1 + (y')^2\right]^{3/2}} = \dfrac{1}{(1 + x^2)^{3/2}}$

At $x = 4$, $K = \dfrac{1}{17^{3/2}}$ and $r = 17^{3/2} = 17\sqrt{17}$.

69. $y = \ln x, x = 1$

$$y' = \frac{1}{x}$$

$$y'' = -\frac{1}{x^2}$$

$$K = \frac{|y''|}{\left[1 + (y')^2\right]^{3/2}} = \frac{1/x^2}{\left[1 + (1/x)^2\right]^{3/2}}$$

At $x = 1, K = \dfrac{1}{2^{3/2}} = \dfrac{1}{2\sqrt{2}} = \dfrac{\sqrt{2}}{4}$ and $r = 2\sqrt{2}$.

71. $\mathbf{F} = ma_N = mk\left(\dfrac{ds}{dt}\right)^2$

$$= \left(\frac{7200 \text{ lb}}{32 \text{ ft/sec}^2}\right)\left(\frac{1}{150 \text{ ft}}\right)\left(\frac{25(5280)\text{ft}}{3600 \text{ sec}}\right)^2$$

$$\approx 2016.67 \text{ pounds}$$

Problem Solving for Chapter 12

1. $x(t) = \displaystyle\int_0^t \cos\left(\frac{\pi u^2}{2}\right) du,\ y(t) = \int_0^t \sin\left(\frac{\pi u^2}{2}\right) du$

$$x'(t) = \cos\left(\frac{\pi t^2}{2}\right), y'(t) = \sin\left(\frac{\pi t^2}{2}\right)$$

(a) $s = \displaystyle\int_0^a \sqrt{x'(t)^2 + y'(t)^2}\, dt = \int_0^a dt = a$

(b) $x''(t) = -\pi t \sin\left(\dfrac{\pi t^2}{2}\right), y''(t) = \pi t \cos\left(\dfrac{\pi t^2}{2}\right)$

$$K = \frac{\left|\pi t \cos^2\left(\frac{\pi t^2}{2}\right) + \pi t \sin^2\left(\frac{\pi t^2}{2}\right)\right|}{1} = \pi t$$

At $t = a, K = \pi a$.

(c) $K = \pi a = \pi$ (length)

3. Bomb: $\mathbf{r}_1(t) = \langle 5000 - 400t, 3200 - 16t^2 \rangle$

Projectile: $\mathbf{r}_2(t) = \langle (v_0 \cos\theta)t, (v_0 \sin\theta)t - 16t^2 \rangle$

At 1600 feet: Bomb:

$3200 - 16t^2 = 1600 \Rightarrow t = 10$ seconds.

Projectile will travel 5 seconds:

$5(v_0 \sin\theta) - 16(25) = 1600$

$\qquad\qquad v_0 \sin\theta = 400$.

Horizontal position:

At $t = 10$, bomb is at $5000 - 400(10) = 1000$.

At $t = 5$, projectile is at $5v_0 \cos\theta$.

So, $v_0 \cos\theta = 200$.

Combining,

$$\frac{v_0 \sin\theta}{v_0 \cos\theta} = \frac{400}{200} \Rightarrow \tan\theta = 2 \Rightarrow \theta \approx 63.43°.$$

$$v_0 = \frac{200}{\cos\theta} \approx 447.2 \text{ ft/sec}$$

5. $x'(\theta) = 1 - \cos\theta, y'(\theta) = \sin\theta, 0 \le \theta \le 2\pi$

$$\sqrt{x'(\theta)^2 y'(\theta)^2} = \sqrt{(1 - \cos\theta)^2 + \sin^2\theta}$$

$$= \sqrt{2 - 2\cos\theta} = \sqrt{4\sin^2\frac{\theta}{2}}$$

$$s(t) = \int_\pi^t 2\sin\frac{\theta}{2}\, d\theta = \left[-4\cos\frac{\theta}{2}\right]_\pi^t = -4\cos\frac{t}{2}$$

$$x''(\theta) = \sin\theta, y''(\theta) = \cos\theta$$

$$K = \frac{|(1 - \cos\theta)\cos\theta - \sin\theta\sin\theta|}{\left(2\sin\frac{\theta}{2}\right)^3}$$

$$= \frac{|\cos\theta - 1|}{8\sin^3\frac{\theta}{2}}$$

$$= \frac{1}{4\sin\frac{\theta}{2}}$$

So, $\rho = \dfrac{1}{K} = 4\sin\dfrac{t}{2}$ and

$$s^2 + \rho^2 = 16\cos^2\left(\frac{t}{2}\right) + 16\sin^2\left(\frac{t}{2}\right) = 16.$$

7. $\|\mathbf{r}(t)\|^2 = \mathbf{r}(t) \cdot \mathbf{r}(t)$

$$\frac{d}{dt}\left(\|\mathbf{r}(t)\|\right)^2 = 2\|\mathbf{r}(t)\|\frac{d}{dt}\|\mathbf{r}(t)\| = \mathbf{r}(t) \cdot \mathbf{r}'(t) + \mathbf{r}'(t) \cdot \mathbf{r}(t) \Rightarrow \frac{d}{dt}\|\mathbf{r}(t)\| = \frac{\mathbf{r}(t) \cdot \mathbf{r}'(t)}{\|\mathbf{r}(t)\|}$$

9. $\mathbf{r}(t) = 4\cos t\,\mathbf{i} + 4\sin t\,\mathbf{j} + 3t\,\mathbf{k}, t = \dfrac{\pi}{2}$

$\mathbf{r}'(t) = -4\sin t\,\mathbf{i} + 4\cos t\,\mathbf{j} + 3\mathbf{k}, \left\|\mathbf{r}'(t)\right\| = 5$

$\mathbf{r}''(t) = -4\cos t\,\mathbf{i} - 4\sin t\,\mathbf{j}$

$\mathbf{T} = -\dfrac{4}{5}\sin t\,\mathbf{i} + \dfrac{4}{5}\cos t\,\mathbf{j} + \dfrac{3}{5}\mathbf{k}$

$\mathbf{T}' = -\dfrac{4}{5}\cos t\,\mathbf{i} - \dfrac{4}{5}\sin t\,\mathbf{j}$

$\mathbf{N} = -\cos t\,\mathbf{i} - \sin t\,\mathbf{j}$

$\mathbf{B} = \mathbf{T}\times\mathbf{N} = \dfrac{3}{5}\sin t\,\mathbf{i} - \dfrac{3}{5}\cos t\,\mathbf{j} + \dfrac{4}{5}\mathbf{k}$

At $t = \dfrac{\pi}{2}$, $\mathbf{T}\!\left(\dfrac{\pi}{2}\right) = -\dfrac{4}{5}\mathbf{i} + \dfrac{3}{5}\mathbf{k}$

$\mathbf{N}\!\left(\dfrac{\pi}{2}\right) = -\mathbf{j}$

$\mathbf{B}\!\left(\dfrac{\pi}{2}\right) = \dfrac{3}{5}\mathbf{i} + \dfrac{4}{5}\mathbf{k}$

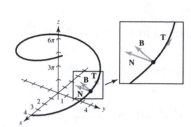

11. (a) $\|\mathbf{B}\| = \|\mathbf{T}\times\mathbf{N}\| = 1$ constant length $\Rightarrow \dfrac{d\mathbf{B}}{ds} \perp \mathbf{B}$

$\dfrac{d\mathbf{B}}{ds} = \dfrac{d}{ds}(\mathbf{T}\times\mathbf{N}) = (\mathbf{T}\times\mathbf{N}') + (\mathbf{T}'\times\mathbf{N})$

$\mathbf{T}\cdot\dfrac{d\mathbf{B}}{ds} = \mathbf{T}\cdot(\mathbf{T}\times\mathbf{N}') + \mathbf{T}\cdot(\mathbf{T}'\times\mathbf{N})$

$\qquad = (\mathbf{T}\times\mathbf{T})\cdot\mathbf{N}' + \mathbf{T}\cdot\left(\mathbf{T}'\times\dfrac{\mathbf{T}'}{\|\mathbf{T}'\|}\right) = 0$

So, $\dfrac{d\mathbf{B}}{ds} \perp \mathbf{B}$ and $\dfrac{d\mathbf{B}}{ds} \perp \mathbf{T} \Rightarrow \dfrac{d\mathbf{B}}{ds} = \tau\mathbf{N}$

for some scalar τ.

(b) $\mathbf{B} = \mathbf{T}\times\mathbf{N}$. Using Section 11.4, exercise 58,

$\mathbf{B}\times\mathbf{N} = (\mathbf{T}\times\mathbf{N})\times\mathbf{N} = -\mathbf{N}\times(\mathbf{T}\times\mathbf{N})$

$\qquad = -\big[(\mathbf{N}\cdot\mathbf{N})\mathbf{T} - (\mathbf{N}\cdot\mathbf{T})\mathbf{N}\big]$

$\qquad = -\mathbf{T}$

$\mathbf{B}\times\mathbf{T} = (\mathbf{T}\times\mathbf{N})\times\mathbf{T} = -\mathbf{T}\times(\mathbf{T}\times\mathbf{N})$

$\qquad = -\big[(\mathbf{T}\cdot\mathbf{N})\mathbf{T} - (\mathbf{T}\cdot\mathbf{T})\mathbf{N}\big]$

$\qquad = \mathbf{N}.$

Now, $K\mathbf{N} = \left\|\dfrac{d\mathbf{T}}{ds}\right\|\dfrac{\mathbf{T}'(s)}{\|\mathbf{T}'(s)\|} = \mathbf{T}'(s) = \dfrac{d\mathbf{T}}{ds}$

Finally,

$\mathbf{N}'(s) = \dfrac{d}{ds}(\mathbf{B}\times\mathbf{T}) = (\mathbf{B}\times\mathbf{T}') + (\mathbf{B}'\times\mathbf{T})$

$\qquad = (\mathbf{B}\times K\mathbf{N}) + (-\tau\mathbf{N}\times\mathbf{T})$

$\qquad = -K\mathbf{T} + \tau\mathbf{B}.$

13. $\mathbf{r}(t) = \langle t\cos \pi t, t\sin \pi t\rangle, 0 \le t \le 2$

(a)

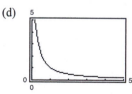

(b) Length $= \displaystyle\int_0^2 \|\mathbf{r}'(t)\|\, dt$

$\qquad = \displaystyle\int_0^2 \sqrt{\pi^2 t^2 + 1}\, dt$

$\qquad \approx 6.766$ (graphing utility)

(c) $K = \dfrac{\pi\left(\pi^2 t^2 + 2\right)}{\left[\pi^2 t^2 + 1\right]^{3/2}}$

$K(0) = 2\pi$

$K(1) = \dfrac{\pi\left(\pi^2 + 2\right)}{\left(\pi^2 + 1\right)^{3/2}} \approx 1.04$

$K(2) \approx 0.51$

(d)

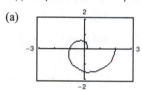

(e) $\displaystyle\lim_{t\to\infty} K = 0$

(f) As $t \to \infty$, the graph spirals outward and the curvature decreases.

C H A P T E R 1 3
Functions of Several Variables

CHAPTER 13
Functions of Several Variables

Section 13.1 Introduction to Functions of Several Variables

1. No, it is not the graph of a function. For some values of x and y (for example, $(x, y) = (0, 0)$), there are 2 z-values.

3. $x^2z + 3y^2 - xy = 10$

$$x^2z = 10 + xy - 3y^2$$

$$z = \frac{10 + xy - 3y^2}{x^2}$$

Yes, z is a function of x and y.

5. $\dfrac{x^2}{4} + \dfrac{y^2}{9} + z^2 = 1$

No, z is not a function of x and y. For example, $(x, y) = (0, 0)$ corresponds to both $z = \pm 1$.

7. $f(x, y) = xy$

(a) $f(3, 2) = 3(2) = 6$

(b) $f(-1, 4) = -1(4) = -4$

(c) $f(30, 5) = 30(5) = 150$

(d) $f(5, y) = 5y$

(e) $f(x, 2) = 2x$

(f) $f(5, t) = 5t$

9. $f(x, y) = xe^y$

(a) $f(5, 0) = 5e^0 = 5$

(b) $f(3, 2) = 3e^2$

(c) $f(2, -1) = 2e^{-1} = \dfrac{2}{e}$

(d) $f(5, y) = 5e^y$

(e) $f(x, 2) = xe^2$

(f) $f(t, t) = te^t$

11. $h(x, y, z) = \dfrac{xy}{z}$

(a) $h(2, 3, 9) = \dfrac{2(3)}{9} = \dfrac{2}{3}$

(b) $h(1, 0, 1) = \dfrac{1(0)}{1} = 0$

(c) $h(-2, 3, 4) = \dfrac{(-2)(3)}{4} = -\dfrac{3}{2}$

(d) $h(5, 4, -6) = \dfrac{5(4)}{-6} = -\dfrac{10}{3}$

13. $f(x, y) = x \sin y$

(a) $f\left(2, \dfrac{\pi}{4}\right) = 2 \sin \dfrac{\pi}{4} = \sqrt{2}$

(b) $f(3, 1) = 3 \sin(1)$

(c) $f\left(-3, \dfrac{\pi}{3}\right) = -3 \sin \dfrac{\pi}{3} = -3\left(\dfrac{\sqrt{3}}{2}\right) = \dfrac{-3\sqrt{3}}{2}$

(d) $f\left(4, \dfrac{\pi}{2}\right) = 4 \sin \dfrac{\pi}{2} = 4$

15. $g(x, y) = \displaystyle\int_x^y (2t - 3)\, dt$

$$= \left[t^2 - 3t\right]_x^y = y^2 - 3y - x^2 + 3x$$

(a) $g(4, 0) = 0 - 16 + 12 = -4$

(b) $g(4, 1) = (1 - 3) - 16 + 12 = -6$

(c) $g\left(4, \tfrac{3}{2}\right) = \left(\tfrac{9}{4} - \tfrac{9}{2}\right) - 16 + 12 = -\tfrac{25}{4}$

(d) $g\left(\tfrac{3}{2}, 0\right) = 0 - \tfrac{9}{4} + \tfrac{9}{2} = \tfrac{9}{4}$

17. $f(x, y) = 2x + y^2$

(a) $\dfrac{f(x + \Delta x, y) - f(x, y)}{\Delta x} = \dfrac{2(x + \Delta x) + y^2 - (2x + y^2)}{\Delta x} = \dfrac{2\Delta x}{\Delta x} = 2,\ \Delta x \neq 0$

(b) $\dfrac{f(x, y + \Delta y) - f(x, y)}{\Delta y} = \dfrac{2x + (y + \Delta y)^2 - 2x - y^2}{\Delta y} = \dfrac{2y\Delta y + (\Delta y^2)}{\Delta y} = 2y + \Delta y,\ \Delta y \neq 0$

653

19. $f(x, y) = x^2 + y^2$

Domain:

$\{(x, y): x \text{ is any real number}, y \text{ is any real number}\}$

Range: $z \geq 0$

21. $g(x, y) = x\sqrt{y}$

Domain: $\{(x, y): y \geq 0\}$

Range: all real numbers

23. $z = \dfrac{x + y}{xy}$

Domain: $\{(x, y): x \neq 0 \text{ and } y \neq 0\}$

Range: all real numbers

25. $f(x, y) = \sqrt{4 - x^2 - y^2}$

Domain: $4 - x^2 - y^2 \geq 0$

$x^2 + y^2 \leq 4$

$\{(x, y): x^2 + y^2 \leq 4\}$

Range: $0 \leq z \leq 2$

27. $f(x, y) = \arccos(x + y)$

Domain: $\{(x, y): -1 \leq x + y \leq 1\}$

Range: $0 \leq z \leq \pi$

29. $f(x, y) = \ln(4 - x - y)$

Domain: $4 - x - y > 0$

$x + y < 4$

$\{(x, y): y < -x + 4\}$

Range: all real numbers

31. $f(x, y) = \dfrac{-4x}{x^2 + y^2 + 1}$

(a) View from the positive x-axis: $(20, 0, 0)$

(b) View where x is negative, y and z are positive: $(-15, 10, 20)$

(c) View from the first octant: $(20, 15, 25)$

(d) View from the line $y = x$ in the xy-plane: $(20, 20, 0)$

33. $f(x, y) = 4$

Plane: $z = 4$

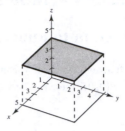

35. $f(x, y) = y^2$

Because the variable x is missing, the surface is a cylinder with rulings parallel to the x-axis. The generating curve is $z = y^2$. The domain is the entire xy-plane and the range is $z \geq 0$.

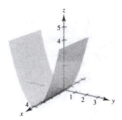

37. $z = -x^2 - y^2$

Paraboloid

Domain: entire xy-plane

Range: $z \leq 0$

39. $f(x, y) = e^{-x}$

Because the variable y is missing, the surface is a cylinder with rulings parallel to the y-axis. The generating curve is $z = e^{-x}$.

The domain is the entire xy-plane and the range is $z > 0$.

41. $z = y^2 - x^2 + 1$

Hyperbolic paraboloid

Domain: entire xy-plane

Range: $-\infty < z < \infty$

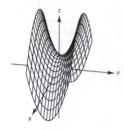

43. $f(x, y) = x^2 e^{(-xy/2)}$

45. $z = e^{1-x^2-y^2}$

Level curves:

$$c = e^{1-x^2-y^2}$$

$$\ln c = 1 - x^2 - y^2$$

$$x^2 + y^2 = 1 - \ln c$$

Circles centered at $(0, 0)$

Matches (c)

46. $z = e^{1-x^2+y^2}$

Level curves:

$$c = e^{1-x^2+y^2}$$

$$\ln c = 1 - x^2 + y^2$$

$$x^2 - y^2 = 1 - \ln c$$

Hyperbolas centered at $(0, 0)$

Matches (d)

47. $z = \ln\left| y - x^2 \right|$

Level curves:

$$c = \ln\left| y - x^2 \right|$$

$$\pm e^c = y - x^2$$

$$y = x^2 \pm e^c$$

Parabolas

Matches (b)

48. $z = \cos\left(\dfrac{x + 2y^2}{4} \right)$

Level curves:

$$c = \cos\left(\dfrac{x^2 + 2y^2}{4} \right)$$

$$\cos^{-1} c = \dfrac{x^2 + 2y^2}{4}$$

$$x^2 + 2y^2 = 4\cos^{-1} c$$

Ellipses

Matches (a)

49. $z = x + y$

Level curves are parallel
lines of the form
$x + y = c.$

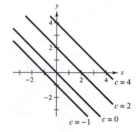

51. $z = x^2 + 4y^2$

The level curves are ellipses of the form

$$x^2 + 4y^2 = c$$

$\left(\text{except } x^2 + 4y^2 = 0 \text{ is the point } (0, 0)\right).$

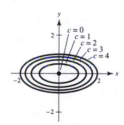

53. $f(x, y) = xy$

The level curves are
hyperbolas of the form
$xy = c.$

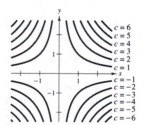

55. $f(x, y) = \dfrac{x}{x^2 + y^2}$

The level curves are of the form

$$c = \dfrac{x}{x^2 + y^2}$$

$$x^2 - \dfrac{x}{c} + y^2 = 0$$

$$\left(x - \dfrac{1}{2c}\right)^2 + y^2 = \left(\dfrac{1}{2c}\right)^2.$$

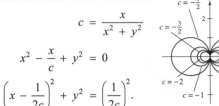

So, the level curves are circles passing through the origin and centered at $(\pm 1/2c, 0)$.

57. $f(x, y) = x^2 - y^2 + 2$

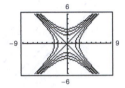

59. $g(x, y) = \dfrac{8}{1 + x^2 + y^2}$

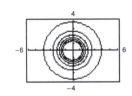

61. The graph of a function of two variables is the set of all points (x, y, z) for which $z = f(x, y)$ and (x, y) is in the domain of f. The graph can be interpreted as a surface in space. Level curves are the scalar fields $f(x, y) = c$, where c is a constant.

63. $f(x, y) = \dfrac{x}{y}$

The level curves are the lines $c = \dfrac{x}{y}$ or $y = \dfrac{1}{c}x$.

These lines all pass through the origin.

65. The surface is sloped like a saddle. The graph is not unique. Any vertical translation would have the same level curves.

One possible function is

$$f(x, y) = |xy|.$$

67. $V(I, R) = 1000\left[\dfrac{1 + 0.06(1 - R)}{1 + I}\right]^{10}$

Tax Rate	Inflation Rate		
	0	0.03	0.05
0	1790.85	1332.56	1099.43
0.28	1526.43	1135.80	937.09
0.35	1466.07	1090.90	900.04

69. $f(x, y, z) = x - y + z, c = 1$

$1 = x - y + z$, Plane

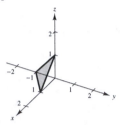

71. $f(x, y, z) = x^2 + y^2 + z^2$

$c = 9$

$9 = x^2 + y^2 + z^2$

Sphere

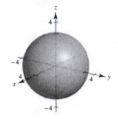

73. $f(x, y, z) = 4x^2 + 4y^2 - z^2$

$c = 0$

$0 = 4x^2 + 4y^2 - z^2$

Elliptic cone

75. $N(d, L) = \left(\dfrac{d - 4}{4}\right)^2 L$

(a) $N(22, 12) = \left(\dfrac{22 - 4}{4}\right)^2 (12) = 243$ board-feet

(b) $N(30, 12) = \left(\dfrac{30 - 4}{4}\right)^2 (12) = 507$ board-feet

77. $T = 600 - 0.75x^2 - 0.75y^2$

The level curves are of the form

$$c = 600 - 0.75x^2 - 0.75y^2$$

$$x^2 + y^2 = \frac{600 - c}{0.75}.$$

The level curves are circles centered at the origin.

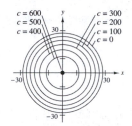

79. $f(x, y) = 100x^{0.6}y^{0.4}$

$$f(2x, 2y) = 100(2x)^{0.6}(2y)^{0.4}$$

$$= 100(2)^{0.6}x^{0.6}(2)^{0.4}y^{0.4}$$

$$= 100(2)^{0.6}(2)^{0.4}x^{0.6}y^{0.4}$$

$$= 2\left[100x^{0.6}y^{0.4}\right] = 2f(x, y)$$

87. False. Let

$$f(x, y) = 2xy$$

$$f(1, 2) = f(2, 1), \text{ but } 1 \neq 2.$$

89. True

91. We claim that $g(x) = f(x, 0)$. First note that $x = y = z = 0$ implies $3f(0, 0) = 0 \Rightarrow f(0, 0) = 0$.

Letting $y = z = 0$ implies $f(x, 0) + f(0, 0) + f(0, x) = 0 \Rightarrow -f(0, x) = f(x, 0)$.

Letting $z = 0$ implies $f(x, y) + f(y, 0) + f(0, x) = 0 \Rightarrow f(x, y) = -f(y, 0) - f(0, x) = f(x, 0) - f(y, 0)$.

Hence, $f(x, y) = g(x) - g(y)$, as desired.

81. $PV = kT$

(a) $26(2000) = k(300) \Rightarrow k = \dfrac{520}{3}$

(b) $P = \dfrac{kT}{V} = \dfrac{520}{3}\left(\dfrac{T}{V}\right)$

The level curves are of the form

$$c = \frac{520}{3}\left(\frac{T}{V}\right), \text{ or } V = \frac{520}{3c}T.$$

These are lines through the origin with slope $\dfrac{520}{3c}$.

83. (a) Highest pressure at C

(b) Lowest pressure at A

(c) Highest wind velocity at B

85. $C = \underbrace{1.20xy}_{\text{base}} + \underbrace{2(0.75)xz}_{\text{front and back}} + \underbrace{2(0.75)yz}_{\text{2 ends}}$

$$= 1.20xy + 1.50(xz + yz)$$

Section 13.2   Limits and Continuity

1. $\lim\limits_{(x, y)\to(1, 0)} x = 1$

$f(x, y) = x, L = 1$

We need to show that for all $\varepsilon > 0$, there exists a δ-neighborhood about $(1, 0)$ such that

$$\left|f(x, y) - L\right| = \left|x - 1\right| < \varepsilon$$

Whenever $(x, y) \neq (1, 0)$ lies in the neighborhood.

From $0 < \sqrt{(x - 1)^2 + (y - 0)^2} < \delta$, it follows that

$$\left|x - 1\right| = \sqrt{(x - 1)^2} \le \sqrt{(x - 1)^2 + (y - 0)^2} < \delta.$$

So, choose $\delta = \varepsilon$ and the limit is verified.

3. $\lim\limits_{(x, y)\to(1, -3)} y = -3. \; f(x, y) = y, \; L = -3$

We need to show that for all $\varepsilon > 0$, there exists a δ-neighborhood about $(1, -3)$ such that

$$\left|f(x, y) - L\right| = \left|y + 3\right| < \varepsilon$$

whenever $(x, y) \neq (1, -3)$ lies in the neighborhood.

From $0 < \sqrt{(x - 1)^2 + (y + 3)^2} < \delta$ it follows that

$$\left|y + 3\right| = \sqrt{(y + 3)^2} \le \sqrt{(x - 1)^2 + (y + 3)^2} < \delta.$$

So, choose $\delta = \varepsilon$ and the limit is verified.

5. $\displaystyle\lim_{(x,y)\to(a,b)}\left[f(x,y)-g(x,y)\right] = \lim_{(x,y)\to(a,b)}f(x,y) - \lim_{(x,y)\to(a,b)}g(x,y) = 4 - 3 = 1$

7. $\displaystyle\lim_{(x,y)\to(a,b)}\left[f(x,y)g(x,y)\right] = \left[\lim_{(x,y)\to(a,b)}f(x,y)\right]\left[\lim_{(x,y)\to(a,b)}g(x,y)\right] = 4(3) = 12$

9. $\displaystyle\lim_{(x,y)\to(2,1)}\left(2x^2+y\right) = 8 + 1 = 9$

Continuous everywhere

11. $\displaystyle\lim_{(x,y)\to(1,2)}e^{xy} = e^{1(2)} = e^2$

Continuous everywhere

13. $\displaystyle\lim_{(x,y)\to(0,2)}\frac{x}{y} = \frac{0}{2} = 0$

Continuous for all $y \neq 0$

15. $\displaystyle\lim_{(x,y)\to(1,1)}\frac{xy}{x^2+y^2} = \frac{1}{2}$

Continuous except at $(0,0)$

17. $\displaystyle\lim_{(x,y)\to(\pi/4,2)}y\cos(xy) = 2\cos\frac{\pi}{2} = 0$

Continuous everywhere

19. $\displaystyle\lim_{(x,y)\to(0,1)}\frac{\arcsin xy}{1-xy} = \frac{\arcsin 0}{1} = 0$

Continuous for $xy \neq 1$, $|xy| \le 1$

21. $\displaystyle\lim_{(x,y,z)\to(1,3,4)}\sqrt{x+y+z} = \sqrt{1+3+4} = 2\sqrt{2}$

Continuous for $x + y + z \ge 0$

23. $\displaystyle\lim_{(x,y)\to(1,1)}\frac{xy-1}{1+xy} = \frac{1-1}{1+1} = 0$

25. $\displaystyle\lim_{(x,y)\to(0,0)}\frac{1}{x+y}$ does not exist

Because the denominator $x + y$ approaches 0 as $(x,y)\to(0,0)$.

27. $\displaystyle\lim_{(x,y)\to(0,0)}\frac{x-y}{\sqrt{x}-\sqrt{y}}$

does not exist because you can't approach $(0,0)$ from negative values of x and y.

29. The limit does not exist because along the line $y = 0$ you have

$$\lim_{(x,y)\to(0,0)}\frac{x+y}{x^2+y} = \lim_{(x,0)\to(0,0)}\frac{x}{x^2} = \lim_{(x,0)\to(0,0)}\frac{1}{x}$$

which does not exist.

31. $\displaystyle\lim_{(x,y)\to(0,0)}\frac{x^2}{\left(x^2+1\right)\left(y^2+1\right)} = \frac{0}{(1)(1)} = 0$

33. The limit does not exist because along the path $x = 0$, $y = 0$, you have

$$\lim_{(x,y,z)\to(0,0,0)}\frac{xy+yz+xz}{x^2+y^2+z^2} = \lim_{(0,0,z)\to(0,0,0)}\frac{0}{z^2} = 0$$

whereas along the path $x = y = z$, you have

$$\lim_{(x,y,z)\to(0,0,0)}\frac{xy+yz+xz}{x^2+y^2+z^2} = \lim_{(x,x,x)\to(0,0,0)}\frac{x^2+x^2+x^2}{x^2+x^2+x^2}$$
$$= 1$$

35. $\displaystyle\lim_{(x,y)\to(0,0)}e^{xy} = 1$

Continuous everywhere

37. $f(x, y) = \dfrac{xy}{x^2 + y^2}$

Continuous except at $(0, 0)$

Path: $y = 0$

(x, y)	$(1, 0)$	$(0.5, 0)$	$(0.1, 0)$	$(0.01, 0)$	$(0.001, 0)$
$f(x, y)$	0	0	0	0	0

Path: $y = x$

(x, y)	$(1, 1)$	$(0.5, 0.5)$	$(0.1, 0.1)$	$(0.01, 0.01)$	$(0.001, 0.001)$
$f(x, y)$	$\frac{1}{2}$	$\frac{1}{2}$	$\frac{1}{2}$	$\frac{1}{2}$	$\frac{1}{2}$

The limit does not exist because along the path $y = 0$ the function equals 0, whereas along the path $y = x$ the function equals $\frac{1}{2}$.

39. $f(x, y) = \dfrac{y}{x^2 + y^2}$

Continuous except at $(0, 0)$

Path: $y = 0$

(x, y)	$(1, 0)$	$(0.5, 0)$	$(0.1, 0)$	$(0.01, 0)$	$(0.001, 0)$
$f(x, y)$	0	0	0	0	0

Path: $y = x$

(x, y)	$(1, 1)$	$(0.5, 0.5)$	$(0.1, 0.1)$	$(0.01, 0.01)$	$(0.001, 0.001)$
$f(x, y)$	$\frac{1}{2}$	1	5	50	500

The limit does not exist because along the path $y = 0$ the function equals 0, whereas along the path $y = x$ the function tends to infinity.

41. $\displaystyle\lim_{(x,y)\to(0,0)} \frac{x^4 - y^4}{x^2 + y^2} = \lim_{(x,y)\to(0,0)} \frac{(x^2 + y^2)(x^2 - y^2)}{x^2 + y^2} = \lim_{(x,y)\to(0,0)} (x^2 - y^2) = 0$

So, f is continuous everywhere, whereas g is continuous everywhere except at $(0, 0)$. g has a removable discontinuity at $(0, 0)$.

43. $\displaystyle\lim_{(x,y)\to(0,0)} \frac{xy^2}{x^2 + y^2} = \lim_{r\to0} \frac{(r\cos\theta)(r^2 \sin^2\theta)}{r^2}$
$= \lim_{r\to0}(r\cos\theta\sin^2\theta) = 0$

45. $\displaystyle\lim_{(x,y)\to(0,0)} \frac{x^2 y^2}{x^2 + y^2} = \lim_{r\to0} \frac{r^4 \cos^2\theta \sin^2\theta}{r^2}$
$= \lim_{r\to0} r^2 \cos^2\theta \sin^2\theta = 0$

47. $\displaystyle\lim_{(x,y)\to(0,0)} \cos(x^2 + y^2) = \lim_{r\to0}\cos(r^2) = \cos(0) = 1$

49. $\sqrt{x^2 + y^2} = r$

$\displaystyle\lim_{(x,y)\to(0,0)} \frac{\sin\sqrt{x^2 + y^2}}{\sqrt{x^2 + y^2}} = \lim_{r\to0^+}\frac{\sin(r)}{r} = 1$

51. $x^2 + y^2 = r^2$

$\displaystyle\lim_{(x,y)\to(0,0)} \frac{1 - \cos(x^2 + y^2)}{x^2 + y^2} = \lim_{x\to0}\frac{1 - \cos(r^2)}{r^2} = 0$

53. $f(x, y, z) = \dfrac{1}{\sqrt{x^2 + y^2 + z^2}}$

Continuous except at $(0, 0, 0)$

55. $f(x, y, z) = \dfrac{\sin z}{e^x + e^y}$

Continuous everywhere

57. For $xy \neq 0$, the function is clearly continuous.

For $xy \neq 0$, let $z = xy$. Then

$$\lim_{z \to 0} \frac{\sin z}{z} = 1$$

implies that f is continuous for all x, y.

59. $f(t) = t^2, \ g(x, y) = 2x - 3y$

$$f(g(x, y)) = f(2x - 3y) = (2x - 3y)^2$$

Continuous everywhere

61. $f(t) = \dfrac{1}{t}, \ g(x, y) = 2x - 3y$

$$f(g(x, y)) = f(2x - 3y) = \frac{1}{2x - 3y}$$

Continuous for all $y \neq \dfrac{2}{3}x$

63. $f(x, y) = x^2 - 4y$

(a) $\displaystyle \lim_{\Delta x \to 0} \frac{f(x + \Delta x, y) - f(x, y)}{\Delta x} = \lim_{\Delta x \to 0} \frac{\left[(x + \Delta x)^2 - 4y \right] - (x^2 - 4y)}{\Delta x} = \lim_{\Delta x \to 0} \frac{2x\Delta x + (\Delta x)^2}{\Delta x} = \lim_{\Delta x \to 0} (2x + \Delta x) = 2x$

(b) $\displaystyle \lim_{\Delta y \to 0} \frac{f(x, y + \Delta y) - f(x, y)}{\Delta y} = \lim_{\Delta y \to 0} \frac{\left[x^2 - 4(y + \Delta y) \right] - (x^2 - 4y)}{\Delta y} = \lim_{\Delta y \to 0} \frac{-4\Delta y}{\Delta y} = \lim_{\Delta y \to 0} (-4) = -4$

65. $f(x, y) = \dfrac{x}{y}$

(a) $\displaystyle \lim_{\Delta x \to 0} \frac{f(x + \Delta x, y) - f(x, y)}{\Delta x} = \lim_{\Delta x \to 0} \frac{\dfrac{x + \Delta x}{y} - \dfrac{x}{y}}{\Delta x} = \lim_{\Delta x \to 0} \frac{\dfrac{\Delta x}{y}}{\Delta x} = \lim_{\Delta x \to 0} \frac{1}{y} = \frac{1}{y}$

(b) $\displaystyle \lim_{\Delta y \to 0} \frac{f(x, y + \Delta y) - f(x, y)}{\Delta y} = \lim_{\Delta y \to 0} \frac{\dfrac{x}{y + \Delta y} - \dfrac{x}{y}}{\Delta y} = \lim_{\Delta y \to 0} \frac{xy - (xy + x\Delta y)}{(y + \Delta y)y\Delta y} = \lim_{\Delta y \to 0} \frac{-x\Delta y}{(y + \Delta y)y\Delta y} = \lim_{\Delta y \to 0} \frac{-x}{(y + \Delta y)y} = \frac{-x}{y^2}$

67. $f(x, y) = 3x + xy - 2y$

(a) $\displaystyle \lim_{\Delta x \to 0} \frac{f(x + \Delta x, y) - f(x, y)}{\Delta x} = \lim_{\Delta x \to 0} \frac{3(x + \Delta x) + (x + \Delta x)y - 2y - (3x + xy - 2y)}{\Delta x}$

$\displaystyle = \lim_{\Delta x \to 0} \frac{3\Delta x + y\Delta x}{\Delta x} = \lim_{\Delta x \to 0} (3 + y) = 3 + y$

(b) $\displaystyle \lim_{\Delta y \to 0} \frac{f(x, y + \Delta y) - f(x, y)}{\Delta y} = \lim_{\Delta y \to 0} \frac{3x + x(y + \Delta y) - 2(y + \Delta y) - (3x + xy - 2y)}{\Delta y}$

$\displaystyle = \lim_{\Delta y \to 0} \frac{x\Delta y - 2\Delta y}{\Delta y} = \lim_{\Delta y \to 0} (x - 2) = x - 2$

69. True. Assuming $f(x, 0)$ exists for $x \neq 0$.

71. False. Let $f(x, y) = \begin{cases} \ln(x^2 + y^2), & (x, y) \neq (0, 0) \\ 0, & x = 0, y = 0 \end{cases}$.

73. $\displaystyle\lim_{(x,y)\to(0,0)} \frac{x^2 + y^2}{xy}$

(a) Along $y = ax$:

$$\lim_{(x,ax)\to(0,0)} \frac{x^2 + (ax)^2}{x(ax)} = \lim_{x\to0} \frac{x^2(1 + a^2)}{ax^2}$$

$$= \frac{1 + a^2}{a}, a \neq 0$$

If $a = 0$, then $y = 0$ and the limit does not exist.

(b) Along $y = x^2$: $\displaystyle\lim_{(x,x^2)\to(0,0)} \frac{x^2 + (x^2)^2}{x(x^2)} = \lim_{x\to0} \frac{1 + x^2}{x}$

Limit does not exist.

(c) No, the limit does not exist. Different paths result in different limits.

75. $\displaystyle\lim_{(x,y,z)\to(0,0,0)} \frac{xyz}{x^2 + y^2 + z^2} = \lim_{\rho\to0^+} \frac{(\rho \sin \phi \cos \theta)(\rho \sin \phi \sin \theta)(\rho \cos \phi)}{\rho^2}$

$$= \lim_{\rho\to0^+} \rho\left[\sin^2 \phi \cos \theta \sin \theta \cos \phi\right] = 0$$

77. As $(x, y) \to (0, 1)$, $x^2 + 1 \to 1$ and $x^2 + (y - 1)^2 \to 0$.

So, $\displaystyle\lim_{(x,y)\to(0,1)} \tan^{-1}\left[\frac{x^2 + 1}{x^2 + (y - 1)^2}\right] = \frac{\pi}{2}$.

79. Because $\displaystyle\lim_{(x,y)\to(a,b)} f(x, y) = L_1$, then for $\varepsilon/2 > 0$, there corresponds $\delta_1 > 0$ such that $\left|f(x, y) - L_1\right| < \varepsilon/2$ whenever

$0 < \sqrt{(x - a)^2 + (y - b)^2} < \delta_1$.

Because $\displaystyle\lim_{(x,y)\to(a,b)} g(x, y) = L_2$, then for $\varepsilon/2 > 0$, there corresponds $\delta_2 > 0$ such that $\left|g(x, y) - L_2\right| < \varepsilon/2$ whenever

$0 < \sqrt{(x - a)^2 + (y - b)^2} < \delta_2$.

Let δ be the smaller of δ_1 and δ_2. By the triangle inequality, whenever $\sqrt{(x - a)^2 + (y - b)^2} < \delta$, we have

$$\left|f(x, y) + g(x, y) - (L_1 + L_2)\right| = \left|(f(x, y) - L_1) + (g(x, y) - L_2)\right| \leq \left|f(x, y) - L_1\right| + \left|g(x, y) - L_2\right| < \frac{\varepsilon}{2} + \frac{\varepsilon}{2} = \varepsilon.$$

So, $\displaystyle\lim_{(x,y)\to(a,b)} \left[f(x, y) + g(x, y)\right] = L_1 + L_2$.

81. See the definition on page 881. Show that the value of $\displaystyle\lim_{(x,y)\to(x_0,y_0)} f(x, y)$ is not the same for two different paths to (x_0, y_0).

83. (a) No. The existence of $f(2, 3)$ has no bearing on the existence of the limit as $(x, y) \to (2, 3)$.

(b) No, $f(2, 3)$ can equal any number, or not even be defined.

Section 13.3   Partial Derivatives

1. No, x only occurs in the numerator.

3. No, y only occurs in the numerator.

5. Yes, x occurs in both the numerator and denominator.

7. $f(x, y) = 2x - 5y + 3$

$f_x(x, y) = 2$

$f_y(x, y) = -5$

9. $f(x, y) = x^2 y^3$

$f_x(x, y) = 2xy^3$

$f_y(x, y) = 3x^2 y^2$

11. $z = x\sqrt{y}$

$\dfrac{\partial z}{\partial x} = \sqrt{y}$

$\dfrac{\partial z}{\partial y} = \dfrac{x}{2\sqrt{y}}$

13. $z = x^2 - 4xy + 3y^2$

$\dfrac{\partial z}{\partial x} = 2x - 4y$

$\dfrac{\partial z}{\partial y} = -4x + 6y$

15. $z = e^{xy}$

$\dfrac{\partial z}{\partial x} = ye^{xy}$

$\dfrac{\partial z}{\partial y} = xe^{xy}$

17. $z = x^2 e^{2y}$

$\dfrac{\partial z}{\partial x} = 2xe^{2y}$

$\dfrac{\partial z}{\partial y} = 2x^2 e^{2y}$

19. $z = \ln \dfrac{x}{y} = \ln x - \ln y$

$\dfrac{\partial z}{\partial x} = \dfrac{1}{x}$

$\dfrac{\partial z}{\partial y} = -\dfrac{1}{y}$

21. $z = \ln(x^2 + y^2)$

$\dfrac{\partial z}{\partial x} = \dfrac{2x}{x^2 + y^2}$

$\dfrac{\partial z}{\partial y} = \dfrac{2y}{x^2 + y^2}$

23. $z = \dfrac{x^2}{2y} + \dfrac{3y^2}{x}$

$\dfrac{\partial z}{\partial x} = \dfrac{2x}{2y} - \dfrac{3y^2}{x^2} = \dfrac{x^3 - 3y^3}{x^2 y}$

$\dfrac{\partial z}{\partial y} = \dfrac{-x^2}{2y^2} + \dfrac{6y}{x} = \dfrac{12y^3 - x^3}{2xy^2}$

25. $h(x, y) = e^{-(x^2+y^2)}$

$h_x(x, y) = -2xe^{-(x^2+y^2)}$

$h_y(x, y) = -2ye^{-(x^2+y^2)}$

27. $f(x, y) = \sqrt{x^2 + y^2}$

$f_x(x, y) = \dfrac{1}{2}(x^2 + y^2)^{-1/2}(2x) = \dfrac{x}{\sqrt{x^2 + y^2}}$

$f_y(x, y) = \dfrac{1}{2}(x^2 + y^2)^{-1/2}(2y) = \dfrac{y}{\sqrt{x^2 + y^2}}$

29. $z = \cos xy$

$\dfrac{\partial z}{\partial x} = -y \sin xy$

$\dfrac{\partial z}{\partial y} = -x \sin xy$

31. $z = \tan(2x - y)$

$\dfrac{\partial z}{\partial x} = 2 \sec^2(2x - y)$

$\dfrac{\partial z}{\partial y} = -\sec^2(2x - y)$

33. $z = e^y \sin xy$

$\dfrac{\partial z}{\partial x} = ye^y \cos xy$

$\dfrac{\partial z}{\partial y} = e^y \sin xy + xe^y \cos x$

$\qquad = e^y(x \cos xy + \sin xy)$

35. $z = \sinh(2x + 3y)$

$\dfrac{\partial z}{\partial x} = 2 \cosh(2x + 3y)$

$\dfrac{\partial z}{\partial y} = 3 \cosh(2x + 3y)$

37. $f(x, y) = \displaystyle\int_x^y (t^2 - 1)\, dt$

$\qquad = \left[\dfrac{t^3}{3} - t\right]_x^y = \left(\dfrac{y^3}{3} - y\right) - \left(\dfrac{x^3}{3} - x\right)$

$f_x(x, y) = -x^2 + 1 = 1 - x^2$

$f_y(x, y) = y^2 - 1$

[You could also use the Second Fundamental Theorem of Calculus.]

39. $f(x, y) = 3x + 2y$

$$\frac{\partial f}{\partial x} = \lim_{\Delta x \to 0} \frac{f(x + \Delta x, y) - f(x, y)}{\Delta x} = \lim_{\Delta x \to 0} \frac{3(x + \Delta x) + 2y - (3x + 2y)}{\Delta x} = \lim_{\Delta x \to 0} \frac{3\Delta x}{\Delta x} = 3$$

$$\frac{\partial f}{\partial y} = \lim_{\Delta y \to 0} \frac{f(x, y + \Delta y) - f(x, y)}{\Delta y} = \lim_{\Delta y \to 0} \frac{3x + 2(y + \Delta y) - (3x + 2y)}{\Delta y} = \lim_{\Delta y \to 0} \frac{2\Delta y}{\Delta y} = 2$$

41. $f(x, y) = \sqrt{x + y}$

$$\frac{\partial f}{\partial x} = \lim_{\Delta x \to 0} \frac{f(x + \Delta x, y) - f(x, y)}{\Delta x}$$

$$= \lim_{\Delta x \to 0} \frac{\sqrt{x + \Delta x + y} - \sqrt{x + y}}{\Delta x}$$

$$= \lim_{\Delta x \to 0} \frac{\left(\sqrt{x + \Delta x + y} - \sqrt{x + y}\right)\left(\sqrt{x + \Delta x + y} + \sqrt{x + y}\right)}{\Delta x\left(\sqrt{x + \Delta x + y} + \sqrt{x + y}\right)} = \lim_{\Delta x \to 0} \frac{1}{\sqrt{x + \Delta x + y} + \sqrt{x + y}} = \frac{1}{2\sqrt{x + y}}$$

$$\frac{\partial f}{\partial y} = \lim_{\Delta y \to 0} \frac{f(x, y + \Delta y) - f(x, y)}{\Delta y} = \lim_{\Delta y \to 0} \frac{\sqrt{x + y + \Delta y} - \sqrt{x + y}}{\Delta y}$$

$$= \lim_{\Delta y \to 0} \frac{\left(\sqrt{x + y + \Delta y} - \sqrt{x + y}\right)\left(\sqrt{x + y + \Delta y} + \sqrt{x + y}\right)}{\Delta y\left(\sqrt{x + y + \Delta y} + \sqrt{x + y}\right)}$$

$$= \lim_{\Delta y \to 0} \frac{1}{\sqrt{x + y + \Delta y} + \sqrt{x + y}} = \frac{1}{2\sqrt{x + y}}$$

43. $f(x, y) = e^y \sin x$

$f_x(x, y) = e^y \cos x$

At $(\pi, 0)$, $f_x(\pi, 0) = -1$.

$f_y(x, y) = e^y \sin x$

At $(\pi, 0)$, $f_y(\pi, 0) = 0$.

45. $f(x, y) = \cos(2x - y)$

$f_x(x, y) = -2 \sin(2x - y)$

At $\left(\frac{\pi}{4}, \frac{\pi}{3}\right)$, $f_x\left(\frac{\pi}{4}, \frac{\pi}{3}\right) = -2 \sin\left(\frac{\pi}{2} - \frac{\pi}{3}\right) = -1$.

$f_y(x, y) = \sin(2x - y)$

At $\left(\frac{\pi}{4}, \frac{\pi}{3}\right)$, $f_y\left(\frac{\pi}{4}, \frac{\pi}{3}\right) = \sin\left(\frac{\pi}{2} - \frac{\pi}{3}\right) = \frac{1}{2}$.

47. $f(x, y) = \arctan \dfrac{y}{x}$

$f_x(x, y) = \dfrac{1}{1 + (y^2/x^2)}\left(-\dfrac{y}{x^2}\right) = \dfrac{-y}{x^2 + y^2}$

At $(2, -2)$: $f_x(2, -2) = \dfrac{1}{4}$

$f_y(x, y) = \dfrac{1}{1 + (y^2/x^2)}\left(\dfrac{1}{x}\right) = \dfrac{x}{x^2 + y^2}$

At $(2, -2)$: $f_y(2, -2) = \dfrac{1}{4}$

49. $f(x, y) = \dfrac{xy}{x - y}$

$f_x(x, y) = \dfrac{y(x - y) - xy}{(x - y)^2} = \dfrac{-y^2}{(x - y)^2}$

At $(2, -2)$: $f_x(2, -2) = -\dfrac{1}{4}$

$f_y(x, y) = \dfrac{x(x - y) + xy}{(x - y)^2} = \dfrac{x^2}{(x - y)^2}$

At $(2, -2)$: $f_y(2, -2) = \dfrac{1}{4}$

51. $g(x, y) = 4 - x^2 - y^2$

$g_x(x, y) = -2x$

At $(1, 1)$: $g_x(1, 1) = -2$

$g_y(x, y) = -2y$

At $(1, 1)$: $g_y(1, 1) = -2$

53. $H(x, y, z) = \sin(x + 2y + 3z)$

$H_x(x, y, z) = \cos(x + 2y + 3z)$

$H_y(x, y, z) = 2 \cos(x + 2y + 3z)$

$H_z(x, y, z) = 3 \cos(x + 2y + 3z)$

55. $w = \sqrt{x^2 + y^2 + z^2}$

$$\frac{\partial w}{\partial x} = \frac{x}{\sqrt{x^2 + y^2 + z^2}}$$

$$\frac{\partial w}{\partial y} = \frac{y}{\sqrt{x^2 + y^2 + z^2}}$$

$$\frac{\partial w}{\partial z} = \frac{z}{\sqrt{x^2 + y^2 + z^2}}$$

57. $F(x, y, z) = \ln\sqrt{x^2 + y^2 + z^2} = \dfrac{1}{2}\ln(x^2 + y^2 + z^2)$

$$F_x(x, y, z) = \frac{x}{x^2 + y^2 + z^2}$$

$$F_y(x, y, z) = \frac{y}{x^2 + y^2 + z^2}$$

$$F_z(x, y, z) = \frac{z}{x^2 + y^2 + z^2}$$

59. $f(x, y, z) = x^3yz^2$

$f_x(x, y, z) = 3x^2yz^2$

$f_x(1, 1, 1) = 3$

$f_y(x, y, z) = x^3z^2$

$f_y(1, 1, 1) = 1$

$f_z(x, y, z) = 2x^3yz$

$f_z(1, 1, 1) = 2$

61. $f(x, y, z) = \dfrac{x}{yz}$

$f_x(x, y, z) = \dfrac{1}{yz}$

$f_x(1, -1, -1) = 1$

$f_y(x, y, z) = \dfrac{-x}{y^2z}$

$f_y(1, -1, -1) = 1$

$f_z(x, y, z) = \dfrac{-x}{yz^2}$

$f_z(1, -1, -1) = 1$

63. $f(x, y, z) = z\sin(x + y)$

$f_x(x, y, z) = z\cos(x + y)$

$f_x\left(0, \dfrac{\pi}{2}, -4\right) = -4\cos\dfrac{\pi}{2} = 0$

$f_y(x, y, z) = z\cos(x + y)$

$f_y\left(0, \dfrac{\pi}{2}, -4\right) = -4\cos\dfrac{\pi}{2} = 0$

$f_z(x, y, z) = \sin(x + y)$

$f_z\left(0, \dfrac{\pi}{2}, -4\right) = \sin\dfrac{\pi}{2} = 1$

65. $f_x(x, y) = 2x + y - 2 = 0$

$f_y(x, y) = x + 2y + 2 = 0$

$2x + y - 2 = 0 \Rightarrow y = 2 - 2x$

$x + 2(2 - 2x) + 2 = 0 \Rightarrow -3x + 6 = 0 \Rightarrow x = 2,$

$y = -2$

Point: $(2, -2)$

67. $f_x(x, y) = 2x + 4y - 4,\ f_y(x, y) = 4x + 2y + 16$

$f_x = f_y = 0:\ 2x + 4y = 4$

$\qquad\qquad\qquad 4x + 2y = -16$

Solving for x and y,

$x = -6$ and $y = 4.$

69. $f_x(x, y) = -\dfrac{1}{x^2} + y,\ f_y(x, y) = -\dfrac{1}{y^2} + x$

$f_x = f_y = 0:\ -\dfrac{1}{x^2} + y = 0$ and $-\dfrac{1}{y^2} + x = 0$

$$y = \frac{1}{x^2}\ \text{and}\ x = \frac{1}{y^2}$$

$y = y^4 \Rightarrow y = 1 = x$

Points: $(1, 1)$

71. $f_x(x, y) = (2x + y)e^{x^2+xy+y^2} = 0$

$f_y(x, y) = (x + 2y)e^{x^2+xy+y^2} = 0$

$2x + y = 0 \Rightarrow y = -2x$

$x + 2(-2x) = 0 \Rightarrow x = 0 \Rightarrow y = 0$

Point: $(0, 0)$

73. $z = 3xy^2$

$$\frac{\partial z}{\partial x} = 3y^2,\ \frac{\partial^2 z}{\partial x^2} = 0,\ \frac{\partial^2 z}{\partial y\partial x} = 6y$$

$$\frac{\partial z}{\partial y} = 6xy,\ \frac{\partial^2 z}{\partial y^2} = 6x,\ \frac{\partial^2 z}{\partial x\partial y} = 6y$$

75. $z = x^2 - 2xy + 3y^2$

$$\frac{\partial z}{\partial x} = 2x - 2y$$

$$\frac{\partial^2 z}{\partial x^2} = 2$$

$$\frac{\partial^2 z}{\partial y \partial x} = -2$$

$$\frac{\partial z}{\partial y} = -2x + 6y$$

$$\frac{\partial^2 z}{\partial y^2} = 6$$

$$\frac{\partial^2 z}{\partial x \partial y} = -2$$

77. $z = \sqrt{x^2 + y^2}$

$$\frac{\partial z}{\partial x} = \frac{x}{\sqrt{x^2 + y^2}}$$

$$\frac{\partial^2 z}{\partial x^2} = \frac{y^2}{\left(x^2 + y^2\right)^{3/2}}$$

$$\frac{\partial^2 z}{\partial y \partial x} = \frac{-xy}{\left(x^2 + y^2\right)^{3/2}}$$

$$\frac{\partial z}{\partial y} = \frac{y}{\sqrt{x^2 + y^2}}$$

$$\frac{\partial^2 z}{\partial y^2} = \frac{x^2}{\left(x^2 + y^2\right)^{3/2}}$$

$$\frac{\partial^2 z}{\partial x \partial y} = \frac{-xy}{\left(x^2 + y^2\right)^{3/2}}$$

79. $z = e^x \tan y$

$$\frac{\partial z}{\partial x} = e^x \tan y$$

$$\frac{\partial^2 z}{\partial x^2} = e^x \tan y$$

$$\frac{\partial^2 z}{\partial y \partial x} = e^x \sec^2 y$$

$$\frac{\partial z}{\partial y} = e^x \sec^2 y$$

$$\frac{\partial^2 z}{\partial y^2} = 2e^x \sec^2 y \tan y$$

$$\frac{\partial^2 z}{\partial x \partial y} = e^x \sec^2 y$$

81. $z = \cos xy$

$$\frac{\partial z}{\partial x} = -y \sin xy, \frac{\partial^2 z}{\partial x^2} = -y^2 \cos xy$$

$$\frac{\partial^2 z}{\partial y \partial x} = -yx \cos xy - \sin xy$$

$$\frac{\partial z}{\partial y} = -x \sin xy, \frac{\partial^2 z}{\partial y^2} = -x^2 \cos xy$$

$$\frac{\partial^2 z}{\partial x \partial y} = -xy \cos xy - \sin xy$$

83. $z = x \sec y$

$$\frac{\partial z}{\partial x} = \sec y$$

$$\frac{\partial^2 z}{\partial x^2} = 0$$

$$\frac{\partial^2 z}{\partial y \partial x} = \sec y \tan y$$

$$\frac{\partial z}{\partial y} = x \sec y \tan y$$

$$\frac{\partial^2 z}{\partial y^2} = x \sec y \left(\sec^2 y + \tan^2 y\right)$$

$$\frac{\partial^2 z}{\partial x \partial y} = \sec y \tan y$$

So, $\dfrac{\partial^2 z}{\partial y \partial x} = \dfrac{\partial^2 z}{\partial x \partial y}$

There are no points for which $z_x = 0 = z_y$, because

$$\frac{\partial z}{\partial x} = \sec y \neq 0.$$

85. $z = \ln\left(\dfrac{x}{x^2 + y^2}\right) = \ln x - \ln\left(x^2 + y^2\right)$

$$\frac{\partial z}{\partial x} = \frac{1}{x} - \frac{2x}{x^2 + y^2} = \frac{y^2 - x^2}{x\left(x^2 + y^2\right)}$$

$$\frac{\partial^2 z}{\partial x^2} = \frac{x^4 - 4x^2y^2 - y^4}{x^2\left(x^2 + y^2\right)^2}$$

$$\frac{\partial^2 z}{\partial y \partial x} = \frac{4xy}{\left(x^2 + y^2\right)^2}$$

$$\frac{\partial z}{\partial y} = -\frac{2y}{x^2 + y^2}$$

$$\frac{\partial^2 z}{\partial y^2} = \frac{2\left(y^2 - x^2\right)}{\left(x^2 + y^2\right)^2}$$

$$\frac{\partial^2 z}{\partial x \partial y} = \frac{4xy}{\left(x^2 + y^2\right)^2}$$

There are no points for which $z_x = z_y = 0$.

87. $f(x, y, z) = xyz$

$f_x(x, y, z) = yz$

$f_y(x, y, z) = xz$

$f_{yy}(x, y, z) = 0$

$f_{xy}(x, y, z) = z$

$f_{yx}(x, y, z) = z$

$f_{yyx}(x, y, z) = 0$

$f_{xyy}(x, y, z) = 0$

$f_{yxy}(x, y, z) = 0$

So, $f_{xyy} = f_{yxy} = f_{yyx} = 0.$

89. $f(x, y, z) = e^{-x} \sin yz$

$f_x(x, y, z) = -e^{-x} \sin yz$

$f_y(x, y, z) = ze^{-x} \cos yz$

$f_{yy}(x, y, z) = -z^2 e^{-x} \sin yz$

$f_{xy}(x, y, z) = -ze^{-x} \cos yz$

$f_{yx}(x, y, z) = -ze^{-x} \cos yz$

$f_{yyx}(x, y, z) = z^2 e^{-x} \sin yz$

$f_{xyy}(x, y, z) = z^2 e^{-x} \sin yz$

$f_{yxy}(x, y, z) = z^2 e^{-x} \sin yz$

So, $f_{xyy} = f_{yxy} = f_{yyz}.$

91. $z = 5xy$

$\dfrac{\partial z}{\partial x} = 5y$

$\dfrac{\partial^2 z}{\partial x^2} = 0$

$\dfrac{\partial z}{\partial y} = 5x$

$\dfrac{\partial^2 z}{\partial y^2} = 0$

So, $\dfrac{\partial^2 z}{\partial x^2} + \dfrac{\partial^2 z}{\partial y^2} = 0 + 0 = 0.$

93. $z = e^x \sin y$

$\dfrac{\partial z}{\partial x} = e^x \sin y$

$\dfrac{\partial^2 z}{\partial x^2} = e^x \sin y$

$\dfrac{\partial z}{\partial y} = e^x \cos y$

$\dfrac{\partial^2 z}{\partial y^2} = -e^x \sin y$

So, $\dfrac{\partial^2 z}{\partial x^2} + \dfrac{\partial^2 z}{\partial y^2} = e^x \sin y - e^x \sin y = 0.$

95. $z = \sin(x - ct)$

$\dfrac{\partial z}{\partial t} = -c \cos(x - ct)$

$\dfrac{\partial^2 z}{\partial t^2} = -c^2 \sin(x - ct)$

$\dfrac{\partial z}{\partial x} = \cos(x - ct)$

$\dfrac{\partial^2 z}{\partial x^2} = -\sin(x - ct)$

So, $\dfrac{\partial^2 z}{\partial t^2} = c^2 \left(\dfrac{\partial^2 z}{\partial x^2} \right).$

97. $z = \ln(x + ct)$

$\dfrac{\partial z}{\partial t} = \dfrac{c}{x + ct}$

$\dfrac{\partial^2 z}{\partial t^2} = \dfrac{-c^2}{(x + ct)^2}$

$\dfrac{\partial z}{\partial x} = \dfrac{1}{x + ct}$

$\dfrac{\partial^2 z}{\partial x^2} = \dfrac{-1}{(x + ct)^2}$

$\dfrac{\partial^2 z}{\partial t^2} = \dfrac{-c^2}{(x + ct)^2} = c^2 \left(\dfrac{\partial^2 z}{\partial x^2} \right)$

99. $z = e^{-t} \cos \dfrac{x}{c}$

$\dfrac{\partial z}{\partial t} = -e^{-t} \cos \dfrac{x}{c}$

$\dfrac{\partial z}{\partial x} = -\dfrac{1}{c} e^{-t} \sin \dfrac{x}{c}$

$\dfrac{\partial^2 z}{\partial x^2} = -\dfrac{1}{c^2} e^{-t} \cos \dfrac{x}{c}$

So, $\dfrac{\partial z}{\partial t} = c^2 \left(\dfrac{\partial^2 z}{\partial x^2} \right).$

101. Yes. The function $f(x, y) = \cos(3x - 2y)$ satisfies both equations.

103. If $z = f(x, y)$, then to find f_x you consider y constant and differentiate with respect to x. Similarly, to find f_y, you consider x constant and differentiate with respect to y.

105. The plane $z = -x + y = f(x, y)$ satisfies

$$\frac{\partial f}{\partial x} < 0 \text{ and } \frac{\partial f}{\partial y} > 0.$$

107. In this case, the mixed partials are equal, $f_{xy} = f_{yx}$.
See Theorem 13.3.

109. $R = 200x_1 + 200x_2 - 4x_1^2 - 8x_1x_2 - 4x_2^2$

(a) $\dfrac{\partial r}{\partial x_1} = 200 - 8x_1 - 8x_2$

At $(x_1, x_2) = (4, 12)$, $\dfrac{\partial R}{\partial x_1} = 200 - 32 - 96 = 72$.

(b) $\dfrac{\partial R}{\partial x_2} = 200 - 8x_1 - 8x_2$

At $(x_1, x_2) = (4, 12)$, $\dfrac{\partial R}{\partial x_2} = 72$.

111. $IQ(M, C) = 100\dfrac{M}{C}$

$IQ_M = \dfrac{100}{C}, IQ_M(12, 10) = 10$

$IQ_c = \dfrac{-100M}{C^2}, IQ_c(12, 10) = -12$

When the chronological age is constant, IQ increases at a rate of 10 points per mental age year.

When the mental age is constant, IQ decreases at a rate of 12 points per chronological age year.

113. An increase in either price will cause a decrease in demand.

115. $T = 500 - 0.6x^2 - 1.5y^2$

$\dfrac{\partial T}{\partial x} = -1.2x, \dfrac{\partial T}{\partial x}(2, 3) = -2.4°/m$

$\dfrac{\partial T}{\partial y} = -3y = \dfrac{\partial T}{\partial y}(2, 3) = -9°/m$

117.
$$PV = \frac{n}{xB}RT$$

$$T = \frac{PV}{\dfrac{n}{xB}R} \Rightarrow \frac{\partial T}{\partial P} = \frac{V}{\dfrac{n}{xB}R}$$

$$P = \frac{\dfrac{n}{xB}RT}{V} \Rightarrow \frac{\partial P}{\partial V} = -\frac{\dfrac{n}{xB}RT}{V^2}$$

$$V = \frac{\dfrac{n}{xB}RT}{P} \Rightarrow \frac{\partial V}{\partial T} = \frac{\dfrac{n}{xB}R}{P}$$

$$\frac{\partial T}{\partial P} \cdot \frac{\partial P}{\partial V} \cdot \frac{\partial V}{\partial T} = \left(\frac{V}{\dfrac{n}{xB}R}\right)\left(-\frac{\dfrac{n}{xB}RT}{V^2}\right)\left(\frac{\dfrac{n}{xB}R}{P}\right)$$

$$= -\frac{\dfrac{n}{xB}RT}{VP} = -\frac{\dfrac{n}{xB}RT}{\dfrac{n}{xB}RT} = -1$$

119. $z = 0.461x + 0.301y - 494$

(a) $\dfrac{\partial z}{\partial x} = 0.461$ $\dfrac{\partial z}{\partial y} = 0.301$

(b) As the expenditures on amusement parks and campgrounds (x) increase, the expenditures on spectator sports (z) increase. As the expenditures on live entertainment (y) increase, the expenditures on spectator sports (z) increase.

121. False
Let $z = x + y + 1$.

123. True

125. $f(x, y) = \begin{cases} \dfrac{xy(x^2 - y^2)}{x^2 + y^2}, & (x, y) \neq (0, 0) \\ 0, & (x, y) = (0, 0) \end{cases}$

(a) $f_x(x, y) = \dfrac{(x^2 + y^2)(3x^2y - y^3) - (x^3y - xy^3)(2x)}{(x^2 + y^2)^2} = \dfrac{y(x^4 + 4x^2y^2 - y^4)}{(x^2 + y^2)^2}$

$f_y(x, y) = \dfrac{(x^2 + y^2)(x^3 - 3xy^2) - (x^3y - xy^3)(2y)}{(x^2 + y^2)^2} = \dfrac{x(x^4 - 4x^2y^2 - y^4)}{(x^2 + y^2)^2}$

(b) $f_x(0, 0) = \lim_{\Delta x \to 0} \dfrac{f(\Delta x, 0) - f(0, 0)}{\Delta x} = \lim_{\Delta x \to 0} \dfrac{0/[(\Delta x)^2] - 0}{\Delta x} = 0$

$f_y(0, 0) = \lim_{\Delta y \to 0} \dfrac{f(0, \Delta y) - f(0, 0)}{\Delta y} = \lim_{\Delta y \to 0} \dfrac{0/[(\Delta y)^2] - 0}{\Delta y} = 0$

(c) $f_{xy}(0, 0) = \dfrac{\partial}{\partial y}\left(\dfrac{\partial f}{\partial x}\right)\bigg|_{(0,0)} = \lim_{\Delta y \to 0} \dfrac{f_x(0, \Delta y) - f_x(0, 0)}{\Delta y} = \lim_{\Delta y \to 0} \dfrac{\Delta y(-(\Delta y)^4)}{((\Delta y)^2)^2(\Delta y)} = \lim_{\Delta y \to 0}(-1) = -1$

$f_{yx}(0, 0) = \dfrac{\partial}{\partial x}\left(\dfrac{\partial f}{\partial y}\right)\bigg|_{(0,0)} = \lim_{\Delta x \to 0} \dfrac{f_y(\Delta x, 0) - f_y(0, 0)}{\Delta x} = \lim_{\Delta x \to 0} \dfrac{\Delta x((\Delta x)^4)}{((\Delta x)^2)^2(\Delta x)} = \lim_{\Delta x \to 0} 1 = 1$

(d) f_{yx} or f_{xy} or both are not continuous at $(0, 0)$.

127. $f(x, y) = (x^2 + y^2)^{2/3}$

For $(x, y) \neq (0, 0)$, $f_x(x, y) = \dfrac{2}{3}(x^2 + y^2)^{-1/3}(2x) = \dfrac{4x}{3(x^2 + y^2)^{1/3}}$.

For $(x, y) = (0, 0)$, use the definition of partial derivative.

$f_x(0, 0) = \lim_{\Delta x \to 0} \dfrac{f(0 + \Delta x) - f(0, 0)}{\Delta x} = \lim_{\Delta x \to 0} \dfrac{(\Delta x)^{4/3}}{\Delta x} = \lim_{\Delta x \to 0}(\Delta x)^{1/3} = 0$

Section 13.4 Differentials

1. $z = 2x^2y^3$

$dz = 4xy^3\, dx + 6x^2y^2\, dy$

3. $z = \dfrac{-1}{x^2 + y^2}$

$dz = \dfrac{2x}{(x^2 + y^2)^2}\, dx + \dfrac{2y}{(x^2 + y^2)^2}\, dy$

$= \dfrac{2}{(x^2 + y^2)^2}(x\, dx + y\, dy)$

5. $z = x\cos y - y\cos x$

$dz = (\cos y + y\sin x)\, dx + (-x\sin y - \cos x)\, dy$

$= (\cos y + y\sin x)\, dx - (x\sin y + \cos x)\, dy$

7. $z = e^x \sin y$

$dz = (e^x \sin y)\, dx + (e^x \cos y)\, dy$

9. $w = 2z^3y \sin x$

$dw = 2z^3y \cos x\, dx + 2z^3 \sin x\, dy + 6z^2y \sin x\, dz$

11. $f(x, y) = 2x - 3y$

(a) $f(2, 1) = 1$

$f(2.1, 1.05) = 1.05$

$\Delta z = f(2.1, 1.05) - f(2, 1) = 0.05$

(b) $dz = 2\, dx - 3\, dy = 2(0.1) - 3(0.05) = 0.05$

13. $f(x, y) = 16 - x^2 - y^2$

   (a) $f(2, 1) = 11$

      $f(2.1, 1.05) = 10.4875$

      $\Delta z = f(2.1, 1.05) - f(2.1) = -0.5125$

   (b) $dz = -2x\, dx - 2y\, dy$

      $= -2(2)(0.1) - 2(1)(0.05) = -0.5$

15. $f(x, y) = ye^x$

   (a) $f(2, 1) = e^2 \approx 7.3891$

      $f(2.1, 1.05) = 1.05e^{2.1} \approx 8.5745$

      $\Delta z = f(2.1, 1.05) - f(2, 1) = 1.1854$

   (b) $dz = ye^x\, dx + e^x\, dy$

      $= e^2(0.1) + e^2(0.05) \approx 1.1084$

17. Let $z = x^2 y$, $x = 2$, $y = 9$, $dx = 0.01$, $dy = 0.02$.

Then: $dz = 2xy\, dx + x^2\, dy$

$(2.01)^2(9.02) - 2^2 \cdot 9 \approx 2(2)(9)(0.01) + 2^2(0.02) = 0.44$

19. Let $z = \sqrt{x^2 + y^2}$, $x = 5$, $y = 3$, $dx = 0.05$, $dy = 0.1$.

Then:

$$dz = \frac{x}{\sqrt{x^2 + y^2}}\, dx + \frac{y}{\sqrt{x^2 + y^2}}\, dy$$

$$\sqrt{(5.05)^2 + (3.1)^2} - \sqrt{5^2 + 3^2} \approx \frac{5}{\sqrt{5^2 + 3^2}}(0.05) + \frac{3}{\sqrt{5^2 + 3^2}}(0.1) = \frac{0.55}{\sqrt{34}} \approx 0.094$$

21. In general, the accuracy worsens as Δx and Δy increase.

23. If $z = f(x, y)$, then $\Delta z \approx dz$ is the propagated error,

and $\dfrac{\Delta z}{z} \approx \dfrac{dz}{z}$ is the relative error.

25. $A = lh$

$dA = l\, dh + h\, dl$

$\Delta A = (1 + dl)(h + dh) - lh$

$= h\, dl + l\, dh + dl\, dh$

$\Delta A - dA = dl\, dh$

27. $V = \dfrac{\pi r^2 h}{3}$, $r = 4$, $h = 8$

$$dV = \frac{2\pi rh}{3}\, dr + \frac{\pi r^2}{3}\, dh = \frac{\pi r}{3}(2h\, dr + r\, dh) = \frac{4\pi}{3}(16\, dr + 4\, dh)$$

$$\Delta V = \frac{\pi}{3}\left[(r + \Delta r)^2(h + \Delta h) - r^2 h\right] = \frac{\pi}{3}\left[(4 + \Delta r)^2(8 + \Delta h) - 128\right]$$

Δr	Δh	dV	ΔV	$\Delta V - dV$
0.1	0.1	8.3776	8.5462	0.1686
0.1	−0.1	5.0265	5.0255	−0.0010
0.001	0.002	0.1005	0.1006	0.0001
−0.0001	0.0002	−0.0034	−0.0034	0.0000

29. $V = xyz$, $dV = yz\, dx + xz\, dy + xy\, dz$

Propagated error $= dV = 5(12)(\pm 0.02) + 8(12)(\pm 0.02) + 8(5)(\pm 0.02)$

$= (60 + 96 + 40)(\pm 0.02) = 196(\pm 0.02) = \pm 3.92\text{ in.}^3$

The measured volume is $V = 8(5)(12) = 480\text{ in.}^3$

Relative error $= \dfrac{\Delta V}{V} \approx \dfrac{dV}{V} = \dfrac{3.92}{480} \approx 0.008167 \approx 0.82\%$

31. $C = 35.74 + 0.6215T - 35.75v^{0.16} + 0.4275Tv^{0.16}$

$\dfrac{\partial C}{\partial T} = 0.6215 + 0.4275v^{0.16}$

$\dfrac{\partial C}{\partial v} = -5.72v^{-0.84} + 0.0684Tv^{-0.84}$

$dC = \dfrac{\partial C}{\partial T}dT + \dfrac{\partial C}{\partial v}dv = \left(0.6215 + 0.4275(23)^{0.16}\right)(\pm 1) + \left(-5.72(23)^{-0.84} + 0.0684(8)(23)^{-0.84}\right)(\pm 3)$

$\quad = \pm 1.3275 \pm 1.1143 = \pm 2.4418 \text{ Maximum propagated error}$

$\dfrac{dC}{C} = \dfrac{2.4418}{-12.6807} \approx 0.19 = 19\% \text{ Maximum relative error}$

33. $P = \dfrac{E^2}{R}, \left|\dfrac{dE}{E}\right| = 3\% = 0.03, \left|\dfrac{dR}{R}\right| = 4\% = 0.04$

$dP = \dfrac{2E}{R}dE - \dfrac{E^2}{R^2}dR$

$\dfrac{dP}{P} = \left[\dfrac{2E}{R}dE - \dfrac{E^2}{R^2}dR\right]\Big/ P = \left[\dfrac{2E}{R}dE - \dfrac{E^2}{R^2}dR\right]\Big/\left(E^2/R\right) = \dfrac{2}{E}dE - \dfrac{1}{R}dR$

Using the worst case scenario, $\dfrac{dE}{E} = 0.03$ and $\dfrac{dR}{R} = -0.04$: $\dfrac{dP}{P} \le 2(0.03) - (-0.04) = 0.10 = 10\%.$

35. (a) $V = \dfrac{1}{2}bhl = \left(18\sin\dfrac{\theta}{2}\right)\left(18\cos\dfrac{\theta}{2}\right)(16)(12) = 31{,}104\sin\theta \text{ in.}^3 = 18\sin\theta \text{ ft}^3$

$\quad$ V is maximum when $\sin\theta = 1$ or $\theta = \pi/2.$

(b) $V = \dfrac{s^2}{2}(\sin\theta)l$

$\quad dV = s(\sin\theta)l\,ds + \dfrac{s^2}{2}l(\cos\theta)\,d\theta + \dfrac{s^2}{2}(\sin\theta)\,dl$

$\quad = 18\left(\sin\dfrac{\pi}{2}\right)(16)(12)\left(\dfrac{1}{2}\right) + \dfrac{18^2}{2}(16)(12)\left(\cos\dfrac{\pi}{2}\right)\left(\dfrac{\pi}{90}\right) + \dfrac{18^2}{2}\left(\sin\dfrac{\pi}{2}\right)\left(\dfrac{1}{2}\right) = 1809 \text{ in.}^3 \approx 1.047 \text{ ft}^3$

37. $L = 0.00021\left(\ln\dfrac{2h}{r} - 0.75\right)$

$dL = 0.00021\left[\dfrac{dh}{h} - \dfrac{dr}{r}\right] = 0.00021\left[\dfrac{(\pm 1/100)}{100} - \dfrac{(\pm 1/16)}{2}\right] \approx (\pm 6.6) \times 10^{-6}$

$L = 0.00021(\ln 100 - 0.75) \pm dL \approx 8.096 \times 10^{-4} \pm 6.6 \times 10^{-6} \text{ micro henrys}$

39. $z = f(x, y) = x^2 - 2x + y$

$\Delta z = f(x + \Delta x, y + \Delta y) - f(x, y) = \left(x^2 + 2x(\Delta x) + (\Delta x)^2 - 2x - 2(\Delta x) + y + (\Delta y)\right) - \left(x^2 - 2x + y\right)$

$\quad = 2x(\Delta x) + (\Delta x)^2 - 2(\Delta x) + (\Delta y) = (2x - 2)\,\Delta x + \Delta y + \Delta x(\Delta x) + 0(\Delta y)$

$\quad = f_x(x, y)\,\Delta x + f_y(x, y)\,\Delta y + \varepsilon_1\Delta x + \varepsilon_2\Delta y \text{ where } \varepsilon_1 = \Delta x \text{ and } \varepsilon_2 = 0.$

As $(\Delta x, \Delta y) \to (0, 0), \varepsilon_1 \to 0$ and $\varepsilon_2 \to 0.$

41. $z = f(x, y) = x^2 y$

$\Delta z = f(x + \Delta x, y + \Delta y) - f(x, y) = \left(x^2 + 2x(\Delta x) + (\Delta x)^2\right)(y + \Delta y) - x^2 y$

$\quad = 2xy(\Delta x) + y(\Delta x)^2 + x^2 \Delta y + 2x(\Delta x)(\Delta y) + (\Delta x)^2 \Delta y = 2xy(\Delta x) + x^2 \Delta y + (y\Delta x)\Delta x + \left[2x\Delta x + (\Delta x)^2\right]\Delta y$

$\quad = f_x(x, y)\,\Delta x + f_y(x, y)\,\Delta y + \varepsilon_1 \Delta x + \varepsilon_2 \Delta y$ where $\varepsilon_1 = y(\Delta x)$ and $\varepsilon_2 = 2x\Delta x + (\Delta x)^2$.

As $(\Delta x, \Delta y) \to (0, 0)$, $\varepsilon_1 \to 0$ and $\varepsilon_2 \to 0$.

43. $f(x, y) = \begin{cases} \dfrac{3x^2 y}{x^4 + y^2}, & (x, y) \neq (0, 0) \\ 0, & (x, y) = (0, 0) \end{cases}$

$f_x(0, 0) = \lim_{\Delta x \to 0} \dfrac{f(\Delta x, 0) - f(0, 0)}{\Delta x} = \lim_{\Delta x \to 0} \dfrac{\frac{0}{(\Delta x)^4} - 0}{\Delta x} = 0$

$f_y(0, 0) = \lim_{\Delta y \to 0} \dfrac{f(0, \Delta y) - f(0, 0)}{\Delta y} = \lim_{\Delta y \to 0} \dfrac{\frac{0}{(\Delta y)^2} - 0}{\Delta y} = 0$

So, the partial derivatives exist at $(0, 0)$.

Along the line $y = x$: $\lim_{(x, y) \to (0, 0)} f(x, y) = \lim_{x \to 0} \dfrac{3x^3}{x^4 + x^2} = \lim_{x \to 0} \dfrac{3x}{x^2 + 1} = 0$

Along the curve $y = x^2$: $\lim_{(x, y) \to (0, 0)} f(x, y) = \dfrac{3x^4}{2x^4} = \dfrac{3}{2}$

f is not continuous at $(0, 0)$. So, f is not differentiable at $(0, 0)$. (See Theorem 12.5)

Section 13.5   Chain Rules for Functions of Several Variables

1. $w = x^2 + y^2$

$x = 2t, \; y = 3t$

$\dfrac{dw}{dt} = \dfrac{\partial w}{\partial x}\dfrac{dx}{dt} + \dfrac{\partial w}{\partial y}\dfrac{dy}{dt} = (2x)(2) + (2y)(3)$

$\quad = 4x + 6y = 8t + 18t = 26t$

When $t = 2$, $\dfrac{dw}{dt} = 26(2) = 52$.

3. $w = x \sin y$

$x = e^t, \; y = \pi - t$

$\dfrac{dw}{dt} = \dfrac{\partial w}{\partial x}\dfrac{dx}{dt} + \dfrac{\partial w}{\partial y}\dfrac{dy}{dt} = \sin y(e^t) + x \cos y(-1)$

$\quad = \sin(\pi - t)e^t - e^t \cos(\pi - t) = e^t \sin t + e^t \cos t$

When $t = 0$, $\dfrac{dw}{dt} = (1)(0) + (1)(1) = 0 + 1 = 1$.

5. $w = xy, \; x = e^t, \; y = e^{-2t}$

(a) $\dfrac{dw}{dt} = \dfrac{\partial w}{\partial x}\dfrac{dx}{dt} + \dfrac{\partial w}{\partial y}\dfrac{dy}{dt}$

$\quad = y(e^t) + x(-2e^{-2t}) = e^{-2t}e^t - e^t 2e^{-2t} = -e^{-t}$

(b) $w = e^t e^{-2t} = e^{-t}$

$\quad \dfrac{dw}{dt} = -e^{-t}$

7. $w = x^2 + y^2 + z^2, \; x = \cos t, \; y = \sin t, \; z = e^t$

(a) $\dfrac{dw}{dt} = \dfrac{\partial w}{\partial x}\dfrac{dx}{dt} + \dfrac{\partial w}{\partial y}\dfrac{dy}{dt} + \dfrac{\partial w}{\partial z}\dfrac{dz}{dt}$

$\quad = 2x(-\sin t) + 2y(\cos t) + 2z(e^t)$

$\quad = -2 \cos t \sin t + 2 \sin t \cos t + 2e^{2t} = 2e^{2t}$

(b) $w = \cos^2 t + \sin^2 t + e^{2t} = 1 + e^{2t}$

$\quad \dfrac{dw}{dt} = 2e^{2t}$

9. $w = xy + xz + yz$, $x = t - 1$, $y = t^2 - 1$, $z = t$

(a) $\dfrac{dw}{dt} = \dfrac{\partial w}{\partial x}\dfrac{dx}{dt} + \dfrac{\partial w}{\partial y}\dfrac{dy}{dt} + \dfrac{\partial w}{\partial z}\dfrac{dz}{dt} = (y + z) + (x + z)(2t) + (x + y)$

$\qquad = (t^2 - 1 + t) + (t - 1 + t)(2t) + (t - 1 + t^2 - 1) = 3(2t^2 - 1)$

(b) $w = (t - 1)(t^2 - 1) + (t - 1)t + (t^2 - 1)t$

$\dfrac{dw}{dt} = 2t(t - 1) + (t^2 - 1) + 2t - 1 + 3t^2 - 1 = 3(2t^2 - 1)$

11. Distance $= f(t) = \sqrt{(x_1 - x_2)^2 + (y_1 - y_2)^2} = \sqrt{(10\cos 2t - 7\cos t)^2 + (6\sin 2t - 4\sin t)^2}$

$f'(t) = \dfrac{1}{2}\left[(10\cos 2t - 7\cos t)^2 + (6\sin 2t - 4\sin t)^2\right]^{-1/2}$

$\qquad \left[\left[2(10\cos 2t - 7\cos t)(-20\sin 2t + 7\sin t)\right] + \left[2(6\sin 2t - 4\sin t)(12\cos 2t - 4\cos t)\right]\right]$

$f'\!\left(\dfrac{\pi}{2}\right) = \dfrac{1}{2}\left[(-10)^2 + 4^2\right]^{-1/2}\left[2(-10)(7)\right] + (2(-4)(-12))\right] = \dfrac{1}{2}(116)^{-1/2}(-44) = \dfrac{-22}{2\sqrt{29}} = \dfrac{-11\sqrt{29}}{29} \approx -2.04$

13. $w = x^2 + y^2$

$x = s + t$, $y = s - t$

$\dfrac{\partial w}{\partial s} = 2x(1) + 2y(1) = 2(s + t) + 2(s - t) = 4s$

$\dfrac{\partial w}{\partial t} = 2x(1) + 2y(-1) = 2(s + t) - 2(s - t) = 4t$

When $s = 1$ and $t = 0$, $\dfrac{\partial w}{\partial s} = 4$ and $\dfrac{\partial w}{\partial t} = 0$.

15. $w = \sin(2x + 3y)$

$x = s + t$

$y = s - t$

$\dfrac{\partial w}{\partial s} = 2\cos(2x + 3y) + 3\cos(2x + 3y)$

$\qquad = 5\cos(2x + 3y) = 5\cos(5s - t)$

$\dfrac{\partial w}{\partial t} = 2\cos(2x + 3y) - 3\cos(2x + 3y)$

$\qquad = -\cos(2x + 3y) = -\cos(5s - t)$

When $s = 0$ and $t = \dfrac{\pi}{2}$, $\dfrac{\partial w}{\partial s} = 0$ and $\dfrac{\partial w}{\partial t} = 0$.

17. (a) $w = xyz$, $x = s + t$, $y = s - t$, $z = st^2$

$\dfrac{\partial w}{\partial s} = yz(1) + xz(1) + xy(t^2)$

$\qquad = (s - t)st^2 + (s + t)st^2 + (s + t)(s - t)t^2 = 2s^2t^2 + s^2t^2 - t^4 = 3s^2t^2 - t^4 = t^2(3s^2 - t^2)$

$\dfrac{\partial w}{\partial t} = yz(1) + xz(-1) + xy(2st) = (s - t)st^2 - (s + t)st^2 + (s + t)(s - t)(2st) = -2st^3 + 2s^3t - 2st^3 = 2s^3t - 4st^3$

$\qquad = 2st(s^2 - 2t^2)$

(b) $w = xyz = (s + t)(s - t)st^2 = (s^2 - t^2)st^2 = s^3t^2 - st^4$

$\dfrac{\partial w}{\partial s} = 3s^2t^2 - t^4 = t^2(3s^2 - t^2)$

$\dfrac{\partial w}{\partial t} = 2s^3t - 4st^3 = 2st(s^2 - 2t^2)$

19. (a) $w = ze^{xy}$, $x = s - t$, $y = s + t$, $z = st$

$$\frac{\partial w}{\partial s} = yze^{xy}(1) + xze^{xy}(1) + e^{xy}(t)$$

$$= e^{(s-t)(s+t)}\left[(s + t)st + (s - t)st + t\right]$$

$$= e^{(s-t)(s+t)}\left[2s^2t + t\right] = te^{s^2-t^2}\left(2s^2 + 1\right)$$

$$\frac{\partial w}{\partial t} = yze^{xy}(-1) + xze^{xy}(1) + e^{xy}(s)$$

$$= e^{(s-t)(s+t)}\left[-(s + t)(st) + (s - t)st + s\right]$$

$$= e^{(s-t)(s+t)}\left[-2st^2 + s\right] = se^{s^2-t^2}\left(1 - 2t^2\right)$$

(b) $w = ze^{xy} = ste^{(s-t)(s+t)} = ste^{s^2-t^2}$

$$\frac{\partial w}{\partial s} = te^{s^2-t^2} + st(2s)e^{s^2-t^2} = te^{s^2-t^2}\left(1 + 2s^2\right)$$

$$\frac{\partial w}{\partial t} = se^{s^2-t^2} + st(-2t)e^{s^2-t^2} = se^{s^2-t^2}\left(1 - 2t^2\right)$$

21. $x^2 - xy + y^2 - x + y = 0$

$$\frac{dy}{dx} = -\frac{F_x(x, y)}{F_y(x, y)} = -\frac{2x - y - 1}{-x + 2y + 1} = \frac{y - 2x + 1}{2y - x + 1}$$

23. $\ln\sqrt{x^2 + y^2} + x + y = 4$

$$\frac{1}{2}\ln(x^2 + y^2) + x + y - 4 = 0$$

$$\frac{dy}{dx} = -\frac{F_x(x, y)}{F_y(x, y)} = -\frac{\dfrac{x}{x^2 + y^2} + 1}{\dfrac{y}{x^2 + y^2} + 1} = -\frac{x + x^2 + y^2}{y + x^2 + y^2}$$

25. $F(x, y, z) = x^2 + y^2 + z^2 - 1$

$$F_x = 2x, \; F_y = 2y, \; F_z = 2z$$

$$\frac{\partial z}{\partial x} = -\frac{F_x}{F_z} = -\frac{x}{z}$$

$$\frac{\partial z}{\partial y} = -\frac{F_y}{F_z} = -\frac{y}{z}$$

27. $F(x, y, z) = x^2 + 2yz + z^2 - 1 = 0$

$$\frac{\partial z}{\partial x} = -\frac{F_x(x, y, z)}{F_z(x, y, z)} = \frac{-2x}{2y + 2z} = \frac{-x}{y + z}$$

$$\frac{\partial z}{\partial y} = -\frac{F_y(x, y, z)}{F_z(x, y, z)} = \frac{-2z}{2y + 2z} = \frac{-z}{y + z}$$

29. $F(x, y, z) = \tan(x + y) + \tan(y + z) - 1$

$$F_x = \sec^2(x + y)$$

$$F_y = \sec^2(x + y) + \sec^2(y + z)$$

$$F_z = \sec^2(y + z)$$

$$\frac{\partial z}{\partial x} = -\frac{F_x}{F_z} = -\frac{\sec^2(x + y)}{\sec^2(y + z)}$$

$$\frac{\partial z}{\partial y} = -\frac{F_y}{F_z} = -\frac{\sec^2(x + y) + \sec^2(y + z)}{\sec^2(y + z)}$$

$$= -\left(\frac{\sec^2(x + y)}{\sec^2(y + z)} + 1\right)$$

31. $F(x, y, z) = e^{xz} + xy = 0$

$$\frac{\partial z}{\partial x} = -\frac{F_x(x, y, z)}{F_z(x, y, z)} = -\frac{ze^{xz} + y}{xe^{xz}}$$

$$\frac{\partial z}{\partial y} = -\frac{F_y(x, y, z)}{F_z(x, y, z)} = \frac{-x}{xe^{xz}} = \frac{-1}{e^{xz}} = -e^{-xz}$$

33. $F(x, y, z, w) = xy + yz - wz + wx - s$

$$F_x = y + w$$

$$F_y = x + z$$

$$F_z = y - w$$

$$F_w = -z + x$$

$$\frac{\partial w}{\partial x} = -\frac{F_x}{F_w} = -\frac{y + w}{-z + x} = \frac{y + w}{z - x}$$

$$\frac{\partial w}{\partial y} = -\frac{F_y}{F_w} = -\frac{x + z}{-z + x} = \frac{x + z}{z - x}$$

$$\frac{\partial w}{\partial z} = -\frac{F_z}{F_w} = -\frac{y - w}{-z + x} = \frac{y - w}{z - x}$$

35. $F(x, y, z, w) = \cos xy + \sin yz + wz - 20$

$$\frac{\partial w}{\partial x} = \frac{-F_x}{F_w} = \frac{y \sin xy}{z}$$

$$\frac{\partial w}{\partial y} = \frac{-F_y}{F_w} = \frac{x \sin xy - z \cos yz}{z}$$

$$\frac{\partial w}{\partial z} = \frac{-F_z}{F_w} = \frac{y \cos zy + w}{z}$$

37. (a) $f(x, y) = \dfrac{xy}{\sqrt{x^2 + y^2}}$

$$f(tx, ty) = \dfrac{(tx)(ty)}{\sqrt{(tx)^2 + (ty)^2}} = t\left(\dfrac{xy}{\sqrt{x^2 + y^2}}\right) = tf(x, y)$$

Degree: 1

(b) $xf_x(x, y) + yf_y(x, y) = x\left(\dfrac{y^3}{\left(x^2 + y^2\right)^{3/2}}\right) + y\left(\dfrac{x^3}{\left(x^2 + y^2\right)^{3/2}}\right) = \dfrac{xy}{\sqrt{x^2 + y^2}} = 1f(x, y)$

39. (a) $f(x, y) = e^{x/y}$

$$f(tx, ty) = e^{tx/ty} = e^{x/y} = f(x, y)$$

Degree: 0

(b) $xf_x(x, y) + yf_y(x, y) = x\left(\dfrac{1}{y}e^{x/y}\right) + y\left(-\dfrac{x}{y^2}e^{x/y}\right) = 0$

41. $\dfrac{dw}{dt} = \dfrac{\partial w}{\partial x}\dfrac{dx}{dt} + \dfrac{\partial w}{\partial y}\dfrac{dy}{dt} = \dfrac{\partial f}{\partial x}\dfrac{dg}{dt} + \dfrac{\partial f}{\partial y}\dfrac{dh}{dt}$

At $t = 2$, $x = 4$, $y = 3$, $f_x(4, 3) = -5$ and

$f_y(4, 3) = 7$.

So, $\dfrac{dw}{dt} = (-5)(-1) + (7)(6) = 47$

43. $\dfrac{dw}{dt} = \dfrac{\partial w}{\partial x}\dfrac{dx}{dt} + \dfrac{\partial w}{\partial y}\dfrac{dy}{dt}$ (Page 907)

45. $\dfrac{dy}{dx} = -\dfrac{f_x(x, y)}{f_y(x, y)}$

$\dfrac{\partial z}{\partial x} = -\dfrac{f_x(x, y, z)}{f_z(x, y, z)}$

$\dfrac{\partial z}{\partial y} = -\dfrac{f_y(x, y, z)}{f_z(x, y, z)}$ (page 912)

47. $V = \pi r^2 h$

$\dfrac{dV}{dt} = \pi\left(2rh\dfrac{dr}{dt} + r^2\dfrac{dh}{dt}\right) = \pi r\left(2h\dfrac{dr}{dt} + r\dfrac{dh}{dt}\right) = \pi(12)\left[2(36)(6) + 12(-4)\right] = 4608\pi \text{ in.}^3/\text{min}$

$S = 2\pi r(r + h)$

$\dfrac{dS}{dt} = 2\pi\left[(2r + h)\dfrac{dr}{dt} + r\dfrac{dh}{dt}\right] = 2\pi\left[(24 + 36)(6) + 12(-4)\right] = 624\pi \text{ in.}^2/\text{min}$

49. $I = \dfrac{1}{2}m\left(r_1^2 + r_2^2\right)$

$\dfrac{dI}{dt} = \dfrac{1}{2}m\left[2r_1\dfrac{dr_1}{dt} + 2r_2\dfrac{dr_2}{dt}\right] = m\left[(6)(2) + (8)(2)\right] = 28m \text{ cm}^2/\text{sec}$

51. $w = f(x, y)$

$x = u - v$

$y = v - u$

$\dfrac{\partial w}{\partial u} = \dfrac{\partial w}{\partial x}\dfrac{dx}{du} + \dfrac{\partial w}{\partial y}\dfrac{dy}{du} = \dfrac{\partial w}{\partial x} - \dfrac{\partial w}{\partial y}$

$\dfrac{\partial w}{\partial v} = \dfrac{\partial w}{\partial x}\dfrac{dx}{dv} + \dfrac{\partial w}{\partial y}\dfrac{dy}{dv} = -\dfrac{\partial w}{\partial x} + \dfrac{\partial w}{\partial y}$

$\dfrac{\partial w}{\partial u} + \dfrac{\partial w}{\partial v} = 0$

53. Given $\dfrac{\partial u}{\partial x} = \dfrac{\partial v}{\partial y}$ and $\dfrac{\partial u}{\partial y} = -\dfrac{\partial v}{\partial x}$, $x = r\cos\theta$ and $y = r\sin\theta$.

$$\frac{\partial u}{\partial r} = \frac{\partial u}{\partial x}\cos\theta + \frac{\partial u}{\partial y}\sin\theta = \frac{\partial v}{\partial y}\cos\theta - \frac{\partial v}{\partial x}\sin\theta$$

$$\frac{\partial v}{\partial \theta} = \frac{\partial v}{\partial x}(-r\sin\theta) + \frac{\partial v}{\partial y}(r\cos\theta) = r\left[\frac{\partial v}{\partial y}\cos\theta - \frac{\partial v}{\partial x}\sin\theta\right]$$

So, $\dfrac{\partial u}{\partial r} = \dfrac{1}{r}\dfrac{\partial v}{\partial \theta}$.

$$\frac{\partial v}{\partial r} = \frac{\partial v}{\partial x}\cos\theta + \frac{\partial v}{\partial y}\sin\theta = -\frac{\partial u}{\partial y}\cos\theta + \frac{\partial u}{\partial x}\sin\theta$$

$$\frac{\partial u}{\partial \theta} = \frac{\partial u}{\partial x}(-r\sin\theta) + \frac{\partial u}{\partial y}(r\cos\theta) = -r\left[-\frac{\partial u}{\partial y}\cos\theta + \frac{\partial u}{\partial x}\sin\theta\right]$$

So, $\dfrac{\partial v}{\partial r} = -\dfrac{1}{r}\dfrac{\partial u}{\partial \theta}$.

55. $g(t) = f(xt, yt) = t^n f(x, y)$

Let $u = xt$, $v = yt$, then

$$g'(t) = \frac{\partial f}{\partial u}\cdot\frac{du}{dt} + \frac{\partial f}{\partial v}\cdot\frac{dv}{dt} = \frac{\partial f}{\partial u}x + \frac{\partial f}{\partial v}y$$

and $g'(t) = nt^{n-1}f(x, y)$.

Now, let $t = 1$ and we have $u = x$, $v = y$. Thus,

$$\frac{\partial f}{\partial x}x + \frac{\partial f}{\partial y}y = nf(x, y).$$

Section 13.6 Directional Derivatives and Gradients

1. $f(x, y) = x^2 + y^2$, $P(1, -2)$, $\theta = \pi/4$

$D_{\mathbf{u}}f(x, y) = f_x(x, y)\cos\theta + f_y(x, y)\sin\theta$

$\qquad\qquad = 2x\cos\theta + 2y\sin\theta$

At $\theta = \pi/4$, $x = 1$, and $y = -2$,

$D_{\mathbf{u}}f(1, -2) = 2(1)\cos\pi/4 + 2(-2)\sin\pi/4$

$\qquad\qquad = \sqrt{2} - 2\sqrt{2} = -\sqrt{2}$.

3. $f(x, y) = \sin(2x + y)$, $P(0, 0)$, $\theta = \pi/3$

$D_{\mathbf{u}}f(x, y) = f_x(x, y)\cos\theta + f_y(x, y)\sin\theta$

$\qquad\qquad = 2\cos(2x + y)\cos\theta + \cos(2x + y)\sin\theta$

At $\theta = \pi/3$ and $x = y = 0$,

$D_{\mathbf{u}}f(0, 0) = 2\cos\pi/3 + \sin\pi/3 = 1 + \sqrt{3}/2$.

5. $f(x, y) = 3x - 4xy + 9y$, $P(1, 2)$, $\mathbf{v} = \dfrac{3}{5}\mathbf{i} + \dfrac{4}{5}\mathbf{j}$

$\mathbf{u} = \dfrac{\mathbf{v}}{\|\mathbf{v}\|} = \dfrac{3}{5}\mathbf{i} + \dfrac{4}{5}\mathbf{j} = \cos\theta\,\mathbf{i} + \sin\theta\,\mathbf{j}$

$D_{\mathbf{u}}f(x, y) = (3 - 4y)\cos\theta + (-4x + 9)\sin\theta$

$D_{\mathbf{u}}f(1, 2) = (3 - 4(2))\dfrac{3}{5} + (-4(1) + 9)\dfrac{4}{5}$

$\qquad\qquad = -3 + 4 = 1$

7. $g(x, y) = \sqrt{x^2 + y^2}$, $P(3, 4)$, $\mathbf{v} = 3\mathbf{i} - 4\mathbf{j}$

$\mathbf{u} = \dfrac{\mathbf{v}}{\|\mathbf{v}\|} = \dfrac{3}{5}\mathbf{i} - \dfrac{4}{5}\mathbf{j}$

$D_{\mathbf{u}}g(x, y) = \dfrac{x}{\sqrt{x^2 + y^2}}\left(\dfrac{3}{5}\right) + \dfrac{y}{\sqrt{x^2 + y^2}}\left(-\dfrac{4}{5}\right)$

$D_{\mathbf{u}}g(3, 4) = \dfrac{3}{5}\left(\dfrac{3}{5}\right) + \dfrac{4}{5}\left(-\dfrac{4}{5}\right) = -\dfrac{7}{25}$

9. $f(x, y) = x^2 + 3y^2$, $P(1, 1)$, $Q(4, 5)$

$$\mathbf{v} = (4 - 1)\mathbf{i} + (5 - 1)\mathbf{j} = 3\mathbf{i} + 4\mathbf{j}$$

$$\mathbf{u} = \frac{\mathbf{v}}{\|\mathbf{v}\|} = \frac{3}{5}\mathbf{i} + \frac{4}{5}\mathbf{j}$$

$$D_{\mathbf{u}} f(x, y) = 2x\left(\frac{3}{5}\right) + 6y\left(\frac{4}{5}\right)$$

$$D_{\mathbf{u}} f(1, 1) = 2\left(\frac{3}{5}\right) + 6\left(\frac{4}{5}\right) = 6$$

11. $f(x, y) = e^y \sin x$, $P(0, 0)$, $Q(2, 1)$

$$\mathbf{v} = (2 - 0)\mathbf{i} + (1 - 0)\mathbf{j}$$

$$\mathbf{v} = 2\mathbf{i} + \mathbf{j}, \mathbf{u} = \frac{\mathbf{v}}{\|\mathbf{v}\|} = \frac{2}{\sqrt{5}}\mathbf{i} + \frac{1}{\sqrt{5}}\mathbf{j}$$

$$D_{\mathbf{u}} f(x, y) = e^y \cos x\left(\frac{2}{\sqrt{5}}\right) + e^y \sin x\left(\frac{1}{\sqrt{5}}\right)$$

$$D_{\mathbf{u}} f(0, 0) = \frac{2}{\sqrt{5}} = \frac{2\sqrt{5}}{5}$$

13. $f(x, y) = 3x + 5y^2 + 1$

$$\nabla f(x, y) = 3\mathbf{i} + 10y\mathbf{j}$$

$$\nabla f(2, 1) = 3\mathbf{i} + 10\mathbf{j}$$

15. $z = \ln(x^2 - y)$

$$\nabla z(x, y) = \frac{2x}{x^2 - y}\mathbf{i} - \frac{1}{x^2 - y}\mathbf{j}$$

$$\nabla z(2, 3) = 4\mathbf{i} - \mathbf{j}$$

17. $w = 3x^2 - 5y^2 + 2z^2$

$$\nabla w(x, y, z) = 6x\mathbf{i} - 10y\mathbf{j} + 4z\mathbf{k}$$

$$\nabla w(1, 1, -2) = 6\mathbf{i} - 10\mathbf{j} - 8\mathbf{k}$$

19. $f(x, y) = xy$

$$\mathbf{v} = \frac{1}{2}(\mathbf{i} + \sqrt{3}\mathbf{j})$$

$$\nabla f(x, y) = y\mathbf{i} + x\mathbf{j}$$

$$\nabla f(0, -2) = -2\mathbf{i}$$

$$\mathbf{u} = \frac{\mathbf{v}}{\|\mathbf{v}\|} = \frac{1}{2}\mathbf{i} + \frac{\sqrt{3}}{2}\mathbf{j}$$

$$D_{\mathbf{u}} f(0, -2) = \nabla f(0, -2) \cdot \mathbf{u} = -1$$

21. $f(x, y, z) = x^2 + y^2 + z^2$

$$\mathbf{v} = \frac{\sqrt{3}}{3}(\mathbf{i} - \mathbf{j} + \mathbf{k})$$

$$\nabla f(x, y, z) = 2x\mathbf{i} + 2y\mathbf{j} + 2z\mathbf{k}$$

$$\nabla f(1, 1, 1) = 2\mathbf{i} + 2\mathbf{j} + 2\mathbf{k}$$

$$\mathbf{u} = \frac{\mathbf{v}}{\|\mathbf{v}\|} = \frac{\sqrt{3}}{3}\mathbf{i} - \frac{\sqrt{3}}{3}\mathbf{j} + \frac{\sqrt{3}}{3}\mathbf{k}$$

$$D_{\mathbf{u}} f(1, 1, 1) = \nabla f(1, 1, 1) \cdot \mathbf{u} = \frac{2}{3}\sqrt{3}$$

23. $\overrightarrow{PQ} = \mathbf{i} + \mathbf{j}$, $\mathbf{u} = \frac{\sqrt{2}}{2}\mathbf{i} + \frac{\sqrt{2}}{2}\mathbf{j}$

$$\nabla g(x, y) = 2x\mathbf{i} + 2y\mathbf{j}, \nabla g(1, 2) = 2\mathbf{i} + 4\mathbf{j}$$

$$D_{\mathbf{u}} g = \nabla g \cdot \mathbf{u} = \sqrt{2} + 2\sqrt{2} = 3\sqrt{2}$$

25. $g(x, y, z) = xye^z$

$$\mathbf{v} = -2\mathbf{i} - 4\mathbf{j}$$

$$\nabla g = ye^z\mathbf{i} + xe^z\mathbf{j} + xye^z\mathbf{k}$$

At $(2, 4, 0)$, $\nabla g = 4\mathbf{i} + 2\mathbf{j} + 8\mathbf{k}$.

$$\mathbf{u} = \frac{\mathbf{v}}{\|\mathbf{v}\|} = -\frac{1}{\sqrt{5}}\mathbf{i} - \frac{2}{\sqrt{5}}\mathbf{j}$$

$$D_{\mathbf{u}} g = \nabla g \cdot \mathbf{u} = -\frac{4}{\sqrt{5}} - \frac{4}{\sqrt{5}} = -\frac{8}{\sqrt{5}}$$

27. $f(x, y) = x^2 + 2xy$

$$\nabla f(x, y) = (2x + 2y)\mathbf{i} + 2x\mathbf{j}$$

$$\nabla f(1, 0) = 2\mathbf{i} + 2\mathbf{j}$$

$$\|\nabla f(1, 0)\| = 2\sqrt{2}$$

29. $h(x, y) = x \tan y$

$$\nabla h(x, y) = \tan y\mathbf{i} + x \sec^2 y\mathbf{j}$$

$$\nabla h\left(2, \frac{\pi}{4}\right) = \mathbf{i} + 4\mathbf{j}$$

$$\left\|\nabla h\left(2, \frac{\pi}{4}\right)\right\| = \sqrt{17}$$

31. $g(x, y) = ye^{-x}$

$$\nabla g(x, y) = -ye^{-x}\mathbf{i} + e^{-x}\mathbf{j}$$

$$\nabla g(0, 5) = -5\mathbf{i} + \mathbf{j}$$

$$\|\nabla g(0, 5)\| = \sqrt{26}$$

33. $f(x, y, z) = \sqrt{x^2 + y^2 + z^2}$

$\nabla f(x, y, z) = \dfrac{1}{\sqrt{x^2 + y^2 + z^2}}(x\mathbf{i} + y\mathbf{j} + z\mathbf{k})$

$\nabla f(1, 4, 2) = \dfrac{1}{\sqrt{21}}(\mathbf{i} + 4\mathbf{j} + 2\mathbf{k})$

$\|\nabla f(1, 4, 2)\| = 1$

35. $w = xy^2z^2$

$\nabla w = y^2z^2\mathbf{i} + 2xyz^2\mathbf{j} + 2xy^2z\mathbf{k}$

$\nabla w(2, 1, 1) = \mathbf{i} + 4\mathbf{j} + 4\mathbf{k}$

$\|\nabla w(2, 1, 1)\| = \sqrt{33}$

For exercises 37–41, $f(x, y) = 3 - \dfrac{x}{3} - \dfrac{y}{2}$ **and**

$D_\mathbf{u} f(x, y) = -\left(\dfrac{1}{3}\right)\cos\theta - \left(\dfrac{1}{2}\right)\sin\theta.$

37. $f(x, y) = 3 - \dfrac{x}{3} - \dfrac{y}{2}$

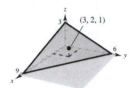

39. (a) $\mathbf{u} = \left(\dfrac{1}{\sqrt{2}}\right)(\mathbf{i} + \mathbf{j})$

$D_\mathbf{u} f = \nabla f \cdot \mathbf{u} = -\left(\dfrac{1}{3}\right)\dfrac{1}{\sqrt{2}} - \left(\dfrac{1}{2}\right)\dfrac{1}{\sqrt{2}} = -\dfrac{5\sqrt{2}}{12}$

(b) $\mathbf{v} = -3\mathbf{i} - 4\mathbf{j}$

$\|\mathbf{v}\| = \sqrt{9 + 16} = 5$

$\mathbf{u} = -\dfrac{3}{5}\mathbf{i} - \dfrac{4}{5}\mathbf{j}$

$D_\mathbf{u} f = \nabla f \cdot \mathbf{u} = \dfrac{1}{5} + \dfrac{2}{5} = \dfrac{3}{5}$

(c) $\mathbf{v} - 3\mathbf{i} + 4\mathbf{j}$

$\|\mathbf{v}\| = \sqrt{9 + 16} = 5$

$\mathbf{u} = -\dfrac{3}{5}\mathbf{i} + \dfrac{4}{5}\mathbf{j}$

$D_\mathbf{u} f = \nabla f \cdot \mathbf{u} = \dfrac{1}{5} - \dfrac{2}{5} = -\dfrac{1}{5}$

(d) $\mathbf{v} = \mathbf{i} + 3\mathbf{j}$

$\|\mathbf{v}\| = \sqrt{10}$

$\mathbf{u} = \dfrac{1}{\sqrt{10}}\mathbf{i} + \dfrac{3}{\sqrt{10}}\mathbf{j}$

$D_\mathbf{u} f = \nabla f \cdot \mathbf{u} = \dfrac{-11}{6\sqrt{10}} = -\dfrac{11\sqrt{10}}{60}$

41. $\|\nabla f\| = \sqrt{\dfrac{1}{9} + \dfrac{1}{4}} = \dfrac{1}{6}\sqrt{13}$

43. (a) In the direction of the vector $-4\mathbf{i} + \mathbf{j}$

(b) $\nabla f = \dfrac{1}{10}(2x - 3y)\mathbf{i} + \dfrac{1}{10}(-3x + 2y)\mathbf{j}$

$\nabla f(1, 2) = \dfrac{1}{10}(-4)\mathbf{i} + \dfrac{1}{10}(1)\mathbf{j} = -\dfrac{2}{5}\mathbf{i} + \dfrac{1}{10}\mathbf{j}$

(Same direction as in part (a))

(c) $-\nabla f = \dfrac{2}{5}\mathbf{i} - \dfrac{1}{10}\mathbf{j}$, the direction opposite that of the gradient

45. $f(x, y) = x^2 - y^2, (4, -3, 7)$

(a)

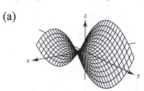

(b) $D_\mathbf{u} f(x, y) = \nabla f(x, y) \cdot \mathbf{u} = 2x\cos\theta - 2y\sin\theta$

$D_\mathbf{u} f(4, -3) = 8\cos\theta + 6\sin\theta$

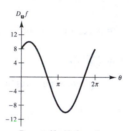

Generated by Mathematica

(c) Zeros: $\theta \approx 2.21, 5.36$

These are the angles θ for which $D_\mathbf{u} f(4, 3)$ equals zero.

(d) $g(\theta) = D_\mathbf{u} f(4, -3) = 8\cos\theta + 6\sin\theta$

$g'(\theta) = -8\sin\theta + 6\cos\theta$

Critical numbers: $\theta \approx 0.64, 3.79$

These are the angels for which $D_\mathbf{u} f(4, -3)$ is a maximum (0.64) and minimum (3.79).

(e) $\|\nabla f(4, -3)\| = \|2(4)\mathbf{i} - 2(-3)\mathbf{j}\| = \sqrt{64 + 36} = 10$, the maximum value of $D_\mathbf{u} f(4, -3)$, at $\theta \approx 0.64$.

(f) $f(x, y) = x^2 - y^2 = 7$

$\nabla f(4, -3) = 8\mathbf{i} + 6\mathbf{j}$ is perpendicular to the level curve at $(4, -3)$.

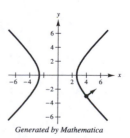

Generated by Mathematica

47. $f(x, y) = 6 - 2x - 3y$

$c = 6, P = (0, 0)$

$\nabla f(x, y) = -2\mathbf{i} - 3\mathbf{j}$

$6 - 2x - 3y = 6$

$\qquad 0 = 2x + 3y$

$\nabla f(0, 0) = -2\mathbf{i} - 3\mathbf{j}$

49. $f(x, y) = xy$

$c = -3, P = (-1, 3)$

$\nabla f(x, y) = y\mathbf{i} + x\mathbf{j}$

$xy = -3$

$\nabla f(-1, 3) = 3\mathbf{i} - \mathbf{j}$

51. $f(x, y) = 4x^2 - y$

(a) $\nabla f(x, y) = 8x\mathbf{i} - \mathbf{j}$

$\nabla f(2, 10) = 16\mathbf{i} - \mathbf{j}$

(b) $\|16\mathbf{i} - \mathbf{j}\| = \sqrt{257}$

$\dfrac{1}{\sqrt{257}}(16\mathbf{i} - \mathbf{j})$ is a unit vector normal to the level

curve $4x^2 - y = 6$ at $(2, 10)$.

(c) The vector $\mathbf{i} + 16\mathbf{j}$ is tangent to the level curve.

Slope $= \dfrac{16}{1} = 16$

$y - 10 = 16(x - 2)$

$\qquad y = 16x - 22$ Tangent line

(d)

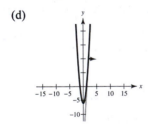

53. $f(x, y) = 3x^2 - 2y^2$

(a) $\nabla f = 6x\mathbf{i} - 4y\mathbf{j}$

$\nabla f(1, 1) = 6\mathbf{i} - 4\mathbf{j}$

(b) $\|\nabla f(1, 1)\| = \sqrt{36 + 16} = 2\sqrt{13}$

$\dfrac{1}{\sqrt{13}}(3\mathbf{i} - 2\mathbf{j})$ is a unit vector normal to the level

curve $3x^2 - 2y^2 = 1$ at $(1, 1)$.

(c) The vector $2\mathbf{i} + 3\mathbf{j}$ is tangent to the level curve.

Slope $= \dfrac{3}{2}$.

$y - 1 = \frac{3}{2}(x - 1)$

$\qquad y = \frac{3}{2}x - \frac{1}{2}$ tangent line

(d)

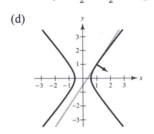

55. See the definition, page 916.

57. See the definition, pages 918 and 919.

59. The gradient vector is normal to the level curves. See Theorem 13.12.

61. $h(x, y) = 5000 - 0.001x^2 - 0.004y^2$

$\nabla h = -0.002x\mathbf{i} - 0.008y\mathbf{j}$

$\nabla h(500, 300) = -\mathbf{i} - 2.4\mathbf{j}$ or

$5\nabla h = -(5\mathbf{i} + 12\mathbf{j})$

63. $T = \dfrac{x}{x^2 + y^2}$

$\nabla T = \dfrac{y^2 - x^2}{\left(x^2 + y^2\right)^2}\mathbf{i} - \dfrac{2xy}{\left(x^2 + y^2\right)^2}\mathbf{j}$

$\nabla T(3, 4) = \dfrac{7}{625}\mathbf{i} - \dfrac{24}{625}\mathbf{j} = \dfrac{1}{625}(7\mathbf{i} - 24\mathbf{j})$

65. $T(x, y) = 80 - 3x^2 - y^2, P(-1, 5)$

$\nabla T(x, y) = -6x\mathbf{i} - 2y\mathbf{j}$

Maximum increase in direction:

$\nabla T(-1, 5) = (-6)(-1)\mathbf{i} - 2(5)\mathbf{j} = 6\mathbf{i} - 10\mathbf{j}$

Maximum rate:

$\|\nabla T(-1, 5)\| = \sqrt{6^2 + (-10)^2} = 2\sqrt{34}$

$\approx 11.66°$ per centimeter

67. $T(x, y) = 400 - 2x^2 - y^2$, $P = (10, 10)$

$$\frac{dx}{dt} = -4x \qquad\qquad \frac{dy}{dt} = -2y$$

$$x(t) = C_1 e^{-4t} \qquad\qquad y(t) = C_2 e^{-2t}$$

$$10 = x(0) = C_1 \qquad 10 = y(0) = C_2$$

$$x(t) = 10e^{-4t} \qquad\qquad y(t) = 10e^{-2t}$$

$$x = \frac{y^2}{10} \qquad\qquad y^2(t) = 100e^{-4t}$$

$$y^2 = 10x$$

69. True

71. True

73. Let $f(x, y, z) = e^x \cos y + \dfrac{z^2}{2} + C$. Then

$$\nabla f(x, y, z) = e^x \cos y\mathbf{i} - e^x \sin y\mathbf{j} + z\mathbf{k}.$$

75. (a) $f(x, y) = \sqrt[3]{xy}$ is the composition of two continuous functions, $h(x, y) = xy$ and $g(z) = z^{1/3}$, and therefore continuous by Theorem 13.2.

(b) $f_x(0, 0) = \lim\limits_{\Delta x \to 0} \dfrac{f(0 + \Delta x, 0) - f(0, 0)}{\Delta x} = \lim\limits_{\Delta x \to 0} \dfrac{(0 \cdot \Delta x)^{1/3} - 0}{\Delta x} = 0$

$f_y(0, 0) = \lim\limits_{\Delta y \to 0} \dfrac{f(0, 0 + \Delta y) - f(0, 0)}{\Delta y} = \lim\limits_{\Delta y \to 0} \dfrac{(0 \cdot \Delta y)^{1/3} - 0}{\Delta y} = 0$

Let $\mathbf{u} = \cos\theta i + \sin\theta\mathbf{j}$, $\theta \neq 0, \dfrac{\pi}{2}, \pi, \dfrac{3\pi}{2}$. Then

$$D_{\mathbf{u}}f(0, 0) = \lim_{t \to 0} \frac{f(0 + t\cos\theta, 0 + t\sin\theta) - f(0, 0)}{t} = \lim_{t \to 0} \frac{\sqrt[3]{t^2 \cos\theta \sin\theta}}{t} = \lim_{t \to 0} \frac{\sqrt[3]{\cos\theta \sin\theta}}{t^{1/3}}, \text{ does not exist.}$$

(c)

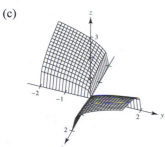

Section 13.7   Tangent Planes and Normal Lines

1. $F(x, y, z) = 3x - 5y + 3z - 15 = 0$

$3x - 5y + 3z = 15$ Plane

3. $F(x, y, z) = 4x^2 + 9y^2 - 4z^2 = 0$

$4x^2 + 9y^2 = 4z^2$ Elliptic cone

5. $F(x, y, z) = 3x + 4y + 12z = 0$

$\nabla F = 3\mathbf{i} + 4\mathbf{j} + 12\mathbf{k}$, $\|\nabla F\| = \sqrt{9 + 16 + 144} = 13$

$\mathbf{n} = \dfrac{\nabla F}{\|\nabla F\|} = \dfrac{3}{13}\mathbf{i} + \dfrac{4}{13}\mathbf{j} + \dfrac{12}{13}\mathbf{k}$

7. $F(x, y, z) = x^2 + 3y + z^3 - 9$

$\nabla F(x, y, z) = 2x\mathbf{i} + 3\mathbf{j} + 3z^2\mathbf{k}$

$\nabla F(2, -1, 2) = 4\mathbf{i} + 3\mathbf{j} + 12\mathbf{k}$

$\mathbf{n} = \dfrac{\nabla F}{\|\nabla F\|} = \dfrac{1}{13}(4\mathbf{i} + 3\mathbf{j} + 12\mathbf{k})$

9. $z = x^2 + y^2 + 3, (2, 1, 8)$

$F(x, y, z) = x^2 + y^2 + 3 - z$

$F_x(x, y, z) = 2x \quad F_y(x, y, z) = 2y \quad F_z(x, y, z) = -1$

$F_x(2, 1, 8) = 4 \qquad F_y(2, 1, 8) = 2 \qquad F_z(2, 1, 8) = -1$

$4(x - 2) + 2(y - 1) - 1(z - 8) = 0$

$4x + 2y - z = 2$

11.
$$z = \sqrt{x^2 + y^2}, (3, 4, 5)$$
$$F(x, y, z) = \sqrt{x^2 + y^2} - z$$

$$F_x(x, y, z) = \frac{x}{\sqrt{x^2 + y^2}} \qquad F_y(x, y, z) = \frac{y}{\sqrt{x^2 + y^2}} \qquad F_z(x, y, z) = -1$$

$$F_x(3, 4, 5) = \frac{3}{5} \qquad\qquad F_y(3, 4, 5) = \frac{4}{5} \qquad\qquad F_z(3, 4, 5) = -1$$

$$\frac{3}{5}(x - 3) + \frac{4}{5}(y - 4) - (z - 5) = 0$$
$$3(x - 3) + 4(y - 4) - 5(z - 5) = 0$$
$$3x + 4y - 5z = 0$$

13.
$$g(x, y) = x^2 + y^2, (1, -1, 2)$$
$$G(x, y, z) = x^2 + y^2 - z$$

$$G_x(x, y, z) = 2x \quad G_y(x, y, z) = 2y \quad G_z(x, y, z) = -1$$

$$G_x(1, -1, 2) = 2 \quad G_y(1, -1, 2) = -2 \quad G_z(1, -1, 2) = -1$$

$$2(x - 1) - 2(y + 1) - 1(z - 2) = 0$$
$$2x - 2y - z = 2$$

15.
$$h(x, y) = \ln\sqrt{x^2 + y^2}, (3, 4, \ln 5)$$
$$H(x, y, z) = \ln\sqrt{x^2 + y^2} - z = \frac{1}{2}\ln(x^2 + y^2) - z$$

$$H_x(x, y, z) = \frac{x}{x^2 + y^2} \qquad H_y(x, y, z) = \frac{y}{x^2 + y^2} \qquad H_z(x, y, z) = -1$$

$$H_x(3, 4, \ln 5) = \frac{3}{25} \qquad\qquad H_y(3, 4, \ln 5) = \frac{4}{25} \qquad\qquad H_z(3, 4, \ln 5) = -1$$

$$\frac{3}{25}(x - 3) + \frac{4}{25}(y - 4) - (z - \ln 5) = 0$$
$$3(x - 3) + 4(y - 4) - 25(z - \ln 5) = 0$$
$$3x + 4y - 25z = 25(1 - \ln 5)$$

17. $x^2 + 4y^2 + z^2 = 36, (2, -2, 4)$
$$F(x, y, z) = x^2 + 4y^2 + z^2 - 36$$

$$F_x(x, y, z) = 2x \qquad F_y(x, y, z) = 8y \qquad F_z(x, y, z) = 2z$$
$$F_x(2, -2, 4) = 4 \qquad F_y(2, -2, 4) = -16 \qquad F_z(2, -2, 4) = 8$$
$$4(x - 2) - 16(y + 2) + 8(z - 4) = 0$$
$$(x - 2) - 4(y + 2) + 2(z - 4) = 0$$
$$x - 4y + 2z = 18$$

19. $xy^2 + 3x - z^2 = 8, (1, -3, 2)$

$\quad F(x, y, z) = xy^2 + 3x - z^2 - 8$

$\quad\quad F_x(x, y, z) = y^2 + 3 \quad\quad F_y(x, y, z) = 2xy \quad\quad F_z(x, y, z) = -2z$

$\quad\quad F_x(1, -3, 2) = 12 \quad\quad\quad F_y(1, -3, 2) = -6 \quad\quad F_z(1, -3, 2) = -4$

$\quad\quad 12(x - 1) - 6(y + 3) - 4(z - 2) = 0$

$\quad\quad\quad\quad\quad\quad\quad\quad 12x - 6y - 4z = 22$

$\quad\quad\quad\quad\quad\quad\quad\quad\quad 6x - 3y - 2z = 11$

21. $x + y + z = 9, (3, 3, 3)$

$\quad F(x, y, z) = x + y + z - 9$

$\quad\quad F_x(x, y, z) = 1 \quad F_y(x, y, z) = 1 \quad F_z(x, y, z) = 1$

$\quad\quad F_x(3, 3, 3) = 1 \quad F_y(3, 3, 3) = 1 \quad F_z(3, 3, 3) = 1$

$\quad\quad (x - 3) + (y - 3) + (z - 3) = 0$

$\quad\quad\quad\quad\quad\quad x + y + z = 9 \ \left(\text{same plane!}\right)$

Direction numbers: 1, 1, 1

Line: $x - 3 = y - 3 = z - 3$

23. $x^2 + y^2 + z = 9, (1, 2, 4)$

$\quad F(x, y, z) = x^2 + y^2 + z - 9$

$\quad\quad F_x(x, y, z) = 2x \quad F_y(x, y, z) = 2y \quad F_z(x, y, z) = 1$

$\quad\quad F_x(1, 2, 4) = 2 \quad\quad F_y(1, 2, 4) = 4 \quad\quad F_z(1, 2, 4) = 1$

Direction numbers: 2, 4, 1

Plane: $2(x - 1) + 4(y - 2) + (z - 4) = 0, \ 2x + 4y + z = 14$

Line: $\dfrac{x - 1}{2} = \dfrac{y - 2}{4} = \dfrac{z - 4}{1}$

25. $z = x^2 - y^2, (3, 2, 5)$

$\quad F(x, y, z) = x^2 - y^2 - z$

$\quad\quad F_x(x, y, z) = 2x \quad F_y(x, y, z) = -2y \quad F_z(x, y, z) = -1$

$\quad\quad F_x(3, 2, 5) = 6 \quad\quad F_y(3, 2, 5) = -4 \quad\quad F_z(3, 2, 5) = -1$

$\quad\quad 6(x - 3) - 4(y - 2) - (z - 5) = 0$

$\quad\quad\quad\quad\quad\quad\quad\quad 6x - 4y - z = 5$

Direction numbers: 6, -4, -1

Line: $\dfrac{x - 3}{6} = \dfrac{y - 2}{-4} = \dfrac{z - 5}{-1}$

27. $xyz = 10, (1, 2, 5)$

$\quad F(x, y, z) = xyz - 10$

$\quad\quad F_x(x, y, z) = yz \quad F_y(x, y, z) = xz \quad F_z(x, y, z) = xy$

$\quad\quad F_x(1, 2, 5) = 10 \quad F_y(1, 2, 5) = 5 \quad\quad F_z(1, 2, 5) = 2$

Direction numbers: 10, 5, 2

Plane: $10(x - 1) + 5(y - 2) + 2(z - 5) = 0, \ 10x + 5y + 2z = 30$

Line: $\dfrac{x - 1}{10} = \dfrac{y - 2}{5} = \dfrac{z - 5}{2}$

29. $z = \arctan\dfrac{y}{x}, \left(1, 1, \dfrac{\pi}{4}\right)$

$F(x, y, z) = \arctan\dfrac{y}{x} - z$

$F_x(x, y, z) = \dfrac{-y}{x^2 + y^2}$ $F_y(x, y, z) = \dfrac{x}{x^2 + y^2}$ $F_z(x, y, z) = -1$

$F_x\left(1, 1, \dfrac{\pi}{4}\right) = -\dfrac{1}{2}$ $F_y\left(1, 1, \dfrac{\pi}{4}\right) = \dfrac{1}{2}$ $F_z\left(1, 1, \dfrac{\pi}{4}\right) = -1$

Direction numbers: 1, –1, 2

Plane: $(x - 1) - (y - 1) + 2\left(z - \dfrac{\pi}{4}\right) = 0, \ x - y + 2z = \dfrac{\pi}{2}$

Line: $\dfrac{x - 1}{1} = \dfrac{y - 1}{-1} = \dfrac{z - (\pi/4)}{2}$

31. $F(x, y, z) = x^2 + y^2 - 2$ $G(x, y, z) = x - z$

$\nabla F(x, y, z) = 2x\mathbf{i} + 2y\mathbf{j}$ $\nabla G(x, y, z) = \mathbf{i} - \mathbf{k}$

$\nabla F(1, 1, 1) = 2\mathbf{i} + 2\mathbf{j}$ $\nabla G(1, 1, 1) = \mathbf{i} - \mathbf{k}$

(a) $\nabla F \times \nabla G = \begin{vmatrix} \mathbf{i} & \mathbf{j} & \mathbf{k} \\ 2 & 2 & 0 \\ 1 & 0 & -1 \end{vmatrix} = -2\mathbf{i} + 2\mathbf{j} - 2\mathbf{k} = -2(\mathbf{i} - \mathbf{j} + \mathbf{k})$

Direction numbers: 1, –1, 1

Line: $x - 1 = \dfrac{y - 1}{-1} = z - 1$

(b) $\cos\theta = \dfrac{|\nabla F \cdot \nabla G|}{\|\nabla F\|\|\nabla G\|} = \dfrac{2}{(2\sqrt{2})\sqrt{2}} = \dfrac{1}{2}$

Not orthogonal

33. $F(x, y, z) = x^2 + z^2 - 25$ $G(x, y, z) = y^2 + z^2 - 25$

$\nabla F = 2x\mathbf{i} + 2z\mathbf{k}$ $\nabla G = 2y\mathbf{j} + 2z\mathbf{k}$

$\nabla F(3, 3, 4) = 6\mathbf{i} + 8\mathbf{k}$ $\nabla G(3, 3, 4) = 6\mathbf{j} + 8\mathbf{k}$

(a) $\nabla F \times \nabla G = \begin{vmatrix} \mathbf{i} & \mathbf{j} & \mathbf{k} \\ 6 & 0 & 8 \\ 0 & 6 & 8 \end{vmatrix} = -48\mathbf{i} - 48\mathbf{j} + 36\mathbf{k} = -12(4\mathbf{i} + 4\mathbf{j} - 3\mathbf{k})$

Direction numbers: 4, 4, –3. $\dfrac{x - 3}{4} = \dfrac{y - 3}{4} = \dfrac{z - 4}{-3}$

(b) $\cos\theta = \dfrac{|\nabla F \cdot \nabla G|}{\|\nabla F\|\|\nabla G\|} = \dfrac{64}{(10)(10)} = \dfrac{16}{25}$; not orthogonal

35. $F(x, y, z) = x^2 + y^2 + z^2 - 14$ $G(x, y, z) = x - y - z$

$\nabla F(x, y, z) = 2x\mathbf{i} + 2y\mathbf{j} + 2z\mathbf{k}$ $\nabla G(x, y, z) = \mathbf{i} - \mathbf{j} - \mathbf{k}$

$\nabla F(3, 1, 2) = 6\mathbf{i} + 2\mathbf{j} + 4\mathbf{k}$ $\nabla G(3, 1, 2) = \mathbf{i} - \mathbf{j} - \mathbf{k}$

(a) $\nabla F \times \nabla G = \begin{vmatrix} \mathbf{i} & \mathbf{j} & \mathbf{k} \\ 6 & 2 & 4 \\ 1 & -1 & -1 \end{vmatrix} = 2\mathbf{i} + 10\mathbf{j} - 8\mathbf{k} = 2[\mathbf{i} + 5\mathbf{j} - 4\mathbf{k}]$

Direction numbers: $1, 5, -4$

Line: $\dfrac{x-3}{1} = \dfrac{y-1}{5} = \dfrac{z-2}{-4}$

(b) $\cos\theta = \dfrac{|\nabla F \cdot \nabla G|}{\|\nabla F\|\|\nabla G\|} = 0 \Rightarrow$ orthogonal

37. $F(x, y, z) = 3x^2 + 2y^2 - z - 15, (2, 2, 5)$

$\nabla F(x, y, z) = 6x\mathbf{i} + 4y\mathbf{j} - \mathbf{k}$

$\nabla F(2, 2, 5) = 12\mathbf{i} + 8\mathbf{j} - \mathbf{k}$

$\cos\theta = \dfrac{|\nabla F(2, 2, 5) \cdot \mathbf{k}|}{\|\nabla F(2, 2, 5)\|} = \dfrac{1}{\sqrt{209}}$

$\theta = \arccos\left(\dfrac{1}{\sqrt{209}}\right) = 86.03°$

39. $F(x, y, z) = x^2 - y^2 + z, (1, 2, 3)$

$\nabla F(x, y, z) = 2x\mathbf{i} - 2y\mathbf{j} + \mathbf{k}$

$\nabla F(1, 2, 3) = 2\mathbf{i} - 4\mathbf{j} + \mathbf{k}$

$\cos\theta = \dfrac{|\nabla F(1, 2, 3) \cdot \mathbf{k}|}{\|\nabla F(1, 2, 3)\|} = \dfrac{1}{\sqrt{21}}$

$\theta = \arccos\dfrac{1}{\sqrt{21}} \approx 77.40°$

41. $F(x, y, z) = 3 - x^2 - y^2 + 6y - z$

$\nabla F(x, y, z) = -2x\mathbf{i} + (-2y + 6)\mathbf{j} - \mathbf{k}$

$-2x = 0, x = 0$

$-2y + 6 = 0, y = 3$

$z = 3 - 0^2 - 3^2 + 6(3) = 12$

$(0, 3, 12)$ (vertex of paraboloid)

43. $F(x, y, z) = x^2 - xy + y^2 - 2x - 2y - z$

$\nabla F(x, y, z) = (2x - y - 2)\mathbf{i} + (-x + 2y - 2)\mathbf{j} - \mathbf{k}$

$2x - y - 2 = 0$

$-x + 2y - 2 = 0$

$y = 2x - 2 \Rightarrow -x + 2(2x - 2) - 2$

$= 3x - 6 = 0 \Rightarrow x = 2$

$y = 2, z = -4$

Point: $(2, 2, -4)$

45. $F(x, y, z) = 5xy - z$

$\nabla F(x, y, z) = 5y\mathbf{i} + 5x\mathbf{j} - \mathbf{k}$

$5y = 0$

$5x = 0$

$x = y = z = 0$

Point: $(0, 0, 0)$

47. $F(x, y, z) = x^2 + 2y^2 + 3z^2 - 3, (-1, 1, 0)$

$F_x(x, y, z) = 2x \quad F_y(x, y, z) = 4y$

$F_x(-1, 1, 0) = -2 \quad F_y(-1, 1, 0) = 4$

$F_z(x, y, z) = 6z$

$F_z(-1, 1, 0) = 0$

$-2(x + 1) + 4(y - 1) + 0(z - 0) = 0$

$\qquad\qquad -2x + 4y = 6$

$\qquad\qquad\quad -x + 2y = 3$

$G(x, y, z) = x^2 + y^2 + z^2 + 6x - 10y + 14, (-1, 1, 0)$

$G_x(x, y, z) = 2x + 6 \quad G_y(x, y, z) = 2y - 10$

$G_x(-1, 1, 0) = 4 \qquad G_y(-1, 1, 0) = -8$

$G_z(x, y, z) = 2z$

$G_z(-1, 1, 0) = 0$

$4(x + 1) - 8(y - 1) + 0(z - 0) = 0$

$\qquad\qquad 4x - 8y + 12 = 0$

$\qquad\qquad\quad -x + 2y = 3$

The tangent planes are the same.

51. $F(x, y, z) = x^2 + 4y^2 + z^2 - 9$

$\nabla F = 2x\mathbf{i} + 8y\mathbf{j} + 2z\mathbf{k}$

This normal vector is parallel to the line with direction number $-4, 8, -2$.

So, $2x = -4t \Rightarrow x = -2t$

$\qquad 8y = 8t \Rightarrow y = t$

$\qquad 2z = -2t \Rightarrow z = -t$

$x^2 + 4y^2 + z^2 - 9 = 4t^2 + 4t^2 + t^2 - 9 = 0 \Rightarrow t = \pm 1$

There are two points on the ellipse where the tangent plane is perpendicular to the line:

$(-2, 1, -1) \ (t = 1)$

$(2, -1, 1) \ \ (t = -1)$

53. $F_x(x_0, y_0, z_0)(x - x_0) + F_y(x_0, y_0, z_0)(y - y_0) + F_z(x_0, y_0, z_0)(z - z_0) = 0$

(Theorem 13.13)

55. Answers will vary.

49. (a) $F(x, y, z) = 2xy^2 - z, F(1, 1, 2) = 2 - 2 = 0$

$G(x, y, z) = 8x^2 - 5y^2 - 8z + 13,$

$G(1, 1, 2) = 8 - 5 - 16 + 13 = 0$

So, $(1, 1, 2)$ lies on both surfaces.

(b) $\nabla F = 2y^2\mathbf{i} + 4xy\mathbf{j} - \mathbf{k}, \nabla F(1, 1, 2) = 2\mathbf{i} + 4\mathbf{j} - \mathbf{k}$

$\nabla G = 16x\mathbf{i} - 10y\mathbf{j} - 8\mathbf{k},$

$\nabla G(1, 1, 2) = 16\mathbf{i} - 10\mathbf{j} - 8\mathbf{k}$

$\nabla F \cdot \nabla G = 2(16) + 4(-10) + (-1)(-8) = 0$

The tangent planes are perpendicular at $(1, 1, 2)$.

57. $z = f(x, y) = \dfrac{4xy}{(x^2 + 1)(y^2 + 1)},\ -2 \le x \le 2, 0 \le y \le 3$

(a) Let $F(x, y, z) = \dfrac{4xy}{(x^2 + 1)(y^2 + 1)} - z$

$\nabla F(x, y, z) = \dfrac{4y}{y^2 + 1}\left(\dfrac{x^2 + 1 - 2x^2}{(x^2 + 1)^2}\right)\mathbf{i} + \dfrac{4x}{x^2 + 1}\left(\dfrac{y^2 + 1 - 2y^2}{(y^2 + 1)^2}\right)\mathbf{j} - \mathbf{k} = \dfrac{4y(1 - x^2)}{(y^2 + 1)(x^2 + 1)^2}\mathbf{i} + \dfrac{4x(1 - y^2)}{(x^2 + 1)(y^2 + 1)^2}\mathbf{j} - \mathbf{k}$

$\nabla F(1, 1, 1) = -\mathbf{k}$

Direction numbers: $0, 0, -1$

Line: $x = 1, y = 1, z = 1 - t$

Tangent plane: $0(x - 1) + 0(y - 1) - 1(z - 1) = 0 \Rightarrow z = 1$

(b) $\nabla F\left(-1, 2, -\dfrac{4}{5}\right) = 0\mathbf{i} + \dfrac{-4(-3)}{(2)(5)^2}\mathbf{j} - \mathbf{k} = \dfrac{6}{25}\mathbf{j} - \mathbf{k}$

Line: $x = -1, y = 2 + \dfrac{6}{25}t, z = -\dfrac{4}{5} - t$

Plane: $0(x + 1) + \dfrac{6}{25}(y - 2) - 1\left(z + \dfrac{4}{5}\right) = 0$

$6y - 12 - 25z - 20 = 0$

$6y - 25z - 32 = 0$

(c)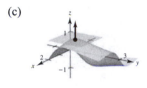

59. $f(x, y) = 6 - x^2 - \dfrac{y^2}{4}, g(x, y) = 2x + y$

(a) $F(x, y, z) = z + x^2 + \dfrac{y^2}{4} - 6 \qquad G(x, y, z) = z - 2x - y$

$\nabla F(x, y, z) = 2x\mathbf{i} + \dfrac{1}{2}y\mathbf{j} + \mathbf{k} \qquad \nabla G(x, y, z) = -2\mathbf{i} - \mathbf{j} + \mathbf{k}$

$\nabla F(1, 2, 4) = 2\mathbf{i} + \mathbf{j} + \mathbf{k} \qquad \nabla G(1, 2, 4) = -2\mathbf{i} - \mathbf{j} + \mathbf{k}$

(b)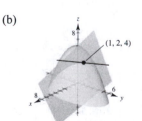

The cross product of these gradients is parallel to the curve of intersection.

$\nabla F(1, 2, 4) \times \nabla G(1, 2, 4) = \begin{vmatrix} \mathbf{i} & \mathbf{j} & \mathbf{k} \\ 2 & 1 & 1 \\ -2 & -1 & 1 \end{vmatrix} = 2\mathbf{i} - 4\mathbf{j}$

Using direction numbers $1, -2, 0$, you get $x = 1 + t, y = 2 - 2t, z = 4$.

$\cos \theta = \dfrac{\nabla F \cdot \nabla G}{\|\nabla F\|\|\nabla G\|} = \dfrac{-4 - 1 + 1}{\sqrt{6}\sqrt{6}} = \dfrac{-4}{6} \Rightarrow \theta \approx 48.2°$

61. $F(x, y, z) = \dfrac{x^2}{a^2} + \dfrac{y^2}{b^2} + \dfrac{z^2}{c^2} - 1$

$F_x(x, y, z) = \dfrac{2x}{a^2}$

$F_y(x, y, z) = \dfrac{2y}{b^2}$

$F_z(x, y, z) = \dfrac{2z}{c^2}$

Plane: $\dfrac{2x_0}{a^2}(x - x_0) + \dfrac{2y_0}{b^2}(y - y_0) + \dfrac{2z_0}{c^2}(z - z_0) = 0$

$\dfrac{x_0 x}{a^2} + \dfrac{y_0 y}{b^2} + \dfrac{z_0 z}{c^2} = \dfrac{x_0^2}{a^2} + \dfrac{y_0^2}{b^2} + \dfrac{z_0^2}{c^2} = 1$

63. $F(x, y, z) = a^2 x^2 + b^2 y^2 - z^2$

$F_x(x, y, z) = 2a^2 x$

$F_y(x, y, z) = 2b^2 y$

$F_z(x, y, z) = -2z$

Plane: $2a^2 x_0(x - x_0) + 2b^2 y_0(y - y_0) - 2z_0(z - z_0) = 0$

$\qquad a^2 x_0 x + b^2 y_0 y - z_0 z = a^2 x_0^2 + b^2 y_0^2 - z_0^2 = 0$

So, the plane passes through the origin.

65. $f(x, y) = e^{x-y}$

$f_x(x, y) = e^{x-y}, \; f_y(x, y) = -e^{x-y}$

$f_{xx}(x, y) = e^{x-y}, \; f_{yy}(x, y) = e^{x-y}, \; f_{xy}(x, y) = -e^{x-y}$

(a) $P_1(x, y) \approx f(0, 0) + f_x(0, 0)x + f_y(0, 0)y = 1 + x - y$

(b) $P_2(x, y) \approx f(0, 0) + f_x(0, 0)x + f_y(0, 0)y + \frac{1}{2}f_{xx}(0, 0)x^2 + f_{xy}(0, 0)xy + \frac{1}{2}f_{yy}(0, 0)y^2 = 1 + x - y + \frac{1}{2}x^2 - xy + \frac{1}{2}y^2$

(c) If $x = 0$, $P_2(0, y) = 1 - y + \frac{1}{2}y^2$. This is the second-degree Taylor polynomial for e^{-y}.

If $y = 0$, $P_2(x, 0) = 1 + x + \frac{1}{2}x^2$. This is the second-degree Taylor polynomial for e^x.

(d)

x	y	$f(x, y)$	$P_1(x, y)$	$P_2(x, y)$
0	0	1	1	1
0	0.1	0.9048	0.9000	0.9050
0.2	0.1	1.1052	1.1000	1.1050
0.2	0.5	0.7408	0.7000	0.7450
1	0.5	1.6487	1.5000	1.6250

(e)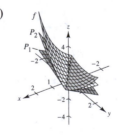

67. Given $z = f(x, y)$, then:

$$F(x, y, z) = f(x, y) - z = 0$$

$$\nabla F(x_0, y_0, z_0) = f_x(x_0, y_0)\mathbf{i} + f_y(x_0, y_0)\mathbf{j} - \mathbf{k}$$

$$\cos\theta = \frac{|\nabla F(x_0, y_0, z_0) \cdot \mathbf{k}|}{\|\nabla F(x_0, y_0, z_0)\|\|\mathbf{k}\|}$$

$$= \frac{|-1|}{\sqrt{[f_x(x_0, y_0)]^2 + [f_y(x_0, y_0)]^2 + (-1)^2}}$$

$$= \frac{1}{\sqrt{[f_x(x_0, y_0)]^2 + [f_y(x_0, y_0)]^2 + 1}}$$

Section 13.8 Extrema of Functions of Two Variables

1. $g(x, y) = (x - 1)^2 + (y - 3)^2 \geq 0$

Relative minimum: $(1, 3, 0)$

Check: $g_x = 2(x - 1) = 0 \Rightarrow x = 1$

$ g_y = 2(y - 3) = 0 \Rightarrow y = 3$

$g_{xx} = 2, g_{yy} = 2, g_{xy} = 0, d = (2)(2) - 0 = 4 > 0$

At critical point $(1, 3)$, $d > 0$ and $g_{xx} > 0 \Rightarrow$ relative minimum at $(1, 3, 0)$.

3. $f(x, y) = \sqrt{x^2 + y^2 + 1} \geq 1$

Relative minimum: $(0, 0, 1)$

Check: $f_x = \dfrac{x}{\sqrt{x^2 + y^2 + 1}} = 0 \Rightarrow x = 0$

$f_y = \dfrac{y}{\sqrt{x^2 + y^2 + 1}} = 0 \Rightarrow y = 0$

$f_{xx} = \dfrac{y^2 + 1}{(x^2 + y^2 + 1)^{3/2}}$

$f_{yy} = \dfrac{x^2 + 1}{(x^2 + y^2 + 1)^{3/2}}$

$f_{xy} = \dfrac{-xy}{(x^2 + y^2 + 1)^{3/2}}$

At the critical point $(0, 0)$, $f_{xx} > 0$ and

$$f_{xx}f_{yy} - (f_{xy})^2 > 0.$$

So, $(0, 0, 1)$ is a relative minimum.

5. $f(x, y) = x^2 + y^2 + 2x - 6y + 6 = (x + 1)^2 + (y - 3)^2 - 4 \geq -4$

Relative minimum: $(-1, 3, -4)$

Check: $f_x = 2x + 2 = 0 \Rightarrow x = -1$

$ f_y = 2y - 6 = 0 \Rightarrow y = 3$

$ f_{xx} = 2, f_{yy} = 2, f_{xy} = 0$

At the critical point $(-1, 3)$, $f_{xx} > 0$ and $f_{xx}f_{yy} - (f_{xy})^2 > 0$. So, $(-1, 3, -4)$ is a relative minimum.

7. $h(x, y) = 80x + 80y - x^2 - y^2$

$\left.\begin{array}{l} h_x = 80 - 2x = 0 \\ h_y = 80 - 2y = 0 \end{array}\right\} x = y = 40$

$h_{xx} = -2, h_{yy} = -2, h_{xy} = 0,$

$d = (-2)(-2) - 0 = 4 > 0$

At the critical point $(40, 40)$, $d > 0$ and

$h_{xx} < 0 \Rightarrow (40, 40, 3200)$ is a relative maximum.

9. $g(x, y) = xy$

$\left.\begin{array}{l} g_x = y \\ g_y = x \end{array}\right\} x = 0$ and $y = 0$

$g_{xx} = 0, g_{yy} = 0, g_{xy} = 1$

At the critical point $(0, 0)$, $g_{xx}g_{yy} - (g_{xy})^2 < 0$.

So, $(0, 0, 0)$ is a saddle point.

11. $f(x, y) = -3x^2 - 2y^2 + 3x - 4y + 5$

$f_x = -6x + 3 = 0$ when $x = \frac{1}{2}$.

$f_y = -4y - 4 = 0$ when $y = -1$.

$f_{xx} = -6, f_{yy} = -4, f_{xy} = 0$

At the critical point $\left(\frac{1}{2}, -1\right)$, $f_{xx} < 0$

and $f_{xx}f_{yy} - \left(f_{xy}\right)^2 > 0$.

So, $\left(\frac{1}{2}, -1, \frac{31}{4}\right)$ is a relative maximum.

13. $f(x, y) = z = x^2 + xy + \frac{1}{2}y^2 - 2x + y$

$\left.\begin{array}{l} f_x = 2x + y - 2 = 0 \\ f_y = x + y + 1 = 0 \end{array}\right\}$ Solving simultaneously yields $x = 3, y = -4$

$f_{xx} = 2, f_{yy} = 1, f_{xy} = 1, d = 2(1) - 1 = 1 > 0$.

At the critical point $(3, -4), d > 0$

and $f_{xx} > 0 \Rightarrow (3, -4, -5)$ is a relative minimum.

15. $f(x, y) = \sqrt{x^2 + y^2}$

$\left.\begin{array}{l} f_x = \dfrac{x}{\sqrt{x^2 + y^2}} = 0 \\[4mm] f_y = \dfrac{y}{\sqrt{x^2 + y^2}} = 0 \end{array}\right\} x = y = 0$

Because $f(x, y) \geq 0$ for all (x, y) and $f(0, 0) = 0, (0, 0, 0)$ is a relative minimum.

17. $f(x, y) = x^2 - xy - y^2 - 3x - y$

$f_x = 2x - y - 3 = 0$

$f_y = -x - 2y - 1 = 0$

Solving simultaneously yields $x = 1, y = -1$.

$f_{xx} = 2, f_{yy} = -2, f_{xy} = -1$

$d = (2)(-2) - (-1)^2 = -5 < 0$

At the critical point $(1, -1), d < 0 \Rightarrow (1, -1, -1)$ is a saddle point.

19. $f(x, y) = e^{-x} \sin y$

$\left.\begin{array}{l} f_x = -e^{-x} \sin y = 0 \\ f_y = e^{-x} \cos y = 0 \end{array}\right\}$ Because $e^{-x} > 0$ for all x and $\sin y$ and $\cos y$ are never both zero for a given value of y, there are no critical points.

21. $z = \dfrac{-4x}{x^2 + y^2 + 1}$

Relative minimum: $(1, 0, -2)$

Relative maximum: $(-1, 0, 2)$

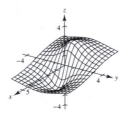

23. $z = \left(x^2 + 4y^2\right)e^{1-x^2-y^2}$

Relative minimum: $(0, 0, 0)$

Relative maxima: $(0, \pm 1, 4)$

Saddle points: $(\pm 1, 0, 1)$

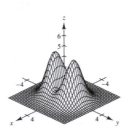

25. $z = \dfrac{(x - y)^4}{x^2 + y^2} \geq 0. z = 0$ if $x = y \neq 0$.

Relative minimum at all points $(x, x), x \neq 0$.

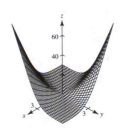

27. $f_{xx}f_{yy} - \left(f_{xy}\right)^2 = (9)(4) - 6^2 = 0$

Insufficient information.

29. $f_{xx}f_{yy} - \left(f_{xy}\right)^2 = (-9)(6) - 10^2 < 0$

f has a saddle point at (x_0, y_0).

31. $d = f_{xx}f_{yy} - f_{xy}^2 = (2)(8) - f_{xy}^2 = 16 - f_{xy}^2 > 0$

$\Rightarrow f_{xy}^2 < 16 \Rightarrow -4 < f_{xy} < 4$

33. $f(x, y) = x^3 + y^3$

(a) $\left. \begin{array}{l} f_x = 3x^2 = 0 \\ f_y = 3y^2 = 0 \end{array} \right\} x = y = 0$

Critical point: $(0, 0)$

(b) $f_{xx} = 6x, f_{yy} = 6y, f_{xy} = 0$

At $(0, 0)$, $f_{xx}f_{yy} - \left(f_{xy}\right)^2 = 0$.

$(0, 0, 0)$ is a saddle point.

(c) Test fails at $(0, 0)$.

(d)

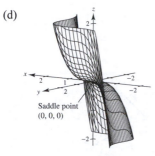

Saddle point
$(0, 0, 0)$

35. $f(x, y) = (x - 1)^2 (y + 4)^2 \ge 0$

(a) $\left. \begin{array}{l} f_x = 2(x - 1)(y + 4)^2 = 0 \\ f_y = 2(x - 1)^2 (y + 4) = 0 \end{array} \right|$ critical points: $(1, a)$ and $(b, -4)$

(b) $f_{xx} = 2(y + 4)^2$

$f_{yy} = 2(x - 1)^2$

$f_{xy} = 4(x - 1)(y + 4)$

At both $(1, a)$ and $(b, -4)$, $f_{xx}f_{yy} - \left(f_{xy}\right)^2 = 0$.

Because $f(x, y) \ge 0$, there are absolute minima at $(1, a, 0)$ and $(b, -4, 0)$.

(c) Test fails at $(1, a)$ and $(b, -4)$.

(d)

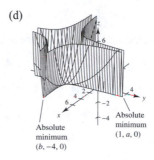

Absolute
minimum
$(b, -4, 0)$

Absolute
minimum
$(1, a, 0)$

37. $f(x, y) = x^{2/3} + y^{2/3} \ge 0$

(a) $\left. \begin{array}{l} f_x = \dfrac{2}{3x^{1/3}} \\ f_y = \dfrac{2}{3y^{1/3}} \end{array} \right|$ f_x and f_y are undefined at $x = 0$ and $y = 0$. Critical point: $(0, 0)$

(b) $f_{xx} = \dfrac{-2}{9x^{4/3}}, f_{yy} = \dfrac{-2}{9y^{4/3}}, f_{xy} = 0$

At $(0, 0)$, $f_{xx}f_{yy} - \left(f_{xy}\right)^2$ is undefined.

$(0, 0, 0)$ is an absolute minimum.

(c) Test fails at $(0, 0)$.

(d)

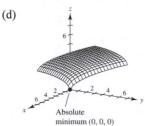

Absolute
minimum $(0, 0, 0)$

39. $f(x, y, z) = x^2 + (y - 3)^2 + (z + 1)^2 \ge 0$

$\left. \begin{array}{l} f_x = 2x = 0 \\ f_y = 2(y - 3) = 0 \\ f_z = 2(z + 1) = 0 \end{array} \right\}$ Solving yields the critical point $(0, 3, -1)$.

Absolute minimum: 0 at $(0, 3, -1)$

41. $f(x, y) = x^2 - 4xy + 5, R = \{(x, y): 1 \leq x \leq 4, 0 \leq y \leq 2\}$

$\left.\begin{array}{l} f_x = 2x - 4y = 0 \\ f_y = -4x = 0 \end{array}\right\} x = y = 0 \quad \text{(not in region R)}$

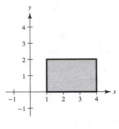

Along $y = 0, 1 \leq x \leq 4: f = x^2 + 5, f(1, 0) = 6, f(4, 0) = 21.$

Along $y = 2, 1 \leq x \leq 4: f = x^2 - 8x + 5, f' = 2x - 8 = 0$

$\qquad f(1, 2) = -2, f(4, 2) = -11.$

Along $x = 1, 0 \leq y \leq 2: f = -4y + 6, f(1, 0) = 6, f(1, 2) = -2.$

Along $x = 4, 0 \leq y \leq 2: f = 21 - 16y, f(4, 0) = 21, f(4, 2) = -11.$

So, the maximum is $(4, 0, 21)$ and the minimum is $(4, 2, -11)$.

43. $f(x, y) = 12 - 3x - 2y$ has no critical points. On the line $y = x + 1, 0 \leq x \leq 1,$

$f(x, y) = f(x) = 12 - 3x - 2(x + 1) = -5x + 10$

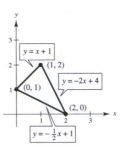

and the maximum is 10, the minimum is 5. On the line $y = -2x + 4, 1 \leq x \leq 2,$

$f(x, y) = f(x) = 12 - 3x - 2(-2x + 4) = x + 4$

and the maximum is 6, the minimum is 5. On the line $y = -\frac{1}{2}x + 1, 0 \leq x \leq 2,$

$f(x, y) = f(x) = 12 - 3x - 2(-\frac{1}{2}x + 1) = -2x + 10$

and the maximum is 10, the minimum is 6.

Absolute maximum: 10 at $(0, 1)$

Absolute minimum: 5 at $(1, 2)$

45. $f(x, y) = 3x^2 + 2y^2 - 4y$

$\left.\begin{array}{l} f_x = 6x = 0 \Rightarrow x = 0 \\ f_y = 4y - 4 = 0 \Rightarrow y = 1 \end{array}\right\} f(0, 1) = -2$

On the line $y = 4, -2 \leq x \leq 2,$

$f(x, y) = f(x) = 3x^2 + 32 - 16 = 3x^2 + 16$

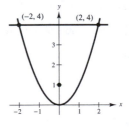

and the maximum is 28, the minimum is 16. On the curve $y = x^2, -2 \leq x \leq 2,$

$f(x, y) = f(x) = 3x^2 + 2(x^2)^2 - 4x^2 = 2x^4 - x^2 = x^2(2x^2 - 1)$

and the maximum is 28, the minimum is $-\frac{1}{8}$.

Absolute maximum: 28 at $(\pm 2, 4)$

Absolute minimum: -2 at $(0, 1)$

47. $f(x, y) = x^2 + 2xy + y^2, R = \{(x, y): |x| \le 2, |y| \le 1\}$

$\left.\begin{array}{l} f_x = 2x + 2y = 0 \\ f_y = 2x + 2y = 0 \end{array}\right\} y = -x$

$f(x, -x) = x^2 - 2x^2 + x^2 = 0$

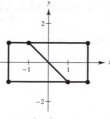

Along $y = 1, -2 \le x \le 2,$

$f = x^2 + 2x + 1, f' = 2x + 2 = 0 \Rightarrow x = -1, f(-2, 1) = 1, f(-1, 1) = 0, f(2, 1) = 9.$

Along $y = -1, -2 \le x \le 2,$

$f = x^2 - 2x + 1, f' = 2x - 2 = 0 \Rightarrow x = 1, f(-2, -1) = 9, f(1, -1) = 0, f(2, -1) = 1.$

Along $x = 2, -1 \le y \le 1, f = 4 + 4y + y^2, f' = 2y + 4 \ne 0.$

Along $x = -2, -1 \le y \le 1, f = 4 - 4y + y^2, f' = 2y - 4 \ne 0.$

So, the maxima are $f(-2, -1) = 9$ and $f(2, 1) = 9,$ and the minima are $f(x, -x) = 0, -1 \le x \le 1.$

49. (a) The function f has a relative minimum at (x_0, y_0) if $f(x, y) \ge f(x_0, y_0)$ for all (x, y) in an open disk containing $(x_0, y_0).$

    (b) The function f has a relative maximum at (x_0, y_0) if $f(x, y) \le f(x_0, y_0)$ for all (x, y) in an open disk containing $(x_0, y_0).$

    (c) The point (x_0, y_0) is a critical point if either

        (1) $f_x(x_0, y_0) = 0$ and $f_y(x_0, y_0) = 0,$ or

        (2) $f_x(x_0, y_0)$ or $f_y(x_0, y_0)$ does not exist.

    (d) A critical point is a saddle point if it is neither a relative minimum nor a relative maximum.

51.

No extrema

53. $f(x, y) = x^2 - y^2, g(x, y) = x^2 + y^2$

    (a) $f_x = 2x = 0, f_y = -2y = 0 \Rightarrow (0, 0)$ is a critical point.

        $g_x = 2x = 0, g_y = 2y = 0 \Rightarrow (0, 0)$ is a critical point.

    (b) $f_{xx} = 2, f_{yy} = -2, f_{xy} = 0$

        $d = 2(-2) - 0 < 0 \Rightarrow (0, 0)$ is a saddle point.

        $g_{xx} = 2, g_{yy} = 2, g_{xy} = 0$

        $d = 2(2) - 0 > 0 \Rightarrow (0, 0)$ is a relative minimum.

55. False.

    Let $f(x, y) = 1 - |x| - |y|.$

    $(0, 0, 1)$ is a relative maximum, but $f_x(0, 0)$ and $f_y(0, 0)$ do not exist.

57. False. Let $f(x, y) = x^2y^2$ (See Example 4 on page 940).

Section 13.9 Applications of Extrema of Functions of Two Variables

1. A point on the plane is given by $(x, y, z) = (x, y, 3 - x + y)$. The square of the distance from $(0, 0, 0)$ to this point is

$$S = x^2 + y^2 + (3 - x + y)^2.$$
$$S_x = 2x - 2(3 - x + y)$$
$$S_y = 2y + 2(3 - x + y)$$

From the equations $S_x = 0$ and $S_y = 0$ we obtain

$$4x - 2y = 6$$
$$-2x + 4y = -6.$$

Solving simultaneously, we have $x = 1, y = -1, z = 1$.

So, the distance is $\sqrt{1^2 + (-1)^2 + 1^2} = \sqrt{3}$.

3. A point on the surface is given by $(x, y, z) = \left(x, y, \sqrt{1 - 2x - 2y}\right)$. The square of the distance from $(-2, -2, 0)$ to a point on the surface is given by

$$S = (x + 2)^2 + (y + 2)^2 + \left(\sqrt{1 - 2x - 2y} - 0\right)^2 = (x + 2)^2 + (y + 2)^2 + 1 - 2x - 2y.$$

$$S_x = 2(x + 2) - 2$$
$$S_y = 2(y + 2) - 2$$

From the equations $S_x = 0$ and $S_y = 0$, we obtain $\left.\begin{array}{l} 2x + 2 = 0 \\ 2y + 2 = 0 \end{array}\right\} \Rightarrow x = y = -1, z = \sqrt{5}.$

So, the distance is $\sqrt{(-1 + 2)^2 + (-1 + 2)^2 + \left(\sqrt{5}\right)^2} = \sqrt{7}$.

5. Let $x, y,$ and z be the numbers. Because $xyz = 27$,

$$z = \frac{27}{xy}.$$

$$S = x + y + z = x + y + \frac{27}{xy}.$$

$$S_x = 1 - \frac{27}{x^2 y} = 0, S_y = 1 - \frac{27}{xy^2} = 0.$$

$$\left.\begin{array}{l} x^2 y = 27 \\ xy^2 = 27 \end{array}\right\} x = y = 3$$

So, $x = y = z = 3$.

7. Let $x, y,$ and z be the numbers and let

$S = x^2 + y^2 + z^2$. Because

$x + y + z = 30$, we have

$$S = x^2 + y^2 + (30 - x - y)^2$$

$$S_x = 2x + 2(30 - x - y)(-1) = 0 \Big] 2x + y = 30$$
$$S_y = 2y + 2(30 - x - y)(-1) = 0 \Big] x + 2y = 30.$$

Solving simultaneously yields $x = 10,$

$y = 10,$ and $z = 10$.

9. The volume is $668.25 = xyz \Rightarrow z = \dfrac{668.25}{xy}$.

$$C = 0.06(2yz + 2xz) + 0.11(xy) = 0.12\left(\dfrac{668.25}{x} + \dfrac{668.25}{y}\right) + 0.11(xy)$$

$$C = \dfrac{80.19}{x} + \dfrac{80.19}{y} + 0.11(xy)$$

$$C_x = \dfrac{-80.19}{x^2} + 0.11y = 0$$

$$C_y = \dfrac{-8.19}{y^2} + 0.11x = 0$$

Solving simultaneously, $x = y = 9$ and $z = 8.25$.

Minimum cost: $\dfrac{80.19}{9} + \dfrac{80.19}{9} + 0.11(xy) = \$26.73$

11. Let x, y, and z be the length, width, and height, respectively and let V_0 be the given volume.

Then $V_0 = xyz$ and $z = V_0/xy$. The surface area is

$$S = 2xy + 2yz + 2xz = 2\left(xy + \dfrac{V_0}{x} + \dfrac{V_0}{y}\right)$$

$$S_x = 2\left(y - \dfrac{V_0}{x^2}\right) = 0 \ \Bigg| \ x^2y - V_0 = 0$$

$$S_y = 2\left(x - \dfrac{V_0}{y^2}\right) = 0 \ \Bigg| \ xy^2 - V_0 = 0.$$

Solving simultaneously yields $x = \sqrt[3]{V_0}$, $y = \sqrt[3]{V_0}$, and $z = \sqrt[3]{V_0}$.

13. $R(x_1, x_2) = -5x_1^2 - 8x_2^2 - 2x_1x_2 + 42x_1 + 102x_2$

$R_{x_1} = -10x_1 - 2x_2 + 42 = 0, 5x_1 + x_2 = 21$

$R_{x_2} = -16x_2 - 2x_1 + 102 = 0, x_1 + 8x_2 = 51$

Solving this system yields $x_1 = 3$ and $x_2 = 6$.

$R_{x_1x_1} = -10$

$R_{x_1x_2} = -2$

$R_{x_2x_2} = -16$

$R_{x_1x_1} < 0$ and $R_{x_1x_1}R_{x_2x_2} - \left(R_{x_1x_2}\right)^2 > 0$

So, revenue is maximized when $x_1 = 3$ and $x_2 = 6$.

15. $P(p, q, r) = 2pq + 2pr + 2qr$.

$p + q + r = 1$ implies that $r = 1 - p - q$.

$$P(p, q) = 2pq + 2p(1 - p - q) + 2q(1 - p - q)$$
$$= 2pq + 2p - 2p^2 - 2pq + 2q - 2pq - 2q^2 = -2pq + 2p + 2q - 2p^2 - 2q^2$$

$$\frac{\partial P}{\partial p} = -2q + 2 - 4p; \frac{\partial P}{\partial q} = -2p + 2 - 4q$$

Solving $\dfrac{\partial P}{\partial p} = \dfrac{\partial P}{\partial q} = 0$ gives $q + 2p = 1$

$$p + 2q = 1$$

and so $p = q = \dfrac{1}{3}$ and $P\left(\dfrac{1}{3}, \dfrac{1}{3}\right) = -2\left(\dfrac{1}{9}\right) + 2\left(\dfrac{1}{3}\right) + 2\left(\dfrac{1}{3}\right) - 2\left(\dfrac{1}{9}\right) - 2\left(\dfrac{1}{9}\right) = \dfrac{6}{9} = \dfrac{2}{3}$.

17. The distance from P to Q is $\sqrt{x^2 + 4}$. The distance from Q to R is $\sqrt{(y - x)^2 + 1}$. The distance from R to S is $10 - y$.

$$C = 3k\sqrt{x^2 + 4} + 2k\sqrt{(y - x)^2 + 1} + k(10 - y)$$

$$C_x = 3k\left(\frac{x}{\sqrt{x^2 + 4}}\right) + 2k\left(\frac{-(y - x)}{\sqrt{(y - x)^2 + 1}}\right) = 0$$

$$C_y = 2k\left(\frac{y - x}{\sqrt{(y - x)^2 + 1}}\right) - k = 0 \Rightarrow \frac{y - x}{\sqrt{(y - x)^2 + 1}} = \frac{1}{2}$$

$$3k\left(\frac{x}{\sqrt{x^2 + 4}}\right) + 2k\left(-\frac{1}{2}\right) = 0$$

$$\frac{x}{\sqrt{x^2 + 4}} = \frac{1}{3}$$

$$3x = \sqrt{x^2 + 4}$$

$$9x^2 = x^2 + 4$$

$$x^2 = \frac{1}{2}$$

$$x = \frac{\sqrt{2}}{2}$$

$$2(y - x) = \sqrt{(y - x)^2 + 1}$$

$$4(y - x)^2 = (y - x)^2 + 1$$

$$(y - x)^2 = \frac{1}{3}$$

$$y = \frac{1}{\sqrt{3}} + \frac{1}{\sqrt{2}} = \frac{2\sqrt{3} + 3\sqrt{2}}{6}$$

So, $x = \dfrac{\sqrt{2}}{2} \approx 0.707$ km and $y = \dfrac{2\sqrt{3} + 3\sqrt{2}}{6} \approx 1.284$ km.

19. Write the equation to be maximized or minimized as a function of two variables. Set the partial derivatives equal to zero (or undefined) to obtain the critical points. Use the Second Partials Test to test for relative extrema using the critical points. Check the boundary points, too.

21. (a)

x	y	xy	x^2
-2	0	0	4
0	1	0	0
2	3	6	4
$\sum x_i = 0$	$\sum y_i = 4$	$\sum x_i y_i = 6$	$\sum x_i^2 = 8$

$$a = \frac{3(6) - 0(4)}{3(8) - 0^2} = \frac{3}{4}, b = \frac{1}{3}\left[4 - \frac{3}{4}(0)\right] = \frac{4}{3}, y = \frac{3}{4}x + \frac{4}{3}$$

(b) $S = \left(-\frac{3}{2} + \frac{4}{3} - 0\right)^2 + \left(\frac{4}{3} - 1\right)^2 + \left(\frac{3}{2} + \frac{4}{3} - 3\right)^2 = \frac{1}{6}$

23. (a)

x	y	xy	x^2
0	4	0	0
1	3	3	1
1	1	1	1
2	0	0	4
$\sum x_i = 4$	$\sum y_i = 8$	$\sum x_i y_i = 4$	$\sum x_i^2 = 6$

$$a = \frac{4(4) - 4(8)}{4(6) - 4^2} = -2, b = \frac{1}{4}\left[8 + 2(4)\right] = 4, y = -2x + 4$$

(b) $S = (4 - 4)^2 + (2 - 3)^2 + (2 - 1)^2 + (0 - 0)^2 = 2$

25. $(0, 0), (1, 1), (3, 4), (4, 2), (5, 5)$

$$\sum x_i = 13, \qquad \sum y_i = 12,$$

$$\sum x_i y_i = 46, \qquad \sum x_i^2 = 51$$

$$a = \frac{5(46) - 13(12)}{5(51) - (13)^2} = \frac{74}{86} = \frac{37}{43}$$

$$b = \frac{1}{5}\left[12 - \frac{37}{43}(13)\right] = \frac{7}{43}$$

$$y = \frac{37}{43}x + \frac{7}{43}$$

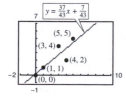

27. $(0, 6), (4, 3), (5, 0), (8, -4), (10, -5)$

$$\sum x_i = 27, \qquad \sum y_i = 0,$$

$$\sum x_i y_i = -70, \qquad \sum x_i^2 = 205$$

$$a = \frac{5(-70) - (27)(0)}{5(205) - (27)^2} = \frac{-350}{296} = -\frac{175}{148}$$

$$b = \frac{1}{5}\left[0 - \left(-\frac{175}{148}\right)(27)\right] = \frac{945}{148}$$

$$y = -\frac{175}{148}x + \frac{945}{148}$$

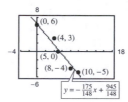

29. (a) Using a graphing utility, $y = 1.6x + 84$.

(b) For each one-year increase in age, the pressure changes by approximately 1.6, the slope of the line.

31. $S(a, b, c) = \sum_{i=1}^{n} \left(y_i - ax_i^2 - bx_i - c \right)^2$

$\dfrac{\partial S}{\partial a} = \sum_{i=1}^{n} -2x_i^2 \left(y_i - ax_i^2 - bx_i - c \right) = 0$

$\dfrac{\partial S}{\partial b} = \sum_{i=1}^{n} -2x_i \left(y_i - ax_i^2 - bx_i - c \right) = 0$

$\dfrac{\partial S}{\partial c} = -2\sum_{i=1}^{n} \left(y_i - ax_i^2 - bx_i - c \right) = 0$

$a\sum_{i=1}^{n} x_i^4 + b\sum_{i=1}^{n} x_i^3 + c\sum_{i=1}^{n} x_i^2 = \sum_{i=1}^{n} x_i^2 y_i$

$a\sum_{i=1}^{n} x_i^3 + b\sum_{i=1}^{n} x_i^2 + c\sum_{i=1}^{n} x_i = \sum_{i=1}^{n} x_i y_i$

$a\sum_{i=1}^{n} x_i^2 + b\sum_{i=1}^{n} x_i + cn = \sum_{i=1}^{n} y_i$

33. $(-2, 0), (-1, 0), (0, 1), (1, 2), (2, 5)$

$\sum x_i = 0$

$\sum y_i = 8$

$\sum x_i^2 = 10$

$\sum x_i^3 = 0$

$\sum x_i^4 = 34$

$\sum x_i y_i = 12$

$\sum x_i^2 y_i = 22$

$34a + 10c = 22, 10b = 12, 10a + 5c = 8$

$a = \dfrac{3}{7}, b = \dfrac{6}{5}, c = \dfrac{26}{35}, y = \dfrac{3}{7}x^2 + \dfrac{6}{5}x + \dfrac{26}{35}$

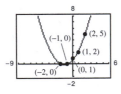

35. $(0, 0), (2, 2), (3, 6), (4, 12)$

$\sum x_i = 9$

$\sum y_i = 20$

$\sum x_i^2 = 29$

$\sum x_i^3 = 99$

$\sum x_i^4 = 353$

$\sum x_i y_i = 70$

$\sum x_i^2 y_i = 254$

$353a + 99b + 29c = 254$

$99a + 29b + 9c = 70$

$29a + 9b + 4c = 20$

$a = 1, b = -1, c = 0, y = x^2 - x$

37. (a) $(0, 0), (2, 15), (4, 30),$

$(6, 50), (8, 65), (10, 70)$

$\sum x_i = 30$

$\sum y_i = 230$

$\sum x_i^2 = 220$

$\sum x_i^3 = 1800$

$\sum x_i^4 = 15{,}664$

$\sum x_i y_i = 1670$

$\sum x_i^2 y_i = 13{,}500$

$15{,}664a + 1800b + 220c = 13{,}500$

$1800a + 220b + 30c = 1670$

$220a + 30b + 6c = 230$

$y = -\dfrac{25}{112}x^2 + \dfrac{541}{56}x - \dfrac{25}{14} \approx -0.22x^2 + 9.66x - 1.79$

(b)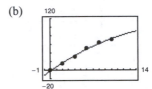

39. (a) $\ln P = -0.1499h + 9.3018$

(b) $\ln P = -0.1499h + 9.3018$

$P = e^{-0.1499h + 9.3018} = 10{,}957.7e^{-0.1499h}$

(c)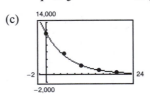

(d) Same answers

41. $S(a, b) = \displaystyle\sum_{i=1}^{n} (ax_i + b - y_i)^2$

$S_a(a, b) = 2a\displaystyle\sum_{i=1}^{n} x_i^2 + 2b\sum_{i=1}^{n} x_i - 2\sum_{i=1}^{n} x_i y_i$

$S_b(a, b) = 2a\displaystyle\sum_{i=1}^{n} x_i + 2nb - 2\sum_{i=1}^{n} y_i$

$S_{aa}(a, b) = 2\displaystyle\sum_{i=1}^{n} x_i^2$

$S_{bb}(a, b) = 2n$

$S_{ab}(a, b) = 2\displaystyle\sum_{i=1}^{n} x_i$

$S_{aa}(a, b) > 0$ as long as $x_i \neq 0$ for all i. (**Note:** If $x_i = 0$ for all i, then $x = 0$ is the least squares regression line.)

$$d = S_{aa}S_{bb} - S_{ab}^2 = 4n\sum_{i=1}^{n} x_i^2 - 4\left(\sum_{i=1}^{n} x_i\right)^2 = 4\left[n\sum_{i=1}^{n} x_i^2 - \left(\sum_{i=1}^{n} x_i\right)^2\right] \geq 0 \text{ since } n\sum_{i=1}^{n} x_i^2 \geq \left(\sum_{i=1}^{n} x_i\right)^2.$$

As long as $d \neq 0$, the given values for a and b yield a minimum.

Section 13.10   Lagrange Multipliers

1. Maximize $f(x, y) = xy$

Constraint: $x + y = 10$

$\nabla f = \lambda \nabla g$

$y\mathbf{i} + x\mathbf{j} = \lambda(\mathbf{i} + \mathbf{j})$

$\left.\begin{array}{c} y = \lambda \\ x = \lambda \\ x + y = 10 \end{array}\right\}$ $x = y = 5$

$f(5, 5) = 25$

3. Minimize $f(x, y) = x^2 + y^2$.

Constraint: $x + 2y - 5 = 0$

$\nabla f = \lambda \nabla g$

$2x\mathbf{i} + 2y\mathbf{j} = \lambda(\mathbf{i} + 2\mathbf{j})$

$\left.\begin{array}{l} 2x = \lambda \\ 2y = 2\lambda \end{array}\right\}$ $\begin{array}{l} x = \lambda/2 \\ y = \lambda \end{array}$

$x + 2y - 5 = 0$

$\dfrac{\lambda}{2} + 2\lambda = 5 \Rightarrow \lambda = 2, x = 1, y = 2$

$f(1, 2) = 5$

5. Maximize $f(x, y) = 2x + 2xy + y$.

Constraint: $2x + y = 100$

$\nabla f = \lambda \nabla g$

$(2 + 2y)\mathbf{i} + (2x + 1)\mathbf{j} = 2\lambda\mathbf{i} + \lambda\mathbf{j}$

$\left.\begin{array}{l} 2 + 2y = 2\lambda \Rightarrow y = \lambda - 1 \\ 2x + 1 = \lambda \Rightarrow x = \dfrac{\lambda - 1}{2} \end{array}\right\}$ $y = 2x$

$2x + y = 100 \Rightarrow 4x = 100$

$x = 25, y = 50$

$f(25, 50) = 2600$

7. Note: $f(x, y) = \sqrt{6 - x^2 - y^2}$ is maximum when

$g(x, y)$ is maximum.

Maximize $g(x, y) = 6 - x^2 - y^2$.

Constraint: $x + y - 2 = 0$

$\left.\begin{array}{l} -2x = \lambda \\ -2y = \lambda \end{array}\right\}$ $x = y$

$x + y = 2 \Rightarrow x = y = 1$

$f(1, 1) = \sqrt{g(1, 1)} = 2$

9. Minimize $f(x, y, z) = x^2 + y^2 + z^2$.

Constraint: $x + y + z - 9 = 0$

$$\left.\begin{array}{l} 2x = \lambda \\ 2y = \lambda \\ 2z = \lambda \end{array}\right\} x = y = z$$

$x + y + z = 9 \Rightarrow x = y = z = 3$

$f(3, 3, 3) = 27$

11. Minimize $f(x, y, z) = x^2 + y^2 + z^2$.

Constraint: $x + y + z = 1$

$$\left.\begin{array}{l} 2x = \lambda \\ 2y = \lambda \\ 2z = \lambda \end{array}\right\} x = y = z$$

$x + y + z = 1 \Rightarrow x = y = z = \frac{1}{3}$

$f\left(\frac{1}{3}, \frac{1}{3}, \frac{1}{3}\right) = \frac{1}{3}$

13. Maximize or minimize $f(x, y) = x^2 + 3xy + y^2$.

Constraint: $x^2 + y^2 \le 1$

Case 1: On the circle $x^2 + y^2 = 1$

$$\left.\begin{array}{l} 2x + 3y = 2x\lambda \\ 3x + 2y = 2y\lambda \end{array}\right\} x^2 = y^2$$

$x^2 + y^2 = 1 \Rightarrow x = \pm\dfrac{\sqrt{2}}{2}, y = \pm\dfrac{\sqrt{2}}{2}$

Maxima: $f\left(\pm\dfrac{\sqrt{2}}{2}, \pm\dfrac{\sqrt{2}}{2}\right) = \dfrac{5}{2}$

Minima: $f\left(\pm\dfrac{\sqrt{2}}{2}, \dfrac{\sqrt{2}}{2}\right) = -\dfrac{1}{2}$

Case 2: Inside the circle

$$\left.\begin{array}{l} f_x = 2x + 3y = 0 \\ f_y = 3x + 2y = 0 \end{array}\right\} x = y = 0$$

$f_{xx} = 2, f_{yy} = 2, f_{xy} = 3, f_{xx}f_{yy} - \left(f_{xy}\right)^2 \le 0$

Saddle point: $f(0, 0) = 0$

By combining these two cases, we have a maximum

of $\dfrac{5}{2}$ at $\left(\pm\dfrac{\sqrt{2}}{2}, \pm\dfrac{\sqrt{2}}{2}\right)$ and a minimum of

$-\dfrac{1}{2}$ at $\left(\pm\dfrac{\sqrt{2}}{2}, \mp\dfrac{\sqrt{2}}{2}\right)$.

15. Maximize $f(x, y, z) = xyz$.

Constraints: $x + y + z = 32$

$x - y + z = 0$

$\nabla f = \lambda \nabla g + \mu \nabla h$

$yz\mathbf{i} + xz\mathbf{j} + xy\mathbf{k} = \lambda(\mathbf{i} + \mathbf{j} + \mathbf{k}) + \mu(\mathbf{i} - \mathbf{j} + \mathbf{k})$

$$\left.\begin{array}{l} yz = \lambda + \mu \\ xz = \lambda - \mu \\ xy = \lambda + \mu \end{array}\right\} yz = xy \Rightarrow x = z$$

$$\left.\begin{array}{l} x + y + z = 32 \\ x - y + z = 0 \end{array}\right\} 2x + 2z = 32 \Rightarrow x = z = 8$$

$y = 16$

$f(8, 16, 8) = 1024$

17. Minimize the square of the distance

$f(x, y) = (x - 0)^2 + (y - 0)^2 = x^2 + y^2$ subject to

the constraint $x + y = 1$.

$$\left.\begin{array}{l} 2x = \lambda \\ 2y = \lambda \end{array}\right\} \quad \left.\begin{array}{l} x = \lambda/2 \\ y = \lambda/2 \end{array}\right\} \Rightarrow x = y$$

$x + y = 1$

$x = y = \dfrac{1}{2}$

The minimum distance is $d = \sqrt{\left(\dfrac{1}{2}\right)^2 + \left(\dfrac{1}{2}\right)^2} = \dfrac{\sqrt{2}}{2}$.

19. Minimize the square of the distance

$f(x, y) = x^2 + (y - 2)^2$

subject to the constraint $x - y = 4$.

$$\left.\begin{array}{l} 2x = \lambda \\ 2(y - 2) = -\lambda \end{array}\right| \quad \begin{array}{l} x = \lambda/2 \\ y = \dfrac{4 - \lambda}{2} \end{array}$$

$x - y = 4$

$\dfrac{\lambda}{2} - \left(\dfrac{4 - \lambda}{2}\right) = 4$

$\lambda = 6$

$x = 3, y = -1$

The minimum distance

is $d = \sqrt{3^2 + (-1 - 2)^2} = 3\sqrt{2}$.

21. Minimize the square of the distance

$f(x, y) = x^2 + (y - 3)^2$ subject to the constraint

$y - x^2 = 0$.

$2x = -2x\lambda$

$2(y - 3) = \lambda$

$y = x^2$

If $x = 0$, $y = 0$, and $f(0, 0) = 9 \Rightarrow$ distance $= 3$.

If $x \neq 0$, $\lambda = -1$, $y = 5/2$, $x = \pm\sqrt{5/2}$

$f(\pm\sqrt{5/2}, 5/2) = 5/2 + \left(\frac{1}{2}\right)^2 = \frac{11}{4} < 3$

The minimum distance is $d = \dfrac{\sqrt{11}}{2}$.

23. Minimize the square of the distance

$f(x, y) = (x - 4)^2 + (y - 4)^2$ subject to the constraint

$x^2 + (y - 1)^2 = 9$.

$2(x - 4) = 2x\lambda$

$2(y - 4) = 2(y - 1)\lambda$

$x^2 + (y - 1)^2 = 9$

Solving these equations, you obtain

$x = 12/5$, $y = 14/5$ and $\lambda = -2/3$.

The minimum distance is

$d = \sqrt{\left(\frac{12}{5} - 4\right)^2 + \left(\frac{14}{5} - 4\right)^2} = \sqrt{\frac{64}{25} + \frac{36}{25}} = 2$.

25. Minimize the square of the distance

$f(x, y, z) = (x - 2)^2 + (y - 1)^2 + (z - 1)^2$

subject to the constraint $x + y + z = 1$.

$\left.\begin{array}{l} 2(x - 2) = \lambda \\ 2(y - 1) = \lambda \\ 2(z - 1) = \lambda \end{array}\right\}$ $y = z$ and $y = x - 1$

$x + y + z = 1 \Rightarrow x + 2(x - 1) = 1$

$\qquad\qquad x = 1, y = z = 0$

The minimum distance is

$d = \sqrt{(1 - 2)^2 + (0 - 1)^2 + (0 - 1)^2} = \sqrt{3}$.

27. Maximize $f(x, y, z) = z$ subject to the constraints

$x^2 + y^2 - z^2 = 0$ and $x + 2z = 4$.

$0 = 2x\lambda + \mu$

$0 = 2y\lambda \Rightarrow y = 0$

$1 = -2z\lambda + 2\mu$

$x^2 + y^2 - z^2 = 0$

$x + 2z = 4 \Rightarrow x = 4 - 2z$

$(4 - 2z)^2 + 0^2 - z^2 = 0$

$3z^2 - 16z + 16 = 0$

$(3z - 4)(z - 4) = 0$

$z = \frac{4}{3}$ or $z = 4$

The maximum value of f occurs when $z = 4$ at the point of $(-4, 0, 4)$.

29. Optimization problems that have restrictions or constraints on the values that can be used to produce the optimal solution are called contrained optimization problems.

31. Minimize $f(x, y, z) = x^2 + y^2 + z^2$.

Constraint: $g(x, y, z) = x - y + z = 3$

$2x = \lambda \Rightarrow x = \lambda/2$

$2y = -\lambda \Rightarrow y = -\lambda/2$

$2z = \lambda \Rightarrow z = \lambda/2$

$x - y + z = 3$

$\frac{\lambda}{2} - \left(-\frac{\lambda}{2}\right) + \frac{\lambda}{2} = 3$

$\frac{3\lambda}{2} = 3$

$\lambda = 2$

$x = 1, y = -1, z = 1$

Minimum distance $= \sqrt{1^2 + (-1)^2 + 1^2} = \sqrt{3}$

33. Minimize $f(x, y, z) = x + y + z$.

Constraint: $g(x, y, z) = xyz = 27$

$\left.\begin{array}{l} 1 = \lambda yz \Rightarrow x = \lambda xyz \\ 1 = \lambda xz \Rightarrow y = \lambda xyz \\ 1 = \lambda xy \Rightarrow z = \lambda xyz \end{array}\right\} \Rightarrow x = y = z$

$xyz = 27$

$x^3 = 27 \Rightarrow x = y = z = 3$

35. Minimize $f(x, y, z) = 0.06(2yz + 2xz) + 0.11(xy)$.

Constraint: $g(x, y, z) = xyz = 668.25$

$0.12z + 0.11y = yz\lambda$

$0.12z + 0.11x = xz\lambda$

$0.12(y + x) = xy\lambda$

$xyz = 668.25$

$0.12xz + 0.11yx = xyz\lambda = 0.12yz + 0.11xy \Rightarrow x = y$

$0.12(2x) = x^2\lambda \Rightarrow \lambda = \dfrac{0.24}{x}$

$0.12z + 0.11x = xz\left(\dfrac{0.24}{x}\right) = 0.24z \Rightarrow z = \dfrac{0.11x}{0.12} = \dfrac{11x}{12}$

$xyz = x^2\left(\dfrac{11}{12}x\right) = 668.25 \Rightarrow x = y = 9, z = \dfrac{33}{4}$

$f\left(9, 9, \dfrac{33}{4}\right) = \$26.73$

37. Maximize $P(p, q, r) = 2pq + 2pr + 2qr$.

Constraint: $g(p, q, r) = p + q + r = 1$

$\left.\begin{array}{r} 2q + 2r = \lambda \\ 2p + 2r = \lambda \\ 2p + 2q = \lambda \end{array}\right\} p = q = r$

$p + q + r = 3p = 1 \Rightarrow p = \frac{1}{3}$ and

$P\left(\frac{1}{3}, \frac{1}{3}, \frac{1}{3}\right) = 3\left(\frac{2}{9}\right) = \frac{2}{3}$.

39. Maximize $V(x, y, z) = (2x)(2y)(2z) = 8xyz$ subject to the constraint $\dfrac{x^2}{a^2} + \dfrac{y^2}{b^2} + \dfrac{z^2}{c^2} = 1$.

$\left.\begin{array}{r} 8yz = \dfrac{2x}{a^2}\lambda \\[2mm] 8xz = \dfrac{2y}{b^2}\lambda \\[2mm] 8xy = \dfrac{2z}{c^2}\lambda \end{array}\right\} \dfrac{x^2}{a^2} = \dfrac{y^2}{b^2} = \dfrac{z^2}{c^2}$

$\dfrac{x^2}{a^2} + \dfrac{y^2}{b^2} + \dfrac{z^2}{c^2} = 1 \Rightarrow \dfrac{3x^2}{a^2} = 1, \dfrac{3y^2}{b^2} = 1, \dfrac{3z^2}{c^2} = 1$

$x = \dfrac{a}{\sqrt{3}}, y = \dfrac{b}{\sqrt{3}}, z = \dfrac{c}{\sqrt{3}}$

So, the dimensions of the box are $\dfrac{2\sqrt{3}a}{3} \times \dfrac{2\sqrt{3}b}{3} \times \dfrac{2\sqrt{3}c}{3}$.

41. Minimize $C(x, y, z) = 5xy + 3(2xz + 2yz + xy)$ subject to the constraint $xyz = 480$.

$$\left.\begin{array}{l} 8y + 6z = yz\lambda \\ 8x + 6z = xz\lambda \\ 6x + 6y = xy\lambda \end{array}\right\} x = y, 4y = 3z$$

$$xyz = 480 \Rightarrow \tfrac{4}{3}y^3 = 480$$

$$x = y = \sqrt[3]{360}, z = \tfrac{4}{3}\sqrt[3]{360}$$

Dimensions: $\sqrt[3]{360} \times \sqrt[3]{360} \times \tfrac{4}{3}\sqrt[3]{360}$ feet.

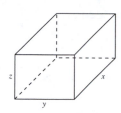

43. Minimize $A(\pi, r) = 2\pi rh + 2\pi r^2$ subject to the constraint $\pi r^2 h = V_0$.

$$\left.\begin{array}{l} 2\pi h + 4\pi r = 2\pi rh\lambda \\ 2\pi r = \pi r^2\lambda \end{array}\right\} h = 2r$$

$$\pi r^2 h = V_0 \Rightarrow 2\pi r^3 = V_0$$

Dimensions: $r = \sqrt[3]{\dfrac{V_0}{2\pi}}$ and $h = 2\sqrt[3]{\dfrac{V_0}{2\pi}}$

45. Using the formula $\text{Time} = \dfrac{\text{Distance}}{\text{Rate}}$, minimize

$$T(x, y) = \frac{\sqrt{d_1^2 + x^2}}{v_1} + \frac{\sqrt{d_2^2 + y^2}}{v_2} \text{ subject to the}$$

constraint $x + y = a$.

$$\left.\begin{array}{l} \dfrac{x}{v_1\sqrt{d_1^2 + x^2}} = \lambda \\[2mm] \dfrac{y}{v_2\sqrt{d_2^2 + y^2}} = \lambda \end{array}\right\} \dfrac{x}{v_1\sqrt{d_1^2 + x^2}} = \dfrac{y}{v_2\sqrt{d_2^2 + y^2}}$$

$$x + y = a$$

Because $\sin \theta_1 = \dfrac{x}{\sqrt{d_1^2 + x^2}}$

and $\sin \theta_2 = \dfrac{y}{\sqrt{d_2^2 + y^2}}$,

we have $\dfrac{x/\sqrt{d_1^2 + x^2}}{v_1} = \dfrac{y/\sqrt{d_2^2 + y^2}}{v_2}$ or

$$\dfrac{\sin \theta_1}{v_1} = \dfrac{\sin \theta_2}{v_2}.$$

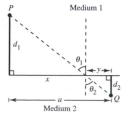

47. Maximize $P(x, y) = 100x^{0.25}y^{0.75}$ subject to the constraint $72x + 60y = 250{,}000$.

$$25x^{-0.75}y^{0.75} = 72\lambda \Rightarrow \left(\frac{y}{x}\right)^{0.75} = \frac{72\lambda}{25}$$

$$75x^{0.25}y^{-0.25} = 60\lambda \Rightarrow \left(\frac{x}{y}\right)^{0.25} = \frac{60\lambda}{75}$$

$$\left(\frac{y}{x}\right)^{0.75}\left(\frac{y}{x}\right)^{0.25} = \left(\frac{72\lambda}{25}\right)\left(\frac{75}{60\lambda}\right)$$

$$\frac{y}{x} = \frac{18}{5}$$

$$y = \frac{18}{5}x$$

$$72x + 60\left(\frac{18}{5}x\right) = 288x = 250{,}000 \Rightarrow x = \frac{15{,}625}{18}$$

$$y = 3125$$

$$P\left(\frac{15625}{18}, 3125\right) \approx 226{,}869$$

49. Minimize $C(x, y) = 72x + 60y$ subject to the constraint $100x^{0.25}y^{0.75} = 50{,}000$.

$$72 = 25x^{-0.75}y^{0.75}\lambda \Rightarrow \left(\frac{y}{x}\right)^{0.75} = \frac{72}{25\lambda}$$

$$60 = 75x^{0.25}y^{-0.25}\lambda \Rightarrow \left(\frac{x}{y}\right)^{0.25} = \frac{60}{75\lambda}$$

$$\left(\frac{y}{x}\right)^{0.75}\left(\frac{y}{x}\right)^{0.25} = \frac{72}{25\lambda} \cdot \frac{75\lambda}{60}$$

$$\frac{y}{x} = \frac{18}{5} \Rightarrow y = \frac{18}{5}x = 3.6x$$

$$100x^{0.25}(3.6x)^{0.75} = 50{,}000$$

$$x = \frac{500}{3.6^{0.75}} \approx 191.3124$$

$$y = 3.6x \approx 688.7247$$

$$C(191.3124, 688.7247) \approx 55{,}097.97$$

51. Let r = radius of cylinder, and h = height of cylinder = height of cone.

$$S = 2\pi rh + 2\pi r\sqrt{h^2 + r^2} = \text{constant surface area}$$

$$V = \pi r^2 h + \frac{2\pi r^2 h}{3} = \frac{5\pi r^2 h}{3} \text{ volume}$$

We maximize $f(r, h) = r^2 h$ subject to $g(r, h) = rh + r\sqrt{h^2 + r^2} = C$.

$$(C - rh)^2 = r^2(h^2 + r^2)$$

$$C^2 - 2Crh = r^4$$

$$h = \frac{C^2 - r^4}{2Cr}$$

$$f(r, h) = F(r) = r^2\left[\frac{C^2 - r^4}{2Cr}\right] = \frac{Cr}{2} - \frac{r^5}{2C}$$

$$F'(r) = \frac{C}{2} - \frac{5r^4}{2C} = 0$$

$$C^2 = 5r^4$$

$$r^2 = \frac{C}{\sqrt{5}}$$

$$F''(r) = \frac{-10r^3}{C}$$

$$h = \frac{C^2 - r^4}{2Cr} = \frac{C^2 - C^2/5}{2C(C^2/5)^{1/4}}$$

$$= \frac{(4/5)C}{2(C^2/5)^{1/4}}$$

$$= \frac{2C}{5r}$$

$$= \frac{2}{5r}\left(\sqrt{5}r^2\right)$$

$$= \frac{2\sqrt{5}}{5}r$$

So, $\dfrac{h}{r} = \dfrac{2\sqrt{5}}{5}$.

By the Second Derivative Test, this is a maximum.

Review Exercises for Chapter 13

1. $f(x, y) = 3x^2 y$

(a) $f(1, 3) = 3(1)^2(3) = 9$

(b) $f(-1, 1) = 3(-1)^2(1) = 3$

(c) $f(-4, 0) = 3(-4)^2(0) = 0$

(d) $f(x, z) = 3x^2(2) = 6x^2$

3. $f(x, y) = \dfrac{\sqrt{x}}{y}$

The domain is $\{(x, y) : x \geq 0, y \neq 0\}$.

The range is all real numbers.

5. $z = 3 - 2x + y$

The level curves are parallel lines of the form
$y = 2x - 3 + c.$

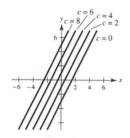

7. $f(x, y) = x^2 + y^2$

(a)

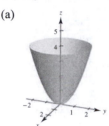

(b) $g(x, y) = f(x, y) + 2$ is a vertical translation of f two units upward.

(c) $g(x, y) = f(x, y - z)$ is a horizontal translation of f two units to the right. The vertex moves from $(0, 0, 0)$ to $(0, 2, 0)$.

(d)

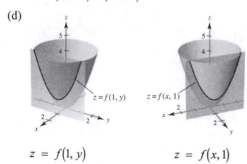

$z = f(1, y)$ $\qquad$ $z = f(x, 1)$

9. $f(x, y, z) = x^2 - y + z^2 = 2$
$$y = x^2 + z^2 - 2$$
Elliptic paraboloid

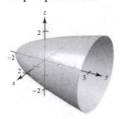

11. $\displaystyle\lim_{(x, y) \to (1, 1)} \dfrac{xy}{x^2 + y^2} = \dfrac{1}{2}$

Continuous except at $(0, 0)$.

13. $\displaystyle\lim_{(x, y) \to (0, 0)} \dfrac{y + xe^{-y^2}}{1 + x^2} = \dfrac{0 + 0}{1 + 0} = 0$

Continuous everywhere.

15. $f(x, y) = 5x^3 + 7y - 3$
$$\dfrac{\partial f}{\partial x} = 15x^2 \qquad \dfrac{\partial f}{\partial y} = 7$$

17. $f(x, y) = e^x \cos y$
$$f_x = e^x \cos y$$
$$f_y = -e^x \sin y$$

19. $f(x, y) = y^3 e^{4x}$
$$\dfrac{\partial f}{\partial x} = 4y^3 e^{4x}$$
$$\dfrac{\partial f}{\partial y} = 3y^2 e^{4x}$$

21. $f(x, y, z) = 2xz^2 + 6xyz - 5xy^3$
$$\dfrac{\partial f}{\partial x} = 2z^2 + 6yz - 5y^3$$
$$\dfrac{\partial f}{\partial y} = 6xz - 15xy^2$$
$$\dfrac{\partial f}{\partial z} = 4xz + 6xy$$

23. $f(x, y) = 3x^2 - xy + 2y^3$
$$f_x = 6x - y$$
$$f_y = -x + 6y^2$$
$$f_{xx} = 6$$
$$f_{yy} = 12y$$
$$f_{xy} = -1$$
$$f_{yx} = -1$$

25. $h(x, y) = x \sin y + y \cos x$

$h_x = \sin y - y \sin x$

$h_y = x \cos y + \cos x$

$h_{xx} = -y \cos x$

$h_{yy} = -x \sin y$

$h_{xy} = \cos y - \sin x$

$h_{yx} = \cos y - \sin x$

27. $z = x^2 \ln(y + 1)$

$\dfrac{\partial z}{\partial x} = 2x \ln(y + 1)$. At $(2, 0, 0)$, $\dfrac{\partial z}{\partial x} = 0$.

Slope in x-direction.

$\dfrac{\partial z}{\partial y} = \dfrac{x^2}{1 + y}$. At $(2, 0, 0)$, $\dfrac{\partial z}{\partial y} = 4$.

Slope in y-direction.

29. $z = x \sin xy$

$dz = \dfrac{\partial z}{\partial x}dx + \dfrac{\partial z}{\partial y}dy = (xy \cos xy + \sin xy)dx + (x^2 \cos xy)dy$

31. $w = 3xy^2 - 2x^3yz^2$

$dw = \dfrac{\partial w}{\partial x}\,dx + \dfrac{\partial w}{\partial y}\,dy + \dfrac{\partial w}{\partial z}\,dz$

$= (3y^2 - 6x^2yz^2)dx + (6xy - 2x^3z^2)dy - 4x^3yz\,dz$

33. $f(x, y) = 4x + 2y$

(a) $f(2, 1) = 4(2) + 2(1) = 10$

$f(2.1, 1.05) = 4(2.1) + 2(1.05) = 10.5$

$\Delta z = 10.5 - 10 = 0.5$

(b) $dz = 4dx + 2dy$

$= 4(0.1) + 2(0.05) = 0.5$

35. $V = \dfrac{1}{3}\pi r^2 h$

$dV = \dfrac{2}{3}\pi rh\,dr + \dfrac{1}{3}\pi r^2\,dh$

$= \dfrac{2}{3}\pi(2)(5)\left(\pm\dfrac{1}{8}\right) + \dfrac{1}{3}\pi(2)^2\left(\pm\dfrac{1}{8}\right)$

$= \pm\dfrac{5}{6}\pi + \dfrac{1}{6}\pi = \pm\pi$ in.^{3} Propogated error

$V = \dfrac{1}{3}\pi(2)^2 5 = \dfrac{20}{3}\pi$ in.^{3}

Relative error $= \dfrac{dV}{V} = \dfrac{\pm\pi}{\left(\dfrac{20}{3}\pi\right)} = \dfrac{3}{20} = 15\%$

37. $w = \ln(x^2 + y)$, $x = 2t$, $y = 4 - t$

(a) Chain Rule: $\dfrac{dw}{dt} = \dfrac{\partial w}{\partial x}\dfrac{dx}{dt} + \dfrac{\partial w}{\partial y}\dfrac{dy}{dt}$

$= \dfrac{2x}{x^2 + y}(2) + \dfrac{1}{x^2 + y}(-1)$

$= \dfrac{8t - 1}{4t^2 + 4 - t}$

(b) Substitution: $w = \ln(x^2 + y) = \ln(4t^2 + 4 - t)$

$\dfrac{dw}{dt} = \dfrac{1}{4t^2 + 4 - t}(8t - 1)$

39. $w = \dfrac{xy}{z}, x = 2r + t, y = rt, z = 2r - t$

    (a) Chain Rule: $\dfrac{\partial w}{\partial r} = \dfrac{\partial w}{\partial x}\dfrac{\partial x}{\partial r} + \dfrac{\partial w}{\partial y}\dfrac{\partial y}{\partial r} + \dfrac{\partial w}{\partial z}\dfrac{\partial z}{\partial r}$

$$= \dfrac{y}{z}(2) + \dfrac{x}{z}(t) - \dfrac{xy}{z^2}(2)$$

$$= \dfrac{2rt}{2r - t} + \dfrac{(2r + t)t}{2r - t} - \dfrac{2(2r + t)(rt)}{(2r - t)^2}$$

$$= \dfrac{4r^2t - 4rt^2 - t^3}{(2r - t)^2}$$

$$\dfrac{\partial w}{\partial t} = \dfrac{\partial w}{\partial x}\dfrac{\partial x}{\partial t} + \dfrac{\partial w}{\partial y}\dfrac{\partial y}{\partial t} + \dfrac{\partial w}{\partial z}\dfrac{\partial z}{\partial t}$$

$$= \dfrac{y}{z}(1) + \dfrac{x}{z}(r) = \dfrac{xy}{z^2}(-1)$$

$$= \dfrac{4r^2t - rt^2 + 4r^3}{(2r - t)^2}$$

    (b) Substitution: $w = \dfrac{xy}{z} = \dfrac{(2r + t)(rt)}{2r - t} = \dfrac{2r^2t + rt^2}{2r - t}$

$$\dfrac{\partial w}{\partial r} = \dfrac{4r^2t - 4rt^2 - t^3}{(2r - t)^2}$$

$$\dfrac{\partial w}{\partial t} = \dfrac{4r^2t - rt^2 + 4r^3}{(2r - t)^2}$$

41. $x^2 + xy + y^2 + yz + z^2 = 0$

$$2x + y + y\dfrac{\partial z}{\partial x} + 2z\dfrac{\partial z}{\partial x} = 0$$

$$\dfrac{\partial z}{\partial x} = \dfrac{-2x - y}{y + 2z}$$

$$x + 2y + y\dfrac{\partial z}{\partial y} + z + 2z\dfrac{\partial z}{\partial y} = 0$$

$$\dfrac{\partial z}{\partial y} = \dfrac{-x - 2y - z}{y + 2z}$$

43. $f(x, y) = x^2y, P(-5, 5), \mathbf{v} = 3\mathbf{i} - 4\mathbf{j}$

$$\mathbf{u} = \dfrac{\mathbf{v}}{\|\mathbf{v}\|} = \dfrac{3}{5}\mathbf{i} - \dfrac{4}{5}\mathbf{j}$$

$$D_{\mathbf{u}}f(x, y) = \dfrac{\partial f}{\partial x}\cos\theta + \dfrac{\partial f}{\partial y}\sin\theta$$

$$= 2xy\cos\theta + x^2\sin\theta$$

$$D_{\mathbf{u}}f(-5, 5) = 2(-5)(5)\left(\dfrac{3}{5}\right) + (-5)^2\left(-\dfrac{4}{5}\right)$$

$$= -30 - 20 = -50$$

45.
$$w = y^2 + xz$$
$$\nabla w = z\mathbf{i} + 2y\mathbf{j} + x\mathbf{k}$$
$$\nabla w(1, 2, 2) = 2\mathbf{i} + 4\mathbf{j} + \mathbf{k}$$
$$\mathbf{u} = \tfrac{1}{3}\mathbf{v} = \tfrac{2}{3}\mathbf{i} - \tfrac{1}{3}\mathbf{j} + \tfrac{2}{3}\mathbf{k}$$
$$D_{\mathbf{u}}w(1, 2, 2) = \nabla w(1, 2, 2)\cdot\mathbf{u} = \tfrac{4}{3} - \tfrac{4}{3} + \tfrac{2}{3} = \tfrac{2}{3}$$

47.
$$z = x^2y$$
$$\nabla z = 2xy\mathbf{i} + x^2\mathbf{j}$$
$$\nabla_z(2, 1) = 4\mathbf{i} + 4\mathbf{j}$$
$$\|\nabla z(2, 1)\| = 4\sqrt{2}$$

49.
$$z = \dfrac{y}{x^2 + y^2}$$
$$\nabla z = -\dfrac{2xy}{\left(x^2 + y^2\right)^2}\mathbf{i} + \dfrac{x^2 - y^2}{\left(x^2 + y^2\right)^2}\mathbf{j}$$
$$\nabla z(1, 1) = -\dfrac{1}{2}\mathbf{i} = \left\langle -\dfrac{1}{2}, 0 \right\rangle$$
$$\|\nabla z(1, 1)\| = \dfrac{1}{2}$$

51. $f(x, y) = 9x^2 - 4y^2, c = 65, P(3, 2)$

(a) $\nabla f(x, y) = 18x\mathbf{i} - 8y\mathbf{j}$

$\nabla f(3, 2) = 54\mathbf{i} - 16\mathbf{j}$

(b) Unit normal: $\dfrac{54\mathbf{i} - 16\mathbf{j}}{\|54\mathbf{i} - 16\mathbf{j}\|} = \dfrac{1}{\sqrt{793}}(27\mathbf{i} - 8\mathbf{j})$

(c) Slope $= \dfrac{27}{8}$.

$y - z = \dfrac{27}{8}(x - 3)$

$y = \dfrac{27}{8}x - \dfrac{65}{8}$ Tangent line

(d)

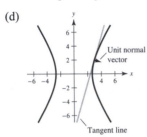

53. $F(x, y, z) = x^2 + y^2 + 2 - z = 0, (1, 3, 12)$

$\nabla F = 2x\mathbf{i} + 2y\mathbf{j} - \mathbf{k}$

$\nabla F(1, 3, 12) = 2\mathbf{i} + 6\mathbf{j} - \mathbf{k}$

Tangent Plane:

$2(x - 1) + 6(y - 3) - (z - 12) = 0$

$2x + 6y - z = 8$

55. $F(x, y, z) = x^2 + y^2 - 4x + 6y + z + 9 = 0$

$\nabla F = (2x - 4)\mathbf{i} + (2y + 6)\mathbf{j} + \mathbf{k}$

$\nabla F(2, -3, 4) = \mathbf{k}$

So, the equation of the tangent plane is

$z - 4 = 0$ or $z = 4$.

57. $F(x, y, z) = x^2y - z = 0$

$\nabla F = 2xy\mathbf{i} + x^2\mathbf{j} - \mathbf{k}$

$\nabla F(2, 1, 4) = 4\mathbf{i} + 4\mathbf{j} - \mathbf{k}$

So, the equation of the tangent plane is

$4(x - 2) + 4(y - 1) - (z - 4) = 0$ or

$4x + 4y - z = 8,$

and the equation of the normal line is

$x = 4t + 2, y = 4t + 1, z = -t + 4.$

Symmetric equations:

$\dfrac{x - 2}{4} = \dfrac{y - 1}{4} = -\dfrac{z - 4}{1}$

59. $f(x, y, z) = x^2 + y^2 + z^2 - 14$

$\nabla f(x, y, z) = 2x\mathbf{i} + 2y\mathbf{j} + 2z\mathbf{k}$

$\nabla f(2, 1, 3) = 4\mathbf{i} + 2\mathbf{j} + 6\mathbf{k}$ Normal vector to plane.

$\cos \theta = \dfrac{|\mathbf{n} \cdot \mathbf{k}|}{\|\mathbf{n}\|} = \dfrac{6}{\sqrt{56}} = \dfrac{3\sqrt{14}}{14}$

$\theta = 36.7°$

61. $f(x, y) = -x^2 - 4y^2 + 8x - 8y - 11$

$f_x = -2x + 8 = 0 \Rightarrow x = 4$

$f_y = -8y - 8 = 0 \Rightarrow y = -1$

$f_{xx} = -2, f_{yy} = -8, f_{xy} = 0$

$f_{xx}f_{yy} - (f_{xy})^2 = (-2)(-8) - 0 = 16 > 0$

So, $(4, -1, 9)$ is a relative minimum.

63. $f(x, y) = 2x^2 + 6xy + 9y^2 + 8x + 14$

$f_x = 4x + 6y + 8 = 0$

$f_y = 6x + 18y = 0, x = -3y$

$4(-3y) + 6y = -8 \Rightarrow y = \frac{4}{3}, x = -4$

$f_{xx} = 4$

$f_{yy} = 18$

$f_{xy} = 6$

$f_{xx}f_{yy} - (f_{xy})^2 = 4(18) - (6)^2 = 36 > 0.$

So, $\left(-4, \frac{4}{3}, -2\right)$ is a relative minimum.

65. $f(x, y) = xy + \dfrac{1}{x} + \dfrac{1}{y}$

$f_x = y - \dfrac{1}{x^2} = 0, x^2y = 1$

$f_y = x - \dfrac{1}{y^2} = 0, xy^2 = 1$

So, $x^2y = xy^2$ or $x = y$ and substitution yields the critical point $(1, 1)$.

$f_{xx} = \dfrac{2}{x^3}$

$f_{xy} = 1$

$f_{yy} = \dfrac{2}{y^3}$

At the critical point $(1, 1)$, $f_{xx} = 2 > 0$ and

$f_{xx}f_{yy} - (f_{xy})^2 = 3 > 0.$

So, $(1, 1, 3)$ is a relative minimum.

67. A point on the plane is given by $(x, y, 4 - x - y)$

The square of the distance from $(2, 1, 4)$ to a point on the plane is

$$S = (x - 2)^2 + (y - 1)^2 + (4 - x - y - 4)^2$$
$$= (x - 2)^2 + (y - 1)^2 + (-x - y)^2.$$

$$S_x = 2(x - 2) - 2(-x - y) = 4x + 2y - 4$$
$$S_y = 2(y - 1) - 2(-x - y) = 2x + 4y - 2$$

$$S_x = S_y = 0 \Rightarrow \begin{cases} 4x + 2y = 4 \\ 2x + 4y = 2 \end{cases} \Rightarrow x = 1, y = 0, z = 3$$

The distance is $\sqrt{(1 - 2)^2 + (0 - 1)^2 + (-1)^2} = \sqrt{3}$.

69. $R = -6x_1^2 - 10x_2^2 - 2x_1x_2 + 32x_1 + 84x_2$

$Rx_1 = -12x_1 - 2x_2 + 32 = 0 \Rightarrow 6x_1 + x_2 = 16$

$Rx_2 = -20x_2 - 2x_1 + 84 = 0 \Rightarrow x_1 + 10x_2 = 42$

Solving this system yields $x_1 = 2$ and $x_2 = 4$.

71. $(0, 4), (1, 5), (3, 6), (6, 8), (8, 10)$

$$\sum x_i = 18 \qquad \sum y_i = 33$$
$$\sum x_i y_i = 151 \qquad \sum x_i^2 = 110$$

$$a = \frac{5(151) - 18(33)}{5(110) - (18)^2} = \frac{161}{226} \approx 0.7124$$

$$b = \frac{1}{5}\left(33 - \frac{161}{226}(18)\right) = \frac{456}{113} \approx 4.0354$$

$$y = \frac{161}{226}x + \frac{456}{113}$$

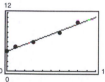

73. $(100, 35), (150, 44), (200, 50), (250, 56)$

(a) Using a graphing utility, you obtain
$$y = 0.138x + 22.1.$$

(b) If $x = 175$, $y = 0.138(175) + 22.1 = 46.25$
bushels per acre.

75. Minimize $f(x, y) = x^2 + y^2$

Constraint: $x + y - 8 = 0$

$\nabla f = \lambda \nabla g$

$2x\mathbf{i} + 2y\mathbf{j} = \lambda(\mathbf{i} + \mathbf{j})$

$$\left. \begin{array}{l} 2x = \lambda \\ 2y = \lambda \end{array} \right\} \quad x = y$$

$x + y - 8 = 2x - 8 = 0 \Rightarrow x = y = 4$

$f(4, 4) = 32$

77. Maximize $f(x, y) = 2x + 3xy + y$

Constraint: $x + 2y = 29$

$\nabla f = \lambda \nabla g$

$$\left. \begin{array}{l} 2 + 3y = \lambda \\ 3x + 1 = 2\lambda \end{array} \right\} \quad 4 + 6y = 3x + 1 \Rightarrow x - 2y = 1$$

$$\left. \begin{array}{l} x - 2y = 1 \\ x + 2y = 29 \end{array} \right\} \quad x = 15, y = 7$$

$f(15, 7) = 2(15) + 3(15)(7) + 7 = 352$

79. Maximize $f(x, y) = 2xy$

Constraint: $2x + y = 12$

$\nabla f = \lambda \nabla g$

$$\left. \begin{array}{l} 2y = 2\lambda \\ 2x = \lambda \end{array} \right\} \quad 4x = 2y \Rightarrow y = 2x$$

$2x + y = 2x + 2x = 12 \Rightarrow x = 3, y = 6$

$f(3, 6) = 2(3)(6) = 36\,0$

81. $PQ = \sqrt{x^2 + 4}$,

$QR = \sqrt{y^2 + 1}$,

$RS = z; x + y + z = 10$

$C = 3\sqrt{x^2 + 4} + 2\sqrt{y^2 + 1} + z$

Constraint: $x + y + z = 10$

$\nabla C = \lambda \nabla g$

$$\frac{3x}{\sqrt{x^2 + 4}}\mathbf{i} + \frac{2y}{\sqrt{y^2 + 1}}\mathbf{j} + \mathbf{k} = \lambda[\mathbf{i} + \mathbf{j} + \mathbf{k}]$$

$3x = \lambda\sqrt{x^2 + 4}$

$2y = \lambda\sqrt{y^2 + 1}$

$1 = \lambda$

$9x^2 = x^2 + 4 \Rightarrow x^2 = \frac{1}{2}$

$4y^2 = y^2 + 1 \Rightarrow y^2 = \frac{1}{3}$

So, $x = \frac{\sqrt{2}}{2} \approx 0.707$ km,

$y = \frac{\sqrt{3}}{3} \approx 0.577$ km,

$z = 10 - \frac{\sqrt{2}}{2} - \frac{\sqrt{3}}{3} \approx 8.716$ km.

Problem Solving for Chapter 13

1. (a) The three sides have lengths 5, 6, and 5.

Thus, $s = \frac{16}{2} = 8$ and $A = \sqrt{8(3)(2)(3)} = 12$.

(b) Let $f(a, b, c) = (\text{area})^2 = s(s - a)(s - b)(s - c)$,

subject to the constraint

$a + b + c = \text{constant (perimeter)}$.

Using Lagrange multipliers,

$-s(s - b)(s - c) = \lambda$

$-s(s - a)(s - c) = \lambda$

$-s(s - a)(s - b) = \lambda$.

From the first 2 equations

$s - b = s - a \Rightarrow a = b$.

Similarly, $b = c$ and hence $a = b = c$ which is an equilateral triangle.

(c) Let $f(a, b, c) = a + b + c$, subject

to $(\text{Area})^2 = s(s - a)(s - b)(s - c)$ constant.

Using Lagrange multipliers,

$1 = -\lambda s(s - b)(s - c)$

$1 = -\lambda s(s - a)(s - c)$

$1 = -\lambda s(s - a)(s - b)$

So, $s - a = s - b \Rightarrow a = b$ and $a = b = c$.

3. (a) $F(x, y, z) = xyz - 1 = 0$

$F_x = yz, F_y = xz, F_z = xy$

Tangent plane:

$y_0 z_0 (x - x_0) + x_0 z_0 (y - y_0) + x_0 y_0 (z - z_0) = 0$

$y_0 z_0 x + x_0 z_0 y + x_0 y_0 z = 3 x_0 y_0 z_0 = 3$

(b) $V = \frac{1}{3}(\text{base})(\text{height})$

$= \frac{1}{3}\left(\frac{1}{2}\frac{3}{y_0 z_0}\frac{3}{x_0 z_0}\right)\left(\frac{3}{x_0 y_0}\right) = \frac{9}{2}$

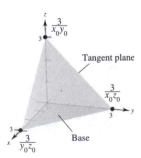

5. (a)

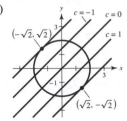

Maximum value of f is $f\left(\sqrt{2}, -\sqrt{2}\right) = 2\sqrt{2}$.

Maximize $f(x, y) = x - y$.

Constraint: $g(x, y) = x^2 + y^2 = 4$

$\nabla f = \lambda \nabla g$: $1 = 2\lambda x$

$-1 = 2\lambda y$

$x^2 + y^2 = 4$

$2\lambda x = -2\lambda y \Rightarrow x = -y$

$2x^2 = 4 \Rightarrow x = \pm\sqrt{2}, y = \mp\sqrt{2}$

$f\left(\sqrt{2}, -\sqrt{2}\right) = 2\sqrt{2}, f\left(-\sqrt{2}, \sqrt{2}\right) = -2\sqrt{2}$

(b) $f(x, y) = x - y$

Constraint: $x^2 + y^2 = 0 \Rightarrow (x, y) = (0, 0)$

Maximum and minimum values are 0.

Lagrange multipliers does not work:

$\left.\begin{array}{r}1 = 2\lambda x \\ -1 = 2\lambda y\end{array}\right\} x = -y = 0$, a contradiction.

Note that $\nabla g(0, 0) = \mathbf{0}$.

7. $H = k(5xy + 6xz + 6yz)$

$z = \frac{1000}{xy} \Rightarrow H = k\left(5xy + \frac{6000}{y} + \frac{6000}{x}\right)$.

$H_x = 5y - \frac{6000}{x^2} = 0 \Rightarrow 5yx^2 = 6000$

By symmetry, $x = y \Rightarrow x^3 = y^3 = 1200$.

So, $x = y = 2\sqrt[3]{150}$ and $z = \frac{5}{3}\sqrt[3]{150}$.

9. (a) $\dfrac{\partial f}{\partial x} = Cax^{a-1}y^{1-a}, \dfrac{\partial f}{\partial y} = C(1-a)x^a y^{-a}$

$x\dfrac{\partial f}{\partial x} + y\dfrac{\partial f}{\partial y} = Cax^a y^{1-a} + C(1-a)x^a y^{1-a}$

$$= \left[Ca + C(1-a)\right]x^a y^{1-a}$$

$$= Cx^a y^{1-a} = f$$

(b) $f(tx, ty) = C(tx)^a (ty)^{1-a} = Ct^a x^a t^{1-a} y^{1-a} = Cx^a y^{1-a}(t) = tf(x, y)$

11. (a) $x = 64(\cos 45°)t = 32\sqrt{2}\,t$

$y = 64(\sin 45°)t - 16t^2 = 32\sqrt{2}\,t - 16t^2$

(b) $\tan \alpha = \dfrac{y}{x + 50}$

$\alpha = \arctan\left(\dfrac{y}{x+50}\right) = \arctan\left(\dfrac{32\sqrt{2}\,t - 16t^2}{32\sqrt{2}\,t + 50}\right)$

(c) $\dfrac{d\alpha}{dt} = \dfrac{1}{1 + \left(\dfrac{32\sqrt{2}\,t - 16t^2}{32\sqrt{2}\,t + 50}\right)^2} \cdot \dfrac{-64\left(8\sqrt{2}\,t^2 + 25t - 25\sqrt{2}\right)}{\left(32\sqrt{2}\,t + 50\right)^2} = \dfrac{-16\left(8\sqrt{2}\,t^2 + 25t - 25\sqrt{2}\right)}{64t^4 - 256\sqrt{2}\,t^3 + 1024t^2 + 800\sqrt{2}\,t + 625}$

(d)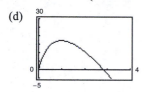

No. The rate of change of α is greatest when the projectile is closest to the camera.

(e) $\dfrac{d\alpha}{dt} = 0$ when

$8\sqrt{2}\,t^2 + 25t - 25\sqrt{2} = 0$

$$t = \dfrac{-25 + \sqrt{25^2 - 4\left(8\sqrt{2}\right)\left(-25\sqrt{2}\right)}}{2\left(8\sqrt{2}\right)} \approx 0.98 \text{ second.}$$

No, the projectile is at its maximum height when $dy/dt = 32\sqrt{2} - 32t = 0$ or $t = \sqrt{2} \approx 1.41$ seconds.

13. (a) There is a minimum at $(0, 0, 0)$, maxima at $(0, \pm1, 2/e)$ and saddle point at $(\pm1, 0, 1/e)$:

$$f_x = \left(x^2 + 2y^2\right)e^{-\left(x^2+y^2\right)}(-2x) + (2x)e^{-\left(x^2+y^2\right)}$$

$$= e^{-\left(x^2+y^2\right)}\left[\left(x^2 + 2y^2\right)(-2x) + 2x\right] = e^{-\left(x^2+y^2\right)}\left[-2x^3 + 4xy^2 + 2x\right] = 0 \Rightarrow x^3 + 2xy^2 - x = 0$$

$$f_y = \left(x^2 + 2y^2\right)e^{-\left(x^2+y^2\right)}(-2y) + (4y)e^{-\left(x^2+y^2\right)}$$

$$= e^{-\left(x^2+y^2\right)}\left[\left(x^2 + 2y^2\right)(-2y) + 4y\right] = e^{-\left(x^2+y^2\right)}\left[-4y^3 - 2x^2y + 4y\right] = 0 \Rightarrow 2y^3 + x^2y - 2y = 0$$

Solving the two equations $x^3 + 2xy^2 - x = 0$ and $2y^3 + x^2y - 2y = 0$, you obtain the following critical points:
$(0, \pm1), (\pm1, 0), (0, 0)$. Using the second derivative test, you obtain the results above.

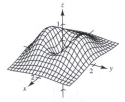

(b) As in part (a), you obtain

$$f_x = e^{-\left(x^2+y^2\right)}\left[2x\left(x^2 - 1 - 2y^2\right)\right]$$

$$f_y = e^{-\left(x^2+y^2\right)}\left[2y\left(2 + x^2 - 2y^2\right)\right]$$

The critical numbers are $(0, 0), (0, \pm1), (\pm1, 0)$.

These yield

$(\pm1, 0, -1/e)$ minima

$(0, \pm1, 2/e)$ maxima

$(0, 0, 0)$ saddle

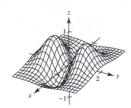

(c) In general, for $\alpha > 0$ you obtain

$(0, 0, 0)$ minimum

$(0, \pm1, \beta/e)$ maxima

$(\pm1, 0, \alpha/e)$ saddle

For $\alpha < 0$, you obtain

$(\pm1, 0, \alpha/e)$ minima

$(0, \pm1, \beta/e)$ maxima

$(0, 0, 0)$ saddle

15. (a)

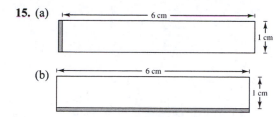

(b)

(c) The height has more effect since the shaded region in (b) is larger than the shaded region in (a).

(d) $A = hl \Rightarrow dA = l\,dh + h\,dl$

If $dl = 0.01$ and $dh = 0$, then $dA = 1(0.01) = 0.01$.

If $dh = 0.01$ and $dl = 0$, then $dA = 6(0.01) = 0.06$.

17. $\dfrac{\partial u}{\partial t} = \dfrac{1}{2}\Big[-\cos(x-t) + \cos(x+t)\Big]$

$\dfrac{\partial^2 u}{\partial t^2} = \dfrac{1}{2}\Big[-\sin(x-t) - \sin(x+t)\Big]$

$\dfrac{\partial u}{\partial x} = \dfrac{1}{2}\Big[\cos(x-t) + \cos(x+t)\Big]$

$\dfrac{\partial^2 u}{\partial x^2} = \dfrac{1}{2}\Big[-\sin(x-t) - \sin(x+t)\Big]$

Then, $\dfrac{\partial^2 u}{\partial t^2} = \dfrac{\partial^2 u}{\partial x^2}$.

19. $w = f(x, y),\ x = r\cos\theta,\ y = r\sin\theta$

$\dfrac{\partial w}{\partial r} = \dfrac{\partial w}{\partial x}\cos\theta + \dfrac{\partial w}{\partial y}\sin\theta$

$\dfrac{\partial w}{\partial\theta} = \dfrac{\partial w}{\partial x}(-r\sin\theta) + \dfrac{\partial w}{\partial y}(r\cos\theta)$

(a)

$r\cos\theta\,\dfrac{\partial w}{\partial r} = \dfrac{\partial w}{\partial x}r\cos^2\theta + \dfrac{\partial w}{\partial y}r\sin\theta\cos\theta$

$-\sin\theta\,\dfrac{\partial w}{\partial\theta} = \dfrac{\partial w}{\partial x}(r\sin^2\theta) - \dfrac{\partial w}{\partial y}r\sin\theta\cos\theta$

$r\cos\theta\,\dfrac{\partial w}{\partial r} - \sin\theta\,\dfrac{\partial w}{\partial\theta} = \dfrac{\partial w}{\partial x}(r\cos^2\theta + r\sin^2\theta)$

$r\dfrac{\partial w}{\partial x} = \dfrac{\partial w}{\partial r}(r\cos\theta) - \dfrac{\partial w}{\partial\theta}\sin\theta$

$\dfrac{\partial w}{\partial x} = \dfrac{\partial w}{\partial r}\cos\theta - \dfrac{\partial w}{\partial\theta}\dfrac{\sin\theta}{r}$ (First Formula)

$r\sin\theta\,\dfrac{\partial w}{\partial r} = \dfrac{\partial w}{\partial x}r\sin\theta\cos\theta + \dfrac{\partial w}{\partial y}r\sin^2\theta$

$\cos\theta\,\dfrac{\partial w}{\partial\theta} = \dfrac{\partial w}{\partial x}(-r\sin\theta\cos\theta) + \dfrac{\partial w}{\partial y}(r\cos^2\theta)$

$r\sin\theta\,\dfrac{\partial w}{\partial r} + \cos\theta\,\dfrac{\partial w}{\partial\theta} = \dfrac{\partial w}{\partial y}(r\sin^2\theta + r\cos^2\theta)$

$r\dfrac{\partial w}{\partial y} = \dfrac{\partial w}{\partial r}r\sin\theta + \dfrac{\partial w}{\partial\theta}\cos\theta$

$\dfrac{\partial w}{\partial y} = \dfrac{\partial w}{\partial r}\sin\theta + \dfrac{\partial w}{\partial\theta}\dfrac{\cos\theta}{r}$ (Second Formula)

(b) $\left(\dfrac{\partial w}{\partial r}\right)^2 + \dfrac{1}{r^2}\left(\dfrac{\partial w}{\partial\theta}\right)^2 = \left(\dfrac{\partial w}{\partial x}\right)^2\cos^2\theta + 2\dfrac{\partial w}{\partial x}\dfrac{\partial w}{\partial y}\sin\theta\cos\theta + \left(\dfrac{\partial w}{\partial y}\right)^2\sin^2\theta + \left(\dfrac{\partial w}{\partial x}\right)^2\sin^2\theta$

$- 2\dfrac{\partial w}{\partial x}\dfrac{\partial w}{\partial y}\sin\theta\cos\theta + \left(\dfrac{\partial w}{\partial y}\right)^2\cos^2\theta = \left(\dfrac{\partial w}{\partial x}\right)^2 + \left(\dfrac{\partial w}{\partial y}\right)^2$

21. $x = r \cos \theta, y = r \sin \theta, z = z$

$$\frac{\partial u}{\partial \theta} = \frac{\partial u}{\partial x}\frac{\partial x}{\partial \theta} + \frac{\partial u}{\partial y}\frac{\partial y}{\partial \theta} + \frac{\partial u}{\partial z}\frac{\partial z}{\partial \theta} = \frac{\partial u}{\partial x}(-r \sin \theta) + \frac{\partial u}{\partial y}r \cos \theta \text{ Similarly,}$$

$$\frac{\partial u}{\partial r} = \frac{\partial u}{\partial x}\cos \theta + \frac{\partial u}{\partial y}\sin \theta.$$

$$\frac{\partial^2 u}{\partial \theta^2} = (-r \sin \theta)\left[\frac{\partial^2 u}{\partial x^2}\frac{\partial x}{\partial \theta} + \frac{\partial^2 u}{\partial x \partial y}\frac{\partial y}{\partial \theta} + \frac{\partial^2 u}{\partial x \partial z}\frac{\partial z}{\partial \theta}\right] - r\frac{\partial u}{\partial x}\cos \theta + (r \cos \theta)\left[\frac{\partial^2 u}{\partial y \partial x}\frac{\partial x}{\partial \theta} + \frac{\partial^2 u}{\partial y^2}\frac{\partial y}{\partial \theta} + \frac{\partial^2 u}{\partial y \partial z}\frac{\partial z}{\partial \theta}\right] - r\frac{\partial u}{\partial y}\sin \theta$$

$$= \frac{\partial^2 u}{\partial x^2}r^2 \sin^2 \theta + \frac{\partial^2 u}{\partial y^2}r^2 \cos^2 \theta - 2\frac{\partial^2 u}{\partial x \partial y}r^2 \sin \theta \cos \theta - \frac{\partial u}{\partial x}r \cos \theta - \frac{\partial u}{\partial y}r \sin \theta$$

Similarly, $\dfrac{\partial^2 u}{\partial r^2} = \dfrac{\partial^2 u}{\partial x^2}\cos^2 \theta + \dfrac{\partial^2 u}{\partial y^2}\sin^2 \theta + 2\dfrac{\partial^2 u}{\partial x \partial y}\cos \theta \sin \theta.$

Now observe that

$$\frac{\partial^2 u}{\partial r^2} + \frac{1}{r}\frac{\partial u}{\partial r} + \frac{1}{r^2}\frac{\partial^2 u}{\partial \theta^2} + \frac{\partial^2 u}{\partial z^2} = \left[\frac{\partial^2 u}{\partial x^2}\cos^2 \theta + \frac{\partial^2 u}{\partial y^2}\sin^2 \theta + 2\frac{\partial^2 u}{\partial x \partial y}\cos \theta \sin \theta\right] + \frac{1}{r}\left[\frac{\partial u}{\partial x}\cos \theta + \frac{\partial u}{\partial y}\sin \theta\right]$$

$$+ \left[\frac{\partial^2 u}{\partial x^2}\sin^2 \theta + \frac{\partial^2 u}{\partial y^2}\cos^2 \theta - 2\frac{\partial^2 u}{\partial x \partial y}\sin \theta \cos \theta - \frac{1}{r}\frac{\partial u}{\partial x}\cos \theta - \frac{1}{r}\frac{\partial u}{\partial y}\sin \theta\right] + \frac{\partial^2 u}{\partial z^2}$$

$$= \frac{\partial^2 u}{\partial x^2} + \frac{\partial^2 u}{\partial y^2} + \frac{\partial^2 u}{\partial z^2}.$$

So, Laplace's equation in cylindrical coordinates, is $\dfrac{\partial^2 u}{\partial r^2} + \dfrac{1}{r}\dfrac{\partial u}{\partial r} + \dfrac{1}{r^2}\dfrac{\partial^2 u}{\partial \theta^2} + \dfrac{\partial^2 u}{\partial z^2} = 0.$

CHAPTER 14
Multiple Integration

C H A P T E R   1 4
Multiple Integration

Section 14.1   Iterated Integrals and Area in the Plane

1. $\int_0^x (x + 2y)\, dy = \left[xy + y^2 \right]_0^x = x^2 + x^2 = 2x^2$

5. $\int_0^{\sqrt{4-x^2}} x^2 y\, dy = \left[\frac{1}{2} x^2 y^2 \right]_0^{\sqrt{4-x^2}} = \frac{4x^2 - x^4}{2}$

3. $\int_1^{2y} \frac{y}{x}\, dx = \left[y \ln x \right]_1^{2y}$
$= y \ln 2y - 0 = y \ln 2y, \, (y > 0)$

7. $\int_{e^y}^y \frac{y \ln x}{x}\, dx = \left[\frac{1}{2} y \ln^2 x \right]_{e^y}^y = \frac{1}{2} y \left[\ln^2 y - \ln^2 e^y \right] = \frac{y}{2} \left[(\ln y)^2 - y^2 \right], \, (y > 0)$

9. $\int_0^{x^3} ye^{-y/x}\, dy = \left[-xye^{-y/x} \right]_0^{x^3} + x \int_0^{x^3} e^{-y/x}\, dy = -x^4 e^{-x^2} - \left[x^2 e^{-y/x} \right]_0^{x^3} = x^2 \left(1 - e^{-x^2} - x^2 e^{-x^2} \right)$
$u = y, \, du = dy, \, dv = e^{-y/x}\, dy, \, v = -xe^{-y/x}$

11. $\int_0^1 \int_0^2 (x + y)\, dy\, dx = \int_0^1 \left[xy + \frac{1}{2} y^2 \right]_0^2 dx = \int_0^1 (2x + 2)\, dx = \left[x^2 + 2x \right]_0^1 = 3$

13. $\int_1^2 \int_0^4 (x^2 - 2y^2)\, dx\, dy = \int_1^2 \left[\frac{x^3}{3} - 2xy^2 \right]_0^4 dy = \int_1^2 \left[\frac{64}{3} - 8y^2 \right] dy = \left[\frac{64}{3} y - \frac{8}{3} y^3 \right]_1^2 = \left(\frac{128}{3} - \frac{64}{3} \right) - \left(\frac{64}{3} - \frac{8}{3} \right) = \frac{8}{3}$

15. $\int_0^{\pi/2} \int_0^1 y \cos x\, dy\, dx = \int_0^{\pi/2} \left[\frac{y^2}{2} \cos x \right]_0^1 dx = \int_0^{\pi/2} \frac{1}{2} \cos x\, dx = \left[\frac{1}{2} \sin x \right]_0^{\pi/2} = \frac{1}{2}$

17. $\int_0^\pi \int_0^{\sin x} (1 + \cos x)\, dy\, dx = \int_0^\pi \left[(y + y \cos x) \right]_0^{\sin x} dx = \int_0^\pi \left[\sin x + \sin x \cos x \right] dx = \left[-\cos x + \frac{1}{2} \sin^2 x \right]_0^\pi = 1 + 1 = 2$

19. $\int_0^1 \int_0^x \sqrt{1 - x^2}\, dy\, dx = \int_0^1 \left[y\sqrt{1 - x^2} \right]_0^x dx = \int_0^1 x\sqrt{1 - x^2}\, dx = \left[-\frac{1}{2} \left(\frac{2}{3} \right) (1 - x^2)^{3/2} \right]_0^1 = \frac{1}{3}$

21. $\int_{-1}^5 \int_0^{3y} \left(3 + x^2 + \frac{1}{4} y^2 \right) dx\, dy = \int_{-1}^5 \left[3x + \frac{x^3}{3} + \frac{1}{4} xy^2 \right]_0^{3y} dy$

$= \int_{-1}^5 \left[9y + 9y^3 + \frac{3}{4} y^3 \right] dy = \int_{-1}^5 \left[9y + \frac{39}{4} y^3 \right] dy = \left[\frac{9}{2} y^2 + \frac{39}{16} y^4 \right]_{-1}^5 = \left(\frac{9}{2}(25) + \frac{39}{16}(625) \right) - \left(\frac{9}{2} + \frac{39}{16} \right) = 1629$

23. $\int_0^1 \int_0^{\sqrt{1-y^2}} (x + y)\, dx\, dy = \int_0^1 \left[\frac{1}{2} x^2 + xy \right]_0^{\sqrt{1-y^2}} dy = \int_0^1 \left[\frac{1}{2}(1 - y^2) + y\sqrt{1 - y^2} \right] dy = \left[\frac{1}{2} y - \frac{1}{6} y^3 - \frac{1}{2} \left(\frac{2}{3} \right) (1 - y^2)^{3/2} \right]_0^1 = \frac{2}{3}$

25. $\int_0^2 \int_0^{\sqrt{4-y^2}} \frac{2}{\sqrt{4 - y^2}}\, dx\, dy = \int_0^2 \left[\frac{2x}{\sqrt{4 - y^2}} \right]_0^{\sqrt{4-y^2}} dy = \int_0^2 2\, dy = \left[2y \right]_0^2 = 4$

27. $\int_0^{\pi/2} \int_0^{2\cos\theta} r\, dr\, d\theta = \int_0^{\pi/2} \left[\frac{r^2}{2} \right]_0^{2\cos\theta} d\theta = \int_0^{\pi/2} 2 \cos^2 \theta\, d\theta = \left[\theta - \frac{1}{2} \sin 2\theta \right]_0^{\pi/2} = \frac{\pi}{2}$

29. $\displaystyle\int_0^{\pi/2}\int_0^{\sin\theta}\theta r\,dr\,d\theta = \int_0^{\pi/2}\left[\theta\frac{r^2}{2}\right]_0^{\sin\theta}d\theta = \int_0^{\pi/2}\frac{1}{2}\theta\sin^2\theta\,d\theta$

$$= \frac{1}{4}\int_0^{\pi/2}(\theta-\theta\cos 2\theta)\,d\theta = \frac{1}{4}\left[\frac{\theta^2}{2}-\left(\frac{1}{4}\cos 2\theta+\frac{\theta}{2}\sin 2\theta\right)\right]_0^{\pi/2} = \frac{\pi^2}{32}+\frac{1}{8}$$

31. $\displaystyle\int_1^\infty\int_0^{1/x}y\,dy\,dx = \int_1^\infty\left[\frac{y^2}{2}\right]_0^{1/x}dx = \frac{1}{2}\int_1^\infty\frac{1}{x^2}\,dx = \left[-\frac{1}{2x}\right]_1^\infty = 0+\frac{1}{2} = \frac{1}{2}$

33. $\displaystyle\int_1^\infty\int_1^\infty\frac{1}{xy}\,dx\,dy = \int_1^\infty\left[\frac{1}{y}\ln x\right]_1^\infty dy = \int_1^\infty\left[\frac{1}{y}(\infty)-\frac{1}{y}(0)\right]dy$

Diverges

35. $A = \displaystyle\int_0^8\int_0^3 dy\,dx = \int_0^8[y]_0^3\,dx = \int_0^8 3\,dx = [3x]_0^8 = 24$

$A = \displaystyle\int_0^3\int_0^8 dx\,dy = \int_0^3[x]_0^8\,dy = \int_0^3 8\,dy = [8y]_0^3 = 24$

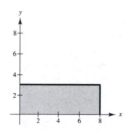

37. $A = \displaystyle\int_0^2\int_0^{4-x^2} dy\,dx = \int_0^2[y]_0^{4-x^2}\,dx = \int_0^2(4-x^2)\,dx = \left[4x-\frac{x^3}{3}\right]_0^2 = \frac{16}{3}$

$A = \displaystyle\int_0^4\int_0^{\sqrt{4-y}} dx\,dy = \int_0^4[x]_0^{\sqrt{4-y}}\,dy = \int_0^4\sqrt{4-y}\,dy = -\int_0^4(4-y)^{1/2}(-1)\,dy = \left[-\frac{2}{3}(4-y)^{3/2}\right]_0^4 = \frac{2}{3}(8) = \frac{16}{3}$

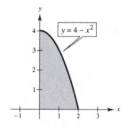

39. $A = \displaystyle\int_0^4\int_0^{(2-\sqrt{x})^2} dy\,dx = \int_0^4[y]_0^{(2-\sqrt{x})^2}\,dx = \int_0^4(4-4\sqrt{x}+x)\,dx = \left[4x-\frac{8}{3}x\sqrt{x}+\frac{x^2}{2}\right]_0^4 = \frac{8}{3}$

$A = \displaystyle\int_0^4\int_0^{(2-\sqrt{y})^2} dx\,dy = \frac{8}{3}$

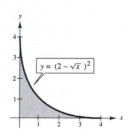

Integration steps are similar to those above.

41. $A = \int_0^3 \int_0^{2x/3} dy\, dx + \int_3^5 \int_0^{5-x} dy\, dx$

$= \int_0^3 \left[y\right]_0^{2x/3} dx + \int_3^5 \left[y\right]_0^{5-x} dx$

$= \int_0^3 \frac{2x}{3}\, dx + \int_3^5 (5-x)\, dx$

$= \left[\frac{1}{3}x^2\right]_0^3 + \left[5x - \frac{1}{2}x^2\right]_3^5 = 5$

$A = \int_0^2 \int_{3y/2}^{5-y} dx\, dy$

$= \int_0^2 \left[x\right]_{3y/2}^{5-y} dy$

$= \int_0^2 \left(5 - y - \frac{3y}{2}\right) dy$

$= \int_0^2 \left(5 - \frac{5y}{2}\right) dy = \left[5y - \frac{5}{4}y^2\right]_0^2 = 5$

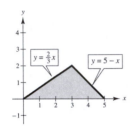

43. $A = \int_{-2}^1 \int_{x+2}^{4-x^2} dy\, dx$

$= \int_{-2}^1 \left[y\right]_{x+2}^{4-x^2} dx$

$= \int_{-2}^1 \left(4 - x^2 - x - 2\right) dx$

$= \int_{-2}^1 \left(2 - x - x^2\right) dx$

$= \left[2x - \frac{1}{2}x^2 - \frac{1}{3}x^3\right]_{-2}^1 = \frac{9}{2}$

$A = \int_0^3 \int_{-\sqrt{4-y}}^{y-2} dx\, dy + 2\int_3^4 \int_0^{\sqrt{4-y}} dx\, dy$

$= \int_0^3 \left[x\right]_{-\sqrt{4-y}}^{y-2} dy + 2\int_3^4 \left[x\right]_0^{\sqrt{4-y}} dy$

$= \int_0^3 \left(y - 2 + \sqrt{4-y}\right) dy + 2\int_3^4 \sqrt{4-y}\, dy$

$= \left[\frac{1}{2}y^2 - 2y - \frac{2}{3}(4-y)^{3/2}\right]_0^3 - \left[\frac{4}{3}(4-y)^{3/2}\right]_3^4 = \frac{9}{2}$

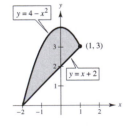

45. $\int_0^4 \int_0^y f(x, y)\, dx\, dy, \, 0 \le x \le y, 0 \le y \le 4$

$= \int_0^4 \int_x^4 f(x, y)\, dy\, dx$

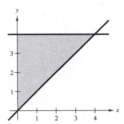

47. $\int_{-2}^2 \int_0^{\sqrt{4-x^2}} f(x, y)\, dy\, dx, \, 0 \le y \le \sqrt{4-x^2}, -2 \le x \le 2$

$= \int_0^2 \int_{-\sqrt{4-y^2}}^{\sqrt{4-y^2}} dx\, dy$

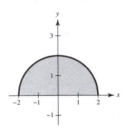

49. $\int_1^{10} \int_0^{\ln y} f(x, y)\, dx\, dy, \, 0 \le x \le \ln y, 1 \le y \le 10$

$= \int_0^{\ln 10} \int_{e^x}^{10} f(x, y)\, dy\, dx$

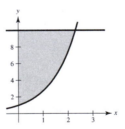

51. $\int_{-1}^1 \int_{x^2}^1 f(x, y)\, dy\, dx, \, x^2 \le y \le 1, -1 \le x \le 1$

$= \int_0^1 \int_{-\sqrt{y}}^{\sqrt{y}} f(x, y)\, dx\, dy$

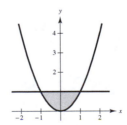

53. $\int_0^1 \int_0^2 dy\, dx = \int_0^2 \int_0^1 dx\, dy = 2$

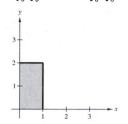

59. $\int_0^2 \int_{x/2}^1 dy\, dx = \int_0^1 \int_0^{2y} dx\, dy = 1$

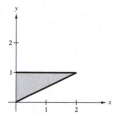

55. $\int_0^1 \int_{-\sqrt{1-y^2}}^{\sqrt{1-y^2}} dx\, dy = \int_{-1}^1 \int_0^{\sqrt{1-x^2}} dy\, dx = \dfrac{\pi}{2}$

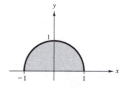

61. $\int_0^1 \int_{y^2}^{\sqrt[3]{y}} dx\, dy = \int_0^1 \int_{x^3}^{\sqrt{x}} dy\, dx = \dfrac{5}{12}$

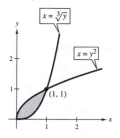

57. $\int_0^2 \int_0^x dy\, dx + \int_2^4 \int_0^{4-x} dy\, dx = \int_0^2 \int_y^{4-y} dx\, dy = 4$

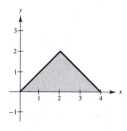

63. The first integral arises using vertical representative rectangles. The second two integrals arise using horizontal representative rectangles.

$$\int_0^5 \int_x^{\sqrt{50-x^2}} x^2 y^2\, dy\, dx = \int_0^5 \left[\tfrac{1}{3}x^2\left(50 - x^2\right)^{3/2} - \tfrac{1}{3}x^5 \right] dx = \frac{15{,}625}{24}\pi$$

$$\int_0^5 \int_0^y x^2 y^2\, dx\, dy + \int_5^{5\sqrt{2}} \int_0^{\sqrt{50-y^2}} x^2 y^2\, dx\, dy = \int_0^5 \tfrac{1}{3}y^5\, dy + \int_5^{5\sqrt{2}} \tfrac{1}{3}\left(50 - y^2\right)^{3/2} y^2\, dy$$

$$= \frac{15{,}625}{18} + \left(\frac{15{,}625}{18}\pi - \frac{15{,}625}{18} \right) = \frac{15{,}625}{24}\pi$$

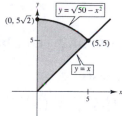

65. $\int_0^2 \int_x^2 x\sqrt{1 + y^3}\, dy\, dx = \int_0^2 \int_0^y x\sqrt{1 + y^3}\, dx\, dy$

$$= \int_0^2 \left[\sqrt{1 + y^3} \cdot \frac{x^2}{2} \right]_0^y dy$$

$$= \frac{1}{2} \int_0^2 \sqrt{1 + y^3}\; y^2\, dy$$

$$= \left[\frac{1}{2} \cdot \frac{1}{3} \cdot \frac{2}{3}\left(1 + y^3\right)^{3/2} \right]_0^2$$

$$= \frac{1}{9}(27) - \frac{1}{9}(1) = \frac{26}{9}$$

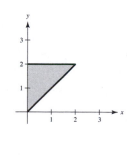

67. $\int_0^1 \int_{2x}^2 4e^{y^2} \, dy \, dx = \int_0^2 \int_0^{y/2} 4e^{y^2} \, dx \, dy$

$$= \int_0^2 \left[4xe^{y^2} \right]_0^{y/2} \, dy = \int_0^2 2ye^{y^2} \, dy$$

$$= \left[e^{y^2} \right]_0^2 = e^4 - 1$$

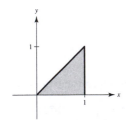

69. $\int_0^1 \int_y^1 \sin(x^2) \, dx \, dy = \int_0^1 \int_0^x \sin(x^2) \, dy \, dx$

$$= \int_0^1 \left[y \sin(x^2) \right]_0^x \, dx$$

$$= \int_0^1 x \sin(x^2) \, dx$$

$$= \left[-\frac{1}{2} \cos(x^2) \right]_0^1$$

$$= -\frac{1}{2} \cos 1 + \frac{1}{2}(1)$$

$$= \frac{1}{2}(1 - \cos 1) \approx 0.2298$$

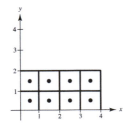

71. $\int_0^2 \int_{x^2}^{2x} (x^3 + 3y^2) \, dy \, dx = \frac{1664}{105} \approx 15.848$

73. $\int_0^4 \int_0^y \frac{2}{(x + 1)(y + 1)} \, dx \, dy = (\ln 5)^2 \approx 2.590$

75. $\int_0^2 \int_0^{4-x^2} e^{xy} \, dy \, dx \approx 20.5648$

77. $\int_0^{2\pi} \int_0^{1+\cos\theta} 6r^2 \cos\theta \, dr \, d\theta = \frac{15\pi}{2}$

79. (a) $x = y^3 \Leftrightarrow y = x^{1/3}$

$$x = 4\sqrt{2y} \Leftrightarrow x^2 = 32y \Leftrightarrow y = \frac{x^2}{32}$$

  (b) $\int_0^8 \int_{x^2/32}^{x^{1/3}} (x^2y - xy^2) \, dy \, dx$

  (c) Both integrals equal $\frac{67,520}{693} \approx 97.43$.

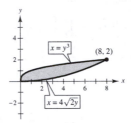

81. An iterated integral is integration of a function of several variables. Integrate with respect to one variable while holding the other variables constant.

83. The region is a rectangle.

85. True

Section 14.2   Double Integrals and Volume

For Exercises 1–3, $\Delta x_i = \Delta y_i = 1$ **and the midpoints of the squares are**

$\left(\frac{1}{2}, \frac{1}{2} \right), \left(\frac{3}{2}, \frac{1}{2} \right), \left(\frac{5}{2}, \frac{1}{2} \right), \left(\frac{7}{2}, \frac{1}{2} \right), \left(\frac{1}{2}, \frac{3}{2} \right), \left(\frac{3}{2}, \frac{3}{2} \right), \left(\frac{5}{2}, \frac{3}{2} \right), \left(\frac{7}{2}, \frac{3}{2} \right).$

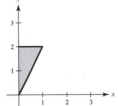

1. $f(x, y) = x + y$

$$\sum_{i=1}^8 f(x_i, y_i) \, \Delta x_i \Delta y_i = 1 + 2 + 3 + 4 + 2 + 3 + 4 + 5 = 24$$

$$\int_0^4 \int_0^2 (x + y) \, dy \, dx = \int_0^4 \left[xy + \frac{y^2}{2} \right]_0^2 \, dx = \int_0^4 (2x + 2) \, dx = \left[x^2 + 2x \right]_0^4 = 24$$

3. $f(x, y) = x^2 + y^2$

$$\sum_{i=1}^{8} f(x_i, y_i)\, \Delta x_i\, \Delta y_i = \frac{2}{4} + \frac{10}{4} + \frac{26}{4} + \frac{50}{4} + \frac{10}{4} + \frac{18}{4} + \frac{34}{4} + \frac{58}{4} = 52$$

$$\int_0^4 \int_0^2 (x^2 + y^2)\, dy\, dx = \int_0^4 \left[x^2 y + \frac{y^3}{3} \right]_0^2 dx = \int_0^4 \left(2x^2 + \frac{8}{3} \right) dx = \left[\frac{2x^3}{3} + \frac{8x}{3} \right]_0^4 = \frac{160}{3}$$

5. $\displaystyle \int_0^2 \int_0^1 (1 + 2x + 2y)\, dy\, dx = \int_0^2 \left[y + 2xy + y^2 \right]_0^1 dx = \int_0^2 (2 + 2x)\, dx = \left[2x + x^2 \right]_0^2 = 8$

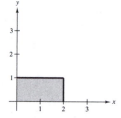

7. $\displaystyle \int_0^6 \int_{y/2}^3 (x + y)\, dx\, dy = \int_0^6 \left[\frac{1}{2}x^2 + xy \right]_{y/2}^3 dy = \int_0^6 \left(\frac{9}{2} + 3y - \frac{5}{8}y^2 \right) dy = \left[\frac{9}{2}y + \frac{3}{2}y^2 - \frac{5}{24}y^3 \right]_0^6 = 36$

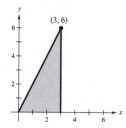

9. $\displaystyle \int_{-a}^a \int_{-\sqrt{a^2-x^2}}^{\sqrt{a^2-x^2}} (x + y)\, dy\, dx = \int_{-a}^a \left[xy + \frac{1}{2}y^2 \right]_{-\sqrt{a^2-x^2}}^{\sqrt{a^2-x^2}} dx = \int_{-a}^a 2x\sqrt{a^2 - x^2}\, dx = \left[-\frac{2}{3}(a^2 - x^2)^{3/2} \right]_{-a}^a = 0$

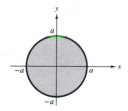

11. $\displaystyle \int_0^5 \int_0^3 xy\, dx\, dy = \int_0^3 \int_0^5 xy\, dy\, dx$

$$= \int_0^3 \left[\tfrac{1}{2}xy^2 \right]_0^5 dx = \tfrac{25}{2} \int_0^3 x\, dx = \left[\tfrac{25}{4}x^2 \right]_0^3 = \tfrac{225}{4}$$

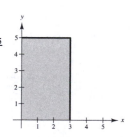

13. $\int_1^2 \int_1^y \dfrac{y}{x^2 + y^2}\, dx\, dy + \int_2^4 \int_{y/2}^2 \dfrac{y}{x^2 + y^2}\, dx\, dy = \int_1^2 \int_x^{2x} \dfrac{y}{x^2 + y^2}\, dy\, dx$

$= \dfrac{1}{2} \int_1^2 \left[\ln(x^2 + y^2) \right]_x^{2x} dx$

$= \dfrac{1}{2} \int_1^2 \left(\ln 5x^2 - \ln 2x^2 \right) dx$

$= \dfrac{1}{2} \ln \dfrac{5}{2} \int_1^2 dx = \left[\dfrac{1}{2}\left(\ln \dfrac{5}{2}\right) x \right]_1^2 = \dfrac{1}{2} \ln \dfrac{5}{2}$

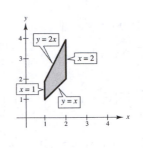

15. $\int_3^4 \int_{4-y}^{\sqrt{4-y}} -2y\, dx\, dy = \int_0^1 \int_{4-x}^{4-x^2} -2y\, dy\, dx$

$= \int_0^1 \left[-y^2 \right]_{4-x}^{4-x^2} dx$

$= -\int_0^1 \left[(4 - x^2)^2 - (4 - x)^2 \right] dx$

$= -\int_0^1 \left[16 - 8x^2 + x^4 - (16 - 8x + x^2) \right] dx$

$= -\left[-3x^3 + \dfrac{x^5}{5} + 4x^2 \right]_0^1 = -\dfrac{6}{5}$

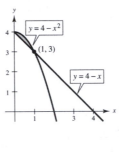

17. $\int_0^4 \int_0^{3x/4} x\, dy\, dx + \int_4^5 \int_0^{\sqrt{25-x^2}} x\, dy\, dx = \int_0^3 \int_{4y/3}^{\sqrt{25-y^2}} x\, dx\, dy$

$= \int_0^3 \left[\dfrac{1}{2}x^2 \right]_{4y/3}^{\sqrt{25-y^2}} dy$

$= \dfrac{25}{18} \int_0^3 \left(9 - y^2 \right) dy = \left[\dfrac{25}{18}\left(9y - \dfrac{1}{3}y^3 \right) \right]_0^3 = 25$

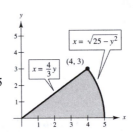

19. $V = \int_0^4 \int_0^2 \dfrac{y}{2}\, dy\, dx = \int_0^4 \left[\dfrac{y^2}{4} \right]_0^2 dx = \int_0^4 dx = 4$

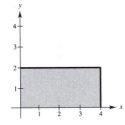

21. $V = \int_0^2 \int_0^y (4 - x - y)\, dx\, dy$

$= \int_0^2 \left[4x - \dfrac{x^2}{2} - xy \right]_0^y dy$

$= \int_0^2 \left(4y - \dfrac{y^2}{2} - y^2 \right) dy$

$= \left[2y^2 - \dfrac{y^3}{6} - \dfrac{y^3}{3} \right]_0^2$

$= 8 - \dfrac{8}{6} - \dfrac{8}{3} = 4$

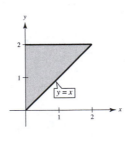

23. $V = \int_0^1 \int_0^y (1 - xy) \, dx \, dy$

$= \int_0^1 \left[x - \dfrac{x^2 y}{2} \right]_0^y dy = \int_0^1 \left(y - \dfrac{y^3}{2} \right) dy = \left[\dfrac{y^2}{2} - \dfrac{y^4}{8} \right]_0^1 = \dfrac{3}{8}$

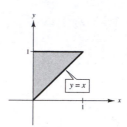

25. $V = \int_0^\infty \int_0^\infty \dfrac{1}{(x+1)^2 (y+1)^2} \, dy \, dx = \int_0^\infty \left[-\dfrac{1}{(x+1)^2(y+1)} \right]_0^\infty dx = \int_0^\infty \dfrac{1}{(x+1)^2} \, dx = \left[-\dfrac{1}{(x+1)} \right]_0^\infty = 1$

27. $V = \int_0^1 \int_0^x xy \, dy \, dx = \int_0^1 \left[\dfrac{1}{2} xy^2 \right]_0^x dx = \dfrac{1}{2} \int_0^1 x^3 \, dx = \left[\dfrac{1}{8} x^4 \right]_0^1 = \dfrac{1}{8}$

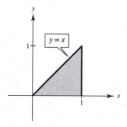

29. Divide the solid into two equal parts.

$V = 2 \int_0^1 \int_0^x \sqrt{1 - x^2} \, dy \, dx = 2 \int_0^1 \left[y\sqrt{1 - x^2} \right]_0^x dx$

$= 2 \int_0^1 x\sqrt{1 - x^2} \, dx = \left[-\dfrac{2}{3}\left(1 - x^2\right)^{3/2} \right]_0^1 = \dfrac{2}{3}$

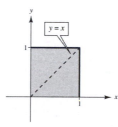

31. $V = \int_0^2 \int_0^{\sqrt{4-x^2}} (x + y) \, dy \, dx = \int_0^2 \left[xy + \dfrac{1}{2} y^2 \right]_0^{\sqrt{4-x^2}} dx$

$= \int_0^2 \left(x\sqrt{4 - x^2} + 2 - \dfrac{1}{2} x^2 \right) dx$

$= \left[-\dfrac{1}{3}\left(4 - x^2\right)^{3/2} + 2x - \dfrac{1}{6} x^3 \right]_0^2 = \dfrac{16}{3}$

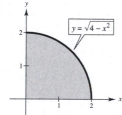

33. $V = 2 \int_0^2 \int_0^{\sqrt{1-(x-1)^2}} \left(\left[4 - x^2 - y^2 \right] - \left[4 - 2x \right] \right) dy \, dx$

$= 2 \int_0^2 \int_0^{\sqrt{1-(x-1)^2}} \left(2x - x^2 - y^2 \right) dy \, dx$

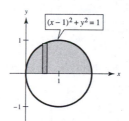

35. $V = 4 \int_0^2 \int_0^{\sqrt{4-x^2}} \left(x^2 + y^2 \right) dy \, dx$

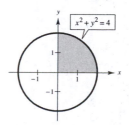

37. $V = \int_0^2 \int_{-\sqrt{2-2(y-1)^2}}^{\sqrt{2-2(y-1)^2}} \left[4y - \left(x^2 + 2y^2\right)\right] dx \, dy$

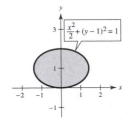

43. f is a continuous function such that
$0 \le f(x, y) \le 1$ over a region R of area 1. Let
$f(m, n) =$ the minimum value of f over R and
$f(M, N) =$ the maximum value of f over R. Then

$$f(m, n)\int_R \int dA \le \int_R \int f(x, y) \, dA \le f(M, N)\int_R \int dA.$$

Because $\int_R \int dA = 1$ and

$0 \le f(m, n) \le f(M, N) \le 1$, you have

$0 \le f(m, n)(1) \le \int_R \int f(x, y) \, dA \le f(M, N)(1) \le 1.$

So, $0 \le \int_R \int f(x, y) \, dA \le 1.$

39. $V = 4\int_0^3 \int_0^{\sqrt{9-x^2}} \left(9 - x^2 - y^2\right) dy \, dx = \dfrac{81\pi}{2}$

41. $V = \int_0^2 \int_0^{-0.5x+1} \dfrac{2}{1 + x^2 + y^2} \, dy \, dx \approx 1.2315$

45. $\int_0^1 \int_{y/2}^{1/2} e^{-x^2} \, dx \, dy = \int_0^{1/2} \int_0^{2x} e^{-x^2} \, dy \, dx$

$= \int_0^{1/2} 2xe^{-x^2} \, dx = \left[-e^{-x^2}\right]_0^{1/2} = -e^{-1/4} + 1 = 1 - e^{-1/4} \approx 0.221$

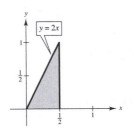

47. $\int_{-2}^{2} \int_{-\sqrt{4-x^2}}^{\sqrt{4-x^2}} \sqrt{4 - y^2} \, dy \, dx = \int_{-2}^{2} \int_{-\sqrt{4-y^2}}^{\sqrt{4-y^2}} \sqrt{4 - y^2} \, dx \, dy = \int_{-2}^{2} \left[x\sqrt{4 - y^2}\right]_{-\sqrt{4-y^2}}^{\sqrt{4-y^2}} dy$

$= \int_{-2}^{2} 2\left(4 - y^2\right) dy = \left[8y - \dfrac{2y^3}{3}\right]_{-2}^{2} = \left(16 - \dfrac{16}{3}\right) - \left(-16 + \dfrac{16}{3}\right)$

$= \dfrac{64}{3}$

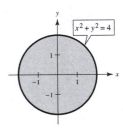

49. $\int_0^1 \int_0^{\arccos y} \sin x\sqrt{1 + \sin^2 x} \, dx \, dy$

$= \int_0^{\pi/2} \int_0^{\cos x} \sin x\sqrt{1 + \sin^2 x} \, dy \, dx$

$= \int_0^{\pi/2} \left(1 + \sin^2 x\right)^{1/2} \sin x \cos x \, dx = \left[\dfrac{1}{2} \cdot \dfrac{2}{3}\left(1 + \sin^2 x\right)^{3/2}\right]_0^{\pi/2} = \dfrac{1}{3}\left[2\sqrt{2} - 1\right]$

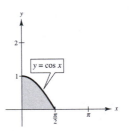

51. Average $= \dfrac{1}{8}\int_0^4 \int_0^2 x \, dy \, dx = \dfrac{1}{8}\int_0^4 2x \, dx = \left[\dfrac{x^2}{8}\right]_0^4 = 2$

53. Average $= \dfrac{1}{4}\int_0^2 \int_0^2 \left(x^2 + y^2\right) dx \, dy$

$= \dfrac{1}{4}\int_0^2 \left[\dfrac{x^3}{3} + xy^2\right]_0^2 dy = \dfrac{1}{4}\int_0^2 \left(\dfrac{8}{3} + 2y^2\right) dy$

$= \left[\dfrac{1}{4}\left(\dfrac{8}{3}y + \dfrac{2}{3}y^3\right)\right]_0^2 = \dfrac{8}{3}$

55. Average $= \dfrac{1}{1/2}\int_0^1 \int_x^1 e^{x+y} \, dy \, dx = 2\int_0^1 e^{x+1} - e^{2x} \, dx$

$= 2\left[e^{x+1} - \dfrac{1}{2}e^{2x}\right]_0^1 = 2\left[e^2 - \dfrac{1}{2}e^2 - e + \dfrac{1}{2}\right]$

$= e^2 - 2e + 1 = (e - 1)^2$

57. Average $= \dfrac{1}{1250} \displaystyle\int_{300}^{325} \int_{200}^{250} 100x^{0.6} y^{0.4} \, dx \, dy$

$= \dfrac{1}{1250} \displaystyle\int_{300}^{325} \left[(100y^{0.4}) \dfrac{x^{1.6}}{1.6} \right]_{200}^{250} dy$

$= \dfrac{128{,}844.1}{1250} \displaystyle\int_{300}^{325} y^{0.4} \, dy$

$= 103.0753 \left[\dfrac{y^{1.4}}{1.4} \right]_{300}^{325} \approx 25{,}645.24$

59. See the definition on page 976.

65. $f(x, y) \geq 0$ for all (x, y) and

$$\int_{-\infty}^{\infty} \int_{-\infty}^{\infty} f(x, y) \, dA = \int_{0}^{3} \int_{3}^{6} \dfrac{1}{27}(9 - x - y) \, dy \, dx = \int_{0}^{3} \dfrac{1}{27} \left[9y - xy - \dfrac{y^2}{2} \right]_{3}^{6} dx = \int_{0}^{3} \left(\dfrac{1}{2} - \dfrac{1}{9}x \right) dx = \left[\dfrac{x}{2} - \dfrac{x^2}{18} \right]_{0}^{3} = 1$$

$$P(0 \leq x \leq 1, 4 \leq y \leq 6) = \int_{0}^{1} \int_{4}^{6} \dfrac{1}{27}(9 - x - y) \, dy \, dx = \int_{0}^{1} \dfrac{2}{27}(4 - x) \, dx = \dfrac{7}{27}.$$

67. $\displaystyle\int_{0}^{4} \int_{0}^{4} f(x, y) \, dy \, dx \approx (32 + 31 + 28 + 23) + (31 + 30 + 27 + 22) + (28 + 27 + 24 + 19) + (23 + 22 + 19 + 14)$

$= 400$

Using the corner of the ith square farthest from the origin, you obtain 272.

69. False

$V = 8 \displaystyle\int_{0}^{1} \int_{0}^{\sqrt{1-y^2}} \sqrt{1 - x^2 - y^2} \, dx \, dy$

71. $z = 9 - x^2 - y^2$ is a paraboloid opening downward with vertex $(0, 0, 9)$. The double integral is maximized if $z \geq 0$. That is,

$R = \{ (x, y) : x^2 + y^2 \leq 9 \}$.

$\left[\text{The maximum value is } \displaystyle\iint_{R} (9 - x^2 - y^2) \, dA = \dfrac{81\pi}{2}. \right]$

61. No, the maximum possible value is $(\text{Area})(6) = 6\pi$.

63. $f(x, y) \geq 0$ for all (x, y) and

$$\int_{-\infty}^{\infty} \int_{-\infty}^{\infty} f(x, y) \, dA = \int_{0}^{5} \int_{0}^{2} \dfrac{1}{10} \, dy \, dx$$

$$= \int_{0}^{5} \dfrac{1}{5} \, dx = 1$$

$$P(0 \leq x \leq 2, 1 \leq y \leq 2) = \int_{0}^{2} \int_{1}^{2} \dfrac{1}{10} \, dy \, dx$$

$$= \int_{0}^{2} \dfrac{1}{10} \, dx = \dfrac{1}{5}.$$

73. Average $= \displaystyle\int_{0}^{1} f(x) \, dx = \int_{0}^{1} \int_{1}^{x} e^{t^2} \, dt \, dx = -\int_{0}^{1} \int_{x}^{1} e^{t^2} \, dt \, dx$

$= -\displaystyle\int_{0}^{1} \int_{0}^{t} e^{t^2} \, dx \, dt = -\int_{0}^{1} t e^{t^2} \, dt$

$= \left[-\dfrac{1}{2} e^{t^2} \right]_{0}^{1} = -\dfrac{1}{2}(e - 1) = \dfrac{1}{2}(1 - e)$

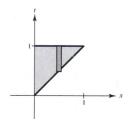

75. Let $I = \displaystyle\int_{0}^{a} \int_{0}^{b} e^{\max\{b^2 x^2, \, a^2 y^2\}} \, dy \, dx$.

Divide the rectangle into two parts by the diagonal line $ay = bx$. On lower triangle, $b^2 x^2 \geq a^2 y^2$ because $y \leq \dfrac{b}{a}x$.

$I = \displaystyle\int_{0}^{a} \int_{0}^{bx/a} e^{b^2 x^2} \, dy \, dx + \int_{0}^{b} \int_{0}^{ay/b} e^{a^2 y^2} \, dx \, dy = \int_{0}^{a} \dfrac{bx}{a} e^{b^2 x^2} \, dx + \int_{0}^{b} \dfrac{ay}{b} e^{a^2 y^2} \, dy$

$= \dfrac{1}{2ab} \left[e^{b^2 x^2} \right]_{0}^{a} + \dfrac{1}{2ab} \left[e^{a^2 y^2} \right]_{0}^{b} = \dfrac{1}{2ab} \left[e^{b^2 a^2} - 1 + e^{a^2 b^2} - 1 \right] = \dfrac{e^{a^2 b^2} - 1}{ab}$

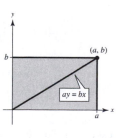

Section 14.3   Change of Variables: Polar Coordinates

1. Rectangular coordinates

3. Polar coordinates

5. $R = \{ (r, \theta) : 0 \leq r \leq 8, 0 \leq \theta \leq \pi \}$

7. $R = \left\{(r, \theta) : 4 \le r \le 8, 0 \le \theta \le \dfrac{\pi}{2}\right\}$

9. $\displaystyle\int_0^{\pi} \int_0^{\cos\theta} r\, dr\, d\theta$

$\displaystyle = \int_0^{\pi} \left[\frac{r^2}{2}\right]_0^{\cos\theta} d\theta$

$\displaystyle = \int_0^{\pi} \frac{1}{2}\cos^2\theta\, d\theta$

$\displaystyle = \int_0^{\pi} \frac{1}{4}(1 + \cos 2\theta)\, d\theta$

$\displaystyle = \frac{1}{4}\left[\theta + \frac{1}{2}\sin 2\theta\right]_0^{\pi} = \frac{\pi}{4}$

11. $\displaystyle\int_0^{2\pi} \int_0^{6} 3r^2 \sin\theta\, dr\, d\theta = \int_0^{2\pi} \left[r^3 \sin\theta\right]_0^{6} d\theta$

$\displaystyle = \int_0^{2\pi} 216 \sin\theta\, d\theta$

$\displaystyle = \left[-216 \cos\theta\right]_0^{2\pi} = 0$

13. $\displaystyle\int_0^{\pi/2} \int_2^{3} \sqrt{9 - r^2}\, r\, dr\, d\theta = \int_0^{\pi/2} \left[-\frac{1}{3}\left(9 - r^2\right)^{3/2}\right]_2^{3} d\theta$

$\displaystyle = \left[\frac{5\sqrt{5}}{3}\theta\right]_0^{\pi/2} = \frac{5\sqrt{5}\pi}{6}$

15. $\displaystyle\int_0^{\pi/2} \int_0^{1+\sin\theta} \theta r\, dr\, d\theta = \int_0^{\pi/2} \left[\frac{\theta r^2}{2}\right]_0^{1+\sin\theta} d\theta$

$\displaystyle = \int_0^{\pi/2} \frac{1}{2}\theta(1 + \sin\theta)^2\, d\theta$

$\displaystyle = \left[\frac{1}{8}\theta^2 + \sin\theta - \theta\cos\theta + \frac{1}{2}\theta\left(-\frac{1}{2}\cos\theta \cdot \sin\theta + \frac{1}{2}\theta\right) + \frac{1}{8}\sin^2\theta\right]_0^{\pi/2}$

$\displaystyle = \frac{3}{32}\pi^2 + \frac{9}{8}$

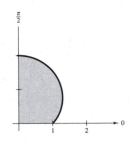

17. $\displaystyle\int_0^{a} \int_0^{\sqrt{a^2-y^2}} y\, dx\, dy = \int_0^{\pi/2} \int_0^{a} r^2 \sin\theta\, dr\, d\theta = \frac{a^3}{3}\int_0^{\pi/2} \sin\theta\, d\theta = \left[\frac{a^3}{3}(-\cos\theta)\right]_0^{\pi/2} = \frac{a^3}{3}$

19. $\displaystyle\int_{-2}^{2} \int_0^{\sqrt{4-x^2}} \left(x^2 + y^2\right) dy\, dx = \int_0^{\pi} \int_0^{2} r^2\, r\, dr\, d\theta = \int_0^{\pi} \left[\frac{r^4}{4}\right]_0^{2} d\theta = \int_0^{\pi} 4\, d\theta = 4\pi$

21. $\displaystyle\int_0^{3} \int_0^{\sqrt{9-x^2}} \left(x^2 + y^2\right)^{3/2} dy\, dx = \int_0^{\pi/2} \int_0^{3} r^4\, dr\, d\theta = \frac{243}{5}\int_0^{\pi/2} d\theta = \frac{243\pi}{10}$

23. $\displaystyle\int_0^{2} \int_0^{\sqrt{2x-x^2}} xy\, dy\, dx = \int_0^{\pi/2} \int_0^{2\cos\theta} r^3 \cos\theta \sin\theta\, dr\, d\theta = 4\int_0^{\pi/2} \cos^5\theta \sin\theta\, d\theta = \left[-\frac{4\cos^6\theta}{6}\right]_0^{\pi/2} = \frac{2}{3}$

25. $\displaystyle\int_{-1}^{1} \int_0^{\sqrt{1-x^2}} \cos\left(x^2 + y^2\right) dy\, dx = \int_0^{\pi} \int_0^{1} \cos\left(r^2\right) r\, dr\, d\theta = \int_0^{\pi} \left[\frac{1}{2}\sin\left(r^2\right)\right]_0^{1} d\theta = \int_0^{\pi} \frac{1}{2}\sin(1)\, d\theta = \frac{\pi}{2}\sin(1) \approx 1.3218$

27. $\int_0^2 \int_0^x \sqrt{x^2 + y^2}\, dy\, dx + \int_2^{2\sqrt{2}} \int_0^{\sqrt{8-x^2}} \sqrt{x^2 + y^2}\, dy\, dx = \int_0^{\pi/4} \int_0^{2\sqrt{2}} r^2\, dr\, d\theta$

$$= \int_0^{\pi/4} \frac{16\sqrt{2}}{3}\, d\theta = \frac{4\sqrt{2}\,\pi}{3}$$

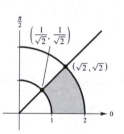

29. $\int_0^2 \int_0^{\sqrt{4-x^2}} (x + y)\, dy\, dx = \int_0^{\pi/2} \int_0^2 (r\cos\theta + r\sin\theta)r\, dr\, d\theta = \int_0^{\pi/2} \int_0^2 (\cos\theta + \sin\theta)r^2\, dr\, d\theta$

$$= \frac{8}{3} \int_0^{\pi/2} (\cos\theta + \sin\theta)\, d\theta = \left[\frac{8}{3}(\sin\theta - \cos\theta)\right]_0^{\pi/2} = \frac{16}{3}$$

31. $\int_0^{1/\sqrt{2}} \int_{\sqrt{1-y^2}}^{\sqrt{4-y^2}} \arctan\frac{y}{x}\, dx\, dy + \int_{1/\sqrt{2}}^{\sqrt{2}} \int_y^{\sqrt{4-y^2}} \arctan\frac{y}{x}\, dx\, dy$

$$= \int_0^{\pi/4} \int_1^2 \theta r\, dr\, d\theta$$

$$= \int_0^{\pi/4} \frac{3}{2}\,\theta\, d\theta = \left[\frac{3\theta^2}{4}\right]_0^{\pi/4} = \frac{3\pi^2}{64}$$

33. $V = \int_0^{\pi/2} \int_0^1 (r\cos\theta)(r\sin\theta)r\, dr\, d\theta$

$$= \frac{1}{2} \int_0^{\pi/2} \int_0^1 r^3 \sin 2\theta\, dr\, d\theta = \frac{1}{8} \int_0^{\pi/2} \sin 2\theta\, d\theta = \left[-\frac{1}{16}\cos 2\theta\right]_0^{\pi/2} = \frac{1}{8}$$

35. $V = \int_0^{2\pi} \int_0^5 r^2\, dr\, d\theta = \int_0^{2\pi} \frac{125}{3}\, d\theta = \frac{250\pi}{3}$

37. $V = 2\int_0^{\pi/2} \int_0^{4\cos\theta} \sqrt{16 - r^2}\, r\, dr\, d\theta = 2\int_0^{\pi/2} \left[-\frac{1}{3}\left(\sqrt{16 - r^2}\right)^3\right]_0^{4\cos\theta} d\theta = -\frac{2}{3} \int_0^{\pi/2} (64\sin^3\theta - 64)\, d\theta$

$$= \frac{128}{3} \int_0^{\pi/2} \left[1 - \sin\theta(1 - \cos^2\theta)\right] d\theta = \frac{128}{3}\left[\theta + \cos\theta - \frac{\cos^3\theta}{3}\right]_0^{\pi/2} = \frac{64}{9}(3\pi - 4)$$

39. $V = \int_0^{2\pi} \int_a^4 \sqrt{16 - r^2}\, r\, dr\, d\theta = \int_0^{2\pi} \left[-\frac{1}{3}\left(\sqrt{16 - r^2}\right)^3\right]_a^4 d\theta = \frac{1}{3}\left(\sqrt{16 - a^2}\right)^3 (2\pi)$

One-half the volume of the hemisphere is $(64\pi)/3$.

$$\frac{2\pi}{3}(16 - a^2)^{3/2} = \frac{64\pi}{3}$$

$$(16 - a^2)^{3/2} = 32$$

$$16 - a^2 = 32^{2/3}$$

$$a^2 = 16 - 32^{2/3} = 16 - 8\sqrt[3]{2}$$

$$a = \sqrt{4\left(4 - 2\sqrt[3]{2}\right)} = 2\sqrt{4 - 2\sqrt[3]{2}} \approx 2.4332$$

41. $A = \int_0^\pi \int_0^{6\cos\theta} r\, dr\, d\theta = \int_0^\pi 18\cos^2\theta\, d\theta = 9\int_0^\pi (1 + \cos 2\theta)\, d\theta = \left[9\left(\theta + \frac{1}{2}\sin 2\theta\right)\right]_0^\pi = 9\pi$

43. $A = \int_0^{2\pi} \int_0^{1+\cos\theta} r \, dr \, d\theta = \frac{1}{2}\int_0^{2\pi} \left(1 + 2\cos\theta + \cos^2\theta\right) d\theta$

$$= \frac{1}{2}\int_0^{2\pi} \left(1 + 2\cos\theta + \frac{1 + \cos 2\theta}{2}\right) d\theta = \frac{1}{2}\left[\theta + 2\sin\theta + \frac{1}{2}\left(\theta + \frac{1}{2}\sin 2\theta\right)\right]_0^{2\pi} = \frac{3\pi}{2}$$

45. $A = 3\int_0^{\pi/3} \int_0^{2\sin 3\theta} r \, dr \, d\theta = \frac{3}{2}\int_0^{\pi/3} 4\sin^2 3\theta \, d\theta = 3\int_0^{\pi/3} (1 - \cos 6\theta)\, d\theta = 3\left[\theta - \frac{1}{6}\sin 6\theta\right]_0^{\pi/3} = \pi$

47. $r = 1 = 2\cos\theta \Rightarrow \theta = \pm\frac{\pi}{3}$

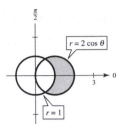

$A = 2\int_0^{\pi/3} \int_1^{2\cos\theta} r \, dr \, d\theta = 2\int_0^{\pi/3} \left[\frac{r^2}{2}\right]_1^{2\cos\theta} d\theta = 2\int_0^{\pi/3} \left(2\cos^2\theta - \frac{1}{2}\right) d\theta$

$\qquad = 2\int_0^{\pi/3} \left(1 + \cos 2\theta - \frac{1}{2}\right) d\theta = 2\left[\frac{1}{2}\theta + \frac{\sin 2\theta}{2}\right]_0^{\pi/3} = 2\left[\frac{\pi}{6} + \frac{\sqrt{3}}{4}\right] = \frac{\pi}{3} + \frac{\sqrt{3}}{2}$

$r = 2\cos\theta$
$r = 1$

49. $r = 3\cos\theta = 1 + \cos\theta \Rightarrow \cos\theta = \frac{1}{2} \Rightarrow \theta = \pm\frac{\pi}{3}$

$A = 2\int_0^{\pi/3} \int_{1+\cos\theta}^{3\cos\theta} r \, dr \, d\theta = 2\int_0^{\pi/3} \left[\frac{r^2}{2}\right]_{1+\cos\theta}^{3\cos\theta} d\theta = \int_0^{\pi/3} \left[9\cos^2\theta - (1 + \cos\theta)^2\right] d\theta$

$\qquad = \int_0^{\pi/3} \left(8\cos^2\theta - 2\cos\theta - 1\right) d\theta = \int_0^{\pi/3} \left[4(1 + \cos 2\theta) - 2\cos\theta - 1\right] d\theta$

$\qquad = \left[3\theta + 2\sin 2\theta - 2\sin\theta\right]_0^{\pi/3} = 3\left(\frac{\pi}{3}\right) + \sqrt{3} - \sqrt{3} = \pi$

$r = 3\cos\theta$
$r = 1 + \cos\theta$

51. $r = 4\sin 3\theta = 2 \Rightarrow \sin 3\theta = \frac{1}{2} \Rightarrow 3\theta = \frac{\pi}{6}, \frac{5\pi}{6} \Rightarrow \theta = \frac{\pi}{18}, \frac{5\pi}{18}$

$A = 3\int_{\pi/18}^{5\pi/18} \int_2^{4\sin 3\theta} r \, dr \, d\theta = 3\int_{\pi/18}^{5\pi/18} \left[\frac{r^2}{2}\right]_2^{4\sin 3\theta} d\theta = \frac{3}{2}\int_{\pi/18}^{5\pi/18} \left[(4\sin 3\theta)^2 - 4\right] d\theta$

$\qquad = \frac{3}{2}\int_{\pi/18}^{5\pi/18} \left[8(1 - \cos 6\theta) - 4\right] d\theta = \frac{3}{2}\left[4\theta - \frac{4}{3}\sin 6\theta\right]_{\pi/18}^{5\pi/18}$

$\qquad = \frac{3}{2}\left[\left(\frac{10}{9}\pi - \frac{4}{3}\left(\frac{-\sqrt{3}}{2}\right)\right) - \left(\frac{2\pi}{9} - \frac{4}{3}\left(\frac{\sqrt{3}}{2}\right)\right)\right] = \frac{4}{3}\pi + 2\sqrt{3}$

$r = 4\sin 3\theta$
$r = 2$

53. Let R be a region bounded by the graphs of $r = g_1(\theta)$ and $r = g_2(\theta)$, and the lines $\theta = a$ and $\theta = b$.

When using polar coordinates to evaluate a double integral over R, R can be partitioned into small polar sectors.

55. r-simple regions have fixed bounds for θ. θ-simple regions have fixed bounds for r.

57. $\int_{-7}^{7} \int_{-\sqrt{49-x^2}}^{\sqrt{49-x^2}} 4000e^{-0.01\left(x^2+y^2\right)} dy \, dx = \int_0^{2\pi} \int_0^7 4000e^{-0.01r^2} r \, dr \, d\theta = \int_0^{2\pi} \left[-200{,}000e^{-0.01r^2}\right]_0^7 d\theta$

$$= 2\pi(-200{,}000)\left(e^{-0.49} - 1\right) = 400{,}000\pi\left(1 - e^{-0.49}\right) \approx 486{,}788$$

59. Total volume $= V = \int_0^{2\pi} \int_0^4 25e^{-r^2/4} r \, dr \, d\theta = \int_0^{2\pi} \left[-50e^{-r^2/4} \right]_0^4 d\theta = \int_0^{2\pi} -50\left(e^{-4} - 1\right) d\theta = \left(1 - e^{-4}\right)100\pi \approx 308.40524$

Let c be the radius of the hole that is removed.

$$\frac{1}{10}V = \int_0^{2\pi} \int_0^c 25e^{-r^2/4} r \, dr \, d\theta = \int_0^{2\pi} \left[-50e^{-r^2/4} \right]_0^c d\theta$$

$$= \int_0^{2\pi} -50\left(e^{-c^2/4} - 1\right) d\theta \Rightarrow 30.84052 = 100\pi\left(1 - e^{-c^2/4}\right)$$

$$\Rightarrow e^{-c^2/4} = 0.90183$$

$$-\frac{c^2}{4} = -0.10333$$

$$c^2 = 0.41331$$

$$c = 0.6429$$

$$\Rightarrow \text{diameter} = 2c = 1.2858$$

61. $\int_{\pi/4}^{\pi/2} \int_0^5 r\sqrt{1 + r^3}\ \sin\sqrt{\theta}\ dr\ d\theta \approx 56.051$

$\left[\textbf{Note:} \text{This integral equals } \left(\int_{\pi/4}^{\pi/2} \sin\sqrt{\theta}\ d\theta \right)\left(\int_0^5 r\sqrt{1 + r^3}\ dr \right). \right]$

63. False

Let $f(r, \theta) = r - 1$ where R is the circular sector $0 \le r \le 6$ and $0 \le \theta \le \pi$. Then,

$\int_R \int (r - 1)\ dA > 0$ but $r - 1 \not> 0$ for all r.

65. (a) $I^2 = \int_{-\infty}^{\infty} \int_{-\infty}^{\infty} e^{-\left(x^2 + y^2\right)/2}\ dA = 4\int_0^{\pi/2} \int_0^{\infty} e^{-r^2/2} r\ dr\ d\theta = 4\int_0^{\pi/2} \left[-e^{-r^2/2} \right]_0^{\infty} d\theta = 4\int_0^{\pi/2} d\theta = 2\pi$

(b) So, $I = \sqrt{2\pi}$.

67. (a) $\int_2^4 \int_{y/\sqrt{3}}^y f\ dx\ dy$

(b) $\int_{2/\sqrt{3}}^2 \int_2^{\sqrt{3}x} f\ dy\ dx$

$+ \int_2^{4/\sqrt{3}} \int_x^{\sqrt{3}x} f\ dy\ dx + \int_{4/\sqrt{3}}^4 \int_x^4 f\ dy\ dx$

(c) $\int_{\pi/4}^{\pi/3} \int_{2\csc\theta}^{4\csc\theta} fr\ dr\ d\theta$

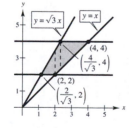

69. $\int_0^{\infty} \int_0^{\infty} ke^{-\left(x^2 + y^2\right)}\ dy\ dx = \int_0^{\pi/2} \int_0^{\infty} ke^{-r^2} r\ dr\ d\theta$

$= \int_0^{\pi/2} \left[-\frac{k}{2}e^{-r^2} \right]_0^{\infty} d\theta$

$= \int_0^{\pi/2} \frac{k}{2}\ d\theta = \frac{k\pi}{4}$

For $f(x, y)$ to be a probability density function,

$$\frac{k\pi}{4} = 1$$

$$k = \frac{4}{\pi}.$$

Section 14.4   Center of Mass and Moments of Inertia

1. $m = \int_0^2 \int_0^2 xy\ dy\ dx = \int_0^2 \left[\frac{xy^2}{2} \right]_0^2 dx = \int_0^2 2x\ dx = \left[x^2 \right]_0^2 = 4$

3. $m = \int_0^{\pi/2} \int_0^1 (r \cos \theta)(r \sin \theta)\, r\, dr\, d\theta = \int_0^{\pi/2} \left[(\cos \theta \sin \theta)\frac{r^4}{4} \right]_0^1 d\theta = \int_0^{\pi/2} \frac{1}{4} \cos \theta \sin \theta\, d\theta = \left[\frac{1}{4} \cdot \frac{\sin^2 \theta}{2} \right]_0^{\pi/2} = \frac{1}{8}$

5. (a) $\quad m = \int_0^a \int_0^a k\, dy\, dx = ka^2$

$\qquad M_x = \int_0^a \int_0^a ky\, dy\, dx = \int_0^a \frac{ka^2}{2}\, dx = \frac{ka^3}{2}$

$\qquad M_y = \int_0^a \int_0^a kx\, dy\, dx = \frac{ka^3}{2}$

$\qquad \bar{x} = \frac{M_y}{m} = \frac{a}{2}, \bar{y} = \frac{M_x}{m} = \frac{a}{2}$

$\qquad (\bar{x}, \bar{y}) = \left(\frac{a}{2}, \frac{a}{2} \right) \qquad$ (center of square)

(b) $\quad m = \int_0^a \int_0^a ky\, dy\, dx = \frac{1}{2} ka^3$

$\qquad M_x = \int_0^a \int_0^a ky^2\, dy\, dx = \frac{1}{3} ka^4$

$\qquad M_y = \int_0^a \int_0^a kyx\, dy\, dx = \frac{1}{4} ka^4$

$\qquad \bar{x} = \frac{M_y}{m} = \frac{a}{2}, \qquad \bar{y} = \frac{M_x}{m} = \frac{2a}{3}$

$\qquad (\bar{x}, \bar{y}) = \left(\frac{a}{2}, \frac{2a}{3} \right)$

(c) $\quad m = \int_0^a \int_0^a kx\, dy\, dx = \frac{1}{2} ka^3$

$\qquad M_x = \int_0^a \int_0^a kxy\, dy\, dx = \frac{1}{4} ka^4$

$\qquad M_y = \int_0^a \int_0^a kx^2\, dy\, dx = \frac{1}{3} ka^3$

$\qquad \bar{x} = \frac{M_y}{m} = \frac{2a}{3}, \qquad \bar{y} = \frac{M_x}{m} = \frac{a}{2}$

$\qquad (\bar{x}, \bar{y}) = \left(\frac{2a}{3}, \frac{a}{2} \right)$

7. (a) $\quad m = \int_0^a \int_0^y k\, dx\, dy = \frac{1}{2} ka^2$

$\qquad M_x = \int_0^a \int_0^y ky\, dx\, dy = \frac{1}{3} ka^3$

$\qquad M_y = \int_0^a \int_0^y kx\, dx\, dy = \frac{1}{6} ka^3$

$\qquad \bar{x} = \frac{M_y}{m} = \frac{a}{3} \qquad \bar{y} = \frac{M_x}{m} = \frac{2a}{3}$

$\qquad (\bar{x}, \bar{y}) = \left(\frac{a}{3}, \frac{2a}{3} \right)$

(b) $\quad m = \int_0^a \int_0^y ky\, dx\, dy = \frac{1}{3} ka^3$

$\qquad M_x = \int_0^a \int_0^y ky^2\, dx\, dy = \frac{1}{4} ka^4$

$\qquad M_y = \int_0^a \int_0^y kxy\, dx\, dy = \frac{1}{8} ka^4$

$\qquad x = \frac{M_y}{m} = \frac{3a}{8}, \qquad \bar{y} = \frac{M_x}{m} = \frac{3a}{4}$

$\qquad (\bar{x}, \bar{y}) = \left(\frac{3a}{8}, \frac{3a}{4} \right)$

(c) $\quad m = \int_0^a \int_0^y kx\, dx\, dy = \frac{1}{6} ka^3$

$\qquad M_x = \int_0^a \int_0^y kxy\, dx\, dy = \frac{1}{8} ka^4$

$\qquad M_y = \int_0^a \int_0^y kx^2\, dx\, dy = \frac{1}{12} ka^4$

$\qquad \bar{x} = \frac{M_y}{m} = \frac{a}{2} \qquad \bar{y} = \frac{M_x}{m} = \frac{3a}{4}$

$\qquad (\bar{x}, \bar{y}) = \left(\frac{a}{2}, \frac{3a}{4} \right)$

9. (a) The x-coordinate changes by 5: $(\bar{x}, \bar{y}) = \left(\frac{a}{2} + 5, \frac{a}{2} \right)$

(b) The x-coordinate changes by 5: $(\bar{x}, \bar{y}) = \left(\frac{a}{2} + 5, \frac{2a}{3} \right)$

(c) $\quad m = \int_5^{a+5} \int_0^a kx\, dy\, dx = \frac{1}{2} ka\left((a + 5)^2 - 25 \right)$

$\qquad M_x = \int_5^{a+5} \int_0^a kxy\, dy\, dx = \frac{1}{4} ka^2\left((a + 5)^2 - 25 \right)$

$\qquad M_y = \int_5^{a+5} \int_0^a kx^2\, dy\, dx = \frac{1}{3} ka\left((a + 5)^3 - 125 \right)$

$\qquad \bar{x} = \frac{M_y}{m} = \frac{2\left[(a + 5)^3 - 125 \right]}{3\left[(a + 5)^2 - 25 \right]} = \frac{2(a^2 + 15a + 75)}{3(a + 10)}$

$\qquad y = \frac{M_x}{m} = \frac{a}{2}$

$\qquad (\bar{x}, \bar{y}) = \left(\frac{2(a^2 + 15a + 75)}{3(a + 10)}, \frac{a}{2} \right)$

11. $m = \int_0^1 \int_0^{\sqrt{x}} ky \, dy \, dx = \frac{1}{4}k$

$M_x = \int_0^1 \int_0^{\sqrt{x}} ky^2 \, dy \, dx = \frac{2}{15}k$

$M_y = \int_0^1 \int_0^{\sqrt{x}} kxy \, dy \, dx = \frac{1}{6}k$

$\bar{x} = \dfrac{M_y}{m} = \dfrac{2}{3}$

$\bar{y} = \dfrac{M_x}{m} = \dfrac{8}{15}$

$(\bar{x}, \bar{y}) = \left(\dfrac{2}{3}, \dfrac{8}{15}\right)$

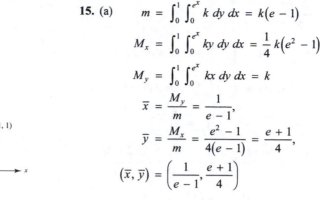

13. $m = \int_1^4 \int_0^{4/x} kx^2 \, dy \, dx = 30k$

$M_x = \int_1^4 \int_0^{4/x} kx^2 y \, dy \, dx = 24k$

$M_y = \int_1^4 \int_0^{4/x} kx^3 \, dy \, dx = 84k$

$\bar{x} = \dfrac{M_y}{m} = \dfrac{84k}{30k} = \dfrac{14}{5}$

$\bar{y} = \dfrac{M_y}{m} = \dfrac{24k}{30k} = \dfrac{4}{5}$

$(\bar{x}, \bar{y}) = \left(\dfrac{14}{5}, \dfrac{4}{5}\right)$

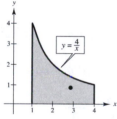

15. (a) $m = \int_0^1 \int_0^{e^x} k \, dy \, dx = k(e - 1)$

$M_x = \int_0^1 \int_0^{e^x} ky \, dy \, dx = \frac{1}{4}k(e^2 - 1)$

$M_y = \int_0^1 \int_0^{e^x} kx \, dy \, dx = k$

$\bar{x} = \dfrac{M_y}{m} = \dfrac{1}{e - 1},$

$\bar{y} = \dfrac{M_x}{m} = \dfrac{e^2 - 1}{4(e - 1)} = \dfrac{e + 1}{4},$

$(\bar{x}, \bar{y}) = \left(\dfrac{1}{e - 1}, \dfrac{e + 1}{4}\right)$

(b) $m = \int_0^1 \int_0^{e^x} ky \, dy \, dx = \dfrac{e^2 - 1}{4}k$

$M_x = \int_0^1 \int_0^{e^x} ky^2 \, dy \, dx = \dfrac{e^3 - 1}{9}k$

$M_y = \int_0^1 \int_0^{e^x} kxy \, dy \, dx = \dfrac{e^2 + 1}{8}k$

$\bar{x} = \dfrac{M_y}{m} = \dfrac{e^2 + 1}{2(e^2 - 1)}, \bar{y} = \dfrac{M_x}{m} = \dfrac{4(e^3 - 1)}{9(e^2 - 1)},$

$(\bar{x}, \bar{y}) = \left(\dfrac{e^2 + 1}{2(e^2 - 1)}, \dfrac{4(e^3 - 1)}{9(e^2 - 1)}\right)$

17. $m = \int_{-2}^2 \int_0^{4-x^2} ky \, dy \, dx = \dfrac{256}{15}k$

$M_x = \int_{-2}^2 \int_0^{4-x^2} ky^2 \, dy \, dx = \dfrac{4096}{105}k$

$\bar{x} = 0 \,(\text{by symmetry})$

$\bar{y} = \dfrac{M_x}{m} = \dfrac{16}{7}$

$(\bar{x}, \bar{y}) = \left(0, \dfrac{16}{7}\right)$

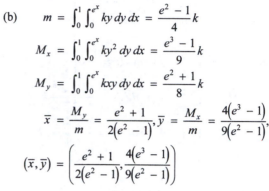

19. $\bar{x} = \dfrac{L}{2} \quad (\text{by symmetry})$

$m = \int_0^L \int_0^{\sin(\pi x/2)} k \, dy \, dx = \dfrac{2kL}{\pi}$

$M_x = \int_0^L \int_0^{\sin(\pi x/L)} ky \, dy \, dx = \dfrac{kL}{4}$

$\bar{y} = \dfrac{M_x}{m} = \dfrac{\pi}{8}$

$(\bar{x}, \bar{y}) = \left(\dfrac{L}{2}, \dfrac{\pi}{8}\right)$

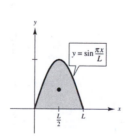

21. $m = \dfrac{\pi a^2 k}{8}$

$M_x = \displaystyle\int_R \int ky \, dA = \int_0^{\pi/4} \int_0^a kr^2 \sin\theta \, dr \, d\theta = \dfrac{ka^3\left(2 - \sqrt{2}\right)}{6}$

$M_y = \displaystyle\int_R \int kx \, dA = \int_0^{\pi/4} \int_0^a kr^2 \cos\theta \, dr \, d\theta = \dfrac{ka^3\sqrt{2}}{6}$

$\overline{x} = \dfrac{M_y}{m} = \dfrac{ka^3\sqrt{2}}{6} \cdot \dfrac{8}{\pi a^2 k} = \dfrac{4a\sqrt{2}}{3\pi}$

$\overline{y} = \dfrac{M_x}{m} = \dfrac{ka^3\left(2 - \sqrt{2}\right)}{6} \cdot \dfrac{8}{\pi a^2 k} = \dfrac{4a\left(2 - \sqrt{2}\right)}{3\pi}$

$(\overline{x}, \overline{y}) = \left(\dfrac{4a\sqrt{2}}{3\pi}, \dfrac{4a\left(2 - \sqrt{2}\right)}{3\pi} \right)$

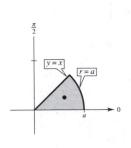

23. $m = \displaystyle\int_0^2 \int_0^{e^{-x}} kxy \, dy \, dx = \dfrac{1 - 5e^{-4}}{8} k$

$M_x = \displaystyle\int_0^2 \int_0^{e^{-x}} kxy^2 \, dy \, dx = \dfrac{1 - 7e^{-6}}{27} k$

$M_y = \displaystyle\int_0^2 \int_0^{e^{-x}} kx^2 y \, dy \, dx = \dfrac{1 - 13e^{-4}}{8} k$

$\overline{x} = \dfrac{M_y}{m} = \dfrac{e^4 - 13}{e^4 - 5}$

$\overline{y} = \dfrac{M_x}{m} = \dfrac{8\left(e^6 - 7\right)}{27\left(e^6 - 5e^2\right)}$

$(\overline{x}, \overline{y}) = \left(\dfrac{e^4 - 13}{e^4 - 5}, \dfrac{8\left(e^6 - 7\right)}{27\left(e^6 - 5e^2\right)} \right)$

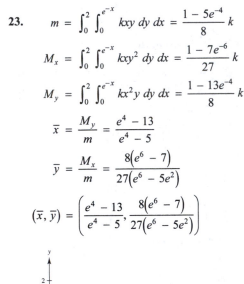

25. $\overline{y} = 0$ by symmetry

$m = \displaystyle\int_R \int k \, dA = \int_{-\pi/6}^{\pi/6} \int_0^{2\cos 3\theta} kr \, dr \, d\theta = \dfrac{k\pi}{3}$

$M_y = \displaystyle\int_R \int kx \, dA$

$= \displaystyle\int_{-\pi/6}^{\pi/6} \int_0^{2\cos 3\theta} kr^2 \cos\theta \, dr \, d\theta$

$= \dfrac{27\sqrt{3}}{40} k \approx 1.17k$

$\overline{x} = \dfrac{M_y}{m} = \dfrac{81\sqrt{3}}{40\pi} \approx 1.12$

$(\overline{x}, \overline{y}) \approx (1.12, 0)$

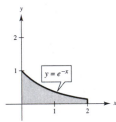

27. $m = bh$

$I_x = \displaystyle\int_0^b \int_0^h y^2 \, dy \, dx = \dfrac{bh^3}{3}$

$I_y = \displaystyle\int_0^b \int_0^h x^2 \, dy \, dx = \dfrac{b^3 h}{3}$

$\overline{\overline{x}} = \sqrt{\dfrac{I_y}{m}} = \sqrt{\dfrac{b^3 h}{3} \cdot \dfrac{1}{bh}} = \sqrt{\dfrac{b^2}{3}} = \dfrac{b}{\sqrt{3}} = \dfrac{\sqrt{3}}{3} b$

$\overline{\overline{y}} = \sqrt{\dfrac{I_x}{m}} = \sqrt{\dfrac{bh^3}{3} \cdot \dfrac{1}{bh}} = \sqrt{\dfrac{h^2}{3}} = \dfrac{h}{\sqrt{3}} = \dfrac{\sqrt{3}}{3} h$

29. $m = \pi a^2$

$I_x = \displaystyle\int_R \int y^2 \, dA = \int_0^{2\pi} \int_0^a r^3 \sin^2\theta \, dr \, d\theta = \dfrac{a^4 \pi}{4}$

$I_y = \displaystyle\int_R \int x^2 \, dA = \int_0^{2\pi} \int_0^a r^3 \cos^2\theta \, dr \, d\theta = \dfrac{a^4 \pi}{4}$

$I_0 = I_x + I_y = \dfrac{a^4 \pi}{4} + \dfrac{a^4 \pi}{4} = \dfrac{a^4 \pi}{2}$

$\overline{\overline{x}} = \overline{\overline{y}} = \sqrt{\dfrac{I_x}{m}} = \sqrt{\dfrac{a^4 \pi}{4} \cdot \dfrac{1}{\pi a^2}} = \dfrac{a}{2}$

31. $m = \dfrac{\pi a^2}{4}$

$I_x = \displaystyle\int_R \int y^2 \, dA = \int_0^{\pi/2} \int_0^a r^3 \sin^2\theta \, dr \, d\theta = \dfrac{\pi a^4}{16}$

$I_y = \displaystyle\int_R \int x^2 \, dA = \int_0^{\pi/2} \int_0^a r^3 \cos^2\theta \, dr \, d\theta = \dfrac{\pi a^4}{16}$

$I_0 = I_x + I_y = \dfrac{\pi a^4}{16} + \dfrac{\pi a^4}{16} = \dfrac{\pi a^4}{8}$

$\overline{\overline{x}} = \overline{\overline{y}} = \sqrt{\dfrac{I_x}{m}} = \sqrt{\dfrac{\pi a^4}{16} \cdot \dfrac{4}{\pi a^2}} = \dfrac{a}{2}$

33. $\rho = kx$

$$m = k\int_0^2 \int_0^{4-x^2} x \, dy \, dx = 4k$$

$$I_x = k\int_0^2 \int_0^{4-x^2} xy^2 \, dy \, dx = \frac{32k}{3}$$

$$I_y = k\int_0^2 \int_0^{4-x^2} x^3 \, dy \, dx = \frac{16k}{3}$$

$$I_0 = I_x + I_y = 16k$$

$$\bar{\bar{x}} = \sqrt{\frac{I_y}{m}} = \sqrt{\frac{16k/3}{4k}} = \sqrt{\frac{4}{3}} = \frac{2}{\sqrt{3}} = \frac{2\sqrt{3}}{3}$$

$$\bar{\bar{y}} = \sqrt{\frac{I_x}{m}} = \sqrt{\frac{32k/3}{4k}} = \sqrt{\frac{8}{3}} = \frac{4}{\sqrt{6}} = \frac{2\sqrt{6}}{3}$$

35. $\rho = kxy$

$$m = \int_0^4 \int_0^{\sqrt{x}} kxy \, dy \, dx = \frac{32k}{3}$$

$$I_x = \int_0^4 \int_0^{\sqrt{x}} kxy^3 \, dy \, dx = 16k$$

$$I_y = \int_0^4 \int_0^{\sqrt{x}} kx^3 y \, dy \, dx = \frac{512k}{5}$$

$$I_0 = I_x + I_y = \frac{592k}{5}$$

$$\bar{\bar{x}} = \sqrt{\frac{I_y}{m}} = \sqrt{\frac{512k}{5} \cdot \frac{3}{32k}} = \sqrt{\frac{48}{5}} = \frac{4\sqrt{15}}{5}$$

$$\bar{\bar{y}} = \sqrt{\frac{I_x}{m}} = \sqrt{\frac{16k}{1} \cdot \frac{3}{32k}} = \sqrt{\frac{3}{2}} = \frac{\sqrt{6}}{2}$$

37. $I = 2k\int_{-b}^{b} \int_0^{\sqrt{b^2-x^2}} (x-a)^2 \, dy \, dx = 2k\int_{-b}^{b} (x-a)^2 \sqrt{b^2 - x^2} \, dx$

$$= 2k\left[\int_{-b}^{b} x^2\sqrt{b^2-x^2} \, dx - 2a\int_{-b}^{b} x\sqrt{b^2-x^2} \, dx + a^2\int_{-b}^{b} \sqrt{b^2-x^2} \, dx\right] = 2k\left[\frac{\pi b^4}{8} + 0 + \frac{\pi a^2 b^2}{2}\right] = \frac{k\pi b^2}{4}(b^2 + 4a^2)$$

39. $I = \int_{-a}^{a} \int_0^{\sqrt{a^2-x^2}} ky(y-a)^2 \, dy \, dx$

$$= \int_{-a}^{a} k\left[\frac{y^4}{4} - \frac{2ay^3}{3} + \frac{a^2 y^2}{2}\right]_0^{\sqrt{a^2-x^2}} dx$$

$$= \int_{-a}^{a} k\left[\frac{1}{4}(a^4 - 2a^2 x^2 + x^4) - \frac{2a}{3}\left(a^2\sqrt{a^2-x^2} - x^2\sqrt{a^2-x^2}\right) + \frac{a^2}{2}(a^2 - x^2)\right] dx$$

$$= k\left[\frac{1}{4}\left(a^4 x - \frac{2a^2 x^3}{3} + \frac{x^5}{5}\right) - \frac{2a}{3}\left[\frac{a^2}{2}\left(x\sqrt{a^2-x^2} + a^2 \arcsin\frac{x}{a}\right)\right.\right.$$

$$\left.\left. - \frac{1}{8}\left(x(2x^2 - a^2)\sqrt{a^2-x^2} + a^4 \arcsin\frac{x}{a}\right)\right] + \frac{a^2}{2}\left(a^2 x - \frac{x^3}{3}\right)\right]_{-a}^{a}$$

$$= 2k\left[\frac{1}{4}\left(a^5 - \frac{2}{3}a^5 + \frac{1}{5}a^5\right) - \frac{2a}{3}\left(\frac{a^4\pi}{4} - \frac{a^4\pi}{16}\right) + \frac{a^2}{2}\left(a^3 - \frac{a^3}{3}\right)\right] = 2k\left(\frac{7a^5}{15} - \frac{a^5\pi}{8}\right) = ka^5\left(\frac{56 - 15\pi}{60}\right)$$

41. $\bar{y} = \dfrac{L}{2}, \; A = bL, \; h = \dfrac{L}{2}$

$$I_{\bar{y}} = \int_0^b \int_0^L \left(y - \frac{L}{2}\right)^2 dy \, dx$$

$$= \int_0^b \left[\frac{[y - (L/2)]^3}{3}\right]_0^L dx = \frac{L^3 b}{12}$$

$$y_a = \bar{y} - \frac{I_{\bar{y}}}{hA} = \frac{L}{2} - \frac{L^3 b/12}{(L/2)(bL)} = \frac{L}{3}$$

43. $\bar{y} = \dfrac{2L}{3}, \; A = \dfrac{bL}{2}, \; h = \dfrac{L}{3}$

$$I_{\bar{y}} = 2\int_0^{b/2} \int_{2Lx/b}^L \left(y - \frac{2L}{3}\right)^2 dy \, dx$$

$$= \frac{2}{3}\int_0^{b/2} \left[\left(y - \frac{2L}{3}\right)^3\right]_{2Lx/b}^L dx$$

$$= \frac{2}{3}\int_0^{b/2} \left[\frac{L^3}{27} - \left(\frac{2Lx}{b} - \frac{2L}{3}\right)^3\right] dx$$

$$= \frac{2}{3}\left[\frac{L^3 x}{27} - \frac{b}{8L}\left(\frac{2Lx}{b} - \frac{2L}{3}\right)^4\right]_0^{b/2} = \frac{L^3 b}{36}$$

$$y_a = \frac{2L}{3} - \frac{L^3 b/36}{L^2 b/6} = \frac{L}{2}$$

45. Let $\rho(x, y)$ be a continuous density function on the planar lamina R.

The movements of mass with respect to the x- and y-axes are

$$M_x = \int_R \int y\rho(x, y)\, dA \text{ and } M_y = \int_R \int x\rho(x, y)\, dA.$$

If m is the mass of the lamina, then the center of mass is

$$(\bar{x}, \bar{y}) = \left(\frac{M_y}{m}, \frac{M_x}{m}\right).$$

47. See the definition on page 999.

49. Orient the xy-coordinate system so that L is along the y-axis and R is the first quadrant. Then the volume of the solid is

$$V = \int_R \int 2\pi x\, dA$$

$$= 2\pi \int_R \int x\, dA$$

$$= 2\pi\left(\frac{\int_R \int x\, dA}{\int_R \int dA}\right)\int_R \int dA$$

$$= 2\pi\bar{x}A.$$

By our positioning, $\bar{x} = r$. So, $V = 2\pi\, rA$.

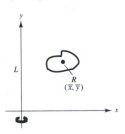

Section 14.5   Surface Area

1. $f(x, y) = 2x + 2y$

$f_x = f_y = 2$

$\sqrt{1 + (f_x)^2 + (f_y)^2} = \sqrt{1 + 4 + 4} = 3$

$S = \int_0^4 \int_0^{4-x} 3\, dy\, dx = 3\int_0^4 (4 - x)\, dx$

$= 3\left[4x - \frac{x^2}{2}\right]_0^4 = 24$

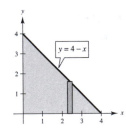

3. $f(x, y) = 7 + 2x + 2y$

$f_x = f_y = 2$

$\sqrt{1 + (f_x)^2 + (f_y)^2} = \sqrt{1 + 4 + 4} = 3$

$S = \int_0^{2\pi} \int_0^2 3\, r\, dr\, d\theta = \int_0^{2\pi} 6\, d\theta = 12\pi$

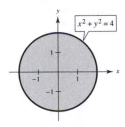

5. $f(x, y) = 9 - x^2$

$f_x = -2x, f_y = 0$

$\sqrt{1 + (f_x)^2 + (f_y)^2} = \sqrt{1 + 4x^2}$

$S = \int_0^2 \int_0^2 \sqrt{1 + 4x^2}\, dy\, dx = 2\int_0^2 \sqrt{1 + 4x^2}\, dx$

$= 2\left[\frac{1}{4}\ln\left(\sqrt{1 + 4x^2} + 2x\right) + \frac{x}{2}\sqrt{1 + 4x^2}\right]_0^2$

$= 2\left[\frac{1}{4}\ln\left(\sqrt{17} + 4\right) + \sqrt{17}\right]$

$= 2\sqrt{17} + \frac{1}{2}\ln\left(4 + \sqrt{17}\right)$

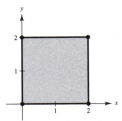

7. $f(x, y) = 3 + x^{3/2}$

$f_x = \dfrac{3}{2}x^{1/2}, f_y = 0$

$\sqrt{1 + (f_x)^2 + (f_y)^2} = \sqrt{1 + \dfrac{9}{4}x} = \dfrac{\sqrt{4 + 9x}}{2}$

$S = \displaystyle\int_0^3 \int_0^4 \dfrac{\sqrt{4 + 9x}}{2} \, dy \, dx = 4\int_0^3 \dfrac{\sqrt{4 + 9x}}{2} \, dx$

$\qquad = \left[\dfrac{4}{27}(4 + 9x)^{3/2}\right]_0^3 = \dfrac{4}{27}\left(31\sqrt{31} - 8\right)$

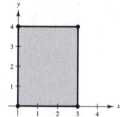

9. $f(x, y) = \ln|\sec x|$

$R = \left\{(x, y)\colon 0 \le x \le \dfrac{\pi}{4}, 0 \le y \le \tan x\right\}$

$f_x = \tan x, f_y = 0$

$\sqrt{1 + (f_x)^2 + (f_y)^2} = \sqrt{1 + \tan^2 x} = \sec x$

$S = \displaystyle\int_0^{\pi/4} \int_0^{\tan x} \sec x \, dy \, dx$

$\quad = \displaystyle\int_0^{\pi/4} \sec x \tan x \, dx$

$\quad = \left[\sec x\right]_0^{\pi/4} = \sqrt{2} - 1$

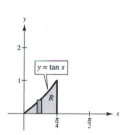

11. $f(x, y) = \sqrt{x^2 + y^2}$

$R = \{(x, y)\colon 0 \le f(x, y) \le 1\}$

$0 \le \sqrt{x^2 + y^2} \le 1, x^2 + y^2 \le 1$

$f_x = \dfrac{x}{\sqrt{x^2 + y^2}}, f_y = \dfrac{y}{\sqrt{x^2 + y^2}}$

$\sqrt{1 + (f_x)^2 + (f_y)^2} = \sqrt{1 + \dfrac{x^2}{x^2 + y^2} + \dfrac{y^2}{x^2 + y^2}} = \sqrt{2}$

$S = \displaystyle\int_{-1}^1 \int_{-\sqrt{1-x^2}}^{\sqrt{1-x^2}} \sqrt{2} \, dy \, dx = \int_0^{2\pi} \int_0^1 \sqrt{2}\,r \, dr \, d\theta = \sqrt{2}\pi$

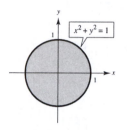

13. $f(x, y) = \sqrt{a^2 - x^2 - y^2}$

$R = \{(x, y)\colon x^2 + y^2 \le b^2, 0 < b < a\}$

$f_x = \dfrac{-x}{\sqrt{a^2 - x^2 - y^2}}, f_y = \dfrac{-y}{\sqrt{a^2 - x^2 - y^2}}$

$\sqrt{1 + (f_x)^2 + (f_y)^2} = \sqrt{1 + \dfrac{x^2}{a^2 - x^2 - y^2} + \dfrac{y^2}{a^2 - x^2 - y^2}} = \dfrac{a}{\sqrt{a^2 - x^2 - y^2}}$

$S = \displaystyle\int_{-b}^b \int_{-\sqrt{b^2-x^2}}^{\sqrt{b^2-x^2}} \dfrac{a}{\sqrt{a^2 - x^2 - y^2}} \, dy \, dx = \int_0^{2\pi} \int_0^b \dfrac{a}{\sqrt{a^2 - r^2}}\,r \, dr \, d\theta = 2\pi a\left(a - \sqrt{a^2 - b^2}\right)$

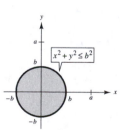

15. $z = 24 - 3x - 2y$

$\sqrt{1 + (f_x)^2 + (f_y)^2} = \sqrt{14}$

$S = \displaystyle\int_0^8 \int_0^{-(3/2)x+12} \sqrt{14} \, dy \, dx = 48\sqrt{14}$

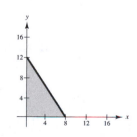

17. $z = \sqrt{25 - x^2 - y^2}$

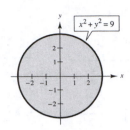

$$\sqrt{1 + (f_x)^2 + (f_y)^2} = \sqrt{1 + \frac{x^2}{25 - x^2 - y^2} + \frac{y^2}{25 - x^2 - y^2}} = \frac{5}{\sqrt{25 - x^2 - y^2}}$$

$$S = 2\int_{-3}^{3}\int_{-\sqrt{9-x^2}}^{\sqrt{9-x^2}} \frac{5}{\sqrt{25 - (x^2 + y^2)}}\, dy\, dx = 2\int_{0}^{2\pi}\int_{0}^{3} \frac{5}{\sqrt{25 - r^2}} r\, dr\, d\theta = 20\pi$$

19. $f(x, y) = 2y + x^2$

$R = $ triangle with vertices $(0, 0), (1, 0), (1, 1)$

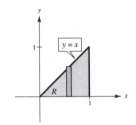

$$\sqrt{1 + (f_x)^2 + (f_y)^2} = \sqrt{5 + 4x^2}$$

$$S = \int_{0}^{1}\int_{0}^{x} \sqrt{5 + 4x^2}\, dy\, dx = \frac{1}{12}\left(27 - 5\sqrt{5}\right)$$

21. $f(x, y) = 9 - x^2 - y^2$

$R = \{(x, y): 0 \le f(x, y)\}$

$0 \le 9 - x^2 - y^2 \Rightarrow x^2 + y^2 \le 9$

$fx = -2x,\ fy = -2y$

$$\sqrt{1 + (f_x)^2 + (f_y)^2} = \sqrt{1 + 4x^2 + 4y^2}$$

$$S = \int_{-3}^{3}\int_{-\sqrt{9-x^2}}^{\sqrt{9-x^2}} \sqrt{1 + 4x^2 + 4y^2}\, dy\, dx$$

$$= \int_{0}^{2\pi}\int_{0}^{3} \sqrt{1 + 4r^2}\, r\, dr\, d\theta$$

$$= \frac{\pi}{6}\left(37\sqrt{37} - 1\right) \approx 117.3187$$

23. $f(x, y) = 4 - x^2 - y^2$

$R = \{(x, y): 0 \le x \le 1, 0 \le y \le 1\}$

$f_x = -2x,\ f_y = -2y$

$$\sqrt{1 + (f_x)^2 + (f_y)^2} = \sqrt{1 + 4x^2 + 4y^2}$$

$$S = \int_{0}^{1}\int_{0}^{1} \sqrt{(1 + 4x^2) + 4y^2}\, dy\, dx \approx 1.8616$$

25. $f(x, y) = e^{xy}$

$R = \{(x, y): 0 \le x \le 4, 0 \le y \le 10\}$

$f_x = ye^{xy},\ f_y = xe^{xy}$

$$\sqrt{1 + (f_x)^2 + (f_y)^2} = \sqrt{1 + y^2e^{2xy} + x^2e^{2xy}} = \sqrt{1 + e^{2xy}(x^2 + y^2)}$$

$$S = \int_{0}^{4}\int_{0}^{10} \sqrt{1 + e^{2xy}(x^2 + y^2)}\, dy\, dx$$

27. $f(x, y) = e^{-x} \sin y$

$f_x = -e^{-x} \sin y,\ f_y = e^{-x} \cos y$

$$\sqrt{1 + (f_x^2) + (f_y^2)} = \sqrt{1 + e^{-2x} \sin^2 y + e^{-2x} \cos^2 y} = \sqrt{1 + e^{-2x}}$$

$$S = \int_{-2}^{2}\int_{-\sqrt{4-x^2}}^{\sqrt{4-x^2}} \sqrt{1 + e^{-2x}}\, dy\, dx$$

29. See the definition on page 1003.

31. No, the surface area is the same.

$$z = f(x, y) \qquad \text{and} \qquad z = f(x, y) + k$$

have the same partial derivatives.

33. (a) $V = \int_R \int f(x, y)$

$= 8 \int_R \int \sqrt{625 - x^2 - y^2} \, dA$ where R is the region in the first quadrant

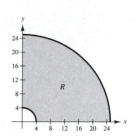

$= 8 \int_0^{\pi/2} \int_4^{25} \sqrt{625 - r^2} \, r \, dr \, d\theta = -4 \int_0^{\pi/2} \left[\frac{2}{3}(625 - r^2)^{3/2} \right]_4^{25} d\theta$

$= -\frac{8}{3}\left[0 - 609\sqrt{609} \right] \cdot \frac{\pi}{2} = 812\pi\sqrt{609} \text{ cm}^3$

(b) $A = \int_R \int \sqrt{1 + (f_x)^2 + (f_y)^2} \, dA = 8 \int_R \int \sqrt{1 + \frac{x^2}{625 - x^2 - y^2} + \frac{y^2}{625 - x^2 - y^2}} \, dA$

$= 8 \int_R \int \frac{25}{\sqrt{625 - x^2 - y^2}} \, dA = 8 \int_0^{\pi/2} \int_4^{25} \frac{25}{\sqrt{625 - r^2}} \, r \, dr \, d\theta$

$= \lim_{b \to 25^-} \left[-200\sqrt{625 - r^2} \right]_4^b \cdot \frac{\pi}{2} = 100\pi\sqrt{609} \text{ cm}^2$

35. $f(x, y) = \sqrt{1 - x^2}; f_x = \frac{-x}{\sqrt{1^2 - x^2}}, f_y = 0$

$S = \int_R \int \sqrt{1 + (f_x)^2 + (f_y)^2} \, dA$

$= 16 \int_0^1 \int_0^x \frac{1}{\sqrt{1 - x^2}} \, dy \, dx$

$= 16 \int_0^1 \frac{x}{\sqrt{1 - x^2}} \, dx = \left[-16(1 - x^2)^{1/2} \right]_0^1 = 16$

Section 14.6   Triple Integrals and Applications

1. $\int_0^3 \int_0^2 \int_0^1 (x + y + z) \, dx \, dz \, dy = \int_0^3 \int_0^2 \left[\frac{x^2}{2} + xy + xz \right]_0^1 dz \, dy = \int_0^3 \int_0^2 \left(\frac{1}{2} + y + z \right) dz \, dy = \int_0^3 \left[\frac{1}{2}z + yz + \frac{z^2}{2} \right]_0^2 dy$

$= \int_0^3 (1 + 2y + 2) \, dy = \left[3y + y^2 \right]_0^3 = 18$

3. $\int_0^1 \int_0^x \int_0^{xy} x \, dz \, dy \, dx = \int_0^1 \int_0^x [xz]_0^{xy} \, dy \, dx = \int_0^1 \int_0^x x^2 y \, dy \, dx = \int_0^1 \left[\frac{x^2 y^2}{2} \right]_0^x dx = \int_0^1 \frac{x^4}{2} \, dx = \left[\frac{x^5}{10} \right]_0^1 = \frac{1}{10}$

5. $\int_1^4 \int_0^1 \int_0^x 2ze^{-x^2} \, dy \, dx \, dz = \int_1^4 \int_0^1 \left[\left(2ze^{-x^2} \right) y \right]_0^x dx \, dz = \int_1^4 \int_0^1 2zxe^{-x^2} \, dx \, dz$

$= \int_1^4 \left[-ze^{-x^2} \right]_0^1 dz = \int_1^4 z(1 - e^{-1}) \, dz = \left[(1 - e^{-1})\frac{z^2}{2} \right]_1^4 = \frac{15}{2}\left(1 - \frac{1}{e} \right)$

7. $\int_0^4 \int_0^{\pi/2} \int_0^{1-x} x \cos y \, dz \, dy \, dx = \int_0^4 \int_0^{\pi/2} \left[(x \cos y)z \right]_0^{1-x} dy \, dx = \int_0^4 \int_0^{\pi/2} x(1 - x) \cos y \, dy \, dx$

$= \int_0^4 \left[x(1 - x) \sin y \right]_0^{\pi/2} dx = \int_0^4 x(1 - x) \, dx = \left[\frac{x^2}{2} - \frac{x^3}{3} \right]_0^4 = 8 - \frac{64}{3} = -\frac{40}{3}$

9. $\int_0^3 \int_{-\sqrt{9-y^2}}^{\sqrt{9-y^2}} \int_0^{y^2} y \, dz \, dx \, dy = \frac{324}{5}$ **11.** $V = \int_0^5 \int_0^{5-x} \int_0^{5-x-y} dz \, dy \, dx$

13. $V = \int_{-\sqrt{6}}^{\sqrt{6}} \int_{-\sqrt{6-x^2}}^{\sqrt{6-x^2}} \int_0^{6-x^2-y^2} dz \, dy \, dx = \int_{-\sqrt{6}}^{\sqrt{6}} \int_{-\sqrt{6-y^2}}^{\sqrt{6-y^2}} \int_0^{6-x^2-y^2} dz \, dx \, dy$

15. $z = \frac{1}{2}(x^2 + y^2) \Rightarrow 2z = x^2 + y^2$

$x^2 + y^2 + z^2 = 2z + z^2 = 80 \Rightarrow z^2 + 2z - 80 = 0 \Rightarrow (z - 8)(z + 10) = 0 \Rightarrow z = 8 \Rightarrow x^2 + y^2 = 2z = 16$

$V = \int_{-4}^{4} \int_{-\sqrt{16-x^2}}^{\sqrt{16-x^2}} \int_{1/2(x^2+y^2)}^{\sqrt{80-x^2-y^2}} dz\, dy\, dx$

17. $V = \int_{-2}^{2} \int_{0}^{4-y^2} \int_{0}^{x} dz\, dx\, dy = \int_{-2}^{2} \int_{0}^{4-y^2} x\, dx\, dy$

$= \frac{1}{2} \int_{-2}^{2} (4 - y^2)^2\, dy = \int_{0}^{2} (16 - 8y^2 + y^4)\, dy = \left[16y - \frac{8}{3}y^3 + \frac{1}{5}y^5\right]_{0}^{2} = \frac{256}{15}$

19. $V = 8\int_{0}^{a} \int_{0}^{\sqrt{a^2-x^2}} \int_{0}^{\sqrt{a^2-x^2-y^2}} dz\, dy\, dx = 8\int_{0}^{a} \int_{0}^{\sqrt{a^2-x^2}} \sqrt{a^2 - x^2 - y^2}\, dy\, dx$

$= 4\int_{0}^{a} \left[y\sqrt{a^2 - x^2 - y^2} + (a^2 - x^2) \arcsin\left(\frac{y}{\sqrt{a^2 - x^2}}\right) \right]_{0}^{\sqrt{a^2-x^2}} dx$

$= 4\left(\frac{\pi}{2}\right) \int_{0}^{a} (a^2 - x^2)\, dx = \left[2\pi\left(a^2 x - \frac{1}{3}x^3\right)\right]_{0}^{a} = \frac{4}{3}\pi a^3$

21. $V = \int_{0}^{2} \int_{0}^{4-x^2} \int_{0}^{4-x^2} dz\, dy\, dx = \int_{0}^{2} (4 - x^2)^2\, dx = \int_{0}^{2} (16 - 8x^2 + x^4)\, dx = \left[16x - \frac{8}{3}x^3 + \frac{1}{5}x^5\right]_{0}^{2} = \frac{256}{15}$

23. $V = \int_{0}^{3} \int_{0}^{2} \int_{2-y}^{4-y^2} dz\, dy\, dx$

$= \int_{0}^{3} \int_{0}^{2} \left[4 - y^2 - 2 + y\right] dy\, dx$

$= \int_{0}^{3} \left[2y - \frac{y^3}{3} + \frac{y^2}{2}\right]_{0}^{2} dx$

$= \int_{0}^{3} \left(4 - \frac{8}{3} + 2\right) dx$

$= \left[\frac{10}{3}x\right]_{0}^{3} = 10$

25.

$\int_{0}^{1} \int_{0}^{1} \int_{-1}^{-\sqrt{z}} dy\, dz\, dx$

27. Plane: $3x + 6y + 4z = 12$

$\int_{0}^{3} \int_{0}^{(12-4z)/3} \int_{0}^{(12-4z-3x)/6} dy\, dx\, dz$

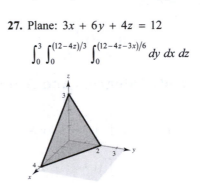

29. Top cylinder: $y^2 + z^2 = 1$

Side plane: $x = y$

$\int_{0}^{1} \int_{0}^{x} \int_{0}^{\sqrt{1-y^2}} dz\, dy\, dx$

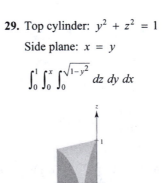

31. $Q = \{(x, y, z): 0 \le x \le 1, 0 \le y \le x, 0 \le z \le 3\}$

$$\iiint_Q xyz \, dV = \int_0^3 \int_0^1 \int_y^1 xyz \, dx \, dy \, dz = \int_0^3 \int_0^1 \int_0^x xyz \, dy \, dx \, dz$$

$$= \int_0^1 \int_0^3 \int_y^1 xyz \, dx \, dz \, dy = \int_0^1 \int_0^3 \int_0^x xyz \, dy \, dz \, dx$$

$$= \int_0^1 \int_y^1 \int_0^3 xyz \, dz \, dx \, dy = \int_0^1 \int_0^x \int_0^3 xyz \, dz \, dy \, dx \left(= \frac{9}{16}\right)$$

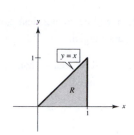

33. $Q = \{(x, y, z): x^2 + y^2 \le 9, 0 \le z \le 4\}$

$$\iiint_Q xyz \, dV = \int_0^4 \int_{-3}^3 \int_{-\sqrt{9-x^2}}^{\sqrt{9-x^2}} xyz \, dy \, dx \, dz = \int_0^4 \int_{-3}^3 \int_{-\sqrt{9-y^2}}^{\sqrt{9-y^2}} xyz \, dx \, dy \, dz$$

$$= \int_{-3}^3 \int_0^4 \int_{-\sqrt{9-y^2}}^{\sqrt{9-y^2}} xyz \, dx \, dz \, dy = \int_{-3}^3 \int_{-\sqrt{9-y^2}}^{\sqrt{9-y^2}} \int_0^4 xyz \, dz \, dx \, dy$$

$$= \int_{-3}^3 \int_0^4 \int_{-\sqrt{9-x^2}}^{\sqrt{9-x^2}} xyz \, dy \, dz \, dx = \int_{-3}^3 \int_{-\sqrt{9-x^2}}^{\sqrt{9-x^2}} \int_0^4 xyz \, dz \, dy \, dx \, (= 0)$$

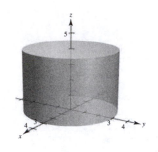

35. $Q = \{(x, y, z): 0 \le y \le 1, 0 \le x \le 1 - y^2, 0 \le z \le 1 - y\}$

$$\int_0^1 \int_0^{1-y^2} \int_0^{1-y} dz \, dx \, dy = \int_0^1 \int_0^{\sqrt{1-x}} \int_0^{1-y} dz \, dy \, dx$$

$$= \int_0^1 \int_0^{2z-z^2} \int_0^{1-z} dy \, dx \, dz + \int_0^1 \int_{2z-z^2}^1 \int_0^{\sqrt{1-x}} dy \, dx \, dz$$

$$= \int_0^1 \int_{1-\sqrt{1-x}}^1 \int_0^{1-z} dy \, dz \, dx + \int_0^1 \int_0^{1-\sqrt{1-x}} \int_0^{\sqrt{1-x}} dy \, dz \, dx$$

$$= \int_0^1 \int_0^{1-y} \int_0^{1-y^2} dx \, dz \, dy = \int_0^1 \int_0^{1-z} \int_0^{1-y^2} dx \, dy \, dz = \frac{5}{12}$$

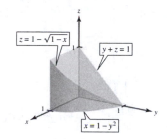

37. $m = k \int_0^6 \int_0^{4-(2x/3)} \int_0^{2-(y/2)-(x/3)} dz \, dy \, dx = 8k$

$M_{yz} = k \int_0^6 \int_0^{4-(2x/3)} \int_0^{2-(y/2)-(x/3)} x \, dz \, dy \, dx = 12k$

$\overline{x} = \dfrac{M_{yz}}{m} = \dfrac{12k}{8k} = \dfrac{3}{2}$

39. $m = k \int_0^4 \int_0^4 \int_0^{4-x} x \, dz \, dy \, dx = k \int_0^4 \int_0^4 x(4 - x) \, dy \, dx$

$ = 4k \int_0^4 (4x - x^2) \, dx = \dfrac{128k}{3}$

$M_{xy} = k \int_0^4 \int_0^4 \int_0^{4-x} xz \, dz \, dy \, dx = k \int_0^4 \int_0^4 x\dfrac{(4 - x)^2}{2} \, dy \, dx$

$\phantom{M_{xy}} = 2k \int_0^4 (16x - 8x^2 + x^3) \, dx = \dfrac{128k}{3}$

$\overline{z} = \dfrac{M_{xy}}{m} = 1$

41. $m = k \int_0^b \int_0^b \int_0^b xy \, dz \, dy \, dx = \dfrac{kb^5}{4}$

$M_{yz} = k \int_0^b \int_0^b \int_0^b x^2 y \, dz \, dy \, dx = \dfrac{kb^6}{6}$

$M_{xz} = k \int_0^b \int_0^b \int_0^b xy^2 \, dz \, dy \, dx = \dfrac{kb^6}{6}$

$M_{xy} = k \int_0^b \int_0^b \int_0^b xyz \, dz \, dy \, dx = \dfrac{kb^6}{8}$

$\overline{x} = \dfrac{M_{yz}}{m} = \dfrac{kb^6/6}{kb^5/4} = \dfrac{2b}{3}$

$\overline{y} = \dfrac{M_{xz}}{m} = \dfrac{kb^6/6}{kb^5/4} = \dfrac{2b}{3}$

$\overline{z} = \dfrac{M_{xy}}{m} = \dfrac{kb^6/8}{kb^5/4} = \dfrac{b}{2}$

43. $\overline{x}$ will be greater than 2, whereas $\overline{y}$ and $\overline{z}$ will be unchanged.

45. $\overline{y}$ will be greater than 0, whereas $\overline{x}$ and $\overline{z}$ will be unchanged.

47. $m = \dfrac{1}{3}k\pi r^2 h$

$\overline{x} = \overline{y} = 0$ by symmetry

$$M_{xy} = 4k\int_0^r \int_0^{\sqrt{r^2-x^2}} \int_{h\sqrt{x^2+y^2}/r}^{h} z \, dz \, dy \, dx$$

$$= \frac{2kh^2}{r^2}\int_0^r \int_0^{\sqrt{r^2-x^2}} \left(r^2 - x^2 - y^2\right) dy \, dx$$

$$= \frac{4kh^2}{3r^2}\int_0^r \left(r^2 - x^2\right)^{3/2} dx = \frac{k\pi r^2 h^2}{4}$$

$$\overline{z} = \frac{M_{xy}}{m} = \frac{k\pi r^2 h^2/4}{k\pi r^2 h/3} = \frac{3h}{4}$$

$$(\overline{x}, \overline{y}, \overline{z}) = \left(0, 0, \frac{3h}{4}\right)$$

49. $m = \dfrac{128k\pi}{3}$

$\overline{x} = \overline{y} = 0$ by symmetry

$z = \sqrt{4^2 - x^2 - y^2}$

$$M_{xy} = 4k\int_0^4 \int_0^{\sqrt{4^2-x^2}} \int_0^{\sqrt{4^2-x^2-y^2}} z \, dz \, dy \, dx$$

$$= 2k\int_0^4 \int_0^{\sqrt{4^2-x^2}} \left(4^2 - x^2 - y^2\right) dy \, dx = 2k\int_0^4 \left[16y - x^2 y - \frac{1}{3}y^3\right]_0^{\sqrt{4^2-x^2}} dx$$

$$= \frac{4k}{3}\int_0^4 \left(4^2 - x^2\right)^{3/2} dx$$

$$= \frac{1024k}{3}\int_0^{\pi/2} \cos^4\theta \, d\theta \quad (\text{let } x = 4\sin\theta)$$

$$= 64\pi k \quad \text{by Wallis's Formula}$$

$$\overline{z} = \frac{M_{xy}}{m} = \frac{64k\pi}{1} \cdot \frac{3}{128k\pi} = \frac{3}{2}$$

$$(\overline{x}, \overline{y}, \overline{z}) = \left(0, 0, \frac{3}{2}\right)$$

51. $f(x, y) = \dfrac{5}{12}y$

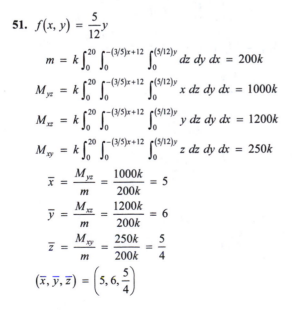

$$m = k\int_0^{20} \int_0^{-(3/5)x+12} \int_0^{(5/12)y} dz \, dy \, dx = 200k$$

$$M_{yz} = k\int_0^{20} \int_0^{-(3/5)x+12} \int_0^{(5/12)y} x \, dz \, dy \, dx = 1000k$$

$$M_{xz} = k\int_0^{20} \int_0^{-(3/5)x+12} \int_0^{(5/12)y} y \, dz \, dy \, dx = 1200k$$

$$M_{xy} = k\int_0^{20} \int_0^{-(3/5)x+12} \int_0^{(5/12)y} z \, dz \, dy \, dx = 250k$$

$$\overline{x} = \frac{M_{yz}}{m} = \frac{1000k}{200k} = 5$$

$$\overline{y} = \frac{M_{xz}}{m} = \frac{1200k}{200k} = 6$$

$$\overline{z} = \frac{M_{xy}}{m} = \frac{250k}{200k} = \frac{5}{4}$$

$$(\overline{x}, \overline{y}, \overline{z}) = \left(5, 6, \frac{5}{4}\right)$$

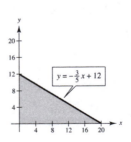

$y = -\dfrac{3}{5}x + 12$

53. (a) $I_x = k \int_0^a \int_0^a \int_0^a \left(y^2 + z^2\right) dx\, dy\, dz = ka \int_0^a \int_0^a \left(y^2 + z^2\right) dy\, dz$

$$= ka \int_0^a \left[\frac{1}{3}y^3 + z^2 y\right]_0^a dz = ka \int_0^a \left(\frac{1}{3}a^3 + az^2\right) dz = \left[ka\left(\frac{1}{3}a^3 z + \frac{1}{3}az^3\right)\right]_0^a = \frac{2ka^5}{3}$$

$I_x = I_y = I_z = \dfrac{2ka^5}{3}$ by symmetry

(b) $I_x = k \int_0^a \int_0^a \int_0^a \left(y^2 + z^2\right) xyz\, dx\, dy\, dz = \dfrac{ka^2}{2} \int_0^a \int_0^a \left(y^3 z + yz^3\right) dy\, dz$

$$= \frac{ka^2}{2} \int_0^a \left[\frac{y^4 z}{4} + \frac{y^2 z^3}{2}\right]_0^a dz = \frac{ka^4}{8} \int_0^a \left(a^2 z + 2z^3\right) dz = \left[\frac{ka^4}{8}\left(\frac{a^2 z^2}{2} + \frac{2z^4}{4}\right)\right]_0^a = \frac{ka^8}{8}$$

$I_x = I_y = I_z = \dfrac{ka^8}{8}$ by symmetry

55. (a) $I_x = k \int_0^4 \int_0^4 \int_0^{4-x} \left(y^2 + z^2\right) dz\, dy\, dx = k \int_0^4 \int_0^4 \left[y^2(4 - x) + \frac{1}{3}(4 - x)^3\right] dy\, dx$

$$= k \int_0^4 \left[\frac{y^3}{3}(4 - x) + \frac{y}{3}(4 - x)^3\right]_0^4 dx = k \int_0^4 \left[\frac{64}{3}(4 - x) + \frac{4}{3}(4 - x)^3\right] dx = k\left[-\frac{32}{3}(4 - x)^2 - \frac{1}{3}(4 - x)^4\right]_0^4 = 256k$$

$I_y = k \int_0^4 \int_0^4 \int_0^{4-x} \left(x^2 + z^2\right) dz\, dy\, dx = k \int_0^4 \int_0^4 \left[x^2(4 - x) + \frac{1}{3}(4 - x)^3\right] dy\, dx$

$$= 4k \int_0^4 \left[4x^2 - x^3 + \frac{1}{3}(4 - x)^3\right] dx = 4k\left[\frac{4}{3}x^3 - \frac{1}{4}x^4 - \frac{1}{12}(4 - x)^4\right]_0^4 = \frac{512k}{3}$$

$I_z = k \int_0^4 \int_0^4 \int_0^{4-x} \left(x^2 + y^2\right) dz\, dy\, dx = k \int_0^4 \int_0^4 \left(x^2 + y^2\right)(4 - x)\, dy\, dx$

$$= k \int_0^4 \left[\left(x^2 y + \frac{y^3}{3}\right)(4 - x)\right]_0^4 dx = k \int_0^4 \left(4x^2 + \frac{64}{3}\right)(4 - x)\, dx = 256k$$

(b) $I_x = k \int_0^4 \int_0^4 \int_0^{4-x} y\left(y^2 + z^2\right) dz\, dy\, dx = k \int_0^4 \int_0^4 \left[y^3(4 - x) + \frac{1}{3}y(4 - x)^3\right] dy\, dx$

$$= k \int_0^4 \left[\frac{y^4}{4}(4 - x) + \frac{y^2}{6}(4 - x)^3\right]_0^4 dx = k \int_0^4 \left[64(4 - x) + \frac{8}{3}(4 - x)^3\right] dx = k\left[-32(4 - x)^2 - \frac{2}{3}(4 - x)^4\right]_0^4 = \frac{2048k}{3}$$

$I_y = k \int_0^4 \int_0^4 \int_0^{4-x} y\left(x^2 + z^2\right) dz\, dy\, dx = k \int_0^4 \int_0^4 \left[x^2 y(4 - x) + \frac{1}{3}y(4 - x)^3\right] dy\, dx$

$$= 8k \int_0^4 \left[4x^2 - x^3 + \frac{1}{3}(4 - x)^3\right] dx = 8k\left[\frac{4}{3}x^3 - \frac{1}{4}x^4 - \frac{1}{12}(4 - x)^4\right]_0^4 = \frac{1024k}{3}$$

$I_z = k \int_0^4 \int_0^4 \int_0^{4-x} y\left(x^2 + y^2\right) dz\, dy\, dx = k \int_0^4 \int_0^4 \left(x^2 y + y^3\right)(4 - x)\, dx$

$$= k \int_0^4 \left[\left(\frac{x^2 y^2}{2} + \frac{y^4}{4}\right)(4 - x)\right]_0^4 dx = k \int_0^4 \left(8x^2 + 64\right)(4 - x)\, dx$$

$$= 8k \int_0^4 \left(32 - 8x + 4x^2 - x^3\right) dx = \left[8k\left(32x - 4x^2 + \frac{4}{3}x^3 - \frac{1}{4}x^4\right)\right]_0^4 = \frac{2048k}{3}$$

57. $I_{xy} = k \int_{-L/2}^{L/2} \int_{-a}^{a} \int_{-\sqrt{a^2-x^2}}^{\sqrt{a^2-x^2}} z^2 \, dz \, dx \, dy = k \int_{-L/2}^{L/2} \int_{-a}^{a} \frac{2}{3}(a^2 - x^2)\sqrt{a^2 - x^2} \, dx \, dy$

$= \frac{2}{3} \int_{-L/2}^{L/2} k \left[\frac{a^2}{2}\left(x\sqrt{a^2 - x^2} + a^2 \arcsin\frac{x}{a} \right) - \frac{1}{8}\left(x(2x^2 - a^2)\sqrt{x^2 - a^2} + a^4 \arcsin\frac{x}{a} \right) \right]_{-a}^{a} dy$

$= \frac{2k}{3} \int_{-L/2}^{L/2} 2\left(\frac{a^4\pi}{4} - \frac{a^4\pi}{16} \right) dy = \frac{a^4\pi L k}{4}$

Because $m = \pi a^2 L k$, $I_{xy} = ma^2/4$.

$I_{xz} = k \int_{-L/2}^{L/2} \int_{-a}^{a} \int_{-\sqrt{a^2-x^2}}^{\sqrt{a^2-x^2}} y^2 \, dz \, dx \, dy = 2k \int_{-L/2}^{L/2} \int_{-a}^{a} y^2 \sqrt{a^2 - x^2} \, dx \, dy$

$= 2k \int_{-L/2}^{L/2} \left[\frac{y^2}{2}\left(x\sqrt{a^2 - x^2} + a^2 \arcsin\frac{x}{a} \right) \right]_{-a}^{a} dy = k\pi a^2 \int_{-L/2}^{L/2} y^2 \, dy = \frac{2k\pi a^2}{3}\left(\frac{L^3}{8} \right) = \frac{1}{12}mL^2$

$I_{yz} = k \int_{-L/2}^{L/2} \int_{-a}^{a} \int_{-\sqrt{a^2-x^2}}^{\sqrt{a^2-x^2}} x^2 \, dz \, dx \, dy = 2k \int_{-L/2}^{L/2} \int_{-a}^{a} x^2 \sqrt{a^2 - x^2} \, dx \, dy$

$= 2k \int_{-L/2}^{L/2} \frac{1}{8}\left[x(2x^2 - a^2)\sqrt{a^2 - x^2} + a^4 \arcsin\frac{x}{a} \right]_{-a}^{a} dy = \frac{ka^4\pi}{4} \int_{-L/2}^{L/2} dy = \frac{ka^4\pi L}{4} = \frac{ma^2}{4}$

$I_x = I_{xy} + I_{xz} = \frac{ma^2}{4} + \frac{mL^2}{12} = \frac{m}{12}(3a^2 + L^2)$

$I_y = I_{xy} + I_{yz} = \frac{ma^2}{4} + \frac{ma^2}{4} = \frac{ma^2}{2}$

$I_z = I_{xz} + I_{yz} = \frac{mL^2}{12} + \frac{ma^2}{4} = \frac{m}{12}(3a^2 + L^2)$

59. $\int_{-1}^{1} \int_{-1}^{1} \int_{0}^{1-x} (x^2 + y^2)\sqrt{x^2 + y^2 + z^2} \, dz \, dy \, dx$

61. $\rho = kz$

(a) $m = \int_{-2}^{2} \int_{-\sqrt{4-x^2}}^{\sqrt{4-x^2}} \int_{0}^{4-x^2-y^2} (kz) \, dz \, dy \, dx \left(= \frac{32k\pi}{3} \right)$

(b) $\bar{x} = \bar{y} = 0$ by symmetry

$\bar{z} = \frac{M_{xy}}{m} = \frac{1}{m} \int_{-2}^{2} \int_{-\sqrt{4-x^2}}^{\sqrt{4-x^2}} \int_{0}^{4-x^2-y^2} kz^2 \, dz \, dy \, dx \, (= 2)$

(c) $I_z = \int_{-2}^{2} \int_{-\sqrt{4-x^2}}^{\sqrt{4-x^2}} \int_{0}^{4-x^2-y^2} (x^2 + y^2)kz \, dz \, dy \, dx \left(= \frac{32k\pi}{3} \right)$

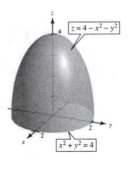

63. $V = 1$ (unit cube)

Average value $= \frac{1}{V} \iiint_Q f(x, y, z) \, dV$

$= \int_{0}^{1} \int_{0}^{1} \int_{0}^{1} (z^2 + 4) \, dx \, dy \, dz$

$= \int_{0}^{1} \int_{0}^{1} (z^2 + 4) \, dy \, dz = \int_{0}^{1} (z^2 + 4) \, dz$

$= \left[\frac{z^3}{3} + 4z \right]_{0}^{1} = \frac{1}{3} + 4 = \frac{13}{3}$

65. $V = \frac{1}{3}$ base $\times$ height

$= \frac{1}{3}\left(\frac{1}{2}(2)(2)\right)(2) = \frac{4}{3}$

$f(x, y, z) = x + y + z$

Plane: $x + y + z = 2$

Average value $= \frac{1}{V}\iiint\limits_{Q} f(x, y, z)\, dV$

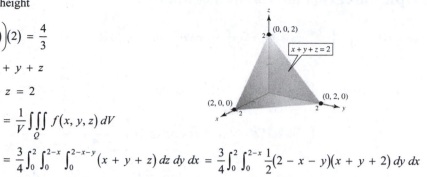

$= \frac{3}{4}\int_0^2 \int_0^{2-x} \int_0^{2-x-y} (x + y + z)\, dz\, dy\, dx = \frac{3}{4}\int_0^2 \int_0^{2-x} \frac{1}{2}(2 - x - y)(x + y + 2)\, dy\, dx$

$= \frac{3}{4}\int_0^2 \frac{1}{6}(x + 4)(x - 2)^2\, dx = \frac{3}{4}(2) = \frac{3}{2}$

67. See the definition, pages 1009 and 1010.

See Theorem 14.4.

69. The region of integration is a cube:

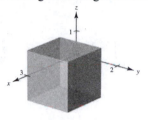

Answer: (b)

71. $1 - 2x^2 - y^2 - 3z^2 \geq 0$

$2x^2 + y^2 + 3z^2 \leq 1$

$Q = \{(x, y, z): 2x^2 + y^2 + 3z^2 \leq 1\}$ ellipsoid

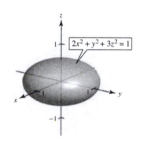

$\int_{-1/\sqrt{2}}^{1/\sqrt{2}} \int_{-\sqrt{1-2x^2}}^{\sqrt{1-2x^2}} \int_{-\sqrt{(1-2x^2-y^2)/3}}^{\sqrt{(1-2x^2-y^2)/3}} \left(1 - 2x^2 - y^2 - 3z^2\right)\, dz\, dy\, dx \approx 0.684$

Exact value: $\dfrac{4\sqrt{6}\pi}{45}$

73. Let $y_k = 1 - x_k$.

$\frac{\pi}{2n}(x_1 + \cdots + x_n) = \frac{\pi}{2n}(n - y_1 - y_2 - \cdots - y_n) = \frac{\pi}{2} - \frac{\pi}{2n}(y_1 + \cdots + y_n)$

So,

$I_1 = \int_0^1 \int_0^1 \cdots \int_0^1 \cos^2\left\{\frac{\pi}{2n}(x_1 + \cdots + x_n)\right\} dx_1\, dx_2 \cdots dx_n$

$= \int_1^0 \int_1^0 \cdots \int_1^0 \sin^2\left\{\frac{\pi}{2n}(y_1 + \cdots + y_n)\right\}(-dy_1)(-dy_2)\cdots(-dy_n) = \int_0^1 \int_0^1 \cdots \int_0^1 \sin^2\left\{\frac{\pi}{2n}(x_1 + \cdots + x_n)\right\} dx_1\, dx_2 \cdots dx_n = I_2$

$I_1 + I_2 = 1 \Rightarrow I_1 = \frac{1}{2}$.

Finally, $\lim\limits_{n \to \infty} I_1 = \frac{1}{2}$.

Section 14.7   Triple Integrals in Other Coordinates

1. $\int_{-1}^{5} \int_{0}^{\pi/2} \int_{0}^{3} r \cos\theta \, dr \, d\theta \, dz = \int_{-1}^{5} \int_{0}^{\pi/2} \frac{9}{2} \cos\theta \, d\theta \, dz = \int_{-1}^{5} \left[\frac{9}{2} \sin\theta \right]_{0}^{\pi/2} dz = \int_{-1}^{5} \frac{9}{2} \, dz = \left[\frac{9}{2} z \right]_{-1}^{5} = \frac{9}{2}(5 - (-1)) = 27$

3. $\int_{0}^{\pi/2} \int_{0}^{2\cos^2\theta} \int_{0}^{4-r^2} r \sin\theta \, dz \, dr \, d\theta = \int_{0}^{\pi/2} \int_{0}^{2\cos^2\theta} r(4 - r^2) \sin\theta \, dr \, d\theta = \int_{0}^{\pi/2} \left[\left(2r^2 - \frac{r^4}{4} \right) \sin\theta \right]_{0}^{2\cos^2\theta} d\theta$

$$= \int_{0}^{\pi/2} \left[8\cos^4\theta - 4\cos^8\theta \right] \sin\theta \, d\theta = \left[-\frac{8\cos^5\theta}{5} + \frac{4\cos^9\theta}{9} \right]_{0}^{\pi/2} = \frac{52}{45}$$

5. $\int_{0}^{2\pi} \int_{0}^{\pi/4} \int_{0}^{\cos\phi} \rho^2 \sin\phi \, d\rho \, d\phi \, d\theta = \frac{1}{3} \int_{0}^{2\pi} \int_{0}^{\pi/4} \cos^3\phi \sin\phi \, d\phi \, d\theta = -\frac{1}{12} \int_{0}^{2\pi} \left[\cos^4\phi \right]_{0}^{\pi/4} d\theta = \frac{\pi}{8}$

7. $\int_{0}^{4} \int_{0}^{z} \int_{0}^{\pi/2} re^r \, d\theta \, dr \, dz = \pi\left(e^4 + 3 \right)$

9. $\int_{0}^{\pi/2} \int_{0}^{3} \int_{0}^{e^{-r^2}} r \, dz \, dr \, d\theta = \int_{0}^{\pi/2} \int_{0}^{3} re^{-r^2} \, dr \, d\theta = \int_{0}^{\pi/2} \left[-\frac{1}{2} e^{-r^2} \right]_{0}^{3} d\theta = \int_{0}^{\pi/2} \frac{1}{2}\left(1 - e^{-9} \right) d\theta = \frac{\pi}{4}\left(1 - e^{-9} \right)$

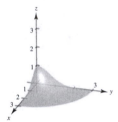

11. $\int_{0}^{2\pi} \int_{\pi/6}^{\pi/2} \int_{0}^{4} \rho^2 \sin\phi \, d\rho \, d\phi \, d\theta = \frac{64}{3} \int_{0}^{2\pi} \int_{\pi/6}^{\pi/2} \sin\phi \, d\phi \, d\theta = \frac{64}{3} \int_{0}^{2\pi} \left[-\cos\phi \right]_{\pi/6}^{\pi/2} d\theta = \frac{32\sqrt{3}}{3} \int_{0}^{2\pi} d\theta = \frac{64\sqrt{3}\pi}{3}$

13. $\int_{0}^{2\pi} \int_{0}^{2} \int_{r^2}^{4} r^2 \cos\theta \, dz \, dr \, d\theta = 0$

$\int_{0}^{2\pi} \int_{0}^{\arctan(1/2)} \int_{0}^{4\sec\phi} \rho^3 \sin^2\phi \cos\theta \, d\rho \, d\phi \, d\theta + \int_{0}^{2\pi} \int_{\arctan(1/2)}^{\pi/2} \int_{0}^{\cot\phi \csc\phi} \rho^3 \sin^2\phi \cos\theta \, d\rho \, d\phi \, d\theta = 0$

15. $\int_{0}^{2\pi} \int_{0}^{a} \int_{a}^{a+\sqrt{a^2-r^2}} r^2 \cos\theta \, dz \, dr \, d\theta = 0$

$\int_{0}^{\pi/4} \int_{0}^{2\pi} \int_{a\sec\phi}^{2a\cos\phi} \rho^3 \sin^2\phi \cos\theta \, d\rho \, d\theta \, d\phi = 0$

17. $V = 4\int_{0}^{\pi/2} \int_{0}^{a\cos\theta} \int_{0}^{\sqrt{a^2-r^2}} r \, dz \, dr \, d\theta = 4\int_{0}^{\pi/2} \int_{0}^{a\cos\theta} r\sqrt{a^2 - r^2} \, dr \, d\theta$

$$= \frac{4}{3} a^3 \int_{0}^{\pi/2} \left(1 - \sin^3\theta \right) d\theta = \frac{4}{3} a^3 \left[\theta + \frac{1}{3} \cos\theta \left(\sin^2\theta + 2 \right) \right]_{0}^{\pi/2} = \frac{4}{3} a^3 \left(\frac{\pi}{2} - \frac{2}{3} \right) = \frac{2a^3}{9}(3\pi - 4)$$

19. In the *xy*-plane, $2x = 2x^2 + 2y^2 \Rightarrow$

$$0 = x^2 - x + y^2 \Rightarrow \left(x^2 - x + 1/4\right) + y^2 = 1/4$$

$$\Rightarrow (x - 1/2)^2 + y^2 = (1/2)^2$$

In polar coordinates, use $r = \cos\theta$ for this circle.

$$V = \int_0^\pi \int_0^{\cos\theta} \int_{2r^2}^{2r\cos\theta} r\,dz\,dr\,d\theta$$

$$= \int_0^\pi \int_0^{\cos\theta} \left(2r^2\cos\theta - 2r^3\right) dr\,d\theta$$

$$= \int_0^\pi \left[\frac{2r^3}{3}\cos\theta - \frac{r^4}{2}\right]_0^{\cos\theta} d\theta$$

$$= \int_0^\pi \left(\frac{2}{3}\cos^4\theta - \frac{\cos^4\theta}{2}\right) d\theta$$

$$= \frac{1}{6}\int_0^\pi \cos^4\theta\,d\theta = \frac{\pi}{16}$$

21. $V = 2\int_0^\pi \int_0^{a\cos\theta} \int_0^{\sqrt{a^2-r^2}} r\,dz\,dr\,d\theta$

$$= 2\int_0^\pi \int_0^{a\cos\theta} r\sqrt{a^2 - r^2}\,dr\,d\theta$$

$$= 2\int_0^\pi \left[-\frac{1}{3}\left(a^2 - r^2\right)^{3/2}\right]_0^{a\cos\theta} d\theta$$

$$= \frac{2a^3}{3}\int_0^\pi \left(1 - \sin^3\theta\right) d\theta$$

$$= \frac{2a^3}{3}\left[\theta + \cos\theta - \frac{\cos^3\theta}{3}\right]_0^\pi = \frac{2a^3}{9}(3\pi - 4)$$

23. $m = \int_0^{2\pi} \int_0^2 \int_0^{9-r\cos\theta-2r\sin\theta} (kr)r\,dz\,dr\,d\theta$

$$= \int_0^{2\pi} \int_0^2 kr^2(9 - r\cos\theta - 2r\sin\theta)\,dr\,d\theta$$

$$= \int_0^{2\pi} k\left[3r^3 - \frac{r^4}{4}\cos\theta - \frac{r^4}{2}\sin\theta\right]_0^2 d\theta$$

$$= \int_0^{2\pi} k[24 - 4\cos\theta - 8\sin\theta]\,d\theta$$

$$= k[24\theta - 4\sin\theta + 8\cos\theta]_0^{2\pi}$$

$$= k[48\pi + 8 - 8] = 48k\pi$$

25. $z = h - \dfrac{h}{r_0}\sqrt{x^2 + y^2} = \dfrac{h}{r_0}(r_0 - r)$

$$V = 4\int_0^{\pi/2} \int_0^{r_0} \int_0^{h(r_0-r)/r_0} r\,dz\,dr\,d\theta$$

$$= \frac{4h}{r_0}\int_0^{\pi/2} \int_0^{r_0} \left(r_0 r - r^2\right) dr\,d\theta$$

$$= \frac{4h}{r_0}\int_0^{\pi/2} \frac{r_0^3}{6}\,d\theta$$

$$= \frac{4h}{r_0}\left(\frac{r_0^3}{6}\right)\left(\frac{\pi}{2}\right) = \frac{1}{3}\pi r_0^2 h$$

27. $\rho = k\sqrt{x^2 + y^2} = kr$

$\bar{x} = \bar{y} = 0$ by symmetry

$$m = 4k\int_0^{\pi/2} \int_0^{r_0} \int_0^{h(r_0-r)/r_0} r^2\,dz\,dr\,d\theta = \frac{1}{6}k\pi r_0^3 h$$

$$M_{xy} = 4k\int_0^{\pi/2} \int_0^{r_0} \int_0^{h(r_0-r)/r_0} r^2 z\,dz\,dr\,d\theta$$

$$= \frac{1}{30}k\pi r_0^3\,h^2$$

$$\bar{z} = \frac{M_{xy}}{m} = \frac{k\pi r_0^3 h^2/30}{k\pi r_0^3 h/6} = \frac{h}{5}$$

$$(\bar{x}, \bar{y}, \bar{z}) = \left(0, 0, \frac{h}{5}\right)$$

29. $I_z = 4k\int_0^{\pi/2} \int_0^{r_0} \int_0^{h(r_0-r)/r_0} r^3\,dz\,dr\,d\theta$

$$= \frac{4kh}{r_0}\int_0^{\pi/2} \int_0^{r_0} \left(r_0 r^3 - r^4\right) dr\,d\theta$$

$$= \frac{4kh}{r_0}\left(\frac{r_0^5}{20}\right)\left(\frac{\pi}{2}\right) = \frac{1}{10}k\pi r_0^4 h$$

Because the mass of the core is $m = kV = k\left(\dfrac{1}{3}\pi r_0^2 h\right)$

from Exercise 25, we have $k = 3m/\pi r_0^2 h$. So,

$$I_z = \frac{1}{10}k\pi r_0^4 h = \frac{1}{10}\left(\frac{3m}{\pi r_0^2 h}\right)\pi r_0^4 h = \frac{3}{10}mr_0^2.$$

31. $m = k\left(\pi b^2 h - \pi a^2 h\right) = k\pi h\left(b^2 - a^2\right)$

$$I_z = 4k\int_0^{\pi/2} \int_a^b \int_0^h r^3\,dz\,dr\,d\theta$$

$$= 4kh\int_0^{\pi/2} \int_a^b r^3\,dr\,d\theta = kh\int_0^{\pi/2} \left(b^4 - a^4\right) d\theta$$

$$= \frac{k\pi\left(b^4 - a^4\right)h}{2} = \frac{k\pi\left(b^2 - a^2\right)\left(b^2 + a^2\right)h}{2}$$

$$= \frac{1}{2}m\left(a^2 + b^2\right)$$

33. $V = \int_0^{2\pi} \int_{\pi/4}^{\pi/2} \int_0^3 \rho^2 \sin\phi \, d\rho \, d\phi \, d\theta$

$= \int_0^{2\pi} \int_{\pi/4}^{\pi/2} 9 \sin\phi \, d\phi \, d\theta$

$= \int_0^{2\pi} \left[-9\cos\phi \right]_{\pi/4}^{\pi/2} d\theta$

$= \int_0^{2\pi} 9\left(\frac{\sqrt{2}}{2} \right) d\theta = 18\pi \left(\frac{\sqrt{2}}{2} \right) = 9\pi\sqrt{2}$

35. $V = \int_0^{2\pi} \int_0^{\pi} \int_0^{4\sin\phi} \rho^2 \sin\phi \, d\rho \, d\phi \, d\theta = 16\pi^2$

37. $m = 8k \int_0^{\pi/2} \int_0^{\pi/2} \int_0^a \rho^3 \sin\phi \, d\rho \, d\theta \, d\phi$

$= 2ka^4 \int_0^{\pi/2} \int_0^{\pi/2} \sin\phi \, d\theta \, d\phi$

$= k\pi a^4 \int_0^{\pi/2} \sin\phi \, d\phi = \left[k\pi a^4 (-\cos\phi) \right]_0^{\pi/2} = k\pi a^4$

39. $m = \frac{2}{3}k\pi r^3$

$\bar{x} = \bar{y} = 0$ by symmetry

$M_{xy} = 4k \int_0^{\pi/2} \int_0^{\pi/2} \int_0^r \rho^3 \cos\phi \sin\phi \, d\rho \, d\theta \, d\phi$

$= \frac{1}{2} kr^4 \int_0^{\pi/2} \int_0^{\pi/2} \sin 2\phi \, d\theta \, d\phi$

$= \frac{kr^4 \pi}{4} \int_0^{\pi/2} \sin 2\phi \, d\phi$

$= \left[-\frac{1}{8} k\pi r^4 \cos 2\phi \right]_0^{\pi/2} = \frac{1}{4} k\pi r^4$

$\bar{z} = \frac{M_{xy}}{m} = \frac{k\pi r^4 / 4}{2k\pi r^3 / 3} = \frac{3r}{8}$

$(\bar{x}, \bar{y}, \bar{z}) = \left(0, 0, \frac{3r}{8} \right)$

41. $I_z = 4k \int_{\pi/4}^{\pi/2} \int_0^{\pi/2} \int_0^{\cos\phi} \rho^4 \sin^3\phi \, d\rho \, d\theta \, d\phi$

$= \frac{4}{5} k \int_{\pi/4}^{\pi/2} \int_0^{\pi/2} \cos^5\phi \sin^3\phi \, d\theta \, d\phi$

$= \frac{2}{5} k\pi \int_{\pi/4}^{\pi/2} \cos^5\phi (1 - \cos^2\phi) \sin\phi \, d\phi$

$= \left[\frac{2}{5} k\pi \left(-\frac{1}{6} \cos^6\phi + \frac{1}{8} \cos^8\phi \right) \right]_{\pi/4}^{\pi/2} = \frac{k\pi}{192}$

43. $x = r\cos\theta \qquad\qquad x^2 + y^2 = r^2$

$y = r\sin\theta \qquad\qquad \tan\theta = \frac{y}{x}$

$z = z \qquad\qquad\qquad z = z$

45. $\int_{\theta_1}^{\theta_2} \int_{g_1(\theta)}^{g_2(\theta)} \int_{h_1(r\cos\theta,\, r\sin\theta)}^{h_2(r\cos\theta,\, r\sin\theta)} f(r\cos\theta,\, r\sin\theta,\, z) r \, dz \, dr \, d\theta$

47. (a) $r = r_0$: right circular cylinder about z-axis

$\theta = \theta_0$: plane parallel to z-axis

$z = z_0$: plane parallel to xy-plane

(b) $\rho = \rho_0$: sphere of radius ρ_0

$\theta = \theta_0$: plane parallel to z-axis

$\phi = \phi_0$: cone

49. $\left(x^2 + y^2 + z^2 + 8 \right)^2 \le 36\left(x^2 + y^2 \right)$

In cylindrical coordinates,

$\left(r^2 + z^2 + 8 \right)^2 \le 36 r^2$

$r^2 + z^2 + 8 \le 6r$

$r^2 - 6r + 9 + z^2 - 1 \le 0$

$(r - 3)^2 + z^2 \le 1.$

This is a torus: rotate $(x - 3)^2 + z^2 = 1$ about the z-axis. By Pappus' Theorem,

$V = 2\pi(3)\pi = 6\pi^2.$

Section 14.8   Change of Variables: Jacobians

1. $x = -\dfrac{1}{2}(u - v)$

$y = \dfrac{1}{2}(u + v)$

$\dfrac{\partial x}{\partial u}\dfrac{\partial y}{\partial v} - \dfrac{\partial y}{\partial u}\dfrac{\partial x}{\partial v} = \left(-\dfrac{1}{2}\right)\left(\dfrac{1}{2}\right) - \left(\dfrac{1}{2}\right)\left(\dfrac{1}{2}\right) = -\dfrac{1}{2}$

3. $x = u - v^2$

$y = u + v$

$\dfrac{\partial x}{\partial u}\dfrac{\partial y}{\partial v} - \dfrac{\partial y}{\partial u}\dfrac{\partial x}{\partial v} = (1)(1) - (1)(-2v) = 1 + 2v$

7. $x = e^u \sin v$

$y = e^u \cos v$

$\dfrac{\partial x}{\partial u}\dfrac{\partial y}{\partial v} - \dfrac{\partial y}{\partial u}\dfrac{\partial x}{\partial v} = \left(e^u \sin v\right)\left(-e^u \sin v\right) - \left(e^u \cos v\right)\left(e^u \cos v\right) = -e^{2u}$

9. $x = 3u + 2v$

$y = 3v$

$v = \dfrac{y}{3}$

$u = \dfrac{x - 2v}{3} = \dfrac{x - 2(y/3)}{3} = \dfrac{x}{3} - \dfrac{2y}{9}$

(x, y)	(u, v)
$(0, 0)$	$(0, 0)$
$(3, 0)$	$(1, 0)$
$(2, 3)$	$(0, 1)$

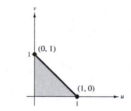

5. $x = u \cos \theta - v \sin \theta$

$y = u \sin \theta + v \cos \theta$

$\dfrac{\partial x}{\partial u}\dfrac{\partial y}{\partial v} - \dfrac{\partial y}{\partial u}\dfrac{\partial x}{\partial v} = \cos^2 \theta + \sin^2 \theta = 1$

11. $x = \frac{1}{2}(u + v)$

$y = \frac{1}{2}(u - v)$

$u = x + y$

$v = x - y$

(x, y)	(u, v)
$\left(\frac{1}{2}, \frac{1}{2}\right)$	$(1, 0)$
$(0, 1)$	$(1, -1)$
$(1, 2)$	$(3, -1)$
$\left(\frac{3}{2}, \frac{3}{2}\right)$	$(3, 0)$

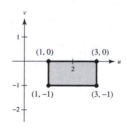

13. $x - 2y = 0$ } $3y = 4$
$x + y = 4$ $\Rightarrow$ $y = \frac{4}{3}, \quad x = \frac{8}{3}$

$x - 2y = -4$ } $3y = 8$
$x + y = 4$ } $y = \frac{8}{3}, \quad x = \frac{4}{3}$

$x - 2y = -4$ } $3y = 5$
$x + y = 1$ } $y = \frac{5}{3}, \quad x = -\frac{2}{3}$

$x - 2y = 0$ } $3y = 1$
$x + y = 1$ } $y = \frac{1}{2}, \quad x = \frac{2}{3}$

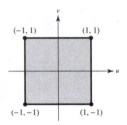

$u = x + y$ } $u - v = 3y \Rightarrow y = \frac{1}{3}(u - v)$
$v = x - 2y$ } $2u + v = 3x \Rightarrow x = \frac{1}{3}(2u + v)$

$\iint\limits_{R} 3 \, xy \, dA = \int_{-2/3}^{2/3} \int_{1-x}^{(x+4)/2} 3 \, xy \, dy \, dx + \int_{2/3}^{4/3} \int_{x/2}^{(x+4)/2} 3 \, xy \, dy \, dx + \int_{4/3}^{8/3} \int_{x/2}^{4-x} 3 \, xy \, dy \, dx = \frac{32}{27} + \frac{164}{27} + \frac{296}{27} = \frac{164}{9}$

15. $x = \frac{1}{2}(u + v)$

$y = \frac{1}{2}(u - v)$

$\frac{\partial x}{\partial u}\frac{\partial y}{\partial v} - \frac{\partial y}{\partial u}\frac{\partial x}{\partial v} = \left(\frac{1}{2}\right)\left(-\frac{1}{2}\right) - \left(\frac{1}{2}\right)\left(\frac{1}{2}\right) = -\frac{1}{2}$

$\iint\limits_{R} 4(x^2 + y^2) \, dA = \int_{-1}^{1} \int_{-1}^{1} 4\left[\frac{1}{4}(u + v)^2 + \frac{1}{4}(u - v)^2\right]\left(\frac{1}{2}\right) dv \, du$

$= \int_{-1}^{1} \int_{-1}^{1} (u^2 + v^2) dv \, du = \int_{-1}^{1} 2\left(u^2 + \frac{1}{3}\right) du = \left[2\left(\frac{u^3}{3} + \frac{u}{3}\right)\right]_{-1}^{1} = \frac{8}{3}$

17. $x = u + v$

$y = u$

$\frac{\partial x}{\partial u}\frac{\partial y}{\partial v} - \frac{\partial y}{\partial u}\frac{\partial x}{\partial v} = (1)(0) - (1)(1) = -1$

$\int_{R}\int y(x - y) \, dA = \int_{0}^{3} \int_{0}^{4} uv(1) \, dv \, du$

$= \int_{0}^{3} 8u \, du = 36$

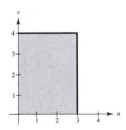

19. $\displaystyle\int_R \int e^{-xy/2}\, dA$

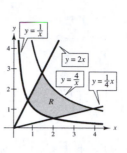

$R: \ y = \dfrac{x}{4}, \ y = 2x, \ y = \dfrac{1}{x}, \ y = \dfrac{4}{x}$

$x = \sqrt{v/u}, \ y = \sqrt{uv} \ \Rightarrow \ u = \dfrac{y}{x}, v = xy$

$$\frac{\partial(x,y)}{\partial(u,v)} = \begin{vmatrix} \dfrac{\partial x}{\partial u} & \dfrac{\partial x}{\partial v} \\[2mm] \dfrac{\partial y}{\partial u} & \dfrac{\partial y}{\partial v} \end{vmatrix} = \begin{vmatrix} -\dfrac{1}{2}\dfrac{v^{1/2}}{u^{3/2}} & \dfrac{1}{2}\dfrac{1}{u^{1/2}v^{1/2}} \\[2mm] \dfrac{1}{2}\dfrac{v^{1/2}}{u^{1/2}} & \dfrac{1}{2}\dfrac{u^{1/2}}{v^{1/2}} \end{vmatrix} = -\frac{1}{4}\left(\frac{1}{u} + \frac{1}{u}\right) = -\frac{1}{2u}$$

Transformed Region:

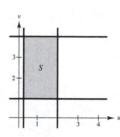

$y = \dfrac{1}{x} \ \Rightarrow \ yx = 1 \ \Rightarrow \ v = 1$

$y = \dfrac{4}{x} \ \Rightarrow \ ux = 4 \ \Rightarrow \ v = 4$

$y = 2x \ \Rightarrow \ \dfrac{y}{x} = 2 \ \Rightarrow \ u = 2$

$y = \dfrac{x}{4} \ \Rightarrow \ \dfrac{y}{x} = \dfrac{1}{4} \ \Rightarrow \ u = \dfrac{1}{4}$

$$\int_R \int e^{-xy/2}\, dA = \int_{1/4}^{2}\int_{1}^{4} e^{-v/2}\left(\frac{1}{2u}\right) dv\, du = -\int_{1/4}^{2}\left[\frac{e^{-v/2}}{u}\right]_1^4 du = -\int_{1/4}^{2}\left(e^{-2} - e^{-1/2}\right)\frac{1}{u}\, du$$

$$= -\left[\left(e^{-2} - e^{-1/2}\right)\ln u\right]_{1/4}^{2} = -\left(e^{-2} - e^{-1/2}\right)\left(\ln 2 - \ln\frac{1}{4}\right) = \left(e^{-1/2} - e^{-2}\right)\ln 8 \approx 0.9798$$

21. $u = x - y = 1, \quad v = x + y = 1$

$u = x - y = -1, \quad v = x + y = 3$

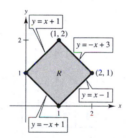

$x = \dfrac{1}{2}(u + v)$

$y = \dfrac{1}{2}(v - u)$

$$\frac{\partial(x,y)}{\partial(u,v)} = \frac{\partial x}{\partial u}\frac{\partial y}{\partial v} - \frac{\partial y}{\partial u}\frac{\partial x}{\partial v} = \frac{1}{2}\left(\frac{1}{2}\right) - \left(-\frac{1}{2}\right)\left(\frac{1}{2}\right) = \frac{1}{2}$$

$$\int_R \int 48\, xy\, dA = \int_1^3\int_{-1}^{1} 48\left(\frac{1}{2}\right)(u+v)\left(\frac{1}{2}\right)(v-u)\left(\frac{1}{2}\right) du\, dv = \int_1^3\int_{-1}^{1} 6\left(v^2 - u^2\right) du\, dv = 6\int_1^3\left[uv^2 - \frac{u^3}{3}\right]_{-1}^{1} dv$$

$$= 6\int_1^3\left(2v^2 - \frac{2}{3}\right) dv = 6\left[\frac{2v^3}{3} - \frac{2}{3}v\right]_1^3 = 6\left[18 - 2 - \frac{2}{3} + \frac{2}{3}\right] = 96$$

23. $u = x + y = 4, \quad v = x - y = 0$

$u = x + y = 8, \quad v = x - y = 4$

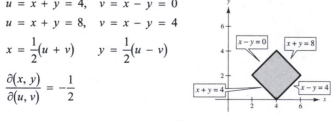

$x = \dfrac{1}{2}(u + v) \qquad y = \dfrac{1}{2}(u - v)$

$$\frac{\partial(x,y)}{\partial(u,v)} = -\frac{1}{2}$$

$$\int_R \int (x+y)e^{x-y}\, dA = \int_4^8\int_0^4 u e^v\left(\frac{1}{2}\right) dv\, du = \frac{1}{2}\int_4^8 u\left(e^4 - 1\right) du = \left[\frac{1}{4}u^2\left(e^4 - 1\right)\right]_4^8 = 12\left(e^4 - 1\right)$$

25. $u = x + 4y = 0,\quad v = x - y = 0$

$\quad u = x + 4y = 5,\quad v = x - y = 5$

$\quad x = \dfrac{1}{5}(u + 4v),\qquad y = \dfrac{1}{5}(u - v)$

$\quad \dfrac{\partial x}{\partial u}\dfrac{\partial y}{\partial v} - \dfrac{\partial y}{\partial u}\dfrac{\partial x}{\partial v} = \left(\dfrac{1}{5}\right)\left(-\dfrac{1}{5}\right) - \left(\dfrac{1}{5}\right)\left(\dfrac{4}{5}\right) = -\dfrac{1}{5}$

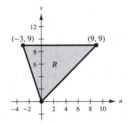

$\quad \displaystyle\int_R\!\!\int \sqrt{(x - y)(x + 4y)}\;dA = \int_0^5\!\!\int_0^5 \sqrt{uv}\left(\dfrac{1}{5}\right)du\,dv = \int_0^5\left[\dfrac{1}{5}\left(\dfrac{2}{3}\right)u^{3/2}\sqrt{v}\right]_0^5 dv = \left[\dfrac{2\sqrt{5}}{3}\left(\dfrac{2}{3}\right)v^{3/2}\right]_0^5 = \dfrac{100}{9}$

27. $u = x + y,\; v = x - y,\; x = \dfrac{1}{2}(u + v),\; y = \dfrac{1}{2}(u - v)$

$\quad \dfrac{\partial x}{\partial u}\dfrac{\partial y}{\partial v} - \dfrac{\partial y}{\partial u}\dfrac{\partial x}{\partial v} = -\dfrac{1}{2}$

$\quad \displaystyle\int_R\!\!\int \sqrt{x + y}\;dA = \int_0^a\!\!\int_{-u}^u \sqrt{u}\left(\dfrac{1}{2}\right)dv\,du = \int_0^a u\sqrt{u}\;du = \left[\dfrac{2}{5}u^{5/2}\right]_0^a = \dfrac{2}{5}a^{5/2}$

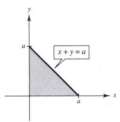 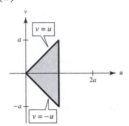

29. $u = 2x - y$

$\quad v = x + y$

$\quad 3x = u + v \Rightarrow x = \tfrac{1}{3}(u + v)$

$\quad$ Then $y = v - x = v - \tfrac{1}{3}(u + v) = \tfrac{1}{3}(2v - u).$

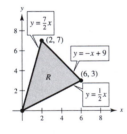

(x, y)	(u, v)
$(0, 0)$	$(0, 0)$
$(6, 3)$	$(9, 9)$
$(2, 7)$	$(-3, 9)$

One side is parallel to the u-axis.

31. $\dfrac{x^2}{a^2} + \dfrac{y^2}{b^2} = 1,\ x = au,\ y = bv$

$\dfrac{(au)^2}{a^2} + \dfrac{(bv)^2}{b^2} = 1$

$u^2 + v^2 = 1$

(a) $\dfrac{x^2}{a^2} + \dfrac{y^2}{b^2} = 1$ $\qquad\qquad\qquad u^2 + v^2 = 1$

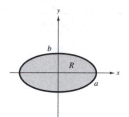   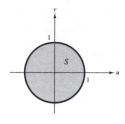

(b) $\dfrac{\partial(x, y)}{\partial(u, v)} = \dfrac{\partial x}{\partial u}\dfrac{\partial y}{\partial v} - \dfrac{\partial y}{\partial u}\dfrac{\partial x}{\partial v} = (a)(b) - (0)(0) = ab$

(c) $A = \displaystyle\int_S \int ab\, dS = ab\left(\pi(1)^2\right) = \pi ab$

33. Jacobian $= \dfrac{\partial(x, y)}{\partial(u, v)} = \dfrac{\partial x}{\partial u}\dfrac{\partial y}{\partial v} - \dfrac{\partial y}{\partial u}\dfrac{\partial x}{\partial v}$

35. $x = u(1 - v),\ y = uv(1 - w),\ z = uvw$

$\dfrac{\partial(x, y, z)}{\partial(u, v, w)} = \begin{vmatrix} 1 - v & -u & 0 \\ v(1 - w) & u(1 - w) & -uv \\ vw & uw & uv \end{vmatrix} = (1 - v)\left[u^2 v(1 - w) + u^2 vw\right] + u\left[uv^2(1 - w) + uv^2 w\right] = (1 - v)\left(u^2 v\right) + u\left(uv^2\right) = u^2 v$

37. $x = \dfrac{1}{2}(u + v),\ y = \dfrac{1}{2}(u - v),\ z = 2uvw$

$\dfrac{\partial(x, y, z)}{\partial(u, v, w)} = \begin{vmatrix} 1/2 & 1/2 & 0 \\ 1/2 & -1/2 & 0 \\ 2vw & 2uw & 2uv \end{vmatrix}$

$= 2uv[-1/4 - 1/4] = -uv$

39. $x = \rho \sin\phi \cos\theta,\ y = \rho \sin\phi \sin\theta,\ z = \rho \cos\phi$

$\dfrac{\partial(x, y, z)}{\partial(\rho, \theta, \phi)} = \begin{vmatrix} \sin\phi \cos\theta & -\rho \sin\phi \sin\theta & \rho \cos\phi \cos\theta \\ \sin\phi \sin\theta & \rho \sin\phi \cos\theta & \rho \cos\phi \sin\theta \\ \cos\phi & 0 & -\rho \sin\phi \end{vmatrix}$

$= \cos\phi\left[-\rho^2 \sin\phi \cos\phi \sin^2\theta - \rho^2 \sin\phi \cos\phi \cos^2\theta\right] - \rho \sin\phi\left[\rho \sin^2\phi \cos^2\theta + \rho \sin^2\phi \sin^2\theta\right]$

$= \cos\phi\left[-\rho^2 \sin\phi \cos\phi\left(\sin^2\theta + \cos^2\theta\right)\right] - \rho \sin\phi\left[\rho \sin^2\phi\left(\cos^2\theta + \sin^2\theta\right)\right]$

$= -\rho^2 \sin\phi \cos^2\phi - \rho^2 \sin^3\phi = -\rho^2 \sin\phi\left(\cos^2\phi + \sin^2\phi\right) = -\rho^2 \sin\phi$

41. Let $u = \dfrac{x}{3}$, $v = y \Rightarrow \dfrac{\partial(x,y)}{\partial(u,v)} = \begin{vmatrix} 3 & 0 \\ 0 & 1 \end{vmatrix} = 3$, $y = \dfrac{x}{2} \Rightarrow v = \dfrac{3u}{2}$.

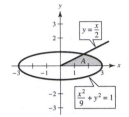 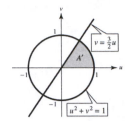 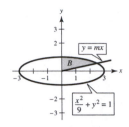 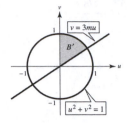

Region A is transformed to region A', and region B is transformed to region B'.

$$A' = B' \Rightarrow \frac{2}{3} = 3m \Rightarrow m = \frac{2}{9}$$

Note: You could also calculate the integrals directly.

Review Exercises for Chapter 14

1. $\displaystyle\int_0^{2x} xy^3 \, dy = \left[\frac{xy^4}{4} \right]_0^{2x} = \frac{x(2x)^4}{4} = 4x^5$

3. $\displaystyle\int_0^1 \int_0^{1+x} (3x + 2y) \, dy \, dx = \int_0^1 \left[3xy + y^2 \right]_0^{1+x} dx$

$$= \int_0^1 (4x^2 + 5x + 1) \, dx$$

$$= \left[\frac{4}{3}x^3 + \frac{5}{2}x^2 + x \right]_0^1$$

$$= \frac{29}{6}$$

5. $\displaystyle\int_0^3 \int_0^{\sqrt{9-x^2}} 4x \, dy \, dx = \int_0^3 4x\sqrt{9 - x^2} \, dx$

$$= \left[-\frac{4}{3}(9 - x^2)^{3/2} \right]_0^3 = 36$$

7. $A = \displaystyle\int_0^1 \int_0^{3-3y} dx \, dy = \int_0^1 (3 - 3y) \, dy$

$$= \left[3y - \frac{3y^2}{2} \right]_0^1$$

$$= \frac{3}{2}$$

9. $A = \displaystyle\int_0^4 \int_x^{2x+2} dy \, dx$

$$= \int_0^4 (x + 2) \, dx$$

$$= \left[\frac{x^2}{2} + 2x \right]_0^4 = 16$$

11.

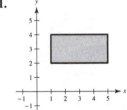

$\displaystyle\int_1^5 \int_2^4 dy \, dx = \int_1^5 \left[y \right]_2^4 dx = \int_1^5 2 \, dx = \left[2x \right]_1^5 = 8$

$\displaystyle\int_2^4 \int_1^5 dx \, dy = \int_2^4 \left[x \right]_1^5 dy = \int_2^4 4 \, dy = \left[4x \right]_2^4 = 8$

13. $\displaystyle\int_0^4 \int_{2x}^8 dy \, dx = \int_0^4 (8 - 2x) \, dx = \left[8x - x^2 \right]_0^4 = 16$

$\displaystyle\int_0^8 \int_0^{y/2} dx \, dy = \int_0^8 \frac{y}{2} \, dy = \left[\frac{y^2}{4} \right]_0^8 = 16$

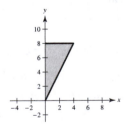

15. $\displaystyle\iint_R 4xy \, dA = \int_0^4 \int_0^2 4xy \, dx \, dy = \int_0^2 \int_0^4 4xy \, dy \, dx$

$\displaystyle\int_0^4 \int_0^2 4xy \, dx \, dy = \int_0^4 \left[2x^2 y \right]_0^2 dy$

$$= \int_0^4 8y \, dy$$

$$= \left[4y^2 \right]_0^4 = 64$$

17. $V = \int_0^3 \int_0^2 (5 - x)\, dy\, dx = \int_0^3 (10 - 2x)\, dx = \left[10x - x^2\right]_0^3 = 30 - 9 = 21$

19. $V = \int_{-1}^1 \int_{-1}^1 \left(4 - x^2 - y^2\right) dy\, dx$

$\quad = \int_{-1}^1 \left[4y - x^2 y - \dfrac{y^3}{3}\right]_{-1}^1 dx$

$\quad = \int_{-1}^1 \left[\left(4 - x^2 - \dfrac{1}{3}\right) - \left(-4 + x^2 - \dfrac{1}{3}\right)\right] dx$

$\quad = \int_{-1}^1 \left[\dfrac{22}{3} - 2x^2\right] dx$

$\quad = \left[\dfrac{22}{3}x - \dfrac{2x^3}{3}\right]_{-1}^1$

$\quad = \dfrac{40}{3}$

Alternate Solution: $V = 4\int_0^1 \int_0^1 \left(4 - x^2 - y^2\right) dy\, dx$

$\qquad\qquad\qquad = 4\int_0^1 \left(4 - x^2 - \dfrac{1}{3}\right) dx$

$\qquad\qquad\qquad = 4\int_0^1 \left(\dfrac{11}{3} - x^2\right) dx = 4\left[\dfrac{11}{3}x - \dfrac{1}{3}x^3\right]_0^1 = \dfrac{40}{3}$

21. Area $R = 16$

$\quad$ Average Value $= \dfrac{1}{16}\int_{-2}^2 \int_{-2}^2 \left(16 - x^2 - y^2\right) dy\, dx$

$\qquad\qquad\qquad\quad = \dfrac{1}{16}\int_{-2}^2 \left[16y - x^2 y - \dfrac{y^3}{3}\right]_{-2}^2 dx$

$\qquad\qquad\qquad\quad = \dfrac{1}{16}\int_{-2}^2 \left[64 - 4x^2 - \dfrac{16}{3}\right] dx$

$\qquad\qquad\qquad\quad = \dfrac{1}{16}\left[64x - \dfrac{4x^3}{3} - \dfrac{16}{3}x\right]_{-2}^2$

$\qquad\qquad\qquad\quad = \dfrac{1}{16}\left[256 - \dfrac{64}{3} - \dfrac{64}{3}\right] = \dfrac{40}{3}$

23. Area $R = 3(5) = 15$

$\quad$ Average temperature $= \dfrac{1}{15}\int_0^3 \int_0^5 \left(40 - 6x^2 - y^2\right) dy\, dx = \dfrac{1}{15}\int_0^3 \left[40y - 6x^2 y - \dfrac{y^3}{3}\right]_0^5 dx$

$\qquad\qquad\qquad\qquad = \dfrac{1}{15}\int_0^3 \left[200 - 30x^2 - \dfrac{125}{3}\right] dx = \dfrac{1}{15}\left[200x - 10x^3 - \dfrac{125x}{3}\right]_0^3 = \dfrac{1}{15}[600 - 270 - 125] = 13\dfrac{2}{3}°\text{C}$

25. $\int_0^h \int_0^x \sqrt{x^2 + y^2}\, dy\, dx = \int_0^{\pi/4} \int_0^{h\sec\theta} r^2\, dr\, d\theta$

$\qquad\qquad\qquad\qquad = \dfrac{h^3}{3}\int_0^{\pi/4} \sec^3\theta\, d\theta = \dfrac{h^3}{6}\left[\sec\theta\tan\theta + \ln|\sec\theta + \tan\theta|\right]_0^{\pi/4} = \dfrac{h^3}{6}\left[\sqrt{2} + \ln\left(\sqrt{2} + 1\right)\right]$

27. $V = \int_0^{\pi/2} \int_0^3 (r \cos \theta)(r \sin \theta)^2 \, r \, dr \, d\theta$

$= \int_0^{\pi/2} \int_0^3 \cos \theta \sin^2 \theta \, r^4 \, dr \, d\theta$

$= \int_0^{\pi/2} \cos \theta \sin^2 \theta \left[\dfrac{r^5}{5} \right]_0^3 d\theta$

$= \dfrac{243}{5} \int_0^{\pi/2} \sin^2 \theta \cos \theta \, d\theta$

$= \dfrac{243}{5} \left[\dfrac{\sin^3 \theta}{3} \right]_0^{\pi/2} = \dfrac{81}{5}$

29. $A = 2 \int_0^\pi \int_0^{2+\cos\theta} r \, dr \, d\theta = \int_0^\pi (2 + \cos \theta)^2 \, d\theta = \int_0^\pi \left[4 + 4\cos\theta + \dfrac{1 + \cos 2\theta}{2} \right] d\theta = \left[4\theta + 4\sin\theta + \dfrac{1}{2}\theta + \dfrac{\sin 2\theta}{4} \right]_0^\pi = \dfrac{9\pi}{2}$

31.

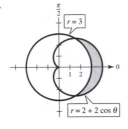

Intersection points: $3 = 2 + 2\cos\theta$

$\cos\theta = \dfrac{1}{2}$

$\theta = \dfrac{\pi}{3}, \dfrac{5\pi}{3}$

$A = 2 \int_0^{\pi/3} \int_3^{2+2\cos\theta} r \, dr \, d\theta$

$= 2 \int_0^{\pi/3} \left[\dfrac{r^2}{2} \right]_3^{2+2\cos\theta} d\theta$

$= 2 \int_0^{\pi/3} \left[\dfrac{(2 + 2\cos\theta)^2}{2} - \dfrac{9}{2} \right] d\theta$

$= \int_0^{\pi/3} \left[4 + 8\cos\theta + 4\cos^2\theta - 9 \right] d\theta$

$= \int_0^{\pi/3} \left[8\cos\theta + 2(1 + \cos 2\theta) - 5 \right] d\theta$

$= \left[8\sin\theta + \sin 2\theta - 3\theta \right]_0^{\pi/3}$

$= \dfrac{8\sqrt{3}}{2} + \dfrac{\sqrt{3}}{2} - \pi = \dfrac{9\sqrt{3}}{2} - \pi$

33. (a) $\left(x^2 + y^2 \right)^2 = 9\left(x^2 - y^2 \right)$

$\left(r^2 \right)^2 = 9\left(r^2 \cos^2 \theta - r^2 \sin^2 \theta \right)$

$r^2 = 9\left(\cos^2 \theta - \sin^2 \theta \right)$

$= 9 \cos 2\theta$

$r = 3\sqrt{\cos 2\theta}$

(b) $A = 4 \int_0^{\pi/4} \int_0^{3\sqrt{\cos 2\theta}} r \, dr \, d\theta = 9$

(c) $V = 4 \int_0^{\pi/4} \int_0^{3\sqrt{\cos 2\theta}} \sqrt{9 - r^2} \, r \, dr \, d\theta \approx 20.392$

35. $m = \int_0^2 \int_0^{x^3} kx \, dy \, dx = \dfrac{32k}{5}$

$M_x = \int_0^2 \int_0^{x^3} kxy \, dy \, dx = 16k$

$M_y = \int_0^2 \int_0^{x^3} kx^2 \, dy \, dx = \dfrac{32k}{3}$

$\bar{x} = \dfrac{M_y}{m} = \dfrac{5}{3}$

$\bar{y} = \dfrac{M_x}{m} = \dfrac{5}{2}$

$(\bar{x}, \bar{y}) = \left(\dfrac{5}{3}, \dfrac{5}{2} \right)$

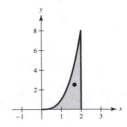

37. $m = k \int_0^1 \int_{2x^3}^{2x} xy \, dy \, dx = \dfrac{k}{4}$

$M_x = k \int_0^1 \int_{2x^3}^{2x} xy^2 \, dy \, dx = \dfrac{16k}{55}$

$M_y = k \int_0^1 \int_{2x^3}^{2x} x^2 y \, dy \, dx = \dfrac{8k}{45}$

$\bar{x} = \dfrac{M_y}{m} = \dfrac{32}{45}$

$\bar{y} = \dfrac{M_x}{m} = \dfrac{64}{55}$

$(\bar{x}, \bar{y}) = \left(\dfrac{32}{45}, \dfrac{64}{55}\right)$

39. $I_x = \int_R \int y^2 \rho(x,y)\, dA = \int_0^a \int_0^b kxy^2 \, dy \, dx = \dfrac{1}{6}kb^3 a^2$

$I_y = \int_R \int x^2 \rho(x,y)\, dA = \int_0^a \int_0^b kx^3 \, dy \, dx = \dfrac{1}{4}kba^4$

$I_0 = I_x + I_y = \dfrac{1}{6}kb^3 a^2 + \dfrac{1}{4}kba^4 = \dfrac{ka^2 b}{12}\left(2b^2 + 3a^2\right)$

$m = \int_R \int \rho(x,y)\, dA = \int_0^a \int_0^b kx \, dy \, dx = \dfrac{1}{2}kba^2$

$\bar{\bar{x}} = \sqrt{\dfrac{I_y}{m}} = \sqrt{\dfrac{(1/4)kba^4}{(1/2)kba^2}} = \sqrt{\dfrac{a^2}{2}} = \dfrac{a\sqrt{2}}{2}$

$\bar{\bar{y}} = \sqrt{\dfrac{I_x}{m}} = \sqrt{\dfrac{(1/6)kb^3 a^2}{(1/2)kba^2}} = \sqrt{\dfrac{b^2}{3}} = \dfrac{b\sqrt{3}}{3}$

43. $f(x,y) = 9 - y^2$

$f_x = 0, \ f_y = -2y$

$S = \int_R \int \sqrt{1 + (f_x)^2 + (f_y)^2}\, dA$

$= \int_0^3 \int_{-y}^{y} \sqrt{1+4y^2}\, dx \, dy = \int_0^3 \left[\sqrt{1+4y^2}\, x\right]_{-y}^{y} dy = \int_0^3 2y\sqrt{1+4y^2}\, dy = \frac{1}{4}\frac{2}{3}\left(1+4y^2\right)^{3/2}\Big]_0^3 = \frac{1}{6}\left[(37)^{3/2} - 1\right]$

41. $f(x,y) = 25 - x^2 - y^2$

$f_x = -2x, \ f_y = -2y$

$S = \int_R \int \sqrt{1 + (f_x)^2 + (f_y)^2}\, dA$

$= \int_R \int \sqrt{1 + 4x^2 + 4y^2}\, dA$

$= 4\int_0^{\pi/2} \int_0^5 \sqrt{1+4r^2}\, r \, dr \, d\theta$

$= \frac{1}{3}\int_0^{\pi/2} \left[\left(1+4r^2\right)^{3/2}\right]_0^5 d\theta$

$= \frac{1}{3}\int_0^{\pi/2} \left[(101)^{3/2} - 1\right] d\theta$

$= \frac{\pi}{6}\left[101\sqrt{101} - 1\right]$

45. (a) $V = \int_0^{50} \int_0^{\sqrt{50^2 - x^2}} \left(20 + \dfrac{xy}{100} - \dfrac{x+y}{5}\right) dy \, dx = \int_0^{50} \left[20\sqrt{50^2 - x^2} + \dfrac{x}{200}(50^2 - x^2) - \dfrac{x}{5}\sqrt{50^2 - x^2} - \dfrac{50^2 - x^2}{10}\right] dy$

$= \left[10\left(x\sqrt{50 - x^2} + 50^2 \arcsin\dfrac{x}{50}\right) + \dfrac{25}{4}x^2 - \dfrac{x^4}{800} + \dfrac{1}{15}(50^2 - x^2)^{3/2} - 250x + \dfrac{x^3}{30}\right]_0^{50} \approx 30{,}415.74 \text{ ft}^3$

(b) $z = 20 + \dfrac{xy}{100}$

$\sqrt{1 + (f_x)^2 + (f_y)^2} = \sqrt{1 + \dfrac{y^2}{100^2} + \dfrac{x^2}{100^2}} = \dfrac{\sqrt{100^2 + x^2 + y^2}}{100}$

$S = \dfrac{1}{100}\int_0^{50} \int_0^{\sqrt{50^2 - x^2}} \sqrt{100^2 + x^2 + y^2}\, dy \, dx = \dfrac{1}{100}\int_0^{\pi/2} \int_0^{50} \sqrt{100^2 + r^2}\, r \, dr \, d\theta \approx 2081.53 \text{ ft}^2$

47. $\int_0^4 \int_0^1 \int_0^2 (2x + y + 4z)\, dy\, dz\, dx = \int_0^4 \int_0^1 \left[2xy + \dfrac{y^2}{2} + 4zy \right]_0^2 dz\, dx$

$$= \int_0^4 \int_0^1 (4x + 2 + 8z)\, dz\, dx$$

$$= \int_0^4 \left[4xz + 2z + 4z^2 \right]_0^1 dx$$

$$= \int_0^4 (4x + 2 + 4)\, dx$$

$$= \left[2x^2 + 6x \right]_0^4 = 56$$

49. $\int_0^a \int_0^b \int_0^c (x^2 + y^2 + z^2)\, dx\, dy\, dz = \int_0^a \int_0^b \left(\tfrac{1}{3}c^3 + cy^2 + cz^2 \right) dy\, dz$

$$= \int_0^a \left(\tfrac{1}{3}bc^3 + \tfrac{1}{3}b^3 c + bcz^2 \right) dz = \tfrac{1}{3}abc^3 + \tfrac{1}{3}ab^3 c + \tfrac{1}{3}a^3 bc = \tfrac{1}{3}abc\left(a^2 + b^2 + c^2 \right)$$

51. $\int_{-1}^1 \int_{-\sqrt{1-x^2}}^{\sqrt{1-x^2}} \int_{-\sqrt{1-x^2-y^2}}^{\sqrt{1-x^2-y^2}} (x^2 + y^2)\, dz\, dy\, dx = \int_0^{2\pi} \int_0^1 \int_{-\sqrt{1-r^2}}^{\sqrt{1-r^2}} r^3\, dz\, dr\, d\theta = \dfrac{8\pi}{15}$

53. $V = \int_0^3 \int_0^4 \int_0^{xy} dz\, dy\, dx$

$$= \int_0^3 \int_0^4 xy\, dy\, dx$$

$$= \int_0^3 \left[\dfrac{xy^2}{2} \right]_0^4 dx$$

$$= \int_0^3 8x\, dx$$

$$= \left[4x^2 \right]_0^3 = 36$$

59. $\int_0^3 \int_0^{\pi/3} \int_0^4 r \cos\theta\, dr\, d\theta\, dz$

$$= \int_0^3 \int_0^{\pi/3} \left[\dfrac{r^2}{2} \cos\theta \right]_0^4 d\theta\, dz$$

$$= \int_0^3 \int_0^{\pi/3} 8 \cos\theta\, d\theta\, dz$$

$$= \int_0^3 \left[8\sin\theta \right]_0^{\pi/3} dz$$

$$= \int_0^3 4\sqrt{3}\, dz = 12\sqrt{3}$$

55. $\int_0^1 \int_0^y \int_0^{\sqrt{1-x^2}} dz\, dx\, dy = \int_0^1 \int_x^1 \int_0^{\sqrt{1-x^2}} dz\, dy\, dx$

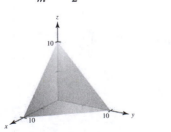

61. $\int_0^{\pi/2} \int_0^{\pi/2} \int_0^2 \rho^2\, d\rho\, d\theta\, d\phi = \int_0^{\pi/2} \int_0^{\pi/2} \left[\dfrac{\rho^3}{3} \right]_0^2 d\theta\, d\phi$

$$= \int_0^{\pi/2} \int_0^{\pi/2} \dfrac{8}{3}\, d\theta\, d\phi$$

$$= \int_0^{\pi/2} \dfrac{8}{3}\left(\dfrac{\pi}{2} \right) d\theta$$

$$= \dfrac{8}{3}\left(\dfrac{\pi}{2} \right)\left(\dfrac{\pi}{2} \right) = \dfrac{2}{3}\pi^2$$

63. $\int_0^{\pi} \int_0^2 \int_0^3 \sqrt{z^2 + 4}\, dz\, dr\, d\theta \approx 48.995$

57. $m = \int_0^{10} \int_0^{10-x} \int_0^{10-x-y} k\, dz\, dy\, dx = \dfrac{500}{3}k$

$Myz = \int_0^{10} \int_0^{10-x} \int_0^{10-x-y} kx\, dz\, dy\, dx = \dfrac{1250}{3}k$

$\bar{x} = \dfrac{Myz}{m} = \dfrac{5}{2}$

65. $z = 8 - x^2 - y^2 = x^2 + y^2$

$$8 = 2(x^2 + y^2)$$

$$x^2 + y^2 = 4$$

$$V = \int_0^{2\pi} \int_0^2 \int_{r^2}^{8-r^2} r \, dz \, dr \, d\theta$$

$$= \int_0^{2\pi} \int_0^2 r\left(8 - r^2 - r^2\right) dr \, d\theta$$

$$= \int_0^{2\pi} \int_0^2 \left(8r - 2r^3\right) dr \, d\theta$$

$$= \int_0^{2\pi} \left[4r^2 - \frac{r^4}{2}\right]_0^2 d\theta$$

$$= \int_0^{2\pi} 8 \, d\theta$$

$$= 16\pi$$

67. $\dfrac{\partial(x, y)}{\partial(u, v)} = \dfrac{\partial x}{\partial u}\dfrac{\partial y}{\partial v} - \dfrac{\partial y}{\partial u}\dfrac{\partial x}{\partial v} = 1(-3) - 2(3) = -9$

69. $\dfrac{\partial(x, y)}{\partial(u, v)} = \dfrac{\partial x}{\partial u}\dfrac{\partial y}{\partial v} - \dfrac{\partial y}{\partial u}\dfrac{\partial x}{\partial v}$

$$= (\sin \theta)(\sin \theta) - (\cos \theta)(\cos \theta)$$

$$= \sin^2 \theta - \cos^2 \theta$$

71. $\dfrac{\partial(x, y)}{\partial(u, v)} = \dfrac{\partial x}{\partial u}\dfrac{\partial y}{\partial v} - \dfrac{\partial x}{\partial v}\dfrac{\partial y}{\partial u} = \dfrac{1}{2}\left(-\dfrac{1}{2}\right) - \dfrac{1}{2}\left(\dfrac{1}{2}\right) = -\dfrac{1}{2}$

$x = \dfrac{1}{2}(u + v),\ y = \dfrac{1}{2}(u - v) \Rightarrow u = x + y,\ v = x - y$

Boundaries in *xy*-plane	Boundaries in *uv*-plane
$x + y = 3$	$u = 3$
$x + y = 5$	$u = 5$
$x - y = -1$	$v = -1$
$x - y = 1$	$v = 1$

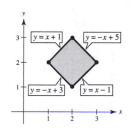

$\displaystyle\int_R \int \ln(x + y) \, dA = \int_3^5 \int_{-1}^1 \ln\left(\frac{1}{2}(u + v) + \frac{1}{2}(u - v)\right)\left(\frac{1}{2}\right) dv \, du = \int_3^5 \int_{-1}^1 \frac{1}{2} \ln u \, dv \, du = \int_3^5 \ln u \, du = \left[u \ln u - u\right]_3^5$

$$= (5 \ln 5 - 5) - (3 \ln u - 3) = 5 \ln 5 - 3 \ln 3 - 2 \approx 2.751$$

73. $\dfrac{\partial(x, y)}{\partial(u, v)} = \dfrac{\partial x}{\partial u}\dfrac{\partial y}{\partial v} - \dfrac{\partial y}{\partial u}\dfrac{\partial x}{\partial v} = 1\left(-\dfrac{1}{3}\right) - \dfrac{1}{3}(0) = -\dfrac{1}{3}$

$x = u,\ y = \dfrac{1}{3}(u - v) \Rightarrow u = x,\ v = x - 3y$

Boundary in *xy*-plane	Boundary in *uv*-plane
$x = 1$	$u = 1$
$x = 4$	$u = 4$
$3y - x = 8$	$v = -8$
$3y - x = 2$	$v = -2$

$\displaystyle\int_R \int \left(xy + x^2\right) dA = \int_1^4 \int_{-2}^{-8} \left[u \frac{1}{3}(u - v) + u^2\right]\left(-\frac{1}{3}\right) dv \, du$

$$= \left(-\frac{1}{3}\right)\int_1^4 \int_{-2}^{-8} \left(\frac{4}{3}u^2 - \frac{1}{3}uv\right) dv \, du = \left(-\frac{1}{3}\right)\int_1^4 \left[\frac{4}{3}u^2 v - \frac{1}{6}uv^2\right]_{-2}^{-8} du$$

$$= \left(-\frac{1}{3}\right)\int_1^4 \left(-8u^2 - 10u\right) du = 81$$

Problem Solving for Chapter 14

1. $V = 16 \int_R \int \sqrt{1 - x^2}\, dA$

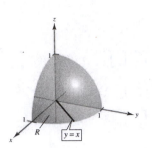

$$= 16 \int_0^{\pi/4} \int_0^1 \sqrt{1 - r^2 \cos^2 \theta}\, r\, dr\, d\theta = -\frac{16}{3} \int_0^{\pi/4} \frac{1}{\cos^2 \theta} \left[\left(1 - \cos^2 \theta \right)^{3/2} - 1 \right] d\theta$$

$$= -\frac{16}{3} \left[\sec \theta + \cos \theta - \tan \theta \right]_0^{\pi/4} = 8\left(2 - \sqrt{2} \right) \approx 4.6863$$

3.

Boundary in xy-plane	Boundary in uv-plane
$y = \sqrt{x}$	$u = 1$
$y = \sqrt{2x}$	$u = 2$
$y = \dfrac{1}{3} x^2$	$v = 3$
$y = \dfrac{1}{4} x^2$	$v = 4$

$$\frac{\partial(x, y)}{\partial(u, v)} = \begin{vmatrix} \dfrac{1}{3}\left(\dfrac{v}{u}\right)^{2/3} & \dfrac{2}{3}\left(\dfrac{u}{v}\right)^{1/3} \\[3mm] \dfrac{2}{3}\left(\dfrac{v}{u}\right)^{1/3} & \dfrac{1}{3}\left(\dfrac{u}{v}\right)^{2/3} \end{vmatrix} = -\frac{1}{3}$$

$$A = \int_R \int 1\, dA = \int_S \int 1 \left| \frac{\partial(x, y)}{\partial(u, v)} \right| dA = \frac{1}{3}$$

5. (a) $\displaystyle \int \frac{du}{a^2 + u^2} = \frac{1}{a} \arctan \frac{u}{a} + c.$ Let $a^2 = 2 - u^2, u = v.$

Then $\displaystyle \int \frac{1}{\left(2 - u^2\right) + v^2}\, dv = \frac{1}{\sqrt{2 - u^2}} \arctan \frac{v}{\sqrt{2 - u^2}} + C.$

(b) $\displaystyle I_1 = \int_0^{\sqrt{2}/2} \left[\frac{2}{\sqrt{2 - u^2}} \arctan \frac{v}{\sqrt{2 - u^2}} \right]_{-u}^{u} du$

$$= \int_0^{\sqrt{2}/2} \frac{2}{\sqrt{2 - u^2}} \left(\arctan \frac{u}{\sqrt{2 - u^2}} - \arctan \frac{-u}{\sqrt{2 - u^2}} \right) du = \int_0^{\sqrt{2}/2} \frac{4}{\sqrt{2 - u^2}} \arctan \frac{u}{\sqrt{2 - u^2}}\, du$$

Let $u = \sqrt{2} \sin \theta, du = \sqrt{2} \cos \theta\, d\theta, 2 - u^2 = 2 - 2 \sin^2 \theta = 2 \cos^2 \theta.$

$$I_1 = 4 \int_0^{\pi/6} \frac{1}{\sqrt{2} \cos \theta} \arctan \left(\frac{\sqrt{2} \sin \theta}{\sqrt{2} \cos \theta} \right) \cdot \sqrt{2} \cos \theta\, d\theta = 4 \int_0^{\pi/6} \arctan(\tan \theta)\, d\theta = \frac{4\theta^2}{2} \Big]_0^{\pi/6} = 2\left(\frac{\pi}{6} \right)^2 = \frac{\pi^2}{18}$$

(c) $\displaystyle I_2 = \int_{\sqrt{2}/2}^{\sqrt{2}} \left[\frac{2}{\sqrt{2 - u^2}} \arctan \frac{v}{\sqrt{2 - u^2}} \right]_{u - \sqrt{2}}^{-u + \sqrt{2}} du$

$$= \int_{\sqrt{2}/2}^{\sqrt{2}} \frac{2}{\sqrt{2 - u^2}} \left[\arctan \left(\frac{-u + \sqrt{2}}{\sqrt{2 - u^2}} \right) - \arctan \left(\frac{u - \sqrt{2}}{\sqrt{2 - u^2}} \right) \right] du = \int_{\sqrt{2}/2}^{\sqrt{2}} \frac{4}{\sqrt{2 - u^2}} \arctan \left(\frac{\sqrt{2} - u}{\sqrt{2 - u^2}} \right) du$$

Let $u = \sqrt{2} \sin \theta.$

$$I_2 = 4 \int_{\pi/6}^{\pi/2} \frac{1}{\sqrt{2} \cos \theta} \arctan \left(\frac{\sqrt{2} - \sqrt{2} \sin \theta}{\sqrt{2} \cos \theta} \right) \cdot \sqrt{2} \cos \theta\, d\theta = 4 \int_{\pi/6}^{\pi/2} \arctan \left(\frac{1 - \sin \theta}{\cos \theta} \right) d\theta$$

(d) $\tan\left(\dfrac{1}{2}\left(\dfrac{\pi}{2} - \theta\right)\right) = \sqrt{\dfrac{1 - \cos((\pi/2) - \theta)}{1 + \cos((\pi/2) - \theta)}} = \sqrt{\dfrac{1 - \sin\theta}{1 + \sin\theta}} = \sqrt{\dfrac{(1 - \sin\theta)^2}{(1 + \sin\theta)(1 - \sin\theta)}} = \sqrt{\dfrac{(1 - \sin\theta)^2}{\cos^2\theta}} = \dfrac{1 - \sin\theta}{\cos\theta}$

(e) $I_2 = 4\displaystyle\int_{\pi/6}^{\pi/2} \arctan\left(\dfrac{1 - \sin\theta}{\cos\theta}\right) d\theta = 4\int_{\pi/6}^{\pi/2} \arctan\left(\tan\left(\dfrac{1}{2}\left(\dfrac{\pi}{2} - \theta\right)\right)\right) d\theta = 4\int_{\pi/6}^{\pi/2} \dfrac{1}{2}\left(\dfrac{\pi}{2} - \theta\right) d\theta = 2\int_{\pi/6}^{\pi/2}\left(\dfrac{\pi}{2} - \theta\right) d\theta$

$= 2\left[\dfrac{\pi}{2}\theta - \dfrac{\theta^2}{2}\right]_{\pi/6}^{\pi/2} = 2\left[\left(\dfrac{\pi^2}{4} - \dfrac{\pi^2}{8}\right) - \left(\dfrac{\pi^2}{12} - \dfrac{\pi^2}{72}\right)\right] = 2\left[\dfrac{18 - 9 - 6 + 1}{72}\pi^2\right] = \dfrac{4}{36}\pi^2 = \dfrac{\pi^2}{9}$

(f) $\dfrac{1}{1 - xy} = 1 + (xy) + (xy)^2 + \cdots \qquad |xy| < 1$

$\displaystyle\int_0^1\int_0^1 \dfrac{1}{1 - xy}\, dx\, dy = \int_0^1\int_0^1 \left[1 + (xy) + (xy)^2 + \cdots\right] dx\, dy = \int_0^1\int_0^1 \sum_{K=0}^{\infty}(xy)^K\, dx\, dy = \sum_{K=0}^{\infty}\int_0^1 \left.\dfrac{x^{K+1}y^K}{K+1}\right|_0^1 dy$

$= \displaystyle\sum_{K=0}^{\infty}\int_0^1 \dfrac{y^K}{K+1}\, dy = \sum_{K=0}^{\infty}\left.\dfrac{y^{K+1}}{(K+1)^2}\right|_0^1 = \sum_{K=0}^{\infty}\dfrac{1}{(K+1)^2} = \sum_{n=1}^{\infty}\dfrac{1}{n^2}$

(g) $u = \dfrac{x + y}{\sqrt{2}},\ v = \dfrac{y - x}{\sqrt{2}}$

$u - v = \dfrac{2x}{\sqrt{2}} \Rightarrow x = \dfrac{u - v}{\sqrt{2}}$

$u + v = \dfrac{2y}{\sqrt{2}} \Rightarrow y = \dfrac{u + v}{\sqrt{2}}$

$\dfrac{\partial(x, y)}{\partial(u, v)} = \begin{vmatrix} 1/\sqrt{2} & -1/\sqrt{2} \\ 1/\sqrt{2} & 1/\sqrt{2} \end{vmatrix} = 1$

R	S
$(0, 0)$	$\leftrightarrow$ $(0, 0)$
$(1, 0)$	$\leftrightarrow$ $\left(\dfrac{1}{\sqrt{2}}, -\dfrac{1}{\sqrt{2}}\right)$
$(0, 1)$	$\leftrightarrow$ $\left(\dfrac{1}{\sqrt{2}}, \dfrac{1}{\sqrt{2}}\right)$
$(1, 1)$	$\leftrightarrow$ $\left(\sqrt{2}, 0\right)$

$\displaystyle\int_0^1\int_0^1 \dfrac{1}{1 - xy}\, dx\, dy = \int_0^{\sqrt{2}/2}\int_{-u}^{u} \dfrac{1}{1 - \dfrac{u^2}{2} + \dfrac{v^2}{2}}\, dv\, du + \int_{\sqrt{2}/2}^{\sqrt{2}}\int_{u - \sqrt{2}}^{-u + \sqrt{2}} \dfrac{1}{1 - \dfrac{u^2}{2} + \dfrac{v^2}{2}}\, dv\, du = I_1 + I_2 = \dfrac{\pi^2}{18} + \dfrac{\pi^2}{9} = \dfrac{\pi^2}{6}$

7. $\displaystyle\int_0^1\int_0^1 \dfrac{x - y}{(x + y)^3}\, dx\, dy = -\dfrac{1}{2}$

$\displaystyle\int_0^1\int_0^1 \dfrac{(x - y)}{(x + y)^3}\, dy\, dx = \dfrac{1}{2}$

The results are not the same. Fubini's Theorem is not valid because *f* is not continuous on the region $0 \le x \le 1, 0 \le y \le 1$.

9. From Exercise 65, Section 14.3,

$$\int_{-\infty}^{\infty} e^{-x^2/2} \, dx = \sqrt{2\pi}.$$

So, $\displaystyle\int_0^{\infty} e^{-x^2/2} \, dx = \frac{\sqrt{2\pi}}{2}$ and $\displaystyle\int_0^{\infty} e^{-x^2} \, dx = \frac{\sqrt{\pi}}{2}$

$$\int_0^{\infty} x^2 e^{-x^2} \, dx = \left[-\frac{1}{2} x e^{-x^2} \right]_0^{\infty} + \frac{1}{2} \int_0^{\infty} e^{-x^2} \, dx$$

$$= \frac{1}{2} \cdot \frac{\sqrt{\pi}}{2} = \frac{\sqrt{\pi}}{4}$$

11. $f(x, y) = \begin{cases} ke^{-(x+y)/a} & x \geq 0, \, y \geq 0 \\ 0 & \text{elsewhere} \end{cases}$

$$\int_{-\infty}^{\infty} \int_{-\infty}^{\infty} f(x, y) \, dA = \int_0^{\infty} \int_0^{\infty} ke^{-(x+y)/a} \, dx \, dy$$

$$= k \int_0^{\infty} e^{-x/a} \, dx \cdot \int_0^{\infty} e^{-y/a} \, dy$$

These two integrals are equal to

$$\int_0^{\infty} e^{-x/a} \, dx = \lim_{b \to \infty} \left[(-a) e^{-x/a} \right]_0^b = a.$$

So, assuming $a, k > 0$, you obtain

$$1 = ka^2 \text{ or } a = \frac{1}{\sqrt{k}}.$$

13. Essay

15. The greater the angle between the given plane and the *xy*-plane, the greater the surface area. So: $z_2 < z_1 < z_4 < z_3$

17. $V = \displaystyle\int_0^3 \int_0^{2x} \int_x^{6-x} dy \, dz \, dx = 18$

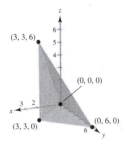

CHAPTER 15
Vector Analysis

C H A P T E R   1 5
Vector Analysis

Section 15.1   Vector Fields

1. All vectors are parallel to x-axis.

Matches (d)

2. All vectors are parallel to y-axis.

Matches (c)

3. Vectors are in rotational pattern.

Matches (a)

4. All vectors point outward.

Matches (b)

5. $\mathbf{F}(x, y) = \mathbf{i} + \mathbf{j}$

$\|\mathbf{F}\| = \sqrt{2}$

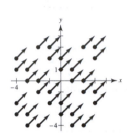

7. $\mathbf{F}(x, y, z) = 3y\mathbf{j}$

$\|\mathbf{F}\| = 3|y| = c$

9. $\mathbf{F}(x, y, z) = \mathbf{i} + \mathbf{j} + \mathbf{k}$

$\|\mathbf{F}\| = \sqrt{3}$

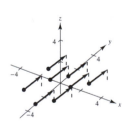

11. $F(x, y) = \dfrac{1}{8}\left(2xy\mathbf{i} + y^2\mathbf{j}\right)$

13. $F(x, y, z) = \dfrac{x\mathbf{i} + y\mathbf{j} + z\mathbf{k}}{\sqrt{x^2 + y^2 + z^2}}$

15. $f(x, y) = x^2 + 2y^2$

$f_x(x, y) = 2x$

$f_y(x, y) = 4y$

$\mathbf{F}(x, y) = 2x\mathbf{i} + 4y\mathbf{j}$

Note that $\nabla f = \mathbf{F}$.

17. $g(x, y) = 5x^2 + 3xy + y^2$

$g_x(x, y) = 10x + 3y$

$g_y(x, y) = 3x + 2y$

$\mathbf{G}(x, y) = (10x + 3y)\mathbf{i} + (3x + 2y)\mathbf{j}$

19. $f(x, y, z) = 6xyz$

$f_x(x, y, z) = 6yz$

$f_y(x, y, z) = 6xz$

$f_z(x, y, z) = 6xy$

$\mathbf{F}(x, y, z) = 6yz\mathbf{i} + 6xz\mathbf{j} + 6xy\mathbf{k}$

21. $g(x, y, z) = z + ye^{x^2}$

$g_x(x, y, z) = 2xye^{x^2}$

$g_y(x, y, z) = e^{x^2}$

$g_z(x, y, z) = 1$

$\mathbf{G}(x, y, z) = 2xye^{x^2}\mathbf{i} + e^{x^2}\mathbf{j} + \mathbf{k}$

23. $h(x, y, z) = xy \ln(x + y)$

$h_x(x, y, z) = y \ln(x + y) + \dfrac{xy}{x + y}$

$h_y(x, y, z) = x \ln(x + y) + \dfrac{xy}{x + y}$

$h_z(x, y, z) = 0$

$\mathbf{H}(x, y, z) = \left[\dfrac{xy}{x + y} + y \ln(x + y)\right]\mathbf{i} + \left[\dfrac{xy}{x + y} + x \ln(x + y)\right]\mathbf{j}$

25. $\mathbf{F}(x, y) = xy^2\mathbf{i} + x^2 y\mathbf{j}$

$M = xy^2$ and $N = x^2 y$ have continuous first partial derivatives.

$\dfrac{\partial N}{\partial x} = 2xy = \dfrac{\partial M}{\partial y} \Rightarrow \mathbf{F}$ conservative

27. $\mathbf{F}(x, y) = \sin y\mathbf{i} + x \cos y\mathbf{j}$

$M = \sin y$ and $N = x \cos y$ have continuous first partial derivatives.

$\dfrac{\partial N}{\partial x} = \cos y = \dfrac{\partial M}{\partial y} \Rightarrow \mathbf{F}$ is conservative.

29. $\mathbf{F}(x, y) = \dfrac{1}{xy}(y\mathbf{i} - x\mathbf{j}) = \dfrac{1}{x}\mathbf{i} - \dfrac{1}{y}\mathbf{j}$

$M = 1/x$ and $N = -1/y$ have continuous first partial derivatives for all $x, y \neq 0$.

$\dfrac{\partial N}{\partial x} = 0 = \dfrac{\partial M}{\partial y} \Rightarrow \mathbf{F}$ is conservative.

31. $M = \dfrac{1}{\sqrt{x^2 + y^2}}, N = \dfrac{1}{\sqrt{x^2 + y^2}}$

$\dfrac{\partial N}{\partial x} = \dfrac{-x}{\left(x^2 + y^2\right)^{3/2}} \neq \dfrac{\partial M}{\partial y} = \dfrac{-y}{\left(x^2 + y^2\right)^{3/2}}$

$\Rightarrow$ Not conservative

33. $\mathbf{F}(x, y) = y\mathbf{i} + x\mathbf{j}$

$\dfrac{\partial}{\partial y}[y] = 1 = \dfrac{\partial}{\partial x}[x] \Rightarrow$ Conservative

$f_x(x, y) = y, f_y(x, y) = x \Rightarrow f(x, y) = xy + k$

35. $\mathbf{F}(x, y) = 2xy\mathbf{i} + x^2\mathbf{j}$

$\dfrac{\partial}{\partial y}[2xy] = 2x$

$\dfrac{\partial}{\partial x}[x^2] = 2x$

Conservative

$f_x(x, y) = 2xy, f_y(x, y) = x^2, f(x, y) = x^2 y + K$

37. $\mathbf{F}(x, y) = 15y^3\mathbf{i} - 5xy^2\mathbf{j}$

$\dfrac{\partial}{\partial y}\left[15y^3\right] = 45y^2 \neq \dfrac{\partial}{\partial x}\left[-5xy^2\right] = -5y^2$

Not conservative

39. $\mathbf{F}(x, y) = \dfrac{2y}{x}\mathbf{i} - \dfrac{x^2}{y^2}\mathbf{j}$

$\dfrac{\partial}{\partial y}\left[\dfrac{2y}{x}\right] = \dfrac{2}{x}$

$\dfrac{\partial}{\partial x}\left[-\dfrac{x^2}{y^2}\right] = -\dfrac{2x}{y^2}$

Not conservative

41. $\mathbf{F}(x, y) = e^x(\cos y\mathbf{i} - \sin y\mathbf{j})$

$\dfrac{\partial}{\partial y}\left[e^x \cos y\right] = -e^x \sin y$

$\dfrac{\partial}{\partial x}\left[-e^x \sin y\right] = -e^x \sin y$

Conservative

$f_x(x, y) = e^x \cos y$

$f_y(x, y) = -e^x \sin y$

$f(x, y) = e^x \cos y + K$

43. $\mathbf{F}(x, y, z) = xyz\,\mathbf{i} + xyz\,\mathbf{j} + xyz\,\mathbf{k}, (2, 1, 3)$

$\text{curl } \mathbf{F} = \begin{vmatrix} \mathbf{i} & \mathbf{j} & \mathbf{k} \\ \dfrac{\partial}{\partial x} & \dfrac{\partial}{\partial y} & \dfrac{\partial}{\partial z} \\ xyz & xyz & xyz \end{vmatrix}$

$= (xz - xy)\mathbf{i} - (yz - xy)\mathbf{j} + (yz - xz)\mathbf{k}$

$\text{curl } \mathbf{F}(2, 1, 3) = (6 - 2)\mathbf{i} - (3 - 2)\mathbf{j} + (3 - 6)\mathbf{k}$

$= 4\mathbf{i} - \mathbf{j} - 3\mathbf{k}$

45. $\mathbf{F}(x, y, z) = e^x \sin y\, \mathbf{i} - e^x \cos y\, \mathbf{j}, \,(0, 0, 1)$

$$\text{curl } \mathbf{F} = \begin{vmatrix} \mathbf{i} & \mathbf{j} & \mathbf{k} \\ \dfrac{\partial}{\partial x} & \dfrac{\partial}{\partial y} & \dfrac{\partial}{\partial z} \\ e^x \sin y & -e^x \cos y & 0 \end{vmatrix} = \left(-e^x \cos y - e^x \cos y\right)\mathbf{k} = -2e^x \cos y\, \mathbf{k}$$

$\text{curl } \mathbf{F}\,(0, 0, 1) = -2\mathbf{k}$

47. $\mathbf{F}(x, y, z) = \arctan\left(\dfrac{x}{y}\right)\mathbf{i} + \ln\sqrt{x^2 + y^2}\,\mathbf{j} + \mathbf{k}$

$$\text{curl } \mathbf{F} = \begin{vmatrix} \mathbf{i} & \mathbf{j} & \mathbf{k} \\ \dfrac{\partial}{\partial x} & \dfrac{\partial}{\partial y} & \dfrac{\partial}{\partial z} \\ \arctan\left(\dfrac{x}{y}\right) & \dfrac{1}{2}\ln(x^2 + y^2) & 1 \end{vmatrix} = \left[\dfrac{x}{x^2 + y^2} - \dfrac{(-x/y^2)}{1 + (x/y)^2}\right]\mathbf{k} = \dfrac{2x}{x^2 + y^2}\mathbf{k}$$

49. $\mathbf{F}(x, y, z) = \sin(x - y)\mathbf{i} + \sin(y - z)\mathbf{j} + \sin(z - x)\mathbf{k}$

$$\text{curl } \mathbf{F} = \begin{vmatrix} \mathbf{i} & \mathbf{j} & \mathbf{k} \\ \dfrac{\partial}{\partial x} & \dfrac{\partial}{\partial y} & \dfrac{\partial}{\partial z} \\ \sin(x - y) & \sin(y - z) & \sin(z - x) \end{vmatrix} = \cos(y - z)\mathbf{i} + \cos(z - x)\mathbf{j} + \cos(x - y)\mathbf{k}$$

51. $\mathbf{F}(x, y, z) = xy^2z^2\mathbf{i} + x^2yz^2\mathbf{j} + x^2y^2z\mathbf{k}$

$$\text{curl } \mathbf{F} = \begin{vmatrix} \mathbf{i} & \mathbf{j} & \mathbf{k} \\ \dfrac{\partial}{\partial x} & \dfrac{\partial}{\partial y} & \dfrac{\partial}{\partial z} \\ xy^2z^2 & x^2yz^2 & x^2y^2z \end{vmatrix} = \mathbf{0}$$

Conservative

$f_x(x, y, z) = xy^2z^2$

$f_y(x, y, z) = x^2yz^2$

$f_z(x, y, z) = x^2y^2z$

$f(x, y, z) = \dfrac{1}{2}x^2y^2z^2 + K$

53. $\mathbf{F}(x, y, z) = \sin z\,\mathbf{i} + \sin x\,\mathbf{j} + \sin y\,\mathbf{k}$

$$\text{curl } \mathbf{F} = \begin{vmatrix} \mathbf{i} & \mathbf{j} & \mathbf{k} \\ \dfrac{\partial}{\partial x} & \dfrac{\partial}{\partial y} & \dfrac{\partial}{\partial z} \\ \sin z & \sin x & \sin y \end{vmatrix}$$

$= \cos y\,\mathbf{i} + \cos z\,\mathbf{j} + \cos x\,\mathbf{k} \neq \mathbf{0}$

Not conservative

55. $F(x, y, z) = \dfrac{z}{y}\mathbf{i} - \dfrac{xz}{y^2}\mathbf{j} + \dfrac{x}{y}\mathbf{k}$

$$\mathbf{curl\ F} = \begin{vmatrix} \mathbf{i} & \mathbf{j} & \mathbf{k} \\ \dfrac{\partial}{\partial x} & \dfrac{\partial}{\partial y} & \dfrac{\partial}{\partial z} \\ \dfrac{z}{y} & -\dfrac{xz}{y^2} & \dfrac{x}{y} \end{vmatrix} = \left(-\dfrac{x}{y^2} + \dfrac{x}{y^2}\right)\mathbf{i} - \left(\dfrac{1}{y} - \dfrac{1}{y}\right)\mathbf{j} + \left(-\dfrac{z}{y^2} + \dfrac{z}{y^2}\right)\mathbf{k} = \mathbf{0}$$

Conservative

$$f_x(x, y, z) = \dfrac{z}{y}$$

$$f_y(x, y, z) = -\dfrac{xz}{y^2}$$

$$f_z(x, y, z) = \dfrac{x}{y}$$

$$f(x, y, z) = \dfrac{xz}{y} + K$$

57. $F(x, y) = x^2\mathbf{i} + 2y^2\mathbf{j}$

$\operatorname{div} F(x, y) = \dfrac{\partial}{\partial x}(x^2) + \dfrac{\partial}{\partial y}(2y^2) = 2x + 4y$

59. $F(x, y, z) = \sin x\mathbf{i} + \cos y\mathbf{j} + z^2\mathbf{k}$

$\operatorname{div} F(x, y, z) = \dfrac{\partial}{\partial x}[\sin x] + \dfrac{\partial}{\partial y}[\cos y] + \dfrac{\partial}{\partial z}[z^2]$

$\qquad\qquad = \cos x - \sin y + 2z$

61. $F(x, y, z) = xyz\mathbf{i} + xy\mathbf{j} + z\mathbf{k}$

$\operatorname{div} F(x, y, z) = yz + x + 1$

$\operatorname{div} F(2, 1, 1) = 1 + 2 + 1 = 4$

63. $F(x, y, z) = e^x \sin y\mathbf{i} - e^x \cos y\mathbf{j} + z^2\mathbf{k}$

$\operatorname{div} F(x, y, z) = e^x \sin y + e^x \sin y + 2z$

$\operatorname{div} F(3, 0, 0) = 0$

65. See the definition, page 1040. Examples include velocity fields, gravitational fields, and magnetic fields.

67. See the definition on page 1046.

69. $F(x, y, z) = \mathbf{i} + 3x\mathbf{j} + 2y\mathbf{k}$

$G(x, y, z) = x\mathbf{i} - y\mathbf{j} + z\mathbf{k}$

$$\mathbf{F \times G} = \begin{vmatrix} \mathbf{i} & \mathbf{j} & \mathbf{k} \\ 1 & 3x & 2y \\ x & -y & z \end{vmatrix}$$

$\qquad = (3xz + 2y^2)\mathbf{i} - (z - 2xy)\mathbf{j} + (-y - 3x^2)\mathbf{k}$

$$\mathbf{curl(F \times G)} = \begin{vmatrix} \mathbf{i} & \mathbf{j} & \mathbf{k} \\ \dfrac{\partial}{\partial x} & \dfrac{\partial}{\partial y} & \dfrac{\partial}{\partial z} \\ 3xz + 2y^2 & -z + 2xy & -y - 3x^2 \end{vmatrix}$$

$\qquad = (-1 + 1)\mathbf{i} - (-6x - 3x)\mathbf{j} + (2y - 4y)\mathbf{k}$

$\qquad = 9x\mathbf{j} - 2y\mathbf{k}$

71. $F(x, y, z) = xyz\mathbf{i} + y\mathbf{j} + z\mathbf{k}$

$$\mathbf{curl\ F} = \begin{vmatrix} \mathbf{i} & \mathbf{j} & \mathbf{k} \\ \dfrac{\partial}{\partial x} & \dfrac{\partial}{\partial y} & \dfrac{\partial}{\partial z} \\ xyz & y & z \end{vmatrix} = xy\mathbf{j} - xz\mathbf{k}$$

$$\mathbf{curl(curl\ F)} = \begin{vmatrix} \mathbf{i} & \mathbf{j} & \mathbf{k} \\ \dfrac{\partial}{\partial x} & \dfrac{\partial}{\partial y} & \dfrac{\partial}{\partial z} \\ 0 & xy & -xz \end{vmatrix} = z\mathbf{j} + y\mathbf{k}$$

73. $\mathbf{F}(x, y, z) = \mathbf{i} + 3x\mathbf{j} + 2y\mathbf{k}$

$\mathbf{G}(x, y, z) = x\mathbf{i} - y\mathbf{j} + z\mathbf{k}$

$$\mathbf{F} \times \mathbf{G} = \begin{vmatrix} \mathbf{i} & \mathbf{j} & \mathbf{k} \\ 1 & 3x & 2y \\ x & -y & z \end{vmatrix}$$

$$= \left(3xz + 2y^2\right)\mathbf{i} - (z - 2xy)\mathbf{j} + \left(-y - 3x^2\right)\mathbf{k}$$

$\text{div}(\mathbf{F} \times \mathbf{G}) = 3z + 2x$

75. $\mathbf{F}(x, y, z) = xyz\mathbf{i} + y\mathbf{j} + z\mathbf{k}$

$$\text{curl } \mathbf{F} = \begin{vmatrix} \mathbf{i} & \mathbf{j} & \mathbf{k} \\ \dfrac{\partial}{\partial x} & \dfrac{\partial}{\partial y} & \dfrac{\partial}{\partial z} \\ xyz & y & z \end{vmatrix} = xy\mathbf{j} - xz\mathbf{k}$$

$\text{div}(\text{curl } \mathbf{F}) = x - x = 0$

77. (a) Let $\mathbf{F} = M\mathbf{i} + N\mathbf{j} + P\mathbf{k}$ and $\mathbf{G} = Q\mathbf{i} + R\mathbf{j} + S\mathbf{k}$ where $M, N, P, Q, R,$ and S have continuous partial derivatives.

$\mathbf{F} + \mathbf{G} = (M + Q)\mathbf{i} + (N + R)\mathbf{j} + (P + S)\mathbf{k}$

$$\text{curl}(\mathbf{F} + \mathbf{G}) = \begin{vmatrix} \mathbf{i} & \mathbf{j} & \mathbf{k} \\ \dfrac{\partial}{\partial x} & \dfrac{\partial}{\partial y} & \dfrac{\partial}{\partial z} \\ M + Q & N + R & P + S \end{vmatrix}$$

$$= \left[\dfrac{\partial}{\partial y}(P + S) - \dfrac{\partial}{\partial z}(N + R)\right]\mathbf{i} - \left[\dfrac{\partial}{\partial x}(P + S) - \dfrac{\partial}{\partial z}(M + Q)\right]\mathbf{j} + \left[\dfrac{\partial}{\partial x}(N + R) - \dfrac{\partial}{\partial y}(M + Q)\right]\mathbf{k}$$

$$= \left(\dfrac{\partial P}{\partial y} - \dfrac{\partial N}{\partial z}\right)\mathbf{i} - \left(\dfrac{\partial P}{\partial x} - \dfrac{\partial M}{\partial z}\right)\mathbf{j} + \left(\dfrac{\partial N}{\partial x} - \dfrac{\partial M}{\partial y}\right)\mathbf{k} + \left(\dfrac{\partial S}{\partial y} - \dfrac{\partial R}{\partial z}\right)\mathbf{i} - \left(\dfrac{\partial S}{\partial x} - \dfrac{\partial Q}{\partial z}\right)\mathbf{j} + \left(\dfrac{\partial R}{\partial x} - \dfrac{\partial Q}{\partial y}\right)\mathbf{k}$$

$$= \text{curl } \mathbf{F} + \text{curl } \mathbf{G}$$

(b) Let $f(x, y, z)$ be a scalar function whose second partial derivatives are continuous.

$$\nabla f = \dfrac{\partial f}{\partial x}\mathbf{i} + \dfrac{\partial f}{\partial y}\mathbf{j} + \dfrac{\partial f}{\partial z}\mathbf{k}$$

$$\text{curl}(\nabla f) = \begin{vmatrix} \mathbf{i} & \mathbf{j} & \mathbf{k} \\ \dfrac{\partial}{\partial x} & \dfrac{\partial}{\partial y} & \dfrac{\partial}{\partial z} \\ \dfrac{\partial f}{\partial x} & \dfrac{\partial f}{\partial y} & \dfrac{\partial f}{\partial z} \end{vmatrix} = \left(\dfrac{\partial^2 f}{\partial y \partial z} - \dfrac{\partial^2 f}{\partial z \partial y}\right)\mathbf{i} - \left(\dfrac{\partial^2 f}{\partial x \partial z} - \dfrac{\partial^2 f}{\partial z \partial x}\right)\mathbf{j} + \left(\dfrac{\partial^2 f}{\partial x \partial y} - \dfrac{\partial^2 f}{\partial y \partial x}\right)\mathbf{k} = \mathbf{0}$$

(c) Let $\mathbf{F} = M\mathbf{i} + N\mathbf{j} + P\mathbf{k}$ and $\mathbf{G} = R\mathbf{i} + S\mathbf{j} + T\mathbf{k}$.

$$\text{div}(\mathbf{F} + \mathbf{G}) = \dfrac{\partial}{\partial x}(M + R) + \dfrac{\partial}{\partial y}(N + S) + \dfrac{\partial}{\partial z}(P + T) \quad = \dfrac{\partial M}{\partial x} + \dfrac{\partial R}{\partial x} + \dfrac{\partial N}{\partial y} + \dfrac{\partial S}{\partial y} + \dfrac{\partial P}{\partial z} + \dfrac{\partial T}{\partial z}$$

$$= \left[\dfrac{\partial M}{\partial x} + \dfrac{\partial N}{\partial y} + \dfrac{\partial P}{\partial z}\right] + \left[\dfrac{\partial R}{\partial x} + \dfrac{\partial S}{\partial y} + \dfrac{\partial T}{\partial z}\right]$$

$$= \text{div } \mathbf{F} + \text{div } \mathbf{G}$$

(d) Let $\mathbf{F} = M\mathbf{i} + N\mathbf{j} + P\mathbf{k}$ and $\mathbf{G} = R\mathbf{i} + S\mathbf{j} + T\mathbf{k}$.

$$\mathbf{F} \times \mathbf{G} = \begin{vmatrix} \mathbf{i} & \mathbf{j} & \mathbf{k} \\ M & N & P \\ R & S & T \end{vmatrix} = (NT - PS)\mathbf{i} - (MT - PR)\mathbf{j} + (MS - NR)\mathbf{k}$$

$$\text{div}(\mathbf{F} \times \mathbf{G}) = \frac{\partial}{\partial x}(NT - PS) + \frac{\partial}{\partial y}(PR - MT) + \frac{\partial}{\partial z}(MS - NR)$$

$$= N\frac{\partial T}{\partial x} + T\frac{\partial N}{\partial x} - P\frac{\partial S}{\partial x} - S\frac{\partial P}{\partial x} + P\frac{\partial R}{\partial y} + R\frac{\partial P}{\partial y} - M\frac{\partial T}{\partial y} - T\frac{\partial M}{\partial y} + M\frac{\partial S}{\partial z} + S\frac{\partial M}{\partial z} - N\frac{\partial R}{\partial z} - R\frac{\partial N}{\partial z}$$

$$= \left[\left(\frac{\partial P}{\partial y} - \frac{\partial N}{\partial z}\right)R + \left(\frac{\partial M}{\partial z} - \frac{\partial P}{\partial x}\right)S + \left(\frac{\partial N}{\partial x} - \frac{\partial M}{\partial y}\right)T\right] - \left[M\left(\frac{\partial T}{\partial y} - \frac{\partial S}{\partial z}\right) + N\left(\frac{\partial R}{\partial z} - \frac{\partial T}{\partial x}\right) + P\left(\frac{\partial S}{\partial x} - \frac{\partial R}{\partial y}\right)\right]$$

$$= (\text{curl } \mathbf{F}) \cdot \mathbf{G} - \mathbf{F} \cdot (\text{curl } \mathbf{G})$$

(e) $\mathbf{F} = M\mathbf{i} + N\mathbf{j} + P\mathbf{k}$

$$\nabla \times \left[\nabla f + (\nabla \times \mathbf{F})\right] = \text{curl}(\nabla f + (\nabla \times \mathbf{F}))$$

$$= \text{curl}(\nabla f) + \text{curl}(\nabla \times \mathbf{F}) \quad \text{(Part (a))}$$

$$= \text{curl}(\nabla \times \mathbf{F}) \quad \text{(Part (b))}$$

$$= \nabla \times (\nabla \times \mathbf{F})$$

(f) Let $\mathbf{F} = M\mathbf{i} + N\mathbf{j} + P\mathbf{k}$.

$$\nabla \times (f\mathbf{F}) = \begin{vmatrix} \mathbf{i} & \mathbf{j} & \mathbf{k} \\ \dfrac{\partial}{\partial x} & \dfrac{\partial}{\partial y} & \dfrac{\partial}{\partial z} \\ fM & fN & fP \end{vmatrix}$$

$$= \left(\frac{\partial f}{\partial y}P + f\frac{\partial P}{\partial y} - \frac{\partial f}{\partial z}N - f\frac{\partial N}{\partial z}\right)\mathbf{i} - \left(\frac{\partial f}{\partial x}P + f\frac{\partial P}{\partial x} - \frac{\partial f}{\partial z}M - f\frac{\partial M}{\partial z}\right)\mathbf{j} + \left(\frac{\partial f}{\partial x}N + f\frac{\partial N}{\partial x} - \frac{\partial f}{\partial y}M - f\frac{\partial M}{\partial y}\right)\mathbf{k}$$

$$= f\left[\left(\frac{\partial P}{\partial y} - \frac{\partial N}{\partial z}\right)\mathbf{i} - \left(\frac{\partial P}{\partial x} - \frac{\partial M}{\partial z}\right)\mathbf{j} + \left(\frac{\partial N}{\partial x} - \frac{\partial M}{\partial y}\right)\mathbf{k}\right] + \begin{vmatrix} \mathbf{i} & \mathbf{j} & \mathbf{k} \\ \dfrac{\partial f}{\partial x} & \dfrac{\partial f}{\partial y} & \dfrac{\partial f}{\partial z} \\ M & N & P \end{vmatrix} = f[\nabla \times \mathbf{F}] + (\nabla f) \times \mathbf{F}$$

(g) Let $\mathbf{F} = M\mathbf{i} + N\mathbf{j} + P\mathbf{k}$, then $f\mathbf{F} = f M\mathbf{i} + f N\mathbf{j} + f P\mathbf{k}$.

$$\text{div}(f\mathbf{F}) = \frac{\partial}{\partial x}(f M) + \frac{\partial}{\partial y}(f N) + \frac{\partial}{\partial z}(f P) = f\frac{\partial M}{\partial x} + M\frac{\partial f}{\partial x} + f\frac{\partial N}{\partial y} + N\frac{\partial f}{\partial y} + f\frac{\partial P}{\partial z} + P\frac{\partial f}{\partial z}$$

$$= f\left(\frac{\partial M}{\partial x} + \frac{\partial N}{\partial y} + \frac{\partial N}{\partial z}\right) + \left(\frac{\partial f}{\partial x}M + \frac{\partial f}{\partial y}N + \frac{\partial f}{\partial z}P\right) = f\,\text{div}\,\mathbf{F} + \nabla f \cdot \mathbf{F}$$

(h) Let $\mathbf{F} = M\mathbf{i} + N\mathbf{j} + P\mathbf{k}$.

$$\text{curl } \mathbf{F} = \left(\frac{\partial P}{\partial y} - \frac{\partial N}{\partial z}\right)\mathbf{i} - \left(\frac{\partial P}{\partial x} - \frac{\partial M}{\partial z}\right)\mathbf{j} + \left(\frac{\partial N}{\partial x} - \frac{\partial M}{\partial y}\right)\mathbf{k}$$

$$\text{div}(\text{curl } \mathbf{F}) = \frac{\partial}{\partial x}\left[\frac{\partial P}{\partial y} - \frac{\partial N}{\partial z}\right] - \frac{\partial}{\partial y}\left[\frac{\partial P}{\partial x} - \frac{\partial M}{\partial z}\right] + \frac{\partial}{\partial z}\left[\frac{\partial N}{\partial x} - \frac{\partial M}{\partial y}\right]$$

$$= \frac{\partial^2 P}{\partial x \partial y} - \frac{\partial^2 N}{\partial x \partial z} - \frac{\partial^2 P}{\partial y \partial x} + \frac{\partial^2 M}{\partial y \partial z} + \frac{\partial^2 N}{\partial z \partial x} - \frac{\partial^2 M}{\partial z \partial y} = 0 \quad \text{(because the mixed partials are equal)}$$

79. True.

$$\|\mathbf{F}(x + y)\| = \sqrt{16x^2 + y^4} \to 0 \text{ as } (x, y) \to (0, 0).$$

81. False. Curl is defined on vector fields, not scalar fields.

83. $\mathbf{F}(x, y) = M(x, y)\mathbf{i} + N(x, y)\mathbf{j} = \dfrac{m}{\left(x^2 + y^2\right)^{5/2}}\left[3xy\mathbf{i} + \left(2y^2 - x^2\right)\mathbf{j}\right]$

$M = \dfrac{3mxy}{\left(x^2 + y^2\right)^{5/2}} = 3mxy\left(x^2 + y^2\right)^{-5/2}$

$\dfrac{\partial M}{\partial y} = 3mxy\left[-\dfrac{5}{2}\left(x^2 + y^2\right)^{-7/2}(2y)\right] + \left(x^2 + y^2\right)^{-5/2}(3mx)$

$\qquad = 3mx\left(x^2 + y^2\right)^{-7/2}\left[-5y^2 + \left(x^2 + y^2\right)\right] = \dfrac{3mx\left(x^2 - 4y^2\right)}{\left(x^2 + y^2\right)^{7/2}}$

$N = \dfrac{m\left(2y^2 - x^2\right)}{\left(x^2 + y^2\right)^{5/2}} = m\left(2y^2 - x^2\right)\left(x^2 + y^2\right)^{-5/2}$

$\dfrac{\partial N}{\partial x} = m\left(2y^2 - x^2\right)\left[-\dfrac{5}{2}\left(x^2 + y^2\right)^{-7/2}(2x)\right] + \left(x^2 + y^2\right)^{-5/2}(-2mx)$

$\qquad = mx\left(x^2 + y^2\right)^{-7/2}\left[\left(2y^2 - x^2\right)(-5) + \left(x^2 + y^2\right)(-2)\right]$

$\qquad = mx\left(x^2 + y^2\right)^{-7/2}\left(3x^2 - 12y^2\right) = \dfrac{3mx\left(x^2 - 4y^2\right)}{\left(x^2 + y^2\right)^{7/2}}$

So, $\dfrac{\partial N}{\partial x} = \dfrac{\partial M}{\partial y}$ and **F** is conservative.

Section 15.2   Line Integrals

1. $\mathbf{r}(t) = \begin{cases} t\mathbf{i} + t\mathbf{j}, & 0 \le t \le 1 \\ (2 - t)\mathbf{i} + \sqrt{2 - t}\,\mathbf{j}, & 1 \le t \le 2 \end{cases}$

3. $\mathbf{r}(t) = \begin{cases} t\mathbf{i}, & 0 \le t \le 3 \\ 3\mathbf{i} + (t - 3)\mathbf{j}, & 3 \le t \le 6 \\ (9 - t)\mathbf{i} + 3\mathbf{j}, & 6 \le t \le 9 \\ (12 - t)\mathbf{j}, & 9 \le t \le 12 \end{cases}$

5.
$\qquad x^2 + y^2 = 9$

$\qquad \dfrac{x^2}{9} + \dfrac{y^2}{9} = 1$

$\qquad \cos^2 t + \sin^2 t = 1$

$\qquad\qquad \cos^2 t = \dfrac{x^2}{9}$

$\qquad\qquad \sin^2 t = \dfrac{y^2}{9}$

$\qquad\qquad\qquad x = 3\cos t$

$\qquad\qquad\qquad y = 3\sin t$

$\qquad\qquad \mathbf{r}(t) = 3\cos t\mathbf{i} + 3\sin t\mathbf{j}$

$\qquad\qquad\qquad 0 \le t \le 2\pi$

7. $\mathbf{r}(t) = 4t\mathbf{i} + 3t\mathbf{j}, \quad 0 \le t \le 1$

$\quad \mathbf{r}'(t) = 4\mathbf{i} + 3\mathbf{j}$

$\quad \displaystyle\int_C xy\,ds = \int_0^1 (4t)(3t)\sqrt{4^2 + 3^2}\,dt$

$\qquad\qquad = \displaystyle\int_0^1 60t^2\,dt = \left[20t^3\right]_0^1 = 20$

9. $\mathbf{r}(t) = \sin t\mathbf{i} + \cos t\mathbf{j} + 2\mathbf{k}, \quad 0 \le t \le \dfrac{\pi}{2}$

$\mathbf{r}'(t) = \cos t\mathbf{i} - \sin t\mathbf{j}$

$\displaystyle\int_C (x^2 + y^2 + z^2)\,ds = \int_0^{\pi/2} (\sin^2 t + \cos^2 t + 4)\sqrt{\cos^2 t + \sin^2 t}\,dt = \int_0^{\pi/2} 5\,dt = \dfrac{5\pi}{2}$

11. (a) $\mathbf{r}(t) = t\mathbf{i} + t\mathbf{j}, \quad 0 \le t \le 1$

(b) $\mathbf{r}'(t) = \mathbf{i} + \mathbf{j}, \|\mathbf{r}'(t)\| = \sqrt{2}$

$\displaystyle\int_C (x^2 + y^2)\,ds = \int_0^1 (t^2 + t^2)\sqrt{2}\,dt = 2\sqrt{2}\left[\dfrac{t^3}{3}\right]_0^1 = \dfrac{2\sqrt{2}}{3}$

13. (a) $\mathbf{r}(t) = \cos t\mathbf{i} + \sin t\mathbf{j}, \quad 0 \le t \le \dfrac{\pi}{2}$

(b) $\displaystyle\int_C (x^2 + y^2)\,ds = \int_0^{\pi/2} \left[\cos^2 t + \sin^2 t\right]\sqrt{(-\sin t)^2 + (\cos t)^2}\,dt = \int_0^{\pi/2} dt = \dfrac{\pi}{2}$

15. (a) $\mathbf{r}(t) = t\mathbf{i}, \quad 0 \le t \le 1$

(b) $\mathbf{r}'(t) = \mathbf{i}, \|\mathbf{r}'(t)\| = 1$

$\displaystyle\int_C (x + 4\sqrt{y})\,ds = \int_0^1 t\,dt = \left[\dfrac{t^2}{2}\right]_0^1 = \dfrac{1}{2}$

17. (a) $\mathbf{r}(t) = \begin{cases} t\mathbf{i}, & 0 \le t \le 1 \\ (2-t)\mathbf{i} + (t-1)\mathbf{j}, & 1 \le t \le 2 \\ (3-t)\mathbf{j}, & 2 \le t \le 3 \end{cases}$

(b) $\displaystyle\int_{C_1} (x + 4\sqrt{y})\,ds = \int_0^1 t\,dt = \dfrac{1}{2}$

$\displaystyle\int_{C_2} (x + 4\sqrt{y})\,ds = \int_1^2 \left[(2-t) + 4\sqrt{t-1}\right]\sqrt{1+1}\,dt = \sqrt{2}\left[2t - \dfrac{t^2}{2} + \dfrac{8}{3}(t-1)^{3/2}\right]_1^2 = \dfrac{19\sqrt{2}}{6}$

$\displaystyle\int_{C_3} (x + 4\sqrt{y})\,ds = \int_2^3 4\sqrt{3-t}\,dt = \left[-\dfrac{8}{3}(3-t)^{3/2}\right]_2^3 = \dfrac{8}{3}$

$\displaystyle\int_C (x + 4\sqrt{y})\,ds = \dfrac{1}{2} + \dfrac{19\sqrt{2}}{6} + \dfrac{8}{3} = \dfrac{19 + 19\sqrt{2}}{6} = \dfrac{19(1 + \sqrt{2})}{6}$

19. (a) $C_1: (0,0,0)$ to $(1,0,0): \mathbf{r}(t) = t\mathbf{i}, 0 \le t \le 1, \mathbf{r}'(t) = \mathbf{i}, \|\mathbf{r}'(t)\| = 1$

$\displaystyle\int_{C_1} (2x + y^2 - z)\,ds = \int_0^1 2t\,dt = t^2\Big]_0^1 = 1$

$C_2: (1,0,0)$ to $(1,0,1): \mathbf{r}(t) = \mathbf{i} + t\mathbf{k}, 0 \le t \le 1, \mathbf{r}'(t) = \mathbf{k}, \|\mathbf{r}'(t)\| = 1$

$\displaystyle\int_{C_2} (2x + y^2 - z)\,ds = \int_0^1 (2 - t)\,dt = \left[2t - \dfrac{t^2}{2}\right]_0^1 = \dfrac{3}{2}$

$C_3: (1,0,1)$ to $(1,1,1): \mathbf{r}(t) = \mathbf{i} + t\mathbf{j} + \mathbf{k}, 0 \le t \le 1, \mathbf{r}'(t) = \mathbf{j}, \|\mathbf{r}'(t)\| = 1$

$\displaystyle\int_{C_3} (2x + y^2 - z)\,ds = \int_0^1 (2 + t^2 - 1)\,dt = \left[t + \dfrac{t^3}{3}\right]_0^1 = \dfrac{4}{3}$

(b) Combining, $\displaystyle\int_C (2x + y^2 - z)\,ds = 1 + \dfrac{3}{2} + \dfrac{4}{3} = \dfrac{23}{6}.$

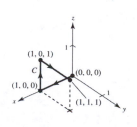

21. $\rho(x, y, z) = \dfrac{1}{2}(x^2 + y^2 + z^2)$

$\quad \mathbf{r}(t) = 2\cos t\mathbf{i} + 2\sin t\mathbf{j} + t\mathbf{k}, 0 \le t \le 4\pi$

$\quad \mathbf{r}'(t) = -2\sin t\mathbf{i} + 2\cos t\mathbf{j} + \mathbf{k}$

$\quad \|\mathbf{r}'(t)\| = \sqrt{4\sin^2 t + 4\cos^2 t + 1} = \sqrt{5}$

$\quad$ Mass $= \displaystyle\int_C \rho(x, y, z)\, ds$

$\qquad = \displaystyle\int_0^{4\pi} \dfrac{1}{2}(4\cos^2 t + 4\sin^2 t + t^2)\sqrt{5}\, dt$

$\qquad = \dfrac{\sqrt{5}}{2}\displaystyle\int_0^{4\pi}(4 + t^2)\, dt = \dfrac{\sqrt{5}}{2}\left[4t + \dfrac{t^3}{3}\right]_0^{4\pi}$

$\qquad = \dfrac{\sqrt{5}}{2}\left[16\pi + \dfrac{64\pi^3}{3}\right] = \dfrac{8\pi\sqrt{5}}{3}(4\pi^2 + 3) \approx 795.7$

23. $\mathbf{r}(t) = \cos t\mathbf{i} + \sin t\mathbf{j}, \; 0 \le t \le \pi$

$\quad \mathbf{r}'(t) = -\sin t\mathbf{i} + \cos t\mathbf{j}, \; \|\mathbf{r}'(t)\| = 1$

$\quad$ Mass $= \displaystyle\int_C \rho(x, y)\, ds = \int_C (x + y + 2)\, ds$

$\qquad = \displaystyle\int_0^\pi (\cos t + \sin t + 2)\, dt$

$\qquad = \left[\sin t - \cos + 2t\right]_0^\pi$

$\qquad = (1 + 2\pi) - (-1) = 2 + 2\pi$

25. $\mathbf{r}(t) = t^2\mathbf{i} + 2t\mathbf{j} + t\mathbf{k}, \; 1 \le t \le 3$

$\quad \mathbf{r}'(t) = 2t\mathbf{i} + 2\mathbf{j} + \mathbf{k}, \; \|\mathbf{r}'(t)\| = \sqrt{4t^2 + 5}$

$\quad$ Mass $= \displaystyle\int_C \rho(x, y, z)\, ds = \int_C kz\, ds$

$\qquad = \displaystyle\int_1^3 kt\sqrt{4t^2 + 5}\, dt$

$\qquad = \dfrac{k(4t^2 + 5)^{3/2}}{12}\Bigg]_1^3$

$\qquad = \dfrac{k}{12}\left[41\sqrt{41} - 27\right]$

27. $\mathbf{F}(x, y) = x\mathbf{i} + y\mathbf{j}$

$\quad C: \mathbf{r}(t) = t\mathbf{i} + t\mathbf{j}, \; 0 \le t \le 1$

$\quad \mathbf{F}(t) = t\mathbf{i} + t\mathbf{j}$

$\quad \mathbf{r}'(t) = \mathbf{i} + \mathbf{j}$

$\quad \displaystyle\int_C \mathbf{F} \cdot d\mathbf{r} = \int_0^1 (t + t)\, dt = \left[t^2\right]_0^1 = 1$

29. $\mathbf{F}(x, y) = 3x\mathbf{i} + 4y\mathbf{j}$

$\quad C: \mathbf{r}(t) = \cos t\mathbf{i} + \sin t\mathbf{j}, \; 0 \le t \le \pi/2$

$\quad \mathbf{F}(t) = 3\cos t\mathbf{i} + 4\sin t\mathbf{j}$

$\quad \mathbf{r}'(t) = -\sin t\mathbf{i} + \cos t\mathbf{j}$

$\quad \displaystyle\int_C \mathbf{F} \cdot d\mathbf{r} = \int_0^{\pi/2}(-3\cos t\sin t + 4\sin t\cos t)\, dt$

$\qquad = \left[\dfrac{\sin^2 t}{2}\right]_0^{\pi/2} = \dfrac{1}{2}$

31. $\mathbf{F}(x, y, z) = xy\mathbf{i} + xz\mathbf{j} + yz\mathbf{k}$

$\quad C: \mathbf{r}(t) = t\mathbf{i} + t^2\mathbf{j} + 2t\mathbf{k}, \; 0 \le t \le 1$

$\quad \mathbf{F}(t) = t^3\mathbf{i} + 2t^2\mathbf{j} + 2t^3\mathbf{k}$

$\quad \mathbf{r}'(t) = \mathbf{i} + 2t\mathbf{j} + 2\mathbf{k}$

$\quad \displaystyle\int_C \mathbf{F} \cdot d\mathbf{r} = \int_0^1 (t^3 + 4t^3 + 4t^3)\, dt = \left[\dfrac{9t^4}{4}\right]_0^1 = \dfrac{9}{4}$

33. $\mathbf{F}(x, y, z) = x^2 z\mathbf{i} + 6y\mathbf{j} + yz^2\mathbf{k}$

$\quad \mathbf{r}(t) = t\mathbf{i} + t^2\mathbf{j} + \ln t\mathbf{k}, \; 1 \le t \le 3$

$\quad \mathbf{F}(t) = t^2\ln t\mathbf{i} + 6t^2\mathbf{j} + t^2\ln^2 t\mathbf{k}$

$\quad d\mathbf{r} = \left(\mathbf{i} + 2t\mathbf{j} + \dfrac{1}{t}\mathbf{k}\right)dt$

$\quad \displaystyle\int_C \mathbf{F} \cdot d\mathbf{r} = \int_1^3 \left[t^2\ln t + 12t^3 + t(\ln t)^2\right]dt \approx 249.49$

35. $\mathbf{F}(x, y) = x\mathbf{i} + 2y\mathbf{j}$

$\quad C: \mathbf{r}(t) = t\mathbf{i} + t^3\mathbf{j}, \; 0 \le t \le 2$

$\quad \mathbf{r}'(t) = \mathbf{i} + 3t^2\mathbf{j}$

$\quad \mathbf{F}(t) = t\mathbf{i} + 2t^3\mathbf{j}$

$\quad$ Work $= \displaystyle\int_C \mathbf{F} \cdot d\mathbf{r} = \int_0^2 (t + 6t^5)\, dt = \left[\dfrac{t^2}{2} + t^6\right]_0^2 = 66$

37. $\mathbf{F}(x, y) = x\mathbf{i} + y\mathbf{j}$

$$C : \mathbf{r}(t) = \begin{cases} t\mathbf{i} & 0 \le t \le 1 \\ (2 - t)\mathbf{i} + (t - 1)\mathbf{j}, & 1 \le t \le 2 \\ (3 - t)\mathbf{j} & 2 \le t \le 3 \end{cases}$$

On C_1, $\mathbf{F}(t) = t\mathbf{i}$, $\mathbf{r}'(t) = \mathbf{i}$

$\text{Work} = \int_{C_1} \mathbf{F} \cdot d\mathbf{r} = \int_0^1 t \, dt = \dfrac{1}{2}$

On C_2, $\mathbf{F}(t) = (2 - t)\mathbf{i} + (t - 1)\mathbf{j}$, $\mathbf{r}'(t) = -\mathbf{i} + \mathbf{j}$

$\begin{aligned} \text{Work} = \int_{C_2} \mathbf{F} \cdot d\mathbf{r} &= \int_1^2 \big[(t - 2) + (t - 1)\big] \, dt \\ &= \Big[t^2 - 3t \Big]_1^2 \\ &= (4 - 6) - (1 - 3) = 0 \end{aligned}$

On C_3, $\mathbf{F}(t) = (3 - t)\mathbf{j}$, $\mathbf{r}'(t) = -\mathbf{j}$

$\begin{aligned} \text{Work} = \int_{C_3} \mathbf{F} \cdot d\mathbf{r} &= \int_2^3 (t - 3) \, dt = \left[\dfrac{t^2}{2} - 3t \right]_2^3 \\ &= \left(\dfrac{9}{2} - 9 \right) - (2 - 6) = -\dfrac{1}{2} \end{aligned}$

$\text{Total work} = \dfrac{1}{2} + 0 - \dfrac{1}{2} = 0$

39. $\mathbf{F}(x, y, z) = x\mathbf{i} + y\mathbf{j} - 5z\mathbf{k}$

$\quad C : \mathbf{r}(t) = 2 \cos t\mathbf{i} + 2 \sin t\mathbf{j} + t\mathbf{k}, \quad 0 \le t \le 2\pi$

$\mathbf{r}'(t) = -2 \sin t\mathbf{i} + 2 \cos t\mathbf{j} + \mathbf{k}$

$\mathbf{F}(t) = 2 \cos t\mathbf{i} + 2 \sin t\mathbf{j} - 5t\mathbf{k}$

$\mathbf{F} \cdot \mathbf{r}' = -5t$

$\text{Work} = \int_C \mathbf{F} \cdot d\mathbf{r} = \int_0^{2\pi} -5t \, dt = -10\pi^2$

41. Because the vector field determined by $\mathbf{F}$ points in the general direction of the path C, $\mathbf{F} \cdot \mathbf{T} > 0$ and work will be positive.

43. Because the vector field determined by $\mathbf{F}$ is perpendicular to the path, work will be 0.

45. $\mathbf{F}(x, y) = x^2\mathbf{i} + xy\mathbf{j}$

(a) $\quad \mathbf{r}_1(t) = 2t\mathbf{i} + (t - 1)\mathbf{j}, \quad 1 \le t \le 3$

$\quad \mathbf{r}_1'(t) = 2\mathbf{i} + \mathbf{j}$

$\quad \mathbf{F}(t) = 4t^2\mathbf{i} + 2t(t - 1)\mathbf{j}$

$\quad \int_{C_1} \mathbf{F} \cdot d\mathbf{r} = \int_1^3 \big(8t^2 + 2t(t - 1)\big) \, dt = \dfrac{236}{3}$

Both paths join $(2, 0)$ and $(6, 2)$. The integrals are negatives of each other because the orientations are different.

(b) $\quad \mathbf{r}_2(t) = 2(3 - t)\mathbf{i} + (2 - t)\mathbf{j}, \quad 0 \le t \le 2$

$\quad \mathbf{r}_2'(t) = -2\mathbf{i} - \mathbf{j}$

$\quad \mathbf{F}(t) = 4(3 - t)^2\mathbf{i} + 2(3 - t)(2 - t)\mathbf{j}$

$\quad \int_{C_2} \mathbf{F} \cdot d\mathbf{r} = \int_0^2 \big[-8(3 - t)^2 - 2(3 - t)(2 - t)\big] \, dt$

$\quad = -\dfrac{236}{3}$

47. $\mathbf{F}(x, y) = y\mathbf{i} - x\mathbf{j}$

$\quad C : \mathbf{r}(t) = t\mathbf{i} - 2t\mathbf{j}$

$\quad \mathbf{r}'(t) = \mathbf{i} - 2\mathbf{j}$

$\quad \mathbf{F}(t) = -2t\mathbf{i} - t\mathbf{j}$

$\quad \mathbf{F} \cdot \mathbf{r}' = -2t + 2t = 0$

So, $\int_C \mathbf{F} \cdot d\mathbf{r} = 0$.

49. $\mathbf{F}(x, y) = (x^3 - 2x^2)\mathbf{i} + \left(x - \dfrac{y}{2} \right)\mathbf{j}$

$\quad C : \mathbf{r}(t) = t\mathbf{i} + t^2\mathbf{j}$

$\quad \mathbf{r}'(t) = \mathbf{i} + 2t\mathbf{j}$

$\quad \mathbf{F}(t) = (t^3 - 2t^2)\mathbf{i} + \left(t - \dfrac{t^2}{2} \right)\mathbf{j}$

$\quad \mathbf{F} \cdot \mathbf{r}' = (t^3 - 2t^2) + 2t\left(t - \dfrac{t^2}{2} \right) = 0$

So, $\int_C \mathbf{F} \cdot d\mathbf{r} = 0$.

51. $x = 2t, \ y = 10t, \ 0 \le t \le 1 \Rightarrow y = 5x$ or $x = \dfrac{y}{5}, \ 0 \le y \le 10$

$\int_C (x + 3y^2) \, dy = \int_0^{10} \left(\dfrac{y}{5} + 3y^2 \right) dy = \left[\dfrac{y^2}{10} + y^3 \right]_0^{10} = 1010$

53. $x = 2t,\ y = 10t,\ 0 \le t \le 1 \Rightarrow x = \dfrac{y}{5},\ 0 \le y \le 10,\ dx = \dfrac{1}{5}\,dy$

$$\int_C xy\,dx + y\,dy = \int_0^{10}\left(\frac{y^2}{25} + y\right)dy = \left[\frac{y^3}{75} + \frac{y^2}{2}\right]_0^{10} = \frac{190}{3}\ \text{or}$$

$y = 5x,\ dy = 5\,dx,\ 0 \le x \le 2$

$$\int_C xy\,dx + y\,dy = \int_0^2 \left(5x^2 + 25x\right)dx = \left[\frac{5x^3}{3} + \frac{25x^2}{2}\right]_0^2 = \frac{190}{3}$$

55. $\mathbf{r}(t) = t\mathbf{i},\ 0 \le t \le 5$

$x(t) = t,\quad y(t) = 0$

$dx = dt,\quad dy = 0$

$$\int_C (2x - y)\,dx + (x + 3y)\,dy = \int_0^5 2t\,dt = 25$$

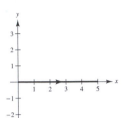

57. $\mathbf{r}(t) = \begin{cases} t\mathbf{i}, & 0 \le t \le 3 \\ 3\mathbf{i} + (t - 3)\mathbf{j}, & 3 \le t \le 6 \end{cases}$

$C_1:\ x(t) = t,\ y(t) = 0,$

$\quad dx = dt,\ dy = 0$

$$\int_{C_1} (2x - y)\,dx + (x + 3y)\,dy = \int_0^3 2t\,dt = 9$$

$C_2:\ x(t) = 3,\ y(t) = t - 3$

$\quad dx = 0,\ dy = dt$

$$\int_{C_2} (2x - y)\,dx + (x + 3y)\,dy = \int_3^6 \left[3 + 3(t - 3)\right]dt = \left[\frac{3t^2}{2} - 6t\right]_3^6 = \frac{45}{2}$$

$$\int_C (2x - y)\,dx + (x + 3y)\,dy = 9 + \frac{45}{2} = \frac{63}{2}$$

59. $x(t) = t,\ y(t) = 1 - t^2,\ 0 \le t \le 1,\ dx = dt,\ dy = -2t\,dt$

$$\int_C (2x - y)\,dx + (x + 3y)\,dy = \int_0^1 \left[\left(2t - 1 + t^2\right) + \left(t + 3 - 3t^2\right)(-2t)\right]dt$$

$$= \int_0^1 \left(6t^3 - t^2 - 4t - 1\right)dt = \left[\frac{3t^4}{2} - \frac{t^3}{3} - 2t^2 - t\right]_0^1 = -\frac{11}{6}$$

61. $x(t) = t,\ y(t) = 2t^2,\ 0 \le t \le 2$

$dx = dt,\ dy = 4t\,dt$

$$\int_C (2x - y)\,dx + (x + 3y)\,dy = \int_0^2 \left(2t - 2t^2\right)dt + \left(t + 6t^2\right)4t\,dt = \int_0^2 \left(24t^3 + 2t^2 + 2t\right)dt = \left[6t^4 + \tfrac{2}{3}t^3 + t^2\right]_0^2 = \frac{316}{3}$$

63. $f(x, y) = h$

C: line from $(0, 0)$ to $(3, 4)$

$\mathbf{r} = 3t\mathbf{i} + 4t\mathbf{j}, \; 0 \le t \le 1$

$\mathbf{r}'(t) = 3\mathbf{i} + 4\mathbf{j}$

$\|\mathbf{r}'(t)\| = 5$

Lateral surface area:

$\int_C f(x, y) \, ds = \int_0^1 5h \, dt = 5h$

65. $f(x, y) = xy$

$C: x^2 + y^2 = 1$ from $(1, 0)$ to $(0, 1)$

$\mathbf{r}(t) = \cos t\mathbf{i} + \sin t\mathbf{j}, \; 0 \le t \le \dfrac{\pi}{2}$

$\mathbf{r}'(t) = -\sin t\mathbf{i} + \cos t\mathbf{j}$

$\|\mathbf{r}'(t)\| = 1$

Lateral surface area:

$\int_C f(x, y) \, ds = \int_0^{\pi/2} \cos t \sin t \, dt = \left[\dfrac{\sin^2 t}{2} \right]_0^{\pi/2} = \dfrac{1}{2}$

67. $f(x, y) = h$

$C: y = 1 - x^2$ from $(1, 0)$ to $(0, 1)$

$\mathbf{r}(t) = (1 - t)\mathbf{i} + \left[1 - (1 - t)^2 \right]\mathbf{j}, \; 0 \le t \le 1$

$\mathbf{r}'(t) = -\mathbf{i} + 2(1 - t)\mathbf{j}$

$\|\mathbf{r}'(t)\| = \sqrt{1 + 4(1 - t)^2}$

Lateral surface area:

$\int_C f(x, y) \, ds = \int_0^1 h\sqrt{1 + 4(1 - t)^2} \, dt = -\dfrac{h}{4}\left[2(1 - t)\sqrt{1 + 4(1 - t)^2} + \ln\left| 2(1 - t) + \sqrt{1 + 4(1 - t)^2} \right| \right]_0^1$

$= \dfrac{h}{4}\left[2\sqrt{5} + \ln\left(2 + \sqrt{5} \right) \right] \approx 1.4789h$

69. $f(x, y) = xy$

$C: y = 1 - x^2$ from $(1, 0)$ to $(0, 1)$

You could parameterize the curve C as in Exercises 67 and 68. Alternatively, let $x = \cos t$, then:

$y = 1 - \cos^2 t = \sin^2 t$

$\mathbf{r}(t) = \cos t\mathbf{i} + \sin^2 t\mathbf{j}, \; 0 \le t \le \dfrac{\pi}{2}$

$\mathbf{r}'(t) = -\sin t\mathbf{i} + 2 \sin t \cos t\mathbf{j}$

$\|\mathbf{r}'(t)\| = \sqrt{\sin^2 t + 4 \sin^2 t \cos^2 t} = \sin t\sqrt{1 + 4\cos^2 t}$

Lateral surface area:

$\int_C f(x, y) \, ds = \int_0^{\pi/2} \cos t \sin^2 t\left(\sin t\sqrt{1 + 4\cos^2 t} \right) dt = \int_0^{\pi/2} \sin^2 t\left[\left(1 + 4\cos^2 t \right)^{1/2} \sin t \cos t \right] dt$

Let $u = \sin^2 t$ and $dv = \left(1 + 4\cos^2 t \right)^{1/2} \sin t \cos t$, then $du = 2 \sin t \cos t \, dt$ and $v = -\dfrac{1}{12}\left(1 + 4\cos^2 t \right)^{3/2}$.

$\int_C f(x, y) \, ds = \left[-\dfrac{1}{12}\sin^2 t\left(1 + 4\cos^2 t \right)^{3/2} \right]_0^{\pi/2} + \dfrac{1}{6}\int_0^{\pi/2} \left(1 + 4\cos^2 t \right)^{3/2} \sin t \cos t \, dt$

$= \left[-\dfrac{1}{12}\sin^2 t\left(1 + 4\cos^2 t \right)^{3/2} - \dfrac{1}{120}\left(1 + 4\cos^2 t \right)^{5/2} \right]_0^{\pi/2} = \left(-\dfrac{1}{12} - \dfrac{1}{120} \right) + \dfrac{1}{120}(5)^{5/2} = \dfrac{1}{120}\left(25\sqrt{5} - 11 \right) \approx 0.3742$

71. (a) $f(x, y) = 1 + y^2$

$\qquad$ $\mathbf{r}(t) = 2 \cos t\mathbf{i} + 2 \sin t\mathbf{j}, \ 0 \le t \le 2\pi$

$\qquad$ $\mathbf{r}'(t) = -2 \sin t\mathbf{i} + 2 \cos t\mathbf{j}$

$\qquad$ $\|\mathbf{r}'(t)\| = 2$

$\qquad$ $S = \int_C f(x, y) \, ds = \int_0^{2\pi} \left(1 + 4 \sin^2 t\right)(2) \, dt = \left[2t + 4(t - \sin t \cos t)\right]_0^{2\pi} = 12\pi \approx 37.70 \text{ cm}^2$

$\quad$ (b) $0.2(12\pi) = \dfrac{12\pi}{5} \approx 7.54 \text{ cm}^3$

$\quad$ (c)

73. $\mathbf{r}(t) = a \cos t\mathbf{i} + a \sin t\mathbf{j}, \ 0 \le t \le 2\pi$

$\qquad$ $\mathbf{r}'(t) = -a \sin t\mathbf{i} + a \cos t\mathbf{j}, \ \|\mathbf{r}'(t)\| = a$

$\qquad$ $I_x = \int_C y^2 \rho(x, y) \, ds = \int_0^{2\pi} \left(a^2 \sin^2 t\right)(1)a \, dt$

$$= a^3 \int_0^{2\pi} \sin^2 t \, dt = a^3 \pi$$

$\qquad$ $I_y = \int_C x^2 \rho(x, y) \, ds = \int_0^{2\pi} \left(a^2 \cos^2 t\right)(1)a \, dt$

$$= a^3 \int_0^{2\pi} \cos^2 t \, dt = a^3 \pi$$

75. (a) Graph of: $\mathbf{r}(t) = 3 \cos t\mathbf{i} + 3 \sin t\mathbf{j} + \left(1 + \sin^2 2t\right)\mathbf{k}, \ 0 \le t \le 2\pi$

$\qquad$ For $y = b$ constant, $3 \sin t = b \Rightarrow \sin t = \dfrac{b}{3}$ and

$\qquad$ $1 + \sin^2 2t = 1 + \left(2 \sin t \cos t\right)^2$

$\qquad\qquad\qquad\qquad = 1 + 4 \sin^2 t \cos^2 t$

$\qquad\qquad\qquad\qquad = 1 + 4 \sin^2 t\left(1 - \sin^2 t\right) = 1 + \dfrac{4}{9}b^2\left(1 - \dfrac{b^2}{9}\right).$

$\quad$ (b) Consider the portion of the surface in the first quadrant. The curve $z = 1 + \sin^2 2t$ is over the curve

$\qquad$ $\mathbf{r}_1(t) = 3 \cos t\mathbf{i} + 3 \sin t\mathbf{j}, 0 \le t \le \pi/2.$ So, the total lateral surface area is

$\qquad$ $4\int_C f(x, y) \, ds = 4\int_0^{\pi/2} \left(1 + \sin^2 2t\right)3 \, dt = 12\left(\dfrac{3\pi}{4}\right) = 9\pi \text{ cm}^2.$

$\quad$ (c) The cross sections parallel to the *xz*-plane are rectangles of height $1 + 4(y/3)^2\left(1 - y^2/9\right)$ and base $2\sqrt{9 - y^2}.$ So,

$\qquad$ Volume $= 2\int_0^3 2\sqrt{9 - y^2}\left(1 + 4\dfrac{y^2}{9}\left(1 - \dfrac{y^2}{9}\right)\right) dy = \dfrac{27\pi}{2} \approx 42.412 \text{ cm}^3.$

77. $r(t) = 3 \sin t\mathbf{i} + 3 \cos t\mathbf{j} + \dfrac{10}{2\pi} t\mathbf{k}, \quad 0 \le t \le 2\pi$

$\mathbf{F} = 175\mathbf{k}$

$d\mathbf{r} = \left(3 \cos t\mathbf{i} - 3 \sin t\mathbf{j} + \dfrac{10}{2\pi}\mathbf{k}\right) dt$

$\displaystyle\int_C \mathbf{F} \cdot d\mathbf{r} = \int_0^{2\pi} \dfrac{1750}{2\pi}\, dt = \left[\dfrac{1750}{2\pi}t\right]_0^{2\pi} = 1750 \text{ ft} \cdot \text{lb}$

79. See the definition of Line Integral, page 1052. See Theorem 15.4.

85. False, the orientations are different.

87. $\mathbf{F}(x, y) = (y - x)\mathbf{i} + xy\mathbf{j}$

$r(t) = kt(1 - t)\mathbf{i} + t\mathbf{j}, \quad 0 \le t \le 1$

$r'(t) = k(1 - 2t)\mathbf{i} + \mathbf{j}$

$\text{Work} = 1 = \displaystyle\int_C \mathbf{F} \cdot d\mathbf{r}$

$= \displaystyle\int_0^1 \Big[\big(t - kt(1 - t)\big)\mathbf{i} + kt^2(1 - t)\mathbf{j}\Big] \cdot \Big[k(1 - 2t)\mathbf{i} + \mathbf{j}\Big] dt$

$= \displaystyle\int_0^1 \Big[\big(t - kt(1 - t)\big)k(1 - 2t) + kt^2(1 - t)\Big] dt$

$= \displaystyle\int_0^1 \big(-2k^2t^3 - kt^3 - kt^2 + 3k^2t^2 - k^2t + kt\big) dt = \dfrac{-k}{12}$

$k = -12$

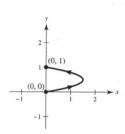

81. The greater the height of the surface over the curve, the greater the lateral surface area. So, $z_3 < z_1 < z_2 < z_4$.

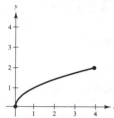

83. False

$\displaystyle\int_C xy\, ds = \sqrt{2} \int_0^1 t^2\, dt$

Section 15.3   Conservative Vector Fields and Independence of Path

1. $\mathbf{F}(x, y) = x^2\mathbf{i} + xy\mathbf{j}$

(a) $r_1(t) = t\mathbf{i} + t^2\mathbf{j}, \quad 0 \le t \le 1$

$r_1'(t) = \mathbf{i} + 2t\mathbf{j}$

$\mathbf{F}(t) = t^2\mathbf{i} + t^3\mathbf{j}$

$\displaystyle\int_C \mathbf{F} \cdot d\mathbf{r} = \int_0^1 \big(t^2 + 2t^4\big) dt = \dfrac{11}{15}$

(b) $r_2(\theta) = \sin \theta\mathbf{i} + \sin^2 \theta\mathbf{j}, \quad 0 \le \theta \le \dfrac{\pi}{2}$

$r_2'(\theta) = \cos \theta\mathbf{i} + 2 \sin \theta \cos \theta\mathbf{j}$

$\mathbf{F}(t) = \sin^2 \theta\mathbf{i} + \sin^3 \theta\mathbf{j}$

$\displaystyle\int_C \mathbf{F} \cdot d\mathbf{r} = \int_0^{\pi/2} \big(\sin^2 \theta \cos \theta + 2 \sin^4 \theta \cos \theta\big) d\theta = \left[\dfrac{\sin^3 \theta}{3} + \dfrac{2 \sin^5 \theta}{5}\right]_0^{\pi/2} = \dfrac{11}{15}$

3. $F(x, y) = y\mathbf{i} - x\mathbf{j}$

(a) $\mathbf{r}_1(\theta) = \sec\theta\mathbf{i} + \tan\theta\mathbf{j}, \ 0 \le \theta \le \dfrac{\pi}{3}$

$\mathbf{r}_1'(\theta) = \sec\theta\tan\theta\mathbf{i} + \sec^2\theta\mathbf{j}$

$F(\theta) = \tan\theta\mathbf{i} - \sec\theta\mathbf{j}$

$$\int_C F \cdot d\mathbf{r} = \int_0^{\pi/3}\left(\sec\theta\tan^2\theta - \sec^3\theta\right)d\theta$$

$$= \int_0^{\pi/3}\left[\sec\theta(\sec^2\theta - 1) - \sec^3\theta\right]d\theta$$

$$= -\int_0^{\pi/3}\sec\theta\,d\theta$$

$$= \left[-\ln|\sec\theta + \tan\theta|\right]_0^{\pi/3}$$

$$= -\ln\left(2 + \sqrt{3}\right) \approx -1.317$$

(b) $\mathbf{r}_2(t) = \sqrt{t + 1}\mathbf{i} + \sqrt{t}\mathbf{j}, \ 0 \le t \le 3$

$\mathbf{r}_2'(t) = \dfrac{1}{2\sqrt{t + 1}}\mathbf{i} + \dfrac{1}{2\sqrt{t}}\mathbf{j}$

$F(t) = \sqrt{t}\mathbf{i} - \sqrt{t + 1}\mathbf{j}$

$$\int_C F \cdot d\mathbf{r} = \int_0^3\left[\frac{\sqrt{t}}{2\sqrt{t + 1}} - \frac{\sqrt{t + 1}}{2\sqrt{t}}\right]dt$$

$$= -\frac{1}{2}\int_0^3\frac{1}{\sqrt{t}\sqrt{t + 1}}\,dt$$

$$= -\frac{1}{2}\int_0^3\frac{1}{\sqrt{t^2 + t + (1/4) - (1/4)}}\,dt$$

$$= -\frac{1}{2}\int_0^3\frac{1}{\sqrt{\left[t + (1/2)\right]^2 - (1/4)}}\,dt$$

$$= \left[-\frac{1}{2}\ln\left|\left(t + \frac{1}{2}\right) + \sqrt{t^2 + t}\right|\right]_0^3$$

$$= -\frac{1}{2}\left[\ln\left(\frac{7}{2} + 2\sqrt{3}\right) - \ln\left(\frac{1}{2}\right)\right]$$

$$= -\frac{1}{2}\ln\left(7 + 4\sqrt{3}\right) \approx -1.317$$

5. $F(x, y) = e^x\sin y\mathbf{i} + e^x\cos y\mathbf{j}$

$\dfrac{\partial N}{\partial x} = e^x\cos y \qquad \dfrac{\partial M}{\partial y} = e^x\cos y$

Because $\dfrac{\partial N}{\partial x} = \dfrac{\partial M}{\partial y}$, F is conservative.

7. $F(x, y) = \dfrac{1}{y}\mathbf{i} + \dfrac{x}{y^2}\mathbf{j}$

$\dfrac{\partial N}{\partial x} = \dfrac{1}{y^2} \qquad \dfrac{\partial M}{\partial y} = -\dfrac{1}{y^2}$

Because $\dfrac{\partial N}{\partial x} \ne \dfrac{\partial M}{\partial y}$, F is not conservative.

9. $F(x, y, z) = y^2z\mathbf{i} + 2xyz\mathbf{j} + xy^2\mathbf{k}$

$\text{curl } F = 0 \Rightarrow F$ is conservative.

11. $F(x, y) = 2xy\mathbf{i} + x^2\mathbf{j}$

(a) $\mathbf{r}_1(t) = t\mathbf{i} + t^2\mathbf{j}, \ 0 \le t \le 1$

$\mathbf{r}_1'(t) = \mathbf{i} + 2t\mathbf{j}$

$F(t) = 2t^3\mathbf{i} + t^2\mathbf{j}$

$$\int_C F \cdot d\mathbf{r} = \int_0^1 4t^3\,dt = 1$$

(b) $\mathbf{r}_2(t) = t\mathbf{i} + t^3\mathbf{j}, \ 0 \le t \le 1$

$\mathbf{r}_2'(t) = \mathbf{i} + 3t^2\mathbf{j}$

$F(t) = 2t^4\mathbf{i} + t^2\mathbf{j}$

$$\int_C F \cdot d\mathbf{r} = \int_0^1 5t^4\,dt = 1$$

13. $F(x, y) = y\mathbf{i} - x\mathbf{j}$

(a) $\mathbf{r}_1(t) = t\mathbf{i} + t\mathbf{j}, \ 0 \le t \le 1$

$\mathbf{r}_1'(t) = \mathbf{i} + \mathbf{j}$

$F(t) = t\mathbf{i} - t\mathbf{j}$

$$\int_C F \cdot d\mathbf{r} = 0$$

(b) $\mathbf{r}_2(t) = t\mathbf{i} + t^2\mathbf{j}, \ 0 \le t \le 1$

$\mathbf{r}_2'(t) = \mathbf{i} + 2t\mathbf{j}$

$F(t) = t^2\mathbf{i} - t\mathbf{j}$

$$\int_C F \cdot d\mathbf{r} = \int_0^1 -t^2\,dt = -\frac{1}{3}$$

(c) $\mathbf{r}_3(t) = t\mathbf{i} + t^3\mathbf{j}, \ 0 \le t \le 1$

$\mathbf{r}_3'(t) = \mathbf{i} + 3t^2\mathbf{j}$

$F(t) = t^3\mathbf{i} - t\mathbf{j}$

$$\int_C F \cdot d\mathbf{r} = \int_0^1 -2t^3\,dt = -\frac{1}{2}$$

15. $\displaystyle\int_C y^2\,dx + 2xy\,dy$

Because

$\partial M/\partial y = \partial N/\partial x = 2y$, $F(x, y) = y^2\mathbf{i} + 2xy\mathbf{j}$ is conservative. The potential function is

$f(x, y) = xy^2 + k$. So, you can use the Fundamental Theorem of Line Integrals.

(a) $\displaystyle\int_C y^2\,dx + 2xy\,dy = \left[x^2y\right]_{(0,0)}^{(4,4)} = 64$

(b) $\displaystyle\int_C y^2\,dx + 2xy\,dy = \left[x^2y\right]_{(-1,0)}^{(1,0)} = 0$

(c) and (d) Because C is a closed curve,
$$\int_C y^2\,dx + 2xy\,dy = 0.$$

17. $\int_C 2xy \, dx + \left(x^2 + y^2\right) dy$

Because $\partial M/\partial y = \partial N/\partial x = 2x$, $\mathbf{F}(x, y) = 2xy\mathbf{i} + \left(x^2 + y^2\right)\mathbf{j}$ is conservative.

The potential function is $f(x, y) = x^2 y + \dfrac{y^3}{3} + k$.

(a) $\int_C 2xy \, dx + \left(x^2 + y^2\right) dy = \left[x^2 y + \dfrac{y^3}{3}\right]_{(5,0)}^{(0,4)} = \dfrac{64}{3}$

(b) $\int_C 2xy \, dx + \left(x^2 + y^2\right) dy = \left[x^2 y + \dfrac{y^3}{3}\right]_{(2,0)}^{(0,4)} = \dfrac{64}{3}$

19. $\mathbf{F}(x, y, z) = yz\mathbf{i} + xz\mathbf{j} + xy\mathbf{k}$

Because $\mathbf{curl}\,\mathbf{F} = \mathbf{0}$, $\mathbf{F}(x, y, z)$ is conservative. The potential function is $f(x, y, z) = xyz + k$.

(a) $\mathbf{r}_1(t) = t\mathbf{i} + 2\mathbf{j} + t\mathbf{k}$, $0 \le t \le 4$

$\int_C \mathbf{F} \cdot d\mathbf{r} = [xyz]_{(0,2,0)}^{(4,2,4)} = 32$

(b) $\mathbf{r}_2(t) = t^2\mathbf{i} + t\mathbf{j} + t^2\mathbf{k}$, $0 \le t \le 2$

$\int_C \mathbf{F} \cdot d\mathbf{r} = [xyz]_{(0,0,0)}^{(4,2,4)} = 32$

21. $\mathbf{F}(x, y, z) = (2y + x)\mathbf{i} + \left(x^2 - z\right)\mathbf{j} + (2y - 4z)\mathbf{k}$

$\mathbf{F}(x, y, z)$ is not conservative.

(a) $\mathbf{r}_1(t) = t\mathbf{i} + t^2\mathbf{j} + \mathbf{k}$, $0 \le t \le 1$

$\mathbf{r}_1'(t) = \mathbf{i} + 2t\mathbf{j}$

$\mathbf{F}(t) = \left(2t^2 + t\right)\mathbf{i} + \left(t^2 - 1\right)\mathbf{j} + \left(2t^2 - 4\right)\mathbf{k}$

$\int_C \mathbf{F} \cdot d\mathbf{r} = \int_0^1 \left(2t^3 + 2t^2 - t\right) dt = \dfrac{2}{3}$

(b) $\mathbf{r}_2(t) = t\mathbf{i} + t\mathbf{j} + (2t - 1)^2\mathbf{k}$, $0 \le t \le 1$

$\mathbf{r}_2'(t) = \mathbf{i} + \mathbf{j} + 4(2t - 1)\mathbf{k}$

$\mathbf{F}(t) = 3t\mathbf{i} + \left[t^2 - (2t - 1)^2\right]\mathbf{j} + \left[2t - 4(2t - 1)^2\right]\mathbf{k}$

$\int_C \mathbf{F} \cdot d\mathbf{r} = \int_0^1 \left[3t + t^2 - (2t - 1)^2 + 8t(2t - 1) - 16(2t - 1)^3\right] dt$

$= \int_0^1 \left[17t^2 - 5t - (2t - 1)^2 - 16(2t - 1)^3\right] dt = \left[\dfrac{17t^3}{3} - \dfrac{5t^2}{2} - \dfrac{(2t - 1)^3}{6} - 2(2t - 1)^4\right]_0^1 = \dfrac{17}{6}$

23. $\mathbf{F}(x, y, z) = e^z(y\mathbf{i} + x\mathbf{j} + xy\mathbf{k})$

$\mathbf{F}(x, y, z)$ is conservative. The potential function is $f(x, y, z) = xye^z + k$.

(a) $\mathbf{r}_1(t) = 4\cos t\mathbf{i} + 4\sin t\mathbf{j} + 3\mathbf{k}$, $0 \le t \le \pi$

$\int_C \mathbf{F} \cdot d\mathbf{r} = \left[xye^z\right]_{(4,0,3)}^{(-4,0,3)} = 0$

(b) $\mathbf{r}_2(t) = (4 - 8t)\mathbf{i} + 3\mathbf{k}$, $0 \le t \le 1$

$\int_C \mathbf{F} \cdot d\mathbf{r} = \left[xye^z\right]_{(4,0,3)}^{(-4,0,3)} = 0$

25. $\int_C (3y\mathbf{i} + 3x\mathbf{j}) \cdot d\mathbf{r} = \left[3xy \right]_{(0,0)}^{(3,8)} = 72$

27. $\int_C \cos x \sin y \, dx + \sin x \cos y \, dy = \left[\sin x \sin y \right]_{(0,-\pi)}^{(3\pi/2, \pi/2)} = -1$

29. $\int_C e^x \sin y \, dx + e^x \cos y \, dy = \left[e^x \sin y \right]_{(0,0)}^{(2\pi, 0)} = 0$

31. $\int_C (z + 2y) \, dx + (2x - z) \, dy + (x - y) \, dz$

    $\mathbf{F}(x, y, z)$ is conservative and the potential function is $f(x, y, z) = xz + 2xy - yz$

    (a) $\left[xz + 2xy - yz \right]_{(0,0,0)}^{(1,1,1)} = 2 - 0 = 2$

    (b) $\left[xz + 2xy - yz \right]_{(0,0,0)}^{(0,0,1)} + \left[xz + 2xy - yz \right]_{(0,0,1)}^{(1,1,1)} = 0 + 2 = 2$

    (c) $\left[xz + 2xy - yz \right]_{(0,0,0)}^{(1,0,0)} + \left[xz + 2xy - yz \right]_{(1,0,0)}^{(1,1,0)} + \left[xz + 2xy - yz \right]_{(1,1,0)}^{(1,1,1)} = 0 + 2 + (2 - 2) = 2$

33. $\int_C -\sin x \, dx + z \, dy + y \, dz = \left[\cos x + yz \right]_{(0,0,0)}^{(\pi/2, 3, 4)} = 12 - 1 = 11$

35. $\mathbf{F}(x, y) = 9x^2 y^2 \mathbf{i} + (6x^3 y - 1) \mathbf{j}$ is conservative.

    Work $= \left[3x^3 y^2 - y \right]_{(0,0)}^{(5,9)} = 30,366$

37. $\mathbf{r}(t) = 2 \cos 2\pi t \mathbf{i} + 2 \sin 2\pi t \mathbf{j}$

    $\mathbf{r}'(t) = -4\pi \sin 2\pi t \mathbf{i} + 4\pi \cos 2\pi t \mathbf{j}$

    $\mathbf{a}(t) = -8\pi^2 \cos 2\pi t \mathbf{i} - 8\pi^2 \sin 2\pi t \mathbf{j}$

    $\mathbf{F}(t) = m\mathbf{a}(t) = \dfrac{1}{32}\mathbf{a}(t) = -\dfrac{\pi^2}{4}(\cos 2\pi t \mathbf{i} + \sin 2\pi t \mathbf{j})$

    $W = \int_C \mathbf{F} \cdot d\mathbf{r} = \int_C -\dfrac{\pi^2}{4}(\cos 2\pi t \mathbf{i} + \sin 2\pi t \mathbf{j}) \cdot 4\pi(-\sin 2\pi t \mathbf{i} + \cos 2\pi t \mathbf{j}) \, dt = -\pi^3 \int_C 0 \, dt = 0$

39. $\mathbf{F} = -175\mathbf{j}$

    (a) $\mathbf{r}(t) = t\mathbf{i} + (50 - t)\mathbf{j}, \quad 0 \le t \le 50$

        $d\mathbf{r} = (\mathbf{i} - \mathbf{j}) \, dt$

        $\int_C \mathbf{F} \cdot d\mathbf{r} = \int_0^{50} 175 \, dt = 8750 \text{ ft} \cdot \text{lbs}$

    (b) $\mathbf{r}(t) = t\mathbf{i} + \frac{1}{50}(50 - t)^2 \mathbf{j}, \quad 0 \le t \le 50$

        $d\mathbf{r} = \mathbf{i} - \frac{1}{25}(50 - t)\mathbf{j}$

        $\int_C \mathbf{F} \cdot d\mathbf{r} = \int_0^{50} (175)\frac{1}{25}(50 - t) \, dt = 7 \left[50t - \dfrac{t^2}{2} \right]_0^{50} = 8750 \text{ ft} \cdot \text{lbs}$

41. See Theorem 15.5.

43. (a) For the circle $\mathbf{r}(t) = a \cos t \mathbf{i} - a \sin t \mathbf{j}, 0 \le t \le 2\pi$, you have $x^2 + y^2 = a^2$, and

        $\int_C \mathbf{F} \cdot d\mathbf{r} = \int_0^{2\pi} \left(\dfrac{-a \sin t}{a^2}\mathbf{i} - \dfrac{a \cos t}{a^2}\mathbf{j} \right) \cdot (-a \sin t \mathbf{i} - a \cos t \mathbf{j}) \, dt = \int_0^{2\pi} (\sin^2 t + \cos^2 t) \, dt = 2\pi.$

    (b) For this curve, the answer is the same, 2π.

    (c) For the opposite overtation, the answer is -2π.

    (d) For the curve away from the origin, the answer is 0.

45. Conservative. $\int_C \mathbf{F} \cdot d\mathbf{r}$ is independent of path.

47. False, it would be true if $\mathbf{F}$ were conservative.

49. True

51. Let

$$\mathbf{F} = M\mathbf{i} + N\mathbf{j} = \frac{\partial f}{\partial y}\mathbf{i} - \frac{\partial f}{\partial x}\mathbf{j}.$$

Then $\dfrac{\partial M}{\partial y} = \dfrac{\partial}{\partial y}\left(\dfrac{\partial f}{\partial y}\right) = \dfrac{\partial^2 f}{\partial y^2}$ and $\dfrac{\partial N}{\partial x} = \dfrac{\partial}{\partial x}\left(-\dfrac{\partial f}{\partial x}\right) = -\dfrac{\partial^2 f}{\partial x^2}$. Because $\dfrac{\partial^2 f}{\partial x^2} + \dfrac{\partial^2 f}{\partial y^2} = 0$ you have $\dfrac{\partial M}{\partial y} = \dfrac{\partial N}{\partial x}$.

So, $\mathbf{F}$ is conservative. Therefore, by Theorem 15.7, you have $\displaystyle\int_C\left(\dfrac{\partial f}{\partial y}\,dx - \dfrac{\partial f}{\partial x}\,dy\right) = \int_C(M\,dx + N\,dy) = \int_C \mathbf{F} \cdot d\mathbf{r} = 0$

for every closed curve in the plane.

53. $\mathbf{F}(x, y) = \dfrac{y}{x^2 + y^2}\mathbf{i} - \dfrac{x}{x^2 + y^2}\mathbf{j}$

   (a) $M = \dfrac{y}{x^2 + y^2}$

   $\dfrac{\partial M}{\partial y} = \dfrac{\left(x^2 + y^2\right)(1) - y(2y)}{\left(x^2 + y^2\right)^2} = \dfrac{x^2 - y^2}{\left(x^2 + y^2\right)^2}$

   $N = -\dfrac{x}{x^2 + y^2}$

   $\dfrac{\partial N}{\partial x} = \dfrac{\left(x^2 + y^2\right)(-1) + x(2x)}{\left(x^2 + y^2\right)^2} = \dfrac{x^2 - y^2}{\left(x^2 + y^2\right)^2}$

   So, $\dfrac{\partial N}{\partial x} = \dfrac{\partial M}{\partial y}$.

   (b) $\mathbf{r}(t) = \cos t\mathbf{i} + \sin t\mathbf{j}, \ 0 \le t \le \pi$

   $\mathbf{F} = \sin t\mathbf{i} - \cos t\mathbf{j}$

   $d\mathbf{r} = (-\sin t\mathbf{i} + \cos t\mathbf{j})\,dt$

   $\displaystyle\int_C \mathbf{F} \cdot d\mathbf{r} = \int_0^\pi \left(-\sin^2 t - \cos^2 t\right)dt = \left[-t\right]_0^\pi = -\pi$

   (c) $\mathbf{r}(t) = \cos t\mathbf{i} - \sin t\mathbf{j}, \ 0 \le t \le \pi$

   $\mathbf{F} = -\sin t\mathbf{i} - \cos t\mathbf{j}$

   $d\mathbf{r} = (-\sin t\mathbf{i} - \cos t\mathbf{j})\,dt$

   $\displaystyle\int_C \mathbf{F} \cdot d\mathbf{r} = \int_0^\pi \left(\sin^2 t + \cos^2 t\right)dt = \left[t\right]_0^\pi = \pi$

   (d) $\mathbf{r}(t) = \cos t\mathbf{i} + \sin t\mathbf{j}, \ 0 \le t \le 2\pi$

   $\mathbf{F} = \sin t\mathbf{i} - \cos t\mathbf{j}$

   $d\mathbf{r} = (-\sin t\mathbf{i} + \cos t\mathbf{j})\,dt$

   $\displaystyle\int_C \mathbf{F} \cdot d\mathbf{r} = \int_0^{2\pi} \left(-\sin^2 t - \cos^2 t\right)dt = \left[-t\right]_0^{2\pi} = -2\pi$

   This does not contradict Theorem 15.7 because $\mathbf{F}$ is not continuous at $(0, 0)$ in R enclosed by curve C.

   (e) $\nabla\left(\arctan\dfrac{x}{y}\right) = \dfrac{1/y}{1 + (x/y)^2}\mathbf{i} + \dfrac{-x/y^2}{1 + (x/y)^2}\mathbf{j} = \dfrac{y}{x^2 + y^2}\mathbf{i} - \dfrac{x}{x^2 + y^2}\mathbf{j} = \mathbf{F}$

Section 15.4   Green's Theorem

1. $r(t) = \begin{cases} t\mathbf{i} + t^2\mathbf{j}, & 0 \le t \le 1 \\ (2-t)\mathbf{i} + (2-t)\mathbf{j}, & 1 \le t \le 2 \end{cases}$

$\int_C y^2\,dx + x^2\,dy = \int_0^1 \left[t^4(dt) + t^2(2t\,dt) \right] + \int_1^2 \left[(2-t)^2(-dt) + (2-t)^2(-dt) \right]$

$= \int_0^1 (t^4 + 2t^3)\,dt + \int_1^2 2(2-t)^2(-dt) = \left[\frac{t^5}{5} + \frac{t^4}{2} \right]_0^1 + \left[\frac{2(2-t)^3}{3} \right]_1^2 = \frac{7}{10} - \frac{2}{3} = \frac{1}{30}$

By Green's Theorem,

$\int_R \int \left(\frac{\partial N}{\partial x} - \frac{\partial M}{\partial y} \right) dA = \int_0^1 \int_{x^2}^x (2x - 2y)\,dy\,dx = \int_0^1 \left[2xy - y^2 \right]_{x^2}^x dx$

$= \int_0^1 (x^2 - 2x^3 + x^4)\,dx = \left[\frac{x^3}{3} - \frac{x^4}{2} + \frac{x^5}{5} \right]_0^1 = \frac{1}{30}$

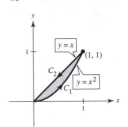

3. $r(t) = \begin{cases} t\mathbf{i} & 0 \le t \le 1 \\ \mathbf{i} + (t-1)\mathbf{j} & 1 \le t \le 2 \\ (3-t)\mathbf{i} + \mathbf{j} & 2 \le t \le 3 \\ (4-t)\mathbf{j} & 3 \le t \le 4 \end{cases}$

$\int_C y^2\,dx + x^2\,dy = \int_0^1 \left[0\,dt + t^2(0) \right] + \int_1^2 \left[(t-1)^2(0) + 1\,dt \right] + \int_2^3 \left[1(-dt) + (3-t)^2(0) \right] + \int_3^4 \left[(4-t)^2(0) + 0(-dt) \right]$

$= \int_1^2 dt + \int_2^3 -dt = 1 - 1 = 0$

By Green's Theorem,

$\int_R \int \left(\frac{\partial N}{\partial x} - \frac{\partial M}{\partial y} \right) dA = \int_0^1 \int_0^1 (2x - 2y)\,dy\,dx$

$= \int_0^1 \left[2xy - y^2 \right]_0^1 dx = \int_0^1 (2x - 1)\,dx = \left[x^2 - x \right]_0^1 = 0$

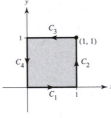

5. $C: x^2 + y^2 = 4$

Let $x = 2\cos t$ and $y = 2\sin t, 0 \le t \le 2\pi$.

$\int_C xe^y\,dx + e^x\,dy = \int_0^{2\pi} \left[2\cos t\, e^{2\sin t}(-2\sin t) + e^{2\cos t}(2\cos t) \right] dt \approx 19.99$

$\int_R \int \left(\frac{\partial N}{\partial x} - \frac{\partial M}{\partial y} \right) dA = \int_{-2}^2 \int_{-\sqrt{4-x^2}}^{\sqrt{4-x^2}} (e^x - xe^y)\,dy\,dx = \int_{-2}^2 \left[2\sqrt{4-x^2}\,e^x - xe^{\sqrt{4-x^2}} + xe^{-\sqrt{4-x^2}} \right] dx \approx 19.99$

In Exercises 7–9, $\frac{\partial N}{\partial x} - \frac{\partial M}{\partial y} = 1.$

7. $\int_C (y - x)\,dx + (2x - y)\,dy = \int_0^3 \int_{x^2-2x}^x dy\,dx$

$= \int_0^3 \left[x - (x^2 - 2x)\,dx \right] = \left[-\frac{x^3}{3} + \frac{3x^2}{2} \right]_0^3 = -9 + \frac{27}{2} = \frac{9}{2}$

9. From the accompanying figure, we see that R is the shaded region. So, Green's Theorem yields

$$\int_C (y - x)\, dx + (2x - y)\, dy = \int_R \int 1\, dA = \text{Area of } R = 6(10) - 2(2) = 56.$$

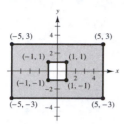

11. $\displaystyle \int_C 2xy\, dx + (x + y)\, dy = \int_R \int \left(\frac{\partial N}{\partial x} - \frac{\partial M}{\partial y} \right) dA$

$$= \int_{-1}^{1} \int_{0}^{1-x^2} (1 - 2x)\, dy\, dx = \int_{-1}^{1} \left[y - 2xy \right]_0^{1-x^2}\, dx = \int_{-1}^{1} \left[(1 - x^2) - 2x(1 - x^2) \right] dx$$

$$= \int_{-1}^{1} \left[1 - x^2 - 2x + 2x^3 \right] dx = \left[x - \frac{x^3}{3} - x^2 + \frac{x^4}{2} \right]_{-1}^{1} = \frac{1}{6} + \frac{7}{6} = \frac{4}{3}$$

13. $\displaystyle \int_C (x^2 - y^2)\, dx + 2xy\, dy = \int_R \int \left(\frac{\partial N}{\partial x} - \frac{\partial M}{\partial y} \right) dA = \int_{-4}^{4} \int_{-\sqrt{16-x^2}}^{\sqrt{16-x^2}} (2y + 2y)\, dy\, dx = \int_{-4}^{4} \left[2y^2 \right]_{-\sqrt{16-x^2}}^{\sqrt{16-x^2}}\, dx = 0$

15. Because $\dfrac{\partial M}{\partial y} = -2e^x \sin 2y = \dfrac{\partial N}{\partial x}$ you have

$$\int_R \int \left(\frac{\partial N}{\partial x} - \frac{\partial M}{\partial y} \right) dA = 0.$$

17. By Green's Theorem,

$$\int_C \cos y\, dx + (xy - x \sin y)\, dy = \int_R \int (y - \sin y + \sin y)\, dA = \int_0^1 \int_x^{\sqrt{x}} y\, dy\, dx = \int_0^1 \left[\frac{y^2}{2} \right]_x^{\sqrt{x}}\, dx$$

$$= \int_0^1 \left(\frac{x}{2} - \frac{x^2}{2} \right) dx = \left[\frac{x^2}{4} - \frac{x^3}{6} \right]_0^1 = \frac{1}{4} - \frac{1}{6} = \frac{1}{12}$$

19. By Green's Theorem,

$$\int_C (x - 3y)\, dx + (x + y)\, dy = \int_R \int (1 + 3)\, dA = 4[\text{Area Large Circle} - \text{Area Small Circle}] = 4[9\pi - \pi] = 32\pi$$

21. $\mathbf{F}(x, y) = xy\mathbf{i} + (x + y)\mathbf{j}$

$C: x^2 + y^2 = 1$

$$\text{Work} = \int_C xy\, dx + (x + y)\, dy = \int_R \int (1 - x)\, dA = \int_0^{2\pi} \int_0^1 (1 - r \cos \theta)\, r\, dr\, d\theta$$

$$= \int_0^{2\pi} \left[\frac{r^2}{2} - \frac{r^2}{2} \cos \theta \right]_0^1 d\theta = \int_0^{2\pi} \frac{1}{2}(1 - \cos \theta)\, d\theta = \left[\frac{1}{2}\theta - \frac{1}{2} \sin \theta \right]_0^{2\pi} = \pi$$

23. $\mathbf{F}(x, y) = (x^{3/2} - 3y)\mathbf{i} + (6x + 5\sqrt{y})\mathbf{j}$

$C:$ boundary of the triangle with vertices $(0, 0), (5, 0), (0, 5)$

$$\text{Work} = \int_C (x^{3/2} - 3y)\, dx + (6x + 5\sqrt{y})\, dy = \int_R \int 9\, dA = 9\left(\tfrac{1}{2}\right)(5)(5) = \frac{225}{2}$$

25. C: let $x = a \cos t$, $y = a \sin t$, $0 \leq t \leq 2\pi$. By Theorem 15.9, you have

$$A = \frac{1}{2}\int_C x\,dy - y\,dx = \frac{1}{2}\int_0^{2\pi}\left[a\cos t(a\cos t) - a\sin t(-a\sin t)\right]dt = \frac{1}{2}\int_0^{2\pi} a^2\,dt = \left[\frac{a^2}{2}t\right]_0^{2\pi} = \pi a^2.$$

27. C_1: $y = x^2 + 1$, $dy = 2x\,dx$

C_2: $y = 5x - 3$, $dy = 5\,dx$

So, by Theorem 15.9 you have

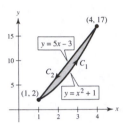

$$A = \frac{1}{2}\int_1^4\left(x(2x) - \left(x^2 + 1\right)\right)dx + \frac{1}{2}\int_4^1\left(x(5) - (5x - 3)\right)dx$$

$$= \frac{1}{2}\left[\frac{x^3}{3} - x\right]_1^4 + \frac{1}{2}[3x]_4^1 = \frac{1}{2}[18] + \frac{1}{2}[-9] = \frac{9}{2}.$$

29. See Theorem 15.8, page 1075.

31. For the moment about the x-axis, $M_x = \int_R \int y\,dA$. Let $N = 0$ and $M = -y^2/2$. By Green's Theorem,

$$M_x = \int_C -\frac{y^2}{2}\,dx = -\frac{1}{2}\int_C y^2\,dx \text{ and } \bar{y} = \frac{M_x}{2A} = -\frac{1}{2A}\int_C y^2\,dx.$$

For the moment about the y-axis, $M_y = \int_R \int x\,dA$. Let $N = x^2/2$ and $M = 0$. By Green's Theorem,

$$M_y = \int_C \frac{x^2}{2}\,dy = \frac{1}{2}\int_C x^2\,dy \text{ and } \bar{x} = \frac{M_y}{2A} = \frac{1}{2A}\int_C x^2\,dy.$$

33. $A = \int_{-2}^2 \left(4 - x^2\right)dx = \left[4x - \frac{x^3}{3}\right]_{-2}^2 = \frac{32}{3}$

$$\bar{x} = \frac{1}{2A}\int_{C_1} x^2\,dy + \frac{1}{2A}\int_{C_2} x^2\,dy$$

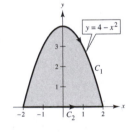

For C_1, $dy = -2x\,dx$ and for C_2, $dy = 0$. So, $\bar{x} = \frac{1}{2(32/3)}\int_2^{-2} x^2(-2x\,dx) = \left[\frac{3}{64}\left(-\frac{x^4}{2}\right)\right]_2^{-2} = 0.$

To calculate $\bar{y}$, note that $y = 0$ along C_2. So,

$$\bar{y} = \frac{-1}{2(32/3)}\int_2^{-2}\left(4 - x^2\right)^2 dx = \frac{3}{64}\int_{-2}^2\left(16 - 8x^2 + x^4\right)dx = \frac{3}{64}\left[16x - \frac{8x^3}{3} + \frac{x^5}{5}\right]_{-2}^2 = \frac{8}{5}.$$

$$(\bar{x}, \bar{y}) = \left(0, \frac{8}{5}\right)$$

35. Because $A = \int_0^1\left(x - x^3\right)dx = \left[\frac{x^2}{2} - \frac{x^4}{4}\right]_0^1 = \frac{1}{4}$, you have $\frac{1}{2A} = 2$. On C_1 you have $y = x^3$, $dy = 3x^2\,dx$ and on C_2 you

have $y = x$, $dy = dx$. So,

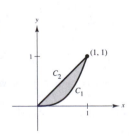

$$\bar{x} = 2\int_C x^2\,dy = 2\int_{C_1} x^2\left(3x^2\,dx\right) + 2\int_{C_2} x^2\,dx = 6\int_0^1 x^4\,dx + 2\int_1^0 x^2\,dx = \frac{6}{5} - \frac{2}{3} = \frac{8}{15}$$

$$\bar{y} = -2\int_C y^2\,dx = -2\int_0^1 x^6\,dx - 2\int_1^0 x^2\,dx = -\frac{2}{7} + \frac{2}{3} = \frac{8}{21}.$$

$$(\bar{x}, \bar{y}) = \left(\frac{8}{15}, \frac{8}{21}\right)$$

37. $A = \dfrac{1}{2}\displaystyle\int_0^{2\pi} a^2(1 - \cos\theta)^2\, d\theta = \dfrac{a^2}{2}\displaystyle\int_0^{2\pi}\left(1 - 2\cos\theta + \dfrac{1}{2} + \dfrac{\cos 2\theta}{2}\right) d\theta = \dfrac{a^2}{2}\left[\dfrac{3\theta}{2} - 2\sin\theta + \dfrac{1}{4}\sin 2\theta\right]_0^{2\pi} = \dfrac{a^2}{2}(3\pi) = \dfrac{3\pi a^2}{2}$

39. In this case the inner loop has domain $\dfrac{2\pi}{3} \le \theta \le \dfrac{4\pi}{3}$. So,

$A = \dfrac{1}{2}\displaystyle\int_{2\pi/3}^{4\pi/3}\left(1 + 4\cos\theta + 4\cos^2\theta\right) d\theta = \dfrac{1}{2}\displaystyle\int_{2\pi/3}^{4\pi/3}\left(3 + 4\cos\theta + 2\cos 2\theta\right) d\theta = \dfrac{1}{2}[3\theta + 4\sin\theta + \sin 2\theta]_{2\pi/3}^{4\pi/3} = \pi - \dfrac{3\sqrt{3}}{2}.$

41. (a) $\displaystyle\int_{C_1} y^3\, dx + \left(27x - x^3\right) dy = \displaystyle\int_R \int \left[(27 - 3x^2) - 3y^2\right] dA$

$\qquad\qquad = \displaystyle\int_0^{2\pi}\int_0^1 (27 - 3r^2)\, r\, dr\, d\theta = \displaystyle\int_0^{2\pi}\left[\dfrac{27r^2}{2} - \dfrac{3r^4}{4}\right]_0^1 d\theta = \displaystyle\int_0^{2\pi}\dfrac{51}{4}\, d\theta = \dfrac{51}{2}\pi$

   (b) You want to find c such that $\displaystyle\int_0^C (27 - 3r^2)r\, dr\, d\theta$ is a maximum:

$\qquad f(c) = \dfrac{27c^2}{2} - \dfrac{3}{4}c^4$

$\qquad f'(c) = 27c - 3c^2 \Rightarrow c = 3$

$\qquad$ Maximum Value: $\displaystyle\int_0^{2\pi}\int_0^3 (27 - 3r^2)r\, dr\, d\theta = \dfrac{243\pi}{2}$

43. $\displaystyle\int_C \left(e^{-x^2/2} - y\right) dx + \left(e^{-y^2/2} + x\right) dy = \displaystyle\int_R\int \left(\dfrac{\partial N}{\partial x} - \dfrac{\partial M}{\partial y}\right) dA$

$\qquad\qquad = \displaystyle\int_R\int (1 - (-1))\, dA$

$\qquad\qquad = 2(\text{area of } R)$

$\qquad\qquad = 2(\pi r^2 - \pi ab)$

$\qquad\qquad = 2(\pi(s^2) - \pi(2)(1))$

$\qquad\qquad = 46\pi$

45. $I = \displaystyle\int_C \dfrac{y\, dx - x\, dy}{x^2 + y^2}$

   (a) Let $\mathbf{F} = \dfrac{y}{x^2 + y^2}\mathbf{i} - \dfrac{x}{x^2 + y^2}\mathbf{j}.$

$\qquad$ $\mathbf{F}$ is conservative because $\dfrac{\partial N}{\partial x} = \dfrac{\partial M}{\partial y} = \dfrac{x^2 - y^2}{\left(x^2 + y^2\right)^2}.$

$\qquad$ $\mathbf{F}$ is defined and has continuous first partials everywhere except at the origin. If C is a circle (a closed path) that does not contain the origin, then

$\qquad$ $\displaystyle\int_C \mathbf{F} \cdot d\mathbf{r} = \displaystyle\int_C M\, dx + N\, dy = \displaystyle\int_R\int \left(\dfrac{\partial N}{\partial x} - \dfrac{\partial M}{\partial y}\right) dA = 0.$

(b) Let $\mathbf{r} = a \cos t\mathbf{i} - a \sin t\mathbf{j}$, $0 \le t \le 2\pi$ be a circle C_1 oriented clockwise inside C (see figure). Introduce line segments C_2 and C_3 as illustrated in Example 6 of this section in the text. For the region inside C and outside C_1, Green's Theorem applies. Note that since C_2 and C_3 have opposite orientations, the line integrals over them cancel. So, $C_4 = C_1 + C_2 + C + C_3$ and

$$\int_{C_4} \mathbf{F} \cdot d\mathbf{r} = \int_{C_1} \mathbf{F} \cdot d\mathbf{r} + \int_C \mathbf{F} \cdot d\mathbf{r} = 0.$$

But,

$$\int_{C_1} \mathbf{F} \cdot d\mathbf{r} = \int_0^{2\pi} \left[\frac{(-a \sin t)(-a \sin t)}{a^2 \cos^2 t + a^2 \sin^2 t} + \frac{(-a \cos t)(-a \cos t)}{a^2 \cos^2 t + a^2 \sin^2 t} \right] dt$$

$$= \int_0^{2\pi} \left(\sin^2 t + \cos^2 t \right) dt = \left[t \right]_0^{2\pi} = 2\pi.$$

Finally, $\int_C \mathbf{F} \cdot d\mathbf{r} = -\int_{C_1} \mathbf{F} \cdot d\mathbf{r} = -2\pi.$

Note: If C were oriented clockwise, then the answer would have been 2π.

47. (a) Let C be the line segment joining (x_1, y_1) and (x_2, y_2).

$$y = \frac{y_2 - y_1}{x_2 - x_1}(x - x_1) + y_1$$

$$dy = \frac{y_2 - y_1}{x_2 - x_1} dx$$

$$\int_C -y\, dx + x\, dy = \int_{x_1}^{x_2} \left[-\frac{y_2 - y_1}{x_2 - x_1}(x - x_1) - y_1 + x\left(\frac{y_2 - y_1}{x_2 - x_1} \right) \right] dx = \int_{x_1}^{x_2} \left[x_1\left(\frac{y_2 - y_1}{x_2 - x_1} \right) - y_1 \right] dx$$

$$= \left[\left[x_1\left(\frac{y_2 - y_1}{x_2 - x_1} \right) - y_1 \right] x \right]_{x_1}^{x_2} = \left[x_1\left(\frac{y_2 - y_1}{x_2 - x_1} \right) - y_1 \right](x_2 - x_1) = x_1(y_2 - y_1) - y_1(x_2 - x_1) = x_1 y_2 - x_2 y_1$$

(b) Let C be the boundary of the region $A = \frac{1}{2}\int_C -y\, dx + x\, dy = \frac{1}{2}\int_R \int (1 - (-1))\, dA = \int_R \int dA.$

So,

$$\int_R \int dA = \frac{1}{2}\left[\int_{C_1} -y\, dx + x\, dy + \int_{C_2} -y\, dx + x\, dy + \cdots + \int_{C_n} -y\, dx + x\, dy \right]$$

where C_1 is the line segment joining (x_1, y_1) and (x_2, y_2), C_2 is the line segment joining (x_2, y_2) and (x_3, y_3), $\cdots$, and C_n is the line segment joining (x_n, y_n) and (x_1, y_1). So,

$$\int_R \int dA = \frac{1}{2}\left[(x_1 y_2 - x_2 y_1) + (x_2 y_3 - x_3 y_2) + \cdots + (x_{n-1} y_n - x_n y_{n-1}) + (x_n y_1 - x_1 y_n) \right].$$

49. Because $\int_C \mathbf{F} \cdot \mathbf{N}\, ds = \int_R \int \operatorname{div} \mathbf{F}\, dA$, then

$$\int_C f D_{\mathbf{N}} g\, ds = \int_C f \nabla g \cdot \mathbf{N}\, ds = \int_R \int \operatorname{div}(f \nabla g)\, dA = \int_R \int \left(f \operatorname{div}(\nabla g) + \nabla f \cdot \nabla g \right) dA = \int_R \int \left(f \nabla^2 g + \nabla f \cdot \nabla g \right) dA.$$

51. $\mathbf{F} = M\mathbf{i} + N\mathbf{j}$

$$\frac{\partial N}{\partial x} = \frac{\partial M}{\partial y} \Rightarrow \frac{\partial N}{\partial x} - \frac{\partial M}{\partial y} = 0$$

$$\int_C \mathbf{F} \cdot d\mathbf{r} = \int_C M\, dx + N\, dy = \int_R \int \left(\frac{\partial N}{\partial x} - \frac{\partial M}{\partial y} \right) dA = \int_R \int (0)\, dA = 0$$

Section 15.5   Parametric Surfaces

1. $\mathbf{r}(u, v) = u\mathbf{i} + v\mathbf{j} + uv\mathbf{k}$

$z = xy$

Matches (e)

2. $\mathbf{r}(u, v) = u \cos v\mathbf{i} + u \sin v\mathbf{j} + u\mathbf{k}$

$x^2 + y^2 = z^2$, cone

Matches (f)

3. $\mathbf{r}(u, v) = u\mathbf{i} + \frac{1}{2}(u + v)\mathbf{j} + v\mathbf{k}$

$2y = x + z$, plane

Matches (b)

4. $\mathbf{r}(u, v) = u\mathbf{i} + \frac{1}{4}v^3\mathbf{j} + v\mathbf{k}$

$4y = z^3$, cylinder

Matches (a)

5. $\mathbf{r}(u, v) = 2 \cos v \cos u\mathbf{i} + 2 \cos v \sin u\mathbf{j} + 2 \sin v\mathbf{k}$

$x^2 + y^2 + z^2 = 4\cos^2 v \cos^2 u + 4 \cos^2 v \sin^2 u + 4 \sin^2 v = 4 \cos^2 v + 4 \sin^2 v = 4$, sphere

Matches (d)

6. $\mathbf{r}(u, v) = 4 \cos u\mathbf{i} + 4 \sin u\mathbf{j} + v\mathbf{k}$

$x^2 + y^2 = 4$, circular cylinder

Matches (c)

7. $\mathbf{r}(u, v) = u\mathbf{i} + v\mathbf{j} + \frac{v}{2}\mathbf{k}$

$y - 2z = 0$

Plane

9. $\mathbf{r}(u, v) = 2 \cos u\mathbf{i} + v\mathbf{j} + 2 \sin u\mathbf{k}$

$x^2 + z^2 = 4$

Cylinder

11. $\mathbf{r}(u, v) = 2u \cos v\mathbf{i} + 2u \sin v\mathbf{j} + u^4\mathbf{k}$,

$0 \le u \le 1, \ 0 \le v \le 2\pi$

$z = \dfrac{(x^2 + y^2)^2}{16}$

13. $\mathbf{r}(u, v) = 2 \sinh u \cos v\mathbf{i} + \sinh u \sin v\mathbf{j} + \cosh u\mathbf{k}$,

$0 \le u \le 2, \ 0 \le v \le 2\pi$

$\dfrac{z^2}{1} - \dfrac{x^2}{4} - \dfrac{y^2}{1} = 1$

15. $\mathbf{r}(u, v) = (u - \sin u) \cos v\mathbf{i} + (1 - \cos u) \sin v\mathbf{j} + u\mathbf{k}$,

$0 \le u \le \pi, \ 0 \le v \le 2\pi$

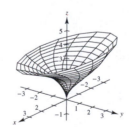

For Exercises 17–19, $\mathbf{r}(u, v) = u\cos v\mathbf{i} + u\sin v\mathbf{j} + u^2\mathbf{k}$, $0 \le u \le 2, 0 \le v \le 2\pi$.

Eliminating the parameter yields $z = x^2 + y^2, 0 \le z \le 4$.

17. $\mathbf{s}(u, v) = u\cos v\mathbf{i} + u\sin v\mathbf{j} - u^2\mathbf{k}$, $0 \le u \le 2$, $0 \le v \le 2\pi$

$z = -(x^2 + y^2)$

The paraboloid is reflected (inverted) through the xy-plane.

19. $\mathbf{s}(u, v) = u\cos v\mathbf{i} + u\sin v\mathbf{j} + u^2\mathbf{k}$, $0 \le u \le 3$, $0 \le v \le 2\pi$

The height of the paraboloid is increased from 4 to 9.

21. $z = y$

$\mathbf{r}(u, v) = u\mathbf{i} + v\mathbf{j} + v\mathbf{k}$

23. $y = \sqrt{4x^2 + 9z^2}$

$\mathbf{r}(x, y) = x\mathbf{i} + \sqrt{4x^2 + 9z^2}\mathbf{j} + z\mathbf{k}$

or,

$\mathbf{r}(u, v) = \frac{1}{2}u\cos v\mathbf{i} + u\mathbf{j} + \frac{1}{3}u\sin v\mathbf{k}$,

$u \ge 0$, $0 \le v \le 2\pi$

25. $x^2 + y^2 = 25$

$\mathbf{r}(u, v) = 5\cos u\mathbf{i} + 5\sin u\mathbf{j} + v\mathbf{k}$

27. $z = x^2$

$\mathbf{r}(u, v) = u\mathbf{i} + v\mathbf{j} + u^2\mathbf{k}$

29. $z = 4$ inside $x^2 + y^2 = 9$.

$\mathbf{r}(u, v) = v\cos u\mathbf{i} + v\sin u\mathbf{j} + 4\mathbf{k}$, $0 \le v \le 3$

31. Function: $y = \dfrac{x}{2}$, $0 \le x \le 6$

Axis of revolution: x-axis

$x = u, y = \dfrac{u}{2}\cos v, z = \dfrac{u}{2}\sin v$

$0 \le u \le 6$, $0 \le v \le 2\pi$

33. Function: $x = \sin z$, $0 \le z \le \pi$

Axis of revolution: z-axis

$x = \sin u\cos v, y = \sin u\sin v, z = u$

$0 \le u \le \pi$, $0 \le v \le 2\pi$

35. $\mathbf{r}(u, v) = (u + v)\mathbf{i} + (u - v)\mathbf{j} + v\mathbf{k}, (1, -1, 1)$

$\mathbf{r}_u(u, v) = \mathbf{i} + \mathbf{j}, \mathbf{r}_v(u, v) = \mathbf{i} - \mathbf{j} + \mathbf{k}$

At $(1, -1, 1)$, $u = 0$ and $v = 1$.

$\mathbf{r}_u(0, 1) = \mathbf{i} + \mathbf{j}, \mathbf{r}_v(0, 1) = \mathbf{i} - \mathbf{j} + \mathbf{k}$

$$\mathbf{N} = \mathbf{r}_u(0, 1) \times \mathbf{r}_v(0, 1) = \begin{vmatrix} \mathbf{i} & \mathbf{j} & \mathbf{k} \\ 1 & 1 & 0 \\ 1 & -1 & 1 \end{vmatrix} = \mathbf{i} - \mathbf{j} - 2\mathbf{k}$$

Tangent plane: $(x - 1) - (y + 1) - 2(z - 1) = 0$

$x - y - 2z = 0$

(The original plane!)

37. $\mathbf{r}(u, v) = 2u\cos v\mathbf{i} + 3u\sin v\mathbf{j} + u^2\mathbf{k}$, $(0, 6, 4)$

$\mathbf{r}_u(u, v) = 2\cos v\mathbf{i} + 3\sin v\mathbf{j} + 2u\mathbf{k}$

$\mathbf{r}_v(u, v) = -2u\sin v\mathbf{i} + 3u\cos v\mathbf{j}$

At $(0, 6, 4)$, $u = 2$ and $v = \pi/2$.

$\mathbf{r}_u\left(2, \dfrac{\pi}{2}\right) = 3\mathbf{j} + 4\mathbf{k}, \mathbf{r}_v\left(2, \dfrac{\pi}{2}\right) = -4\mathbf{i}$

$$\mathbf{N} = \mathbf{r}_u\left(2, \dfrac{\pi}{2}\right) \times \mathbf{r}_v\left(2, \dfrac{\pi}{2}\right) = \begin{vmatrix} \mathbf{i} & \mathbf{j} & \mathbf{k} \\ 0 & 3 & 4 \\ -4 & 0 & 0 \end{vmatrix} = -16\mathbf{j} + 12\mathbf{k}$$

Direction numbers: $0, 4, -3$

Tangent plane: $4(y - 6) - 3(z - 4) = 0$

$4y - 3z = 12$

39. $\mathbf{r}(u, v) = 4u\mathbf{i} - v\mathbf{j} + v\mathbf{k}, \quad 0 \le u \le 2, 0 \le v \le 1$

$\mathbf{r}_u(u, v) = 4\mathbf{i}, \mathbf{r}_v(u, v) = -\mathbf{j} + \mathbf{k}$

$\mathbf{r}_u \times \mathbf{r}_v = \begin{vmatrix} \mathbf{i} & \mathbf{j} & \mathbf{k} \\ 4 & 0 & 0 \\ 0 & -1 & 1 \end{vmatrix} = -4\mathbf{j} - 4\mathbf{k}$

$\|\mathbf{r}_u \times \mathbf{r}_v\| = \sqrt{16 + 16} = 4\sqrt{2}$

$A = \int_0^1 \int_0^2 4\sqrt{2} \, du \, dv = 4\sqrt{2}(2)(1) = 8\sqrt{2}$

41. $\mathbf{r}(u, v) = a \cos u\mathbf{i} + a \sin u\mathbf{j} + v\mathbf{k}, \quad 0 \le u \le 2\pi, 0 \le v \le b$

$\mathbf{r}_u(u, v) = -a \sin u\mathbf{i} + a \cos u\mathbf{j}$

$\mathbf{r}_v(u, v) = \mathbf{k}$

$\mathbf{r}_u \times \mathbf{r}_v = \begin{vmatrix} \mathbf{i} & \mathbf{j} & \mathbf{k} \\ -a \sin u & a \cos u & 0 \\ 0 & 0 & 1 \end{vmatrix} = a \cos u\mathbf{i} + a \sin u\mathbf{j}$

$\|\mathbf{r}_u \times \mathbf{r}_v\| = a$

$A = \int_0^b \int_0^{2\pi} a \, du \, dv = 2\pi ab$

43. $\mathbf{r}(u, v) = au \cos v\mathbf{i} + au \sin v\mathbf{j} + u\mathbf{k}, \quad 0 \le u \le b, \quad 0 \le v \le 2\pi$

$\mathbf{r}_u(u, v) = a \cos v\mathbf{i} + a \sin v\mathbf{j} + \mathbf{k}$

$\mathbf{r}_v(u, v) = -au \sin v\mathbf{i} + au \cos v\mathbf{j}$

$\mathbf{r}_u \times \mathbf{r}_v = \begin{vmatrix} \mathbf{i} & \mathbf{j} & \mathbf{k} \\ a \cos v & a \sin v & 1 \\ -au \sin v & au \cos v & 0 \end{vmatrix} = -au \cos v\mathbf{i} - au \sin v\mathbf{j} + a^2 u\mathbf{k}$

$\|\mathbf{r}_u \times \mathbf{r}_v\| = au\sqrt{1 + a^2}$

$A = \int_0^{2\pi} \int_0^b a\sqrt{1 + a^2} \, u \, du \, dv = \pi ab^2 \sqrt{1 + a^2}$

45. $\mathbf{r}(u, v) = \sqrt{u} \cos v\mathbf{i} + \sqrt{u} \sin v\mathbf{j} + u\mathbf{k}, 0 \le u \le 4, 0 \le v \le 2\pi$

$\mathbf{r}_u(u, v) = \dfrac{\cos v}{2\sqrt{u}}\mathbf{i} + \dfrac{\sin v}{2\sqrt{u}}\mathbf{j} + \mathbf{k}$

$\mathbf{r}_v(u, v) = -\sqrt{u} \sin v\mathbf{i} + \sqrt{u} \cos v\mathbf{j}$

$\mathbf{r}_u \times \mathbf{r}_v = \begin{vmatrix} \mathbf{i} & \mathbf{j} & \mathbf{k} \\ \dfrac{\cos v}{2\sqrt{u}} & \dfrac{\sin v}{2\sqrt{u}} & 1 \\ -\sqrt{u} \sin v & \sqrt{u} \cos v & 0 \end{vmatrix} = -\sqrt{u} \cos v\mathbf{i} - \sqrt{u} \sin v\mathbf{j} + \dfrac{1}{2}\mathbf{k}$

$\|\mathbf{r}_u \times \mathbf{r}_v\| = \sqrt{u + \dfrac{1}{4}}$

$A = \int_0^{2\pi} \int_0^4 \sqrt{u + \dfrac{1}{4}} \, du \, dv = \dfrac{\pi}{6}\left(17\sqrt{17} - 1\right) \approx 36.177$

47. See the definition, page 1084.

49. Function: $z = x$

Axis of revolution: z-axis

$x = u \cos v,\ y = u \sin v,\ z = u$

$\mathbf{r}(u, v) = u \cos v\mathbf{i} + u \sin v\mathbf{j} + u\mathbf{k}$

$u \le 0, \quad 0 \le v \le 2\pi$

51. $\mathbf{r}(u, v) = a \sin^3 u \cos^3 v\mathbf{i} + a \sin^3 u \sin^3 v\mathbf{j} + a \cos^3 u\mathbf{k}$

$0 \le u \le \pi, \quad 0 \le v \le 2\pi$

$x = a \sin^3 u \cos^3 v \Rightarrow x^{2/3} = a^{2/3} \sin^2 u \cos^2 v$

$y = a \sin^3 u \sin^3 v \Rightarrow y^{2/3} = a^{2/3} \sin^2 u \sin^2 v$

$z = a \cos^3 u \Rightarrow z^{2/3} = a^{2/3} \cos^2 u$

$x^{2/3} + y^{2/3} + z^{2/3} = a^{2/3}\left[\sin^2 u \cos^2 v + \sin^2 u \sin^2 v + \cos^2 u\right] = a^{2/3}\left[\sin^2 u + \cos^2 u\right] = a^{2/3}$

53. (a) $\mathbf{r}(u, v) = (4 + \cos v) \cos u\mathbf{i} +$
$(4 + \cos v) \sin u\mathbf{j} + \sin v\mathbf{k},$
$0 \le u \le 2\pi, 0 \le v \le 2\pi$

(b) $\mathbf{r}(u, v) = (4 + 2 \cos v) \cos u\mathbf{i} +$
$(4 + 2 \cos v) \sin u\mathbf{j} + 2 \sin v\mathbf{k},$
$0 \le u \le 2\pi, 0 \le v \le 2\pi$

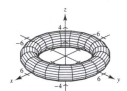

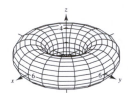

(c) $\mathbf{r}(u, v) = (8 + \cos v) \cos u\mathbf{i} +$
$(8 + \cos v) \sin u\mathbf{j} + \sin v\mathbf{k},$
$0 \le u \le 2\pi, 0 \le v \le 2\pi$

(d) $\mathbf{r}(u, v) = (8 + 3 \cos v) \cos u\mathbf{i} +$
$(8 + 3 \cos v) \sin u\mathbf{j} + 3 \sin v\mathbf{k},$
$0 \le u \le 2\pi, 0 \le v \le 2\pi$

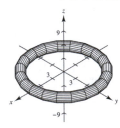

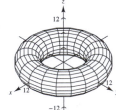

The radius of the generating circle that is revolved about the z-axis is b, and its center is a units from the axis of revolution.

55. $\mathbf{r}(u, v) = 20 \sin u \cos v\mathbf{i} + 20 \sin u \sin v\mathbf{j} + 20 \cos u\mathbf{k},\ 0 \le u \le \pi/3, \quad 0 \le v \le 2\pi$

$\mathbf{r}_u = 20 \cos u \cos v\mathbf{i} + 20 \cos u \sin v\mathbf{j} - 20 \sin u\mathbf{k}$

$\mathbf{r}_v = -20 \sin u \sin v\mathbf{i} + 20 \sin u \cos v\mathbf{j}$

$$\mathbf{r}_u \times \mathbf{r}_v = \begin{vmatrix} \mathbf{i} & \mathbf{j} & \mathbf{k} \\ 20 \cos u \cos v & 20 \cos u \sin v & -20 \sin u \\ -20 \sin u \sin v & 20 \sin u \cos v & 0 \end{vmatrix}$$

$= 400 \sin^2 u \cos v\mathbf{i} + 400 \sin^2 u \sin v\mathbf{j} + 400(\cos u \sin u \cos^2 v + \cos u \sin u \sin^2 v)\mathbf{k}$

$= 400\left[\sin^2 u \cos v\mathbf{i} + \sin^2 u \sin v\mathbf{j} + \cos u \sin u\mathbf{k}\right]$

$\|\mathbf{r}_u \times \mathbf{r}_v\| = 400\sqrt{\sin^4 u \cos^2 v + \sin^4 u \sin^2 v + \cos^2 u \sin^2 u} = 400\sqrt{\sin^4 u + \cos^2 u \sin^2 u} = 400\sqrt{\sin^2 u} = 400 \sin u$

$S = \int_S \int dS = \int_0^{2\pi} \int_0^{\pi/3} 400 \sin u\ du\ dv = \int_0^{2\pi} \left[-400 \cos u\right]_0^{\pi/3} dv = \int_0^{2\pi} 200\ dv = 400\pi\ \text{m}^2$

57. $\mathbf{r}(u, v) = u \cos v\mathbf{i} + u \sin v\mathbf{j} + 2v\mathbf{k}, \quad 0 \le u \le 3, 0 \le v \le 2\pi$

$\mathbf{r}_u(u, v) = \cos v\mathbf{i} + \sin v\mathbf{j}$

$\mathbf{r}_v(u, v) = -u \sin v\mathbf{i} + u \cos v\mathbf{j} + 2\mathbf{k}$

$\mathbf{r}_u \times \mathbf{r}_v = \begin{vmatrix} \mathbf{i} & \mathbf{j} & \mathbf{k} \\ \cos v & \sin v & 0 \\ -u \sin v & u \cos v & 2 \end{vmatrix} = 2 \sin v\mathbf{i} - 2 \cos v\mathbf{j} + u\mathbf{k}$

$\|\mathbf{r}_u \times \mathbf{r}_v\| = \sqrt{4 + u^2}$

$A = \int_0^{2\pi} \int_0^3 \sqrt{4 + u^2} \; du \; dv = \pi\left[3\sqrt{13} + 4\ln\left(\dfrac{3 + \sqrt{13}}{2}\right)\right]$

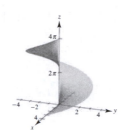

59. Answers will vary.

Section 15.6   Surface Integrals

1. $S: z = 4 - x, \quad 0 \le x \le 4, \quad 0 \le y \le 3, \quad \dfrac{\partial z}{\partial x} = -1, \quad \dfrac{\partial z}{\partial y} = 0$

$\displaystyle\int_S\!\!\int (x - 2y + z)\, dS = \int_0^4 \int_0^3 (x - 2y + 4 - x)\sqrt{1 + (-1)^2 + 0^2}\; dy\, dx = \sqrt{2}\int_0^4 \int_0^3 (4 - 2y)\, dy\, dx = \sqrt{2}\int_0^4 3\, dx = 12\sqrt{2}$

3. $S: z = 2, \quad x^2 + y^2 \le 1, \dfrac{\partial z}{\partial x} = \dfrac{\partial z}{\partial y} = 0$

$\displaystyle\int_S\!\!\int (x - 2y + z)\, dS = \int_{-1}^{1} \int_{-\sqrt{1-x^2}}^{\sqrt{1-x^2}} (x - 2y + 2)\sqrt{1 + 0^2 + 0^2}\; dy\, dx = \int_0^{2\pi} \int_0^1 (r \cos\theta - 2r\sin\theta + 2)\, r\, dr\, d\theta$

$= \int_0^{2\pi}\left[\dfrac{1}{3}\cos\theta - \dfrac{2}{3}\sin\theta + 1\right]d\theta = \left[\dfrac{1}{3}\sin\theta + \dfrac{2}{3}\cos\theta + \theta\right]_0^{2\pi} = \dfrac{2}{3} + 2\pi - \dfrac{2}{3} = 2\pi$

5. $S: z = 3 - x - y \quad \text{(first octant)}, \quad \dfrac{\partial z}{\partial x} = -1, \quad \dfrac{\partial z}{\partial y} = -1$

$\displaystyle\int_S\!\!\int xy\, dS = \int_0^3 \int_0^{3-x} xy\sqrt{1 + (-1)^2 + (-1)^2}\; dy\, dx = \sqrt{3}\int_0^3\left[x\dfrac{y^2}{2}\right]_0^{3-x}$

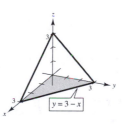

$y = 3 - x$

$= \dfrac{\sqrt{3}}{2}\int_0^3 x(3 - x)^2\, dx = \dfrac{\sqrt{3}}{2}\left[\dfrac{x^4}{4} - 2x^3 + \dfrac{9x^2}{2}\right]_0^3 = \dfrac{\sqrt{3}}{2}\left[\dfrac{27}{4}\right] = \dfrac{27\sqrt{3}}{8}$

7. $S: z = 9 - x^2, \quad 0 \le x \le 2, \quad 0 \le y \le x,$

$\dfrac{\partial z}{\partial x} = -2x, \dfrac{\partial z}{\partial y} = 0$

$\displaystyle\int_S\!\!\int xy\, dS = \int_0^2 \int_y^2 xy\sqrt{1 + 4x^2}\; dx\, dy = \dfrac{391\sqrt{17} + 1}{240}$

9. $S: z = 10 - x^2 - y^2, \quad 0 \le x \le 2, \quad 0 \le y \le 2$

$\displaystyle\int_S\!\!\int (x^2 - 2xy)\, dS = \int_0^2 \int_0^2 (x^2 - 2xy)\sqrt{1 + 4x^2 + 4y^2}\; dy\, dx \approx -11.47$

11. $S: 2x + 3y + 6z = 12$ (first octant) $\Rightarrow z = 2 - \frac{1}{3}x - \frac{1}{2}y$

$\rho(x, y, z) = x^2 + y^2$

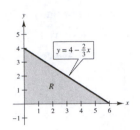

$m = \int_R \int (x^2 + y^2)\sqrt{1 + \left(-\frac{1}{3}\right)^2 + \left(-\frac{1}{2}\right)^2}\, dA = \frac{7}{6}\int_0^6 \int_0^{4-(2x/3)} (x^2 + y^2)\, dy\, dx$

$= \frac{7}{6}\int_0^6 \left[x^2\left(4 - \frac{2}{3}x\right) + \frac{1}{3}\left(4 - \frac{2}{3}x\right)^3\right] dx = \frac{7}{6}\left[\frac{4}{3}x^3 - \frac{1}{6}x^4 - \frac{1}{8}\left(4 - \frac{2}{3}x\right)^4\right]_0^6 = \frac{364}{3}$

13. $S: \mathbf{r}(u, v) = u\mathbf{i} + v\mathbf{j} + 2v\mathbf{k}, \quad 0 \le u \le 1, \quad 0 \le v \le 2$

$\mathbf{r}_u = \mathbf{i}, \quad \mathbf{r}_v = \mathbf{j} + 2\mathbf{k}$

$\mathbf{r}_u \times \mathbf{r}_v = \begin{vmatrix} \mathbf{i} & \mathbf{j} & \mathbf{k} \\ 1 & 0 & 0 \\ 0 & 1 & 2 \end{vmatrix} = -2\mathbf{j} + \mathbf{k}$

$\|\mathbf{r}_u \times \mathbf{r}_v\| = \sqrt{5}$

$\int_S \int (y + 5)\, dS = \int_0^2 \int_0^1 (v + 5)\sqrt{5}\, du\, dv = \int_0^2 (v + 5)\sqrt{5}\, dv = \sqrt{5}\left[\frac{v^2}{2} + 5v\right]_0^2 = 12\sqrt{5}$

15. $S: \mathbf{r}(u, v) = 2\cos u\mathbf{i} + 2\sin u\mathbf{j} + v\mathbf{k}, \quad 0 \le u \le \pi/2, \quad 0 \le v \le 1$

$\mathbf{r}_u = -2\sin u\mathbf{i} + 2\cos u\mathbf{j}, \quad \mathbf{r}_v = \mathbf{k}$

$\mathbf{r}_u \times \mathbf{r}_v = \begin{vmatrix} \mathbf{i} & \mathbf{j} & \mathbf{k} \\ -2\sin u & 2\cos u & 0 \\ 0 & 0 & 1 \end{vmatrix} = 2\cos u\mathbf{i} + 2\sin u\mathbf{j}$

$\|\mathbf{r}_u \times \mathbf{r}_v\| = \sqrt{4\cos^2 u + 4\sin^2 u} = 2$

$\int_S \int (x + y)\, dS = \int_0^1 \int_0^{\pi/2} (2\cos u + 2\sin u)2\, du\, dv = 4\int_0^1 [\sin u - \cos u]_0^{\pi/2}\, dv = 4\int_0^1 2\, dv = 8$

17. $f(x, y, z) = x^2 + y^2 + z^2$

$S: z = x + y, \quad x^2 + y^2 \le 1, \quad \dfrac{\partial z}{\partial x} = \dfrac{\partial z}{\partial y} = 1$

$\int_S \int f(x, y, z)\, dS = \int_{-1}^1 \int_{-\sqrt{1-x^2}}^{\sqrt{1-x^2}} \left[x^2 + y^2 + (x + y)^2\right]\sqrt{1 + 1^2 + 1^2}\, dy\, dx$

$= \sqrt{3}\int_{-1}^1 \int_{-\sqrt{1-x^2}}^{\sqrt{1-x^2}} \left[2x^2 + 2y^2 + 2xy\right] dy\, dx = \sqrt{3}\int_0^{2\pi} \int_0^1 (2r^2 + 2r\cos\theta\, r\sin\theta)\, r\, dr\, d\theta$

$= 2\sqrt{3}\int_0^{2\pi} \left[\frac{r^4}{4} + \frac{r^4}{4}\cos\theta\sin\theta\right]_0^1 d\theta = \frac{\sqrt{3}}{2}\int_0^{2\pi} (1 + \cos\theta\sin\theta)\, d\theta = \frac{\sqrt{3}}{2}\left[\theta + \frac{\sin^2\theta}{2}\right]_0^{2\pi} = \sqrt{3}\pi$

19. $f(x, y, z) = \sqrt{x^2 + y^2 + z^2}$

$S: z = \sqrt{x^2 + y^2}, x^2 + y^2 \le 4$

$\int_S \int f(x, y, z)\, dS = \int_{-2}^2 \int_{-\sqrt{4-x^2}}^{\sqrt{4-x^2}} \sqrt{x^2 + y^2 + \left(\sqrt{x^2 + y^2}\right)^2}\sqrt{1 + \left(\frac{x}{\sqrt{x^2 + y^2}}\right)^2 + \left(\frac{y}{\sqrt{x^2 + y^2}}\right)^2}\, dy\, dx$

$= \sqrt{2}\int_{-2}^2 \int_{-\sqrt{4-x^2}}^{\sqrt{4-x^2}} \sqrt{x^2 + y^2}\sqrt{\frac{x^2 + y^2 + x^2 + y^2}{x^2 + y^2}}\, dy\, dx$

$= 2\int_{-2}^2 \int_{-\sqrt{4-x^2}}^{\sqrt{4-x^2}} \sqrt{x^2 + y^2}\, dy\, dx = 2\int_0^{2\pi} \int_0^2 r^2\, dr\, d\theta = 2\int_0^{2\pi} \left[\frac{r^3}{3}\right]_0^2 d\theta = \left[\frac{16}{3}\theta\right]_0^{2\pi} = \frac{32\pi}{3}$

21. $f(x, y, z) = x^2 + y^2 + z^2$

   $S: x^2 + y^2 = 9, \ 0 \le x \le 3, \ 0 \le y \le 3, \ 0 \le z \le 9$

   Project the solid onto the yz-plane; $x = \sqrt{9 - y^2}, \ 0 \le y \le 3, \ 0 \le z \le 9$.

   $$\int_S \int f(x, y, z) \, dS = \int_0^3 \int_0^9 \left[(9 - y^2) + y^2 + z^2 \right] \sqrt{1 + \left(\frac{-y}{\sqrt{9 - y^2}} \right)^2 + (0)^2} \, dz \, dy$$

   $$= \int_0^3 \int_0^9 (9 + z^2) \frac{3}{\sqrt{9 - y^2}} \, dz \, dy = \int_0^3 \left[\frac{3}{\sqrt{9 - y^2}} \left(9z + \frac{z^3}{3} \right) \right]_0^9 \, dy$$

   $$= 324 \int_0^3 \frac{3}{\sqrt{9 - y^2}} \, dy = \left[972 \arcsin\left(\frac{y}{3} \right) \right]_0^3 = 972\left(\frac{\pi}{2} - 0 \right) = 486\pi$$

23. $\mathbf{F}(x, y, z) = 3z\mathbf{i} - 4\mathbf{j} + y\mathbf{k}$

   $S: z = 1 - x - y$ (first octant)

   $G(x, y, z) = x + y + z - 1$

   $\nabla G(x, y, z) = \mathbf{i} + \mathbf{j} + \mathbf{k}$

   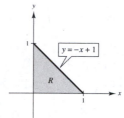

   $$\int_S \int \mathbf{F} \cdot \mathbf{N} \, dS = \int_R \int \mathbf{F} \cdot \nabla G \, dA = \int_0^1 \int_0^{1-x} (3z - 4 + y) \, dy \, dx$$

   $$= \int_0^1 \int_0^{1-x} \left[3(1 - x - y) - 4 + y \right] \, dy \, dx$$

   $$= \int_0^1 \int_0^{1-x} (-1 - 3x - 2y) \, dy \, dx = \int_0^1 \left[-y - 3xy - y^2 \right]_0^{1-x} \, dx$$

   $$= -\int_0^1 \left[(1 - x) + 3x(1 - x) + (1 - x)^2 \right] \, dx = -\int_0^1 (2 - 2x^2) \, dx = -\frac{4}{3}$$

25. $\mathbf{F}(x, y, z) = x\mathbf{i} + y\mathbf{j} + z\mathbf{k}$

   $S: z = 1 - x^2 - y^2, \quad z \ge 0$

   $G(x, y, z) = x^2 + y^2 + z - 1$

   $\nabla G(x, y, z) = 2x\mathbf{i} + 2y\mathbf{j} + \mathbf{k}$

   $$\int_S \int \mathbf{F} \cdot \mathbf{N} \, dS = \int_R \int \mathbf{F} \cdot \nabla G \, dA$$

   $$= \int_R \int (2x^2 + 2y^2 + z) \, dA$$

   $$= \int_R \int \left(2x^2 + 2y^2 + (1 - x^2 - y^2) \right) \, dA$$

   $$= \int_R \int (1 + x^2 + y^2) \, dA$$

   $$= \int_0^{2\pi} \int_0^1 (r^2 + 1) r \, dr \, d\theta$$

   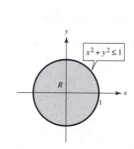

   $$= \int_0^{2\pi} \left[\frac{r^4}{4} + \frac{r^2}{2} \right]_0^1 \, d\theta = \int_0^{2\pi} \frac{3}{4} \, d\theta = \frac{3\pi}{2}$$

27. $\mathbf{F}(x, y, z) = 4\mathbf{i} - 3\mathbf{j} + 5\mathbf{k}$

$S\!: z = x^2 + y^2,\ x^2 + y^2 \le 4$

$G(x, y, z) = -x^2 - y^2 + z$

$\nabla G(x, y, z) = -2x\mathbf{i} - 2y\mathbf{j} + \mathbf{k}$

$$\int_S \int \mathbf{F} \cdot \mathbf{N}\, dS = \int_R \int \mathbf{F} \cdot \nabla G\, dA = \int_R \int (-8x + 6y + 5)\, dA$$

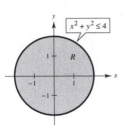

$$= \int_0^{2\pi} \int_0^2 \left[-8r\cos\theta + 6r\sin\theta + 5\right] r\, dr\, d\theta$$

$$= \int_0^{2\pi} \left[-\tfrac{8}{3}r^3 \cos\theta + 2r^3 \sin\theta + \tfrac{5}{2}r^2\right]_0^2 d\theta$$

$$= \int_0^{2\pi} \left[-\tfrac{64}{3}\cos\theta + 16\sin\theta + 10\right] d\theta$$

$$= \left[-\tfrac{64}{3}\sin\theta - 16\cos\theta + 10\theta\right]_0^{2\pi} = 20\pi$$

29. $\mathbf{F}(x, y, z) = (x + y)\mathbf{i} + y\mathbf{j} + z\mathbf{k}$

$S\!: z = 16 - x^2 - y^2,\ \ z = 0$

$G(x, y, z) = z + x^2 + y^2 - 16$

$\nabla G(x, y, z) = 2x\mathbf{i} + 2y\mathbf{j} + \mathbf{k}$

$\mathbf{F} \cdot \nabla G = 2x(x + y) + 2y^2 + z = 2x^2 + 2xy + 2y^2 + 16 - x^2 - y^2 = x^2 + y^2 + 2xy + 16$

$$\int_S \int \mathbf{F} \cdot \mathbf{N}\, dS = \int_R \int \mathbf{F} \cdot \nabla G\, dA$$

$$= \int_0^{2\pi} \int_0^4 (r^2 + 2r^2 \cos\theta \sin\theta + 16)\, r\, dr\, d\theta$$

$$= \int_0^{2\pi} \left[\frac{r^4}{4} + \frac{r^4}{2}\cos\theta \sin\theta + 8r^2\right]_0^4 d\theta = \int_0^{2\pi}[192 + 128\cos\theta\sin\theta]\,d\theta = \left[192 + 64\sin^2\theta\right]_0^{2\pi} = 384\pi$$

(The flux across the bottom $z = 0$ is 0.)

31. $\mathbf{E} = yz\mathbf{i} + xz\mathbf{j} + xy\mathbf{k}$

$S\!: z = \sqrt{1 - x^2 - y^2}$

$$\int_S \int \mathbf{E} \cdot \mathbf{N}\, dS = \int_R \int \mathbf{E} \cdot \left(-g_x(x, y)\mathbf{i} - g_y(x, y)\mathbf{j} + \mathbf{k}\right) dA$$

$$= \int_R \int (yz\mathbf{i} + xz\mathbf{j} + xy\mathbf{k}) \cdot \left(\frac{x}{\sqrt{1 - x^2 - y^2}}\mathbf{i} + \frac{y}{\sqrt{1 - x^2 - y^2}}\mathbf{j} + \mathbf{k}\right) dA$$

$$= \int_R \int \left(\frac{2xyz}{\sqrt{1 - x^2 - y^2}} + xy\right) dA = \int_R \int 3xy\, dA = \int_{-1}^1 \int_{-\sqrt{1-x^2}}^{\sqrt{1-x^2}} 3xy\, dy\, dx = 0$$

33. $z = \sqrt{x^2 + y^2},\, 0 \le z \le a$

$$m = \int_S \int k\, dS = k \int_R \int \sqrt{1 + \left(\frac{x}{\sqrt{x^2 + y^2}}\right)^2 + \left(\frac{y}{\sqrt{x^2 + y^2}}\right)^2}\, dA = k\int_R \int \sqrt{2}\, dA = \sqrt{2}\, k\pi a^2$$

$$I_z = \int_S \int k(x^2 + y^2)\, dS = \int_R \int k(x^2 + y^2)\sqrt{2}\, dA = \sqrt{2}\,k\int_0^{2\pi}\int_0^a r^3\, dr\, d\theta = \frac{\sqrt{2}\,ka^4}{4}(2\pi) = \frac{\sqrt{2}\,k\pi a^4}{2} = \frac{a^2}{2}\left(\sqrt{2}\,k\pi a^2\right) = \frac{a^2 m}{2}$$

35. $x^2 + y^2 = a^2, 0 \le z \le h$

$\rho(x, y, z) = 1$

$y = \pm\sqrt{a^2 - x^2}$

Project the solid onto the *xz*-plane.

$$I_z = 4\int_S \int (x^2 + y^2)(1)\, dS = 4\int_0^h \int_0^a \left[x^2 + (a^2 - x^2)\right]\sqrt{1 + \left(\frac{-x}{\sqrt{a^2 - x^2}}\right)^2 + (0)^2}\, dx\, dz$$

$$= 4a^3 \int_0^h \int_0^a \frac{1}{\sqrt{a^2 - x^2}}\, dx\, dz = 4a^3 \int_0^h \left[\arcsin\frac{x}{a}\right]_0^a dz = 4a^3\left(\frac{\pi}{2}\right)(h) = 2\pi a^3 h$$

37. $S: z = 16 - x^2 - y^2, z \ge 0$

$\mathbf{F}(x, y, z) = 0.5z\mathbf{k}$

$$\int_S \int \rho \mathbf{F} \cdot \mathbf{N}\, dS = \int_R \int \rho \mathbf{F} \cdot \left(-g_x(x, y)\mathbf{i} - g_y(x, y)\mathbf{j} + \mathbf{k}\right) dA = \int_R \int 0.5\rho z\mathbf{k} \cdot (2x\mathbf{i} + 2y\mathbf{j} + \mathbf{k})\, dA$$

$$= \int_R \int 0.5\rho z\, dA = \int_R \int 0.5\rho(16 - x^2 - y^2)\, dA$$

$$= 0.5\rho \int_0^{2\pi} \int_0^4 (16 - r^2)r\, dr\, d\theta = 0.5\rho \int_0^{2\pi} 64\, d\theta = 64\pi\rho$$

39. The surface integral of f over a surface S, where S is given by $z = g(x, y)$, is defined as

$$\int_S \int f(x, y, z)\, dS = \lim_{\|\Delta\| \to 0} \sum_{i=1}^n f(x_i, y_i, z_i)\Delta S_i. \text{ (page 1112)}$$

See Theorem 15.10, page 1094.

41. See the definition, page 1100.

See Theorem 15.11, page 1100.

43. (a)

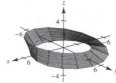

(b) If a normal vector at a point P on the surface is moved around the Möbius strip once, it will point in the opposite direction.

(c) $\mathbf{r}(u, 0) = 4\cos(2u)\mathbf{i} + 4\sin(2u)\mathbf{j}$

This is circle.

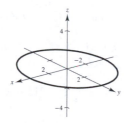

(d) (construction)

(e) You obtain a strip with a double twist and twice as long as the original Möbius strip.

Section 15.7 Divergence Theorem

1. Surface Integral: There are six surfaces to the cube, each with $dS = \sqrt{1}\, dA$.

$z = 0, \quad \mathbf{N} = -\mathbf{k}, \quad \mathbf{F} \cdot \mathbf{N} = -z^2, \quad \int_{S_1} \int 0\, dA = 0$

$z = a, \quad \mathbf{N} = \mathbf{k}, \quad \mathbf{F} \cdot \mathbf{N} = z^2, \quad \int_{S_2} \int a^2\, dA = \int_0^a \int_0^a a^2\, dx\, dy = a^4$

$x = 0, \quad \mathbf{N} = -\mathbf{i}, \quad \mathbf{F} \cdot \mathbf{N} = -2x, \quad \int_{S_3} \int 0\, dA = 0$

$x = a, \quad \mathbf{N} = \mathbf{i}, \quad \mathbf{F} \cdot \mathbf{N} = 2x, \quad \int_{S_4} \int 2a\, dy\, dz = \int_0^a \int_0^a 2a\, dy\, dz = 2a^3$

$y = 0, \quad \mathbf{N} = -\mathbf{j}, \quad \mathbf{F} \cdot \mathbf{N} = 2y, \quad \int_{S_5} \int 0\, dA = 0$

$y = a, \quad \mathbf{N} = \mathbf{j}, \quad \mathbf{F} \cdot \mathbf{N} = -2y, \quad \int_{S_6} \int -2a\, dA = \int_0^a \int_0^a -2a\, dz\, dx = -2a^3$

So, $\int_s \int \mathbf{F} \cdot \mathbf{N}\, dS = a^4 + 2a^3 - 2a^3 = a^4$.

Divergence Theorem: Because div $\mathbf{F} = 2z$, the Divergence Theorem yields

$$\iiint_Q \text{div } \mathbf{F}\, dV = \int_0^a \int_0^a \int_0^a 2z\, dz\, dy\, dx = \int_0^a \int_0^a a^2\, dy\, dx = a^4.$$

3. Surface Integral: There are four surfaces to this solid.

$z = 0, \quad \mathbf{N} = -\mathbf{k}, \quad \mathbf{F} \cdot \mathbf{N} = -z$

$\int_{S_1} \int 0\, dS = 0$

$y = 0, \quad \mathbf{N} = -\mathbf{j}, \quad \mathbf{F} \cdot \mathbf{N} = 2y - z, \quad dS = dA = dx\, dz$

$\int_{S_2} \int -z\, dS = \int_0^6 \int_0^{6-z} -z\, dx\, dz = \int_0^6 (z^2 - 6z)\, dz = -36$

$x = 0, \quad \mathbf{N} = -\mathbf{i}, \quad \mathbf{F} \cdot \mathbf{N} = y - 2x, \quad dS = dA = dz\, dy$

$\int_{S_3} \int y\, dS = \int_0^3 \int_0^{6-2y} y\, dz\, dy = \int_0^3 (6y - 2y^2)\, dy = 9$

$x + 2y + z = 6, \mathbf{N} = \dfrac{\mathbf{i} + 2\mathbf{j} + \mathbf{k}}{\sqrt{6}}, \mathbf{F} \cdot \mathbf{N} = \dfrac{2x - 5y + 3z}{\sqrt{6}}, dS = \sqrt{6}\, dA$

$\int_{S_4} \int (2x - 5y + 3z)\, dz\, dy = \int_0^3 \int_0^{6-2y} (18 - x - 11y)\, dx\, dy = \int_0^3 (90 - 90y + 20y^2)\, dy = 45$

So, $\int_s \int \mathbf{F} \cdot \mathbf{N}\, dS = 0 - 36 + 9 + 45 = 18$.

Divergence Theorem: Because div $\mathbf{F} = 1$, you have

$$\iiint_Q dV = (\text{Volume of solid}) = \frac{1}{3}(\text{Area of base}) \times (\text{Height}) = \frac{1}{3}(9)(6) = 18.$$

5. $F(x, y, z) = xz\mathbf{i} + yz\mathbf{j} + 2z^2\mathbf{k}$

Surface Integral: There are two surfaces.

Bottom: $z = 0$, $\mathbf{N} = -\mathbf{k}$, $\mathbf{F} \cdot \mathbf{N} = -2z^2$

$$\int_{S_1}\int \mathbf{F} \cdot \mathbf{N}\, dS = \int_R\int -2z^2\, dA = \iint 0\, dA = 0$$

Side: Outward unit normal is

$$\mathbf{N} = \frac{2x\mathbf{i} + 2y\mathbf{j} + \mathbf{k}}{\sqrt{4x^2 + 4y^2 + 1}}$$

$$\mathbf{F} \cdot \mathbf{N} = \frac{1}{\sqrt{4x^2 + 4y^2 + 1}}\left[2x^2z + 2y^2z + 2z^2\right]$$

$$\int_{S_2}\int \mathbf{F} \cdot \mathbf{N}\, dS = \int_{S_2}\int \left[2(x^2 + y^2)z + 2z^2\right] dA$$

$$= \int_0^{2\pi}\int_0^1 \left[2r^2(1 - r^2) + 2(1 - r^2)^2\right] r\, dr\, d\theta = \int_0^{2\pi}\int_0^1 (2r - 2r^3)\, dr\, d\theta = \int_0^{2\pi}\frac{1}{2}\, d\theta = \pi$$

Divergence Theorem: div $\mathbf{F} = z + z + 4z = 6z$

$$\iiint_Q \text{div } \mathbf{F}\, dV = \int_0^{2\pi}\int_0^1\int_0^{1-r^2} 6z\, r\, dz\, dr\, d\theta$$

$$= \int_0^{2\pi}\int_0^1 3(1 - r^2)^2\, r\, dr\, d\theta = \int_0^{2\pi}\int_0^1 (3 - 6r^2 + 3r^4)\, r\, dr\, d\theta = \int_0^{2\pi}\left[\frac{3}{2} - \frac{3}{2} + \frac{1}{2}\right] d\theta = \pi$$

7. Because div $\mathbf{F} = 2x + 2y + 2z$, you have

$$\iiint_Q \text{div } \mathbf{F}\, dV = \int_0^a\int_0^a\int_0^a (2x + 2y + 2z)\, dz\, dy\, dx$$

$$= \int_0^a\int_0^a (2ax + 2ay + a^2)\, dy\, dx = \int_0^a (2a^2x + 2a^3)\, dx = \left[a^2x^2 + 2a^3x\right]_0^a = 3a^4.$$

9. Because div $\mathbf{F} = 2x - 2x + 2xyz = 2xyz$,

$$\iiint_Q \text{div } \mathbf{F}\, dV = \iiint_Q 2xyz\, dV = \int_0^a\int_0^{2\pi}\int_0^{\pi/2} 2(\rho \sin\phi \cos\theta)(\rho \sin\phi \sin\theta)(\rho \cos\phi)\rho^2 \sin\phi\, d\phi\, d\theta\, d\rho$$

$$= \int_0^a\int_0^{2\pi}\int_0^{\pi/2} 2\rho^5(\sin\theta \cos\theta)(\sin^3\phi \cos\phi)\, d\phi\, d\theta\, d\rho$$

$$= \int_0^a\int_0^{2\pi}\frac{1}{2}\rho^5 \sin\theta \cos\theta\, d\theta\, d\rho = \int_0^a\left[\left(\frac{\rho^5}{2}\right)\frac{\sin^2\theta}{2}\right]_0^{2\pi} d\rho = 0.$$

11. Because div $\mathbf{F} = 3$, you have

$$\iiint_Q 3\, dV = 3\,(\text{Volume of Sphere}) = 3\left[\frac{4}{3}\pi(3^3)\right] = 108\pi.$$

13. Because div $\mathbf{F} = 1 + 2y - 1 = 2y$, you have

$$\iiint_Q 2y\, dV = \int_0^7\int_{-5}^5\int_{-\sqrt{25-y^2}}^{\sqrt{25-y^2}} 2y\, dx\, dy\, dz = \int_0^7\int_{-5}^5 4y\sqrt{25 - y^2}\, dy\, dz = \int_0^7 \left[\frac{-4}{3}(25 - y^2)^{3/2}\right]_{-5}^5 dz = 0.$$

15. Because div $\mathbf{F} = e^z + e^z + e^z = 3e^z$, you have

$$\iiint_Q 3e^z\, dV = \int_0^6\int_0^4\int_0^{4-y} 3e^z\, dz\, dy\, dx = \int_0^6\int_0^4 3\left[e^{4-y} - 1\right] dy\, dx = \int_0^6 3(e^4 - 5)\, dx = 18(e^4 - 5).$$

17. Using the Divergence Theorem, you have

$$\int_S \int \text{curl } \mathbf{F} \cdot \mathbf{N} \, dS = \iiint_Q \text{div} \left(\text{curl } \mathbf{F} \right) dV$$

$$\text{curl } \mathbf{F}(x,y,z) = \begin{vmatrix} \mathbf{i} & \mathbf{j} & \mathbf{k} \\ \dfrac{\partial}{\partial x} & \dfrac{\partial}{\partial y} & \dfrac{\partial}{\partial z} \\ 4xy + z^2 & 2x^2 + 6yz & 2xz \end{vmatrix} = -6y\mathbf{i} - (2z - 2z)\mathbf{j} + (4x - 4x)\mathbf{k} = -6y\mathbf{i}$$

$$\text{div} \left(\text{curl } \mathbf{F} \right) = 0.$$

So, $\iiint_Q \text{div} \left(\text{curl } \mathbf{F} \right) dV = 0.$

19. See Theorem 15.12.

21. Using the Divergence Theorem, you have $\int_S \int \text{curl } \mathbf{F} \cdot \mathbf{N} \, dS = \iiint_Q \text{div} \left(\text{curl } \mathbf{F} \right) dV.$ Let

$$\mathbf{F}(x, y, z) = M\mathbf{i} + N\mathbf{j} + P\mathbf{k}$$

$$\text{curl } \mathbf{F} = \left(\dfrac{\partial P}{\partial y} - \dfrac{\partial N}{\partial z} \right)\mathbf{i} - \left(\dfrac{\partial P}{\partial x} - \dfrac{\partial M}{\partial z} \right)\mathbf{j} + \left(\dfrac{\partial N}{\partial x} - \dfrac{\partial M}{\partial y} \right)\mathbf{k}$$

$$\text{div} \left(\text{curl } \mathbf{F} \right) = \dfrac{\partial^2 P}{\partial x \partial y} - \dfrac{\partial^2 N}{\partial x \partial z} - \dfrac{\partial^2 P}{\partial y \partial x} + \dfrac{\partial^2 M}{\partial y \partial z} + \dfrac{\partial^2 N}{\partial z \partial x} - \dfrac{\partial^2 M}{\partial z \partial y} = 0.$$

So, $\int_S \int \text{curl } \mathbf{F} \cdot \mathbf{N} \, dS = \iiint_Q 0 \, dV = 0.$

23. (a) Using the triple integral to find volume, you need $\mathbf{F}$ so that

$$\text{div } \mathbf{F} = \dfrac{\partial M}{\partial x} + \dfrac{\partial N}{\partial y} + \dfrac{\partial P}{\partial z} = 1.$$

So, you could have $\mathbf{F} = x\mathbf{i}, \ \mathbf{F} = y\mathbf{j}, \ \text{or} \ \mathbf{F} = z\mathbf{k}.$

For $dA = dy \, dz$ consider $\mathbf{F} = x\mathbf{i}, x = f(y, z)$, then $\mathbf{N} = \dfrac{\mathbf{i} + f_y\mathbf{j} + f_z\mathbf{k}}{\sqrt{1 + f_y^2 + f_z^2}}$ and $dS = \sqrt{1 + f_y^2 + f_z^2} \, dy \, dz.$

For $dA = dz \, dx$ consider $\mathbf{F} = y\mathbf{j}, y = f(x, z)$, then $\mathbf{N} = \dfrac{f_x\mathbf{i} + \mathbf{j} + f_z\mathbf{k}}{\sqrt{1 + f_x^2 + f_z^2}}$ and $dS = \sqrt{1 + f_x^2 + f_z^2} \, dz \, dx.$

For $dA = dx \, dy$ consider $\mathbf{F} = z\mathbf{k}, z = f(x, y)$, then $\mathbf{N} = \dfrac{f_x\mathbf{i} + f_y\mathbf{j} + \mathbf{k}}{\sqrt{1 + f_x^2 + f_y^2}}$ and $dS = \sqrt{1 + f_x^2 + f_y^2} \, dx \, dy.$

Correspondingly, you then have $V = \int_S \int \mathbf{F} \cdot \mathbf{N} \, dS = \int_S \int x \, dy \, dz = \int_S \int y \, dz \, dx = \int_S \int z \, dx \, dy.$

(b) $v = \int_0^a \int_0^a x \, dy \, dz = \int_0^a \int_0^a a \, dy \, dz = \int_0^a a^2 \, dz = a^3$

Similarly, $\int_0^a \int_0^a y \, dz \, dx = \int_0^a \int_0^a z \, dx \, dy = a^3.$

25. If $\mathbf{F}(x, y, z) = x\mathbf{i} + y\mathbf{j} + z\mathbf{k}$, then $\text{div } \mathbf{F} = 3.$

$$\int_S \int \mathbf{F} \cdot \mathbf{N} \, dS = \iiint_Q \text{div } \mathbf{F} \, dV = \iiint_Q 3 \, dV = 3V.$$

27. $\int_S \int f D_\mathbf{N} g \, dS = \int_S \int f \nabla g \cdot \mathbf{N} \, dS = \iiint_Q \text{div}(f \nabla g) \, dV = \iiint_Q (f \, \text{div} \nabla g + \nabla f \cdot \nabla g) \, dV = \iiint_Q (f \nabla^2 g + \nabla f \cdot \nabla g) \, dV$

Section 15.8   Stokes's Theorem

1. $F(x, y, z) = (2y - z)\mathbf{i} + e^z\mathbf{j} + xyz\mathbf{k}$

$$\text{curl } \mathbf{F} = \begin{vmatrix} \mathbf{i} & \mathbf{j} & \mathbf{k} \\ \dfrac{\partial}{\partial x} & \dfrac{\partial}{\partial y} & \dfrac{\partial}{\partial z} \\ 2y - z & e^z & xyz \end{vmatrix}$$

$$= (xz - e^z)\mathbf{i} - (yz + 1)\mathbf{j} - 2\mathbf{k}$$

3. $F(x, y, z) = e^{x^2 + y^2}\mathbf{i} + e^{y^2 + z^2}\mathbf{j} + xyz\mathbf{k}$

$$\text{curl } \mathbf{F} = \begin{vmatrix} \mathbf{i} & \mathbf{j} & \mathbf{k} \\ \dfrac{\partial}{\partial x} & \dfrac{\partial}{\partial y} & \dfrac{\partial}{\partial z} \\ e^{x^2 + y^2} & e^{y^2 + z^2} & xyz \end{vmatrix}$$

$$= \left(xz - 2ze^{y^2 + z^2}\right)\mathbf{i} - yz\mathbf{j} - 2ye^{x^2 + y^2}\mathbf{k}$$

$$= z\left(x - 2e^{y^2 + z^2}\right)\mathbf{i} - yz\mathbf{j} - 2ye^{x^2 + y^2}\mathbf{k}$$

5. $C: x^2 + y^2 = 9, \quad z = 0, dz = 0$

Line Integral:

$$\int_C \mathbf{F} \cdot d\mathbf{r} = \int_C -y\, dx + x\, dy$$

$x = 3\cos t, dx = -3\sin t\, dt, y = 3\sin t, dy = 3\cos t\, dt$

$$\int_C \mathbf{F} \cdot d\mathbf{r} = \int_0^{2\pi} \left[(-3\sin t)(-3\sin t) + (3\cos t)(3\cos t)\right] dt$$

$$= \int_0^{2\pi} 9\, dt = 18\pi$$

Double Integral: $g(x, y) = 9 - x^2 - y^2, g_x = -2x, g_y = -2y$

$\text{curl } \mathbf{F} = 2\mathbf{k}$

$$\int_S \int \text{curl } \mathbf{F} \cdot \mathbf{N}\, dS = \int_R \int 2\, dA = 2(\text{area circle}) = 18\pi$$

7. Line Integral:

From the figure you see that

$C_1: z = 0, dz = 0$

$C_2: x = 0, dx = 0$

$C_3: y = 0, dy = 0$

$$\int_C \mathbf{F} \cdot d\mathbf{r} = \int_C xyz\, dx + y\, dy + z\, dz = \int_{C_1} y\, dy + \int_{C_2} y\, dy + z\, dz + \int_{C_3} z\, dz = \int_0^2 y\, dy + \int_2^0 y\, dy + \int_0^{12} z\, dz + \int_{12}^0 z\, dz = 0$$

Double Integral: $\text{curl } \mathbf{F} = xy\mathbf{j} - xz\mathbf{k}$

Letting $z = 12 - 6x - 6y = g(x, y), g_x = -6 = g_y$.

$$\int_S \int (\text{curl } \mathbf{F}) \cdot \mathbf{N}\, dS = \int_R \int (\text{curl } \mathbf{F}) \cdot [6\mathbf{i} + 6\mathbf{j} + \mathbf{k}]\, dA = \int_R \int (6xy - xz)\, dA$$

$$= \int_0^2 \int_0^{2-x} \left[6xy - x(12 - 6x - 6y)\right] dy\, dx = \int_0^2 \int_0^{2-x} \left(12xy - 12x + 6x^2\right) dy\, dx$$

$$= \int_0^2 \left[6xy^2 - 12xy + 6x^2y\right]_0^{2-x} dx = 0$$

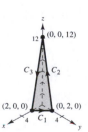

9. These three points have equation:

$x + y + z = 2$.

Normal vector: $\mathbf{N} = \mathbf{i} + \mathbf{j} + \mathbf{k}$

$\text{curl } \mathbf{F} = -3\mathbf{i} - \mathbf{j} - 2\mathbf{k}$

$\int_S \int \text{curl } \mathbf{F} \cdot \mathbf{N} \, dS = \int_R \int (-6) \, dA = -6(\text{area of triangle in } xy\text{-plane})$

$$= -6(2) = -12$$

11. $\text{curl } \mathbf{F} = \begin{vmatrix} \mathbf{i} & \mathbf{j} & \mathbf{k} \\ \dfrac{\partial}{\partial x} & \dfrac{\partial}{\partial y} & \dfrac{\partial}{\partial z} \\ z^2 & 2x & y^2 \end{vmatrix} = 2y\mathbf{i} + 2z\mathbf{j} + 2\mathbf{k}$

$z = G(x, y) = 1 - x^2 - y^2, G_x = -2x, G_y = -2y$

$\int_S \int \text{curl } \mathbf{F} \cdot \mathbf{N} \, dS = \int_R \int (2y\mathbf{i} + 2z\mathbf{j} + 2\mathbf{k}) \cdot (2x\mathbf{i} + 2y\mathbf{j} + \mathbf{k}) \, dA = \int_R \int \left[4xy + 4y(1 - x^2 - y^2) + 2 \right] dA$

$$= \int_{-1}^1 \int_{-\sqrt{1-x^2}}^{\sqrt{1-x^2}} \left[4xy + 4y - 4x^2y - 4y^3 + 2 \right] dy \, dx$$

$$= \int_{-1}^1 4\sqrt{1 - x^2} \, dx = 2\left[\arcsin x + x\sqrt{1 - x^2} \right]_{-1}^1 = 2\pi$$

13. $\text{curl } \mathbf{F} = \begin{vmatrix} \mathbf{i} & \mathbf{j} & \mathbf{k} \\ \dfrac{\partial}{\partial x} & \dfrac{\partial}{\partial y} & \dfrac{\partial}{\partial z} \\ z^2 & y & z \end{vmatrix} = 2z\mathbf{j}$

$z = G(x, y) = \sqrt{4 - x^2 - y^2}, G_x = \dfrac{-x}{\sqrt{4 - x^2 - y^2}}, G_y = \dfrac{-y}{\sqrt{4 - x^2 - y^2}}$

$\int_S \int \text{curl } \mathbf{F} \cdot \mathbf{N} = \int_R \int (2z\mathbf{j}) \cdot \left(\dfrac{x}{\sqrt{4 - x^2 - y^2}}\mathbf{i} + \dfrac{y}{\sqrt{4 - x^2 - y^2}}\mathbf{j} + \mathbf{k} \right) dA$

$$= \int_R \int \dfrac{2yz}{\sqrt{4 - x^2 - y^2}} \, dA = \int_R \int \dfrac{2y\sqrt{4 - x^2 - y^2}}{\sqrt{4 - x^2 - y^2}} \, dA = \int_{-2}^2 \int_{-\sqrt{4-x^2}}^{\sqrt{4-x^2}} 2y \, dy \, dx = 0$$

15. $\mathbf{F}(x, y, z) = -\ln\sqrt{x^2 + y^2}\,\mathbf{i} + \arctan\dfrac{x}{y}\mathbf{j} + \mathbf{k}$

$\text{curl } \mathbf{F} = \begin{vmatrix} \mathbf{i} & \mathbf{j} & \mathbf{k} \\ \dfrac{\partial}{\partial x} & \dfrac{\partial}{\partial y} & \dfrac{\partial}{\partial z} \\ -1/2 \ln(x^2 + y^2) & \arctan x/y & 1 \end{vmatrix} = \left[\dfrac{(1/y)}{1 + (x^2/y^2)} + \dfrac{y}{x^2 + y^2} \right] \mathbf{k} = \left[\dfrac{2y}{x^2 + y^2} \right] \mathbf{k}$

$S: z = 9 - 2x - 3y$ over one petal of $r = 2 \sin 2\theta$ in the first octant.

$G(x, y, z) = 2x + 3y + z - 9$

$\nabla G(x, y, z) = 2\mathbf{i} + 3\mathbf{j} + \mathbf{k}$

$\int_S \int (\text{curl } \mathbf{F}) \cdot \mathbf{N} \, dS = \int_R \int \dfrac{2y}{x^2 + y^2} \, dA = \int_0^{\pi/2} \int_0^{2\sin 2\theta} \dfrac{2r \sin \theta}{r^2} r \, dr \, d\theta$

$$= \int_0^{\pi/2} \int_0^{4\sin\theta\cos\theta} 2 \sin \theta \, dr \, d\theta = \int_0^{\pi/2} 8 \sin^2 \theta \cos \theta \, d\theta = \left[\dfrac{8 \sin^3 \theta}{3} \right]_0^{\pi/2} = \dfrac{8}{3}$$

17. $\text{curl } \mathbf{F} = xy\mathbf{j} - xz\mathbf{k}$

$z = G(x, y) = x^2, G_x = 2x, G_y = 0$

$\displaystyle \int_S \int \text{curl } \mathbf{F} \cdot \mathbf{N} = \int_R \int (xy\mathbf{j} - xz\mathbf{k}) \cdot (2x\mathbf{i} - \mathbf{k}) \, dA = \int_R \int xz \, dA = \int_0^a \int_0^a x(x^2) \, dy \, dx = \dfrac{a^5}{4}$

19. $\mathbf{F}(x, y, z) = \mathbf{i} + \mathbf{j} - 2\mathbf{k}$

$\text{curl } \mathbf{F} = \begin{vmatrix} \mathbf{i} & \mathbf{j} & \mathbf{k} \\ \dfrac{\partial}{\partial x} & \dfrac{\partial}{\partial y} & \dfrac{\partial}{\partial z} \\ 1 & 1 & -2 \end{vmatrix} = \mathbf{0}$

Letting $\mathbf{N} = \mathbf{k}$, you have $\displaystyle \int_S \int (\text{curl } \mathbf{F}) \cdot \mathbf{N} \, dS = 0$.

21. See Theorem 15.13.

23. Let $\mathbf{C} = a\mathbf{i} + b\mathbf{j} + c\mathbf{k}$, then $\dfrac{1}{2} \int_C (\mathbf{C} \times \mathbf{r}) \cdot d\mathbf{r} = \dfrac{1}{2} \int_S \int \text{curl}(\mathbf{C} \times \mathbf{r}) \cdot \mathbf{N} \, dS = \dfrac{1}{2} \int_S \int 2\mathbf{C} \cdot \mathbf{N} \, dS = \int_S \int \mathbf{C} \cdot \mathbf{N} \, dS$

because $\mathbf{C} \times \mathbf{r} = \begin{vmatrix} \mathbf{i} & \mathbf{j} & \mathbf{k} \\ a & b & c \\ x & y & z \end{vmatrix} = (bz - cy)\mathbf{i} - (az - cx)\mathbf{j} + (ay - bx)\mathbf{k}$

and $\text{curl}(\mathbf{C} \times \mathbf{r}) = \begin{vmatrix} \mathbf{i} & \mathbf{j} & \mathbf{k} \\ \dfrac{\partial}{\partial x} & \dfrac{\partial}{\partial y} & \dfrac{\partial}{\partial z} \\ bz - cy & cx - az & ay - bx \end{vmatrix} = 2(a\mathbf{i} + b\mathbf{j} + c\mathbf{k}) = 2\mathbf{C}.$

25. Let S be the upper portion of the ellipsoid

$x^2 + 4y^2 + z^2 = 4, z \geq 0$

Let $C: \mathbf{r}(t) = \langle 2 \cos t, \sin t, 0 \rangle, 0 \leq t \leq 2\pi$, be the boundary of S.

If $\mathbf{F} = \langle M, N, P \rangle$ exists, then

$0 = \displaystyle \int_S \int (\text{curl } \mathbf{F}) \cdot \mathbf{N} \, dS$ (by (i))

$= \displaystyle \int_C \mathbf{F} \cdot d\mathbf{r}$ (Stokes's Theorem)

$= \displaystyle \int_C \mathbf{G} \cdot d\mathbf{r}$ (by (iii))

$= \displaystyle \int_0^{2\pi} \left\langle \dfrac{-\sin t}{4}, \dfrac{2 \cos t}{4}, 0 \right\rangle \cdot \langle -2 \sin t, \cos t, 0 \rangle \, dt = \dfrac{1}{4} \int_0^{2\pi} (2 \sin^2 t + 2 \cos^2 t) \, dt = \pi$

So, there is no such $\mathbf{F}$.

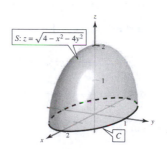

Review Exercises for Chapter 15

1. $\mathbf{F}(x, y, z) = x\mathbf{i} + \mathbf{j} + 2\mathbf{k}$

$\|\mathbf{F}\| = \sqrt{x^2 + 1^2 + 2^2} = \sqrt{x^2 + 5}$

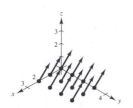

3. $f(x, y, z) = 2x^2 + xy + z^2$

$\mathbf{F}(x, y, z) = \nabla f = (4x + y)\mathbf{i} + x\mathbf{j} + 2z\mathbf{k}$

5. Because $\partial M / \partial y = -1/x^2 = \partial N / \partial x$, $\mathbf{F}$ is conservative.

From $M = \partial U / \partial x = -y/x^2$ and

$N = \partial U / \partial y = 1/x$, partial integration yields

$U = (y/x) + h(y)$ and $U = (y/x) + g(x)$ which

suggests that $U(x, y) = (y/x) + C$.

7. Because $\dfrac{\partial M}{\partial y} = 2xy$ and $\dfrac{\partial N}{\partial x} = 2xy$, **F** is conservative.

From $M = \dfrac{\partial U}{\partial x} = xy^2 - x^2$ and

$N = \dfrac{\partial U}{\partial y} = x^2 y + y^2$, partial integration yields

$$U = \frac{1}{2}x^2 y^2 - \frac{x^3}{3} + h(y)$$

and

$$U = \frac{1}{2}x^2 y^2 + \frac{y^3}{3} + g(x).$$

So, $h(y) = y^3/3$ and $g(x) = -x^3/3$. So,

$$U(x, y) = \frac{1}{2}x^2 y^2 - \frac{x^3}{3} + \frac{y^3}{3} + C.$$

9. Because $\dfrac{\partial M}{\partial y} = 8xy$ and $\dfrac{\partial N}{\partial x} = 4x$, $\dfrac{\partial M}{\partial y} \neq \dfrac{\partial N}{\partial x}$, so **F** is not conservative.

11. Because

$$\frac{\partial M}{\partial y} = \frac{-1}{y^2 z} = \frac{\partial N}{\partial x}, \frac{\partial M}{\partial z} = \frac{-1}{yz^2} = \frac{\partial P}{\partial x},$$

$$\frac{\partial N}{\partial z} = \frac{x}{y^2 z^2} = \frac{\partial P}{\partial y},$$

F is conservative. From

$$M = \frac{\partial U}{\partial x} = \frac{1}{yz}, \ N = \frac{\partial U}{\partial y} = \frac{-x}{y^2 z}, \ P = \frac{\partial U}{\partial z} = \frac{-x}{yz^2}$$

you obtain

$$U = \frac{x}{yz} + f(y, z), \ U = \frac{x}{yz} + g(x, z),$$

$$U = \frac{x}{yz} + h(x, y) \Rightarrow f(x, y \, z) = \frac{x}{yz} + K.$$

13. Because $\mathbf{F}(x, y, z) = x^2 \mathbf{i} + xy^2 \mathbf{j} + x^2 z \mathbf{k}$:

(a) $\operatorname{div} \mathbf{F} = 2x + 2xy + x^2$

(b) $\operatorname{curl} \mathbf{F} = \begin{vmatrix} \mathbf{i} & \mathbf{j} & \mathbf{k} \\ \dfrac{\partial}{\partial x} & \dfrac{\partial}{\partial y} & \dfrac{\partial}{\partial z} \\ x^2 & xy^2 & x^2 z \end{vmatrix} = -(2xz)\mathbf{j} + y^2 \mathbf{k}$

15. Because $\mathbf{F} = (\cos y + y \cos x)\mathbf{i} + (\sin x - x \sin y)\mathbf{j} + xyz\,\mathbf{k}$:

(a) $\operatorname{div} \mathbf{F} = -y \sin x - x \cos y + xy$

(b) $\operatorname{curl} \mathbf{F} = xz\mathbf{i} - yz\mathbf{j} + (\cos x - \sin y + \sin y - \cos x)\mathbf{k} = xz\mathbf{i} - yz\mathbf{j}$

17. Because $\mathbf{F} = \arcsin x\mathbf{i} + xy^2\mathbf{j} + yz^2\mathbf{k}$:

(a) $\operatorname{div} \mathbf{F} = \dfrac{1}{\sqrt{1 - x^2}} + 2xy + 2yz$

(b) $\operatorname{curl} \mathbf{F} = z^2\mathbf{i} + y^2\mathbf{k}$

19. Because $\mathbf{F} = \ln(x^2 + y^2)\mathbf{i} + \ln(x^2 + y^2)\mathbf{j} + z\mathbf{k}$:

(a) $\operatorname{div} \mathbf{F} = \dfrac{2x}{x^2 + y^2} + \dfrac{2y}{x^2 + y^2} + 1 = \dfrac{2x + 2y}{x^2 + y^2} + 1$

(b) $\operatorname{curl} \mathbf{F} = \dfrac{2x - 2y}{x^2 + y^2}\mathbf{k}$

21. (a) Let $x = 3t, y = 4t, \quad 0 \le t \le 1$,

then $ds = \sqrt{9 + 16}\, dt = 5\, dt$.

$$\int_C (x^2 + y^2)\, ds = \int_0^1 (9t^2 + 16t^2)5\, dt$$

$$= \left[125 \frac{t^3}{3}\right]_0^1 = \frac{125}{3}$$

(b) Let $x = \cos t, y = \sin t, \quad 0 \le t \le 2\pi$,

then $ds = \sqrt{(-\sin t)^2 + (\cos t)^2} = 1$.

$$\int_C (x^2 + y^2)\, ds = \int_0^{2\pi} dt = 2\pi$$

23. $x = 1 - \sin t, y = 1 - \cos t, 0 \le t \le 2\pi$

$$\frac{dx}{dt} = -\cos t, \frac{dy}{dt} = \sin t, ds = \sqrt{(-\cos t)^2 + (\sin t)^2}\, dt = dt$$

$$\int_C (x^2 + y^2)\, ds = \int_0^{2\pi} \left[(1 - \sin t)^2 + (1 - \cos t)^2\right] dt = \int_0^{2\pi} \left[1 - 2\sin t + \sin^2 t + 1 - 2\cos t + \cos^2 t\right] dt$$

$$= \int_0^{2\pi} \left[3 - 2\sin t - 2\cos t\right] dt = \left[3t + 2\cos t - 2\sin t\right]_0^{2\pi} = 6\pi$$

25. (a) Let $x = 3t$, $y = -3t$, $\quad 0 \le t \le 1$.

$$\int_C (2x - y)\,dx + (x + 2y)\,dy = \int_0^1 \left[(6t + 3t)3 + (3t - 6t)(-3)\right] dt = \int_0^1 (27t + 9t)\,dt = 18t^2 \Big]_0^1 = 18$$

(b) Let $x = 3\cos t$, $y = 3\sin t$, $dx = -3\sin t\,dt$, $dy = 3\cos t\,dt$, $0 \le t \le 2\pi$.

$$\int_C (2x - y)\,dx + (x + 2y)\,dy = \int_0^{2\pi} \left[(6\cos t - 3\sin t)(-3\sin t) + (3\cos t + 6\sin t)(3\cos t)\right] dt = \int_0^{2\pi} 9\,dt = 18\pi$$

27. $\int_C (2x + y)\,ds, \mathbf{r}(t) = a\cos^3 t\,\mathbf{i} + a\sin^3 t\,\mathbf{j}, 0 \le t \le \dfrac{\pi}{2}$

$x'(t) = -3a \cdot \cos^2 t \sin t$

$y'(t) = 3a \cdot \sin^2 t \cos t$

$$\int_C (2x + y)\,ds = \int_0^{\pi/2} \left(2(a \cdot \cos^3 t) + a \cdot \sin^3 t\right)\sqrt{x'(t)^2 + y'(t)^2}\,dt = \frac{9a^2}{5}$$

29. $f(x, y) = 3 + \sin(x + y)$

$C: y = 2x$ from $(0, 0)$ to $(2, 4)$

$\mathbf{r}(t) = t\mathbf{i} + 2t\mathbf{j}, 0 \le t \le 2$

$\mathbf{r}'(t) = \mathbf{i} + 2\mathbf{j}$

$\|\mathbf{r}'(t)\| = \sqrt{5}$

Lateral surface area:

$$\int_C f(x, y)\,ds = \int_0^2 \left[3 + \sin(t + 2t)\right]\sqrt{5}\,dt$$

$$= \sqrt{5}\int_0^2 \left[3 + \sin 3t\right] dt$$

$$= \sqrt{5}\left[3t - \frac{1}{3}\cos 3t\right]_0^2$$

$$= \sqrt{5}\left[6 - \frac{1}{3}\cos 6 + \frac{1}{3}\right]$$

$$= \frac{\sqrt{5}}{3}(19 - \cos 6) \approx 13.446$$

31. $\mathbf{F}(x, y) = xy\mathbf{i} + 2xy\mathbf{j}$

$\mathbf{r}(t) = t^2\mathbf{i} + t^2\mathbf{j}, \ 0 \le t \le 1$

$\mathbf{r}'(t) = 2t\mathbf{i} + 2t\mathbf{j}$

$$\int_C \mathbf{F} \cdot d\mathbf{r} = \int_0^1 \left[t^2(t^2)(2t) + 2(t^2)(t^2)(2t)\right] dt$$

$$= \int_0^1 6t^5\,dt = t^6 \Big]_0^1 = 1$$

33. $d\mathbf{r} = \left[(-2\sin t)\mathbf{i} + (2\cos t)\mathbf{j} + \mathbf{k}\right] dt$

$\mathbf{F} = (2\cos t)\mathbf{i} + (2\sin t)\mathbf{j} + t\mathbf{k}, 0 \le t \le 2\pi$

$$\int_C \mathbf{F} \cdot d\mathbf{r} = \int_0^{2\pi} t\,dt = 2\pi^2$$

35. $\mathbf{F}(x, y, z) = (y + z)\mathbf{i} + (x + z)\mathbf{j} + (x + y)\mathbf{k}$

Curve of intersection: $x = t$, $y = t$, $z = t^2 + t^2 = 2t^2$

$\mathbf{r}(t) = t\mathbf{i} + t\mathbf{j} + 2t^2\mathbf{k}, \quad 0 \le t \le 2$

$\mathbf{r}'(t) = \mathbf{i} + \mathbf{j} + 4t\mathbf{k}$

$$\int_C \mathbf{F} \cdot d\mathbf{r} = \int_0^2 \left[(t + 2t^2) + (t + 2t^2) + (2t)(4t)\right] dt = \int_0^2 \left[12t^2 + 2t\right] dt = \left[4t^3 + t^2\right]_0^2 = 36$$

37. For $y = x^2$, $\mathbf{r}_1(t) = t\mathbf{i} + t^2\mathbf{j}, 0 \le t \le 2$

For $y = 2x$, $\mathbf{r}_2(t) = (2 - t)\mathbf{i} + (4 - 2t)\mathbf{j}, 0 \le t \le 2$

$$\int_C xy\,dx + (x^2 + y^2)\,dy = \int_{C_1} xy\,dx + (x^2 + y^2)\,dy + \int_{C_2} xy\,dx + (x^2 + y^2)\,dy$$

$$= \frac{100}{3} + (-32) = \frac{4}{3}$$

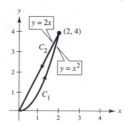

39. $\mathbf{F} = x\mathbf{i} - \sqrt{y}\mathbf{j}$ is conservative.

$$\text{Work} = \left[\frac{1}{2}x^2 - \frac{2}{3}y^{3/2} \right]_{(0,0)}^{(4,8)} = \frac{1}{2}(16) - \left(\frac{2}{3}\right)8^{3/2} = \frac{8}{3}\left(3 - 4\sqrt{2}\right)$$

41. $\displaystyle\int_c 2xyz \, dx + x^2z \, dy + x^2y \, dz = \left[x^2yz \right]_{(0,0,0)}^{(1,3,2)} = 6$

43. (a) $\displaystyle\int_C y^2 \, dx + 2xy \, dy = \int_0^1 \left[(1 + t)^2(3) + 2(1 + 3t)(1 + t) \right] dt$

$$= \int_0^1 3(t^2 + 2t + 1) + 2(3t^2 + 4t + 1) \, dt = \int_0^1 \left(9t^2 + 14t + 5\right) dt = \left[3t^2 + 7t^2 + 5t \right]_0^1 = 15$$

(b) $\displaystyle\int_C y^2 \, dx + 2xy \, dy = \int_1^4 \left[t(1) + 2(t)(\sqrt{t})\frac{1}{2\sqrt{t}} \right] dt = \int_1^4 \left[(t + t) \right] dt = \left[t^2 \right]_1^4 = 15$

(c) $\mathbf{F}(x, y) = y^2\mathbf{i} + 2xy \, \mathbf{j} = \nabla f$ where $f(x, y) = xy^2$.

So, $\displaystyle\int_C \mathbf{F} \cdot d\mathbf{r} = 4(2)^2 - 1(1)^2 = 15.$

45. $\displaystyle\int_C y \, dx + 2x \, dy = \int_0^1 \int_0^1 \left(\frac{\partial N}{\partial x} - \frac{\partial M}{\partial y} \right) dy \, dx$

$$= \int_0^1 \int_0^1 (2 - 1) \, dy \, dx = 1$$

47. $\displaystyle\int_C xy^2 \, dx + x^2y \, dy = \int_R \int \left(\frac{\partial N}{\partial x} - \frac{\partial M}{\partial y} \right) dA$

$$= \int_R \int (2xy - 2xy) \, dA = 0$$

49. $\displaystyle\int_C xy \, dx + x^2 \, dy = \int_R \int \left(\frac{\partial N}{\partial x} - \frac{\partial M}{\partial y} \right) dA$

$$= \int_{-1}^1 \int_{x^2}^1 (2x - x) \, dy \, dx$$

$$= \int_{-1}^1 [xy]_{x^2}^1 = dx$$

$$= \int_{-1}^1 (x - x^3) \, dx$$

$$= \left[\frac{x^2}{2} - \frac{x^4}{4} \right]_{-1}^1 = 0$$

51. $\mathbf{r}(u, v) = \sec u \cos v\mathbf{i} + (1 + 2 \tan u)\sin v\mathbf{j} + 2u\mathbf{k}$

$$0 \le u \le \frac{\pi}{3}, \quad 0 \le v \le 2\pi$$

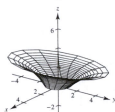

53. (a)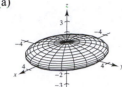

(b)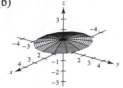

(c)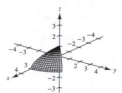

(d)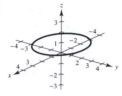

The space curve is a circle: $\mathbf{r}\left(u, \dfrac{\pi}{4}\right) = \dfrac{3\sqrt{2}}{2}\cos u\mathbf{i} + \dfrac{3\sqrt{2}}{2}\sin u\mathbf{j} + \dfrac{\sqrt{2}}{2}\mathbf{k}$

(e) $\mathbf{r}_u = -3\cos v\sin u\,\mathbf{i} + 3\cos v\cos u\,\mathbf{j}$

$\mathbf{r}_v = -3\sin v\sin u\,\mathbf{i} - 3\sin v\sin u\,\mathbf{j} + \cos v\,\mathbf{k}$

$$\mathbf{r}_u \times \mathbf{r}_v = \begin{vmatrix} \mathbf{i} & \mathbf{j} & \mathbf{k} \\ -3\cos v\sin u & 3\cos v\cos u & 0 \\ -3\sin v\sin u & 3\sin v\sin u & \cos v \end{vmatrix}$$

$= \left(3\cos^2 v\cos u\right)\mathbf{i} + \left(3\cos^2 v\sin u\right)\mathbf{j} + \left(9\cos v\sin v\sin^2 u + 9\cos v\sin v\cos^2 u\right)\mathbf{k}$

$= \left(3\cos^2 v\cos u\right)\mathbf{i} + \left(3\cos^2 v\sin u\right)\mathbf{j} + \left(9\cos v\sin v\right)\mathbf{k}$

$\|\mathbf{r}_u \times \mathbf{r}_v\| = \sqrt{9\cos^4 v\cos^2 u + 9\cos^4 v\sin^2 u + 81\cos^2 v\sin^2 v} = \sqrt{9\cos^4 v + 81\cos^2 v\sin^2 v}$

Using a Symbolic integration utility, $\displaystyle\int_{\pi/4}^{\pi/2}\int_0^{2\pi}\|\mathbf{r}_u \times \mathbf{r}_v\|\,dv\,du \approx 14.44$.

(f) Similarly, $\displaystyle\int_0^{\pi/4}\int_0^{\pi/2}\|\mathbf{r}_u \times \mathbf{r}_v\|\,dv\,du \approx 4.27$.

55. $S: \mathbf{r}(u, v) = u\cos v\mathbf{i} + u\sin v\mathbf{j} + (u - 1)(2 - u)\mathbf{k}, \quad 0 \le u \le 2, 0 \le v \le 2\pi$

$\mathbf{r}_u(u, v) = \cos v\mathbf{i} + \sin v\mathbf{j} + (3 - 2u)\mathbf{k}$

$\mathbf{r}_v(u, v) = -u\sin v\mathbf{i} + u\cos v\mathbf{j}$

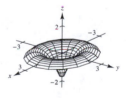

$$\mathbf{r}_u \times \mathbf{r}_v = \begin{vmatrix} \mathbf{i} & \mathbf{j} & \mathbf{k} \\ \cos v & \sin v & 3 - 2u \\ -u\sin v & u\cos v & 0 \end{vmatrix} = (2u - 3)u\cos v\mathbf{i} + (2u - 3)u\sin v\mathbf{j} + u\mathbf{k}$$

$\|\mathbf{r}_u \times \mathbf{r}_v\| = u\sqrt{(2u - 3)^2 + 1}$

$\displaystyle\int_S\!\!\int (x + y)\,dS = \int_0^{2\pi}\int_0^2 (u\cos v + u\sin v)u\sqrt{(2u - 3)^2 + 1}\,du\,dv = \int_0^2\int_0^{2\pi}(\cos v + \sin v)u^2\sqrt{(2u - 3)^2 + 1}\,dv\,du = 0$

57. $\mathbf{F}(x, y, z) = x^2\mathbf{i} + xy\mathbf{j} + z\mathbf{k}$

Q: solid region bounded by the coordinates planes and the plane $2x + 3y + 4z = 12$

Surface Integral: There are four surfaces for this solid.

$z = 0,\ \mathbf{N} = -\mathbf{k}, \mathbf{F} \cdot \mathbf{N} = -z,\ \int_{S_1} \int 0\, dS = 0$

$y = 0,\ \mathbf{N} = -\mathbf{j}, \mathbf{F} \cdot \mathbf{N} = -xy,\ \int_{S_2} \int 0\, dS = 0$

$x = 0,\ \mathbf{N} = -\mathbf{i}, \mathbf{F} \cdot \mathbf{N} = -x^2,\ \int_{S_3} \int 0\, dS = 0$

$2x + 3y + 4z = 12,\ \mathbf{N} = \dfrac{2\mathbf{i} + 3\mathbf{j} + 4\mathbf{k}}{\sqrt{29}},\ dS = \sqrt{1 + \left(\dfrac{1}{4}\right) + \left(\dfrac{9}{16}\right)}\, dA = \dfrac{\sqrt{29}}{4}\, dA$

$\displaystyle \int_{S_4} \int \mathbf{F} \cdot \mathbf{N}\, dS = \frac{1}{4} \int_R \int \left(2x^2 + 3xy + 4z\right) dA$

$\displaystyle \qquad = \frac{1}{4} \int_0^6 \int_0^{4-(2x/3)} \left(2x^2 + 3xy + 12 - 2x - 3y\right) dy\, dx$

$\displaystyle \qquad = \frac{1}{4} \int_0^6 \left[2x^2\left(\frac{12 - 2x}{3}\right) + \frac{3x}{2}\left(\frac{12 - 2x}{3}\right)^2 + 12\left(\frac{12 - 2x}{3}\right) - 2x\left(\frac{12 - 2x}{3}\right) - \frac{3}{2}\left(\frac{12 - 2x}{3}\right)^2 \right] dx$

$\displaystyle \qquad = \frac{1}{6} \int_0^6 \left(-x^3 + x^2 + 24x + 36\right) dx$

$\displaystyle \qquad = \frac{1}{6}\left[-\frac{x^4}{4} + \frac{x^3}{3} + 12x^2 + 36x \right]_0^6 = 66$

Divergence Theorem: Because div $\mathbf{F} = 2x + x + 1 = 3x + 1$, Divergence Theorem yields

$\displaystyle \iiint_Q \text{div } \mathbf{F}\, dV = \int_0^6 \int_0^{(12-2x)/3} \int_0^{(12-2x-3y)/4} \left(3x + 1\right) dz\, dy\, dx$

$\displaystyle \qquad = \int_0^6 \int_0^{(12-2x)/3} (3x + 1)\left(\frac{12 - 2x - 3y}{4}\right) dy\, dx$

$\displaystyle \qquad = \frac{1}{4} \int_0^6 (3x + 1)\left(12y - 2xy - \frac{3}{2}y^2 \right)_0^{(12-2x)/3} dx$

$\displaystyle \qquad = \frac{1}{4} \int_0^6 (3x + 1)\left[4(12 - 2x) - 2x\left(\frac{12 - 2x}{3}\right) - \frac{3}{2}\left(\frac{12 - 2x}{3}\right)^2 \right] dx$

$\displaystyle \qquad = \frac{1}{4} \int_0^6 \frac{2}{3}\left(3x^3 - 35x^2 + 96x + 36 \right) dx$

$\displaystyle \qquad = \frac{1}{6}\left[\frac{3x^4}{4} - \frac{35x^3}{3} + 48x^2 + 36x \right] = 66.$

59. $\mathbf{F}(x, y, z) = (\cos y + y \cos x)\mathbf{i} + (\sin x - x \sin y)\mathbf{j} + xyz\mathbf{k}$

S: portion of $z = y^2$ over the square in the xy-plane with vertices $(0, 0), (a, 0), (a, a), (0, a)$

Line Integral: Using the line integral you have:

C_1: $y = 0, \ dy = 0$

C_2: $x = 0, \ dx = 0, \ z = y^2, \ dz = 2y \, dy$

C_3: $y = a, \ dy = 0, \ z = a^2, \ dz = 0$

C_4: $x = a, \ dx = 0, \ z = y^2, \ dz = 2y \, dy$

$$\int_C \mathbf{F} \cdot d\mathbf{r} = \int_C (\cos y + y \cos x) \, dx + (\sin x - x \sin y) \, dy + xyz \, dz$$

$$= \int_{C_1} dx + \int_{C_2} 0 + \int_{C_3} (\cos a + a \cos x) \, dx + \int_{C_4} (\sin a - a \sin y) \, dy + ay^3 (2y \, dy)$$

$$= \int_0^a dx + \int_a^0 (\cos a + a \cos x) \, dx + \int_0^a (\sin a - a \sin y) \, dy + \int_0^a 2ay^4 \, dy$$

$$= a + \left[x \cos a + a \sin x \right]_a^0 + \left[y \sin a + a \cos y \right]_0^a + \left[2a \frac{y^5}{5} \right]_0^a$$

$$= a - a \cos a - a \sin a + a \sin a + a \cos a - a + \frac{2a^6}{5} = \frac{2a^6}{5}$$

Double Integral: Considering $f(x, y, z) = z - y^2$, you have:

$$\mathbf{N} = \frac{\nabla f}{\|\nabla f\|} = \frac{-2y\mathbf{j} + \mathbf{k}}{\sqrt{1 + 4y^2}}, \ dS = \sqrt{1 + 4y^2} \, dA, \text{ and } \mathbf{curl} \ \mathbf{F} = xz\mathbf{i} - yz\mathbf{j}.$$

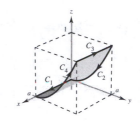

So, $\displaystyle \int_S \int (\mathbf{curl} \ \mathbf{F}) \cdot \mathbf{N} \, dS = \int_0^a \int_0^a 2y^2 z \, dy \, dx = \int_0^a \int_0^a 2y^4 z \, dy \, dx = \int_0^a \frac{2a^5}{5} \, dx = \frac{2a^6}{5}.$

61. If $\mathbf{curl} \, (\mathbf{F}) = x\mathbf{i} + y\mathbf{j} + z\mathbf{k}$, then $\text{div}(\mathbf{curl} \ \mathbf{F}) = 1 + 1 + 1 = 3$, contradicting Theorem 15.3.

Problem Solving for Chapter 15

1. (a) $\displaystyle \nabla T = \frac{-25}{\left(x^2 + y^2 + z^2\right)^{3/2}} \left[x\mathbf{i} + y\mathbf{j} + z\mathbf{k} \right]$

$$\mathbf{N} = x\mathbf{i} + \sqrt{1 - x^2} \, \mathbf{k}$$

$$dS = \frac{1}{\sqrt{1 - x^2}} \, dA$$

$$\text{Flux} = \int_S \int -k\nabla T \cdot \mathbf{N} \, dS = 25k \int_R \int \left[\frac{x^2}{\left(x^2 + y^2 + z^2\right)^{3/2} \left(1 - x^2\right)^{1/2}} + \frac{z}{\left(x^2 + y^2 + z^2\right)^{3/2}} \right] dA$$

$$= 25k \int_{-1/2}^{1/2} \int_0^1 \left[\frac{x^2}{\left(x^2 + y^2 + z^2\right)^{3/2} \left(1 - x^2\right)^{1/2}} + \frac{1 - x^2}{\left(x^2 + y^2 + z^2\right)^{3/2} \left(1 - x^2\right)^{1/2}} \right] dy \, dx$$

$$= 25k \int_{-1/2}^{1/2} \int_0^1 \frac{1}{\left(1 + y^2\right)^{3/2} \left(1 - x^2\right)^{1/2}} \, dy \, dx = 25k \int_0^1 \frac{1}{\left(1 + y^2\right)^{3/2}} \, dy \int_{-1/2}^{1/2} \frac{1}{\left(1 - x^2\right)^{1/2}} \, dx = 25k \left(\frac{\sqrt{2}}{2} \right) \left(\frac{\pi}{3} \right) = 25k \frac{\sqrt{2}\pi}{6}$$

(b) $\mathbf{r}(u,v) = \langle \cos u, v, \sin u \rangle$

$\mathbf{r}_u = \langle -\sin u, 0, \cos u \rangle, \ \mathbf{r}_v = \langle 0, 1, 0 \rangle$

$\mathbf{r}_u \times \mathbf{r}_v = \langle -\cos u, 0, \sin u \rangle$

$$\nabla T = \frac{-25}{\left(x^2 + y^2 + z^2\right)^{3/2}}[x\mathbf{i} + y\mathbf{j} + z\mathbf{k}] = \frac{-25}{\left(v^2 + 1\right)^{3/2}}[\cos u\mathbf{i} + v\mathbf{j} + \sin u\mathbf{k}]$$

$$\nabla T \cdot (\mathbf{r}_u \times \mathbf{r}_v) = \frac{-25}{\left(v^2 + 1\right)^{3/2}}\left(-\cos^2 u - \sin^2 u\right) = \frac{25}{\left(v^2 + 1\right)^{3/2}}$$

$$\text{Flux} = \int_0^1 \int_{\pi/3}^{2\pi/3} \frac{25}{\left(v^2 + 1\right)^{3/2}} \, du \, dv = 25k\frac{\sqrt{2}\pi}{6}$$

3. $\mathbf{r}(t) = \langle 3\cos t, 3\sin t, 2t \rangle$

$\mathbf{r}'(t) = \langle -3\sin t, 3\cos t, 2t \rangle, \|\mathbf{r}'(t)\| = \sqrt{13}$

$$I_x = \int_C \left(y^2 + z^2\right)\rho \, ds = \int_0^{2\pi} \left(9\sin^2 t + 4t^2\right)\sqrt{13} \, dt = \frac{1}{3}\sqrt{13}\pi\left(32\pi^2 + 27\right)$$

$$I_y = \int_C \left(x^2 + z^2\right)\rho \, ds = \int_0^{2\pi} \left(9\cos^2 t + 4t^2\right)\sqrt{13} \, dt = \frac{1}{3}\sqrt{13}\pi\left(32\pi^2 + 27\right)$$

$$I_z = \int_C \left(x^2 + y^2\right)\rho \, ds = \int_0^{2\pi} \left(9\cos^2 t + 9\sin^2 t\right)\sqrt{13} \, dt = 18\pi\sqrt{13}$$

5. (a) $\ln f = \frac{1}{2}\ln\left(x^2 + y^2 + z^2\right)$

$$\nabla(\ln f) = \frac{x}{x^2 + y^2 + z^2}\mathbf{i} + \frac{y}{x^2 + y^2 + z^2}\mathbf{j} + \frac{z}{x^2 + y^2 + z^2}\mathbf{k} = \frac{x\mathbf{i} + y\mathbf{j} + z\mathbf{k}}{x^2 + y^2 + z^2} = \frac{\mathbf{F}}{f^2}$$

(b) $\dfrac{1}{f} = \dfrac{1}{\sqrt{x^2 + y^2 + z^2}}$

$$\nabla\left(\frac{1}{f}\right) = \frac{-x}{\left(x^2 + y^2 + z^2\right)^{3/2}}\mathbf{i} + \frac{-y}{\left(x^2 + y^2 + z^2\right)^{3/2}}\mathbf{j} + \frac{-z}{\left(x^2 + y^2 + z^2\right)^{3/2}}\mathbf{k} = \frac{-(x\mathbf{i} + y\mathbf{j} + z\mathbf{k})}{\left(\sqrt{x^2 + y^2 + z^2}\right)^3} = \frac{\mathbf{F}}{f^3}$$

(c) $f^n = \left(\sqrt{x^2 + y^2 + z^2}\right)^n$

$$\nabla f^n = n\left(\sqrt{x^2 + y^2 + z^2}\right)^{n-1}\frac{x}{\sqrt{x^2 + y^2 + z^2}}\mathbf{i} + n\left(\sqrt{x^2 + y^2 + z^2}\right)^{n-1}\frac{y}{\sqrt{x^2 + y^2 + z^2}}\mathbf{j}$$

$$+ \, n\left(\sqrt{x^2 + y^2 + z^2}\right)^{n-1}\frac{z}{\sqrt{x^2 + y^2 + z^2}}\mathbf{k}$$

$$= n\left(\sqrt{x^2 + y^2 + z^2}\right)^{n-2}(x\mathbf{i} + y\mathbf{j} + z\mathbf{k})$$

$$= nf^{n-2}\mathbf{F}$$

(d) $w = \dfrac{1}{f} = \dfrac{1}{\sqrt{x^2 + y^2 + z^2}}$ $\qquad \dfrac{\partial^2 w}{dx^2} = \dfrac{2x^2 - y^2 - z^2}{\left(x^2 + y^2 + z^2\right)^{5/2}}$

$\dfrac{\partial w}{dx} = -\dfrac{x}{\left(x^2 + y^2 + z^2\right)^{3/2}}$ $\qquad \dfrac{\partial^2 w}{dy^2} = \dfrac{2y^2 - x^2 - z^2}{\left(x^2 + y^2 + z^2\right)^{5/2}}$

$\dfrac{\partial w}{dy} = -\dfrac{y}{\left(x^2 + y^2 + z^2\right)^{3/2}}$ $\qquad \dfrac{\partial^2 w}{dz^2} = \dfrac{2z^2 - x^2 - y^2}{\left(x^2 + y^2 + z^2\right)^{5/2}}$

$\dfrac{\partial w}{dz} = -\dfrac{z}{\left(x^2 + y^2 + z^2\right)^{3/2}}$ $\qquad \nabla^2 w = \dfrac{\partial^2 w}{dx^2} + \dfrac{\partial^2 w}{dy^2} + \dfrac{\partial^2 w}{dz^2} = 0$

Therefore $w = \dfrac{1}{f}$ is harmonic.

7. $\dfrac{1}{2}\displaystyle\int_C x\,dy - y\,dx = \dfrac{1}{2}\displaystyle\int_0^{2\pi}\left[a(\theta - \sin\theta)(a\sin\theta)\,d\theta - a(1 - \cos\theta)(a(1 - \cos\theta))\,d\theta\right]$

$\qquad = \dfrac{1}{2}a^2\displaystyle\int_0^{2\pi}\left[\theta\sin\theta = \sin^2\theta - 1 + 2\cos\theta - \cos^2\theta\right]d\theta = \dfrac{1}{2}a^2\displaystyle\int_0^{2\pi}\left(\theta\sin\theta + 2\cos\theta - 2\right)d\theta = -3\pi a^2$

So, the area is $3\pi a^2$.

9. (a) $\mathbf{r}(t) = t\mathbf{j}, 0 \le t \le 1$

$\mathbf{r}'(t) = \mathbf{j}$

$W = \displaystyle\int_C \mathbf{F} \cdot d\mathbf{r} = \int_0^1 (t\mathbf{i} + \mathbf{j}) \cdot \mathbf{j}\,dt = \int_0^1 dt = 1$

(b) $\mathbf{r}(t) = \left(t - t^2\right)\mathbf{i} + t\mathbf{j}, 0 \le t \le 1$

$\mathbf{r}'(t) = (1 - 2t)\mathbf{i} + \mathbf{j}$

$W = \mathbf{F} \cdot d\mathbf{r} = \displaystyle\int_0^1 \left(\left(2t - t^2\right)\mathbf{i} + \left[\left(t - t^2\right)^2 + 1\right]\mathbf{j}\right) \cdot \left((1 - 2t)\mathbf{i} + \mathbf{j}\right)dt$

$\qquad = \displaystyle\int_0^1\left[(1 - 2t)\left(2t - t^2\right) + \left(t^4 - 2t^3 + t^2 + 1\right)\right]dt = \int_0^1\left(t^4 - 4t^2 + 2t + 1\right)dt = \dfrac{13}{15}$

(c) $\mathbf{r}(t) = c\left(t - t^2\right)\mathbf{i} + t\mathbf{j}, 0 \le t \le 1$

$\mathbf{r}'(t) = c(1 - 2t)\mathbf{i} + \mathbf{j}$

$\mathbf{F} \cdot d\mathbf{r} = \left(c\left(t - t^2\right) + t\right)\left(c(1 - 2t)\right) + \left(c^2\left(t - t^2\right)^2 + 1\right)(1)$

$\qquad = c^2 t^4 - 2c^2 t^3 + c^2 t - 2ct^2 + ct + 1$

$W = \displaystyle\int_C \mathbf{F} \cdot d\mathbf{r} = \dfrac{1}{30}c^2 - \dfrac{1}{6}c + 1$

$\dfrac{dW}{dc} = \dfrac{1}{15}c - \dfrac{1}{6} = 0 \Rightarrow c = \dfrac{5}{2}$

$\dfrac{d^2 W}{dc^2} = \dfrac{1}{15} > 0$ $c = \dfrac{5}{2}$ minimum.

11. $\mathbf{v} \times \mathbf{r} = \langle a_1, a_2, a_3\rangle \times \langle x, y, z\rangle$

$\qquad = \langle a_2 z - a_3 y, -a_1 z + a_3 x, a_1 y - a_2 x\rangle$

$\mathbf{curl}(\mathbf{v} \times \mathbf{r}) = \langle 2a_1, 2a_2, 2a_3\rangle = 2\mathbf{v}$

By Stokes's Theorem,

$\displaystyle\int_C (\mathbf{v} \times \mathbf{r})\,d\mathbf{r} = \int_S\int \mathbf{curl}(\mathbf{v} \times \mathbf{r}) \cdot \mathbf{N}\,dS = \int_S\int 2\mathbf{v} \cdot \mathbf{N}\,dS.$

13. (a) (i) $\int_C f\nabla g \cdot d\mathbf{r} = \int_S \int \mathbf{curl}[f\nabla g] \cdot \mathbf{N}\, dS$ (Stokes's Theorem)

$$f\nabla g = f\frac{\partial g}{\partial x}\mathbf{i} + f\frac{\partial g}{\partial y}\mathbf{j} + f\frac{\partial g}{\partial z}\mathbf{k}$$

$$\mathbf{curl}\,(f\nabla g) = \begin{vmatrix} \mathbf{i} & \mathbf{j} & \mathbf{k} \\ \dfrac{\partial}{\partial x} & \dfrac{\partial}{\partial y} & \dfrac{\partial}{\partial z} \\ f(\partial g/\partial x) & f(\partial g/\partial y) & f(\partial g/\partial z) \end{vmatrix}$$

$$= \left[\left[f\left(\frac{\partial^2 g}{\partial y \partial z}\right) + \left(\frac{\partial f}{\partial y}\right)\left(\frac{\partial g}{\partial z}\right)\right] - \left[f\left(\frac{\partial^2 g}{\partial z \partial y}\right) + \left(\frac{\partial f}{\partial z}\right)\left(\frac{\partial g}{\partial y}\right)\right]\right]\mathbf{i}$$

$$- \left[\left[f\left(\frac{\partial^2 g}{\partial x \partial z}\right) + \left(\frac{\partial f}{\partial x}\right)\left(\frac{\partial g}{\partial z}\right)\right] - \left[f\left(\frac{\partial^2 g}{\partial z \partial x}\right) + \left(\frac{\partial f}{\partial z}\right)\left(\frac{\partial g}{\partial x}\right)\right]\right]\mathbf{j}$$

$$+ \left[\left[f\left(\frac{\partial^2 g}{\partial x \partial y}\right) + \left(\frac{\partial f}{\partial x}\right)\left(\frac{\partial g}{\partial y}\right)\right] - \left[f\left(\frac{\partial^2 g}{\partial y \partial x}\right) + \left(\frac{\partial f}{\partial y}\right)\left(\frac{\partial g}{\partial x}\right)\right]\right]\mathbf{k}$$

$$= \left[\left(\frac{\partial f}{\partial y}\right)\left(\frac{\partial g}{\partial z}\right) - \left(\frac{\partial f}{\partial z}\right)\left(\frac{\partial g}{\partial y}\right)\right]\mathbf{i} - \left[\left(\frac{\partial f}{\partial x}\right)\left(\frac{\partial g}{\partial z}\right) - \left(\frac{\partial f}{\partial z}\right)\left(\frac{\partial g}{\partial x}\right)\right]\mathbf{j} + \left[\left(\frac{\partial f}{\partial x}\right)\left(\frac{\partial g}{\partial y}\right) - \left(\frac{\partial f}{\partial y}\right)\left(\frac{\partial g}{\partial x}\right)\right]\mathbf{k}$$

$$= \begin{vmatrix} \mathbf{i} & \mathbf{j} & \mathbf{k} \\ \dfrac{\partial f}{\partial x} & \dfrac{\partial f}{\partial y} & \dfrac{\partial f}{\partial z} \\ \dfrac{\partial g}{\partial x} & \dfrac{\partial g}{\partial y} & \dfrac{\partial g}{\partial z} \end{vmatrix} = \nabla f \times \nabla g$$

So, $\int_C f\nabla g \cdot d\mathbf{r} = \int_S \int \mathbf{curl}[f\nabla g] \cdot \mathbf{N}\, dS = \int_S \int [\nabla f \times \nabla g] \cdot \mathbf{N}\, dS.$

(ii) $\quad \int_C (f\nabla f) \cdot d\mathbf{r} = \int_S \int (\nabla f \times \nabla f) \cdot \mathbf{N}\, dS$ (using part a)

$$= 0 \text{ because } \nabla f \times \nabla f = \mathbf{0}.$$

(iii) $\int_C (f\nabla g + g\nabla f) \cdot d\mathbf{r} = \int_C (f\nabla g) \cdot d\mathbf{r} + \int_C (g\nabla f) \cdot d\mathbf{r}$

$$= \int_S \int (\nabla f \times \nabla g) \cdot \mathbf{N}\, dS + \int_S \int (\nabla g \times \nabla f) \cdot \mathbf{N}\, dS \text{ (using part a)}$$

$$= \int_S \int (\nabla f \times \nabla g) \cdot \mathbf{N}\, dS + \int_S \int -(\nabla f \times \nabla g) \cdot \mathbf{N}\, dS = 0$$

(b) $f(x, y, z) = xyz, \ g(x, y, z) = z, \ S\!: z = \sqrt{4 - x^2 - y^2}$

(i) $\nabla g(x, y, z) = \mathbf{k}$

$f(x, y, z)\nabla g(x, y, z) = xyz\mathbf{k}$

$\mathbf{r}(t) = 2\cos t\,\mathbf{i} + 2\sin t\,\mathbf{j} + 0\mathbf{k}, \ 0 \le t \le 2\pi$

$\int_C [f(x, y, z)\nabla g(x, y, z)] \cdot d\mathbf{r} = 0$

(ii) $\nabla f(x, y, z) = yz\mathbf{i} + xz\mathbf{j} + xy\mathbf{k}$

$\nabla g(x, y, z) = \mathbf{k}$

$$\nabla f \times \nabla g = \begin{vmatrix} \mathbf{i} & \mathbf{j} & \mathbf{k} \\ yz & xz & xy \\ 0 & 0 & 1 \end{vmatrix} = xz\mathbf{i} - yz\mathbf{j}$$

$$\mathbf{N} = \frac{x}{\sqrt{4 - x^2 - y^2}}\mathbf{i} + \frac{y}{\sqrt{4 - x^2 - y^2}}\mathbf{j} + \mathbf{k}$$

$$dS = \sqrt{1 + \left(\frac{-x}{\sqrt{4 - x^2 - y^2}}\right)^2 + \left(\frac{-y}{\sqrt{4 - x^2 - y^2}}\right)^2}\, dA = \frac{2}{\sqrt{4 - x^2 - y^2}}\, dA$$

$$\int_S \int \left[\nabla f(x, y, z) \times \nabla g(x, y, z)\right] \cdot \mathbf{N}\, dS = \int_S \int \left[\frac{x^2 z}{\sqrt{4 - x^2 - y^2}} - \frac{y^2 z}{\sqrt{4 - x^2 - y^2}}\right] \frac{2}{\sqrt{4 - x^2 - y^2}}\, dA$$

$$= \int_S \int \frac{2(x^2 - y^2)}{\sqrt{4 - x^2 - y^2}}\, dA$$

$$= \int_0^2 \int_0^{2\pi} \frac{2r^2(\cos^2\theta - \sin^2\theta)}{\sqrt{4 - r^2}} r\, d\theta\, dr = \int_0^2 \left[\frac{2r^3}{\sqrt{4 - r^2}}\left(\frac{1}{2}\sin 2\theta\right)\right]_0^{2\pi}\, dr = 0$$

C H A P T E R   1 6
Additional Topics in Differential Equations

CHAPTER 16
Additional Topics in Differential Equations

Section 16.1   Exact First-Order Equations

1. $\left(2x + xy^2\right) dx + \left(3 + x^2y\right) dy = 0$

$\dfrac{\partial M}{\partial y} = 2xy$

$\dfrac{\partial N}{\partial x} = 2xy$

$\dfrac{\partial M}{\partial y} = \dfrac{\partial N}{\partial x}$ Exact

3. $x \sin y \, dx + x \cos y \, dy = 0$

$\dfrac{\partial M}{\partial y} = x \cos y$

$\dfrac{\partial N}{\partial x} = \cos y$

$\dfrac{\partial M}{\partial y} \neq \dfrac{\partial N}{\partial x}$ Not exact

5. $\left(2x - 3y\right) dx + \left(2y - 3x\right) dy = 0$

$\dfrac{\partial M}{\partial y} = -3 = \dfrac{\partial N}{\partial x}$ Exact

$f(x, y) = \int M(x, y) \, dx$

$\qquad = \int \left(2x - 3y\right) dx$

$\qquad = x^2 - 3xy + g(y)$

$f_y(x, y) = -3x + g'(y)$

$\qquad = 2y - 3x \Rightarrow g'(y) = 2y$

$\Rightarrow g(y) = y^2 + C_1$

$f(x, y) = x^2 - 3xy + y^2 + C_1$

$x^2 - 3xy + y^2 = C$

7. $\left(3y^2 + 10xy^2\right) dx + \left(6xy - 2 + 10x^2y\right) = 0$

$\dfrac{\partial M}{\partial y} = 6y + 20xy = \dfrac{\partial N}{\partial x}$ Exact

$f(x, y) = \int M(x, y) \, dx = \int \left(3y^2 + 10xy^2\right) dx$

$\qquad = 3xy^2 + 5x^2y^2 + g(y)$

$f_y(x, y) = 6xy + 10x^2y + g'(y) = 6xy - 2 + 10x^2y$

$\qquad\qquad \Rightarrow g'(y) = -2 \Rightarrow g(y) = -2y + C_1$

$f(x, y) = 3xy^2 + 5x^2y^2 - 2y + C_1$

$3xy^2 + 5x^2y^2 - 2y = C$

9. $\left(4x^3 - 6xy^2\right) dx + \left(4y^3 - 6xy\right) dy = 0$

$\dfrac{\partial M}{\partial y} = -12xy$

$\dfrac{\partial N}{\partial x} = -6y$

Not exact

11. $\dfrac{-y}{x^2 + y^2} dx + \dfrac{x}{x^2 + y^2} dy = 0$

$\dfrac{\partial M}{\partial y} = \dfrac{y^2 - x^2}{\left(x^2 + y^2\right)^2} = \dfrac{\partial N}{\partial x}$ Exact

$f(x, y) = \int M(x, y) \, dx = -\arctan\left(\dfrac{x}{y}\right) + g(y)$

$f_y(x, y) = \dfrac{x}{x^2 + y^2} + g'(y)$

$\qquad = \dfrac{x}{x^2 + y^2} \Rightarrow g'(y) = 0 \Rightarrow g(y) = C_1$

$f(x, y) = -\arctan\left(\dfrac{x}{y}\right) + C_1$

$\arctan\left(\dfrac{x}{y}\right) = C$

13. $\left(\dfrac{y}{x - y}\right)^2 dx + \left(\dfrac{x}{x - y}\right)^2 dy = 0$

$\dfrac{\partial M}{\partial y} = \dfrac{2xy}{\left(x - y\right)^3}$

$\dfrac{\partial N}{\partial x} = \dfrac{-2xy}{\left(x - y\right)^3}$

Not exact

15. (a) and (c)

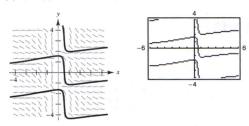

(b) $(2x \tan y + 5)\, dx + (x^2 \sec^2 y)\, dy = 0, \ y\left(\dfrac{1}{2}\right) = \dfrac{\pi}{4}$

$$\dfrac{\partial M}{\partial y} = 2x \sec^2 y = \dfrac{\partial N}{\partial x} \Rightarrow \text{Exact}$$

$$f(x, y) = \int M(x, y)\, dx = \int (2x \tan y + 5)\, dx$$

$$= x^2 \tan y + 5x + g(y)$$

$$f_y(x, y) = x^2 \sec^2 y + g'(y)$$

$$= x^2 \sec^2 y$$

$$\Rightarrow g'(y) = 0 \Rightarrow g(y) = C$$

$$f(x, y) = x^2 \tan y + 5x = C$$

$$f\left(\dfrac{1}{2}, \dfrac{\pi}{4}\right) = \dfrac{1}{4} + \dfrac{5}{2} = \dfrac{11}{4} = C$$

Answer: $x^2 \tan y + 5x = \dfrac{11}{4}$

17. $\dfrac{y}{x-1}\, dx + \left[\ln(x-1) + 2y\right] dy = 0$

$$\dfrac{\partial M}{\partial y} = \dfrac{1}{x-1} = \dfrac{\partial N}{\partial x} \text{ Exact}$$

$$f(x, y) = \int M(x, y)\, dx = y \ln(x-1) + g(y)$$

$$f_y(x, y) = \ln(x-1) + g'(y)$$

$$\Rightarrow g'(y) = 2y \Rightarrow g(y) = y^2 + C_1$$

$$f(x, y) = y \ln(x-1) + y^2 + C_1$$

$$y \ln(x-1) + y^2 = C$$

$$y(2) = 4: 4 \ln(2-1) + 16 = C \Rightarrow C = 16$$

Solution: $y \ln(x-1) + y^2 = 16$

19. $\left(e^{3x} \sin 3y\right) dx + \left(e^{3x} \cos 3y\right) dy = 0$

$$\dfrac{\partial M}{\partial y} = 3e^{3x} \cos 3y = \dfrac{\partial N}{\partial x} \text{ Exact}$$

$$f(x, y) = \int M(x, y)\, dx$$

$$= \int e^{3x} \sin 3y\, dx = \dfrac{1}{3} e^{3x} \sin 3y + g(y)$$

$$f_y(x, y) = e^{3x} \cos 3y + g'(y)$$

$$\Rightarrow g'(y) = 0 \Rightarrow g(y) = C_1$$

$$f(x, y) = \dfrac{1}{3} e^{3x} \sin 3y + C_1$$

$$e^{3x} \sin 3y = C$$

$$y(0) = \pi: C = 0$$

Solution: $e^{3x} \sin 3y = 0$

21. $\left(2xy - 9x^2\right) dx + \left(2y + x^2 + 1\right) dy = 0$

$$\dfrac{\partial M}{\partial y} = 2x = \dfrac{\partial N}{\partial x} \text{ Exact}$$

$$f(x, y) = \int M(x, y)\, dx = \int (2xy - 9x^2)\, dx$$

$$= x^2 y - 3x^3 + g(y)$$

$$f_y(x, y) = x^2 + g'(y) = 2y + x^2 + 1$$

$$\Rightarrow g'(y) = 2y + 1$$

$$\Rightarrow g(y) = y^2 + y + C_1$$

$$f(x, y) = x^2 y - 3x^2 + y^2 + y + C_1$$

$$x^2 y - 3x^2 + y^2 + y = C$$

$$y(0) = -3: 9 - 3 = 6 = C$$

Solution: $x^2 y - 3x^2 + y^2 + y = 6$

23. $y\, dx - \left(x + 6y^2\right) dy = 0$

$$\dfrac{(\partial N / \partial x) - (\partial M / \partial y)}{M} = -\dfrac{2}{y} = k(y)$$

Integrating factor: $e^{\int k(y)\, dy} = e^{\ln y^{-2}} = \dfrac{1}{y^2}$

Exact equation: $\dfrac{1}{y}\, dx - \left(\dfrac{x}{y^2} + 6\right) dy = 0$

$$f(x, y) = \dfrac{x}{y} + g(y)$$

$$g'(y) = -6$$

$$g(y) = -6y + C_1$$

$$\dfrac{x}{y} - 6y = C$$

25. $\left(5x^2 - y\right)dx + x\,dy = 0$

$$\frac{(\partial M/\partial y) - (\partial N/\partial x)}{N} = -\frac{2}{x} = h(x)$$

Integrating factor: $e^{\int h(x)\,dx} = e^{\ln x^{-2}} = \frac{1}{x^2}$

Exact equation: $\left(5 - \frac{y}{x^2}\right)dx + \frac{1}{x}\,dy = 0$

$f(x, y) = 5x + \frac{y}{x} + g(y)$

$g'(y) = 0$

$g(y) = C_1$

$5x + \frac{y}{x} = C$

27. $(x + y)\,dx + (\tan x)\,dy = 0$

$$\frac{(\partial M/\partial y) - (\partial N/\partial x)}{N} = -\tan x = h(x)$$

Integrating factor: $e^{\int h(x)\,dx} = e^{\ln \cos x} = \cos x$

Exact equation: $(x + y)\cos x\,dx + \sin x\,dy = 0$

$f(x, y) = x\sin x + \cos x + y\sin x + g(y)$

$g'(y) = 0$

$g(y) = C_1$

$x\sin x + \cos x + y\sin x = C$

29. $y^2\,dx + (xy - 1)\,dy = 0$

$$\frac{(\partial N/\partial x) - (\partial M/\partial y)}{M} = -\frac{1}{y} = k(y)$$

Integrating factor: $e^{\int k(y)\,dy} = e^{\ln(1/y)} = \frac{1}{y}$

Exact equation: $y\,dx + \left(x - \frac{1}{y}\right)dy = 0$

$f(x, y) = xy + g(y)$

$g'(y) = -\frac{1}{y}$

$g(y) = -\ln|y| + C_1$

$xy - \ln|y| = C$

31. $2y\,dx + \left(x - \sin\sqrt{y}\right)dy = 0$

$$\frac{(\partial N/\partial x) - (\partial M/\partial y)}{M} = \frac{-1}{2y} = k(y)$$

Integrating factor: $e^{\int k(y)\,dy} = e^{\ln(1/\sqrt{y})} = \frac{1}{\sqrt{y}}$

Exact equation: $2\sqrt{y}\,dy + \left(\frac{x}{\sqrt{y}} - \frac{\sin\sqrt{y}}{\sqrt{y}}\right)dy = 0$

$f(x, y) = 2\sqrt{y}\,x + g(y)$

$g'(y) = -\frac{\sin\sqrt{y}}{\sqrt{y}}$

$g(y) = 2\cos\sqrt{y} + C_1$

$\sqrt{y}\,x + \cos\sqrt{y} = C$

33. $\left(4x^2y + 2y^2\right)dx + \left(3x^3 + 4xy\right)dy = 0$

Integrating factor: xy^2

Exact equation:

$$\left(4x^3y^3 + 2xy^4\right)dy + \left(3x^4y^2 + 4x^2y^3\right)dy = 0$$

$f(x, y) = x^4y^3 + x^2y^4 + g(y)$

$g'(y) = 0$

$g(y) = C_1$

$x^4y^3 + x^2y^4 = C$

35. $\left(-y^5 + x^2y\right)dx + \left(2xy^4 - 2x^3\right)dy = 0$

Integrating factor: $x^{-2}y^{-3}$

Exact equation:

$$\left(-\frac{y^2}{x^2} + \frac{1}{y^2}\right)dx + \left(2\frac{y}{x} - 2\frac{x}{y^3}\right)dy = 0$$

$f(x, y) = \frac{y^2}{x} + \frac{x}{y^2} + g(y)$

$g'(y) = 0$

$g(y) = C_1$

$\frac{y^2}{x} + \frac{x}{y^2} = C$

37. $y \, dx - x \, dy = 0$

(a) $\dfrac{1}{x^2}, \dfrac{y}{x^2} \, dx - \dfrac{1}{x} \, dy = 0, \dfrac{\partial M}{\partial y} = \dfrac{1}{x^2} = \dfrac{\partial N}{\partial x}$

(b) $\dfrac{1}{y^2}, \dfrac{1}{y} \, dx - \dfrac{x}{y^2} \, dy = 0, \dfrac{\partial M}{\partial y} = \dfrac{-1}{y^2} = \dfrac{\partial N}{\partial x}$

(c) $\dfrac{1}{xy}, \dfrac{1}{x} \, dx - \dfrac{1}{y} \, dy = 0, \dfrac{\partial M}{\partial y} = 0 = \dfrac{\partial N}{\partial x}$

(d) $\dfrac{1}{x^2 + y^2}, \dfrac{y}{x^2 + y^2} \, dx - \dfrac{x}{x^2 + y^2} \, dy = 0,$

$\dfrac{\partial M}{\partial y} = \dfrac{x^2 - y^2}{\left(x^2 + y^2\right)^2} = \dfrac{\partial N}{\partial x}$

39. $\mathbf{F}(x, y) = \dfrac{y}{\sqrt{x^2 + y^2}} \mathbf{i} - \dfrac{x}{\sqrt{x^2 + y^2}} \mathbf{j}$

$\dfrac{dy}{dx} = -\dfrac{x}{y}$

$y \, dy + x \, dx = 0$

$y^2 + x^2 = C$

Family of circles

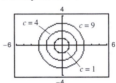

41. $\mathbf{F}(x, y) = 4x^2 y \mathbf{i} - \left(2xy^2 + \dfrac{x}{y^2}\right) \mathbf{j}$

$\dfrac{dy}{dx} = \dfrac{-y}{2x} - \dfrac{1}{4xy^3}$

$\dfrac{8y^3}{2y^4 + 1} \, dy = -\dfrac{2}{x} \, dx$

$\ln\left(2y^4 + 1\right) = \ln\left(\dfrac{1}{x^2}\right) + \ln C$

$2y^4 + 1 = \dfrac{C}{x^2}$

$2x^2 y^4 + x^2 = C$

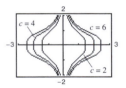

43. $E(x) = \dfrac{20x - y}{2y - 10x} = \dfrac{x \, dy}{y \, dx}$

$\left(20xy - y^2\right) dx + \left(10x^2 - 2xy\right) dy = 0$

$\dfrac{\partial M}{\partial y} = 20x - 2y = \dfrac{\partial N}{\partial x}$

$f(x, y) = 10x^2 y - xy^2 + g(y)$

$g'(y) = 0$

$g(y) = C_1$

$10x^2 y - xy^2 = K$

Initial condition: $C(100) = 500, 100 \le x, K = 25{,}000{,}000$

$$10x^2 y - xy^2 = 25{,}000{,}000$$

$xy^2 - 10x^2 y + 25{,}000{,}000 = 0$ Quadratic Formula

$$y = \dfrac{10x^2 + \sqrt{100x^4 - 4x(25{,}000{,}000)}}{2x} = \dfrac{5\left(x^2 + \sqrt{x^4 - 1{,}000{,}000x}\right)}{x}$$

45. $\dfrac{dy}{dx} = \dfrac{-2xy}{x^2 + y^2}$

$2xy\,dx + \left(x^2 + y^2\right)dy = 0$

$\dfrac{\partial M}{\partial y} = 2x = \dfrac{\partial N}{\partial x}$

$f(x, y) = x^2 y + g(y)$

$g'(y) = y^2$

$g(y) = \dfrac{y^3}{3} + C_1$

$3x^2 y + y^3 = C$

Initial condition: $y(0) = 2, 8 = C$

Particular solution: $3x^2 y + y^3 = 8$

47. (a) $y(4) \approx 0.5231$

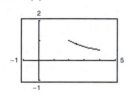

(b) $\dfrac{dy}{dx} = \dfrac{-xy}{x^2 + y^2}$

$xy\,dx + \left(x^2 + y^2\right)dy = 0$

$\dfrac{1}{M}\left[N_x - M_y\right] = \dfrac{1}{xy}[2x - x] = \dfrac{1}{y}$ function of y alone.

Integrating factor: $e^{\int (1/y)\,dy} = e^{\ln y} = y$

$xy^2\,dx + \left(x^2 y + y^3\right)dy = 0$

$f(x, y) = \int xy^2\,dx = \dfrac{x^2 y^2}{2} + g(y)$

$f_y(x, y) = x^2 y + g'(y) \Rightarrow g(y) = \dfrac{y^4}{4} + C_1$

$f(x, y) = \dfrac{x^2 y^2}{2} + \dfrac{y^4}{4} = C$

Initial condition: $y(2) = 1, \dfrac{4}{2} + \dfrac{1}{4} = \dfrac{9}{4} = C$

Particular solution: $\dfrac{x^2 y^2}{2} + \dfrac{y^4}{4} = \dfrac{9}{4}$ or

$2x^2 y^2 + y^4 = 9.$

For $x = 4$, $32y^2 + y^4 = 9 \Rightarrow y(4) = 0.528$

(c)

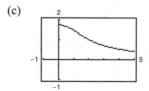

49. (a) $y(4) \approx 0.408$

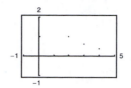

(b) $\dfrac{dy}{dx} = \dfrac{-xy}{x^2 + y^2}$

$xy\,dx + \left(x^2 + y^2\right)dy = 0$

$\dfrac{1}{M}\left[N_x - M_y\right] = \dfrac{1}{xy}[2x - x]$

$= \dfrac{1}{y}$ function of y alone.

Integrating factor: $e^{\int 1/y\,dy} = e^{\ln y} = y$

$xy^2\,dx + \left(x^2 y + y^3\right)dy = 0$

$f(x, y) = \int xy^2\,dx = \dfrac{x^2 y^2}{2} + g(y)$

$f_y(x, y) = x^2 y + g'(y) \Rightarrow g(y) = \dfrac{y^4}{4} + C_1$

$f(x, y) = \dfrac{x^2 y^2}{2} + \dfrac{y^4}{4} = C$

Initial condition: $y(2) = 1, \dfrac{4}{2} + \dfrac{1}{4} = \dfrac{9}{4} = C$

Particular solution: $\dfrac{x^2 y^2}{2} + \dfrac{y^4}{4} = \dfrac{9}{4}$ or

$2x^2 y^2 + y^4 = 9.$

For $x = 4$, $32y^2 + y^4 = 9 \Rightarrow y(4) = 0.528$

(c)

The solution is less accurate. For Exercise 47, Euler's Method gives $y(4) \approx 0.523$, whereas in Exercise 49, you obtain $y(4) \approx 0.408$. The errors are $0.528 - 0.523 = 0.005$ and $0.528 - 0.408 = 0.120$.

51. If M and N have continuous partial derivatives on an open disc R, then $M(x, y)\,dx + N(x, y)\,dy = 0$ is exact if and only if $\dfrac{\partial M}{\partial y} = \dfrac{\partial N}{\partial x}$.

53. False

$$\frac{\partial M}{\partial y} = 2x \text{ and } \frac{\partial N}{\partial x} = -2x$$

55. True

$$\frac{\partial}{\partial y}\big[f(x) + M\big] = \frac{\partial M}{\partial y} \text{ and } \frac{\partial}{\partial x}\big[g(y) + N\big] = \frac{\partial N}{\partial x}$$

57. $M = xy^2 + kx^2y + x^3$, $N = x^3 + x^2y + y^2$

$$\frac{\partial M}{\partial y} = 2xy + kx^2, \quad \frac{\partial N}{\partial x} = 3x^2 + 2xy$$

$$\frac{\partial M}{\partial y} = \frac{\partial N}{\partial x} \Rightarrow k = 3$$

59. $M = g(y)\sin x$, $N = y^2 f(x)$

$$\frac{\partial M}{\partial y} = g'(y)\sin x, \quad \frac{\partial N}{\partial x} = y^2 f'(x)$$

$$\frac{\partial M}{\partial y} = \frac{\partial N}{\partial x}: \ g'(y)\sin x = f'(x)y^2$$

$$g'(y) = y^2 \Rightarrow g(y) = \frac{y^3}{3} + C_1$$

$$f'(x) = \sin x \Rightarrow f(x) = -\cos x + C_2$$

Section 16.2   Second-Order Homogeneous Linear Equations

1.
$$y = C_1 e^{-3x} + C_2 xe^{-3x}$$
$$y' = -3C_1 e^{-3x} + C_2 e^{-3x} - 3C_2 xe^{-3x}$$
$$y'' = 9C_1 e^{-3x} - 6C_2 e^{-3x} + 9C_2 xe^{-3x}$$
$$y'' + 6y' + 9y = \big(9C_1 e^{-3x} - 6C_2 e^{-3x} + 9C_2 xe^{-3x}\big) + \big(-18C_1 e^{-3x} + 6C_2 e^{-3x} - 18C_2 xe^{-3x}\big) + \big(9C_1 e^{-3x} + 9C_2 xe^{-3x}\big) = 0$$

y approaches zero as $x \to \infty$.

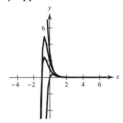

3.
$$y = C_1 \cos 2x + C_2 \sin 2x$$
$$y' = -2C_1 \sin 2x + 2C_2 \cos 2x$$
$$y'' = -4C_1 \cos 2x - 4C_2 \sin 2x = -4y$$
$$y'' + 4y = -4y + 4y = 0$$

The graphs are basically the same shape, with left and right shifts and varying ranges.

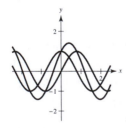

5. $y'' - y' = 0$

Characteristic equation: $m^2 - m = 0$

Roots: $m = 0, 1$

$$y = C_1 + C_2 e^x$$

7. $y'' - y' - 6y = 0$

Characteristic equation: $m^2 - m - 6 = 0$

Roots: $m = 3, -2$

$$y = C_1 e^{3x} + C_2 e^{-2x}$$

9. $2y'' + 3y' - 2y = 0$

Characteristic equation: $2m^2 + 3m - 2 = 0$

Roots: $m = \frac{1}{2}, -2$

$$y = C_1 e^{(1/2)x} + C_2 e^{-2x}$$

11. $y'' + 6y' + 9y = 0$

Characteristic equation: $m^2 + 6m + 9 = 0$

Roots: $m = -3, -3$

$$y = C_1 e^{-3x} + C_2 xe^{-3x}$$

13. $16y'' - 8y' + y = 0$

Characteristic equation: $16m^2 - 8m + 1 = 0$

Roots: $m = \frac{1}{4}, \frac{1}{4}$

$$y = C_1 e^{(1/4)x} + C_2 xe^{(1/4)x}$$

15. $y'' + y = 0$

Characteristic equation: $m^2 + 1 = 0$

Roots: $m = -i, i$

$y = C_1 \cos x + C_2 \sin x$

17. $y'' - 9y = 0$

Characteristic equation: $m^2 - 9 = 0$

Roots: $m = -3, 3$

$y = C_1 e^{3x} + C_2 e^{-3x}$

19. $y'' - 2y' + 4y = 0$

Characteristic equation: $m^2 - 2m + 4 = 0$

Roots: $m = 1 - \sqrt{3}i, 1 + \sqrt{3}i$

$y = e^x \left(C_1 \cos \sqrt{3}x + C_2 \sin \sqrt{3}x \right)$

21. $y'' - 3y' + y = 0$

Characteristic equation: $m^2 - 3m + 1 = 0$

Roots: $m = \dfrac{3 - \sqrt{5}}{2}, \dfrac{3 + \sqrt{5}}{2}$

$y = C_1 e^{[(3+\sqrt{5})/2]x} + C_2 e^{[(3-\sqrt{5})/2]x}$

23. $9y'' - 12y' + 11y = 0$

Characteristic equation: $9m^2 - 12m + 11 = 0$

Roots: $m = \dfrac{2 + \sqrt{7}i}{3}, \dfrac{2 - \sqrt{7}i}{3}$

$y = e^{(2/3)x} \left[C_1 \cos\left(\dfrac{\sqrt{7}}{3}x\right) + C_2 \sin\left(\dfrac{\sqrt{7}}{3}x\right) \right]$

25. $y^{(4)} - y = 0$

Characteristic equation: $m^4 - 1 = 0$

Roots: $m = -1, 1, -i, i$

$y = C_1 e^x + C_2 e^{-x} + C_3 \cos x + C_4 \sin x$

27. $y''' - 6y'' + 11y' - 6y = 0$

Characteristic equation: $m^3 - 6m^2 + 11m - 6 = 0$

Roots: $m = 1, 2, 3$

$y = C_1 e^x + C_2 e^{2x} + C_3 e^{3x}$

29. $y''' - 3y'' + 7y' - 5y = 0$

Characteristic equation: $m^3 - 3m^2 + 7m - 5 = 0$

Roots: $m = 1, 1 - 2i, 1 + 2i$

$y = C_1 e^x + e^x \left(C_2 \cos 2x + C_3 \sin 2x \right)$

31. $y'' + 100y = 0$

$y = C_1 \cos 10x + C_2 \sin 10x$

$y' = -10C_1 \sin 10x + 10C_2 \cos 10x$

(a) $y(0) = 2$: $2 = C_1$

$y'(0) = 0$: $0 = 10C_2 \Rightarrow C_2 = 0$

Particular solution: $y = 2 \cos 10x$

(b) $y(0) = 0$: $0 = C_1$

$y'(0) = 2$: $2 = 10C_2 \Rightarrow C_2 = \frac{1}{5}$

Particular solution: $y = \frac{1}{5} \sin 10x$

(c) $y(0) = -1$: $-1 = C_1$

$y'(0) = 3$: $3 = 10C_2 \Rightarrow C_2 = \frac{3}{10}$

Particular solution: $y = -\cos 10x + \frac{3}{10} \sin 10x$

33. $y'' - y' - 30y = 0, \ y(0) = 1, \ y'(0) = -4$

Characteristic equation: $m^2 - m - 30 = 0$

Roots: $m = 6, -5$

$y = C_1 e^{6x} + C_2 e^{-5x}, \ y' = 6C_1 e^{6x} - 5C_2 e^{-5x}$

Initial conditions:

$y(0) = 1, \ y'(0) = -4, \ 1 = C_1 + C_2, \ -4 = 6C_1 - 5C_2$

Solving simultaneously: $C_1 = \dfrac{1}{11}, C_2 = \dfrac{10}{11}$

Particular solution: $y = \dfrac{1}{11}\left(e^{6x} + 10e^{-5x} \right)$

35. $y'' + 16y = 0, \ y(0) = 0, \ y'(0) = 2$

Characteristic equation: $m^2 + 16 = 0$

Roots: $m = \pm 4i$

$y = C_1 \cos 4x + C_2 \sin 4x$

$y' = -4C_1 \sin 4x + 4C_2 \cos 4x$

Initial conditions: $y(0) = 0 = C_1$

$y'(0) = 2 = 4C_2 \Rightarrow C_2 = \frac{1}{2}$

Particular solution: $y = \frac{1}{2} \sin 4x$

37. $9y'' - 6y' + y = 0$, $y(0) = 2$, $y'(0) = 1$

Characteristic equation: $9m^2 - 6m + 1 = 0$

Roots: $m = \frac{1}{3}, \frac{1}{3}$

$y = C_1 e^{(1/3)x} + C_2 x e^{(1/3)x}$

$y' = \frac{1}{3} C_1 e^{(1/3)x} + \frac{1}{3} C_2 x e^{(1/3)x} + C_2 e^{(1/3)x}$

Initial conditions: $y(0) = 2$, $y'(0) = 1$

$\left. \begin{array}{l} C_1 = 2 \\ \frac{1}{3} C_1 + C_2 = 1 \end{array} \right\} \Rightarrow C_1 = 2, C_2 = \frac{1}{3}$

Particular solution: $y = 2e^{x/3} + \frac{1}{3} x e^{x/3}$

39. $y'' - 4y' + 3y = 0$, $y(0) = 1$, $y(1) = 3$

Characteristic equation: $m^2 - 4m + 3 = 0$

Roots: $m = 1, 3$

$y = C_1 e^x + C_2 e^{3x}$

$y(0) = 1$: $C_1 + C_2 = 1$

$y(1) = 3$: $C_1 e + C_2 e^3 = 3$

Solving simultaneously, $C_1 = \dfrac{e^3 - 3}{e^3 - e}, C_2 = \dfrac{3 - e}{e^3 - e}$

Solution: $y = \dfrac{e^3 - 3}{e^3 - e} e^x + \dfrac{3 - e}{e^3 - e} e^{3x}$

41. $y'' + 9y = 0$, $y(0) = 3$, $y(\pi) = 5$

Characteristic equation: $m^2 + 9m = 0$

Roots: $m = \pm 3i$

$y = C_1 \cos 3x + C_2 \sin 3x$

$y(0) = 3$: $C_1 = 3$

$y(\pi) = 5$: $-C_1 = 5$

No solution

43. $4y'' - 28y' + 49y = 0$, $y(0) = 2$, $y(1) = -1$

Characteristic equation: $4m^2 - 28m + 49 = 0$

Roots: $m = \dfrac{7}{2}, \dfrac{7}{2}$

$y = C_1 e^{(7/2)x} + C_2 x e^{(7/2)x}$

$y(0) = 2$: $C_1 = 2$

$y(1) = -1$: $C_1 e^{7/2} + C_2 e^{7/2} = -1 \Rightarrow C_2 = \dfrac{-1 - 2e^{7/2}}{e^{7/2}}$

Solution: $y = 2e^{(7/2)x} + \left(\dfrac{-1 - 2e^{7/2}}{e^{7/2}} \right) x e^{(7/2)x}$

45. Answers will vary. See Theorem 16.4.

47. The motion of a spring in a shock absorber is damped.

49. $y'' + 9y = 0$

Undamped vibration

Period: $\dfrac{2\pi}{3}$

Matches (b)

50. $y'' + 25y = 0$

Undamped vibration

Period: $\dfrac{2\pi}{5}$

Matches (d)

51. $y'' + 2y' + 10y = 0$

Damped vibration

Matches (c)

52. $y'' + y' + \frac{37}{4} y = 0$

Damped vibration

Matches (a)

53. By Hooke's Law, $F = kx$

$$k = \frac{F}{x} = \frac{32}{2/3} = 48.$$

Also, $F = ma$, and $m = \dfrac{F}{a} = \dfrac{32}{32} = 1.$

So, $y = \dfrac{1}{2} \cos\left(4\sqrt{3}\, t\right)$

55. $y = C_1 \cos\left(\sqrt{k/m}\, t\right) + C_2 \sin\left(\sqrt{k/m}\, t\right),$

$\sqrt{k/m} = \sqrt{48} = 4\sqrt{3}$

Initial conditions: $y(0) = -\dfrac{2}{3}$, $y'(0) = \dfrac{1}{2}$

$y = C_1 \cos\left(4\sqrt{3}\, t\right) + C_2 \sin\left(4\sqrt{3}\, t\right)$

$y(0) = C_1 = -\dfrac{2}{3}$

$y'(t) = -4\sqrt{3}\, C_1 \sin\left(4\sqrt{3}\, t\right) + 4\sqrt{3}\, C_2 \cos\left(4\sqrt{3}\, t\right)$

$y'(0) = 4\sqrt{3}\, C_2 = \dfrac{1}{2} \Rightarrow C_2 = \dfrac{1}{8\sqrt{3}} = \dfrac{\sqrt{3}}{24}$

$y(t) = -\dfrac{2}{3} \cos\left(4\sqrt{3}\, t\right) + \dfrac{\sqrt{3}}{24} \sin\left(4\sqrt{3}\, t\right)$

57. By Hooke's Law, $32 = k(2/3)$, so $k = 48$. Moreover, because the weight w is given by mg, it follows that $m = w/g = 32/32 = 1$. Also, the damping force is given by $(-1/8)(dy/dt)$. So, the differential equation for the oscillations of the weight is

$$m\left(\frac{d^2y}{dt^2}\right) = -\frac{1}{8}\left(\frac{dy}{dt}\right) - 48y$$

$$m\left(\frac{d^2y}{dt^2}\right) + \frac{1}{8}\left(\frac{dy}{dt}\right) + 48y = 0.$$

In this case the characteristic equation is $8m^2 + m + 384 = 0$ with complex roots $m = (-1/16) \pm \left(\sqrt{12,287}/16\right)i$.

So, the general solution is $y(t) = e^{-t/16}\left(C_1 \cos\dfrac{\sqrt{12,287}t}{16} + C_2 \sin\dfrac{\sqrt{12,287}t}{16}\right)$.

Using the initial conditions, you have $y(0) = C_1 = \dfrac{1}{2}$

$$y'(t) = e^{-t/16}\left[\left(-\frac{\sqrt{12,287}}{16}C_1 - \frac{C_2}{16}\right)\sin\frac{\sqrt{12,287}t}{16} + \left(\frac{\sqrt{12,287}}{16}C_2 - \frac{C_1}{16}\right)\cos\frac{\sqrt{12,287}t}{16}\right]$$

$$y'(0) = \frac{\sqrt{12,287}}{16}C_2 - \frac{C_1}{16} = 0 \Rightarrow C_2 = \frac{\sqrt{12,287}}{24,574}$$

and the particular solution is

$$y(t) = \frac{e^{-t/16}}{2}\left(\cos\frac{\sqrt{12,287}t}{16} + \frac{\sqrt{12,287}}{12,287}\sin\frac{\sqrt{12,287}t}{16}\right).$$

59. Because $m = -a/2$ is a double root of the characteristic equation, you have

$$\left(m + \frac{a}{2}\right)^2 = m^2 + am + \frac{a^2}{4} = 0$$

and the differential equation is $y'' + ay' + \left(a^2/4\right)y = 0$. The solution is

$$y = (C_1 + C_2x)e^{-(a/2)x}$$

$$y' = \left(-\frac{C_1a}{2} + C_2 - \frac{C_2a}{2}x\right)e^{-(a/2)x}$$

$$y'' = \left(\frac{C_1a^2}{4} - aC_2 + \frac{C_2a^2}{4}x\right)e^{-(a/2)x}$$

$$y'' + ay' + \frac{a^2}{4}y = e^{-(a/2)x}\left[\left(\frac{C_1a^2}{4} - C_2a + \frac{C_2a^2}{4}x\right) + \left(-\frac{C_1a^2}{2} + C_2a - \frac{C_2a^2}{2}x\right) + \left(\frac{C_1a^2}{4} + \frac{C_2a^2}{4}x\right)\right] = 0.$$

61. False. The general solution is $y = C_1e^{3x} + C_2xe^{3x}$.

63. True

65. $y_1 = e^{ax}$, $y_2 = e^{bx}$, $a \neq b$

$$W(y_1, y_2) = \begin{vmatrix} e^{ax} & e^{bx} \\ ae^{ax} & be^{bx} \end{vmatrix}$$

$$= (b - a)e^{ax+bx} \neq 0 \text{ for any value of } x.$$

67. $y_1 = e^{ax}\sin bx$, $y_2 = e^{ax}\cos bx$, $b \neq 0$

$$W(y_1, y_2) = \begin{vmatrix} e^{ax}\sin bx & e^{ax}\cos bx \\ ae^{ax}\sin bx + be^{ax}\cos bx & ae^{ax}\cos bx - be^{ax}\sin bx \end{vmatrix}$$

$$= -be^{2ax}\sin^2 bx - be^{2ax}\cos^2 bx = -be^{2ax} \neq 0 \text{ for any value of } x.$$

Section 16.3 Second-Order Nonhomogeneous Linear Equations

1. $y = 2e^{2x} - 2\cos x$

$y' = 4e^{2x} + 2\sin x$

$y'' = 8e^{2x} + 2\cos x$

$y'' + y = \left(8e^{2x} + 2\cos x\right) + \left(2e^{2x} - \cos x\right) = 10e^{2x}$

3. $y = 3\sin x - \cos x \ln|\sec x + \tan x|$

$y' = 3\cos x - 1 + \sin x \ln|\sec x + \tan x|$

$y'' = -3\sin x + \tan x + \cos x \ln|\sec x + \tan x|$

$y'' + y = \left(-3\sin x + \tan x + \cos x \ln|\sec x \tan x|\right) + \left(3\sin x - \cos x \ln|\sec x + \tan x|\right) = \tan x$

5. $y'' + 7y' + 12y = 3x + 1$

$y'' + 7y' + 12y = 0$

$m^2 - 7m + 12 = (m - 3)(m - 4) = 0$ when $m = 3, 4$

$y_h = C_1 e^{3x} + C_2 e^{4x}$

$y_p = A_0 + A_1 x$

$y_p' = A_1, \; y_p'' = 0$

$y_p'' + 7y_p' + 12y_p = 7A_1 + 12(A_0 + A_1 x) = 3x + 1$

$\left.\begin{array}{r} 12A_1 = 3 \\ 7A_1 + 12A_0 = 1 \end{array}\right\} \Rightarrow A_1 = \frac{1}{4}, A_0 = -\frac{1}{16}$

Solution: $y_p = -\frac{1}{16} + \frac{1}{4}x$

7. $y'' - 8y' + 16y = e^{3x}$

$y'' - 8y' + 16y = 0$

$m^2 - 8m + 16 = (m - 4)^2 = 0$ when $m = 4$

$y_h = C_1 e^{4x} + C_2 x e^{4x}$

$y_p = Ae^{3x}, \; y_p' = 3Ae^{3x}, \; y_p'' = 9Ae^{3x}$

$y_p'' - 8y_p' + 16y_p = 9Ae^{3x} - 8\left(3Ae^{3x}\right) + 16\left(Ae^{3x}\right)$
$\qquad = e^{3x}$

$9A - 24A + 16A = 1 \Rightarrow A = 1$

Solution: $y_p = e^{3x}$

9. $y'' - 2y' - 15y = \sin x$

$y'' - 2y' - 15y = 0$

$m^2 - 2m - 15 = (m - 5)(m + 3) = 0 \Rightarrow m = 5, -3$

$y_p = A\sin x + B\cos x$

$y_p' = A\cos x - B\sin x$

$y_p'' = -A\sin x - B\cos x$

$y_p'' - 2y_p' - 15y_p = (-A\sin x - B\cos x) - 2(A\cos x - B\sin x) - 15(A\sin x + B\cos x) = \sin x$

$(-A + 2B - 15A)\sin x + (-B - 2A - 15B)\cos x = \sin x$

$\left.\begin{array}{r} -16A + 2B = 1 \\ -2A - 16B = 0 \end{array}\right\} \Rightarrow A = -\frac{4}{65}, B = \frac{1}{130}$

Solution: $y_p = -\dfrac{4}{65}\sin x + \dfrac{1}{130}\cos x$

11. $y'' - 3y' + 2y = 2x$

$y'' - 3y' + 2y = 0$

$m^2 - 3m + 2 = 0$ when $m = 1, 2.$

$y_h = C_1 e^x + C_2 e^{2x}$

$y_p = A_0 + A_1 x$

$y_p' = A_1$

$y_p'' = 0$

$y_p'' - 3y_p' + 2y_p = (2A_0 - 3A_1) + 2A_1 x = 2x$

$\left. \begin{array}{l} 2A_0 - 3A_1 = 0 \\ 2A_1 = 2 \end{array} \right\} A_1 = 1, A_0 = \frac{3}{2}$

$y = C_1 e^x + C_2 e^{2x} + x + \frac{3}{2}$

13. $y'' + 2y' = 2e^x$

$y'' + 2y' = 0$

$m^2 + 2m = 0$ when $m = 0, -2.$

$y_h = C_1 + C_2 e^{-2x}$

$y_p = Ae^x = y_p' = y_p''$

$y_p'' + 2y_p' = 3Ae^x = 2e^x$ or $A = \frac{2}{3}$

$y = C_1 + C_2 e^{-2x} + \frac{2}{3} e^x$

15. $y'' - 10y' + 25y = 5 + 6e^x$

$y'' - 10y' + 25y = 0$

$m^2 - 10m + 25 = 0$ when $m = 5, 5.$

$y_h = C_1 e^{5x} + C_2 x e^{5x}$

$y_p = A_0 + A_1 e^x$

$y_p' = y_p'' = A_1 e^x$

$y_p'' - 10y_p' + 25y_p = 25A_0 + 16A_1 e^x = 5 + 6e^x$

or $A_0 = \frac{1}{5}, A_1 = \frac{3}{8}$

$y = (C_1 + C_2 x)e^{5x} + \frac{3}{8} e^x + \frac{1}{5}$

17. $y'' + 9y = \sin 3x$

$y'' + 9y = 0$

$m^2 + 9 = 0$ when $m = -3i, 3i.$

$y_h = C_1 \cos 3x + C_2 \sin 3x$

$y_p = A_0 \sin 3x + A_1 x \sin 3x + A_2 \cos 3x + A_3 x \cos 3x$

$y_p'' = (-9A_0 - 6A_3) \sin 3x - 9A_1 x \sin 3x$

$\quad + (6A_1 - 9A_2) \cos 3x - 9A_3 x \cos 3x$

$y_p'' + 9y_p = -6A_3 \sin 3x + 6A_1 \cos 3x = \sin 3x,$

$A_1 = 0, A_3 = -\frac{1}{6}$

$y = \left(C_1 - \frac{1}{6}x\right) \cos 3x + C_2 \sin 3x$

19. $y'' + y = x^3, y(0) = 1, y'(0) = 0$

$y'' + y = 0$

$m^2 + 1 = 0$ when $m = i, -i.$

$y_h = C_1 \cos x + C_2 \sin x$

$y_p = A_0 + A_1 x + A_2 x^2 + A_3 x^3$

$y_p' = A_1 + 2A_2 x + 3A_3 x^2$

$y_p'' = 2A_2 + 6A_3 x$

$y_p'' + y_p = A_3 x^3 + A_2 x^2 + (A_1 + 6A_3)x + (A_0 + 2A_2)$

$\quad = x^3$

or $A_3 = 1, A_2 = 0, A_1 = -6, A_0 = 0$

$y = C_1 \cos x + C_2 \sin x + x^3 - 6x$

$y' = -C_1 \sin x + C_2 \cos x + 3x^2 - 6$

Initial conditions:

$y(0) = 1, y'(0) = 0, 1 = C_1, 0 = C_2 - 6, C_2 = 6$

Particular solution: $y = \cos x + 6 \sin x + x^3 - 6x$

21. $y'' + y' = 2 \sin x, y(0) = 0, y'(0) = -3$

$y'' + y' = 0$

$m^2 + m = 0$ when $m = 0, -1.$

$y_h = C_1 + C_2 e^{-x}$

$y_p = A \cos x + B \sin x$

$y_p' = -A \sin x + B \cos x$

$y_p'' = -A \cos x - B \sin x$

$y_p'' + y_p' = (-A + B) \cos x + (-A - B) \sin x = 2 \sin x$

$\left. \begin{array}{l} -A + B = 0 \\ -A - B = 2 \end{array} \right\} A = -1, B = -1$

$y = C_1 + C_2 e^x - (\cos x + \sin x)$

$y' = -C_2 e^{-x} - (-\sin x + \cos x)$

Initial conditions: $y(0) = 0, y'(0) = -3,$

$\quad 0 = C_1 + C_2 - 1, -3 = -C_2 - 1,$

$\quad C_2 = 2, C_1 = -1$

Particular solution: $y = -1 + 2e^{-x} - (\cos x + \sin x)$

23. $y' - 4y = xe^x - xe^{4x}$, $y(0) = \frac{1}{3}$

$y' - 4y = 0$

$m - 4 = 0$ when $m = 4$.

$y_h = Ce^{4x}$

$y_p = (A_0 + A_1 x)e^x + (A_2 x + A_3 x^2)e^{4x}$

$y_p' = (A_0 + A_1 x)e^x + A_1 e^x$
$\qquad + 4(A_2 x + A_3 x^2)e^{4x} + (A_2 + 2A_3 x)e^{4x}$

$y_p' - 4y_p = (-3A_0 - 3A_1 x)e^x + A_1 e^x + A_2 e^{4x}$
$\qquad\qquad + 2A_3 x e^{4x} = xe^x - xe^{4x}$

$A_0 = -\frac{1}{9}, A_1 = -\frac{1}{3}, A_2 = 0, A_3 = -\frac{1}{2}$

$y = \left(C - \frac{1}{2}x^2\right)e^{4x} - \frac{1}{9}(1 + 3x)e^x$

Initial conditions: $y(0) = \frac{1}{3}$, $\frac{1}{3} = C - \frac{1}{9}$, $C = \frac{4}{9}$

Particular solution: $y = \left(\frac{4}{9} - \frac{1}{2}x^2\right)e^{4x} - \frac{1}{9}(1 + 3x)e^x$

25. $y'' + y = \sec x$

$y'' + y = 0$

$m^2 + 1 = 0$ when $m = -i, i$.

$y_h = C_1 \cos x + C_2 \sin x$

$y_p = v_1 \cos x + v_2 \sin x$

$v_1' \cos x + v_2' \sin x = 0$

$v_1'(-\sin x) + v_2'(\cos x) = \sec x$

$v_1' = \dfrac{\begin{vmatrix} 0 & \sin x \\ \sec x & \cos x \end{vmatrix}}{\begin{vmatrix} \cos x & \sin x \\ -\sin x & \cos x \end{vmatrix}} = -\tan x$

$v_1 = \int -\tan x \, dx = \ln|\cos x|$

$v_2' = \dfrac{\begin{vmatrix} \cos x & 0 \\ -\sin x & \sec x \end{vmatrix}}{\begin{vmatrix} \cos x & \sin x \\ -\sin x & \cos x \end{vmatrix}} = 1$

$v_2 = \int dx = x$

$y = (C_1 + \ln|\cos x|)\cos x + (C_2 + x)\sin x$

27. $y'' + 4y = \csc 2x$

$y'' + 4y = 0$

$m^2 + 4 = 0$ when $m = -2i, 2i$.

$y_h = C_1 \cos 2x + C_2 \sin 2x$

$y_p = v_1 \cos 2x + v_2 \sin 2x = 0$

$v_1' \cos 2x + v_2' \sin 2x = 0$

$v_1'(-2\sin 2x) + v_2'(2\cos 2x) = \csc 2x$

$v_1' = \dfrac{\begin{vmatrix} 0 & \sin 2x \\ \csc 2x & 2\cos 2x \end{vmatrix}}{\begin{vmatrix} \cos 2x & \sin 2x \\ -2\sin 2x & 2\cos 2x \end{vmatrix}} = -\dfrac{1}{2}$

$v_1 = \int -\dfrac{1}{2} \, dx = -\dfrac{1}{2}x$

$v_2' = \dfrac{\begin{vmatrix} \cos 2x & 0 \\ -2\sin 2x & \csc 2x \end{vmatrix}}{\begin{vmatrix} \cos 2x & \sin 2x \\ -2\sin 2x & 2\cos 2x \end{vmatrix}} = \dfrac{1}{2}\cot 2x$

$v_2 = \int \dfrac{1}{2}\cot 2x \, dx = \dfrac{1}{4}\ln|\sin 2x|$

$y = \left(C_1 - \dfrac{1}{2}x\right)\cos 2x + \left(C_2 + \dfrac{1}{4}\ln|\sin 2x|\right)\sin 2x$

29. $y'' - 2y' + y = e^x \ln x$

$y'' - 2y' + y = 0$

$m^2 - 2m + 1 = 0$ when $m = 1, 1$.

$y_h = (C_1 + C_2 x)e^x$

$y_p = (v_1 + v_2 x)e^x$

$v_1' e^x + v_2' x e^x = 0$

$v_1' e^x + v_2'(x + 1)e^x = e^x \ln x$

$v_1' = -x \ln x$

$v_1 = \int -x \ln x \, dx = -\dfrac{x^2}{2}\ln x + \dfrac{x^2}{4}$

$v_2' = \ln x$

$v_2 = \int \ln x \, dx = x \ln x - x$

$y = (C_1 + C_2 x)e^x + \dfrac{x^2 e^x}{4}(\ln x^2 - 3)$

31. (a) $y'' - y' - 12y = 0$

$m^2 - m - 12 = (m - 4)(m + 3) = 0 \Rightarrow m = 4, -3$

Let $y_p = Ax^2 + Bx + C$. This is a generalized form of $F(x) = x^2$.

(b) $y'' - y' - 12y = 0$

$m^2 - m - 12 = (m - 4)(m + 3) = 0 \Rightarrow m = 4, -3$

Because $y_h = C_1 e^{4x} + C_2 e^{-3x}$, let

$y_p = Axe^{4x}$.

33. $q'' + 10q' + 25q = 6 \sin 5t, q(0) = 0, q'(0) = 0$

$m^2 + 10m + 25 = 0$ when $m = -5, -5$.

$q_h = (C_1 + C_2 t)e^{-5t}$

$q_p = A \cos 5t + B \sin 5t$

$q_p' = -5A \sin 5t + 5B \cos 5t$

$q_p'' = -25A \cos 5t - 25B \sin 5t$

$q_p'' + 10q_p' + 25q_p = 50B \cos 5t - 50A \sin 5t$

$\qquad\qquad = 6 \sin 5t, A = -\tfrac{3}{25}, B = 0$

$q = (C_1 + C_2 t)e^{-5t} - \tfrac{3}{25} \cos 5t$

Initial conditions:

$q(0) = 0, q'(0) = 0, C_1 - \tfrac{3}{25} = 0, -5C_1 + C_2 = 0,$

$C_1 = \tfrac{3}{25}, C_2 = \tfrac{3}{5}$

Particular solution: $q = \tfrac{3}{25}\left(e^{-5t} + 5te^{-5t} - \cos 5t\right)$

35. $\tfrac{24}{32} y'' + 48y = \tfrac{24}{32}(48 \sin 4t), y(0) = \tfrac{1}{4}, y'(0) = 0$

$\tfrac{24}{32} m^2 + 48 = 0$ when $m = \pm 8i$.

$y_h = C_1 \cos 8t + C_2 \sin 8t$

$y_p = A \sin 4t + B \cos 4t$

$y_p' = 4A \cos 4t - 4B \sin 4t$

$y_p'' = -16A \sin 4t - 16B \cos 4t$

$\tfrac{24}{32} y_p'' + 48y_p = 36A \sin 4t + 36B \cos 4t$

$\qquad\qquad = \tfrac{24}{32}(48 \sin 4t), B = 0, A = 1$

$y = y_h + y_p = C_1 \cos 8t + C_2 \sin 8t + \sin 4t$

Initial conditions: $y(0) = \tfrac{1}{4}, y'(0) = 0, \tfrac{1}{4} = C_1,$

$\qquad\qquad 0 = 8C_2 + 4 \Rightarrow C_2 = -\tfrac{1}{2}$

Particular solution: $y = \tfrac{1}{4} \cos 8t - \tfrac{1}{2} \sin 8t + \sin 4t$

37. $\tfrac{2}{32} y'' + y' + 4y = \tfrac{2}{32}(4 \sin 8t), y(0) = \tfrac{1}{4}, y'(0) = -3$

$\tfrac{1}{16} m^2 + m + 4 = 0$

when $m = -8, -8$.

$y_h = (C_1 + C_2 t)e^{-8t}$

$y_p = A \sin 8t + B \cos 8t$

$y_p' = 8A \cos 8t - 8B \sin 8t$

$y_p'' = -64A \sin 8t - 64B \cos 8t$

$\tfrac{2}{32} y_p'' + y_p' + 4y_p = -8B \sin 8t + 8A \cos 8t$

$\qquad\qquad = \tfrac{2}{32}(4 \sin 8t) - 8B$

$\qquad\qquad = \tfrac{1}{4} \Rightarrow B = -\tfrac{1}{32}, 8A = 0 \Rightarrow A = 0$

Initial conditions:

$y(0) = \tfrac{1}{4}, y'(0) = -3, \tfrac{1}{4} = C_1 - \tfrac{1}{32} \Rightarrow C_1 = \tfrac{9}{32},$

$-3 = -8C_1 + C_2 \Rightarrow C_2 = -\tfrac{3}{4}$

Particular solution: $y = \left(\tfrac{9}{32} - \tfrac{3}{4} t\right)e^{-8t} - \tfrac{1}{32} \cos 8t$

39. In Exercise 35,

$y_h = \dfrac{1}{4} \cos 8t - \dfrac{1}{2} \sin 8t - \dfrac{\sqrt{5}}{4} \sin\left[8t + \pi + \arctan\left(-\dfrac{1}{2}\right)\right] = \dfrac{\sqrt{5}}{4} \sin\left(8t + \pi - \arctan \dfrac{1}{2}\right) \approx \dfrac{\sqrt{5}}{4} \sin(8t + 2.6779).$

41. (a) $\frac{4}{32}y'' + \frac{25}{2}y = 0$

$y = C_1 \cos 10x + C_2 \sin 10x$

$y(0) = \frac{1}{2}: \frac{1}{2} = C_1$

$y'(0) = -4: -4 = 10C_2 \Rightarrow C_2 = -\frac{2}{5}$

$y = \frac{1}{2}\cos 10x - \frac{2}{5}\sin 10x$

The motion is undamped.

(b) If $b > 0$, the motion is damped.

(c) If $b > \frac{5}{2}$, the solution to the differential equation is of the form $y = C_1 e^{m_1 x} + C_2 e^{m_2 x}$.

There would be no oscillations in this case.

43. True. $y_p = -e^{2x}\cos e^{-x}$

$y_p' = e^{2x}\sin e^{-x}(-e^{-x}) - 2e^{2x}\cos e^{-x} = -e^x \sin e^{-x} - 2e^{2x}\cos e^{-x}$

$y_p'' = \left[-e^x \cos e^{-x}(-e^{-x}) - e^x \sin e^{-x}\right] + \left[2e^{2x}\sin e^{-x}(-e^{-x}) - 4e^{2x}\cos e^{-x}\right]$

So,

$y_p'' - 3y_p' + 2y_p = \left[\cos e^{-x} - e^x \sin e^{-x} - 2e^x \sin e^{-x} - 4e^{2x}\cos e^{-x}\right] - 3\left[-e^x \sin e^{-x} - 2e^{2x}\cos e^{-x}\right] - 2e^{2x}\cos e^{-x}$

$\qquad = \left[-e^x - 2e^x + 3e^x\right]\sin e^{-x} + \left[1 - 4e^{2x} + 6e^{2x} - 2e^{2x}\right]\cos e^{-x} = \cos e^{-x}$.

45. $y'' - 2y' + y = 2e^x$

$m^2 - 2m + 1 = 0 \Rightarrow m = 1, 1$

$y_h = C_1 e^x + C_2 x e^x$, $y_p = x^2 e^x$, particular solution

General solution: $f(x) = (C_1 + C_2 x)e^x + x^2 e^x = (C_1 + C_2 x + x^2)e^x$

$f'(x) = (C_2 + 2x + C_1 + C_2 x + x^2)e^x = (x^2 + (C_2 + 2)x + (C_1 + C_2))e^x$

(a) No. If $f(x) > 0$ for all x, then $x^2 + C_2 x + C_1 > 0 \Leftrightarrow C_2^2 - 4C_1 < 0$ for all x.

So, let $C_1 = C_2 = 1$. Then $f'(x) = (x^2 + 3x + 2)e^x$ and $f'(-\frac{3}{2}) = -\frac{1}{4} < 0$.

(b) Yes. If $f'(x) > 0$ for all x, then

$(C_2 + 2)^2 - 4(C_1 + C_2) < 0$

$\Rightarrow C_2^2 - 4C_1 + 4 < 0$

$\qquad C_2^2 - 4C_1 < -4$

$\qquad C_2^2 - 4C_1 < 0$

$\Rightarrow f(x) > 0$ for all x.

Section 16.4   Series Solutions of Differential Equations

1. $y' - y = 0.$ Letting $y = \sum_{n=0}^{\infty} a_n x^n$:

$$y' - y = \sum_{n=1}^{\infty} na_n x^{n-1} - \sum_{n=0}^{\infty} a_n x^n = \sum_{n=0}^{\infty} (n+1)a_{n+1}x^n - \sum_{n=0}^{\infty} a_n x^n = 0$$

$$(n+1)a_{n+1} = a_n$$

$$a_{n+1} = \frac{a_n}{n+1}$$

$$a_1 = a_0, a_2 = \frac{a_1}{2} = \frac{a_0}{2}, a_3 = \frac{a_2}{3} = \frac{a_0}{1 \cdot 2 \cdot 3}, \ldots, a_n = \frac{a_0}{n!}$$

$$y = \sum_{n=0}^{\infty} \frac{a_0}{n!} x^n = a_0 e^x$$

Check: By separation of variables, you have:

$$\int \frac{dy}{y} = \int dx$$

$$\ln y = x + C_1$$

$$y = Ce^x$$

3. $y'' - 9y = 0.$ Letting $y = \sum_{n=0}^{\infty} a_n x^n$:

$$y'' - 9y = \sum_{n=2}^{\infty} n(n-1)a_n x^{n-2} - 9\sum_{n=0}^{\infty} a_n x^n = \sum_{n=0}^{\infty} (n+2)(n+1)a_{n+2}x^n - \sum_{n=0}^{\infty} 9a_n x^n = 0$$

$$(n+2)(n+1)a_{n+2} = 9a_n$$

$$a_{n+2} = \frac{9a_n}{(n+2)(n+1)}$$

$$a_0 = a_0 \qquad\qquad a_1 = a_1$$

$$a_2 = \frac{9a_0}{2} \qquad\qquad a_3 = \frac{9a_1}{3 \cdot 2}$$

$$a_4 = \frac{9a_2}{4 \cdot 3} = \frac{9^2 a_0}{4 \cdot 3 \cdot 2 \cdot 1} \qquad a_5 = \frac{9a_3}{5 \cdot 4} = \frac{9^2 a_1}{5 \cdot 4 \cdot 3 \cdot 2 \cdot 1}$$

$$\vdots \qquad\qquad\qquad \vdots$$

$$a_{2n} = \frac{9^n a_0}{(2n)!} \qquad\qquad a_{2n+1} = \frac{9^n a_1}{(2n+1)!}$$

$$y = \sum_{n=0}^{\infty} \frac{9^n a_0}{(2n)!} x^{2n} + \sum_{n=0}^{\infty} \frac{9^n a_1}{(2n+1)!} x^{2n+1} = a_0 \sum_{n=0}^{\infty} \frac{(3x)^{2n}}{(2n)!} + \frac{a_1}{3} \sum_{n=0}^{\infty} \frac{(3x)^{2n+1}}{(2n+1)!} = C_0 \sum_{n=0}^{\infty} \frac{(3x)^n}{n!} + C_1 \sum_{n=0}^{\infty} \frac{(-3x)^n}{n!}$$

$$= C_0 e^{3x} + C_1 e^{-3x} \text{ where } C_0 + C_1 = a_0 \text{ and } C_0 - C_1 = \frac{a_1}{3}.$$

Check: $y'' - 9y = 0$ is a second-order homogeneous linear equation.

$$m^2 - 9 = 0 \Rightarrow m_1 = 3 \text{ and } m_2 = -3$$

$$y = C_1 e^{3x} + C_2 e^{-3x}$$

5. $y'' + 4y = 0$. Letting $y = \sum\limits_{n=0}^{\infty} a_n x^n$:

$$y'' + 4y = \sum_{n=2}^{\infty} n(n-1)a_n x^{n-2} + 4\sum_{n=0}^{\infty} a_n x^n = \sum_{n=0}^{\infty} (n+2)(n+1)a_{n+2}x^n + \sum_{n=0}^{\infty} 4a_n x^n = 0$$

$$(n+2)(n+1)a_{n+2} = -4a_n$$

$$a_{n+2} = \frac{-4a_n}{(n+2)(n+1)}$$

$a_0 = a_0$ $\qquad\qquad\qquad\qquad$ $a_1 = a_1$

$a_2 = \dfrac{-4a_0}{2}$ $\qquad\qquad\qquad$ $a_3 = \dfrac{-4a_1}{3 \cdot 2}$

$a_4 = \dfrac{-4a_2}{4 \cdot 3} = \dfrac{(-4)^2 a_0}{4!}$ $\qquad\qquad$ $a_5 = \dfrac{-4a_3}{5 \cdot 4} = \dfrac{(-4)^2 a_1}{5!}$

$\qquad\vdots$ $\qquad\qquad\qquad\qquad\qquad$ $\vdots$

$a_{2n} = \dfrac{(-1)^n 4^n}{(2n)!} a_0$ $\qquad\qquad$ $a_{2n+1} = \dfrac{(-1)^n 4^n}{(2n+1)!} a_1$

$$y = \sum_{n=0}^{\infty} \frac{(-1)^n 4^n a_0}{(2n)!} x^{2n} + \sum_{n=0}^{\infty} \frac{(-1)^n 4^n a_1}{(2n+1)!} x^{2n+1} = a_0 \sum_{n=0}^{\infty} \frac{(-1)^n (2x)^{2n}}{(2n)!} + \frac{a_1}{2} \sum_{n=0}^{\infty} \frac{(-1)^n (2x)^{2n+1}}{(2n+1)!} = C_0 \cos 2x + C_1 \sin 2x$$

Check: $y'' + 4y = 0$ is a second-order homogeneous linear equation.

$m^2 + 4 = 0 \Rightarrow m = \pm 2i$

$y = C_1 \cos 2x + C_2 \sin 2x$

7. $y' + 3xy = 0$. Letting $y = \sum\limits_{n=0}^{\infty} a_n x^n$:

$$y' + 3xy = \sum_{n=1}^{\infty} na_n x^{n-1} + \sum_{n=0}^{\infty} 3a_n x^{n+1} = 0$$

$$\sum_{n=-1}^{\infty} (n+2)a_{n+2}x^{n+1} = \sum_{n=0}^{\infty} -3a_n x^{n+1} \Rightarrow a_1 = 0 \text{ and } a_{n+2} = \frac{-3a_n}{n+2}$$

$a_0 = a_0$ $\qquad\qquad\qquad\qquad\qquad\qquad$ $a_1 = 0$

$a_2 = -\dfrac{3a_0}{2}$ $\qquad\qquad\qquad\qquad\qquad$ $a_3 = -\dfrac{3a_1}{3} = 0$

$a_4 = -\dfrac{3}{4}\left(-\dfrac{3a_0}{2}\right) = \dfrac{3^2}{2^3}a_0$ $\qquad\qquad$ $a_5 = -\dfrac{3}{5}\left(-\dfrac{3a_1}{3}\right) = 0$

$a_6 = -\dfrac{3}{6}\left(-\dfrac{3^2}{2^3}a_0\right) = -\dfrac{3^3 a_0}{2^3(3 \cdot 2)}$ $\qquad\qquad$ $a_7 = -\dfrac{3}{7}\left(\dfrac{3^2 a_1}{3 \cdot 5}\right) = 0$

$a_8 = -\dfrac{3}{8}\left(-\dfrac{3^3 a_0}{2^3(3 \cdot 2)}\right) = \dfrac{3^4 a_0}{2^4(4 \cdot 3 \cdot 2)}$ $\qquad\qquad$ $a_9 = -\dfrac{3}{9}\left(-\dfrac{3^3 a_1}{3 \cdot 5 \cdot 7}\right) = 0$

$$y = a_0 \sum_{n=0}^{\infty} \frac{(-3)^n x^{2n}}{2^n n!}$$

$$\lim_{n \to \infty} \left|\frac{u_{n+1}}{u_n}\right| = \lim_{n \to \infty} \left|\frac{(-3)^{n+1} x^{2n+2}}{2^{n+1}(n+1)!} \cdot \frac{2^n n!}{(-3)^n x^{2n}}\right| = \lim_{n \to \infty} \frac{3x^2}{2(n+1)} = 0$$

The interval of convergence for the solution is $(-\infty, \infty)$.

9. $y'' - xy' = 0$. Letting $y = \displaystyle\sum_{n=0}^{\infty} a_n x^n$:

$$y'' - xy' = \sum_{n=2}^{\infty} n(n-1)a_n x^{n-2} - x\sum_{n=1}^{\infty} na_n x^{n-1} = 0$$

$$\sum_{n=2}^{\infty} n(n-1)a_n x^{n-2} = \sum_{n=0}^{\infty} na_n x^n$$

$$\sum_{n=0}^{\infty} (n+2)(n+1)a_{n+2} x^n = \sum_{n=0}^{\infty} na_n x^n$$

$$a_{n+2} = \frac{na_n}{(n+2)(n+1)}$$

$$a_0 = a_0 \qquad\qquad a_1 = a_1$$

$$a_2 = 0 \qquad\qquad a_3 = \frac{a_1}{3\cdot 2}$$

There are no even powered terms. $a_5 = \dfrac{3a_3}{5\cdot 4} = \dfrac{3a_1}{5!}$

$$a_7 = \frac{5a_5}{7\cdot 6} = \frac{5\cdot 3a_1}{7!}$$

$$y = a_0 + a_1\sum_{n=0}^{\infty} \frac{1\cdot 3\cdot 5\cdot 7\cdots(2n-1)x^{2n+1}}{(2n+1)!} = a_0 + a_1\sum_{n=0}^{\infty} \frac{(2n)!x^{2n+1}}{2^n n!(2n+1)!} = a_0 + a_1\sum_{n=0}^{\infty} \frac{x^{2n+1}}{2^n n!(2n+1)}$$

$$\lim_{n\to\infty}\left|\frac{u_{n+1}}{u_n}\right| = \lim_{n\to\infty}\left|\frac{x^{2n+3}}{2^{n+1}(n+1)!(2n+3)}\cdot\frac{2^n n!(2n+1)}{x^{2n+1}}\right| = \lim_{n\to\infty}\frac{(2n+1)x^2}{2(n+1)(2n+3)} = 0$$

Interval of convergence: $(-\infty, \infty)$

11. $(x^2+4)y'' + y = 0$. Letting $y = \displaystyle\sum_{n=0}^{\infty} a_n x^n$:

$$(x^2+4)y'' + y = \sum_{n=2}^{\infty} n(n-1)a_n x^n + 4\sum_{n=2}^{\infty} n(n-1)a_n x^{n-2} + \sum_{n=0}^{\infty} a_n x^n$$

$$= \sum_{n=0}^{\infty} (n^2 - n + 1)a_n x^n + \sum_{n=0}^{\infty} 4(n+2)(n+1)a_{n+2} x^n = 0$$

$$a_{n+2} = \frac{-(n^2 - n + 1)a_n}{4(n+2)(n+1)}$$

$$a_0 = a_0 \qquad\qquad\qquad a_1 = a_1$$

$$a_2 = \frac{-a_0}{4(2)(1)} = \frac{-a_0}{8} \qquad\qquad a_3 = \frac{-a_1}{4(3)(2)} = \frac{-a_1}{24}$$

$$a_4 = \frac{-3a_2}{4(4)(3)} = \frac{a_0}{128} \qquad\qquad a_5 = \frac{-7a_3}{4(5)(4)} = \frac{7a_1}{1920}$$

$$y = a_0\left(1 - \frac{x^2}{8} + \frac{x^4}{128} - \cdots\right) + a_1\left(x - \frac{x^3}{24} + \frac{7x^5}{1920} - \cdots\right)$$

13. $y' + (2x - 1)y = 0, \; y(0) = 2$

$$y' = (1 - 2x)y \qquad\qquad y'(0) = 0$$
$$y'' = (1 - 2x)y' - 2y \qquad y''(0) = -2$$
$$y''' = (1 - 2x)y'' - 4y' \qquad y'''(0) = -10$$
$$y^4 = (1 - 2x)y''' - 6y'' \qquad y^{(4)}(0) = 2$$
$$\vdots \qquad\qquad\qquad \vdots$$

$$y(x) = 2 + \frac{2}{1!}x - \frac{2}{2!}x^2 - \frac{10}{3!}x^3 + \frac{2}{4!}x^4 + \cdots$$

Using the first five terms of the series, $y\left(\frac{1}{2}\right) = \frac{163}{64} \approx 2.547$.

Using Euler's Method with $\Delta x = 0.1$ you have $y' = (1 - 2x)y$.

i	x_i	y_i
0	0	2
1	0.1	2.2
2	0.2	2.376
3	0.3	2.51856
4	0.4	2.61930
5	0.5	2.67169

15. Given a differential equation, assume that the solution is of the form $y = \sum\limits_{n=0}^{\infty} a_n x^n$. Then substitute y and its derivatives into the differential equation. You should then be able to determine the coefficients $a_0, a_1, \dots$.

17. (a) From Exercise 9, the general solution is

$$y = a_0 + a_1 \sum_{n=0}^{\infty} \frac{x^{2n+1}}{2^n n!(2n + 1)}.$$

$$y(0) = 0 \Rightarrow a_0 = 0$$

$$y' = a_1 \sum_{n=0}^{\infty} \frac{(2n + 1)x^{2n}}{2^n n!(2n + 1)} = a_1 \sum_{n=0}^{\infty} \frac{x^{2n}}{2^n n!}$$

$$y'(0) = 2 = a_1$$

$$y = 2 \sum_{n=0}^{\infty} \frac{x^{2n+1}}{2^n n!(2n + 1)}$$

(b) $P_3(x) = 2\left[x + \dfrac{x^3}{2 \cdot 3}\right] = 2x + \dfrac{x^3}{3}$

$P_5(x) = 2x + \dfrac{x^3}{3} + 2\dfrac{x^5}{4 \cdot 2 \cdot 5} = 2x + \dfrac{x^3}{3} + \dfrac{x^5}{20}$

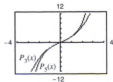

(c) The solution is symmetric about the origin.

19. $y'' - 2xy = 0, \; y(0) = 1, \; y'(0) = -3$

$$y'' = 2xy \qquad\qquad y''(0) = 0$$
$$y''' = 2(xy' + y) \qquad y'''(0) = 2$$
$$y^{(4)} = 2(xy'' + 2y') \qquad y^{(4)}(0) = -12$$
$$y^{(5)} = 2(xy''' + 3y'') \qquad y^{(5)}(0) = 0$$
$$y^{(6)} = 2\left(xy^{(4)} + 4y'''\right) \qquad y^{(6)}(0) = 16$$
$$y^{(7)} = 2\left(xy^{(5)} + 5y^{(4)}\right) \qquad y^{(7)}(0) = -120$$
$$\vdots \qquad\qquad\qquad \vdots$$

$$y \approx 1 - \frac{3}{1!}x + \frac{2}{3!}x^3 - \frac{12}{4!}x^4 + \frac{16}{6!}x^6 - \frac{120}{7!}x^7$$

Using the first six terms of the series, $y\left(\dfrac{1}{4}\right) \approx 0.253$.

21. $y'' + x^2 y' - (\cos x) y = 0$, $y(0) = 3$, $y'(0) = 2$

$y'' = -x^2 y' + (\cos x) y$ $\qquad\qquad\qquad$ $y''(0) = 3$

$y''' = -2x^2 y' - x^2 y'' - (\sin x) y + (\cos x) y'$ $\qquad$ $y'''(0) = 2$

$y \approx 3 + \dfrac{2}{1!} x + \dfrac{3}{2!} x^2 + \dfrac{2}{3!} x^3$

Using the first four terms of the series, $y\left(\dfrac{1}{3}\right) \approx 3.846$.

23. $f(x) = e^x$, $f'(x) = e^x$, $y' - y = 0$. Assume $y = \displaystyle\sum_{n=0}^{\infty} a_n x^n$, then:

$$y' = \sum_{n=1}^{\infty} n a_n x^{n-1}$$

$$\sum_{n=1}^{\infty} n a_n x^{n-1} = \sum_{n=0}^{\infty} a_n x^n$$

$$\sum_{n=0}^{\infty} (n+1) a_{n+1} x^n = \sum_{n=0}^{\infty} a_n x^n$$

$$a_{n+1} = \frac{a_n}{n+1}, \quad n \geq 0$$

$n = 0$, $\qquad\qquad$ $a_1 = a_0$

$n = 1$, $\qquad\qquad$ $a_2 = \dfrac{a_1}{2} = \dfrac{a_0}{2}$

$n = 2$, $\qquad\qquad$ $a_3 = \dfrac{a_2}{3} = \dfrac{a_0}{2(3)}$

$n = 3$, $\qquad\qquad$ $a_4 = \dfrac{a_3}{4} = \dfrac{a_0}{2(3)(4)}$

$n = 4$, $\qquad\qquad$ $a_5 = \dfrac{a_4}{5} = \dfrac{a_0}{2(3)(4)(5)}$

$$\vdots$$

$$a_{n+1} = \frac{a_0}{(n+1)!} \;\Rightarrow\; a_n = \frac{a_0}{n!}$$

$y = a_0 \displaystyle\sum_{n=0}^{\infty} \dfrac{x^n}{n!}$ which converges on $(-\infty, \infty)$. When $a_0 = 1$, you have the Maclaurin Series for $f(x) = e^x$.

25.
$$f(x) = \arctan x$$

$$f'(x) = \frac{1}{1 + x^2}$$

$$f''(x) = \frac{-2x}{\left(1 + x^2\right)^2}$$

$$y'' = \frac{-2x}{1 + x^2}y'$$

$$\left(1 + x^2\right)y'' + 2xy' = 0$$

Assume $y = \displaystyle\sum_{n=0}^{\infty} a_n x^n$, then:

$$y' = \sum_{n=1}^{\infty} n a_n x^{n-1}$$

$$y'' = \sum_{n=2}^{\infty} n(n-1) a_n x^{n-2}$$

$$\left(1 + x^2\right)y'' + 2xy' = \sum_{n=2}^{\infty} n(n-1)a_n x^{n-2} + \sum_{n=0}^{\infty} n(n-1)a_n x^n + \sum_{n=0}^{\infty} 2n a_n x^n = 0$$

$$\sum_{n=2}^{\infty} n(n-1)a_n x^{n-2} = -\sum_{n=0}^{\infty} n(n-1)a_n x^n - \sum_{n=0}^{\infty} 2n a_n x^n$$

$$\sum_{n=0}^{\infty} (n+2)(n+1)a_{n+2} x^n = -\sum_{n=0}^{\infty} n(n+1)a_n x^n$$

$$(n+2)(n+1)a_{n+2} = -n(n+1)a_n$$

$$a_{n+2} = -\frac{n}{n+2}a_n, n \geq 0$$

$n = 0 \Rightarrow a_2 = 0 \Rightarrow$ all the even-powered terms have a coefficient of 0.

$n = 1,$
$$a_3 = -\frac{1}{3}a_1$$

$n = 3,$
$$a_5 = -\frac{3}{5}a_3 = \frac{1}{5}a_1$$

$n = 5,$
$$a_7 = -\frac{5}{7}a_5 = -\frac{1}{7}a_1$$

$n = 7,$
$$a_9 = -\frac{7}{9}a_7 = \frac{1}{9}a_1$$

$$\vdots$$

$$a_{2n+1} = \frac{(-1)^n a_1}{2n+1}$$

$$y = a_1 \sum_{n=0}^{\infty} \frac{(-1)^n x^{2n+1}}{2n+1}$$ which converges on $(-1, 1)$. When $a_1 = 1$, you have the Maclaurin Series for $f(x) = \arctan x$.

27. $y'' - xy = 0.$ Let $y = \displaystyle\sum_{n=0}^{\infty} a_n x^n.$

$$y'' - xy = \sum_{n=2}^{\infty} n(n-1)a_n x^{n-2} - x\sum_{n=0}^{\infty} a_n x^n = \sum_{n=-1}^{\infty} (n+3)(n+2)a_{n+3}x^{n+1} - \sum_{n=0}^{\infty} a_n x^{n+1} = 0$$

$$2a_2 + \sum_{n=0}^{\infty} \left[(n+3)(n+2)a_{n+3} - a_n\right]x^{n+1} = 0$$

So, $a_2 = 0$ and $a_{n+3} = \dfrac{a_n}{(n+3)(n+2)}$ for $n = 0, 1, 2, \ldots$

The constants a_0 and a_1 are arbitrary.

$a_0 = a_0$ $a_1 = a_1$

$a_3 = \dfrac{a_0}{3 \cdot 2}$ $a_4 = \dfrac{a_1}{4 \cdot 3}$

$a_6 = \dfrac{a_3}{6 \cdot 5} = \dfrac{a_0}{6 \cdot 5 \cdot 3 \cdot 2}$ $a_7 = \dfrac{a_4}{7 \cdot 6} = \dfrac{a_1}{7 \cdot 6 \cdot 4 \cdot 3}$

So, $y = a_0 + a_1 x + \dfrac{a_0}{6}x^3 + \dfrac{a_1}{12}x^4 + \dfrac{a_0}{180}x^6 + \dfrac{a_1}{504}x^7.$

Review Exercises for Chapter 16

1. $\left(y + x^3 + xy^2\right)dx - x\,dy = 0$

$\dfrac{\partial M}{\partial y} = 1 + 2xy \neq \dfrac{\partial N}{\partial x} = -1$

Not exact

3. $(10x + 8y + 2)\,dx + (8x + 5y + 2)\,dy = 0$

Exact: $\dfrac{\partial M}{\partial y} = 8 = \dfrac{\partial N}{\partial x}$

$f(x, y) = \displaystyle\int(10x + 8y + 2)\,dx = 5x^2 + 8xy + 2x + g(y)$

$f_y(x, y) = 8x + g'(y) = 8x + 5y + 2$

$g'(y) = 5y + 2$

$g(y) = \dfrac{5}{2}y^2 + 2y + C_1$

$f(x, y) = 5x^2 + 8xy + 2x + \dfrac{5}{2}y^2 + 2y + C_1$

$5x^2 + 8xy + 2x + \dfrac{5}{2}y^2 + 2y = C$

5. $(x - y - 5)\,dx - (x + 3y - z)\,dy = 0$

$\dfrac{\partial M}{\partial y} = -1 = \dfrac{\partial N}{\partial x}$ Exact

$f(x, y) = \displaystyle\int(x - y - 5)\,dx = \dfrac{x^2}{2} - xy - 5x + g(y)$

$f_y(x, y) = -x + g'(y) = -x - 3y + 2$

$g'(y) = -3y + 2$

$g(y) = \dfrac{-3}{2}y^2 + 2y + C_1$

$\dfrac{x^2}{2} - xy - 5x - \dfrac{3}{2}y^2 + 2y + C_1 = 0$

$x^2 - 2xy - 10x - 3y^2 + 4y = C$

7. $\dfrac{x}{y}\,dx - \dfrac{x}{y^2}\,dy = 0$

$\dfrac{\partial M}{\partial y} = \dfrac{-x}{y^2} \neq \dfrac{\partial N}{\partial x} = \dfrac{-1}{y^2}$

Not exact

9. (a)

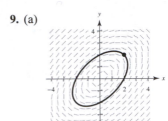

(b) $(2x - y)\,dx + (2y - x)\,dy = 0$

$$\frac{\partial M}{\partial y} = -1 = \frac{\partial N}{\partial x} \text{ Exact}$$

$$f(x, y) = \int (2x - y)\,dx = x^2 - xy + g(y)$$

$$f_y(x, y) = -x + g'(y) = 2y - x$$

$$g'(y) = 2y$$

$$g(y) = y^2 + C_1$$

$$x^2 - xy + y^2 = C$$

$$y(2) = 2 : 4 - 4 + 4 = 4 = C$$

Particular solution: $x^2 - xy + y^2 = 4$

(c)

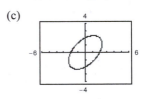

11. $(2x + y - 3)\,dx + (x - 3y + 1)\,dy = 0$

Exact: $\dfrac{\partial M}{\partial y} = 1 = \dfrac{\partial N}{\partial x}$

$$f(x, y) = \int (2x + y - 3)\,dx$$

$$= x^2 + xy - 3x + g(y)$$

$$f_y(x, y) = x + g'(y)$$

$$= x - 3y + 1$$

$$g'(y) = -3y + 1$$

$$g(y) = -\frac{3}{2}y^2 + y + C_1$$

$$f(x, y) = x^2 + xy - 3x$$

$$-\frac{3}{2}y^2 + y + C_1$$

$$2x^2 + 2xy - 6x - 3y^2 + 2y = C$$

Initial condition:

$$y(2) = 0$$

$$8 + 0 - 12 - 0 + 0 = C \Rightarrow C = -4$$

Particular solution:

$$2x^2 + 2xy - 6x - 3y^2 + 2y = -4$$

13. $\left(3x^2 - y^2\right)dx + 2xy\,dy = 0$

$$\frac{(\partial M/\partial y) - (\partial N/\partial x)}{N} = \frac{-2y - 2y}{2xy} = -\frac{2}{x} = h(x)$$

Integrating factor: $e^{\int h(x)\,dx} = e^{\ln x^{-2}} = \dfrac{1}{x^2}$

Exact equation: $\left(3 - \dfrac{y^2}{x^2}\right)dx + \dfrac{2y}{x}\,dy = 0$

$$f(x, y) = \int \left(3 - \frac{y^2}{x^2}\right)dx = 3x + \frac{y^2}{x} + g(y)$$

$$f_y(x, y) = \frac{2y}{x} + g'(y) = \frac{2y}{x}$$

$$g'(y) = 0 \Rightarrow g(y) = C_1$$

$$3x + \frac{y^2}{x} = C$$

15. $dx + \left(3x - e^{-2y}\right)dy = 0$

$$\frac{(\partial N/\partial x) - (\partial M/\partial y)}{M} = \frac{3 - 0}{1} = 3 = k(y)$$

Integrating factor: $e^{\int k(y)\,dy} = e^{3y}$

Exact equation: $e^{3y}\,dx + \left(3xe^{3y} - e^y\right)dy = 0$

$$f(x, y) = \int e^{3y}\,dx = xe^{3y} + g(y)$$

$$f_y(x, y) = 3xe^{3y} + g'(y) = 3xe^{3y} - e^y$$

$$g'(y) = -e^y$$

$$g(y) = -e^y + C_1$$

$$xe^{3y} - e^y = C$$

17. $y = C_1 e^{2x} + C_2 e^{-2x}$

$$y' = 2C_1 e^{2x} - 2C_2 e^{-2x}$$

$$y'' = 4C_1 e^{2x} + 4C_2 e^{-2x}$$

$$y'' - 4y = 4C_1 e^{2x} + 4C_2 e^{-2x}$$

$$- 4\left(C_1 e^{2x} + C_2 e^{-2x}\right) = 0$$

19. $y'' - y' - 2y = 0$

$$m^2 - m - 2 = (m - 2)(m + 1) = 0, m = 2, -1$$

$$y = C_1 e^{2x} + C_2 e^{-x}$$

$$y' = 2C_1 e^{2x} - C_2 e^{-x}$$

$$y(0) = 0 = C_1 + C_2$$

$$y'(0) = 3 = 2C_1 - C_2$$

Adding these equations, $3 = 3C_1 \Rightarrow C_1 = 1$ and $C_2 = -1$.

$$y = e^{2x} - e^{-x}$$

21. $y'' + 2y' - 3y = 0$

$m^2 + 2m - 3 = (m + 3)(m - 1) = 0 \Rightarrow m = -3, 1$

$y = C_1 e^{-3x} + C_2 e^x$

$y' = -3C_1 e^{-3x} + C_2 e^x$

$y(0) = 2 = C_1 + C_2$

$y'(0) = 0 = -3C_1 + C_2$

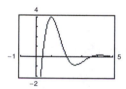

Subtracting these equations, $2 = 4C_1 \Rightarrow C_1 = \frac{1}{2}$ and $C_2 = \frac{3}{2}$.

$y = \frac{3}{2} e^x + \frac{1}{2} e^{-3x}$

23. $y'' + 2y' + 5y = 0$

$m^2 + 2m + 5 = 0 \Rightarrow m = \dfrac{-2 \pm \sqrt{4 - 20}}{2} = -1 \pm 2i$

$y = e^{-x}(C_1 \cos 2x + C_2 \sin 2x)$

$y(1) = 4 = e^{-1}(C_1 \cos 2 + C_2 \sin 2)$

$y(2) = 0 = e^{-2}(C_1 \cos 4 + C_2 \sin 4)$

Solving this system, you obtain $C_1 = -9.0496$, $C_2 = 7.8161$.

$y = e^{-x}(-9.0496 \cos 2x + 7.8161 \sin 2x)$

25. No, it is not homogeneous because of the nonzero term $\sin x$.

27. $y'' + y = x^3 + x$

$m^2 + 1 = 0$ when $m = -i, i$.

$y_h = C_1 \cos x + C_2 \sin x$

$y_p = A_0 + A_1 x + A_2 x^2 + A_3 x^3$

$y_p' = A_1 + 2A_2 x + 3A_3 x^2$

$y_p'' = 2A_2 + 6A_3 x$

$y_p'' + y_p = (A_0 + 2A_2) + (A_1 + 6A_3)x + A_2 x^2 + A_3 x^3$

$= x^3 + x$

$A_0 = 0, A_1 = -5, A_2 = 0, A_3 = 1$

$y = C_1 \cos x + C_2 \sin x - 5x + x^3$

29. $y'' + y = 2 \cos x$

$m^2 + 1 = 0$ when $m = -i, i$.

$y_h = C_1 \cos x + C_2 \sin x$

$y_p = Ax \cos x + Bx \sin x$

$y_p' = (Bx + A) \cos x + (B - Ax) \sin x$

$y_p'' = (2B - Ax) \cos x + (-Bx - 2A) \sin x$

$y_p'' + y_p = 2B \cos x - 2A \sin x = 2 \cos x$

$A = 0, B = 1$

$y = C_1 \cos x + (C_2 + x) \sin x$

31. $y'' - 2y' + y = 2xe^x$

$m^2 - 2m + 1 = 0$ when $m = 1, 1$.

$y_h = (C_1 + C_2 x)e^x$

$y_p = (v_1 + v_2 x)e^x$

$v_1' e^x + v_2' xe^x = 0$

$v_1' e^x + v_2'(x + 1)e^x = 2xe^x$

$v_1' = -2x^2$

$v_1 = \int -2x^2 \, dx = -\frac{2}{3}x^3$

$v_2' = 2x$

$v_2 = \int 2x \, dx = x^2$

$y = \left(C_1 + C_2 x + \frac{1}{3}x^3\right)e^x$

33. $y'' + y' - 6y = 54, y(0) = 2, y'(0) = 0$

$m^2 - m - 6 = 0$

$(m - 3)(m + 2) = 0$

$m_1 = 3, m_2 = -2$

$y_h = C_1 e^{3x} + C_2 e^{-2x}$

$y_p = -9$ by inspection

$y = y_h + y_p = C_1 e^{3x} + C_2 e^{-2x} - 9$

Initial conditions:

$y(0) = 2: 2 = C_1 + C_2 - 9 \Rightarrow C_1 + C_2 = 11$

$y'(0) = 0: 0 = 3C_1 - 2C_2 \Rightarrow C_1 = \frac{22}{5}, C_2 = \frac{33}{5}$

$y = \frac{11}{5}\left(2e^{3x} + 3e^{-2x}\right) - 9$

35. $y'' + 4y = \cos x$

$m^2 + 4 = 0 \Rightarrow m = \pm 2i$

$y_h = C_1 \cos 2x + C_2 \sin 2x$

$y_p = A \cos x + B \sin x$

$y_p' = -A \sin x + B \cos x$

$y_p'' = -A \cos x - B \sin x$

$y_p'' + 4y_p = (-A \cos x - B \sin x) + 4(A \cos x + B \sin x) = \cos x$

$3A \cos x + 3B \sin x = \cos x \Rightarrow A = \frac{1}{3}$ and $B = 0$

$y_p = \frac{1}{3} \cos x$

$y = y_h + y_p = C_1 \cos 2x + C_2 \sin 2x + \frac{1}{3} \cos x$

Initial conditions: $\quad y(0) = 6 : 6 = C_1 + \frac{1}{3} \Rightarrow C_1 = \frac{17}{3}$

$\qquad\qquad\qquad\qquad y'(0) = -6 : -6 = 2C_2 \Rightarrow C_2 = -3$

Particular solution: $\quad y = \frac{17}{3} \cos 2x - 3 \sin 2x + \frac{1}{3} \cos x$

37. $y'' - y' - 2y = 1 + xe^{-x}, y(0) = 1, y'(0) = 3$

$m^2 - m - 2 = (m - 2)(m + 1) = 0 \Rightarrow m = 2, -1$

$y_h = C_1 e^{2x} + C_2 e^{-x}$

$y_p = A + (Bx + Cx^2)e^{-x}$

$y_p' = -(Bx + Cx^2)e^{-x} + (B + 2Cx)e^{-x} = (B + (2C - B)x - Cx^2)e^{-x}$

$y_p'' = -(B + (2C - B)x - Cx^2)e^{-x} + (2C - B - 2Cx)e^{-x} = (Cx^2 + (B - 4C)x + 2C - 2B)e^{-x}$

$y_p'' - y_p' - 2y_p = (2C - 2B + (-4C + B)x + Cx^2)e^{-x} - (B + (2C - B)x - Cx^2)e^{-x} - 2(A + (Bx + Cx^2)e^{-x})$

$\qquad\qquad = -2A + (-6Cx + 2C - 3B)e^{-x} = 1 + xe^{-x} \Rightarrow A = -\frac{1}{2}, -6C = 1$ and $2C - 3B = 0$.

So, $C = -\frac{1}{6}$ and $B = -\frac{1}{9}$.

$y = y_h + y_p = C_1 e^{2x} + C_2 e^{-x} - \frac{1}{2} + \left(-\frac{1}{9}x - \frac{1}{6}x^2\right)e^{-x}$

Initial conditions: $y(0) = 1 = C_1 + C_2 - \frac{1}{2} \Rightarrow C_1 + C_2 = \frac{3}{2}$

$\qquad\qquad\qquad\quad y'(0) = 3 = 2C_1 - C_2 - \frac{1}{9} \Rightarrow 2C_1 - C_2 = \frac{28}{9}$

Adding, $3C_1 = \frac{83}{18} \Rightarrow C_1 = \frac{83}{54}$.

So, $C_2 = -\frac{1}{27}$.

Particular solution: $y = \frac{83}{54}e^{2x} - \frac{1}{27}e^{-x} - \frac{1}{2} - \left(\frac{1}{9} + \frac{1}{6}x\right)xe^{-x}$

39. By Hooke's Law, $F = kx, k = F/x = 64/(4/3) = 48$. Also, $F = ma$ and $m = F/a = 64/32 = 2$. So,

$\dfrac{d^2 y}{dt^2} + \left(\dfrac{48}{2}\right)y = 0$

$y = C_2 \cos\left(2\sqrt{6}\,t\right) + C_2 \sin\left(2\sqrt{6}\,t\right)$.

Because $y(0) = \frac{1}{2}$ you have $C_1 = \frac{1}{2}$ and $y'(0) = 0$ yields $C_2 = 0$. So, $y = \frac{1}{2}\cos\left(2\sqrt{6}\,t\right)$.

41. (a) (i) $y = \dfrac{1}{2}\cos 2t + \dfrac{12\pi}{\pi^2 - 4}\sin 2t + \dfrac{24}{4 - \pi^2}\sin \pi t$ (ii) $y = \dfrac{1}{2}\left[\left(1 - 6\sqrt{2}\,t\right)\cos\left(2\sqrt{2}\,t\right) + 3\sin\left(2\sqrt{2}\,t\right)\right]$

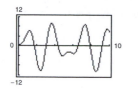 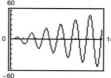

(iii) $y = \dfrac{e^{-t/5}}{398}\left[199\cos\dfrac{\sqrt{199}\,t}{5} + \sqrt{199}\sin\dfrac{\sqrt{199}\,t}{5}\right]$ (iv) $y = \dfrac{1}{2}e^{-2t}(\cos 2t + \sin 2t)$

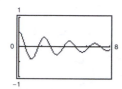 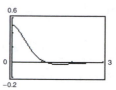

(b) The object comes to rest more quickly. It may not even oscillate, as in part (iv).

(c) It would oscillate more rapidly.

(d) Part (ii). The amplitude becomes increasingly large.

43. (a) $y_p{}'' = -A\sin x$ and $3y_p = 3A\sin x$.

So, $y_p{}'' + 3y_p = -A\sin x + 3A\sin x = 2A\sin x = 12\sin x$

(b) $y_p = \dfrac{5}{2}\cos x$

(c) If $y_p = A\cos x + B\sin x$, then $y_p{}'' = -A\cos x - B\sin x$, and solving for A and B would be more difficult.

45. $(x - 4)y' + y = 0.$ Letting $y = \displaystyle\sum_{n=0}^{\infty} a_n x^n$:

$$xy' - 4y' + y = \sum_{n=0}^{\infty} na_n x^n - 4\sum_{n=1}^{\infty} na_n x^{n-1} + \sum_{n=0}^{\infty} a_n x^n$$

$$= \sum_{n=0}^{\infty}(n+1)a_n x^n - \sum_{n=1}^{\infty} 4na_n x^{n-1} = \sum_{n=0}^{\infty}(n+1)a_n x^n - \sum_{n=-1}^{\infty} 4(n+1)a_{n+1}x^n = 0$$

$$(n+1)a_n = 4(n+1)a_{n+1}$$

$$a_{n+1} = \frac{1}{4}a_n$$

$$a_0 = a_0,\ a_1 = \frac{1}{4}a_0,\ a_2 = \frac{1}{4}a_1 = \frac{1}{4^2}a_0,\ \dots,\ a_n = \frac{1}{4^n}a_0$$

$$y = a_0\sum_{n=0}^{\infty}\frac{x^n}{4^n}$$

47. $y'' + y' - e^x y = 0,\ y(0) = 2,\ y'(0) = 0$

$y'' = -y' + e^x y$ $\qquad\qquad\qquad y''(0) = 2$

$y''' = -y'' + e^x(y + y')$ $\qquad\qquad y'''(0) = -2 + 2 = 0$

$y^{(4)} = -y''' + e^x(y + 2y' + y'')$ $\qquad y^{(4)}(0) = 4$

$y^{(5)} = -y^{(4)} + e^x(y + 3y' + 3y'' + y''')$ $\qquad y^{(5)}(0) = -4 + 8 = 4$

$y \approx y(0) + y'(0)x + \dfrac{y''(0)}{2!}x^2 + \dfrac{y'''(0)}{3!}x^3 + \dfrac{y^{(4)}(0)}{4!}x^4 + \dfrac{y^{(5)}(0)}{5!}x^5 = 2 + x^2 + \dfrac{1}{6}x^4 + \dfrac{1}{30}x^5$

Using the first four terms of the series, $y\left(\dfrac{1}{4}\right) \approx 2.063$.

Problem Solving for Chapter 16

1. $\left(3x^2 + kxy^2\right)dx - \left(5x^2y + ky^2\right)dy = 0$

$\dfrac{\partial M}{\partial y} = 2kxy$

$\dfrac{\partial N}{\partial x} = -10xy$

$\dfrac{\partial M}{\partial y} = \dfrac{\partial N}{\partial x} \Rightarrow k = -5$

$\left(3x^2 - 5xy^2\right)dx - \left(5x^2y - 5y^2\right)dy = 0$ Exact

$f(x, y) = \int \left(3x^2 - 5xy^2\right)dx = x^3 - \dfrac{5}{2}x^2y^2 + g(y)$

$f_y(x, y) = -5x^2y + g'(y) = -5x^2y + 5y^2$

$g'(y) = 5y^2 \Rightarrow g(y) = \dfrac{5}{3}y^3 + C_1$

$x^3 - \dfrac{5}{2}x^2y^2 + \dfrac{5}{3}y^3 = C_2$

$6x^3 - 15x^2y^2 + 10y^3 = C$

3. $y'' - a^2y = 0,\ y > 0$

$m^2 - a^2 = (m + a)(m - a) = 0 \Rightarrow m = \pm a$

$y = B_1e^{ax} + B_2e^{-ax} = \dfrac{C_1 + C_2}{2}e^{ax} + \dfrac{C_1 - C_2}{2}e^{-ax}$

$= C_1\left(\dfrac{e^{ax} + e^{-ax}}{2}\right) + C_2\left(\dfrac{e^{ax} - e^{-ax}}{2}\right)$

$= C_1 \cosh ax + C_2 \sinh ax$

5. The general solution to $y'' + ay' + by = 0$ is

$y = B_1e^{(r+s)x} + B_2e^{(r-s)x}$.

Let $C_1 = B_1 + B_2$ and $C_2 = B_1 - B_2$.

Then $B_1 = \dfrac{C_1 + C_2}{2}$ and $B_2 = \dfrac{C_1 - C_2}{2}$.

So $y = \left(\dfrac{C_1 + C_2}{2}\right)e^{(r+s)x} + \left(\dfrac{C_1 - C_2}{2}\right)e^{(r-s)x}$

$= e^{rx}\left[C_1\left(\dfrac{e^{sx} + e^{-sx}}{2}\right) + C_2\left(\dfrac{e^{sx} - e^{-sx}}{2}\right)\right]$

$= e^{rx}\left[C_1 \cosh sx + C_2 \sinh sx\right]$.

7. $y'' + ay = 0,\ y(0) = y(L) = 0$

(a) If $a = 0$, $y'' = 0 \Rightarrow y = cx + d$. $y(0) = 0 = d$
    and $y(L) = 0 = cL \Rightarrow c = 0$. So $y = 0$ is the
    solution.

(b) If $a < 0$, $y'' + ay = 0$ has characteristic equation

$m^2 + a = 0 \Rightarrow m = \pm\sqrt{-a}$.

$y = C_1e^{\sqrt{-a}x} + C_2e^{-\sqrt{-a}x}$

$y(0) = 0 = C_1 + C_2 \Rightarrow -C_1 = C_2$

$y(L) = 0 = C_1e^{\sqrt{-a}L} + C_2e^{-\sqrt{-a}L}$

$\qquad = C_1e^{\sqrt{-a}L} - C_1e^{-\sqrt{-a}L}$

$\qquad = 2C_1\left(\dfrac{e^{\sqrt{-a}L} - e^{-\sqrt{-a}L}}{2}\right)$

$\qquad = 2C_1 \sinh\left(\sqrt{-a}\,L\right) \Rightarrow C_1 = 0 = C_2$

So, $y = 0$ is the only solution.

(c) For $a > 0$:

$m^2 + a = 0 \Rightarrow m = \pm\sqrt{a}\,i$

$y = C_1 \cos\left(\sqrt{a}\,x\right) + C_2 \sin\left(\sqrt{a}\,x\right)$.

$y(0) = 0 = C_1$

$y = C_2 \sin\left(\sqrt{a}\,x\right)$

$y(L) = 0 = C_2 \sin\left(\sqrt{a}\,L\right)$

So $\sqrt{a}\,L = n\pi$

$a = \left(\dfrac{n\pi}{L}\right)^2,\ n$ an integer.

9. $\dfrac{d^2\theta}{dt^2} + \dfrac{g}{L}\theta = 0, \dfrac{g}{L} > 0$

(a) $\theta(t) = C_1 \sin\left(\sqrt{\dfrac{g}{L}}\,t\right) + C_2 \cos\left(\sqrt{\dfrac{g}{L}}\,t\right)$

  Let ϕ be given by $\tan\left(\sqrt{\dfrac{g}{L}}\,\phi\right) = -\dfrac{C_1}{C_2}, -\dfrac{\pi}{2} < \phi < \dfrac{\pi}{2}$.

  Then $C_2 \sin\left(\sqrt{\dfrac{g}{L}}\,\phi\right) = -C_1 \cos\left(\sqrt{\dfrac{g}{L}}\,\phi\right)$.

  Let $A = \dfrac{C_2}{\cos\left(\sqrt{\dfrac{g}{L}}\,\phi\right)} = -\dfrac{C_1}{\sin\left(\sqrt{\dfrac{g}{L}}\,\phi\right)}$

  $\theta(t) = C_1 \sin\left(\sqrt{\dfrac{g}{L}}\,t\right) + C_2 \cos\left(\sqrt{\dfrac{g}{L}}\,t\right) = -A \sin\left(\sqrt{\dfrac{g}{L}}\,\phi\right)\sin\left(\sqrt{\dfrac{g}{L}}\,t\right) + A \cos\left(\sqrt{\dfrac{g}{L}}\,\phi\right)\cos\left(\sqrt{\dfrac{g}{L}}\,t\right) = A \cos\left[\sqrt{\dfrac{g}{L}}(t + \phi)\right]$

(b) $\theta(t) = A \cos\left[\sqrt{\dfrac{g}{L}}(t + \phi)\right], g = 9.8, L = 0.25$

  $\theta(0) = A \cos\left[\sqrt{39.2}\,\phi\right] = 0.1$

  $\theta'(t) = -A\sqrt{\dfrac{g}{L}}\,\sin\left[\sqrt{\dfrac{g}{4}}(t + \phi)\right]$

  $\theta'(0) = -A\sqrt{39.2}\,\sin\left[\sqrt{39.2}\,\phi\right] = 0.5$

  Dividing, $\tan\left[\sqrt{39.2}\,\phi\right] = \dfrac{-5}{\sqrt{39.2}} \Rightarrow \phi \approx -0.1076 \Rightarrow A \approx 0.128$.

  $\theta(t) = 0.128 \cos\left[\sqrt{39.2}(t - 0.108)\right]$

(c) Period $= \dfrac{2\pi}{\sqrt{39.2}} \approx 1$ sec

(d) Maximum is 0.128.

(e) $\theta(t) = 0$ at $t \approx 0.359$ sec, and at $t \approx 0.860$ sec.

(f) $\theta'(0.359) \approx -0.801, \theta'(0.860) \approx 0.801$

11. $y'' + 8y' + 16y = 0$, $y(0) = 1$, $y'(0) = 1$

(a) $\lambda = 4$, $\omega = 4$, $\lambda^2 - \omega^2 = 0$, critically damped

(b) $m_1 = m_2 = -4$

$$y = (C_1 + C_2 t)e^{-4t},$$
$$y' = -4(C_1 + C_2 t)e^{-4t} + C_2 e^{-4t}$$

$$y(0) = 1 = C_1$$
$$y'(0) = 1 = -4 + C_2 \Rightarrow C_2 = 5$$
$$y = (1 + 5t)e^{-4t}$$

(c)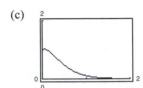

The solution tends to zero quickly.

13. $y'' + 20y' + 64y = 0$, $y(0) = 2$, $y'(0) = -20$

(a) $\lambda = 10$, $\omega = 8$, $\lambda^2 - \omega^2 = 36 > 0$, overdamped

(b) $m_1 = -10 + 6 = -4$, $m_2 = -10 - 6 = -16$

$$y = C_1 e^{-4t} + C_2 e^{-16t}$$
$$y(0) = 2 = C_1 + C_2$$
$$y'(t) = -4C_1 e^{-4t} - 16C_2 e^{-16t}$$
$$y'(0) = -20 = -4C_1 - 16C_2$$

$$\left.\begin{array}{r} C_1 + C_2 = 2 \\ -C_1 - 4C_2 = -5 \end{array}\right\} C_1 = 1,\ C_2 = 1$$

$$y = e^{-4t} + e^{-16t}$$

(c)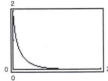

The solution tends to zero quickly.

15. Airy's Equation: $y'' - xy = 0$

$$y'' - xy + y - y = y'' - (x-1)y - y = 0$$

Let $y = \sum_{n=0}^{\infty} a_n(x-1)^n$, $y' = \sum_{n=1}^{\infty} na_n(x-1)^{n-1}$, $y'' = \sum_{n=2}^{\infty} n(n-1)a_n(x-1)^{n-2}$.

$$y'' - (x-1)y - y = 0$$

$$\sum_{n=2}^{\infty} n(n-1)a_n(x-1)^{n-2} - (x-1)\sum_{n=0}^{\infty} a_n(x-1)^n - \sum_{n=0}^{\infty} a_n(x-1)^n = 0$$

$$\sum_{n=-1}^{\infty} (n+3)(n+2)a_{n+3}(x-1)^{n+1} - \sum_{n=0}^{\infty} a_n(x-1)^{n+1} - \sum_{n=-1}^{\infty} a_{n+1}(x-1)^{n+1} = 0$$

$$(2a_2 - a_0) + \sum_{n=0}^{\infty} \left[(n+3)(n+2)a_{n+3} - a_n - a_{n+1} \right](x-1)^{n+1} = 0$$

$$2a_2 - a_0 = 0 \Rightarrow a_2 = \frac{1}{2}a_0; \; a_0, a_1 \text{ arbitrary}$$

In general, $a_{n+3} = \dfrac{a_n + a_{n+1}}{(n+3)(n+2)}$.

$$a_3 = \frac{a_0 + a_1}{6}$$

$$a_4 = \frac{a_1 + a_2}{12} = \frac{a_1 + \left(\frac{1}{2}a_0\right)}{12} = \frac{2a_1 + a_0}{24}$$

$$a_5 = \frac{a_2 + a_3}{20} = \frac{\frac{1}{2}a_0 + \frac{a_0 + a_1}{6}}{20} = \frac{4a_0 + a_1}{120}$$

$$a_6 = \frac{a_3 + a_4}{30} = \frac{\left(\frac{a_0 + a_1}{6}\right) + \left(\frac{2a_1 + a_0}{24}\right)}{30} = \frac{5a_0 + 6a_1}{720}$$

$$a_7 = \frac{a_4 + a_5}{42} = \frac{\left(\frac{2a_1 + a_0}{24}\right) + \left(\frac{4a_0 + a_1}{120}\right)}{42} = \frac{9a_0 + 11a_1}{5040}$$

So, the first eight terms are

$$y = a_0 + a_1(x-1) + \frac{a_0}{2}(x-1)^2 + \frac{a_0 + a_1}{6}(x-1)^3 + \frac{2a_1 + a_0}{24}(x-1)^4 + \frac{4a_0 + a_1}{120}(x-1)^5$$

$$+ \frac{5a_0 + 6a_1}{720}(x-1)^6 + \frac{9a_0 + 11a_1}{5040}(x-1)^7.$$

17. $x^2 y'' + xy' + x^2 y = 0$ Bessell equation of order zero

(a) Let $y = \sum_{n=0}^{\infty} a_n x^n$, $y' = \sum_{n=1}^{\infty} n a_n x^{n-1}$, $y'' = \sum_{n=2}^{\infty} n(n-1) a_n x^{n-2}$.

$$x^2 y'' + xy' + x^2 y = 0$$

$$x^2 \sum_{n=2}^{\infty} n(n-1) a_n x^{n-2} + x \sum_{n=1}^{\infty} n a_n x^{n-1} + x^2 \sum_{n=0}^{\infty} a_n x^n = 0$$

$$\sum_{n=2}^{\infty} n(n-1) a_n x^n + \sum_{n=1}^{\infty} n a_n x^n + \sum_{n=0}^{\infty} a_n x^{n+2} = 0$$

$$\sum_{n=0}^{\infty} (n+2)(n+1) a_{n+2} x^{n+2} + \sum_{n=-1}^{\infty} (n+2) a_{n+2} x^{n+2} + \sum_{n=0}^{\infty} a_n x^{n+2} = 0$$

$$a_1 x + \sum_{n=0}^{\infty} \left[(n+2)(n+1) a_{n+2} + (n+2) a_{n+2} + a_n \right] x^{n+2} = 0$$

$$a_1 = 0 \text{ and } a_{n+2} = \frac{-a_n}{(n+2)^2}.$$

All odd terms a_i are 0.

$$a_2 = \frac{-a_0}{2^2}$$

$$a_4 = \frac{-a_2}{4^2} = a_0 \frac{1}{2^2 \cdot 4^2} = \frac{a_0}{2^4 (1 \cdot 2)^2}$$

$$a_6 = \frac{-a_4}{6^2} = -a_0 \frac{1}{2^2 \cdot 4^2 \cdot 6^2} = \frac{-a_0}{2^6 (3!)^2}$$

$$y = a_0 \sum_{n=0}^{\infty} \frac{(-1)^n x^{2n}}{2^{2n} (n!)^2}$$

(b) This is the same function (assuming $a_0 = 1$).

19. (a) Let $y = \sum\limits_{n=0}^{\infty} a_n x^n$, $y' = \sum\limits_{n=1}^{\infty} n a_n x^{n-1}$, $y'' = \sum\limits_{n=2}^{\infty} n(n-1) a_n x^{n-2}$.

$$y'' - 2xy' + 8y = 0$$

$$\sum_{n=2}^{\infty} n(n-1) a_n x^{n-2} - 2x \sum_{n=1}^{\infty} n a_n x^{n-1} + 8 \sum_{n=0}^{\infty} a_n x^n = 0$$

$$\sum_{n=0}^{\infty} (n+2)(n+1) a_{n+2} x^n - \sum_{n=0}^{\infty} 2n a_n x^n + \sum_{n=0}^{\infty} 8 a_n x^n = 0$$

$$\sum_{n=0}^{\infty} \left[(n+2)(n+1) a_{n+2} - 2n a_n + 8 a_n \right] x^n = 0$$

$$a_{n+2} = \frac{2(n-4)}{(n+2)(n+1)} a_n$$

$$a_4 = 16 = \frac{2(-2)}{4(3)} a_2 = -\frac{1}{3} a_2 \Rightarrow a_2 = -48$$

$$a_2 = -48 = \frac{2(-4)}{2} a_0 = -4 a_0 \Rightarrow a_0 = 12$$

$$H_4(x) = 16x^4 - 48x^2 + 12$$

(b) $H_0(x) = \dfrac{(2x)^0}{0!} = 1$

$H_1(x) = \dfrac{(2x)^1}{1!} = 2x$

$H_2(x) = \sum\limits_{n=0}^{1} \dfrac{(-1)^n 2! (2x)^{2-2n}}{n!(2-2n)!} = \dfrac{2(2x)^2}{2!} - \dfrac{2}{1} = 4x^2 - 2$

$H_3(x) = \sum\limits_{n=0}^{1} \dfrac{(-1)^n 3! (2x)^{3-2n}}{n!(3-2n)!} = \dfrac{3!(2x)^3}{3!} - \dfrac{3!(2x)^1}{1} = 8x^3 - 12x$

$H_4(x) = \sum\limits_{n=0}^{2} \dfrac{(-1)^n 4! (2x)^{4-2n}}{n!(4-2n)!} = \dfrac{4!(2x)^4}{4!} - \dfrac{4!(2x)^2}{2!} + \dfrac{4!}{2!} = 16x^4 - 48x^2 + 12$

Transmission and Population Genetics

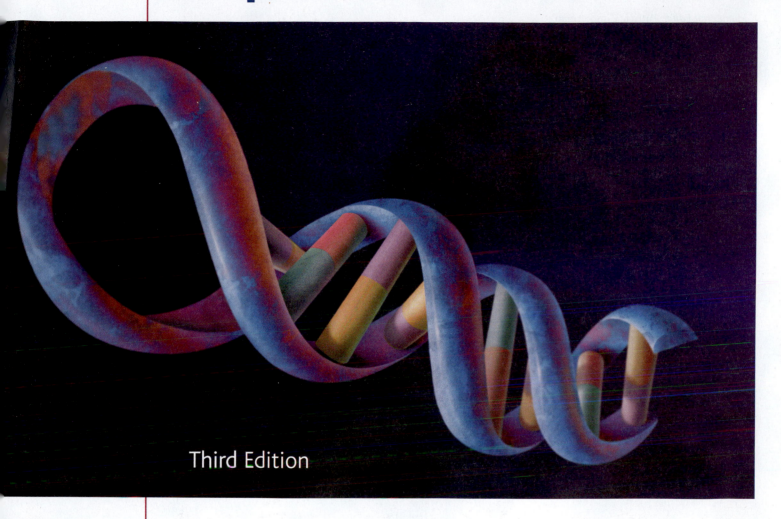

Third Edition

Benjamin A. Pierce

Southwestern University

W. H. Freeman and Company • New York

Acquisitions Editor: Jerry Correa

Development Editor: Lisa Samols

Media Editors: Patrick Shriner, Marc Mazzoni

Supplements Editors: Hannah Thonet, Beth McHenry

Project Editor: Leigh Renhard

Manuscript Editor: Patricia Zimmerman

Associate Director of Marketing: Debbie Clare

Publisher: Sara Tenney

Text Designer: Marsha Cohen/Parallelogram Graphics

Cover Illustration: Rosetta Genomics

Illustrations: Dragonfly Media Group

Senior Illustration Coordinator: Bill Page

Photo Editors: Ted Szczepanski, Christine Buese

Photo Researcher: Patty Catuera

Production Coordinator: Susan Wein

Composition: Black Dot Group

Printing and Binding: RR Donnelley

Library of Congress Control Number: 2007924869

ISBN-13: 978-1-4292-1118-5
ISBN-10: 1-4292-1118-0

Printed in the United States of America
First printing

W. H. Freeman and Company
Madison Avenue
York, NY 10010
ills, Basingstoke RG21 6XS, England

an.com

To my parents, Rush and Amanda Pierce;

my children, Sarah and Michael Pierce;

and my genetic partner, friend,
and soul mate for 27 years, Marlene Tyrrell

Brief Contents

Contents

CHAPTER 7

Linkage, Recombination, and Eukaryotic Gene Mapping / 160

CHAPTER 8

Bacterial and Viral Genetic Systems / 200

Letter from the Author

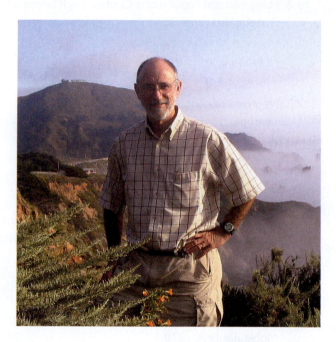

A former genetics student—a user of this book—once told me that, when she read *Genetics: A Conceptual Approach*, she could hear my voice. My goal as the author of your textbook is to share my knowledge and experience as we journey together through genetics.

In this book, I've tried to share some of what I've learned in my 27 years of teaching genetics. One of my strengths is helping students create a mental map of genetics—one that shows where we've been, where we'll go, and how we'll get there. I provide advice and encouragement at places that are often rough spots and I tell stories of the people, places, and experiments of genetics—past and present—to keep the subject relevant, interesting, and alive. My goal is to help you learn the necessary details, concepts, and problem-solving skills, while encouraging you to see the overarching beauty of the discipline.

At Southwestern University, my office door is always open, and my students often drop by to share their own approaches to learning, things that they have read about genetics, and their experiences, concerns, and triumphs. I learn as much from my students as they learn from me, and I would love to learn from you—by email (pierceb@southwestern.edu), by telephone (512-863-1974), or in person (Southwestern University, Georgetown, Texas).

Ben Pierce
PROFESSOR OF BIOLOGY AND
HOLDER OF THE LILLIAN NELSON CHAIR
SOUTHWESTERN UNIVERSITY

Preface

*T*ransmission and Population Genetics was developed to address the needs of introductory genetics courses that focus solely on those topics. In response to instructors' comments about their course, this edition includes a chapter on gene mutations and DNA repair (chapter 10). Like its parent text, *Genetics* 3e, this edition of *Transmission and Population Genetics* builds on features that contributed to the success of the preceding editions: an informal writing style; accessible and instructive illustrations; an emphasis on problem solving; and, most importantly, a strong focus on concepts and connections. To bring these key concepts into sharper focus, I've selectively reduced the amount of detail in this edition and made it easier for students to make meaningful connections among the different areas of genetics.

CONCEPTS

In an inversion, a segment of a chromosome is inverted. Inversions cause breaks in some genes and may move others to new locations. In heterozygotes for a chromosome inversion, the homologous chromosomes form a loop in prophase I of meiosis. When crossing over takes place within the inverted region, nonviable gametes are usually produced, resulting in a depression in observed recombination frequencies.

✔ CONCEPT CHECK 3

A dicentric chromosome is produced when crossing over takes place in an individual heterozygous for which type of chromosome rearrangement?

a. Duplication
b. Deletion
c. Paracentric inversion
d. Pericentric inversion

Instructive Art Program

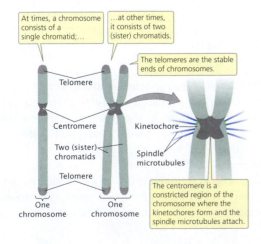

At times, a chromosome consists of a single chromatid;…

…at other times, it consists of two (sister) chromatids.

The telomeres are the stable ends of chromosomes.

Telomere

Centromere

Two (sister) chromatids

Telomere

One chromosome

One chromosome

Kinetochore

Spindle microtubules

The centromere is a constricted region of the chromosome where the kinetochores form and the spindle microtubules attach.

Hallmark Features

■ **Key Concepts and Connections** Throughout the book, I've included pedagogical devices to help students focus on the major concepts of each topic.

• *Concept boxes* summarize the important take-home messages and key points of the chapter. New *Concept Check* questions—some open ended, some multiple choice—allow students to assess their understanding of the take-home message of the preceding section. Answers to the Concept Checks are included in the end-of-chapter material.

• *Connecting Concepts* sections help students see how key ideas within a chapter relate to one another. These sections integrate preceding discussions, showing how processes are similar, where they differ, and how one informs the other. After reading Connecting Concepts sections, students will better understand how newly learned concepts fit into the bigger picture of genetics. All of the key concepts in the chapter are listed at the end of the chapter in the *Concepts Summary*.

■ **Accessibility** I have intentionally used a friendly and conversational writing style so that students will find the book inviting and informative. The **introductory story** at the beginning of every chapter draws students into the material. These stories highlight the relevance of genetics to the student's daily life and feature new research in genetics, the genetic basis of human disease, hereditary oddities, and other interesting topics. More than 50% of the introductory stories are new to this edition.

■ **Clear, Simple Illustration Program** The attractive and instructive illustration program continues to play a pivotal role in reinforcing the key concepts presented in each chapter. Because many students are visual learners, I've worked closely with the illustrators to make sure that the main point of each illustration is easily identified and understood. Color is used in many

New Introductory Stories

6 | Pedigree Analysis, Applications, and Genetic Testing

HUTCHINSON–GILFORD SYNDROME AND THE SECRET OF AGING

The Spanish explorer Juan Ponce de León searched in vain for the fabled fountain of youth, the healing waters of which were reported to prevent old age. In the course of his explorations, Ponce de León discovered Florida, but he never located the source of perpetual youth. Like Ponce de León, scientists today seek to forestall aging, not in the waters of a mythical spring, but rather in understanding the molecular changes associated with the aging process. Curiously, their search has led to a small group of special children, who experience, not perpetual youth, but alarmingly premature old age.

These children were first noticed by an English physician, Jonathan Hutchinson, who published in 1886 a report of a young patient only 6 years old with the appearance of an 80-year-old man; the patient was bald with aged facial features and stiffness of joints. A few years later, another physician, Hastings Gilford, observed an additional case of the disorder, giving it the name progeria, derived from the Greek meaning "prematurely old."

This exceedingly rare disease, known today as Hutchinson–Gilford progeria syndrome (HGPS), affects about 50 children worldwide and is characterized by the same features described by Hutchinson and Gilford more than 100 years ago. Children with HGPS appear healthy at birth but, by 2 years of age, usually begin to show features of accelerated aging, including cessation of growth, hair loss, aged skin, stiffness of joints, heart disease, and osteoporosis. By the age of 13, many have died, usually from coronary artery disease. HGPS is clearly genetic in origin and is usually inherited as an autosomal dominant trait.

In 1998, Leslie Gordon and Scott Berns's 2-year-old child Sam was diagnosed with HGPS. Gordon and Berns—both physicians—begin to seek information about the disease but, to their dismay, virtually none existed. Shocked, Gordon and Berns established the Progeria Research Foundation, and Gordon focused her career on directing the

Megan, a 5-year-old with Hutchinson–Gilford progeria syndrome. Today, researchers seek to understand aging from the study of children with this rare autosomal dominant disorder that causes premature aging. [Courtesy of John Hurley for Progeria Research Foundation (progeriaresearch.org).]

New In-Text Worked Problems

Worked Problem

Now that we understand the pattern of X-linked inheritance, let's apply our knowledge to answer a specific question in regard to X-linked inheritance of color blindness in humans.

Betty has normal vision, but her mother is color blind. Bill is color blind. If Bill and Betty marry and have a child together, what is the probability that the child will be color blind?

• Solution

Because color blindness is an X-linked recessive characteristic, Betty's color-blind mother must be homozygous for the color-blind allele (X^cX^c). Females inherit one X chromosome from each of their parents; so Betty must have inherited a color-blind allele from her mother. Because Betty has normal color vision, she must have inherited an allele for normal vision (X^+) from her father; thus Betty is heterozygous (X^+X^c). Bill is color blind. Because males are hemizygous for X-linked alleles, he must be (X^cY). A mating between Betty and Bill is represented as:

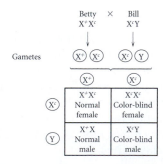

Betty $\times$ Bill
X^+X^c X^cY

Gametes X^+ X^c X^c Y

	X^+	X^c
X^c	X^+X^c Normal female	X^cX^c Color-blind female
Y	X^+X Normal male	X^cY Color-blind male

Thus, $\frac{1}{4}$ of the children are expected to be female with normal color vision, $\frac{1}{4}$ female with color blindness, $\frac{1}{4}$ male with normal color vision, and $\frac{1}{4}$ male with color blindness.

? Get some additional practice with X-linked inheritance by working Problem 23 at the end of this chapter.

---

illlustrations to help students orient themselves as they study experiments and processes. Most include balloon descriptions that walk students step-by-step through a process or point out important features of a structure or experiment. Each of the 19 "experiment" illustrations found throughout the book facilitates student understanding of the experimental process by posing a question, describing experimental methodology, presenting results, and drawing a conclusion that reinforces the major concept being discussed.

■ **Emphasis on Problem Solving** Problem solving is among the most difficult skills for a student to learn, but, is essential to the mastery of genetics. In this new edition, I've increased the number of in-text worked problems, which walk students through a key problem and review important strategies for students to consider when tackling a problem of a similar type. This edition continues to provide extensive problem sets, broken down into three categories: comprehension questions; application questions and problems; and challenge questions. Also, new to this edition are data-analysis problems based on real data from the scientific literature. More than 25% of the problems are new in this edition.

New to the Third Edition

■ **Increased Clarity and Accessibility** To focus on essential concepts, instructors have been asked to help in eliminating details that were either infrequently taught or less central to the course. On the basis of this feedback, the text was carefully edited. As a result, the material is more streamlined and better focused on core topics.

■ **New Introductory Stories** A favorite feature of the earlier editions, a brief story introduces each chapter and is meant to be a glimpse of the material in action. These stories are intended to make genetics more relevant to the student and to draw students into the chapters. Half of the introductory stories in this edition are new and include topics such as "Cuénot's Odd Yellow Mice," "Hutchinson–Gilford Syndrome and the Secret of Aging," "A Fly without a Heart" and "Taster Genes in spitting Apes".

■ **Concept Check Questions** The Concepts boxes help students identify and review the core concepts in each chapter. I've enhanced this feature by including Concept Check questions in many of the Concepts boxes. By answering an open-ended or multiple-choice question, students test their knowledge and understanding of the material. The questions and their answers (found in the end-of-chapter material) provide immediate feedback to the student on their progress.

■ **New In-Text Worked Problems** Many genetics students struggle with problem-solving. The in-text worked problems carefully guide students through each step in finding the solutions to the types of problems that they often find difficult. These problems are placed strategically throughout the chapters so that students get help with problem-solving in the context of learning the material. In the third edition, I've increased the number of in-text worked problems by one third. I've also added a reference at the end of each Worked Problem that gives students an opportunity to check their understanding by solving a similar problem at the end of the chapter.

■ ![DATA ANALYSIS icon] **Data-Analysis Problems** New problems at the end of each chapter feature real data from the scientific literature. Students have an opportunity to apply their problem-solving skills in the context of real experiments. These problems are easily identified by an icon placed next to each of them.

■ **New End-of-Chapter Questions and Problems** The third edition builds on the versatile problem sets from the preceding editions by a 25% increase in the number of new questions. The end-of-chapter problem sections include new problems that emphasize data analysis and feature real data from current experiments. These sections are now organized by **section heading**, enabling students to easily identify specific sections that they would like to review and practice.

■ **Updated Coverage** The third edition addresses the rapidly changing nature of the field and provides the most-up-to date coverage of the following key topics.

  • chromosome segregation (Chapter 2)
  • mechanism of X-inactivation (Chapter 4)
  • genes that make us human: a comparison of chimpanzee and human genes (Chapter 6)
  • contingency chi-square test for independent assortment (Chapter 7)
  • variation in recombination rates (Chapter 7)
  • HIV virus evolution (Chapter 8)
  • chromosome variation and evolution (Chapter 9)
  • the genetics of speciation (Chapter 13)
  • tracking human migratory patterns throughout history (Chapter 13)

eBook

The new eBook, based on the full version of *Genetics* 3e, can be customized to
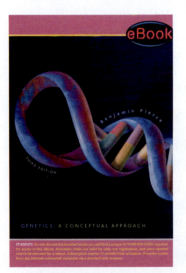
include only the chapters that appear in *Transmission and Population Genetics*. The eBook is a completely integrated electronic version of the textbook available as a stand-alone text at a reduced price or packaged with new copies of the book at a nominal cost. The online **eBook** contains everything found in the printed book plus:

■ Instant navigation to any section or subsection of the book or to any printed book page number

■ Electronic study tools including bookmarks, highlighting, and note-taking capabilities, as well as instant search of any term, pop-up of key-term definitions, and a spoken glossary

■ Access to all the fully integrated media resources created specifically for this book, including the **Interactive Animations, Podcasts, Problem-Solving Videos** as well as **Mining Genomes** bioinformatic tutorials and **Concept Check Questions.**

Instructor Media and Supplements

Instructors are provided with a comprehensive set of teaching tools, carefully developed to support lecture and individual teaching styles. These resources are

Updated Coverage: HIV Virus Evolution

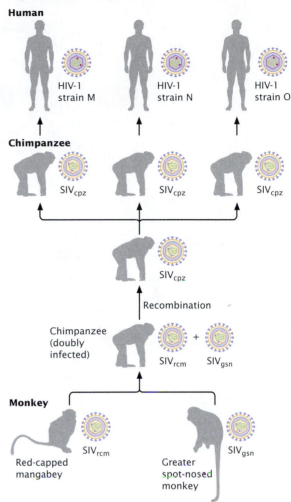

Fully Optimized JPEGS, with and without labeling

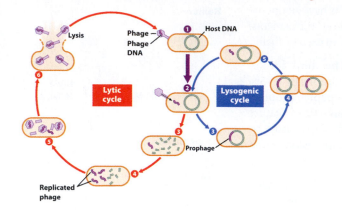

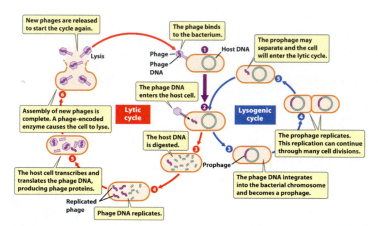

made available to adopters using the printed textbook or the **eBook** or both. The **Instructor Media** described below, along with the **Solutions** to all textbook problems, can be found on the **Instructors' Resource CD/DVD** (IRCD) as well as on the instructor side of the **Book Companion Website** (www.whfreeman.com/pierceTAP3e).

- **Fully Optimized JPEGs** of every illustration, photograph, and table in the text in labeled and unlabeled versions. Type size, configuration, and color saturation of every image has been individually treated, and the resulting files have been reviewed by instructors of this course and tested in a large lecture hall to ensure maximum clarity and visibility. These JPEGs are offered separately as well as in **PowerPoint Presentations** for each chapter.

- The most popular images in the textbook are also available as **Overhead Transparencies**, which also have been prepared for maximum visibility.

- **New Lecture-Connection PowerPoint Presentations** for each chapter have been developed to minimize preparation time for new users of the book. These files offer suggested lectures including key illustrations and summaries that instructors can adapt to their teaching styles.

- **New Layered PowerPoint Presentations** provide step-by-step depictions of key processes, sequences, and concepts in a visual format to allow instructors to present complex ideas in clear, manageable chunks.

- Special lecture-ready versions of the **Interactive Podcasts/Animations** (listed on page xv) have been created and preloaded into PowerPoint in both PC and Macintosh versions to save instructors time.

- **New Clicker Questions**, by Steven Gorsich, Central Michigan University, allow instructors to integrate active learning in the classroom and to assess student understanding of key concepts during lectures. Available in Microsoft Word and PowerPoint, questions are based on the Concept Check questions featured in the textbook.

- The **Test Bank**, prepared by Gregory Copenhaver, University of North Carolina at Chapel Hill, Rodney Mauricio, University of Georgia, and Ravishankar Palanivelu, University of Arizona, contains 50 questions per chapter, including multiple-choice, true–false, and short-answer questions. The Test Bank, available on the Instructor's CD-ROM and on the Book Companion Website, consists of chapter-by-chapter Word files that are easy to download, edit, and print.

- **Blackboard** and **WebCT** cartridges are available and include the Test Bank and end-of-chapter questions in multiple-choice and fill-in-the blank-format.

Student Media and Supplements

Students are provided with media designed to help them grasp genetic concepts and improve their problem-solving ability.

- **Solutions and Problem-Solving MegaManual** by Jung Choi, Georgia Institute of Technology, and Mark McCallum, Pfeiffer University. This manual contains complete answers and worked-out solutions to all questions and problems in the textbook. The MegaManual has been reviewed extensively by instructors and has been found to be an indispensable tool for success by students.

- **New Analytical Problem-Solving Videos** provide a fresh focus on crucial objectives of this course: increasing student problem-solving proficiency by developing their analytical skills and ability to successfully apply strategies. These videos developed by Susan Elrod, California Polytechnic State University, San Luis Obispo, have a student and an instructor interacting in an office-hours setting. The student is instructed in applying the concepts of the course in the solving of key end-of-chapter problems in Chapters 3, 4, 5, 6, and 7. These videos are available on the **eBook**.

- **Interactive Podcasts/Animations** can be downloaded to MP3 players and consist of 9 tutorials that illuminate important concepts in genetics. They help students understand key processes in genetics by outlining these processes in a step-by-step manner. The tutorials are available on the **eBook** and the **Book Companion Website** (www.whfreeman.com/pierceTAP3e). The animated tutorials are:

  2.1 Cell Cycle and Mitosis
  2.2 Meiosis
  2.3 Genetic Variation in Meiosis
  2.4 Genetic Crosses
  4.1 X-Linked Inheritance
  7.1 Determining Gene Order
  8.1 Bacterial Conjugation
  10.1 DNA Mutations
  13.1 The Hardy–Weinberg Law

- **Updated Mining Genomes** by Mark S. Wilson, Humboldt State University, consists of 10 interactive tutorials that guide students through live analyses and searches of online databases, giving them experience in using the most-current tools of bioinformatics. **Mining Genomes** is available on the **eBook** and on the **Book Companion Website**. A print component allows students to review the information that they have learned through the online version

- The **New Genetics Ethics, Science, and Technology** boxes and the **Suggested Readings** included in the first and second editions of the textbook are now available online (as are the Suggested Readings for the third edition).

Also available for students is *Genetics: A Conceptual Approach*, Third Edition, in hardcover and paperback for the introductory genetics course covering transmission, molecular and population genetics.

Interactive Animations

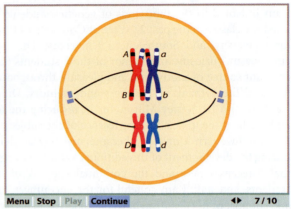

| Menu | Stop | Play | Continue | | ◄► | 7/10 |

In crossing over, the DNA strands of paired, nonsister chromatids of homologous chromosomes undergo breakage. The breaks are repaired in suce a way that segments of nonsister chromatids are exchanged. After crossing over, the two sister chromatids are no longer identical.

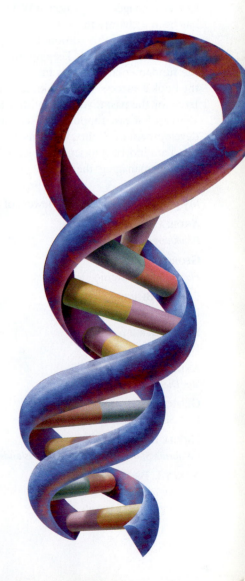

Acknowledgments

I am indebted to the thousands of genetics students who have filled my classes in the past 27 years, at Connecticut College, Baylor University, and Southwestern University. The intelligence, enthusiasm, curiosity, and humor of these students have been a constant source of motivation and pleasure throughout my professional life. I thank my teachers and mentors, Dr. Raymond Canham and Dr. Jeffrey Mitton, for introducing me to genetics and teaching me to be a lifelong learner of the subject.

Southwestern University became my new academic home during the development of this edition; I value the warm welcome that I received here and the continual support of its faculty, students, and staff. I am grateful for the encouragement, advice, and support of my department chair, Rebecca Sheller, who helped immensely with my transition to Southwestern. I am grateful to James Hunt, Provost and Dean of Brown College at Southwestern University; Jim has been a good friend, supporter, and mentor throughout the past 2 years. I value the friendship, collegiality, and support provided by my colleagues in the Department of Biology at Southwestern.

Modern science textbooks are a team effort, and I have been blessed with an outstanding team. Acquisitions editor Jerry Correa managed this project at W. H. Freeman and Company; he has been a champion of the book and contributed creatively to many aspects of this edition. Developmental editor Lisa Samols did an outstanding job, keeping me on schedule and focused on the fundamental elements that made the preceding editions of the book a success. I have enjoyed working closely with Lisa and Jerry for the past 2 years. Publisher Sara Tenney has been a key source of ideas, support, and encouragement throughout the development of all three editions of the book.

I am indebted to senior project editor Georgia Lee Hadler for expertly managing the book's production. As in the first two editions, Patricia Zimmerman was a fantastic manuscript editor, turning my awkward sentences into acceptable prose and contributing many valuable editorial suggestions. I thank Craig Durant at Dragonfly Media Group for creating the new and revised illustrations in this edition, continuing the book's outstanding illustration program. Additional thanks go to Susan Wein at W. H. Freeman and Denise Kadlubowski at Black Dot Group for ably coordinating the composition and manufacturing phases of production, and to Debbie Clare, for directing the marketing program. Vicki Tomaselli developed the book's updated design and created the beautiful and unique cover. I thank Ted Szczepanski, Christine Buese, and Patty Catuera, for photo research, and Bill Page, for coordinating the illustration program. Patrick Shriner and Marc Mazzoni managed multimedia, and Hannah Thonet and Beth McHenry managed the supplements. I am particularly grateful to Jung H. Choi and Mark McCallum, for writing solutions to many new end-of-chapter problems, and to Gregory P. Copenhaver, Rodney Mauricio, and Ravishankar Palanivelu, for writing the Test Bank. I thank Mark Samols for his review and advice on RNAi and other recent developments in molecular genetics.

I extend special thanks to the W. H. Freeman sales representatives, regional managers, and regional sales specialists. I have enjoyed getting to know and communicating with many of them. Ultimately, it is their hard work and good service that have made *Genetics: A Conceptual Approach* a success.

A large number of colleagues served as reviewers and class testers of the textbook, kindly imparting to me their technical expertise and teaching experience. Their assistance is gratefully acknowledged; any remaining errors are entirely my responsibility.

Throughout three editions of this book, my family—Marlene, Sarah, and Michael—have been my primary source of inspiration, encouragement, and support.

My gratitude goes to the reviewers of the third edition:

ANDREA BAILEY, PH.D.
Brookhaven College

GEORGE W. BATES
Florida State University

EDWARD BERGER
Dartmouth University

DANIEL BERGEY
Black Hills State University

ANDREW J. BOHONAK
San Diego State University

GREGORY C. BOOTON
Ohio State University

NICOLE BOURNIAS
California State University, Channel Islands

NANCY L. BROOKER, PH.D.
Professor of Genetics,
Pittsburg State University

ROBB T. BRUMFIELD
Louisiana State University

J. AARON CASSILL
University of Texas, San Antonio

HENRY C. CHANG
Purdue University

CAROL J. CHIHARA, PH.D.
University of San Francisco

HUI-MIN CHUNG
University of West Florida

MARY C. COLAVITO
Santa Monica College

DEBORAH A. EASTMAN
Connecticut College

BERT ELY
University of South Carolina

F. LES ERICKSON
Salisbury State University

ROBERT FARRELL, PH.D.
Penn State University

WAYNE C. FORRESTER
Indiana University

ROBERT G. FOWLER
San Jose State University

LAURA L. FROST, PH.D.
Point Park University

JACK R. GIRTON
Iowa State University

ELLIOT S. GOLDSTEIN, PH.D.
Arizona State University

JESSICA L. GOLDSTEIN
Barnard College

STEVEN W. GORSICH
Central Michigan University

PATRICK GUILFOILE, PH.D.
Bemidji State University

ASHLEY A. HAGLER, M.S.
University of North Carolina, Charlotte

ROBERT D. HINRICHSEN
Indian University of Pennsylvania

STAN HOEGERMAN
College of William and Mary

MARGARET HOLLINGSWORTH
State University of New York, Buffalo

CHERYL L. JORCYK, PH.D.
Boise State University

ANTHONY KERN
Northland College

MARGARET J. KOVACH
University of Tennessee, Chattanooga

BRIAN KREISER
University of Southern Mississippi

CATHERINE B. KUNST
University of Denver

MARY ROSE LAMB
University of Puget Sound

PATRICK H. MASSON
Professor of Genetics,
University of Wisconsin, Madison

SHAWN MEAGHER, PH.D.
Western Illinois University

MARCIE H. MOEHNKE
Baylor University

JESSICA L. MOORE
University of South Florida

NANCY MORVILLO, PH.D.
Florida Southern College

HARRY NICKLA
Creighton University

ANN V. PATERSON
Williams Baptist College

GREG PODGORSKI
Utah State University

CATHERINE A. REINKE
Northwestern University

KATHERINE T. SCHMEIDLER
Irvine Valley Community College

JON SCHNORR
Pacific University

STEPHANIE C. SCHROEDER, PH.D.
Webster University

RODNEY J. SCOTT, PH.D.
Wheaton College

BARKUR S. SHASTRY, PH.D.
Oakland University

DR. WENDY A. SHUTTLEWORTH
Lewis-Clark State College

THOMAS SMITH
Southern Arkansas University

WALTER SOTERO-ESTEVA
University of Central Florida

DOUGLAS THROWER
University of California, Santa Barbara

DOROTHY E. TUTHILL
University of Wyoming

TZVI TZFIRA
University of Michigan

NANETTE VAN LOON
Borough of Manhattan Community College

ERIK VOLLBRECHT
Iowa State University

DANIEL WANG
University of Miami

YI-HONG WANG
Penn State University, Erie-Behrend College

WILLIAM R. WELLNITZ
Augusta State University

CINDY L. WHITE, PH.D.
University of Colorado

DR. KATHLEEN WOOD
University of Mary Hardin-Baylor

JIANZHI ZHANG
University of Michigan, Ann Arbor

I wish to thank the reviewers and class testers of the first and second editions:

First edition

Joan Abramowitz, *University of Houston, Downtown;* James Allan, *Edinburgh University;* Fred Allendorf, *University of Montana;* James O. Allen, *Florida International University;* Ross Anderson, *Master's College;* Pablo Arenaz, *University of Texas, El Paso;* Alan Atherly, *Iowa State University;* Charles Atkins, *Ohio University;* Vance Baird, *Clemson University;* Stephanie Baker, *Erskine College;* Phillip T. Barnes, *Connecticut College;* George Bates, *Florida State University;* Amy Bejsovec, *Duke University;* John Belote, *Syracuse University;* David Benner, *East Tennessee State University;* Spencer Benson, *University of Maryland, College Park;* Dan Bergey, *Montana State University;* Andrew Bohonak, *San Diego State University;* J. Hoyt Bowers, *Wayland Baptist University;* Jane Bradley, *Des Moines Area Community College;* Elizabeth Bryda, *Marshall University;* Michael Buratovich, *Spring Arbor University;* Peter Burgers, *Washington University in St. Louis;* Jeffrey Byrd, *St.*

Mary's College of Maryland; Patrick Calie, *Eastern Kentucky University, Richmond;* Arthur Champlin, *Colby College;* Lee Anne Chaney, *Whitworth College;* Bruce Chase, *University of Nebraska, Omaha;* Christian Chavret, *Indiana University, Kokomo;* Carol Chihara, *University of San Francisco;* Joseph Chinnici, *VA Commonwealth University;* Jung H. Choi, *Georgia Institute of Technology;* Craig Coates, *Texas A & M University;* John Condie, *Fort Lewis College;* Patricia Conklin, *State University of New York College, Cortland;* Victor Corces, *Johns Hopkins University;* Victor Cox, *Parkland College;* Drew Cressman, *Sarah Lawrence College;* David Crowley, *Mercer Community College;* Laszlo Csonka, *Purdue University;* Michael R. Culbertson, *University of Wisconsin, Madison;* Mike Dalbey, *University of California, Santa Cruz;* Alix Darden, *The Citadel;* Terry Davin, *Penn Valley Community College;* Sandra Davis, *University of Louisiana, Monroe;* Thomas Davis, *University of New Hampshire;* Ann Marie Davison, *Kwantlen University College;* Steven Denison, *Eckerd College;* Carter Denniston, *University of Wisconsin, Madison;* Myra Derbyshire, *Mount Saint Mary's College;* Andrew A. Dewees,

Sam Houston State University; AnnMarie DiLorenzo, *Montclair State University;* Judith A. Dilts, *William Jewell College;* Stephen DiNardo, *University of Pennsylvania;* Linda Dixon, *University of Colorado, Denver;* Frank Doe, *University of Dallas;* Diana Downs, *University of Wisconsin, Madison;* Richard Duhrkopf, *Baylor University;* Maureen Dunbar, *Penn State, Berks College;* Lynn Ebersole, *Northern Kentucky University;* Larry Eckroat, *Penn State, Erie;* Johnny El-Rady, *University of South Florida;* Ted English, *North Carolina State University;* Scott Erdman, *Syracuse University;* Asim Esen, *Virginia Polytechnic Institute and State University;* Paul Evans, *Brigham Young University;* Elsa Q. Falls, *Randolph-Macon College;* Robert Farrell, *Penn State, York;* Kathleen M. Fisher, *San Diego State University;* Valerie Flechtner, *John Carroll University;* Mary Flynn, *MCP Hahnemann University;* Thomas Fogle, *Saint Mary's College;* Rosemary Ford, *Washington College;* Laurie Freeman, *Fulton-Montgomery Community College;* Julia Frugoli, *Clemson University;* Maria Galb-Meagher, *University of Florida;* Arupa Ganguly, *University of Pennsylvania;* Dan Garza, *Novartis Pharmaceutical Company;* Ivan Gepner, *Monmouth University;* Elliot S. Goldstein, *Arizona State University;* Paul Goldstein, *University of Texas, El Paso;* Javier Gonzalez, *University of Texas, Brownsville;* Deborah Good, *University of Massachusetts;* Harvey F. Good, *University of Le Verne;* Myron Goodman, *University of Southern California;* Richard L. Gourse, *University of Wisconsin, Madison;* Michael W. Gray, *Dalhousie University;* Michael Grotewiel, *Michigan State University;* James Haber, *Brandeis University;* Randall Harris, *William Carey College;* Stan Hattman, *University of Rochester;* Martha Haviland, *Rutgers University;* John Hays, *Oregon State University;* James L. Hayward, *Andrews University;* Donna Hazelwood, *Dakota State University;* Kaius Helenurm, *University of South Dakota;* Curtis Henderson, *MacMurray College;* Jerald Hendrix, *Kennesaw State Universtiy;* Richard P. Hershberger, *Carlow College;* Karen Hicks, *Kenyon College;* Jerry Higginbotham, *Transylvania University;* David Hillis, *University of Texas;* Alan Hinnebusch, *National Institutes of Health;* Deborah Hinson, *Dallas Baptist University;* Margaret Hollingsworth, *State University of New York, Buffalo;* George Hudock, *Indiana University;* Kim Hunter, *Salisbury State University;* David Hyde, *University of Notre Dame;* Colleen Jacks, *Gustavus Adolphus College;* William Jackson, *University of South Carolina, Aiken;* Tony Jilek, *University of Wisconsin, River Falls;* Rick Johns, *Northern Illinois University;* Casonya M. Johnson, *Morgan State University;* Hugh Johnson, *Southern Arkansas University;* J. Spencer Johnston, *Texas A&M University;* Greg Jones, *Santa Fe Community College;* Michael Jones, *Indiana University;* Gregg Jongeward, *University of the Pacific;* Chris Kapicka, *Northwest Nazarene University;* Clifford Kiel, *University of Delaware;* Hai Kinal, *Springfield College;* Olga Ruiz Kopp, *University of Tennessee;* Gae Kovalik, *University of Texas of the Permian Basin;* Rhonda J. Kuykindoll, *Rust College;* Trip Lamb, *East Carolina University;* Franz Lang, *Université de Montréal;* Allan Larsen, *Washington University, St. Louis;* Chris Lawrence, *University of Rochester;* Alicia Lesnikowska, *Georgia Southwestern State University;* Ricki Lewis, *University of Albany;* Alice Lindgren, *Bemidji State University;* Malcolm Lippert, *Saint Michael's College;* John Locke, *University of Alberta;* Pat Lord, *Wake Forest University;* Charles Louis, *Georgia State University;* Carl Luciano, In*diana University of Pennsylvania;* Paul Lurquin, *Washington State University;* Aldons Lusis, *University of California, Los Angeles;* Peter Luykx, *University of Miami;* J. David MacDonald, *Wichita State University;* Paul Mangum, *Midland College;* Michael Markovitz, *Midland Lutheran College;* Marcie Marston, *Roger Williams University;* Alfred Martin, *Benedictine University;* Robert Martinez, *Quinnipiac College;* David Matthes, *San Jose State University;* Joyce Maxwell, *California Institute of Technology;* Mark McCallum, *Pfeiffer University;* Jim McGivern, *Gannon University;* Denise McGuire, *Saint Cloud State University;* Lauren McIntyre, *Purdue University;* Sandra Michael, *Binghamton University;* Gail Miller, *York College;* John Mishler, *Delaware Valley College;* Jeffrey Mitton, *University of Colorado, Boulder;* Aaron Moe, *Concordia University;* Mary Montgomery, *Macalaster College;* Ammini Moorthy, *Wagner College;* Nancy Morvillo, *Florida Southern College;* Mary Murnik, *Ferris State College;* Jennifer Myka, *Brescia University;* Elbert Myles, *Tennessee State University;* William Nelson, *Morgan State University;* Bryan Ness, *Pacific Union College;* John Newell, *Shaw University;* Jeffrey Newman, *Lycoming College;* Harry Nickla, *Creighton University;* Timothy Nilsen, *Center for RNA Molecular Biology;* Donna Nofziger-Plank, *Pepperdine University;* Marcia O'Connell, *State University of New York, Stony Brook;* Ronald Ostrowski, *University of North Carolina, Charlotte;* Tony Palombella, *Longwood College;* Louise Paquin, *Western Maryland College;* Ann Paterson, *Williams Baptist College;* Jackie Peltier, *Houston Baptist University;* Dorene Petrosky, *Delaware State University;* Lynn Petrullo, *College of New Rochelle;* Bernie Possidente, *Skidmore College;* Daphne Preuss, *University of Chicago;* Jim Price, *Utah Valley State College;* Frank Pugh, *Pennsylvania State University;* Mary Puterbaugh, *University of Pittsburgh, Bradford;* Todd Rainey, *Bluffton College;* Wendy Raymond, *Williams College;* Peggy Redshaw, *Austin College;* William S. Reznikoff, *University of Wisconsin, Madison;* R. H. Richardson, *University of Texas, Austin;* Todd Rimkus, *Marymount College;* John Ringo, *University of Maine;* Tara Robinson, *Oregon State Univesity;* Torbert R. Rocheford, *University of Illinois;* Stephen Roof, *Fairmont State College;* Jim Sanders, *Texas A&M University;* Mark Sanders, *University of California, Davis;* Andrew Scala, *Dutchess Community College;* Joanne Scalzitti, *University of Nebraska, Kearney;* Marilyn Schendel, *Carroll College;* Jennifer Schisa, *Central Michigan University;* Kathy Schmeidler, *Irvine Valley College;* Malcolm Schug, *University of North Carolina, Greensboro;* Terry Schwaner, *Southern Utah University;* Ralph Seelke, *University of Wisconsin, Superior;* Jeanine Seguin, *Keuka College;* Jeff Sekelsky, *University of North Carolina;* David Sherratt, *University of Oxford;* DeWayne Shoemaker, *University of Wisconsin, Madison;* Nancy N. Shontz, *Grand Valley State University;* J. Kenneth Shull, *Appalachian State University;* Laura Sigismondi, *University of Rio Grande;* Pat Singer, *Simpson College;* Laurie Smith, *University of California, San Diego;* Ken Spitze, *University of Illinois, Urbana-Champaign;* Martha Stauderman, *University of San Diego;* Todd Steck, *University of North Carolina, Charlotte;* Anne

Stone, *University of New Mexico;* Susan Strome, *Indiana University;* Heidi Super, *Minot State University;* Chris Tachibana, *University of Washington;* Jennifer Thomson, *University of Cape Town;* Grant Thorsett, *Willamette University;* Tammy Tobin-Janzen, *Susquehanna University;* John Tonzetich, *Bucknell University;* Andrew Travers, *Medical Research Council, Cambridge, England;* Carol Trent, *Western Washington University;* Callie Vanderbilt, *San Juan College;* Albrecht Von Arnim, *University of Tennessee;* Alan Waldman, *University of South Carolina;* Melinda Wales, *Texas A&M University;* Nancy Walker, *Clemson University;* David Weber, *Illinois State University;* Karen Weiler, *Idaho State University;* William Wellnitz, *Augusta State University;* Michelle Whaley; *University of Notre Dame;* Matt White, *Ohio University;* Betsy Wilson, *University of North Carolina, Asheville;* Mark Wilson, *Humboldt State University;* Tom Wiltshire, *Culver-Stockton College;* Mark Winey, *University of Colorado;* David Wing, *Shepherd College;* Darla Wise, *Concord College;* D. S. Wofford, *University of Florida;* Bonnie Wood, *University of Maine, Presque Isle;* Kathleen Wood, *University of Mary Hardin-Baylor;* David Woods, *Rhodes University;* Michael Wooten, *Auburn Univesity;* Debbie Wygal, *College of Saint Catherine;* Krassimir (Joseph) Yankulov, *University of Guelph;* Jeff Young, *Western Washington University;* Malcolm Zellars, *Georgia State University;* Hong Zhang, *Texas Tech University;* Suzanne Ziesmann, *Sweet Briar College.*

Second edition

Lucy Andrews, *University of Alabama, Birmingham;* John Belote, *Syracuse University;* Susan Bergeson, *University of Texas, Austin;* Andrew Bohanak, *San Diego State University;* Nancy Brooker, *Pittsburg State University;* Mary Bryk, *Texas A&M University;*

Mary Colavito, *Santa Monica College;* Georgina Cornwall, *Metropolitan State College of Denver;* Susan Elrod, *California Polytechnic State University, San Luis Obispo;* Les Erickson, *Salisbury State University;* Robert Farrell, *Pennsylvania State University, York;* Julia Frugoli, *Clemson University;* Paul Gauss, *Western State College;* Elliot Goldstein, *Arizona State University;* Paul Goldstein, *University of Texas, El Paso;* Peter Hart, *University of Massachusetts, Dartmouth;* Kaius Helenurm, *University of South Dakota;* Leslie Hickok, *University of Tennessee, Knoxville;* Bruce Hofkin, *University of New Mexico;* Margaret Hollingsworth, *State University of New York, Buffalo;* Richard Imberski, *University of Maryland, College Park;* Ikhide Imumorin, *Valdosta State University;* Mitrick Johns, *Northern Illinois State University;* Lee Johnson, *The Ohio State University;* Spencer J. Johnson, *Texas A&M University;* Chris Korey, *College of Charleston;* Margaret Kovach, *University of Tennessee, Chattanooga;* Patrick H. Masson, *University of Wisconsin, Madison;* Mary Montgomery, *Macalester College;* James Morris, *Clemson University;* David Muir, *Stanford University;* John Nambu, *University of Massachusetts, Amherst;* John Osterman, *University of Nebraska, Lincoln;* Francisco Pelegri, *University of Wisconsin, Madison;* Michael Polymenis, *Texas A&M University;* Dennis Ray, *University of Arizona;* Leslie Rye, *University of Guelph;* Stephanie Schroeder, *Webster University;* E. Shokraii, *Virginia Polytechnic Institute and State University;* Ken Shull, *Appalachian State University;* Raymond St. Leger, *University of Maryland, College Park;* Jim Straka, *Macalester College;* Martin Tracey, *Florida International University;* Alan Waldman, *University of South Carolina, Columbia;* Fred Whipple, *California State University, Fullerton;* Michael Windelspecht, *Appalachian State University;* Andrew J. Wood, *Southern Illinois University*

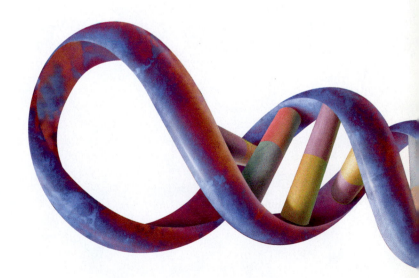

1 | Introduction to Genetics

ALBINISM IN THE HOPIS

Hopi bowl, early twentieth century. Albinism, a genetic condition, arises with high frequency among the Hopi people and occupies a special place in the Hopi culture. *[The Newark Museum/Art Resource, NY.]*

Rising a thousand feet above the desert floor, Black Mesa dominates the horizon of the Enchanted Desert and provides a familiar landmark for travelers passing through northeastern Arizona. Black Mesa is not only a prominent geological feature; more significantly, it is the ancestral home of the Hopi Native Americans. Fingers of the mesa reach out into the desert, and alongside or on top of each finger is a Hopi village. Most of the villages are quite small, filled with only a few dozen inhabitants, but they are incredibly old. One village, Oraibi, has existed on Black Mesa since 1150 A.D. and is the oldest continually occupied settlement in North America.

In 1900, Aleš Hrdliĕka, an anthropologist and physician working for the American Museum of Natural History, visited the Hopi villages of Black Mesa and reported a startling discovery. Among the Hopis were 11 white people—not Caucasians, but actually white Hopi Native Americans. These persons had a genetic condition known as albinism (**Figure 1.1**).

Albinism is caused by a defect in one of the enzymes required to produce melanin, the pigment that darkens our skin, hair, and eyes. People with albinism don't produce melanin or they produce only small amounts of it and, consequently, have white hair, light skin, and no pigment in the irises of their eyes. Melanin normally protects the DNA of skin cells from the damaging effects of ultraviolet radiation in sunlight, and melanin's presence in the developing eye is essential for proper eyesight.

The genetic basis of albinism was first described by Archibald Garrod, who recognized in 1908 that the condition was inherited as an autosomal recessive trait, meaning that a person must receive two copies of an albino mutation—one from each parent—to have albinism. In recent years, the molecular natures of the mutations that lead to albinism have been elucidated. Albinism in humans is caused by defects in any one of four different genes that control the synthesis and storage of melanin; many different types of mutations can occur at each gene, any one of which may lead to albinism. The form of albinism found in the Hopis is most likely oculocutaneous albinism type 2, due to a defect in the *OCA* gene on chromosome 15.

1.1 Albinism among the Hopi Native Americans. In this photograph, taken about 1900, the Hopi girl in the center has albinism. *[The Field Museum/Charles Carpenter.]*

The Hopis are not unique in having albinos among the members of their tribe. Albinism is found in almost all human ethnic groups and is described in ancient writings; it has probably been present since humankind's beginnings. What is unique about the Hopis is the high frequency of albinism. In most human groups, albinism is rare, present in only about 1 in 20,000 persons. In the villages on Black Mesa, it reaches a frequency of 1 in 200, a hundred times as frequent as in most other populations.

Why is albinism so frequent among the Hopi Native Americans? The answer to this question is not completely known, but geneticists who have studied albinism in the Hopis speculate that the high frequency of the albino gene is related to the special place that albinism occupied in the Hopi culture. For much of their history, the Hopis considered members of their tribe with albinism to be important and special. People with albinism were considered pretty, clean, and intelligent. Having a number of people with albinism in one's village was considered a good sign, a symbol that the people of the village contained particularly pure Hopi blood. Albinos performed in Hopi ceremonies and assumed positions of leadership within the tribe, often becoming chiefs, healers, and religious leaders.

Hopi albinos were also given special treatment in everyday activities. For example, the Hopis farmed small garden plots at the foot of Black Mesa for centuries. Every day throughout the growing season, the men of the tribe trek to the base of Black Mesa and spend much of the day in the bright southwestern sunlight tending their corn and vegetables. With little or no melanin pigment in their skin, people with albinism are extremely susceptible to sunburn and have increased incidences of skin cancer when exposed to the sun. Furthermore, many don't see well in bright sunlight. But the male Hopis with albinism were excused from this normal male labor and allowed to remain behind in the village with the women of the tribe, performing other duties.

Geneticists have suggested that these special considerations given to albino members of the tribe are partly responsible for the high frequency of albinism among the Hopis. Throughout the growing season, the albino men were the only male members of the tribe in the village during the day with all the women and, thus, they enjoyed a mating advantage, which helped to spread their albino genes. In addition, the special considerations given to albino Hopis allowed them to avoid the detrimental effects of albinism—increased skin cancer and poor eyesight. The small size of the Hopi tribe probably also played a role by allowing chance to increase the frequency of the albino gene. Regardless of the factors that led to the high frequency of albinism, the Hopis clearly had great respect and appreciation for the members of their tribe who possessed this particular trait. Unfortunately, people with genetic conditions in other societies are more often subject to discrimination and prejudice.

Genetics is one of the frontiers of modern science. Pick up almost any major newspaper or news magazine and chances are that you will see something related to genetics: the discovery of cancer-causing genes; the use of gene therapy to treat diseases; or reports of possible hereditary influences on intelligence, personality, and sexual orientation. These findings often have significant economic and ethical implications, making the study of genetics relevant, timely, and interesting.

This chapter introduces you to genetics and reviews some concepts that you may have encountered briefly in a preceding biology course. We begin by considering the importance of genetics to each of us, to society at large, and to students of biology. We then turn to the history of genet-

ics, how the field as a whole developed. The final part of the chapter reviews some fundamental terms and principles of genetics that are used throughout the book.

1.1 Genetics Is Important to Individuals, to Society, and to the Study of Biology

Albinism among the Hopis illustrates the important role that genes play in our lives. This one genetic defect, among the 20,000 to 25,000 genes that humans possess, completely changes the life of a Hopi who possesses it. It alters his or her occupation, role in Hopi society, and relations with other

(a)   **(b)**

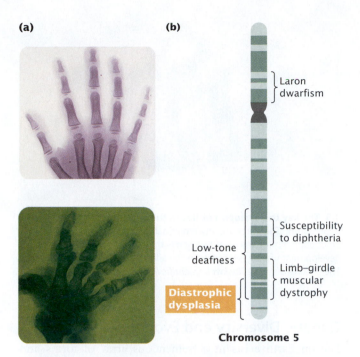

Laron
dwarfism

Low-tone
deafness

Susceptibility
to diphtheria

Limb–girdle
muscular
dystrophy

Diastrophic
dysplasia

Chromosome 5

1.2 Genes influence susceptibility to many diseases and disorders. (a) An X-ray of the hand of a person suffering from diastrophic dysplasia (bottom), a hereditary growth disorder that results in curved bones, short limbs, and hand deformities, compared with an X-ray of a normal hand (top). (b) This disorder is due to a defect in a gene on chromosome 5. Braces indicate regions on chromosome 5 where genes giving rise to other disorders are located. *[Part a: (top) Biophoto Associates/Science Source/Photo Researchers; (bottom) courtesy of Eric Lander, Whitehead Institute, MIT.]*

(a)

(b)

1.3 In the Green Revolution, genetic techniques were used to develop new high-yielding strains of crops. (a) Norman Borlaug, a leader in the development of new strains of wheat that led to the Green Revolution. Borlaug was awarded the Nobel Peace Prize in 1970. (b) Modern, high-yielding rice plant (left) and traditional rice plant (right). *[Part a: UPI/Corbis-Bettman. Part b: IRRI.]*

members of the tribe. We all possess genes that influence our lives in significant ways. Genes affect our height, weight, hair color, and skin pigmentation. They influence our susceptibility to many diseases and disorders (**Figure 1.2**) and even contribute to our intelligence and personality. Genes are fundamental to who and what we are.

Although the science of genetics is relatively new compared with many other sciences, people have understood the hereditary nature of traits and have practiced genetics for thousands of years. The rise of agriculture began when people started to apply genetic principles to the domestication of plants and animals. Today, the major crops and animals used in agriculture have undergone extensive genetic alterations to greatly increase their yields and provide many desirable traits, such as disease and pest resistance, special nutritional qualities, and characteristics that facilitate harvest. The Green Revolution, which expanded food production throughout the world in the 1950s and 1960s, relied heavily on the application of genetics (**Figure 1.3**). Today, genetically engineered corn, soybeans, and other crops constitute a significant proportion of all the food produced worldwide.

The pharmaceutical industry is another area in which genetics plays an important role. Numerous drugs and food

additives are synthesized by fungi and bacteria that have been genetically manipulated to make them efficient producers of these substances. The biotechnology industry employs molecular genetic techniques to develop and mass-produce substances of commercial value. Growth hormone, insulin, and clotting factor are now produced commercially by genetically engineered bacteria (**Figure 1.4**). Techniques of molecular genetics have also been used to produce bacteria that remove minerals from ore, break down toxic chemicals, and inhibit damaging frost formation on crop plants.

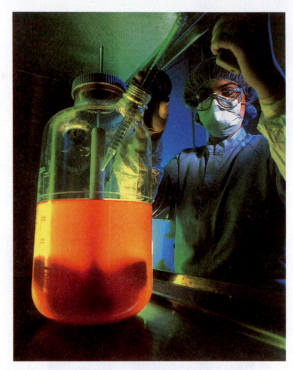

1.4 The biotechnology industry uses molecular genetic methods to produce substances of economic value. *[James Holmes/Celltech Ltd./Science Photo Library/Photo Researchers.]*

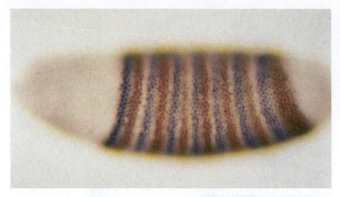

1.5 The key to development lies in the regulation of gene expression. This early fruit-fly embryo illustrates the localized production of proteins from two genes that determine the development of body segments in the adult fly. *[From Peter Lawrence, The Making of a Fly (Blackwell Scientific Publications, 1992).]*

Genetics plays a critical role in medicine. Physicians recognize that many diseases and disorders have a hereditary component, including genetic disorders such as sickle-cell anemia and Huntington disease as well as many common diseases such as asthma, diabetes, and hypertension. Advances in molecular genetics have resulted not only in important insights into the nature of cancer but also in the development of many diagnostic tests. Gene therapy—the direct alteration of genes to treat human diseases—has now been carried out on thousands of patients.

The Role of Genetics in Biology

Although an understanding of genetics is important to all people, it is critical to the student of biology. Genetics provides one of biology's unifying principles: all organisms use genetic systems that have a number of features in common. Genetics also undergirds the study of many other biological disciplines. Evolution, for example, is genetic change taking place through time; so the study of evolution requires an understanding of genetics. Developmental biology relies heavily on genetics: tissues and organs form through the regulated expression of genes (**Figure 1.5**). Even such fields as taxonomy, ecology, and animal behavior are making increasing use of genetic methods. The study of almost any field of biology or medicine is incomplete without a thorough understanding of genes and genetic methods.

Genetic Diversity and Evolution

Life on Earth exists in a tremendous array of forms and features that occupy almost every conceivable environment. Life is also characterized by adaptation: many organisms are exquisitely suited to the environment in which they are found. The history of life is a chronicle of new forms of life emerging, old forms disappearing, and existing forms changing.

Despite their tremendous diversity, living organisms have an important feature in common: all use similar genetic systems. A complete set of genetic instructions for any organism is its **genome,** and all genomes are encoded in nucleic acids—either DNA or RNA. The coding system for genomic information also is common to all life: genetic instructions are in the same format and, with rare exceptions, the code words are identical. Likewise, the process by which genetic information is copied and decoded is remarkably similar for all forms of life. These common features of heredity suggest that all life on Earth evolved from the same primordial ancestor that arose between 3.5 billion and 4 billion years ago. Biologist Richard Dawkins describes life as a river of DNA that runs through time, connecting all organisms past and present.

That all organisms have similar genetic systems means that the study of one organism's genes reveals principles that apply to other organisms. Investigations of how bacterial DNA is copied (replicated), for example, provide information that applies to the replication of human DNA. It also means that genes will function in foreign cells, which makes genetic engineering possible. Unfortunately, these similar genetic systems are also the basis for diseases such as AIDS (acquired immune deficiency syndrome), in which viral genes are able to function—sometimes with alarming efficiency—in human cells.

Life's diversity and adaptation are products of evolution, which is simply genetic change through time. Evolution is a two-step process: first, genetic variants arise randomly and, then, the proportion of particular variants increases or decreases. Genetic variation is therefore the foundation of all evolutionary change and is ultimately the basis of all life as we know it. Genetics, the study of genetic variation, is critical to understanding the past, present, and future of life.

CONCEPTS

Heredity affects many of our physical features as well as our susceptibility to many diseases and disorders. Genetics contributes to advances in agriculture, pharmaceuticals, and medicine and is fundamental to modern biology. All organisms use similar genetic systems, and genetic variation is the foundation of the diversity of all life.

✔ CONCEPT CHECK 1

What are some of the implications of all organisms having similar genetic systems?

a. That all life forms are genetically related

b. That research findings on one organism's gene function can often be applied to other organisms

c. That genes from one organism can often exist and thrive in another organism

d. All of the above

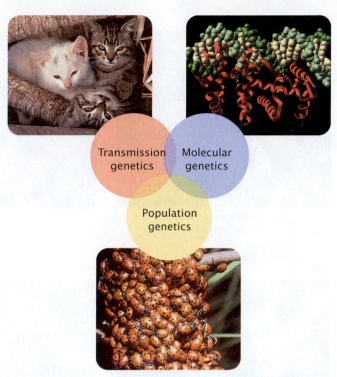

1.6 Genetics can be subdivided into three interrelated fields.
[Top left: Alan Carey/Photo Researchers. Top right: Mona file M0214602tif. Bottom: J. Alcock/Visuals Unlimited.]

Divisions of Genetics

Traditionally, the study of genetics has been divided into three major subdisciplines: transmission genetics, molecular genetics, and population genetics (**Figure 1.6**). Also known as classical genetics, **transmission genetics** encompasses the basic principles of heredity and how traits are passed from one generation to the next. This area addresses the relation between chromosomes and heredity, the arrangement of genes on chromosomes, and gene mapping. Here, the focus is on the individual organism—how an individual organism inherits its genetic makeup and how it passes its genes to the next generation.

Molecular genetics concerns the chemical nature of the gene itself: how genetic information is encoded, replicated, and expressed. It includes the cellular processes of replication, transcription, and translation—by which genetic information is transferred from one molecule to another—and gene regulation—the processes that control the expression of genetic information. The focus in molecular genetics is the gene—its structure, organization, and function.

Population genetics explores the genetic composition of groups of individual members of the same species (populations) and how that composition changes over time and geographic space. Because evolution is genetic change, population genetics is fundamentally the study of evolution. The focus of population genetics is the group of genes found in a population.

Division of the study of genetics into these three groups is convenient and traditional, but we should recognize that the fields overlap and that each major subdivision can be further divided into a number of more specialized fields, such as chromosomal genetics, biochemical genetics, quantitative genetics, and so forth. Alternatively, genetics can be subdivided by organism (fruit fly, corn, or bacterial genetics), and each of these organisms can be studied at the level of transmission, molecular, and population genetics. Modern genetics is an extremely broad field, encompassing many interrelated subdisciplines and specializations.

Model Genetic Organisms

Through the years, genetic studies have been conducted on thousands of different species, including almost all major groups of bacteria, fungi, protists, plants, and animals. Nevertheless, a few species have emerged as **model genetic organisms**—organisms having characteristics that make them particularly useful for genetic analysis and about which a tremendous amount of genetic information has accumulated. Six model organisms that have been the subject of intensive genetic study are: *Drosophila melanogaster*, the fruit fly; *Escherichia coli*, a bacterium present in the gut of humans and other mammals; *Caenorhabditis elegans*, a nematode worm; *Arabidopsis thaliana*, the thale cress plant; *Mus musculus*, the house mouse; and *Saccharomyces*

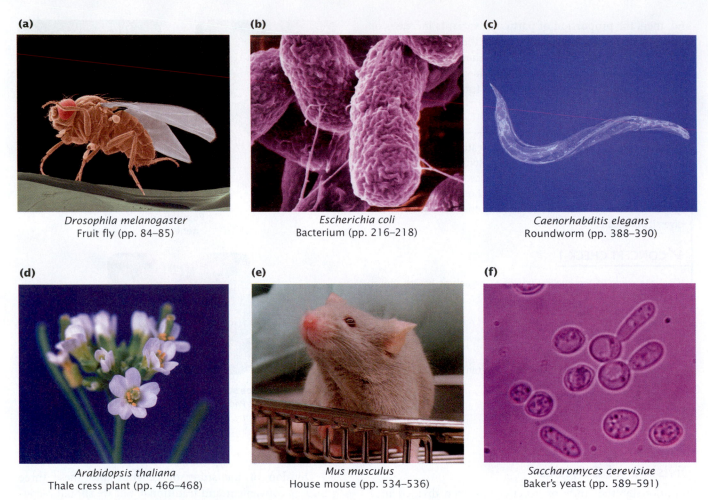

(a)
Drosophila melanogaster
Fruit fly (pp. 84–85)

(b)
Escherichia coli
Bacterium (pp. 216–218)

(c)
Caenorhabditis elegans
Roundworm (pp. 388–390)

(d)
Arabidopsis thaliana
Thale cress plant (pp. 466–468)

(e)
Mus musculus
House mouse (pp. 534–536)

(f)
Saccharomyces cerevisiae
Baker's yeast (pp. 589–591)

1.7 Model genetic organisms are species having features that make them useful for genetic analysis. *[Part a: SPL/Photo Researchers. Part b: Gary Gaugler/Visuals Unlimited. Part c: Natalie Pujol/Visuals Unlimited. Part d: Peggy Greb/ARS. Part e: Joel Page/AP. Part f: T. E. Adams/Visuals Unlimited.]*

cerevisiae, baker's yeast (**Figure 1.7**). These species are the organisms of choice for many genetic researchers, and their genomes were sequenced as a part of the Human Genome Project.

At first glance, this group of lowly and sometimes despised creatures might seem unlikely candidates for model organisms. However, all possess life cycles and traits that make them particularly suitable for genetic study, including a short generation time, manageable numbers of progeny, adaptability to a laboratory environment, and the ability to be housed and propagated inexpensively. The life cycles, genomic characteristics, and features that make these model organisms useful for genetic studies are included in special model-organism illustrations that appear throughout this book. Other species that are frequently the subject of genetic research and are also considered genetic models include bread mold (*Neurospora crassa*), corn (*Zea mays*), zebrafish (*Danio rerio*), and clawed frog (*Xenopus laevis*). Although not generally considered a genetic model, humans also have been subjected to intensive genetic scrutiny; special techniques for the genetic analysis of humans are discussed in Chapter 6.

The value of model genetic organisms is illustrated by the use of zebrafish to identify genes that affect skin pigmentation in humans. For many years, geneticists have recognized that differences in pigmentation among human ethnic groups (**Figure 1.8a**) are genetic, but the genes causing these differences were largely unknown. Zebrafish have recently become an important model in genetic studies because they are small vertebrates that produce many offspring and are easy to rear in the laboratory. The zebrafish *golden* mutant, caused by a recessive mutation, has light pigmentation due to the presence of fewer, smaller, and less-dense pigment-containing structures called melanosomes in its cells (**Figure 1.8b**). Light skin in humans is similarly due to fewer and less-dense melanosomes in pigment-containing cells.

Keith Cheng and his colleagues at Pennsylvania State University College of Medicine hypothesized that light skin in humans might result from a mutation that is similar to the

(a)

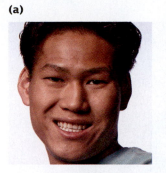

1.8 The zebrafish, a genetic model organism, has been instrumental in helping to identify genes encoding pigmentation differences among humans. (a) Human ethnic groups differ in degree of skin pigmentation. (b) The zebrafish *golden* mutation is caused by a gene that controls the amount of melanin pigment in melanosomes. *[Part a: PhotoDisc. Part b: K. Cheng / J. Gittlen, Cancer Research Foundation, Pennsylvania State College of Medicine.]*

(b)

Normal zebrafish   *Golden* mutant

golden mutation in zebrafish. Taking advantage of the ease with which zebrafish can be manipulated in the laboratory, they isolated and sequenced the gene responsible for the *golden* mutation and found that it encodes a protein that takes part in calcium uptake by melanosomes. They then searched a database of all known human genes and found a similar gene called *SLC24A5,* which encodes the same function in human cells. When they examined human populations, they found that light-skinned Europeans typically possessed one form of this gene, whereas darker-skinned Africans, Eastern Asians, and Native Americans usually possessed a different form of the gene. Many other genes also affect pigmentation in humans, as illustrated by mutations in the *OCA* gene that produce albinism among the Hopi Indians (discussed in the introduction to this chapter). Nevertheless, *SLC24A5* appears to be responsible for 24% to 38% of the differences in pigmentation between Africans and Europeans. This example illustrates the power of model organisms in genetic research.

CONCEPTS

The three major divisions of genetics are transmission genetics, molecular genetics, and population genetics. Transmission genetics examines the principles of heredity; molecular genetics deals with the gene and the cellular processes by which genetic information is transferred and expressed; population genetics concerns the genetic composition of groups of organisms and how that composition changes over time and geographic space. Model genetic organisms are species that have received special emphasis in genetic research; they have characteristics that make them useful for genetic analysis.

✔ CONCEPT CHECK 2

Would the horse make a good model genetic organism? Why or why not?

1.2 Humans Have Been Using Genetics for Thousands of Years

Although the science of genetics is young—almost entirely a product of the past 100 years or so—people have been using genetic principles for thousands of years.

The Early Use and Understanding of Heredity

The first evidence that people understood and applied the principles of heredity in earlier times is found in the domestication of plants and animals, which began between approximately 10,000 and 12,000 years ago. The world's first agriculture is thought to have developed in the Middle East, in what is now Turkey, Iraq, Iran, Syria, Jordan, and Israel, where domesticated plants and animals were major dietary components of many populations by 10,000 years ago. The first domesticated organisms included wheat, peas, lentils, barley, dogs, goats, and sheep (**Figure 1.9a**). By 4000 years ago, sophisticated genetic techniques were already in use in the Middle East. Assyrians and Babylonians developed several hundred varieties of date palms that differed in fruit size, color, taste, and time of ripening (**Figure 1.9b**). Other crops and domesticated animals were developed by cultures in Asia, Africa, and the Americas in the same period.

CONCEPTS

Humans first applied genetics to the domestication of plants and animals between approximately 10,000 and 12,000 years ago. This domestication led to the development of agriculture and fixed human settlements.

(a)   **(b)**

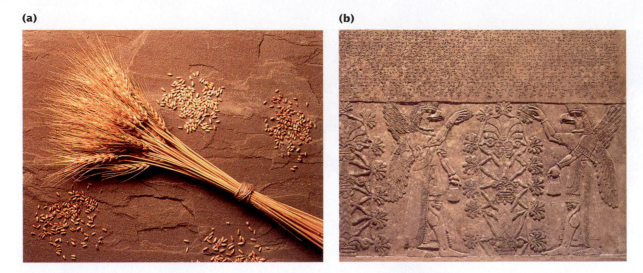

1.9 Ancient peoples practiced genetic techniques in agriculture. (a) Modern wheat, with larger and more numerous seeds that do not scatter before harvest, was produced by interbreeding at least three different wild species. (b) Assyrian bas-relief sculpture showing artificial pollination of date palms at the time of King Assurnasirpalli II, who reigned from 883 to 859 B.C. *[Part a: Scott Bauer/ARS/USDA. Part b: The Metropolitan Museum of Art, gift of John D. Rockefeller, Jr., 1932. (32.143.3) Photograph © 1996 Metropolitan Museum of Art.]*

Ancient writings demonstrate that early humans were also aware of their own heredity. Hindu sacred writings dating to 2000 years ago attribute many traits to the father and suggest that differences between siblings can be accounted for by effects from the mother. The Talmud, the Jewish book of religious laws based on oral traditions dating back thousands of years, presents an uncannily accurate understanding of the inheritance of hemophilia. It directs that, if a woman bears two sons who die of bleeding after circumcision, any additional sons that she bears should not be circumcised; nor should the sons of her sisters be circumcised, although the sons of her brothers should. This advice accurately corresponds to the X-linked pattern of inheritance of hemophilia (discussed further in Chapter 6).

The ancient Greeks gave careful consideration to human reproduction and heredity. The Greek physician Alcmaeon (circa 520 B.C.) conducted dissections of animals and proposed that the brain was not only the principal site of perception, but also the origin of semen. This proposal sparked a long philosophical debate about where semen was produced and its role in heredity. The debate culminated in the concept of **pangenesis,** in which specific particles, later called gemmules, carry information from various parts of the body to the reproductive organs, from which they are passed to the embryo at the moment of conception (**Figure 1.10a**). Although incorrect, the concept of pangenesis was highly influential and persisted until the late 1800s.

Pangenesis led the ancient Greeks to propose the notion of the **inheritance of acquired characteristics,** in which traits acquired in one's lifetime become incorporated into one's hereditary information and are passed on to offspring; for example, people who developed musical ability through diligent study would produce children who are innately endowed with musical ability. The notion of the inheritance of acquired characteristics also is no longer accepted, but it remained popular through the twentieth century.

The Greek philosopher Aristotle (384–322 B.C.) was keenly interested in heredity. He rejected the concepts of both pangenesis and the inheritance of acquired characteristics, pointing out that people sometimes resemble past ancestors more than their parents and that acquired characteristics such as mutilated body parts are not passed on. Aristotle believed that both males and females made contributions to the offspring and that there was a struggle of sorts between male and female contributions.

Although the ancient Romans contributed little to an understanding of human heredity, they successfully developed a number of techniques for animal and plant breeding; the techniques were based on trial and error rather than any general concept of heredity. Little new information was added to the understanding of genetics in the next 1000 years.

Dutch eyeglass makers began to put together simple microscopes in the late 1500s, enabling Robert Hooke (1635–1703) to discover cells in 1665. Microscopes provided naturalists with new and exciting vistas on life, and perhaps it was excessive enthusiasm for this new world of the very small that gave rise to the idea of **preformationism.** According to preformationism, inside the egg or sperm there exists a tiny miniature adult, a *homunculus,* which simply enlarges during development. Ovists argued that the homunculus resides in the egg, whereas spermists insisted that it is in the sperm

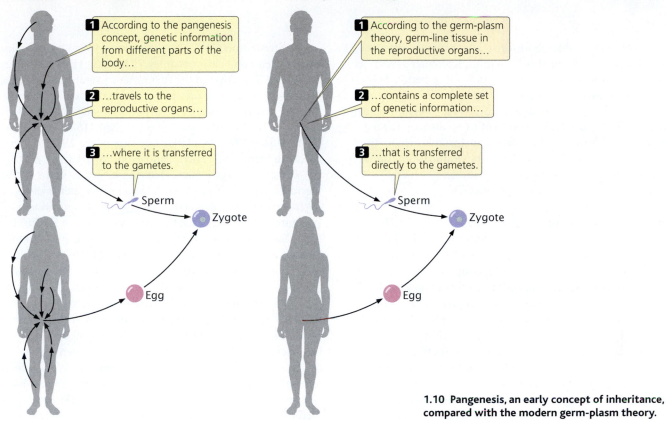

(a) Pangenesis concept

1 According to the pangenesis concept, genetic information from different parts of the body...

2 ...travels to the reproductive organs...

3 ...where it is transferred to the gametes.

Sperm

Zygote

Egg

(b) Germ-plasm theory

1 According to the germ-plasm theory, germ-line tissue in the reproductive organs...

2 ...contains a complete set of genetic information...

3 ...that is transferred directly to the gametes.

Sperm

Zygote

Egg

1.10 Pangenesis, an early concept of inheritance, compared with the modern germ-plasm theory.

(**Figure 1.11**). Preformationism meant that all traits would be inherited from only one parent—from the father if the homunculus was in the sperm or from the mother if it was in the egg. Although many observations suggested that offspring possess a mixture of traits from both parents, preformationism remained a popular concept throughout much of the seventeenth and eighteenth centuries.

Another early notion of heredity was **blending inheritance,** which proposed that offspring are a blend, or mixture, of parental traits. This idea suggested that the genetic material itself blends, much as blue and yellow pigments blend to make green paint. Once blended, genetic differences could not be separated out in future generations, just as green paint cannot be separated out into blue and yellow pigments. Some traits do *appear* to exhibit blending inheritance; however, we realize today that individual genes do not blend.

The Rise of the Science of Genetics

In 1676, Nehemiah Grew (1641–1712) reported that plants reproduce sexually by using pollen from the male sex cells. With this information, a number of botanists began to experiment with crossing plants and creating hybrids. Foremost among these early plant breeders was Joseph Gottlieb Kölreuter (1733–1806), who carried out numerous crosses and studied pollen under the microscope. He observed that many hybrids were intermediate between the parental varieties. Because he crossed plants that differed in many traits,

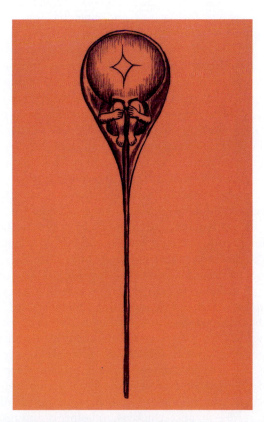

1.11 Preformationism was a popular idea of inheritance in the seventeenth and eighteenth centuries. Shown here is a drawing of a homunculus inside a sperm. *[Science VU/Visuals Unlimited.]*

Kölreuter was unable to discern any general pattern of inheritance. In spite of this limitation, Kölreuter's work set the foundation for the modern study of genetics. Subsequent to his work, a number of other botanists began to experiment with hybridization, including Gregor Mendel (1822–1884; **Figure 1.12**), who went on to discover the basic principles of heredity. Mendel's conclusions, which were unappreciated for 35 years, laid the foundation for our modern understanding of heredity, and he is generally recognized today as the father of genetics.

Developments in cytology (the study of cells) in the 1800s had a strong influence on genetics. Robert Brown (1773–1858) described the cell nucleus in 1833. Building on the work of others, Matthias Jacob Schleiden (1804–1881) and Theodor Schwann (1810–1882) proposed the concept of the **cell theory** in 1839. According to this theory, all life is composed of cells, cells arise only from preexisting cells, and the cell is the fundamental unit of structure and function in living organisms. Biologists began to examine cells to see how traits were transmitted in the course of cell division.

Charles Darwin (1809–1882), one of the most influential biologists of the nineteenth century, put forth the theory of evolution through natural selection and published his ideas in *On the Origin of Species* in 1859. Darwin recognized that heredity was fundamental to evolution, and he conducted extensive genetic crosses with pigeons and other organisms. However, he never understood the nature of inheritance, and this lack of understanding was a major omission in his theory of evolution.

In the last half of the nineteenth century, the invention of the microtome (for cutting thin sections of tissue for microscopic examination) and the development of improved histological stains stimulated a flurry of cytological research. Several cytologists demonstrated that the nucleus had a role in fertilization. Walther Flemming (1843–1905) observed the division of chromosomes in 1879 and published a superb description of mitosis. By 1885, it was generally recognized that the nucleus contained the hereditary information.

Near the close of the nineteenth century, August Weismann (1834–1914) finally laid to rest the notion of the inheritance of acquired characteristics. He cut off the tails of mice for 22 consecutive generations and showed that the tail length in descendants remained stubbornly long. Weismann proposed the **germ-plasm theory,** which holds that the cells in the reproductive organs carry a complete set of genetic information that is passed to the egg and sperm (see Figure 1.10b).

The year 1900 was a watershed in the history of genetics. Gregor Mendel's pivotal 1866 publication on experiments with pea plants, which revealed the principles of heredity, was rediscovered, as discussed in more detail in Chapter 3. The significance of his conclusions was recognized, and other biologists immediately began to conduct similar genetic studies on mice, chickens, and other organisms. The results of these investigations showed that many traits indeed follow Mendel's rules.

Walter Sutton (1877–1916) proposed in 1902 that genes are located on chromosomes. Thomas Hunt Morgan (1866–1945) discovered the first genetic mutant of fruit flies in 1910 and used fruit flies to unravel many details of transmission genetics. Ronald A. Fisher (1890–1962), John B. S. Haldane (1892–1964), and Sewall Wright (1889–1988) laid the foundation for population genetics in the 1930s by synthesizing Mendelian genetics and evolutionary theory.

Geneticists began to use bacteria and viruses in the 1940s; the rapid reproduction and simple genetic systems of these organisms allowed detailed study of the organization and structure of genes. At about this same time, evidence accumulated that DNA was the repository of genetic information. James Watson (b. 1928) and Francis Crick (1916–2004), along with Maurice Wilkins (1916–2004) and Rosalind Franklin (1920–1958), described the three-dimensional structure of DNA in 1953, ushering in the era of molecular genetics.

By 1966, the chemical structure of DNA and the system by which it determines the amino acid sequence of proteins had been worked out. Advances in molecular genetics led to the first recombinant DNA experiments in 1973, which touched off another revolution in genetic research. Walter Gilbert (b. 1932) and Frederick Sanger (b. 1918) developed methods for sequencing DNA in 1977. The polymerase chain reaction, a technique for quickly amplifying tiny amounts of DNA, was developed by Kary Mullis (b. 1944)

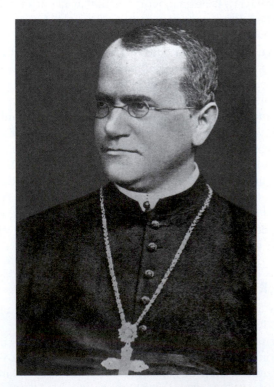

1.12 Gregor Mendel was the founder of modern genetics. Mendel first discovered the principles of heredity by crossing different varieties of pea plants and analyzing the pattern of transmission of traits in subsequent generations. *[Hulton Archive/Getty Images.]*

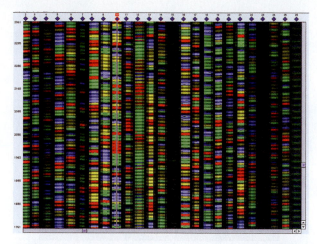

1.13 The human genome was completely sequenced in 2003.
Each of the colored bars represents one nucleotide base in the DNA.

and others in 1983. In 1990, gene therapy was used for the first time to treat human genetic disease in the United States, and the Human Genome Project was launched. By 1995, the first complete DNA sequence of a free-living organism—the bacterium *Haemophilus influenzae*—was determined, and the first complete sequence of a eukaryotic organism (yeast) was reported a year later. A rough draft of the human genome sequence was reported in 2000, with the sequence essentially completed in 2003, ushering in a new era in genetics (**Figure 1.13**). Today, the genomes of numerous organisms are being sequenced, analyzed, and compared.

The Future of Genetics

Numerous advances in genetics are being made today, and genetics is at the forefront of biological research. For example, the information content of genetics is increasing at a rapid pace, as the genome sequences of many organisms are added to DNA databases every year. New details about gene structure and function are continually expanding our knowledge of how genetic information is encoded and how it specifies phenotypic traits.

Information about sequence differences among individual organisms is a source of new insights about evolution and helps to locate genes that affect complex traits such as hypertension in humans and weight gain in cattle. In recent years, our understanding of the role of RNA in many cellular processes has expanded greatly; RNA has a role in many aspects of gene function. New genetic microchips that simultaneously analyze thousands of RNA molecules are providing information about the activity of thousands of genes in a given cell, allowing a detailed picture of how cells respond to external signals, environmental stresses, and disease states such as cancer. In the emerging field of proteomics, powerful computer programs are being used to model the structure and function of proteins from

DNA sequence information. All of this information provides us with a better understanding of numerous biological processes and evolutionary relationships. The flood of new genetic information requires the continuous development of sophisticated computer programs to store, retrieve, compare, and analyze genetic data and has given rise to the field of bioinformatics, a merging of molecular biology and computer science.

In the future, the focus of DNA-sequencing efforts will shift from the genomes of different species to individual differences within species. In the not too distant future, each person may possess a copy of his or her entire genome sequence, which can be used to assess the risk of acquiring various diseases and to tailor their treatment should they arise. The use of genetics in the agricultural, chemical, and health-care fields will continue to expand. This ever-widening scope of genetics will raise significant ethical, social, and economic issues.

This brief overview of the history of genetics is not intended to be comprehensive; rather it is designed to provide a sense of the accelerating pace of advances in genetics. In the chapters to come, we will learn more about the experiments and the scientists who helped shape the discipline of genetics.

CONCEPTS

Developments in plant hybridization and cytology in the eighteenth and nineteenth centuries laid the foundation for the field of genetics today. After Mendel's work was rediscovered in 1900, the science of genetics developed rapidly and today is one of the most active areas of science.

✔ CONCEPT CHECK 3

How did developments in cytology in the nineteenth century contribute to our modern understanding of genetics?

1.3 A Few Fundamental Concepts Are Important for the Start of Our Journey into Genetics

Undoubtedly, you learned some genetic principles in other biology classes. Let's take a few moments to review some of the fundamental genetic concepts.

Cells are of two basic types: eukaryotic and prokaryotic. Structurally, cells consist of two basic types, although, evolutionarily, the story is more complex (see Chapter 2). Prokaryotic cells lack a nuclear membrane and possess no membrane-bounded cell organelles, whereas eukaryotic cells are more complex, possessing a nucleus and membrane-bounded organelles such as chloroplasts and mitochondria.

The gene is the fundamental unit of heredity.   The precise way in which a gene is defined often varies, depending on the biological context. At the simplest level, we can think of a gene as a unit of information that encodes a genetic characteristic. We will enlarge this definition as we learn more about what genes are and how they function.

Genes come in multiple forms called alleles.   A gene that specifies a characteristic may exist in several forms, called alleles. For example, a gene for coat color in cats may exist in an allele that encodes black fur or an allele that encodes orange fur.

Genes confer phenotypes.   One of the most important concepts in genetics is the distinction between traits and genes. Traits are not inherited directly. Rather, genes are inherited and, along with environmental factors, determine the expression of traits. The genetic information that an individual organism possesses is its genotype; the trait is its phenotype. For example, the A blood type is a phenotype; the genetic information that encodes the blood-type-A antigen is the genotype.

Genetic information is carried in DNA and RNA.   Genetic information is encoded in the molecular structure of nucleic acids, which come in two types: deoxyribonucleic acid (DNA) and ribonucleic acid (RNA). Nucleic acids are polymers consisting of repeating units called nucleotides; each nucleotide consists of a sugar, a phosphate, and a nitrogenous base. The nitrogenous bases in DNA are of four types (abbreviated A, C, G, and T), and the sequence of these bases encodes genetic information. DNA consists of two complementary nucleotide strands. Most organisms carry their genetic information in DNA, but a few viruses carry it in RNA. The four nitrogenous bases of RNA are abbreviated A, C, G, and U.

Genes are located on chromosomes.   The vehicles of genetic information within a cell are chromosomes (**Figure 1.14**), which consist of DNA and associated proteins. The cells of each species have a characteristic number of chromosomes; for example, bacterial cells normally possess a single chromosome; human cells possess 46; pigeon cells possess 80. Each chromosome carries a large number of genes.

Chromosomes separate through the processes of mitosis and meiosis.   The processes of mitosis and

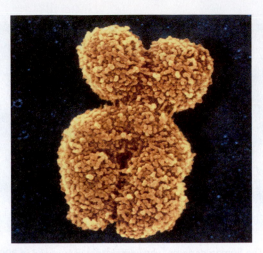

1.14 Genes are carried on chromosomes. [Biophoto Associates/Science Source/Photo Researchers.]

meiosis ensure that each daughter cell receives a complete set of an organism's chromosomes. Mitosis is the separation of replicated chromosomes in the division of somatic (nonsex) cells. Meiosis is the pairing and separation of replicated chromosomes in the division of sex cells to produce gametes (reproductive cells).

Genetic information is transferred from DNA to RNA to protein.   Many genes encode traits by specifying the structure of proteins. Genetic information is first transcribed from DNA into RNA, and then RNA is translated into the amino acid sequence of a protein.

Mutations are permanent, heritable changes in genetic information.   Gene mutations affect the genetic information of only a single gene; chromosome mutations alter the number or the structure of chromosomes and therefore usually affect many genes.

Some traits are affected by multiple factors.   Some traits are influenced by multiple genes that interact in complex ways with environmental factors. Human height, for example, is affected by hundreds of genes as well as environmental factors such as nutrition.

Evolution is genetic change.   Evolution can be viewed as a two-step process: first, genetic variation arises and, second, some genetic variants increase in frequency, whereas other variants decrease in frequency.

CONCEPTS SUMMARY

- Genetics is central to the life of every person: it influences a person's physical features, susceptibility to numerous diseases, personality, and intelligence.

- Genetics plays important roles in agriculture, the pharmaceutical industry, and medicine. It is central to the study of biology.

- All organisms use similar genetic systems. Genetic variation is the foundation of evolution and is critical to understanding all life.

- The study of genetics can be divided into transmission genetics, molecular genetics, and population genetics.

- Model genetic organisms are species having characteristics that make them particularly amenable to genetic analysis and about which much genetic information exists.

- The use of genetics by humans began with the domestication of plants and animals.

- The ancient Greeks developed the concepts of pangenesis and the inheritance of acquired characteristics. Ancient Romans developed practical measures for the breeding of plants and animals.

- Preformationism suggested that a person inherits all of his or her traits from one parent. Blending inheritance proposed that offspring possess a mixture of the parental traits.

- By studying the offspring of crosses between varieties of peas, Gregor Mendel discovered the principles of heredity. Developments in cytology in the nineteenth century led to the understanding that the cell nucleus is the site of heredity.

- In 1900, Mendel's principles of heredity were rediscovered. Population genetics was established in the early 1930s, followed closely by biochemical genetics and bacterial and viral genetics. The structure of DNA was discovered in 1953, stimulating the rise of molecular genetics.

- Cells are of two basic types: prokaryotic and eukaryotic.

- The genes that determine a trait are termed the genotype; the trait that they produce is the phenotype.

- Genes are located on chromosomes, which are made up of nucleic acids and proteins and are partitioned into daughter cells through the process of mitosis or meiosis.

- Genetic information is expressed through the transfer of information from DNA to RNA to proteins.

- Evolution requires genetic change in populations.

IMPORTANT TERMS

genome (p. 4)
transmission genetics (p. 5)
molecular genetics (p. 5)
population genetics (p. 5)

model genetic organism (p. 5)
pangenesis (p. 8)
inheritance of acquired
 characteristics (p. 8)

preformationism (p. 8)
blending inheritance (p. 9)
cell theory (p. 10)
germ-plasm theory (p. 10)

ANSWERS TO CONCEPT CHECKS

1. d

2. No, because horses are expensive to house, feed, and propagate, they have too few progeny, and their generation time is too long.

3. Developments in cytology in the 1800s led to the identification of parts of the cell, including the cell nucleus and chromosomes. The cell theory focused the attention of biologists on the cell, which eventually led to the conclusion that the nucleus contains the hereditary information.

COMPREHENSION QUESTIONS

Answers to questions and problems preceded by an asterisk can be found at the end of the book.

Section 1.1

*1. How does the Hopi culture contribute to the high incidence of albinism among members of the Hopi tribe?

2. Outline some of the ways in which genetics is important to each of us.

*3. Give at least three examples of the role of genetics in society today.

4. Briefly explain why genetics is crucial to modern biology.

*5. List the three traditional subdisciplines of genetics and summarize what each covers.

6. What are some characteristics of model genetic organisms that make them useful for genetic studies?

Section 1.2

7. When and where did agriculture first arise? What role did genetics play in the development of the first domesticated plants and animals?

*8. Outline the notion of pangenesis and explain how it differs from the germ-plasm theory.

9. What does the concept of the inheritance of acquired characteristics propose and how is it related to the notion of pangenesis?

*10. What is preformationism? What did it have to say about how traits are inherited?

11. Define blending inheritance and contrast it with preformationism.

12. How did developments in botany in the seventeenth and eighteenth centuries contribute to the rise of modern genetics?

*13. Who first discovered the basic principles that laid the foundation for our modern understanding of heredity?

14. List some advances in genetics that have been made in the twentieth century.

Section 1.3

15. What are the two basic cell types (from a structural perspective) and how do they differ?

*16. Outline the relations between genes, DNA, and chromosomes.

APPLICATION QUESTIONS AND PROBLEMS

Section 1.1

17. What is the relation between genetics and evolution?

*18. For each of the following genetic topics, indicate whether it focuses on transmission genetics, molecular genetics, or population genetics.

 a. Analysis of pedigrees to determine the probability of someone inheriting a trait

 b. Study of the genetic history of people on a small island to determine why a genetic form of asthma is so prevalent on the island

 c. The influence of nonrandom mating on the distribution of genotypes among a group of animals

 d. Examination of the nucleotide sequences found at the ends of chromosomes

 e. Mechanisms that ensure a high degree of accuracy during DNA replication

 f. Study of how the inheritance of traits encoded by genes on sex chromosomes (sex-linked traits) differs from the inheritance of traits encoded by genes on nonsex chromosomes (autosomal traits)

Section 1.2

*19. Genetics is said to be both a very old science and a very young science. Explain what is meant by this statement.

20. Match the description (*a* through *d*) with the correct theory or concept listed below.

 Preformationism   Germ-plasm theory
 Pangenesis   Inheritance of acquired
                          characteristics

 a. Each reproductive cell contains a complete set of genetic information.

 b. All traits are inherited from one parent.

 c. Genetic information may be altered by the use of a characteristic.

 d. Cells of different tissues contain different genetic information.

*21. Compare and contrast the following ideas about inheritance.

 a. Pangenesis and germ-plasm theory

 b. Preformationism and blending inheritance

 c. The inheritance of acquired characteristics and our modern theory of heredity

Section 1.3

*22. Compare and contrast the following terms:

 a. Eukaryotic and prokaryotic cells

 b. Gene and allele

 c. Genotype and phenotype

 d. DNA and RNA

 e. DNA and chromosome

CHALLENGE QUESTIONS

Section 1.1

23. We now know as much or more about the genetics of humans as we know about that of any other organism, and humans are the focus of many genetic studies. Do you think humans should be considered a model genetic organism? Why or why not?

24. Describe some of the ways in which your own genetic makeup affects you as a person. Be as specific as you can.

25. Describe at least one trait that appears to run in your family (appears in multiple members of the family). Do you think this trait runs in your family because it is an inherited trait or because is caused by environmental factors that are common to family members? How might you distinguish between these possibilities?

Section 1.3

*26. Suppose that life exists elsewhere in the universe. All life must contain some type of genetic information, but alien genomes might not consist of nucleic acids and have the same features as those found in the genomes of life on Earth. What do you think might be the common features of all genomes, no matter where they exist?

27. Pick one of the following ethical or social issues and give your opinion on the issue. For background information, you might read one of the articles on ethics listed and marked with an asterisk in the Suggested Readings section for Chapter 1 at www.whfreeman.com/pierce.

　　a. Should a person's genetic makeup be used in determining his or her eligibility for life insurance?

　　b. Should biotechnology companies be able to patent newly sequenced genes?

　　c. Should gene therapy be used on people?

　　d. Should genetic testing be made available for inherited conditions for which there is no treatment or cure?

　　e. Should governments outlaw the cloning of people?

*28. Suppose that you could undergo genetic testing at age 18 for susceptibility to a genetic disease that would not appear until middle age and has no available treatment.

　　a. What would be some of the possible reasons for having such a genetic test and some of the possible reasons for not having the test?

　　b. Would you personally want to be tested? Explain your reasoning.

2 | Chromosomes and Cellular Reproduction

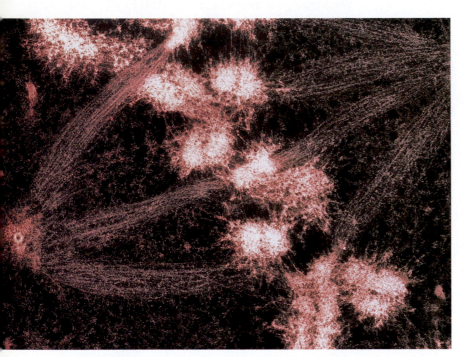

Chromosomes in mitosis, the process whereby each new cell receives a complete copy of the genetic material. *[Photograph by Conly L. Reider/ Biological Photo Service.]*

THE BLIND MEN'S RIDDLE

In a well-known riddle, two blind men by chance enter a department store at the same time, go to the same counter, and both order five pairs of socks, each pair of different color. The sales clerk is so befuddled by this strange coincidence that he places all ten pairs (two black pairs, two blue pairs, two gray pairs, two brown pairs, and two green pairs) into a single shopping bag and gives the bag with all ten pairs to one blind man and an empty bag to the other. The two blind men happen to meet on the street outside, where they discover that one of their bags contains all ten pairs of socks. How do the blind men, without seeing and without any outside help, sort out the socks so that each man goes home with exactly five pairs of different colored socks? Can you come up with a solution to the riddle?

By an interesting coincidence, cells have the same dilemma as that of the blind men in the riddle. Most organisms possess two sets of genetic information, one set inherited from each parent. Before cell division, the DNA in each chromosome replicates; after replication, there are two copies—called sister chromatids—of each chromosome. At the end of cell division, it is critical that each new cell receives a complete copy of the genetic material, just as each blind man needed to go home with a complete set of socks.

The solution to the riddle is simple. Socks are sold as pairs; the two socks of a pair are typically connected by a thread. As a pair is removed from the bag, the men each grasp a different sock of the pair and pull in opposite directions. When the socks are pulled tight, it is easy for one of the men to take a pocket knife and cut the thread connecting the pair. Each man then deposits his single sock in his own bag. At the end of the process, each man's bag will contain exactly two black socks, two blue socks, two gray socks, two brown socks, and two green socks.*

Remarkably, cells employ a similar solution for separating their chromosomes into new daughter cells. As we will learn in this chapter, the replicated chromosomes line up at the center of a cell undergoing division and, like the socks in the riddle, the sister chromatids of each chromosome are pulled in opposite directions. Like the thread connecting two socks of a pair, a molecule called cohesin holds the sister chromatids together until severed by a molecular knife called separase. The two resulting chromosomes separate and the cell divides, ensuring that a complete set of chromosomes is deposited in each cell.

*This analogy is adapted from K. Nasmyth. Disseminating the genome: joining, resolving, and separating sister chromatids, during mitosis and meiosis. *Annual Review of Genetics* 35:673–745, 2001.

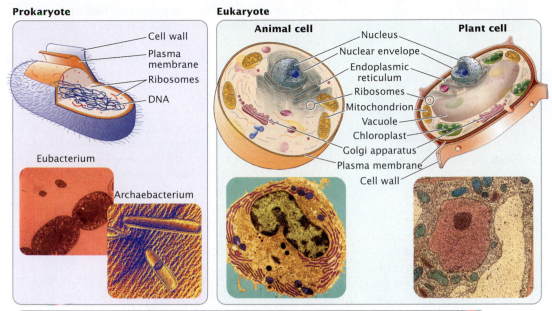

Prokaryote

- Cell wall
- Plasma membrane
- Ribosomes
- DNA

Eubacterium

Archaebacterium

Eukaryote

Animal cell   **Plant cell**

- Nucleus
- Nuclear envelope
- Endoplasmic reticulum
- Ribosomes
- Mitochondrion
- Vacuole
- Chloroplast
- Golgi apparatus
- Plasma membrane
- Cell wall

	Prokaryotic cells	Eukaryotic cells
Nucleus	Absent	Present
Cell diameter	Relatively small, from 1 to 10 μm	Relatively large, from 10 to 100 μm
Genome	Usually one circular DNA molecule	Multiple linear DNA molecules
DNA	Not complexed with histones in eubacteria; some histones in archaea	Complexed with histones
Amount of DNA	Relatively small	Relatively large
Membrane-bounded organelles	Absent	Present
Cytoskeleton	Absent	Present

2.1 Prokaryotic and eukaryotic cells differ in structure.
[Photographs (left to right) by T. J. Beveridge/Visuals Unlimited; W. Baumeister/Science Photo Library/Photo Researchers; G. Murti/Phototake; Biophoto Associates/Photo Researchers.]

In this analogy, the blind men and cells differ in one critical regard: if the blind men make a mistake, one man ends up with an extra sock and the other is a sock short, but no great harm results. The same cannot be said for human cells. Errors in chromosome separation, producing cells with too many or too few chromosomes, are frequently catastrophic, leading to cancer, reproductive failure, or—sometimes—a child with severe handicaps.

This chapter explores the process of cell reproduction and how a complete set of genetic information is transmitted to new cells. In prokaryotic cells, reproduction is simple, because prokaryotic cells possess a single chromosome. In eukaryotic cells, multiple chromosomes must be copied and distributed to each of the new cells, and so cell reproduction is more complex. Cell division in eukaryotes takes place through mitosis and meiosis, processes that serve as the foundation for much of genetics.

Grasping mitosis and meiosis requires more than simply memorizing the sequences of events that take place in each stage, although these events are important. The key is to understand how genetic information is apportioned in the course of cell reproduction through a dynamic interplay of DNA synthesis, chromosome movement, and cell division. These processes bring about the transmission of genetic information and are the basis of similarities and differences between parents and progeny.

2.1 Prokaryotic and Eukaryotic Cells Differ in a Number of Genetic Characteristics

Biologists traditionally classify all living organisms into two major groups, the *prokaryotes* and the *eukaryotes* (**Figure 2.1**). A **prokaryote** is a unicellular organism with a relatively simple cell structure. A **eukaryote** has a compartmentalized cell structure having components bounded by intracellular membranes; eukaryotes may be unicellular or multicellular.

Research indicates that a division of life into two major groups, the prokaryotes and eukaryotes, is not so simple. Although similar in cell structure, prokaryotes include at least two fundamentally distinct types of bacteria: the **eubacteria** (true bacteria) and the **archaea** (ancient bacteria). An examination of equivalent DNA sequences reveals that eubacteria and archaea are as distantly related to one another as they are to the eukaryotes. Although eubacteria and archaea are similar

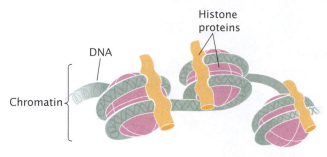

2.2 In eukaryotic cells, DNA is complexed with histone proteins to form chromatin.

in cell structure, some genetic processes in archaea (such as transcription) are more similar to those in eukaryotes, and the archaea are actually closer evolutionarily to eukaryotes than to eubacteria. Thus, from an evolutionary perspective, there are three major groups of organisms: eubacteria, archaea, and eukaryotes. In this book, the prokaryotic–eukaryotic distinction will be made frequently, but important eubacterial–archaeal differences also will be noted.

From the perspective of genetics, a major difference between prokaryotic and eukaryotic cells is that a eukaryote has a *nuclear envelope*, which surrounds the genetic material to form a **nucleus** and separates the DNA from the other cellular contents. In prokaryotic cells, the genetic material is in close contact with other components of the cell—a property that has important consequences for the way in which genes are controlled.

Another fundamental difference between prokaryotes and eukaryotes lies in the packaging of their DNA. In eukaryotes, DNA is closely associated with a special class of proteins, the **histones,** to form tightly packed chromosomes. This complex of DNA and histone proteins is termed **chromatin,** which is the stuff of eukaryotic chromosomes (**Figure 2.2**). Histone proteins limit the accessibility of enzymes and other proteins that copy and read the DNA, but they enable the DNA to fit into the nucleus. Eukaryotic DNA must separate from the histones before the genetic information in the DNA can be accessed. Archaea also have some histone proteins that complex with DNA, but the structure of their chromatin is different from that found in eukaryotes. However, eubacteria do not possess histones; so their DNA does not exist in the highly ordered, tightly packed arrangement found in eukaryotic cells (**Figure 2.3**). The copying and reading of DNA are therefore simpler processes in eubacteria.

Genes of prokaryotic cells are generally on a single, circular molecule of DNA—the chromosome of a prokaryotic cell. In eukaryotic cells, genes are located on multiple, usually linear DNA molecules (multiple chromosomes). Eukaryotic cells therefore require mechanisms that ensure that a copy of each chromosome is faithfully transmitted to each new cell. This generalization—a single, circular chromosome in prokaryotes and multiple, linear chromosomes in eukaryotes—is not always true. A few bacteria have more

than one chromosome, and important bacterial genes are frequently found on other DNA molecules called *plasmids* (see Chapter 8). Furthermore, in some eukaryotes, a few genes are located on circular DNA molecules found outside the nucleus.

CONCEPTS

Organisms are classified as prokaryotes or eukaryotes, and prokaryotes consist of archaea and eubacteria. A prokaryote is a unicellular organism that lacks a nucleus, its DNA is not complexed to histone proteins, and its genome is usually a single chromosome. Eukaryotes are either unicellular or multicellular, their cells possess a nucleus, their DNA is complexed to histone proteins, and their genomes consist of multiple chromosomes.

✔ CONCEPT CHECK 1

List several characteristics that eubacteria and archaea have in common and that distinguish them from eukaryotes.

(a)

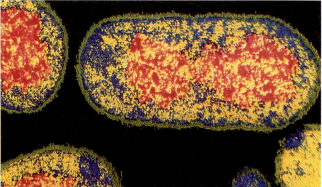

(b)

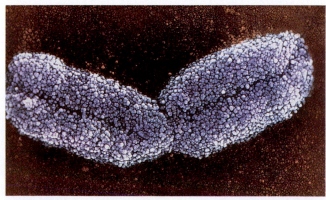

2.3 Prokaryotic and eukaryotic DNA compared. (a) Prokaryotic DNA (shown in red) is neither surrounded by a nuclear membrane nor complexed with histone proteins. (b) Eukaryotic DNA is complexed to histone proteins to form chromosomes (one of which is shown) that are located in the nucleus. [Part a: A. B. Dowsett/Science Photo Library/Photo Researchers. Part b: Biophoto Associates/Photo Researchers.]

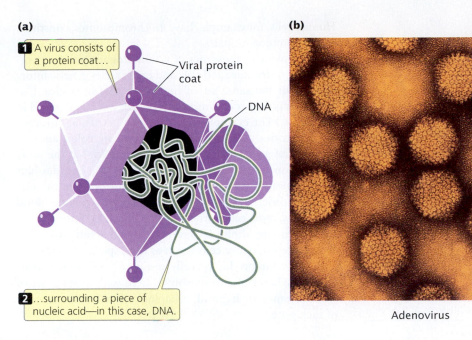

(a)

1 A virus consists of a protein coat…

Viral protein coat

DNA

2 …surrounding a piece of nucleic acid—in this case, DNA.

(b)

Adenovirus

2.4 A virus is a simple replicative structure consisting of protein and nucleic acid. *[Micrograph by Hans Gelderblom/Visuals Unlimited.]*

Viruses are simple structures composed of an outer protein coat surrounding nucleic acid (either DNA or RNA; **Figure 2.4**). Viruses are neither cells nor primitive forms of life: they can reproduce only within host cells, which means that they must have evolved after, rather than before, cells evolved. In addition, viruses are not an evolutionarily distinct group but are most closely related to their hosts—the genes of a plant virus are more similar to those in a plant cell than to those in animal viruses, which suggests that viruses evolved from their hosts, rather than from other viruses. The close relationship between the genes of virus and host makes viruses useful for studying the genetics of host organisms.

2.2 Cell Reproduction Requires the Copying of the Genetic Material, Separation of the Copies, and Cell Division

For any cell to reproduce successfully, three fundamental events must take place: (1) its genetic information must be copied, (2) the copies of genetic information must be separated from each other, and (3) the cell must divide. All cellular reproduction includes these three events, but the processes that lead to these events differ in prokaryotic and eukaryotic cells because of their structural differences.

Prokaryotic Cell Reproduction

When prokaryotic cells reproduce, the circular chromosome of the bacterium is replicated (**Figure 2.5**). Replication usually begins at a specific place on the bacterial chromosome, called the origin of replication. In a process that is not fully understood, the origins of the two newly replicated chromosomes move away from each other and toward opposite ends of the cell. In at least some bacteria, proteins bind near the replication origins and anchor the new chromosomes to the plasma membrane at opposite ends of the cell. Finally, a new cell wall forms between the two chromosomes, producing two cells, each with an identical copy of the chromosome. Under optimal conditions, some bacterial cells divide every 20 minutes. At this rate, a single bacterial cell could produce a billion descendants in a mere 10 hours.

Eukaryotic Cell Reproduction

Like prokaryotic cell reproduction, eukaryotic cell reproduction requires the processes of DNA replication, copy separation, and division of the cytoplasm. However, the presence of multiple DNA molecules requires a more complex mechanism to ensure that exactly one copy of each molecule ends up in each of the new cells.

Eukaryotic chromosomes are separated from the cytoplasm by the nuclear envelope. The nucleus was once thought to be a fluid-filled bag in which the chromosomes floated, but we now know that the nucleus has a highly organized internal scaffolding called the *nuclear matrix*. This matrix consists of a network of protein fibers that maintains precise spatial relations among the nuclear components and takes part in DNA replication, the expression of genes, and the modification of gene products before they leave the nucleus. We will now take a closer look at the structure of eukaryotic chromosomes.

Eukaryotic chromosomes Each eukaryotic species has a characteristic number of chromosomes per cell: potatoes have 48 chromosomes, fruit flies have 8, and humans have 46. There appears to be no special significance between the complexity of an organism and its number of chromosomes per cell.

(a)

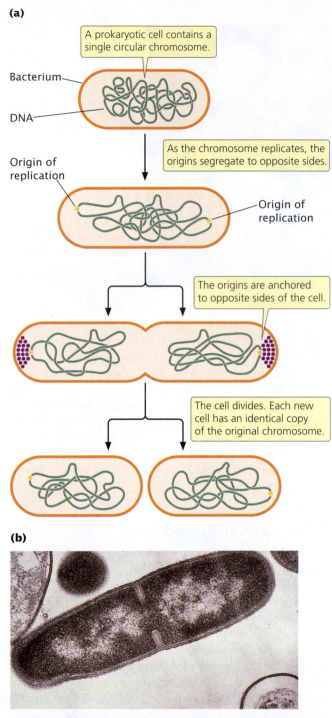

A prokaryotic cell contains a single circular chromosome.

Bacterium

DNA

Origin of replication

As the chromosome replicates, the origins segregate to opposite sides.

Origin of replication

The origins are anchored to opposite sides of the cell.

The cell divides. Each new cell has an identical copy of the original chromosome.

(b)

2.5 Prokaryotic cells reproduce by simple division. *[Micrograph by Lee D. Simon/Photo Researchers.]*

In most eukaryotic cells, there are two sets of chromosomes. The presence of two sets is a consequence of sexual reproduction: one set is inherited from the male parent and the other from the female parent. Each chromosome in one set has a corresponding chromosome in the other set, together constituting a homologous pair (Figure 2.6).

Human cells, for example, have 46 chromosomes, constituting 23 homologous pairs.

The two chromosomes of a **homologous pair** are usually alike in structure and size, and each carries genetic information for the same set of hereditary characteristics. (An exception is the sex chromosomes, which will be discussed in Chapter 4.) For example, if a gene on a particular chromosome encodes a characteristic such as hair color, another copy of the gene (each copy is called an *allele*) at the same position on that chromosome's homolog *also* encodes hair color. However, these two alleles need not be identical: one might encode red hair and the other might encode blond hair. Thus, most cells carry two sets of genetic information; these cells are **diploid.** But not all eukaryotic cells are diploid: reproductive cells (such as eggs, sperm, and spores) and even nonreproductive cells in some organisms may contain a single set of chromosomes. Cells with a single set of chromosomes are **haploid.** A haploid cell has only one copy of each gene.

CONCEPTS

Cells reproduce by copying and separating their genetic information and then dividing. Because eukaryotes possess multiple chromosomes, mechanisms exist to ensure that each new cell receives one copy of each chromosome. Most eukaryotic cells are diploid, and their two chromosome sets can be arranged in homologous pairs. Haploid cells contain a single set of chromosomes.

✔ CONCEPT CHECK 2

Diploid cells have

a. two chromosomes

b. two sets of chromosomes

c. one set of chromosomes

d. two pairs of homologous chromosomes

Chromosome structure   The chromosomes of eukaryotic cells are larger and more complex than those found in prokaryotes, but each unreplicated chromosome nevertheless consists of a single molecule of DNA. Although linear, the DNA molecules in eukaryotic chromosomes are highly folded and condensed; if stretched out, some human chromosomes would be several centimeters long—thousands of times as long as the span of a typical nucleus. To package such a tremendous length of DNA into this small volume, each DNA molecule is coiled again and again and tightly packed around histone proteins, forming a rod-shaped chromosome. Most of the time, the chromosomes are thin and difficult to observe but, before cell division, they condense further into thick, readily observed

(a)

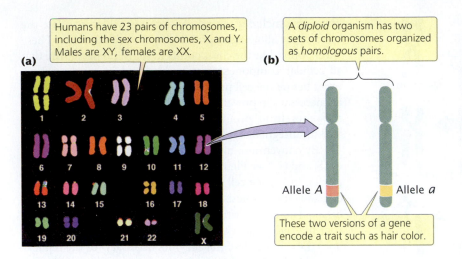

Humans have 23 pairs of chromosomes, including the sex chromosomes, X and Y. Males are XY, females are XX.

(b)

A *diploid* organism has two sets of chromosomes organized as *homologous* pairs.

Allele *A*   Allele *a*

These two versions of a gene encode a trait such as hair color.

2.6 Diploid eukaryotic cells have two sets of chromosomes. (a) A set of chromosomes from a female human cell. Each pair of chromosomes is hybridized to a uniquely colored probe, giving it a distinct color. (b) The chromosomes are present in homologous pairs, which consist of chromosomes that are alike in size and structure and carry information for the same characteristics. [Part a: Courtesy of Dr. Thomas Ried and Dr. Evelin Schrock.]

structures; it is at this stage that chromosomes are usually studied.

A functional chromosome has three essential elements: a centromere, a pair of telomeres, and origins of replication. The *centromere* is the attachment point for *spindle microtubules*, which are the filaments responsible for moving chromosomes during cell division (**Figure 2.7**). The centromere appears as a constricted region. Before cell division, a protein complex called the *kinetochore* assembles on the centromere; later spindle microtubules attach to the kinetochore. Chromosomes lacking a centromere cannot be drawn into the newly formed nuclei; these chromosomes are lost, often with catastrophic consequences to the cell. On the basis of the location of the centromere, chromosomes are classified into four types: metacentric, submetacentric, acrocentric, and telocentric (**Figure 2.8**). One of the two arms of a chromosome (the short arm of a submetacentric or acrocentric chromosome) is designated by the letter p and the other arm is designated by q.

Telomeres are the natural ends, the tips, of a linear chromosome (see Figure 2.7); they serve to stabilize the chromosome ends. If a chromosome breaks, producing new ends, these ends have a tendency to stick together, and the chromosome is degraded at the newly broken ends. Telomeres provide chromosome stability. The results of research suggest that telomeres also participate in limiting cell division and may play important roles in aging and cancer.

Origins of replication are the sites where DNA synthesis begins; they are not easily observed by microscopy. In preparation for cell division, each chromosome replicates, making a copy of itself, as already mentioned. These two initially identical copies, called **sister chromatids,** are held together at the centromere (see Figure 2.7). Each sister chromatid consists of a single molecule of DNA.

CONCEPTS

Sister chromatids are copies of a chromosome held together at the centromere. Functional chromosomes contain centromeres, telomeres, and origins of replication. The kinetochore is the point of attachment for the spindle microtubules; telomeres are the stabilizing ends of a chromosome; origins of replication are sites where DNA synthesis begins.

✔ CONCEPT CHECK 3

What are three essential elements required for a chromosome to function?

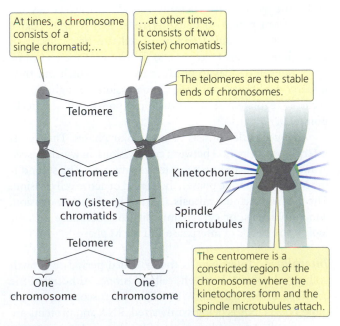

At times, a chromosome consists of a single chromatid;...

...at other times, it consists of two (sister) chromatids.

The telomeres are the stable ends of chromosomes.

Telomere

Centromere   Kinetochore

Two (sister) chromatids   Spindle microtubules

Telomere

One chromosome   One chromosome

The centromere is a constricted region of the chromosome where the kinetochores form and the spindle microtubules attach.

2.7 Each eukaryotic chromosome has a centromere and telomeres.

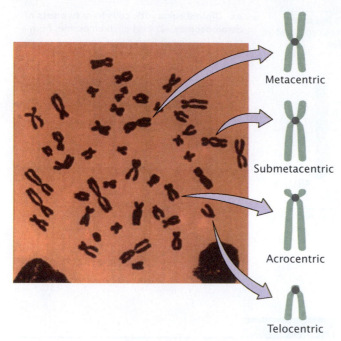

Metacentric

Submetacentric

Acrocentric

Telocentric

2.8 Eukaryotic chromosomes exist in four major types based on the position of the centromere. [*Micrograph by L. Lisco, D. W. Fawcett/Visuals Unlimited.*]

The Cell Cycle and Mitosis

The **cell cycle** is the life story of a cell, the stages through which it passes from one division to the next (**Figure 2.9**). This process is critical to genetics because, through the cell cycle, the genetic instructions for all characteristics are passed from parent to daughter cells. A new cycle begins after a cell has divided and produced two new cells. Each new cell metabolizes, grows, and develops. At the end of its cycle, the cell divides to produce two cells, which can then undergo additional cell cycles. Progression through the cell cycle is regulated at key transition points called **checkpoints.**

The cell cycle consists of two major phases. The first is **interphase,** the period between cell divisions, in which the cell grows, develops, and prepares for cell division. The second is the **M phase** (mitotic phase), the period of active cell division. The M phase includes **mitosis,** the process of nuclear division, and **cytokinesis,** or cytoplasmic division. Let's take a closer look at the details of interphase and the M phase.

Interphase   Interphase is the extended period of growth and development between cell divisions. Although little activity can be observed with a light microscope, the cell is quite busy: DNA is being synthesized, RNA and proteins are being produced, and hundreds of biochemical reactions are taking place. In addition to growth and development, inter-

phase also includes several checkpoints, which regulate the cell cycle by allowing or prohibiting the cell's division. These checkpoints, like the checkpoints in the M phase, ensure that all cellular components are present and in good working order before the cell proceeds to the next stage. Checkpoints are necessary to prevent cells with damaged or missing chromosomes from proliferating. Defects in checkpoints can lead to unregulated cell growth, as is seen in some cancers.

By convention, interphase is divided into three phases: G_1, S, and G_2 (see Figure 2.9). Interphase begins with G_1 (for gap 1). In G_1, the cell grows, and proteins necessary for cell division are synthesized; this phase typically lasts several hours. There is a critical point termed the G_1/S *checkpoint* near the end of G_1. The G_1/S checkpoint holds the cell in G_1 until the cell has all of the enzymes necessary for the replication of DNA. After this checkpoint has been passed, the cell is committed to divide.

Before reaching the G_1/S checkpoint, cells may exit from the active cell cycle in response to regulatory signals and pass into a nondividing phase called G_0, which is a stable state during which cells usually maintain a constant size. They can remain in G_0 for an extended period of time, even indefinitely, or they can reenter G_1 and the active cell cycle. Many cells never enter G_0; rather, they cycle continuously.

After G_1, the cell enters the *S* phase (for DNA synthesis), in which each chromosome duplicates. Although the cell is committed to divide after the G_1/S checkpoint has been passed, DNA synthesis must take place before the cell can proceed to mitosis. If DNA synthesis is blocked (by drugs or by a mutation), the cell will not be able to undergo mitosis. Before the S phase, each chromosome is composed of one chromatid; after the S phase, each chromosome is composed of two chromatids (see Figure 2.7).

After the S phase, the cell enters G_2 (gap 2). In this phase, several additional biochemical events necessary for cell division take place. The important G_2/M *checkpoint* is reached near the end of G_2. This checkpoint is passed only if the cell's DNA is undamaged. Damaged DNA can inhibit the activation of some proteins that are necessary for mitosis to take place. After the G_2/M checkpoint has been passed, the cell is ready to divide and enters the M phase. Although the length of interphase varies from cell type to cell type, a typical dividing mammalian cell spends about 10 hours in G_1, 9 hours in S, and 4 hours in G_2 (see Figure 2.9).

Throughout interphase, the chromosomes are in a relaxed, but by no means uncoiled, state, and individual chromosomes cannot be seen with the use of a microscope. This condition changes dramatically when interphase draws to a close and the cell enters the M phase.

M phase   The M phase is the part of the cell cycle in which the copies of the cell's chromosomes (sister chromatids) separate and the cell undergoes division. The separation of sister chromatids in the M phase is a critical process that results in a complete set of genetic information for each of the resulting

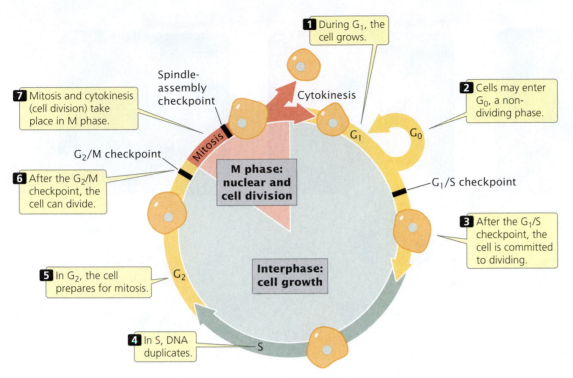

1 During G₁, the cell grows.

2 Cells may enter G₀, a non-dividing phase.

7 Mitosis and cytokinesis (cell division) take place in M phase.

Spindle-assembly checkpoint

Cytokinesis

6 After the G₂/M checkpoint, the cell can divide.

G₂/M checkpoint

G₁/S checkpoint

3 After the G₁/S checkpoint, the cell is committed to dividing.

5 In G₂, the cell prepares for mitosis.

4 In S, DNA duplicates.

M phase: nuclear and cell division

Interphase: cell growth

2.9 The cell cycle consists of interphase and M phase.

cells. Biologists usually divide the M phase into six stages: the five stages of mitosis (prophase, prometaphase, metaphase, anaphase, and telophase), illustrated in **Figure 2.10**, and cytokinesis. It's important to keep in mind that the M phase is a continuous process, and its separation into these six stages is somewhat arbitrary.

During interphase, the chromosomes are relaxed and are visible only as diffuse chromatin, but they condense during **prophase,** becoming visible under a light microscope. Each chromosome possesses two chromatids because the chromosome was duplicated in the preceding S phase. The *mitotic spindle,* an organized array of microtubules that move the chromosomes in mitosis, forms. In animal cells, the spindle grows out from a pair of *centrosomes* that migrate to opposite sides of the cell. Within each centrosome is a special organelle, the *centriole,* which also is composed of microtubules. Some plant cells do not have centrosomes or centrioles, but they do have mitotic spindles.

Disintegration of the nuclear membrane marks the start of **prometaphase.** Spindle microtubules, which until now have been outside the nucleus, enter the nuclear region. The ends of certain microtubules make contact with the chromosomes. For each chromosome, a microtubule from one of the centrosomes anchors to the kinetochore of *one* of the sister chromatids; a microtubule from the opposite centrosome then attaches to the other sister chromatid, and so the chromosome is anchored to both of the centrosomes. The microtubules lengthen and shorten, pushing and pulling the chromosomes about. Some microtubules extend from each

centrosome toward the center of the spindle but do not attach to a chromosome.

During **metaphase,** the chromosomes become arranged in a single plane, the *metaphase plate,* between the two centrosomes. The centrosomes, now at opposite ends of the cell with microtubules radiating outward and meeting in the middle of the cell, center at the spindle poles. A *spindle-assembly checkpoint* ensures that each chromosome is aligned on the metaphase plate and attached to spindle fibers from opposite poles.

Anaphase begins when the sister chromatids separate and move toward opposite spindle poles. The microtubules that connect the chromosomes to the spindle poles are composed of subunits of a protein called tubulin (**Figure 2.11**). Chromosome movement is due to the disassembly of tubulin molecules at both the kinetochore end (called the + end) and the spindle end (called the − end) of the spindle fiber. Special proteins called molecular motors disassemble tubulin molecules from the spindle and generate forces that pull the chromosome toward the spindle pole.

After the chromatids have separated, each is considered a separate chromosome. **Telophase** is marked by the arrival of the chromosomes at the spindle poles. The nuclear membrane re-forms around each set of chromosomes, producing two separate nuclei within the cell. The chromosomes relax and lengthen, once again disappearing from view. In many cells, division of the cytoplasm (cytokinesis) is simultaneous with telophase. The major features of the cell cycle are summarized in **Table 2.1**.

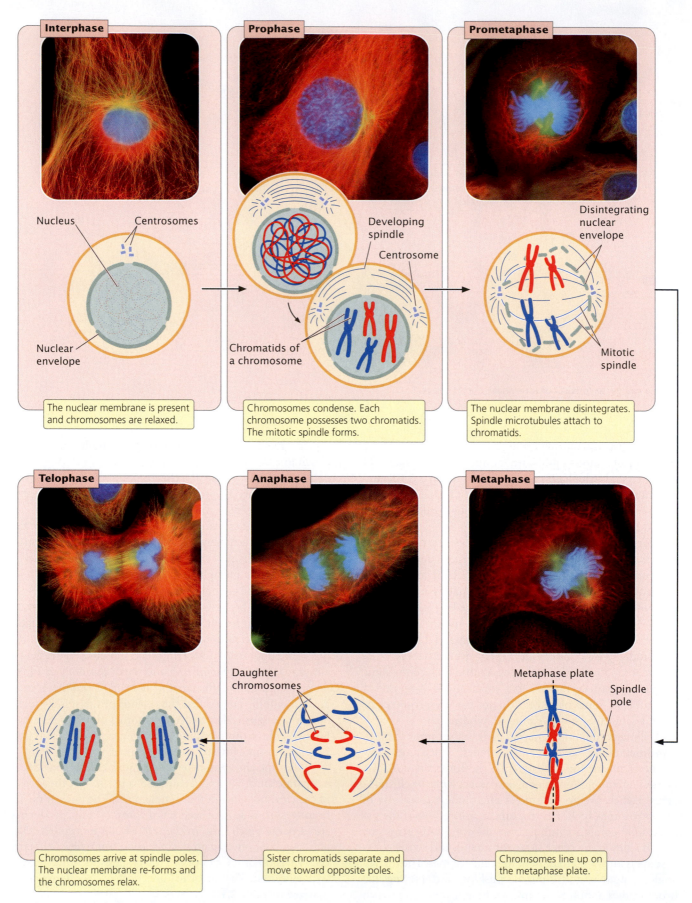

2.10 The cell cycle is divided into stages. [Photographs by Conly L. Rieder/Biological Photo Service.]

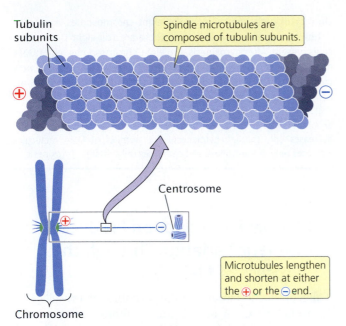

Tubulin subunits

Spindle microtubules are composed of tubulin subunits.

Centrosome

Microtubules lengthen and shorten at either the ⊕ or the ⊖ end.

Chromosome

2.11 Microtubules are composed of tubulin subunits. Each microtubule has a positively charged (+) end at the kinetochore and a negatively charged (−) end at the centrosome.

synthesis in the S phase creates an exact copy of each DNA molecule, giving rise to two genetically identical sister chromatids. Mitosis then ensures that one chromatid from each replicated chromosome passes into each new cell.

Another genetically important result of the cell cycle is that each of the cells produced contains a full complement of chromosomes—there is no net reduction or increase in chromosome number. Each cell also contains approximately half the cytoplasm and organelle content of the original parental cell, but no precise mechanism analogous to mitosis ensures that organelles are evenly divided. Consequently, not all cells resulting from the cell cycle are identical in their cytoplasmic content.

Genetic Consequences of the Cell Cycle

What are the genetically important results of the cell cycle? From a single cell, the cell cycle produces two cells that contain the same genetic instructions. These two cells are genetically identical with each other and with the cell that gave rise to them. They are genetically identical because DNA

CONCEPTS

The active cell-cycle phases are interphase and the M phase. Interphase consists of G_1, S, and G_2. In G_1, the cell grows and prepares for cell division; in the S phase, DNA synthesis takes place; in G_2, other biochemical events necessary for cell division take place. Some cells enter a quiescent phase called G_0. The M phase includes mitosis and cytokinesis and is divided into prophase, prometaphase, metaphase, anaphase, and telophase.

✔ CONCEPT CHECK 4

Which is the correct order of stages in the cell cycle?

a. G_1, S, prophase, metaphase, anaphase
b. S, G_1, prophase, metaphase, anaphase
c. Prophase, S, G_1, metaphase, anaphase
d. S, G_1, anaphase, prophase, metaphase

Table 2.1	Features of the cell cycle
Stage	**Major Features**
G_0 phase	Stable, nondividing period of variable length.
Interphase	
G_1 phase	Growth and development of the cell; G_1/S checkpoint.
S phase	Synthesis of DNA.
G_2 phase	Preparation for division; G_2/M checkpoint.
M phase	
Prophase	Chromosomes condense and mitotic spindle forms.
Prometaphase	Nuclear envelope disintegrates, and spindle microtubules anchor to kinetochores.
Metaphase	Chromosomes align on the spindle-assembly checkpoint.
Anaphase	Sister chromatids separate, becoming individual chromosomes that migrate toward spindle poles.
Telophase	Chromosomes arrive at spindle poles, the nuclear envelope re-forms, and the condensed chromosomes relax.
Cytokinesis	Cytoplasm divides; cell wall forms in plant cells.

CONNECTING CONCEPTS

Counting Chromosomes and DNA Molecules

The relations among chromosomes, chromatids, and DNA molecules frequently cause confusion. At certain times, chromosomes are unreplicated; at other times, each possesses two chromatids (see Figure 2.7). Chromosomes sometimes consist of a single DNA molecule; at other times, they consist of two DNA molecules. How can we keep track of the number of these structures in the cell cycle?

There are two simple rules for counting chromosomes and DNA molecules: (1) to determine the number of chromosomes, count the number of functional centromeres; (2) to determine the number of DNA molecules, count the number of chromatids. Let's examine a hypothetical cell as it passes through the cell cycle (**Figure 2.12**). At the beginning of G_1, this diploid cell has two complete sets of chromosomes, inherited from its parent cell. Each chromosome consists of a single chromatid—a single DNA molecule—and so there are four DNA molecules in the cell during G_1. In the S phase, each DNA molecule is copied. The two resulting DNA molecules combine with histones and other proteins to form sister chromatids. Although the amount of DNA doubles in the S phase, the number of chromosomes remains the same, because the two sister chromatids are tethered together and share a single functional centromere. At the end of the S phase, this cell still contains four chromosomes, each with two chromatids; so there are eight DNA molecules present.

Through prophase, prometaphase, and metaphase, the cell has four chromosomes and eight DNA molecules. At anaphase, however, the sister chromatids separate. Each now has its own functional centromere, and so each is considered a separate chromosome.

Until cytokinesis, the cell contains eight chromosomes, each consisting of a single chromatid; thus, there are still eight DNA molecules present. After cytokinesis, the eight chromosomes (eight DNA molecules) are distributed equally between two cells; so each new cell contains four chromosomes and four DNA molecules, the number present at the beginning of the cell cycle.

In summary, the number of chromosomes increases briefly only in anaphase, when the two chromatids of a chromosome separate, and decreases only through cytokinesis. The number of DNA molecules increases only in the S phase and decreases only through cytokinesis.

2.3 Sexual Reproduction Produces Genetic Variation Through the Process of Meiosis

If all reproduction were accomplished through the cell cycle, life would be quite dull, because mitosis produces only genetically identical progeny. With only mitosis, you, your children, your parents, your brothers and sisters, your cousins, and many people you don't even know would be clones—copies of one another. Only the occasional mutation would introduce any genetic variability. All organisms reproduced in this way for the first 2 billion years of Earth's existence (and it is the way in which some organisms still reproduce today). Then, some 1.5 billion to 2 billion years ago, something remarkable evolved: cells that produce genetically variable offspring through sexual reproduction.

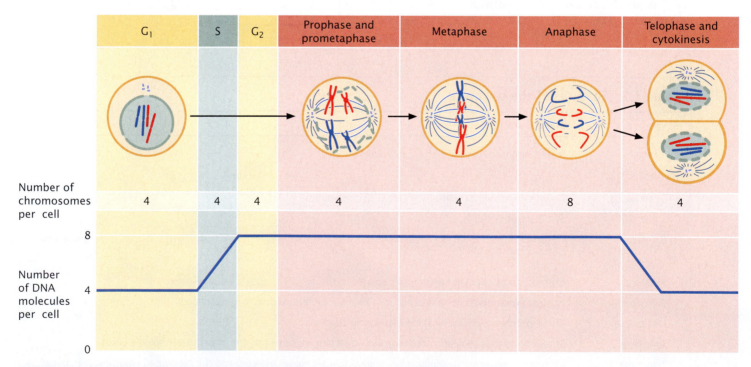

2.12 The number of chromosomes and the number of DNA molecules change in the course of the cell cycle. The number of chromosomes per cell equals the number of functional centromeres, and the number of DNA molecules per cell equals the number of chromatids.

The evolution of sexual reproduction is one of the most significant events in the history of life. As will be discussed in Chapters 11 and 12, the pace of evolution depends on the amount of genetic variation present. By shuffling the genetic information from two parents, sexual reproduction greatly increases the amount of genetic variation and allows for accelerated evolution. Most of the tremendous diversity of life on Earth is a direct result of sexual reproduction.

Sexual reproduction consists of two processes. The first is **meiosis,** which leads to *gametes* in which chromosome number is reduced by half. The second process is **fertilization,** in which two haploid gametes fuse and restore chromosome number to its original diploid value.

Meiosis

The words *mitosis* and *meiosis* are sometimes confused. They sound a bit alike, and both refer to chromosome division and cytokinesis. But don't be deceived. The outcomes of mitosis and meiosis are radically different, and several unique events that have important genetic consequences take place only in meiosis.

How does meiosis differ from mitosis? Mitosis consists of a single nuclear division and is usually accompanied by a single cell division. Meiosis, on the other hand, consists of two divisions. After mitosis, chromosome number in newly formed cells is the same as that in the original cell, whereas meiosis causes chromosome number in the newly formed cells to be reduced by half. Finally, mitosis produces genetically identical cells, whereas meiosis produces genetically variable cells. Let's see how these differences arise.

Like mitosis, meiosis is preceded by an interphase stage that includes G_1, S, and G_2 phases. Meiosis consists of two distinct processes: *meiosis I* and *meiosis II,* each of which includes a cell division. The first division, which comes at the end of meiosis I, is termed the reduction division because the number of chromosomes per cell is reduced by half (**Figure 2.13**). The second division, which comes at the end of meiosis II, is sometimes termed the equational division. The events of meiosis II are similar to those of mitosis. However, meiosis II differs from mitosis in that chromosome number has already been halved in meiosis I, and the cell does not begin with the same number of chromosomes as it does in mitosis (see Figure 2.13).

The stages of meiosis are outlined in **Figure 2.14.** During interphase, the chromosomes are relaxed and visible as diffuse chromatin. **Prophase I** is a lengthy stage, divided into five substages (**Figure 2.15**). In *leptotene,* the chromosomes contract and become visible. In *zygotene,* the chromosomes continue to condense; homologous chromosomes pair up and begin **synapsis,** a very close pairing association. Each homologous pair of synapsed chromosomes consists of four chromatids called a **bivalent** or **tetrad.** In *pachytene,* the chromosomes become shorter and thicker, and a three-part *synaptonemal complex* develops between homologous chromosomes. **Crossing over** takes place, in which homologous chromosomes exchange genetic information. The centromeres of the paired chromosomes move apart in *diplotene;* the two homologs remain attached at each *chiasma* (plural, *chiasmata*), which is the result of crossing over. In *diakinesis,* chromosome condensation continues, and the chiasmata move toward the ends of the chromosomes as the strands slip apart; so the homologs remain paired only at the tips. Near the end of prophase I, the nuclear membrane breaks down and the spindle forms.

Metaphase I is initiated when homologous pairs of chromosomes align along the metaphase plate (see Figure 2.14). A microtubule from one pole attaches to one chromosome of a homologous pair, and a microtubule from the other pole attaches to the other member of the pair. **Anaphase I** is marked by the separation of homologous chromosomes. The two chromosomes of a homologous pair are pulled toward opposite poles. Although the homologous chromosomes separate, the sister chromatids remain attached and travel together. In **telophase I,** the chromosomes arrive at the spindle poles and the cytoplasm divides.

The period between meiosis I and meiosis II is **interkinesis,** in which the nuclear membrane re-forms around the chromosomes clustered at each pole, the spindle breaks down, and the chromosomes relax. These cells then pass through **prophase II,** in which the events of interkinesis are reversed: the chromosomes recondense, the spindle re-forms, and the nuclear envelope once again breaks down. In

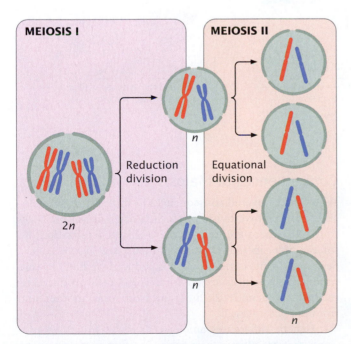

2.13 Meiosis includes two cell divisions. In this illustration, the original cell is $2n = 4$. After two meiotic divisions, each resulting cell is $1n = 2$.

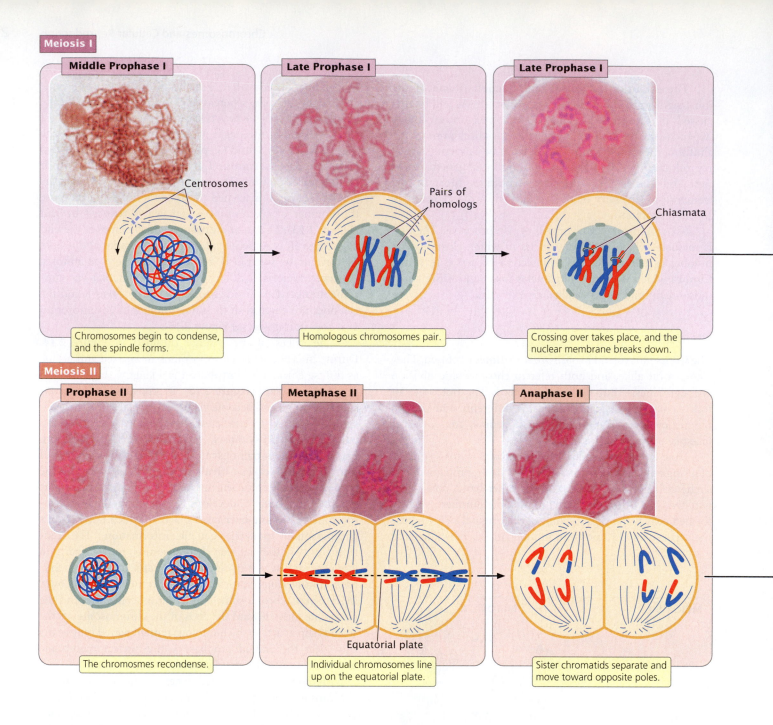

Meiosis I

Middle Prophase I

Centrosomes

Chromosomes begin to condense, and the spindle forms.

Late Prophase I

Pairs of homologs

Homologous chromosomes pair.

Late Prophase I

Chiasmata

Crossing over takes place, and the nuclear membrane breaks down.

Meiosis II

Prophase II

The chromosmes recondense.

Metaphase II

Equatorial plate

Individual chromosomes line up on the equatorial plate.

Anaphase II

Sister chromatids separate and move toward opposite poles.

interkinesis in some types of cells, the chromosomes remain condensed, and the spindle does not break down. These cells move directly from cytokinesis into **metaphase II,** which is similar to metaphase of mitosis: the individual chromosomes line up on the metaphase plate, with the sister chromatids facing opposite poles.

In **anaphase II,** the kinetochores of the sister chromatids separate and the chromatids are pulled to opposite poles. Each chromatid is now a distinct chromosome. In **telophase II,** the chromosomes arrive at the spindle poles, a nuclear envelope re-forms around the chromosomes, and the cytoplasm divides. The chromosomes relax and are no longer visible. The major events of meiosis are summarized in **Table 2.2.**

Consequences of Meiosis

What are the overall consequences of meiosis? First, meiosis comprises two divisions; so each original cell produces four cells (there are exceptions to this generalization, as, for example, in many female animals; see Figure 2.20b). Second, chromosome number is reduced by half; so cells produced by meiosis are haploid. Third, cells produced by meiosis are genetically different from one another and from the parental cell.

Genetic differences among cells result from two processes that are unique to meiosis. The first is crossing over, which takes place in prophase I. Crossing over refers to the exchange of genes between nonsister chromatids (chromatids

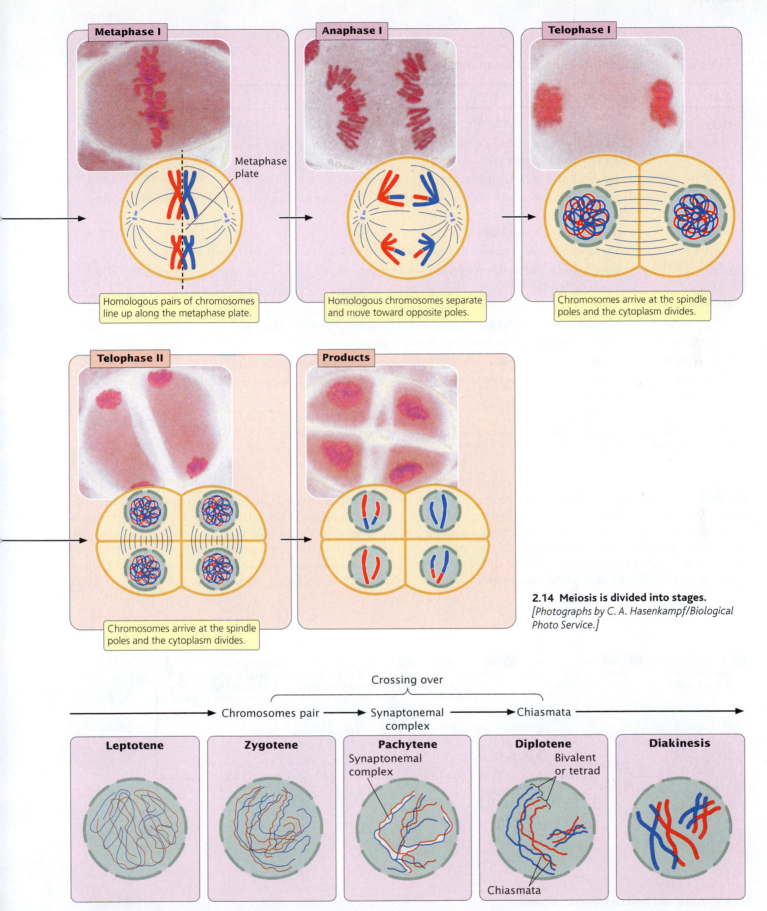

Metaphase I

Metaphase plate

Homologous pairs of chromosomes line up along the metaphase plate.

Anaphase I

Homologous chromosomes separate and move toward opposite poles.

Telophase I

Chromosomes arrive at the spindle poles and the cytoplasm divides.

Telophase II

Chromosomes arrive at the spindle poles and the cytoplasm divides.

Products

2.14 Meiosis is divided into stages.
[Photographs by C. A. Hasenkampf/Biological Photo Service.]

Crossing over

Chromosomes pair → Synaptonemal complex → Chiasmata

Leptotene **Zygotene** **Pachytene** Synaptonemal complex **Diplotene** Bivalent or tetrad **Diakinesis**

Chiasmata

2.15 Crossing over takes place in prophase I. In yeast, rough pairing of chromosomes begins in leptotene and continues in zygotene. The synaptonemal complex forms in pachytene. Crossing over is initiated in zygotene, before the synaptonemal complex develops, and is not completed until near the end of prophase I.

Stage	Major Events
Meiosis I	
Prophase I	Chromosomes condense, homologous chromosomes synapse, crossing over takes place, nuclear envelope breaks down, and mitotic spindle forms.
Metaphase I	Homologous pairs of chromosomes line up on the metaphase plate.
Anaphase I	The two chromosomes (each with two chromatids) of each homologous pair separate and move toward opposite poles.
Telophase I	Chromosomes arrive at the spindle poles.
Cytokinesis	The cytoplasm divides to produce two cells, each having half the original number of chromosomes.
Interkinesis	In some types of cells, the spindle breaks down, chromosomes relax, and a nuclear envelope re-forms, but no DNA synthesis takes place.
Meiosis II	
Prophase II*	Chromosomes condense, the spindle forms, and the nuclear envelope disintegrates.
Metaphase II	Individual chromosomes line up on the metaphase plate.
Anaphase II	Sister chromatids separate and move as individual chromosomes toward the spindle poles.
Telophase II	Chromosomes arrive at the spindle poles; the spindle breaks down and a nuclear envelope re-forms.
Cytokinesis	The cytoplasm divides.

Table 2.2   Major events in each stage of meiosis

*Only in cells in which the spindle has broken down, chromosomes have relaxed, and the nuclear envelope has re-formed in telophase I. Other types of cells proceed directly to metaphase II after cytokinesis.

from different homologous chromosomes). At one time, this process was thought to take place in pachytene, and the synaptonemal complex was believed to be a requirement for crossing over. However, evidence from yeast now suggests that the situation is more complex, as shown in Figure 2.15. Crossing over is initiated in zygotene, before the synaptone-

mal complex develops, and is not completed until near the end of prophase I. In other organisms, recombination is initiated after the formation of the synaptonemal complex and, in yet others, there is no synaptonemal complex.

After crossing over has taken place, the sister chromatids may no longer be identical. Crossing over is the

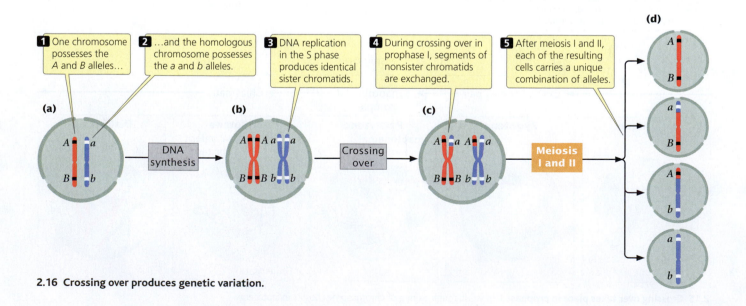

1 One chromosome possesses the *A* and *B* alleles...

2 ...and the homologous chromosome possesses the *a* and *b* alleles.

3 DNA replication in the S phase produces identical sister chromatids.

4 During crossing over in prophase I, segments of nonsister chromatids are exchanged.

5 After meiosis I and II, each of the resulting cells carries a unique combination of alleles.

2.16 Crossing over produces genetic variation.

basis for intrachromosomal **recombination,** creating new combinations of alleles on a chromatid. To see how crossing over produces genetic variation, consider two pairs of alleles, which we will abbreviate *Aa* and *Bb*. Assume that one chromosome possesses the *A* and *B* alleles and its homolog possesses the *a* and *b* alleles (**Figure 2.16a**). When DNA is replicated in the S phase, each chromosome duplicates, and so the resulting sister chromatids are identical (**Figure 2.16b**).

In the process of crossing over, there are breaks in the DNA strands and the breaks are repaired in such a way that segments of nonsister chromatids are exchanged (**Figure 2.16c**). The important thing here is that, after crossing over has taken place, the two sister chromatids are no longer identical —one chromatid has alleles *A* and *B*, whereas its sister chromatid (the chromatid that underwent crossing over) has alleles *a* and *B*. Likewise, one chromatid of the other chromosome

has alleles *a* and *b*, and the other has alleles *A* and *b*. Each of the four chromatids now carries a unique combination of alleles: *A B*, *a B*, *A b*, and *a b*. Eventually, the two homologous chromosomes separate, each going into a different cell. In meiosis II, the two chromatids of each chromosome separate, and thus each of the four cells resulting from meiosis carries a different combination of alleles (**Figure 2.16d**).

The second process of meiosis that contributes to genetic variation is the random distribution of chromosomes in anaphase I of meiosis after their random alignment in metaphase I. To illustrate this process, consider a cell with three pairs of chromosomes, I, II, and III (**Figure 2.17a**). One chromosome of each pair is maternal in origin (I_m, II_m, and III_m); the other is paternal in origin (I_p, II_p, and III_p). The chromosome pairs line up in the center of the cell in metaphase I; and, in anaphase I, the chromosomes of each homologous pair separate.

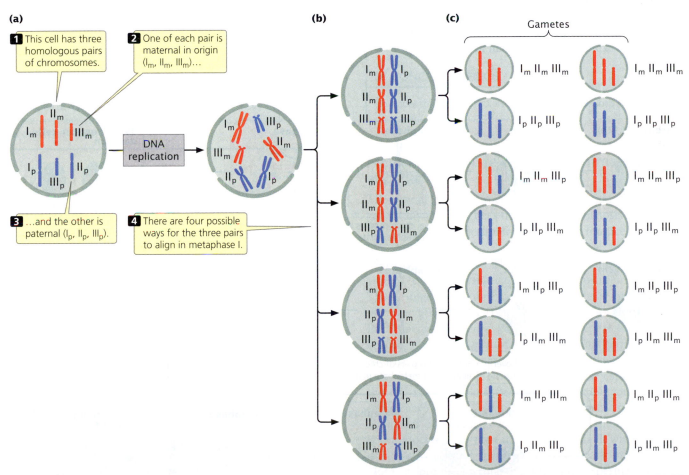

2.17 Genetic variation is produced through the random distribution of chromosomes in meiosis. In this example, the cell possesses three homologous pairs of chromosomes.

Conclusion: Eight different combinations of chromosomes in the gametes are possible, depending on how the chromosomes align and separate in meiosis I and II.

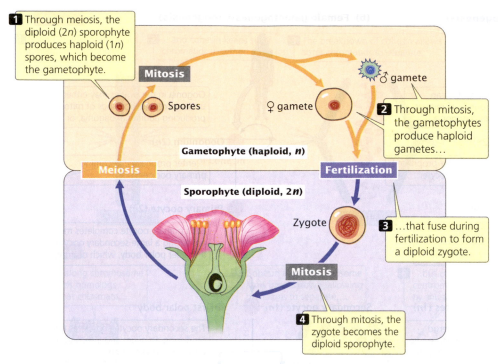

1 Through meiosis, the diploid (2*n*) sporophyte produces haploid (1*n*) spores, which become the gametophyte.

Mitosis

Spores

♀ gamete

♂ gamete

2 Through mitosis, the gametophytes produce haploid gametes…

Gametophyte (haploid, *n*)

Meiosis

Fertilization

Sporophyte (diploid, 2*n*)

Zygote

3 …that fuse during fertilization to form a diploid zygote.

Mitosis

4 Through mitosis, the zygote becomes the diploid sporophyte.

2.21 Plants alternate between diploid and haploid life stages (female, ♀ ; male, ♂).

microsporocytes, each of which undergoes meiosis to produce four haploid **microspores** (**Figure 2.22a**). Each microspore divides mitotically, producing an immature pollen grain consisting of two haploid nuclei. One of these nuclei, called the tube nucleus, directs the growth of a pollen tube. The other, termed the generative nucleus, divides mitotically to produce two sperm cells. The pollen grain, with its two haploid nuclei, is the male gametophyte.

The female part of the flower, the ovary, contains diploid cells called **megasporocytes,** each of which undergoes meiosis to produce four haploid **megaspores** (**Figure 2.22b**), only one of which survives. The nucleus of the surviving megaspore divides mitotically three times, producing a total of eight haploid nuclei that make up the female gametophyte, the embryo sac. Division of the cytoplasm then produces separate cells, one of which becomes the *egg*.

When the plant flowers, the stamens open and release pollen grains. Pollen lands on a flower's stigma—a sticky platform that sits on top of a long stalk called the style. At the base of the style is the ovary. If a pollen grain germinates, it grows a tube down the style into the ovary. The two sperm cells pass down this tube and enter the embryo sac (**Figure 2.22c**). One of the sperm cells fertilizes the egg cell, producing a diploid zygote, which develops into an embryo. The other sperm cell fuses with two nuclei enclosed in a single cell, giving rise to a 3*n* (triploid) endosperm, which stores food that will be used later by the embryonic plant. These two fertilization events are termed *double fertilization.*

CONCEPTS

In the stamen of a flowering plant, meiosis produces haploid microspores that divide mitotically to produce haploid sperm in a pollen grain. Within the ovary, meiosis produces four haploid megaspores, only one of which divides mitotically three times to produce eight haploid nuclei. After pollination, one sperm fertilizes the egg cell, producing a diploid zygote; the other fuses with two nuclei to form the endosperm.

✔ CONCEPT CHECK 8

Which structure is diploid?

a. Microspore
c. Megaspore

b. Egg
d. Microsporocyte

We have now examined the place of meiosis in the sexual cycle of two organisms, a typical multicellular animal and a flowering plant. These cycles are just two of the many variations found among eukaryotic organisms. Although the cellular events that produce reproductive cells in plants and animals differ in the number of cell divisions, the number of haploid gametes produced, and the relative size of the final products, the overall result is the same: meiosis gives rise to haploid, genetically variable cells that then fuse during fertilization to produce diploid progeny.

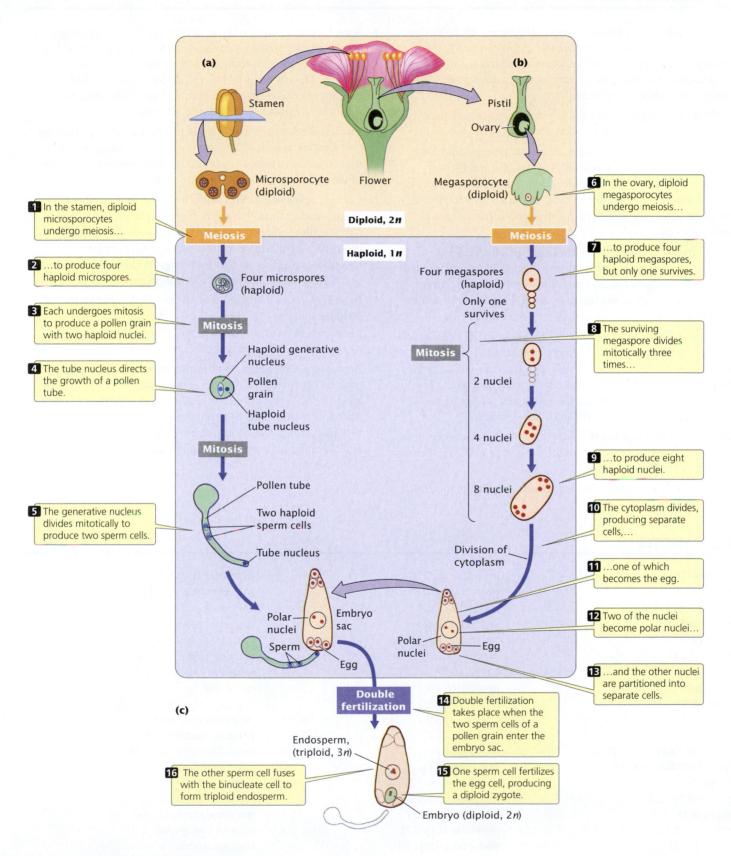

2.22 Sexual reproduction in flowering plants.

CONCEPTS SUMMARY

- A prokaryotic cell possesses a simple structure, with no nuclear envelope and usually a single, circular chromosome. A eukaryotic cell possesses a more complex structure, with a nucleus and multiple linear chromosomes consisting of DNA complexed to histone proteins.

- Cell reproduction requires the copying of the genetic material, separation of the copies, and cell division.

- In a prokaryotic cell, the single chromosome replicates, each copy moves toward opposite sides of the cell, and the cell divides. In eukaryotic cells, reproduction is more complex than in prokaryotic cells, requiring mitosis and meiosis to ensure that a complete set of genetic information is transferred to each new cell.

- In eukaryotic cells, chromosomes are typically found in homologous pairs. Each functional chromosome consists of a centromere, telomeres, and multiple origins of replication. After a chromosome has been copied, the two copies remain attached at the centromere, forming sister chromatids.

- The cell cycle consists of the stages through which a eukaryotic cell passes between cell divisions. It consists of (1) interphase, in which the cell grows and prepares for division and (2) the M phase, in which nuclear and cell division take place. The M phase consists of (1) mitosis, the process of nuclear division, and (2) cytokinesis, the division of the cytoplasm.

- Progression through the cell cycle is controlled at checkpoints, which regulate the cell cycle by allowing or prohibiting the cell to proceed to the next stage.

- Mitosis usually results in the production of two genetically identical cells.

- Sexual reproduction produces genetically variable progeny and allows for accelerated evolution. It includes meiosis, in which haploid sex cells are produced, and fertilization, the fusion of sex cells. Meiosis includes two cell divisions. In meiosis I, crossing over takes place and homologous chromosomes separate. In meiosis II, chromatids separate.

- The usual result of meiosis is the production of four haploid cells that are genetically variable. Genetic variation in meiosis is produced by crossing over and by the random distribution of maternal and paternal chromosomes.

- Cohesin holds sister chromatids together. In metaphase of mitosis and in metaphase II of meiosis, the breakdown of cohesin allows sister chromatids to separate. In meiosis I, centromeric cohesin remains intact and keeps sister chromatids together so that homologous chromosomes, but not sister chromatids, separate in anaphase I.

- In animals, a diploid spermatogonium undergoes meiosis to produce four haploid sperm cells. A diploid oogonium undergoes meiosis to produce one large haploid ovum and one or more smaller polar bodies.

- In plants, a diploid microsporocyte in the stamen undergoes meiosis to produce four pollen grains, each with two haploid sperm cells. In the ovary, a diploid megasporocyte undergoes meiosis to produce eight haploid nuclei, one of which forms the egg. During pollination, one sperm fertilizes the egg cell and the other fuses with two haploid nuclei to form a $3n$ endosperm.

IMPORTANT TERMS

prokaryote (p. 17)
eukaryote (p. 17)
eubacteria (p. 17)
archaea (p. 17)
nucleus (p. 18)
histone (p. 18)
chromatin (p. 18)
homologous pair (p. 20)
diploid (p. 20)
haploid (p. 20)
telomere (p. 21)
origin of replication (p. 21)
sister chromatid (p. 21)
cell cycle (p. 22)
checkpoint (p. 22)
interphase (p. 22)
M phase (p. 22)
mitosis (p. 22)
cytokinesis (p. 22)

prophase (p. 23)
prometaphase (p. 23)
metaphase (p. 23)
anaphase (p. 23)
telophase (p. 23)
meiosis (p. 27)
fertilization (p. 27)
prophase I (p. 27)
synapsis (p. 27)
bivalent (p. 27)
tetrad (p. 27)
crossing over (p. 27)
metaphase I (p. 27)
anaphase I (p. 27)
telophase I (p. 27)
interkinesis (p. 27)
prophase II (p. 27)
metaphase II (p. 28)
anaphase II (p. 28)

telophase II (p. 28)
recombination (p. 31)
cohesin (p. 33)
spermatogenesis (p. 33)
spermatogonium (p. 33)
primary spermatocyte (p. 33)
secondary spermatocyte (p. 33)
spermatid (p. 33)
oogenesis (p. 33)
oogonium (p. 34)
primary oocyte (p. 34)
secondary oocyte (p. 34)
first polar body (p. 34)
ovum (p. 34)
second polar body (p. 34)
microsporocyte (p. 36)
microspore (p. 36)
megasporocyte (p. 36)
megaspore (p. 36)

ANSWERS TO CONCEPT CHECKS

1. Eubacteria and archaea are prokaryotes. They differ from eukaryotes in possessing no nucleus, a genome that usually consists of a single, circular chromosome, and a small amount of DNA.

2. b

3. A centromere, a pair of telomeres, and an origin of replication

4. a

5. d

6. During anaphase I, shugoshin protects cohesin at the centromeres from the action of separase; so cohesin remains intact and the sister chromatids remain together. Subsequently, shugoshin breaks down; so centromeric cohesin is cleaved in anaphase II and the chromatids separate.

7. d

8. d

WORKED PROBLEMS

1. A student examines a thin section of an onion-root tip and records the number of cells that are in each stage of the cell cycle. She observes 94 cells in interphase, 14 cells in prophase, 3 cells in prometaphase, 3 cells in metaphase, 5 cells in anaphase, and 1 cell in telophase. If the complete cell cycle in an onion-root tip requires 22 hours, what is the average duration of each stage in the cycle? Assume that all cells are in the active cell cycle (not G_0).

Solution

This problem is solved in two steps. First, we calculate the proportions of cells in each stage of the cell cycle, which correspond to the amount of time that an average cell spends in each stage. For example, if cells spend 90% of their time in interphase, then, at any given moment, 90% of the cells will be in interphase. The second step is to convert the proportions into lengths of time, which is done by multiplying the proportions by the total time of the cell cycle (22 hours).

Step 1. Calculate the proportion of cells at each stage.
The proportion of cells at each stage is equal to the number of cells found in that stage divided by the total number of cells examined:

Interphase	$^{94}/_{120} = 0.783$
Prophase	$^{14}/_{120} = 0.117$
Prometaphase	$^{3}/_{120} = 0.025$
Metaphase	$^{3}/_{120} = 0.025$
Anaphase	$^{5}/_{120} = 0.042$
Telophase	$^{1}/_{120} = 0.008$

We can check our calculations by making sure that the proportions sum to 1.0, which they do.

Step 2. Determine the average duration of each stage.
To determine the average duration of each stage, multiply the proportion of cells in each stage by the time required for the entire cell cycle:

Interphase	0.783×22 hours $= 17.23$ hours
Prophase	0.117×22 hours $= 2.57$ hours
Prometaphase	0.025×22 hours $= 0.55$ hour
Metaphase	0.025×22 hours $= 0.55$ hour
Anaphase	0.042×22 hours $= 0.92$ hour
Telophase	0.008×22 hours $= 0.18$ hour

2. A cell in G_1 of interphase has 8 chromosomes. How many chromosomes and how many DNA molecules will be found per cell as this cell progresses through the following stages: G_2, metaphase of mitosis, anaphase of mitosis, after cytokinesis in mitosis, metaphase I of meiosis, metaphase II of meiosis, and after cytokinesis of meiosis II?

Solution

Remember the rules about counting chromosomes and DNA molecules: (1) to determine the number of chromosomes, count the functional centromeres; (2) to determine the number of DNA molecules, count the chromatids. Think carefully about when and how the numbers of chromosomes and DNA molecules change in the course of mitosis and meiosis.

The number of DNA molecules increases only in the S phase, when DNA replicates; the number of DNA molecules decreases only when the cell divides. Chromosome number increases only when sister chromatids separate in anaphase of mitosis and in anaphase II of meiosis (homologous chromosomes, not chromatids, separate in anaphase I of meiosis). Chromosome number, like the number of DNA molecules, is reduced only by cell division.

Let's now apply these principles to the problem. A cell in G_1 has 8 chromosomes, each consisting of a single chromatid; so 8 DNA molecules are present in G_1. DNA replicates in the S phase; so, in G_2, 16 DNA molecules are present per cell. However, the two copies of each DNA molecule remain attached at the centromere; so there are still only 8 chromosomes present. As the cell passes through prophase and metaphase of the cell cycle, the

*30. The fruit fly *Drosophila melanogaster* has four pairs of chromosomes, whereas the house fly *Musca domestica* has six pairs of chromosomes. Other things being equal, in which species would you expect to see more genetic variation among the progeny of a cross? Explain your answer.

*31. A cell has two pairs of submetacentric chromosomes, which we will call chromosomes I_a, I_b, II_a, and II_b (chromosomes I_a and I_b are homologs, and chromosomes II_a and II_b are homologs). Allele M is located on the long arm of chromosome I_a, and allele m is located at the same position on chromosome I_b. Allele P is located on the short arm of chromosome I_a, and allele p is located at the same position on chromosome I_b. Allele R is located on chromosome II_a and allele r is located at the same position on chromosome II_b.

a. Draw these chromosomes, identifying genes M, m, P, p, R, and r, as they might appear in metaphase I of meiosis. Assume that there is no crossing over.

b. Taking into consideration the random separation of chromosomes in anaphase I, draw the chromosomes (with genes identified) present in all possible types of gametes that might result from this cell's undergoing meiosis. Assume that there is no crossing over.

32. A horse has 64 chromosomes and a donkey has 62 chromosomes. A cross between a female horse and a male donkey produces a mule, which is usually sterile. How many chromosomes does a mule have? Can you think of any reasons for the fact that most mules are sterile?

33. Normal somatic cells of horses have 64 chromosomes ($2n = 64$). How many chromosomes and DNA molecules will be present in the following types of horse cells?

Cell type	Number of chromosomes	Number of DNA molecules
a. Spermatogonium	_____	_____
b. First polar body	_____	_____
c. Primary oocyte	_____	_____
d. Secondary spermatocyte	_____	_____

*34. A primary oocyte divides to give rise to a secondary oocyte and a first polar body. The secondary oocyte then divides to give rise to an ovum and a second polar body.

a. Is the genetic information found in the first polar body identical with that found in the secondary oocyte? Explain your answer.

b. Is the genetic information found in the second polar body identical with that in the ovum? Explain your reasoning.

CHALLENGE QUESTIONS

Section 2.3

35. From 80% to 90% of the most common human chromosome abnormalities arise because the chromosomes fail to divide properly in oogenesis. Can you think of a reason why failure of chromosome division might be more common in female gametogenesis than in male gametogenesis?

36. On average, what proportion of the genome in the following pairs of humans would be exactly the same if no crossing over occurred? (For the purposes of this question only, we will ignore the special case of the X and Y sex chromosomes and assume that all genes are located on nonsex chromosomes.)

a. Father and child

b. Mother and child

c. Two full siblings (offspring that have the same two biological parents)

d. Half siblings (offspring that have only one biological parent in common)

e. Uncle and niece

f. Grandparent and grandchild

*37. Female bees are diploid, and male bees are haploid. The haploid males produce sperm and can successfully mate with diploid females. Fertilized eggs develop into females and unfertilized eggs develop into males. How do you think the process of sperm production in male bees differs from sperm production in other animals?

3 Basic Principles of Heredity

Red hair is caused by recessive mutations at the melanocortin 1 receptor gene. *Lady Lilith,* 1868, by Dante Charles Gabriel Rosetti. Oil on canvas. *[© Delaware Art Museum, Wilmington, USA/Samuel and Mary R. Bancroft Memorial/The Bridgeman Art Library.]*

THE GENETICS OF RED HAIR

Whether because of its exotic hue or its novelty, red hair has long been a subject of fascination for historians, poets, artists, and scientists. Greek historians made special note of the fact that Boudica, the Celtic queen who led a revolt against the Roman Empire, possessed a "great mass of red hair." Early Christian artists frequently portrayed Mary Magdalene as a striking red head (though there is no mention of her red hair in the Bible), and the famous artist Botticelli painted the goddess Venus as a red-haired beauty in his masterpiece *The Birth of Venus.* Queen Elizabeth I of England possessed curly red hair; during her reign, red hair was quite fashionable in London society.

The color of our hair is caused largely by a pigment called melanin that comes in two primary forms: eumelanin, which is black or brown, and pheomelanin, which is red or yellow. The color of a person's hair is determined by two factors: (1) the amount of melanin produced (more melanin causes darker hair; less melanin causes lighter hair) and (2) the relative amounts of eumelanin compared with pheomelanin (more eumelanin produces black or brown hair; more pheomelanin produces red or yellow hair). The color of our hair is not just an academic curiosity; melanin protects against the harmful effects of sunlight, and people with red hair are usually fair skinned and particularly susceptible to skin cancer.

The inheritance of red hair has long been a subject of scientific debate. In 1909, Charles and Gertrude Davenport speculated on the inheritance of hair color in humans. Charles Davenport was an early enthusiast of genetics, particularly of inheritance in humans, and was the first director of the Biological Laboratory in Cold Spring Harbor, New York. He later became a leading proponent of eugenics, a movement—now discredited—that advocated improvement of the human race through genetics. The Davenports' study was based on family histories sent in by untrained amateurs and was methodologically flawed, but their results suggested that red hair is recessive to black and brown, meaning that a person must inherit two copies of a red-hair gene—one from each parent—to have red hair. Subsequent research contradicted this initial conclusion, suggesting that red hair is inherited instead as a dominant trait and

43

that a person will have red hair even if possessing only a single red-hair gene. Controversy over whether red hair color is dominant or recessive, or even dependent on combinations of several different genes, continued for many years.

In 1993, scientists who were investigating a gene that affects the color of fur in mice discovered that the gene encodes the melanocortin-1 receptor. This receptor, when activated, increases the production of black eumelanin and decreases the production of red pheomelanin, resulting in black or brown fur. Shortly thereafter, the same melanocortin-1 receptor gene (*MC1R*) was located on human chromosome 16, cloned, and sequenced. When this gene is mutated in humans, red hair results. Most people with red hair carry two defective copies of the *MC1R* gene, which means that the trait is recessive (as originally proposed by the Davenports back in 1909). However, from 10% to 20% of red heads possess only a single mutant copy of *MC1R*, muddling the recessive interpretation of red hair (the people with a single mutant copy of the gene tend to have lighter red hair than those who harbor two mutant copies). The type and frequency of mutations at the *MC1R* gene vary widely among human populations, accounting for ethnic differences in the preponderance of red hair: Among those of African and Asian descent, mutations for red hair are uncommon, whereas almost 40% of the people from the northern part of the United Kingdom carry at least one mutant gene for red hair.

This chapter is about the principles of heredity: how genes—like the one for the melanocortin-1 receptor—are passed from generation to generation and how factors such as dominance influence that inheritance. The principles of heredity were first put forth by Gregor Mendel, and so we begin this chapter by examining Mendel's scientific achievements. We then turn to simple genetic crosses, those in which a single characteristic is examined. We will learn some techniques for predicting the outcome of genetic crosses and then turn to crosses in which two or more characteristics are examined. We will see how the principles applied to simple genetic crosses and the ratios of offspring that they produce serve as the key for understanding more complicated crosses. The chapter ends with a discussion of statistical tests for analyzing crosses.

Throughout this chapter, a number of concepts are interwoven: Mendel's principles of segregation and independent assortment, probability, and the behavior of chromosomes. These concepts might at first appear to be unrelated, but they are actually different views of the same phenomenon, because the genes that undergo segregation and independent assortment are located on chromosomes. The principal aim of this chapter is to examine these different views and to clarify their relations.

3.1 Gregor Mendel Discovered the Basic Principles of Heredity

In 1909, when the Davenports speculated about the inheritance of red hair, the basic principles of heredity were just becoming widely known among biologists. Surprisingly, these principles had been discovered some 44 years earlier by Johann Gregor Mendel (1822–1884).

Mendel was born in what is now part of the Czech Republic. Although his parents were simple farmers with little money, he was able to achieve a sound education and was admitted to the Augustinian monastery in Brno in September 1843. After graduating from seminary, Mendel was ordained a priest and appointed to a teaching position in a local school. He excelled at teaching, and the abbot of the monastery recommended him for further study at the University of Vienna, which he attended from 1851 to 1853. There, Mendel enrolled in the newly opened Physics Institute and took courses in mathematics, chemistry, entomology, paleontology, botany, and plant physiology. It was probably there that Mendel acquired knowledge of the scientific method, which he later applied so successfully to his genetics experiments. After 2 years of study in Vienna, Mendel returned to Brno, where he taught school and began his experimental work with pea plants. He conducted breeding experiments from 1856 to 1863 and presented his results publicly at meetings of the Brno Natural Science Society in 1865. Mendel's paper from these lectures was published in 1866. In spite of widespread interest in heredity, the effect of his research on the scientific community was minimal. At the time, no one seemed to have noticed that Mendel had discovered the basic principles of inheritance.

In 1868, Mendel was elected abbot of his monastery, and increasing administrative duties brought an end to his teaching and eventually to his genetics experiments. He died at the age of 61 on January 6, 1884, unrecognized for his contribution to genetics.

The significance of Mendel's discovery was unappreciated until 1900, when three botanists—Hugo de Vries, Erich von Tschermak, and Karl Correns—began independently conducting similar experiments with plants and arrived at

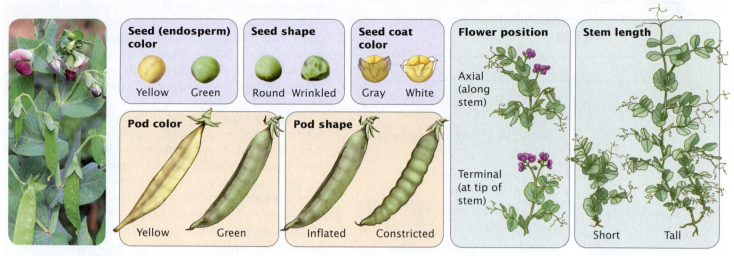

3.1 Mendel used the pea plant *Pisum sativum* in his studies of heredity. He examined seven characteristics that appeared in the seeds and in plants grown from the seeds. *[Photograph by Wally Eberhart/Visuals Unlimited.]*

conclusions similar to those of Mendel. Coming across Mendel's paper, they interpreted their results in accord with his principles and drew attention to his pioneering work.

Mendel's Success

Mendel's approach to the study of heredity was effective for several reasons. Foremost was his choice of experimental subject, the pea plant *Pisum sativum* (**Figure 3.1**), which offered clear advantages for genetic investigation. The plant is easy to cultivate, and Mendel had the monastery garden and greenhouse at his disposal. Compared with some other plants, peas grow relatively rapidly, completing an entire generation in a single growing season. By today's standards, one generation per year seems frightfully slow—fruit flies complete a generation in 2 weeks and bacteria in 20 minutes—but Mendel was under no pressure to publish quickly and was able to follow the inheritance of individual characteristics for several generations. Had he chosen to work on an organism with a longer generation time—horses, for example—he might never have discovered the basis of inheritance. Pea plants also produce many offspring—their seeds—which allowed Mendel to detect meaningful mathematical ratios in the traits that he observed in the progeny.

The large number of varieties of peas that were available to Mendel also was crucial, because these varieties differed in various traits and were genetically pure. Mendel was therefore able to begin with plants of variable, known genetic makeup.

Much of Mendel's success can be attributed to the seven characteristics that he chose for study (see Figure 3.1). He avoided characteristics that display a range of variation; instead, he focused his attention on those that exist in two easily differentiated forms, such as white versus gray seed coats, round versus wrinkled seeds, and inflated versus constricted pods.

Finally, Mendel was successful because he adopted an experimental approach and interpreted his results by using mathematics. Unlike many earlier investigators who just described the *results* of crosses, Mendel formulated *hypotheses* based on his initial observations and then conducted additional crosses to test his hypotheses. He kept careful records of the numbers of progeny possessing each type of trait and computed ratios of the different types. He paid close attention to detail, was adept at seeing patterns in detail, and was patient and thorough, conducting his experiments for 10 years before attempting to write up his results.

CONCEPTS

Gregor Mendel put forth the basic principles of inheritance, publishing his findings in 1866. The significance of his work did not become widely appreciated until 1900.

✔ CONCEPT CHECK 1

Which of the following factors did not contribute to Mendel's success in his study of heredity?

a. His use of the pea plant

b. His study of plant chromosomes

c. His adoption of an experimental approach

d. His use of mathematics

Genetic Terminology

Before we examine Mendel's crosses and the conclusions that he drew from them, it will be helpful to review some terms commonly used in genetics (**Table 3.1**). The term *gene* is a word that Mendel never knew. It was not coined until 1909, when Danish geneticist Wilhelm Johannsen first used

Table 3.1	Summary of important genetic terms
Term	Definition
Gene	A genetic factor (region of DNA) that helps determine a characteristic
Allele	One of two or more alternate forms of a gene
Locus	Specific place on a chromosome occupied by an allele
Genotype	Set of alleles possessed by an individual organism
Heterozygote	An individual organism possessing two different alleles at a locus
Homozygote	An individual organism possessing two of the same alleles at a locus
Phenotype or trait	The appearance or manifestation of a character
Character or characteristic	An attribute or feature

it. The definition of a gene varies with the context of its use, and so its definition will change as we explore different aspects of heredity. For our present use in the context of genetic crosses, we will define a **gene** as an inherited factor that determines a characteristic.

Genes frequently come in different versions called **alleles** (**Figure 3.2**). In Mendel's crosses, seed shape was determined by a gene that exists as two different alleles: one allele encodes round seeds and the other encodes wrinkled seeds. All alleles for any particular gene will be found at a specific place on a chromosome called the **locus** for that gene. (The plural of locus is loci; it's bad form in genetics—and incorrect—to

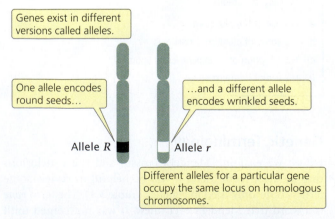

Genes exist in different versions called alleles.

One allele encodes round seeds...

...and a different allele encodes wrinkled seeds.

Allele R   Allele r

Different alleles for a particular gene occupy the same locus on homologous chromosomes.

3.2 At each locus, a diploid organism possesses two alleles located on different homologous chromosomes.

speak of locuses.) Thus, there is a specific place—a locus—on a chromosome in pea plants where the shape of seeds is determined. This locus might be occupied by an allele for round seeds or one for wrinkled seeds. We will use the term *allele* when referring to a specific version of a gene; we will use the term *gene* to refer more generally to any allele at a locus.

The **genotype** is the set of alleles that an individual organism possesses. A diploid organism that possesses two identical alleles is **homozygous** for that locus. One that possesses two different alleles is **heterozygous** for the locus.

Another important term is **phenotype,** which is the manifestation or appearance of a characteristic. A phenotype can refer to any type of characteristic—physical, physiological, biochemical, or behavioral. Thus, the condition of having round seeds is a phenotype, a body weight of 50 kilograms (50 kg) is a phenotype, and having sickle-cell anemia is a phenotype. In this book, the term *characteristic* or *character* refers to a general feature such as eye color; the term *trait* or *phenotype* refers to specific manifestations of that feature, such as blue or brown eyes.

A given phenotype arises from a genotype that develops within a particular environment. The genotype determines the potential for development; it sets certain limits, or boundaries, on that development. How the phenotype develops within those limits is determined by the effects of other genes and of environmental factors, and the balance between these influences varies from character to character. For some characters, the differences between phenotypes are determined largely by differences in genotype; in other words, the genetic limits for that phenotype are narrow. Seed shape in Mendel's peas is a good example of a characteristic for which the genetic limits are narrow and the phenotypic differences are largely genetic. For other characters, environmental differences are more important; in this case, the limits imposed by the genotype are broad. The height reached by an oak tree at maturity is a phenotype that is strongly influenced by environmental factors, such as the availability of water, sunlight, and nutrients. Nevertheless, the tree's genotype still imposes some limits on its height: an oak tree will never grow to be 300 meters (300 m) tall no matter how much sunlight, water, and fertilizer are provided. Thus, even the height of an oak tree is determined to some degree by genes. For many characteristics, both genes and environment are important in determining phenotypic differences.

An obvious but important concept is that only the genotype is inherited. Although the phenotype is determined, at least to some extent, by genotype, organisms do not transmit their phenotypes to the next generation. The distinction between genotype and phenotype is one of the most important principles of modern genetics. The next section describes Mendel's careful observation of phenotypes through several generations of breeding experiments. These experiments allowed him to deduce not only the genotypes of the individual plants, but also the rules governing their inheritance.

CONCEPTS

Each phenotype results from a genotype developing within a specific environment. The genotype, not the phenotype, is inherited.

✔ **CONCEPT CHECK 2**

Distinguish among the following terms: locus, allele, genotype.

3.2 Monohybrid Crosses Reveal the Principle of Segregation and the Concept of Dominance

Mendel started with 34 varieties of peas and spent 2 years selecting those varieties that he would use in his experiments. He verified that each variety was genetically pure (homozygous for each of the traits that he chose to study) by growing the plants for two generations and confirming that all offspring were the same as their parents. He then carried out a number of crosses between the different varieties. Although peas are normally self-fertilizing (each plant crosses with itself), Mendel conducted crosses between different plants by opening the buds before the anthers were fully developed, removing the anthers, and then dusting the stigma with pollen from a different plant (**Figure 3.3**).

Mendel began by studying **monohybrid crosses**—those between parents that differed in a single characteristic. In one experiment, Mendel crossed a pea plant homozygous for round seeds with one that was homozygous for wrinkled seeds (see Figure 3.3). This first generation of a cross is the **P (parental) generation.**

After crossing the two varieties in the P generation, Mendel observed the offspring that resulted from the cross. In regard to seed characteristics, such as seed shape, the phenotype develops as soon as the seed matures, because the seed traits are determined by the newly formed embryo within the seed. For characters associated with the plant itself, such as stem length, the phenotype doesn't develop until the plant grows from the seed; for these characters, Mendel had to wait until the following spring, plant the seeds, and then observe the phenotypes on the plants that germinated.

The offspring from the parents in the P generation are the F_1 (**filial 1**) **generation.** When Mendel examined the F_1 generation of this cross, he found that they expressed only one of the phenotypes present in the parental generation: all the F_1 seeds were round. Mendel carried out 60 such crosses and always obtained this result. He also conducted **reciprocal crosses:** in one cross, pollen (the male gamete) was taken from a plant with round seeds and, in its reciprocal cross, pollen was taken from a plant with wrinkled seeds. Reciprocal crosses gave the same result: all the F_1 were round.

Mendel wasn't content with examining only the seeds arising from these monohybrid crosses. The following spring, he planted the F_1 seeds, cultivated the plants that

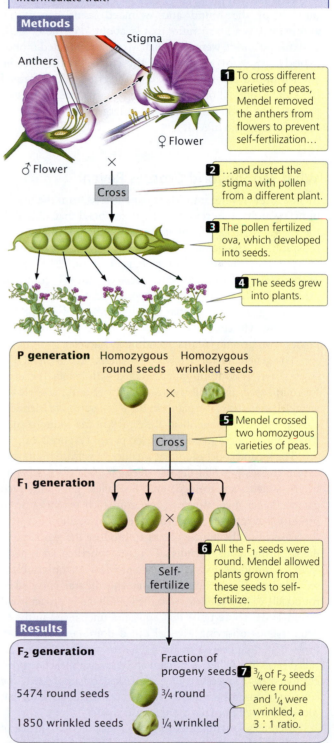

Experiment

Question: When peas with two different traits—round and wrinkled seeds—are crossed, will their progeny exhibit one of those traits, both of those traits, or a "blended" intermediate trait?

Methods

Stigma

Anthers

♂ Flower

♀ Flower

Cross

1 To cross different varieties of peas, Mendel removed the anthers from flowers to prevent self-fertilization…

2 …and dusted the stigma with pollen from a different plant.

3 The pollen fertilized ova, which developed into seeds.

4 The seeds grew into plants.

P generation   Homozygous round seeds   Homozygous wrinkled seeds

×

Cross

5 Mendel crossed two homozygous varieties of peas.

F₁ generation

×

Self-fertilize

6 All the F_1 seeds were round. Mendel allowed plants grown from these seeds to self-fertilize.

Results

F₂ generation

Fraction of progeny seeds

5474 round seeds   ¾ round

1850 wrinkled seeds   ¼ wrinkled

7 ¾ of F_2 seeds were round and ¼ were wrinkled, a 3 : 1 ratio.

Conclusion: The traits of the parent plants do not blend. Although F_1 plants display the phenotype of one parent, both traits are passed to F_2 progeny in a 3 : 1 ratio.

3.3 Mendel conducted monohybrid crosses.

germinated from them, and allowed the plants to self-fertilize, producing a second generation—the F_2 (**filial 2**) **generation**. Both of the traits from the P generation emerged in the F_2 generation; Mendel counted 5474 round seeds and 1850 wrinkled seeds in the F_2 (see Figure 3.3). He noticed that the number of the round and wrinkled seeds constituted approximately a 3 to 1 ratio; that is, about $\frac{3}{4}$ of the F_2 seeds were round and $\frac{1}{4}$ were wrinkled. Mendel conducted monohybrid crosses for all seven of the characteristics that he studied in pea plants and, in all of the crosses, he obtained the same result: all of the F_1 resembled only one of the two parents, but both parental traits emerged in the F_2 in an approximate ratio of 3 : 1.

What Monohybrid Crosses Reveal

Mendel drew several important conclusions from the results of his monohybrid crosses. First, he reasoned that, although the F_1 plants display the phenotype of only one parent, they must inherit genetic factors from both parents because they transmit both phenotypes to the F_2 generation. The presence of both round and wrinkled seeds in the F_2 could be explained only if the F_1 plants possessed both round and wrinkled genetic factors that they had inherited from the P generation. He concluded that each plant must therefore possess two genetic factors encoding a character.

The genetic factors (now called alleles) that Mendel discovered are, by convention, designated with letters; the allele for round seeds is usually represented by R, and the allele for wrinkled seeds by r. The plants in the P generation of Mendel's cross possessed two identical alleles: RR in the round-seeded parent and rr in the wrinkled-seeded parent (**Figure 3.4a**).

The second conclusion that Mendel drew from his monohybrid crosses was that the two alleles in each plant separate when gametes are formed, and one allele goes into each gamete. When two gametes (one from each parent) fuse to produce a zygote, the allele from the male parent unites with the allele from the female parent to produce the genotype of the offspring. Thus, Mendel's F_1 plants inherited an R allele from the round-seeded plant and an r allele from the wrinkled-seeded plant (**Figure 3.4b**). However, only the trait encoded by round allele (R) was *observed* in the F_1—all the F_1 progeny had round seeds. Those traits that appeared unchanged in the F_1 heterozygous offspring Mendel called **dominant,** and those traits that disappeared in the F_1 heterozygous offspring he called **recessive.** When dominant and recessive alleles are present together, the recessive allele is masked, or suppressed. The concept of dominance was the third important conclusion that Mendel derived from his monohybrid crosses.

Mendel's fourth conclusion was that the two alleles of an individual plant separate with equal probability into the gametes. When plants of the F_1 (with genotype Rr) produced gametes, half of the gametes received the R allele for round seeds and half received the r allele for wrinkled seeds. The gametes then paired randomly to produce the following

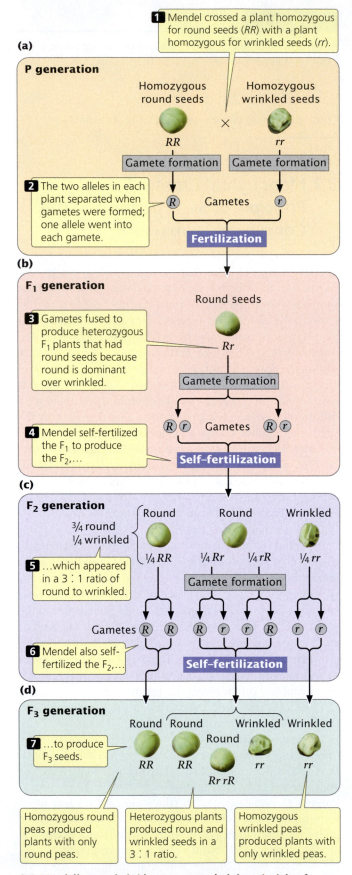

3.4 Mendel's monohybrid crosses revealed the principle of segregation and the concept of dominance.

genotypes in equal proportions among the F_2: *RR, Rr, rR, rr* (**Figure 3.4c**). Because round (*R*) is dominant over wrinkled (*r*), there were three round progeny in the F_2 (*RR, Rr, rR*) for every one wrinkled progeny (*rr*) in the F_2. This 3 : 1 ratio of round to wrinkled progeny that Mendel observed in the F_2 could occur only if the two alleles of a genotype separated into the gametes with equal probability.

The conclusions that Mendel developed about inheritance from his monohybrid crosses have been further developed and formalized into the principle of segregation and the concept of dominance. The **principle of segregation** (Mendel's first law) states that each individual diploid organism possesses two alleles for any particular characteristic. These two alleles segregate (separate) when gametes are formed, and one allele goes into each gamete. Furthermore, the two alleles segregate into gametes in equal proportions. The **concept of dominance** states that, when two different alleles are present in a genotype, only the trait encoded by one of them—the "dominant" allele—is observed in the phenotype.

Mendel confirmed these principles by allowing his F_2 plants to self-fertilize and produce an F_3 generation. He found that the F_2 plants grown from the wrinkled seeds—those displaying the recessive trait (*rr*)—produced an F_3 in which all plants produced wrinkled seeds. Because his wrinkled-seeded plants were homozygous for wrinkled alleles (*rr*), they could pass on only wrinkled alleles to their progeny (**Figure 3.4d**).

The F_2 plants grown from round seeds—the dominant trait—fell into two types (see Figure 3.4c). On self-fertilization, about $2/3$ of the F_2 plants produced both round and wrinkled seeds in the F_3 generation. These F_2 plants were heterozygous (*Rr*); so they produced $1/4$ *RR* (round), $1/2$ *Rr* (round), and $1/4$ *rr* (wrinkled) seeds, giving a 3 : 1 ratio of round to wrinkled in the F_3. About $1/3$ of the F_2 plants were of the second type; they produced only the dominant round-seeded trait in the F_3. These F_2 plants were homozygous for the round allele (*RR*) and could thus produce only round offspring in the F_3 generation. Mendel planted the seeds obtained in the F_3 and carried these plants through three more rounds of self-fertilization. In each generation, $2/3$ of the round-seeded plants produced round and wrinkled offspring, whereas $1/3$ produced only round offspring. These results are entirely consistent with the principle of segregation.

CONCEPTS

The principle of segregation states that each individual organism possesses two alleles that can encode a characteristic. These alleles segregate when gametes are formed, and one allele goes into each gamete. The concept of dominance states that, when the two alleles of a genotype are different, only the trait encoded by one of them—the "dominant" allele—is observed.

✔ **CONCEPT CHECK 3**

How did Mendel know that each of his pea plants carried two alleles encoding a characteristic?

CONNECTING CONCEPTS

Relating Genetic Crosses to Meiosis

We have now seen how the results of monohybrid crosses are explained by Mendel's principle of segregation. Many students find that they enjoy working genetic crosses but are frustrated by the abstract nature of the symbols. Perhaps you feel the same at this point. You may be asking, "What do these symbols really represent? What does the genotype *RR* mean in regard to the biology of the organism?" The answers to these questions lie in relating the abstract symbols of crosses to the structure and behavior of chromosomes, the repositories of genetic information (see Chapter 2).

In 1900, when Mendel's work was rediscovered and biologists began to apply his principles of heredity, the relation between genes and chromosomes was still unclear. The theory that genes are located on chromosomes (the **chromosome theory of heredity**) was developed in the early 1900s by Walter Sutton, then a graduate student at Columbia University. Through the careful study of meiosis in insects, Sutton documented the fact that each homologous pair of chromosomes consists of one maternal chromosome and one paternal chromosome. Showing that these pairs segregate independently into gametes in meiosis, he concluded that this process is the biological basis for Mendel's principles of heredity. German cytologist and embryologist Theodor Boveri came to similar conclusions at about the same time.

Sutton knew that diploid cells have two sets of chromosomes. Each chromosome has a pairing partner, its homologous chromosome. One chromosome of each homologous pair is inherited from the mother and the other is inherited from the father. Similarly, diploid cells possess two alleles at each locus, and these alleles constitute the genotype for that locus. The principle of segregation indicates that one allele of the genotype is inherited from each parent.

This similarity between the number of chromosomes and the number of alleles is not accidental—the two alleles of a genotype are located on homologous chromosomes. The symbols used in genetic crosses, such as *R* and *r*, are just shorthand notations for particular sequences of DNA in the chromosomes that encode particular phenotypes. The two alleles of a genotype are found on different but homologous chromosomes. In the S phase of meiotic interphase, each chromosome replicates, producing two copies of each allele, one on each chromatid (**Figure 3.5a**). The homologous chromosomes segregate in anaphase I, thereby separating the two different alleles (**Figure 3.5b and c**). This chromosome segregation is the basis of the principle of segregation. In anaphase II of meiosis, the two chromatids of each replicated chromosome separate; so each gamete resulting from meiosis carries only a single allele at each locus, as Mendel's principle of segregation predicts.

If crossing over has taken place in prophase I of meiosis, then the two chromatids of each replicated chromosome are no longer identical, and the segregation of different alleles takes place at anaphase I and anaphase II (see Figure 3.5c). However, Mendel didn't know anything about chromosomes; he formulated his principles of heredity entirely on the basis of the results of the crosses that he carried out. Nevertheless, we should not forget that these principles work because they are based on the behavior of actual chromosomes in meiosis.

(a)

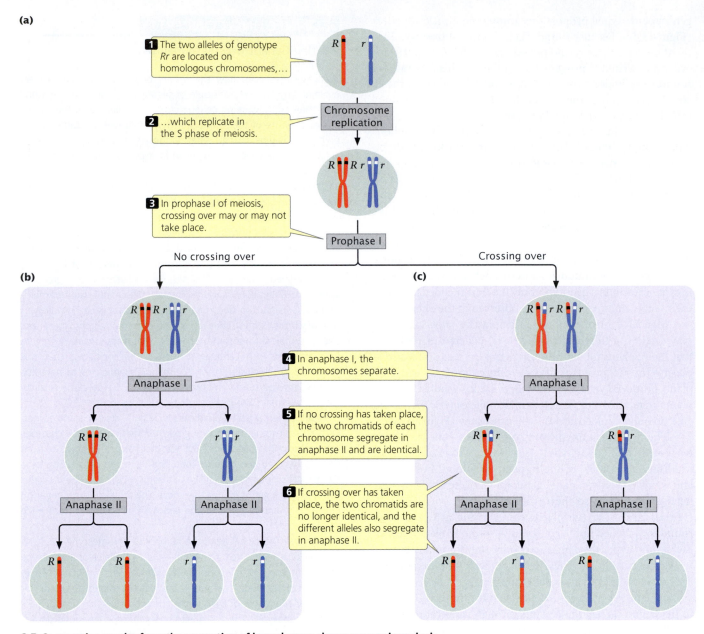

1 The two alleles of genotype *Rr* are located on homologous chromosomes,…

2 …which replicate in the S phase of meiosis.

Chromosome replication

3 In prophase I of meiosis, crossing over may or may not take place.

Prophase I

No crossing over   Crossing over

(b)   **(c)**

4 In anaphase I, the chromosomes separate.

Anaphase I   Anaphase I

5 If no crossing has taken place, the two chromatids of each chromosome segregate in anaphase II and are identical.

Anaphase II   Anaphase II

6 If crossing over has taken place, the two chromatids are no longer identical, and the different alleles also segregate in anaphase II.

Anaphase II   Anaphase II

3.5 Segregation results from the separation of homologous chromosomes in meiosis.

Predicting the Outcomes of Genetic Crosses

One of Mendel's goals in conducting his experiments on pea plants was to develop a way to predict the outcome of crosses between plants with different phenotypes. In this section, we will first learn a simple, shorthand method for predicting outcomes of genetic crosses (the Punnett square), and then we will learn how to use probability to predict the results of crosses.

The Punnett square   The Punnett square was developed by English geneticist Reginald C. Punnett in 1917. To illustrate the Punnett square, let's examine another cross that Mendel carried out. By crossing two varieties of peas that differed in height, Mendel established that tall (*T*) was dom-

inant over short (*t*). He tested his theory concerning the inheritance of dominant traits by crossing an F_1 tall plant that was heterozygous (*Tt*) with the short homozygous parental variety (*tt*). This type of cross, between an F_1 genotype and either of the parental genotypes, is called a **backcross.**

To predict the types of offspring that result from this backcross, we first determine which gametes will be produced by each parent (**Figure 3.6a**). The principle of segregation tells us that the two alleles in each parent separate, and one allele passes to each gamete. All gametes from the homozygous *tt* short plant will receive a single short (*t*) allele. The tall plant in this cross is heterozygous (*Tt*); so 50% of its gametes will receive a tall allele (*T*) and the other 50% will receive a short allele (*t*).

A **Punnett square** is constructed by drawing a grid, putting the gametes produced by one parent along the upper edge and the gametes produced by the other parent down the left side (**Figure 3.6b**). Each cell (a block within the Punnett square) contains an allele from each of the corresponding gametes, generating the genotype of the progeny produced by fusion of those gametes. In the upper left-hand cell of the Punnett square in Figure 3.6b, a gamete containing T from the tall plant unites with a gamete containing t from the short plant, giving the genotype of the progeny (Tt). It is useful to write the phenotype expressed by each genotype; here the progeny will be tall, because the tall allele is dominant over the short allele. This process is repeated for all the cells in the Punnett square.

By simply counting, we can determine the types of progeny produced and their ratios. In Figure 3.6b, two cells contain tall (Tt) progeny and two cells contain short (tt)

progeny; so the genotypic ratio expected for this cross is 2 Tt to 2 tt (a 1 : 1 ratio). Another way to express this result is to say that we expect $\frac{1}{2}$ of the progeny to have genotype Tt (and phenotype tall) and $\frac{1}{2}$ of the progeny to have genotype tt (and phenotype short). In this cross, the genotypic ratio and the phenotypic ratio are the same, but this outcome need not be the case. Try completing a Punnett square for the cross in which the F_1 round-seeded plants in Figure 3.4 undergo self-fertilization (you should obtain a phenotypic ratio of 3 round to 1 wrinkled and a genotypic ratio of 1 RR to 2 Rr to 1 rr).

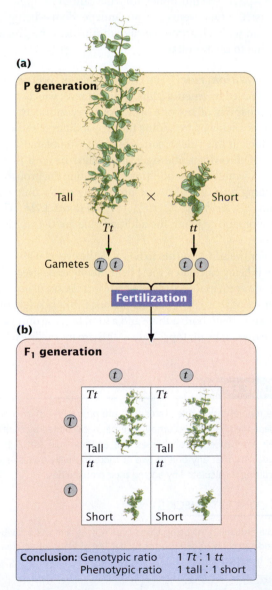

(a)

P generation

Tall   ×   Short

Tt tt

Gametes T t t t

Fertilization

(b)

F₁ generation

	t	t
T	Tt Tall	Tt Tall
t	tt Short	tt Short

Conclusion: Genotypic ratio   1 Tt : 1 tt
Phenotypic ratio   1 tall : 1 short

3.6 The Punnett square can be used to determine the results of a genetic cross.

CONCEPTS

The Punnett square is a shorthand method of predicting the genotypic and phenotypic ratios of progeny from a genetic cross.

✔ CONCEPT CHECK 4

If an F_1 plant depicted in Figure 3.4 is backcrossed to the parent with round seeds, what proportion of the progeny will have wrinkled seeds? (Use a Punnett square.)

a. $\frac{3}{4}$ c. $\frac{1}{4}$

b. $\frac{1}{2}$ d. 0

Probability as a tool in genetics   Another method for determining the outcome of a genetic cross is to use the rules of probability, as Mendel did with his crosses. **Probability** expresses the likelihood of the occurrence of a particular event. It is the number of times that a particular event occurs, divided by the number of all possible outcomes. For example, a deck of 52 cards contains only one king of hearts. The probability of drawing one card from the deck at random and obtaining the king of hearts is $\frac{1}{52}$, because there is only one card that is the king of hearts (one event) and there are 52 cards that can be drawn from the deck (52 possible outcomes). The probability of drawing a card and obtaining an ace is $\frac{4}{52}$, because there are four cards that are aces (four events) and 52 cards (possible outcomes). Probability can be expressed either as a fraction ($\frac{4}{52}$ in this case) or as a decimal number (0.077 in this case).

The probability of a particular event may be determined by knowing something about *how* the event occurs or *how often* it occurs. We know, for example, that the probability of rolling a six-sided die and getting a four is $\frac{1}{6}$, because the die has six sides and any one side is equally likely to end up on top. So, in this case, understanding the nature of the event—the shape of the thrown die—allows us to determine the probability. In other cases, we determine the probability of an event by making a large number of observations. When a weather forecaster says that there is a 40% chance of rain on a particular day, this probability was obtained by observing a large number of days with similar atmospheric conditions and finding that it rains on 40% of those days. In this case, the probability has been determined empirically (by observation).

The multiplication rule Two rules of probability are useful for predicting the ratios of offspring produced in genetic crosses. The first is the **multiplication rule,** which states that the probability of two or more independent events occurring together is calculated by multiplying their independent probabilities.

To illustrate the use of the multiplication rule, let's again consider the roll of a die. The probability of rolling one die and obtaining a four is $\frac{1}{6}$. To calculate the probability of rolling a die twice and obtaining 2 fours, we can apply the

(a) The multiplication rule

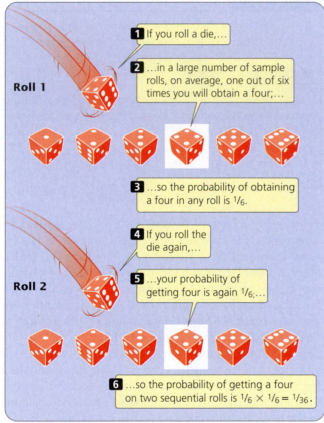

(b) The addition rule

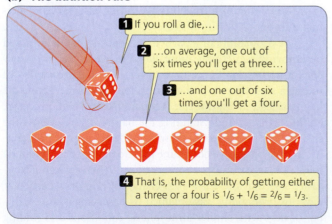

3.7 The multiplication and addition rules can be used to determine the probability of combinations of events.

multiplication rule. The probability of obtaining a four on the first roll is $\frac{1}{6}$ and the probability of obtaining a four on the second roll is $\frac{1}{6}$; so the probability of rolling a four on both is $\frac{1}{6} \times \frac{1}{6} = \frac{1}{36}$ (**Figure 3.7a**). The key indicator for applying the multiplication rule is the word *and;* in the example just considered, we wanted to know the probability of obtaining a four on the first roll *and* a four on the second roll.

For the multiplication rule to be valid, the events whose joint probability is being calculated must be independent—the outcome of one event must not influence the outcome of the other. For example, the number that comes up on one roll of the die has no influence on the number that comes up on the other roll; so these events are independent. However, if we wanted to know the probability of being hit on the head with a hammer and going to the hospital on the same day, we could not simply multiply the probability of being hit on the head with a hammer by the probability of going to the hospital. The multiplication rule cannot be applied here, because the two events are not independent—being hit on the head with a hammer certainly influences the probability of going to the hospital.

The addition rule The second rule of probability frequently used in genetics is the **addition rule,** which states that the probability of any one of two or more mutually exclusive events is calculated by adding the probabilities of these events. Let's look at this rule in concrete terms. To obtain the probability of throwing a die once and rolling *either* a three *or* a four, we would use the addition rule, adding the probability of obtaining a three ($\frac{1}{6}$) to the probability of obtaining a four (again, $\frac{1}{6}$), or $\frac{1}{6} + \frac{1}{6} = \frac{2}{6} = \frac{1}{3}$ (**Figure 3.7b**). The key indicators for applying the addition rule are the words *either* and *or*.

For the addition rule to be valid, the events whose probability is being calculated must be mutually exclusive, meaning that one event excludes the possibility of the occurrence of the other event. For example, you cannot throw a single die just once and obtain both a three and a four, because only one side of the die can be on top. These events are mutually exclusive.

CONCEPTS

The multiplication rule states that the probability of two or more independent events occurring together is calculated by multiplying their independent probabilities. The addition rule states that the probability that any one of two or more mutually exclusive events occurring is calculated by adding their probabilities.

✔ CONCEPT CHECK 5

If the probability of being blood-type A is $\frac{1}{8}$ and the probability of blood-type O is $\frac{1}{2}$, what is the probability of being either blood-type A or blood-type O?

a. $\frac{5}{8}$ c. $\frac{1}{8}$

b. $\frac{1}{2}$ d. $\frac{1}{16}$

The application of probability to genetic crosses   The multiplication and addition rules of probability can be used in place of the Punnett square to predict the ratios of progeny expected from a genetic cross. Let's first consider a cross between two pea plants heterozygous for the locus that determines height, $Tt \times Tt$. Half of the gametes produced by each plant have a T allele, and the other half have a t allele; so the probability for each type of gamete is $\frac{1}{2}$.

The gametes from the two parents can combine in four different ways to produce offspring. Using the multiplication rule, we can determine the probability of each possible type. To calculate the probability of obtaining TT progeny, for example, we multiply the probability of receiving a T allele from the first parent ($\frac{1}{2}$) times the probability of receiving a T allele from the second parent ($\frac{1}{2}$). The multiplication rule should be used here because we need the probability of receiving a T allele from the first parent *and* a T allele from the second parent—two independent events. The four types of progeny from this cross and their associated probabilities are:

TT (T gamete and T gamete)	$\frac{1}{2} \times \frac{1}{2} = \frac{1}{4}$	tall
Tt (T gamete and t gamete)	$\frac{1}{2} \times \frac{1}{2} = \frac{1}{4}$	tall
tT (t gamete and T gamete)	$\frac{1}{2} \times \frac{1}{2} = \frac{1}{4}$	tall
tt (t gamete and t gamete)	$\frac{1}{2} \times \frac{1}{2} = \frac{1}{4}$	short

Notice that there are two ways for heterozygous progeny to be produced: a heterozygote can either receive a T allele from the first parent and a t allele from the second or receive a t allele from the first parent and a T allele from the second.

After determining the probabilities of obtaining each type of progeny, we can use the addition rule to determine the overall phenotypic ratios. Because of dominance, a tall plant can have genotype TT, Tt, or tT; so, using the addition rule, we find the probability of tall progeny to be $\frac{1}{4} + \frac{1}{4} + \frac{1}{4} = \frac{3}{4}$. Because only one genotype encodes short (tt), the probability of short progeny is simply $\frac{1}{4}$.

Two methods have now been introduced to solve genetic crosses: the Punnett square and the probability method. At this point, you may be asking, "Why bother with probability rules and calculations? The Punnett square is easier to understand and just as quick." For simple monohybrid crosses, the Punnett square is simpler than the probability method and is just as easy to use. However, for tackling more-complex crosses concerning genes at two or more loci, the probability method is both clearer and quicker than the Punnett square.

The binomial expansion and probability   When probability is used, it is important to recognize that there may be several different ways in which a set of events can occur. Consider two parents who are both heterozygous for albinism, a recessive condition in humans that causes reduced pigmentation in the skin, hair, and eyes (**Figure 3.8**; see also the intro-

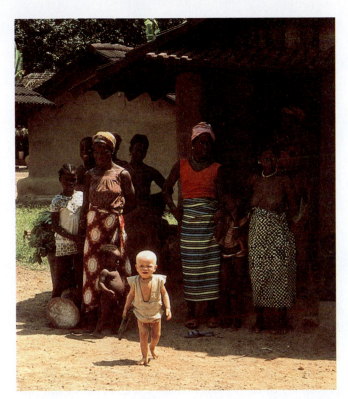

3.8 Albinism in human beings is usually inherited as a recessive trait. [*Richard Dranitzke/SS/Photo Researchers.*]

duction to Chapter 1). When two parents heterozygous for albinism mate ($Aa \times Aa$), the probability of their having a child with albinism (aa) is $\frac{1}{4}$ and the probability of having a child with normal pigmentation (AA or Aa) is $\frac{3}{4}$. Suppose we want to know the probability of this couple having three children, all three with albinism. In this case, there is only one way in which they can have three children with albinism—their first child has albinism *and* their second child has albinism *and* their third child has albinism. Here, we simply apply the multiplication rule: $\frac{1}{4} \times \frac{1}{4} \times \frac{1}{4} = \frac{1}{64}$.

Suppose we now ask, What is the probability of this couple having three children, one with albinism and two with normal pigmentation. This situation is more complicated. The first child might have albinism, whereas the second and third are unaffected; the probability of this sequence of events is $\frac{1}{4} \times \frac{3}{4} \times \frac{3}{4} = \frac{9}{64}$. Alternatively, the first and third child might have normal pigmentation, whereas the second has albinism; the probability of this sequence is $\frac{3}{4} \times \frac{1}{4} \times \frac{3}{4} = \frac{9}{64}$. Finally, the first two children might have normal pigmentation and the third albinism; the probability of this sequence is $\frac{3}{4} \times \frac{3}{4} \times \frac{1}{4} = \frac{9}{64}$. Because *either* the first sequence *or* the second sequence *or* the third sequence produces one child with albinism and two with normal pigmentation, we apply the addition rule and add the probabilities: $\frac{9}{64} + \frac{9}{64} + \frac{9}{64} = \frac{27}{64}$.

If we want to know the probability of this couple having five children, two with albinism and three with normal pigmentation, figuring out *all* the different combinations of children and their probabilities becomes more difficult. This task is made easier if we apply the binomial expansion.

The binomial takes the form $(p + q)^n$, where p equals the probability of one event, q equals the probability of the alternative event, and n equals the number of times the event occurs. For figuring the probability of two out of five children with albinism:

p = the probability of a child having albinism ($\frac{1}{4}$)

q = the probability of a child having normal pigmentation ($\frac{3}{4}$)

The binomial for this situation is $(p + q)^5$ because there are five children in the family ($n = 5$). The expansion is:

$$(p + q)^5 = p^5 + 5p^4q + 10p^3q^2 + 10p^2q^3 + 5pq^4 + q^5$$

The first term in the expansion (p^5) equals the probability of having five children all with albinism, because p is the probability of albinism. The second term ($5p^4q$) equals the probability of having four children with albinism and one with normal pigmentation, the third term ($10p^3q^2$) equals the probability of having three children with albinism and two with normal pigmentation, and so forth.

To obtain the probability of any combination of events, we insert the values of p and q; so the probability of having two out of five children with albinism is:

$$10p^2q^3 = 10(\tfrac{1}{4})^2(\tfrac{3}{4})^3 = \tfrac{270}{1024} = 0.26$$

We could easily figure out the probability of any desired combination of albinism and pigmentation among five children by using the other terms in the expansion.

How did we expand the binomial in this example? In general, the expansion of any binomial $(p + q)^n$ consists of a series of $n + 1$ terms. In the preceding example, $n = 5$; so there are $5 + 1 = 6$ terms: p^5, $5p^4q$, $10p^3q^2$, $10p^2q^3$, $5pq^4$, and q^5. To write out the terms, first figure out their exponents. The exponent of p in the first term always begins with the power to which the binomial is raised, or n. In our example, n equals 5, so our first term is p^5. The exponent of p decreases by one in each successive term; so the exponent of p is 4 in the second term (p^4), 3 in the third term (p^3), and so forth. The exponent of q is 0 (no q) in the first term and increases by 1 in each successive term, increasing from 0 to 5 in our example.

Next, determine the coefficient of each term. The coefficient of the first term is always 1; so, in our example, the first term is $1p^5$, or just p^5. The coefficient of the second term is always the same as the power to which the binomial is raised; in our example, this coefficient is 5 and the term is $5p^4q$. For the coefficient of the third term, look back at the preceding term; multiply the coefficient of the preceding term (5 in our example) by the exponent of p in that term (4) and then divide by the number of that term (second term, or 2). So the coefficient of the third term in our example is $(5 \times 4)/2 = 20/2 = 10$ and the term is $10p^3q^2$. Follow this procedure for each successive term.

Another way to determine the probability of any particular combination of events is to use the following formula:

$$P = \frac{n!}{s!t!} p^s q^t$$

where P equals the overall probability of event X with probability p occurring s times and event Y with probability q occurring t times. For our albinism example, event X would be the occurrence of a child with albinism ($\frac{1}{4}$) and event Y the occurrence of a child with normal pigmentation ($\frac{3}{4}$); s would equal the number of children with albinism (2) and t the number of children with normal pigmentation (3). The ! symbol stands for factorial, and it means the product of all the integers from n to 1. In this example, $n = 5$; so $n! = 5 \times 4 \times 3 \times 2 \times 1$. Applying this formula to obtain the probability of two out of five children having albinism, we obtain:

$$P = \frac{5!}{2!3!} (\tfrac{1}{4})^2(\tfrac{3}{4})^3$$

$$= \frac{5 \times 4 \times 3 \times 2 \times 1}{2 \times 1 \times 3 \times 2 \times 1} (\tfrac{1}{4})^2(\tfrac{3}{4})^3 = 0.26$$

This value is the same as that obtained with the binomial expansion.

The Testcross

A useful tool for analyzing genetic crosses is the **testcross,** in which one individual of unknown genotype is crossed with another individual with a homozygous recessive genotype for the trait in question. Figure 3.6 illustrates a testcross (as well as a backcross). A testcross tests, or reveals, the genotype of the first individual.

Suppose you were given a tall pea plant with no information about its parents. Because tallness is a dominant trait in peas, your plant could be either homozygous (TT) or heterozygous (Tt), but you would not know which. You could determine its genotype by performing a testcross. If the plant were homozygous (TT), a testcross would produce all tall progeny ($TT \times tt \rightarrow$ all Tt); if the plant were heterozygous (Tt), the testcross would produce half tall progeny and half short progeny ($Tt \times tt \rightarrow \frac{1}{2}\ Tt$ and $\frac{1}{2}\ tt$). When a testcross is performed, any recessive allele in the unknown genotype is expressed in the progeny, because it will be paired with a recessive allele from the homozygous recessive parent.

Incomplete Dominance

We have seen that, in a cross between two individuals heterozygous for a dominant trait ($Aa \times Aa$), the offspring have a $\frac{3}{4}$ probability of exhibiting the dominant trait and a $\frac{1}{4}$ probability of exhibiting the recessive trait. We also examined the outcome of a cross between an individual heterozygous for a dominant trait and an individual homozygous for a recessive trait ($Aa \times aa$); in this case, $\frac{1}{2}$ of the offspring exhibit the dominant trait and $\frac{1}{2}$ exhibit the recessive trait. Later in the chapter, we will see how probabilities for such individual traits can be combined to determine the overall probability of offspring with combinations of two or more different traits. But, before exploring the inheritance of multiple traits, we must consider an additional phenotypic ratio that is obtained when dominance is lacking.

All of the seven characters in pea plants that Mendel chose to study extensively exhibited dominance and produced a $3:1$ phenotypic ratio in the progeny. However, Mendel did realize that not all characters have traits that exhibit dominance. He conducted some crosses concerning the length of time that pea plants take to flower. When he crossed two homozygous varieties that differed in their flowering time by an average of 20 days, the length of time taken by the F_1 plants to flower was intermediate between those of the two parents. When the heterozygote has a phenotype intermediate between the phenotypes of the two homozygotes, the trait is said to display **incomplete dominance.**

Incomplete dominance is also exhibited in the fruit color of eggplants. When a homozygous plant that produces purple fruit (PP) is crossed with a homozygous plant that produces white fruit (pp), all the heterozygous F_1 (Pp) produce violet fruit (**Figure 3.9a**). When the F_1 are crossed with each other, $\frac{1}{4}$ of the F_2 are purple (PP), $\frac{1}{2}$ are violet (Pp), and $\frac{1}{4}$ are white (pp), as shown in **Figure 3.9b**. This $1:2:1$ ratio is different from the $3:1$ ratio that we would observe if eggplant fruit color exhibited dominance. When a trait displays incomplete dominance, the genotypic ratios and phenotypic ratios of the offspring are the *same,* because each genotype has its own phenotype. It is impossible to obtain eggplants that are pure breeding for violet fruit, because all plants with violet fruit are heterozygous.

Another example of incomplete dominance is feather color in chickens. A cross between a homozygous black

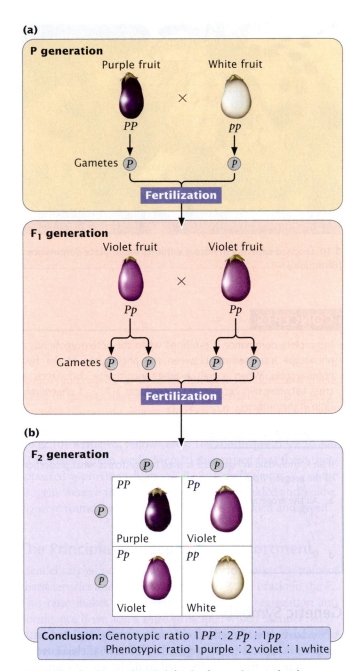

(a)

P generation

Purple fruit   ×   White fruit

PP pp

Gametes P p

Fertilization

F_1 generation

Violet fruit   ×   Violet fruit

Pp Pp

Gametes P p P p

Fertilization

(b)

F_2 generation

	P	p
P	PP Purple	Pp Violet
p	Pp Violet	pp White

Conclusion: Genotypic ratio $1\,PP : 2\,Pp : 1\,pp$
Phenotypic ratio 1 purple $: 2$ violet $: 1$ white

3.9 Fruit color in eggplant is inherited as an incompletely dominant characteristic.

chicken and a homozygous white chicken produces F_1 chickens that are gray. If these gray F_1 are intercrossed, they produce F_2 birds in a ratio of 1 black : 2 gray : 1 white. Leopard white spotting in horses is incompletely dominant over unspotted horses: LL horses are white with numerous dark spots, heterozygous Ll horses have fewer spots, and ll horses have no spots (**Figure 3.10**). The concept of dominance and some of its variations are discussed further in Chapter 5.

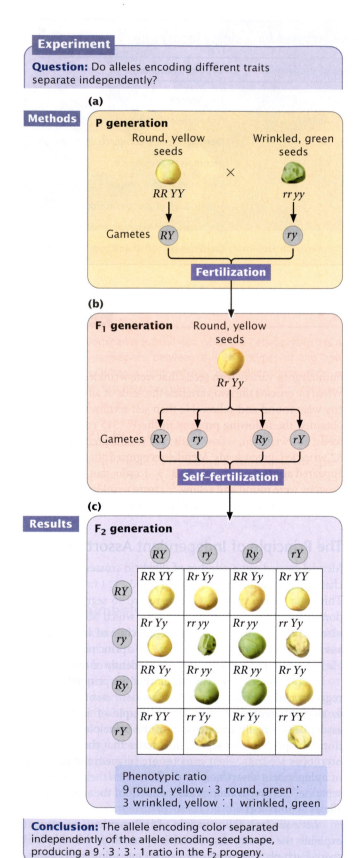

Question: Do alleles encoding different traits separate independently?

(a)

Methods

P generation

Round, yellow seeds   Wrinkled, green seeds

RR YY   ×   *rr yy*

Gametes *RY*   *ry*

Fertilization

(b)

F₁ generation Round, yellow seeds

Rr Yy

Gametes *RY*   *ry*   *Ry*   *rY*

Self-fertilization

(c)

Results

F₂ generation

	RY	*ry*	*Ry*	*rY*
RY	*RR YY*	*Rr Yy*	*RR Yy*	*Rr YY*
ry	*Rr Yy*	*rr yy*	*Rr yy*	*rr Yy*
Ry	*RR Yy*	*Rr yy*	*RR yy*	*Rr Yy*
rY	*Rr YY*	*rr Yy*	*Rr Yy*	*rr YY*

Phenotypic ratio
9 round, yellow : 3 round, green :
3 wrinkled, yellow : 1 wrinkled, green

Conclusion: The allele encoding color separated independently of the allele encoding seed shape, producing a 9 : 3 : 3 : 1 ratio in the F₂ progeny.

3.11 Mendel's dihybrid crosses revealed the principle of independent assortment.

allele for each locus passes to each gamete. The gametes produced by the round, yellow parent therefore contain alleles *RY*, whereas the gametes produced by the wrinkled, green parent contain alleles *ry*. These two types of gametes unite to produce the F₁, all with genotype *Rr Yy*. Because round is dominant over wrinkled and yellow is dominant over green, the phenotype of the F₁ will be round and yellow.

When Mendel self-fertilized the F₁ plants to produce the F₂, the alleles for each locus separated, with one allele going into each gamete. This event is where the principle of independent assortment becomes important. Each pair of alleles can separate in two ways: (1) *R* separates with *Y* and *r* separates with *y* to produce gametes *RY* and *ry* or (2) *R* separates with *y* and *r* separates with *Y* to produce gametes *Ry* and *rY*. The principle of independent assortment tells us that the alleles at each locus separate independently; thus, both kinds of separation occur equally and all four type of gametes (*RY*, *ry*, *Ry*, and *rY*) are produced in equal proportions (**Figure 3.11b**). When these four types of gametes are combined to produce the F₂ generation, the progeny consist of ⁹⁄₁₆ round and yellow, ³⁄₁₆ wrinkled and yellow, ³⁄₁₆ round and green, and ¹⁄₁₆ wrinkled and green, resulting in a 9 : 3 : 3 : 1 phenotypic ratio (**Figure 3.11c**).

Relating the Principle of Independent Assortment to Meiosis

An important qualification of the principle of independent assortment is that it applies to characters encoded by loci located on different chromosomes because, like the principle of segregation, it is based wholly on the behavior of chromosomes during meiosis. Each pair of homologous chromosomes separates independently of all other pairs in anaphase I of meiosis (see Figure 2.17); so genes located on different pairs of homologs will assort independently. Genes that happen to be located on the same chromosome will travel together during anaphase I of meiosis and will arrive at the same destination—within the same gamete (unless crossing over takes place). Genes located on the same chromosome therefore do not assort independently (unless they are located sufficiently far apart that crossing over takes place every meiotic division, as will be discussed fully in Chapter 7).

CONCEPTS

The principle of independent assortment states that genes encoding different characteristics separate independently of one another when gametes are formed, owing to the independent separation of homologous pairs of chromosomes in meiosis. Genes located close together on the same chromosome do not, however, assort independently.

✔ **CONCEPT CHECK 7**

How are the principles of segregation and independent assortment related and how are they different?

Applying Probability and the Branch Diagram to Dihybrid Crosses

When the genes at two loci separate independently, a dihybrid cross can be understood as two monohybrid crosses. Let's examine Mendel's dihybrid cross ($Rr\,Yy \times Rr\,Yy$) by considering each characteristic separately (**Figure 3.12a**). If we consider only the shape of the seeds, the cross was $Rr \times Rr$, which yields a 3 : 1 phenotypic ratio ($\frac{3}{4}$ round and $\frac{1}{4}$ wrinkled progeny, see Table 3.2). Next consider the other characteristic, the color of the seed. The cross was $Yy \times Yy$, which produces a 3 : 1 phenotypic ratio ($\frac{3}{4}$ yellow and $\frac{1}{4}$ green progeny).

We can now combine these monohybrid ratios by using the multiplication rule to obtain the proportion of progeny with different combinations of seed shape and color. The

proportion of progeny with round and yellow seeds is $\frac{3}{4}$ (the probability of round) $\times \frac{3}{4}$ (the probability of yellow) $= \frac{9}{16}$. The proportion of progeny with round and green seeds is $\frac{3}{4} \times \frac{1}{4} = \frac{3}{16}$; the proportion of progeny with wrinkled and yellow seeds is $\frac{1}{4} \times \frac{3}{4} = \frac{3}{16}$; and the proportion of progeny with wrinkled and green seeds is $\frac{1}{4} \times \frac{1}{4} = \frac{1}{16}$.

Branch diagrams are a convenient way of organizing all the combinations of characteristics (**Figure 3.12b**). In the first column, list the proportions of the phenotypes for one character (here, $\frac{3}{4}$ round and $\frac{1}{4}$ wrinkled). In the second column, list the proportions of the phenotypes for the second character ($\frac{3}{4}$ yellow and $\frac{1}{4}$ green) twice, next to each of the phenotypes in the first column: put $\frac{3}{4}$ yellow and $\frac{1}{4}$ green next to the round phenotype and again next to the wrinkled phenotype. Draw lines between the phenotypes in the first column and each of the phenotypes in the second column. Now follow each branch of the diagram, multiplying the probabilities for each trait along that branch. One branch leads from round to yellow, yielding round and yellow progeny. Another branch leads from round to green, yielding round and green progeny, and so forth. We calculate the probability of progeny with a particular combination of traits by using the multiplication rule: the probability of round ($\frac{3}{4}$) and yellow ($\frac{3}{4}$) seeds is $\frac{3}{4} \times \frac{3}{4} = \frac{9}{16}$. The advantage of the branch diagram is that it helps keep track of all the potential combinations of traits that may appear in the progeny. It can be used to determine phenotypic or genotypic ratios for any number of characteristics.

Using probability is much faster than using the Punnett square for crosses that include multiple loci. Genotypic and phenotypic ratios can be quickly worked out by combining, with the multiplication rule, the simple ratios in Tables 3.2 and 3.3. The probability method is particularly efficient if we need the probability of only a *particular* phenotype or genotype among the progeny of a cross. Suppose we needed to know the probability of obtaining the genotype $Rr\,yy$ in the F_2 of the dihybrid cross in Figure 3.11. The probability of obtaining the Rr genotype in a cross of $Rr \times Rr$ is $\frac{1}{2}$ and that of obtaining yy progeny in a cross of $Yy \times Yy$ is $\frac{1}{4}$ (see Table 3.3). Using the multiplication rule, we find the probability of $Rr\,yy$ to be $\frac{1}{2} \times \frac{1}{4} = \frac{1}{8}$.

To illustrate the advantage of the probability method, consider the cross $Aa\,Bb\,cc\,Dd\,Ee \times Aa\,Bb\,Cc\,dd\,Ee$. Suppose we wanted to know the probability of obtaining offspring with the genotype $aa\,bb\,cc\,dd\,ee$. If we used a Punnett square to determine this probability, we might be working on the solution for months. However, we can quickly figure the probability of obtaining this one genotype by breaking this cross into a series of single-locus crosses:

Progeny cross	Genotype	Probability
$Aa \times Aa$	aa	$\frac{1}{4}$
$Bb \times Bb$	bb	$\frac{1}{4}$
$cc \times Cc$	cc	$\frac{1}{2}$
$Dd \times dd$	dd	$\frac{1}{2}$
$Ee \times Ee$	ee	$\frac{1}{4}$

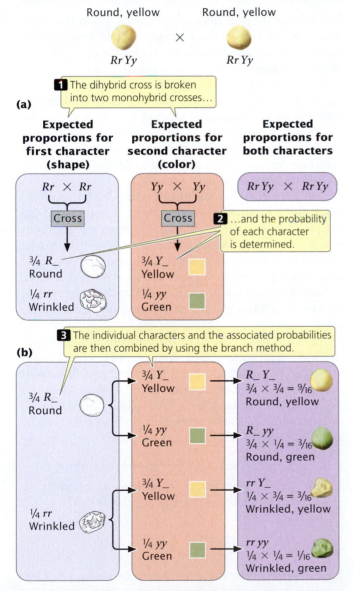

3.12 A branch diagram can be used to determine the phenotypes and expected proportions of offspring from a dihybrid cross ($Rr\,Yy \times Rr\,Yy$).

The probability of an offspring from this cross having genotype *aa bb cc dd ee* is now easily obtained by using the multiplication rule: $\frac{1}{4} \times \frac{1}{4} \times \frac{1}{2} \times \frac{1}{2} \times \frac{1}{4} = \frac{1}{256}$. This calculation assumes that genes at these five loci all assort independently.

CONCEPTS

A cross including several characteristics can be worked by breaking the cross down into single-locus crosses and using the multiplication rule to determine the proportions of combinations of characteristics (provided the genes assort independently).

The Dihybrid Testcross

Let's practice using the branch diagram by determining the types and proportions of phenotypes in a dihybrid testcross

Round, yellow   Wrinkled, green

×

Rr Yy   *rr yy*

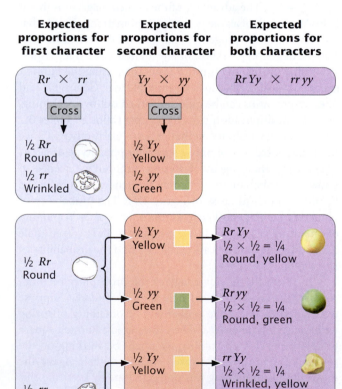

3.13 A branch diagram can be used to determine the phenotypes and expected proportions of offspring from a dihybrid testcross (*Rr Yy* × *rr yy*).

between the round and yellow F_1 plants (*Rr Yy*) that Mendel obtained in his dihybrid cross and the wrinkled and green plants (*rr yy*) (**Figure 3.13**). Break the cross down into a series of single-locus crosses. The cross *Rr* × *rr* yields $\frac{1}{2}$ round (*Rr*) progeny and $\frac{1}{2}$ wrinkled (*rr*) progeny. The cross *Yy* × *yy* yields $\frac{1}{2}$ yellow (*Yy*) progeny and $\frac{1}{2}$ green (*yy*) progeny. Using the multiplication rule, we find the proportion of round and yellow progeny to be $\frac{1}{2}$ (the probability of round) × $\frac{1}{2}$ (the probability of yellow) = $\frac{1}{4}$. Four combinations of traits with the following proportions appear in the offspring: $\frac{1}{4}$ *Rr Yy*, round yellow; $\frac{1}{4}$ *Rr yy*, round green; $\frac{1}{4}$ *rr Yy*, wrinkled yellow; and $\frac{1}{4}$ *rr yy*, wrinkled green.

Worked Problem

Not only are the principles of segregation and independent assortment important because they explain how heredity works, but they also provide the means for predicting the outcome of genetic crosses. This predictive power has made genetics a powerful tool in agriculture and other fields, and the ability to apply the principles of heredity is an important skill for all students of genetics. Practice with genetic problems is essential for mastering the basic principles of heredity—no amount of reading and memorization can substitute for the experience gained by deriving solutions to specific problems in genetics.

Students may have difficulty with genetics problems when they are unsure of where to begin or how to organize the problem and plan a solution. In genetics, every problem is different, and so no common series of steps can be applied to all genetics problems. Logic and common sense must be used to analyze a problem and arrive at a solution. Nevertheless, certain steps can facilitate the process, and solving the following problem will serve to illustrate these steps.

In mice, black coat color (*B*) is dominant over brown (*b*), and a solid pattern (*S*) is dominant over white spotted (*s*). Color and spotting are controlled by genes that assort independently. A homozygous black, spotted mouse is crossed with a homozygous brown, solid mouse. All the F_1 mice are black and solid. A testcross is then carried out by mating the F_1 mice with brown, spotted mice.

a. Give the genotypes of the parents and the F_1 mice.

b. Give the genotypes and phenotypes, along with their expected ratios, of the progeny expected from the testcross.

• **Solution**

Step 1:   Determine the questions to be answered. What question or questions is the problem asking? Is it asking for genotypes, genotypic ratios, or phenotypic ratios? This problem asks you to provide the *genotypes* of the parents and the F_1, the *expected genotypes* and *phenotypes* of the progeny of the testcross, and their *expected proportions*.

Step 2:   Write down the basic information given in the problem. This problem provides important information about the dominance relations of the characters and about the mice being crossed. Black is dominant over brown, and solid is dominant over white spotted. Furthermore, the genes for the two characters assort independently. In this problem, symbols are provided for the different alleles (B for black, b for brown, S for solid, and s for spotted); had these symbols not been provided, you would need to choose symbols to represent these alleles. It is useful to record these symbols at the beginning of the solution:

B — black S — solid
b — brown s — white spotted

Next, write out the crosses given in the problem.

P   Homozygous   ×   Homozygous
           black, spotted   brown, solid

↓

F_1 Black, solid

Testcross   Black, solid   ×   Brown, spotted

Step 3:   Write down any genetic information that can be determined from the phenotypes alone. From the phenotypes and the statement that they are homozygous, you know that the P-generation mice must be $BB\,ss$ and $bb\,SS$. The F_1 mice are black and solid, both dominant traits, and so the F_1 mice must possess at least one black allele (B) and one solid allele (S). At this point, you cannot be certain about the other alleles; so represent the genotype of the F_1 as $B?\,S?$. The brown, spotted mice in the testcross must be $bb\,ss$, because both brown and spotted are recessive traits that will be expressed only if two recessive alleles are present. Record these genotypes on the crosses that you wrote out in step 2:

P   Homozygous   ×   Homozygous
           black, spotted   brown, solid
           $BB\,ss$ $bb\,SS$

↓

F_1 Black, solid
                              $B?\,S?$

Testcross   Black, solid   ×   Brown, spotted
                   $B?\,S?$ $bb\,ss$

Step 4:   Break the problem down into smaller parts. First, determine the genotype of the F_1. After this genotype has been determined, you can predict the results of the testcross and determine the genotypes and phenotypes of the progeny from the testcross. Second, because this cross includes two independently assorting loci, it can be conveniently broken down into two single-locus crosses: one for coat color and the other for spotting. Third, use a branch diagram to determine the proportion of progeny of the testcross with different combinations of the two traits.

Step 5:   Work the different parts of the problem. Start by determining the genotype of the F_1 progeny. Mendel's first law indicates that the two alleles at a locus separate, one going into each gamete. Thus, the gametes produced by the black, spotted parent contain $B\,s$ and the gametes produced by the brown, solid parent contain $b\,S$, which combine to produce F_1 progeny with the genotype $Bb\,Ss$:

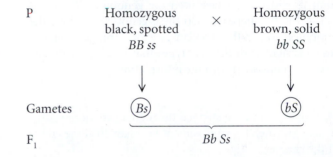

P   Homozygous   ×   Homozygous
           black, spotted   brown, solid
           $BB\,ss$ $bb\,SS$

Gametes Bs bS

F_1 $Bb\,Ss$

Use the F_1 genotype to work the testcross ($Bb\,Ss \times bb\,ss$), breaking it into two single-locus crosses. First, consider the cross for coat color: $Bb \times bb$. Any cross between a heterozygote and a homozygous recessive genotype produces a $1:1$ phenotypic ratio of progeny (see Table 3.2):

$Bb \times bb$

↓

$\frac{1}{2}\,Bb$ black
$\frac{1}{2}\,bb$ brown

Next, do the cross for spotting: $Ss \times ss$. This cross also is between a heterozygote and a homozygous recessive genotype and will produce $\frac{1}{2}$ solid (Ss) and $\frac{1}{2}$ spotted (ss) progeny (see Table 3.2).

$Ss \times ss$

↓

$\frac{1}{2}\,Ss$ solid
$\frac{1}{2}\,ss$ spotted

Finally, determine the proportions of progeny with combinations of these characters by using the branch diagram.

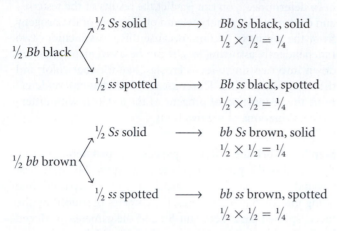

Step 6:   Check all work. As a last step, reread the problem, checking to see if your answers are consistent with the information provided. You have used the genotypes *BB ss* and *bb SS* in the P generation. Do these genotypes encode the phenotypes given in the problem? Are the F_1 progeny phenotypes consistent with the genotypes that you assigned? The answers are consistent with the information.

> Now that we have stepped through a genetics problem together, try your hand at Problem 30 at the end of the chapter.

3.4 Observed Ratios of Progeny May Deviate from Expected Ratios by Chance

When two individual organisms of known genotype are crossed, we expect certain ratios of genotypes and phenotypes in the progeny; these expected ratios are based on the Mendelian principles of segregation, independent assortment, and dominance. The ratios of genotypes and phenotypes *actually* observed among the progeny, however, may deviate from these expectations.

For example, in German cockroaches, brown body color (*Y*) is dominant over yellow body color (*y*). If we cross a brown, heterozygous cockroach (*Yy*) with a yellow cockroach (*yy*), we expect a 1 : 1 ratio of brown (*Yy*) and yellow (*yy*) progeny. Among 40 progeny, we would therefore expect to see 20 brown and 20 yellow offspring. However, the observed numbers might deviate from these expected values; we might in fact see 22 brown and 18 yellow progeny.

Chance plays a critical role in genetic crosses, just as it does in flipping a coin. When you flip a coin, you expect a 1 : 1 ratio—½ heads and ½ tails. If you flip a coin 1000 times, the proportion of heads and tails obtained would probably be very close to that expected 1 : 1 ratio. However, if you flip the coin 10 times, the ratio of heads to tails might be quite different from 1 : 1. You could easily get 6 heads and 4 tails, or 3 heads and 7 tails, just by chance. You might even get 10 heads and 0 tails. The same thing happens in genetic crosses. We may expect 20 brown and 20 yellow cockroaches, but 22 brown and 18 yellow progeny *could* arise as a result of chance.

The Goodness-of-Fit Chi-Square Test

If you expected a 1 : 1 ratio of brown and yellow cockroaches but the cross produced 22 brown and 18 yellow, you probably wouldn't be too surprised even though it wasn't a perfect 1 : 1 ratio. In this case, it seems reasonable to assume that chance produced the deviation between the expected and the observed results. But, if you observed 25 brown and 15 yellow, would the ratio still be 1 : 1? Something other than chance might have caused the deviation. Perhaps the inheritance of this character is more complicated than was assumed or perhaps some of the yellow progeny died before they were counted. Clearly, we need some means of evaluating how likely it is that chance is responsible for the deviation between the observed and the expected numbers.

To evaluate the role of chance in producing deviations between observed and expected values, a statistical test called the **goodness-of-fit chi-square test** is used. This test provides information about how well observed values fit expected values. Before we learn how to calculate the chi square, it is important to understand what this test does and does not indicate about a genetic cross.

The chi-square test cannot tell us whether a genetic cross has been correctly carried out, whether the results are correct, or whether we have chosen the correct genetic explanation for the results. What it does indicate is the *probability* that the difference between the observed and the expected values is due to chance. In other words, it indicates the likelihood that chance alone could produce the deviation between the expected and the observed values.

If we expected 20 brown and 20 yellow progeny from a genetic cross, the chi-square test gives the probability that we might observe 25 brown and 15 yellow progeny simply owing to chance deviations from the expected 20 : 20 ratio. This hypothesis, that chance alone is responsible for any deviations between observed and expected values, is sometimes called the null hypothesis. When the probability calculated from the chi-square test is high, we assume that chance alone produced the difference (the null hypothesis is true). When the probability is low, we assume that some factor other than chance—some significant factor—produced the deviation (the null hypothesis is false).

To use the goodness-of-fit chi-square test, we first determine the expected results. The chi-square test must always be applied to numbers of progeny, not to proportions or percentages. Let's consider a locus for coat color in domestic cats, for which black color (*B*) is dominant over gray (*b*). If we crossed two heterozygous black cats (*Bb* × *Bb*), we would expect a 3 : 1 ratio of black and gray kittens. A series of such crosses yields a total of 50 kittens—30 black and 20 gray. These numbers are our *observed* values. We can obtain the expected numbers by multiplying the *expected* proportions by the total number of observed progeny. In this case, the expected number of black kittens is $\frac{3}{4} \times 50 = 37.5$ and the expected number of gray kittens is $\frac{1}{4} \times 50 = 12.5$. The chi-square ($\chi^2$) value is calculated by using the following formula:

$$\chi^2 = \Sigma \frac{(\text{observed} - \text{expected})^2}{\text{expected}}$$

where Σ means the sum. We calculate the sum of all the squared differences between observed and expected and divide by the expected values. To calculate the chi-square value for our black and gray kittens, we would first subtract the number of *expected* black kittens from the number of *observed* black kittens ($30 - 37.5 = -7.5$) and square this value: $-7.5^2 = 56.25$. We then divide this result by the expected number of black kittens, $56.25/37.5 = 1.5$. We repeat the calculations on the number of expected gray kittens:

$(20 - 12.5)^2/12.5 = 4.5$. To obtain the overall chi-square value, we sum the (observed − expected)^2/expected values: $1.5 + 4.5 = 6.0$.

The next step is to determine the probability associated with this calculated chi-square value, which is the probability that the deviation between the observed and the expected results could be due to chance. This step requires us to compare the calculated chi-square value (6.0) with theoretical values that have the same degrees of freedom in a chi-square table. The degrees of freedom represent the number of ways in which the expected classes are free to vary. For a goodness-of-fit chi-square test, the degrees of freedom are equal to $n - 1$, where n is the number of different expected phenotypes. In our example, there are two expected phenotypes (black and gray); so $n = 2$, and the degree of freedom equals $2 - 1 = 1$.

Now that we have our calculated chi-square value and have figured out the associated degrees of freedom, we are ready to obtain the probability from a chi-square table (**Table 3.4**). The degrees of freedom are given in the left-hand column of the table and the probabilities are given at the top; within the body of the table are chi-square values associated with these probabilities. First, find the row for the appropriate degrees of freedom; for our example with 1 degree of freedom, it is the first row of the table. Find

Table 3.4   Critical values of the χ^2 distribution

df					P				
	0.995	0.975	0.9	0.5	0.1	0.05	0.025	0.01	0.005
1	0.000	0.000	0.016	0.455	2.706	3.841	5.024	6.635	7.879
2	0.010	0.051	0.211	1.386	4.605	5.991	7.378	9.210	10.597
3	0.072	0.216	0.584	2.366	6.251	7.815	9.348	11.345	12.838
4	0.207	0.484	1.064	3.357	7.779	9.488	11.143	13.277	14.860
5	0.412	0.831	1.610	4.351	9.236	11.070	12.832	15.086	16.750
6	0.676	1.237	2.204	5.348	10.645	12.592	14.449	16.812	18.548
7	0.989	1.690	2.833	6.346	12.017	14.067	16.013	18.475	20.278
8	1.344	2.180	3.490	7.344	13.362	15.507	17.535	20.090	21.955
9	1.735	2.700	4.168	8.343	14.684	16.919	19.023	21.666	23.589
10	2.156	3.247	4.865	9.342	15.987	18.307	20.483	23.209	25.188
11	2.603	3.816	5.578	10.341	17.275	19.675	21.920	24.725	26.757
12	3.074	4.404	6.304	11.340	18.549	21.026	23.337	26.217	28.300
13	3.565	5.009	7.042	12.340	19.812	22.362	24.736	27.688	29.819
14	4.075	5.629	7.790	13.339	21.064	23.685	26.119	29.141	31.319
15	4.601	6.262	8.547	14.339	22.307	24.996	27.488	30.578	32.801

P, probability; df, degrees of freedom.

4.1 The sex chromosomes of males (Y, at the left) and females (X, at the right) differ in size and shape. [Biophoto Associates/Photo Researchers.]

male, with two Z chromosomes, but they develop female traits and reproduce as fully functional females.

Why do the *Wolbachia* bacteria go to the trouble of converting male isopods into females? The answer lies in the way in which the bacteria are transmitted. *Wolbachia* are found in the cytoplasm of isopod cells and are transmitted from one isopod to another strictly through the cytoplasm of an isopod egg; because sperm contain little or no cytoplasm, the bacteria are not transmitted by males. Thus, bacteria that end up inside a male isopod are at a dead end. Natural selection, acting on the bacteria, favors any trait that gets the bacteria into a female and has led to the bacteria's ability to convert males into females. Natural selection acting on the isopod, however, favors a 50 : 50 sex ratio (as stated earlier) and, in an evolutionary sense, the isopods have fought back by evolving an autosomal dominant gene that can sometimes override the feminizing effects of the bacteria. In this fascinating situation, the isopods and bacteria are locked in an evolutionary tug of war, with the effect of selection on the bacteria favoring more female isopods and that on the isopods favoring equal numbers of males and females. Sex in rolly-pollies is just one example of the varied ways in which sex is determined and influences inheritance.

In Chapter 3, we studied Mendel's principles of segregation and independent assortment and saw how these principles explain much about the nature of inheritance. After Mendel's principles were rediscovered in 1900, biologists began to conduct genetic studies on a wide array of different organisms. As they applied Mendel's principles more widely, exceptions were observed, and it became necessary to devise extensions to his basic principles of heredity. In this chapter, we explore one of the major extensions to Mendel's principles: the inheritance of characteristics encoded by genes located on the sex chromosomes, which often differ in males and females (**Figure 4.1**). These characteristics and the genes that produce them are referred to as sex linked. To understand the inheritance of sex-linked characteristics, we must first know how sex is determined—why some members of a species are male and others are female. Sex determination is the focus of the first part of the chapter. The second part examines how characteristics encoded by genes on the sex chromosomes are inherited. In Chapter 5, we will explore some additional ways in which sex and inheritance interact.

As we consider sex determination and sex-linked characteristics, it will be helpful to think about two important principles. First, there are several different mechanisms of sex determination and, ultimately, the mechanism of sex determination controls the inheritance of sex-linked characteristics. Second, like other pairs of chromosomes, the X and Y sex chromosomes may pair in the course of meiosis and segregate, but, throughout most of their length, they are not homologous (their gene sequences don't encode the same characteristics): most genes on the X chromosome are different from genes on the Y chromosome. Consequently, males and females do not possess the same number of alleles at sex-linked loci. This difference in the number of sex-linked alleles produces distinct patterns of inheritance in males and females.

4.1 Sex Is Determined by a Number of Different Mechanisms

Sexual reproduction is the formation of offspring that are genetically distinct from their parents; most often, two parents contribute genes to their offspring and the genes are assorted into new combinations through meiosis. Among most eukaryotes, sexual reproduction consists of two processes that lead to an alternation of haploid and diploid cells: meiosis produces haploid gametes (or spores in plants), and fertilization produces diploid zygotes (**Figure 4.2**).

The term **sex** refers to sexual phenotype. Most organisms have only two sexual phenotypes: male and female. The fundamental difference between males and females is gamete size: males produce small gametes; females produce relatively larger gametes (**Figure 4.3**).

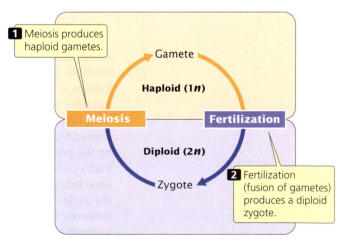

1 Meiosis produces haploid gametes.

Gamete

Haploid (1*n*)

Meiosis

Fertilization

Diploid (2*n*)

Zygote

2 Fertilization (fusion of gametes) produces a diploid zygote.

4.2 In most eukaryotic organisms, sexual reproduction consists of an alternation of haploid (1*n*) and diploid (2*n*) cells.

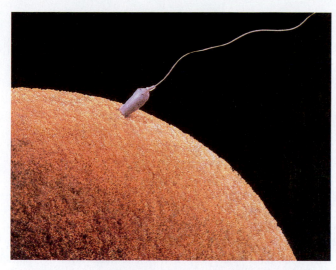

4.3 Male and female gametes (sperm and egg, respectively) differ in size. In this photograph, a human sperm (with flagellum) penetrates a human egg cell. *[Francis Leroy, Biocosmos/Science Photo Library/Photo Researchers.]*

The mechanism by which sex is established is termed **sex determination.** We define the sex of an individual organism in reference to its phenotype. Sometimes an individual organism has chromosomes or genes that are normally associated with one sex but a morphology corresponding to the opposite sex. For instance, the cells of female humans normally have two X chromosomes, and the cells of males have one X chromosome and one Y chromosome. A few rare persons have male anatomy, although their cells each contain two X chromosomes. Even though these people are genetically female, we refer to them as male because their sexual phenotype is male. (As we will see later in the chapter, these XX males usually have a small piece of the Y chromosome, which is attached to another chromosome.)

CONCEPTS

In sexual reproduction, parents contribute genes to produce an offspring that is genetically distinct from both parents. In most eukaryotes, sexual reproduction consists of meiosis, which produces haploid gametes (or spores), and fertilization, which produces a diploid zygote.

✔ CONCEPT CHECK 1

What process causes the genetic variation seen in offspring produced by sexual reproduction?

There are many ways in which sex differences arise. In some species, both sexes are present in the same organism, a condition termed **hermaphroditism;** organisms that bear both male and female reproductive structures are said to be **monoecious** (meaning "one house"). Species in which the organism has either male or female reproductive structures

are said to be **dioecious** (meaning "two houses"). Humans are dioecious. Among dioecious species, sex may be determined chromosomally, genetically, or environmentally.

Chromosomal Sex-Determining Systems

The chromosome theory of inheritance (discussed in Chapter 3) states that genes are located on chromosomes, which serve as the vehicles for gene segregation in meiosis. Definitive proof of this theory was provided by the discovery that the sex of certain insects is determined by the presence or absence of particular chromosomes.

In 1891, Hermann Henking noticed a peculiar structure in the nuclei of cells from male insects. Understanding neither its function nor its relation to sex, he called this structure the X body. Later, Clarence E. McClung studied the X body in grasshoppers and recognized that it was a chromosome. McClung called it the accessory chromosome, but eventually it became known as the X chromosome, from Henking's original designation. McClung observed that the cells of female grasshoppers had one more chromosome than the number of chromosomes in cells of male grasshoppers, and he concluded that accessory chromosomes played a role in sex determination. In 1905, Nettie Stevens and Edmund Wilson demonstrated that, in grasshoppers and other insects, the cells of females have two X chromosomes, whereas the cells of males have a single X. In some insects, they counted the same number of chromosomes in cells of males and females but saw that one chromosome pair was different: two X chromosomes were found in female cells, whereas a single X chromosome plus a smaller chromosome, which they called Y, was found in male cells.

Stevens and Wilson also showed that the X and Y chromosomes separate into different cells in sperm formation; half of the sperm receive an X chromosome and half receive a Y. All egg cells produced by the female in meiosis receive one X chromosome. A sperm containing a Y chromosome unites with an X-bearing egg to produce an XY male, whereas a sperm containing an X chromosome unites with an X-bearing egg to produce an XX female. This accounts for the 50 : 50 sex ratio observed in most dioecious organisms (**Figure 4.4**). Because sex is inherited like other genetically determined characteristics, Stevens and Wilson's discovery that sex is associated with the inheritance of a particular chromosome also demonstrated that genes are on chromosomes.

As Stevens and Wilson found for insects, sex in many organisms is determined by a pair of chromosomes, the **sex chromosomes,** which differ between males and females. The nonsex chromosomes, which are the same for males and females, are called **autosomes.** We think of sex in these organisms as being determined by the presence of the sex chromosomes, but, in fact, the individual genes located on the sex chromosomes are usually responsible for the sexual phenotypes.

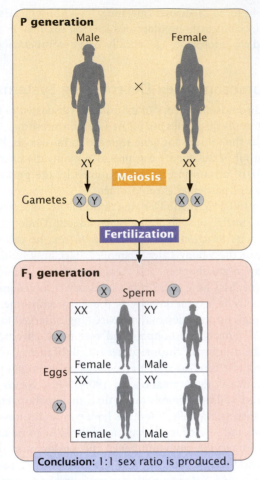

4.4 Inheritance of sex in organisms with X and Y chromosomes results in equal numbers of male and female offspring.

XX-XO sex determination   The mechanism of sex determination in the grasshoppers studied by McClung is one of the simplest mechanisms of chromosomal sex determination and is called the XX-XO system. In this system, females have two X chromosomes (XX), and males possess a single X chromosome (XO). There is no O chromosome; the letter O signifies the absence of a sex chromosome.

In meiosis in females, the two X chromosomes pair and then separate, with one X chromosome entering each haploid egg. In males, the single X chromosome segregates in meiosis to half the sperm cells—the other half receive no sex chromosome. Because males produce two different types of gametes with respect to the sex chromosomes, they are said to be the **heterogametic sex.** Females, which produce gametes that are all the same with respect to the sex chromosomes, are the **homogametic sex.** In the XX-XO system, the sex of an individual organism is therefore determined by which type of male gamete fertilizes the egg. X-bearing sperm unite with X-bearing eggs to produce XX zygotes, which eventually develop as females. Sperm lacking an X chromosome unite with X-bearing eggs to produce XO zygotes, which develop into males.

XX-XY sex determination   In many species, the cells of males and females have the same number of chromosomes, but the cells of females have two X chromosomes (XX) and the cells of males have a single X chromosome and a smaller sex chromosome called the Y chromosome (XY). In humans and many other organisms, the Y chromosome is acrocentric (**Figure 4.5**), not Y shaped as is commonly assumed. In this type of sex-determining system, the male is the heterogametic sex—half of his gametes have an X chromosome and half have a Y chromosome. The female is the homogametic sex—all her egg cells contain a single X chromosome. Many organisms, including some plants, insects, and reptiles, and all mammals (including humans), have the XX-XY sex-determining system. Other organisms have odd variations of the XX-XY system of sex determination, including the duck-billed platypus, in which females have five pairs of X chromosomes and males have five pairs of X and Y chromosomes.

Although the X and Y chromosomes are not generally homologous, they do pair and segregate into different cells in meiosis. They can pair because these chromosomes are homologous at small regions called the **pseudoautosomal regions** (see Figure 4.5), in which they carry the same genes. In humans, there are pseudoautosomal regions at both tips of the X and Y chromosomes.

ZZ-ZW sex determination   In this system, the female is heterogametic and the male is homogametic. To prevent confusion with the XX-XY system, the sex chromosomes in this system are labeled Z and W, but the chromosomes do not resemble Zs and Ws. Females in this system are ZW; after meiosis, half of the eggs have a Z chromosome and the other half have a W chromosome. Males are ZZ; all sperm contain a single Z chromosome. The ZZ-ZW system is found in birds, snakes, butterflies, some amphibians, and some fishes. It is also found in some isopods, including those mentioned in the introduction to this chapter.

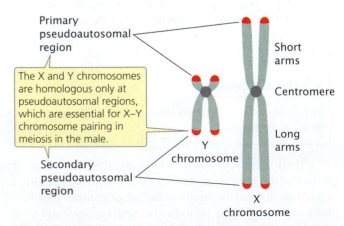

4.5 The X and Y chromosomes in humans differ in size and genetic content. They are homologous only at the pseudoautosomal regions.

The discovery that the presence or absence of particular chromosomes determines sex in insects provided proof that genes are located on chromosomes. In XX-XO sex determination, the male is XO and heterogametic, and the female is XX and homogametic. In XX-XY sex determination, the male is XY and the female is XX; in this system, the male is heterogametic. In ZZ-ZW sex determination, the female is ZW and the male is ZZ; in this system, the female is the heterogametic sex.

✔ CONCEPT CHECK 2

How does the heterogametic sex differ from the homogametic sex?

a. The heterogametic sex is male; the homogametic sex is female.

b. Gametes of the heterogametic sex have different sex chromosomes; gametes of homogametic sex have the same sex chromosome.

c. Gametes of the heterogametic sex all contain a Y chromosome, gametes of the homogametic sex all contain an X chromosome.

Haplodiploidy Some insects in the order Hymenoptera (bees, wasps, and ants) have no sex chromosomes; instead, sex is based on the number of chromosome sets found in the nucleus of each cell. Males develop from unfertilized eggs, and females develop from fertilized eggs. The cells of male hymenopterans possess only a single set of chromosomes (they are haploid) inherited from the mother. In contrast, the cells of females possess two sets of chromosomes (they are diploid), one set inherited from the mother and the other set from the father (**Figure 4.6**).

The haploidiploid method of sex determination produces some odd genetic relationships. When both parents are diploid, siblings on average have half their genes in common because they have a 50% chance of receiving the same allele from each parent. In insects with haplodiploid sex determination, however, males produce sperm by mitosis (they are already haploid), and so all offspring receive the same set of paternal genes. The diploid females produce eggs by normal meiosis. Therefore, sisters have a 50% chance of receiving the same allele from their mother and a 100% chance of receiving the same allele from their father; the average relatedness between sisters is therefore 75%. Brothers have a 50% chance of receiving the same copy of each of their mother's two alleles at any particular locus; so their average relatedness is only 50%. The greater genetic relatedness among female siblings in insects with haplodiploid sex determination may contribute to the high degree of social cooperation that exists among females (the workers) of these insects.

Some insects possess haplodiploid sex determination, in which males develop from unfertilized eggs and are haploid; females develop from fertilized eggs and are diploid.

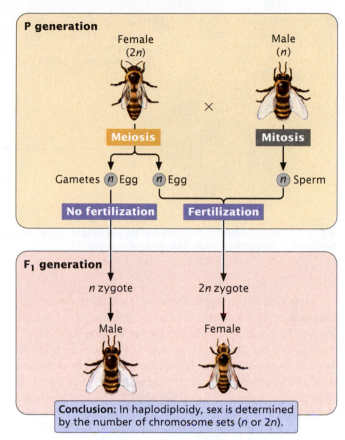

P generation

Female (2*n*)   ×   Male (*n*)

Meiosis   Mitosis

Gametes *n* Egg   *n* Egg   *n* Sperm

No fertilization   Fertilization

F₁ generation

n zygote   2*n* zygote

Male   Female

Conclusion: In haplodiploidy, sex is determined by the number of chromosome sets (*n* or 2*n*).

4.6 In insects with haplodiploidy, males develop from unfertilized eggs and are haploid; females develop from fertilized eggs and are diploid.

Genic Sex-Determining Systems

In some plants and protozoans, sex is genetically determined, but there are no obvious differences in the chromosomes of males and females: there are no sex chromosomes. These organisms have **genic sex determination**; genotypes at one or more loci determine the sex of an individual plant or protozoan.

It is important to understand that, even in chromosomal sex-determining systems, sex is actually determined by individual genes. For example, in mammals, a gene (*SRY*, discussed later in this chapter) located on the Y chromosome determines the male phenotype. In both genic sex determination and chromosomal sex determination, sex is controlled by individual genes; the difference is that, with chromosomal sex determination, the chromosomes that carry those genes look different in males and females.

Environmental Sex Determination

Genes have had a role in all of the examples of sex determination discussed thus far, but sex is determined fully or in part by environmental factors in a number of organisms.

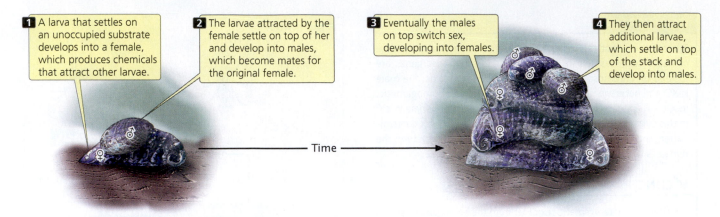

1 A larva that settles on an unoccupied substrate develops into a female, which produces chemicals that attract other larvae.

2 The larvae attracted by the female settle on top of her and develop into males, which become mates for the original female.

3 Eventually the males on top switch sex, developing into females.

4 They then attract additional larvae, which settle on top of the stack and develop into males.

Time

4.7 In *Crepidula fornicata*, the common slipper limpet, sex is determined by an environmental factor—the limpet's position in a stack of limpets.

A fascinating example of environmental sex determination is seen in the marine mollusk *Crepidula fornicata*, also known as the common slipper limpet (**Figure 4.7**). Slipper limpets live in stacks, one on top of another. Each limpet begins life as a swimming larva. The first larva to settle on a solid, unoccupied substrate develops into a female limpet. It then produces chemicals that attract other larvae, which settle on top of it. These larvae develop into males, which then serve as mates for the limpet below. After a period of time, the males on top develop into females and, in turn, attract additional larvae that settle on top of the stack, develop into males, and serve as mates for the limpets under them. Limpets can form stacks of a dozen or more animals; the uppermost animals are always male. This type of sexual development is called **sequential hermaphroditism;** each individual animal can be both male and female, although not at the same time. In *Crepidula fornicata*, sex is determined environmentally by the limpet's position in the stack.

Environmental factors are also important in determining sex in many reptiles. Although most snakes and lizards have sex chromosomes, the sexual phenotype of many turtles, crocodiles, and alligators is affected by temperature during embryonic development. In turtles, for example, warm temperatures produce females during certain times of the year, whereas cool temperatures produce males. In alligators, the reverse is true.

Now that we have surveyed some of the different ways that sex can be determined, we will examine in detail one mechanism (the XX, XY system). Both fruit flies and humans possess XX-XY sex determination but, as we will see, the way in which the X and Y chromosomes determine sex in these two organisms is quite different.

CONCEPTS

In genic sex determination, sex is determined by genes at one or more loci, but there are no obvious differences in the chromosomes of males and females. In environmental sex determination, sex is determined fully or in part by environmental factors.

✔ CONCEPT CHECK 3

How do chromosomal, genic, and environmental sex determination differ?

Sex Determination in *Drosophila melanogaster*

The fruit fly *Drosophila melanogaster* has eight chromosomes: three pairs of autosomes and one pair of sex chromosomes. Thus, it has two haploid sets of autosomes and two sex chromosomes, one set of autosomes and one sex chromosome inherited from each parent. Normally, females have two X chromosomes and males have an X chromosome and a Y chromosome. However, the presence of the Y chromosome does not determine maleness in *Drosophila*; instead, each fly's sex is determined by a balance between genes on the autosomes and genes on the X chromosome. This type of sex determination is called the **genic balance system.** In this system, a number of different genes influence sexual development. The X chromosome contains genes with female-producing effects, whereas the autosomes contain genes with male-producing effects. Consequently, a fly's sex is determined by the **X : A ratio,** the number of X chromosomes divided by the number of haploid sets of autosomal chromosomes.

An X : A ratio of 1.0 produces a female fly; an X : A ratio of 0.5 produces a male. If the X : A ratio is less than 0.5, a male phenotype is produced, but the fly is weak and sterile—such flies are sometimes called metamales. An X : A ratio between 1.0 and 0.5 produces an intersex fly, with a mixture of male and female characteristics. If the X : A ratio is greater than 1.0, a female phenotype is produced, but this fly (called a metafemale) has serious developmental problems and many never complete development. **Table 4.1** presents some different chromosome complements in *Drosophila* and their associated sexual phenotypes. Normal females have two X

Table 4.1	Chromosome complements and sexual phenotypes in *Drosophila*		
Sex-Chromosome Complement	Haploid Sets of Autosomes	X : A Ratio	Sexual Phenotype
XX	AA	1.0	Female
XY	AA	0.5	Male
XO	AA	0.5	Male
XXY	AA	1.0	Female
XXX	AA	1.5	Metafemale
XXXY	AA	1.5	Metafemale
XX	AAA	0.67	Intersex
XO	AAA	0.33	Metamale
XXXX	AAA	1.3	Metafemale

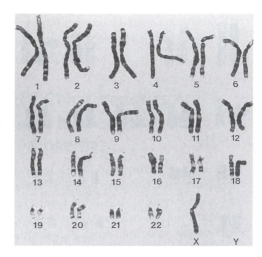

4.8 Persons with Turner syndrome have a single X chromosome in their cells. Chromosomes of a person with Turner syndrome. *[Department of Clinical Cytogenetics, Addenbrookes Hospital/Science Photo Library/Photo Reseachers.]*

chromosomes and two sets of autosomes (XX, AA), and so their X : A ratio is 1.0. Males, on the other hand, normally have a single X and two sets of autosomes (XY, AA), and so their X : A ratio is 0.5. Flies with XXY sex chromosomes and two sets of autosomes (an X : A ratio of 1.0) develop as fully fertile females, in spite of the presence of a Y chromosome. Flies with only a single X and two sets of autosomes (XO, AA, for an X : A ratio of 0.5) develop as males, although they are sterile. These observations confirm that the Y chromosome does not determine sex in *Drosophila*.

CONCEPTS

The sexual phenotype of a fruit fly is determined by the ratio of the number of X chromosomes to the number of haploid sets of autosomal chromosomes (the X : A ratio).

✔ CONCEPT CHECK 4

What will be the sexual phenotype of a fruit fly with XXYYY sex chromosomes and two sets of autosomes?

a. Male c. Intersex

b. Female d. Metamale

Sex Determination in Humans

Humans, like *Drosophila*, have XX-XY sex determination, but, in humans, the presence of a gene (*SRY*) on the Y chromosome determines maleness. The phenotypes that result from abnormal numbers of sex chromosomes, which arise when the sex chromosomes do not segregate properly in meiosis or mitosis, illustrate the importance of the Y chromosome in human sex determination.

Turner syndrome Persons who have **Turner syndrome** are female and often have underdeveloped secondary sex characteristics. This syndrome is seen in 1 of 3000 female births. Affected women are frequently short and have a low hairline, a relatively broad chest, and folds of skin on the neck. Their intelligence is usually normal. Most women who have Turner syndrome are sterile. In 1959, Charles Ford used new techniques to study human chromosomes and discovered that cells from a 14-year-old girl with Turner syndrome had only a single X chromosome (**Figure 4.8**); this chromosome complement is usually referred to as XO.

There are no known cases in which a person is missing both X chromosomes, an indication that at least one X chromosome is necessary for human development. Presumably, embryos missing both Xs are spontaneously aborted in the early stages of development.

Klinefelter syndrome Persons who have **Klinefelter syndrome,** which occurs with a frequency of about 1 in 1000 male births, have cells with one or more Y chromosomes and multiple X chromosomes. The cells of most males having this condition are XXY (**Figure 4.9**), but cells of a few Klinefelter males are XXXY, XXXXY, or XXYY. Persons with this condition are male, frequently with small testes and reduced facial and pubic hair. They are often taller than normal and sterile; most have normal intelligence.

Poly-X females In about 1 in 1000 female births, the infant's cells possess three X chromosomes, a condition often referred to as **triplo-X syndrome.** These persons have no distinctive features other than a tendency to be tall and thin. Although a few are sterile, many menstruate regularly and are fertile. The incidence of mental retardation among triple-X females is slightly greater than that in the general population, but most XXX females have normal intelligence. Much rarer

males ($Z^{Ca+}Z^{ca}$), $^1/_4$ are blue females (Z^{Ca+}W), $^1/_4$ are cameo males ($Z^{ca}Z^{ca}$), and $^1/_4$ are cameo females (Z^{ca}W). The reciprocal cross of a cameo female with a homozygous blue male produces an F_1 generation in which all offspring are blue and an F_2 consisting of $^1/_2$ blue males ($Z^{Ca+}Z^{Ca+}$ and $Z^{Ca+}Z^{ca}$), $^1/_4$ blue females (Z^{Ca+}W), and $^1/_4$ cameo females (Z^{ca}W).

In organisms with ZZ-ZW sex determination, the female always inherits her W chromosome from her mother, and she inherits her Z chromosome, along with any Z-linked alleles, from her father. In this system, the male inherits Z chromosomes, along with any Z-linked alleles, from both the mother and the father. This pattern of inheritance is the reverse of that of X-linked alleles in organisms with XX-XY sex determination.

Y-Linked Characteristics

Y-linked traits exhibit a distinct pattern of inheritance and are present only in males, because only males possess a Y chromosome. All male offspring of a male with a Y-linked trait will display the trait, because every male inherits the Y chromosome from his father.

Characteristics of the human Y chromosome

The genetic sequence of most of the human Y chromosome has now been determined as part of the Human Genome Project. This work reveals that about two-thirds of the Y chromosome is heterochromatin, consisting of short DNA sequences that are repeated many times and contain no active genes. The other third consists of euchromatin, but there are few genes present. Outside of the pseudoautosomal region, only a little more than 150 genes have been identified on the human Y chromosome, compared with thousands on most chromosomes, and only about half of those identified on the Y chromosome encode proteins. The function of most Y-linked genes is poorly understood; many appear to influence male sexual development and fertility. Some are expressed throughout the body, but many are expressed predominately or exclusively in the testes.

A surprising feature revealed by sequencing is the presence of eight massive palindromic sequences on the Y chromosome. A palindrome is defined as a word, such as "rotator," or sentence that reads the same backward and forward. A palindromic sequence in DNA reads the same on both strands of the double helix, creating two nearly identical copies stretching out from a central point, such as:

Arm 1

5′–TGGGAG . . . CTCCCA–3′
3′–ACCCTC . . . GAGGGT–5′

Arm 2

Thus, a palindromic sequence in DNA appears twice, very much like the two copies of a DNA sequence that are found on two homologous chromosomes. As already indicated, the X and Y chromosomes are not homologous at the vast majority of their sequences, and most of the Y chromosome does not undergo crossing over with the X chromosome. Evidence suggests that X–Y recombination has been replaced by Y–Y recombination between the two arms of the palindromes, which may help to maintain the sequences and functions of genes on the Y chromosomes.

CONCEPTS

Y-linked characteristics exhibit a distinct pattern of inheritance: they are present only in males, and all male offspring of a male with a Y-linked trait inherit the trait.

✔ CONCEPT CHECK 9

What unusual feature of the Y chromosome allows some recombination among the genes found on it?

Use of Y-linked genetic markers

DNA sequences in the Y chromosome undergo mutation with the passage of time and vary among individual males. Like Y-linked traits, these variants—called genetic markers—are passed from father to son and can be used to study male ancestry. Although the markers themselves do not encode any physical traits, they can be detected with the use of molecular methods. Much of the Y chromosome is nonfunctional; so mutations readily accumulate. Many of these mutations are unique; they arise only once and are passed down through the generations without undergoing recombination. Individual males possessing the same set of mutations are therefore related, and the distribution of these genetic markers on Y chromosomes provides clues about genetic relationships of present-day people.

Y-linked markers have been used to study the offspring of Thomas Jefferson, principal author of the Declaration of Independence and third president of the United States. In 1802, Jefferson was accused by a political enemy of fathering a child by his slave Sally Hemings, but the evidence was circumstantial. Hemings, who worked in the Jefferson household and accompanied Jefferson on a trip to Paris, had five children. Jefferson was accused of fathering the first child, Tom, but rumors about the paternity of the other children circulated as well. Hemings's last child, Eston, bore a striking resemblance to Jefferson, and her fourth child, Madison, testified late in life that Jefferson was the father of all of Hemings's children. Ancestors of Hemings's children maintained that they were descendants of the Jefferson line, but some Jefferson descendants refused to recognize their claim.

To resolve this long-standing controversy, geneticists examined markers from the Y chromosomes of male-line descendants of Hemings's first son (Thomas Woodson), her last son (Eston Hemings), and a paternal uncle of Thomas Jefferson with whom Jefferson had Y chromosomes in common. (Descendants of Jefferson's uncle were used because Jefferson himself had no verified male descendants.) Geneticists determined that Jefferson possessed a rare and distinctive set of genetic markers on his Y chromosome. The same markers were also found on the Y chromosomes of the male-line descendants of Eston Hemings. The probability of such a match arising by chance is less than 1%. (The markers were not found on the Y chromosomes of the descendants of Thomas Woodson.) Together with the circumstantial historical evidence, these matching markers strongly suggest that Jefferson fathered Eston Hemings but not Thomas Woodson.

Another study utilizing Y-linked genetic markers focused on the origins of the Lemba, an African tribe comprising 50,000 people who reside in South Africa and parts of Zimbabwe. Members of the Lemba tribe practice some traditions, including circumcision and food taboos, that superficially resemble those of Jewish people, and Lemba oral tradition suggests that the tribe originally came from the Middle East by boat. Today, most Lemba belong to Christian churches, are Muslims, or claim to be Lemba in religion. Their religious practices have little in common with Judaism and, with the exception of their oral tradition and a few cultural practices, there is little to suggest a Jewish origin.

To reveal the genetic origin of the Lemba, scientists examined genetic markers on the Y chromosomes of males in several populations, including the Lemba in Africa, Bantu (another South African tribe), two groups from Yemen, and several groups of Jews. This analysis revealed that Y chromosomes in the Lemba were of two types: those of Bantu origin and those similar to chromosomes found in Jewish and Yemen populations. Most importantly, members of one Lemba clan carried a large number of Y chromosomes that had a rare combination of alleles also found on the Y chromosomes of members of the Jewish priesthood. This set of alleles is thought to be an important indicator of Judaic origin. These findings are consistent with the Lemba oral tradition and strongly suggest a genetic contribution from Jewish populations.

CONNECTING CONCEPTS

Recognizing Sex-Linked Inheritance

What features should we look for to identify a trait as sex linked? A common misconception is that any genetic characteristic in which the phenotypes of males and females differ must be sex linked. In fact, the expression of many *autosomal* characteristics differs between males and females. The genes that encode these characteristics are the same in both sexes, but their expression is influenced by sex hormones. The different sex hormones of males and females cause the same genes to generate different phenotypes in males and females.

Another misconception is that any characteristic that is found more frequently in one sex is sex linked. A number of autosomal traits are expressed more commonly in one sex than in the other. These traits are said to be sex influenced. Some autosomal traits are expressed in only one sex; these traits are said to be sex limited. Both sex-influenced and sex-limited characteristics will be discussed in more detail in Chapter 5.

Several features of sex-linked characteristics make them easy to recognize. Y-linked traits are found only in males, but this fact does not guarantee that a trait is Y linked, because some autosomal characteristics are expressed only in males. A Y-linked trait is unique, however, in that all the male offspring of an affected male will express the father's phenotype, and a Y-linked trait can be inherited only from the father's side of the family. Thus, a Y-linked trait can be inherited only from the paternal grandfather (the father's father), never from the maternal grandfather (the mother's father).

X-linked characteristics also exhibit a distinctive pattern of inheritance. X linkage is a possible explanation when the results of reciprocal crosses differ. If a characteristic is X linked, a cross between an affected male and an unaffected female will not give the same results as a cross between an affected female and an unaffected male. For almost all autosomal characteristics, the results of reciprocal crosses are the same. We should not conclude, however, that, when the reciprocal crosses give different results, the characteristic is X linked. Other sex-associated forms of inheritance, discussed in Chapter 5, also produce different results in reciprocal crosses. The key to recognizing X-linked inheritance is to remember that a male always inherits his X chromosome from his mother, not from his father. Thus, an X-linked characteristic is not passed directly from father to son; if a male clearly inherits a characteristic from his father—and the mother is not heterozygous—it cannot be X linked.

CONCEPTS SUMMARY

- Sexual reproduction is the production of offspring that are genetically distinct from their parents. Most organisms have two sexual phenotypes—males and females. Males produce small gametes; females produce large gametes.

- The mechanism by which sex is specified is termed sex determination. Sex may be determined by differences in specific chromosomes, ploidy level, genotypes, or environment.

- Sex chromosomes differ in number and appearance between males and females. The homogametic sex produces gametes that are all identical with regard to sex chromosomes; the heterogametic sex produces gametes that differ in their sex-chromosome composition.

- In the XX-XO system, females possess two X chromosomes, and males possess a single X chromosome. In the XX-XY system, females possess two X chromosomes, and males possess a single X and a single Y chromosome. In the ZZ-ZW system of sex determination, males possess two Z chromosomes and females possess a Z and a W chromosome.

- In some organisms, ploidy level determines sex; males develop from unfertilized eggs (and are haploid) and females develop from fertilized eggs (and are diploid). Other organisms have genic sex determination, in which genotypes at one or more loci determine the sex of an individual organism. Still others have environmental sex determination.

- In *Drosophila melanogaster*, sex is determined by a balance between genes on the X chromosomes and genes on the autosomes, the X : A ratio.

- In humans, sex is ultimately determined by the presence or absence of the *SRY* gene located on the Y chromosome.

- Sex-linked characteristics are determined by genes on the sex chromosomes; X-linked characteristics are encoded by genes on the X chromosome, and Y-linked characteristics are encoded by genes on the Y chromosome.

- A female inherits X-linked alleles from both parents; a male inherits X-linked alleles from his female parent only.

- The fruit fly *Drosophila melanogaster* has a number of characteristics that make it an ideal model organism for genetic studies, including a short generation time, large numbers of progeny, small size, ease of rearing, and a small genome.

- Dosage compensation equalizes the amount of protein produced by X-linked genes in males and females. In placental mammals, one of the two X chromosomes in females normally becomes inactivated. Which X chromosome is inactivated is random and varies from cell to cell. Some X-linked genes excape X inactivation, and other X-linked genes may be inactivated in some females but not in others. X inactivation is controlled by the *Xist* gene.

- Y-linked characteristics are found only in males and are passed from father to all sons.

IMPORTANT TERMS

sex (p. 74)
sex determination (p. 75)
hermaphroditism (p. 75)
monoecious organism(p. 75
dioecious organism (p. 75)
sex chromosome (p. 75)
autosome (p. 75)
heterogametic sex (p. 76)
homogametic sex (p. 76)
pseudoautosomal region (p. 76)

genic sex determination (p. 77)
sequential hermaphroditism (p. 78)
genic balance system (p. 78)
X : A ratio (p. 78)
Turner syndrome (p. 79)
Klinefelter syndrome (p. 79)
triplo-X syndrome (p. 79)
sex-determining region Y (*SRY*)
 gene (p. 80)
sex-linked characteristic (p. 81)

X-linked characteristic (p. 81)
Y-linked characteristic (p. 81)
hemizygosity (p. 82)
nondisjunction (p. 83)
dosage compensation (p. 87)
Barr body (p. 87)
Lyon hypothesis (p. 88)

ANSWERS TO CONCEPT CHECKS

1. Meiosis

2. b

3. In chromosomal sex determination, males and females have chromosomes that are distinguishable. In genic sex determination, sex is determined by genes but the chromosomes of males and females are indistinguishable. In environmental sex determination, sex is determined by environmental effects.

4. b

5. a

6. c

7. All male offspring will have hemophilia, and all female offspring will not have hemophilia; so the overall probability of hemophilia in the offspring is $\frac{1}{2}$.

8. Two Barr bodies. A Barr body is an inactivated X chromosome.

9. Eight large palindromes that allow crossing over within the Y chromosome

WORKED PROBLEMS

1. A fruit fly has XXXYY sex chromosomes; all the autosomal chromosomes are normal. What sexual phenotype will this fly have?

• **Solution**

Sex in fruit flies is determined by the X : A ratio—the ratio of the number of X chromosomes to the number of haploid autosomal sets. An X : A ratio of 1.0 produces a female fly; an X : A

ratio of 0.5 produces a male. If the X : A ratio is greater than 1.0, the fly is a metafemale; if it is less than 0.5, the fly is a metamale; if the X : A ratio is between 1.0 and 0.5, the fly is an intersex.

This fly has three X chromosomes and normal autosomes. Normal diploid flies have two autosomal sets of chromosomes; so the X : A ratio in this case is $\frac{3}{2}$, or 1.5. Thus, this fly is a metafemale.

2. Chickens, like all birds, have ZZ-ZW sex determination. The bar-feathered phenotype in chickens results from a Z-linked allele that is dominant over the allele for nonbar feathers. A barred female is crossed with a nonbarred male. The F_1 from this cross are intercrossed to produce the F_2. What will the phenotypes and their proportions be in the F_1 and F_2 progeny?

• Solution

With the ZZ-ZW system of sex determination, females are the heterogametic sex, possessing a Z chromosome and a W chromosome; males are the homogametic sex, with two Z chromosomes. In this problem, the barred female is hemizygous for the bar phenotype ($Z^B W$). Because bar is dominant over nonbar, the nonbarred male must be homozygous for nonbar ($Z^b Z^b$). Crossing these two chickens, we obtain:

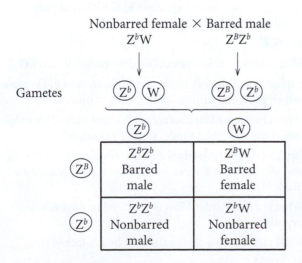

Thus, all the males in the F_1 will be barred ($Z^B Z^b$), and all the females will be nonbarred ($Z^b W$).

We now cross the F_1 to produce the F_2:

So, $\frac{1}{4}$ of the F_2 are barred males, $\frac{1}{4}$ are nonbarred males, $\frac{1}{4}$ are barred females, and $\frac{1}{4}$ are nonbarred females.

3. In *Drosophila melanogaster,* forked bristles are caused by an allele (X^f) that is X linked and recessive to an allele for normal bristles (X^+). Brown eyes are caused by an allele (b) that is autosomal and recessive to an allele for red eyes (b^+). A female fly that is homozygous for normal bristles and red eyes mates with a male fly that has forked bristles and brown eyes. The F_1 are intercrossed to produce the F_2. What will the phenotypes and proportions of the F_2 flies be from this cross?

• Solution

This problem is best worked by breaking the cross down into two separate crosses, one for the X-linked genes that determine the type of bristles and one for the autosomal genes that determine eye color.

Let's begin with the autosomal characteristics. A female fly that is homozygous for red eyes ($b^+ b^+$) is crossed with a male with brown eyes. Because brown eyes are recessive, the male fly must be homozygous for the brown-eyed allele (bb). All of the offspring of this cross will be heterozygous ($b^+ b$) and will have red eyes:

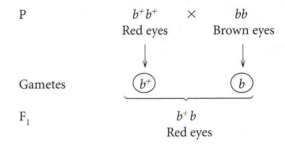

The F_1 are then intercrossed to produce the F_2. Whenever two individual organisms heterozygous for an autosomal recessive characteristic are crossed, $\frac{3}{4}$ of the offspring will have the dominant trait and $\frac{1}{4}$ will have the recessive trait; thus, $\frac{3}{4}$ of the F_2 flies will have red eyes and $\frac{1}{4}$ will have brown eyes:

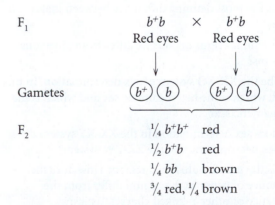

Next, we work out the results for the X-linked characteristic. A female that is homozygous for normal bristles ($X^+ X^+$) is crossed with a male that has forked bristles ($X^f Y$). The female F_1 from this cross are heterozygous ($X^+ X^f$), receiving an X chromosome with a normal-bristle allele from their mother (X^+) and an X chromosome with a forked-bristle allele (X^f) from their father. The male

F_1 are hemizygous (X^+Y), receiving an X chromosome with a normal-bristle allele from their mother (X^+) and a Y chromosome from their father:

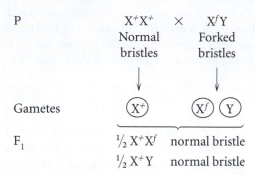

P X^+X^+ × X^fY
           Normal   Forked
           bristles   bristles

Gametes X^+ X^f Y

F_1 $\frac{1}{2}\,X^+X^f$ normal bristle
             $\frac{1}{2}\,X^+Y$ normal bristle

When these F_1 are intercrossed, $\frac{1}{2}$ of the F_2 will be normal-bristle females, $\frac{1}{4}$ will be normal-bristle males, and $\frac{1}{4}$ will be forked-bristle males:

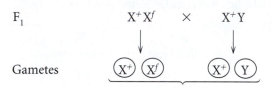

F_1 X^+X^f × X^+Y

Gametes X^+ X^f X^+ Y

	X^+	X^f
X^+	X^+X^+ Normal female	X^+X^f Normal female
Y	X^+Y Normal male	X^fY Forked-bristle male

F_2

$\frac{1}{2}$ normal female, $\frac{1}{4}$ normal male, $\frac{1}{4}$ forked-bristle male

To obtain the phenotypic ratio in the F_2, we now combine these two crosses by using the multiplication rule of probability and the branch diagram:

Eye color	Bristle and sex	F_2 phenotype	Probability
red ($\frac{3}{4}$)	normal female ($\frac{1}{2}$)	red normal female	$\frac{3}{4} \times \frac{1}{2} = \frac{3}{8}$ = $\frac{6}{8}$
	normal male ($\frac{1}{4}$)	red normal male	$\frac{3}{4} \times \frac{1}{4} = \frac{3}{16}$
	forked-bristle male ($\frac{1}{4}$)	red forked male	$\frac{3}{4} \times \frac{1}{4} = \frac{3}{16}$
brown ($\frac{1}{4}$)	normal female ($\frac{1}{2}$)	brown normal female	$\frac{1}{4} \times \frac{1}{2} = \frac{1}{8}$ = $\frac{2}{16}$
	normal male ($\frac{1}{4}$)	brown normal male	$\frac{1}{4} \times \frac{1}{4} = \frac{1}{16}$
	forked-bristle male ($\frac{1}{4}$)	brown forked male	$\frac{1}{4} \times \frac{1}{4} = \frac{1}{16}$

COMPREHENSION QUESTIONS

Section 4.1

*1. What is the most defining difference between males and females?

2. How do monoecious organisms differ from dioecious organisms?

3. Describe the XX-XO system of sex determination. In this system, which is the heterogametic sex and which is the homogametic sex?

4. How does sex determination in the XX-XY system differ from sex determination in the ZZ-ZW system?

*5. What is the pseudoautosomal region? How does the inheritance of genes in this region differ from the inheritance of other Y-linked characteristics?

*6. How is sex determined in insects with haplodiploid sex determination?

7. What is meant by genic sex determination?

8. How does sex determination in *Drosophila* differ from sex determination in humans?

9. Give the typical sex chromosomes found in the cells of people with Turner syndrome, Klinefelter syndrome, and androgen-insensitivity syndrome, as well as in poly-X females.

Section 4.2

*10. What characteristics are exhibited by an X-linked trait?

11. Explain how Bridges's study of nondisjunction in *Drosophila* helped prove the chromosome theory of inheritance.

12. What are some of its characteristics that make *Drosophila melanogaster* a good model genetic organism?

13. Explain why tortoiseshell cats are almost always female and why they have a patchy distribution of orange and black fur.

14. What is a Barr body? How is it related to the Lyon hypothesis?

*15. What characteristics are exhibited by a Y-linked trait?

APPLICATION QUESTIONS AND PROBLEMS

Section 4.1

*16. What is the sexual phenotype of fruit flies having the following chromosomes?

	Sex chromosomes	Autosomal chromosomes
a.	XX	all normal
b.	XY	all normal
c.	XO	all normal
d.	XXY	all normal
e.	XYY	all normal
f.	XXYY	all normal
g.	XXX	all normal
h.	XX	four haploid sets
i.	XXX	four haploid sets
j.	XXX	three haploid sets
k.	X	three haploid sets
l.	XY	three haploid sets
m.	XX	three haploid sets

17. What will be the phenotypic sex of a human with the following gene or chromosomes or both?

   a. XY with the *SRY* gene deleted

   b. XY with the *SRY* gene located on an autosomal chromosome

   c. XX with a copy of *SRY* gene on an autosomal chromosome

   d. XO with a copy of *SRY* gene on an autosome

   e. XXY with the *SRY* gene deleted

   g. XXYY with one copy of the *SRY* gene deleted

18. A normal female *Drosophila* produces abnormal eggs that contain all (a complete diploid set) of her chromosomes. She mates with a normal male *Drosophila* that produces normal sperm. What will be the sex of the progeny from this cross?

19. In certain salamanders, the sex of a genetic female can be altered, making her into a functional male; these salamanders are called sex-reversed males. When a sex-reversed male is mated with a normal female, approximately $\frac{2}{3}$ of the offspring are female and $\frac{1}{3}$ are male. How is sex determined in these salamanders? Explain the results of this cross.

20. In some mites, males pass genes to their grandsons, but they never pass genes to male offspring. Explain.

*21. In organisms with the ZZ-ZW sex-determining system, from which of the following possibilities can a female inherit her Z chromosome?

	Yes	No
Her mother's mother	_____	_____
Her mother's father	_____	_____
Her father's mother	_____	_____
Her father's father	_____	_____

Section 4.2

22. When Bridges crossed white-eyed females with red-eyed males, he obtained a few red-eyed males and white-eyed females (see Figure 4.13). What types of offspring would be produced if these red-eyed males and white-eyed females were crossed with each other?

*23. Joe has classic hemophilia, an X-linked recessive disease. Could Joe have inherited the gene for this disease from the following persons?

	Yes	No
a. His mother's mother	_____	_____
b. His mother's father	_____	_____
c. His father's mother	_____	_____
d. His father's father	_____	_____

*24. In *Drosophila,* yellow body is due to an X-linked gene that is recessive to the gene for gray body.

   a. A homozygous gray female is crossed with a yellow male. The F_1 are intercrossed to produce F_2. Give the genotypes and phenotypes, along with the expected proportions, of the F_1 and F_2 progeny.

   b. A yellow female is crossed with a gray male. The F_1 are intercrossed to produce the F_2. Give the genotypes and phenotypes, along with the expected proportions, of the F_1 and F_2 progeny.

   c. A yellow female is crossed with a gray male. The F_1 females are backcrossed with gray males. Give the genotypes and phenotypes, along with the expected proportions, of the F_2 progeny.

   d. If the F_2 flies in part *b* mate randomly, what are the expected phenotypic proportions of flies in the F_3?

*25. Red–green color blindness in humans is due to an X-linked recessive gene. Both John and Cathy have normal color vision. After 10 years of marriage to John, Cathy gave birth to a color-blind daughter. John filed for divorce, claiming that he is not the father of the child. Is John justified in his claim of nonpaternity? Explain why. If Cathy had given birth to a color-blind son, would John be justified in claiming nonpaternity?

26. Red–green color blindness in humans is due to an X-linked recessive gene. A woman whose father is color blind possesses one eye with normal color vision and one eye with color blindness.

   a. Propose an explanation for this woman's vision pattern. Assume that no new mutations have spontaneously arisen.

   b. Would it be possible for a man to have one eye with normal color vision and one eye with color blindness?

*27. Bob has XXY chromosomes (Klinefelter syndrome) and is color blind. His mother and father have normal color

buildup causes the formation of thick mucus and produces the symptoms of the disease.

Most people have two copies of the normal allele for CFTR and produce only functional CFTR protein. Those with cystic fibrosis possess two copies of the mutated CFTR allele and produce only the defective CFTR protein. Heterozygotes, having one normal and one defective CFTR allele, produce both functional and defective CFTR protein. Thus, at the molecular level, the alleles for normal and defective CFTR are codominant, because both alleles are expressed in the heterozygote. However, because one functional allele produces enough functional CFTR protein to allow normal chloride ion transport, the heterozygote exhibits no adverse effects, and the mutated CFTR allele appears to be recessive at the physiological level. The type of dominance expressed by an allele, as illustrated in this example, is a function of the phenotypic aspect of the allele that is observed.

In summary, several important characteristics of dominance should be emphasized. First, dominance is a result of interactions between genes at the same locus; in other words, dominance is *allelic* interaction. Second, dominance does not alter the way in which the genes are inherited; it only influences the way in which they are expressed as a phenotype. The allelic interaction that characterizes dominance is therefore interaction between the *products* of the genes. Finally, dominance is frequently "in the eye of the beholder," meaning that the classification of dominance depends on the level at which the phenotype is examined. As seen for cystic fibrosis, an allele may exhibit codominance at one level and be recessive at another level.

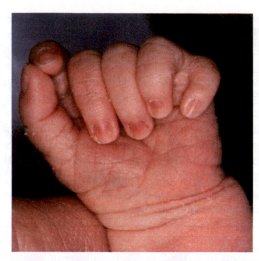

5.3 Human polydactyly (extra digits) exhibits incomplete penetrance and variable expressivity. [SPL/Photo Researchers.]

CONCEPTS

Dominance entails interactions between genes at the same locus (allelic genes) and is an aspect of the phenotype; dominance does not affect the way in which genes are inherited. The type of dominance exhibited by a characteristic frequently depends on the level of the phenotype examined.

✔CONCEPT CHECK 1

How do complete dominance, incomplete dominance, and codominance differ?

5.2 Penetrance and Expressivity Describe How Genes Are Expressed As Phenotype

In the genetic crosses presented thus far, we have considered only the interactions of alleles and have assumed that every individual organism having a particular genotype expresses the expected phenotype. We assumed, for example, that the genotype Rr always produces round seeds and that the genotype rr always produces wrinkled seeds. For some characters, however, such an assumption is incorrect: the genotype does not always produce the expected phenotype, a phenomenon termed **incomplete penetrance.**

Incomplete penetrance is seen in human polydactyly, the condition of having extra fingers and toes (**Figure 5.3**). There are several different forms of human polydactyly, but the trait is usually caused by a dominant allele. Occasionally, people possess the allele for polydactyly (as evidenced by the fact that their children inherit the polydactyly) but nevertheless have a normal number of fingers and toes. In these cases, the gene for polydactyly is not fully penetrant. **Penetrance** is defined as the percentage of individuals having a particular genotype that express the expected phenotype. For example, if we examined 42 people having an allele for polydactyly and found that only 38 of them were polydactylous, the penetrance would be $^{38}/_{42} = 0.90$ (90%).

A related concept is that of **expressivity,** the degree to which a character is expressed. In addition to incomplete penetrance, polydactyly exhibits variable expressivity. Some polydactylous persons possess extra fingers and toes that are fully functional, whereas others possess only a small tag of extra skin.

Incomplete penetrance and variable expressivity are due to the effects of other genes and to environmental factors that can alter or completely suppress the effect of a particular gene. For example, a gene may encode an enzyme that produces a particular phenotype only within a limited temperature range. At higher or lower temperatures, the enzyme does not function and the phenotype is not expressed; the allele encoding such an enzyme is therefore penetrant only within a particular temperature range. Many characters exhibit incomplete penetrance and variable expressivity; thus the mere presence of a gene does not guarantee its expression.

CONCEPTS

Penetrance is the percentage of individuals having a particular genotype that express the associated phenotype. Expressivity is the degree to which a trait is expressed. Incomplete penetrance and variable expressivity result from the influence of other genes and environmental factors on the phenotype.

✔ CONCEPT CHECK 2

Assume that long fingers are inherited as a recessive trait with 80% penetrance. Two people heterozygous for long fingers mate. What is the probability that their first child will have long fingers?

5.3 Lethal Alleles May Alter Phenotypic Ratios

As described in the introduction to this chapter, Lucien Cuénot reported the first case of a lethal allele, the allele for yellow coat color in mice (see Figure 5.1). A **lethal allele** causes death at an early stage of development—often before birth—and so some genotypes may not appear among the progeny.

Another example of a lethal allele, originally described by Erwin Baur in 1907, is found in snapdragons. The *aurea* strain in these plants has yellow leaves. When two plants with yellow leaves are crossed, $\frac{2}{3}$ of the progeny have yellow leaves and $\frac{1}{3}$ have green leaves. When green is crossed with green, all the progeny have green leaves; however, when yellow is crossed with green, $\frac{1}{2}$ of the progeny have green leaves and $\frac{1}{2}$ have yellow leaves, confirming that all yellow-leaved snapdragons are heterozygous. A 2 : 1 ratio is almost always produced by a recessive lethal allele; so observing this ratio among the progeny of a cross between individuals with the same phenotype is a strong clue that one of the alleles is lethal.

In this example, like that of yellow coat color in mice, the lethal allele is recessive because it causes death only in homozygotes. Unlike its effect on *survival,* the effect of the allele on *color* is dominant; in both mice and snapdragons, a single copy of the allele in the heterozygote produces a yellow color. It illustrates the point made earlier that the type of dominance depends on the aspect of the phenotype examined.

Lethal alleles also can be dominant; in this case, homozygotes and heterozygotes for the allele die. Truly dominant lethal alleles cannot be transmitted unless they are expressed after the onset of reproduction, as in Huntington disease.

CONCEPTS

A lethal allele causes death, frequently at an early developmental stage, and so one or more genotypes are missing from the progeny of a cross. Lethal alleles modify the ratio of progeny resulting from a cross.

✔ CONCEPT CHECK 3

A cross between two green corn plants yields $\frac{2}{3}$ progeny that are green and $\frac{1}{3}$ progeny that are white. What is the genotype of the green progeny?

5.4 Multiple Alleles at a Locus Create a Greater Variety of Genotypes and Phenotypes Than Do Two Alleles

Most of the genetic systems that we have examined so far consist of two alleles. In Mendel's peas, for instance, one allele encoded round seeds and another encoded wrinkled seeds; in cats, one allele produced a black coat and another produced a gray coat. For some loci, more than two alleles are present within a group of individuals—the locus has **multiple alleles.** (Multiple alleles may also be referred to as an *allelic series.*) Although there may be more than two alleles present within a *group,* the genotype of each individual diploid organism still consists of only two alleles. The inheritance of characteristics encoded by multiple alleles is no different from the inheritance of characteristics encoded by two alleles, except that a greater variety of genotypes and phenotypes are possible.

Duck-Feather Patterns

An example of multiple alleles is at a locus that determines the feather pattern of mallard ducks. One allele, M, produces the wild-type *mallard* pattern. A second allele, M^R, produces a different pattern called *restricted,* and a third allele, m^d, produces a pattern termed *dusky.* In this allelic series, restricted is dominant over mallard and dusky, and mallard is dominant over dusky: $M^R > M > m^d$. The six genotypes possible with these three alleles and their resulting phenotypes are:

Genotype	Phenotype
$M^R M^R$	restricted
$M^R M$	restricted
$M^R m^d$	restricted
MM	mallard
$M m^d$	mallard
$m^d m^d$	dusky

In general, the number of genotypes possible will be $[n(n + 1)]/2$, where n equals the number of different alleles

(a)

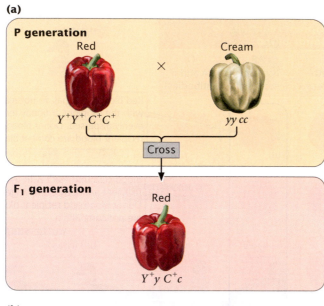

P generation

Red × Cream

$Y^+Y^+ C^+C^+$ $yy\, cc$

Cross

F₁ generation

Red

$Y^+y\, C^+c$

(b)

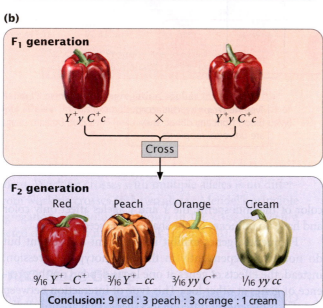

F₁ generation

$Y^+y\, C^+c$ × $Y^+y\, C^+c$

Cross

F₂ generation

Red	Peach	Orange	Cream
$\frac{9}{16}\, Y^+\_ C^+\_$	$\frac{3}{16}\, Y^+\_ cc$	$\frac{3}{16}\, yy\, C^+$	$\frac{1}{16}\, yy\, cc$

Conclusion: 9 red : 3 peach : 3 orange : 1 cream

5.6 Gene interaction in which two loci determine a single characteristic, fruit color, in the pepper *Capsicum annuum*.

Gene Interaction That Produces Novel Phenotypes

Let's first examine gene interaction in which genes at two loci interact to produce a single characteristic. Fruit color in the pepper *Capsicum annuum* is determined in this way. Certain types of peppers produce fruits in one of four colors: red, peach, orange (sometimes called yellow), and cream (or white). If a homozygous plant with red peppers is crossed with a homozygous plant with cream peppers, all the F₁ plants have red peppers (**Figure 5.6a**). When the F₁ are crossed with one another, the F₂ are in a ratio of

9 red : 3 peach : 3 orange : 1 cream (**Figure 5.6b**). This dihybrid ratio (see Chapter 3) is produced by a cross between two plants that are both heterozygous for two loci ($Y^+y\, C^+c \times Y^+y\, C^+c$). In this example, the Y locus and the C locus interact to produce a single phenotype—the color of the pepper:

Genotype	Phenotype
$Y^+\_ C^+\_$	red
$Y^+\_ cc$	peach
$yy\, C^+\_$	orange
$yy\, cc$	cream

Color in peppers of *Capsicum annuum* results from the relative amounts of red and yellow carotenoids, compounds that are synthesized in a complex biochemical pathway (**Figure 5.7**). The Y locus encodes one enzyme (the first step in the pathway outlined in Figure 5.7), and the C locus encodes a different enzyme (the last step in the pathway). When different loci influence different steps in a common biochemical pathway, gene interaction often arises because the product of one enzyme affects the substrate of another enzyme.

To illustrate how Mendel's rules of heredity can be used to understand the inheritance of characteristics determined

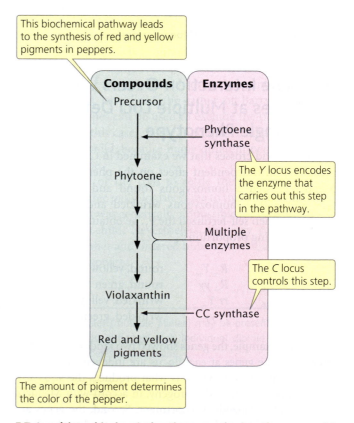

This biochemical pathway leads to the synthesis of red and yellow pigments in peppers.

Compounds   **Enzymes**

Precursor

Phytoene synthase

The Y locus encodes the enzyme that carries out this step in the pathway.

Phytoene

Multiple enzymes

The C locus controls this step.

Violaxanthin

CC synthase

Red and yellow pigments

The amount of pigment determines the color of the pepper.

5.7 A multistep biochemical pathway synthesizes the carotenoid pigments responsible for color variation in peppers.

by gene interaction, let's consider a testcross between an F$_1$ plant from the cross in Figure 5.6 ($Y^+y\ C^+c$) and a plant with cream peppers ($yy\ cc$). As outlined in Chapter 3 for independent loci, we can work this cross by breaking it down into two simple crosses. At the first locus, the heterozygote Y^+y is crossed with the homozygote yy; this cross produces $\frac{1}{2}\ Y^+y$ and $\frac{1}{2}\ yy$ progeny. Similarly, at the second locus, the heterozygous genotype C^+c is crossed with the homozygous genotype cc, producing $\frac{1}{2}\ C^+c$ and $\frac{1}{2}\ cc$ progeny. In accord with Mendel's principle of independent assortment, these single-locus ratios can be combined by using the multiplication rule: the probability of obtaining the genotype $Y^+y\ C^+c$ is the probability of Y^+y ($\frac{1}{2}$) multiplied by the probability of C^+c ($\frac{1}{2}$), or $\frac{1}{4}$. The probability of each progeny genotype resulting from the testcross is:

Progeny genotype	Probability at each locus	Overall probability	Phenotype
$Y^+y\ C^+c$	$\frac{1}{2} \times \frac{1}{2} =$	$\frac{1}{4}$	red peppers
$Y^+y\ cc$	$\frac{1}{2} \times \frac{1}{2} =$	$\frac{1}{4}$	peach peppers
$yy\ C^+c$	$\frac{1}{2} \times \frac{1}{2} =$	$\frac{1}{4}$	orange peppers
$yy\ cc$	$\frac{1}{2} \times \frac{1}{2} =$	$\frac{1}{4}$	cream peppers

When you work problems with gene interaction, it is especially important to determine the probabilities of single-locus genotypes and to multiply the probabilities of *genotypes*, not phenotypes, because the phenotypes cannot be determined without considering the effects of the genotypes at all the contributing loci.

Gene Interaction with Epistasis

Sometimes the effect of gene interaction is that one gene masks (hides) the effect of another gene at a different locus, a phenomenon known as **epistasis**. This phenomenon is similar to dominance, except that dominance entails the masking of genes at the *same* locus (allelic genes). In epistasis, the gene that does the masking is called an **epistatic gene**; the gene whose effect is masked is a **hypostatic gene**. Epistatic genes may be recessive or dominant in their effects.

Recessive epistasis Recessive epistasis is seen in the genes that determine coat color in Labrador retrievers. These dogs may be black, brown, or yellow; their different coat colors are determined by interactions between genes at two loci (although a number of other loci also help to determine coat color; see pp. 113–114). One locus determines the type of pigment produced by the skin cells: a dominant allele B encodes black pigment, whereas a recessive allele b encodes brown pigment. Alleles at a second locus affect the *deposition* of the pigment in the shaft of the hair; allele E allows dark pigment (black or brown) to be deposited, whereas a recessive allele e prevents the deposition of dark pigment, causing the hair to be yellow. The presence of genotype ee at the second locus therefore masks the expression of the black and brown alleles at the first locus. The genotypes that determine coat color and their phenotypes are:

Genotype	Phenotype
$B\_\ E\_$	black
$bb\ E\_$	brown (frequently called chocolate)
$B\_\ ee$	yellow
$bb\ ee$	yellow

If we cross a black Labrador homozygous for the dominant alleles with a yellow Labrador homozygous for the recessive alleles and then intercross the F$_1$, we obtain progeny in the F$_2$ in a 9 : 3 : 4 ratio:

P $BB\ EE \times bb\ ee$
           Black   Yellow

F$_1$ $Bb\ Ee$
                     Black
                     Intercross

F$_2$ $\frac{9}{16}\ B\_\ E\_$ black
           $\frac{3}{16}\ bb\ E\_$ brown
           $\frac{3}{16}\ B\_\ ee$ yellow $\Big\}\ \frac{4}{16}$ yellow
           $\frac{1}{16}\ bb\ ee$ yellow

Notice that yellow dogs can carry alleles for either black or brown pigment, but these alleles are not expressed in their coat color.

In this example of gene interaction, allele e is epistatic to B and b, because e masks the expression of the alleles for black and brown pigments, and alleles B and b are hypostatic to e. In this case, e is a recessive epistatic allele, because two copies of e must be present to mask the expression of the black and brown pigments.

Another example of a recessive epistatic gene is the Bombay phenotype, which suppresses the expression of alleles at the ABO locus. As mentioned earlier in the chapter, the alleles at the ABO locus encode antigens on the red blood cells; the antigens consist of short chains of carbohydrates embedded in the membranes of red blood cells. The difference between the A and the B antigens is a function of chemical differences in the terminal sugar of the chain. The I^A and I^B alleles actually encode different enzymes, which add sugars designated A or B to the ends of the carbohydrate chains

5.8 Expression of the ABO antigens depend on alleles at the H locus. The H locus encodes a precursor to the antigens called compound H. Alleles at the ABO locus determine which types of terminal sugars are added to compound H.

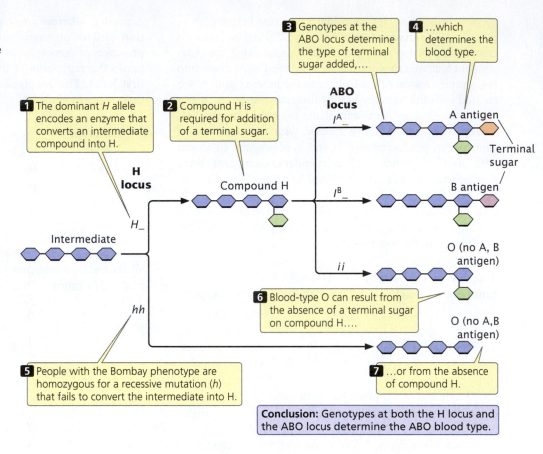

3 Genotypes at the ABO locus determine the type of terminal sugar added,...

4 ...which determines the blood type.

1 The dominant *H* allele encodes an enzyme that converts an intermediate compound into H.

2 Compound H is required for addition of a terminal sugar.

H locus

ABO locus

I^A_

A antigen

Terminal sugar

*H*_

Compound H

I^B_

B antigen

Intermediate

ii

O (no A, B antigen)

6 Blood-type O can result from the absence of a terminal sugar on compound H....

hh

O (no A, B antigen)

5 People with the Bombay phenotype are homozygous for a recessive mutation (*h*) that fails to convert the intermediate into H.

7 ...or from the absence of compound H.

Conclusion: Genotypes at both the H locus and the ABO locus determine the ABO blood type.

(**Figure 5.8**). The common substrate on which these enzymes act is a molecule called H. The enzyme encoded by the *i* allele apparently either adds no sugar to H or no functional enzyme is specified.

In most people, a dominant allele (*H*) at the H locus encodes an enzyme that makes H, but people with the Bombay phenotype are homozygous for a recessive mutation (*h*) that encodes a defective enzyme. The defective enzyme is incapable of making H and, because H is not produced, no ABO antigens are synthesized. Thus, the expression of the alleles at the ABO locus depends on the genotype at the H locus.

Genotype	H present	ABO phenotype
$H\_ I^A I^A$, $H\_ I^A i$	Yes	A
$H\_ I^B I^B$, $H\_ I^B i$	Yes	B
$H\_ I^A I^B$	Yes	AB
$H\_ ii$	Yes	O
$hh\ I^A I^A$, $hh\ I^A i$, $hh\ I^B I^B$, $hh\ I^B i$, $hh\ I^A I^B$, and $hh\ ii$	No	O

In this example, the alleles at the ABO locus are hypostatic to the recessive *h* allele.

The Bombay phenotype provides us with a good opportunity for considering how epistasis often arises when genes affect a series of steps in a biochemical pathway. The ABO antigens are produced in a multistep biochemical pathway (see Figure 5.8), which depends on enzymes that make H and other enzymes that convert H into the A or B antigen. Note that blood-type O may arise in one of two ways: (1) from failure to add a terminal sugar to compound H (genotype $H\_ ii$) or (2) from failure to produce compound H (genotype $hh\ \_$). Many cases of epistasis arise in this way. A gene (such as *h*) that has an effect on an early step in a biochemical pathway will be epistatic to genes (such as I^A and I^B) that affect subsequent steps, because the effects of the genes in the later step depend on the product of the earlier reaction.

Dominant epistasis Dominant epistasis is seen in the interaction of two loci that determine fruit color in summer squash, which is commonly found in one of three colors: yellow, white, or green. When a homozygous plant that produces white squash is crossed with a homozygous plant that

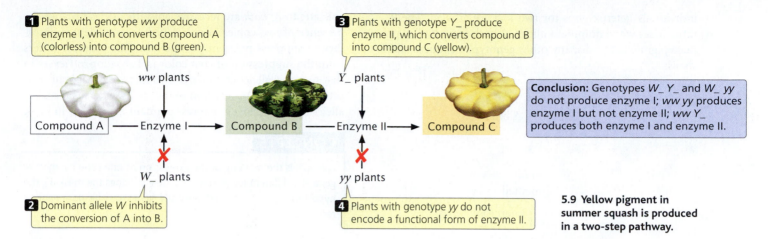

1 Plants with genotype *ww* produce enzyme I, which converts compound A (colorless) into compound B (green).

3 Plants with genotype *Y_* produce enzyme II, which converts compound B into compound C (yellow).

ww plants

*Y_* plants

| Compound A | ── Enzyme I ──► | Compound B | ── Enzyme II ──► | Compound C |

Conclusion: Genotypes *W_ Y_* and *W_ yy* do not produce enzyme I; *ww yy* produces enzyme I but not enzyme II; *ww Y_* produces both enzyme I and enzyme II.

*W_* plants

yy plants

2 Dominant allele *W* inhibits the conversion of A into B.

4 Plants with genotype *yy* do not encode a functional form of enzyme II.

5.9 Yellow pigment in summer squash is produced in a two-step pathway.

produces green squash and the F_1 plants are crossed with each other, the following results are obtained:

P   Plants with   Plants with
          white squash   ×   green squash

F_1 Plants with
                     white squash

                     Intercross

F_2 $^{12}/_{16}$ plants with white squash
          $^{3}/_{16}$ plants with yellow squash
          $^{1}/_{16}$ plants with green squash

How can gene interaction explain these results?

In the F_2, $^{12}/_{16}$, or $^{3}/_{4}$, of the plants produce white squash and $^{3}/_{16} + ^{1}/_{16} = ^{4}/_{16} = ^{1}/_{4}$ of the plants produce squash having color. This outcome is the familiar 3 : 1 ratio produced by a cross between two heterozygotes, which suggests that a dominant allele at one locus inhibits the production of pigment, resulting in white progeny. If we use the symbol *W* to represent the dominant allele that inhibits pigment production, then genotype *W_* inhibits pigment production and produces white squash, whereas *ww* allows pigment and results in colored squash.

Among those *ww* F_2 plants with pigmented fruit, we observe $^{3}/_{16}$ yellow and $^{1}/_{16}$ green (a 3 : 1 ratio). In this outcome, a second locus determines the type of pigment produced in the squash, with yellow (*Y_*) dominant over green (*yy*). This locus is expressed only in *ww* plants, which lack the dominant inhibitory allele *W*. We can assign the genotype *ww Y_* to plants that produce yellow squash and the genotype *ww yy* to plants that produce green squash. The genotypes and their associated phenotypes are:

*W_ Y_*	white squash
W_ yy	white squash
*ww Y_*	yellow squash
ww yy	green squash

Allele *W* is epistatic to *Y* and *y*: it suppresses the expression of these pigment-producing genes. Allele *W* is a dominant epistatic allele because, in contrast with *e* in Labrador retriever coat color and with *h* in the Bombay phenotype, a single copy of the allele is sufficient to inhibit pigment production.

Yellow pigment in the squash is most likely produced in a two-step biochemical pathway (**Figure 5.9**). A colorless (white) compound (designated A in Figure 5.9) is converted by enzyme I into green compound B, which is then converted into compound C by enzyme II. Compound C is the yellow pigment in the fruit. Plants with the genotype *ww* produce enzyme I and may be green or yellow, depending on whether enzyme II is present. When allele *Y* is present at a second locus, enzyme II is produced and compound B is converted into compound C, producing a yellow fruit. When two copies of allele *y*, which does not encode a functional form of enzyme II, are present, squash remain green. The presence of *W* at the first locus inhibits the conversion of compound A into compound B; plants with genotype *W_* do not make compound B and their fruit remains white, regardless of which alleles are present at the second locus.

Duplicate recessive epistasis Let's consider one more detailed example of epistasis. Albinism is the absence of pigment and is a common genetic trait in many plants and animals. Pigment is almost always produced through a multistep biochemical pathway; thus, albinism may entail gene interaction. Robert T. Dillon and Amy R. Wethington found that albinism in the common freshwater snail *Physa heterostroha* can result from the presence of either of two recessive alleles at two different loci. Inseminated snails were collected from a natural population and placed in cups of water, where they laid eggs. Some of the eggs hatched into albino snails. When two albino snails were crossed, all of the F_1 were pigmented. When the F_1 were intercrossed, the F_2 consisted of $^{9}/_{16}$ pigmented snails and $^{7}/_{16}$ albino snails. How did this 9 : 7 ratio arise?

The 9 : 7 ratio seen in the F_2 snails can be understood as a modification of the 9 : 3 : 3 : 1 ratio obtained when two individuals heterozygous for two loci are crossed. The 9 : 7 ratio arises when dominant alleles at both loci ($A\_ B\_$) produce pigmented snails; any other genotype produces albino snails:

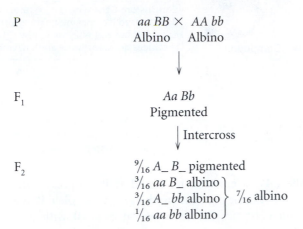

P $aa\ BB \times\ AA\ bb$
           Albino   Albino

F$_1$ $Aa\ Bb$
                 Pigmented

                 ↓ Intercross

F$_2$ $^9/_{16}\ A\_ B\_$ pigmented
            $^3/_{16}\ aa\ B\_$ albino
            $^3/_{16}\ A\_ bb$ albino } $^7/_{16}$ albino
            $^1/_{16}\ aa\ bb$ albino

The 9 : 7 ratio in these snails is probably produced by a two-step pathway of pigment production (**Figure 5.10**). Pigment (compound C) is produced only after compound A has been converted into compound B by enzyme I and after compound B has been converted into compound C by enzyme II. At least one dominant allele A at the first locus is required to produce enzyme I; similarly, at least one dominant allele B at the second locus is required to produce enzyme II. Albinism arises from the absence of compound C, which may happen in one of three ways. First, two recessive alleles at the first locus (genotype $aa\ B\_$) may prevent the production of enzyme I, and so compound B is never produced. Second, two recessive alleles at the second locus (genotype $A\_ bb$) may prevent the production of enzyme II. In this case, compound B is never converted into compound C. Third, two recessive alleles may be present at both loci ($aa\ bb$), causing the absence of both enzyme I and enzyme II. In this example of gene interaction, a is epistatic to B, and b is

epistatic to A; *both* are recessive epistatic alleles because the presence of two copies of either allele a or allele b is necessary to suppress pigment production. This example differs from the suppression of coat color in Labrador retrievers in that recessive alleles at either of two loci are capable of suppressing pigment production in the snails, whereas recessive alleles at a single locus suppress pigment expression in Labs.

CONCEPTS

Epistasis is the masking of the expression of one gene by another gene at a different locus. The epistatic gene does the masking; the hypostatic gene is masked. Epistatic genes can be dominant or recessive.

✔CONCEPT CHECK 6

A number of all-white cats are crossed and they produce the following types of progeny: $^{12}/_{16}$ all-white, $^3/_{16}$ black, and $^1/_{16}$ gray. Give the genotypes of the progeny. Which gene is epistatic?

CONNECTING CONCEPTS

Interpreting Ratios Produced by Gene Interaction

A number of modified ratios that result from gene interaction are shown in **Table 5.2**. Each of these examples represents a modification of the basic 9 : 3 : 3 : 1 dihybrid ratio. In interpreting the genetic basis of modified ratios, we should keep several points in mind. First, the inheritance of the genes producing these characteristics is no different from the inheritance of genes encoding simple genetic characters. Mendel's principles of segregation and independent assortment still apply; each individual possesses two alleles at each locus, which separate in meiosis, and genes at the different loci assort independently. The only difference is in how the *products* of the genotypes interact to produce the phenotype. Thus, we cannot consider the expression of genes at each locus separately; instead, we must take into consideration how the genes at different loci interact.

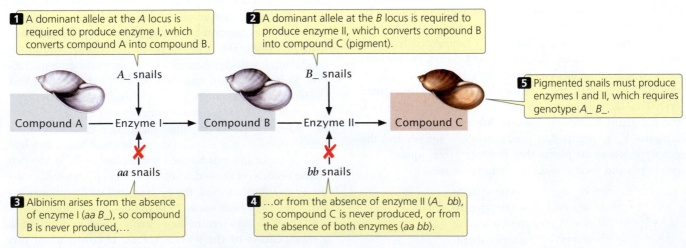

1 A dominant allele at the A locus is required to produce enzyme I, which converts compound A into compound B.

2 A dominant allele at the B locus is required to produce enzyme II, which converts compound B into compound C (pigment).

$A\_$ snails

$B\_$ snails

5 Pigmented snails must produce enzymes I and II, which requires genotype $A\_ B\_$.

Compound A ——Enzyme I——➤ Compound B ——Enzyme II——➤ Compound C

aa snails

bb snails

3 Albinism arises from the absence of enzyme I ($aa\ B\_$), so compound B is never produced,…

4 …or from the absence of enzyme II ($A\_ bb$), so compound C is never produced, or from the absence of both enzymes ($aa\ bb$).

5.10 Pigment is produced in a two-step pathway in snails.

Table 5.2 Modified dihybrid phenotypic ratios due to gene interaction

Ratio*	Genotype				Type of Interaction	Example
	*A_ B_*	*A_ bb*	*aa B_*	*aa bb*		
9 : 3 : 3 : 1	9	3	3	1	None	Seed shape and endosperm color in peas
9 : 3 : 4	9	3	4		Recessive epistasis	Coat color in Labrador retrievers
12 : 3 : 1	12		3	1	Dominant epistasis	Color in squash
9 : 7	9	7			Duplicate recessive epistasis	Albinism in snails
9 : 6 : 1	9	6		1	Duplicate interaction	—
15 : 1	15			1	Duplicate dominant epistasis	—
13 : 3	13		3		Dominant and recessive epistasis	—

*Each ratio is produced by a dihybrid cross (*Aa Bb* × *Aa Bb*). Shaded bars represent combinations of genotypes that give the same phenotype.

A second point is that, in the examples that we have considered, the phenotypic proportions were always in sixteenths because, in all the crosses, pairs of alleles segregated at two independently assorting loci. The probability of inheriting one of the two alleles at a locus is $\frac{1}{2}$. Because there are two loci, each with two alleles, the probability of inheriting any particular combination of genes is $(\frac{1}{2})^4 = \frac{1}{16}$. For a trihybrid cross, the progeny proportions should be in sixty-fourths, because $(\frac{1}{2})^6 = \frac{1}{64}$. In general, the progeny proportions should be in fractions of $(\frac{1}{2})^{2n}$, where n equals the number of loci with two alleles segregating in the cross.

Crosses rarely produce exactly 16 progeny; therefore, modifications of a dihybrid ratio are not always obvious. Modified dihybrid ratios are more easily seen if the number of individuals of each phenotype is expressed in sixteenths:

$$\frac{x}{16} = \frac{\text{number of progeny with a phenotype}}{\text{total number of progeny}}$$

where $x/16$ equals the proportion of progeny with a particular phenotype. If we solve for x (the proportion of the particular phenotype in sixteenths), we have:

$$x = \frac{\text{number of progeny with a phenotype} \times 16}{\text{total number of progeny}}$$

For example, suppose we cross two homozygotes, interbreed the F_1, and obtain 63 red, 21 brown, and 28 white F_2 individuals. Using the preceding formula, we find the phenotypic ratio in the F_2 to be: red $= (63 \times 16)/112 = 9$; brown $= (21 \times 16)/112 = 3$; and white $= (28 \times 16)/112 = 4$. The phenotypic ratio is 9 : 3 : 4.

A final point to consider is how to assign genotypes to the phenotypes in modified ratios that result from gene interaction. Don't try to *memorize* the genotypes associated with all the modified ratios in Table 5.2. Instead, practice relating modified ratios to known ratios, such as the 9 : 3 : 3 : 1 dihybrid ratio. Suppose we obtain $\frac{15}{16}$ green progeny and $\frac{1}{16}$ white progeny in a cross between two plants. If we compare this 15 : 1 ratio with the standard 9 : 3 : 3 : 1 dihybrid ratio, we see that $\frac{9}{16} + \frac{3}{16} + \frac{3}{16}$ equals $\frac{15}{16}$. All the genotypes associated with these proportions in the dihybrid cross (*A_ B_, A_ bb*, and *aa B_*) must give the same phenotype, the green progeny. Genotype *aa bb* makes up $\frac{1}{16}$ of the progeny in a dihybrid cross, the white progeny in this cross.

In assigning genotypes to phenotypes in modified ratios, students sometimes become confused about which letters to assign to which phenotype. Suppose we obtain the following phenotypic ratio: $\frac{9}{16}$ black : $\frac{3}{16}$ brown : $\frac{4}{16}$ white. Which genotype do we assign to the brown progeny, *A_ bb* or *aa B_*? Either answer is correct, because the letters are just arbitrary symbols for the genetic information. The important thing to realize about this ratio is that the brown phenotype arises when two recessive alleles are present at one locus.

Worked Problem

A homozygous strain of yellow corn is crossed with a homozygous strain of purple corn. The F_1 are intercrossed, producing an ear of corn with 119 purple kernels and 89 yellow kernels (the progeny). What is the genotype of the yellow kernels?

(a)

(b)

(c)

5.11 Coat color in dogs is determined by interactions between genes at a number of loci. (a) Most Labrador retrievers are genotype A^sA^s SS, varying only at the B and E loci. (b) Most beagles are genotype a^sa^s BB s^ps^p. (c) Dalmations are genotype A^sA^s EE s^ws^w, varying at the B locus, which makes the dogs black ($B\_$) or brown (bb). [Part a: Kent and Donna Dannen. Part b: Tara Darling. Part c: PhotoDisc.]

3. **Extension (E) locus.** Four alleles at this locus determine where the genotype at the A locus is expressed. For example, if a dog has the A^s allele (solid black) at the A locus, then black pigment will either be extended throughout the coat or be restricted to some areas, depending on the alleles present at the E locus. Areas where the A locus is not expressed may appear as yellow, red, or tan, depending on the presence of particular genes at other loci. When A^s is present at the A locus, the four alleles at the E locus have the following effects:

E^m Black mask with a tan coat.
E The A locus expressed throughout (solid black).
e^{br} Brindle, in which black and yellow are in layers to give a tiger-striped appearance.
e No black in the coat, but the nose and eyes may be black.

The dominance relations among these alleles are poorly known.

4. **Spotting (S) locus.** Alleles at this locus determine whether white spots will be present. There are four common alleles:

S No spots.
s^i Irish spotting; numerous white spots.
s^p Piebald spotting; various amounts of white.
s^w Extreme white piebald; almost all white.

Allele S is completely dominant over alleles s^i, s^p, and s^w; alleles s^i and s^p are dominant over allele s^w ($S > s^i$, $s^p > s^w$). The relation between s^i and s^p is poorly defined; indeed, they may not be separate alleles. Genes at other poorly known loci also modify spotting patterns.

To illustrate how genes at these loci interact in determining a dog's coat color, let's consider a few examples:

Labrador retriever   Labrador retrievers (**Figure 5.11a**) may be black, brown, or yellow. Most are homozygous A^sA^s SS; thus, they vary only at the B and E loci. The

A^s allele allows dark pigment to be expressed; whether a dog is black depends on which genes are present at the B and E loci. As discussed earlier in the chapter, all black Labradors must carry at least one B allele and one E allele ($B\_$ $E\_$). Brown dogs are homozygous bb and have at least one E allele (bb $E\_$). Yellow dogs are a result of the presence of ee ($B\_$ ee or bb ee). Labrador retrievers are homozygous for the S allele, which produces a solid color; the few white spots that appear in some dogs of this breed are due to other modifying genes.

Beagle   Most beagles (**Figure 5.11b**) are homozygous a^sa^s BB s^ps^p, although other alleles at these loci are occasionally present. The a^s allele produces the saddle markings—dark back and sides, with tan head and legs—that are characteristic of the breed. Allele B allows black to be produced, but its distribution is limited by the a^s allele. Most beagles are $E\_$, but the genotype ee does occasionally arise, leading to a few all-tan beagles. White spotting in beagles is due to the s^p allele.

Dalmatian   Dalmatians (**Figure 5.11c**) have an interesting genetic makeup. Most are homozygous A^sA^s EE s^ws^w; so they vary only at the B locus. Notice that these dogs possess genotype A^sA^s EE, which allows for a solid coat that would be black, if genotype $B\_$ is present, or brown (called liver), if genotype bb is present. However, the presence of the s^w allele produces a white coat, masking the expression of the solid color. The dog's color appears only in the pigmented spots, which are due to the presence of an allele at yet another locus that allows the color to penetrate in a limited number of spots. **Table 5.3** gives the common genotypes of other breeds of dogs.

5.6 Sex Influences the Inheritance and Expression of Genes in a Variety of Ways

In Chapter 4, we considered characteristics encoded by genes located on the sex chromosomes (sex-linked traits) and how their inheritance differs from the inheritance of

Table 5.3   Common genotypes in different breeds of dogs

Breed	Usual Homozygous Genes[*]	Other Genes Present Within the Breed
Basset hound	$BB\ EE$	$a^y, a^t\quad S, s^p, s^i$
Beagle	$a^s a^s\ BB\ s^p s^p$	E, e
English bulldog	BB	$A^s, a^y, a^t\quad E^m, E, e^{br}\quad S, s^i, s^p, s^w$
Chihuahua	$A^s, a^y, a^s, a^t\quad B, b\quad E^m, E, e^{br}, e\quad S, s^i, s^p, s^w$
Collie	$BB\ EE$	$a^y, a^t\quad s^i, s^w$
Dalmatian	$A^s A^s\ EE\ s^w s^w$	B, b
Doberman	$a^t a^t\ EE\ SS$	B, b
German shepherd	$BB\ SS$	$a^y, a, a^s, a^t\quad E^m, E, e$
Golden retriever	$A^s A^s\ BB\ SS$	E, e
Greyhound	BB	$A^s, a^y\quad E, e^{br}, e\quad S, s^p, s^w, s^i$
Irish setter	$BB\ ee\ SS$	A, a^t
Labrador retriever	$A^s A^s\ SS$	$B, b\quad E, e$
Poodle	SS	$A^s, a^t\quad B, b\quad E, e$
Rottweiler	$a^t a^t\ BB\ EE\ SS$	
St. Bernard	$a^y a^y\ BB$	$E^m, E\quad s^i, s^p, s^w$

[*]Most dogs in the breed are homozygous for these genes; a few individual dogs may possess other alleles at these loci.

Source: Data from M. B. Willis, *Genetics of the Dog* (London: Witherby, 1989).

traits encoded by autosomal genes. X-linked traits, for example, are passed from father to daughter, but never from father to son, and Y-linked traits are passed from father to all sons. Now, we will examine additional influences of sex, including the effect of the sex of an individual on the expression of genes on autosomal chromosomes, on characteristics determined by genes located in the cytoplasm, and on characteristics for which the genotype of only the maternal parent determines the phenotype of the offspring. Finally, we'll look at situations in which the expression of genes on autosomal chromosomes is affected by the sex of the parent from whom they are inherited.

Sex-Influenced and Sex-Limited Characteristics

Sex-influenced characteristics are determined by autosomal genes and are inherited according to Mendel's principles, but they are expressed differently in males and females. In this case, a particular trait is more readily expressed in one sex; in other words, the trait has higher penetrance in one of the sexes.

For example, the presence of a beard on some goats is determined by an autosomal gene (B^b) that is dominant in males and recessive in females. In males, a single allele is required for the expression of this trait: both the homozygote ($B^b B^b$) and the heterozygote ($B^b B^+$) have beards, whereas the $B^+ B^+$ male is beardless. In contrast, females require two alleles in order for this trait to be expressed: the homozygote $B^b B^b$ has a beard, whereas the heterozygote ($B^b B^+$) and the other homozygote ($B^+ B^+$) are beardless. The key to understanding the expression of the bearded gene is to look at the heterozygote. In males (for which the presence of a beard is dominant), the heterozygous genotype produces a beard but, in females (for which the presence of a beard is recessive and its absence is dominant), the heterozygous genotype produces a goat without a beard.

Figure 5.12a illustrates a cross between a beardless male ($B^+ B^+$) and a bearded female ($B^b B^b$). The alleles separate into gametes according to Mendel's principle of segregation, and all the F_1 are heterozygous ($B^+ B^b$). Because the trait is dominant in males and recessive in females, all the F_1 males will be bearded, and all the F_1 females will be beardless. When the F_1 are crossed with one another, $1/4$ of the F_2 progeny are $B^b B^b$, $1/2$ are $B^b B^+$, and $1/4$ are $B^+ B^+$ (**Figure 5.12b**). Because male heterozygotes are bearded, $3/4$ of the males in the F_2 possess beards; because female heterozygotes are beardless, only $1/4$ of the females in the F_2 are bearded.

An extreme form of sex-influenced inheritance, a **sex-limited characteristic** is encoded by autosomal genes that

(a)

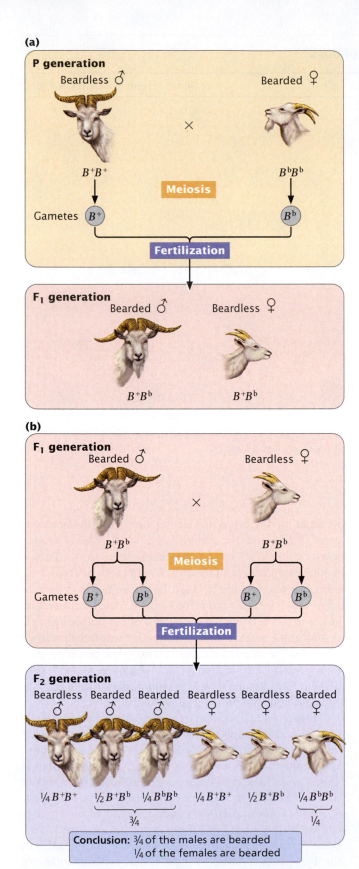

are expressed in only one sex; the trait has zero penetrance in the other sex. In domestic chickens, some males display a plumage pattern called cock feathering (**Figure 5.13a**). Other males and all females display a pattern called hen feathering (**Figure 5.13b and c**). Cock feathering is an autosomal recessive trait that is sex limited to males. Because the trait is autosomal, the genotypes of males and females are the same, but the phenotypes produced by these genotypes differ in males and females:

Genotype	Male phenotype	Female phenotype
HH	hen feathering	hen feathering
Hh	hen feathering	hen feathering
hh	cock feathering	hen feathering

An example of a sex-limited characteristic in humans is male-limited precocious puberty. There are several types of precocious puberty in humans, most of which are not genetic. Male-limited precocious puberty, however, results from an autosomal dominant allele (*P*) that is expressed only in males; females with the gene are normal in phenotype. Males with precocious puberty undergo puberty at an early age, usually before the age of 4. At this time, the penis enlarges, the voice deepens, and pubic hair develops. There is no impairment of sexual function; affected males are fully fertile. Most are short as adults, because the long bones stop growing after puberty.

Because the trait is rare, affected males are usually heterozygous (*Pp*). A male with precocious puberty who mates with a woman who has no family history of this condition will transmit the allele for precocious puberty to $1/2$ of their children (**Figure 5.14a**), but it will be expressed only in the sons. If one of the heterozygous daughters (*Pp*) mates with a male who has normal puberty (*pp*), $1/2$ of their sons will exhibit precocious puberty (**Figure 5.14b**). Thus a sex-limited characteristic can be inherited from either parent, although the trait appears in only one sex.

The results of molecular studies reveal that the underlying genetic defect in male-limited precocious puberty affects the receptor for luteinizing hormone (LH). This hormone normally attaches to receptors found on certain cells of the testes and stimulates these cells to produce testosterone. During normal puberty in males, high levels of LH stimulate the increased production of testosterone, which, in turn, stimulates the anatomical and physiological changes associated with puberty. The *P* allele for precocious puberty encodes a defective LH receptor, which stimulates testosterone production even in the absence of LH. Boys with this allele produce high levels of testosterone at an early age, when levels of LH are low. Defective LH receptors are also found in females who carry the precocious-puberty gene, but their presence does not result in precocious puberty, because additional hormones are required along with LH to induce puberty in girls.

5.12 Genes that encode sex-influenced traits are inherited according to Mendel's principles but are expressed differently in males and females.

5.13 A sex-limited characteristic is encoded by autosomal genes that are expressed in only one sex. An example is cock feathering in chickens, an autosomal recessive trait that is limited to males. (a) Cock-feathered male. (b) Hen-feathered female. (c) Hen-feathered male. *[Larry Lefever/Grant Heilman Photography.]*

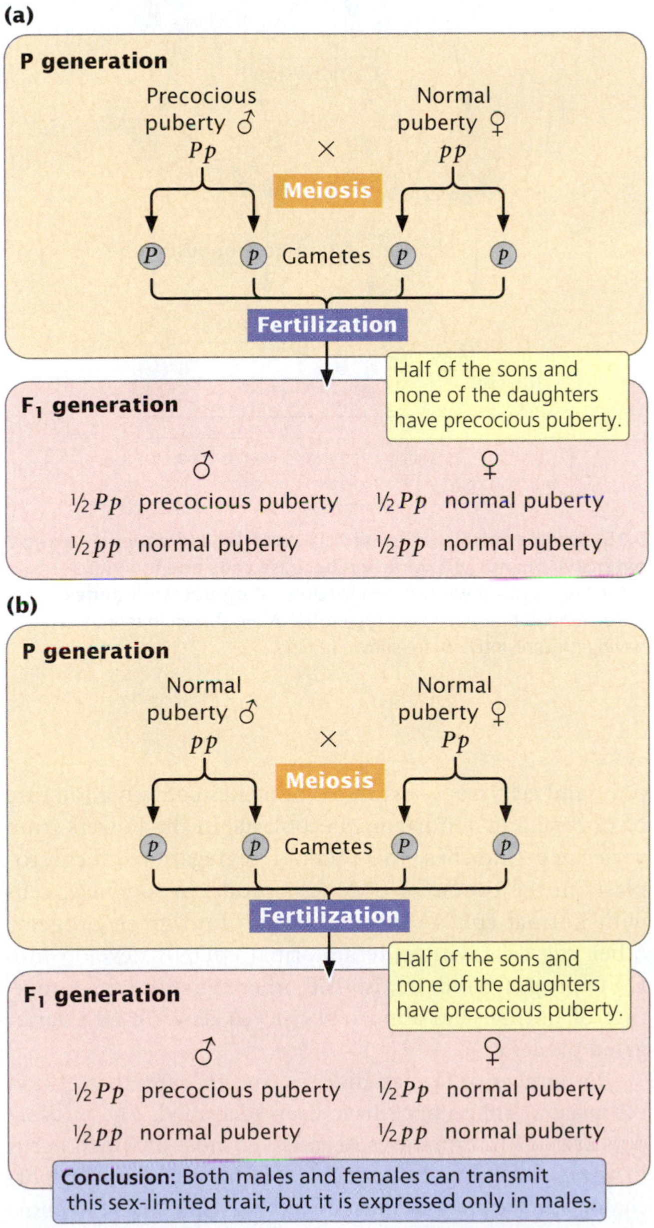
(a)
P generation
Precocious puberty ♂
Pp
×
Normal puberty ♀
pp
Meiosis
P
p
Gametes
p
p
Fertilization
F1 generation
Half of the sons and none of the daughters have precocious puberty.
♂
½ Pp precocious puberty
½ pp normal puberty
♀
½ Pp normal puberty
½ pp normal puberty
(b)
P generation
Normal puberty ♂
pp
×
Normal puberty ♀
Pp
Meiosis
p
p
Gametes
P
p
Fertilization
F1 generation
Half of the sons and none of the daughters have precocious puberty.
♂
½ Pp precocious puberty
½ pp normal puberty
♀
½ Pp normal puberty
½ pp normal puberty
Conclusion: Both males and females can transmit this sex-limited trait, but it is expressed only in males.

5.14 Sex-limited characteristics are inherited according to Mendel's principles. Precocious puberty is an autosomal dominant trait that is limited to males.

CONCEPTS

Sex-influenced characteristics are encoded by autosomal genes that are more readily expressed in one sex. Sex-limited characteristics are encoded by autosomal genes whose expression is limited to one sex.

✔ CONCEPT CHECK 8

How do sex-influenced and sex-limited traits differ from sex-linked traits?

Cytoplasmic Inheritance

Mendel's principles of segregation and independent assortment are based on the assumption that genes are located on chromosomes in the nucleus of the cell. For most genetic characteristics, this assumption is valid, and Mendel's principles allow us to predict the types of offspring that will be produced in a genetic cross. However, not all the genetic material of a cell is found in the nucleus; some characteristics are encoded by genes located in the cytoplasm. These characteristics exhibit **cytoplasmic inheritance.**

A few organelles, notably chloroplasts and mitochondria, contain DNA. Each human mitochondrion contains about 15,000 nucleotides of DNA, encoding 37 genes. Compared with that of nuclear DNA, which contains some 3 billion nucleotides encoding perhaps 25,000 genes, the amount of mitochondrial DNA (mtDNA) is very small; nevertheless, mitochondrial and chloroplast genes encode some important characteristics. In this chapter we will focus on *patterns* of cytoplasmic inheritance.

Cytoplasmic inheritance differs from the inheritance of characteristics encoded by nuclear genes in several important respects. A zygote inherits nuclear genes from both parents; but, typically, all its cytoplasmic organelles, and thus all its cytoplasmic genes, come from only one of the gametes, usually the egg. A sperm generally contributes only a set of nuclear genes from the male parent. In a few organisms, cytoplasmic genes are inherited from the male parent or from both parents; however, for most organisms, all the cytoplasm is inherited from the egg. In this case, cytoplasmically inherited traits are present in both males and females and are passed from mother to offspring, never from father to offspring. Reciprocal crosses, therefore, give different results when cytoplasmic genes encode a trait.

Cytoplasmically inherited characteristics frequently exhibit extensive phenotypic variation, because no mechanism analogous to mitosis or meiosis ensures that cytoplasmic genes are evenly distributed in cell division. Thus, different cells and individual offspring will contain various proportions of cytoplasmic genes.

Consider mitochondrial genes. Most cells contain thousands of mitochondria, and each mitochondrion contains from 2 to 10 copies of mtDNA. Suppose that half of the mitochondria in a cell contain a normal wild-type copy of mtDNA and the other half contain a mutated copy (**Figure 5.15**). In cell division, the mitochondria segregate into progeny cells at random. Just by chance, one cell may receive mostly mutated mtDNA and another cell may receive mostly wild-type mtDNA. In this way, different progeny from the same mother and even cells within an individual offspring may vary in their phenotypes. Traits encoded by chloroplast DNA (cpDNA) are similarly variable.

In 1909, cytoplasmic inheritance was recognized by Carl Correns as one of the first exceptions to Mendel's principles. Correns, one of the biologists who rediscovered Mendel's work, studied the inheritance of leaf variegation in the four-o'clock plant, *Mirabilis jalapa*. Correns found that the leaves and shoots of one variety of four-o'clock were variegated, displaying a mixture of green and white splotches. He also noted that some branches of the variegated strain had all-green leaves; other branches had all-white leaves. Each branch produced flowers; so Correns was able to cross flowers from variegated, green, and white branches in all combinations (**Figure 5.16**). The seeds from green branches always gave rise to green progeny, no matter whether the pollen was from a green, white, or variegated branch. Similarly, flowers on white branches always produced white progeny. Flowers on the variegated branches gave rise to green, white, and variegated progeny, in no particular ratio.

Correns's crosses demonstrated cytoplasmic inheritance of variegation in the four-o'clocks. The phenotypes of the offspring were determined entirely by the maternal parent, never by the paternal parent (the source of the pollen). Furthermore, the production of all three phenotypes by flowers on variegated branches is consistent with cytoplasmic inheritance. Variegation in these plants is caused by a defective gene in the cpDNA, which results in a failure to produce the green pigment chlorophyll. Cells from green branches contain normal chloroplasts only, cells from white branches contain abnormal chloroplasts

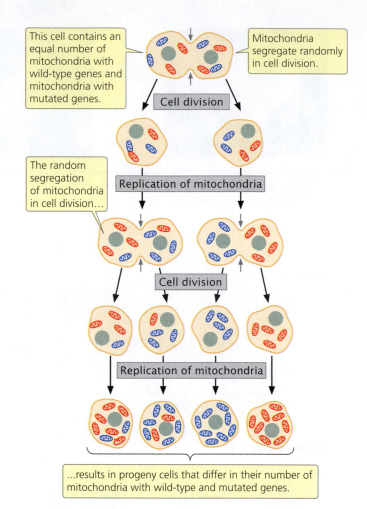

5.15 **Cytoplasmically inherited characteristics frequently exhibit extensive phenotypic variation because cells and individual offspring contain various proportions of cytoplasmic genes.** Mitochondria that have wild-type mtDNA are shown in red; those having mutant mtDNA are shown in blue.

only, and cells from variegated branches contain a mixture of normal and abnormal chloroplasts. In the flowers from variegated branches, the random segregation of chloroplasts in the course of oogenesis produces some egg cells with normal cpDNA, which develop into green progeny; other egg cells with only abnormal cpDNA develop into white progeny; and, finally, still other egg cells with a mixture of normal and abnormal cpDNA develop into variegated progeny.

A number of human diseases (mostly rare) that exhibit cytoplasmic inheritance have been identified. These disorders arise from mutations in mtDNA, most of which occur in genes encoding components of the electron-transport chain, which generates most of the ATP (adenosine triphosphate) in aerobic cellular respiration. One such disease is Leber hereditary optic neuropathy (LHON). Patients who

Experiment

Question: How is stem and leaf color inherited in the four-o'clock plant?

Methods
Cross flowers from white, green, and variegated plants in all combinations.

Pollen plant (♂)

Seed plant (♀)	Pollen ← White	Pollen ← Green	Pollen ← Variegated
Results			
White	White	White	White
Green	Green	Green	Green
Variegated	White / Green / Variegated	White / Green / Variegated	White / Green / Variegated

Conclusion: The phenotype of the progeny is determined by the phenotype of the branch from which the seed originated, not from the branch on which the pollen originated. Stem and leaf color exhibits cytoplasmic inheritance.

5.16 Crosses for leaf type in four-o'clocks illustrate cytoplasmic inheritance.

have this disorder experience rapid loss of vision in both eyes, resulting from the death of cells in the optic nerve. This loss of vision typically occurs in early adulthood (usually between the ages of 20 and 24), but it can occur any time after adolescence. There is much clinical variability in the severity of the disease, even within the same family. Leber hereditary optic neuropathy exhibits maternal inheritance: the trait is always passed from mother to child.

Genetic Maternal Effect

A genetic phenomenon that is sometimes confused with cytoplasmic inheritance is **genetic maternal effect,** in which the phenotype of the offspring is determined by the genotype of the mother. In cytoplasmic inheritance, the genes for a characteristic are inherited from only one parent, usually the mother. In genetic maternal effect, the genes are inherited from both parents, but the offspring's phenotype is determined not by its own genotype but by the genotype of its mother.

Genetic maternal effect frequently arises when substances present in the cytoplasm of an egg (encoded by the mother's nuclear genes) are pivotal in early development. An excellent example is the shell coiling of the snail *Limnaea peregra* (**Figure 5.17**). In most snails of this species, the shell coils to the right, which is termed dextral coiling. However, some snails possess a left-coiling shell, exhibiting sinistral coiling. The direction of coiling is determined by a pair of alleles; the allele for dextral (s^+) is dominant over the allele for sinistral (s). However, the direction of coiling is determined not by that snail's own genotype, but by the genotype of its *mother*. The direction of coiling is affected by the way in which the cytoplasm divides soon after fertilization, which in turn is determined by a substance produced by the mother and passed to the offspring in the cytoplasm of the egg.

If a male homozygous for dextral alleles (s^+s^+) is crossed with a female homozygous for sinistral alleles (ss), all of the F_1 are heterozygous (s^+s) and have a sinistral shell, because the genotype of the mother (ss) encodes sinistral coiling (Figure 5.17). If these F_1 snails are self-fertilized, the genotypic ratio of the F_2 is $1\ s^+s^+ : 2\ s^+s : 1\ ss$.

Notice that that the phenotype of all the F_2 snails is dextral coiled, regardless of their genotypes. The F_2 offspring are dextral coiled because the genotype of their mother (s^+s) encodes a right-coiling shell and determines their phenotype. With genetic maternal effect, the phenotype of the progeny is not necessarily the same as the phenotype of the mother, because the progeny's phenotype is determined by the mother's *genotype,* not her phenotype. Neither the male parent's nor the offspring's own genotype has any role in the offspring's phenotype. However, a male does influence the phenotype of the F_2 generation: by contributing to the genotypes of his daughters, he affects the phenotypes of their offspring. Genes that exhibit genetic maternal effect are therefore transmitted through males to future generations. In contrast, genes that exhibit cytoplasmic inheritance are always transmitted through only one of the sexes (usually the female).

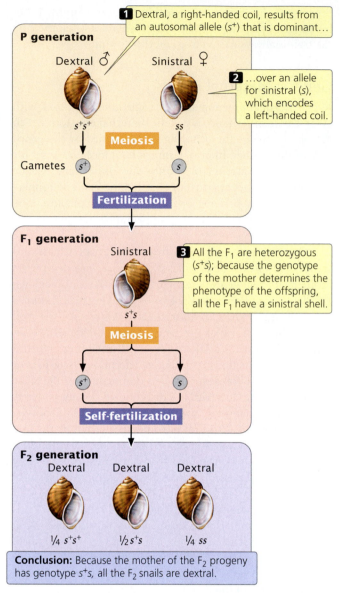

1. Dextral, a right-handed coil, results from an autosomal allele (s^+) that is dominant…

P generation

Dextral ♂   Sinistral ♀

2. …over an allele for sinistral (*s*), which encodes a left-handed coil.

s^+s^+ ss

Meiosis

Gametes s^+ s

Fertilization

F_1 generation

Sinistral

3. All the F_1 are heterozygous (s^+s); because the genotype of the mother determines the phenotype of the offspring, all the F_1 have a sinistral shell.

s^+s

Meiosis

s^+ s

Self-fertilization

F_2 generation

Dextral   Dextral   Dextral

¼ s^+s^+ ½ s^+s ¼ ss

Conclusion: Because the mother of the F_2 progeny has genotype s^+s, all the F_2 snails are dextral.

5.17 In genetic maternal effect, the genotype of the maternal parent determines the phenotype of the offspring. The shell coiling of a snail is a trait that exhibits genetic maternal effect.

CONCEPTS

Characteristics exhibiting cytoplasmic inheritance are encoded by genes in the cytoplasm and are usually inherited from one parent, most commonly the mother. In genetic maternal effect, the genotype of the mother determines the phenotype of the offspring.

✔ CONCEPT CHECK 9

How might you determine whether a particular trait is due to cytoplasmic inheritance or to genetic maternal effect?

Genomic Imprinting

A basic tenet of Mendelian genetics is that the parental origin of a gene does not affect its expression and, therefore, reciprocal crosses give identical results. We have seen that there are some genetic characteristics—those encoded by X-linked genes and cytoplasmic genes—for which reciprocal crosses do not give the same results. In these cases, males and females do not contribute the same genetic material to the offspring. With regard to autosomal genes, males and females contribute the same number of genes, and paternal and maternal genes have long been assumed to have equal effects. However, the expression of some genes is significantly affected by their parental origin. This phenomenon, the differential expression of genetic material depending on whether it is inherited from the male or female parent, is called **genomic imprinting.**

A gene that exhibits genomic imprinting in both mice and humans is *Igf2*, which encodes a protein called insulin-like growth factor II (Igf 2). Offspring inherit one *Igf2* allele from their mother and one from their father. The paternal copy of *Igf2* is actively expressed in the fetus and placenta, but the maternal copy is completely silent (**Figure 5.18**). Both male and female offspring possess *Igf2* genes; the key to whether the gene is expressed is the sex of the parent transmitting the gene. In the present example, the gene is expressed only when it is transmitted by a male parent. In other genomically imprinted traits, the trait is expressed only when the gene is transmitted by the female parent. In a way that is not completely understood, the paternal *Igf 2* allele (but not the maternal allele) promotes placental and fetal growth; when the paternal copy of *Igf 2* is deleted in mice, a small placenta and low-birth-weight offspring result.

Genomic imprinting has been implicated in several human disorders, including Prader–Willi and Angelman syndromes. Children with Prader–Willi syndrome have small hands and feet, short stature, poor sexual development, and mental retardation. These children are small at birth and suckle poorly; but, as toddlers, they develop voracious appetites and frequently become obese. Many persons with Prader–Willi syndrome are missing a small region on the long arm of chromosome 15. The deletion of this region is always inherited from the *father*. Thus, children with Prader–Willi syndrome lack a paternal copy of genes on the long arm of chromosome 15.

The deletion of this same region of chromosome 15 can also be inherited from the *mother*, but this inheritance results in a completely different set of symptoms, producing Angelman syndrome. Children with Angelman syndrome exhibit frequent laughter, uncontrolled muscle movement, a large mouth, and unusual seizures. They are missing a maternal copy of genes on the long arm of chromosome 15. For normal development to take place, copies of this region of chromosome 15 from both male and female parents are apparently required.

(a)

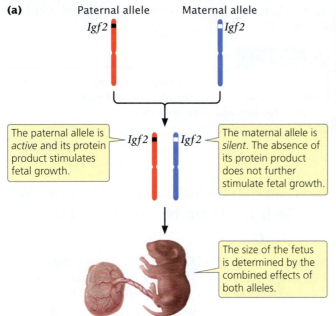

Paternal allele   Maternal allele
Igf2   Igf2

The paternal allele is *active* and its protein product stimulates fetal growth.

Igf2   Igf2

The maternal allele is *silent*. The absence of its protein product does not further stimulate fetal growth.

The size of the fetus is determined by the combined effects of both alleles.

(b)

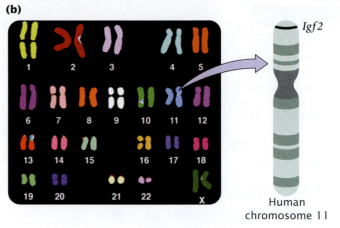

Igf2

Human chromosome 11

5.18 Genomic imprinting of the *Igf2* gene in mice and humans affects fetal growth. (a) The paternal *Igf2* allele is active in the fetus and placenta, whereas the maternal allele is silent. (b) The human *Igf2* locus is on the short arm of chromosome 11; the locus in mice is on chromosome 7. [*Courtesy of Dr. Thomas Ried and Dr. Evelin Schrock.*]

Many imprinted genes in mammals are associated with fetal growth. Imprinting has also been reported in plants, with differential expression of paternal and maternal genes in the endosperm which, like the placenta in mammals, provides nutrients for the growth of the embryo. The mechanism of imprinting is still under active investigation, but the methylation of DNA—the addition of methyl (CH_3) groups to DNA nucleotides (see Chapters 10 and 16)—is essential to the process. In mammals, methylation is erased in the germ cells each generation and then reestablished in the course of gamete formation, with sperm and eggs undergoing different levels of methylation, which then causes the differential expression of male and female alleles in the offspring.

Genomic imprinting is just one form of a phenomenon known as **epigenetics.** Most traits are encoded by genetic information that resides in the sequence of nucleotide bases of the DNA, the so-called genetic code. However, some traits may be caused by other alterations to the DNA that affect the way the DNA sequences are expressed. These changes are often stable and heritable in the sense that they are passed from one cell to another. In genomic imprinting, whether the gene passes through the egg or sperm determines how much methylation takes place. The methylation remains on the DNA as it is passed from cell to cell through mitosis and ultimately determines whether the gene is expressed in the offspring. However, the methylation can be reversed when the DNA passes through a gamete. These types of reversible changes to DNA that influence the expression of traits are termed epigenetic marks. Because

they appear to be responsible for some human diseases, including cancer, epigenetic marks are receiving increased attention by geneticists. Some of the ways in which sex interacts with heredity are summarized in **Table 5.4**.

Table 5.4   Sex influences on heredity

Genetic Phenomenon	Phenotype determined by
Sex-linked characteristic	Genes located on the sex chromosome
Sex-influenced characteristic	Genes on autosomal chromosomes that are more readily expressed in one sex
Sex-limited characteristic	Autosomal genes whose expression is limited to one sex
Genetic maternal effect	Nuclear genotype of the maternal parent
Cytoplasmic inheritance	Cytoplasmic genes, which are usually inherited entirely from only one parent
Genomic imprinting	Genes whose expression is affected by the sex of the transmitting parent

5.7 Anticipation Is the Stronger or Earlier Expression of Traits in Succeeding Generations

Another genetic phenomenon that is not explained by Mendel's principles is **anticipation,** in which a genetic trait becomes more strongly expressed or is expressed at an earlier age as it is passed from generation to generation. In the early 1900s, several physicians observed that many patients with moderate to severe myotonic dystrophy—an autosomal dominant muscle disorder—had ancestors who were only mildly affected by the disease. These observations led to the concept of anticipation. However, the concept quickly fell out of favor with geneticists because there was no obvious mechanism to explain it; traditional genetics held that genes are passed unaltered from parents to offspring. Geneticists tended to attribute anticipation to observational bias.

Research has now reestablished anticipation as a legitimate genetic phenomenon. The mutation causing myotonic dystrophy consists of an unstable region of DNA that can increase or decrease in size as the gene is passed from generation to generation. The age of onset and the severity of the disease are correlated with the size of the unstable region; an increase in the size of the region through generations produces anticipation. The phenomenon has now been implicated in a number of genetic diseases. We will examine these interesting types of mutations in more detail in Chapter 10.

5.8 The Expression of a Genotype May Be Influenced by Environmental Effects

In Chapter 3, we learned that each phenotype is the result of a genotype developing within a specific environment; the genotype sets the potential for development, but how the phenotype actually develops within the limits imposed by the genotype depends on environmental effects. Stated another way, each genotype may produce several different phenotypes, depending on the environmental conditions in which development takes place. For example, a fruit fly homozygous for the vestigial mutation (*vg vg*) develops reduced wings when raised at a temperature below 29°C, but the same genotype develops much longer wings when raised at 31°C. The range of phenotypes (in this case, wing length) produced by a genotype in different environments is called the **norm of reaction** (**Figure 5.19**).

For most of the characteristics discussed so far, the effect of the environment on the phenotype has been slight. Mendel's peas with genotype *yy*, for example, developed green endosperm regardless of the environment in which they were raised. Similarly, persons with genotype $I^A I^A$ have the A antigen on their red blood cells regardless of their diet, socioeconomic status, or family environment. For other phenotypes, however, environmental effects play a more important role.

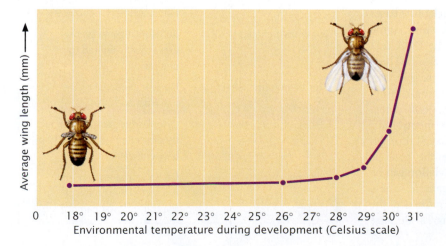

5.19 Norm of reaction is the range of phenotypes produced by a genotype in different environments. This norm of reaction is for individual *Drosophila* flies homozygous for vestigial wings. [*Data from M. H. Harnly,* Journal of Experimental Zoology *56:363–379, 1936.*]

Reared at 20°C or less

Reared at temperatures above 30°C

5.20 The expression of some genotypes depends on specific environments. The expression of a temperature-sensitive allele, *himalayan*, is shown in rabbits reared at different temperatures.

Environmental Effects on Gene Expression

The expression of some genotypes critically depends on the presence of a specific environment. For example, the *himalayan* allele in rabbits produces dark fur at the extremities of the body—on the nose, ears, and feet (**Figure 5.20**). The dark pigment develops, however, only when the rabbit is reared at a temperature of 25°C or less; if a Himalayan rabbit is reared at 30°C, no dark patches develop. The expression of the *himalayan* allele is thus temperature dependent; an enzyme necessary for the production of dark pigment is inactivated at higher temperatures. The pigment is restricted to the nose, feet, and ears of a Himalayan rabbit because the animal's core body temperature is normally above 25°C and the enzyme is functional only in the cells of the relatively cool extremities. The *himalayan* allele is an example of a **temperature-sensitive allele,** an allele whose product is functional only at certain temperatures.

Some types of albinism in plants are temperature dependent. In barley, an autosomal recessive allele inhibits chlorophyll production, producing albinism when the plant is grown at a temperature below 7°C. At temperatures above 18°C, a plant homozygous for the albino allele develops normal chlorophyll and is green. Similarly, vestigial wings in *Drosophila melanogaster* is caused by a temperature-dependent mutation (see Figure 5.19).

Environmental factors also play an important role in the expression of a number of human genetic diseases. Glucose-6-phosphate dehydrogenase is an enzyme that helps to supply energy to the cell. In humans, there are a number of genetic variants of glucose-6-phosphate dehydrogenase, some of which destroy red blood cells when the body is stressed by infection or by the ingestion of certain drugs or foods. The symptoms of the genetic disease, called glucose-6-phosphate dehydrogenase deficiency, appear only in the presence of these specific environmental factors.

Another genetic disease, phenylketonuria (PKU), is due to an autosomal recessive allele that causes mental retardation. The disorder arises from a defect in an enzyme that normally metabolizes the amino acid phenylalanine. When this enzyme is defective, phenylalanine is not metabolized, and its buildup causes brain damage in children. A simple environmental change, putting an affected child on a low-phenylalanine diet, prevents retardation. Phenylketonuria is discussed in more detail in Chapter 6.

These examples illustrate the point that genes and their products do not act in isolation; rather, they frequently interact with environmental factors. Occasionally, environmental factors alone can produce a phenotype that is the same as the phenotype produced by a genotype; this phenotype is called a **phenocopy.** In fruit flies, for example, the autosomal recessive mutation *eyeless* produces greatly reduced eyes. The eyeless phenotype can also be produced by exposing the larvae of normal flies to sodium metaborate.

CONCEPTS

The expression of many genes is modified by the environment. The range of phenotypes produced by a genotype in different environments is called the norm of reaction. A phenocopy is a trait produced by environmental effects that mimics the phenotype produced by a genotype.

✔ CONCEPT CHECK 11

How can you determine whether a phenotype such as reduced eyes in fruit flies is due to a recessive mutation or is a phenocopy?

The Inheritance of Continuous Characteristics

So far, we've dealt primarily with characteristics that have only a few distinct phenotypes. In Mendel's peas, for example, the seeds were either smooth or wrinkled, yellow or green; the coats of dogs were black, brown, or yellow; blood types were of four distinct types, A, B, AB, or O. Such characteristics, which have a few easily distinguished phenotypes, are called **discontinuous characteristics.**

Not all characteristics exhibit discontinuous phenotypes. Human height is an example of such a characteristic; people do not come in just a few distinct heights but, rather, display a continuum of heights. Indeed, there are so many possible phenotypes of human height that we must use a measurement to describe a person's height. Characteristics that exhibit a continuous distribution of phenotypes are termed **continuous characteristics.** Because such characteristics have many possible phenotypes and must be described in quantitative terms, continuous characteristics are also called **quantitative characteristics.**

Continuous characteristics frequently arise because genes at many loci interact to produce the phenotypes. When a single locus with two alleles encodes a characteristic, there are three genotypes possible: *AA, Aa,* and *aa.* With two loci, each with two alleles, there are $3^2 = 9$ genotypes possible. The number of genotypes encoding a characteristic is 3^n, where n equals the number of loci with two alleles that influence the characteristic. For example, when a characteristic is determined by eight loci, each with two alleles, there are $3^8 = 6561$ different genotypes possible for this characteristic. If each genotype produces a different phenotype, many phenotypes will be possible. The slight differences between the phenotypes will be indistinguishable, and the characteristic will appear continuous. Characteristics encoded by genes at many loci are called **polygenic characteristics.**

The converse of polygeny is **pleiotropy,** in which one gene affects multiple characteristics. Many genes exhibit pleiotropy. Phenylketonuria, mentioned earlier, results from a recessive allele; persons homozygous for this allele, if untreated, exhibit mental retardation, blue eyes, and light skin color.

Frequently, the phenotypes of continuous characteristics are also influenced by environmental factors. Each genotype is capable of producing a range of phenotypes: it has a broad norm of reaction. In this situation, the particular phenotype that results depends on both the genotype and the environmental conditions in which the genotype develops. For example, only three genotypes may encode a characteristic, but, because each genotype has a broad norm of reaction, the phenotype of the characteristic exhibits a continuous distribution. Many continuous characteristics are both polygenic and influenced by environmental factors; such characteristics are called **multifactorial characteristics** because many factors help determine the phenotype.

The inheritance of continuous characteristics may appear to be complex, but the alleles at each locus follow Mendel's principles and are inherited in the same way as alleles encoding simple, discontinuous characteristics. However, because many genes participate, because environmental factors influence the phenotype, and because the phenotypes do not sort out into a few distinct types, we cannot observe the distinct ratios that have allowed us to interpret the genetic basis of discontinuous characteristics. To analyze continuous characteristics, we must employ special statistical tools, as will be discussed in Chapter 11.

CONCEPTS

Discontinuous characteristics exhibit a few distinct phenotypes; continuous characteristics exhibit a range of phenotypes. A continuous characteristic is frequently produced when genes at many loci and environmental factors combine to determine a phenotype.

✔ **CONCEPT CHECK 12**

What is the difference between polygeny and pleiotropy?

CONCEPTS SUMMARY

- Dominance always refers to genes at the same locus (allelic genes) and can be understood in regard to how the phenotype of the heterozygote relates to the phenotypes of the homozygotes.

- Dominance is complete when a heterozygote has the same phenotype as a homozygote, is incomplete when the heterozygote has a phenotype intermediate between those of two parental homozygotes, and is codominant when the heterozygote exhibits traits of both parental homozygotes.

- The type of dominance does not affect the inheritance of an allele; it does affect the phenotypic expression of the allele. The classification of dominance may depend on the level of the phenotype examined.

- Penetrance is the percentage of individuals having a particular genotype that exhibit the expected phenotype. Expressivity is the degree to which a character is expressed.

- Lethal alleles cause the death of an individual possessing them, usually at an early stage of development, and may alter phenotypic ratios.

- Multiple alleles refer to the presence of more than two alleles at a locus within a group. Their presence increases the number of genotypes and phenotypes possible.

- Gene interaction refers to the interaction between genes at different loci to produce a single phenotype. An epistatic gene at one locus suppresses or masks the expression of hypostatic genes at other loci. Gene interaction frequently produces phenotypic ratios that are modifications of dihybrid ratios.

- Sex-influenced characteristics are encoded by autosomal genes that are expressed more readily in one sex. Sex-limited characteristics are encoded by autosomal genes expressed in only one sex.

- In cytoplasmic inheritance, the genes for the characteristic are found in the organelles and are usually inherited from a single (usually maternal) parent. Genetic maternal effect is present when an offspring inherits genes from both parents, but the nuclear genes of the mother determine the offspring's phenotype.
- Genomic imprinting refers to characteristics encoded by autosomal genes whose expression is affected by the sex of the parent transmitting the genes.

- Anticipation refers to a genetic trait that is more strongly expressed or is expressed at an earlier age in succeeding generations.
- Phenotypes are often modified by environmental effects. A phenocopy is a phenotype produced by an environmental effect that mimics a phenotype produced by a genotype.
- Continuous characteristics are those that exhibit a wide range of phenotypes; they are frequently produced by the combined effects of many genes and environmental effects.

IMPORTANT TERMS

codominance (p. 101)
incomplete penetrance (p. 102)
penetrance (p. 102)
expressivity (p. 102)
lethal allele (p. 103)
multiple alleles (p. 103)
gene interaction (p. 105)
epistasis (p. 107)
epistatic gene (p. 107)
hypostatic gene (p. 107)

complementation test (p. 113)
complementation (p. 113)
sex-influenced characteristic (p. 115)
sex-limited characteristic (p. 115)
cytoplasmic inheritance (p. 117)
genetic maternal effect (p. 119)
genomic imprinting (p. 120)
epigenetics (p. 121)
anticipation (p. 122)
norm of reaction (p. 122)

temperature-sensitive allele (p. 123)
phenocopy (p. 123)
discontinuous characteristic (p. 123)
continuous characteristic (p. 124)
quantitative characteristic (p. 124)
polygenic characteristic (p. 124)
pleiotropy (p. 124)
multifactorial characteristic (p. 124)

ANSWERS TO CONCEPT CHECKS

1. With complete dominance, the heterozygote expresses the same phenotype as that of one of the homozygotes. With incomplete dominance, the heterozygote has a phenotype that is intermediate between the two homozygotes. And, with codominance, the heterozygote has a phenotype that simultaneously expresses the phenotypes of both homozygotes.

2. The cross is $Ll \times Ll$, where l is an allele for long fingers and L is an allele for normal fingers. The probability that the child will possess the genotype for long fingers (ll) is $\frac{1}{4}$, or 0.25. The trait has a penetrance of 80%, which indicates that a person with the genotype for long fingers has a probability of 0.8 of actually having long fingers. The probability that the child will have long fingers is found by multiplying the probability of the genotype by the probability that a person with that genotype will express the trait: $0.25 \times 0.8 = 0.2$.

3. A 2 : 1 ratio is usually due to a lethal gene. The cross is $Ww \times Ww \rightarrow \frac{1}{4} WW$, $\frac{1}{2} Ww$, and $\frac{1}{4} ww$. One of the homozygotes dies, yielding $\frac{2}{3} Ww$ (green) and $\frac{1}{3} ww$ (white). The green progeny are therefore heterozygous (Ww).

4. People with blood-type A can be I^AI^A or I^Ai. People with blood-type B can be either I^BI^B or I^Bi. The types of matings possible between a man with blood-type A and a woman with blood-type B, along with the offspring that each mating would produce, are as follows:

Possible matings	Offspring
$I^AI^A \times I^BI^B$	I^AI^B
$I^AI^A \times I^Bi$	I^AI^B, I^Ai
$I^Ai \times I^BI^B$	I^AI^B, I^Bi
$I^Ai \times I^Bi$	I^AI^B, I^Ai, I^Bi, ii

Thus, the offspring could have blood-types AB, A, B, or O.

5. Gene interaction is interaction between genes at different loci. Dominance is interaction between alleles at a single locus.

6. The 12 all-white : 3 black : 1 gray ratio is a modification of the 9 : 3 : 3 : 1 ratio produced in a cross between two double heterozygotes:

$$Ww\,Gg \times Ww\,Gg$$
$$\downarrow$$

$\frac{9}{16}\ W\_\ G\_$	all-white
$\frac{3}{16}\ W\_\ gg$	all-white
$\frac{3}{16}\ ww\ G\_$	black
$\frac{1}{16}\ ww\ gg$	gray

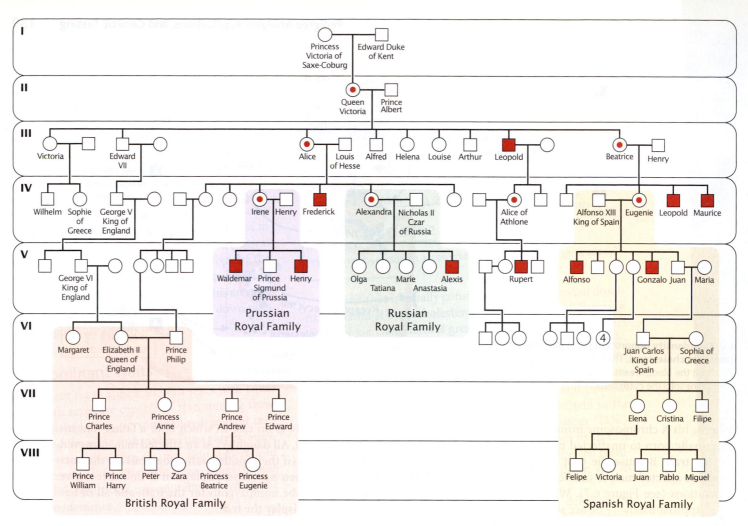

6.8 Classic hemophilia is inherited as an X-linked recessive trait. This pedigree is of hemophilia in the royal families of Europe.

producing pain, swelling, and erosion of the bone. Fortunately, bleeding in people with hemophilia A can now be controlled by administering concentrated doses of factor VIII.

X-Linked Dominant Traits

X-linked dominant traits appear in males and females, although they often affect more females than males. Each person with an X-linked dominant trait must have an affected parent (unless the person possesses a new mutation or the trait has reduced penetrance). X-linked dominant traits do not skip generations (**Figure 6.9**); affected men pass the trait on to all their daughters and none of their sons, as is seen in the children of I-1 in Figure 6.9. In contrast, affected women (if heterozygous) pass the trait on to about $\frac{1}{2}$ of their sons and about $\frac{1}{2}$ of their daughters, as seen in the children of III-6 in the pedigree. As with X-linked recessive traits, a male inherits an X-linked dominant trait only from his mother; the trait is not passed from father to son. A female, on the other hand, inherits an X chromosome from both her mother and her father; so females can receive an X-linked trait from either parent.

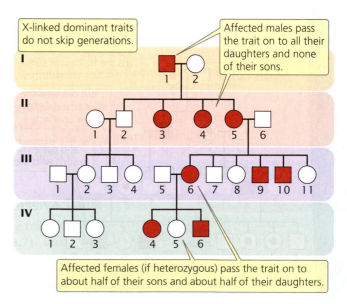

6.9 X-linked dominant traits affect both males and females. An affected male must have an affected mother.

CONCEPTS

X-linked dominant traits affect both males and females. Affected males must have affected mothers (unless the males possess a new mutation), and they pass the trait on to all their daughters.

✔ CONCEPT CHECK 4

A male affected with an X-linked dominant trait will have what proportion of offspring affected with the trait?

a. ½ sons and ½ daughters   c. All daughters and no sons

b. All sons and no daughters   d. ¾ daughters and ¼ sons

An example of an X-linked dominant trait in humans is hypophosphatemia, or familial vitamin-D-resistant rickets. People with this trait have features that superficially resemble those produced by rickets: bone deformities, stiff spines and joints, bowed legs, and mild growth deficiencies. This disorder, however, is resistant to treatment with vitamin D, which normally cures rickets. X-linked hypophosphatemia results from the defective transport of phosphate, especially in cells of the kidneys. People with this disorder excrete large amounts of phosphate in their urine, resulting in low levels of phosphate in the blood and reduced deposition of minerals in the bone. As is common with X-linked dominant traits, males with hypophosphatemia are often more severely affected than females.

Y-Linked Traits

Y-linked traits exhibit a specific, easily recognized pattern of inheritance. Only males are affected, and the trait is passed from father to son. If a man is affected, all his male offspring also should be affected, as is the case for I-1, II-4, II-6, III-6, and III-10 of the pedigree in **Figure 6.10**. Y-linked traits do

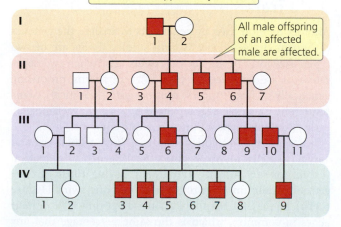

6.10 Y-linked traits appear only in males and are passed from a father to all his sons.

not skip generations. As mentioned in Chapter 4, little genetic information is found on the human Y chromosome. Maleness is one of the few traits in humans that has been shown to be Y linked.

CONCEPTS

Y-linked traits appear only in males and are passed from a father to all his sons.

✔ CONCEPT CHECK 5

What features of a pedigree would distinguish between a Y-linked trait and a trait that was rare, autosomal dominant, and sex limited to males?

The major characteristics of autosomal recessive, autosomal dominant, X-linked recessive, X-linked dominant, and Y-linked traits are summarized in **Table 6.1**.

Worked Problem

The following pedigree represents the inheritance of a rare disorder in an extended family. What is the most likely mode of inheritance for this disease? (Assume that the trait is fully penetrant.)

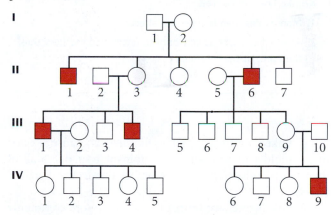

• Solution

To answer this question, we should consider each mode of inheritance and determine which, if any, we can eliminate. The trait appears only in males, and so autosomal dominant and autosomal recessive modes of inheritance are unlikely, because traits with these modes appear equally in males and females. Additionally, autosomal dominance can be eliminated because some affected persons do not have an affected parent.

The trait is observed only among males in this pedigree, which might suggest Y-linked inheritance. However, for a Y-linked trait, affected men should pass the trait on to all their sons, which is not the case here; II-6 is an affected man who has four unaffected male offspring. We can eliminate Y-linked inheritance.

Table 6.1 Pedigree characteristics of autosomal recessive, autosomal dominant, X-linked recessive, X-linked dominant, and Y-linked traits

Autosomal recessive trait

1. Appears in both sexes with equal frequency.
2. Tends to skip generations.
3. Affected offspring are usually born to unaffected parents.
4. When both parents are heterozygous, approximately $1/4$ of the offspring will be affected.
5. Appears more frequently among the children of consanguine marriages.

Autosomal dominant trait

1. Appears in both sexes with equal frequency.
2. Both sexes transmit the trait to their offspring.
3. Does not skip generations.
4. Affected offspring must have an affected parent, unless they possess a new mutation.
5. When one parent is affected (heterozygous) and the other parent is unaffected, approximately $1/2$ of the offspring will be affected.
6. Unaffected parents do not transmit the trait.

X-linked recessive trait

1. More males than females are affected.
2. Affected sons are usually born to unaffected mothers; thus, the trait skips generations.

3. Approximately $1/2$ of a carrier (heterozygous) mother's sons are affected.
4. Never passed from father to son.
5. All daughters of affected fathers are carriers.

X-linked dominant trait

1. Both males and females are affected; often more females than males are affected.
2. Does not skip generations. Affected sons must have an affected mother; affected daughters must have either an affected mother or an affected father.
3. Affected fathers will pass the trait on to all their daughters.
4. Affected mothers (if heterozygous) will pass the trait on to $1/2$ of their sons and $1/2$ of their daughters.

Y-linked trait

1. Only males are affected.
2. Passed from father to all sons.
3. Does not skip generations.

X-linked dominance can be eliminated because affected men should pass an X-linked dominant trait on to all of their female offspring, and II-6 has an unaffected daughter (III-9).

X-linked recessive traits often appear more commonly in males, and affected males are usually born to unaffected female carriers; the pedigree shows this pattern of inheritance. For an X-linked trait, about half the sons of a heterozygous carrier mother should be affected. II-3 and III-9 are suspected carriers, and about $1/2$ of their male children (three of five) are affected. Another important characteristic of an X-linked recessive trait is that it is not passed on from father to son. We observe no father-to-son transmission in this pedigree. X-linked recessive is therefore the most likely mode of inheritance.

> ❓ For additional practice, try to determine the mode of inheritance for the pedigrees in Problem 21 at the end of the chapter.

6.4 The Study of Twins Can Be Used to Assess the Importance of Genes and Environment on Variation in a Trait

Another method used by geneticists to analyze the genetics of human characteristics is twin studies. Twins are of two types: **dizygotic** (nonidentical) **twins** arise when two separate eggs are fertilized by two different sperm, producing genetically distinct zygotes; **monozygotic** (identical) **twins** result when a single egg, fertilized by a single sperm, splits early in development into two separate embryos.

Because monozygotic twins arise from a single egg and sperm (a single, "mono," zygote), they're genetically identical (except for rare somatic mutations), having 100% of their genes in common (**Figure 6.11a**). Dizygotic twins (**Figure 6.11b**), on the other hand, have on average only 50% of their genes in common, which is the same percentage that any pair of siblings has in common. Like other siblings, dizygotic twins may be of the same sex or of different sexes. The only

(a)

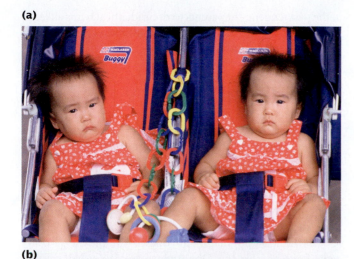

(b)

6.11 Monozygotic twins (a) are identical; dizygotic twins (b) are nonidentical. *[Part a: Joe Carini/Index Stock Imagery/PictureQuest. Part b: Courtesy of Randi Rossignol.]*

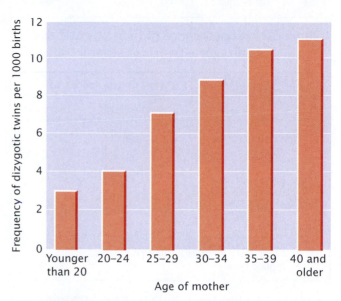

6.12 Older women tend to have more dizygotic twins than do younger women. Relation between the rate of dizygotic twinning and maternal age. *[Data from J. Yerushalmy and S. E. Sheeras,* Human Biology *12:95–113, 1940.]*

✔ **CONCEPT CHECK 6**

Why are monozygotic twins genetically identical, whereas dizygotic twins have only $\frac{1}{2}$ of their genes in common on average?

a. Monozygotic twins tend to look more similar.

b. Monozygotic twins are always the same sex.

c. Dizygotic twins occur more frequently with older mothers.

d. Monozygotic twins develop from a single embryo, whereas dizygotic twins develop from two embryos.

---

difference between dizygotic twins and other siblings is that dizygotic twins are the same age and shared the same uterine environment.

The frequency with which dizygotic twins are born varies among populations. Among North American Caucasians, about 7 dizygotic twin pairs are born per 1000 births but, among Japanese, the rate is only about 3 pairs per 1000 births; among Nigerians, about 40 dizygotic twin pairs are born per 1000 births. The rate of dizygotic twinning also varies with maternal age (**Figure 6.12**), and dizygotic twinning tends to run in families. In contrast, monozygotic twinning is relatively constant. The frequency of monozygotic twinning in most ethnic groups is about 4 twin pairs per 1000 births, and there is little tendency for monozygotic twins to run in families.

CONCEPTS

Dizygotic twins develop from two eggs fertilized by two separate sperm; they have, on average, 50% of their genes in common. Monozygotic twins develop from a single egg, fertilized by a single sperm, that splits into two embryos; they have 100% percent of their genes in common.

Concordance

Comparisons of dizygotic and monozygotic twins can be used to assess the importance of genetic and environmental factors in producing differences in a characteristic. This assessment is often made by calculating the concordance for a trait. If both members of a twin pair have a trait, the twins are said to be *concordant;* if only one member of the pair has the trait, the twins are said to be *discordant.* **Concordance** is the percentage of twin pairs that are concordant for a trait. Because identical twins have 100% of their genes in common and dizygotic twins have on average only 50% in common, genetically influenced traits should exhibit higher concordance in monozygotic twins. For instance, when one member of a monozygotic twin pair has asthma, the other twin of the pair has asthma about 48% of the time; so the monozygotic concordance for asthma is 48%. However, when a dizygotic twin has asthma, the other twin has asthma only 19% of the time (19% dizygotic concordance). The higher concordance in the monozygotic twins suggests that genes influence asthma, a finding supported by the results of

Table 6.2   Concordance of monozygotic and dizygotic twins for several traits

Trait	Concordance (%)	
	Monozygotic	Dizygotic
(1) Heart attack (males)	39	26
(2) Heart attack (females)	44	14
(3) Bronchial asthma	47	24
(4) Cancer (all sites)	12	15
(5) Epilepsy	59	19
(6) Rheumatoid arthritis	32	6
(7) Multiple sclerosis	28	5

Sources: (1 and 2) B. Havald and M. Hauge, U.S. Public Health Service Publication 1103 (1963), pp. 61–67. (3, 4, and 5) B. Havald and M. Hauge, *Genetics and the Epidemiology of Chronic Diseases* (U.S. Department of Health, Education, and Welfare, 1965). (6) J. S. Lawrence, *Annals of Rheumatic Diseases* 26:357–379, 1970. (7) G. C. Ebers et al., *American Journal of Human Genetics* 36:495, 1984.

other family studies of this disease. Concordance values for several human traits and diseases are listed in **Table 6.2.**

The hallmark of a genetic influence on a particular trait is higher concordance in monozygotic twins compared with concordance in dizygotic twins. High concordance in monozygotic twins by itself does not signal a genetic influence. Twins normally share the same environment—they are raised in the same home, have the same friends, attend the same school—and so high concordance may be due to common genes or to common environment. If the high concordance is due to environmental factors, then dizygotic twins, who also share the same environment, should have just as high a concordance as that of monozygotic twins. When genes influence the trait, however, monozygotic twin pairs should exhibit higher concordance than dizygotic twin pairs, because monozygotic twins have a greater percentage of genes in common. It is important to note that any discordance among monozygotic twins must be due to environmental factors, because monozygotic twins are genetically identical.

The use of twins in genetic research rests on the important assumption that, when there is greater concordance in monozygotic twins than in dizygotic twins, it is because monozygotic twins are more similar in their genes and not because they have experienced a more similar environment. The degree of environmental similarity between monozygotic twins and dizygotic twins is assumed to be the same. This assumption may not always be correct, particularly for human behaviors. Because they look alike, identical twins may be treated more similarly by parents, teachers, and peers than are nonidentical twins. Evidence of this similar treatment is seen in the past tendency of parents to dress identical twins alike. In spite of this potential complication, twin studies have played a pivotal role in the study of human genetics.

Twin Studies and Obesity

To illustrate the use of twins in genetic research, let's consider a genetic study of obesity. Obesity is an excess of body fat, often loosely defined as being more than 20% over the ideal weight for height. (Technically, obesity is defined as having a body-mass index (BMI) greater than 30, where BMI is body weight adjusted for height.) Obesity is a serious public-health problem that has reached epidemic proportions in many developed countries. In the United States, 33% of the adult population are obese—double the percentages of just 20 years ago. Obesity increases the risk of a number of medical conditions, including diabetes, gallbladder disease, stroke, high blood pressure, some cancers, and heart disease. An estimated 112,000 people in the United States alone die each year of obesity-related diseases.

Obesity is clearly familial (tends to run in families): when both parents are obese, 80% of their children also will become obese; when both parents are not overweight, only 15% of their children will eventually become obese. The familial nature of obesity could result from genes that influence body weight; alternatively, it could be entirely environmental, resulting from the fact that family members usually have similar diets and exercise habits.

A number of genetic studies have examined twins in an effort to untangle the genetic and environmental contributions to obesity. The largest twin study of obesity was conducted on more than 4000 pairs of twins taken from the National Academy of Sciences National Research Council twin registry. This registry is a database of almost 16,000 male twin pairs, born between 1917 and 1927, who served in the U.S. armed forces in World War II or the Korean War. Albert Stunkard and his colleagues obtained weight and height for each of the twins from medical records compiled at the time of their induction into the armed forces. Equivalent data were again collected in 1967, when the men were 40 to 50 years old. The researchers then computed how overweight each man was at induction and at middle age in 1967. Concordance values for monozygotic and dizygotic twins were then computed for several weight categories (**Table 6.3**).

In each weight category, concordance was significantly higher in monozygotic twins than in dizygotic twins at induction and in middle age 25 years later. The researchers concluded that, in the group being studied, body weight appeared to be strongly influenced by genetic factors. Using statistics that are beyond the scope of this discussion, the researchers further concluded that genetics accounted for 77% of variation in body weight at induction and for 84% at middle age in 1967. (Because a characteristic such as body weight changes in a lifetime, the effects of genes on the characteristic may vary with age.)

Table 6.3	Concordance values for body weight among monozygotic twins (MZ) and dizygotic twins (DZ) at induction in the armed services and at follow-up

| | Concordance (%) | | | |
| Percent Overweight* | At Induction | | At Follow-up in 1967 | |
	MZ	DZ	MZ	DZ
15	61	31	68	49
20	57	27	60	40
25	46	24	54	26
30	51	19	47	16
35	44	12	43	9
40	44	0	36	6

*Percent overweight was determined by comparing each man's actual weight with a standard recommended weight for his height.

Source: After A. J. Stunkard, T. T. Foch, and Z. Hrubec, A twin study of human obesity, *Journal of the American Medical Association* 256:52, 1986.

Findings from this study show that genes influence variation in body weight, yet genes *alone* do not cause obesity. In less-affluent societies, obesity is rare, and no one can become overweight unless caloric intake exceeds energy expenditure. A person does not inherit obesity; rather, the person inherits a predisposition toward a particular body weight; geneticists say that some people are genetically more *at risk* for obesity than others.

How genes affect the risk of obesity is not yet completely understood. In 1994, scientists at Rockefeller University isolated a gene that causes an inherited form of obesity in mice (**Figure 6.13**). This gene encodes a protein called leptin, named after the Greek word for "thin." Leptin is produced by fat tissue and decreases appetite by affecting the hypothalamus, a part of the brain. A decrease in body fat leads to decreased leptin, which stimulates appetite; an increase in body fat leads to increased levels of leptin, which reduces appetite. Obese mice possess two mutated copies of the gene for leptin and produce no functional leptin; giving leptin to these mice promotes weight loss.

The discovery of the gene that encodes leptin raised hopes that obesity in humans might be influenced by defects in the same gene and that the administration of leptin might be an effective treatment for obesity. A few rare obese people are genetically deficient in leptin but, unfortunately, most overweight people actually have elevated levels of leptin and appear to be resistant to its effects. Findings from further studies have revealed that the genetic and hormonal control

6.13 Obesity in some mice is due to a defect in the gene that encodes the protein leptin. Obese mouse on the left compared with normal-sized mouse on the right. *[Remi Banali/Liaison/Getty.]*

of body weight is quite complex; other genes that also cause obesity in mice and humans have been identified, and the molecular basis of weight control is still being investigated.

CONCEPTS

Higher concordance in monozygotic twins compared with that in dizygotic twins indicates that genetic factors play a role in determining differences in a trait. Low concordance in monozygotic twins indicates that environmental factors play a significant role.

✔ CONCEPT CHECK 7

A trait exhibits 100% concordance in both monozygotic and dizygotic twins. What conclusion can you draw about the role of genetic factors in determining differences in the trait?

a. Genetic factors are extremely important.

b. Genetic factors are somewhat important.

c. Genetic factors are unimportant.

d. Both genetic and environmental factors are important.

6.5 Adoption Studies Are Another Technique for Examining the Effects of Genes and Environment on Variation in Traits

A third technique used by geneticists to analyze human inheritance is the study of adopted people. This approach is one of the most powerful for distinguishing the effects of genes and environment on characteristics.

For a variety of reasons, many children each year are separated from their biological parents soon after birth and adopted by adults with whom they have no genetic relationship. These adopted persons have no more genes in common with their adoptive parents than do two randomly chosen

Experiment

Question: Is body-mass index (BMI) influenced by genetic factors?

Methods

Compare the BMIs of adopted children with those of their adoptive and biological parents.

Results

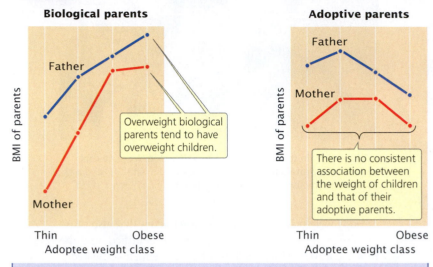

Conclusion: Genetic factors influence body-mass index.

✔ CONCEPT CHECK 8

What assumptions underlie the use of adoption studies in genetics?

a. Adoptees have no contact with their biological parents after birth.

b. The foster parents and biological parents are not related.

c. The environments of biological and adopted parents are independent.

d. All of the above.

Adoption Studies and Obesity

Like twin studies, adoption studies have played an important role in demonstrating that obesity has a genetic influence. In 1986, geneticists published the results of a study of 540 people who had been adopted in Denmark between 1924 and 1947. The geneticists obtained information concerning the adult body weight and height of the adopted persons, along with the adult weight and height of their biological parents and their unrelated adoptive parents.

Geneticists used the body-mass index to analyze the relation between the weight of the adopted persons and that of their parents. On the basis of the BMI, sex, and age, the adopted persons were divided into four weight classes: thin, median weight, overweight, and obese. A strong relation was found between the weight classification of the adopted persons and the BMIs of their biological parents: obese adoptees tended to have heavier biological parents, whereas thin adoptees tended to have lighter biological parents (**Figure 6.14**). Because the only connection between the adoptees and their biological parents was the genes that they have in common, the investigators concluded that genetic factors influence adult body weight. There was no clear relation between the weight classification of adoptees and the BMIs of their adoptive parents (see Figure 6.14), suggesting that the rearing environment has little effect on adult body weight.

6.6 Genetic Counseling Provides Information to Those Concerned about Genetic Diseases and Traits

Our knowledge of human genetic diseases and disorders has expanded rapidly in recent years. The *Online Mendelian Inheritance in Man* now lists more than 16,000 human genetic diseases, disorders, and traits that have a simple genetic basis. Research has provided a great deal of information about the inheritance, chromosomal location, biochemical basis, and symptoms of many of these genetic traits, diseases, and disorders. This information is often useful to people who have a genetic condition.

persons; however, they do share an environment with their adoptive parents. In contrast, the adopted persons have 50% of their genes in common with each of their biological parents but do not share the same environment with them. If adopted persons and their adoptive parents show similarities in a characteristic, these similarities can be attributed to environmental factors. If, on the other hand, adopted persons and their biological parents show similarities, these similarities are likely to be due to genetic factors. Comparisons of adopted persons with their adoptive parents and with their biological parents can therefore help to define the roles of genetic and environmental factors in the determination of human variation.

Adoption studies assume that the environments of biological and adoptive families are independent (i.e., not more alike than would be expected by chance). This assumption may not always be correct, because adoption agencies carefully choose adoptive parents and may select a family that resembles the biological family. Thus, some of the similarity between adopted persons and their biological parents may be due to these similar environments and not due to common genetic factors. In addition, offspring and the biological mother share the same environment during prenatal development.

CONCEPTS

Similarities between adopted persons and their genetically unrelated adoptive parents indicate that environmental factors affect a particular characteristic; similarities between adopted persons and their biological parents indicate that genetic factors influence the characteristic.

Table 6.4	**Common reasons for seeking genetic counseling**

1. A person knows of a genetic disease in the family.

2. A couple has given birth to a child with a genetic disease, birth defect, or chromosomal abnormality.

3. A couple has a child who is mentally retarded or has a close relative who is mentally retarded.

4. An older woman becomes pregnant or wants to become pregnant. There is disagreement about the age at which a prospective mother who has no other risk factor should seek genetic counseling; many experts suggest that any prospective mother age 35 or older should seek genetic counseling.

5. Husband and wife are closely related (e.g., first cousins).

6. A couple experiences difficulties achieving a successful pregnancy.

7. A pregnant woman is concerned about exposure to an environmental substance (drug, chemical, or virus) that causes birth defects.

8. A couple needs assistance in interpreting the results of a prenatal or other test.

9. Both prospective parents are known carriers for a recessive genetic disease.

Genetic counseling is a field that provides information to patients and others who are concerned about hereditary conditions. It is an educational process that helps patients and family members deal with many aspects of a genetic condition including a diagnosis, information about symptoms and treatment, and information about the mode of inheritance. Genetic counseling also helps the patient and family cope with the psychological and physical stress that may be associated with their disorder. Clearly, all of these considerations cannot be handled by a single person; so most genetic counseling is done by a team that can include counselors, physicians, medical geneticists, and laboratory personnel. **Table 6.4** lists some common reasons for seeking genetic counseling.

Genetic counseling usually begins with a diagnosis of the condition. On the bases of a physical examination, biochemical tests, DNA testing, chromosome analysis, family history, and other information, a physician determines the cause of the condition. An accurate diagnosis is critical, because treatment and the probability of passing the condition on may vary, depending on the diagnosis. For example, there are a number of different types of dwarfism, which may be caused by chromosome abnormalities, single-gene mutations, hormonal imbalances, or environmental factors. People who have dwarfism resulting from an autosomal dominant gene have a 50% chance of passing the condition

on to their children, whereas people who have dwarfism caused by a rare recessive gene have a low likelihood of passing the trait on to their children.

When the nature of the condition is known, a genetic counselor meets with the patient and members of the patient's family and explains the diagnosis. A family pedigree may be constructed, and the probability of transmitting the condition to future generations can be calculated for different family members. The counselor helps the family interpret the genetic risks and explains various available reproductive options, including prenatal diagnosis, artificial insemination, and in vitro fertilization. A family's decision about future pregnancies frequently depends on the magnitude of the genetic risk, the severity and effects of the condition, the importance of having children, and religious and cultural views. Throughout the process, a good genetic counselor uses *nondirected* counseling, which means that he or she provides information and facilitates discussion but does not bring his or her own opinion and values into the discussion. The goal of nondirected counseling is for the family to reach its own decision on the basis of the best available information.

Genetic conditions are often perceived differently from other diseases and medical problems, because genetic conditions are intrinsic to the individual person and can be passed on to children. Such perceptions may produce feelings of guilt about past reproductive choices and intense personal dilemmas about future choices. Genetic counselors are trained to help patients and their families recognize and cope with these feelings.

CONCEPTS

Genetic counseling is an educational process that provides patients and their families with information about a genetic condition, its medical implications, the probabilities that others in the family may have the disease, and reproductive options. It also helps patients and their families cope with the psychological and physical stress associated with a genetic condition.

✔ CONCEPT CHECK 9

After a person has been diagnosed with a genetic disease, what kinds of help can a genetic counselor provide?

6.7 Genetic Testing Provides Information about the Potential for Inheriting or Developing a Genetic Condition

The ultimate goal of genetic testing is to recognize the potential for a genetic condition at an early stage. In some cases, genetic testing allows people to make informed

Table 6.5 Examples of genetic diseases and disorders that can be detected prenatally and the techniques used in their detection

Disorder	Method of Detection
Chromosome abnormalities	Examination of a karyotype from cells obtained by amniocentesis or chorionic villus sampling
Cleft lip and palate	Ultrasound
Cystic fibrosis	DNA analysis of cells obtained by amniocentesis or chorionic villus sampling
Dwarfism	Ultrasound or X-ray; some forms can be detected by DNA analysis of cells obtained by amniocentesis or chorionic villus sampling
Hemophilia	Fetal blood sampling* or DNA analysis of cells obtained by amniocentesis or chorionic villus sampling
Lesch–Nyhan syndrome	Biochemical tests on cells obtained by amniocentesis or chorionic villus sampling
Neural-tube defects	Initial screening with maternal blood test, followed by biochemical tests on amniotic fluid obtained by amniocentesis or by the detection of birth defects with the use of ultrasound
Osteogenesis imperfecta (brittle bones)	Ultrasound or X-ray
Phenylketonuria	DNA analysis of cells obtained by amniocentesis or chorionic villus sampling
Sickle-cell anemia	Fetal blood sampling* or DNA analysis of cells obtained by amniocentesis or chorionic villus sampling
Tay–Sachs disease	Biochemical tests on cells obtained by amniocentesis or chorionic villus sampling

*A sample of fetal blood is obtained by inserting a needle into the umbilical cord.

choices about reproduction. In other cases, genetic testing allows early intervention that may lessen or even prevent the development of the condition. For those who know that they are at risk for a genetic condition, genetic testing may help alleviate anxiety associated with the uncertainty of their situation. Genetic testing includes prenatal testing and postnatal testing.

Prenatal Genetic Testing

Prenatal genetic tests are those that are conducted before birth and now include procedures for diagnosing several hundred genetic diseases and disorders (**Table 6.5**). The major purpose of prenatal tests is to provide families with the information that they need to make choices during pregnancies and, in some cases, to prepare for the birth of a child with a genetic condition. The Human Genome Project has accelerated the rate at which new genes are being isolated and new genetic tests are being developed. In spite of these advances, prenatal tests are still not available for many common genetic diseases, and no test can guarantee that a "perfect" child will be born. Several approaches to prenatal diagnosis are described in the following sections.

Ultrasonography Some genetic conditions can be detected through direct visualization of the fetus. Such visualization is most commonly done with the use of **ultrasonography**—usually referred to as ultrasound. In this technique, high-frequency sound is beamed into the uterus;

when the sound waves encounter dense tissue, they bounce back and are transformed into a picture (**Figure 6.15**). The size of the fetus can be determined, as can genetic conditions such as neural-tube defects (defects in the development of the spinal column and the skull) and skeletal abnormalities.

Amniocentesis Most prenatal testing requires fetal tissue, which can be obtained in several ways. The most widely used

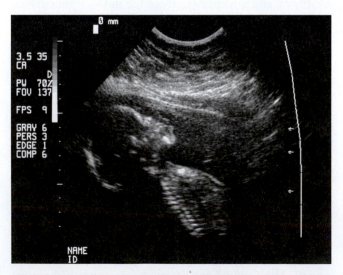

6.15 Ultrasonography can be used to detect some genetic disorders in a fetus and to locate the fetus during amniocentesis and chorionic villus sampling. [PhotoDisc.]

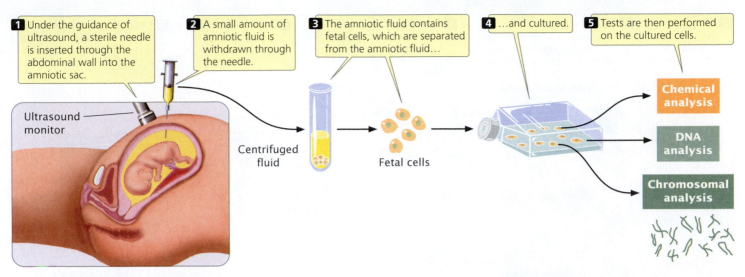

1 | Under the guidance of ultrasound, a sterile needle is inserted through the abdominal wall into the amniotic sac.

2 | A small amount of amniotic fluid is withdrawn through the needle.

3 | The amniotic fluid contains fetal cells, which are separated from the amniotic fluid…

4 | …and cultured.

5 | Tests are then performed on the cultured cells.

Ultrasound monitor

Centrifuged fluid

Fetal cells

Chemical analysis

DNA analysis

Chromosomal analysis

6.16 Amniocentesis is a procedure for obtaining fetal cells for genetic testing.

method is **amniocentesis,** a procedure for obtaining a sample of amniotic fluid from a pregnant woman (**Figure 6.16**). Amniotic fluid—the substance that fills the amniotic sac and surrounds the developing fetus—contains fetal cells that can be used for genetic testing.

Amniocentesis is routinely performed as an outpatient procedure either with or without the use of a local anesthetic. First, ultrasonography is used to locate the position of the fetus in the uterus. Next, a long, sterile needle is inserted through the abdominal wall into the amniotic sac, and a small amount of amniotic fluid is withdrawn through the needle. Fetal cells are separated from the amniotic fluid and placed in a culture medium that stimulates them to grow and divide. Genetic tests are then performed on the cultured cells. Complications with amniocentesis (mostly miscarriage) are uncommon, arising in only about 1 in 400 procedures.

Chorionic villus sampling   A major disadvantage of amniocentesis is that it is routinely performed at about the 15th to 18th week of a pregnancy (although many obstetricians now successfully perform amniocentesis several weeks earlier). The cells obtained by amniocentesis must then be cultured before genetic tests can be performed, requiring yet more time. For these reasons, genetic information about the fetus may not be available until the 17th or 18th week of pregnancy. By this stage, abortion carries a risk of complications and is even more stressful for the parents. **Chorionic villus sampling** (CVS) can be performed earlier (between the 10th and 12th weeks of pregnancy) and collects a larger amount of fetal tissue, which eliminates the necessity of culturing the cells.

In CVS, a catheter—a soft plastic tube—is inserted into the vagina (**Figure 6.17**) and, with the use of ultrasonography for guidance, is pushed through the cervix into the uterus. The tip of the tube is placed into contact with the chorion, the outer layer of the placenta. Suction is then

applied, and a small piece of the chorion is removed. Although the chorion is composed of fetal cells, it is a part of the placenta that is expelled from the uterus after birth; so the tissue that is removed is not actually from the fetus. This tissue contains millions of actively dividing cells that can be used directly in many genetic tests. Chorionic villus sampling has a somewhat higher risk of complication than that of amniocentesis; the results of several studies suggest that this procedure may increase the incidence of limb defects in the fetus when performed earlier than 10 weeks of gestation.

Fetal cells obtained by amniocentesis or by CVS can be used to prepare a **karyotype,** which is a picture of a complete set of metaphase chromosomes. Karyotypes can be studied for chromosome abnormalities (see Chapter 9). Biochemical analyses can be conducted on fetal cells to determine the presence of particular metabolic products of genes. For genetic diseases in which the DNA sequence of the causative gene has been determined, the DNA sequence (DNA testing; see Chapter 10) can be examined for defective alleles.

Maternal blood tests   Some genetic conditions can be detected by performing a blood test on the mother (referred to as **maternal blood testing**). For instance, α-fetoprotein is normally produced by the fetus during development and is present in the fetal blood, the amniotic fluid, and the mother's blood during pregnancy. The level of α-fetoprotein is significantly higher than normal when the fetus has a neural-tube defect or one of several other disorders. Some chromosome abnormalities produce lower-than-normal levels of α-fetoprotein. Measuring the amount of α-fetoprotein in the mother's blood gives an indication of these conditions. However, because other factors affect the amount of α-fetoprotein in maternal blood, a high or low level by itself does not necessarily indicate a problem. Thus, when a blood test

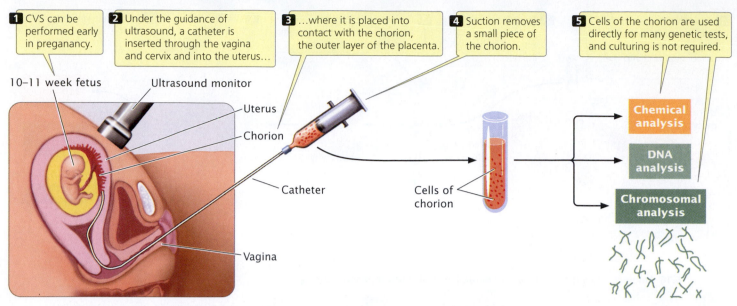

1 CVS can be performed early in preganancy.

2 Under the guidance of ultrasound, a catheter is inserted through the vagina and cervix and into the uterus…

3 …where it is placed into contact with the chorion, the outer layer of the placenta.

4 Suction removes a small piece of the chorion.

5 Cells of the chorion are used directly for many genetic tests, and culturing is not required.

10–11 week fetus   Ultrasound monitor

Uterus

Chorion

Catheter

Vagina

Cells of chorion

Chemical analysis

DNA analysis

Chromosomal analysis

6.17 Chorionic villus sampling (CVS) is another procedure for obtaining fetal cells for genetic testing.

indicates that the amount of α-fetoprotein is abnormal, follow-up tests (additional α-fetoprotein determinations, ultrasound, amniocentesis, or all three) are usually conducted.

Noninvasive fetal diagnosis Prenatal tests that utilize only maternal blood are highly desirable because they are noninvasive and pose no risk to the fetus. During pregnancy, a few fetal cells are released into the mother's circulatory system, where they mix and circulate with her blood. Recent advances have made it possible to separate fetal cells from a maternal blood sample (a procedure called **fetal cell sorting**). With the use of lasers and automated cell-sorting machines, fetal cells can be detected and separated from maternal blood cells. The fetal cells obtained can be cultured for chromosome analysis or used as a source of fetal DNA for molecular testing (see Chapter 10). Maternal blood also contains low levels of free-floating fetal DNA, which also can be tested for mutations. Although still experimental, noninvasive fetal diagnosis has now been used to detect Down syndrome and diseases such as cystic fibrosis and thalassemia (a blood disorder). A current limitation is that the procedure is not as accurate as amniocentesis or CVS, because enough fetal cells for genetic testing cannot always be obtained. Additionally, only mutated genes from the father can be detected, because there is no way to completely separate maternal DNA from the fetal cells. Thus, if the mother carries a copy of the mutation, determining whether the fetus also carries the gene is impossible.

Preimplantation genetic diagnosis Prenatal genetic tests provide today's prospective parents with increasing amounts of information about the health of their future

children. New reproductive technologies provide couples with options for using this information. One of these technologies is in vitro fertilization. In this procedure, hormones are used to induce ovulation. The ovulated eggs are surgically removed from the surface of the ovary, placed in a laboratory dish, and fertilized with sperm. The resulting embryo is then implanted in the uterus. Thousands of babies resulting from in vitro fertilization have now been born.

Genetic testing can be combined with in vitro fertilization to allow the implantation of embryos that are free of a specific genetic defect. Called **preimplantation genetic diagnosis** (PGD), this technique allows people who carry a genetic defect to avoid producing a child with the disorder. For example, when a woman is a carrier of an X-linked recessive disease, approximately half of her sons are expected to have the disease. Through in vitro fertilization and preimplantation testing, an embryo without the disorder can be selected for implantation in her uterus.

The procedure begins with the production of several single-celled embryos through in vitro fertilization. The embryos are allowed to divide several times until they reach the 8- or 16-cell stage. At this point, one cell is removed from each embryo and tested for the genetic abnormality. Removing a single cell at this early stage does not harm the embryo. After determination of which embryos are free of the disorder, a healthy embryo is selected and implanted in the woman's uterus.

Preimplantation genetic diagnosis requires the ability to conduct a genetic test on a single cell. Such testing is possible with the use of the polymerase chain reaction through which minute quantities of DNA can be amplified (replicated) quickly. After amplification of the cell's DNA, the DNA sequence is examined. Although relatively new compared with other techniques, preimplantation diagnosis has been used successfully in more than 1000 births. Its use raises a number of ethical concerns, because it provides a means of actively selecting for or against certain genetic traits.

Postnatal Genetic Testing

Postnatal testing is conducted after birth and includes newborn screening, heterozygote screening, and presymptomatic diagnosis.

Newborn screening Testing for genetic disorders in newborn infants is called **newborn screening.** Most states in the United States and many other countries require that newborn infants be tested for phenylketonuria and galactosemia. These metabolic diseases are caused by autosomal recessive alleles; if not treated at an early age, they can result in mental retardation. But early intervention, through the administration of a modified diet, prevents retardation. Testing is done by analyzing a drop of an infant's blood collected soon after birth. Because of widespread screening, the frequency of mental retardation due to these genetic conditions has dropped tremendously. Screening newborns for additional genetic diseases that benefit from treatment, such as sickle-cell anemia and hypothyroidism, also is common.

Heterozygote screening Testing members of a population to identify heterozygous carriers of recessive disease-causing alleles—members who are healthy but have the potential to produce children with a particular disease—is termed **heterozygote screening.**

Testing for Tay–Sachs disease is a successful example of heterozygote screening. In the general population of North America, the frequency of Tay–Sachs disease is only about 1 person in 360,000. Among Ashkenazi Jews (descendants of Jewish people who settled in eastern and central Europe), the frequency is 100 times as great. A simple blood test is used to identify Ashkenazi Jews who carry the allele for Tay–Sachs disease. If a man and woman are both heterozygotes, approximately one in four of their children is expected to have Tay–Sachs disease. A prenatal test for the Tay–Sachs allele also is available. Screening programs have led to a significant decline in the number of children of Ashkenazi ancestry born with Tay–Sachs disease (now fewer than 10 children per year in the United States).

Presymptomatic testing Evaluating healthy people to determine whether they have inherited a disease-causing allele is known as **presymptomatic genetic testing.** For example, presymptomatic testing is available for members of families that have an autosomal dominant form of breast cancer. In this case, early identification of the disease-causing allele allows for closer surveillance and the early detection of tumors. Presymptomatic testing is also available for some genetic diseases for which no treatment is available, such as Huntington disease, an autosomal dominant disease that leads to slow physical and mental deterioration in middle age. Presymptomatic testing for untreatable conditions raises a number of social and ethical questions (see Chapter 10).

CONCEPTS

Genetic testing is used to screen newborns for genetic diseases, detect persons who are heterozygous for recessive diseases, detect disease-causing alleles in those who have not yet developed symptoms of the disease, and detect defective alleles in unborn babies. Preimplantation genetic diagnosis combined with in vitro fertilization allows for the selection of embryos that are free from specific genetic diseases.

✔ **CONCEPT CHECK 10**

How does preimplantation genetic diagnosis differ from prenatal genetic testing?

6.8 Comparison of Human and Chimpanzee Genomes is Helping to Reveal Genes That Make Humans Unique

Chimpanzees are our closest living relatives, yet we differ from chimps in a huge number of anatomical, behavioral, social, and intellectual skills. The perceived large degree of difference between chimpanzees and humans is manifested by the placement of these two species in entirely different primate families (humans in Hominidae and chimpanzees in Pongidae).

In spite of the large phenotypic gulf between humans and chimpanzees, findings in early DNA-hybridization studies suggested that they differed by only about 1% of their DNA sequences. Recent sequencing of human and chimpanzee genomes has provided more-precise estimates of the genetic differences that separate these species: about 1% of the two genomes differ in base sequences and about 3% differ in regard to deletions and insertions. Thus, more than 95% of the DNA of humans and chimpanzees is identical. But, clearly, humans are not chimpanzees. What, then, are the genetic differences that make us distinctly human? What are the genes that give us our human qualities?

Geneticists are now identifying genes that contribute to human uniqueness and potentially played an important role in the evolution of modern humans. In many cases, these genes have been identified through the study of mutations that cause abnormalities in our human traits, such as brain size or language ability. The recent availability of complete genome sequences for humans and chimpanzees is facilitating the search for genes that give humans their distinctive traits.

One set of genes that potentially contribute to human uniqueness regulates brain size. Mutations at six loci, know as *microcephalin 1* through *microcephalin 6* (*MCPH1–MCPH6*), cause microcephaly, a condition in which the brain is severely reduced in overall size but brain structure is not affected. The observation that mutations in *microcephalin* drastically affect brain size has led to the suggestion that, in the course of human evolution, selection for alleles encoding large brains at one or more of these genes might have led to enlarged brain size in humans.

Geneticists have studied variation in the *microcephalin* genes of different primates and have come to the conclusion that strong selection occurred in the recent past for the sequences of the *microcephalin* genes that are currently found in humans.

Another candidate gene for helping to make humans unique is called *FOXP2*, which is required for human speech. Mutations at the *FOXP2* gene often cause speech and language disabilities. Evolutionary studies of sequence variation in the *FOXP2* gene of mice, humans, and other primates reveal evidence of past selection and suggest that humans acquired their version of the gene (which enables human speech) no more than 200,000 years ago, about the time at which modern humans emerged.

CONCEPTS SUMMARY

- Constraints on the genetic study of human traits include the inability to conduct controlled crosses, long generation time, small family size, and the difficulty of separating genetic and environmental influences. Pedigrees are often used to study the inheritance of traits in humans.

- Autosomal recessive traits typically appear with equal frequency in both sexes and tend to skip generations. When both parents are heterozygous for a particular autosomal recessive trait, approximately $\frac{1}{4}$ of their offspring will have the trait. Recessive traits are more likely to appear in families with consanguinity (mating between closely related persons).

- Autosomal dominant traits usually appear equally in both sexes and do not skip generations. When one parent is affected and heterozygous for an autosomal dominant trait, approximately $\frac{1}{2}$ of the offspring will have the trait. Unaffected people do not normally transmit an autosomal dominant trait to their offspring.

- X-linked recessive traits appear more frequently in males than in females. When a woman is a heterozygous carrier for an X-linked recessive trait and a man is unaffected, approximately $\frac{1}{2}$ of their sons will have the trait and $\frac{1}{2}$ of their daughters will be unaffected carriers. X-linked traits are not passed on from father to son.

- X-linked dominant traits appear in males and females, but more frequently in females. They do not skip generations.

Affected men pass an X-linked dominant trait on to all of their daughters but none of their sons. Heterozygous women pass the trait on to $\frac{1}{2}$ of their sons and $\frac{1}{2}$ of their daughters.

- Y-linked traits appear only in males and are passed on from father to all sons.

- A trait's higher concordance in monozygotic than in dizygotic twins indicates a genetic influence on the trait; less than 100% concordance in monozygotic twins indicates environmental influences on the trait.

- Similarities between adopted children and their biological parents indicate the importance of genetic factors in the expression of a trait; similarities between adopted children and their genetically unrelated adoptive parents indicate the influence of environmental factors.

- Genetic counseling provides information and support to people concerned about hereditary conditions in their families.

- Genetic testing includes prenatal diagnosis, screening for disease-causing alleles in newborns, the detection of people heterozygous for recessive alleles, and presymptomatic testing for the presence of a disease-causing allele in at risk people.

- Genetic research has identified a number of genes that may contribute to human uniqueness, including genes that influence brain size and speech.

IMPORTANT TERMS

pedigree (p. 136)
proband (p. 136)
consanguinity (p. 137)

dizygotic twins (p. 142)
monozygotic twins (p. 142)
concordance (p. 143)

genetic counseling (p. 147)
ultrasonography (p. 148)
amniocentesis (p. 149)

chorionic villus sampling (CVS) (p. 149)
karyotype (p. 150)
maternal blood testing (p. 150)

fetal cell sorting (p. 150)
preimplantation genetic diagnosis (PGD)
 (p. 150)

newborn screening (p. 151)
heterozygote screening (p. 151)
presymptomatic genetic testing (p. 151)

ANSWERS TO CONCEPT CHECKS

1. b

2. It might skip generations when a new mutation arises or the trait has reduced penetrance.

3. If X-linked recessive, the trait will not be passed from father to son.

4. c

5. If the trait were Y linked, an affected male would pass it on to all his sons, whereas, if the trait were autosomal and sex limited, affected heterozygous males would pass it on to only $\frac{1}{2}$ of their sons on average.

6. d

7. c

8. d

9. A genetic counselor can help the patient and family better understand the genetic disease, including its diagnosis, symptoms, and treatment, as well as its mode of inheritance and the risk of other family members having the disease or passing it on to their offspring. The counselor can also help the patient and family members cope with the physical and psychological stress that may be associated with the disease.

10. Preimplantation genetic diagnosis determines the presence of disease-causing genes in an embryo at an early stage, before it is implanted in the uterus and initiates pregnancy. Prenatal genetic diagnosis determines the presence of disease-causing genes or chromosomes in a developing fetus.

WORKED PROBLEMS

1. Joanna has "short fingers" (brachydactyly). She has two older brothers who are identical twins; both have short fingers. Joanna's two younger sisters have normal fingers. Joanna's mother has normal fingers, and her father has short fingers. Joanna's paternal grandmother (her father's mother) has short fingers; her paternal grandfather (her father's father), who is now deceased, had normal fingers. Both of Joanna's maternal grandparents (her mother's parents) have normal fingers. Joanna marries Tom, who has normal fingers; they adopt a son named Bill who has normal fingers. Bill's biological parents both have normal fingers. After adopting Bill, Joanna and Tom produce two children: an older daughter with short fingers and a younger son with normal fingers.

   a. Using standard symbols and labels, draw a pedigree illustrating the inheritance of short fingers in Joanna's family.

   b. What is the most likely mode of inheritance for short fingers in this family?

   c. If Joanna and Tom have another biological child, what is the probability (based on your answer to part *b*) that this child will have short fingers?

• **Solution**

   a. In the pedigree for the family, identify persons with the trait (short fingers) by filled circles (females) and filled

squares (males). Connect Joanna's identical twin brothers to the line above by drawing diagonal lines that have a horizontal line between them. Enclose the adopted child of Joanna and Tom in brackets; connect him to his biological parents by drawing a diagonal line and to his biological parents by a dashed line.

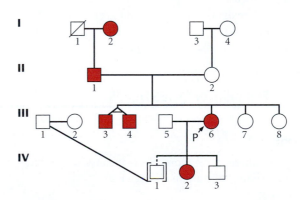

   b. The most likely mode of inheritance for short fingers in this family is autosomal dominant. The trait appears equally in males and females and does not skip generations. When one parent has the trait, it appears in approximately half of that parent's sons and daughters, although the number of children in the families is small. We can eliminate Y-linked inheritance because the trait is found in females as well as males. If short fingers were X-linked recessive, females with the trait would be expected to pass the trait on to all their sons, but Joanna (III-6), who has short fingers, produced a son with normal fingers. For X-linked dominant traits, affected men should pass the trait on to all their daughters; because male II-1 has short fingers and produced two daughters without short fingers (III-7 and III-8), we know that the trait cannot be X-linked dominant. It is unlikely that the trait is autosomal recessive, because it does not skip generations and approximately half of the children of affected parents have the trait.

c. If having short fingers is autosomal dominant, Tom must be homozygous (bb) because he has normal fingers. Joanna must be heterozygous (Bb) because she and Tom have produced both short- and normal- fingered offspring. In a cross between a heterozygote and homozygote, half of the progeny are expected to be heterozygous and half homozygous ($Bb \times bb$: $\frac{1}{2} Bb$, $\frac{1}{2} bb$); so the probability that Joanna's and Tom's next biological child will have short fingers is $\frac{1}{2}$.

2. Concordance values for a series of traits were measured in monozygotic twins and dizygotic twins; the results are shown in the following table. For each trait, indicate whether the rates of concordance suggest genetic influences, environmental influences, or both. Explain your reasoning.

Characteristic	Concordance (%)	
	Monozygotic	Dizygotic
a. ABO blood type	100	65
b. Diabetes	85	36
c. Coffee drinking	80	80
d. Smoking	75	42
e. Schizophrenia	53	16

• Solution

a. The concordance for ABO blood type in the monozygotic twins is 100%. This high concordance in monozygotic twins does not, by itself, indicate a genetic basis for the trait. An important indicator of a genetic influence on the trait is lower concordance in dizygotic twins. Because concordance for ABO blood type is substantially lower in the dizygotic twins, we would be safe in concluding that genes play a role in determining differences in ABO blood types.

b. The concordance for diabetes is substantially higher in monozygotic twins than in dizygotic twins; therefore, we can conclude that genetic factors play some role in susceptibility to diabetes. The fact that monozygotic twins show a concordance less than 100% suggests that environmental factors also play a role.

c. Both monozygotic and dizygotic twins exhibit the same high concordance for coffee drinking; so we can conclude that there is little genetic influence on coffee drinking. The fact that monozygotic twins show a concordance less than 100% suggests that environmental factors play a role.

d. The concordance for smoking is lower in dizygotic twins than in monozygotic twins; so genetic factors appear to influence the tendency to smoke. The fact that monozygotic twins show a concordance less than 100% suggests that environmental factors also play a role.

e. Monozygotic twins exhibit substantially higher concordance for schizophrenia than do dizygotic twins; so we can conclude that genetic factors influence this psychiatric disorder. Because the concordance of monozygotic twins is substantially less than 100%, we can also conclude that environmental factors play a role in the disorder as well.

COMPREHENSION QUESTIONS

Section 6.1

*1. What three factors complicate the task of studying the inheritance of human characteristics?

Section 6.2

2. Who is the proband in a pedigree? Is the proband always found in the last generation of the pedigree? Why or why not?

Section 6.3

*3. For each of the following modes of inheritance, describe the features that will be exhibited in a pedigree in which the trait is present: autosomal recessive, autosomal dominant, X-linked recessive, X-linked dominant, and Y-linked inheritance.

4. How does the pedigree of an autosomal recessive trait differ from the pedigree of an X-linked recessive trait?

5. Other than the fact that a Y-linked trait appears only in males, how does the pedigree of a Y-linked trait differ from the pedigree of an autosomal dominant trait?

Section 6.4

*6. What are the two types of twins and how do they arise?

7. Explain how a comparison of concordance in monozygotic and dizygotic twins can be used to determine the extent to which the expression of a trait is influenced by genes or by environmental factors.

Section 6.5

8. How are adoption studies used to separate the effects of genes and environment in the study of human characteristics?

Section 6.6

*9. What is genetic counseling?

10. Give at least four different reasons that a person might seek genetic counseling.

Section 6.7

11. Briefly define newborn screening, heterozygote screening, presymptomatic testing, and prenatal diagnosis.

12. Compare the advantages and disadvantages of aminocentesis versus chorionic villus sampling for prenatal diagnosis.

13. What is preimplantation genetic diagnosis?

14. How does heterozygote screening differ from presymptomatic genetic testing?

Section 6.8

15. Briefly describe some of the recently discovered genes that contribute to human uniqueness and the importance that they may have had in human evolution.

APPLICATION QUESTIONS AND PROBLEMS

Section 6.1

16. If humans have characteristics that make them unsuitable for genetic analysis, such as long generation time, small family size, and uncontrolled crosses, why do geneticists study humans? Give several reasons why humans have been the focus of so much genetic study.

Section 6.2

*17. Joe is color blind. Both his mother and his father have normal vision, but his mother's father (Joe's maternal grandfather) is color blind. All Joe's other grandparents have normal color vision. Joe has three sisters—Patty, Betsy, and Lora—all with normal color vision. Joe's oldest sister, Patty, is married to a man with normal color vision; they have two children, a 9-year-old color-blind boy and a 4-year-old girl with normal color vision.

 a. Using standard symbols and labels, draw a pedigree of Joe's family.

 b. What is the most likely mode of inheritance for color blindness in Joe's family?

 c. If Joe marries a woman who has no family history of color blindness, what is the probability that their first child will be a color-blind boy?

 d. If Joe marries a woman who is a carrier of the color-blind allele, what is the probability that their first child will be a color-blind boy?

 e. If Patty and her husband have another child, what is the probability that the child will be a color-blind boy?

Section 6.3

18. ▲DATA ANALYSIS Many studies have suggested a strong genetic predisposition to migraine headaches, but the mode

of inheritance is not clear. L. Russo and colleagues examined migraine headaches in several families, two of which are shown below (L. Russo et al. 2005. *American Journal of Human Genetics* 76:327–333). What is the most likely mode of inheritance for migraine headaches in these families? Explain your reasoning.

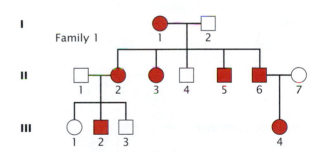

Family 1

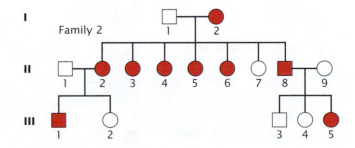

Family 2

19. ▲DATA ANALYSIS Dent disease is a rare disorder of the kidney, in which there is impaired reabsorption of filtered solutes and progressive renal failure. R. R. Hoopes and colleagues studied mutations associated with Dent disease in the

following family (R. R. Hoopes et al. 2005. *American Journal of Human Genetics* 76:260–267).

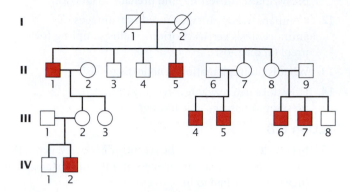

a. On the basis of this pedigree, what is the most likely mode of inheritance for the disease? Explain your reasoning.

b. From your answer to part *a*, give the most likely genotypes for all persons in the pedigree.

20. A man with a specific unusual genetic trait marries an unaffected woman and they have four children. Pedigrees of this family are shown in parts *a* through *e*, but the presence or absence of the trait in the children is not indicated. For each type of inheritance, indicate how many children of each sex are expected to express the trait by filling in the appropriate circles and squares. Assume that the trait is rare and fully penetrant.

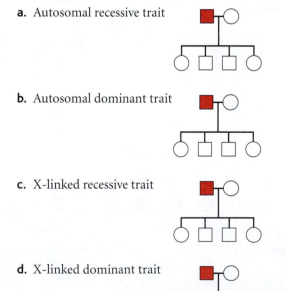

a. Autosomal recessive trait

b. Autosomal dominant trait

c. X-linked recessive trait

d. X-linked dominant trait

e. Y-linked trait

*21. For each of the following pedigrees, give the most likely mode of inheritance, assuming that the trait is rare. Carefully explain your reasoning.

a.

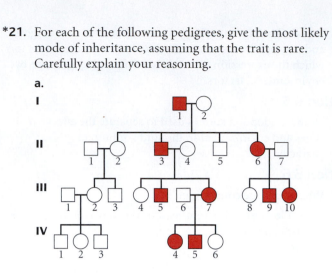

b.

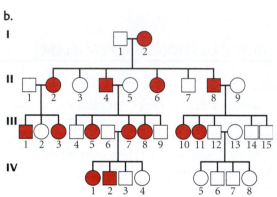

c.

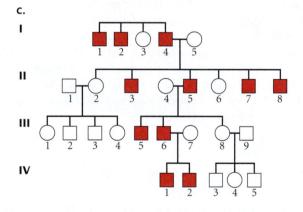

d.

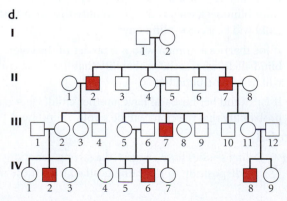

e.

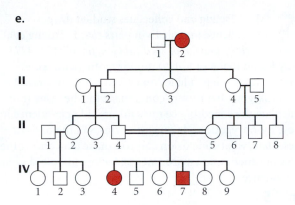

c. If III-2 and III-7 were to mate, what is the probability that one of their children would have Nance–Horan syndrome?

24. **DATA ANALYSIS** The following pedigree illustrates the inheritance of ringed hair, a condition in which each hair is differentiated into light and dark zones. What mode or modes of inheritance are possible for the ringed-hair trait in this family?

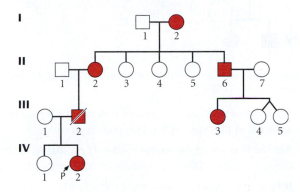

(Pedigree adapted from L. M. Ashley and R. S. Jacques. 1950. *Journal of Heredity* 41:83.)

22. The trait represented in the following pedigree is expressed only in the males of the family. Is the trait Y linked? Why or why not? If you believe the trait is not Y linked, propose an alternate explanation for its inheritance.

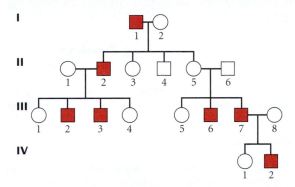

25. **DATA ANALYSIS** Ectodactyly is a rare condition in which the fingers are absent and the hand is split. This condition is usually inherited as an autosomal dominant trait. Ademar Freire-Maia reported the appearance of ectodactyly in a family in São Paulo, Brazil, whose pedigree is shown here. Is this pedigree consistent with autosomal dominant inheritance? If not, what mode of inheritance is most likely? Explain your reasoning.

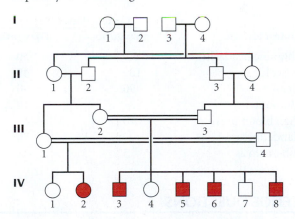

(Pedigree adapted from A. Freire-Maia. 1971. *Journal of Heredity* 62:53.)

*23. **DATA ANALYSIS** The following pedigree illustrates the inheritance of Nance–Horan syndrome, a rare genetic condition in which affected persons have cataracts and abnormally shaped teeth.

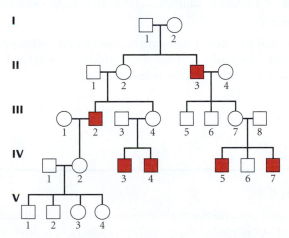

(Pedigree adapted from D. Stambolian, R. A. Lewis, K. Buetow, A. Bond, and R. Nussbaum. 1990. *American Journal of Human Genetics* 47:15.)

a. On the basis of this pedigree, what do you think is the most likely mode of inheritance for Nance–Horan syndrome?

b. If couple III-7 and III-8 have another child, what is the probability that the child will have Nance–Horan syndrome?

26. **DATA ANALYSIS** The complete absence of one or more teeth (tooth agenesis) is a common trait in humans—indeed, more than 20% of humans lack one or more of their third molars. However, more-severe absence of teeth, defined as missing six or more teeth, is less common and frequently an inherited condition. L. Lammi and colleagues examined tooth agenesis in the Finnish family shown in the pedigree on the following page (L. Lammi. 2004. *American Journal of Human Genetics* 74:1043–1050).

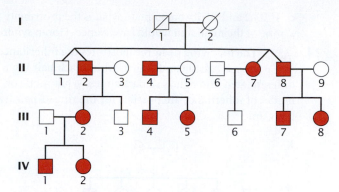

(Pedigree adapted from L. Lammi. 2004. *American Journal of Human Genetics* 74:1043–1050.)

a. What is the most likely mode of inheritance for tooth agenesis in this family? Explain your reasoning.

b. Are the two sets of twins in this family monozygotic or dizygotic twins? What is the basis of your answer?

c. If IV-2 married a man who has a full set of teeth, what is the probability that their child would have tooth agenesis?

d. If III-2 and III-7 married and had a child, what is the probability that their child would have tooth agenesis?

Section 6.4

*27. A geneticist studies a series of characteristics in monozygotic twins and dizygotic twins, obtaining the following concordances. For each characteristic, indicate whether the rates of concordance suggest genetic influences, environmental influences, or both. Explain your reasoning.

Characteristic	Concordance (%)	
	Monozygotic	Dizygotic
Migraine headaches	60	30
Eye color	100	40
Measles	90	90
Clubfoot	30	10
High blood pressure	70	40
Handedness	70	70
Tuberculosis	5	5

28. DATA ANALYSIS M. T. Tsuang and colleagues studied drug dependence in male twin pairs (M. T. Tsuang et al. 1996. *American Journal of Medical Genetics* 67:473–477). They found that 4 out of 30 monozygotic twins were concordant for dependence on opioid drugs, whereas 1 out of 34 dizygotic twins were concordant for the same trait. Calculate the concordance rates for opioid dependence in these monozygotic and dizygotic twins. On the basis of these data, what conclusion can you make concerning the roles of genetic and environmental factors in opioid dependence?

Section 6.5

29. DATA ANALYSIS In a study of schizophrenia (a mental disorder including disorganization of thought and withdrawal from reality), researchers looked at the prevalence of the disorder in the biological and adoptive parents of people who were adopted as children; they found the following results:

	Prevalence of schizophrenia	
Adopted persons	Biological parents	Adoptive parents
With schizophrenia	12	2
Without schizophrenia	6	4

(Source: S. S. Kety et al. 1978. The biological and adoptive families of adopted individuals who become schizophrenic: prevalence of mental illness and other characteristics, in *The Nature of Schizophrenia: New Approaches to Research and Treatment*, L. C. Wynne, R. L. Cromwell, and S. Matthysse, Eds. New York: Wiley, 1978, pp. 25–37.)

What can you conclude from these results concerning the role of genetics in schizophrenia? Explain your reasoning.

Section 6.7

30. What, if any, ethical issues might arise from the widespread use of noninvasive fetal diagnosis, which can be carried out much earlier than amniocentesis or CVS?

CHALLENGE QUESTIONS

Section 6.1

31. Many genetic studies, particularly those of recessive traits, have focused on small isolated human populations, such as those on islands. Suggest one or more advantages that isolated populations might have for the study of recessive traits.

Section 6.3

32. Draw a pedigree that represents an autosomal dominant trait, sex limited to males, and that excludes the possibility that the trait is Y linked.

33. DATA ANALYSIS A. C. Stevenson and E. A. Cheeseman studied deafness in a family in Northern Ireland and

recorded the following pedigree (A. C. Stevenson and E. A. Cheeseman. 1956. *Annals of Human Genetics* 20:177–231).

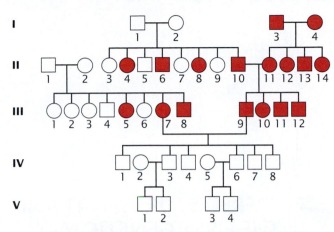

(Pedigree adapted from A. C. Stevenson and E. A. Cheeseman. 1956. *Annals of Human Genetics* 20:177–231.)

a. If you consider only generations I through III, what is the most likely mode of inheritance for this type of deafness?

b. Provide a possible explanation for the cross between III-7 and III-9 and the results for generations IV through V.

Section 6.4

34. Dizygotic twinning often runs in families and its frequency varies among ethnic groups, whereas monozygotic twinning rarely runs in families and its frequency is quite constant among ethnic groups. These observations have been interpreted as evidence for a genetic basis for variation in dizygotic twinning but for little genetic basis for variation in monozygotic twinning. Can you suggest a possible reason for these differences in the genetic tendencies of dizygotic and monozygotic twinning?

7 Linkage, Recombination, and Eukaryotic Gene Mapping

Alfred Henry Sturtevant, an early geneticist, developed the first genetic map.
[*Institute Archives, California Institute of Technology.*]

ALFRED STURTEVANT AND THE FIRST GENETIC MAP

In 1909, Thomas Hunt Morgan taught the introduction to zoology class at Columbia University. Seated in the lecture hall were sophomore Alfred Henry Sturtevant and freshman Calvin Bridges. Sturtevant and Bridges were excited by Morgan's teaching style and intrigued by his interest in biological problems. They asked Morgan if they could work in his laboratory and, the following year, both young men were given desks in the "Fly Room," Morgan's research laboratory where the study of *Drosophila* genetics was in its infancy (see pp. 84–85 in Chapter 4). Sturtevant, Bridges, and Morgan's other research students virtually lived in the laboratory, raising fruit flies, designing experiments, and discussing their results.

In the course of their research, Morgan and his students observed that some pairs of genes did not segregate randomly according to Mendel's principle of independent assortment but instead tended to be inherited together. Morgan suggested that possibly the genes were located on the same chromosome and thus traveled together during meiosis. He further proposed that closely linked genes—those that are rarely shuffled by recombination—lie close together on the same chromosome, whereas loosely linked genes—those more frequently shuffled by recombination—lie farther apart.

One day in 1911, Sturtevant and Morgan were discussing independent assortment when, suddenly, Sturtevant had a flash of inspiration: variation in the strength of linkage indicated how genes are positioned along a chromosome, providing a way of mapping genes. Sturtevant went home and, neglecting his undergraduate homework, spent most of the night working out the first genetic map (**Figure 7.1**). Sturtevant's first chromosome map was remarkably accurate, and it established the basic methodology used today for mapping genes.

Sturtevant went on to become a leading geneticist. His research included gene mapping and basic mechanisms of inheritance in *Drosophila*, cytology, embryology, and evolution. Sturtevant's career was deeply influenced by his early years in the Fly Room, where Morgan's unique personality and the close quarters combined to stimulate intellectual excitement and the free exchange of ideas.

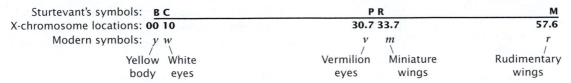

Sturtevant's symbols:	**B C**		**P R**		**M**
X-chromosome locations:	**00 10**		**30.7 33.7**		**57.6**
Modern symbols:	*y w*		*v*	*m*	*r*

Yellow   White   Vermilion   Miniature   Rudimentary
body   eyes   eyes   wings   wings

7.1 Sturtevant's map included five genes on the X chromosome of *Drosophila*. The genes are yellow body (*y*), white eyes (*w*), vermilion eyes (*v*), miniature wings (*m*), and rudimentary wings (*r*).

This chapter explores the inheritance of genes located on the same chromosome. These linked genes do not strictly obey Mendel's principle of independent assortment; rather, they tend to be inherited together. This tendency requires a new approach to understanding their inheritance and predicting the types of offspring produced. A critical piece of information necessary for predicting the results of these crosses is the arrangement of the genes on the chromosomes; thus, it will be necessary to think about the relation between genes and chromosomes. A key to understanding the inheritance of linked genes is to make the conceptual connection between the genotypes in a cross and the behavior of chromosomes in meiosis.

We will begin our exploration of linkage by first comparing the inheritance of two linked genes with the inheritance of two genes that assort independently. We will then examine how crossing over breaks up linked genes. This knowledge of linkage and recombination will be used for predicting the results of genetic crosses in which genes are linked and for mapping genes. Later in the chapter, we will focus on physical methods of determining the chromosomal locations of genes. The final section examines variation in rates of recombination.

7.1 Linked Genes Do Not Assort Independently

Chapter 3 introduced Mendel's principles of segregation and independent assortment. Let's take a moment to review these two important concepts. The principle of segregation states that each individual diploid organism possesses two alleles at a locus that separate in meiosis, with one allele going into each gamete. The principle of independent assortment provides additional information about the process of segregation: it tells us that, in the process of separation, the two alleles at a locus act independently of alleles at other loci.

The independent separation of alleles results in *recombination*, the sorting of alleles into new combinations. Consider a cross between individuals homozygous for two different pairs of alleles: *AA BB* × *aa bb*. The first parent, *AA BB*, produces gametes with alleles *A B*, and the second parent, *aa bb*, produces gametes with the alleles *a b*, resulting in F₁ progeny with genotype *Aa Bb* (**Figure 7.2**). Recombination means that, when one of the F₁ progeny reproduces, the combination of

alleles in its gametes may differ from the combinations in the gametes from its parents. In other words, the F₁ may produce gametes with alleles *A b* or *a B* in addition to gametes with *A B* or *a b*.

Mendel derived his principles of segregation and independent assortment by observing the progeny of genetic crosses, but he had no idea of what biological processes produced these phenomena. In 1903, Walter Sutton proposed a biological basis for Mendel's principles, called the chromosome theory of heredity (see Chapter 3). This theory holds that genes are found on chromosomes. Let's restate Mendel's two principles in relation to the chromosome theory of heredity. The principle of segregation states that a diploid organism possesses two alleles for a trait, each of which is located at the same position, or locus, on each of the two homologous chromosomes. These chromosomes segregate in meiosis, with each gamete receiving one homolog. The principle of independent assortment states that, in meiosis, each pair of homologous chromosomes assorts independently of other homologous pairs. With this new perspective,

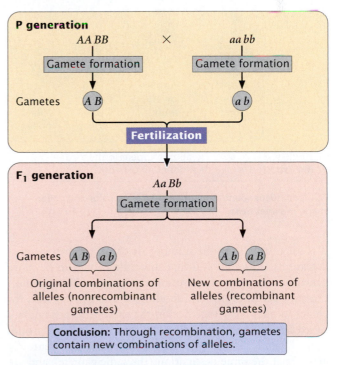

7.2 Recombination is the sorting of alleles into new combinations.

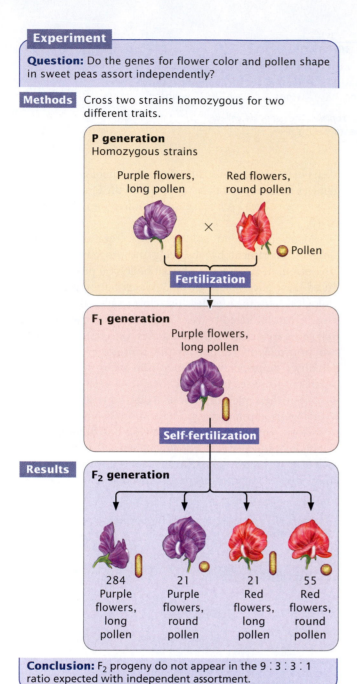

Experiment

Question: Do the genes for flower color and pollen shape in sweet peas assort independently?

Methods Cross two strains homozygous for two different traits.

P generation
Homozygous strains

Purple flowers, long pollen   Red flowers, round pollen

×   Pollen

Fertilization

F₁ generation

Purple flowers, long pollen

Self-fertilization

Results

F₂ generation

| 284 Purple flowers, long pollen | 21 Purple flowers, round pollen | 21 Red flowers, long pollen | 55 Red flowers, round pollen |

Conclusion: F₂ progeny do not appear in the 9 : 3 : 3 : 1 ratio expected with independent assortment.

7.3 Nonindependent assortment of flower color and pollen shape in sweet peas.

it is easy to see that the number of chromosomes in most organisms is limited and that there are certain to be more genes than chromosomes; so some genes must be present on the same chromosome and should not assort independently. Genes located close together on the same chromosome are called **linked genes** and belong to the same **linkage group.** Linked genes travel together during meiosis, eventually arriving at the same destination (the same gamete), and are not expected to assort independently.

All of the characteristics examined by Mendel in peas did display independent assortment and, after the rediscovery of Mendel's work, the first genetic characteristics studied in other organisms also seemed to assort independently. How could genes be carried on a limited number of chromosomes and yet assort independently?

The apparent inconsistency between the principle of independent assortment and the chromosome theory of heredity soon disappeared as biologists began finding genetic characteristics that did not assort independently. One of the first cases was reported in sweet peas by William Bateson, Edith Rebecca Saunders, and Reginald C. Punnett in 1905. They crossed a homozygous strain of peas having purple flowers and long pollen grains with a homozygous strain having red flowers and round pollen grains. All the F_1 had purple flowers and long pollen grains, indicating that purple was dominant over red and long was dominant over round. When they intercrossed the F_1, the resulting F_2 progeny did not appear in the 9 : 3 : 3 : 1 ratio expected with independent assortment (**Figure 7.3**). An excess of F_2 plants had purple flowers and long pollen or red flowers and round pollen (the parental phenotypes). Although Bateson, Saunders, and Punnett were unable to explain these results, we now know that the two loci that they examined lie close together on the same chromosome and therefore do not assort independently.

7.2 Linked Genes Segregate Together and Crossing Over Produces Recombination Between Them

Genes that are close together on the same chromosome usually segregate as a unit and are therefore inherited together. However, genes occasionally switch from one homologous chromosome to the other through the process of crossing over (see Chapter 2), as illustrated in **Figure 7.4.** Crossing over results in recombination; it breaks up the associations of genes that are close together on the same chromosome. Linkage and crossing over can be seen as processes that have opposite effects: linkage keeps particular genes together, and crossing over mixes them up. In Chapter 5 we considered a number of exceptions and extensions to Mendel's principles of heredity. The concept of linked genes adds a further complication to interpretations of the results of genetic crosses. However, with an understanding of how linkage affects heredity, we can analyze crosses for linked genes and successfully predict the types of progeny that will be produced.

Notation for Crosses with Linkage

In analyzing crosses with linked genes, we must know not only the genotypes of the individuals crossed, but also the arrangement of the genes on the chromosomes. To keep

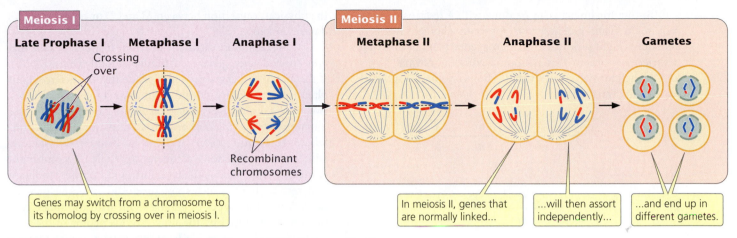

Meiosis I

Late Prophase I   **Metaphase I**   **Anaphase I**

Crossing over

Recombinant chromosomes

> Genes may switch from a chromosome to its homolog by crossing over in meiosis I.

Meiosis II

Metaphase II   **Anaphase II**   **Gametes**

> In meiosis II, genes that are normally linked...

> ...will then assort independently...

> ...and end up in different gametes.

7.4 Crossing over takes place in meiosis and is responsible for recombination.

track of this arrangement, we introduce a new system of notation for presenting crosses with linked genes. Consider a cross between an individual homozygous for dominant alleles at two linked loci and another individual homozygous for recessive alleles at those loci (*AA BB* × *aa bb*). For linked genes, it's necessary to write out the specific alleles as they are arranged on each of the homologous chromosomes:

$$\frac{A \qquad B}{A \qquad B} \times \frac{a \qquad b}{a \qquad b}$$

In this notation, each line represents one of the two homologous chromosomes. Inheriting one chromosome from each parent, the F$_1$ progeny will have the following genotype:

$$\frac{A \qquad B}{a \qquad b}$$

Here, the importance of designating the alleles on each chromosome is clear. One chromosome has the two dominant alleles *A* and *B*, whereas the homologous chromosome has the two recessive alleles *a* and *b*. The notation can be simplified by drawing only a single line, with the understanding that genes located on the same side of the line lie on the same chromosome:

$$\frac{A \qquad B}{a \qquad b}$$

This notation can be simplified further by separating the alleles on each chromosome with a slash: *AB/ab*.

Remember that the two alleles at a locus are always located on different homologous chromosomes and therefore must lie on opposite sides of the line. Consequently, we would *never* write the genotypes as

$$\frac{A \qquad a}{B \qquad b}$$

because the alleles *A* and *a* can *never* be on the same chromosome.

It is also important to always keep the same order of the genes on both sides of the line; thus, we should *never* write

$$\frac{A \qquad B}{b \qquad a}$$

because it would imply that alleles *A* and *b* are allelic (at the same locus).

Complete Linkage Compared with Independent Assortment

We will first consider what happens to genes that exhibit complete linkage, meaning that they are located very close together on the same chromosome and do not exhibit crossing over. Genes are rarely completely linked but, by assuming that no crossing over occurs, we can see the effect of linkage more clearly. We will then consider what happens when genes assort independently. Finally, we will consider the results obtained if the genes are linked but exhibit some crossing over.

A testcross reveals the effects of linkage. For example, if a heterozygous individual is test-crossed with a homozygous recessive individual (*Aa Bb* × *aa bb*), the alleles that are present in the gametes contributed by the heterozygous parent will be expressed in the phenotype of the offspring, because the homozygous parent could not contribute dominant alleles that might mask them. Consequently, traits that appear in the progeny reveal which alleles were transmitted by the heterozygous parent.

Consider a pair of linked genes in tomato plants. One pair affects the type of leaf: an allele for mottled leaves (*m*) is recessive to an allele that produces normal leaves (*M*). Nearby on the same chromosome is another locus that determines the height of the plant: an allele for dwarf (*d*) is recessive to an allele for tall (*D*).

Testing for linkage can be done with a testcross, which requires a plant heterozygous for both characteristics. A geneticist might produce this heterozygous plant by crossing a variety of tomato that is homozygous for normal leaves and tall height with a variety that is homozygous for mottled leaves and dwarf height:

$$\text{P} \qquad \frac{M \qquad D}{M \qquad D} \times \frac{m \qquad d}{m \qquad d}$$

$$\downarrow$$

$$\text{F}_1 \qquad \frac{M \qquad D}{m \qquad d}$$

The geneticist would then use these F_1 heterozygotes in a testcross, crossing them with plants homozygous for mottled leaves and dwarf height:

$$\frac{M \qquad D}{m \qquad d} \times \frac{m \qquad d}{m \qquad d}$$

The results of this testcross are diagrammed in **Figure 7.5a**. The heterozygote produces two types of gametes: some with the $\underline{M \quad D}$ chromosome and others with the $\underline{m \quad d}$ chromosome. Because no crossing over occurs, these gametes are the only types produced by the heterozygote. Notice that these gametes contain only combinations of alleles that were present in the original parents: either the allele for normal leaves together with the allele for tall height (M and D) or the allele for mottled leaves together with the allele for dwarf height (m and d). Gametes that contain only original combinations of alleles present in the parents are **nonrecombinant gametes,** or *parental* gametes.

The homozygous parent in the testcross produces only one type of gamete; it contains chromosome $\underline{m \quad d}$ and pairs with one of the two gametes generated by the heterozygous parent (see Figure 7.5a). Two types of progeny result: half have normal leaves and are tall:

$$\frac{M \qquad D}{m \qquad d}$$

and half have mottled leaves and are dwarf:

$$\frac{m \qquad d}{m \qquad d}$$

These progeny display the original combinations of traits present in the P generation and are **nonrecombinant progeny,** or *parental* progeny. No new combinations of the two traits, such as normal leaves with dwarf or mottled leaves with tall, appear in the offspring, because the genes affecting the two traits are completely linked and are inherited together. New combinations of traits could arise only if

the physical connection between M and D or between m and d were broken.

These results are distinctly different from the results that are expected when genes assort independently (**Figure 7.5b**). If the M and D loci assorted independently, the heterozygous plant ($Mm\ Dd$) would produce four types of gametes: two nonrecombinant gametes containing the original combinations of alleles ($M\ D$ and $m\ d$) and two gametes containing new combinations of alleles ($M\ d$ and $m\ D$). Gametes with new combinations of alleles are called **recombinant gametes.** With independent assortment, nonrecombinant and recombinant gametes are produced in equal proportions. These four types of gametes join with the single type of gamete produced by the homozygous parent of the testcross to produce four kinds of progeny in equal proportions (see Figure 7.5b). The progeny with new combinations of traits formed from recombinant gametes are termed **recombinant progeny.**

In summary, a testcross in which one of the plants is heterozygous for two completely linked genes yields two types of progeny, each type displaying one of the original combinations of traits present in the P generation. Independent assortment, in contrast, produces progeny in a $1:1:1:1$ ratio. That is, there are four types of progeny—two types of recombinant progeny and two types of nonrecombinant progeny in equal proportions.

Crossing Over with Linked Genes

Usually, there is some crossing over between genes that lie on the same chromosome, producing new combinations of traits. Genes that exhibit crossing over are incompletely linked. Let's see how it occurs.

Theory   The effect of crossing over on the inheritance of two linked genes is shown in **Figure 7.6**. Crossing over, which takes place in prophase I of meiosis, is the exchange of genetic material between nonsister chromatids (see Figures 2.16 and 2.18). After a single crossover has taken place, the two chromatids that did not participate in crossing over are unchanged; gametes that receive these chromatids are nonrecombinants. The other two chromatids, which did participate in crossing over, now contain new combinations of alleles; gametes that receive these chromatids are recombinants. For each meiosis in which a single crossover takes place, then, two nonrecombinant gametes and two recombinant gametes will be produced. This result is the same as that produced by independent assortment (see Figure 7.5b); so, when crossing over between two loci takes place in every meiosis, it is impossible to determine whether the genes are on the same chromosome and crossing over took place or whether the genes are on different chromosomes.

For closely linked genes, crossing over does not take place in every meiosis. In meioses in which there is no

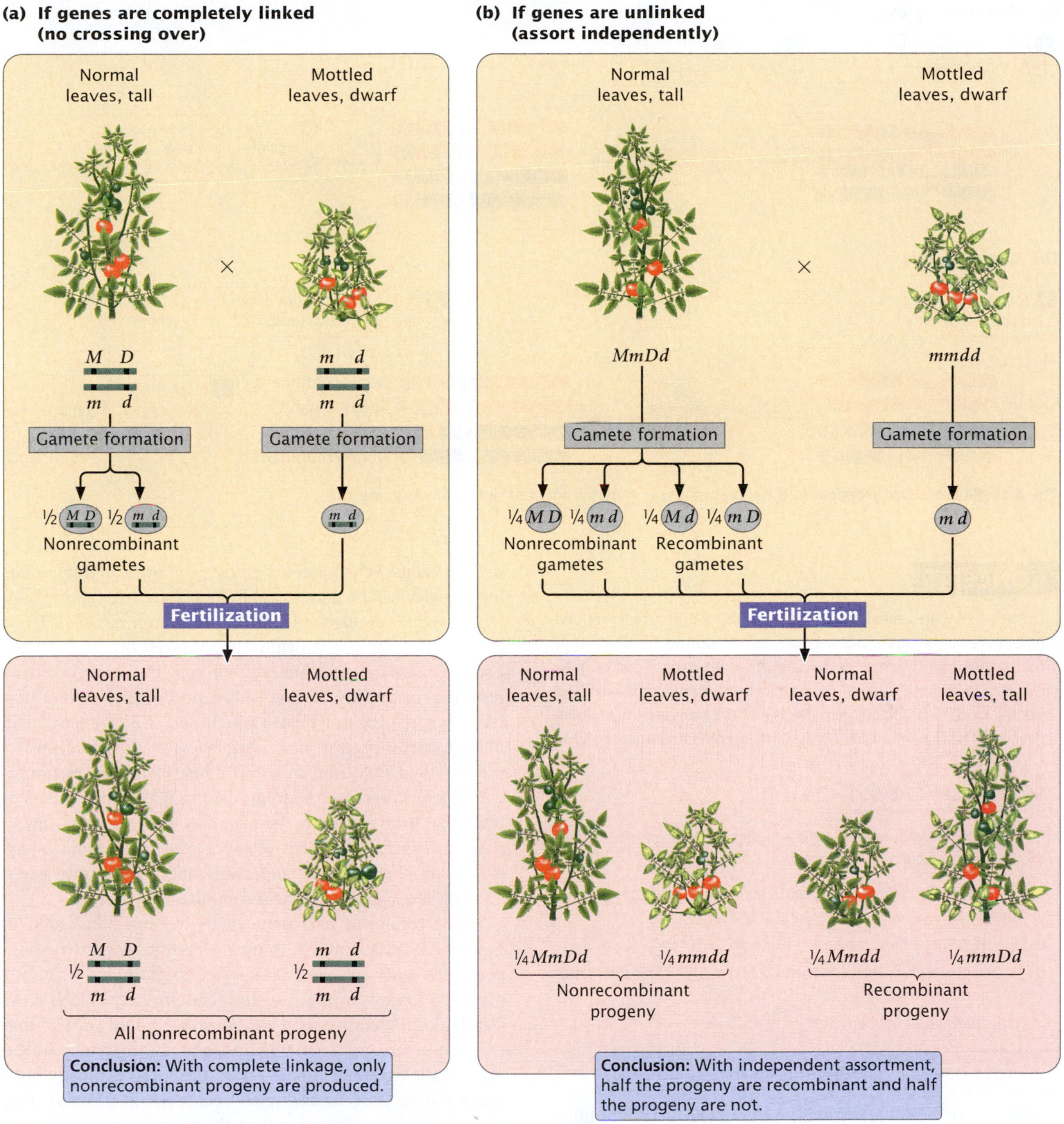

**(a) If genes are completely linked
(no crossing over)**

Normal
leaves, tall
×
Mottled
leaves, dwarf

M   D
m   d

m   d
m   d

Gamete formation

Gamete formation

½ M D   ½ m d
Nonrecombinant
gametes

m d

Fertilization

Normal
leaves, tall

Mottled
leaves, dwarf

½
M   D
m   d

½
m   d
m   d

All nonrecombinant progeny

Conclusion: With complete linkage, only
nonrecombinant progeny are produced.

**(b) If genes are unlinked
(assort independently)**

Normal
leaves, tall
×
Mottled
leaves, dwarf

MmDd

mmdd

Gamete formation

Gamete formation

¼ M D   ¼ m d   ¼ M d   ¼ m D
Nonrecombinant   Recombinant
gametes   gametes

m d

Fertilization

Normal
leaves, tall

Mottled
leaves, dwarf

Normal
leaves, dwarf

Mottled
leaves, tall

¼ MmDd   ¼ mmdd
Nonrecombinant
progeny

¼ Mmdd   ¼ mmDd
Recombinant
progeny

Conclusion: With independent assortment,
half the progeny are recombinant and half
the progeny are not.

7.5 A testcross reveals the effects of linkage. Results of a testcross for two loci in tomatoes that
determine leaf type and plant height.

crossing over, only nonrecombinant gametes are produced. In meioses in which there is a single crossover, half the gametes are recombinants and half are nonrecombinants (because a single crossover affects only two of the four chromatids); so the total percentage of recombinant gametes is always half the percentage of meioses in which

crossing over takes place. Even if crossing over between two genes takes place in every meiosis, only 50% of the resulting gametes will be recombinants. Thus, the frequency of recombinant gametes is always half the frequency of crossing over, and the maximum proportion of recombinant gametes is 50%.

(a) No crossing over

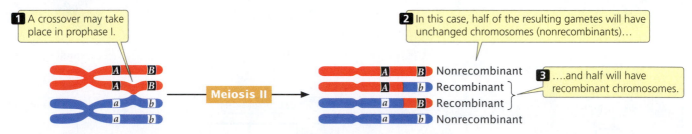

1 Homologous chromosomes pair in prophase I.

2 If no crossing over takes place,…

Meiosis II

3 …all resulting chromosomes in gametes have original allele combinations and are nonrecombinants.

(b) Crossing over

1 A crossover may take place in prophase I.

2 In this case, half of the resulting gametes will have unchanged chromosomes (nonrecombinants)…

Meiosis II

Nonrecombinant
Recombinant
Recombinant
Nonrecombinant

3 ….and half will have recombinant chromosomes.

7.6 A single crossover produces half nonrecombinant gametes and half recombinant gametes.

CONCEPTS

Linkage between genes causes them to be inherited together and reduces recombination; crossing over breaks up the associations of such genes. In a testcross for two linked genes, each crossover produces two recombinant gametes and two nonrecombinants. The frequency of recombinant gametes is half the frequency of crossing over, and the maximum frequency of recombinant gametes is 50%.

✔ CONCEPT CHECK 1

For single crossovers, the frequency of recombinant gametes is half the frequency of crossing over because

a. a test cross between a homozygote and heterozygote produces $1/2$ heterozygous and $1/2$ homozygous progeny.

b. the frequency of recombination is always 50%.

c. each crossover takes place between only two of the four chromatids of a homologous pair.

d. crossovers occur in about 50% of meioses.

Application   Let's apply what we have learned about linkage and recombination to a cross between tomato plants that differ in the genes that encode leaf type and plant height. Assume now that these genes are linked and that some crossing over takes place between them. Suppose a geneticist carried out the testcross outlined earlier:

$$\frac{M \qquad D}{m \qquad d} \times \frac{m \qquad d}{m \qquad d}$$

When crossing over takes place between the genes for leaf type and height, two of the four gametes produced will be

recombinants. When there is no crossing over, all four resulting gametes will be nonrecombinants. Thus, over all meioses, the majority of gametes will be nonrecombinants. These gametes then unite with gametes produced by the homozygous recessive parent, which contain only the recessive alleles, resulting in mostly nonrecombinant progeny and a few recombinant progeny (**Figure 7.7**). In this cross, we see that 55 of the testcross progeny have normal leaves and are tall and 53 have mottled leaves and are dwarf. These plants are the noncombinant progeny, containing the original combinations of traits that were present in the parents. Of the 123 progeny, 15 have new combinations of traits that were not seen in the parents: 8 are normal leaved and dwarf, and 7 are mottle leaved and tall. These plants are the recombinant progeny.

The results of a cross such as the one illustrated in Figure 7.7 reveal several things. A testcross for two independently assorting genes is expected to produce a 1 : 1 : 1 : 1 phenotypic ratio in the progeny. The progeny of this cross clearly do not exhibit such a ratio; so we might suspect that the genes are not assorting independently. When linked genes undergo some crossing over, the result is mostly nonrecombinant progeny and fewer recombinant progeny. This result is what we observe among the progeny of the testcross illustrated in Figure 7.7; so we conclude that the two genes show evidence of linkage with some crossing over.

Calculating Recombination Frequency

The percentage of recombinant progeny produced in a cross is called the **recombination frequency,** which is calculated as follows:

$$\text{recombinant frequency} = \frac{\text{number of recombinant progeny}}{\text{total number of progeny}} \times 100\%$$

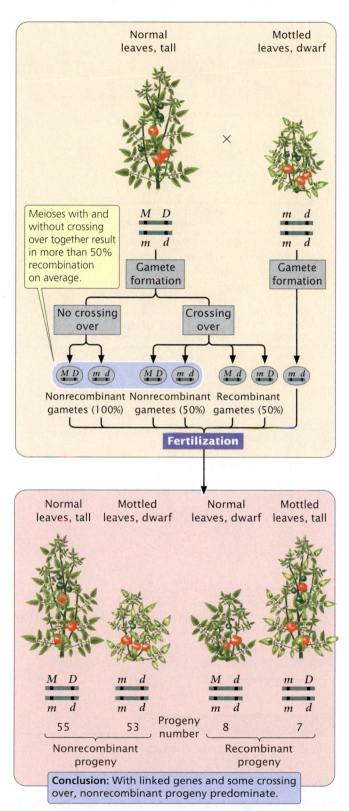

7.7 Crossing over between linked genes produces nonrecombinant and recombinant offspring. In this testcross, genes are linked and there is some crossing over.

In the testcross shown in Figure 7.7, 15 progeny exhibit new combinations of traits; so the recombination frequency is:

$$\frac{8 + 7}{55 + 53 + 8 + 7} \times 100\% = \frac{15}{123} \times 100\% = 12.2\%$$

Thus, 12.2% of the progeny exhibit new combinations of traits resulting from crossing over. The recombination frequency can also be expressed as a decimal fraction (0.122).

Coupling and Repulsion

In crosses for linked genes, the arrangement of alleles on the homologous chromosomes is critical in determining the outcome of the cross. For example, consider the inheritance of two genes in the Australian blowfly, *Lucilia cuprina*. In this species, one locus determines the color of the thorax: a purple thorax (p) is recessive to the normal green thorax (p^+). A second locus determines the color of the puparium: a black puparium (b) is recessive to the normal brown puparium (b^+). These loci are located close together on the second chromosome. Suppose we test cross a fly that is heterozygous at both loci with a fly that is homozygous recessive at both. Because these genes are linked, there are two possible arrangements on the chromosomes of the heterozygous progeny fly. The dominant alleles for green thorax (p^+) and brown puparium (b^+) might reside on the same chromosome, and the recessive alleles for purple thorax (p) and black puparium (b) might reside on the other homologous chromosome:

$$\frac{p^+ \qquad b^+}{p \qquad b}$$

This arrangement, in which wild-type alleles are found on one chromosome and mutant alleles are found on the other chromosome, is referred to as the **coupling** or **cis configuration**. Alternatively, one chromosome might bear the alleles for green thorax (p^+) and black puparium (b), and the other chromosome would carry the alleles for purple thorax (p) and brown puparium (b^+):

$$\frac{p^+ \qquad b}{p \qquad b^+}$$

This arrangement, in which each chromosome contains one wild-type and one mutant allele, is called the **repulsion** or **trans configuration**. Whether the alleles in the heterozygous parent are in coupling or repulsion determines which phenotypes will be most common among the progeny of a testcross.

When the alleles are in the coupling configuration, the most numerous progeny types are those with green thorax and brown puparium and those with purple thorax and black puparium (**Figure 7.8a**); but, when the alleles of the heterozygous parent are in repulsion, the most numerous progeny types are those with green thorax and black puparium and

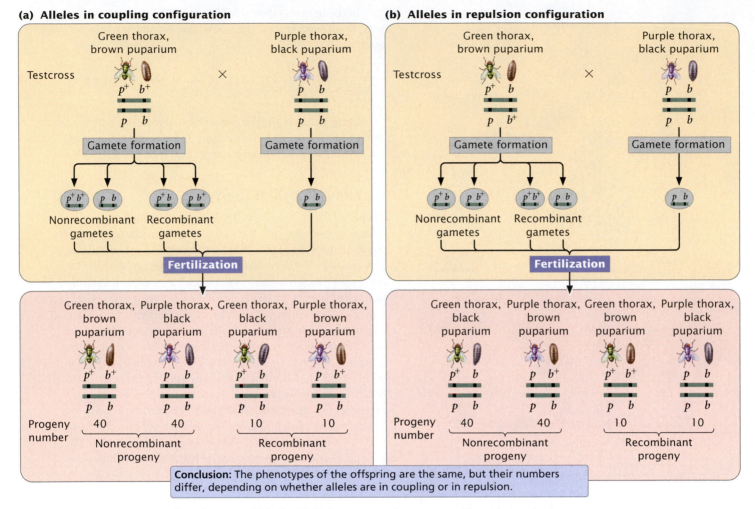

(a) Alleles in coupling configuration

(b) Alleles in repulsion configuration

Conclusion: The phenotypes of the offspring are the same, but their numbers differ, depending on whether alleles are in coupling or in repulsion.

7.8 The arrangement (coupling or repulsion) of linked genes on a chromosome affects the results of a testcross. Linked loci in the Australian blowfly, *Lucilia cuprina,* determine the color of the thorax and that of the puparium.

those with purple thorax and brown puparium (**Figure 7.8b**). Notice that the genotypes of the parents in Figure 7.8a and b are the same ($p^+p\ b^+b \times pp\ bb$) and that the dramatic difference in the phenotypic ratios of the progeny in the two crosses results entirely from the configuration—coupling or repulsion—of the chromosomes. It is essential to know the arrangement of the alleles on the chromosomes to accurately predict the outcome of crosses in which genes are linked.

CONCEPTS

In a cross, the arrangement of linked alleles on the chromosomes is critical for determining the outcome. When two wild-type alleles are on one homologous chromosome and two mutant alleles are on the other, they are in the coupling configuration; when each chromosome contains one wild-type allele and one mutant allele, the alleles are in repulsion.

✔ CONCEPT CHECK 2

The following testcross produces the progeny shown: *Aa Bb* × *aa bb* → 10 *Aa Bb*, 40 *Aa bb*, 40 *aa Bb*, 10 *aa bb*. What is the percent recombination between the *A* and *B* loci? Were the genes in the *Aa Bb* parent in coupling or in repulsion?

CONNECTING CONCEPTS

Relating Independent Assortment, Linkage, and Crossing Over

We have now considered three situations concerning genes at different loci. First, the genes may be located on different chromosomes; in this case, they exhibit independent assortment and combine randomly when gametes are formed. An individual heterozygous at two loci (*Aa Bb*) produces four types of gametes (*A B, a b, A b,* and *a B*) in equal proportions: two types of nonrecombinants and two types of recombinants.

Second, the genes may be completely linked—meaning that they're on the same chromosome and lie so close together that crossing over between them is rare. In this case, the genes do not recombine. An individual heterozygous for two closely linked genes in the coupling configuration

$$\frac{A \qquad B}{a \qquad b}$$

produces only the nonrecombinant gametes containing alleles *A B* or *a b*. The alleles do not assort into new combinations such as *A b* or *a B*.

The third situation, incomplete linkage, is intermediate between the two extremes of independent assortment and complete linkage. Here, the genes are physically linked on the same chromosome, which prevents independent assortment. However, occasional crossovers break up the linkage and allow the genes to recombine. With incomplete linkage, an individual heterozygous at two loci produces four types of gametes—two types of recombinants and two types of nonrecombinants—but the nonrecombinants are produced more frequently than the recombinants because crossing over does not take place in every meiosis.

Earlier in the chapter, the term recombination was defined as the sorting of alleles into new combinations. We can now distinguish between two types of recombination that differ in the mechanism that generates these new combinations of alleles.

Interchromosomal recombination is between genes on *different* chromosomes. It arises from independent assortment—the random segregation of chromosomes in anaphase I of meiosis. This is the kind of recombination that Mendel discovered while studying dihybrid crosses. **Intrachromosomal recombination** is between genes located on the *same* chromosome. It arises from crossing over—the exchange of genetic material in prophase I of meiosis. Both types of recombination produce new allele combinations in the gametes; so they cannot be distinguished by examining the types of gametes produced. Nevertheless, they can often be distinguished by the *frequencies* of types of gametes: interchromosomal recombination produces 50% nonrecombinant gametes and 50% recombinant gametes, whereas intrachromosomal recombination frequently produces fewer than 50% recombinant gametes. However, when the genes are very far apart on the same chromosome, they assort independently, as if they were on different chromosomes. In this case, intrachromosomal recombination also produces 50% recombinant gametes. Intrachromosomal recombination of genes that lie far apart on the same chromosome and interchromosomal recombination are genetically indistinguishable.

Evidence for the Physical Basis of Recombination

Walter Sutton's chromosome theory of inheritance, which stated that genes are physically located on chromosomes, was supported by Nettie Stevens and Edmund Wilson's discovery that sex was associated with a specific chromosome in insects (pp. 75–77 in Chapter 4) and Calvin Bridges's demonstration that nondisjunction of X chromosomes was related to the inheritance of eye color in *Drosophila*

7.9 Barbara McClintock (left) and Harriet Creighton (right) provided evidence that genes are located on chromosomes. [*Karl Maramorosch/Courtesy of Cold Spring Harbor Laboratory Archives.*]

(pp. 81–83). Further evidence for the chromosome theory of heredity came in 1931, when Harriet Creighton and Barbara McClintock (**Figure 7.9**) obtained evidence that intrachromosomal recombination was the result of physical exchange between chromosomes. Creighton and McClintock discovered a strain of corn that had an abnormal chromosome 9, containing a densely staining knob at one end and a small piece of another chromosome attached to the other end. This aberrant chromosome allowed them to visually distinguish the two members of a homologous pair.

They studied the inheritance of two traits in corn determined by genes on chromosome 9. At one locus, a dominant allele (*C*) produced colored kernels, whereas a recessive allele (*c*) produced colorless kernels. At a second, linked locus, a dominant allele (*Wx*) produced starchy kernels, whereas a recessive allele (*wx*) produced waxy kernels. Creighton and McClintock obtained a plant that was heterozygous at both loci in repulsion, with the alleles for colored and waxy on the aberrant chromosome and the alleles for colorless and starchy on a normal chromosome:

They crossed this heterozygous plant with a plant that was homozygous for colorless and heterozygous for waxy:

$$\frac{C \qquad wx}{c \qquad Wx} \times \frac{c \qquad Wx}{c \qquad wx}$$

This cross will produce different combinations of traits in the progeny, but the only way that colorless and waxy progeny can arise is through crossing over in the doubly heterozygous parent:

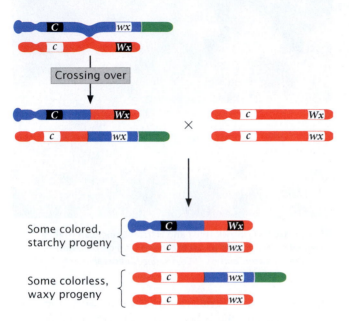

Note: Not all progeny genotypes are shown.

Notice that, if crossing over entails physical exchange between the chromosomes, then the colorless, waxy progeny resulting from recombination should have a chromosome with an extra piece but not a knob. Furthermore, some of the colored, starchy progeny should possess a knob but not the extra piece. This outcome is precisely what Creighton and McClintock observed, confirming the chromosomal theory of inheritance. Curt Stern provided a similar demonstration by using chromosomal markers in *Drosophila* at about the same time.

Predicting the Outcomes of Crosses with Linked Genes

Knowing the arrangement of alleles on a chromosome allows us to predict the types of progeny that will result from a cross entailing linked genes and to determine which of these types will be the most numerous. Determining the *proportions* of the types of offspring requires an additional piece of information—the recombination frequency. The recombination frequency provides us with information about how often the alleles in the gametes appear in new combinations and allows us to predict the proportions of offspring phenotypes that will result from a specific cross with linked genes.

In cucumbers, smooth fruit (*t*) is recessive to warty fruit (*T*) and glossy fruit (*d*) is recessive to dull fruit (*D*). Geneticists have determined that these two genes exhibit a recombination frequency of 16%. Suppose we cross a plant

homozygous for warty and dull fruit with a plant homozygous for smooth and glossy fruit and then carry out a testcross by using the F_1:

$$\frac{T \quad D}{t \quad d} \times \frac{t \quad d}{t \quad d}$$

What types and proportions of progeny will result from this testcross?

Four types of gametes will be produced by the heterozygous parent, as shown in **Figure 7.10**: two types of nonrecombinant gametes ($\underline{T \quad D}$ and $\underline{t \quad d}$) and two types of recombinant gametes ($\underline{T \quad d}$ and $\underline{t \quad D}$). The recombination frequency tells us that 16% of the gametes produced by the heterozygous parent will be recombinant. Because there are two types of recombinant gametes, each should arise with a frequency of $^{16\%}/_2 = 8\%$. This frequency can also be represented as a probability of 0.08. All the other gametes will be nonrecombinants; so they should arise with a frequency of $100\% - 16\% = 84\%$. Because there are two types of nonrecombinant gametes, each should arise with a frequency of $^{84\%}/_2 = 42\%$ (or 0.42). The other parent in the testcross is homozygous and therefore produces only a single type of gamete ($\underline{t \quad d}$) with a frequency of 100% (or 1.00).

The progeny of the cross result from the union of two gametes, producing four types of progeny (see Figure 7.10). The expected proportion of each type can be determined by using the multiplication rule, multiplying together the probability of each uniting gamete. Testcross progeny with warty and dull fruit

$$\frac{T \quad D}{t \quad d}$$

appear with a frequency of 0.42 (the probability of inheriting a gamete with chromosome $\underline{T \quad D}$ from the heterozygous parent) $\times$ 1.00 (the probability of inheriting a gamete with chromosome $\underline{t \quad d}$ from the recessive parent) = 0.42. The proportions of the other types of F_2 progeny can be calculated in a similar manner (see Figure 7.10). This method can be used for predicting the outcome of any cross with linked genes for which the recombination frequency is known.

Testing for Independent Assortment

In some crosses, the genes are obviously linked because there are clearly more nonrecombinant progeny than recombinant progeny. In other crosses, the difference between independent assortment and linkage is not so obvious. For example, suppose we did a testcross for two pairs of genes, such as *Aa Bb* × *aa bb*, and observed the following numbers of progeny: 54 *Aa Bb*, 56 *aa bb*, 42 *Aa bb*, and 48 *aa Bb*. Is this outcome the 1:1:1:1 ratio we would expect if *A* and *B* assorted independently? Not exactly, but it's pretty close. Perhaps these genes assorted independently and chance produced the slight deviations between the observed numbers and the expected 1:1:1:1 ratio. Alternatively, the genes might be linked, with considerable

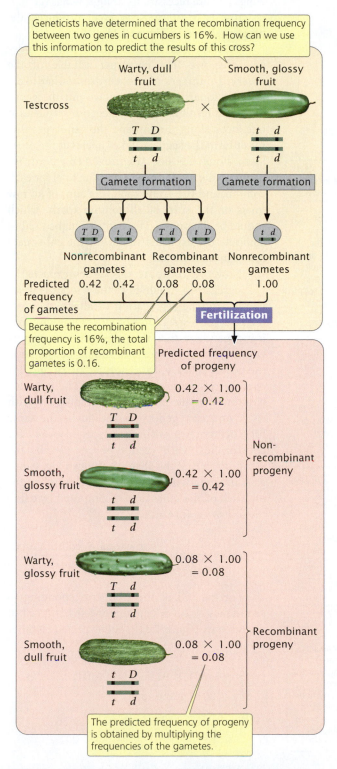

Geneticists have determined that the recombination frequency between two genes in cucumbers is 16%. How can we use this information to predict the results of this cross?

Testcross

Warty, dull fruit   ×   Smooth, glossy fruit

T D   t d
t d   t d

Gamete formation   Gamete formation

T D   t d   T d   t D   t d

Nonrecombinant   Recombinant   Nonrecombinant
gametes   gametes   gametes

Predicted   0.42   0.42   0.08   0.08   1.00
frequency
of gametes

Because the recombination frequency is 16%, the total proportion of recombinant gametes is 0.16.

Fertilization

Predicted frequency of progeny

Warty, dull fruit   0.42 × 1.00 = 0.42

T D
t d

Smooth, glossy fruit   0.42 × 1.00 = 0.42

t d
t d

Non-recombinant progeny

Warty, glossy fruit   0.08 × 1.00 = 0.08

T d
t d

Smooth, dull fruit   0.08 × 1.00 = 0.08

t D
t d

Recombinant progeny

The predicted frequency of progeny is obtained by multiplying the frequencies of the gametes.

7.10 The recombination frequency allows a prediction of the proportions of offspring expected for a cross entailing linked genes.

crossing over taking place between them, and so the number of nonrecombinants is only slightly greater than the number of recombinants. How do we distinguish between the role of chance and the role of linkage in producing deviations from the results expected with independent assortment?

We encountered a similar problem in crosses in which genes were unlinked—the problem of distinguishing between deviations due to chance and those due to other factors. We addressed this problem (in Chapter 3) with the goodness-of-fit chi-square test, which helps us evaluate the likelihood that chance alone is responsible for deviations between the numbers of progeny that we observed and the numbers that we expected by applying the principles of inheritance. Here, we are interested in a different question: Is the inheritance of alleles at one locus independent of the inheritance of alleles at a second locus? If the answer to this question is yes, then the genes are assorting independently; if the answer is no, then the genes are probably linked.

A possible way to test for independent assortment is to calculate the expected probability of each progeny type, assuming independent assortment, and then use the goodness-of-fit chi-square test to evaluate whether the observed numbers deviate significantly from the expected numbers. With independent assortment, we expect $\frac{1}{4}$ of each phenotype: $\frac{1}{4}$ *Aa Bb*, $\frac{1}{4}$ *aa bb*, $\frac{1}{4}$ *Aa bb*, and $\frac{1}{4}$ *aa Bb*. This expected probability of each genotype is based on the multiplication rule of probability, which we considered in Chapter 3. For example, if the probability of *Aa* is $\frac{1}{2}$ and the probability of *Bb* is $\frac{1}{2}$, then the probability of *Aa Bb* is $\frac{1}{2} \times \frac{1}{2} = \frac{1}{4}$. In this calculation, we are making two assumptions: (1) the probability of each single-locus genotype is $\frac{1}{2}$, and (2) genotypes at the two loci are inherited independently ($\frac{1}{2} \times \frac{1}{2} = \frac{1}{4}$).

One problem with this approach is that a significant chi-square test can result from a violation of either assumption. If the genes are linked, then the inheritance of genotypes at the two loci are not independent (assumption 2), and we will get a significant deviation between observed and expected numbers. But we can also get a significant deviation if the probability of each single-locus genotype is not $\frac{1}{2}$ (assumption 1), even when the genotypes are assorting independently. We may obtain a significant deviation, for example, if individuals with one genotype have a lower probability of surviving or the penetrance of a genotype is not 100%. We could test both assumptions by conducting a series of chi-square tests, first testing the inheritance of genotypes at each locus separately (assumption 1) and then testing for independent assortment (assumption 2). However, a faster method is to test for independence in genotypes with a *chi-square test of independence*.

The chi-square test of independence allows us to evaluate whether the segregation of alleles at one locus is independent of the segregation of alleles at another locus, without making any assumption about the probability of single-locus genotypes. To illustrate this analysis, we will examine results from a cross between German cockroaches, in which yellow body (y) is recessive to brown body (y^+) and curved wings (cv) are recessive to straight wings (cv^+). A testcross ($y^+y\ cv^+cv \times yy\ cvcv$) produced the progeny shown in **Figure 7.11a**.

To carry out the chi-square test of independence, we first construct a table of the observed numbers of progeny, somewhat like a Punnett square, except, here, we put the genotypes that result from the segregation of alleles at one locus along the top and the genotypes that result from the segregation of alleles at the other locus along the side (**Figure 7.11b**). Next, we compute the total for each row, the total for each column, and the grand total (the sum of all row totals or the sum of all column totals, which should be the same). These totals will be used to compute the expected values for the chi-square test of independence.

Now, we compute the values expected if the segregation of alleles at the y locus is independent of the segregation of alleles at the cv locus. If the segregation of alleles at each locus is independent, then the proportion of progeny with y^+y and yy genotypes should be the same for cockroaches with genotype cv^+cv and for cockroaches with genotype $cvcv$. The converse is also true; the proportions of progeny with cv^+cv and $cvcv$ genotypes should be the same for cockroaches with genotype y^+y and for cockroaches with genotype yy. With the assumption that the alleles at the two loci segregate independently, the expected number for each cell of the table can be computed by using the following formula:

$$\text{expected number} = \frac{\text{row total} \times \text{column total}}{\text{grand total}}$$

For the cell of the table corresponding to genotype $y^+y\ cv^+cv$ (the upper-left-hand cell of the table in Figure 7.11b) the expected number is:

$$\frac{96\ (\text{row total}) \times 91\ (\text{column total})}{201\ (\text{grand total})} =$$

$$\frac{8736}{201} = 43.46$$

With the use of this method, the expected numbers for each cell are given in **Figure 7.11c**.

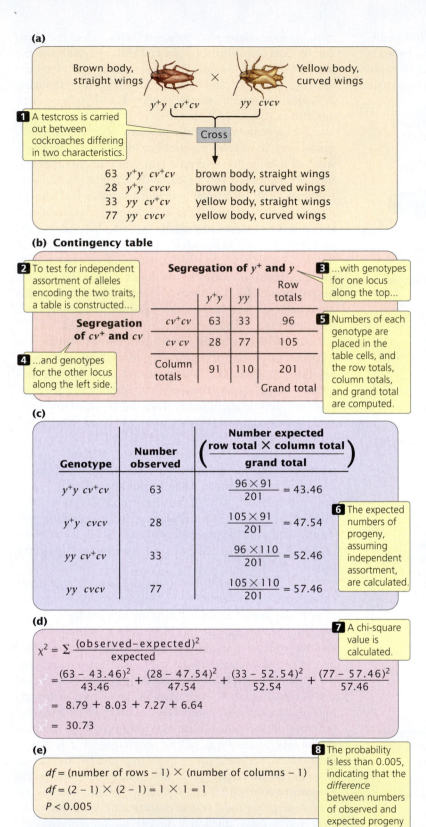

(a)

Brown body, straight wings × Yellow body, curved wings

$y^+y\ cv^+cv$ $yy\ cvcv$

1 A testcross is carried out between cockroaches differing in two characteristics.

Cross

63	$y^+y\ cv^+cv$	brown body, straight wings
28	$y^+y\ cvcv$	brown body, curved wings
33	$yy\ cv^+cv$	yellow body, straight wings
77	$yy\ cvcv$	yellow body, curved wings

(b) Contingency table

2 To test for independent assortment of alleles encoding the two traits, a table is constructed...

Segregation of y^+ and y

3 ...with genotypes for one locus along the top...

	y^+y	yy	Row totals
Segregation of cv^+ and cv cv^+cv	63	33	96
$cv\ cv$	28	77	105
Column totals	91	110	201

Grand total

4 ...and genotypes for the other locus along the left side.

5 Numbers of each genotype are placed in the table cells, and the row totals, column totals, and grand total are computed.

(c)

Genotype	Number observed	Number expected $\left(\dfrac{\text{row total} \times \text{column total}}{\text{grand total}}\right)$
$y^+y\ cv^+cv$	63	$\dfrac{96 \times 91}{201} = 43.46$
$y^+y\ cvcv$	28	$\dfrac{105 \times 91}{201} = 47.54$
$yy\ cv^+cv$	33	$\dfrac{96 \times 110}{201} = 52.46$
$yy\ cvcv$	77	$\dfrac{105 \times 110}{201} = 57.46$

6 The expected numbers of progeny, assuming independent assortment, are calculated.

(d)

7 A chi-square value is calculated.

$$\chi^2 = \sum \frac{(\text{observed} - \text{expected})^2}{\text{expected}}$$

$$\chi^2 = \frac{(63 - 43.46)^2}{43.46} + \frac{(28 - 47.54)^2}{47.54} + \frac{(33 - 52.54)^2}{52.54} + \frac{(77 - 57.46)^2}{57.46}$$

$$\chi^2 = 8.79 + 8.03 + 7.27 + 6.64$$

$$\chi^2 = 30.73$$

(e)

8 The probability is less than 0.005, indicating that the *difference* between numbers of observed and expected progeny is probably not due to chance.

$df = (\text{number of rows} - 1) \times (\text{number of columns} - 1)$

$df = (2 - 1) \times (2 - 1) = 1 \times 1 = 1$

$P < 0.005$

Conclusion: The genes for body color and type of wing are not assorting independently and must be linked.

7.11 A chi-square test of independence can be used to determine if genes at two loci are assorting independently.

We now calculate a chi-square value by using the same formula that we used for the goodness-of-fit chi-square test in Chapter 3:

$$\chi^2 = \Sigma \, \frac{(\text{observed} - \text{expected})^2}{\text{expected}}$$

With the observed and expected numbers of cockroaches from the testcross, the calculated chi-square value is 30.73 (**Figure 7.11d**).

To determine the probability associated with this chi-square value, we need the degrees of freedom. Recall from Chapter 3 that the degrees of freedom are the number of ways in which the observed classes are free to vary from the expected values. In general, for the chi-square test of independence, the degrees of freedom equal the number of rows in the table minus 1 multiplied by the number of columns in the table minus 1 (**Figure 7.11e**), or

$$df = (\text{number of rows} - 1) \times (\text{number of columns} - 1)$$

In our example, there were two rows and two columns, and so the degrees of freedom are:

$$df = (2 - 1) \times (2 - 1) = 1 \times 1 = 1$$

Therefore, our calculated chi-square value is 30.73, with 1 degree of freedom. We can use Table 3.4 to find the associated probability. Looking at Table 3.4, we find our calculated chi-square value is larger than the largest chi-square value given for 1 degree of freedom, which has a probability of 0.005. Thus, our calculated chi-square value has a probability less than 0.005. This very small probability indicates that the genotypes are not in the proportions that we would expect if independent assortment were taking place. Our conclusion, then, is that these genes are not assorting independently and must be linked. As is the case for the goodness-of-fit chi-square test, geneticists generally consider that any chi-square value for the test of independence with a probability less than 0.05 is significantly different from the expected values and is therefore evidence that the genes are not assorting independently.

Gene Mapping with Recombination Frequencies

Morgan and his students developed the idea that physical distances between genes on a chromosome are related to the rates of recombination. They hypothesized that crossover events occur more or less at random up and down the chromosome and that two genes that lie far apart are more likely to undergo a crossover than are two genes that lie close together. They proposed that recombination frequencies could provide a convenient way to determine the order of

genes along a chromosome and would give estimates of the relative distances between the genes. Chromosome maps calculated by using the genetic phenomenon of recombination are called **genetic maps.** In contrast, chromosome maps calculated by using physical distances along the chromosome (often expressed as numbers of base pairs) are called **physical maps.**

Distances on genetic maps are measured in **map units** (abbreviated m.u.); one map unit equals 1% recombination. Map units are also called **centiMorgans** (cM), in honor of Thomas Hunt Morgan; 100 centiMorgans equals one **Morgan.** Genetic distances measured with recombination rates are approximately additive: if the distance from gene A to gene B is 5 m.u., the distance from gene B to gene C is 10 m.u., and the distance from gene A to gene C is 15 m.u., then gene B must be located between genes A and C. On the basis of the map distances just given, we can draw a simple genetic map for genes A, B, and C, as shown here:

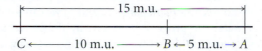

We could just as plausibly draw this map with C on the left and A on the right:

$$C \longleftarrow \text{10 m.u.} \longrightarrow B \longleftarrow \text{5 m.u.} \longrightarrow A$$

Both maps are correct and equivalent because, with information about the relative positions of only three genes, the most that we can determine is which gene lies in the middle. If we obtained distances to an additional gene, then we could position A and C relative to that gene. An additional gene D, examined through genetic crosses, might yield the following recombination frequencies:

Gene pair	Recombination frequency (%)
A and D	8
B and D	13
C and D	23

Notice that C and D exhibit the greatest amount of recombination; therefore, C and D must be farthest apart, with genes A and B between them. Using the recombination frequencies and remembering that 1 m.u. = 1% recombination, we can now add D to our map:

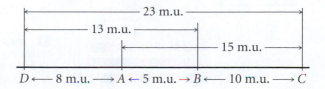

By doing a series of crosses between pairs of genes, we can construct genetic maps showing the linkage arrangements of a number of genes.

Two points should be emphasized about constructing chromosome maps from recombination frequencies. First, recall that we cannot distinguish between genes on different chromosomes and genes located far apart on the same chromosome. If genes exhibit 50% recombination, the most that can be said about them is that they belong to different groups of linked genes (different linkage groups), either on different chromosomes or far apart on the same chromosome.

The second point is that a testcross for two genes that are relatively far apart on the same chromosome tends to underestimate the true physical distance, because the cross does not reveal double crossovers that might take place between the two genes (**Figure 7.12**). A double crossover arises when two separate crossover events take place between two loci. (For now, we will consider only double crossovers that occur between two of the four chromatids of a homologous pair—a two-strand double crossover. Double crossovers entailing three and four chromatids will be considered later.) Whereas a single crossover produces combinations of alleles that were not present on the original parental chromosomes, a second crossover between the same two genes reverses the effects of the first, thus restoring the original parental combination of alleles (see Figure 7.12). Two-strand double crossovers produce only nonrecombinant gametes, and so we cannot distinguish between the progeny produced by double crossovers and the progeny produced when there is no crossing over at all. As we shall see in the next section, we can detect double crossovers if we examine a third gene that lies between the two crossovers. Because double crossovers between two genes go undetected, map distances will be underestimated whenever double crossovers take place. Double crossovers are more frequent between genes that are far apart; therefore genetic maps based on short distances are usually more accurate than those based on longer distances.

CONCEPTS

A genetic map provides the order of the genes on a chromosome and the approximate distances from one gene to another based on recombination frequencies. In genetic maps, 1% recombination equals 1 map unit, or 1 centiMorgan. Double crossovers between two genes go undetected; so map distances between distant genes tend to underestimate genetic distances.

✔ CONCEPT CHECK 3

How does a genetic map differ from a physical map?

Constructing a Genetic Map with Two-Point Testcrosses

Genetic maps can be constructed by conducting a series of testcrosses between pairs of genes and examining the recombination frequencies between them. A testcross between two genes is called a **two-point testcross** or a two-point cross for short. Suppose that we carried out a series of two-point crosses for four genes, *a, b, c,* and *d,* and obtained the following recombination frequencies:

Gene loci in testcross	Recombination frequency (%)
a and *b*	50
a and *c*	50
a and *d*	50
b and *c*	20
b and *d*	10
c and *d*	28

We can begin constructing a genetic map for these genes by considering the recombination frequencies for each pair of genes. The recombination frequency between *a* and *b* is 50%, which is the recombination frequency expected with

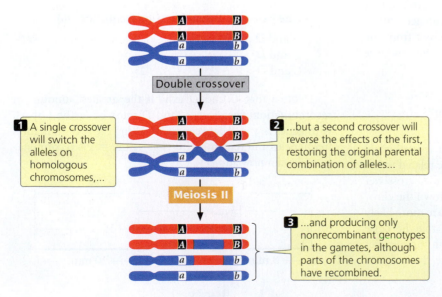

1 A single crossover will switch the alleles on homologous chromosomes,...

2 ...but a second crossover will reverse the effects of the first, restoring the original parental combination of alleles...

Double crossover

Meiosis II

3 ...and producing only nonrecombinant genotypes in the gametes, although parts of the chromosomes have recombined.

7.12 A two-strand double crossover between two linked genes produces only nonrecombinant gametes.

independent assortment. Therefore, genes *a* and *b* may either be on different chromosomes or be very far apart on the same chromosome; so we will place them in different linkage groups with the understanding that they may or may not be on the same chromosome:

Linkage group 1

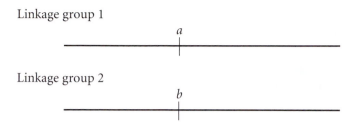

Linkage group 2

The recombination frequency between *a* and *c* is 50%, indicating that they, too, are in different linkage groups. The recombination frequency between *b* and *c* is 20%; so these genes are linked and separated by 20 map units:

Linkage group 1

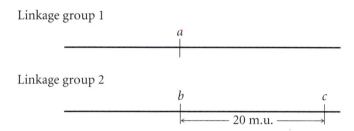

Linkage group 2

The recombination frequency between *a* and *d* is 50%, indicating that these genes belong to different linkage groups, whereas genes *b* and *d* are linked, with a recombination frequency of 10%. To decide whether gene *d* is 10 m.u. to the left or to the right of gene *b*, we must consult the *c*-to-*d* distance. If gene *d* is 10 m.u. to the left of gene *b*, then the distance between *d* and *c* should be 20 m.u. + 10 m.u. = 30 m.u. This distance will be only approximate because any double crossovers between the two genes will be missed and the map distance will be underestimated. If, on the other hand, gene *d* lies to the right of gene *b*, then the distance between gene *d* and gene *c* will be much shorter, approximately 20 m.u. − 10 m.u. = 10 m.u.

By examining the recombination frequency between *c* and *d*, we can distinguish between these two possibilities. The recombination frequency between *c* and *d* is 28%; so gene *d* must lie to the left of gene *b*. Notice that the sum of the recombination between *d* and *b* (10%) and between *b* and *c* (20%) is greater than the recombination between *d* and *c* (28%). As already discussed, this discrepancy arises because double crossovers between the two outer genes go undetected, causing an underestimation of the true map distance. The genetic map of these genes is now complete:

Linkage group 1

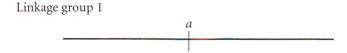

Linkage group 2

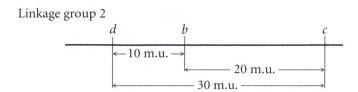

7.3 A Three-Point Testcross Can Be Used to Map Three Linked Genes

Genetic maps can be constructed from a series of testcrosses for pairs of genes, but this approach is not particularly efficient, because numerous two-point crosses must be carried out to establish the order of the genes and because double crossovers are missed. A more efficient mapping technique is a testcross for three genes—a **three-point testcross,** or three-point cross. With a three-point cross, the order of the three genes can be established in a single set of progeny and some double crossovers can usually be detected, providing more accurate map distances.

Consider what happens when crossing over takes place among three hypothetical linked genes. **Figure 7.13** illustrates a pair of homologous chromosomes from an individual that is heterozygous at three loci (*Aa Ba Cc*). Notice that the genes are in the coupling configuration; that is, all the dominant alleles are on one chromosome (*A   B   C*) and all the recessive alleles are on the other chromosome (*a   b   c*). Three types of crossover events can take place between these three genes: two types of single crossovers (see Figure 7.13a and b) and a double crossover (see Figure 7.13c). In each type of crossover, two of the resulting chromosomes are recombinants and two are nonrecombinants.

Notice that, in the recombinant chromosomes resulting from the double crossover, the outer two alleles are the same as in the nonrecombinants, but the middle allele is different. This result provides us with an important clue about the order of the genes. In progeny that result from a double crossover, only the middle allele should differ from the alleles present in the nonrecombinant progeny.

Constructing a Genetic Map with the Three-Point Testcross

To examine gene mapping with a three-point testcross, we will consider three recessive mutations in the fruit fly *Drosophila melanogaster*. In this species, scarlet eyes (*st*) are recessive to red eyes (*st*⁺), ebony body color (*e*) is recessive to gray body color (*e*⁺), and spineless (*ss*)—that is, the presence of small bristles—is recessive to normal bristles (*ss*⁺). All three mutations are linked and located on the third chromosome.

We will refer to these three loci as *st*, *e*, and *ss*, but keep in mind that either the recessive alleles (*st*, *e*, and *ss*) or the dominant alleles (*st*⁺, *e*⁺, and *ss*⁺) may be present at each

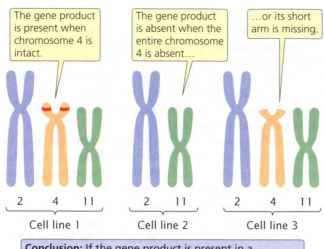

The gene product is present when chromosome 4 is intact.

The gene product is absent when the entire chromosome 4 is absent...

...or its short arm is missing.

2   4   11   2   11   2   4   11
Cell line 1   Cell line 2   Cell line 3

Conclusion: If the gene product is present in a cell line with an intact chromosome but missing from a line with a chromosome deletion, the gene for that product must be located in the deleted region.

7.23 Genes can be localized to a specific part of a chromosome by using somatic-cell hybridization.

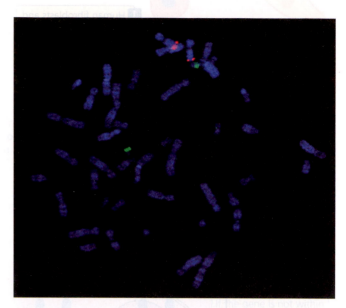

7.24 In situ hybridization is another technique for determining the chromosomal location of a gene. The red fluorescence is produced by a probe for sequences on chromosome 9; the green fluorescence is produced by a probe for sequences on chromosome 22. [*Genetic image courtesy of Applied Imaging Corp.*]

• Solution

First, identify the cell lines that are positive for the protein (human haptoglobin) and determine the chromosomes that they have in common. Lines B and C produce human haptoglobin; the only chromosomes they have in common are chromosomes 1 and 16. Next, examine all the cell lines that possess either chromosomes 1 and 16 and determine whether they produce haptoglobin. Chromosome 1 is found in cell lines A, B, C, and D. If the gene for human haptoglobin were found on chromosome 1, human haptoglobin would be present in all of these cell lines. However, lines A and D do not produce human haptoglobin; so the gene cannot be on chromosome 1. Chromosome 16 is found only in cell lines B and C, and only these lines produce human haptoglobin; so the gene for human haptoglobin lies on chromosome 16.

❓ For more practice with somatic cell hybridizations, work Problem 33 at the end of this chapter.

Physical Chromosome Mapping Through Molecular Analysis

So far, we have explored methods to indirectly determine the chromosomal location of a gene by deletion mapping or by looking for gene products. Now, researchers have the information and technology to actually see where a gene lies. In situ hybridization is a method for determining the chromosomal location of a particular gene through molecular analysis. This method requires creating a probe for the gene—a probe that is radioactive or fluoresces under ultraviolet light. The probe binds to the DNA sequence of the gene on the chromosome. The presence of radioactivity or fluorescence from the bound probe reveals the location of the gene on a particular chromosome (**Figure 7.24**).

In addition to allowing us to see where a gene is located on a chromosome, modern laboratory techniques now allow researchers to identify the precise location of a gene at the nucleotide level. For example, with DNA sequencing, physical distances between genes can be determined in numbers of base pairs.

CONCEPTS

Physical-mapping methods determine the physical locations of genes on chromosomes and include deletion mapping, somatic-cell hybridization, in situ hybridization, and direct DNA sequencing.

7.5 Recombination Rates Exhibit Extensive Variation

In recent years, geneticists have studied variation in rates of recombination and found that levels of recombination vary widely—among species, among and along chromosomes of a single species, and even between males and females of the same species. For example, about twice as much recombination takes place in humans as in the mouse and the rat. Within the human genome, recombination varies among chromosomes, with chromosomes 21 and 22 having the highest rates and chromosomes 2 and 4 having the lowest rates. Researchers have also detected differences between male and female humans in rates of recombination: the autosomal chromosomes of females undergo about 50% more recombination than the autosomal chromosomes of males.

Geneticists have found numerous recombination hotspots, where recombination is at least 10 times as high as the average elsewhere in the genome. The human genome may contain an estimated 25,000 to 50,000 such recombination hotspots. For humans, recombination hotspots tend to be found near, but not within, active genes. Recombination hotspots have been detected in the genomes of other organisms as well. In comparing recombination hotspots in the genomes of humans and chimpanzees (our closest living relative), geneticists have determined that, although the DNA sequences of humans and chimpanzees are extremely similar, their recombination hotspots are at entirely different places.

CONCEPTS

Rates of recombination vary among species, among and along chromosomes, and even between males and females.

CONCEPTS SUMMARY

- Linked genes do not assort independently. In a testcross for two completely linked genes (no crossing over), only nonrecombinant progeny are produced. When two genes assort independently, recombinant progeny and nonrecombinant progeny are produced in equal proportions. When two genes are linked with some crossing over between them, more nonrecombinant progeny than recombinant progeny are produced.

- Recombination frequency is calculated by summing the number of recombinant progeny, dividing by the total number of progeny produced in the cross, and multiplying by 100%. The recombination frequency is half the frequency of crossing over, and the maximum frequency of recombinant gametes is 50%.

- Coupling and repulsion refer to the arrangement of alleles on a chromosome. Whether genes are in coupling configuration or in repulsion determines which combination of phenotypes will be most frequent in the progeny of a testcross.

- Interchromosomal recombination takes place among genes located on different chromosomes through the random segregation of chromosomes in meiosis. Intrachromosomal recombination takes place among genes located on the same chromosome through crossing over.

- A chi-square test of independence can be used to determine if genes are linked.

- Recombination rates can be used to determine the relative order of genes and distances between them on a chromosome. One percent recombination equals one map unit. Maps based on recombination rates are called genetic maps; maps based on physical distances are called physical maps.

- Genetic maps can be constructed by examining recombination rates from a series of two-point crosses or by examining the progeny of a three-point testcross.

- Some multiple crossovers go undetected; thus, genetic maps based on recombination rates underestimate the true physical distances between genes.

- Human genes can be mapped by examining the cosegregation of traits in pedigrees.

- A lod score is the logarithm of the ratio of the probability of obtaining the observed progeny with the assumption of linkage to the probability of obtaining the observed progeny with the assumption of independent assortment. A lod score of 3 or higher is usually considered evidence for linkage.

- Molecular techniques that allow the detection of variable differences in DNA sequence have greatly facilitated gene mapping.

- Nucleotide sequencing is another method of physically mapping genes.

- Rates of recombination vary widely, differing among species, among and along chromosomes within a single species, and even between males and females of the same species.

IMPORTANT TERMS

linked genes (p. 162)
linkage group (p. 162)
nonrecombinant (parental) gamete (p. 164)
nonrecombinant (parental) progeny (p. 164)
recombinant gamete (p. 164)
recombinant progeny (p. 164)
recombination frequency (p. 166)
coupling (cis) configuration (p. 167)
repulsion (trans) configuration (p. 167)

interchromosomal recombination (p. 169)
intrachromosomal recombination (p. 169)
genetic map (p. 173)
physical map (p. 173)
map unit (m.u.) (p. 173)
centiMorgan (p. 173)
Morgan (p. 173)
two-point testcross (p. 174)
three-point testcross (p. 175)

interference (p. 180)
coefficient of coincidence (p. 181)
mapping function (p. 183)
lod (logarithm of odds) score (p. 185)
genetic marker (p. 185)
deletion mapping (p. 185)
somatic-cell hybridization (p. 186)
cell line (p. 186)
heterokaryon (p. 186)

ANSWERS TO CONCEPT CHECKS

1. c

2. 20%, in repulsion

3. Genetic maps are based on rates of recombination; physical maps are based on physical distances.

4. $\dfrac{m^+ p^+ s^+}{m\ p\ s}\ \dfrac{m^+ p\ s}{m\ p\ s}\ \dfrac{m\ p^+ s^+}{m\ p\ s}\ \dfrac{m^+ p^+ s}{m\ p\ s}\ \dfrac{m\ p^+ s^+}{m\ p\ s}\ \dfrac{m^+ p\ s^+}{m\ p\ s}\ \dfrac{m\ p^+ s}{m\ p\ s}\ \dfrac{m\ p\ s}{m\ p\ s}$

5. The c locus

6. b

WORKED PROBLEMS

1. In guinea pigs, white coat (w) is recessive to black coat (W) and wavy hair (v) is recessive to straight hair (V). A breeder crosses a guinea pig that is homozygous for white coat and wavy hair with a guinea pig that is black with straight hair. The F_1 are then crossed with guinea pigs having white coats and wavy hair in a series of testcrosses. The following progeny are produced from these testcrosses:

black, straight	30
black, wavy	10
white, straight	12
white, wavy	31
Total	83

   a. Are the genes that determine coat color and hair type assorting independently? Carry out chi-square tests to test your hypothesis.

   b. If the genes are not assorting independently, what is the recombination frequency between them?

• **Solution**

   a. Assuming independent assortment, outline the crosses conducted by the breeder:

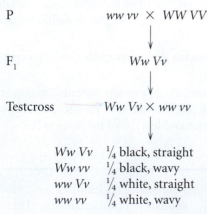

   P $ww\ vv \times WW\ VV$

   F_1 $Ww\ Vv$

   Testcross $Ww\ Vv \times ww\ vv$

$Ww\ Vv$	$\frac{1}{4}$ black, straight
$Ww\ vv$	$\frac{1}{4}$ black, wavy
$ww\ Vv$	$\frac{1}{4}$ white, straight
$ww\ vv$	$\frac{1}{4}$ white, wavy

Because a total of 83 progeny were produced in the testcrosses, we expect $\frac{1}{4} \times 83 = 20.75$ of each. The observed numbers of progeny from the testcross (30, 10, 12, 31) do not appear to fit the expected numbers (20.75, 20.75, 20.75, 20.75) well; so independent assortment may not have taken place.

   To test the hypothesis, carry out a chi-square test of independence. Construct a table, with the genotypes of the first locus along the top and the genotypes of the second locus along the side. Compute the totals for the rows and columns and the grand total.

	Ww	ww	Row totals
Vv	30	12	42
vv	10	31	41
Column totals	40	43	83 ← Grand total

The expected value for each cell of the table is calculated with the formula:

$$\text{expected number} = \frac{\text{row total} \times \text{column total}}{\text{grand total}}$$

Using this formula, we find the expected values (given in parentheses) to be:

	Ww	ww	Row totals
Vv	30 (20.24)	12 (21.76)	42
vv	10 (19.76)	31 (21.24)	41
Column totals	40	43	83 ← Grand total

Using these observed and expected numbers, we find the calculated chi-square value to be:

$$\chi^2 = \Sigma \frac{(\text{observed} - \text{expected})^2}{\text{expected}}$$

$$= \frac{(30 - 20.24)^2}{20.24} + \frac{(10 - 19.76)^2}{19.76} + \frac{(12 - 21.76)^2}{21.76}$$

$$+ \frac{(31 - 21.24)^2}{21.24}$$

$$= 4.71 + 4.82 + 4.38 + 4.48$$

$$= 18.39$$

The degrees of freedom for the chi-square test of independence are $df = $ (number of rows $- 1$) $\times$ (number of columns $- 1$). There are two rows and two columns, so the degrees of freedom are:

$$df = (2 - 1) \times (2 - 1) = 1 \times 1 = 1$$

In Table 3.4, the probability associated with a chi-square value of 18.39 and 1 degree of freedom is less than 0.005, indicating that chance is very unlikely to be responsible for the differences between the observed numbers and the numbers expected with independent assortment. The genes for coat color and hair type have therefore not assorted independently.

b. To determine the recombination frequencies, identify the recombinant progeny. Using the notation for linked genes, write the crosses:

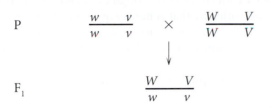

	$\dfrac{W}{w}\quad\dfrac{V}{v}$	30 black, straight (nonrecombinant progeny)

$\dfrac{w}{w}\quad\dfrac{v}{v}$ — 31 white, wavy (nonrecombinant progeny)

$\dfrac{W}{w}\quad\dfrac{v}{v}$ — 10 black, wavy (recombinant progeny)

$\dfrac{w}{w}\quad\dfrac{V}{v}$ — 12 white, straight (recombinant progeny)

The recombination frequency is:

$$\frac{\text{number of recombinant progeny}}{\text{total number progeny}} \times 100\%$$

or

recombinant frequency =

$$\frac{10 + 12}{30 + 31 + 10 + 12} \times 100\% = \frac{22}{83} \times 100\% = 26.5$$

2. A series of two-point crosses were carried out among seven loci (a, b, c, d, e, f, and g), producing the following recombination frequencies. Using these recombination frequencies, map the seven loci, showing their linkage groups, the order of the loci in each linkage group, and the distances between the loci of each group:

Loci	Recombination frequency (%)	Loci	Recombination frequency (%)
a and b	10	c and d	50
a and c	50	c and e	8
a and d	14	c and f	50
a and e	50	c and g	12
a and f	50	d and e	50
a and g	50	d and f	50
b and c	50	d and g	50
b and d	4	e and f	50
b and e	50	e and g	18
b and f	50	f and g	50
b and g	50		

• Solution

To work this problem, remember that 1% recombination equals 1 map unit and a recombination frequency of 50% means that genes at the two loci are assorting independently (located in different linkage groups).

The recombination frequency between a and b is 10%; so these two loci are in the same linkage group, approximately 10 m.u. apart.

Linkage group 1

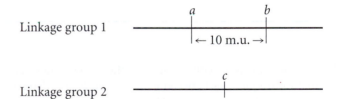

The recombination frequency between a and c is 50%; so c must lie in a second linkage group.

The recombination frequency between a and d is 14%; so d is located in linkage group 1. Is locus d 14 m.u. to the right or to the left of gene a? If d is 14 m.u. to the left of a, then the b-to-d distance should be 10 m.u. $+$ 14 m.u. $=$ 24 m.u. On the other hand, if d is to the right of a, then the distance between b and d should be 14 m.u. $-$ 10 m.u. $=$ 4 m.u. The b–d recombination frequency is 4%; so d is 14 m.u. to the right of a. The updated map is:

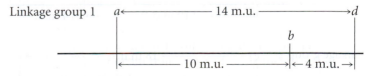

Linkage group 2

Mutation *a* is expressed in flies with deletions 4, 5, and 6 but not in flies with other deletions; so *a* must be in the area that is unique to deletions 4, 5, and 6:

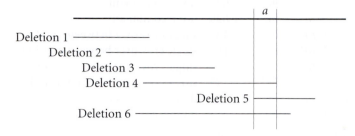

Mutation *b* is expressed only when deletion 1 is present; so *b* must be located in the region of the chromosome covered by deletion 1 and none of the other deletions:

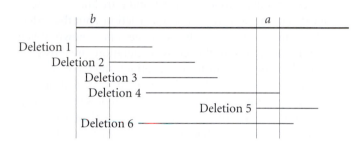

Using this procedure, we can map the remaining mutations. For each mutation, we look for the areas of overlap among deletions that express the mutations and exclude any areas of overlap that are covered by other deletions that do not express the mutation:

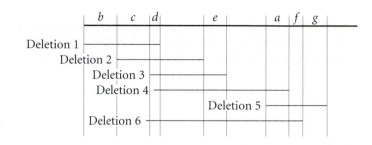

COMPREHENSION QUESTIONS

Section 7.1

*1. What does the term recombination mean? What are two causes of recombination?

Section 7.2

*2. In a testcross for two genes, what types of gametes are produced with (a) complete linkage, (b) independent assortment, and (c) incomplete linkage?

3. What effect does crossing over have on linkage?

4. Why is the frequency of recombinant gametes always half the frequency of crossing over?

*5. What is the difference between genes in coupling configuration and genes in repulsion? What effect does the arrangement of linked genes (whether they are in coupling configuration or in repulsion) have on the results of a cross?

6. How would you test to see if two genes are linked?

7. What is the difference between a genetic map and a physical map?

*8. Why do calculated recombination frequencies between pairs of loci that are located far apart underestimate the true genetic distances between loci?

Section 7.3

9. Explain how to determine which of three linked loci is the middle locus from the progeny of a three-point testcross.

*10. What does the interference tell us about the effect of one crossover on another?

Section 7.4

11. What is a lod score and how is it calculated?

12. List some of the methods for physically mapping genes and explain how they are used to position genes on chromosomes.

APPLICATION QUESTIONS AND PROBLEMS

Section 7.2

13. In the snail *Cepaea nemoralis*, an autosomal allele causing a banded shell (B^B) is recessive to the allele for an unbanded

shell (B^O). Genes at a different locus determine the background color of the shell; here, yellow (C^Y) is recessive to brown (C^{Bw}). A banded, yellow snail is crossed with a homozygous brown, unbanded snail. The F_1 are then crossed with banded, yellow snails (a testcross).

a. What will be the results of the testcross if the loci that control banding and color are linked with no crossing over?

b. What will be the results of the testcross if the loci assort independently?

c. What will be the results of the testcross if the loci are linked and 20 m.u. apart?

14. In silkmoths (*Bombyx mori*), red eyes (*re*) and white-banded wing (*wb*) are encoded by two mutant alleles that are recessive to those that produce wild-type traits (*re*$^+$ and *wb*$^+$); these two genes are on the same chromosome. A moth homozygous for red eyes and white-banded wings is crossed with a moth homozygous for the wild-type traits. The F_1 have normal eyes and normal wings. The F_1 are crossed with moths that have red eyes and white-banded wings in a testcross. The progeny of this testcross are:

wild-type eyes, wild-type wings	418
red eyes, wild-type wings	19
wild-type eyes, white-banded wings	16
red eyes, white-banded wings	426

a. What phenotypic proportions would be expected if the genes for red eyes and for white-banded wings were located on different chromosomes?

b. What is the genetic distance between the genes for red eyes and for white-banded wings?

*15. A geneticist discovers a new mutation in *Drosophila melanogaster* that causes the flies to shake and quiver. She calls this mutation spastic (*sps*) and determines that spastic is due to an autosomal recessive gene. She wants to determine if the gene encoding spastic is linked to the recessive gene for vestigial wings (*vg*). She crosses a fly homozygous for spastic and vestigial traits with a fly homozygous for the wild-type traits and then uses the resulting F_1 females in a testcross. She obtains the following flies from this testcross.

vg$^+$	*sps*$^+$	230
vg	*sps*	224
vg	*sps*$^+$	97
vg$^+$	*sps*	99
Total		650

Are the genes that cause vestigial wings and the spastic mutation linked? Do a chi-square test of independence to determine if the genes have assorted independently.

16. In cucumbers, heart-shaped leaves (*hl*) are recessive to normal leaves (*Hl*) and having numerous fruit spines (*ns*) is recessive to having few fruit spines (*Ns*). The genes for

leaf shape and for number of spines are located on the same chromosome; findings from mapping experiments indicate that they are 32.6 m.u. apart. A cucumber plant having heart-shaped leaves and numerous spines is crossed with a plant that is homozygous for normal leaves and few spines. The F_1 are crossed with plants that have heart-shaped leaves and numerous spines. What phenotypes and proportions are expected in the progeny of this cross?

*17. In tomatoes, tall (*D*) is dominant over dwarf (*d*) and smooth fruit (*P*) is dominant over pubescent fruit (*p*), which is covered with fine hairs. A farmer has two tall and smooth tomato plants, which we will call plant A and plant B. The farmer crosses plants A and B with the same dwarf and pubescent plant and obtains the following numbers of progeny:

	Progeny of	
	Plant A	Plant B
Dd Pp	122	2
Dd pp	6	82
dd Pp	4	82
dd pp	124	4

a. What are the genotypes of plant A and plant B?

b. Are the loci that determine height of the plant and pubescence linked? If so, what is the map distance between them?

c. Explain why different proportions of progeny are produced when plant A and plant B are crossed with the same dwarf pubescent plant.

18. Alleles *A* and *a* are at a locus that is located on the same chromosome as is a locus with alleles *B* and *b*. *Aa Bb* is crossed with *aa bb* and the following progeny are produced:

Aa Bb	5
Aa bb	45
aa Bb	45
aa bb	5

What conclusion can be made about the arrangement of the genes on the chromosome in the *Aa Bb* parent?

19. **DATA ANALYSIS** Daniel McDonald and Nancy Peer determined that eyespot (a clear spot in the center of the eye) in flour beetles is caused by an X-linked gene (*es*) that is recessive to the allele for the absence of eyespot (*es*$^+$). They conducted a series of crosses to determine the distance between the gene for eyespot and a dominant X-linked gene for stripped (*St*), which acted as a recessive lethal (is lethal when homozygous in females or hemizygous in males). The cross on the next page was carried out (D. J. McDonald and N. J. Peer. 1961. *Journal of Heredity* 52:261–264).

$$\female \; \frac{es^+}{es} \; \frac{St}{St^+} \times \frac{es}{Y} \; \frac{St^+}{} \; \male$$

$$\downarrow$$

$$\frac{es^+}{es} \; \frac{St}{St^+} \qquad 1630$$

$$\frac{es}{es} \; \frac{St^+}{St^+} \qquad 1665$$

$$\frac{es}{es} \; \frac{St}{St^+} \qquad 935$$

$$\frac{es^+}{es} \; \frac{St^+}{St^+} \qquad 1005$$

$$\frac{es}{Y} \; \frac{St^+}{} \qquad 1661$$

$$\frac{es^+}{Y} \; \frac{St^+}{} \qquad 1024$$

a. Which progeny are the recombinants and which progeny are the nonrecombinants?

b. Calculate the recombination frequency between *es* and *St*.

c. Are some potential genotypes missing among the progeny of the cross? If so, which ones and why?

20. In tomatoes, dwarf (*d*) is recessive to tall (*D*) and opaque (light green) leaves (*op*) are recessive to green leaves (*Op*). The loci that determine height and leaf color are linked and separated by a distance of 7 m.u. For each of the following crosses, determine the phenotypes and proportions of progeny produced.

a. $\dfrac{D \quad Op}{d \quad op} \times \dfrac{d \quad op}{d \quad op}$

b. $\dfrac{D \quad op}{d \quad Op} \times \dfrac{d \quad op}{d \quad op}$

c. $\dfrac{D \quad Op}{d \quad op} \times \dfrac{D \quad Op}{d \quad op}$

d. $\dfrac{D \quad op}{d \quad Op} \times \dfrac{D \quad op}{d \quad Op}$

21. In German cockroaches, bulging eyes (*bu*) are recessive to normal eyes (*bu⁺*) and curved wings (*cv*) are recessive to straight wings (*cv⁺*). Both traits are encoded by autosomal genes that are linked. A cockroach has genotype $bu^+bu \; cv^+cv$ and the genes are in repulsion. Which of the following sets of genes will be found in the most-common gametes produced by this cockroach?

a. $bu^+ \; cv^+$ **d.** $cv^+ \; cv$

b. $bu \; cv$ **e.** $bu \; cv^+$

c. $bu^+ \; bu$

Explain your answer.

*22. In *Drosophila melanogaster,* ebony body (*e*) and rough eyes (*ro*) are encoded by autosomal recessive genes found on chromosome 3; they are separated by 20 m.u. The gene that encodes forked bristles (*f*) is X-linked recessive and assorts independently of *e* and *ro*. Give the phenotypes of progeny and their expected proportions when a female of each of the following genotypes is test-crossed with a male.

a. $\dfrac{e^+ \quad ro^+}{e \quad ro} \; \dfrac{f^+}{f}$

b. $\dfrac{e^+ \quad ro}{e \quad ro^+} \; \dfrac{f^+}{f}$

23. DATA ANALYSIS   Honeybees have haplodiploid sex determination: females are diploid, developing from fertilized eggs, whereas males are haploid, developing from unfertilized eggs (see Chapter 4). Otto Mackensen studied linkage relations among eight mutations in honeybees (O. Mackensen. 1958. *Journal of Heredity* 49:99–102). The following table gives the results of two of MacKensen's crosses including three recessive mutations: *cd* (cordovan body color), *h* (hairless), and *ch* (chartreuse eye color).

Queen genotype	Phenotypes of drone (male) progeny
$\dfrac{cd \quad h^+}{cd^+ \quad h}$	294 cordovan, 236 hairless, 262 cordovan and hairless, 289 wild .type
$\dfrac{h \quad ch^+}{h^+ \quad ch}$	3131 hairless, 3064 chartreuse, 96 chartreuse and hairless, 132 wild type

a. Only the genotype of the queen is given. Why is the genotype of the male parent not needed for mapping these genes? Would the genotype of the male parent be required if we examined female progeny instead of male progeny?

b. Determine the nonrecombinant and recombinant progeny for each cross and calculate the map distances between *cd, h,* and *ch*. Draw a linkage map illustrating the linkage arrangements among these three enes.

*24. A series of two-point crosses were carried out among seven loci (*a, b, c, d, e, f,* and *g*), producing the following recombination frequencies, which continues on page 197. Map the seven loci, showing their linkage groups, the order of the loci in each linkage group, and the distances between the loci of each group.

Loci	Percent recombination	Loci	Percent recombination
a and *b*	50	*b* and *c*	10
a and *c*	50	*b* and *d*	50
a and *d*	12	*b* and *e*	18
a and *e*	50	*b* and *f*	50
a and *f*	50	*b* and *g*	50
a and *g*	4		

Loci	Percent recombination	Loci	Percent recombination
c and d	50	d and f	50
c and e	26	d and g	8
c and f	50	e and f	50
c and g	50	e and g	50
d and e	50	f and g	50

25. R. W. Allard and W. M. Clement determined recombination rates for a series of genes in lima beans (R. W. Allard and W. M. Clement. 1959. *Journal of Heredity* 50:63–67). The following table lists paired recombination rates for eight of the loci (D, Wl, R, S, L_1, Ms, C, and G) that they mapped. On the basis of these data, draw a series of genetic maps for the different linkage groups of the genes, indicating the distances between the genes. Keep in mind that these rates are estimates of the true recombination rates and that some error is associated with each estimate. An asterisk beside a recombination frequency indicates that the recombination frequency is significantly different from 50%.

Recombination Rates (%) among Seven Loci in Lima Beans

	Wl	R	S	L_1	Ms	C	G
D	2.1*	39.3*	52.4	48.1	53.1	51.4	49.8
Wl		38.0*	47.3	47.7	48.8	50.3	50.4
R			51.9	52.7	54.6	49.3	52.6
S				26.9*	54.9	52.0	48.0
L_1					48.2	45.3	50.4
Ms						14.7*	43.1
C							52.0

*Significantly different from 50%.

Section 7.3

26. Raymond Popp studied linkage among genes for pink eye (p), shaker-1 (sh-1), and hemoglobin (Hb) in mice (R. A. Popp. 1962. *Journal of Heredity* 53:73–80). He performed a series of test crosses, in which mice heterozygous for pink eye, shaker-1, and hemoglobin 1 and 2 were crossed with mice that were homozygous for pink eye, shaker-1 and hemoglobin 2.

$$\frac{P\ Sh\text{-}1\ Hb^1}{p\ sh\text{-}1\ Hb^2} \times \frac{p\ sh\text{-}1\ Hb^2}{p\ sh\text{-}1\ Hb^2}$$

The following progeny were produced.

Progeny genotype	Number
$\dfrac{p\ sh\text{-}1\ Hb^2}{p\ sh\text{-}1\ Hb^2}$	274
$\dfrac{P\ Sh\text{-}1\ Hb^1}{p\ sh\text{-}1\ Hb^2}$	320
$\dfrac{P\ sh\text{-}1\ Hb^2}{p\ sh\text{-}1\ Hb^2}$	57
$\dfrac{p\ Sh\text{-}1\ Hb^1}{p\ sh\text{-}1\ Hb^2}$	45
$\dfrac{p\ Sh\text{-}1\ Hb^2}{p\ sh\text{-}1\ Hb^2}$	6
$\dfrac{p\ sh\text{-}1\ Hb^1}{p\ sh\text{-}1\ Hb^2}$	5
$\dfrac{p\ Sh\text{-}1\ Hb^2}{p\ sh\text{-}1\ Hb^2}$	0
$\dfrac{P\ sh\text{-}1\ Hb^1}{p\ sh\text{-}1\ Hb^2}$	1
Total	708

a. Determine the order of these genes on the chromosome.

b. Calculate the map distances between the genes.

c. Determine the coefficient of coincidence and the interference among these genes.

*27. Waxy endosperm (wx), shrunken endosperm (sh), and yellow seedling (v) are encoded by three recessive genes in corn that are linked on chromosome 5. A corn plant homozygous for all three recessive alleles is crossed with a plant homozygous for all the dominant alleles. The resulting F_1 are then crossed with a plant homozygous for the recessive alleles in a three-point testcross. The progeny of the testcross are:

wx	sh	V	87
Wx	Sh	v	94
Wx	Sh	V	3,479
wx	sh	v	3,478
Wx	sh	V	1,515
wx	Sh	v	1,531
wx	Sh	V	292
Wx	sh	v	280
Total			10,756

a. Determine the order of these genes on the chromosome.

b. Calculate the map distances between the genes.

c. Determine the coefficient of coincidence and the interference among these genes.

28. Priscilla Lane and Margaret Green studied the linkage relations of three genes affecting coat color in mice: mahogany (mg), agouti (a), and ragged (Ra). They carried out a series of three-point crosses, mating mice that were heterozygous at all three loci with mice that were homozygous for the recessive alleles at these loci (P. W. Lane and M. C. Green. 1960. *Journal of Heredity*

51:228–230). The following table lists the results of the testcrosses.

Phenotype			Number
a	*Rg*	+	1
+	+	*mg*	1
a	+	+	15
+	*Rg*	*mg*	9
+	+	+	16
a	*Ra*	*mg*	36
a	+	*mg*	76
+	*Ra*	+	69
Total			213

a. Determine the order of the loci that encode mahogany, agouti, and ragged on the chromosome, the map distances between them, and the interference and coefficient of coincidence for these genes.

b. Draw a picture of the two chromosomes in the triply heterozygous mice used in the testcrosses, indicating which of the alleles are present on each chromosome.

29. Fine spines (*s*), smooth fruit (*tu*), and uniform fruit color (*u*) are three recessive traits in cucumbers, the genes of which are linked on the same chromosome. A cucumber plant heterozygous for all three traits is used in a testcross, and the following progeny are produced from this testcross:

S	*U*	*Tu*	2
s	*u*	*Tu*	70
S	*u*	*Tu*	21
s	*u*	*tu*	4
S	*U*	*tu*	82
s	*U*	*tu*	21
s	*U*	*Tu*	13
S	*u*	*tu*	17
Total			230

a. Determine the order of these genes on the chromosome.

b. Calculate the map distances between the genes.

c. Determine the coefficient of coincidence and the interference among these genes.

d. List the genes found on each chromosome in the parents used in the testcross.

*30. In *Drosophila melanogaster*, black body (*b*) is recessive to gray body (*b*⁺), purple eyes (*pr*) are recessive to red eyes (*pr*⁺), and vestigial wings (*vg*) are recessive to normal wings (*vg*⁺). The loci encoding these traits are linked, with the following map distances:

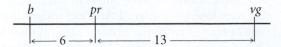

The interference among these genes is 0.5. A fly with a black body, purple eyes, and vestigial wings is crossed with a fly homozygous for a gray body, red eyes, and normal wings. The female progeny are then crossed with males that have a black body, purple eyes, and vestigial wings. If 1000 progeny are produced from this testcross, what will be the phenotypes and proportions of the progeny?

31. A group of geneticists are interested in identifying genes that may play a role in susceptibility to asthma. They study the inheritance of genetic markers in a series of families that have two or more asthmatic children. They find an association between the presence or absence of asthma and a genetic marker on the short arm of chromosome 20 and calculate a lod score of 2 for this association. What does this lod score indicate about genes that may influence asthma?

Section 7.4

*32. The locations of six deletions have been mapped to the *Drosophila* chromosome, as shown in the following deletion map. Recessive mutations *a, b, c, d, e,* and *f* are known to be located in the same region as the deletions, but the order of the mutations on the chromosome is not known.

Chromosome

Deletion 1 ————————
    Deletion 2 ——
        Deletion 3 ————————————
        Deletion 4 ——————————
            Deletion 5 ————————
                Deletion 6 ————————————

When flies homozygous for the recessive mutations are crossed with flies homozygous for the deletions, the following results are obtained, in which "m" represents a mutant phenotype and a plus sign (+) represents the wild type. On the basis of these data, determine the relative order of the seven mutant genes on the chromosome:

Deletion	Mutations					
	a	*b*	*c*	*d*	*e*	*f*
1	m	+	m	+	+	m
2	m	+	+	+	+	+
3	+	m	m	m	m	+
4	+	+	m	m	m	+
5	+	+	+	m	m	+
6	+	m	+	m	+	+

33. A panel of cell lines was created from human–mouse somatic-cell fusions. Each line was examined for the presence of human chromosomes and for the production of an enzyme. The following results were obtained:

Cell line	Enzyme	Human chromosomes											
		1	2	3	4	5	6	7	8	9	10	17	22
A	−	+	−	−	−	+	−	−	−	−	−	+	−
B	+	+	+	−	−	−	−	−	+	−	−	+	+
C	−	+	−	−	−	+	−	−	−	−	−	−	+
D	−	−	−	−	+	−	−	−	−	−	−	−	−
E	+	+	−	−	−	−	−	−	+	−	+	+	−

On the basis of these results, which chromosome has the gene that encodes the enzyme?

*34. A panel of cell lines was created from human–mouse somatic-cell fusions. Each line was examined for the presence of human chromosomes and for the production of three enzymes. The following results were obtained.

Cell line	Enzyme			Human chromosomes									
	1	2	3	4	8	9	12	15	16	17	22	X
A	+	−	+	−	−	+	−	+	+	−	−	+
B	+	−	−	−	−	+	−	−	+	+	−	−
C	−	+	+	+	−	−	−	−	−	+	−	+
D	−	+	+	+	+	−	−	−	+	−	−	+

On the basis of these results, give the chromosome location of enzyme 1, enzyme 2, and enzyme 3.

CHALLENGE QUESTION

Section 7.5

35. [DATA ANALYSIS] Transferrin is a blood protein that is encoded by the transferrin locus (*Trf*). In house mice the two alleles at this locus (*Trf*^{a} and *Trf*^{b}) are codominant and encode three types of transferrin:

Genotype	Phenotype
Trf^a/*Trf*^{a}	Trf-a
Trf^a/*Trf*^{b}	Trf-ab
Trf^b/*Trf*^{b}	Trf-b

The dilution locus, found on the same chromosome, determines whether the color of a mouse is diluted or full; an allele for dilution (*d*) is recessive to an allele for full color (*d*$^+$):

Genotype	Phenotype
d$^+$*d*$^+$	*d*$^+$ (full color)
d$^+$*d*	*d*$^+$ (full color)
dd	*d* (dilution)

Donald Shreffler conducted a series of crosses to determine the map distance between the tranferrin locus and the dilution locus (D. C. Shreffler. 1963 *Journal of Heredity* 54:127–129). The table at right presents a series of crosses carried out by Shreffler and the progeny resulting from these crosses.

a. Calculate the recombinant frequency between the *Trf* and the *d* loci by using the pooled data from all the crosses.

b. Which crosses represent recombination in male gamete formation and which crosses represent recombination in female gamete formation?

c. On the basis of your answer to part *b*, calculate the frequency of recombination among male parents and female parents separately.

d. Are the rates of recombination in males and females the same? If not, what might produce the difference?

Cross	♂	♀	d^+ Trf-ab	d^+ Trf-b	d Trf-ab	d Trf-b	Total
1	$\dfrac{d^+\ Trf^a}{d\ Trf^b}$	$\dfrac{d\ Trf^b}{d\ Trf^b}$	32	3	6	21	62
2	$\dfrac{d\ Trf^b}{d\ Trf^b}$	$\dfrac{d^+\ Trf^a}{d\ Trf^b}$	16	0	2	20	38
3	$\dfrac{d^+\ Trf^a}{d\ Trf^b}$	$\dfrac{d\ Trf^b}{d\ Trf^b}$	35	9	4	30	78
4	$\dfrac{d\ Trf^b}{d\ Trf^b}$	$\dfrac{d^+\ Trf^a}{d\ Trf^b}$	21	3	2	19	45
5	$\dfrac{d^+\ Trf^b}{d\ Trf^a}$	$\dfrac{d\ Trf^b}{d\ Trf^b}$	8	29	22	5	64
6	$\dfrac{d\ Trf^b}{d\ Trf^b}$	$\dfrac{d^+\ Trf^b}{d\ Trf^a}$	4	14	11	0	29

(table header: Progeny phenotypes)

8 | Bacterial and Viral Genetic Systems

Past human migrations can be charted by examining the present-day genetic diversity of the bacterium *Helicobacter pylori,* which resides in the human stomach and causes peptic ulcers. This bacterium has been transported throughout the world in the guts of its human hosts. *[© 2004 Gwendolyn Knight Lawrence/Artists Rights Society (ARS), New York/The Phillips Collection, Washington, D.C.]*

GUTSY TRAVELERS

Peptic ulcers are tissue-damaging sores of the stomach and upper intestinal tract that affect 25 million Americans and, in serious cases, lead to life-threatening blood loss. For many years, peptic ulcers were attributed to stress and spicy foods, and ulcers were treated by encouraging changes in diet and life style, as well as by drugs that limited acid production by the stomach. Although these treatments often brought short-term relief, peptic ulcers in many patients returned and proved to be a recurring problem.

In 1982, physicians Barry Marshall and Robin Warren made a startling proposal. They suggested that most peptic ulcers are actually caused by a bacterium, *Helicobacter pylori* (**Figure 8.1**), which is able to tolerate the acidic environment of the stomach. Treatment for ulcers has now changed from an adjustment in life style to the administration of antibiotics, which has proved to be effective in eliminating the presence of *H. pylori* and permanently curing the disease. For their discovery, Marshall and Warren were awarded the Nobel Prize in medicine or physiology in 2005.

Interestingly, about half of the world's population is infected with *H. pylori,* but only a few people suffer from peptic ulcers. Thus, infection alone cannot be responsible for peptic ulcers, and other factors, including stress, diet, genetic differences among *H. pylori* strains, and even the genetic constitution of the host, are thought to play important roles in whether peptic ulcers arise.

Most people become infected with *H. pylori* in infancy or early childhood and remain infected for life. The source of the infection is usually other family members. A strain of *H. pylori* from one family can be differentiated from a strain from another family and so, like a surname, provides an accurate record of familial connections. Geneticists are now using this property of *H. pylori* to trace historical migrations of human populations.

In 2003, a team of geneticists led by Mark Achtman at the Max Planck Institute for Infection Biology in Berlin examined DNA sequences in eight genes of *H. pylori* bacteria collected from humans throughout the world. They observed that the bacterial sequences clustered into four major groups—two from Africa, one from east Asia, and one from Europe—which corresponded well to the human populations from which the bacteria were isolated. Further analysis revealed affinities among human populations within and between the groups. For example, bacteria from the Maoris (a group of native people from New Zealand) were closely related to those from Polynesia, corroborating other evidence that Maoris originated from South Pacific islanders who migrated to New Zealand several thousand years ago. Similarly, bacteria from Native American clustered with the East Asian bacterial strains, concurring with the Asian origin of Native Americans. The genes of the bacteria also fit together with more recent human migrations: African strains were found in high frequency among African Americans in Louisiana and Tennessee, and European strains of the bacteria were found among people in Singapore, South Africa, and North America. The results of these studies reveal that *H. pylori* travels the world in the guts of its human hosts and can be used to help resolve the details of past human migrations.

Just as scientists are recognizing the value of *H. pylori* in studies of human evolution and history, the bacteria seem to be disappearing from human guts, particularly those of people in developed countries. In developing countries, from 70% to 100% of children are infected with *H. pylori*, but fewer than 10% of children in the United States have the bacteria. The cause for this bacterial decline is unknown—widespread use of antibiotics and better hygiene are suspected—but it has led to a dramatic decrease in the incidence of peptic ulcers and stomach cancer (which is also associated with the presence of the bacteria) in developed countries.

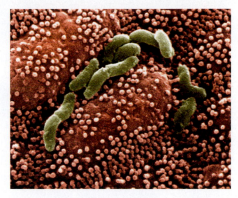

8.1 *Helicobacter pylori* is the bacterium that causes peptic ulcers. *[Veronika Burmeister/Visuals Unlimited.]*

In this chapter, we will examine some of the mechanisms by which bacteria like *H. pylori* exchange and recombine their genes. Since the 1940s, the genetic systems of bacteria and viruses have contributed to the discovery of many important concepts in genetics. The study of molecular genetics initially focused almost entirely on their genes; today, bacteria and viruses are still essential tools for probing the nature of genes in more-complex organisms, in part because they possess a number of characteristics that make them suitable for genetic studies (**Table 8.1**).

The genetic systems of bacteria and viruses are also studied because these organisms play important roles in human society. As illustrated by *H. pylori*, many bacteria are an important part of human ecology. They have been harnessed to produce a number of economically important substances, and they are of immense medical significance, causing many human diseases. In this chapter, we focus on several unique aspects of bacterial and viral genetic systems.

Important processes of gene transfer and recombination, like those that contribute to the genetic structure of *H. pylori*, will be described, and we will see how these processes can be used to map bacterial and viral genes.

8.1 Genetic Analysis of Bacteria Requires Special Approaches and Methods

Heredity in bacteria is fundamentally similar to heredity in more-complex organisms, but the bacterial haploid genome and the small size of bacteria (which makes observation of their phenotypes difficult) require different approaches and methods. First, we will consider how bacteria are studied and, then, we will examine several processes that transfer genes from one bacterium to another.

Techniques for the Study of Bacteria

Microbiologists have defined the nutritional needs of a number of bacteria and developed culture media for growing them in the laboratory. Culture media typically contain a carbon source, essential elements such as nitrogen and phosphorus, certain vitamins, and other required ions and nutrients. Wild-type (*prototrophic*) bacteria can use these simple ingredients to synthesize all the compounds that they need for growth and reproduction. A medium that contains only the nutrients required by prototrophic bacteria is termed **minimal medium**. Mutant strains called *auxotrophs* lack one or more enzymes necessary for metabolizing nutrients or synthesizing essential molecules and will grow only on medium supplemented with one or more nutrients. For example, auxotrophic strains that are unable to synthesize the amino acid leucine will not grow on minimal medium but *will* grow on medium to which leucine has been added. **Complete medium** contains all the substances, such as the amino acid leucine, required by bacteria for growth and reproduction.

Table 8.1	Advantages of using bacteria and viruses for genetic studies

1. Reproduction is rapid.
2. Many progeny are produced.
3. Haploid genome allows all mutations to be expressed directly.
4. Asexual reproduction simplifies the isolation of genetically pure strains.
5. Growth in the laboratory is easy and requires little space.
6. Genomes are small.
7. Techniques are available for isolating and manipulating their genes.
8. They have medical importance.
9. They can be genetically engineered to produce substances of commercial value.

(a)   **(b)**

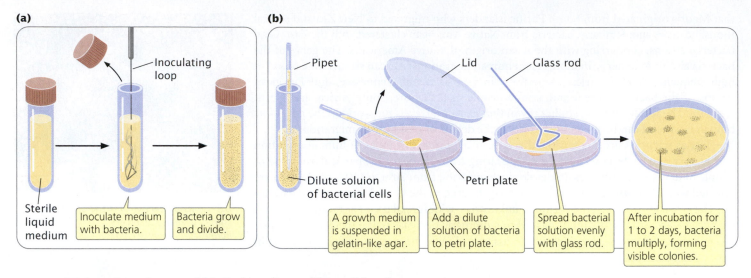

8.2 Bacteria can be grown (a) in liquid medium or (b) on solid medium.

Cultures of bacteria are often grown in test tubes that contain sterile liquid medium (**Figure 8.2a**). A few bacteria are added to a tube, and they grow and divide until all the nutrients are used up or—more commonly—until the concentration of their waste products becomes toxic. Bacteria are also grown in petri plates (**Figure 8.2b**). Growth medium suspended in agar is poured into the bottom half of the petri plate, providing a solid, gel-like base for bacterial growth. The chief advantage of this method is that it allows one to isolate and count bacteria, which individually are too small to see without a microscope. In a process called plating, a dilute solution of bacteria is spread over the surface of an agar-filled petri plate. As each bacterium grows and divides, it gives rise to a visible clump of genetically identical cells (a **colony**). Genetically pure strains of the bacteria can be isolated by collecting bacteria from a single colony and transferring them to a new test tube or petri plate.

Because individual bacteria are too small to be seen directly, it is often easier to study phenotypes that affect the appearance of the colony (**Figure 8.3**) or can be detected by simple chemical tests. Auxotrophs are commonly studied phenotypes. Suppose we want to detect auxotrophs that cannot synthesize leucine (*leu⁻* mutants). We first spread the bacteria on a petri plate containing medium that includes leucine; both prototrophs that have the *leu⁺* allele and auxotrophs that have

leu⁻ alleles will grow on it (**Figure 8.4**). Next, using a technique called replica plating, we transfer a few cells from each of the colonies on the original plate to two new replica plates: one plate contains medium to which leucine has been added; the other plate contains selective medium—that is, a medium in this case lacking leucine. The *leu⁺* bacteria will grow on both media, but the *leu⁻* mutants will grow only on the medium supplemented by leucine, because they cannot synthesize their own leucine. Any colony that grows on medium that contains leucine but not on medium that lacks leucine consists of *leu⁻* bacteria. The auxotrophs that grow on the supplemented medium can then be cultured for further study.

The Bacterial Genome

Bacteria are unicellular organisms that lack a nuclear membrane. Most bacterial genomes consist of a circular chromosome that contains a single DNA molecule several million base pairs (bp) in length (**Figure 8.5**). For example, the genome of *E. coli* has approximately 4.6 million base pairs of DNA. However, some bacteria (such as *Vibrio cholerae*, which causes cholera) contain multiple chromosomes, and a few even have linear chromosomes. Most bacterial chromosomes are organized efficiently, with little DNA between genes.

(a)   **(b)**

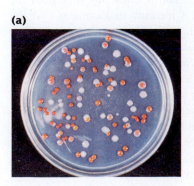   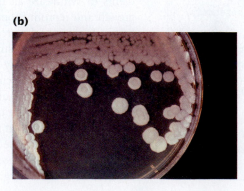

8.3 Bacteria have a variety of phenotypes.
(a) *Serratia marcescens* with color variation.
(b) *Bacillus cereus*. [Part a: Dr. E. Bottone/Peter Arnold. Part b: Biophoto Associates/Photo Researchers.]

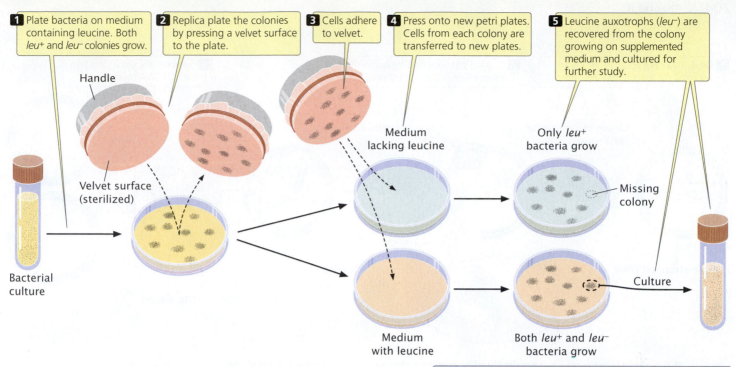

1 Plate bacteria on medium containing leucine. Both *leu⁺* and *leu⁻* colonies grow.

2 Replica plate the colonies by pressing a velvet surface to the plate.

3 Cells adhere to velvet.

4 Press onto new petri plates. Cells from each colony are transferred to new plates.

5 Leucine auxotrophs (*leu⁻*) are recovered from the colony growing on supplemented medium and cultured for further study.

Handle

Velvet surface (sterilized)

Bacterial culture

Medium lacking leucine

Only *leu⁺* bacteria grow

Missing colony

Medium with leucine

Both *leu⁺* and *leu⁻* bacteria grow

Culture

8.4 Mutant bacterial strains can be isolated on the basis of their nutritional requirements.

Conclusion: A colony that grows only on the supplemented medium has a mutation in a gene that encodes the synthesis of an essential nutrient.

Plasmids

In addition to having a chromosome, many bacteria possess **plasmids**—small, circular DNA molecules (**Figure 8.6**). Some plasmids are present in many copies per cell, whereas others are present in only one or two copies. In general, plasmids carry genes that are not essential to bacterial function but that may play an important role in the life cycle and growth of their bacterial hosts. Some plasmids promote mating between bacteria; others contain genes that kill other bacteria. Of great importance, plasmids are used extensively in genetic engineering, and some of them play a role in the spread of antibiotic resistance among bacteria.

Most plasmids are circular and several thousand base pairs in length, although plasmids consisting of several hundred thousand base pairs also have been found. Possessing its own origin of replication, a plasmid replicates independently

8.5 Most bacterial cells possess a single, circular chromosome, shown here emerging from a ruptured bacterial cell. [*David L. Nelson and Michael M. Cox*, Lehninger Principles of Biochemistry, *4th ed. (New York: Worth Publishers, 2004), from Huntington Potter and David Dressler, Harvard Medical School, Department of Neurobiology.*]

8.6 Many bacteria contain plasmids—small, circular molecules of DNA. [*Professor Stanley N. Cohen/Photo Researchers.*]

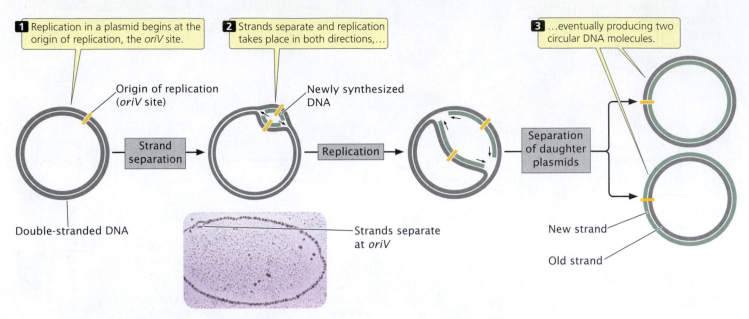

1 Replication in a plasmid begins at the origin of replication, the *oriV* site.

2 Strands separate and replication takes place in both directions,…

3 …eventually producing two circular DNA molecules.

Origin of replication (*oriV* site)

Newly synthesized DNA

Strand separation

Replication

Separation of daughter plasmids

Double-stranded DNA

Strands separate at *oriV*

New strand

Old strand

8.7 A plasmid replicates independently of its bacterial chromosome. Replication begins at the origin of replication (*oriV*) and continues around the circle. In this diagram, replication is taking place in both directions; in some plasmids, replication is in one direction only.

of the bacterial chromosome. Replication proceeds from the origin in one or two directions until the entire plasmid is copied. In **Figure 8.7,** the origin of replication is *oriV*. A few plasmids have multiple replication origins.

Episomes are plasmids that are capable of freely replicating and able to integrate into the bacterial chromosomes.

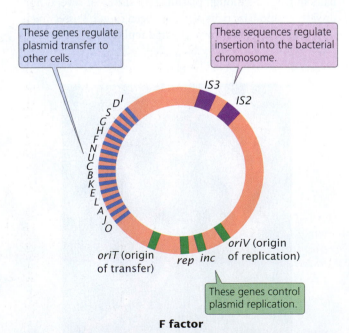

These genes regulate plasmid transfer to other cells.

These sequences regulate insertion into the bacterial chromosome.

IS3

IS2

D
S
G
H
F
N
U
C
B
K
E
L
A
J
O

oriT (origin of transfer)

rep *inc*

oriV (origin of replication)

These genes control plasmid replication.

F factor

8.8 The F factor, a circular episome of *E. coli*, contains a number of genes that regulate transfer into the bacterial cell, replication, and insertion into the bacterial chromosome. Replication is initiated at *oriV*. Insertion sequences (see Chapter 11) *IS3* and *IS2* control insertion into the bacterial chromosome and excision from it.

The **F** (fertility) **factor** of *E. coli* (**Figure 8.8**) is an episome that controls mating and gene exchange between *E. coli* cells, as will be discussed in the next section.

CONCEPTS

Bacteria can be studied in the laboratory by growing them on defined liquid or solid medium. A typical bacterial genome consists of a single circular chromosome that contains several million base pairs. Some bacterial genes may be present on plasmids, which are small, circular DNA molecules that replicate independently of the bacterial chromosome.

✔ CONCEPT CHECK 1

Which is true of plasmids?

a. They are composed of RNA.

b. They normally exist outside of bacterial cells.

c. They possess only a single strand of DNA.

d. They replicate independently of the bacterial chromosome.

Gene Transfer in Bacteria

Bacteria exchange genetic material by three different mechanisms, all entailing some type of DNA transfer and recombination between the transferred DNA and the bacterial chromosome.

1. **Conjugation** takes place when genetic material passes directly from one bacterium to another (**Figure 8.9a**). In conjugation, two bacteria lie close together and a

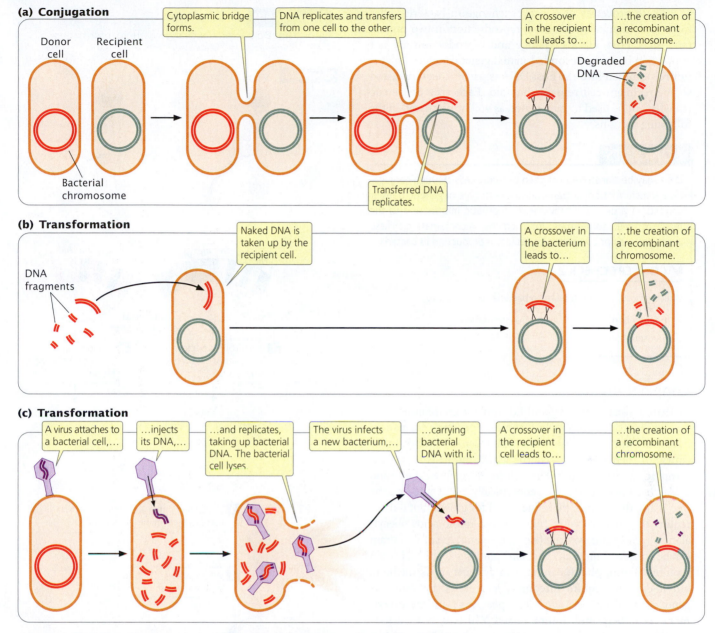

(a) Conjugation

Donor cell   Recipient cell

Bacterial chromosome

Cytoplasmic bridge forms.

DNA replicates and transfers from one cell to the other.

Transferred DNA replicates.

A crossover in the recipient cell leads to…

…the creation of a recombinant chromosome.

Degraded DNA

(b) Transformation

DNA fragments

Naked DNA is taken up by the recipient cell.

A crossover in the bacterium leads to…

…the creation of a recombinant chromosome.

(c) Transformation

A virus attaches to a bacterial cell,…

…injects its DNA,…

…and replicates, taking up bacterial DNA. The bacterial cell lyses.

The virus infects a new bacterium,…

…carrying bacterial DNA with it.

A crossover in the recipient cell leads to…

…the creation of a recombinant chromosome.

8.9 Conjugation, transformation, and transduction are three processes of gene transfer in bacteria. For the transferred DNA to be stably inherited, all three processes require the transferred DNA to undergo recombination with the bacterial chromosome.

connection forms between them. A plasmid or a part of the bacterial chromosome passes from one cell (the donor) to the other (the recipient). Subsequent to conjugation, crossing over may take place between homologous sequences in the transferred DNA and the chromosome of the recipient cell. In conjugation, DNA is transferred only from donor to recipient, with no reciprocal exchange of genetic material.

2. **Transformation** takes place when a bacterium takes up DNA from the medium in which it is growing (**Figure 8.9b**). After transformation, recombination may take place between the introduced genes and those of the bacterial chromosome.

3. **Transduction** takes place when bacterial viruses (bacteriophages) carry DNA from one bacterium to another (**Figure 8.9c**). Inside the bacterium, the newly introduced DNA may undergo recombination with the bacterial chromosome.

Not all bacterial species exhibit all three types of genetic transfer. Conjugation takes place more frequently in some species than in others. Transformation takes place to a limited extent in many species of bacteria, but laboratory techniques increase the rate of DNA uptake. Most bacteriophages have a limited host range; so transduction is normally between bacteria of the same or closely related species only.

These processes of genetic exchange in bacteria differ from diploid eukaryotic sexual reproduction in two important ways. First, DNA exchange and reproduction are not coupled in bacteria. Second, donated genetic material that is not recombined into the host DNA is usually degraded, and so the recipient cell remains haploid. Each type of genetic transfer can be used to map genes, as will be discussed in the following sections.

CONCEPTS

DNA may be transferred between bacterial cells through conjugation, transformation, or transduction. Each type of genetic transfer consists of a one-way movement of genetic information to the recipient cell, sometimes followed by recombination. These processes are not connected to cellular reproduction in bacteria.

✔ CONCEPT CHECK 2

Which process of DNA transfer in bacteria requires a virus?

a. Conjugation   c. Transformation

b. Transduction   d. All of the above

Conjugation

In 1946, Joshua Lederberg and Edward Tatum demonstrated that bacteria can transfer and recombine genetic information, paving the way for the use of bacteria in genetic studies. In the course of their research, Lederberg and Tatum studied auxotrophic strains of *E. coli*. The Y10 strain required the amino acids threonine (and was genotypically *thr⁻*) and leucine (*leu⁻*) and the vitamin thiamine (*thi⁻*) for growth but did not require the vitamin biotin (*bio⁺*) or the amino acids phenylalanine (*phe⁺*) and cysteine (*cys⁺*); the genotype of this strain can be written as *thr⁻ leu⁻ thi⁻ bio⁺ phe⁺ cys⁺*. The Y24 strain required biotin, phenylalanine, and cysteine in its medium, but it did not require threonine, leucine, or thiamine; its genotype was *thr⁺ leu⁺ thi⁺ bio⁻ phe⁻ cys⁻*. In one experiment, Lederberg and Tatum mixed Y10 and Y24 bacteria together and plated them on minimal medium (**Figure 8.10**). Each strain was also plated separately on minimal medium.

Alone, neither Y10 nor Y24 grew on minimal medium. Strain Y10 was unable to grow, because it required threonine, leucine, and thiamine, which were absent in the minimal medium; strain Y24 was unable to grow, because it required biotin, phenylalanine, and cysteine, which also were absent from the minimal medium. When Lederberg and Tatum mixed the two strains, however, a few colonies did grow on the minimal medium. These prototrophic bacteria must have had genotype *thr⁺ leu⁺ thi⁺ bio⁺ phe⁺ cys⁺*. Where had they come from?

If mutations were responsible for the prototrophic colonies, then some colonies should also have grown on the plates containing Y10 or Y24 alone, but no bacteria grew on these plates. Multiple simultaneous mutations (*thr⁻ → thr⁺*, *leu⁻ → leu⁺*, and *thi⁻ → thi⁺* in strain Y10 or *bio⁻ → bio⁺*,

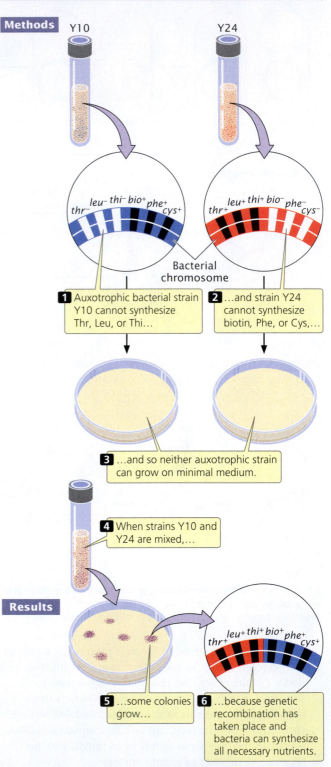

Experiment

Question: Do bacteria exchange genetic information?

Methods   Y10   Y24

leu⁻ thi⁻ bio⁺ phe⁺
thr⁻   *cys⁺*

leu⁺ thi⁺ bio⁻ phe⁻
thr⁺   *cys⁻*

Bacterial chromosome

1 Auxotrophic bacterial strain Y10 cannot synthesize Thr, Leu, or Thi…

2 …and strain Y24 cannot synthesize biotin, Phe, or Cys,…

3 …and so neither auxotrophic strain can grow on minimal medium.

4 When strains Y10 and Y24 are mixed,…

Results

leu⁺ thi⁺ bio⁺ phe⁺
thr⁺   *cys⁺*

5 …some colonies grow…

6 …because genetic recombination has taken place and bacteria can synthesize all necessary nutrients.

Conclusion: Yes, genetic exchange and recombination took place between the two mutant strains.

8.10 Lederberg and Tatum's experiment demonstrated that bacteria undergo genetic exchange.

$phe^- \rightarrow phe^+$, and $cys^- \rightarrow cys^+$ in strain Y24) would have been required for either strain to become prototrophic by mutation, which was very improbable. Lederberg and Tatum concluded that some type of genetic transfer and recombination had taken place:

Auxotrophic strain

Y10 $thr^-\ leu^-\ thi^-\ bio^+\ phe^+\ cys^+$

Y24 $thr^+\ leu^+\ thi^+\ bio^-\ phe^-\ cys^-$

$thr^-\ leu^-\ thi^- \quad bio^+\ phe^+\ cys^+$
$thr^+\ leu^+\ thi^+ \quad bio^-\ phe^-\ cys^-$

$thr^-\ leu^-\ thi^-\ bio^-\ phe^-\ cys^-$

Prototrophic strain $thr^+\ leu^+\ thi^+\ bio^+\ phe^+\ cys^+$

What they did not know was *how* it had taken place.

To study this problem, Bernard Davis constructed a U-shaped tube (**Figure 8.11**) that was divided into two compartments by a filter having fine pores. This filter allowed liquid medium to pass from one side of the tube to the other, but the pores of the filter were too small to allow the passage of bacteria. Two auxotrophic strains of bacteria were placed on opposite sides of the filter, and suction was applied alternately to the ends of the U-tube, causing the medium to flow back and forth between the two compartments. Despite hours of incubation in the U-tube, bacteria plated out on minimal medium did not grow; there had been no genetic exchange between the strains. The exchange of bacterial genes clearly required direct contact, or conjugation, between the bacterial cells.

F^+ and F^- cells

In most bacteria, conjugation depends on a fertility (F) factor that is present in the donor cell and absent in the recipient cell. Cells that contain F are referred to as F^+, and cells lacking F are F^-.

The F factor contains an origin of replication and a number of genes required for conjugation (see Figure 8.8). For example, some of these genes encode sex **pili** (singular, pilus), slender extensions of the cell membrane. A cell containing F produces the sex pili, one of which makes contact with a receptor on an F^- cell (**Figure 8.12**) and pulls the two cells together. DNA is then transferred from the F^+ cell to the F^- cell. Conjugation can take place only between a cell that possesses F and a cell that lacks F.

In most cases, the only genes transferred during conjugation between an F^+ and F^- cell are those on the F factor (**Figure 8.13a and b**). Transfer is initiated when one of the DNA strands on the F factor is nicked at an origin (*oriT*). One end of the nicked DNA separates from the circle and passes into the recipient cell (**Figure 8.13c**). Replication

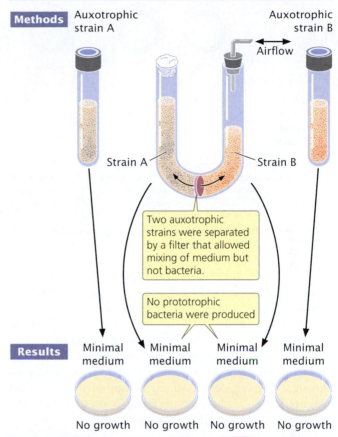

Experiment

Question: How did the genetic exchange seen in Lederberg and Tatum's experiment take place?

Methods   Auxotrophic strain A   Auxotrophic strain B

Airflow

Strain A   Strain B

Two auxotrophic strains were separated by a filter that allowed mixing of medium but not bacteria.

No prototrophic bacteria were produced

Results   Minimal medium   Minimal medium   Minimal medium   Minimal medium

No growth   No growth   No growth   No growth

Conclusion: Genetic exchange requires direct contact between bacterial cells.

8.11 Davis's U-tube experiment.

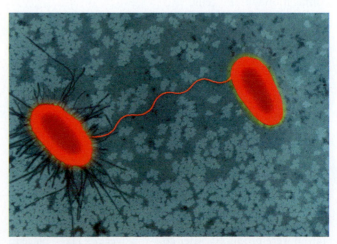

8.12 A sex pilus connects F^+ and F^- cells during bacterial conjugation. *E. coli* cells in conjugation. [Dr. Dennis Kunkel/Phototake.]

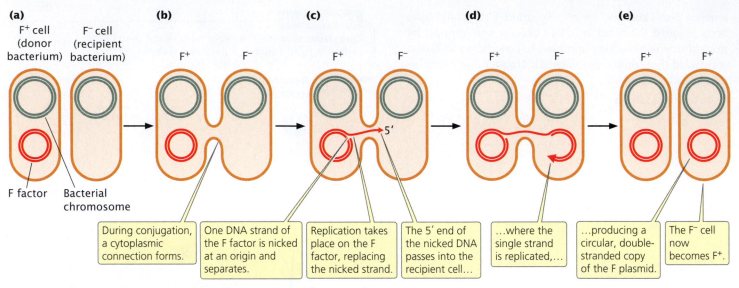

(a)
F⁺ cell   F⁻ cell
(donor   (recipient
bacterium)   bacterium)

(b)   F⁺   F⁻

(c)   F⁺   F⁻

(d)   F⁺   F⁻

(e)   F⁺   F⁺

F factor   Bacterial
            chromosome

| During conjugation, a cytoplasmic connection forms. | One DNA strand of the F factor is nicked at an origin and separates. | Replication takes place on the F factor, replacing the nicked strand. | The 5′ end of the nicked DNA passes into the recipient cell… | …where the single strand is replicated,… | …producing a circular, double-stranded copy of the F plasmid. | The F⁻ cell now becomes F⁺. |

8.13 The F factor is transferred during conjugation between an F⁺ and F⁻ cell.

takes place on the nicked strand, proceeding around the circular plasmid in the F⁺ cell and replacing the transferred strand (**Figure 8.13d**). Because the plasmid in the F⁺ cell is always nicked at the *oriT* site, this site always enters the recipient cell first, followed by the rest of the plasmid. Thus, the transfer of genetic material has a defined direction. Inside the recipient cell, the single strand is replicated, producing a circular, double-stranded copy of the F plasmid (**Figure 8.13e**). If the entire F factor is transferred to the recipient F⁻ cell, that cell becomes an F⁺ cell.

Hfr cells Conjugation transfers genetic material in the F plasmid from F⁺ to F⁻ cells but does not account for the transfer of chromosomal genes observed by Lederberg and Tatum. In Hfr (high-frequency) strains, the F factor is integrated into the bacterial chromosome (**Figure 8.14**). Hfr cells behave as F⁺ cells, forming sex pili and undergoing conjugation with F⁻ cells.

In conjugation between Hfr and F⁻ cells (**Figure 8.15a**), the integrated F factor is nicked, and the end of the nicked strand moves into the F⁻ cell (**Figure 8.15b**), just as it does in conjugation between F⁺ and F⁻ cells. Because, in an Hfr cell, the F factor has been integrated into the bacterial chromosome, the chromosome follows it into the recipient cell. How much of the bacterial chromosome is transferred depends on the length of time that the two cells remain in conjugation.

Inside the recipient cell, the donor DNA strand is replicated (**Figure 8.15c**), and crossing over between it and the original chromosome of the F⁻ cell (**Figure 8.15d**) may take place. This gene transfer between Hfr and F⁻ cells is how the recombinant prototrophic cells observed by Lederberg and Tatum were produced. After crossing over has taken place in the recipient cell, the donated chromosome is degraded, and the recombinant recipient chromosome remains (**Figure**

8.15e), to be replicated and passed on to later generations by binary fission.

In a mating of Hfr × F⁻, the F⁻ cell almost never becomes F⁺ or Hfr, because the F factor is nicked in the middle in the initiation of strand transfer, placing part of F at the beginning and part at the end of the strand to be transferred. To become F⁺ or Hfr, the recipient cell must receive the entire F factor, requiring the entire bacterial chromosome to be transferred. This event happens rarely, because most conjugating cells break apart before the entire chromosome has been transferred.

The F plasmid in F⁺ cells integrates into the bacterial chromosome, causing an F⁺ cell to become Hfr, at a frequency of only about 1/10,000. This low frequency accounts for the low rate of recombination observed by Lederberg and Tatum in their F⁺ cells. The F factor is excised from the bacterial chromosome at a similarly low rate, causing a few Hfr cells to become F⁺.

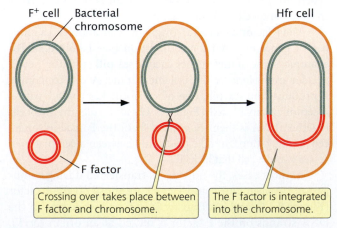

F⁺ cell   Bacterial
           chromosome

                                Hfr cell

F factor

| Crossing over takes place between F factor and chromosome. | The F factor is integrated into the chromosome. |

8.14 The F factor is integrated into the bacterial chromosome in an Hfr cell.

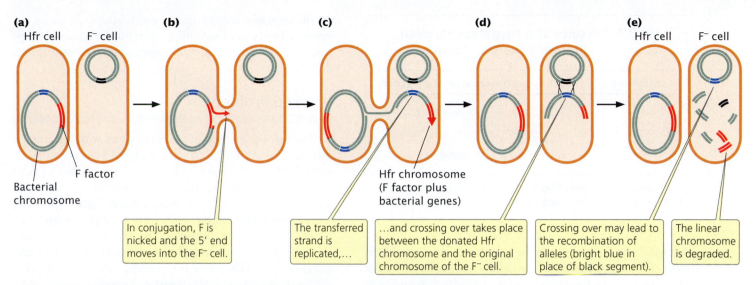

In conjugation, F is nicked and the 5′ end moves into the F⁻ cell.

The transferred strand is replicated,…

…and crossing over takes place between the donated Hfr chromosome and the original chromosome of the F⁻ cell.

Crossing over may lead to the recombination of alleles (bright blue in place of black segment).

The linear chromosome is degraded.

Hfr chromosome (F factor plus bacterial genes)

Bacterial chromosome

F factor

8.15 Bacterial genes may be transferred from an Hfr cell to an F⁻ cell in conjugation. In an Hfr cell, the F factor has been integrated into the bacterial chromosome.

F′ cells When an F factor does excise from the bacterial chromosome, a small amount of the bacterial chromosome may be removed with it, and these chromosomal genes will then be carried with the F plasmid (**Figure 8.16**). Cells containing an F plasmid with some bacterial genes are called F prime (F′). For example, if an F factor integrates into a chromosome adjacent to the *lac* genes (genes that enable a cell to metabolize the sugar lactose), the F factor may pick up *lac* genes when it excises, becoming F′*lac*. F′ cells can conjugate with F⁻ cells, given that F′ cells possess the F plasmid with all the genetic information necessary for conjugation and gene transfer. Characteristics of different mating types of *E. coli* (cells with different types of F) are summarized in **Table 8.2.**

During conjugation between an F′*lac* cell and an F⁻ cell, the F plasmid is transferred to the F⁻ cell, which means that any genes on the F plasmid, including those from the

Table 8.2   Characteristics of *E. coli* cells with different types of F factor

Type	F Factor Characteristics	Role in Conjugation
F⁺	Present as separate circular DNA	Donor
F⁻	Absent	Recipient
Hfr	Present, integrated into bacterial chromosome	High-frequency donor
F′	Present as separate circular DNA, carrying some bacterial genes	Donor

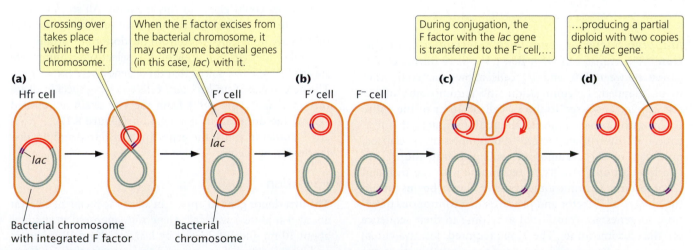

Crossing over takes place within the Hfr chromosome.

When the F factor excises from the bacterial chromosome, it may carry some bacterial genes (in this case, *lac*) with it.

During conjugation, the F factor with the *lac* gene is transferred to the F⁻ cell,…

…producing a partial diploid with two copies of the *lac* gene.

Hfr cell

Bacterial chromosome with integrated F factor

Bacterial chromosome

8.16 An Hfr cell may be converted into an F′ cell when the F factor excises from the bacterial chromosome and carries bacterial genes with it. Conjugation produces a partial diploid.

Table 8.3	Results of conjugation between cells with different F factors

Conjugating	Cell Types Present after Conjugation
F⁺ × F⁻	Two F⁺ cells (F⁻ cell becomes F⁺)
Hfr × F⁻	One Hfr cell and one F⁻ (no change)*
F′ × F⁺	Two F′ cells (F⁻ cell becomes F′)

*Rarely, the F⁻ cell becomes F⁺ in an Hfr × F⁻ conjugation if the entire chromosome is transferred during conjugation.

bacterial chromosome, may be transferred to F⁻ recipient cells. This process is called sexduction. It produces partial diploids, or *merozygotes,* which are cells with two copies of some genes, one on the bacterial chromosome and one on the newly introduced F plasmid. The outcomes of conjugation between different mating types of *E. coli* are summarized in **Table 8.3.**

CONCEPTS

Conjugation in *E. coli* is controlled by an episome called the F factor. Cells containing F (F⁺ cells) are donors during gene transfer; cells without F (F⁻ cells) are recipients. Hfr cells possess F integrated into the bacterial chromosome; they donate DNA to F⁻ cells at a high frequency. F′ cells contain a copy of F with some bacterial genes.

✔ CONCEPT CHECK 3

Conjugation between an F⁺ and an F⁻ cell usually results in

a. two F⁺ cells.

b. two F⁻ cells.

c. an F⁺ and an F⁻ cell.

d. an Hfr cell and an F⁺ cell.

Mapping bacterial genes with interrupted conjugation   The transfer of DNA that takes place during conjugation between Hfr and F⁻ cells allows bacterial genes to be mapped. In conjugation, the chromosome of the Hfr cell is transferred to the F⁻ cell. Transfer of the entire *E. coli* chromosome requires about 100 minutes; if conjugation is interrupted before 100 minutes have elapsed, only part of the chromosome will pass into the F⁻ cell and have an opportunity to recombine with the recipient chromosome. Chromosome transfer always begins within the integrated F factor and proceeds in a continuous direction; so genes are transferred according to their sequence on the chromosome. The time required for individual genes to be transferred indicates their relative positions on the chromosome. In most genetic maps, distances are expressed as percent recombination; but, in bacterial maps constructed with interrupted conjugation, the basic unit of distance is a minute.

Worked Problem

To illustrate the method of mapping genes with interrupted conjugation, let's look at a cross analyzed by François Jacob and Elie Wollman, who first developed this method of gene mapping (**Figure 8.17a**). They used donor Hfr cells that were sensitive to the antibiotic streptomycin (genotype *str*ˢ), resistant to sodium azide (*azi*ʳ) and infection by bacteriophage T1 (*ton*ʳ), prototrophic for threonine (*thr*⁺) and leucine (*leu*⁺), and able to break down lactose (*lac*⁺) and galactose (*gal*⁺). They used F⁻ recipient cells that were resistant to streptomycin (*str*ʳ), sensitive to sodium azide (*azi*ˢ) and to infection by bacteriophage T1 (*ton*ˢ), auxotrophic for threonine (*thr*⁻) and leucine (*leu*⁻), and unable to break down lactose (*lac*⁻) and galactose (*gal*⁻). Thus, the genotypes of the donor and recipient cells were:

Donor Hfr cells: *str*ˢ *leu*⁺ *thr*⁺ *azi*ʳ *ton*ʳ *lac*⁺ *gal*⁺

Recipient F⁻ cells: *str*ʳ *leu*⁻ *thr*⁻ *azi*ˢ *ton*ˢ *lac*⁻ *gal*⁻

The two strains were mixed in nutrient medium and allowed to conjugate. After a few minutes, the medium was diluted to prevent any new pairings. At regular intervals, a sample of cells was removed and agitated vigorously in a kitchen blender to halt all conjugation and DNA transfer. The cells were plated on a selective medium that contained streptomycin and lacked leucine and threonine. The original donor cells were streptomycin sensitive (*str*ˢ) and would not grow on this medium. The F⁻ recipient cells were auxotrophic for leucine and threonine, and they also failed to grow on this medium. Only cells that underwent conjugation and received at least the *leu*⁺ and *thr*⁺ genes from the Hfr donors could grow on this medium. All *str*ʳ *leu*⁺ *thr*⁺ cells were then tested for the presence of other genes that might have been transferred from the donor Hfr strain.

All of the cells that grew on this selective medium were *str*ʳ *leu*⁺ *thr*⁺; so we know that these genes were transferred. The percentage of *str*ʳ *leu*⁺ *thr*⁺ cells receiving specific alleles (*azi*ʳ, *ton*ʳ, *leu*⁺, and *gal*⁺) from the Hfr strain are plotted against the duration of conjugation (**Figure 8.17b**). What are the order in which the genes are transferred and the distances among them?

• Solution

The first donor gene to appear in all of the recipient cells (at about 9 minutes) was *azi*ʳ. Gene *ton*ʳ appeared next (after about 10 minutes), followed by *lac*⁺ (at about 18 minutes) and by *gal*⁺ (after 25 minutes). These transfer times indicate

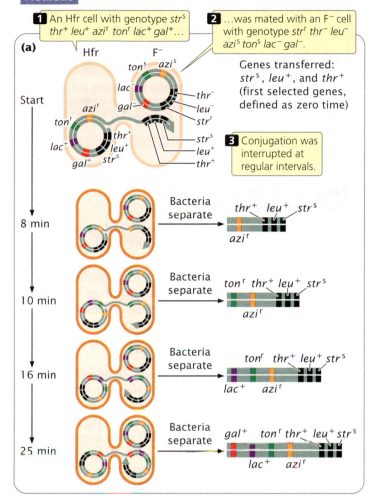

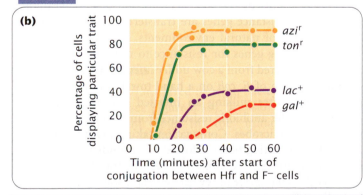

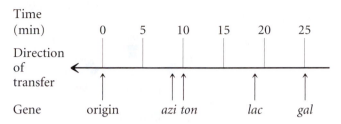

Experiment

Question: How can interrupted conjugation be used to map bacterial genes?

Methods

1 An Hfr cell with genotype *str*^s *thr*^+ *leu*^+ *azi*^r *ton*^r *lac*^+ *gal*^+...

2 ...was mated with an F^− cell with genotype *str*^r *thr*^− *leu*^− *azi*^s *ton*^s *lac*^− *gal*^−.

(a)

Hfr   F^−

Start

Genes transferred: *str*^s, *leu*^+, and *thr*^+ (first selected genes, defined as zero time)

3 Conjugation was interrupted at regular intervals.

8 min

Bacteria separate   *thr*^+ *leu*^+ *str*^s
                            azi^r

10 min

Bacteria separate   *ton*^r *thr*^+ *leu*^+ *str*^s
                            azi^r

16 min

Bacteria separate   *ton*^r *thr*^+ *leu*^+ *str*^s
                            lac^+ *azi*^r

25 min

Bacteria separate   *gal*^+ *ton*^r *thr*^+ *leu*^+ *str*^s
                            lac^+ *azi*^r

Results

(b)

azi^r
ton^r
lac^+
gal^+

Percentage of cells displaying particular trait

Time (minutes) after start of conjugation between Hfr and F^− cells

Conclusion: The transfer times indicate the order and relative distances between genes and can be used to construct a genetic map.

8.17 Jacob and Wollman used interrupted conjugation to map bacterial genes.

the order of gene transfer and the relative distances among the genes (see Figure 8.17b).

Time (min)   0   5   10   15   20   25

Direction of transfer ←

Gene   origin   *azi ton*   *lac*   *gal*

Notice that the frequency of gene transfer from donor to recipient cells decreased for the more distant genes. For example, about 90% of the recipients received the *azi*^r allele, but only about 30% received the *gal*^+ allele. The lower percentage for *gal*^+ is due to the fact that some conjugating cells spontaneously broke apart before they were disrupted by the blender. The probability of spontaneous disruption increases with time; so fewer cells had an opportunity to receive genes that were transferred later.

> **?** For additional practice mapping bacterial genes with interrupted conjugation, try Problem 19 at the end of the chapter.

Directional transfer and mapping Different Hfr strains have the F factor integrated into the bacterial chromosome at different sites and in different orientations. Gene transfer always begins within F, and the orientation and position of F determine the direction and starting point of gene transfer. **Figure 8.18a** shows that, in strain Hfr1, F is integrated between *leu* and *azi;* the orientation of F at this site dictates that gene transfer will proceed in a counterclockwise direction around the circular chromosome. Genes from this strain will be transferred in the order of:

← *leu–thr–thi–his–gal–lac–pro–azi*

In strain Hfr5, F is integrated between the *thi* and the *his* genes (**Figure 8.18b**) and in the opposite orientation. Here gene transfer will proceed in a clockwise direction:

← *thi–thr–leu–azi–pro–lac–gal–his*

Although the starting point and direction of transfer may differ between two strains, the relative distance in time between any two pairs of genes is constant.

organisms. Methods have been developed to introduce specific mutations within *E. coli* genes, and so genetic analysis no longer depends on the isolation of randomly occurring mutations. New DNA sequences produced by recombinant DNA can be introduced by transformation into special strains of *E. coli* that are particularly efficient (competent) at taking up DNA.

Because of its powerful advantages as a model genetic organism, *E. coli* has played a leading role in many fundamental discoveries in genetics, including elucidation of the genetic code, probing the nature of replication, and working out the basic mechanisms of gene regulation. ■

8.2 Viruses Are Simple Replicating Systems Amenable to Genetic Analysis

All organisms—plants, animals, fungi, and bacteria—are infected by viruses. A **virus** is a simple replicating structure made up of nucleic acid surrounded by a protein coat (see Figure 2.4). Viruses come in a great variety of shapes and sizes (**Figure 8.23**). Some have DNA as their genetic material, whereas others have RNA; the nucleic acid may be double stranded or single stranded, linear or circular. Not surprisingly, viruses reproduce in a number of different ways.

Bacteriophages (phages) have played a central role in genetic research since the late 1940s. They are ideal for many types of genetic research because they have small and easily manageable genomes, reproduce rapidly, and produce large numbers of progeny. Bacteriophages have two alternative life cycles: the lytic and the lysogenic cycles. In the lytic cycle, a phage attaches to a receptor on the bacterial cell wall and injects its DNA into the cell (**Figure 8.24**). Inside the host cell, the phage DNA is replicated, transcribed, and translated, producing more phage DNA and phage proteins. New phage particles are assembled from these components. The phages then produce an enzyme that breaks open the host cell, releasing the new phages. **Virulent phages** reproduce strictly through the lytic cycle and always kill their host cells.

Temperate phages can undergo either the lytic or the lysogenic cycle. The lysogenic cycle begins like the lytic cycle (see Figure 8.24) but, inside the cell, the phage DNA integrates into the bacterial chromosome, where it remains as an inactive **prophage.** The prophage is replicated along with the bacterial DNA and is passed on when the bacterium divides. Certain stimuli can cause the prophage to dissociate from the bacterial chromosome and enter into the lytic cycle, producing new phage particles and lysing the cell.

Techniques for the Study of Bacteriophages

Viruses reproduce only within host cells; so bacteriophages must be cultured in bacterial cells. To do so, phages and bacteria are mixed together and plated on solid medium on a petri plate. A high concentration of bacteria is used so that the colonies grow into one another and produce a continuous layer of bacteria, or "lawn," on the agar. An individual phage infects a single bacterial cell and goes through its lytic cycle. Many new phages are released from the lysed cell and infect additional cells; the cycle is then repeated. The bacteria grow on solid medium; so the diffusion of the phages is restricted, and only nearby cells are infected. After several rounds of phage reproduction, a clear patch of lysed cells, or **plaque,** appears on the plate (**Figure 8.25**). Each plaque represents a single phage that multiplied and lysed many cells. Plating a known volume of a dilute solution of phages on a bacterial lawn and counting the number of plaques that appear can be used to determine the original concentration of phage in the solution.

> ### CONCEPTS
>
> Viral genomes may be DNA or RNA, circular or linear, and double or single stranded. Bacteriophages are used in many types of genetic research.

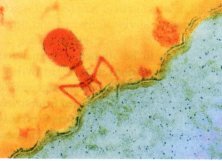
(a)

(b)

8.23 Viruses come in different structures and sizes. (a) T4 bacteriophage (bright orange). (b) Influenza A virus (green structures). *[Left: Biozentrum, University of Basel/Photo Researchers. Right: Eye of Science/Photo Researchers.]*

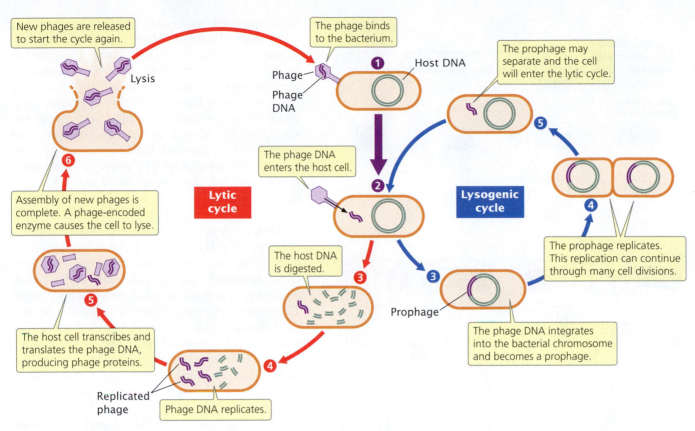

New phages are released to start the cycle again.

Lysis

The phage binds to the bacterium.

1 Host DNA

Phage

Phage DNA

The prophage may separate and the cell will enter the lytic cycle.

5

6

Assembly of new phages is complete. A phage-encoded enzyme causes the cell to lyse.

The phage DNA enters the host cell.

Lytic cycle

2

Lysogenic cycle

The prophage replicates. This replication can continue through many cell divisions.

4

The host DNA is digested.

3

3

Prophage

The phage DNA integrates into the bacterial chromosome and becomes a prophage.

5

The host cell transcribes and translates the phage DNA, producing phage proteins.

Replicated phage

4

Phage DNA replicates.

8.24 Bacteriophages have two alternative life cycles: lytic and lysogenic.

✔ **CONCEPT CHECK 6**

In which bacteriophage life cycle does the phage DNA become incorporated into the bacterial chromosome?

a. Lytic

c. Both lytic and lysogenic

b. Lysogenic

d. Neither lytic or lysogenic

Transduction: Using Phages to Map Bacterial Genes

In the discussion of bacterial genetics, three mechanisms of gene transfer were identified: conjugation, transformation, and transduction (see Figure 8.9). Let's take a closer look at transduction, in which genes are transferred between bacteria by viruses. In **generalized transduction,** any gene may be transferred. In **specialized transduction,** only a few genes are transferred.

Generalized transduction Joshua Lederberg and Norton Zinder discovered generalized transduction in 1952. They were trying to produce recombination in the bac-

terium *Salmonella typhimurium* by conjugation. They mixed a strain of *S. typhimurium* that was phe^+ trp^+ tyr^+ met^- his^- with a strain that was phe^- trp^- tyr^- met^+ his^+ (**Figure 8.26**) and plated them on minimal medium. A few prototrophic recombinants (phe^+ trp^+ tyr^+ met^+

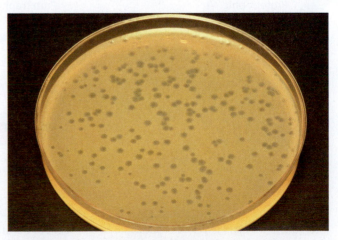

8.25 Plaques are clear patches of lysed cells on a lawn of bacteria. [Carolina Biological/Visuals Unlimited.]

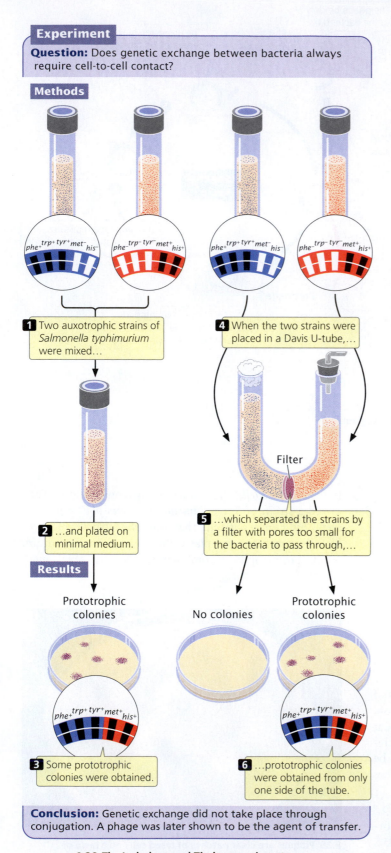

Experiment

Question: Does genetic exchange between bacteria always require cell-to-cell contact?

Methods

phe$^+$*trp*$^+$*tyr*$^+$*met*$^-$*his*$^-$ *phe*$^-$*trp*$^-$*tyr*$^-$*met*$^+$*his*$^+$ *phe*$^+$*trp*$^+$*tyr*$^+$*met*$^-$*his*$^-$ *phe*$^-$*trp*$^-$*tyr*$^-$*met*$^+$*his*$^+$

1 Two auxotrophic strains of *Salmonella typhimurium* were mixed…

4 When the two strains were placed in a Davis U-tube,…

Filter

2 …and plated on minimal medium.

5 …which separated the strains by a filter with pores too small for the bacteria to pass through,…

Results

Prototrophic colonies   No colonies   Prototrophic colonies

phe$^+$*trp*$^+$*tyr*$^+$*met*$^+$*his*$^+$ *phe*$^+$*trp*$^+$*tyr*$^+$*met*$^+$*his*$^+$

3 Some prototrophic colonies were obtained.

6 …prototrophic colonies were obtained from only one side of the tube.

Conclusion: Genetic exchange did not take place through conjugation. A phage was later shown to be the agent of transfer.

8.26 The Lederberg and Zinder experiment.

his$^+$) appeared, suggesting that conjugation had taken place. However, when they tested the two strains in a U-shaped tube similar to the one used by Davis, some *phe*$^+$ *trp*$^+$ *tyr*$^+$ *met*$^+$ *his*$^+$ prototrophs were obtained on one side of the tube (compare Figure 8.26 with Figure 8.11). This apparatus separated the two strains by a filter with pores too small for the passage of bacteria; so how were genes being transferred between bacteria in the absence of conjugation? The results of subsequent studies revealed that the agent of transfer was a bacteriophage.

In the lytic cycle of phage reproduction, the bacterial chromosome is broken into random fragments (**Figure 8.27**). For some types of bacteriophage, a piece of the bacterial chromosome instead of phage DNA occasionally gets packaged into a phage coat; these phage particles are called **transducing phages.** The transducing phage infects a new cell, releasing the bacterial DNA, and the introduced genes may then become integrated into the bacterial chromosome by a double crossover. Bacterial genes can, by this process, be moved from one bacterial strain to another, producing recombinant bacteria called **transductants.**

Not all phages are capable of transduction, a rare event that requires (1) that the phage degrade the bacterial chromosome; (2) that the process of packaging DNA into the phage protein not be specific for phage DNA; and (3) that the bacterial genes transferred by the virus recombine with the chromosome in the recipient cell.

Because of the limited size of a phage particle, only about 1% of the bacterial chromosome can be transduced. Only genes located close together on the bacterial chromosome will be transferred together, or **cotransduced.** The overall rate of transduction ranges from only about 1 in 100,000 to 1 in 1,000,000. Because the chance of a cell being transduced by two separate phages is exceedingly small, any cotransduced genes are usually located close together on the bacterial chromosome. Thus, rates of cotransduction, like rates of cotransformation, give an indication of the physical distances between genes on a bacterial chromosome.

To map genes by using transduction, two bacterial strains with different alleles at several loci are used. The donor strain is infected with phages (**Figure 8.28**), which reproduce within the cell. When the phages have lysed the donor cells, a suspension of the progeny phages is mixed with a recipient strain of bacteria, which is then plated on several different kinds of media to determine the phenotypes of the transducing progeny phages.

CONCEPTS

In transduction, bacterial genes become packaged into a viral coat, are transferred to another bacterium by the virus, and become incorporated into the bacterial chromosome by crossing over. Bacterial genes can be mapped with the use of generalized transduction.

✔ CONCEPT CHECK 7

In gene mapping experiments using generalized transduction, bacterial genes that are cotransduced are

a. far apart on the bacterial chromosome.

b. on different bacterial chromosomes.

c. close together on the bacterial chromosome.

d. on a plasmid.

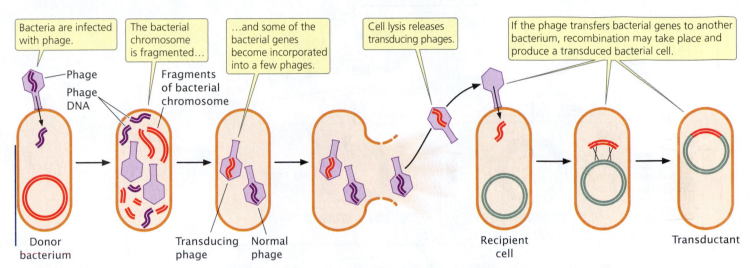

Bacteria are infected with phage.

The bacterial chromosome is fragmented…

…and some of the bacterial genes become incorporated into a few phages.

Cell lysis releases transducing phages.

If the phage transfers bacterial genes to another bacterium, recombination may take place and produce a transduced bacterial cell.

Phage
Phage DNA
Fragments of bacterial chromosome

Donor bacterium   Transducing phage   Normal phage   Recipient cell   Transductant

8.27 Genes can be transferred from one bacterium to another through generalized transduction.

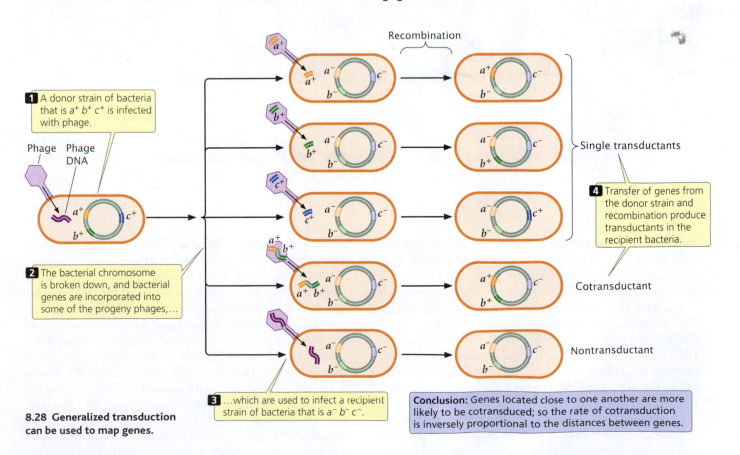

Recombination

1 A donor strain of bacteria that is $a^+ b^+ c^+$ is infected with phage.

Phage   Phage DNA

2 The bacterial chromosome is broken down, and bacterial genes are incorporated into some of the progeny phages,…

3 …which are used to infect a recipient strain of bacteria that is $a^- b^- c^-$.

4 Transfer of genes from the donor strain and recombination produce transductants in the recipient bacteria.

Single transductants

Cotransductant

Nontransductant

Conclusion: Genes located close to one another are more likely to be cotransduced; so the rate of cotransduction is inversely proportional to the distances between genes.

8.28 Generalized transduction can be used to map genes.

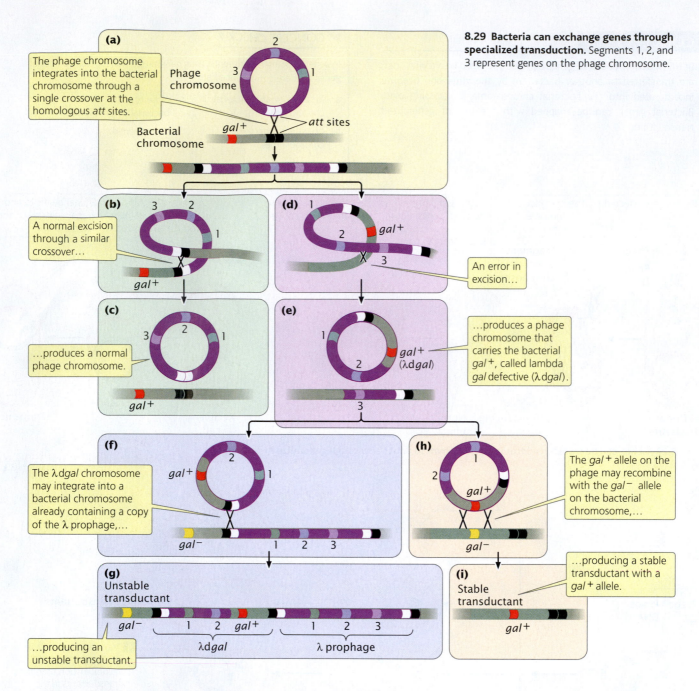

8.29 Bacteria can exchange genes through specialized transduction. Segments 1, 2, and 3 represent genes on the phage chromosome.

(a) The phage chromosome integrates into the bacterial chromosome through a single crossover at the homologous *att* sites.

Phage chromosome

Bacterial chromosome

gal+   *att* sites

(b) A normal excision through a similar crossover...

gal+

(c) ...produces a normal phage chromosome.

gal+

(d)

gal+

An error in excision...

(e) ...produces a phage chromosome that carries the bacterial *gal*+, called lambda *gal* defective (λd*gal*).

gal+ (λd*gal*)

(f) The λd*gal* chromosome may integrate into a bacterial chromosome already containing a copy of the λ prophage,...

gal+

gal−

(g) Unstable transductant

gal−   1   2   *gal*+   1   2   3

λd*gal*   λ prophage

...producing an unstable transductant.

(h) The *gal*+ allele on the phage may recombine with the *gal*− allele on the bacterial chromosome,...

gal+

gal−

(i) Stable transductant

...producing a stable transductant with a *gal*+ allele.

gal+

Specialized transduction Like generalized transduction, specialized transduction requires gene transfer from one bacterium to another through phages, but, here, only genes near particular sites on the bacterial chromosome are transferred. This process requires lysogenic bacteriophages. The prophage may imperfectly excise from the bacterial chromosome, carrying with it a small part of the bacterial DNA adjacent to the site of prophage integration. A phage carrying this DNA will then inject it into another bacterial cell in the next round of infection. This process resembles the situation in F′ cells, in which the F plasmid carries genes from one bacterium into another (see Figure 8.16).

One of the best-studied examples of specialized transduction is in bacteriophage lambda (λ), which integrates into the

E. coli chromosome at the **attachment** (*att*) **site.** The phage DNA contains a site similar to the *att* site; a single crossover integrates the phage DNA into the bacterial chromosome (**Figure 8.29a**). The λ prophage is excised through a similar crossover that reverses the process (**Figure 8.29b and c**).

An error in excision may cause genes on either side of the bacterial *att* site to be excised along with some of the phage DNA (**Figure 8.29d and e**). In *E. coli*, these genes are usually the *gal* (galactose fermentation) and *bio* (biotin biosynthesis) genes. When a transducing phage carrying the *gal* gene infects another bacterium, the gene may integrate into the bacterial chromosome along with the prophage (**Figure 8.29f**), giving the bacterial chromosome two copies of the *gal* gene (**Figure 8.29g**). These transductants are unstable, often reverting back to the wild type, because the prophage DNA may excise from the chromosome, carrying the introduced gene with it. Stable transductants are produced when the *gal* gene in the phage is exchanged for the *gal* gene in the bacterial chromosome through a double crossover (**Figure 8.29h and i**).

CONCEPTS

Specialized transduction transfers only those bacterial genes located near the site of prophage insertion.

CONNECTING CONCEPTS

Three Methods for Mapping Bacterial Genes

Three methods of mapping bacterial genes have now been outlined: (1) interrupted conjugation; (2) transformation; and (3) transduction. These methods have important similarities and differences.

Mapping with interrupted conjugation is based on the time required for genes to be transferred from one bacterium to another by means of cell-to-cell contact. The key to this technique is that the bacterial chromosome itself is transferred, and the order of genes and the time required for their transfer provide information about the positions of the genes on the chromosome. In contrast with other mapping methods, the distance between genes is measured not in recombination frequencies but in units of time required for genes to be transferred. Here, the basic unit of conjugation mapping is a minute.

In gene mapping with transformation, DNA from the donor strain is isolated, broken up, and mixed with the recipient strain. Some fragments pass into the recipient cells, where the transformed DNA may recombine with the bacterial chromosome. The unit of transfer here is a random fragment of the chromosome. Loci that are close together on the donor chromosome tend to be on the same DNA fragment; so the rates of cotransformation provide information about the relative positions of genes on the chromosome.

Transduction mapping also relies on the transfer of genes between bacteria that differ in two or more traits, but, here, the vehicle of gene transfer is a bacteriophage. In a number of respects, transduction mapping is similar to transformation mapping. Small fragments of DNA are carried by the phage from donor to recipient bacteria, and the rates of cotransduction, like the rates of cotransformation, provide information about the relative distances between the genes.

All of the methods use a common strategy for mapping bacterial genes. The movement of genes from donor to recipient is detected by using strains that differ in two or more traits, and the transfer of one gene relative to the transfer of others is examined. Additionally, all three methods rely on recombination between the transferred DNA and the bacterial chromosome. In mapping with interrupted conjugation, the relative order and timing of gene transfer provide the information necessary to map the genes; in transformation and transduction, the rate of cotransfer provides this information.

In conclusion, the same basic strategies are used for mapping with interrupted conjugation, transformation, and transduction. The methods differ principally in their mechanisms of transfer: in conjugation mapping, DNA is transferred though contact between bacteria; in transformation, DNA is transferred as small naked fragments; and, in transduction, DNA is transferred by bacteriophages.

Gene Mapping in Phages

Mapping genes in the bacteriophages themselves requires the application of the same principles as those applied to mapping genes in eukaryotic organisms (see Chapter 7). Crosses are made between viruses that differ in two or more genes, and recombinant progeny phages are identified and counted. The proportion of recombinant progeny is then used to estimate the distances between the genes and their linear order on the chromosome.

In 1949, Alfred Hershey and Raquel Rotman examined rates of recombination in the T2 bacteriophage, which has single-stranded DNA. They studied recombination between genes in two strains that differed in plaque appearance and host range (the bacterial strains that the phages could infect). One strain was able to infect and lyse type B *E. coli* cells but not B/2 cells (wild type with normal host range, h^+) and produced an abnormal plaque that was large with distinct borders (r^-). The other strain was able to infect and lyse *both* B *and* B/2 cells (mutant host range, h^-) and produced wild-type plaques that were small with fuzzy borders (r^+).

Hershey and Rotman crossed the $h^+\ r^-$ and $h^-\ r^+$ strains of T2 by infecting type B *E. coli* cells with a mixture of the two strains. They used a high concentration of phages so that most cells could be simultaneously infected by both strains (**Figure 8.30**). Homologous recombination occasionally took place between the chromosomes of the different strains, producing $h^+\ r^+$ and $h^-\ r^-$ chromosomes, which were then packaged into new phage particles. When the cells lysed, the recombinant phages were released, along with the nonrecombinant $h^+\ r^-$ phages and $h^-\ r^+$ phages.

Hershey and Rotman diluted and plated the progeny phages on a bacterial lawn that consisted of a *mixture* of B and B/2 cells. Phages carrying the h^+ allele (which conferred the ability to infect only B cells) produced a cloudy plaque because the B/2 cells did not lyse. Phages carrying the h^- allele produced a clear plaque because all the cells within the plaque were lysed. The r^+ phages produced small plaques, whereas the r^- phages produced large plaques. The genotypes of these progeny phages could therefore be determined by the appearance of the plaque (see Figure 8.30 and **Table 8.4**).

In this type of phage cross, the recombination frequency (RF) between the two genes can be calculated by using the following formula:

$$RF = \frac{\text{recombinant plaques}}{\text{total plaques}}$$

In Hershey and Rotman's cross, the recombinant plaques were $h^+\ r^+$ and $h^-\ r^-$; so the recombination frequency was

$$RF = \frac{(h^+\ r^+) + (h^-\ r^-)}{\text{total plaques}}$$

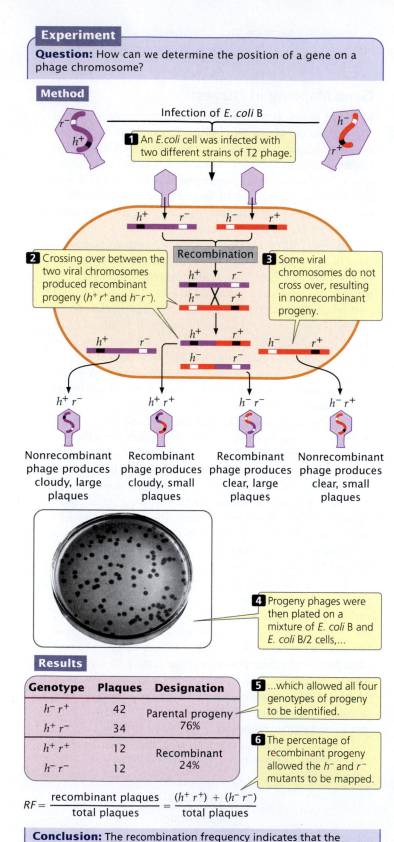

Method

Infection of *E. coli* B

1 An *E. coli* cell was infected with two different strains of T2 phage.

h^+ r^- h^- r^+

Recombination

2 Crossing over between the two viral chromosomes produced recombinant progeny ($h^+ r^+$ and $h^- r^-$).

h^+ r^-
h^- r^+

3 Some viral chromosomes do not cross over, resulting in nonrecombinant progeny.

h^+ r^- h^+ r^+ h^- r^+
                     h^- r^-

$h^+ r^-$ $h^+ r^+$ $h^- r^-$ $h^- r^+$

Nonrecombinant phage produces cloudy, large plaques

Recombinant phage produces cloudy, small plaques

Recombinant phage produces clear, large plaques

Nonrecombinant phage produces clear, small plaques

4 Progeny phages were then plated on a mixture of *E. coli* B and *E. coli* B/2 cells,...

Results

Genotype	Plaques	Designation
$h^- r^+$	42	Parental progeny 76%
$h^+ r^-$	34	
$h^+ r^+$	12	Recombinant 24%
$h^- r^-$	12	

5 ...which allowed all four genotypes of progeny to be identified.

6 The percentage of recombinant progeny allowed the h^- and r^- mutants to be mapped.

$$RF = \frac{\text{recombinant plaques}}{\text{total plaques}} = \frac{(h^+ \, r^+) + (h^- \, r^-)}{\text{total plaques}}$$

Conclusion: The recombination frequency indicates that the distance between *h* and *r* genes is 24%.

8.30 Hershey and Rotman developed a technique for mapping viral genes. *[Photograph from G. S. Stent,* Molecular Biology of Bacterial Viruses. *© 1963 by W. H. Freeman and Company.]*

Recombination frequencies can be used to determine the distances between genes and their order on the phage chromosome, just as recombination frequencies are used to map genes in eukaryotes.

CONCEPTS

To map phage genes, bacterial cells are infected with viruses that differ in two or more genes. Recombinant plaques are counted, and rates of recombination are used to determine the linear order of the genes on the chromosome and the distances between them.

Fine-Structure Analysis of Bacteriophage Genes

In the 1950s and 1960s, Seymour Benzer conducted a series of experiments to examine the structure of a gene. Because no molecular techniques were available at the time for directly examining nucleotide sequences, Benzer was forced to infer gene structure from analyses of mutations and their effects. The results of his studies showed that different mutational sites *within* a single gene could be mapped (referred to as **intragenic mapping**) by using techniques similar to those described for mapping bacterial genes by transduction. Different sites within a single gene are very close together; so recombination between them takes place at a very low frequency. Because large numbers of progeny are required to detect these recombination events, Benzer used the bacteriophage T4, which reproduces rapidly and produces large numbers of progeny.

Benzer's mapping techniques Wild-type T4 phages normally produce small plaques with rough edges when grown on a lawn of *E. coli* bacteria. Certain mutants, called *r* for rapid lysis, produce larger plaques with sharply defined edges. Benzer isolated phages with a number of different *r* mutations, concentrating on one particular subgroup called *rII* mutants.

Wild-type T4 phages produce typical plaques (**Figure 8.31**) on *E. coli* strains B and K. In contrast, the *rII* mutants produce *r* plaques on strain B and do not form plaques at all on

Table 8.4	Progeny phages produced from $h^- r^+ \times h^+ r^-$	

Phenotype	Genotype
Clear and small	$h^- r^+$
Cloudy and large	$h^+ r^-$
Cloudy and small	$h^+ r^+$
Clear and large	$h^- r^-$

8.31 T4 phage *rII* mutants produce distinct plaques when grown on *E. coli* B cells. (Upper image) Plaque produced by wild-type phage. (Lower image) Plaque produced by *rII* mutant. [*Dr. D. P. Snustad, College of Biological Sciences, University of Minnesota.*]

strain K. Benzer recognized the *r* mutants by their distinctive plaques when grown on *E. coli* B. He then collected lysate from these plaques and used it to infect *E. coli* K. Phages that did not produce plaques on *E. coli* K were defined as the *rII* type.

Benzer collected thousands of *rII* mutations. He simultaneously infected bacterial cells with two different mutants and looked for recombinant progeny (**Figure 8.32**). Consider two *rII* mutations, a^- and b^- (their wild-type alleles are a^+ and b^+). Benzer infected *E. coli* B cells with two different strains of phages, one a^- b^+ and the other a^+ b^- (see Figure 8.32, step 3). While reproducing within the B cells, a few phages of the two strains recombined (see Figure 8.32, step 4). A single crossover produces two recombinant chromosomes; one with genotype a^+ b^+ and the other with genotype a^- b^-:

Phage 1 a^- b^+

Phage 2 a^+ b^-

$\downarrow$

a^- b^+

a^+ b^-

$\downarrow$

a^- b^-

a^+ b^+

The resulting recombinant chromosomes, along with the nonrecombinant (parental) chromosomes, were incorporated into progeny phages, which were then used to infect *E. coli* K cells. The resulting plaques were examined to determine the genotype of the infecting phage and map the *rII* mutants (see Figure 8.32, step 5).

The *rII* mutants did not grow on *E. coli* K (see Figure 8.32, step 2), but wild-type phages grew; so progeny phages

with the recombinant genotype a^+ b^+ produced plaques on *E. coli* K. Each recombination event produces equal numbers of double mutants (a^- b^-) and wild-type chromosomes (a^+ b^+); so the number of recombinant progeny should be twice the number of wild-type plaques that appeared on *E. coli* K. The recombination frequency between the two *rII* mutants would be:

$$RF = \frac{2 \times \text{number of plaques on } E.\ coli\ \text{K}}{\text{total number of plaques on } E.\ coli\ \text{B}}$$

Benzer was able to detect a single recombinant among billions of progeny phages, allowing very low rates of recombination to be detected. Recombination frequencies are proportional to physical distances along the chromosome (see p. 173 in Chapter 7), revealing the positions of the different mutations within the *rII* region of the phage chromosome. In this way, Benzer eventually mapped more than 2400 *rII* mutations, many corresponding to single base pairs in the viral DNA. His work provided the first molecular view of a gene.

Complementation experiments   Benzer's mapping experiments demonstrated that some *rII* mutations were very closely linked. This finding raised the question of whether they were at the same locus or at different loci. To determine whether different *rII* mutations belonged to different functional loci, Benzer used the complementation (cis–trans) test (see p. 113 in Chapter 5).

To carry out the complementation test in bacteriophage, Benzer infected cells of *E. coli* K with large numbers of two mutant strains of phage (**Figure 8.33**, step 1) so that cells would become doubly infected with both strains. Consider two *rII* mutations: r_{101}^- and r_{104}^-. Cells infected with both mutants:

$$\frac{r_{101}^- \qquad r_{104}^+}{r_{101}^+ \qquad r_{104}^-}$$

were effectively heterozygous for the phage genes, with the mutations in the trans configuration (see Figure 8.33, step 2). In the complementation testing, the phenotypes of progeny phages were examined on the K strain, rather than the B strain as illustrated in Figure 8.32.

If the r_{101}^- and r_{104}^- mutations occur at different functional loci that encode different proteins, then, in bacterial cells infected by both mutants, the wild-type sequences on the chromosome opposite each mutation will overcome the effects of the recessive mutations; the phages will produce normal plaques on *E. coli* K cells (Figure 8.33, steps 3, 4, and 5). If, on the other hand, the mutations occur at the same locus, no functional protein is produced by either chromosome, and no plaques develop in the *E. coli* K cells (Figure 8.33, steps 6, 7, and 8). Thus, the absence of plaques indicates that the two mutations occur at the same locus. Benzer

Experiment

Question: How can *rII* phage mutants be mapped and what can they reveal about the structure of the gene?

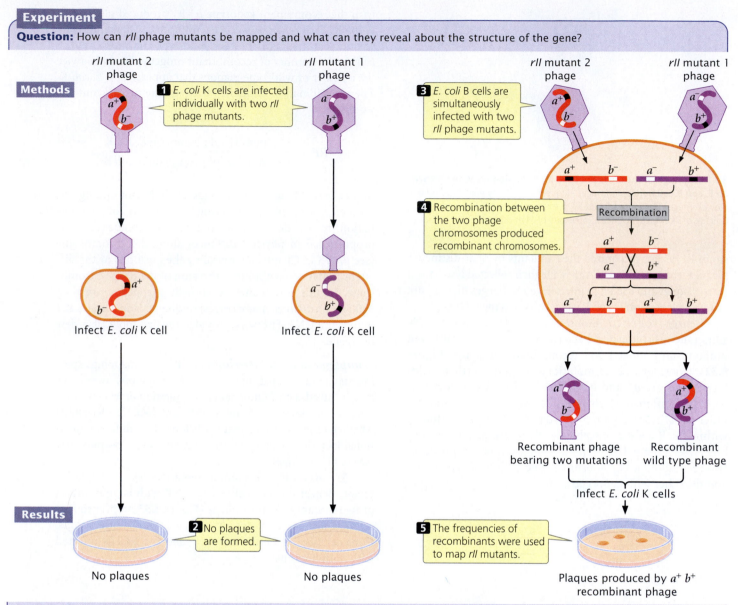

Conclusion: Mapping more than 2400 *rII* mutants provided information about the internal structure of a gene at the base-pair level—the first view of the molecular structure of a gene.

8.32 Benzer developed a procedure for mapping *rII* mutants. Two different *rII* mutants ($a^-\ b^+$ and $a^+\ b^-$) are isolated on *E. coli* B cells. Only the $a^+\ b^+$ recombinant can grow on *E. coli* K, allowing these recombinants to be identified. *rIIA* and *rIIB* refer to different parts of the gene.

coined the term *cistron* to designate a functional gene defined by the complementation test.

In the complementation test, the cis heterozygote is used as a control. Benzer simultaneously infected bacteria with wild-type phage ($r_{101}^+\ \ r_{104}^+$) and with phage carrying both mutations ($r_{101}^-\ \ r_{104}^-$). This test also produced cells that were heterozygous and in cis configuration for the phage genes:

$$\frac{r_{101}^+ \quad r_{104}^+}{r_{101}^- \quad r_{104}^-}$$

Regardless of whether the r_{101}^- and r_{104}^- mutations are in the same functional unit, these cells contain a copy of the wild-type phage chromosome ($r_{101}^+\ \ r_{104}^+$) and will produce normal plaques in *E. coli* K.

Benzer carried out complementation testing on many pairs of *rII* mutants. He found that the *rII* region consists of two loci, designated *rIIA* and *rIIB*. Mutations belonging to the *rIIA* and *rIIB* groups complemented each other, but mutations in the *rIIA* group did not complement others in *rIIA*; nor did mutations in the *rIIB* group complement others in *rIIB*.

8.33 Complementation tests are used to determine whether different mutations are at the same functional gene.

CONCEPTS

In a series of experiments with the bacteriophage T4, Seymour Benzer showed that recombination could take place within a single gene and created the first molecular map of a gene. Benzer used the complementation test to distinguish between functional genes (loci).

✔ CONCEPT CHECK 8

In complementation tests, Benzer simultaneously infected *E. coli* cells with two phages, each of which carried a different mutation. What conclusion did he make when the progeny phage produced normal plaques?

a. The mutations occurred at the same locus.

b. The mutations occurred at different loci.

c. The mutations occurred close together on the chromosome.

d. The genes were in the cis configuration.

At the time of Benzer's research, the relation between genes and DNA structure was unknown. A gene was defined as a functional unit of heredity that encoded a phenotype. Many geneticists believed that genes were indivisible and that recombination could not take place within them. Benzer demonstrated that intragenic recombination did indeed take place (although at a very low rate) and gave geneticists their first glimpse at the structure of an individual gene.

Overlapping Genes

The first viral genome to be completely sequenced, that of bacteriophage φX174, revealed surprising information: the nucleotide sequences of several genes overlapped. This genome encodes nine proteins (**Figure 8.34**). Two of the genes are nested within other genes; in both cases, the same DNA sequence encodes two different proteins by using different reading frames. In five of the φX174 genes, the initiation codon of one gene overlaps the termination codon of another.

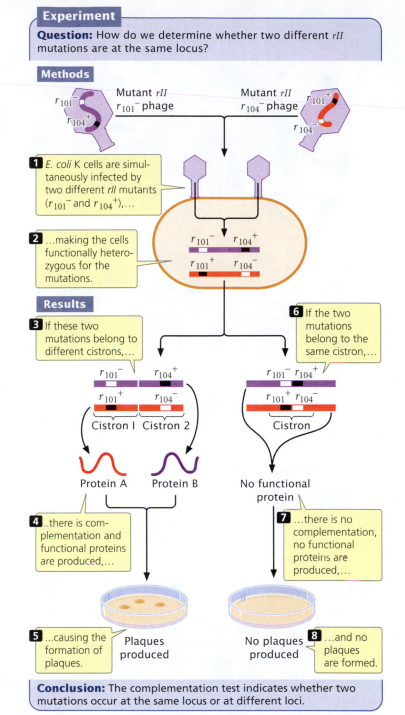

Experiment

Question: How do we determine whether two different *rII* mutations are at the same locus?

Methods

Mutant *rII* r_{101}^- phage   Mutant *rII* r_{104}^- phage

1 *E. coli* K cells are simultaneously infected by two different *rII* mutants (r_{101}^- and r_{104}^+),…

2 …making the cells functionally heterozygous for the mutations.

Results

3 If these two mutations belong to different cistrons,…

6 If the two mutations belong to the same cistron,…

Cistron I   Cistron 2   Cistron

Protein A   Protein B   No functional protein

4 …there is complementation and functional proteins are produced,…

7 …there is no complementation, no functional proteins are produced,…

5 …causing the formation of plaques.   Plaques produced   No plaques produced   8 …and no plaques are formed.

Conclusion: The complementation test indicates whether two mutations occur at the same locus or at different loci.

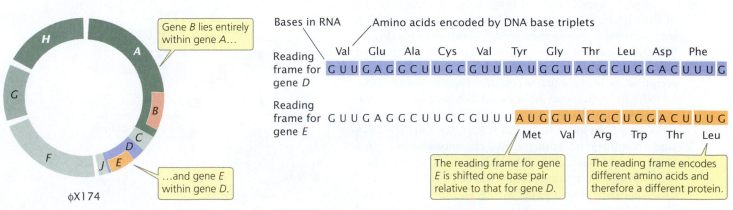

Gene *B* lies entirely within gene *A*…

…and gene *E* within gene *D*.

φX174

Bases in RNA   Amino acids encoded by DNA base triplets

Reading frame for gene *D*

Val	Glu	Ala	Cys	Val	Tyr	Gly	Thr	Leu	Asp	Phe

GUU GAG GCU UGC GUU UAU GGU ACG CUG GAC UUU G

Reading frame for gene *E*

GUU GAG GCU UGC GUU UAUGGU ACG CUG GAC UUU G

Met Val Arg Trp Thr Leu

The reading frame for gene *E* is shifted one base pair relative to that for gene *D*.

The reading frame encodes different amino acids and therefore a different protein.

8.34 The genome of bacteriophage φX174 contains overlapping genes. The genome contains nine genes (*A* through *J*).

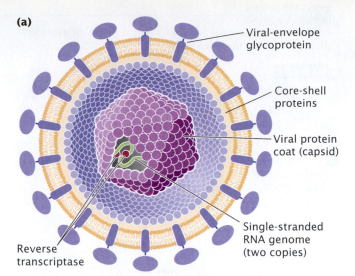

(a)

Viral-envelope glycoprotein

Core-shell proteins

Viral protein coat (capsid)

Single-stranded RNA genome (two copies)

Reverse transcriptase

8.35 A retrovirus uses reverse transcription to incorporate its RNA into the host DNA. (a) Structure of a typical retrovirus. Two copies of the single-stranded RNA genome and the reverse transcriptase enzyme are shown enclosed within a protein capsid. The capsid is surrounded by a viral envelope that is studded with viral glycoproteins. (b) The retrovirus life cycle.

The results of subsequent studies revealed that overlapping genes are found in a number of viruses and bacteria. Viral genome size is strictly limited by the capacity of the viral protein coat; so there is strong selective pressure for the economic use of the DNA.

CONCEPTS

Some viruses contain overlapping genes, in which the same base sequence specifies more than one protein.

RNA Viruses

Viral genomes may be encoded in either DNA or RNA, as stated earlier. RNA is the genetic material of some medically important human viruses, including those that cause influenza, common colds, polio, and AIDS. Almost all viruses that infect plants have RNA genomes. The medical and economic importance of RNA viruses has encouraged their study.

RNA viruses capable of integrating into the genomes of their hosts, much as temperate phages insert themselves into bacterial chromosomes, are called **retroviruses** (**Figure 8.35a**). Because the retroviral genome is RNA, whereas that of the host is DNA, a retrovirus must produce **reverse transcriptase,** an enzyme that synthesizes complementary DNA (cDNA) from either an RNA or a DNA template. A retrovirus uses reverse transcriptase to make a double-stranded DNA copy from its single-stranded RNA genome. The DNA copy then integrates into the host chromosome to form a **provirus,** which is replicated by host enzymes when the host chromosome is duplicated (**Figure 8.35b**).

When conditions are appropriate, the provirus undergoes transcription to produce numerous copies of the original

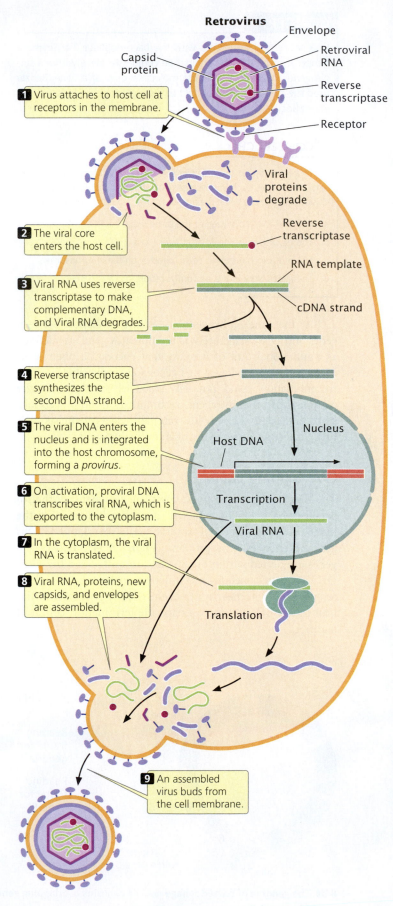

(b)

Retrovirus

Envelope

Capsid protein

Retroviral RNA

Reverse transcriptase

Receptor

1 Virus attaches to host cell at receptors in the membrane.

Viral proteins degrade

2 The viral core enters the host cell.

Reverse transcriptase

RNA template

3 Viral RNA uses reverse transcriptase to make complementary DNA, and Viral RNA degrades.

cDNA strand

4 Reverse transcriptase synthesizes the second DNA strand.

5 The viral DNA enters the nucleus and is integrated into the host chromosome, forming a *provirus*.

Nucleus

Host DNA

6 On activation, proviral DNA transcribes viral RNA, which is exported to the cytoplasm.

Transcription

Viral RNA

7 In the cytoplasm, the viral RNA is translated.

8 Viral RNA, proteins, new capsids, and envelopes are assembled.

Translation

9 An assembled virus buds from the cell membrane.

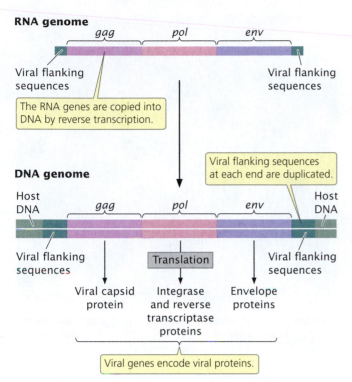

8.36 The typical genome of a retrovirus contains *gag, pol,* and *env* genes.

RNA genome. This RNA encodes viral proteins and serves as genomic RNA for new viral particles. As these viruses escape the cell, they collect patches of the cell membrane to use as their envelopes.

All known retroviral genomes have in common three genes: *gag, pol,* and *env* (**Figure 8.36**), each encoding a precursor protein that is cleaved into two or more functional proteins. The *gag* gene encodes the three or four proteins that make up the viral capsid. The *pol* gene encodes reverse transcriptase and an enzyme, called **integrase,** that inserts the viral DNA into the host chromosome. The *env* gene codes for the glycoproteins that appear on the viral envelope that surrounds the viral capsid.

Some retroviruses contain **oncogenes** that may stimulate cell division and cause the formation of tumors. The first retrovirus to be isolated, the Rous sarcoma virus, was originally recognized by its ability to produce connective-tissue tumors (sarcomas) in chickens.

Human Immunodeficiency Virus and AIDS

The human immunodeficiency virus (HIV) causes acquired immune deficiency syndrome (AIDS), a disease that killed 2.8 million people in 2005 alone. AIDS was first recognized in 1982, when a number of homosexual males in the United States began to exhibit symptoms of a new immune-system-deficiency disease. In that year, Robert Gallo proposed that AIDS was caused by a retrovirus. Between 1983 and 1984, as the AIDS epidemic became widespread, the HIV retro-

virus was isolated from AIDS patients. AIDS is now known to be caused by two different immunodeficiency viruses, HIV-1 and HIV-2, which together have infected more than 65 million people worldwide. Of those infected, 25 million have died. Most cases of AIDS are caused by HIV-1, which now has a global distribution; HIV-2 is primarily found in western Africa.

HIV illustrates the importance of genetic recombination in viral evolution. Studies of the DNA sequences of HIV and other retroviruses reveal that HIV-1 is closely related to the simian immunodeficiency virus found in chimpanzees (SIVcpz). Many wild chimpanzees in Africa are infected with SIVcpz, although it doesn't cause AIDS-like symptoms in chimps. SIVcpz is itself a hybrid that resulted from recombination between a retrovirus found in the red-capped mangabey (a monkey) and a retrovirus found in the greater spot-nosed monkey (**Figure 8.37**).

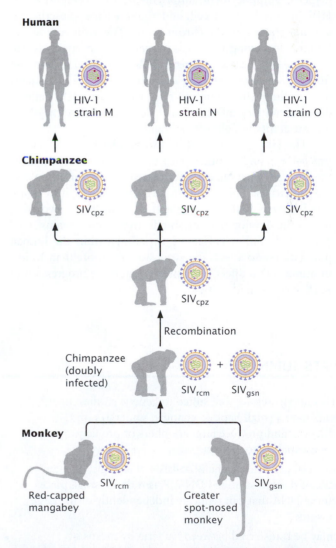

8.37 HIV-1 evolved from a similar virus (SIVcpz) found in chimpanzees and was transmitted to humans. SIVcpz arose from recombination taking place between retroviruses in red-capped mangabeys and greater spot-nosed monkeys.

Apparently, one or more chimpanzees became infected with both viruses; recombination between the viruses produced SIVcpz, which was then transmitted to humans through contact with infected chimpanzees. In humans, SIVcpz underwent significant evolution to become HIV-1, which then spread throughout the world to produce the AIDS epidemic. Several independent transfers of SIVcpz to humans gave rise to different strains of HIV-1. HIV-2 evolved from a different retrovirus, SIVsm, found in sooty mangabeys.

HIV is transmitted by sexual contact between humans and through any type of blood-to-blood contact, such as that caused by the sharing of dirty needles by drug addicts. Until screening tests could identify HIV-infected blood, transfusions and clotting factors used by hemophiliacs also were sources of infection.

HIV principally attacks a class of blood cells called helper T lymphocytes or, simply, helper T cells (**Figure 8.38**). HIV enters a helper T cell, undergoes reverse transcription, and integrates into the chromosome. The virus reproduces rapidly, destroying the T cell as new virus particles escape from the cell. Because helper T cells are central to immune function and are destroyed in the infection, AIDS patients have a diminished immune response; most AIDS patients die of secondary infections that develop because they have lost the ability to fight off pathogens.

The HIV genome is 9749 nucleotides long and carries *gag*, *pol*, *env*, and six other genes that regulate the life cycle of the virus. HIV's reverse transcriptase is very error prone, giving the virus a high mutation rate and allowing it to evolve rapidly, even within a single host. This rapid evolution makes the development of an effective vaccine against HIV particularly difficult. Genetic variation within the human population also affects the virus. To date, more than 10 loci in humans that affect HIV infection and the progression of AIDS have been identified.

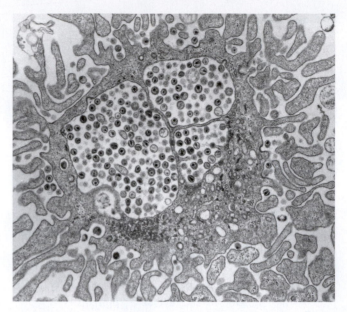

8.38 HIV principally attacks T lymphocytes. Electron micrograph showing a T cell infected with HIV, visible as small circles with dark centers. [*Courtesy of Dr. Hans Gelderblom.*]

CONCEPTS

A retrovirus is an RNA virus that integrates into its host's chromosome by making a DNA copy of its RNA genome through the process of reverse transcription. Human immunodeficiency virus, the causative agent of AIDS, is a retrovirus. It evolved from related retroviruses found in other primates.

✔ **CONCEPT CHECK 9**

What enzyme is used by a retrovirus to make a DNA copy of its genome?

CONCEPTS SUMMARY

- Bacteria and viruses are well suited to genetic studies: they are small, have a small haploid genome, undergo rapid reproduction, and produce large numbers of progeny through asexual reproduction.

- The bacterial genome normally consists of a single, circular molecule of double-stranded DNA. Plasmids are small pieces of bacterial DNA that can replicate independently of the large chromosome.

- DNA may be transferred between bacteria by means of conjugation, transformation, and transduction.

- Conjugation is the union of two bacterial cells and the transfer of genetic material between them. It is controlled by an episome called F. The rate at which individual genes are transferred during conjugation provides information about the order of the genes and the distances between them on the bacterial chromosome.

- Bacteria take up DNA from the environment through the process of transformation. Frequencies of the cotransformation of genes provide information about the physical distances between chromosomal genes.

- Complete DNA sequences of a number of bacterial species have been determined. This sequence information indicates that horizontal gene transfer—the movement of DNA between species—is common in bacteria.

- The bacterium *Escherichia coli* is an important model genetic organism that has the advantages of small size, rapid reproduction, and a small genome.

- Viruses are replicating structures with DNA or RNA genomes that may be double stranded or single stranded, linear or circular.

- Bacterial genes become incorporated into phage coats and are transferred to other bacteria by phages through the process of transformation. Rates of cotransduction can be used to map bacterial genes. In specialized transduction, DNA near the site of phage integration on the bacterial chromosome is transferred from one bacterium to another.

- Phage genes can be mapped by infecting bacterial cells with two different phage strains and counting the number of recombinant plaques produced by the progeny phages.

- Benzer mapped a large number of mutations that occurred within the *rII* region of phage T4. The results of his complementation studies demonstrated that the *rII* region consists of two functional units that he called cistrons. He showed that intragenic recombination takes place.

- A number of viruses have RNA genomes. Retroviruses encode a reverse transcriptase enzyme used to make a DNA copy of the viral genome, which then integrates into the host genome as a provirus. HIV is a retrovirus that is the causative agent for AIDS.

IMPORTANT TERMS

minimal medium (p. 201)
complete medium (p. 201)
colony (p. 202)
plasmid (p. 203)
episome (p. 204)
F factor (p. 204)
conjugation (p. 204)
transformation (p. 205)
transduction (p. 205)
pili (singular, pilus) (p. 207)
competent cell (p. 213)

transformant (p. 214)
cotransformation (p. 214)
horizontal gene transfer (p. 215)
virus (p. 218)
virulent phage (p. 218)
temperate phage (p. 218)
prophage (p. 218)
plaque (p. 218)
generalized transduction (p. 219)
specialized transduction (p. 219)
transducing phage (p. 220)

transductant (p. 220)
cotransduction (p. 220)
attachment site (p. 222)
intragenic mapping (p. 224)
retrovirus (p. 228)
reverse transcriptase (p. 228)
provirus (p. 228)
integrase (p. 229)
oncogene (p. 229)

ANSWERS TO CONCEPT CHECKS

1. d
2. b
3. a
4. *gal*
5. *his* and *leu*

6. b
7. c
8. b
9. Reverse transcriptase

WORKED PROBLEMS

1. DNA from a strain of bacteria with genotype $a^+ b^+ c^+ d^+ e^+$ was isolated and used to transform a strain of bacteria that was $a^- b^- c^- d^- e^-$. The transformed cells were tested for the presence of donated genes. The following genes were cotransformed:

a^+ and d^+ c^+ and d^+
b^+ and e^+ c^+ and e^+

What is the order of genes *a, b, c, d,* and *e* on the bacterial chromosome?

- **Solution**

The rate at which genes are cotransformed is inversely proportional to the distance between them: genes that are close together are frequently cotransformed, whereas genes that are far apart are rarely cotransformed. In this transformation experiment, gene c^+ is cotransformed with both genes e^+ and d^+, but genes e^+ and d^+ are not cotransformed; therefore the *c* locus must be between the *d* and *e* loci:

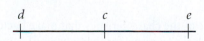

Gene e^+ is also cotransformed with gene b^+; so the e and b loci must be located close together. Locus b could be on either side of locus e. To determine whether locus b is on the same side of e as locus c, we look to see whether genes b^+ and c^+ are cotransformed. They are not; so locus b must be on the opposite side of e from c:

Gene a^+ is cotransformed with gene d^+; so they must be located close together. If locus a were located on the same side of d as locus c, then genes a^+ and c^+ would be cotransformed. Because these genes display no cotransformation, locus a must be on the opposite side of locus d:

2. Consider three genes in *E. coli*: thr^+ (the ability to synthesize threonine), ara^+ (the ability to metabolize arabinose), and leu^+ (the ability to synthesize leucine). All three of these genes are close together on the *E. coli* chromosome. Phages are grown in a thr^+ ara^+ leu^+ strain of bacteria (the donor strain). The phage lysate is collected and used to infect a strain of bacteria that is thr^- ara^- leu^-. The recipient bacteria are then tested on medium lacking leucine. Bacteria that grow and form colonies on this medium (leu^+ transductants) are then replica plated onto medium lacking threonine and onto medium lacking arabinose to see which are thr^+ and which are ara^+.

    Another group of recipient bacteria are tested on medium lacking threonine. Bacteria that grow and form colonies on this medium (thr^+ transductants) are then replica plated onto medium lacking leucine and onto medium lacking arabinose to see which are ara^+ and which are leu^+. Results from these experiments are as follows:

Selected marker	Cells with cotransduced genes (%)
leu^+	3 thr^+
	76 ara^+
thr^+	3 leu^+
	0 ara^+

How are the loci arranged on the chromosome?

• Solution

Notice that, when we select for leu^+ (the top half of the table), most of the selected cells are also ara^+. This finding indicates that the leu and ara genes are located close together, because they are usually cotransduced. In contrast, thr^+ is only rarely cotransduced with leu^+, indicating that leu and thr are much farther apart. On the basis of these observations, we know that leu and ara are closer together than are leu and thr, but we don't yet know the order of three genes—whether thr is on the same side of ara as leu or on the opposite side, as shown here:

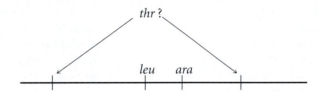

We can determine the position of thr with respect to the other two genes by looking at the cotransduction frequencies when thr^+ is selected (the bottom half of the preceding table). Notice that, although the cotransduction frequency for thr and leu also is 3%, no thr^+ ara^+ cotransductants are observed. This finding indicates that thr is closer to leu than to ara, and therefore thr must be to the left of leu, as shown here:

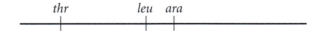

COMPREHENSION QUESTIONS

Section 8.1

1. Explain how auxotrophic bacteria are isolated.

2. Briefly explain the differences between F⁺, F⁻, Hfr, and F′ cells.

*3. What types of matings are possible between F⁺, F⁻, Hfr, and F′ cells? What outcomes do these matings produce? What is the role of F factor in conjugation?

*4. Explain how interrupted conjugation, transformation, and transduction can be used to map bacterial genes. How are these methods similar and how are they different?

5. What is horizontal gene transfer and how might it take place?

Section 8.2

*6. List some of the characteristics that make bacteria and viruses ideal organisms for many types of genetic studies.

7. What types of genomes do viruses have?

8. Briefly describe the differences between the lytic cycle of virulent phages and the lysogenic cycle of temperate phages.

9. Briefly explain how genes in phages are mapped.

*10. How does specialized transduction differ from generalized transduction?

*11. Briefly explain the method used by Benzer to determine whether two different mutations occurred at the same locus.

*12. Explain how a retrovirus, which has an RNA genome, is able to integrate its genetic material into that of a host having a DNA genome.

13. Briefly describe the genetic structure of a typical retrovirus.

14. What are the evolutionary origins of HIV-1 and HIV-2?

APPLICATION QUESTIONS AND PROBLEMS

Section 8.1

*15. John Smith is a pig farmer. For the past 5 years, Smith has been adding vitamins and low doses of antibiotics to his pig food; he says that these supplements enhance the growth of the pigs. Within the past year, however, several of his pigs died from infections of common bacteria, which failed to respond to large doses of antibiotics. Can you offer an explanation for the increased rate of mortality due to infection in Smith's pigs? What advice might you offer Smith to prevent this problem in the future?

16. Rarely, the conjugation of Hfr and F⁻ cells produces two Hfr cells. Explain how this event occurs.

17. [DATA ANALYSIS] Austin Taylor and Edward Adelberg isolated some new strains of Hfr cells that they then used to map several genes in *E. coli* by using interrupted conjugation (A. L. Taylor and E. A. Adelberg. 1960. *Genetics* 45:1233–1243). In one experiment, they mixed cells of Hfr strain AB-312, which were xyl^+ mtl^+ mal^+ met^+ and sensitive to phage T6, with F⁻ strain AB-531, which was xyl^- mtl^- mal^- met^- and resistant to phage T6. The cells were allowed to undergo conjugation. At regular intervals, the researchers removed a sample of cells and interrupted conjugation by killing the Hfr cells with phage T6. The F⁻ cells, which were resistant to phage T6, survived and were then tested for the presence of genes transferred from the Hfr strain. The results of this experiment are shown in the accompanying graph. On the basis of these data, give the order of the *xyl*, *mtl*, *mal*, and *met* genes on the bacterial chromosome and indicate the minimum distances between them.

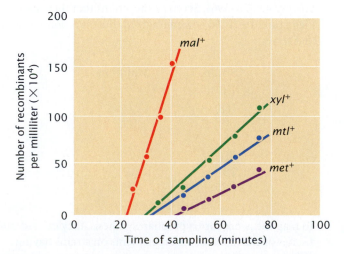

*18. A series of Hfr strains that have genotype m^+ n^+ o^+ p^+ q^+ r^+ are mixed with an F⁻ strain that has genotype m^- n^- o^- p^- q^- r^-. Conjugation is interrupted at regular intervals and the order of the appearance of genes from the Hfr strain is determined in the recipient cells. The order of gene transfer for each Hfr strain is:

Hfr5	m^+ q^+ p^+ n^+ r^+ o^+
Hfr4	n^+ r^+ o^+ m^+ q^+ p^+
Hfr1	o^+ m^+ q^+ p^+ n^+ r^+
Hfr9	q^+ m^+ o^+ r^+ n^+ p^+

What is the order of genes on the circular bacterial chromosome? For each Hfr strain, give the location of the F factor in the chromosome and its polarity.

*19. Crosses of three different Hfr strains with separate samples of an F⁻ strain are carried out, and the following mapping data are provided from studies of interrupted conjugation:

Appearance of genes in F⁻ cells

Hfr1:	Genes	b^+	d^+	c^+	f^+	g^+
	Time	3	5	16	27	59
Hfr2:	Genes	e^+	f^+	c^+	d^+	b^+
	Time	6	24	35	46	48
Hfr3:	Genes	d^+	c^+	f^+	e^+	g^+
	Time	4	15	26	44	58

Construct a genetic map for these genes, indicating their order on the bacterial chromosome and the distances between them.

20. DNA from a strain of *Bacillus subtilis* with the genotype trp^+ tyr^+ was used to transform a recipient strain with the genotype trp^- tyr^-. The following numbers of transformed cells were recovered:

Genotype	Number of transformed cells
trp^+ tyr^-	154
trp^- tyr^+	312
trp^+ tyr^+	354

What do these results suggest about the linkage of the *trp* and *tyr* genes?

21. DNA from a strain of *Bacillus subtilis* with genotype a^+ b^+ c^+ d^+ e^+ is used to transform a strain with genotype a^- b^- c^- d^- e^-. Pairs of genes are checked for cotransformation and the following results are obtained:

Pair of genes	Cotransformation	Pair of genes	Cotransformation
a^+ and b^+	no	b^+ and d^+	no
a^+ and c^+	no	b^+ and e^+	yes
a^+ and d^+	yes	c^+ and d^+	no
a^+ and e^+	yes	c^+ and e^+	yes
b^+ and c^+	yes	d^+ and e^+	no

On the basis of these results, what is the order of the genes on the bacterial chromosome?

22. DNA from a bacterial strain that is his^+ leu^+ lac^+ is used to transform a strain that is his^- leu^- lac^-. The following percentages of cells were transformed:

Donor strain	Recipient strain	Genotype of transformed cells	Percentage
his^+ leu^+ lac^+	his^- leu^- lac^-	his^+ leu^+ lac^+	0.02
		his^+ leu^+ lac^-	0.00
		his^+ leu^- lac^+	2.00
		his^+ leu^- lac^-	4.00
		his^- leu^+ lac^+	0.10
		his^- leu^- lac^+	3.00
		his^- leu^+ lac^-	1.50

a. What conclusions can you make about that order of these three genes on the chromosome?

b. Which two genes are closest?

23. [DATA ANALYSIS] Rollin Hotchkiss and Julius Marmur studied transformation in the bacterium *Streptococcus pneumoniae* (R. D. Hotchkiss and J. Marmur. 1954. *Proceedings of the National Academy of Sciences* 40:55–60). They examined four mutations in this bacterium: penicillin resistance (P), streptomycin resistance (S), sulfanilamide resistance (F), and the ability to utilize mannitol (M). They extracted DNA from strains of bacteria with different combinations of different mutations and used this DNA to transform wild-type bacterial cells (P^+ S^+ F^+ M^+). The results from one of their transformation experiments are shown here.

Donor DNA	Recipient DNA	Transformants	Percentage of all cells
M S F	M^+ S^+ F^+	M^+ S F^+	4.0
		M^+ S^+ F	4.0
		M S^+ F^+	2.6
		M S F^+	0.41
		M^+ S F	0.22
		M S^+ F	0.0058
		M S F	0.0071

a. Hotchkiss and Marmur noted that the percentage of cotransformation was higher than would be expected on a random basis. For example, the results show that the 2.6% of the cells were transformed into M and 4% were transformed into S. If the M and S traits were inherited independently, the expected probability of cotransformation of M and S (M S) would be 0.026 × 0.04 = 0.001, or 0.1%. However, they observed 0.41% M S cotransformants, four times more than they expected. What accounts for the relatively high

frequency of cotransformation of the traits they observed?

b. On the basis of the results, what conclusion can you make about the order of the M, S, and F genes on the bacterial chromosome?

c. Why is the rate of cotransformation for all three genes (M S F) almost the same as the cotransformation of M F alone?

24. [DATA ANALYSIS] In the course of a study on the effects of the mechanical shearing of DNA, Eugene Nester, A. T. Ganesan, and Joshua Lederberg studied the transfer, by transformation, of sheared DNA from a wild type strain of *Bacillus subtilis* (his_2^+ aro_3^+ try_2^+ aro_1^+ tyr_1^+ aro_2^+) to strains of bacteria carrying a series of mutations (E. W. Nester, A. T. Ganesan, and J. Lederberg. 1963. *Proceedings of the National Academy of Sciences* 49:61–68). They reported the following rates of cotransformation between his_2^+ and the other genes (expressed as cotransfer rate), shown here.

Genes	Rate of cotransfer
his_2^+ and aro_3^+	0.015
his_2^+ and try_2^+	0.10
his_2^+ and aro_1^+	0.12
his_2^+ and tyr_1^+	0.23
his_2^+ and aro_2^+	0.05

On the basis of these data, which gene is farthest from his_2^+? Which gene is closest?

25. [DATA ANALYSIS] C. Anagnostopoulos and I. P. Crawford isolated and studied a series of mutations that affected several steps in the biochemical pathway leading to tryptophan in the bacterium *Bacillus subtilis* (C. Anagnostopoulos and I. P. Crawford. 1961. *Proceedings of the National Academy of Sciences* 47:378–390). Seven of the strains that they used in their study are listed here, along with the mutation found in that strain.

Strain	Mutation
T3	T^-
168	I^-
168PT	I^-
TI	I^-
TII	I^-
T8	A^-
H25	H^-

To map the genes for tryptophan synthesis, they carried out a series of transformation experiments on strains having different mutations and determined the percentage of recombinants among the transformed bacteria. Their results were as follows:

Recipient	Donor	Percent recombinants
T3	168PT	12.7
T3	T11	11.8
T3	T8	43.5
T3	H25	28.6
168	H25	44.9
TII	H25	41.4
TI	H25	31.3
T8	H25	67.4
H25	T3	19.0
H25	TII	26.3
H25	TI	13.4
H25	T8	45.0

On the basis of these two-point recombination frequencies, determine the order of the genes and the distances between them. Where more than one cross was completed for a pair of genes, average the recombination rates from the different crosses. Draw a map of the genes on the chromosome.

Section 8.2

26. Two mutations that affect plaque morphology in phages (a^- and b^-) have been isolated. Phages carrying both mutations ($a^- b^-$) are mixed with wild-type phages ($a^+ b^+$) and added to a culture of bacterial cells. Subsequent to infection and lysis, samples of the phage lysate are collected and cultured on bacterial cells. The following numbers of plaques are observed:

Plaque phenotype	Number
$a^+ b^+$	2043
$a^+ b^-$	320
$a^- b^+$	357
$a^- b^-$	2134

What is the frequency of recombination between the a and b genes?

27. **DATA ANALYSIS** T. Miyake and M. Demerec examined proline-requiring mutations in the bacterium *Salmonella typhimurium* (T. Miyake and M. Demerec. 1960. *Genetics* 45:755–762). On the basis of complementation studies, they found four proline auxotrophs: *proA*, *proB*, *proC*, and *proD*. To determine if *proA*, *proB*, *proC*, and *proD* loci were located close together on the bacterial chromosome, they conducted a transduction experiment. Bacterial strains that were *proC⁺* and had mutations at *proA*, *proB*, or *proD*, were used as donors. The donors were infected with bacteriophages, and progeny phages were allowed to infect recipient bacteria with genotype *proC⁻ proA⁺ proB⁺ proD⁺*. The bacteria were then plated on a selective medium that allowed only *proC⁺* bacteria to grow. The following results were obtained:

Donor genotype	Transductant genotype	Number
$proC^+\ proA^-\ proB^+\ proD^+$	$proC^+\ proA^+\ proB^+\ proD^+$	2765
	$proC^+\ proA^-\ proB^+\ proD^+$	3
$proC^+\ proA^+\ proB^-\ proD^+$	$proC^+\ proA^+\ proB^+\ proD^+$	1838
	$proC^+\ proA^+\ proB^-\ proD^+$	2
$proC^+\ proA^+\ proB^+\ proD^-$	$proC^+\ proA^+\ proB^+\ proD^+$	1166
	$proC^+\ proA^+\ proB^+\ proD^-$	0

a. Why are there no *proC⁻* genotypes among the transductants?

b. Which genotypes represent single transductants and which represent cotransductants?

c. Is there evidence that *proA*, *proB*, and *proD* are located close to *proC*? Explain your answer.

*28. A geneticist isolates two mutations in a bacteriophage. One mutation causes clear plaques (c), and the other produces minute plaques (m). Previous mapping experiments have established that the genes responsible for these two mutations are 8 m.u. apart. The geneticist mixes phages with genotype $c^+ m^+$ and genotype $c^- m^-$ and uses the mixture to infect bacterial cells. She collects the progeny phages and cultures a sample of them on plated bacteria. A total of 1000 plaques are observed. What numbers of the different types of plaques ($c^+ m^+, c^- m^-, c^+ m^-, c^- m^+$) should she expect to see?

29. The geneticist carries out the same experiment described in Problem 28, but this time she mixes phages with genotypes $c^+ m^-$ and $c^- m^+$. What results are expected from *this* cross?

*30. A geneticist isolates two r mutants (r_{13} and r_2) that cause rapid lysis. He carries out the following crosses and counts the number of plaques listed here:

Genotype of parental phage	Progeny	Number of plaques
$h^+\ r_{13}^-\ \times\ h^-\ r_{13}^+$	$h^+\ r_{13}^+$	1
	$h^-\ r_{13}^+$	104
	$h^+\ r_{13}^-$	110
	$h^-\ r_{13}^-$	2
Total		216
$h^+\ r_2^-\ \times\ h^-\ r_2^+$	$h^+\ r_2^+$	6
	$h^-\ r_2^+$	86
	$h^+\ r_2^-$	81
	$h^-\ r_2^-$	7
Total		180

a. Calculate the recombination frequencies between r_2 and h and between r_{13} and h.

b. Draw all possible linkage maps for these three genes.

*31. *E. coli* cells are simultaneously infected with two strains of phage λ. One strain has a mutant host range, is temperature sensitive, and produces clear plaques (genotype is *h st c*); another strain carries the wild-type alleles (genotype is *h⁺ st⁺ c⁺*). Progeny phages are collected from the lysed cells and are plated on bacteria. The genotypes of the progeny phages are given here:

Progeny phage genotype	Number of plaques
h⁺ c⁺ st⁺	321
h c st	338
h⁺ c st	26
h c⁺ st⁺	30
h⁺ c st⁺	106
h c⁺ st	110
h⁺ c⁺ st	5
h c st⁺	6

a. Determine the order of the three genes on the phage chromosome.

b. Determine the map distances between the genes.

c. Determine the coefficient of coincidence and the interference (see pp. 179–181 in Chapter 7).

32. A donor strain of bacteria with genes *a⁺ b⁺ c⁺* is infected with phages to map the donor chromosome with generalized transduction. The phage lysate from the bacterial cells is collected and used to infect a second strain of bacteria that are *a⁻ b⁻ c⁻*. Bacteria with the *a⁺* gene are selected, and the percentage of cells with cotransduced *b⁺* and *c⁺* genes are recorded.

Donor	Recipient	Selected gene	Cells with cotransduced gene (%)
a⁺ b⁺ c⁺	*a⁻ b⁻ c⁻*	*a⁺*	25 *b⁺*
		a⁺	3 *c⁺*

Is the *b* or *c* gene closer to *a*? Explain your reasoning.

33. A donor strain of bacteria with genotype *leu⁺ gal⁻ pro⁺* is infected with phages. The phage lysate from the bacterial cells is collected and used to infect a second strain of bacteria that are *leu⁻ gal⁺ pro⁻*. The second strain is selected for *leu⁺*, and the following cotransduction data are obtained:

Donor	Recipient	Selected gene	Cells with cotransduced gene (%)
leu⁺ gal⁻ pro⁺	*leu⁻ gal⁺ pro⁻*	*leu⁺*	47 *pro⁺*
		leu⁺	26 *gal⁻*

Which genes are closest, *leu* and *gal* or *leu* and *pro*?

34. A geneticist isolates two new mutations, called *rII*ₓ and *rII*ᵧ, from the *rII* region of bacteriophage T4. *E. coli* B cells are simultaneously infected with phages carrying the *rII*ₓ mutation and with phages carrying the *rII*ᵧ mutation. After the cells have lysed, samples of the phage lysate are collected. One sample is grown on *E. coli* K cells and a second sample on *E. coli* B cells. There are 8322 plaques on *E. coli* B and 3 plaques on *E. coli* K. What is the recombination frequency between these two mutations?

35. A geneticist is working with a new bacteriophage called phage Y3 that infects *E. coli*. He has isolated eight mutant phages that fail to produce plaques when grown on *E. coli* strain K. To determine whether these mutations occur at the same functional gene, he simultaneously infects *E. coli* K cells with paired combinations of the mutants and looks to see whether plaques are formed. He obtains the following results. (A plus sign means that plaques were formed on *E. coli* K; a minus sign means that no plaques were formed on *E. coli* K.)

Mutant	1	2	3	4	5	6	7	8
1								
2	+							
3	+	+						
4	+	−	+					
5	−	+	+	+				
6	−	+	+	+	−			
7	+	−	+	−	+	+		
8	−	+	+	+	−	−	+	

a. To how many functional genes (cistrons) do these mutations belong?

b. Which mutations belong to the same functional gene?

CHALLENGE QUESTIONS

Section 8.1

36. As a summer project, a microbiology student independently isolates two mutations in *E. coli* that are auxotrophic for glycine (*gly⁻*). The student wants to know whether these two mutants are at the same functional unit. Outline a procedure that the student could use to determine whether these two *gly⁻* mutations occur within the same functional unit.

37. A group of genetics students mix two auxotrophic strains of bacteria: one is *leu⁺ trp⁺ his⁻ met⁻* and the other is *leu⁻ trp⁻ his⁺ met⁺*. After mixing the two strains, they plate the bacteria on minimal medium and observe a few prototrophic colonies (*leu⁺ trp⁺ his⁺ met⁺*). They assume that some gene transfer has taken place between the two strains. How can they determine whether the transfer of genes is due to conjugation, transduction, or transformation?

9 Chromosome Variation

Down syndrome is caused by the presence of three copies of one or more genes located on chromosome 21. [Stockbyte.]

TRISOMY 21 AND THE DOWN-SYNDROME CRITICAL REGION

In 1866, John Langdon Down, physician and medical superintendent of the Earlswood Asylum in Surrey, England, noticed a remarkable resemblance among a number of his mentally retarded patients: all of them possessed a broad, flat face, a thick tongue, a small nose, and oval-shaped eyes. Their features were so similar, in fact, that he felt that they might easily be mistaken as children of the same family. Down did not understand the cause of their retardation, but his original description faithfully records the physical characteristics of this most common genetic form of mental retardation. In his honor, the disorder is today known as Down syndrome.

As early as the 1930s, geneticists suggested that Down syndrome might be due to a chromosome abnormality, but not until 1959 did researchers firmly establish the cause of Down syndrome: most people with the disorder have three copies of chromosome 21, a condition known as trisomy 21. In a few rare cases, people having the disorder are trisomic for smaller parts of chromosome 21. By comparing the parts of chromosome 21 that these people have in common, geneticists have established that a specific segment—called the Down-syndrome critical region or DSCR—probably contains one or more genes responsible for the features of Down syndrome. Sequencing of the human genome has established that the DSCR consists of a little more than 5 million base pairs and only 33 genes.

Research by several groups in 2006 was a source of insight into the roles of two genes in the DSCR. Mice that have deficiencies in the regulatory pathway controlled by two genes, called *DSCR1* and *DYRK1A,* found in the DSCR exhibit many of the features seen in Down syndrome. Mice genetically engineered to express *DYRK1A* and *DSCR1* at high levels have abnormal heart development, similar to congenital heart problems seen in many people with Down syndrome.

In spite of this exciting finding, the genetics of Down syndrome appears to be more complex than formerly thought. Geneticist Lisa Olson and her colleagues had conducted an earlier study on mice to determine if genes within the DSCR are solely responsible for Down syndrome. Mouse breeders have developed several strains of mice that are trisomic for most of the genes found on human chromosome 21 (the equivalent mouse genes are actually found on mouse chromosome 16). These mice display many of the same anatomical features found in people with Down syndrome, as well as altered behavior, and they are considered an animal model for Down syndrome. Olson and her colleagues carefully created mice that were trisomic for genes in the DSCR but possessed the normal two copies of other genes found on human chromosome 21. If one or more of the genes in the DSCR are indeed responsible for Down syndrome, these mice should exhibit the features of Down syndrome. Surprisingly, the engineered mice exhibited none of the anatomical features of Down syndrome, demonstrating that three copies of genes in the DSCR are not the sole cause of the features of Down syndrome, at least in mice. Another study examined a human gene called *APP*, which lies outside of the DSCR. This gene appears to be responsible for at least some of the Alzheimer-like features observed in older Down-syndrome people.

Taken together, findings from these studies suggest that Down syndrome is not due to a single gene but is instead caused by complex interactions among multiple genes that are affected when an extra copy of chromosome 21 is present. Research on Down syndrome illustrates the principle that chromosome abnormalities often affect many genes that interact in complex ways.

Most species have a characteristic number of chromosomes, each with a distinct size and structure, and all the tissues of an organism (except for gametes) generally have the same set of chromosomes. Nevertheless, variations in chromosome number—such as the extra chromosome 21 that leads to Down syndrome—do periodically arise. Variations may also arise in chromosome structure: individual chromosomes may lose or gain parts and the order of genes within a chromosome may become altered. These variations in the number and structure of chromosomes are termed **chromosome mutations,** and they frequently play an important role in evolution.

We begin this chapter by briefly reviewing some basic concepts of chromosome structure, which we learned in Chapter 2. We then consider the different types of chromosome mutations, their definitions, features, phenotypic effects, and influence on evolution.

9.1 Chromosome Mutations Include Rearrangements, Aneuploids, and Polyploids

Before we consider the different types of chromosome mutations, their effects, and how they arise, we will review the basics of chromosome structure.

Chromosome Morphology

Each functional chromosome has a centromere, where spindle fibers attach, and two telomeres that stabilize the chromosome (see Figure 2.7). Chromosomes are classified into four basic types:

1. **Metacentric.** The centromere is located approximately in the middle, and so the chromosome has two arms of equal length.

2. **Submetacentric.** The centromere is displaced toward one end, creating a long arm and a short arm. (On human chromosomes, the short arm is designated by the letter p and the long arm by the letter q.)

3. **Acrocentric.** The centromere is near one end, producing a long arm and a knob, or satellite, at the other end.

4. **Telocentric.** The centromere is at or very near the end of the chromosome (see Figure 2.8).

The complete set of chromosomes possessed by an organism is called its *karyotype* and is usually presented as a picture of metaphase chromosomes lined up in descending order of their size (**Figure 9.1**). Karyotypes are prepared from actively dividing cells, such as white blood cells, bone-marrow cells, or cells from meristematic tissues of plants. After treatment with a chemical (such as colchicine) that prevents them from entering anaphase, the cells are chemi-

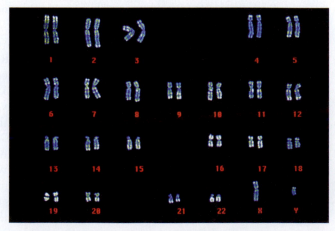

9.1 A human karyotype consists of 46 chromosomes. A karyotype for a male is shown here; a karyotype for a female would have two X chromosomes. *[ISM/Phototake.]*

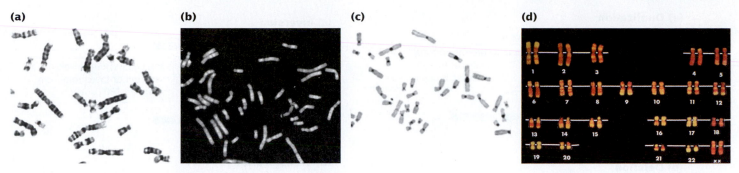

(a)   **(b)**   **(c)**   **(d)**

9.2 Chromosome banding is revealed by special staining techniques. (a) G banding. (b) Q banding. (c) C banding. (d) R banding. *[Part a: Leonard Lessin/Peter Arnold. Parts b and c: University of Washington Pathology Department. Part d: Dr. Ram Verma/Phototake.]*

cally preserved, spread on a microscope slide, stained, and photographed. The photograph is then enlarged, and the individual chromosomes are cut out and arranged in a karyotype. For human chromosomes, karyotypes are often routinely prepared by automated machines, which scan a slide with a video camera attached to a microscope, looking for chromosome spreads. When a spread has been located, the camera takes a picture of the chromosomes, the image is digitized, and the chromosomes are sorted and arranged electronically by a computer.

Preparation and staining techniques help to distinguish among chromosomes of similar size and shape. For instance, the staining of chromosomes with a special dye called Giemsa reveals G bands, which distinguish areas of DNA that are rich in adenine–thymine (A–T) base pairs (**Figure 9.2a**). Q bands (**Figure 9.2b**) are revealed by staining chromosomes with quinacrine mustard and viewing the chromosomes under ultraviolet light; variation in the brightness of Q

bands results from differences in the relative amounts of cytosine–guanine (C–G) and adenine–thymine base pairs. Other techniques reveal C bands (**Figure 9.2c**), which are regions of DNA occupied by centromeric heterochromatin, and R bands (**Figure 9.2d**), which are rich in cytosine–guanine base pairs.

Types of Chromosome Mutations

Chromosome mutations can be grouped into three basic categories: chromosome rearrangements, aneuploids, and polyploids (**Figure 9.3**). Chromosome rearrangements alter the *structure* of chromosomes; for example, a piece of a chromosome might be duplicated, deleted, or inverted. In aneuploidy, the *number* of chromosomes is altered: one or more individual chromosomes are added or deleted. In polyploidy, one or more complete *sets* of chromosomes are added. Some organisms (such as yeast) possess a single chromosome set (1*n*) for

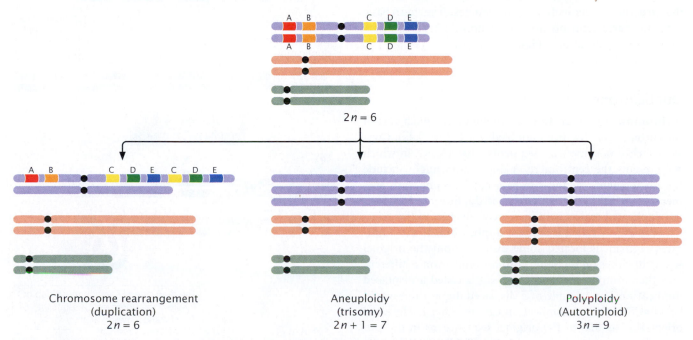

9.3 Chromosome mutations consist of chromosome rearrangements, aneuploids, and polyploids. Duplications, trisomy, and autotriploids are examples of each category of mutation.

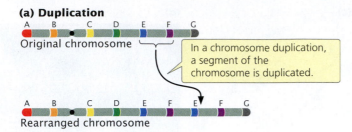

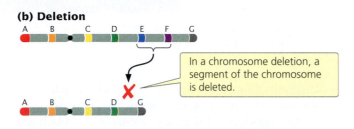

9.4 The four basic types of chromosome rearrangements are duplication, deletion, inversion, and translocation.

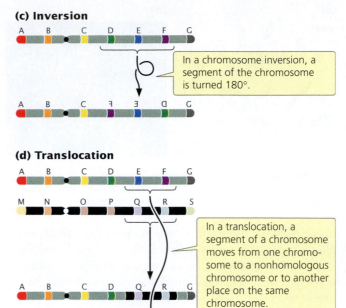

most of their life cycles and are referred to as haploid, whereas others possess two chromosome sets and are referred to as diploid ($2n$). A polyploid is any organism that has more than two sets of chromosomes ($3n$, $4n$, $5n$, or more).

9.2 Chromosome Rearrangements Alter Chromosome Structure

Chromosome rearrangements are mutations that change the structure of individual chromosomes. The four basic types of rearrangements are duplications, deletions, inversions, and translocations (**Figure 9.4**).

Duplications

A **chromosome duplication** is a mutation in which part of the chromosome has been doubled (see Figure 9.4a). Consider a chromosome with segments AB•CDEFG, in which • represents the centromere. A duplication might include the EF segments, giving rise to a chromosome with segments AB•CDEFEFG. This type of duplication, in which the duplicated region is immediately adjacent to the original segment, is called a **tandem duplication.** If the duplicated segment is located some distance from the original segment, either on the same chromosome or on a different one, the chromosome rearrangement is called a **displaced duplication.** An example of a displaced duplication would be AB•CDEFGEF. A duplication can be either in the same orientation as that of the original sequence, as in the two preceding examples, or inverted: AB•CDEFFEG. When the duplication is inverted, it is called a **reverse duplication.**

An individual homozygous for a duplication carries the duplication (the mutated sequence) on both homologous chromosomes, and an individual heterozygous for a duplication has one unmutated chromosome and one chromosome with the duplication. In the heterozygotes (**Figure 9.5a**),

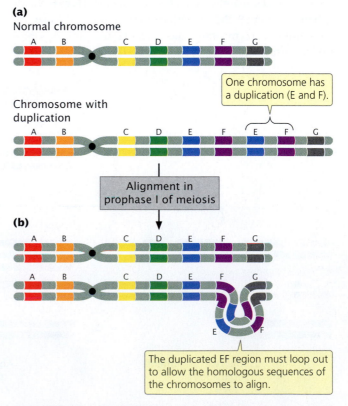

9.5 In an individual heterozygous for a duplication, the duplicated chromosome loops out during pairing in prophase I.

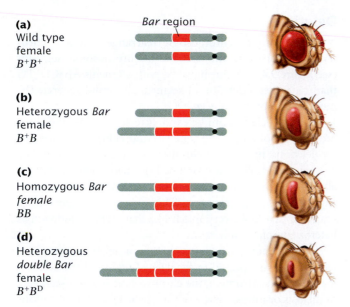

(a)
Wild type
female
B^+B^+

Bar region

(b)
Heterozygous *Bar*
female
B^+B

(c)
Homozygous *Bar*
female
BB

(d)
Heterozygous
double Bar
female
B^+B^D

9.6 The Bar phenotype in *Drosophila melanogaster* results from an X-linked duplication. (a) Wild-type fruit flies have normal-size eyes. (b) Flies heterozygous and (c) homozygous for the *Bar* mutation have smaller, bar-shaped eyes. (d) Flies with *double Bar* have three copies of the duplication and much smaller bar-shaped eyes.

problems arise in chromosome pairing at prophase I of meiosis, because the two chromosomes are not homologous throughout their length. The pairing and synapsis of homologous regions require that one or both chromosomes loop and twist so that these regions are able to line up (**Figure 9.5b**). The appearance of this characteristic loop structure in meiosis is one way to detect duplications.

Duplications may have major effects on the phenotype. Among fruit flies (*Drosophila melanogaster*), for example, a fly having a *Bar* mutation has a reduced number of facets in the eye, making the eye smaller and bar shaped instead of oval (**Figure 9.6**). The *Bar* mutation results from a small duplication on the X chromosome that is inherited as an incompletely dominant, X-linked trait: heterozygous female flies have somewhat smaller eyes (the number of facets is reduced; see Figure 9.6b), whereas, in homozygous female and hemizygous male flies, the number of facets is greatly reduced (see Figure 9.6c). Occasionally, a fly carries three

copies of the *Bar* duplication on its X chromosome; for flies carrying such mutations, which are termed *double Bar*, the number of facets is extremely reduced (see Figure 9.6d). The *Bar* mutation arises from unequal crossing over, a duplication-generating process (**Figure 9.7**; see also Figure 18.14).

How does a chromosome duplication alter the phenotype? After all, gene sequences are not altered by duplications, and no genetic information is missing; the only change is the presence of additional copies of normal sequences. The answer to this question is not well understood, but the effects are most likely due to imbalances in the amounts of gene products (abnormal gene dosage). The amount of a particular protein synthesized by a cell is often directly related to the number of copies of its corresponding gene: an individual organism with three functional copies of a gene often produces 1.5 times as much of the protein encoded by that gene as that produced by an individual with two copies. Because developmental processes require the interaction of many proteins, they may critically depend on the relative amounts of the proteins. If the amount of one protein increases while the amounts of others remain constant, problems can result (**Figure 9.8**). Although duplications can have severe consequences when the precise balance of a gene product is critical to cell function, duplications have arisen frequently throughout the evolution of many eukaryotic organisms and are a source of new genes that may provide novel functions. For example, humans have a series of genes that encode different globin chains, some of which function as an oxygen carrier during adult stages and others that function during embryonic and fetal development. All of these globin genes arose from an original ancestral gene that underwent a series of duplications. Human phenotypes associated with some duplications are summarized in **Table 9.1**.

CONCEPTS

A chromosome duplication is a mutation that doubles part of a chromosome. In individuals heterozygous for a chromosome duplication, the duplicated region of the chromosome loops out when homologous chromosomes pair in prophase I of meiosis. Duplications often have major effects on the phenotype, possibly by altering gene dosage.

Wild-type chromosomes

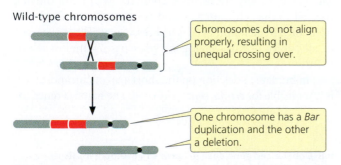

Chromosomes do not align properly, resulting in unequal crossing over.

One chromosome has a *Bar* duplication and the other a deletion.

Bar chromosomes

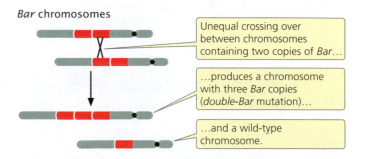

Unequal crossing over between chromosomes containing two copies of *Bar*...

...produces a chromosome with three *Bar* copies (*double-Bar* mutation)...

...and a wild-type chromosome.

9.7 Unequal crossing over produces *Bar* and *double-Bar* mutations.

Chromosome duplications often result in abnormal phenotypes because

a. developmental processes depend on the relative amounts of proteins encoded by different genes.

b. extra copies of the genes within the duplicated region do not pair in meiosis.

c. the chromosome is more likely to break when it loops in meiosis.

d. extra DNA must be replicated, which slows down cell division.

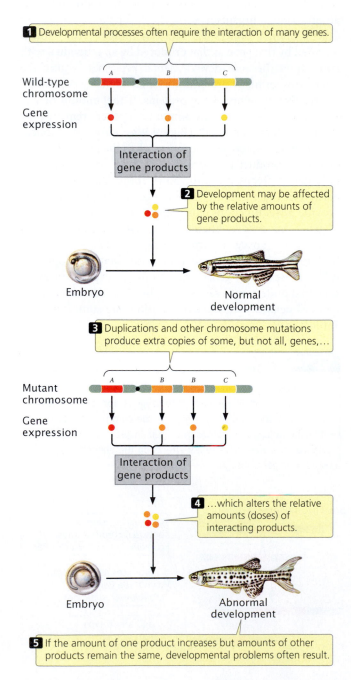

1 Developmental processes often require the interaction of many genes.

Wild-type chromosome

Gene expression

Interaction of gene products

2 Development may be affected by the relative amounts of gene products.

Embryo

Normal development

3 Duplications and other chromosome mutations produce extra copies of some, but not all, genes,…

Mutant chromosome

Gene expression

Interaction of gene products

4 …which alters the relative amounts (doses) of interacting products.

Embryo

Abnormal development

5 If the amount of one product increases but amounts of other products remain the same, developmental problems often result.

Deletions

A second type of chromosome rearrangement is a **chromosome deletion,** the loss of a chromosome segment (see Figure 9.4b). A chromosome with segments AB•CDEFG that undergoes a deletion of segment EF would generate the mutated chromosome AB•CDG.

A large deletion can be easily detected because the chromosome is noticeably shortened. In individuals heterozygous for deletions, the normal chromosome must loop during the pairing of homologs in prophase I of meiosis (**Figure 9.9**) to allow the homologous regions of the two chromosomes to align and undergo synapsis. This looping out generates a structure that looks very much like that seen for individuals heterozygous for duplications.

The phenotypic consequences of a deletion depend on which genes are located in the deleted region. If the deletion includes the centromere, the chromosome will not segregate in meiosis or mitosis and will usually be lost. Many deletions are lethal in the homozygous state because all copies of any essential genes located in the deleted region are missing. Even individuals heterozygous for a deletion may have multiple defects for three reasons.

First, the heterozygous condition may produce imbalances in the amounts of gene products, similar to the imbalances produced by extra gene copies. Second, recessive mutations on the homologous chromosome lacking the deletion may be expressed when the wild-type allele has been deleted (and is no longer present to mask the recessive allele's expression). The expression of a recessive mutation is referred to as **pseudodominance,** and it is an indication that one of the homologous chromosomes has a deletion. Third, some genes must be present in two copies for normal function. When a single copy of a gene is not sufficient to produce a wild-type phenotype, it is said to be a **haploinsufficient gene.** *Notch* is a series of X-linked wing mutations in *Drosophila* that often result from chromosome deletions. *Notch* deletions behave as dominant mutations: when heterozygous for the *Notch* deletion, a fly has wings that are notched at the tips and along the edges (**Figure 9.10**). The *Notch* locus is therefore haploinsufficient. Females that are homozygous for a *Notch* deletion (or males that are hemizygous) die early in embryonic development. The *Notch* gene encodes a receptor that normally transmits signals received from outside the cell to the cell's interior and is important in fly development. The deletion acts as a recessive lethal because loss of all copies of the *Notch* gene prevents normal development.

In humans, a deletion on the short arm of chromosome 5 is responsible for *cri-du-chat* syndrome. The name (French for

9.8 Unbalanced gene dosage leads to developmental abnormalities.

Table 9.1 Effects of some human chromosome rearrangements

Type of Rearrangement	Chromosome	Disorder	Symptoms
Duplication	4, short arm	—	Small head, short neck, low hairline, growth and mental retardation
Duplication	4, long arm	—	Small head, sloping forehead, hand abnormalities
Duplication	7, long arm	—	Delayed development, asymmetry of the head, fuzzy scalp, small nose, low-set ears
Duplication	9, short arm	—	Characteristic face, variable mental retardation, high and broad forehead, hand abnormalities
Deletion	5, short arm	*Cri-du-chat* syndrome	Small head, distinctive cry, widely spaced eyes, round face, mental retardation
Deletion	4, short arm	Wolf–Hirschhorn syndrome	Small head with high forehead, wide nose, cleft lip and palate, severe mental retardation
Deletion	4, long arm	—	Small head, from mild to moderate mental retardation, cleft lip and palate, hand and foot abnormalities
Deletion	7, long arm	Williams–Beuren syndrome	Facial features, heart defects, mental impairment
Deletion	15, long arm	Prader–Willi syndrome	Feeding difficulty at early age, but becoming obese after 1 year of age, from mild to moderate mental retardation
Deletion	18, short arm	—	Round face, large low-set ears, from mild to moderate mental retardation
Deletion	18, long arm	—	Distinctive mouth shape, small hands, small head, mental retardation

"cry of the cat") derives from the peculiar, catlike cry of infants with this syndrome. A child who is heterozygous for this deletion has a small head, widely spaced eyes, and a round face and is mentally retarded. Deletion of part of the short arm of chromosome 4 results in another human disorder—Wolf–Hirschhorn syndrome, which is characterized by seizures and severe mental and growth retardation. A deletion of a tiny segment of chromosome 7 causes haploinsufficiency of the gene encoding elastin and a few other genes and leads to a condition known as Williams–Beuren syndrome (WBS), characterized by distinctive facial features, heart defects, high blood pressure, and cognitive impairments.

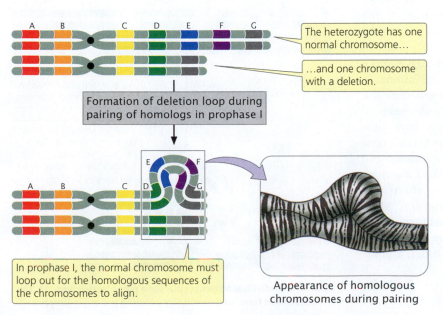

The heterozygote has one normal chromosome…

…and one chromosome with a deletion.

Formation of deletion loop during pairing of homologs in prophase I

In prophase I, the normal chromosome must loop out for the homologous sequences of the chromosomes to align.

Appearance of homologous chromosomes during pairing

9.9 In an individual heterozygous for a deletion, the normal chromosome loops out during chromosome pairing in prophase I.

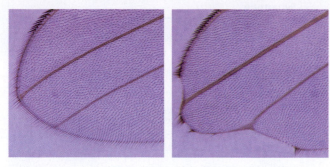

9.10 The Notch phenotype is produced by a chromosome deletion that includes the *Notch* gene. (Left) Normal wing veination. (Right) Wing veination produced by *Notch* mutation. [Spyros Artavanis-Tsakonas, Kenji Matsuno, and Mark E. Fortini.]

CONCEPTS

A chromosomal deletion is a mutation in which a part of a chromosome is lost. In individuals heterozygous for a deletion, the normal chromosome loops out during prophase I of meiosis. Deletions cause recessive genes on the homologous chromosome to be expressed and may cause imbalances in gene products.

✔ CONCEPT CHECK 2

What is pseudodominance and how is it produced by a chromosome deletion?

Inversions

A third type of chromosome rearrangement is a **chromosome inversion,** in which a chromosome segment is inverted—turned 180 degrees (see Figure 9.4c). If a chromosome originally had segments AB•CDEFG, then chromosome AB•CFEDG represents an inversion that includes segments DEF. For an inversion to take place, the chromosome must break in two places. Inversions that do not include the centromere, such as AB•CFEDG, are termed **paracentric inversions** (*para* meaning "next to"), whereas inversions that include the centromere, such as ADC•BEFG, are termed **pericentric inversions** (*peri* meaning "around").

Individual organisms with inversions have neither lost nor gained any genetic material; just the gene order has been altered. Nevertheless, these mutations often have pronounced phenotypic effects. An inversion may break a gene into two parts, with one part moving to a new location and destroying the function of that gene. Even when the chromosome breaks are between genes, phenotypic effects may arise from the inverted gene order in an inversion. Many genes are regulated in a position-dependent manner; if their positions are altered by an inversion, they may be expressed at inappropriate times or in inappropriate tissues, an outcome referred to as a **position effect.**

When an individual is homozygous for a particular inversion, no special problems arise in meiosis, and the two

homologous chromosomes can pair and separate normally. When an individual is heterozygous for an inversion, however, the gene order of the two homologs differs, and the homologous sequences can align and pair only if the two chromosomes form an inversion loop (**Figure 9.11**).

Individuals heterozygous for inversions also exhibit reduced recombination among genes located in the inverted region. The frequency of crossing over within the inversion is not actually diminished but, when crossing over does take place, the result is a tendency to produce gametes that are not viable and thus no recombinant progeny are observed. Let's see why this result occurs.

Figure 9.12 illustrates the results of crossing over within a paracentric inversion. The individual is heterozygous for an inversion (see Figure 9.12a), with one wild-type, unmutated chromosome (AB•CDEFG) and one inverted chromosome (AB•EDCFG). In prophase I of meiosis, an inversion loop forms, allowing the homologous sequences to pair up (see Figure 9.12b). If a single crossover takes place in the inverted region (between segments C and D in Figure 9.12), an unusual structure results (see Figure 9.12c). The two outer chromatids, which did not participate in crossing over, contain original, nonrecombinant gene sequences. The two inner chromatids, which did cross over, are highly abnormal: each has two copies of some genes and no copies of others. Furthermore, one of the four chromatids now has two centromeres and is said to be a **dicentric chromatid;** the other lacks a centromere and is an **acentric chromatid.**

In anaphase I of meiosis, the centromeres are pulled toward opposite poles and the two homologous chromosomes separate. This action stretches the dicentric chromatid across the center of the nucleus, forming a structure called a

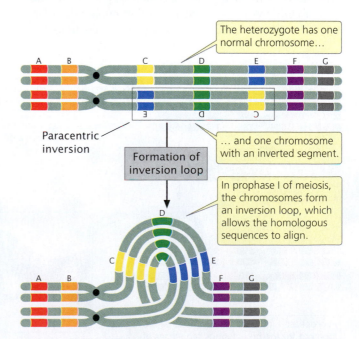

9.11 In an individual heterozygous for a paracentric inversion, the chromosomes form an inversion loop during pairing in prophase I.

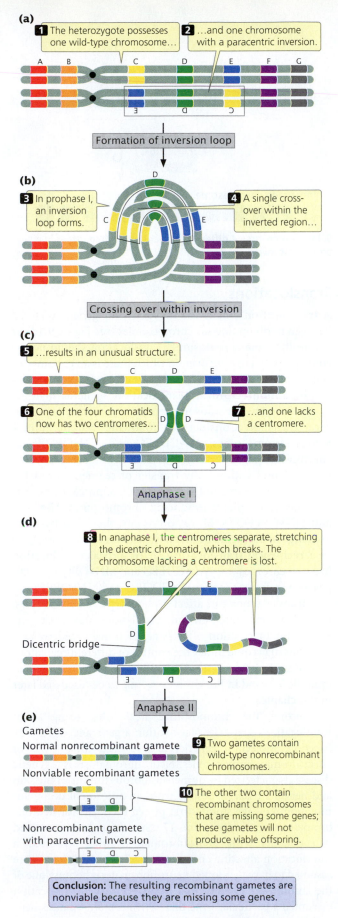

(a)

1 The heterozygote possesses one wild-type chromosome...

2 ...and one chromosome with a paracentric inversion.

A B C D E F G

Formation of inversion loop

(b)

3 In prophase I, an inversion loop forms.

4 A single crossover within the inverted region...

Crossing over within inversion

(c)

5 ...results in an unusual structure.

C D E

6 One of the four chromatids now has two centromeres...

7 ...and one lacks a centromere.

Anaphase I

(d)

8 In anaphase I, the centromeres separate, stretching the dicentric chromatid, which breaks. The chromosome lacking a centromere is lost.

C D E

Dicentric bridge

Anaphase II

(e)

Gametes

Normal nonrecombinant gamete

9 Two gametes contain wild-type nonrecombinant chromosomes.

Nonviable recombinant gametes

10 The other two contain recombinant chromosomes that are missing some genes; these gametes will not produce viable offspring.

Nonrecombinant gamete with paracentric inversion

Conclusion: The resulting recombinant gametes are nonviable because they are missing some genes.

9.12 In a heterozygous individual, a single crossover within a paracentric inversion leads to abnormal gametes.

dicentric bridge (see Figure 9.12d). Eventually, the dicentric bridge breaks, as the two centromeres are pulled farther apart. Spindle fibers do not attach to the acentric fragment, and so this fragment does not segregate into a nucleus in meiosis and is usually lost.

In the second division of meiosis, the chromatids separate and four gametes are produced (see Figure 9.12e). Two of the gametes contain the original, nonrecombinant chromosomes (AB•CDEFG and AB•EDCFG). The other two gametes contain recombinant chromosomes that are missing some genes; these gametes will not produce viable offspring. Thus, no recombinant progeny result when crossing over takes place within a paracentric inversion. The key is to recognize that crossing over still takes place, but, when it does so, the resulting recombinant gametes are not viable; so no recombinant progeny are observed.

Recombination is also reduced within a pericentric inversion (**Figure 9.13**). No dicentric bridges or acentric fragments are produced, but the recombinant chromosomes have too many copies of some genes and no copies of others; so gametes that receive the recombinant chromosomes cannot produce viable progeny.

Figures 9.12 and 9.13 illustrate the results of single crossovers within inversions. Double crossovers, in which both crossovers are on the same two strands (two-strand double crossovers), result in functional recombinant chromosomes. (Try drawing out the results of a double crossover.) Thus, even though the overall rate of recombination is reduced within an inversion, some viable recombinant progeny may still be produced through two-strand double crossovers.

Inversion heterozygotes are common in many organisms, including a number of plants, some species of *Drosophila*, mosquitoes, and grasshoppers. Inversions may have played an important role in human evolution: G-banding patterns reveal that several human chromosomes differ from those of chimpanzees by only a pericentric inversion (**Figure 9.14**).

CONCEPTS

In an inversion, a segment of a chromosome is inverted. Inversions cause breaks in some genes and may move others to new locations. In heterozygotes for a chromosome inversion, the homologous chromosomes form a loop in prophase I of meiosis. When crossing over takes place within the inverted region, nonviable gametes are usually produced, resulting in a depression in observed recombination frequencies.

✔ CONCEPT CHECK 3

A dicentric chromosome is produced when crossing over takes place in an individual heterozygous for which type of chromosome rearrangement?

a. Duplication

b. Deletion

c. Paracentric inversion

d. Pericentric inversion

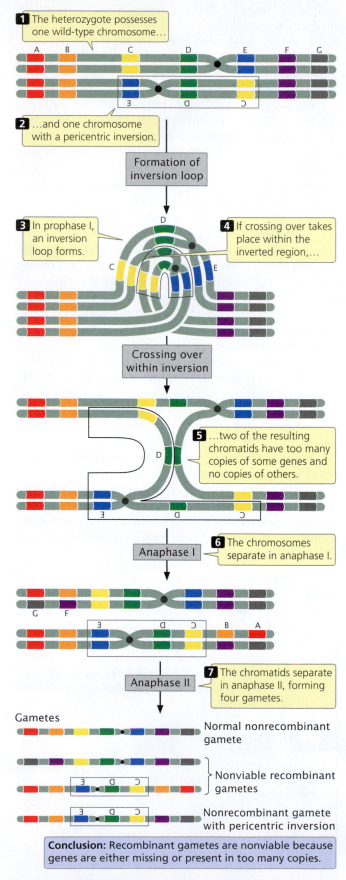

1 The heterozygote possesses one wild-type chromosome…

2 …and one chromosome with a pericentric inversion.

Formation of inversion loop

3 In prophase I, an inversion loop forms.

4 If crossing over takes place within the inverted region,…

Crossing over within inversion

5 …two of the resulting chromatids have too many copies of some genes and no copies of others.

Anaphase I

6 The chromosomes separate in anaphase I.

Anaphase II

7 The chromatids separate in anaphase II, forming four gametes.

Gametes

Normal nonrecombinant gamete

Nonviable recombinant gametes

Nonrecombinant gamete with pericentric inversion

Conclusion: Recombinant gametes are nonviable because genes are either missing or present in too many copies.

9.13 In a heterozygous individual, a single crossover within a pericentric inversion leads to abnormal gametes.

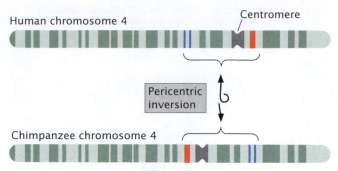

Human chromosome 4

Centromere

Pericentric inversion

Chimpanzee chromosome 4

9.14 Chromosome 4 differs in humans and chimpanzees in a pericentric inversion.

Translocations

A **translocation** entails the movement of genetic material between nonhomologous chromosomes (see Figure 9.4d) or within the same chromosome. Translocation should not be confused with crossing over, in which there is an exchange of genetic material between *homologous* chromosomes.

In a **nonreciprocal translocation,** genetic material moves from one chromosome to another without any reciprocal exchange. Consider the following two nonhomologous chromosomes: AB•CDEFG and MN•OPQRS. If chromosome segment EF moves from the first chromosome to the second without any transfer of segments from the second chromosome to the first, a nonreciprocal translocation has taken place, producing chromosomes AB•CDG and MN•OPEFQRS. More commonly, there is a two-way exchange of segments between the chromosomes, resulting in a **reciprocal translocation.** A reciprocal translocation between chromosomes AB•CDEFG and MN•OPQRS might give rise to chromosomes AB•CDQRG and MN•OPEFS.

Translocations can affect a phenotype in several ways. First, they can create new linkage relations that affect gene expression (a position effect): genes translocated to new locations may come under the control of different regulatory sequences or other genes that affect their expression—an example is found in Burkitt lymphoma, to be discussed later in this chapter.

Second, the chromosomal breaks that bring about translocations may take place within a gene and disrupt its function. Molecular geneticists have used these types of effects to map human genes. Neurofibromatosis is a genetic disease characterized by numerous fibrous tumors of the skin and nervous tissue; it results from an autosomal dominant mutation. Linkage studies first placed the locus for neurofibromatosis on chromosome 17. Geneticists later identified two patients with neurofibromatosis who possessed a translocation affecting chromosome 17. These patients were assumed to have developed neurofibromatosis because one of the chromosome breaks that occurred in the translocation disrupted a particular gene that, when mutated, causes neurofibromatosis. DNA from the regions around the breaks was sequenced and eventually led to the identification of the gene responsible for neurofibromatosis.

Deletions frequently accompany translocations. In a **Robertsonian translocation,** for example, the long arms of two acrocentric chromosomes become joined to a common centromere through a translocation, generating a metacentric chromosome with two long arms and another chromosome with two very short arms (**Figure 9.15**). The smaller chromosome often fails to segregate, leading to an overall reduction in chromosome number. As we will see, Robertsonian translocations are the cause of some cases of Down syndrome, a chromosome disorder discussed in the introduction to the chapter.

The effects of a translocation on chromosome segregation in meiosis depend on the nature of the translocation. Let's consider what happens in an individual heterozygous for a reciprocal translocation. Suppose that the original chromosomes were AB•CDEFG and MN•OPQRS (designated N_1 and N_2, respectively) and that a reciprocal translocation takes place, producing chromosomes AB•CDQRS and MN•OPEFG (designated T_1 and T_2, respectively). An individual heterozygous for this translocation would possess one normal copy of each chromosome and one translocated copy (**Figure 9.16a**). Each of these chromosomes contains segments that are homologous to two other chromosomes. When the homologous sequences pair in prophase I of meiosis, crosslike configurations consisting of all four chromosomes (**Figure 9.16b**) form.

Notice that N_1 and T_1 have homologous centromeres (in both chromosomes, the centromere is between segments B and C); similarly, N_2 and T_2 have homologous centromeres (between segments N and O). Normally, homologous centromeres separate and move toward opposite poles in anaphase I of meiosis. With a reciprocal translocation, the chromosomes may segregate in three different ways. In **alternate segregation** (**Figure 9.16c**), N_1 and N_2 move toward one pole and T_1 and T_2 move toward the opposite pole. In **adjacent-1 segregation,** N_1 and T_2 move toward one pole and T_1 and N_2 move toward the other pole. In both alternate

and adjacent-1 segregation, homologous centromeres segregate toward opposite poles. **Adjacent-2 segregation,** in which N_1 and T_1 move toward one pole and T_2 and N_2 move toward the other, is rare.

The products of the three segregation patterns are illustrated in **Figure 9.16d.** As you can see, the gametes produced by alternate segregation possess one complete set of the chromosome segments. These gametes are therefore functional and can produce viable progeny. In contrast, gametes produced by adjacent-1 and adjacent-2 segregation are not viable, because some chromosome segments are present in two copies, whereas others are missing. Because adjacent-2 segregation is rare, most gametes are produced by alternate or adjacent-1 segregation. Therefore, approximately half of the gametes from an individual heterozygous for a reciprocal translocation are expected to be functional.

Translocations can play an important role in the evolution of karyotypes. Chimpanzees, gorillas, and orangutans all have 48 chromosomes, whereas humans have 46. Human chromosome 2 is a large, metacentric chromosome with G-banding patterns that match those found on two different acrocentric chromosomes of the apes (**Figure 9.17**). Apparently, a Robertsonian translocation took place in a human ancestor, creating a large metacentric chromosome from the two long arms of the ancestral acrocentric chromosomes and a small chromosome consisting of the two short arms. The small chromosome was subsequently lost, leading to the reduced human chromosome number.

CONCEPTS

In translocations, parts of chromosomes move to other, nonhomologous chromosomes or to other regions of the same chromosome. Translocations may affect the phenotype by causing genes to move to new locations, where they come under the influence of new regulatory sequences, or by breaking genes and disrupting their function.

✔ CONCEPT CHECK 4

What is the outcome of a Robertsonian translocation?

a. Two acrocentric chromosomes

b. One metacentric chromosome and one chromosome with two very short arms

c. One metacentric and one acrocentric chromosome

d. Two metacentric chromosomes

Fragile Sites

Chromosomes of cells grown in culture sometimes develop constrictions or gaps at particular locations called **fragile sites** (**Figure 9.18**), because they are prone to breakage under certain conditions. A number of fragile sites have been identified on human chromosomes. One of the most intensively studied is a fragile site on the human X chromosome, a site associated

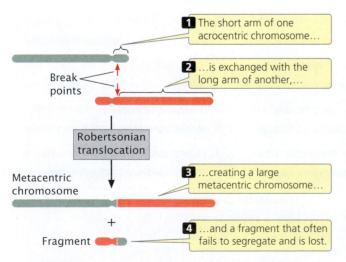

1 The short arm of one acrocentric chromosome…

2 …is exchanged with the long arm of another,…

Break points

Robertsonian translocation

Metacentric chromosome

3 …creating a large metacentric chromosome…

+

Fragment

4 …and a fragment that often fails to segregate and is lost.

9.15 In a Robertsonian translocation, the short arm of one acrocentric chromosome is exchanged with the long arm of another.

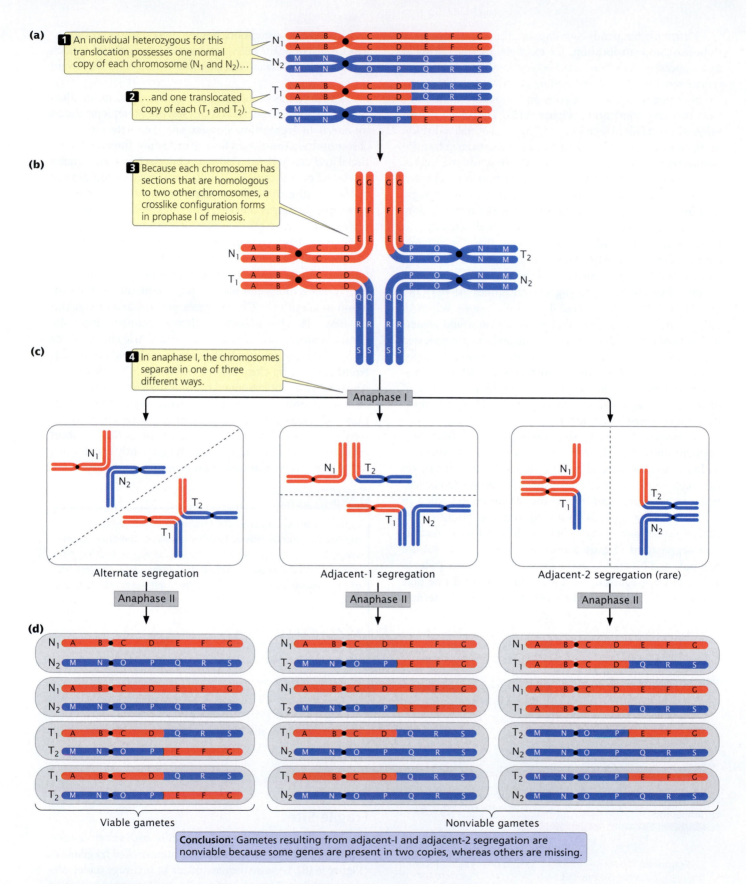

(a)

1 An individual heterozygous for this translocation possesses one normal copy of each chromosome (N₁ and N₂)...

2 ...and one translocated copy of each (T₁ and T₂).

(b)

3 Because each chromosome has sections that are homologous to two other chromosomes, a crosslike configuration forms in prophase I of meiosis.

(c)

4 In anaphase I, the chromosomes separate in one of three different ways.

Anaphase I

Alternate segregation

Adjacent-1 segregation

Adjacent-2 segregation (rare)

Anaphase II   Anaphase II   Anaphase II

(d)

Viable gametes   Nonviable gametes

Conclusion: Gametes resulting from adjacent-I and adjacent-2 segregation are nonviable because some genes are present in two copies, whereas others are missing.

9.16 In an individual heterozygous for a reciprocal translocation, crosslike structures form in homologous pairing.

Human chromosome 2

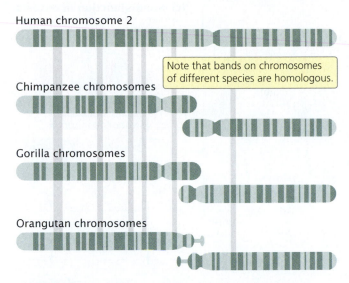

Note that bands on chromosomes of different species are homologous.

Chimpanzee chromosomes

Gorilla chromosomes

Orangutan chromosomes

9.17 Human chromosome 2 contains a Robertsonian translocation that is not present in chimpanzees, gorillas, or orangutans. G-banding reveals that a Robertsonian translocation in a human ancestor switched the long and short arms of the two acrocentric chromosomes that are still found in the other three primates. This translocation created the large metacentric human chromosome 2.

with mental retardation known as the fragile-X syndrome. Exhibiting X-linked inheritance and arising with a frequency of about 1 in 1250 male births, fragile-X syndrome has been shown to result from an increase in the number of repeats of a CGG trinucleotide (see Chapter 10). However, other common fragile sites do not consist of trinucleotide repeats, and their nature is still incompletely understood.

9.3 Aneuploidy Is an Increase or Decrease in the Number of Individual Chromosomes

In addition to chromosome rearrangements, chromosome mutations include changes in the number of chromosomes. Variations in chromosome number can be classified into two basic types: **aneuploidy,** which is a change in the number of individual chromosomes, and **polyploidy,** which is a change in the number of chromosome sets.

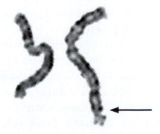

9.18 Fragile sites are chromosomal regions susceptible to breakage under certain conditions. Shown here is a fragile site on human chromosome X. [*University of Wisconsin Cytogenic Services Laboratory.*]

Aneuploidy can arise in several ways. First, a chromosome may be lost in the course of mitosis or meiosis if, for example, its centromere is deleted. Loss of the centromere prevents the spindle fibers from attaching; so the chromosome fails to move to the spindle pole and does not become incorporated into a nucleus after cell division. Second, the small chromosome generated by a Robertsonian translocation may be lost in mitosis or meiosis. Third, aneuploids may arise through nondisjunction, the failure of homologous chromosomes or sister chromatids to separate in meiosis or mitosis (see p. 82–84 in Chapter 4). Nondisjunction leads to some gametes or cells that contain an extra chromosome and others that are missing a chromosome (**Figure 9.19**).

Types of Aneuploidy

We will consider four types of common aneuploid conditions in diploid individuals: nullisomy, monosomy, trisomy, and tetrasomy.

1. **Nullisomy** is the loss of both members of a homologous pair of chromosomes. It is represented as $2n - 2$, where n refers to the haploid number of chromosomes. Thus, among humans, who normally possess $2n = 46$ chromosomes, a nullisomic person has 44 chromosomes.

2. **Monosomy** is the loss of a single chromosome, represented as $2n - 1$. A monosomic person has 45 chromosomes.

3. **Trisomy** is the gain of a single chromosome, represented as $2n + 1$. A trisomic person has 47 chromosomes. The gain of a chromosome means that there are three homologous copies of one chromosome. Most cases of Down syndrome, discussed in the introduction to the chapter, result from trisomy of chromosome 21.

4. **Tetrasomy** is the gain of two homologous chromosomes, represented as $2n + 2$. A tetrasomic person has 48 chromosomes. Tetrasomy is not the gain of *any* two extra chromosomes, but rather the gain of two homologous chromosomes; so there will be four homologous copies of a particular chromosome.

More than one aneuploid mutation may occur in the same individual organism. An individual that has an extra copy of two different (nonhomologous) chromosomes is referred to as being double trisomic and represented as $2n + 1 + 1$. Similarly, a double monosomic has two fewer nonhomologous chromosomes ($2n - 1 - 1$), and a double tetrasomic has two extra pairs of homologous chromosomes ($2n + 2 + 2$).

Effects of Aneuploidy

One of the first aneuploids to be recognized was a fruit fly with a single X chromosome and no Y chromosome discovered by Calvin Bridges in 1913 (see pp. 83–84 in Chapter 4). Another early study of aneuploidy focused on mutants in

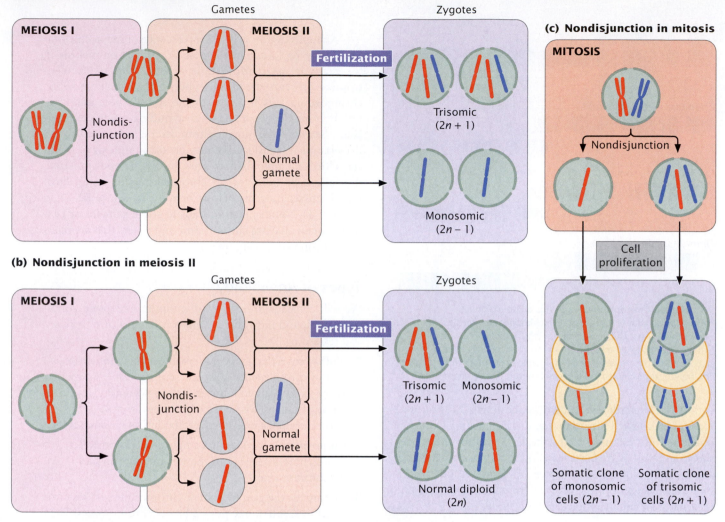

(a) Nondisjunction in meiosis I

Gametes

Zygotes

(c) Nondisjunction in mitosis

MEIOSIS I

MEIOSIS II

Fertilization

MITOSIS

Nondis-junction

Normal gamete

Trisomic (2n + 1)

Monosomic (2n − 1)

Nondisjunction

Cell proliferation

(b) Nondisjunction in meiosis II

Gametes

Zygotes

MEIOSIS I

MEIOSIS II

Fertilization

Nondis-junction

Normal gamete

Trisomic (2n + 1)

Monosomic (2n − 1)

Normal diploid (2n)

Somatic clone of monosomic cells (2n − 1)

Somatic clone of trisomic cells (2n + 1)

9.19 Aneuploids can be produced through nondisjunction in meiosis I, meiosis II, and mitosis.
The gametes that result from meioses with nondisjunction combine with a gamete (with blue chromosome) that results from normal meiosis to produce the zygotes. (a) Nondisjunction in meiosis I. (b) Nondisjunction in meiosis II. (c) Nondisjunction in mitosis.

the Jimson weed, *Datura stramonium*. A. Francis Blakeslee began breeding this plant in 1913, and he observed that crosses with several Jimson mutants produced unusual ratios of progeny. For example, the *globe* mutant (producing a seedcase globular in shape) was dominant but was inherited primarily from the female parent. When plants having the *globe* mutation were self-fertilized, only 25% of the progeny had the globe phenotype, an unusual ratio for a dominant trait. Blakeslee isolated 12 different mutants (**Figure 9.20**) that exhibited peculiar patterns of inheritance. Eventually, John Belling demonstrated that these 12 mutants are in fact trisomics. *Datura stramonium* has 12 pairs of chromosomes (2n = 24), and each of the 12 mutants is trisomic for a different chromosome pair. The aneuploid nature of the mutants explained the unusual ratios that Blakeslee had observed in the progeny. Many of the extra chromosomes in the trisomics were lost in meiosis; so fewer than 50% of the gametes carried the extra chromosome, and the proportion of trisomics in the progeny was low. Furthermore, the pollen containing an extra

chromosome was not as successful in fertilization, and trisomic zygotes were less viable.

Aneuploidy usually alters the phenotype drastically. In most animals and many plants, aneuploid mutations are lethal. Because aneuploidy affects the number of gene copies but not their nucleotide sequences, the effects of aneuploidy are most likely due to abnormal gene dosage. Aneuploidy alters the dosage for some, but not all, genes, disrupting the relative concentrations of gene products and often interfering with normal development.

A major exception to the relation between gene number and protein dosage pertains to genes on the mammalian X chromosome. In mammals, X-chromosome inactivation ensures that males (who have a single X chromosome) and females (who have two X chromosomes) receive the same functional dosage for X-linked genes (see pp. 87–89 in Chapter 4 for further discussion of X-chromosome inactivation). Extra X chromosomes in mammals are inactivated; so we might expect that aneuploidy of the sex chromosomes would be less detrimental in these animals. Indeed, it is the

Seed cases

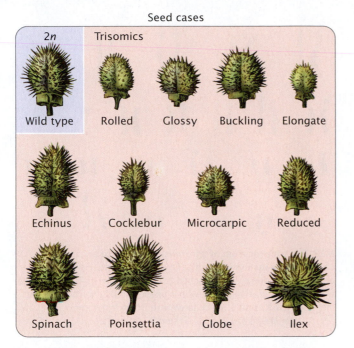

9.20 Mutant capsules in Jimson weed (Datura stramonium) result from different trisomies. Each type of capsule is a phenotype that is trisomic for a different chromosome.

case for mice and humans, for whom aneuploids of the sex chromosomes are the most common form of aneuploidy seen in living organisms. Y-chromosome aneuploids are probably common because there is so little information on the Y-chromosome.

CONCEPTS

Aneuploidy, the loss or gain of one or more individual chromosomes, may arise from the loss of a chromosome subsequent to translocation or from nondisjunction in meiosis or mitosis. It disrupts gene dosage and often has severe phenotypic effects.

✔ CONCEPT CHECK 5

A diploid organism has $2n = 36$ chromosomes. How many chromosomes will be found in a trisomic member of this species?

Aneuploidy in Humans

For unknown reasons, an incredibly high percentage of all human embryos that are conceived possess chromosome abnormalities. Findings from studies of women who are attempting pregnancy suggest that more than 30% of all conceptions spontaneously abort (miscarry), usually so early in development that the mother is not even aware of her pregnancy. Chromosome defects are present in at least 50% of spontaneously aborted human fetuses, with aneuploidy accounting for most of them. This rate of chromosome abnormality in humans is higher than in other organisms

that have been studied; in mice, for example, aneuploidy is found in no more than 2% of fertilized eggs. Aneuploidy in humans usually produces such serious developmental problems that spontaneous abortion results. Only about 2% of all fetuses with a chromosome defect survive to birth.

Sex-chromosome aneuploids   The most common aneuploidy seen in living humans has to do with the sex chromosomes. As is true of all mammals, aneuploidy of the human sex chromosomes is better tolerated than aneuploidy of autosomal chromosomes. Both Turner syndrome and Klinefelter syndrome (see Figures 4.8 and 4.9) result from aneuploidy of the sex chromosomes.

Autosomal aneuploids   Autosomal aneuploids resulting in live births are less common than sex-chromosome aneuploids in humans, probably because there is no mechanism of dosage compensation for autosomal chromosomes. Most autosomal aneuploids are spontaneously aborted, with the exception of aneuploids of some of the small autosomes such as chromosome 21. Because these chromosomes are small and carry fewer genes, the presence of extra copies is less detrimental than it is for larger chromosomes. For example, the most common autosomal aneuploidy in humans is **trisomy 21,** also called **Down syndrome** (discussed in the introduction to the chapter). The number of genes on different human chromosomes is not precisely known at the present time, but DNA sequence data indicate that chromosome 21 has fewer genes than any other autosome, with only about 230 genes of a total of 20,000 to 25,000 for the entire genome.

The incidence of Down syndrome in the United States is similar to that of the world, about 1 in 700 human births, although the incidence increases among children born to older mothers. Approximately 92% of those who have Down syndrome have three full copies of chromosome 21 (and therefore a total of 47 chromosomes), a condition termed **primary Down syndrome** (**Figure 9.21**). Primary Down syndrome usually arises from random nondisjunction in egg formation: about 75% of the nondisjunction events that cause Down syndrome are maternal in origin, most arising in meiosis I. Most children with Down syndrome are born to normal parents, and the failure of the chromosomes to divide has little hereditary tendency. A couple who has conceived one child with primary Down syndrome has only a slightly higher risk of conceiving a second child with Down syndrome (compared with other couples of similar age who have not had any Down-syndrome children). Similarly, the couple's relatives are not more likely to have a child with primary Down syndrome.

About 4% of people with Down syndrome have 46 chromosomes, but an extra copy of part of chromosome 21 is attached to another chromosome through a translocation (**Figure 9.22**). This condition is termed **familial Down syndrome** because it has a tendency to run in families. The phenotypic characteristics of familial Down syndrome are the same as those for primary Down syndrome.

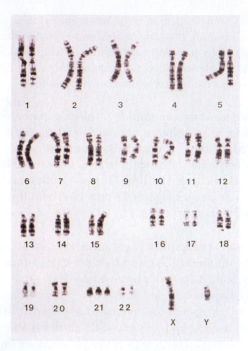

9.21 Primary Down syndrome is caused by the presence of three copies of chromosome 21. Karyotype of a person who has primary Down syndrome. *[L. Willatt, East Anglian Regional Genetics Service/Science Photo Library/Photo Researchers.]*

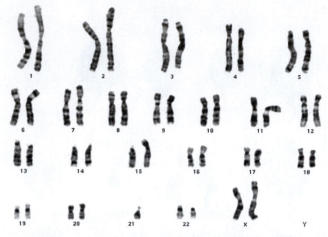

9.22 The translocation of chromosome 21 onto another chromosome results in familial Down syndrome. Here, the long arm of chromosome 21 is attached to chromosome 15. This karyotype is from a translocation carrier, who is phenotypically normal but is at increased risk for producing children with Down syndrome. *[Dr. Dorothy Warburton, HICCC, Columbia University.]*

Familial Down syndrome arises in offspring whose parents are carriers of chromosomes that have undergone a Robertsonian translocation, most commonly between chromosome 21 and chromosome 14: the long arm of 21 and the short arm of 14 exchange places. This exchange produces a chromosome that includes the long arms of chromosomes 14 and 21, and a very small chromosome that consists of the short arms of chromosomes 21 and 14. The small chromosome is generally lost after several cell divisions.

Persons with the translocation, called **translocation carriers,** do not have Down syndrome. Although they possess only 45 chromosomes, their phenotypes are normal because they have two copies of the long arms of chromosomes 14 and 21, and apparently the short arms of these chromosomes (which are lost) carry no essential genetic information. Although translocation carriers are completely healthy, they have an increased chance of producing children with Down syndrome.

When a translocation carrier produces gametes, the translocation chromosome may segregate in three different ways. First, it may separate from the normal chromosomes 14 and 21 in anaphase I of meiosis (**Figure 9.23a**). In this type of segregation, half of the gametes will have the translocation chromosome and no other copies of chromosomes 21 and 14; the fusion of such a gamete with a normal gamete will give rise to a translocation carrier. The other half of the gametes produced by this first type of segregation will be normal, each with a single copy of chromosomes 21 and 14, and will result in normal offspring.

Alternatively, the translocation chromosome may separate from chromosome 14 and pass into the same cell with the normal chromosome 21 (**Figure 9.23b**). This type of segregation produces abnormal gametes only; half will have two functional copies of chromosome 21 (one normal and one attached to chromosome 14) and the other half will lack chromosome 21. The gametes with the two functional copies of chromosome 21 will produce children with familial Down syndrome; the gametes lacking chromosome 21 will result in zygotes with monosomy 21 and will be spontaneously aborted.

In the third type of segregation, the translocation chromosome and the normal copy of chromosome 14 segregate together, and the normal chromosome 21 segregates by itself (**Figure 9.23c**). This pattern is presumably rare, because the two centromeres are both derived from chromosome 14 and usually separate from each other. In any case, all the gametes produced by this process are abnormal: half result in monosomy 14 and the other half result in trisomy 14—all are spontaneously aborted. Thus, only three of the six types of gametes that can be produced by a translocation carrier will result in the birth of a baby and, theoretically, these gametes should arise with equal frequency. One-third of the offspring of a translocation carrier should be translocation carriers like their parent, one-third should have familial Down syndrome, and one-third should be normal. In reality, however, fewer than one-third of the children born to translocation carriers have Down syndrome, which suggests that some of the embryos with Down syndrome are spontaneously aborted.

Few autosomal aneuploids besides trisomy 21 result in human live births. **Trisomy 18,** also known as **Edward syndrome,** arises with a frequency of approximately 1 in 8000 live births. Babies with Edward syndrome are severely retarded

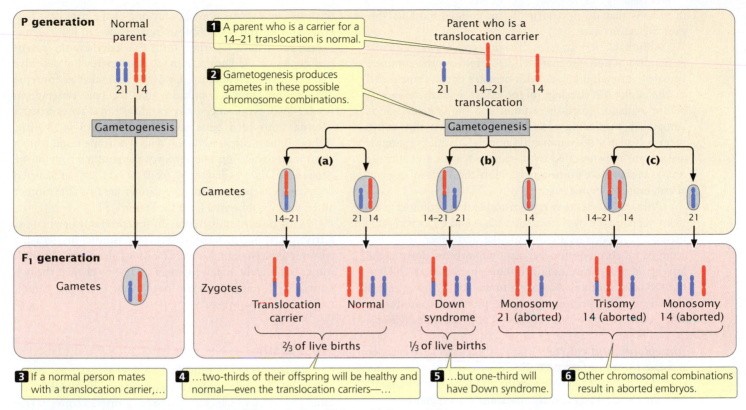

9.23 Translocation carriers are at increased risk for producing children with Down syndrome.

and have low-set ears, a short neck, deformed feet, clenched fingers, heart problems, and other disabilities. Few live for more than a year after birth. **Trisomy 13** has a frequency of about 1 in 15,000 live births and produces features that are collectively known as **Patau syndrome.** Characteristics of this condition include severe mental retardation, a small head, sloping forehead, small eyes, cleft lip and palate, extra fingers and toes, and numerous other problems. About half of children with trisomy 13 die within the first month of life, and 95% die by the age of 3. Rarer still is **trisomy 8,** which arises with a frequency ranging from about 1 in 25,000 to 1 in 50,000 live births. This aneuploid is characterized by mental retardation, contracted fingers and toes, low-set malformed ears, and a prominent forehead. Many who have this condition have normal life expectancy.

Aneuploidy and maternal age   Most cases of Down syndrome and other types of aneuploidy in humans arise from maternal nondisjunction, and the frequency of aneuploidy correlates with maternal age (**Figure 9.24**). Why maternal age is associated with nondisjunction is not known for certain, but the results of recent studies indicate a strong correlation between nondisjunction and aberrant meiotic recombination. Most chromosomes that failed to separate in meiosis I do not show any evidence of having recombined with one another. Conversely, chromosomes that failed to separate in meiosis II often show evidence of recombination

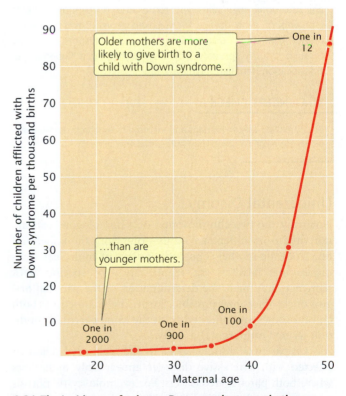

9.24 The incidence of primary Down syndrome and other aneuploids increases with maternal age.

in regions that do not normally recombine, most notably near the centromere.

Although aberrant recombination appears to play a role in nondisjunction, the maternal-age effect is more complex. Female mammals are born with primary oocytes suspended in the diplotene substage of prophase I of meiosis. Just before ovulation, meiosis resumes and the first division is completed, producing a secondary oocyte. At this point, meiosis is suspended again and remains so until the secondary oocyte is penetrated by a sperm. The second meiotic division takes place immediately before the nuclei of egg and sperm unite to form a zygote.

Primary oocytes may remain suspended in diplotene for many years before ovulation takes place and meiosis recommences. Components of the spindle and other structures required for chromosome segregation may break down in the long arrest of meiosis, leading to more aneuploidy in children born to older mothers. According to this theory, no age effect is seen in males, because sperm are produced continuously after puberty with no long suspension of the meiotic divisions.

Aneuploidy and cancer Many tumor cells have extra or missing chromosomes or both; some types of tumors are consistently associated with specific chromosome mutations, including aneuploidy and chromosome rearrangements.

CONCEPTS

In humans, sex-chromosome aneuploids are more common than are autosomal aneuploids. X-chromosome inactivation prevents problems of gene dosage for X-linked genes. Down syndrome results from three functional copies of chromosome 21, either through trisomy (primary Down syndrome) or a Robertsonian translocation (familial Down syndrome).

✔ CONCEPT CHECK 6

Briefly explain why, in humans and mammals, sex-chromosome aneuploids are more common than autosomal aneuploids?

Uniparental Disomy

Normally, the two chromosomes of a homologous pair are inherited from different parents—one from the father and one from the mother. The development of molecular techniques that facilitate the identification of specific DNA sequences has made the determination of the parental origins of chromosomes possible. Surprisingly, sometimes both chromosomes are inherited from the same parent, a condition termed **uniparental disomy.**

Uniparental disomy violates the rule that children affected with a recessive disorder appear only in families where both parents are carriers. For example, cystic fibrosis is an autosomal recessive disease; typically, both parents of an affected child are heterozygous for the cystic fibrosis mutation on chromosome 7. However, for a small proportion

of people with cystic fibrosis, only one of the parents is heterozygous for the cystic fibrosis gene. How can it be? These people must have inherited from the heterozygous parent two copies of the chromosome 7 that carries the defective cystic fibrosis allele and no copy of the normal allele from the other parent. Uniparental disomy has also been observed in Prader–Willi syndrome, a rare condition that arises when a paternal copy of a gene on chromosome 15 is missing. Although most cases of Prader–Willi syndrome result from a chromosome deletion that removes the paternal copy of the gene (see p. 120 in Chapter 5), from 20% to 30% arise when both copies of chromosome 15 are inherited from the mother and no copy is inherited from the father.

Many cases of uniparental disomy probably originate as a trisomy. Although most autosomal trisomies are lethal, a trisomic embryo can survive if one of the three chromosomes is lost early in development. If, just by chance, the two remaining chromosomes are both from the same parent, uniparental disomy results.

Mosaicism

Nondisjunction in a mitotic division may generate patches of cells in which every cell has a chromosome abnormality and other patches in which every cell has a normal karyotype. This type of nondisjunction leads to regions of tissue with different chromosome constitutions, a condition known as **mosaicism.** Growing evidence suggests that mosaicism is common.

Only about 50% of those diagnosed with Turner syndrome have the 45,X karyotype (presence of a single X chromosome) in all their cells; most others are mosaics, possessing some 45,X cells and some normal 46,XX cells. A few may even be mosaics for two or more types of abnormal karyotypes. The 45,X/46,XX mosaic usually arises when an X chromosome is lost soon after fertilization in an XX embryo.

Fruit flies that are XX/XO mosaics (O designates the absence of a homologous chromosome; XO means the cell has a single X chromosome and no Y chromosome) develop a mixture of male and female traits, because the presence of two X chromosomes in fruit flies produces female traits and the presence of a single X chromosome produces male traits (**Figure 9.25**). In fruit flies, sex is determined independently in each cell in the course of development. Those cells that are XX express female traits; those that are XO express male traits. Such sexual mosaics are called **gynandromorphs.** Normally, X-linked recessive genes are masked in heterozygous females but, in XX/XO mosaics, any X-linked recessive genes present in the cells with a single X chromosome will be expressed.

CONCEPTS

In uniparental disomy, an individual has two copies of a chromosome from one parent and no copy from the other. Uniparental disomy may arise when a trisomic embryo loses one of the triplicate chromosomes early in development. In mosaicism, different cells within the same individual have different chromosome constitutions.

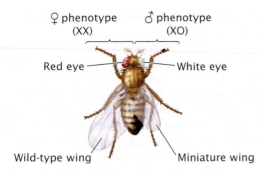

♀ phenotype (XX)   ♂ phenotype (XO)

Red eye —   — White eye

Wild-type wing   Miniature wing

9.25 Mosaicism for the sex chromosomes produces a gynandromorph. This XX/XO gynandromorph fruit fly carries one wild-type X chromosome and one X chromosome with recessive alleles for white eyes and miniature wings. The left side of the fly has a normal female phenotype, because the cells are XX and the recessive alleles on one X chromosome are masked by the presence of wild-type alleles on the other. The right side of the fly has a male phenotype with white eyes and miniature wing, because the cells are missing the wild-type X chromosome (are XO), allowing the white and miniature alleles to be expressed.

9.4 Polyploidy Is the Presence of More Than Two Sets of Chromosomes

Most eukaryotic organisms are diploid ($2n$) for most of their life cycles, possessing two sets of chromosomes. Occasionally, whole sets of chromosomes fail to separate in meiosis or mitosis, leading to polyploidy, the presence of more than two genomic sets of chromosomes. Polyploids include triploids ($3n$), tetraploids ($4n$), pentaploids ($5n$), and even higher numbers of chromosome sets.

Polyploidy is common in plants and is a major mechanism by which new plant species have evolved. Approximately 40% of all flowering-plant species and from 70% to 80% of grasses are polyploids. They include a number of agriculturally important plants, such as wheat, oats, cotton, potatoes, and sugar cane. Polyploidy is less common in animals but is found in some invertebrates, fishes, salamanders, frogs, and lizards. No naturally occurring, viable polyploids are known in birds, but at least one polyploid mammal—a rat in Argentina—has been reported.

We will consider two major types of polyploidy: **autopolyploidy,** in which all chromosome sets are from a single species; and **allopolyploidy,** in which chromosome sets are from two or more species.

Autopolyploidy

Autopolyploidy occurs when accidents of meiosis or mitosis produce extra sets of chromosomes, all derived from a single species. Nondisjunction of all chromosomes in mitosis in an early $2n$ embryo, for example, doubles the chromosome number and produces an autotetraploid ($4n$) (**Figure 9.26a**). An autotriploid ($3n$) may arise when nondisjunction in meiosis produces a diploid gamete that then fuses with a normal haploid gamete to produce a triploid zygote (**Figure 9.26b**). Alternatively, triploids may arise from a cross between an

(a) Autopolyploidy through mitosis

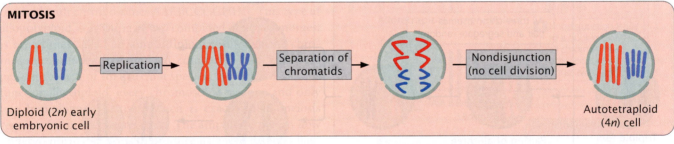

MITOSIS

Diploid ($2n$) early embryonic cell → Replication → Separation of chromatids → Nondisjunction (no cell division) → Autotetraploid ($4n$) cell

(b) Autopolyploidy through meiosis

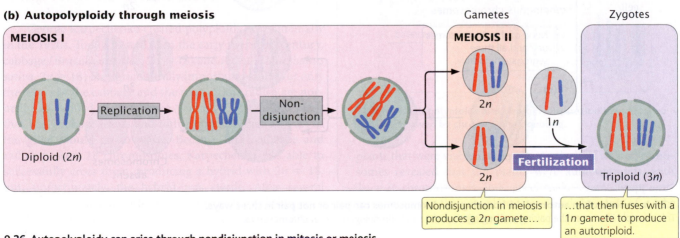

MEIOSIS I   Gametes   Zygotes

Diploid ($2n$) → Replication → Nondisjunction → MEIOSIS II → $2n$ / $2n$ / $1n$ → Fertilization → Triploid ($3n$)

Nondisjunction in meiosis I produces a $2n$ gamete… …that then fuses with a $1n$ gamete to produce an autotriploid.

9.26 Autopolyploidy can arise through nondisjunction in mitosis or meiosis.

Worked Problem

Species I has $2n = 14$ and species II has $2n = 20$. Give all possible chromosome numbers that may be found in the following individuals.

a. An autotriploid of species I

b. An autotetraploid of species II

c. An allotriploid formed from species I and species II

d. An allotetraploid formed from species I and species II

• Solution

The haploid number of chromosomes (n) for species I is 7 and for species II is 10.

a. A triploid individual is $3n$. A common mistake is to assume that $3n$ means three times as many chromosomes as in a normal individual, but remember that normal individuals are $2n$. Because n for species I is 7 and all genomes of an autopolyploid are from the same species, $3n = 3 \times 7 = 21$.

b. A autotetraploid is $4n$ with all genomes from the same species. The n for species II is 10, so $4n = 4 \times 10 = 40$.

c. A triploid individual is $3n$. By definition, an allopolyploid must have genomes from two different species. An allotriploid could have $1n$ from species I and $2n$ from species II or $(1 \times 7) + (2 \times 10) = 27$. Alternatively, it might have $2n$ from species I and $1n$ from species II, or $(2 \times 7) + (1 \times 10) = 24$. Thus, the number of chromosomes in an allotriploid could be 24 or 27.

d. A tetraploid is $4n$. By definition, an allotetraploid must have genomes from at least two different species. An allotetraploid could have $3n$ from species I and $1n$ from species II or $(3 \times 7) + (1 \times 10) = 31$; or $2n$ from species I and $2n$ from species II or $(2 \times 7) + (2 \times 10) = 34$; or $1n$ from species I and $3n$ from species II or $(1 \times 7) + (3 \times 10) = 37$. Thus, the number of chromosomes could be 31, 34, or 37.

> ❓ For additional practice, try Problem 36 at the end of this chapter.

The Significance of Polyploidy

In many organisms, cell volume is correlated with nuclear volume, which, in turn, is determined by genome size. Thus, the increase in chromosome number in polyploidy is often associated with an increase in cell size, and many polyploids are physically larger than diploids. Breeders have used this effect to produce plants with larger leaves, flowers, fruits, and seeds. The hexaploid ($6n = 42$) genome of wheat probably contains chromosomes derived from three different

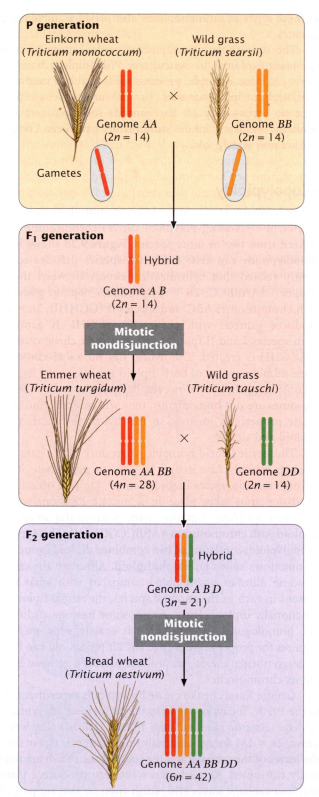

9.29 Modern bread wheat, *Triticum aestivum*, is a hexaploid with genes derived from three different species. Two diploid species, *T. monococcum* ($n = 14$) and probably *T. searsii* ($n = 14$), originally crossed to produce a diploid hybrid ($2n = 14$) that underwent chromosome doubling to create *T. turgidum* ($4n = 28$). A cross between *T. turgidum* and *T. tauschi* ($2n = 14$) produced a triploid hybrid ($3n = 21$) that then underwent chromosome doubling to produce *T. aestivum*, which is a hexaploid ($6n = 42$).

wild species (**Figure 9.29**). Many other cultivated plants also are polyploid (**Table 9.2**).

Polyploidy is less common in animals than in plants for several reasons. As discussed, allopolyploids require hybridization between different species, which happens less frequently in animals than in plants. Animal behavior often prevents interbreeding among species, and the complexity of animal development causes most interspecific hybrids to be nonviable. Many of the polyploid animals that do arise are in groups that reproduce through parthenogenesis (a type of reproduction in which the animal develops from an unfertilized egg). Thus, asexual reproduction may facilitate the development of polyploids, perhaps because the perpetuation of hybrids through asexual reproduction provides greater opportunities for nondisjunction than does sexual reproduction. Only a few human polyploid babies have been reported, and most died within a few days of birth. Polyploidy—usually triploidy—is seen in about 10% of all spontaneously aborted human fetuses. Different types of chromosome mutations are summarized in **Table 9.3**.

Table 9.2   Examples of polyploid crop plants

Plant	Type of Polyploidy	Ploidy	Chromosome Number
Potato	Autopolyploid	$4n$	48
Banana	Autopolyploid	$3n$	33
Peanut	Autopolyploid	$4n$	40
Sweet potato	Autopolyploid	$6n$	90
Tobacco	Allopolyploid	$4n$	48
Cotton	Allopolyploid	$4n$	52
Wheat	Allopolyploid	$6n$	42
Oats	Allopolyploid	$6n$	42
Sugar cane	Allopolyploid	$8n$	80
Strawberry	Allopolyploid	$8n$	56

Source: After F. C. Elliot, *Plant Breeding and Cytogenetics* (New York: McGraw-Hill, 1958).

Table 9.3   Different types of chromosome mutations

Chromosome Mutation	Definition
Chromosome rearrangement	Change in chromosome structure
Chromosome duplication	Duplication of a chromosome segment
Chromosome deletion	Deletion of a chromosome segment
Inversion	Chromosome segment inverted 180 degrees
Paracentric inversion	Inversion that does not include the centromere in the inverted region
Pericentric inversion	Inversion that includes the centromere in the inverted region
Translocation	Movement of a chromosome segment to a nonhomologous chromosome or to another region of the same chromosome
Nonreciprocal translocation	Movement of a chromosome segment to a nonhomologous chromosome or to another region of the same chromosome without reciprocal exchange
Reciprocal translocation	Exchange between segments of nonhomologous chromosomes or between regions of the same chromosome
Aneuploidy	Change in number of individual chromosomes
Nullisomy	Loss of both members of a homologous pair
Monosomy	Loss of one member of a homologous pair
Trisomy	Gain of one chromosome, resulting in three homologous chromosomes
Tetrasomy	Gain of two homologous chromosomes, resulting in four homologous chromosomes
Polyploidy	Addition of entire chromosome sets
Autopolyploidy	Polyploidy in which extra chromosome sets are derived from the same species
Allopolyploidy	Polyploidy in which extra chromosome sets are derived from two or more species

CONCEPTS

Polyploidy is the presence of extra chromosome sets: autopolyploids possess extra chromosome sets from the same species; allopolyploids possess extra chromosome sets from two or more species. Problems in chromosome pairing and segregation often lead to sterility in autopolyploids, but many allopolyploids are fertile.

✔ **CONCEPT CHECK 7**

Species A has $2n = 16$ chromosomes and species B has $2n = 14$. How many chromosomes would be found in an allotriploid of these two species?

a. 21 or 24   c. 22 or 23
b. 42 or 48   d. 45

9.5 Chromosome Variation Plays an Important Role in Evolution

Chromosome variations are potentially important in evolution and, within a number of different groups of organisms, have clearly played a significant role in past evolution. Chromosome duplications provide one way in which new genes may evolve. In many cases, existing copies of a gene are not free to vary, because they encode a product that is essential to development or function. However, after a chromosome undergoes duplication, extra copies of genes within the duplicated region are present. The original copy can provide the essential function while an extra copy from the duplication is free to undergo mutation and change. Over evolutionary time, the extra copy may acquire enough mutations to assume a new function that benefits the organism.

Inversions also can play important evolutionary roles by suppressing recombination among a set of genes. As we have seen, crossing over within an inversion in an individual heterozygous for a pericentric or paracentric inversion leads to unbalanced gametes and no recombinant progeny. This suppression of recombination allows particular sets of co-adapted alleles that function well together to remain intact, unshuffled by recombination.

Polyploidy, particularly allopolyploidy, often gives rise to new species and has been particularly important in the evolution of flowering plants. Occasional genome doubling through polyploidy has been a major contributor to evolutionary success in animal groups. For example, *Saccharomyces cerevisiae* (yeast) is a tetraploid, having undergone whole-genome duplication about 100 million years ago. The vertebrate genome has duplicated twice, once in the common ancestor to jawed vertebrates and again in the ancestor of fishes. Certain groups of vertebrates, such as some frogs and some fishes, have undergone additional polyploidy. Cereal plants have undergone several genome duplication events.

CONCEPTS SUMMARY

- Three basic types of chromosome mutations are: (1) chromosome rearrangements, which are changes in the structure of chromosomes; (2) aneuploidy, which is an increase or decrease in chromosome number; and (3) polyploidy, which is the presence of extra chromosome sets.

- Chromosome rearrangements include duplications, deletions, inversions, and translocations.

- In individuals heterozygous for a duplication, the duplicated region will form a loop when homologous chromosomes pair in meiosis. Duplications often have pronounced effects on the phenotype owing to unbalanced gene dosage.

- In individuals heterozygous for a deletion, one of the chromosomes will loop out during pairing in meiosis. Deletions may cause recessive alleles to be expressed.

- Pericentric inversions include the centromere; paracentric inversions do not. In individuals heterozygous for an inversion, the homologous chromosomes form inversion loops in meiosis, with reduced recombination taking place within the inverted region.

- In translocation heterozygotes, the chromosomes form crosslike structures in meiosis, and the segregation of chromosomes produces unbalanced gametes.

- Fragile sites are constrictions or gaps that appear at particular regions on the chromosomes of cells grown in culture and are prone to breakage under certain conditions.

- Nullisomy is the loss of two homologous chromosomes; monosomy is the loss of one homologous chromosome; trisomy is the addition of one homologous chromosome; tetrasomy is the addition of two homologous chromosomes.

- Aneuploidy usually causes drastic phenotypic effects because it leads to unbalanced gene dosage.

- Primary Down syndrome is caused by the presence of three full copies of chromosome 21, whereas familial Down syndrome is caused by the presence of two normal copies of chromosome 21 and a third copy that is attached to another chromosome through a translocation.

- Uniparental disomy is the presence of two copies of a chromosome from one parent and no copy from the other. Mosaicism is caused by nondisjunction in an early mitotic division that leads to different chromosome constitutions in different cells of a single individual.

- All the chromosomes in an autopolyploid derive from one species; chromosomes in an allopolyploid come from two or more species.

- Chromosome variations have played an important role in the evolution of many groups of organisms.

IMPORTANT TERMS

chromosome mutation (p. 238)
metacentric chromosome (p.238)
submetacentric chromosome (p. 238)
acrocentric chromosome (p. 238)
telocentric chromosome (p. 238)
chromosome rearrangement (p. 240)
chromosome duplication (p. 240)
tandem duplication (p. 240)
displaced duplication (p. 240)
reverse duplication (p. 240)
chromosome deletion (p. 242)
pseudodominance (p. 242)
haploinsufficient gene (p. 242)
chromosome inversion (p. 244)
paracentric inversion (p. 244)
pericentric inversion (p. 244)

position effect (p. 244)
dicentric chromatid (p. 244)
acentric chromatid (p. 244)
dicentric bridge (p. 245)
translocation (p. 246)
nonreciprocal translocation (p. 246)
reciprocal translocation (p. 246)
Robertsonian translocation (p. 247)
alternate segregation (p. 247)
adjacent-1 segregation (p. 247)
adjacent-2 segregation (p. 247)
fragile site (p. 247)
aneuploidy (p. 249)
polyploidy (p. 249)
nullisomy (p. 249)
monosomy (p. 249)

trisomy (p. 249)
tetrasomy (p. 249)
Down syndrome (trisomy 21) (p. 251)
primary Down syndrome (p. 251)
familial Down syndrome (p. 251)
translocation carrier (p. 252)
Edward syndrome (trisomy 18) (p. 252)
Patau syndrome (trisomy 13) (p. 253)
trisomy 8 (p. 253)
uniparental disomy (p. 254)
mosaicism (p. 254)
gynandromorph (p. 254)
autopolyploidy (p. 255)
allopolyploidy (p. 255)
unbalanced gametes (p. 256)
amphidiploidy (p 257)

ANSWERS TO CONCEPT CHECKS

1. a
2. Pseudodominance is the expression of a recessive mutation. It is produced when the wild-type allele in a heterozygous individual is absent due to a deletion on one chromosome.
3. c
4. b
5. 37

6. Dosage compensation prevents the expression of additional copies of X-linked genes in mammals, and there is little information on the Y chromosome; so extra copies of the X and Y chromosomes do not have major effects on development. In contrast, there is no mechanism of dosage compensation for autosomes, and so extra copies of autosomal genes are expressed, upsetting development and causing the spontaneous abortion of aneuploid embryos.

7. c

WORKED PROBLEMS

1. A chromosome has the following segments, where • represents the centromere.

$$\underline{A \quad B \quad C \quad D \quad E \bullet F \quad G}$$

What types of chromosome mutations are required to change this chromosome into each of the following chromosomes? (In some cases, more than one chromosome mutation may be required.)

a. $\underline{A \quad B \quad E \bullet F \quad G}$
b. $\underline{A \quad E \quad D \quad C \quad B \bullet F \quad G}$
c. $\underline{A \quad B \quad A \quad B \quad C \quad D \quad E \bullet F \quad G}$
d. $\underline{A \quad F \bullet E \quad D \quad C \quad B \quad G}$
e. $\underline{A \quad B \quad C \quad D \quad E \quad E \quad D \quad C \bullet F \quad G}$

• **Solution**

The types of chromosome mutations are identified by comparing the mutated chromosome with the original, wild-type chromosome.

a. The mutated chromosome ($\underline{A \quad B \quad E \bullet F \quad G}$) is missing segment $\underline{C \quad D}$; so this mutation is a deletion.

b. The mutated chromosome ($\underline{A \quad E \quad D \quad C \quad B \bullet F \quad G}$) has one and only one copy of all the gene segments, but segment $\underline{B \quad C \quad D \quad E}$ has been inverted 180 degrees. Because the centromere has not changed location and is not in the inverted region, this chromosome mutation is a paracentric inversion.

c. The mutated chromosome ($\underline{A \quad B \quad A \quad B \quad C \quad D \quad E \bullet F \quad G}$) is longer than normal, and we see that segment $\underline{A \quad B}$ has been duplicated. This mutation is a tandem duplication.

d. The mutated chromosome ($\underline{A \quad F \bullet E \quad D \quad C \quad B \quad G}$) is normal length, but the gene order and the location of the centromere have changed; this mutation is therefore a pericentric inversion of region ($\underline{B \quad C \quad D \quad E \bullet F}$).

e. The mutated chromosome
(A B C D E E D C • F G) contains a
duplication (C D E) that is also inverted; so this
chromosome has undergone a duplication and a
paracentric inversion.

2. Species I is diploid ($2n = 4$) with chromosomes AABB;
related species II is diploid ($2n = 6$) with chromosomes
MMNNOO. Give the chromosomes that would be
found in individuals with the following chromosome
mutations.

   a. Autotriploidy in species I

   b. Allotetraploidy including species I and II

   c. Monosomy in species I

   d. Trisomy in species II for chromosome M

   e. Tetrasomy in species I for chromosome A

   f. Allotriploidy including species I and II

   g. Nullisomy in species II for chromosome N

• Solution

To work this problem, we should first determine the haploid
genome complement for each species. For species I, $n = 2$ with
chromosomes AB and, for species II, $n = 3$ with chromosomes
MNO.

   a. An autotriploid is $3n$, with all the chromosomes coming
from a single species; so an autotriploid of species I
would have chromosomes AAABBB ($3n = 6$).

b. An allotetraploid is $4n$, with the chromosomes coming
from more than one species. An allotetraploid could
consist of $2n$ from species I and $2n$ from species II,
giving the allotetraploid ($4n = 2 + 2 + 3 + 3 = 10$)
chromosomes AABBMMNNOO. An allotetraploid
could also possess $3n$ from species I and $1n$ from species
II ($4n = 2 + 2 + 2 + 3 = 9$; AAABBBMNO) or $1n$
from species I and $3n$ from species II ($4n = 2 + 3 + 3
+ 3 = 11$; ABMMMNNNOOO).

c. A monosomic is missing a single chromosome; so a
monosomic for species I would be $2n - 1 = 4 - 1 = 3$.
The monosomy might include either of the two
chromosome pairs, with chromosomes ABB or
AAB.

d. Trisomy requires an extra chromosome; so a trisomic of
species II for chromosome M would be $2n + 1 = 6 + 1
= 7$ (MMMNNOO).

e. A tetrasomic has two extra homologous chromosomes;
so a tetrasomic of species I for chromosome A would be
$2n + 2 = 4 + 2 = 6$ (AAAABB).

f. An allotriploid is $3n$ with the chromosomes coming
from two different species; so an allotriploid could be $3n$
$= 2 + 2 + 3 = 7$ (AABBMNO) or $3n = 2 + 3 + 3 = 8$
(ABMMNNOO).

g. A nullisomic is missing both chromosomes of a
homologous pair; so a nullisomic of species II for
chromosome N would be $2n - 2 = 6 - 2 = 4$
(MMOO).

COMPREHENSION QUESTIONS

Section 9.1

*1. List the different types of chromosome mutations and
define each one.

Section 9.2

*2. Why do extra copies of genes sometimes cause drastic
phenotypic effects?

3. Draw a pair of chromosomes as they would appear during
synapsis in prophase I of meiosis in an individual
heterozygous for a chromosome duplication.

4. What is haploinsufficiency?

*5. What is the difference between a paracentric and a
pericentric inversion?

6. How do inversions cause phenotypic effects?

7. Explain, with the aid of a drawing, how a dicentric bridge is
produced when crossing over takes place in an individual
heterozygous for a paracentric inversion.

8. Explain why recombination is suppressed in individuals
heterozygous for paracentric and pericentric inversions.

*9. How do translocations produce phenotypic effects?

10. Sketch the chromosome pairing and the different
segregation patterns that can arise in an individual
heterozygous for a reciprocal translocation.

11. What is a Robertsonian translocation?

Section 9.3

12. List four major types of aneuploidy.

*13. What is the difference between primary Down syndrome
and familial Down syndrome? How does each type arise?

*14. What is uniparental disomy and how does it arise?

15. What is mosaicism and how does it arise?

Section 9.4

*16. What is the difference between autopolyploidy and
allopolyploidy? How does each arise?

17. Explain why autopolyploids are usually sterile, whereas
allopolyploids are often fertile.

APPLICATION QUESTIONS AND PROBLEMS

Section 9.1

*18. Which types of chromosome mutations

   a. increase the amount of genetic material in a particular chromosome?

   b. increase the amount of genetic material in all chromosomes?

   c. decrease the amount of genetic material in a particular chromosome?

   d. change the position of DNA sequences in a single chromosome without changing the amount of genetic material?

   e. move DNA from one chromosome to a nonhomologous chromosome?

Section 9.2

*19. A chromosome has the following segments, where • represents the centromere:

   A B • C D E F G

   What types of chromosome mutations are required to change this chromosome into each of the following chromosomes? (In some cases, more than one chromosome mutation may be required.)

   a. A B A B • C D E F G
   b. A B • C D E A B F G
   c. A B • C F E D G
   d. A • C D E F G
   e. A B • C D E
   f. A B • E D C F G
   g. C • B A D E F G
   h. A B • C F E D F E D G
   i. A B • C D E F C D F E G

20. A chromosome initially has the following segments:

   A B • C D E F G

   Draw the chromosome, identifying its segments, that would result from each of the following mutations.

   a. Tandem duplication of DEF

   b. Displaced duplication of DEF

   c. Deletion of FG

   d. Paracentric inversion that includes DEFG

   e. Pericentric inversion of BCDE

21. The following diagram represents two nonhomologous chromosomes:

   A B • C D E F G
   R S • T U V W X

What type of chromosome mutation would produce each of the following chromosomes?

   a. A B • C D
      R S • T U V W X E F G
   b. A U V B • C D E F G
      R S • T W X
   c. A B • T U V F G
      R S • C D E W X
   d. A B • C W G
      R S • T U V D E F X

*22. The *Notch* mutation is a deletion on the X chromosome of *Drosophila melanogaster.* Female flies heterozygous for *Notch* have an indentation on the margins of their wings; *Notch* is lethal in the homozygous and hemizygous conditions. The *Notch* deletion covers the region of the X chromosome that contains the locus for white eyes, an X-linked recessive trait. Give the phenotypes and proportions of progeny produced in the following crosses.

   a. A red-eyed, Notch female is mated with a white-eyed male.

   b. A white-eyed, Notch female is mated with a red-eyed male.

   c. A white-eyed, Notch female is mated with a white-eyed male.

23. The green-nose fly normally has six chromosomes, two metacentric and four acrocentric. A geneticist examines the chromosomes of an odd-looking green-nose fly and discovers that it has only five chromosomes; three of them are metacentric and two are acrocentric. Explain how this change in chromosome number might have taken place.

*24. A wild-type chromosome has the following segments:

   A B C • D E F G H I

   An individual is heterozygous for the following chromosome mutations. For each mutation, sketch how the wild-type and mutated chromosomes would pair in prophase I of meiosis, showing all chromosome strands.

   a. A B C • D E F D E F G H I
   b. A B C • D H I
   c. A B C • D G F E H I
   d. A B E D • C F G H I

25. Draw the chromatids that would result from a two-strand double crossover between E and F in Problem 24.

26. **DATA ANALYSIS** As discussed in this chapter, crossing over within a pericentric inversion produces chromosomes that have extra copies of some genes and no copies of other genes. The fertilization of gametes containing such

duplication/deficient chromosomes often results in children with syndromes characterized by developmental delay, mental retardation, abnormal development of organ systems, and early death. Using a special two-color FISH analysis that revealed the presence of crossing over within pericentric inversions, Maarit Jaarola and colleagues examined individual sperm cells of a male who was heterozygous for a pericentric inversion on chromosome 8 and determined that crossing over took place within the pericentric inversion in 26% of the meiotic divisions (M. Jaarola, R. H. Martin, and T. Ashley. 1998. *American Journal of Human Genetics* 63:218–224).

Assume that you are a genetic counselor and that a couple seeks genetic counseling from you. Both the man and the woman are phenotypically normal, but the woman is heterozygous for a pericentric inversion on chromosome 8. The man is karyotypically normal. What is the probability that this couple will produce a child with a debilitating syndrome as the result of crossing over within the pericentric inversion?

*27. An individual heterozygous for a reciprocal translocation possesses the following chromosomes:

A B • C D E F G

A B • C D V W X

R S • T U E F G

R S • T U V W X

a. Draw the pairing arrangement of these chromosomes in prophase I of meiosis.

b. Diagram the alternate, adjacent-1, and adjacent-2 segregation patterns in anaphase I of meiosis.

c. Give the products that result from alternate, adjacent-1, and adjacent-2 segregation.

Section 9.33

28. Red–green color blindness is a human X-linked recessive disorder. A young man with a 47,XXY karyotype (Klinefelter syndrome) is color blind. His 46,XY brother also is color blind. Both parents have normal color vision. Where did the nondisjunction occur that gave rise to the young man with Klinefelter syndrome?

29. [DATA ANALYSIS] Junctional epidermolysis bullosa (JEB) is a severe skin disorder that results in blisters over the entire body. The disorder is caused by autosomal recessive mutations at any one of three loci that help to encode laminin 5, a major component in the dermal–epidermal basement membrane. Leena Pulkkinen and colleagues described a male newborn who was born with JEB and died at 2 months of age (L. Pulkkinen et al. 1997. *American Journal of Human Genetics* 61:611–619); the child had

healthy unrelated parents. Chromosome analysis revealed that the infant had 46 normal-appearing chromosomes. Analysis of DNA showed that his mother was heterozygous for a JEB-causing allele at the *LAMB3* locus, which is on chromosome 1. The father had two normal alleles at this locus. DNA fingerprinting demonstrated that the male assumed to be the father had, in fact, conceived the child.

a. Assuming that no new mutations occurred in this family, explain the presence of an autosomal recessive disease in the child when the mother is heterozygous and the father is homozygous normal.

b. How might you go about proving your explanation? Assume that a number of genetic markers are available for each chromosome.

30. Some people with Turner syndrome are 45,X/46,XY mosaics. Explain how this mosaicism could arise.

*31. Bill and Betty have had two children with Down syndrome. Bill's brother has Down syndrome and his sister has two children with Down syndrome. On the basis of these observations, which of the following statements is most likely correct? Explain your reasoning.

a. Bill has 47 chromosomes.

b. Betty has 47 chromosomes.

c. Bill and Betty's children each have 47 chromosomes.

d. Bill's sister has 45 chromosomes.

e. Bill has 46 chromosomes.

f. Betty has 45 chromosomes.

g. Bill's brother has 45 chromosomes.

*32. In mammals, sex-chromosome aneuploids are more common than autosomal aneuploids but, in fish, sex-chromosome aneuploids and autosomal aneuploids are found with equal frequency. Offer an explanation for these differences in mammals and fish.

*33. A young couple is planning to have children. Knowing that there have been a substantial number of stillbirths, miscarriages, and fertility problems on the husband's side of the family, they see a genetic counselor. A chromosome analysis reveals that, whereas the woman has a normal karyotype, the man possesses only 45 chromosomes and is a carrier of a Robertsonian translocation between chromosomes 22 and 13.

a. List all the different types of gametes that might be produced by the man.

b. What types of zygotes will develop when each of gametes produced by the man fuses with a normal gamete produced by the woman?

c. If trisomies and monosomies entailing chromosomes 13 and 22 are lethal, what proportion of the surviving offspring will be carriers of the translocation?

Section 9.4

*34. Species I has $2n = 16$ chromosomes. How many chromosomes will be found per cell in each of the following mutants in this species?

a. Monosomic   e. Double monosomic

b. Autotriploid   f. Nullisomic

c. Autotetraploid   g. Autopentaploid

d. Trisomic   h. Tetrasomic

35. Species I is diploid ($2n = 8$) with chromosomes AABBCCDD; related species II is diploid ($2n = 8$) with chromosomes MMNNOOPP. What types of chromosome mutations do individual organisms with the following sets of chromosomes have?

a. AAABBCCDD   e. AAABBCCDDD

b. MMNNOOOOPP   f. AABBDD

c. AABBCDD   g. AABBCCDDMMNNOOPP

d. AAABBBCCCDDD   h. AABBCCDDMNOP

36. Species I has $2n = 8$ chromosomes and species II has $2n = 14$ chromosomes. What would be the expected chromosome numbers in individual organisms with the following chromosome mutations? Give all possible answers.

a. Allotriploidy including species I and II

b. Autotetraploidy in species II

c. Trisomy in species I

d. Monosomy in species II

e. Tetrasomy in species I

f. Allotetraploidy including species I and II

37. Consider a diploid cell that has $2n = 4$ chromosomes—one pair of metacentric chromosomes and one pair of acrocentric chromosomes. Suppose this cell undergoes nondisjunction giving rise to an autotriploid cell ($3n$). The triploid cell then undergoes meiosis. Draw the different types of gametes that may result from meiosis in the triploid cell, showing the chromosomes present in each type. To distinguish between the different metacentric and acrocentric chromosomes, use a different color to draw each metacentric chromosome; similarly, use a different color to draw each acrocentric chromosome. (Hint: See Figure 9.27).

38. *DATA ANALYSIS* *Nicotiana glutinosa* ($2n = 24$) and *N. tabacum* ($2n = 48$) are two closely related plant that can be intercrossed, but the F_1 hybrid plants that result are usually sterile. In 1925, Roy Clausen and Thomas Goodspeed crossed *N. glutinosa* and *N. tabacum*, and obtained one fertile F_1 plant (R. E. Clausen and T. H. Goodspeed. 1925 *Genetics* 10:278–284). They were able to self-pollinate the flowers of this plant to produce an F_2 generation. Surprisingly, the F_2 plants were fully fertile and produced viable seed. When Clausen and Goodspeed examined the chromosomes of the F_2 plants, they observed 36 pairs of chromosomes in metaphase I and 36 individual chromosomes in metaphase II. Explain the origin of the F_2 plants obtained by Clausen and Goodspeed and the numbers of chromosomes observed.

39. What would be the chromosome number of progeny resulting from the following crosses in wheat (see Figure 9.29)? What type of polyploid (allotriploid, allotetraploid, etc.) would result from each cross?

a. Einkorn wheat and Emmer wheat

b. Bread wheat and Emmer wheat

c. Einkorn wheat and bread wheat

40. *DATA ANALYSIS* Karl and Hally Sax crossed *Aegilops cylindrical* ($2n = 28$), a wild grass found in the Mediterranean region, with *Triticum vulgare* ($2n = 42$), a type of wheat (K. Sax and H. J. Sax. 1924. *Genetics* 9:454–464). The resulting F_1 plants from this cross had 35 chromosomes. Examination of metaphase I in the F_1 plants revealed the presence of 7 pairs of chromosomes (bivalents) and 21 unpaired chromosomes (univalents).

a. If the unpaired chromosomes segregate randomly, what possible chromosome numbers will appear in the gametes of the F_1 plants?

b. What does the appearance of the bivalents in the F_1 hybrids suggest about the origin of *Triticum vulgare* wheat?

CHALLENGE QUESTIONS

Section 9.3

41. Red–green color blindness is a human X-linked recessive disorder. Jill has normal color vision, but her father is color blind. Jill marries Tom, who also has normal color vision.

Jill and Tom have a daughter who has Turner syndrome and is color blind.

a. How did the daughter inherit color blindness?

b. Did the daughter inherit her X chromosome from Jill or from Tom?

42. *DATA ANALYSIS* Progeny of triploid tomato plants often contain parts of an extra chromosome, in addition to the normal complement of 24 chromosomes (J. W. Lesley and M. M. Lesley. 1929. *Genetics* 14:321–336). Mutants with a part of an extra chromosome are referred to as secondaries. James and Margaret Lesley observed that secondaries arise from triploid ($3n$), trisomic ($3n + 1$), and double trisomic ($3n + 1 + 1$) parents, but never from diploids ($2n$). Suggest one or more possible reasons that secondaries arise from parents that have unpaired chromosomes but not from parents that are normal diploids.

43. DATA ANALYSIS Mules result from a cross between a horse ($2n = 64$) and a donkey ($2n = 62$), have 63 chromosomes and are almost always sterile. However, in the summer of 1985, a female mule named Krause who was pastured with a male donkey gave birth to a newborn foal (O. A. Ryder et al. 1985. *Journal of Heredity* 76:379–381). Blood tests established that the male foal, appropriately named Blue Moon, was the offspring of Krause and that Krause was indeed a mule. Both Blue Moon and Krause were fathered by the same donkey (see the illustration). The foal, like his mother, had 63 chromosomes—half of them horse chromosomes and the other half donkey chromosomes. Analyses of genetic markers showed that, remarkably, Blue Moon seemed to have inherited a complete set of horse chromosomes from his mother, instead of a random mixture of horse and donkey chromosomes that would be expected with normal meiosis. Thus, Blue Moon and Krause were not only mother and son, but also brother and sister.

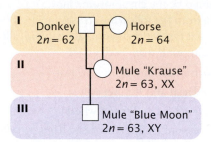

a. With the use of a diagram, show how, if Blue Moon inherited only horse chromosomes from his mother,

Blue Moon and Krause are both mother and son as well as brother and sister.

b. Although rare, additional cases of fertile mules giving births to offspring have been reported. In these cases, when a female mule mates with a male horse, the offspring is horselike in appearance but, when a female mule mates with a male donkey, the offspring is mulelike in appearance. Is this observation consistent with the idea that the offspring of fertile female mules inherit only a set of horse chromosomes from their mule mothers? Explain your reasoning.

c. Can you suggest a possible mechanism for how the offspring of fertile female mules might pass on a complete set of horse chromosomes to their offspring?

Section 9.5

44. Humans and many other complex organisms are diploid, possessing two sets of genes, one inherited from the mother and one from the father. However, a number of eukaryotic organisms spend most of their life cycles in a haploid state. Many of these eukaryotes, such as *Neurospora* and yeast, still undergo meiosis and sexual reproduction, but most of the cells that make up the organism are haploid.

Considering that haploid organisms are fully capable of sexual reproduction and generating genetic variation, why are most complex eukaryotes diploid? In other words, what might be the evolutionary advantage of existing in a diploid state instead of a haploid state? And why might a few organisms, such as *Neurospora* and yeast, exist as haploids?

Gene Mutations and DNA Repair

People with a mutation in the *tinman* gene, named after Tin Man in *The Wizard of Oz*, often have congenital heart defects. *[Mary Evans Picture Library/Alamy.]*

A FLY WITHOUT A HEART

The heart of a fruit fly is a simple organ, an open-ended tube that rhythmically contracts, pumping fluid—rather inefficiently—around the body of the fly. Although simple and inelegant, the fruit fly's heart is nevertheless essential, at least for a fruit fly. Remarkably, a few rare mutant fruit flies never develop a heart and die (not surprisingly) at an early embryonic stage. Geneticist Rolf Bodmer analyzed these mutants in the 1980s and made an important discovery—a gene that specifies the development of a heart. He named the gene *tinman*, after the character in *The Wizard of Oz* who also lacked a heart. Bodmer's research revealed that *tinman* encodes a transcription factor, which binds to DNA and turns on other genes that are essential for the normal development of a heart. In the mutant flies studied by Bodmer, this gene was lacking, the transcription factor was never produced, and the heart never developed. Findings from subsequent research revealed the existence of a human gene (called *Nkx2.5*) with a sequence similar to that of *tinman*, but the function of the human gene was unknown.

Then, in the 1990s, physicians Jonathan and Christine Seidman began studying people born with abnormal hearts, such as those with a hole in the septum that separates the chambers on the left and right sides of the heart. Such heart defects can cause abnormal blood flow through the heart, causing the heart to work harder than normal and the mixing of oxygenated and deoxyganted blood. Congenital heart defects are not uncommon; they're found in about 1 of every 125 babies. Some of the defects heal on their own, but others require corrective surgery. Although surgery is often successful in reversing congenital heart problems, many of these patients begin to have irregular heartbeats in their 20s and 30s. The Seidmans and their colleagues found several families in which congenital heart defects and irregular heart beats were inherited together in an autosomal dominant fashion. Detailed molecular analysis of one of these families revealed that the gene responsible for the heart problems was located on chromosome 5, at a spot where the human *tinman* gene (*Nkx2.5*) had been previously mapped. All members of this family who inherited the heart defects also inherited a mutation in the *tinman* gene. Subsequent studies with additional patients found that many people with congenital heart defects have a mutation in the *tinman* gene. The human version of this gene, like its counterpart in flies, encodes a transcription factor that controls heart development. Despite tremendous differences in size, anatomy, and physiology, humans and flies use the same gene to make a heart.

The story of *tinman* illustrates the central importance of studying mutations: the analysis of mutants is often a source of key insights into important biological processes. This chapter focuses on gene mutations—how errors arise in genetic instructions and how those errors are studied. We begin with a brief examination of the different types of mutations, including their phenotypic effects, how they can be suppressed, and mutation rates. The next section explores how mutations spontaneously arise in the course of replication and afterward, as well as how chemicals and radiation induce mutations. We then consider the analysis of mutations. Finally, we take a look at DNA repair and some of the diseases that arise when DNA repair is defective.

gous to figuring out how an automobile works by breaking different parts of a car and observing the effects; for example, smash the radiator and the engine overheats, revealing that the radiator cools the engine. Using mutations to disrupt function can likewise can be a source of insight into biological processes. For example, geneticists have begun to unravel the molecular details of development by studying mutations, such as *tinman*, that interrupt various embryonic stages in *Drosophila*. Although this method of breaking "parts" to determine their function might seem like a crude approach to understanding a system, it is actually very powerful and has been used extensively in biochemistry, developmental biology, physiology, and behavioral science (but this method is *not* recommended for learning how your car works).

10.1 Mutations Are Inherited Alterations in the DNA Sequence

DNA is a highly stable molecule that replicates with amazing accuracy, but changes in DNA structure and errors of replication do occur. A **mutation** is defined as an inherited change in genetic information; the descendants may be cells or organisms.

CONCEPTS

Mutations are heritable changes in DNA. They are essential to the study of genetics and are useful in many other biological fields.

✔ CONCEPT CHECK 1

How are mutations used to help in understanding basic biological processes?

The Importance of Mutations

Mutations are both the sustainer of life and the cause of great suffering. On the one hand, mutation is the source of all genetic variation, the raw material of evolution. On the other hand, many mutations have detrimental effects, and mutation is the source of many human diseases and disorders.

Much of the study of genetics focuses on how variants produced by mutation are inherited; genetic crosses are meaningless if all individual members of a species are identically homozygous for the same alleles. Much of Gregor Mendel's success in unraveling the principles of inheritance can be traced to his use of carefully selected variants of the garden pea; similarly, Thomas Hunt Morgan and his students discovered many basic principles of genetics by analyzing mutant fruit flies.

Mutations are also useful for probing fundamental biological processes. Finding or creating mutations that affect different components of a biological system and studying their effects can often lead to an understanding of the system. This method, referred to as genetic dissection, is analo-

Categories of Mutations

In multicellular organisms, we can distinguish between two broad categories of mutations: somatic mutations and germ-line mutations. **Somatic mutations** arise in somatic tissues, which do not produce gametes (**Figure 10.1**). When a somatic cell with a mutation divides (mitosis), the mutation is passed on to the daughter cells, leading to a population of genetically identical cells (a clone). The earlier in development that a somatic mutation occurs, the larger the clone of cells within that individual organism that will contain the mutation.

Because of the huge number of cells present in a typical eukaryotic organism, somatic mutations are numerous. For example, there are about 10^{14} cells in the human body. Typically, a mutation arises once in every million cell divisions, and so hundreds of millions of somatic mutations must arise in each person. Many somatic mutations have no obvious effect on the phenotype of the organism, because the function of the mutant cell (even the cell itself) is

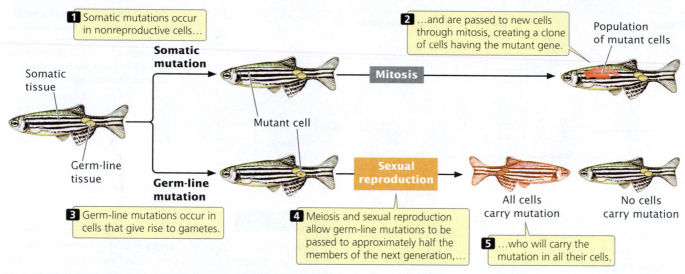

1 Somatic mutations occur in nonreproductive cells...

2 ...and are passed to new cells through mitosis, creating a clone of cells having the mutant gene.

Population of mutant cells

Somatic mutation

Somatic tissue

Mutant cell

Mitosis

Germ-line tissue

Germ-line mutation

3 Germ-line mutations occur in cells that give rise to gametes.

Sexual reproduction

4 Meiosis and sexual reproduction allow germ-line mutations to be passed to approximately half the members of the next generation,...

All cells carry mutation

No cells carry mutation

5 ...who will carry the mutation in all their cells.

10.1 The two basic classes of mutations are somatic mutations and germ-line mutations.

replaced by that of normal cells. However, cells with a somatic mutation that stimulates cell division can increase in number and spread; this type of mutation can give rise to cells with a selective advantage and is the basis for all cancers.

Germ-line mutations arise in cells that ultimately produce gametes. A germ-line mutation can be passed to future generations, producing individual organisms that carry the mutation in all their somatic and germ-line cells (see Figure 10.1). When we speak of mutations in multicellular organisms, we're usually talking about germ-line mutations.

Historically, mutations have been partitioned into those that affect a single gene, called *gene mutations,* and those that affect the number or structure of chromosomes, called *chromosome mutations.* This distinction arose because chromosome mutations could be observed directly, by looking at chromosomes with a microscope, whereas gene mutations could be detected only by observing their phenotypic effects. Now, with the development of DNA sequencing, gene mutations and chromosome mutations are distinguished somewhat arbitrarily on the basis of the size of the DNA lesion. Nevertheless, it is useful to use the term *chromosome mutation* for a large-scale genetic alteration that affects chromosome structure or the number of chromosomes and to use the term **gene mutation** for a relatively small DNA lesion that affects a single gene. This chapter focuses on gene mutations; chromosome mutations were discussed in Chapter 9.

Types of Gene Mutations

There are a number of ways to classify gene mutations. Some classification schemes are based on the nature of the phenotypic effect, others are based on the causative agent of the mutation, and still others focus on the molecular nature of the defect. Here, we will categorize mutations primarily on

the basis of their molecular nature, but we will also encounter some terms that relate the causes and the phenotypic effects of mutations.

Base substitutions The simplest type of gene mutation is a **base substitution,** the alteration of a single nucleotide in the DNA (**Figure 10.2a**). Base substitutions are of two types. In a **transition,** a purine is replaced by a different purine or, alternatively, a pyrimidine is replaced by a different pyrimidine (**Figure 10.3**). In a **transversion,** a purine is replaced by a pyrimidine or a pyrimidine is replaced by a purine. The number of possible transversions (see Figure 10.3) is twice the number of possible transitions, but transitions arise more frequently.

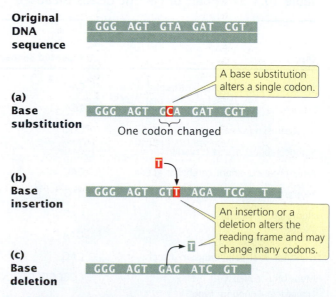

Original DNA sequence

GGG AGT GTA GAT CGT

(a) Base substitution

A base substitution alters a single codon.

GGG AGT GCA GAT CGT

One codon changed

(b) Base insertion

T

GGG AGT GTT AGA TCG T

An insertion or a deletion alters the reading frame and may change many codons.

(c) Base deletion

T

GGG AGT GAG ATC GT

10.2 Three basic types of gene mutations are base substitutions, insertions, and deletions.

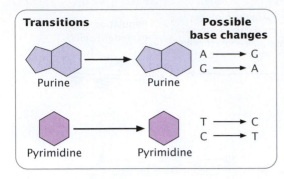

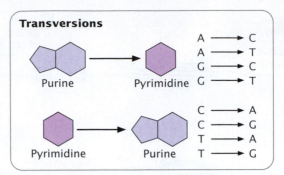

10.3 A transition is the substitution of a purine for a purine or of a pyrimidine for a pyrimidine; a transversion is the substitution of a pyrimidine for a purine or of a purine for a pyrimidine.

Insertions and deletions   The second major class of gene mutations contains **insertions** and **deletions**—the addition or the removal, respectively, of one or more nucleotide pairs (**Figure 10.2b and c**). Although base substitutions are often assumed to be the most common type of mutation, molecular analysis has revealed that insertions and deletions are more frequent. Insertions and deletions within sequences that encode proteins may lead to **frameshift mutations,** changes in the reading frame of the gene. Frameshift mutations usually alter all amino acids encoded by nucleotides following the mutation, and so they generally have drastic effects on the phenotype. Not all insertions and deletions lead to frameshifts, however; insertions and deletions consisting of any multiple of three nucleotides will leave the reading frame intact, although the addition or removal of one or more amino acids may still affect the phenotype. These mutations are called **in-frame insertions** and **deletions,** respectively.

CONCEPTS

Gene mutations consist of changes in a single gene and can be base substitutions (a single pair of nucleotides is altered) or insertions or deletions (nucleotides are added or removed). A base substitution can be a transition (substitution of like bases) or a transversion (substitution of unlike bases). Insertions and deletions often lead to a change in the reading frame of a gene.

✔ CONCEPT CHECK 2

Which of the following changes is a transition base substitution?

a. Adenine is replaced by thymine.

b. Cytosine is replaced by adenine.

c. Guanine is replaced by adenine.

d. Three nucleotide pairs are inserted into DNA.

Table 10.1 Examples of genetic diseases caused by expanding trinucleotide repeats

		Number of Copies of Repeat	
Disease	Repeated Sequence	Normal Range	Disease Range
Spinal and bulbar muscular atrophy	CAG	11–33	40–62
Fragile-X syndrome	CGG	6–54	50–1500
Jacobsen syndrome	CGG	11	100–1000
Spinocerebellar ataxia (several types)	CAG	4–44	21–130
Autosomal dominant cerebellar ataxia	CAG	7–19	37–220
Myotonic dystrophy	CTG	5–37	44–3000
Huntington disease	CAG	9–37	37–121
Friedreich ataxia	GAA	6–29	200–900
Dentatorubral-pallidoluysian atrophy	CAG	7–25	49–75
Myoclonus epilepsy of the Unverricht-Lundborg type*	CCCCGCCCCGCG	2–3	12–13

*Technically not a trinucleotide repeat but does entail a multiple of three nucleotides that expands and contracts in similar fashion to trinucleotide repeats.

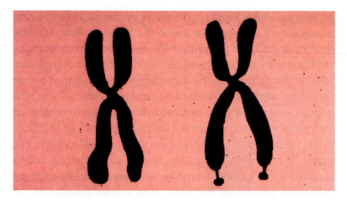

10.4 The fragile-X chromosome is associated with a characteristic constriction (fragile site) on the long arm. [Visuals Unlimited.]

Expanding trinucleotide repeats Mutations in which the number of copies of a trinucleotide (a set of three nucleotides) increase in number are called **expanding trinucleotide repeats.** This type of mutation was first observed in 1991 in a gene called *FMR-1*, which causes fragile-X syndrome, the most common hereditary cause of mental retardation. The disorder is so named because, in specially treated cells from persons having the condition, the tip of each long arm of the X chromosome is attached by only a slender thread (**Figure 10.4**). The normal *FMR-1* allele (not containing the mutation) has 60 or fewer copies of the trinucleotide CGG but, in persons with fragile-X syndrome, the allele may harbor hundreds or even thousands of copies.

Expanding trinucleotide repeats have been found in several other human diseases (**Table 10.1**). The number of copies of the trinucleotide repeat often correlates with the severity or age of onset of the disease. The number of copies of the repeat also corresponds to the instability of trinucleotide repeats: when more repeats are present, the probability of expansion to even more repeats increases. This association between the number of copies of trinucleotide repeats, the severity of the disease, and the probability of expansion leads to a phenomenon known as anticipation (see p. 122 in Chapter 5), in which diseases caused by trinucleotide-repeat expansions become more severe in each generation. Less commonly, the number of trinucleotide repeats may decrease within a family.

How an increase in the number of trinucleotides produces disease symptoms is not yet understood. In several of the diseases (e.g., Huntington disease), the trinucleotide expands within the coding part of a gene, producing a toxic protein that has extra glutamine residues (the amino acid encoded by CAG). In other diseases (e.g., fragile-X syndrome and myotonic dystrophy), the repeat is outside the coding region of the gene and therefore must have some other mode of action. At least one disease (a rare type of epilepsy) has now been associated with an expanding repeat of a 12-bp sequence. Although this repeat is not a trinucleotide, it is included as a type of expanding trinucleotide because its repeat is a multiple of three.

The mechanism that leads to the expansion of trinucleotide repeats also is not completely understood. A possi-

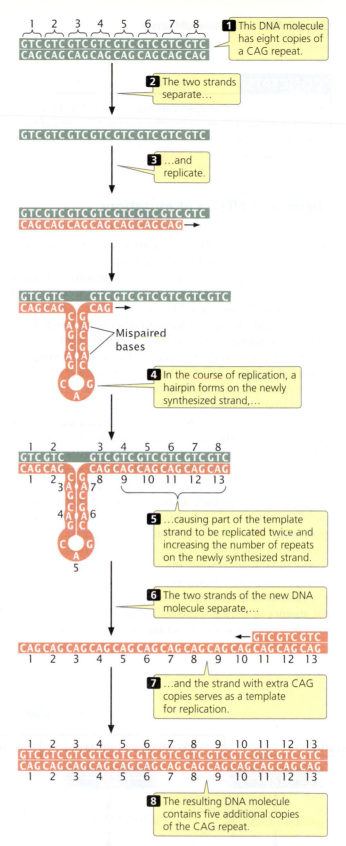

10.5 The number of copies of a trinucleotide may increase in replication owing to the formation of hairpins.

ble source of expansion is the formation of hairpins and other special DNA structures, which can cause nucleotides in the template strand to be replicated twice, thus increasing the number of repeats on the newly synthesized strand (**Figure 10.5**).

mutation. These suppressors sometimes work by changing the way that the mRNA is translated. In the example illustrated in **Figure 10.9a,** the original DNA sequence is AAC (UUG in the mRNA) and specifies leucine. This sequence mutates to ATC (UAG in mRNA), a termination codon (**Figure 10.9b**). The ATC nonsense mutation could be suppressed by a second mutation in a different gene that encodes a tRNA; this second mutation would result in a codon capable of pairing with the UAG termination codon (**Figure 10.9c**). For example, the gene that encodes the tRNA for tyrosine (tRNA^{Tyr}), which has the anticodon AUA, might be mutated to have the anticodon AUC, which will then pair with the UAG stop codon. Instead of translation

terminating at the UAG codon, tyrosine would be inserted into the protein and a full-length protein would be produced, although tyrosine would now substitute for leucine. The effect of this change would depend on the role of this amino acid in the overall structure of the protein, but the effect of the suppressor mutation is likely to be less detrimental than the effect of the nonsense mutation, which would halt translation prematurely.

Because cells in many organisms have multiple copies of tRNA genes, other unmutated copies of tRNA^{Tyr} would remain available to recognize tyrosine codons in the transcripts of the mutant gene in question and other genes being expressed concurrently. However, we might expect that the

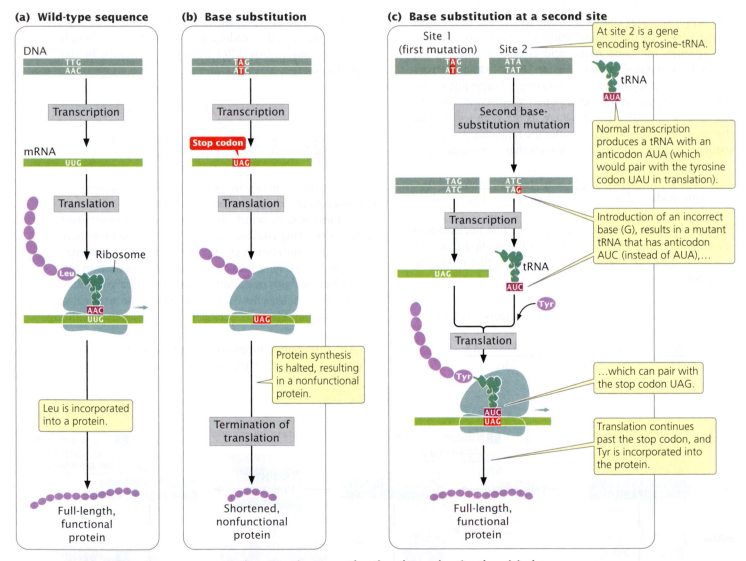

10.9 An intergenic suppressor mutation occurs in a gene other than the one bearing the original mutation. (a) The wild-type sequence produces a full-length, functional protein. (b) A base substitution at a site in the same gene produces a premature stop codon, resulting in a shortened, nonfunctional protein. (c) A base substitution at a site in another gene, which in this case encodes tRNA, alters the anticodon of tRNA^{Tyr} so that tRNA^{Tyr} can pair with the stop codon produced by the original mutation, allowing tyrosine to be incorporated into the protein and translation to continue.

Table 10.2 Characteristics of different types of mutations

Type of Mutation	Definition
Base substitution	Changes the base of a single DNA nucleotide
Transition	Base substitution in which a purine replaces a purine or a pyrimidine replaces a pyrimidine
Transversion	Base substitution in which a purine replaces a pyrimidine or a pyrimidine replaces a purine
Insertion	Addition of one or more nucleotides
Deletion	Deletion of one or more nucleotides
Frameshift mutation	Insertion or deletion that alters the reading frame of a gene
In-frame deletion or insertion	Deletion or insertion of a multiple of three nucleotides that does not alter the reading frame
Expanding trinucleotide repeats	Repeated sequence of three nucleotides (trinucleotide) in which the number of copies of the trinucleotide increases
Forward mutation	Changes the wild-type phenotype to a mutant phenotype
Reverse mutation	Changes a mutant phenotype back to the wild-type phenotype
Missense mutation	Changes a sense codon into a different sense codon, resulting in the incorporation of a different amino acid in the protein
Nonsense mutation	Changes a sense codon into a nonsense codon, causing premature termination of translation
Silent mutation	Changes a sense codon into a synonymous codon, leaving unchanged the amino acid sequence of the protein
Neutral mutation	Changes the amino acid sequence of a protein without altering its ability to function
Loss-of-function mutation	Causes a complete or partial loss of function
Gain-of-function mutation	Causes the appearance of a new trait or function or causes the appearance of a trait in inappropriate tissue or at an inappropriate time
Lethal mutation	Causes premature death
Suppressor mutation	Suppresses the effect of an earlier mutation at a different site
Intragenic suppressor mutation	Suppresses the effect of an earlier mutation within the same gene
Intergenic suppressor mutation	Suppresses the effect of an earlier mutation in another gene

tRNAs that have undergone a suppressor mutation would also suppress the normal termination codons at the ends of coding sequences, resulting in the production of longer-than-normal proteins, but this event does not usually take place.

Intergenic suppressors can also work through genic interactions (see pp. 105–111 in Chapter 5). Polypeptide chains that are produced by two genes may interact to produce a functional protein. A mutation in one gene may alter the encoded polypeptide so that the interaction between the two polypeptides is destroyed, in which case a functional protein is not produced. A suppressor mutation in the second gene may produce a compensatory change in its polypeptide, therefore restoring the original interaction.

Characteristics of some of the different types of mutations are summarized in **Table 10.2**.

CONCEPTS

A suppressor mutation overrides the effect of an earlier mutation at a different site. An intragenic suppressor mutation occurs within the *same* gene as that containing the original mutation, whereas an intergenic suppressor mutation occurs in a *different* gene.

✔ CONCEPT CHECK 3

How is a suppressor mutation different from a reverse mutation?

Worked Problem

A gene encodes a protein with the following amino acid sequence:

Met-Arg-Cys-Ile-Lys-Arg

A mutation of a single nucleotide alters the amino acid sequence to:

Met-Asp-Ala-Tyr-Lys-Gly-Glu-Ala-Pro-Val

A second single-nucleotide mutation occurs in the same gene and suppresses the effects of the first mutation (an intragenic suppressor). With the original mutation and the intragenic suppressor present, the protein has the following amino acid sequence:

Met-Asp-Gly-Ile-Lys-Arg

What is the nature and location of the first mutation and of the intragenic suppressor mutation?

• Solution

The first mutation alters the reading frame, because all amino acids after Met are changed. Insertions and deletions affect the reading frame; so the original mutation consists of a single-nucleotide insertion or deletion in the second codon. The intragenic suppressor restores the reading frame; so the intragenic suppressor also is most likely a single-nucleotide insertion or deletion: if the first mutation is an insertion, the suppressor must be a deletion; if the first mutation is a deletion, then the suppressor must be an insertion. Notice that the protein produced by the suppressor still differs from the original protein at the second and third amino acids, but the second amino acid produced by the suppressor is the same as that in the protein produced by the original mutation. Thus, the suppressor mutation must have occurred in the third codon, because the suppressor does not alter the second amino acid.

> ❓ For more practice with analyzing mutations, try working Problem 22 at the end of the chapter.

Mutation Rates

The frequency with which a wild-type allele at a locus changes into a mutant allele is referred to as the **mutation rate** and is generally expressed as the number of mutations per biological unit, which may be mutations per cell divi-sion, per gamete, or per round of replication. For example, achondroplasia is a type of hereditary dwarfism in humans that results from a dominant mutation. On average, about four achondroplasia mutations arise in every 100,000 gametes, and so the mutation rate is $^4/_{100,000}$, or 0.00004 mutations per gamete. The mutation rate provides information about how often a mutation arises.

Calculations of mutation rates are affected by three factors. First, they depend on the frequency with which a change takes place in DNA. A change in the DNA can arise from spontaneous molecular changes in DNA or it can be induced by chemical or physical agents in the environment.

The second factor influencing the mutation rate is the probability that, when a change takes place, it will be repaired. Most cells possess a number of mechanisms for repairing altered DNA, and so most alterations are corrected before they are replicated. If these repair systems are effective, mutation rates will be low; if they are faulty, mutation rates will be elevated. Some mutations increase the overall rate of mutation at other genes; these mutations usually occur in genes that encode components of the replication machinery or DNA repair enzymes.

The third factor is the probability that a mutation will be recognized and recorded. When DNA is sequenced, all mutations are potentially detectable. In practice, however, mutations are usually detected by their phenotypic effects. Some mutations may appear to arise at a higher rate simply because they are easier to detect.

Mutation rates vary among genes and species (**Table 10.3**), but we can draw several general conclusions about mutation rates. First, spontaneous mutation rates are low for all organisms studied. Typical mutation rates for bacterial genes range from about 1 to 100 mutations per 10 billion cells (from 1×10^{-8} to 1×10^{-10}). The mutation rates for most eukaryotic genes are a bit higher, from about 1 to 10 mutations per million gametes (from 1×10^{-5} to 1×10^{-6}). These higher values in eukaryotes may be due to the fact that the rates are calculated per gamete, and several cell divisions are required to produce a gamete, whereas mutation rates in prokaryotic cells are calculated *per cell division*.

The differences in mutation rates among species may be due to differing abilities to repair mutations, unequal exposures to mutagens, or biological differences in rates of spontaneously arising mutations. Even within a single species, spontaneous rates of mutation vary among genes. The reason for this variation is not entirely understood, but some regions of DNA are known to be more susceptible to mutation than others.

Several recent studies have measured mutation rates directly by sequencing genes of organisms before and after a number of generations. These new studies suggest that mutation rates are often higher than those previously measured on the basis of changes in phenotype. In one study, geneticists sequenced randomly chosen stretches of DNA in

Table 10.3 Mutation rates of different genes in different organisms

Organism	Mutation	Rate	Unit
Bacteriophage T2	Lysis inhibition	1×10^{-8}	Per replication
	Host range	3×10^{-9}	
Escherichia coli	Lactose fermentation	2×10^{-7}	Per cell division
	Histidine requirement	2×10^{-8}	
Neurospora crassa	Inositol requirement	8×10^{-8}	Per asexual spore
	Adenine requirement	4×10^{-8}	
Corn	Kernel color	2.2×10^{-6}	Per gamete
Drosophila	Eye color	4×10^{-5}	Per gamete
	Allozymes	5.14×10^{-6}	
Mouse	Albino coat color	4.5×10^{-5}	Per gamete
	Dilution coat color	3×10^{-5}	
Human	Huntington disease	1×10^{-6}	Per gamete
	Achondroplasia	1×10^{-5}	
	Neurofibromatosis (Michigan)	1×10^{-4}	
	Hemophilia A (Finland)	3.2×10^{-5}	
	Duchenne muscular dystrophy (Wisconsin)	9.2×10^{-5}	

the nematode worm *Caenorhabditis elegans* and found about 2.1 mutations per genome per generation, which was 10 times as high as previous estimates based on phenotypic changes. The researchers found that about half of the mutations were insertions and deletions.

As will be discussed in Chapters 11 through 13, evolutionary change that brings about adaptation to new environments depends critically on the presence of genetic variation. New genetic variants arise primarily through mutation. For many years, genetic variation was assumed to arise randomly and at rates that are independent of the need for adaptation. However, some evidence now suggests that stressful environments—where adaptation may be necessary to survive—can induce more mutations in bacteria, a process that has been termed **adaptive mutation.** The idea of adaptive mutation has been intensely debated; critics counter that most mutations are expected to be deleterious, and so increased mutagenesis would likely be harmful most of the time.

Research findings have shown that mutation rates in bacteria collected from the wild do increase in stressful environments, such as those in which nutrients are limited. There is considerable disagreement about whether (1) the increased mutagenesis is a genetic strategy selected to increase the probability of evolving beneficial traits in the stressful environment or (2) mutagenesis is merely an accidental consequence of error-prone DNA repair mechanisms that are more active in stressful environments (see Section 10.4 on DNA repair).

CONCEPTS

Mutation rate is the frequency with which a specific mutation arises. Rates of mutations are generally low and are affected by environmental and genetic factors.

✔ **CONCEPT CHECK 4**

What three factors affect mutation rates?

10.2 Mutations Are Potentially Caused by a Number of Different Natural and Unnatural Factors

Mutations result from both internal and external factors. Those that are a result of natural changes in DNA structure are termed **spontaneous mutations,** whereas those that result from changes caused by environmental chemicals or radiation are **induced mutations.**

Spontaneous Replication Errors

Replication is amazingly accurate: less than one error in a billion nucleotides arises in the course of DNA synthesis. However, spontaneous replication errors do occasionally occur.

10.19 Chemicals may alter DNA bases.

	Original base	Mutagen	Modified base	Pairing partner	Type of mutation
(a)	Guanine	EMS Alkylation	O^6-Ethylguanine	Thymine	CG → TA
(b)	Cytosine	Nitrous acid (HNO₂) Deamination	Uracil	Adenine	CG → TA
(c)	Cytosine	Hydroxylamine (NH₂OH) Hydroxylation	Hydroxylamino-cytosine	Adenine	CG → TA

Through mispairing, 5-bromouracil can also be incorporated into a newly synthesized DNA strand opposite guanine. In the next round of replication 5-bromouracil pairs with adenine, leading to another transition (G·C→G·5BU→A·5BU→A·T).

Another mutagenic chemical is 2-aminopurine (2AP), which is a base analog of adenine. Normally, 2-aminopurine base pairs with thymine, but it may mispair with cytosine, causing a transition mutation (T·A→T·2AP→C·2AP→C·G). Alternatively, 2-aminopurine may be incorporated through mispairing into the newly synthesized DNA opposite cytosine and then later pair with thymine, leading to a C·G→C·2AP→T·2AP→T·A transition.

Thus, both 5-bromouracil and 2-aminopurine can produce transition mutations. In the laboratory, mutations caused by base analogs can be reversed by treatment with the same analog or by treatment with a different analog.

Alkylating agents Alkylating agents are chemicals that donate alkyl groups, such as methyl (CH_3) and ethyl (CH_3–CH_2) groups, to nucleotide bases. For example, ethylmethylsulfonate (EMS) adds an ethyl group to guanine, producing O^6-ethylguanine, which pairs with thymine (**Figure 10.19a**). Thus, EMS produces C·G→T·A transitions. EMS is also capable of adding an ethyl group to thymine, producing 4-ethylthymine, which then pairs with guanine,

leading to a T·A→C·G transition. Because EMS produces both C·G→T·A and T·A→C·G transitions, mutations produced by EMS can be reversed by additional treatment with EMS. Mustard gas is another alkylating agent.

Deamination In addition to its spontaneous occurrence (see Figure 10.16), deamination can be induced by some chemicals. For instance, nitrous acid deaminates cytosine, creating uracil, which in the next round of replication pairs with adenine (**Figure 10.19b**), producing a C·G→T·A transition mutation. Nitrous acid changes adenine into hypoxanthine, which pairs with cytosine, leading to a T·A→C·G transition. Nitrous acid also deaminates guanine, producing xanthine, which pairs with cytosine just as guanine does; however, xanthine can also pair with thymine, leading to a C·G→T·A transition. Nitrous acid produces exclusively transition mutations and, because both C·G→T·A and T·A→C·G transitions are produced, these mutations can be reversed with nitrous acid.

Hydroxylamine Hydroxylamine is a very specific base-modifying mutagen that adds a hydroxyl group to cytosine, converting it into hydroxylaminocytosine (**Figure 10.19c**). This conversion increases the frequency of a rare tautomer that pairs with adenine instead of guanine and leads to C·G→T·A transitions. Because hydroxylamine acts only on

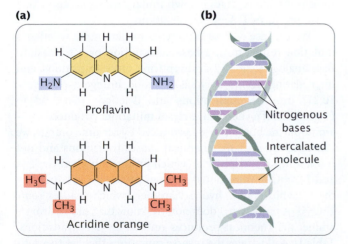

10.20 Oxidative radicals convert guanine into 8-oxy-7, 8-dihydrodeoxyguanine, which frequently mispairs with adenine instead of cytosine, producing a C·G→T·A transversion.

cytosine, it will not generate T·A→C·G transitions; thus, hydroxylamine will not reverse the mutations that it produces.

Oxidative reactions   Reactive forms of oxygen (including superoxide radicals, hydrogen peroxide, and hydroxyl radicals) are produced in the course of normal aerobic metabolism, as well as by radiation, ozone, peroxides, and certain drugs. These reactive forms of oxygen damage DNA and induce mutations by bringing about chemical changes to DNA. For example, oxidation converts guanine into 8-oxy-7,8-dihydrodeoxyguanine (**Figure 10.20**), which frequently mispairs with adenine instead of cytosine, causing a G·C→T·A transversion mutation.

Intercalating agents   Proflavin, acridine orange, ethidium bromide, and dioxin are **intercalating agents** (**Figure 10.21a**), which produce mutations by sandwiching themselves (intercalating) between adjacent bases in DNA, distorting the

(a)   **(b)**

Proflavin

Acridine orange

Nitrogenous bases

Intercalated molecule

10.21 Intercalating agents such as proflavin and acridine orange insert themselves between adjacent bases in DNA, distorting the three-dimensional structure of the helix and causing single-nucleotide insertions and deletions in replication.

three-dimensional structure of the helix and causing single-nucleotide insertions and deletions in replication (**Figure 10.21b**). These insertions and deletions frequently produce frameshift mutations, and so the mutagenic effects of intercalating agents are often severe. Because intercalating agents generate both additions and deletions, they can reverse the effects of their own mutations.

CONCEPTS

Chemicals can produce mutations by a number of mechanisms. Base analogs are inserted into DNA and frequently pair with the wrong base. Alkylating agents, deaminating chemicals, hydroxylamine, and oxidative radicals change the structure of DNA bases, thereby altering their pairing properties. Intercalating agents wedge between the bases and cause single-base insertions and deletions in replication.

✔ CONCEPT CHECK 6

Base analogs are mutagenic because of which characteristic?

a. They produce changes in DNA polymerase that cause it to malfunction.
b. They distort the structure of DNA.
c. They are similar in structure to the normal bases.
d. They chemically modify the normal bases.

Radiation

In 1927, Hermann Muller demonstrated that mutations in fruit flies could be induced by X-rays. The results of subsequent studies showed that X-rays greatly increase mutation rates in all organisms. Because of their high energies, X-rays, gamma rays, and cosmic rays are all capable of penetrating tissues and damaging DNA. These forms of radiation, called ionizing radiation, dislodge electrons from the atoms that they encounter, changing stable molecules into free radicals and reactive ions, which then alter the structures of bases and break phosphodiester bonds in DNA. Ionizing radiation also frequently results in double-strand breaks in DNA. Attempts to repair these breaks can produce chromosome mutations (discussed in Chapter 9).

Ultraviolet light has less energy than that of ionizing radiation and does not eject electrons and cause ionization but is nevertheless highly mutagenic. Purine and pyrimidine bases readily absorb UV light, resulting in the formation of chemical bonds between adjacent pyrimidine molecules on the same strand of DNA and in the creation of **pyrimidine dimers** (**Figure 10.22a**). Pyrimidine dimers consisting of two thymine bases (called thymine dimers) are most frequent, but cytosine dimers and thymine–cytosine dimers also can form. Dimers distort the configuration of DNA (**Figure 10.22b**) and often block replication. Most pyrimidine dimers

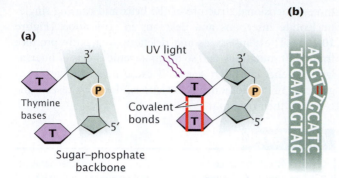

10.22 Pyrimidine dimers result from ultraviolet light. (a) Formation of thymine dimer. (b) Distorted DNA.

CONCEPTS

Ionizing radiation such as X-rays and gamma rays damage DNA by dislodging electrons from atoms; these electrons then break phosphodiester bonds and alter the structure of bases. Ultraviolet light causes mutations primarily by producing pyrimidine dimers that disrupt replication and transcription. The SOS system enables bacteria to overcome replication blocks but introduces mistakes in replication.

are immediately repaired by mechanisms discussed later in this chapter, but some escape repair and inhibit replication and transcription.

When pyrimidine dimers block replication, cell division is inhibited and the cell usually dies; for this reason, UV light kills bacteria and is an effective sterilizing agent. For a mutation—a hereditary error in the genetic instructions—to occur, the replication block must be overcome. Bacteria can sometimes circumvent replication blocks produced by pyrimidine dimers and other types of DNA damage by means of the **SOS system.** This system allows replication blocks to be overcome but, in the process, makes numerous mistakes and greatly increases the rate of mutation. Indeed, the very reason that replication can proceed in the presence of a block is that the enzymes in the SOS system do not strictly adhere to the base-pairing rules. The trade-off is that replication may continue and the cell survives, but only by sacrificing the normal accuracy of DNA synthesis.

The SOS system is complex, including the products of at least 25 genes. A protein called RecA binds to the damaged DNA at the blocked replication fork and becomes activated. This activation promotes the binding of a protein called LexA, which is a repressor of the SOS system. The activated RecA complex induces LexA to undergo self-cleavage, destroying its repressive activity. This inactivation enables other SOS genes to be expressed, and the products of these genes allow replication of the damaged DNA to proceed. The SOS system allows bases to be inserted into a new DNA strand in the absence of bases on the template strand, but these insertions result in numerous errors in the base sequence.

Eukaryotic cells have a specialized DNA polymerase called polymerase η (eta) that bypasses pyrimidine dimers. Polymerase η preferentially inserts AA opposite a pyrimidine dimer. This strategy seems to be reasonable because about two-thirds of pyrimidine dimers are thymine dimers. However, the insertion of AA opposite a CT dimer results in a C·G→A·T transversion. Polymerase η is therefore said to be an error-prone polymerase.

10.3 Mutations Are the Focus of Intense Study by Geneticists

Because mutations often have detrimental effects, they are often studied by geneticists. These studies have included the analysis of reverse mutations, which are often sources of important insight into how mutations cause DNA damage; the development of tests to determine the mutagenic properties of chemical compounds; and the investigation of human populations tragically exposed to high levels of radiation.

The Analysis of Reverse Mutations

The study of reverse mutations (reversions) can provide useful information about how mutagens alter DNA structure. For example, any mutagen that produces both A·T→G·C and G·C→A·T transitions should be able to reverse its own mutations. However, if the mutagen produces only G·C→A·T transitions, then reversion by the same mutagen is not possible. Hydroxylamine (see Figure 10.19c) exhibits this type of one-way mutagenic activity; it causes G·C→A·T transitions but is incapable of reversing the mutations that it produces; so we know that it does not produce A·T→G·C transitions. Ethylmethylsulfonate (see Figure 10.19a), on the other hand, produces C·G→T·A transitions and reverses its own mutations; so we know that it also produces T·A→C·G transitions.

We can use reverse mutations to determine whether a mutation results from a base substitution or a frameshift. Base analogs such as 2-aminopurine cause transitions, and intercalating agents such as acridine orange (see Figure 10.21) produce duplications and deletions that lead to frameshifts. If a chemical reverses mutations produced by 2-aminopurine but not those produced by acridine orange, we can conclude that the chemical causes transitions and not frameshifts. If nitrous acid (which produces both C·G→T·A and T·A→C·G transitions) reverses mutations produced by the chemical but hydroxylamine (which causes *only* C·G→T·A transitions) does not, we know that, like hydroxylamine, the chemical produces only C·G→T·A transitions. **Table 10.4** illustrates the reverse mutations that are theoretically possible among several mutagenic agents. The actual ability of mutagens to produce reversals is more complex than is suggested by Table 10.4 and depends on environmental conditions and the organism tested.

Table 10.4 Theoretical reverse mutations possible by various mutagenic agents

Type of Mutagen	Mutation	Reversal of Mutation by					
		5-Bromo-uracil	2-Amino-purine	Ethylmethyl-sulfonate	Nitrous Acid	Hydroxyl-amine	Acridine Orange
5-Bromouracil	C·G↔T·A	+	+	+	+	+/−	−
2-Aminopurine	C·G↔T·A	+	+	+	+	+/−	−
Nitrous acid	C·G↔T·A	+	+	+	+	+/−	−
Ethylmethyl-sulfonate	C·G↔T·A	+	+	+	+	+/−	−
Hydroxylamine	C·G↔T·A	+	+	+	+	−	−
Acridine orange	Frameshift	−	−	−	−	−	+

Note: A plus (+) sign indicates that reverse mutations occur, a minus (−) sign indicates that reverse mutations do not occur, and +/− indicates that only some mutations are reversed. Not all reverse mutations are equally likely.

CONCEPTS

The study of the ability of mutagenic agents to produce reverse mutations provides important information about how mutagens alter DNA.

✔ CONCEPT CHECK 7

Mutations produced by chemical X223 are reversed by an intercalating agent but not by a base analog, which suggests that the mutations produced by X223 are

a. frameshifts (insertions and deletions).　　c. transversions.

b. transitions.　　d. all of the above.

Detecting Mutations with the Ames Test

People in industrial societies are surrounded by a multitude of artificially produced chemicals: more than 50,000 different chemicals are in commercial and industrial use today, and from 500 to 1000 new chemicals are introduced each year. Some of these chemicals are potential carcinogens and may cause harm to humans. One method for testing the cancer-causing potential of chemicals is to administer them to laboratory animals (rats or mice) and compare the incidence of cancer in the treated animals with that of control animals. These tests are unfortunately time consuming and expensive. Furthermore, the ability of a substance to cause cancer in rodents is not always indicative of its effect on humans. After all, we aren't rats!

In 1974, Bruce Ames developed a simple test for evaluating the potential of chemicals to cause cancer. The **Ames test** is based on the principle that both cancer and mutations result from damage to DNA, and the results of experiments have demonstrated that 90% of known carcinogens are also mutagens. Ames proposed that mutagenesis in bacteria could serve as an indicator of carcinogenesis in humans.

The Ames test uses four auxotrophic strains of the bacterium *Salmonella typhimurium* that have defects in the lipopolysaccharide coat, which normally protects the bacteria from chemicals in the environment. Furthermore, the DNA repair system in these strains has been inactivated, enhancing their susceptibility to mutagens.

One of the four auxotrophic strains used in the Ames test detects base-pair substitutions; the other three detect different types of frameshift mutations. Each strain carries a *his*⁻ mutation, which renders it unable to synthesize the amino acid histidine, and the bacteria are plated onto medium that lacks histidine (**Figure 10.23**). Only bacteria that have undergone a reverse mutation of the histidine gene (*his*⁻→*his*⁺) are able to synthesize histidine and grow on the medium. Different dilutions of a chemical to be tested are added to plates inoculated with the bacteria, and the number of mutant bacterial colonies that appear on each plate is compared with the number that appear on control plates with no chemical (i.e., that arose through spontaneous mutation). Any chemical that significantly increases the number of colonies appearing on a treated plate is mutagenic and is probably also carcinogenic.

Some compounds are not active carcinogens but can be converted into cancer-causing compounds in the body. To make the Ames test sensitive for such *potential* carcinogens, a compound to be tested is first incubated in mammalian liver extract that contains metabolic enzymes.

The Ames test has been applied to thousands of chemicals and commercial products. An early demonstration of its usefulness was the discovery, in 1975, that many hair dyes sold in the United States contained compounds that were mutagenic to bacteria. These compounds were then removed from most hair dyes.

CONCEPTS

The Ames test uses *his*⁻ strains of bacteria to test chemicals for their ability to produce *his*⁻→*his*⁺ mutations. Because mutagenic activity and carcinogenic potential are closely correlated, the Ames test is widely used to screen chemicals for their cancer-causing potential.

Question: How can chemicals be quickly screened for their for their ability to cause cancer?

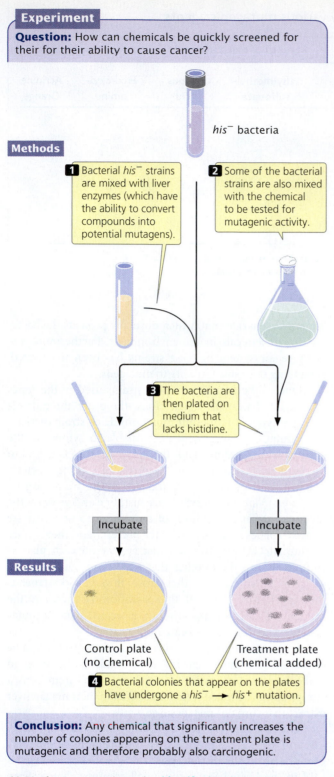

his⁻ bacteria

Methods

1 Bacterial *his⁻* strains are mixed with liver enzymes (which have the ability to convert compounds into potential mutagens).

2 Some of the bacterial strains are also mixed with the chemical to be tested for mutagenic activity.

3 The bacteria are then plated on medium that lacks histidine.

Incubate   Incubate

Results

Control plate (no chemical)   Treatment plate (chemical added)

4 Bacterial colonies that appear on the plates have undergone a *his⁻* → *his⁺* mutation.

Conclusion: Any chemical that significantly increases the number of colonies appearing on the treatment plate is mutagenic and therefore probably also carcinogenic.

10.23 The Ames test is used to identify chemical mutagens.

Radiation Exposure in Humans

People are routinely exposed to low levels of radiation from cosmic, medical, and environmental sources, but there have also been tragic events that produced exposures of much higher degree.

On August 6, 1945, a high-flying American airplane dropped a single atomic bomb on the city of Hiroshima,

Japan. The explosion devastated an area of the city measuring 4.5 square miles, killed from 90,000 to 140,000 people, and injured almost as many (**Figure 10.24**). Three days later, the United States dropped an atomic bomb on the city of Nagasaki, this time destroying an area measuring 1.5 square miles and killing between 60,000 and 80,000 people. Huge amounts of radiation were released during these explosions, and many people were exposed.

After the war, a joint Japanese–U.S. effort was made to study the biological effects of radiation exposure on the survivors of the atomic blasts and their children. Somatic mutations were examined by studying radiation sickness and cancer among the survivors; germ-line mutations were assessed by looking at birth defects, chromosome abnormalities, and gene mutations in children born to people that had been exposed to radiation.

Geneticist James Neel and his colleagues examined almost 19,000 children of people who were within 2000 meters (1.2 miles) of the center of the atomic blast at Hiroshima or Nagasaki, along with a similar number of children whose parents did not receive radiation exposure. Radiation doses were estimated for a child's parents on the basis of careful assessment of the parents' location, posture, and position at the time of the blast. A blood sample was collected from each child, and gel electrophoresis was used to investigate amino acid substitutions in 28 proteins. When rare variants were detected, blood samples from the child's parents also were analyzed to establish whether the variant was inherited or a new mutation.

Of a total of 289,868 genes examined by Neel and his colleagues, only one mutation was found in the children of exposed parents; no mutations were found in the control group. From these findings, a mutation rate of 3.4×10^{-6} was estimated for the children whose parents were exposed to the blast, which is within the range of spontaneous mutation rates observed for other eukaryotes. Neel and his colleagues also examined the frequency of chromosome mutations, sex ratios of children born to exposed parents, and frequencies of chromosome aneuploidy. There was no evidence in any of these assays for increased mutations among the children of the people who were exposed to radiation from the atomic explosions, suggesting that germ-line mutations were not elevated.

Animal studies clearly show that radiation causes germ-line mutations; so why was there no apparent increase in germ-line mutations among the inhabitants of Hiroshima and Nagasaki? The exposed parents did exhibit an increased incidence of leukemia and other types of cancers; so somatic mutations were clearly induced. The answer to the question is not known, but the lack of germ-line mutations may be due to the fact that those persons who received the largest radiation doses died soon after the blasts.

The Techa River in southern Russia is another place where people have been tragically exposed to high levels of radiation. The Mayak nuclear facility, produced plutonium for

10.24 Hiroshima was destroyed by an atomic bomb on August 6, 1945. The atomic explosion produced many somatic mutations among the survivors. *[Stanley Troutman/AP.]*

nuclear warheads in the early days of the Cold War. Between 1949 and 1956, this plant dumped some 76 million cubic meters of radioactive sludge into the Techa River. People downstream used the river for drinking water and crop irrigation; some received radiation doses 1700 times the annual amount considered safe by today's standards. Radiation in the area was further elevated by a series of nuclear accidents at the Mayak plant; the worst was an explosion of a radioactive liquid storage tank in 1957, which showered radiation over a 27,000-square-kilometer (10,425-square-mile) area.

Although Soviet authorities suppressed information about the radiation problems along the Techa until the 1990s, Russian physicians led by Mira Kossenko quietly began studying cancer and other radiation-related illnesses among the inhabitants in the 1960s. They found that the overall incidence of cancer was elevated among people who lived on the banks of the Techa River.

Most data on radiation exposure in humans are from the intensive study of the survivors of the atomic bombing of Hiroshima and Nagasaki. However, the inhabitants of Hiroshima and Nagasaki were exposed in one intense burst of radiation, and these data may not be appropriate for understanding the effects of long-term low-dose radiation. Today, U.S. and Russian scientists are studying the people of the Techa River region in an attempt to better understand the effects of chronic radiation exposure on human populations.

10.4 A Number of Pathways Repair Changes in DNA

The integrity of DNA is under constant assault from radiation, chemical mutagens, and spontaneously arising changes. In spite of this onslaught of damaging agents, the rate of mutation remains remarkably low, thanks to the efficiency with which DNA is repaired. Less than one in a thousand

DNA lesions is estimated to become a mutation; all the others are corrected.

There are a number of complex pathways for repairing DNA, but several general statements can be made about DNA repair. First, most DNA repair mechanisms require two nucleotide strands of DNA because most replace whole nucleotides, and a template strand is needed to specify the base sequence.

A second general feature of DNA repair is redundancy, meaning that many types of DNA damage can be corrected by more than one pathway of repair. This redundancy testifies to the extreme importance of DNA repair to the survival of the cell: it ensures that almost all mistakes are corrected. If a mistake escapes one repair system, it's likely to be repaired by another system.

We will consider four general mechanisms of DNA repair: mismatch repair, direct repair, base-excision repair, and nucleotide-excision repair (**Table 10.5**).

Table 10.5 Summary of common DNA repair mechanisms

Repair System	Type of Damage Repaired
Mismatch	Replication errors, including mispaired bases and strand slippage
Direct	Pyrimidine dimers; other specific types of alterations
Base excision	Abnormal bases, modified bases, and pyrimidine dimers
Nucleotide excision	DNA damage that distorts the double helix, including abnormal bases, modified bases, and pyrimidine dimers

siblings) should resemble one another more than distantly related individuals (such as cousins). Comparing individuals with different degrees of relatedness, then, provides information about the extent to which genes influence a characteristic.

In 2003, geneticists used a combination of quantitative genetics and molecular techniques to identify and isolate chromosomal regions that play an important role in determining increased muscle mass in pigs. Chromosome regions containing genes that influence a quantitative trait are termed **quantitative trait loci** (QTLs). To locate QTLs affecting muscle mass, the geneticists started with crosses between European wild boars and Large White domestic pigs. The alleles from some of the domestic pigs in these crosses markedly increased muscle mass and back-fat thickness in the offspring, indicating that the domestic pigs possess genes that stimulated muscle growth.

The geneticists then used molecular markers to map the position of the QTLs that influence muscle mass. They were able to narrow their search for the location of one important QTL to a 250,000-bp interval on pig chromosome 2. This region is known to contain several genes, including one for insulin-like growth factor 2 (IGF2). Because IGF2 is known to stimulate muscle mass in mammals, the gene immediately attracted their attention. By sequencing the *IGF2* gene of the more-muscled pigs and comparing their sequences with those from less-muscled pigs, the geneticists were able to demonstrate that a change in a single nucleotide, from a G to an A, added 3% to 5% more meat to a pig. Interestingly, the nucleotide change is not in a part of the gene that encodes the protein, but instead is in an intron. Findings from further research revealed that this substitution increases the expression of *IGF2* mRNA threefold in muscle cells. The increased levels of *IGF2* mRNA result in more insulin-like growth factor 2, which stimulates muscle growth and results in more-muscled, leaner pigs. This study demonstrates the power of quantitative genetics coupled with modern molecular techniques to identify and exploit genetic variation that influences economically important characteristics such as muscle mass in pigs.

This chapter is about the genetic analysis of complex characteristics such as muscle mass. We begin by considering the differences between quantitative and qualitative characteristics and why the expression of some characteristics varies continuously. We'll see how quantitative characteristics are often influenced by many genes, each of which has a small effect on the phenotype. Next, we will examine statistical procedures for describing and analyzing quantitative characteristics. We will consider the question of how much of phenotypic variation can be attributed to genetic and environmental influences and will conclude by looking at the effects of selection on quantitative characteristics. It's important to recognize that the methods of quantitative genetics are not designed to identify individual genes and genotypes. Rather, the focus is on statistical predictions based on groups of individuals.

11.1 Quantitative Characteristics Vary Continuously and Many Are Influenced by Alleles at Multiple Loci

Qualitative, or discontinuous, characteristics possess only a few distinct phenotypes (**Figure 11.1a**); these characteristics are the types studied by Mendel and have been the focus of

our attention thus far. However, many characteristics vary continuously along a scale of measurement with many overlapping phenotypes (**Figure 11.1b**). They are referred to as *continuous characteristics;* they are also called *quantitative characteristics* because any individual's phenotype must be described with a quantitative measurement. Quantitative characteristics might include height, weight, and blood pressure in humans, growth rate in mice, seed weight in plants, and milk production in cattle.

Quantitative characteristics arise from two phenomena. First, many are polygenic: they are influenced by genes at many loci. If many loci take part, many genotypes are possible, each producing a slightly different phenotype. Second, quantitative characteristics often arise when environmental factors affect the phenotype, because environmental differences result in a single genotype producing a range of phenotypes. Most continuously varying characteristics are *both* polygenic *and* influenced by environmental factors, and these characteristics are said to be multifactorial.

The Relation Between Genotype and Phenotype

For many discontinuous characteristics, the relation between genotype and phenotype is straightforward. Each genotype produces a single phenotype, and most pheno-

(a) Discontinuous characteristic

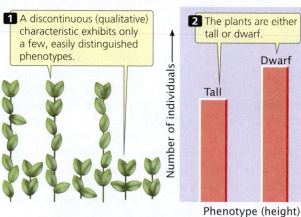

1 A discontinuous (qualitative) characteristic exhibits only a few, easily distinguished phenotypes.

2 The plants are either tall or dwarf.

(b) Continuous characteristic

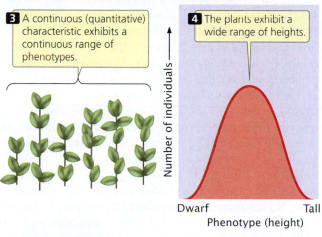

3 A continuous (quantitative) characteristic exhibits a continuous range of phenotypes.

4 The plants exhibit a wide range of heights.

11.1 Discontinuous and continuous characteristics differ in the number of phenotypes exhibited.

loci are taken into account, there are a total of $3^3 = 27$ possible multilocus genotypes (A^+A^+ B^+B^+ C^+C^+, A^+A^- B^+B^+ C^+C^+, etc.). Although there are 27 genotypes, they produce only seven phenotypes (10 cm, 11 cm, 12 cm, 13 cm, 14 cm, 15 cm, and 16 cm in height). Some of the genotypes produce the same phenotype (**Table 11.1**); for example, genotypes A^+A^- B^-B^- C^-C^-, A^-A^- B^+B^- C^-C^-, and A^-A^- B^-B^- C^+C^- all have one gene that encodes a plant hormone. These genotypes produce one dose of the hormone and a plant that is 11 cm tall. Even in this simple example of only three loci, the relation between genotype and phenotype is quite complex. The more loci encoding a characteristic, the greater the complexity.

Table 11.1 Hypothetical example of plant height determined by pairs of alleles at each of three loci

Plant Genotype	Doses of Hormone	Height (cm)
A^-A^- B^-B^- C^-C^-	0	10
A^+A^- B^-B^- C^-C^-	1	11
A^-A^- B^+B^- C^-C^-		
A^-A^- B^-B^- C^-C^+		
A^+A^+ B^-B^- C^-C^-	2	12
A^-A^- B^+B^+ C^-C^-		
A^-A^- B^-B^- C^+C^+		
A^+A^- B^+B^- C^-C^-		
A^+A^- B^-B^- C^+C^-		
A^-A^- B^+B^- C^+C^-		
A^+A^+ B^+B^- C^-C^-	3	13
A^+A^+ B^-B^- C^+C^-		
A^+A^- B^+B^+ C^-C^-		
A^-A^- B^+B^+ C^+C^-		
A^+A^- B^-B^- C^+C^+		
A^-A^- B^+B^- C^+C^+		
A^+A^- B^+B^- C^+C^-		
A^+A^+ B^+B^+ C^-C^-	4	14
A^+A^+ B^+B^- C^+C^-		
A^+A^- B^+B^+ C^+C^-		
A^-A^- B^+B^+ C^+C^+		
A^+A^+ B^-B^- C^+C^+		
A^+A^- B^+B^- C^+C^+		
A^+A^+ B^+B^+ C^+C^-	5	15
A^+A^- B^+B^+ C^+C^+		
A^+A^+ B^+B^- C^+C^+		
A^+A^+ B^+B^+ C^+C^+	6	16

Note: Each + allele contributes 1 cm in height above a baseline of 10 cm.

types are encoded by a single genotype. Dominance and epistasis may allow two or three genotypes to produce the same phenotype, but the relation remains simple. This simple relation between genotype and phenotype allowed Mendel to decipher the basic rules of inheritance from his crosses with pea plants; it also permits us both to predict the outcome of genetic crosses and to assign genotypes to individuals.

For quantitative characteristics, the relation between genotype and phenotype is often more complex. If the characteristic is polygenic, many different genotypes are possible, several of which may produce the same phenotype. For instance, consider a plant whose height is determined by three loci (A, B, and C), each of which has two alleles. Assume that one allele at each locus (A^+, B^+, and C^+) encodes a plant hormone that causes the plant to grow 1 cm above its baseline height of 10 cm. The other allele at each locus (A^-, B^-, and C^-) does not encode a plant hormone and thus does not contribute to additional height. If we consider only the two alleles at a single locus, 3 genotypes are possible (A^+A^+, A^+A^-, and A^-A^-). If all three

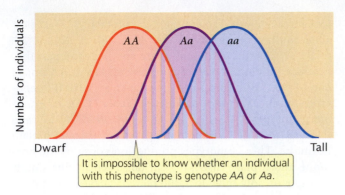

It is impossible to know whether an individual with this phenotype is genotype *AA* or *Aa*.

11.2 For a quantitative characteristic, each genotype may produce a range of possible phenotypes. In this hypothetical example, the phenotypes produced by genotypes *AA*, *Aa*, and *aa* overlap.

The influence of environment on a characteristic also can complicate the relation between genotype and phenotype. Because of environmental effects, the same genotype may produce a range of potential phenotypes (the norm of reaction; see p. 122 in Chapter 5). The phenotypic ranges of different genotypes may overlap, making it difficult to know whether individuals differ in phenotype because of genetic or environmental differences (**Figure 11.2**).

In summary, the simple relation between genotype and phenotype that exists for many qualitative (discontinuous) characteristics is absent in quantitative characteristics, and it is impossible to assign a genotype to an individual on the basis of its phenotype alone. The methods used for analyzing qualitative characteristics (examining the phenotypic ratios of progeny from a genetic cross) will not work with quantitative characteristics. Our goal remains the same: we wish to make predictions about the phenotypes of offspring produced in a genetic cross. We may also want to know how much of the variation in a characteristic results from genetic differences and how much results from environmental differences. To answer these questions, we must turn to statistical methods that allow us to make predictions about the inheritance of phenotypes in the absence of information about the underlying genotypes.

Types of Quantitative Characteristics

Before we look more closely at polygenic characteristics and relevant statistical methods, we need to more clearly define what is meant by a quantitative characteristic. Thus far, we have considered only quantitative characteristics that vary continuously in a population. A *continuous characteristic* can theoretically assume any value between two extremes; the number of phenotypes is limited only by our ability to precisely measure the phenotype. Human height is a continuous characteristic because, within certain limits, people can theoretically have any height. Although the number of phenotypes possible with a continuous characteristic is infinite,

we often group similar phenotypes together for convenience; we may say that two people are both 5 feet 11 inches tall, but careful measurement may show that one is slightly taller than the other.

Some characteristics are not continuous but are nevertheless considered quantitative because they are determined by multiple genetic and environmental factors. **Meristic characteristics,** for instance, are measured in whole numbers. An example is litter size: a female mouse may have 4, 5, or 6 pups but not 4.13 pups. A meristic characteristic has a limited number of distinct phenotypes, but the underlying determination of the characteristic may still be quantitative. These characteristics must therefore be analyzed with the same techniques that we use to study continuous quantitative characteristics.

Another type of quantitative characteristic is a **threshold characteristic,** which is simply present or absent. For example, the presence of some diseases can be considered a threshold characteristic. Although threshold characteristics exhibit only two phenotypes, they are considered quantitative because they, too, are determined by multiple genetic and environmental factors. The expression of the characteristic depends on an underlying susceptibility (usually referred to as liability or risk) that varies continuously. When the susceptibility is larger than a threshold value, a specific trait is expressed (**Figure 11.3**). Diseases are often threshold characteristics because many factors, both genetic and environmental, contribute to disease susceptibility. If enough of the susceptibility factors are present, the disease develops; otherwise, it is absent. Although we focus on the genetics of continuous characteristics in this chapter, the same principles apply to many meristic and threshold characteristics.

It is important to point out that just because a characteristic can be measured on a continuous scale does not mean that it exhibits quantitative variation. One of the characteristics studied by Mendel was height of the pea plant, which can be described by measuring the length of the plant's stem. However, Mendel's particular plants exhibited only two distinct phenotypes (some were tall and others short), and these differences were determined by alleles at a

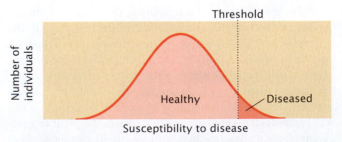

11.3 Threshold characteristics display only two possible phenotypes—the trait is either present or absent—but they are quantitative because the underlying susceptibility to the characteristic varies continuously. When the susceptibility exceeds a threshold value, the characteristic is expressed.

single locus. The differences that Mendel studied were therefore discontinuous in nature.

CONCEPTS

Characteristics whose phenotypes vary continuously are called quantitative characteristics. For most quantitative characteristics, the relation between genotype and phenotype is complex. Some characteristics whose phenotypes do not vary continuously also are considered quantitative because they are influenced by multiple genes and environmental factors.

Polygenic Inheritance

After the rediscovery of Mendel's work in 1900, questions soon arose about the inheritance of continuously varying characteristics. These characteristics had already been the focus of a group of biologists and statisticians, led by Francis Galton, who used statistical procedures to examine the inheritance of quantitative characteristics such as human height and intelligence. The results of these studies showed that quantitative characteristics are inherited, although the mechanism of inheritance was not yet known. Some biometricians argued that the inheritance of quantitative characteristics could not be explained by Mendelian principles, whereas others felt that Mendel's principles acting on numerous genes (polygenes) could adequately account for the inheritance of quantitative characteristics.

This conflict began to be resolved by the work of Wilhelm Johannsen, who showed that continuous variation in the weight of beans was influenced by both genetic and environmental factors. George Udny Yule, a mathematician, proposed in 1906 that several genes acting together could produce continuous characteristics. This hypothesis was later confirmed by Herman Nilsson-Ehle, working on wheat and tobacco, and by Edward East, working on corn. The argument was finally laid to rest in 1918, when Ronald Fisher demonstrated that the inheritance of quantitative characteristics could indeed be explained by the cumulative effects of many genes, each following Mendel's rules.

Kernel Color in Wheat

To illustrate how multiple genes acting on a characteristic can produce a continuous range of phenotypes, let us examine one of the first demonstrations of polygenic inheritance. Nilsson-Ehle studied kernel color in wheat and found that the intensity of red pigmentation was determined by three unlinked loci, each of which had two alleles.

Nilsson-Ehle obtained several homozygous varieties of wheat that differed in color. Like Mendel, he performed crosses between these homozygous varieties and studied the ratios of phenotypes in the progeny. In one experiment, he crossed a variety of wheat that possessed white kernels with a variety that possessed purple (very dark red) kernels and obtained the following results:

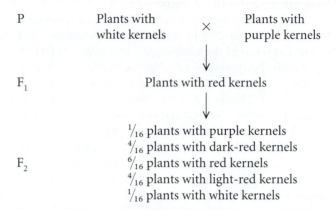

Nilsson-Ehle interpreted this phenotypic ratio as the result of the segregation of alleles at two loci. (Although he found alleles at three loci that affect kernel color, the two varieties used in this cross differed at only two of the loci.) He proposed that there were two alleles at each locus: one that produced red pigment and another that produced no pigment. We'll designate the alleles that encoded pigment A^+ and B^+ and the alleles that encoded no pigment A^- and B^-. Nilsson-Ehle recognized that the effects of the genes were additive. Each gene seemed to contribute equally to color; so the overall phenotype could be determined by adding the effects of all the genes, as shown in the following table.

Genotype	Doses of pigment	Phenotype
$A^+A^+\ B^+B^+$	4	purple
$A^+A^+\ B^+B^-$ $A^+A^-\ B^+B^+$	3	dark red
$A^+A^+\ B^-B^-$ $A^-A^-\ B^+B^+$ $A^+A^-\ B^+B^-$	2	red
$A^+A^-\ B^-B^-$ $A^-A^-\ B^+B^-$	1	light red
$A^-A^-\ B^-B^-$	0	white

Notice that the purple and white phenotypes are each encoded by a single genotype, but other phenotypes may result from several different genotypes.

From these results, we see that five phenotypes are possible when alleles at two loci influence the phenotype and the effects of the genes are additive. When alleles at more than two loci influence the phenotype, more phenotypes are possible, and the color would appear to vary continuously between white and purple. If environmental factors had influenced the characteristic, individuals of the same genotype would vary somewhat in color, making it even more difficult to distinguish between discrete phenotypic classes. Luckily, environment played little role in determining kernel color in Nilsson-Ehle's crosses, and only a few loci encoded color; so Nilsson-Ehle was able to distinguish among the different phenotypic classes. This ability allowed him to see the Mendelian nature of the characteristic.

Let's now see how Mendel's principles explain the ratio obtained by Nilsson-Ehle in his F_2 progeny. Remember that Nilsson-Ehle crossed the homozygous purple variety ($A^+A^+ B^+B^+$) with the homozygous white variety ($A^-A^-\ B^-B^-$), producing F_1 progeny that were heterozygous at both loci ($A^+A^-\ B^+B^-$). This is a dihybrid cross, like those that we worked in Chapter 3, except that both loci encode the same trait. All the F_1 plants possessed two pigment-producing alleles that allowed two doses of color to make red kernels. The types and proportions of progeny expected in the F_2 can be found by applying Mendel's principles of segregation and independent assortment.

Let's first examine the effects of each locus separately. At the first locus, two heterozygous F_1s are crossed ($A^+A^- \times A^+A^-$). As we learned in Chapter 3, when two heterozygotes are crossed, we expect progeny in the proportions $\frac{1}{4} A^+A^+$, $\frac{1}{2} A^+A^-$, and $\frac{1}{4} A^-A^-$. At the second locus, two heterozygotes also are crossed, and, again, we expect progeny in the proportions $\frac{1}{4} B^+B^+$, $\frac{1}{2} B^+B^-$, and $\frac{1}{4} B^-B^-$.

To obtain the probability of combinations of genes at both loci, we must use the multiplication rule of probability (see Chapter 3), which is based on Mendel's principle of independent assortment. The expected proportion of F_2 progeny with genotype $A^+A^+\ B^+B^+$ is the product of the probability of obtaining genotype A^+A^+ ($\frac{1}{4}$) and the probability of obtaining genotype B^+B^+ ($\frac{1}{4}$), or $\frac{1}{4} \times \frac{1}{4} = \frac{1}{16}$ (**Figure 11.4**). The probabilities of each of the phenotypes can then be obtained by adding the probabilities of all the genotypes that produce that phenotype. For example, the red phenotype is produced by three genotypes:

Genotype	Probability
$A^+A^+\ B^-B^-$	$\frac{1}{16}$
$A^-A^-\ B^+B^+$	$\frac{1}{16}$
$A^+A^-\ B^+B^-$	$\frac{1}{4}$

Thus, the overall probability of obtaining red kernels in the F_2 progeny is $\frac{1}{16} + \frac{1}{16} + \frac{1}{4} = \frac{6}{16}$. Figure 11.4 shows that the phenotypic ratio expected in the F_2 is $\frac{1}{16}$ purple, $\frac{4}{16}$ dark red, $\frac{6}{16}$ red, $\frac{4}{16}$ light red, and $\frac{1}{16}$ white. This phenotypic ratio is precisely what Nilsson-Ehle observed in his F_2 progeny, demonstrating that the inheritance of a continu-

11.4 Nilsson-Ehle demonstrated that kernel color in wheat is inherited according to Mendelian principles. He crossed two varieties of wheat that differed in pairs of alleles at two loci affecting kernel color. A purple strain ($A^+A^+\ B^+B^+$) was crossed with a white strain ($A^-A^-\ B^-B^-$), and the F_1 was intercrossed to produce F_2 progeny. The ratio of phenotypes in the F_2 can be determined by breaking the dihybrid cross into two simple single-locus crosses and combining the results by using the multiplication rule.

Question: How is a continous trait, such as kernel color in wheat, inherited?

Methods   Cross wheat having white kernels and wheat having purple kernels. Intercross the F_1 to produce F_2.

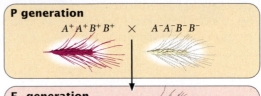

P generation

$$A^+A^+B^+B^+ \quad \times \quad A^-A^-B^-B^-$$

Results

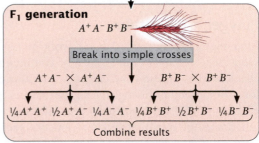

F_1 generation

$$A^+A^-B^+B^-$$

Break into simple crosses

$$A^+A^- \times A^+A^- \qquad B^+B^- \times B^+B^-$$

$$\frac{1}{4}A^+A^+\ \ \frac{1}{2}A^+A^-\ \ \frac{1}{4}A^-A^- \qquad \frac{1}{4}B^+B^+\ \ \frac{1}{2}B^+B^-\ \ \frac{1}{4}B^-B^-$$

Combine results

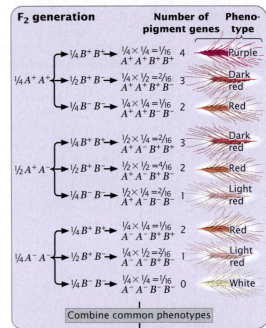

F_2 generation		Number of pigment genes	Phenotype

$\frac{1}{4}A^+A^+$
- $\frac{1}{4}B^+B^+ \rightarrow \frac{1}{4} \times \frac{1}{4} = \frac{1}{16}$ $A^+A^+B^+B^+$ 4 Purple
- $\frac{1}{2}B^+B^- \rightarrow \frac{1}{4} \times \frac{1}{2} = \frac{2}{16}$ $A^+A^+B^+B^-$ 3 Dark red
- $\frac{1}{4}B^-B^- \rightarrow \frac{1}{4} \times \frac{1}{4} = \frac{1}{16}$ $A^+A^+B^-B^-$ 2 Red

$\frac{1}{2}A^+A^-$
- $\frac{1}{4}B^+B^+ \rightarrow \frac{1}{2} \times \frac{1}{4} = \frac{2}{16}$ $A^+A^-B^+B^+$ 3 Dark red
- $\frac{1}{2}B^+B^- \rightarrow \frac{1}{2} \times \frac{1}{2} = \frac{4}{16}$ $A^+A^-B^+B^-$ 2 Red
- $\frac{1}{4}B^-B^- \rightarrow \frac{1}{2} \times \frac{1}{4} = \frac{2}{16}$ $A^+A^-B^-B^-$ 1 Light red

$\frac{1}{4}A^-A^-$
- $\frac{1}{4}B^+B^+ \rightarrow \frac{1}{4} \times \frac{1}{4} = \frac{1}{16}$ $A^-A^-B^+B^+$ 2 Red
- $\frac{1}{2}B^+B^- \rightarrow \frac{1}{4} \times \frac{1}{2} = \frac{2}{16}$ $A^-A^-B^+B^-$ 1 Light red
- $\frac{1}{4}B^-B^- \rightarrow \frac{1}{4} \times \frac{1}{4} = \frac{1}{16}$ $A^-A^-B^-B^-$ 0 White

Combine common phenotypes

F_2 ratio		
Frequency	Number of pigment genes	Phenotype
$\frac{1}{16}$	4	Purple
$\frac{4}{16}$	3	Dark red
$\frac{6}{16}$	2	Red
$\frac{4}{16}$	1	Light red
$\frac{1}{16}$	0	White

Conclusion: Kernel color in wheat is inherited according to Mendel's principles acting on alleles at two loci.

ously varying characteristic such as kernel color is indeed according to Mendel's basic principles.

Nilsson-Ehle's crosses demonstrated that the difference between the inheritance of genes influencing quantitative characteristics and the inheritance of genes influencing discontinuous characteristics is in the *number* of loci that determine the characteristic. When multiple loci affect a character, more genotypes are possible; so the relation between the genotype and the phenotype is less obvious. As the number of loci affecting a character increases, the number of phenotypic classes in the F_2 increases (**Figure 11.5**).

Several conditions of Nilsson-Ehle's crosses greatly simplified the polygenic inheritance of kernel color and made it possible for him to recognize the Mendelian nature of the characteristic. First, genes affecting color segregated at only two or three loci. If genes at many loci had been segregating, he would have had difficulty in distinguishing the phenotypic classes. Second, the genes affecting kernel color had strictly additive effects, making the relation between genotype and phenotype simple. Third, environment played almost no role in the phenotype; had environmental factors modified the phenotypes, distinguishing between the five phenotypic classes would have been difficult. Finally, the loci

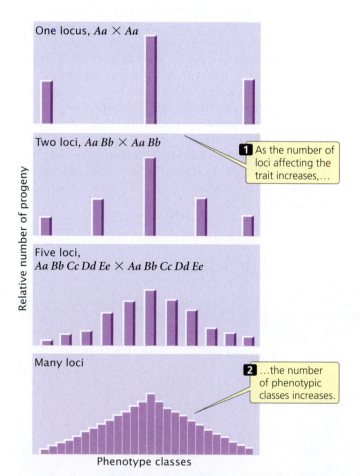

11.5 The results of crossing individuals heterozygous for different numbers of loci affecting a characteristic.

that Nilsson-Ehle studied were not linked; so the genes assorted independently. Nilsson-Ehle was fortunate: for many polygenic characteristics, these simplifying conditions are not present and Mendelian inheritance of these characteristics is not obvious.

Determining Gene Number for a Polygenic Characteristic

When two individuals homozygous for different alleles at a single locus are crossed ($A^1A^1 \times A^2A^2$) and the resulting F_1 are interbred ($A^1A^2 \times A^1A^2$), one-fourth of the F_2 should be homozygous like each of the original parents. If the original parents are homozygous for different alleles at two loci, as are those in Nilsson-Ehle's crosses, then $\frac{1}{4} \times \frac{1}{4} = \frac{1}{16}$ of the F_2 should resemble one of the original homozygous parents. Generally, $(\frac{1}{4})^n$ will be the number of individuals in the F_2 progeny that should resemble each of the original homozygous parents, where n equals the number of loci with a segregating pair of alleles that affects the characteristic. This equation provides us with a possible means of determining the number of loci influencing a quantitative characteristic.

To illustrate the use of this equation, assume that we cross two different homozygous varieties of pea plants that differ in height by 16 cm, interbreed the F_1, and find that approximately $\frac{1}{256}$ of the F_2 are similar to one of the original homozygous parental varieties. This outcome would suggest that four loci with segregating pairs of alleles [$\frac{1}{256} = (\frac{1}{4})^4$] are responsible for the height difference between the two varieties. Because the two homozygous strains differ in height by 16 cm and there are four loci each of which has two alleles (eight alleles in all), each of the alleles contributes 16 cm/8 = 2 cm in height.

This method for determining the number of loci affecting phenotypic differences requires the use of homozygous strains, which may be difficult to obtain in some organisms. It also assumes that all the genes influencing the characteristic have equal effects, that their effects are additive, and that the loci are unlinked. For many polygenic characteristics, these assumptions are not valid, and so this method of determining the number of genes affecting a characteristic has limited application.

CONCEPTS

The principles that determine the inheritance of quantitative characteristics are the same as the principles that determine the inheritance of discontinuous characteristics, but more genes take part in the determination of quantitative characteristics.

✔ CONCEPT CHECK 1

Briefly explain how the number of genes influencing a polygenic trait can be determined?

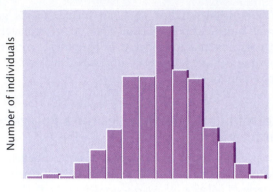

11.6 A frequency distribution is a graph that displays the number or proportion of different phenotypes. Phenotypic values are plotted on the horizontal axis, and the numbers (or proportions) of individuals in each class are plotted on the vertical axis.

11.2 Statistical Methods Are Required for Analyzing Quantitative Characteristics

Because quantitative characteristics are described by a measurement and are influenced by multiple factors, their inheritance must be analyzed statistically. This section will explain the basic concepts of statistics that are used to analyze quantitative characteristics.

Distributions

Understanding the genetic basis of any characteristic begins with a description of the numbers and kinds of phenotypes present in a group of individuals. Phenotypic variation in a group, such as the progeny of a cross, can be conveniently represented by a **frequency distribution,** which is a graph of the frequencies (numbers or proportions) of the different phenotypes (**Figure 11.6**). In a typical frequency distribution, the phenotypic classes are plotted on the horizontal (x) axis, and the numbers (or proportions) of individuals in each class are plotted on the vertical (y) axis. A frequency distribution is a concise method of summarizing all phenotypes of a quantitative characteristic.

Connecting the points of a frequency distribution with a line creates a curve that is characteristic of the distribution. Many quantitative characteristics exhibit a symmetrical (bell-shaped) curve called a **normal distribution** (**Figure 11.7a**). Normal distributions arise when a large number of independent factors contribute to a measurement, as is often the case in quantitative characteristics. Two other common types of distributions (skewed and bimodal) are illustrated in **Figure 11.7b** and **c**.

Samples and Populations

Biologists frequently need to describe the distribution of phenotypes exhibited by a group of individuals. We might want to describe the height of students at the University of Texas (UT), but there are more than 40,000 students at UT, and measuring every one of them would be impractical. Scientists are constantly confronted with this problem: the group of interest, called the **population,** is too large for a complete census. One solution is to measure a smaller collection of individuals, called a **sample,** and use measurements made on the sample to describe the population.

To provide an accurate description of the population, a good sample must have several characteristics. First, it must be representative of the whole population. If our sample consisted entirely of members of the UT basketball team, for instance, we would probably overestimate the true height of the students. One way to ensure that a sample is representative of the population is to select the members of the sample randomly. Second, the sample must be large enough that

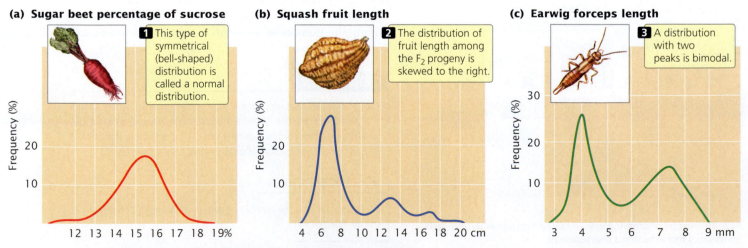

(a) Sugar beet percentage of sucrose

1 This type of symmetrical (bell-shaped) distribution is called a normal distribution.

(b) Squash fruit length

2 The distribution of fruit length among the F₂ progeny is skewed to the right.

(c) Earwig forceps length

3 A distribution with two peaks is bimodal.

11.7 Distributions of phenotypes can assume several different shapes.

chance differences between individuals in the sample and the overall population do not distort the estimate of the population measurements. If we measured only three students at UT and just by chance all three were short, we would underestimate the true height of the student population. Statistics can provide information about how much confidence to expect from estimates based on random samples.

CONCEPTS

In statistics, the population is the group of interest; a sample is a subset of the population. The sample should be representative of the population and large enough to minimize chance differences between the population and the sample.

✔ CONCEPT CHECK 2

A geneticist is interested in whether asthma is caused by a mutation in the *DS112* gene. The geneticist collects DNA from 120 people with asthma and 100 healthy people and sequences the DNA. She finds that 35 of the people with asthma and none of the healthy people have a mutation in the *DS112* gene. What is the population in this study?

a. The 120 people with asthma

b. The 100 healthy people

c. The 35 people with a mutation in their gene

d. All people with asthma

The Mean

The **mean,** also called the average, provides information about the center of the distribution. If we measured the heights of 10-year-old and 18-year-old boys and plotted a frequency distribution for each group, we would find that both distributions are normal, but the two distributions would be centered at different heights, and this difference would be indicated in their different means (**Figure 11.8**).

Suppose we have five measurements of height in centimeters: 160, 161, 167, 164, and 165. If we represent a group of measurements as x_1, x_2, x_3, and so forth, then the mean ($\bar{x}$) is calculated by adding all the individual measurements and dividing by the total number of measurements in the sample (n):

$$\bar{x} = \frac{x_1 + x_2 + x_3 + \cdots + x_n}{n} \tag{11.1}$$

In our example, $x_1 = 160$, $x_2 = 161$, $x_3 = 167$, and so forth. The mean height ($\bar{x}$) equals:

$$\bar{x} = \frac{160 + 161 + 167 + 164 + 165}{5} = \frac{817}{5} = 163.4$$

A shorthand way to represent this formula is

$$\bar{x} = \frac{\Sigma x_i}{n} \tag{11.2}$$

or

$$\bar{x} = \frac{1}{n}\Sigma x_i \tag{11.3}$$

where the symbol Σ means "the summation of" and x_i represents individual x values.

The Variance and Standard Deviation

A statistic that provides key information about a distribution is the **variance,** which indicates the variability of a group of measurements, or how spread out the distribution is. Distributions may have the same mean but different variances (**Figure 11.9**). The larger the variance, the greater the spread of measurements in a distribution about its mean.

The variance (s^2) is defined as the average squared deviation from the mean:

$$s^2 = \frac{\Sigma(x_i - \bar{x})^2}{n - 1} \tag{11.4}$$

To calculate the variance, we (1) subtract the mean from each measurement and square the value obtained, (2) add all the squared deviations, and (3) divide this sum by the number of original measurements minus 1.

Another statistic that is closely related to the variance is the **standard deviation** (s), which is defined as the square root of the variance:

$$s = \sqrt{s^2} \tag{11.5}$$

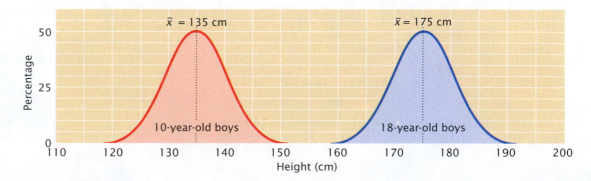

11.8 The mean provides information about the center of a distribution. Both distributions of heights of 10-year-old and 18-year-old boys are normal, but they have different locations along a continuum of height, which makes their means different.

$\bar{x} = 135$ cm $\bar{x} = 175$ cm

10-year-old boys   18-year-old boys

Percentage

Height (cm)

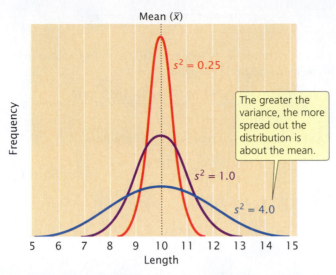

11.9 The variance provides information about the variability of a group of phenotypes. Shown here are three distributions with the same mean but different variances.

Whereas the variance is expressed in units squared, the standard deviation is in the same units as the original measurements; so the standard deviation is often preferred for describing the variability of a measurement.

A normal distribution is symmetrical; so the mean and standard deviation are sufficient to describe its shape. The mean plus or minus one standard deviation ($\bar{x} \pm s$) includes approximately 66% of the measurements in a normal distribution; the mean plus or minus two standard deviations ($\bar{x} \pm 2s$) includes approximately 95% of the measurements, and the mean plus or minus three standard deviations ($\bar{x} \pm 3s$) includes approximately 99% of the measurements (**Figure 11.10**). Thus, only 1% of a normally distributed population lies outside the range of $\bar{x} \pm 3s$.

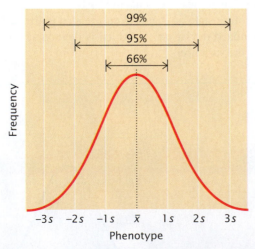

11.10 The proportions of a normal distribution occupied by plus or minus one, two, and three standard deviations from the mean.

CONCEPTS

The mean and variance describe a distribution of measurements: the mean provides information about the location of the center of a distribution, and the variance provides information about its variability.

✔ CONCEPT CHECK 3

The measurements of a distribution with a higher ___ will be more spread out.

a. Mean   c. Standard deviation

b. Variance   d. Both b and c above

Correlation

The mean and the variance can be used to describe an individual characteristic, but geneticists are frequently interested in more than one characteristic. Often, two or more characteristics vary together. For instance, both the number and the weight of eggs produced by hens are important to the poultry industry. These two characteristics are not independent of each other. There is an inverse relation between egg number and weight: hens that lay more eggs produce smaller eggs. This kind of relation between two characteristics is called a **correlation.** When two characteristics are correlated, a change in one characteristic is likely to be associated with a change in the other.

Correlations between characteristics are measured by a **correlation coefficient** (designated r), which measures the strength of their association. Consider two characteristics, such as human height (x) and arm length (y). To determine how these characteristics are correlated, we first obtain the covariance (cov) of x and y:

$$\text{cov}_{xy} = \frac{\Sigma(x_i - \bar{x})(y_i - \bar{y})}{n - 1} \tag{11.6}$$

The covariance is computed by (1) taking an x value for an individual and subtracting it from the mean of x ($\bar{x}$); (2) taking the y value for the same individual and subtracting it from the mean of y ($\bar{y}$); (3) multiplying the results of these two subtractions; (4) adding the results for all the xy pairs; and (5) dividing this sum by $n - 1$ (where n equals the number of xy pairs).

The correlation coefficient (r) is obtained by dividing the covariance of x and y by the product of the standard deviations of x and y:

$$r = \frac{\text{cov}_{xy}}{s_x s_y} \tag{11.7}$$

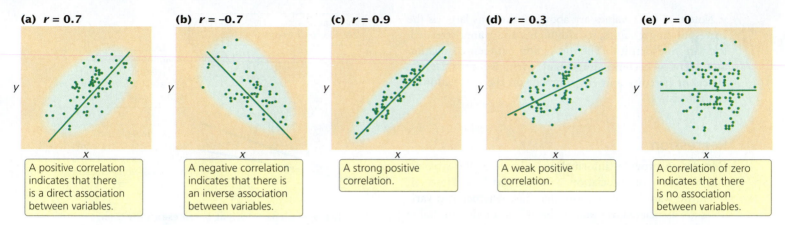

(a) *r* = 0.7

y

x

A positive correlation indicates that there is a direct association between variables.

(b) *r* = −0.7

y

x

A negative correlation indicates that there is an inverse association between variables.

(c) *r* = 0.9

y

x

A strong positive correlation.

(d) *r* = 0.3

y

x

A weak positive correlation.

(e) *r* = 0

y

x

A correlation of zero indicates that there is no association between variables.

11.11 The correlation coefficient describes the relation between two or more variables.

A correlation coefficient can theoretically range from −1 to +1. A positive value indicates that there is a direct association between the variables (**Figure 11.11a**): as one variable increases, the other variable also tends to increase. A positive correlation exists for human height and weight: tall people tend to weigh more. A negative correlation coefficient indicates that there is an inverse relation between the two variables (**Figure 11.11b**): as one variable increases, the other tends to decrease (as is the case for egg number and egg weight in chickens).

The absolute value of the correlation coefficient (the size of the coefficient, ignoring its sign) provides information about the strength of association between the variables. A coefficient of −1 or +1 indicates a perfect correlation between the variables, meaning that a change in x is always accompanied by a proportional change in y. Correlation coefficients close to −1 or close to +1 indicate a strong association between the variables: a change in x is almost always associated with a proportional increase in y, as seen in **Figure 11.11c**. On the other hand, a correlation coefficient closer to 0 indicates a weak correlation: a change in x is associated with a change in y but not always (**Figure 11.11d**). A correlation of 0 indicates that there is no association between variables (**Figure 11.11e**).

A correlation coefficient can be computed for two variables measured for the same individual, such as height (x) and weight (y). A correlation coefficient can also be computed for a single variable measured for pairs of individuals. For example, we can calculate for fish the correlation between the number of vertebrae of a parent (x) and the number of vertebrae of its offspring (y), as shown in **Figure 11.12**. This approach is often used in quantitative genetics.

A correlation between two variables indicates only that the variables are associated; it does not imply a cause-and-effect relation. Correlation also does not mean that the values of two variables are the same; it means only that a change in one variable is associated with a proportional change in the other variable. For example, the x and y variables in the

following list are almost perfectly correlated, with a correlation coefficient of 0.99.

	x value	*y* value
	12	123
	14	140
	10	110
	6	61
	3	32
Average:	9	90

A high correlation is found between these x and y variables; larger values of x are always associated with larger values of

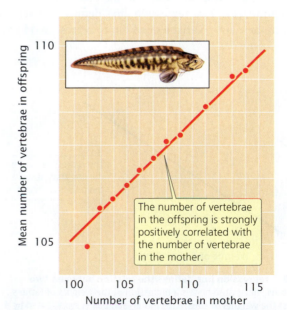

The number of vertebrae in the offspring is strongly positively correlated with the number of vertebrae in the mother.

11.12 A correlation coefficient can be computed for a single variable measured for pairs of individuals. Here, the numbers of vertebrae in mothers and offspring of the fish *Zoarces viviparus* are compared.

y. Note that the *y* values are about 10 times as large as the corresponding *x* values; so, although *x* and *y* are correlated, they are not identical. The distinction between correlation and identity becomes important when we consider the effects of heredity and environment on the correlation of characteristics.

Regression

Correlation provides information only about the strength and direction of association between variables. However, often, we want to know more than just whether two variables are associated; we want to be able to predict the value of one variable, given a value of the other.

A positive correlation exists between the body weight of parents and the body weight of their offspring; this correlation exists in part because genes influence body weight, and parents and children have genes in common. Because of this association between parental and offspring phenotypes, we can predict the weight of an individual on the basis of the weights of its parents. This type of statistical prediction is called **regression.** This technique plays an important role in quantitative genetics because it allows us to predict the characteristics of offspring from a given mating, even without knowledge of the genotypes that encode the characteristic.

Regression can be understood by plotting a series of *x* and *y* values. **Figure 11.13** illustrates the relation between the weight of a father (*x*) and the weight of his son (*y*). Each father–son pair is represented by a point on the graph. The overall relation between these two variables is depicted by the regression line, which is the line that best fits all the points on the graph (deviations of the points from the line are minimized). The regression line defines the relation between the *x* and *y* variables and can be represented by

$$y = a + bx \qquad (11.8)$$

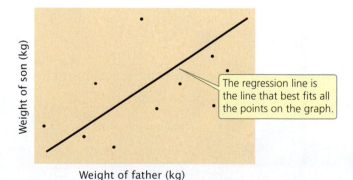

The regression line is the line that best fits all the points on the graph.

11.13 A regression line defines the relation between two variables. Illustrated here is a regression of the weights of fathers against the weights of sons. Each father–son pair is represented by a point on the graph: the *x* value of a point is the father's weight and the *y* value of the point is the son's weight.

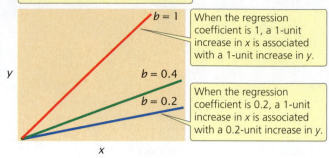

The regression coefficient represents the slope of the regression line.

When the regression coefficient is 1, a 1-unit increase in *x* is associated with a 1-unit increase in *y*.

When the regression coefficient is 0.2, a 1-unit increase in *x* is associated with a 0.2-unit increase in *y*.

11.14 The regression coefficient, *b*, represents the change in *y* per unit change in *x*. Shown here are regression lines with different regression coefficients.

In Equation 11.8, *x* and *y* represent the *x* and *y* variables (in this case, the father's weight and the son's weight, respectively). The variable *a* is the *y* intercept of the line, which is the expected value of *y* when *x* is 0. Variable *b* is the slope of the regression line, also called the **regression coefficient.**

Trying to position a regression line by eye is not only very difficult but also inaccurate when there are many points scattered over a wide area. Fortunately, the regression coefficient and the *y* intercept can be obtained mathematically. The regression coefficient (*b*) can be computed from the covariance of *x* and *y* (cov_{xy}) and the variance of *x* (*s*) by

$$b = \frac{\text{cov}_{xy}}{s_x^2} \qquad (11.9)$$

The regression coefficient indicates how much *y* increases, on average, per increase in *x*. Several regression lines with different regression coefficients are illustrated in **Figure 11.14.** Notice that as the regression coefficient increases, the slope of the regression line increases.

After the regression coefficient has been calculated, the *y* intercept can be calculated by substituting the regression coefficient and the mean values of *x* and *y* into the following equation:

$$a = \bar{y} - b\bar{x} \qquad (11.10)$$

The regression equation (*y* = *a* + *bx*, Equation 11.8) can then be used to predict the value of any *y* given the value of *x*.

CONCEPTS

A correlation coefficient measures the strength of association between two variables. The sign (positive or negative) indicates the direction of the correlation; the absolute value measures the strength of the association. Regression is used to predict the value of one variable on the basis of the value of a correlated variable.

✔ CONCEPT CHECK 4

In Lubbock, Texas, rainfall and temperature exhibit a significant correlation of −0.7. Which conclusion is correct?

a. There is usually more rainfall when the temperature is high.

b. There is usually more rainfall when the temperature is low.

c. Rainfall is equally likely when the temperature is high or low.

Worked Problem

Body weights of 11 female fishes and the numbers of eggs that they produce are:

Weight (mg) x	Eggs (thousands) y
14	61
17	37
24	65
25	69
27	54
33	93
34	87
37	89
40	100
41	90
42	97

What are the correlation coefficient and the regression coefficient for body weight and egg number in these 11 fishes?

• Solution

The computations needed to answer this question are given in the table at the bottom of the page. To calculate the correlation and regression coefficients, we first obtain the sum of all the x_i values (Σx_i) and the sum of all the y_i values (Σy_i); these sums are shown in the last row of the table. We can calculate the means of the two variables by dividing the sums by the number of measurements, which is 11:

$$\bar{x} = \frac{\Sigma x_i}{n} = \frac{334}{11} = 30.36$$

$$\bar{y} = \frac{\Sigma x_i}{n} = \frac{842}{11} = 76.55$$

After the means have been calculated, the deviations of each value from the mean are computed; these deviations are shown in columns B and E of the table. The deviations are then squared (columns C and F) and summed (last row of columns C and F). Next, the products of the deviation of the x values and the deviation of the y values $[(x_i - \bar{x})(y_i - \bar{y})]$ are calculated; these products are shown in column G, and their sum is shown in the last row of column G.

To calculate the covariance, we use Equation 11.6:

$$\text{cov}_{xy} = \frac{\Sigma(x_i - \bar{x})(y_i - \bar{y})}{n-1} = \frac{1743.84}{10} = 174.38$$

A Weight (mg) x	B $x_i - \bar{x}$	C $(x_i - \bar{x})^2$	D Eggs (thousands) y	E $y_i - \bar{y}$	F $(y_i - \bar{y})^2$	G $(x_i - \bar{x})(y_i - \bar{y})$
14	−16.36	267.65	61	−15.55	241.80	254.40
17	−13.36	178.49	37	−39.55	1564.20	528.39
24	−6.36	40.45	65	−11.55	133.40	73.46
25	−5.36	28.73	69	−7.55	57.00	40.47
27	−3.36	11.29	54	−22.55	508.50	75.77
33	2.64	6.97	93	16.45	270.60	43.43
34	3.64	13.25	87	10.45	109.20	38.04
37	6.64	44.09	89	12.45	155.00	82.67
40	9.64	92.93	100	23.45	549.90	226.06
41	10.64	113.21	90	13.45	180.90	143.11
42	11.64	135.49	97	20.45	418.20	238.04
$\Sigma x_i = 334$		$\Sigma(x - \bar{x})^2 = 932.55$	$\Sigma y_i = 842$		$\Sigma(y - \bar{y})^2 = 4188.70$	$\Sigma(x_i - \bar{x})(y_i - \bar{y}) = 1743.84$

Source: R. R. Sokal and F. J. Rohlf, *Biometry*, 2d ed. (San Francisco: W. H. Freeman and Company, 1981).

To calculate the covariance and the regression requires the variances and standard deviations of x and y:

$$s_x^2 = \frac{\Sigma(x_i - \bar{x})^2}{n - 1} = \frac{932.55}{10} = 93.26$$

$$s_x = \sqrt{s_x^2} = \sqrt{93.26} = 9.66$$

$$s_y^2 = \frac{\Sigma(y_i - \bar{y})^2}{n - 1} = \frac{4188.70}{10} = 418.87$$

$$s_y = \sqrt{s_y^2} = \sqrt{418.87} = 20.47$$

We can now compute the correlation and regression coefficients as shown here.

Correlation coefficient:

$$r = \frac{\text{cov}_{xy}}{s_x s_y} = \frac{174.38}{9.66 \times 20.47} = 0.88$$

Regression coefficient:

$$b = \frac{\text{cov}_{xy}}{s_x^2} = \frac{174.38}{93.26} = 1.87$$

? Practice your understanding of correlation and regression by working Problem 26 at the end of the chapter.

Applying Statistics to the Study of a Polygenic Characteristic

Edward East carried out one early statistical study of polygenic inheritance on the length of flowers in tobacco (*Nicotiana longiflora*). He obtained two varieties of tobacco that differed in flower length: one variety had a mean flower length of 40.5 mm, and the other had a mean flower length of 93.3 mm (**Figure 11.15**). These two varieties had been inbred for many generations and were homozygous at all loci contributing to flower length. Thus, there was no genetic variation in the original parental strains; the small differences in flower length within each strain were due to environmental effects on flower length.

When East crossed the two strains, he found that flower length in the F_1 was about halfway between that in the two parents (see Figure 11.15), as would be expected if the genes determining the differences in the two strains were additive in their effects. The variance of flower length in the F_1 was similar to that seen in the parents, because all the F_1 had the same genotype, as did each parental strain (the F_1 were all heterozygous at the genes that differed between the two parental varieties).

East then interbred the F_1 to produce F_2 progeny. The mean flower length of the F_2 was similar to that of the F_1, but

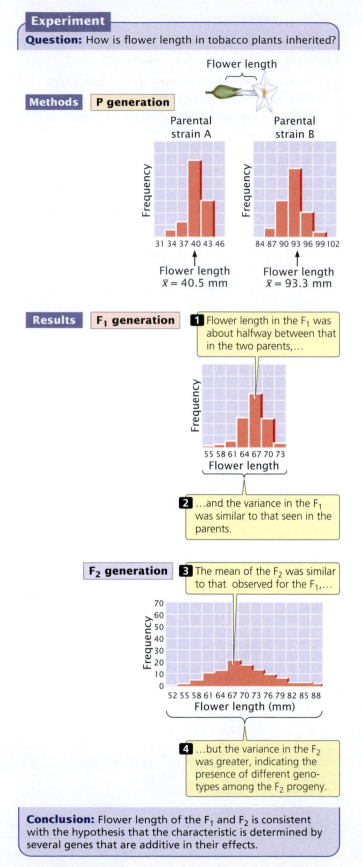

Experiment

Question: How is flower length in tobacco plants inherited?

Flower length

Methods **P generation**

Parental strain A — Frequency vs. Flower length (31 34 37 40 43 46) — Flower length $\bar{x} = 40.5$ mm

Parental strain B — Frequency vs. Flower length (84 87 90 93 96 99 102) — Flower length $\bar{x} = 93.3$ mm

Results **F_1 generation**

1 Flower length in the F_1 was about halfway between that in the two parents,...

Frequency vs. Flower length (55 58 61 64 67 70 73)

2 ...and the variance in the F_1 was similar to that seen in the parents.

F_2 generation

3 The mean of the F_2 was similar to that observed for the F_1,...

Frequency (0 10 20 30 40 50 60 70) vs. Flower length (mm) (52 55 58 61 64 67 70 73 76 79 82 85 88)

4 ...but the variance in the F_2 was greater, indicating the presence of different genotypes among the F_2 progeny.

Conclusion: Flower length of the F_1 and F_2 is consistent with the hypothesis that the characteristic is determined by several genes that are additive in their effects.

11.15 Edward East conducted an early statistical study of the inheritance of flower length in tobacco.

the variance of the F_2 was much greater (see Figure 11.15). This greater variability indicates that not all of the F_2 progeny had the same genotype.

East selected some F_2 plants and interbred them to produce F_3 progeny. He found that flower length of the F_3 depended on flower length in the plants selected as their parents. This finding demonstrated that flower-length differences in the F_2 were partly genetic and were therefore passed to the next generation. None of the 444 F_2 plants that East raised exhibited flower lengths similar to those of the two parental strains. This result suggested that more than four loci with pairs of alleles affected flower length in his varieties, because four allelic pairs are expected to produce 1 of 256 progeny $[(\frac{1}{4})^4 = \frac{1}{256}]$ having one or the other of the original parental phenotypes.

11.3 Heritability Is Used to Estimate the Proportion of Variation in a Trait That Is Genetic

In addition to being polygenic, quantitative characteristics are frequently influenced by environmental factors. It is often useful to know how much of the variation in a quantitative characteristic is due to genetic differences and how much is due to environmental differences. The proportion of the total phenotypic variation that is due to genetic differences is known as the **heritability.**

Consider a dairy farmer who owns several hundred milk cows. The farmer notices that some cows consistently produce more milk than others. The nature of these differences is important to the profitability of his dairy operation. If the differences in milk production are largely genetic in origin, then the farmer may be able to boost milk production by selectively breeding the cows that produce the most milk. On the other hand, if the differences are largely environmental in origin, selective breeding will have little effect on milk production, and the farmer might better boost milk production by adjusting the environmental factors associated with higher milk production. To determine the extent of genetic and environmental influences on variation in a characteristic, phenotypic variation in the characteristic must be partitioned into components attributable to different factors.

Phenotypic Variance

To determine how much of phenotypic differences in a population is due to genetic and environmental factors, we must first have some quantitative measure of the phenotype under consideration. Consider a population of wild plants that differ in size. We could collect a representative sample of plants from the population, weigh each plant in the sample, and calculate the mean and variance of plant weight. This **phenotypic variance** is represented by V_P.

Components of phenotypic variance   First, some of the phenotypic variance may be due to differences in genotypes among individual members of the population. These differences are termed the **genetic variance** and are represented by V_G.

Second, some of the differences in phenotype may be due to environmental differences among the plants; these differences are termed the **environmental variance,** V_E. Environmental variance includes differences in environmental factors such as the amount of light or water that the plant receives; it also includes random differences in development that cannot be attributed to any specific factor. Any variation in phenotype that is not inherited is, by definition, a part of the environmental variance.

Third, **genetic–environmental interaction variance** (V_{GE}) arises when the effect of a gene depends on the specific environment in which it is found. An example is shown in **Figure 11.16.** In a dry environment, genotype AA produces a plant that averages 12 g in weight, and genotype aa produces a smaller plant that averages 10 g. In a wet environment, genotype aa produces the larger plant, averaging 24 g in weight, whereas genotype AA produces a plant that averages 20 g. In this example, there are clearly differences in the two environments: both genotypes produce heavier plants in the wet environment. There are also differences in the weights of the two genotypes, but the relative performances of the genotypes depend on whether the plants are grown in a wet or a dry environment. In this case, the influences on phenotype cannot

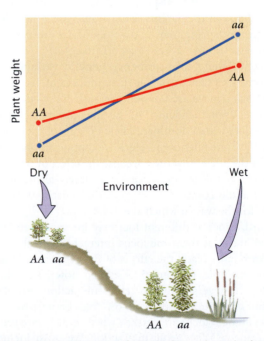

11.16 Genetic–environmental interaction variance is obtained when the effect of a gene depends on the specific environment in which it is found. In this example, the genotype affects plant weight, but the environmental conditions determine which genotype produces the heavier plant.

be neatly allocated into genetic and environmental components, because the expression of the genotype depends on the environment in which the plant grows. The phenotypic variance must therefore include a component that accounts for the way in which genetic and environmental factors interact.

In summary, the total phenotypic variance can be apportioned into three components:

$$V_P = V_G + V_E + V_{GE} \qquad (11.11)$$

Components of genetic variance   Genetic variance can be further subdivided into components consisting of different types of genetic effects. First, **additive genetic variance** (V_A) comprises the additive effects of genes on the phenotype, which can be summed to determine the overall effect on the phenotype. For example, suppose that, in a plant, allele A^1 contributes 2 g in weight and allele A^2 contributes 4 g. If the alleles are strictly additive, then the genotypes would have the following weights:

$$A^1A^1 = 2 + 2 = 4\,\text{g}$$
$$A^1A^2 = 2 + 4 = 6\,\text{g}$$
$$A^2A^2 = 4 + 4 = 8\,\text{g}$$

The genes that Nilsson-Ehle studied, which affected kernel color in wheat, were additive in this way. The additive genetic variance primarily determines the resemblance between parents and offspring. For example, if all of the phenotypic variance is due to additive genetic variance, then the phenotypes of the offspring will be exactly intermediate between those of the parents, but, if some genes have dominance, then offspring may be phenotypically different from both parents (i.e., $Aa \times Aa \rightarrow aa$ offspring).

Second, there is **dominance genetic variance** (V_D) when some genes have a dominance component. In this case, the alleles at a locus are not additive; rather, the effect of an allele depends on the identity of the other allele at that locus. For example, with a dominant allele (T), genotypes TT and Tt have the same phenotype. Here, we cannot simply add the effects of the alleles together, because the effect of the small t allele is masked by the presence of the large T allele. Instead, we must add a component (V_D) to the genetic variance to account for the way in which alleles interact.

Third, genes at different loci may interact in the same way that alleles at the same locus interact. When this genic interaction takes place, the effects of the genes are not additive. For example, Chapter 5 described how coat color in Labrador retrievers exhibits genic interaction; genotypes $BB\ ee$ and $bb\ ee$ both produce yellow dogs, because the effect of alleles at the B locus are masked when ee alleles are present at the E locus. With genic interaction, we must include a third component, called **genic interaction variance** (V_I), to the genetic variance:

$$V_G = V_A + V_D + V_I \qquad (11.12)$$

Summary equation   We can now integrate these components into one equation to represent all the potential contributions to the phenotypic variance:

$$V_P = V_A + V_D + V_I + V_E + V_{GE} \qquad (11.13)$$

This equation provides us with a model that describes the potential causes of differences that we observe among individual phenotypes. It's important to note that this model deals strictly with the observable *differences* (variance) in phenotypes among individual members of a population; it says nothing about the absolute value of the characteristic or about the underlying genotypes that produce these differences.

Types of Heritability

The model of phenotypic variance that we've just developed can be used to address the question of how much of the phenotypic variance in a characteristic is due to genetic differences. **Broad-sense heritability** (H^2) represents the proportion of phenotypic variance that is due to genetic variance and is calculated by dividing the genetic variance by the phenotypic variance:

$$\text{broad-sense heritability} = H^2 = \frac{V_G}{V_P} \qquad (11.14)$$

The symbol H^2 represents broad-sense heritability because it is a measure of variance, which is in units squared.

Broad-sense heritability can potentially range from 0 to 1. A value of 0 indicates that none of the phenotypic variance results from differences in genotype and all of the differences in phenotype result from environmental variation. A value of 1 indicates that all of the phenotypic variance results from differences in genotypes. A heritability value between 0 and 1 indicates that both genetic and environmental factors influence the phenotypic variance.

Often, we are more interested in the proportion of the phenotypic variance that results from the additive genetic variance because, as mentioned earlier, the additive genetic variance primarily determines the resemblance between parents and offspring. **Narrow-sense heritability** (h^2) is equal to the additive genetic variance divided by the phenotypic variance:

$$\text{narrow-sense heritability} = h^2 = \frac{V_A}{V_P} \qquad (11.15)$$

Calculating Heritability

Having considered the components that contribute to phenotypic variance and having developed a general concept of heritability, we can ask, How do we go about estimating these different components and calculating heritability?

There are several ways to measure the heritability of a characteristic. They include eliminating one or more variance components, comparing the resemblance of parents and offspring, comparing the phenotypic variances of individuals with different degrees of relatedness, and measuring the response to selection. The mathematical theory that underlies these calculations of heritability is complex and beyond the scope of this book. Nevertheless, we can develop a general understanding of how heritability is measured.

Heritability by elimination of variance components
One way of calculating the broad-sense heritability is to eliminate one of the variance components. We have seen that $V_P = V_G + V_E + V_{GE}$. If we eliminate all environmental variance ($V_E = 0$), then $V_{GE} = 0$ (because, if either V_G or V_E is zero, no genetic–environmental interaction can take place), and $V_P = V_G$. In theory, we might make V_E equal to 0 by ensuring that all individuals were raised in exactly the same environment but, in practice, it is virtually impossible. Instead, we could make V_G equal to 0 by raising genetically identical individuals, causing V_P to be equal to V_E. In a typical experiment, we might raise cloned or highly inbred, identically homozygous individuals in a defined environment and measure their phenotypic variance to estimate V_E. We could then raise a group of genetically variable individuals and measure their phenotypic variance (V_P). Using V_E calculated on the genetically identical individuals, we could obtain the genetic variance of the variable individuals by subtraction:

$$V_{\text{G (of genetically varying individuals)}} =$$
$$V_{\text{P (of genetically varying individuals)}} \tag{11.16}$$
$$- V_{\text{E (of genetically identical individuals)}}$$

The broad-sense heritability of the genetically variable individuals would then be calculated as follows:

$$H^2 = \frac{V_{\text{G (of genetically varying individuals)}}}{V_{\text{P (of genetically varying individuals)}}} \tag{11.17}$$

Sewall Wright used this method to estimate the heritability of white spotting in guinea pigs. He first measured the phenotypic variance for white spotting in a genetically variable population and found that $V_P = 573$. Then he inbred the guinea pigs for many generations so that they were essentially homozygous and genetically identical. When he measured their phenotypic variance in white spotting, he obtained V_P equal to 340. Because $V_G = 0$ in this group, their $V_P = V_E$. Wright assumed this value of environmental variance for the original (genetically variable) population and estimated their genetic variance:

$$V_P - V_E = V_G$$
$$573 - 340 = 233$$

He then estimated the broad-sense heritability from the genetic and phenotypic variance:

$$H^2 = \frac{V_G}{V_P}$$
$$H^2 = \frac{233}{573} = 0.41$$

This value implies that 41% of the variation in spotting of guinea pigs in Wright's population was due to differences in genotype.

Estimating heritability by using this method assumes that the environmental variance of genetically identical individuals is the same as the environmental variance of the genetically variable individuals, which may not be true. Additionally, this approach can be applied only to organisms for which it is possible to create genetically identical individuals.

Heritability by parent–offspring regression
Another method for estimating heritability is to compare the phenotypes of parents and offspring. When genetic differences are responsible for phenotypic variance, offspring should resemble their parents more than they resemble unrelated individuals, because offspring and parents have some genes in common that help determine their phenotype. Correlation and regression can be used to analyze the association of phenotypes in different individuals.

To calculate the narrow-sense heritability in this way, we first measure the characteristic on a series of parents and offspring. The data are arranged into families, and the mean parental phenotype is plotted against the mean offspring phenotype (**Figure 11.17**). Each data point in the graph represents one family; the value on the x (horizontal) axis is the mean phenotypic value of the parents in a family, and the value on the y (vertical) axis is the mean phenotypic value of the offspring for the family.

Let's assume that there is no narrow-sense heritability for the characteristic ($h^2 = 0$), meaning that genetic differences do not contribute to the phenotypic differences among individuals. In this case, offspring will be no more similar to their parents than they are to unrelated individuals, and the data points will be scattered randomly, generating a regression coefficient of zero (see Figure 11.17a). Next, let's assume that all of the phenotypic differences are due to additive genetic differences ($h^2 = 1.0$). In this case, the mean phenotype of the offspring will be equal to the mean phenotype of the parents, and the regression coefficient will be 1 (see Figure 11.17b). If genes and environment both contribute to the differences in phenotype, both heritability and the regression coefficient will lie between 0 and 1 (see Figure 11.17c). The regression coefficient therefore provides information about the magnitude of the heritability.

A complex mathematical proof (which we will not go into here) demonstrates that, in a regression of the mean phenotype of the offspring against the mean phenotype of

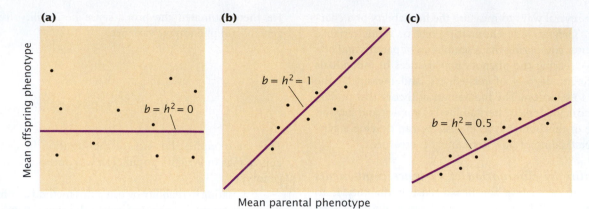

11.17 The narrow-sense heritability, h^2, equals the regression coefficient, b, in a regression of the mean phenotype of the offspring against the mean phenotype of the parents. (a) There is no relation between the parental phenotype and the offspring phenotype. (b) The offspring phenotype is the same as the parental phenotypes. (c) Both genes and environment contribute to the differences in phenotype.

the parents, narrow-sense heritability (h^2) equals the regression coefficient (b):

$$h^2 = b_{\text{(regression of mean offspring against mean of both parents)}}$$ (11.18)

An example of calculating heritability by regression of the phenotypes of parents and offspring is illustrated in **Figure 11.18.** In a regression of the mean offspring phenotype against the phenotype of only one parent, the narrow-sense heritability equals twice the regression coefficient:

$$h^2 = 2b_{\text{(regression of mean offspring against mean of one parent)}}$$ (11.19)

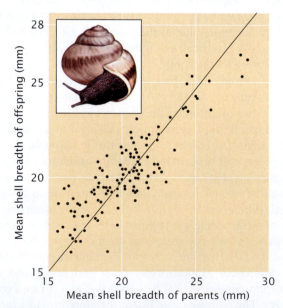

11.18 The heritability of shell breadth in snails can be determined by regression of the phenotype of offspring against the mean phenotype of the parents. The regression coefficient, which equals the heritability, is 0.70. *[From L. M. Cook, Evolution 19:86–94, 1965.]*

With only one parent, the heritability is twice the regression coefficient because only half the genes of the offspring come from one parent; thus, we must double the regression coefficient to obtain the full heritability.

Heritability and degrees of relatedness A third method for calculating heritability is to compare the phenotypes of individuals having different degrees of relatedness. This method is based on the concept that the more closely related two individuals are, the more genes they have in common.

Monozygotic (identical) twins have 100% of their genes in common, whereas dizygotic (nonidentical) twins have, on average, 50% of their genes in common. If genes are important in determining variability in a characteristic, then monozygotic twins should be more similar in a particular characteristic than dizygotic twins. By using correlation to compare the phenotypes of monozygotic and dizygotic twins, we can estimate broad-sense heritability. A rough estimate of the broad-sense heritability can be obtained by taking twice the difference of the correlation coefficients for a quantitative characteristic in monozygotic and dizygotic twins:

$$H^2 = 2(r_{\text{MZ}} - r_{\text{DZ}})$$ (11.20)

where r_{MZ} equals the correlation coefficient among monozygotic twins and r_{DZ} equals the correlation coefficient among dizygotic twins. For example, suppose we found the correlation of height among the two members of monozygotic twin pairs (r_{MZ}) to be 0.9 and the correlation of height among the two members of dizygotic twins (r_{DZ}) to be 0.5. The broad sense heritability for height would be $H^2 = 2(0.9 - 0.5)$ $= 2(0.4) = 0.8$. This calculation assumes that the two individuals of a monozygotic twin pair experience environments that are no more similar to each other than those experienced by the two individuals of a dizygotic twin pair. This assumption is often not met when twins have been reared together.

Narrow-sense heritability also can be estimated by comparing the phenotypic variances for a characteristic in full sibs (siblings who have both parents in common as well as 50% of their genes on the average) and half sibs (who have only one parent in common and thus 25% of their genes on the average).

All estimates of heritability depend on the assumption that the environments of related individuals are not more similar than those of unrelated individuals. This assumption is difficult to meet in human studies, because related people are usually reared together. Heritability estimates for humans should therefore always be viewed with caution.

CONCEPTS

Broad-sense heritability is the proportion of phenotypic variance that is due to genetic variance. Narrow-sense heritability is the proportion of phenotypic variance that is due to additive genetic variance. Heritability can be measured by eliminating one of the variance components, by analyzing parent–offspring regression, or by comparing individuals having different degrees of relatedness.

✔ CONCEPT CHECK 5

If the environmental variance (V_E) increases and all other variance components remain the same, what will the effect be?

a. Broad-sense heritability will decrease.

b. Broad-sense heritability will increase.

c. Narrow-sense heritability will increase.

d. Broad-sense heritability will increase, but narrow-sense heritability will decrease.

The Limitations of Heritability

Knowledge of heritability has great practical value, because it allows us to statistically predict the phenotypes of offspring on the basis of their parent's phenotype. It also provides useful information about how characteristics will respond to selection (see Section 11.4). In spite of its importance, heritability is frequently misunderstood. Heritability does not provide information about an individual's genes or the environmental factors that control the development of a characteristic, and it says nothing about the nature of differences between groups. This section outlines some limitations and common misconceptions concerning broad- and narrow-sense heritability.

Heritability does not indicate the degree to which a characteristic is genetically determined   Heritability is the proportion of the phenotypic variance that is due to genetic variance; it says nothing about the degree to which

genes determine a characteristic. Heritability indicates only the degree to which genes determine *variation* in a characteristic. The determination of a characteristic and the determination of variation in a characteristic are two very different things.

Consider polydactyly (the presence of extra digits) in rabbits, which can be caused either by environmental factors or by a dominant gene. Suppose we have a group of rabbits all homozygous for a gene that produces the usual numbers of digits. None of the rabbits in this group carries a gene for polydactyly, but a few of the rabbits are polydactylous because of environmental factors. Broad-sense heritability for polydactyly in this group is zero, because there is no genetic variation for polydactyly; all of the variation is due to environmental factors. However, it would be incorrect for us to conclude that genes play no role in determining the number of digits in rabbits. Indeed, we know that there are specific genes that can produce extra digits. Heritability indicates nothing about whether genes control the development of a characteristic; it provides information only about causes of the variation in a characteristic within a defined group.

An individual does not have heritability   Broad- and narrow-sense heritabilities are statistical values based on the genetic and phenotypic variances found in a *group* of individuals. Heritability cannot be calculated for an individual, and heritability has no meaning for a specific individual. Suppose we calculate the narrow-sense heritability of adult body weight for the students in a biology class and obtain a value of 0.6. We could conclude that 60% of the variation in adult body weight among the students in this class is determined by additive genetic variation. We should not, however, conclude that 60% of any particular student's body weight is due to additive genes.

There is no universal heritability for a characteristic   The value of heritability for a characteristic is specific for a given population in a given environment. Recall that broad-sense heritability is genetic variance divided by phenotypic variance. Genetic variance depends on which genes are present, which often differs between populations. In the example of polydactyly in rabbits, there were no genes for polydactyly in the group; so the heritability of the characteristic was zero. A different group of rabbits might contain many genes for polydactyly, and the heritability of the characteristic might then be high.

Environmental differences may affect heritability, because V_P is composed of both genetic and environmental variance. When the environmental differences that affect a characteristic differ between two groups, the heritabilities for the two groups also will often differ.

Because heritability is specific to a defined population in a given environment, it is important not to extrapolate heritabilities from one population to another. For example, human height is determined by environmental factors (such as nutrition and health) and by genes. If we measured the heritability of height in a developed country, we might obtain a value of 0.8, indicating that the variation in height in this population is largely genetic. This population has a high heritability because most people have adequate nutrition and health care (V_E is low); so most of the phenotypic variation in height is genetically determined. It would be incorrect for us to assume that height has a high heritability in all human populations. In developing countries, there may be more variation in a range of environmental factors; some people may enjoy good nutrition and health, whereas others may have a diet deficient in protein and suffer from diseases that affect stature. If we measured the heritability of height in such a country, we would undoubtedly obtain a lower value than we observed in the developed country, because there is more environmental variation and the genetic variance in height constitutes a smaller proportion of the phenotypic variation, making the heritability lower. The important point to remember is that heritability must be calculated separately for each population and each environment.

Even when heritability is high, environmental factors may influence a characteristic   High heritability does not mean that environmental factors cannot influence the expression of a characteristic. High heritability indicates only that the environmental variation to which the population is *currently* exposed is not responsible for variation in the characteristic. Let's look again at human height. In most developed countries, heritability of human height is high, indicating that genetic differences are responsible for most of the variation in height. It would be wrong for us to conclude that human height cannot be changed by alteration of the environment. Indeed, height decreased in several European cities during World War II owing to hunger and disease, and height can be increased dramatically by the administration of growth hormone to children. The absence of environmental variation in a characteristic does not mean that the characteristic will not respond to environmental change.

Heritabilities indicate nothing about the nature of population differences in a characteristic   A common misconception about heritability is that it provides information about population differences in a characteristic. Heritability is specific for a given population in a given environment, and so it cannot be used to draw conclusions about why populations differ in a characteristic.

Suppose we measured heritability for human height in two groups. One group is from a small town in a developed country, where everyone consumes a high-protein diet. Because there is little variation in the environmental factors that affect human height and there is some genetic variation, the heritability of height in this group is high. The second group comprises the inhabitants of a single village in a developing country. The consumption of protein by these people is only 25% of that consumed by those in the first group; so their average adult height is several centimeters less than that in the developed country. Again, there is little variation in the environmental factors that determine height in this group, because everyone in the village eats the same types of food and is exposed to the same diseases. Because there is little environmental variation and there is some genetic variation, the heritability of height in this group also is high.

Thus, the heritability of height in both groups is high, and the average height in the two groups is considerably different. We might be tempted to conclude that the difference in height between the two groups is genetically based—that the people in the developed country are genetically taller than the people in the developing country. This conclusion is obviously wrong, however, because the differences in height are due largely to diet—an environmental factor. Heritability provides no information about the causes of differences between populations.

These limitations of heritability have often been ignored, particularly in arguments about possible social implications of genetic differences between humans. Soon after Mendel's principles of heredity were rediscovered, some geneticists began to claim that many human behavioral characteristics are determined entirely by genes. This claim led to debates about whether characteristics such as human intelligence are determined by genes or environment. Many of the early claims of genetically based human behavior were based on poor research; unfortunately, the results of these studies were often accepted at face value and led to a number of eugenic laws that discriminated against certain groups of people. Today, geneticists recognize that many behavioral characteristics are influenced by a complex interaction of genes and environment, and separating genetic effects from those of the environment is very difficult.

The results of a number of modern studies indicate that human intelligence as measured by IQ and other intelligence tests has a moderately high heritability (usually from 0.4 to 0.8). On the basis of this observation, some people have argued that intelligence is innate and that enhanced educational opportunities cannot boost intelligence. This argument is based on the misconception that, when heritability is high, changing the environment will not alter the characteristic. In addition, because heritabilities of intelligence range from 0.4 to 0.8, a considerable amount of the variance in intelligence originates from environmental differences.

Another argument based on a misconception about heritability is that ethnic differences in measures of intelligence are genetically based. Because the results of some genetic studies show that IQ has moderately high heritability and because other studies find differences in the average IQ of ethnic groups, some people have suggested that ethnic differences in IQ are genetically based. As in the example of the effects of diet on human height, heritability provides no information about causes of differences among groups; it indicates only the degree to which phenotypic variance within a single group is genetically based. High heritability for a characteristic does not mean that phenotypic differences between ethnic groups are genetic. We should also remember that separating genetic and environmental effects in humans is very difficult; so heritability estimates themselves may be unreliable.

CONCEPTS

Heritability provides information only about the degree to which *variation* in a characteristic is genetically determined. There is no universal heritability for a characteristic; heritability is specific for a given population in a specific environment. Environmental factors can potentially affect characteristics with high heritability, and heritability says nothing about the nature of population differences in a characteristic.

✔ CONCEPT CHECK 6

Suppose that you just learned that the narrow-sense heritability of blood pressure measured among a group of African Americans in Detroit, Michigan, is 0.40. What does this heritability tell us about genetic and environmental contributions to blood pressure?

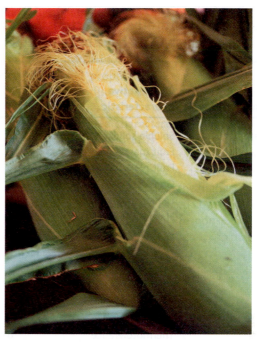

11.19 The availability of molecular markers makes the mapping of QTLs possible in many organisms. QTL mapping is used to identify genes that affect yield in corn and other agriculturally important plants. [Brand X Pictures.]

Locating Genes That Affect Quantitative Characteristics

The statistical methods described for use in analyzing quantitative characteristics can be used both to make predictions about the average phenotype expected in offspring and to estimate the overall contribution of genes to variation in the characteristic. These methods do not, however, allow us to identify and determine the influence of individual genes that affect quantitative characteristics. As discussed in the introduction to this chapter, chromosome regions with genes that control polygenic characteristics are referred to as quantitative trait loci. Although quantitative genetics has made important contributions to basic biology and to plant and animal breeding, the past inability to identify QTLs and measure their individual effects severely limited the application of quantitative genetic methods.

Mapping QTLs   In recent years, numerous genetic markers have been identified and mapped with the use of molecular techniques, making it possible to identify QTLs by linkage analysis. The underlying idea is simple: if the inheritance of a genetic marker is associated consistently with the inheritance of a particular characteristic (such as increased height), then that marker must be linked to a QTL that affects height. The key is to have enough genetic markers so that QTLs can be detected throughout the genome. With the introduction of restriction fragment length polymorphisms,

microsatellite variations, and single-nucleotide polymorphisms, variable markers are now available for mapping QTLs in a number of different organisms (**Figure 11.19**).

A common procedure for mapping QTLs is to cross two homozygous strains that differ in alleles at many loci. The resulting F_1 progeny are then intercrossed or backcrossed to allow the genes to recombine through independent assortment and crossing over. Genes on different chromosomes and genes that are far apart on the same chromosome will recombine freely; genes that are closely linked will be inherited together. The offspring are measured for one or more quantitative characteristics; at the same time, they are genotyped for numerous genetic markers that span the genome. Any correlation between the inheritance of a particular marker allele and a quantitative phenotype indicates that a QTL is linked to that marker. If enough markers are used, the detection of all the QTLs affecting a characteristic is theoretically possible. It is important to recognize that a QTL is not a gene, but, rather, a map location for a chromosome region that is associated with that trait. After a QTL has been identified, it can be studied for the presence of specific genes that influence the quantitative trait. This approach has been used to detect genes affecting various characteristics in several plant and animal species (**Table 11.2**).

Applications of QTL mapping   The number of genes affecting a quantitative characteristic can be estimated by

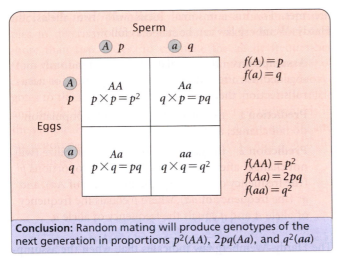

Sperm

$f(A) = p$
$f(a) = q$

$f(AA) = p^2$
$f(Aa) = 2pq$
$f(aa) = q^2$

Conclusion: Random mating will produce genotypes of the next generation in proportions $p^2(AA)$, $2pq(Aa)$, and $q^2(aa)$

12.2 Random mating produces genotypes in the proportions p^2, $2pq$, and q^2.

those in the gametes. If mating is random (one of the assumptions of the Hardy–Weinberg law), the gametes will come together in random combinations, which can be represented by a Punnett square as shown in **Figure 12.2**.

The multiplication rule of probability can be used to determine the probability of various gametes pairing. For example, the probability of a sperm containing allele A is p and the probability of an egg containing allele A is p. Applying the multiplicative rule, we find that the probability that these two gametes will combine to produce an AA homozygote is $p \times p = p^2$. Similarly, the probability of a sperm containing allele a combining with an egg containing allele a to produce an aa homozygote is $q \times q = q^2$. An Aa heterozygote can be produced in one of two ways: (1) a sperm containing allele A may combine with an egg containing allele a ($p \times q$) or (2) an egg containing allele A may combine with a sperm containing allele a ($p \times q$). Thus, the probability of alleles A and a combining to produce an Aa heterozygote is $2pq$. In summary, whenever the frequencies of alleles in a population are p and q, the frequencies of the genotypes in the next generation will be p^2, $2pq$, and q^2.

Closer Examination of the Assumptions of the Hardy–Weinberg Law

Before we consider the implications of the Hardy–Weinberg law, we need to take a closer look at the three assumptions that it makes about a population. First, it assumes that the population is large. How big is "large"? Theoretically, the Hardy–Weinberg law requires that a population be infinitely large in size, but this requirement is obviously unrealistic. In practice, many large populations are in the predicated Hardy–Weinberg proportions, and significant deviations arise only when population size is rather small. Later in the chapter, we will examine the effects of small population size on allelic frequencies.

The second assumption of the Hardy–Weinberg law is that members of the population mate randomly, which means that each genotype mates relative to its frequency. For example, suppose that three genotypes are present in a population in the following proportions: $f(AA) = 0.6$, $f(Aa) = 0.3$, and $f(aa) = 0.1$. With random mating, the frequency of mating between two AA homozygotes ($AA \times AA$) will be equal to the multiplication of their frequencies: $0.6 \times 0.6 = 0.36$, whereas the frequency of mating between two aa homozygotes ($aa \times aa$) will be only $0.1 \times 0.1 = 0.01$.

The third assumption of the Hardy–Weinberg law is that the allelic frequencies of the population are not affected by natural selection, migration, and mutation. Although mutation occurs in every population, its rate is so low that it has little short-term effect on the predictions of the Hardy–Weinberg law (although it may largely shape allelic frequencies over long periods of time when no other forces are acting). Although natural selection and migration are significant factors in real populations, we must remember that the purpose of the Hardy–Weinberg law is to examine only the effect of reproduction on the gene pool. When this effect is known, the effects of other factors (such as migration and natural selection) can be examined.

A final point is that the assumptions of the Hardy–Weinberg law apply to a single locus. No real population mates randomly for all traits; and a population is not completely free of natural selection for all traits. The Hardy–Weinberg law, however, does not require random mating and the absence of selection, migration, and mutation for all traits; it requires these conditions only for the locus under consideration. A population may be in Hardy–Weinberg equilibrium for one locus but not for others.

Implications of the Hardy–Weinberg Law

The Hardy–Weinberg law has several important implications for the genetic structure of a population. One implication is that a population cannot evolve if it meets the Hardy–Weinberg assumptions, because evolution consists of change in the allelic frequencies of a population. Therefore the Hardy–Weinberg law tells us that reproduction alone will not bring about evolution. Other processes such as natural selection, mutation, migration, or chance are required for populations to evolve.

A second important implication is that, when a population is in Hardy–Weinberg equilibrium, the genotypic frequencies are determined by the allelic frequencies. For a locus with two alleles, the frequency of the heterozygote is greatest when allelic frequencies are between 0.33 and 0.66 and is at a maximum when allelic frequencies are each 0.5 (**Figure 12.3**). The heterozygote frequency also never exceeds 0.5 when the population is in Hardy–Weinberg equilibrium. Furthermore, when the frequency of one allele is low, homozygotes for that allele will be rare, and most of the copies of a rare allele will be present in heterozygotes. As

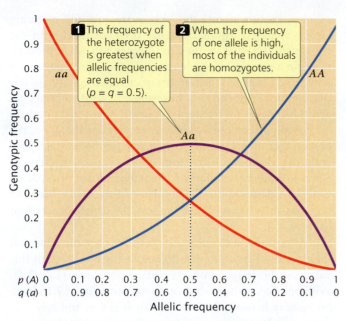

12.3 When a population is in Hardy–Weinberg equilibrium, the proportions of genotypes are determined by the frequencies of alleles.

Table 12.1 Extensions of the Hardy–Weinberg law

Situation	Allelic Frequencies	Genotypic Frequencies
Three alleles	$f(A^1) = p$ $f(A^2) = q$ $f(A^3) = r$	$f(A^1A^1) = p^2$ $f(A^1A^2) = 2pq$ $f(A^2A^2) = q^2$ $f(A^1A^3) = 2pr$ $f(A^2A^3) = 2qr$ $f(A^3A^3) = r^2$
X-linked alleles	$f(X^1) = p$ $f(X^2) = q$	$f(X^1X^1 \text{ female}) = p^2$ $f(X^1X^2 \text{ female}) = 2pq$ $f(X^2X^2 \text{ female}) = q^2$ $f(X^1Y \text{ male}) = p$ $f(X^2Y \text{ male}) = q$

Note: For X-linked female genotypes, the frequencies are the proportions among all females; for X-linked male genotypes, the frequencies are the proportions among all males.

you can see from Figure 12.3, when the frequency of allele a is 0.2, the frequency of the aa homozygote is only 0.04 (q^2), but the frequency of Aa heterozygotes is 0.32 ($2pq$); 80% of the a alleles are in heterozygotes.

A third implication of the Hardy–Weinberg law is that a single generation of random mating produces the equilibrium frequencies of p^2, $2pq$, and q^2. The fact that genotypes are in Hardy–Weinberg proportions does not prove that the population is free from natural selection, mutation, and migration. It means only that these forces have not acted since the last time random mating took place.

Extensions of the Hardy–Weinberg Law

The Hardy–Weinberg expected proportions can also be applied to multiple alleles and X-linked alleles (**Table 12.1**). With multiple alleles, the genotypic frequencies expected at equilibrium are the square of the allelic frequencies. For an autosomal locus with three alleles, the equilibrium genotypic frequencies will $(p + q + r)^2 = p^2 + 2pq + q^2 + 2pr + 2qr + r^2$. For an X-linked locus with two alleles, X^A and X^a, the equilibrium frequencies of the female genotypes are $(p + q)^2 = p^2 + 2pq + q^2$, where p^2 is the frequency of X^AX^A, $2pq$ is the frequency of X^AX^a, and q^2 is the frequency of X^aX^a. Males have only a single X-linked allele, and so the frequencies of the male genotypes are p (frequency of X^AY) and q (frequency of X^aY). These proportions are those of the genotypes among males and females rather than the proportions among the entire population. Thus, p^2 is the expected proportion of females with the genotype X^AX^A; if females make up 50% of the

population, then the expected proportion of this genotype in the entire population is $0.5 \times p^2$.

The frequency of an X-linked recessive trait among males is q, whereas the frequency among females is q^2. When an X-linked allele is uncommon, the trait will therefore be much more frequent in males than in females. Consider hemophilia A, a clotting disorder caused by an X-linked recessive allele with a frequency (q) of approximately 1 in 10,000, or 0.0001. At Hardy–Weinberg equilibrium, this frequency will also be the frequency of the disease among males. The frequency of the disease among females, however, will be $q^2 = (0.0001)^2 = 0.00000001$, which is only 1 in 10 million. Hemophilia is 1000 times as frequent in males as in females.

Testing for Hardy–Weinberg Proportions

To determine if a population's genotypes are in Hardy–Weinberg equilibrium, the genotypic proportions expected under the Hardy–Weinberg law must be compared with the observed genotypic frequencies. To do so, we first calculate the allelic frequencies, then find the expected genotypic frequencies by using the square of the allelic frequencies, and, finally, compare the observed and expected genotypic frequencies by using a chi-square test.

Worked Problem

Jeffrey Mitton and his colleagues found three genotypes (R^2R^2, R^2R^3, and R^3R^3) at a locus encoding the enzyme peroxidase in ponderosa pine trees growing at Glacier Lake, Colorado. The observed numbers of these genotypes were:

Genotypes	Number observed
R^2R^2	135
R^2R^3	44
R^3R^3	11

Are the ponderosa pine trees at Glacier Lake in Hardy–Weinberg equilibrium at the peroxidase locus?

• Solution

If the frequency of the R^2 allele equals p and the frequency of the R^3 allele equals q, the frequency of the R^2 allele is

$$p = f(R^2) = \frac{(2n_{R^2R^2}) + (n_{R^2R^3})}{2N} = \frac{2(135) + 44}{2(190)} = 0.826$$

The frequency of the R^3 allele is obtained by subtraction:

$$q = f(R^3) = 1 - p = 0.174$$

The frequencies of the genotypes expected under Hardy–Weinberg equilibrium are then calculated by using p^2, $2pq$, and q^2:

$$R^2R^2 = p^2 = (0.826)^2 = 0.683$$
$$R^2R^3 = 2pq = 2(0.826)(0.174) = 0.287$$
$$R^3R^3 = q^2 = (0.174)^2 = 0.03$$

Multiplying each of these expected genotypic frequencies by the total number of observed genotypes in the sample (190), we obtain the numbers expected for each genotype:

$$R^2R^2 = 0.0683 \times 190 = 129.8$$
$$R^2R^3 = 0.287 \times 190 = 54.5$$
$$R^3R^3 = 0.03 \times 190 = 5.7$$

By comparing these expected numbers with the observed numbers of each genotype, we see that there are more R^2R^2 homozygotes and fewer R^2R^3 heterozygotes and R^3R^3 homozygotes in the population than we expect at equilibrium.

A goodness-of-fit chi-square test is used to determine whether the differences between the observed and the expected numbers of each genotype are due to chance:

$$\chi^2 = \Sigma \frac{(\text{observed} - \text{expected})^2}{\text{expected}}$$

$$= \frac{(135 - 129.8)^2}{129.8} + \frac{(44 - 54.5)^2}{54.5} + \frac{(11 - 5.7)^2}{5.7}$$

$$= 0.21 + 2.02 + 4.93 = 7.16$$

The calculated chi-square value is 7.16; to obtain the probability associated with this chi-square value, we determine the appropriate degrees of freedom.

Up to this point, the chi-square test for assessing Hardy–Weinberg equilibrium has been identical with the chi-square tests that we used in Chapter 3 to assess progeny ratios in a genetic cross, where the degrees of freedom were $n - 1$ and n equaled the number of expected genotypes. For the Hardy–Weinberg test, however, we must subtract an additional degree of freedom, because the expected numbers are based on the observed allelic frequencies; therefore, the observed numbers are not completely free to vary. In general, the degrees of freedom for a chi-square test of Hardy–Weinberg equilibrium equal the number of expected genotypic classes minus the number of associated alleles. For this particular Hardy–Weinberg test, the degree of freedom is $3 - 2 = 1$.

After we have calculated both the chi-square value and the degrees of freedom, the probability associated with this value can be sought in a chi-square table (see Table 3.4). With one degree of freedom, a chi-square value of 7.16 has a probability between 0.01 and 0.001. It is very unlikely that the peroxidase genotypes observed at Glacier Lake are in Hardy–Weinberg proportions.

? For additional practice, determine whether the genotypic frequencies in Problem 20 at the end of the chapter are in Hardy–Weinberg equilibrium.

CONCEPTS

The observed number of genotypes in a population can be compared with the Hardy–Weinberg expected proportions by using a goodness-of-fit chi-square test.

✔ CONCEPT CHECK 3

What is the expected frequency of heterozygotes in a population with allelic frequencies x and y that is in Hardy–Weinberg equilibrium?

a. $x + y$ c. $2xy$

b. xy d. $(x - y)^2$

Estimating Allelic Frequencies with the Hardy–Weinberg Law

A practical use of the Hardy–Weinberg law is that it allows us to calculate allelic frequencies when dominance is present. For example, cystic fibrosis is an autosomal recessive disorder characterized by respiratory infections, incomplete digestion, and abnormal sweating (see pp. 101–102 in Chapter 5). Among North American Caucasians, the incidence of the disease is approximately 1 person in 2000. The formula for calculating allelic frequency (see Equation 12.3) requires that we know the numbers of homozygotes and heterozygotes, but cystic fibrosis is a recessive disease and so we cannot easily distinguish between homozygous normal persons and heterozygous carriers. Although molecular tests are available for identifying heterozygous carriers of the cystic fibrosis gene, the low frequency of the disease makes widespread screening impractical. In such situations, the Hardy–Weinberg law can be used to estimate the allelic frequencies.

If we assume that a population is in Hardy–Weinberg equilibrium with regard to this locus, then the frequency of the recessive genotype (aa) will be q^2, and the allelic frequency is the square root of the genotypic frequency:

$$q = \sqrt{f(aa)} \qquad (12.9)$$

If the frequency of cystic fibrosis in North American Caucasians is approximately 1 in 2000, or 0.0005, then $q = \sqrt{0.0005} = 0.02$. Thus, about 2% of the alleles in the Caucasian population encode cystic fibrosis. We can calculate the frequency of the normal allele by subtracting: $p = 1 - q = 1 - 0.02 = 0.98$. After we have calculated p and q, we can use the Hardy–Weinberg law to determine the frequencies of homozygous normal people and heterozygous carriers of the gene:

$$f(AA) = p^2 = (0.98)^2 = 0.960$$
$$f(Aa) = 2pq = 2(0.02)(0.98) = 0.0392$$

Thus, about 4% (1 of 25) of Caucasians are heterozygous carriers of the allele that causes cystic fibrosis.

CONCEPTS

Although allelic frequencies cannot be calculated directly for traits that exhibit dominance, the Hardy–Weinberg law can be used to estimate the allelic frequencies if the population is in Hardy–Weinberg equilibrium for that locus. The frequency of the recessive allele will be equal to the square root of the frequency of the recessive trait.

✔ CONCEPT CHECK 4

In cats, all-white color is dominant over not all-white. In a population of 100 cats, 19 are all-white cats. Assuming that the population is in Hardy–Weinberg equilibrium, what is the frequency of the all-white allele in this population?

12.3 Nonrandom Mating Affects the Genotypic Frequencies of a Population

An assumption of the Hardy–Weinberg law is that mating is random with respect to genotype. Nonrandom mating affects the way in which alleles combine to form genotypes and alters the genotypic frequencies of a population.

We can distinguish between two types of nonrandom mating. **Positive assortative mating** refers to a tendency for like individuals to mate. For example, humans exhibit positive assortative mating for height: tall people mate preferentially with other tall people; short people mate preferentially with other short people. **Negative assortative mating** refers to a tendency for unlike individuals to mate. If people engaged in negative assortative mating for height, tall and short people would preferentially mate. Assortative mating is usually for a particular trait and will affect only those genes that encode the trait (and genes closely linked to them).

One form of nonrandom mating is **inbreeding,** which is preferential mating between related individuals. Inbreeding is actually positive assortative mating for relatedness, but it differs from other types of assortative mating because it affects all genes, not just those that determine the trait for which the mating preference exists. Inbreeding causes a departure from the Hardy–Weinberg equilibrium frequencies of p^2, $2pq$, and q^2. More specifically, it leads to an increase in the proportion of homozygotes and a decrease in the proportion of heterozygotes in a population. **Outcrossing** is the avoidance of mating between related individuals.

In a diploid organism, a homozygous individual has two copies of the same allele. These two copies may be the same in *state*, which means that the two alleles are alike in structure and function but do not have a common origin. Alternatively, the two alleles in a homozygous individual may be the same because they are identical by *descent;* that is, the copies are descended from a single allele that was present in an ancestor (**Figure 12.4**). If we go back far enough in time, many alleles are likely to be identical by descent but, for calculating the effects of inbreeding, we consider identity by descent by going back only a few generations.

Inbreeding is usually measured by the **inbreeding coefficient,** designated F, which is a measure of the probability that two alleles are identical by descent. Inbreeding coefficients can range from 0 to 1. A value of 0 indicates that mating in a large population is random; a value of 1 indicates that all alleles are identical by descent. Inbreeding coefficients can be calculated from analyses of pedigrees or they can be determined from the reduction in the heterozygosity of a population. Although we will not go into the details of how F is calculated, an understanding of how inbreeding affects genotypic frequencies is important.

When inbreeding takes place, the proportion of heterozygotes decreases by $2Fpq$, and half of this value (Fpq) is *added* to the proportion of each homozygote each generation. The frequencies of the genotypes will then be

$$f(AA) = p^2 + Fpq$$
$$f(Aa) = 2pq - 2Fpq \qquad (12.10)$$
$$f(aa) = q^2 + Fpq$$

Consider a population that reproduces by self-fertilization (so $F = 1$). We will assume that this population begins with genotypic frequencies in Hardy–Weinberg proportions

(a) Alleles identical by descent

(b) Alleles identical by state

12.4 Individuals may be homozygous by descent or by state. Inbreeding is a measure of the probability that two alleles are identical by descent.

> These two copies of the A^1 allele are descended from the same copy in a common ancestor; so they are identical by descent.

> These two copies of the A^1 allele are the same in structure and function, but are descended from two different copies in ancestors; so they are identical in state.

$(p^2, 2pq,$ and $q^2)$. With selfing, each homozygote produces progeny only of the same homozygous genotype ($AA \times AA$ produces all AA; and $aa \times aa$ produces all aa), whereas only half the progeny of a heterozygote will be like the parent ($Aa \times Aa$ produces $\frac{1}{4}$ AA, $\frac{1}{2}$ Aa, and $\frac{1}{4}$ aa). Selfing therefore reduces the proportion of heterozygotes in the population by half with each generation, until all genotypes in the population are homozygous (**Table 12.2** and **Figure 12.5**).

For most outcrossing species, close inbreeding is harmful because it increases the proportion of homozygotes and thereby boosts the probability that deleterious and lethal recessive alleles will combine to produce homozygotes with a harmful trait. Assume that a recessive allele (a) that causes a genetic disease has a frequency (q) of 0.01. If the population mates randomly ($F = 0$), the frequency of individuals affected with the disease (aa) will be $q^2 = 0.01^2 = 0.0001$; so only 1 in 10,000 individuals will have the disease. However, if $F = 0.25$ (the equivalent of brother–sister mating), then the expected frequency of the homozygote genotype is $q^2 + Fpq = (0.01)^2 + (0.25)(0.99)(0.01) = 0.0026$; thus, the genetic disease is 26 times as frequent at this level of inbreeding. This increased appearance of lethal and deleterious traits with inbreeding is termed **inbreeding depression;** the more intense the inbreeding, the more severe the inbreeding depression.

Table 12.2 Generational increase in frequency of homozygotes in a self-fertilizing population starting with $p = q = 0.5$

	Genotypic Frequencies		
Generation	AA	Aa	aa
1	$\frac{1}{4}$	$\frac{1}{2}$	$\frac{1}{4}$
2	$\frac{1}{4} + \frac{1}{8} = \frac{3}{8}$	$\frac{1}{4}$	$\frac{1}{4} + \frac{1}{8} = \frac{3}{8}$
3	$\frac{3}{8} + \frac{1}{16} = \frac{7}{16}$	$\frac{1}{8}$	$\frac{3}{8} + \frac{1}{16} = \frac{7}{16}$
4	$\frac{7}{16} + \frac{1}{32} = \frac{15}{32}$	$\frac{1}{16}$	$\frac{7}{16} + \frac{1}{32} = \frac{15}{32}$
n	$\dfrac{1 - (\frac{1}{2})^n}{2}$	$(\frac{1}{2})^n$	$\dfrac{1 - (\frac{1}{2})^n}{2}$
	$\frac{1}{2}$	0	$\frac{1}{2}$

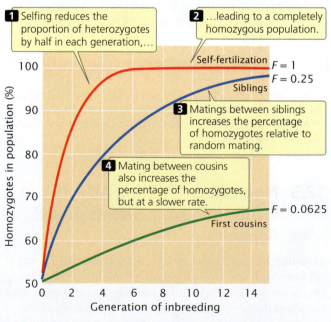

1 Selfing reduces the proportion of heterozygotes by half in each generation,…

2 …leading to a completely homozygous population.

Self-fertilization $F = 1$

$F = 0.25$

Siblings

3 Matings between siblings increases the percentage of homozygotes relative to random mating.

4 Mating between cousins also increases the percentage of homozygotes, but at a slower rate.

$F = 0.0625$

First cousins

12.5 Inbreeding increases the percentage of homozygous individuals in a population.

Table 12.3 Effects of inbreeding on Japanese children

Genetic Relationship of Parents	F	Mortality of Children (through 12 years of age)
Unrelated	0	0.082
Second cousins	0.016 ($\frac{1}{64}$)	0.108
First cousins	0.0625 ($\frac{1}{16}$)	0.114

Source: After D. L. Hartl and A. G. Clark, *Principles of Population Genetics,* 2d ed. (Sunderland, Mass.: Sinauer, 1989), Table 2. Original data from W. J. Schull and J. V. Neel, *The Effects of Inbreeding on Japanese Children* (New York: Harper & Row, 1965).

The harmful effects of inbreeding have been recognized by people for thousands of years and may be the basis of cultural taboos against mating between close relatives. William Schull and James Neel found that, for each 10% increase in *F*, the mean IQ of Japanese children dropped six points. Child mortality also increases with close inbreeding (**Table 12.3**); children of first cousins have a 40% increase in mortality over that seen among the children of unrelated people. Inbreeding also has deleterious effects on crops (**Figure 12.6**) and domestic animals.

Inbreeding depression is most often studied in humans, as well as in plants and animals reared in captivity, but the negative effects of inbreeding may be more severe in natural populations. Julie Jimenez and her colleagues collected wild mice from a natural population in Illinois and bred them in the laboratory for three to four generations. Laboratory matings were chosen so that some mice had no inbreeding,

whereas others had an inbreeding coefficient of 0.25. When both types of mice were released back into the wild, the weekly survival of the inbred mice was only 56% of that of the noninbred mice. Inbred male mice also continously lost body weight after release into the wild, whereas noninbred male mice initially lost weight but then regained it within a few days after release.

In spite of the fact that inbreeding is generally harmful for outcrossing species, a number of plants and animals regularly inbreed and are successful (**Figure 12.7**). Inbreeding is commonly used to produce domesticated plants and animals having desirable traits. As stated earlier, inbreeding increases homozygosity, and eventually all individuals in the population become homozygous for the same allele. If a species undergoes inbreeding for a number of generations, many deleterious recessive alleles are weeded out by natural or artificial selection so that the population becomes homozygous for beneficial alleles. In this way, the harmful effects of inbreeding may eventually be eliminated, leaving a population that is homozygous for beneficial traits.

CONCEPTS

Nonrandom mating alters the frequencies of the genotypes but not the frequencies of the alleles. Inbreeding is preferential mating between related individuals. With inbreeding, the frequency of homozygotes increases, whereas the frequency of heterozygotes decreases.

✔ CONCEPT CHECK 5

What is the effect of outcrossing on a population?

a. Allelic frequencies change.
b. There will be more heterozygotes than predicted by the Hardy–Weinberg law.
c. There will be fewer heterozygotes than predicted by the Hardy–Weinberg law.
d. Genotypic frequencies will equal those predicted by the Hardy–Weinberg law.

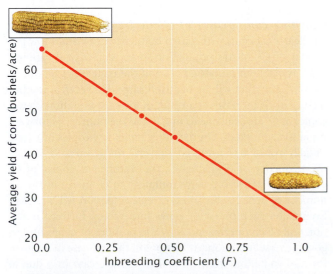

12.6 Inbreeding often has deleterious effects on crops. As inbreeding increases, the average yield of corn, for example, decreases.

12.7 Although inbreeding is generally harmful, a number of inbreeding organisms are successful. Shown here is the terrestrial slug *Arion circumscriptos,* an inbreeding species that causes damage in greenhouses and flower gardens. *[William Leonard/DRK Photo.]*

12.4 Several Evolutionary Forces Potentially Cause Changes in Allelic Frequencies

The Hardy–Weinberg law indicates that allelic frequencies do not change as a result of reproduction; thus, other processes must cause alleles to increase or decrease in frequency. Processes that bring about change in allelic frequency include mutation, migration, genetic drift (random effects due to small population size), and natural selection.

Mutation

Before evolution can take place, genetic variation must exist within a population; consequently, all evolution depends on processes that generate genetic variation. Although new *combinations* of existing genes may arise through recombination in meiosis, all genetic variants ultimately arise through mutation.

The effect of mutation on allelic frequencies   Mutation can influence the rate at which one genetic variant increases at the expense of another. Consider a single locus in a population of 25 diploid individuals. Each individual possesses two alleles at the locus under consideration; so the gene pool of the population consists of 50 allelic copies. Let us assume that there are two different alleles, designated G^1 and G^2 with frequencies p and q, respectively. If there are 45 copies of G^1 and 5 copies of G^2 in the population, $p = 0.90$ and $q = 0.10$. Now suppose that a mutation changes a G^1 allele into a G^2 allele. After this mutation, there are 44 copies of G^1 and 6 copies of G^2, and the frequency of G^2 has increased from 0.10 to 0.12. Mutation has changed the allelic frequency.

If copies of G^1 continue to mutate to G^2, the frequency of G^2 will increase and the frequency of G^1 will decrease (**Figure 12.8**). The amount that G^2 will change (Δq) as a result of mutation depends on: (1) the rate of G^1-to-G^2 mutation (μ); and (2) p, the frequency of G^1 in the population. When p is large, there are many copies of G^1 available to mutate to G^2, and the amount of change will be relatively large. As more mutations occur and p decreases, there will be fewer copies of G^1 available to mutate to G^2. The change in G^2 as a result of mutation equals the mutation rate times the allelic frequency:

$$\Delta q = \mu p \qquad (12.11)$$

As the frequency of p decreases as a result of mutation, the change in frequency due to mutation will be less and less.

So far, we have considered only the effects of $G^1 \rightarrow G^2$ forward mutations. Reverse $G^2 \rightarrow G^1$ mutations also occur at rate ν, which will probably be different from the forward mutation rate, μ. Whenever a reverse mutation occurs, the frequency of G^2 decreases and the frequency of G^1 increases (see Figure 12.8). The rate of change due to reverse mutations equals the reverse mutation rate times the allelic fre-

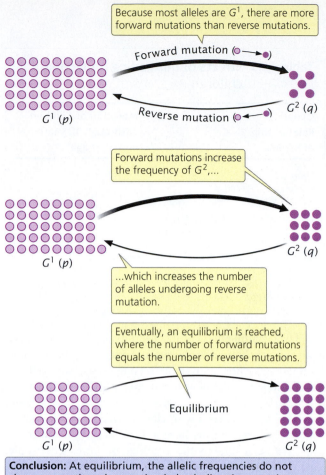

Because most alleles are G^1, there are more forward mutations than reverse mutations.

Forward mutation (○ ——▶ ●)

Reverse mutation (○ ◀—— ●)

G^1 (p) G^2 (q)

Forward mutations increase the frequency of G^2,...

G^1 (p) G^2 (q)

...which increases the number of alleles undergoing reverse mutation.

Eventually, an equilibrium is reached, where the number of forward mutations equals the number of reverse mutations.

Equilibrium

G^1 (p) G^2 (q)

Conclusion: At equilibrium, the allelic frequencies do not change even though mutation in both directions continues.

12.8 Recurrent mutation changes allelic frequencies. Forward and reserve mutations eventually lead to a stable equilibrium.

quency of G^2 ($\Delta q = \nu q$). The overall change in allelic frequency is a balance between the opposing forces of forward mutation and reverse mutation:

$$\Delta q = \mu p - \nu q \qquad (12.12)$$

Reaching equilibrium of allelic frequencies   Consider an allele that begins with a high frequency of G^1 and a low frequency of G^2. In this population, many copies of G^1 are initially available to mutate to G^2, and the increase in G^2 due to forward mutation will be relatively large. However, as the frequency of G^2 increases as a result of forward mutations, fewer copies of G^1 are available to mutate; so the number of forward mutations decreases. On the other hand, few copies of G^2 are initially available to undergo a reverse mutation to G^1 but, as the frequency of G^2 increases, the number of copies of G^2 available to undergo reverse mutation to G^1 increases; so the number of genes undergoing reverse mutation will increase. Eventually, the number of genes undergoing forward mutation will be counterbalanced by the number of genes undergoing reverse mutation. At this point, the increase in q due to forward mutation will be equal to the decrease in q due to reverse mutation, and there will be no net change in allelic frequency ($\Delta q = 0$), in spite of the fact that forward and reverse mutations continue to occur. The point at which there is no change in the allelic frequency of a population is referred to as **equilibrium** (see Figure 12.8). At equilibrium, the frequency of G^2 ($\hat{q}$) will be

$$\hat{q} = \frac{\mu}{\mu + \nu} \tag{12.13}$$

This final equation tells us that the allelic frequency at equilibrium is determined solely by the forward (μ) and reverse (ν) mutation rates.

Summary of effects   When the only evolutionary force acting on a population is mutation, allelic frequencies change with the passage of time because some alleles mutate into others. Eventually, these allelic frequencies reach equilibrium and are determined only by the forward and reverse mutation rates. When the allelic frequencies reach equilibrium, the Hardy–Weinberg law tells us that genotypic frequencies also will remain the same.

The mutation rates for most genes are low; so change in allelic frequency due to mutation in one generation is very small, and long periods of time are required for a population to reach mutational equilibrium. For example, if the forward and reverse mutation rates for alleles at a locus are 1×10^{-5} and 0.3×10^{-5} per generation, respectively (rates that have actually been measured at several loci in mice), and the allelic frequencies are $p = 0.9$ and $q = 0.1$, then the net change in allelic frequency per generation due to mutation is

$$\begin{aligned}
\Delta q &= \mu p - \nu q \\
&= (1 \times 10^{-5})(0.9) - (0.3 \times 10^{-5})(0.1) \\
&= 8.7 \times 10^{-6} = 0.0000087
\end{aligned}$$

Therefore, change due to mutation in a single generation is extremely small and, as the frequency of p drops as a result of mutation, the amount of change will become even smaller (**Figure 12.9**). The effect of typical mutation rates on Hardy–Weinberg equilibrium is negligible, and many generations are required for a population to reach mutational equilibrium. Nevertheless, if mutation is the only force acting on a population for long periods of time, mutation rates will determine allelic frequencies.

CONCEPTS

Recurrent mutation causes changes in the frequencies of alleles. At equilibrium, the allelic frequencies are determined by the forward and reverse mutation rates. Because mutation rates are low, the effect of mutation per generation is very small.

✔ **CONCEPT CHECK 6**

What will be the equilibrium frequency of an allele if its forward and reverse mutation rates are 0.6×10^{-6} and 0.2×10^{-6}, respectively? Assume that no other evolutionary forces are present.

a. 0.80   c. 0.25

b. 0.75   d. 0.12

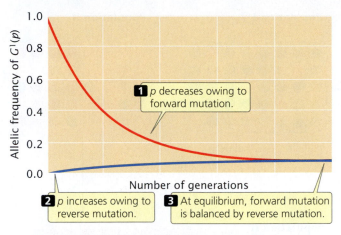

12.9 Change due to recurrent mutation slows as the frequency of p drops. Allelic frequencies are approaching mutational equilibrium at typical low mutation rates. The allelic frequency of G^1 decreases as a result of forward ($G^1 \rightarrow G^2$) mutation at rate m (0.0001) and increases as a result of reverse ($G^2 \rightarrow G^1$) mutation at rate n (0.00001). Owing to the low rate of mutations, eventual equilibrium takes many generations to be reached.

Migration

Another process that may bring about change in the allelic frequencies is the influx of genes from other populations, commonly called **migration** or **gene flow**. One of the assumptions of the Hardy–Weinberg law is that migration does not take place, but many natural populations do experience migration from other populations. The overall effect of migration is twofold: (1) it prevents populations from becoming genetically different from one another and (2) it increases genetic variation within populations.

The effect of migration on allelic frequencies   Let us consider the effects of migration by looking at a simple, unidirectional model of migration between two populations that differ in the frequency of an allele a. Say the frequency of this allele in population I is q_{I} and in population II is q_{II} (**Figure 12.10**). In each generation, a representative sample of the individuals in population I migrates to population II and reproduces, adding its genes to population II's gene pool. Migration is only from population I to population II (is unidirectional), and all the conditions of the Hardy–Weinberg law apply, except the absence of migration.

After migration, population II consists of two types of individuals. Some are migrants; they make up proportion m of population II, and they carry genes from population I; so the frequency of allele a in the migrants is q_{I}. The other individuals in population II are the original residents. If the migrants make up proportion m of population II, then the residents make up $1 - m$; because the residents originated in population II, the frequency of allele a in this group is q_{II}. After migration, the frequency of allele a in the merged population II (q'_{II}) is

$$q'_{\text{II}} = q_{\text{I}}(m) + q_{\text{II}}(1 - m) \tag{12.14}$$

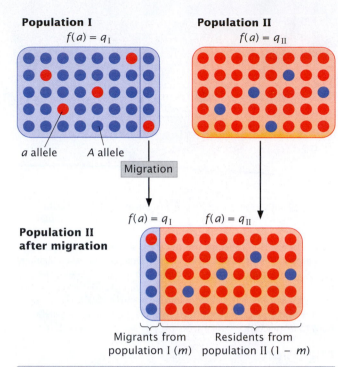

Population I
$f(a) = q_I$

a allele   *A* allele

Population II
$f(a) = q_{II}$

Migration

$f(a) = q_I$ $f(a) = q_{II}$

Population II after migration

Migrants from population I (*m*)   Residents from population II (1 − *m*)

Conclusion: The frequency of allele *a* in population II after migration is $q'_{II} = q_I m + q_{II}(1 − m)$.

12.10 The amount of change in allelic frequency due to migration between populations depends on the difference in allelic frequency and the extent of migration. Shown here is a model of the effect of unidirectional migration on allelic frequencies. The frequency of allele *a* in the source population (population I) is q_I. The frequency of this allele in the recipient population (population II) is q_{II}. Each generation, a random sample of individuals migrate from population I to population II. After migration, population II consists of migrants and residents. The migrants constitute proportion *m* and have a frequency of *q* equal to q_I; the residents constitute proportion 1 − *m* and have a frequency of *q* equal to q_{II}.

where $q_I(m)$ is the contribution to *q* made by the copies of allele *a* in the migrants and $q_{II}(1 − m)$ is the contribution to *q* made by copies of allele *a* in the residents. The change in the allelic frequency due to migration (Δq) will be

$$\Delta q = m(q_I − q_{II}) \tag{12.15}$$

Equation 12.15 summarizes the factors that determine the amount of change in allelic frequency due to migration. The amount of change in *q* is directly proportional to the migration (*m*); as the amount of migration increases, the change in allelic frequency increases. The magnitude of change is also affected by the differences in allelic frequencies of the two populations ($q_I − q_{II}$); when the difference is large, the change in allelic frequency will be large.

With each generation of migration, the frequencies of the two populations become more and more similar until, eventually, the allelic frequency of population II equals that of population I. When $q_I − q_{II} = 0$, there will be no further change in the allelic frequency of population II, in spite of

the fact that migration continues. If migration between two populations takes place for a number of generations with no other evolutionary forces present, an equilibrium is reached at which the allelic frequency of the recipient population equals that of the source population.

The simple model of unidirectional migration between two populations just outlined can be expanded to accommodate multidirectional migration between several populations.

The overall effect of migration Migration has two major effects. First, it causes the gene pools of populations to become more similar. Later, we will see how genetic drift and natural selection lead to genetic differences between populations; migration counteracts this tendency and tends to keep populations homogeneous in their allelic frequencies. Second, migration adds genetic variation to populations. Different alleles may arise in different populations owing to rare mutational events, and these alleles can be spread to new populations by migration, increasing the genetic variation within the recipient population.

CONCEPTS

Migration causes changes in the allelic frequency of a population by introducing alleles from other populations. The magnitude of change due to migration depends on both the extent of migration and the difference in allelic frequencies between the source and the recipient populations. Migration decreases genetic differences between populations and increases genetic variation within populations.

✔ **CONCEPT CHECK 7**

Each generation, 10 random individuals migrate from population A to population B. What will happen to allelic frequency *q* as a result of migration when *q* is equal in populations A and B?

a. *q* in A will decrease.
b. *q* in B will increase.
c. *q* will not change in either A or B.
d. *q* in B will become q^2.

Genetic Drift

The Hardy–Weinberg law assumes random mating in an infinitely large population; only when population size is infinite will the gametes carry genes that perfectly represent the parental gene pool. But no real population is infinitely large, and when population size is limited, the gametes that unite to form individuals of the next generation carry a sample of alleles present in the parental gene pool. Just by chance, the composition of this sample will often deviate from that of the parental gene pool, and this deviation may cause allelic frequencies to change. The smaller the gametic sample, the greater the chance that its composition will deviate from that of the entire gene pool.

The role of chance in altering allelic frequencies is analogous to the flip of a coin. Each time we flip a coin, we have a 50% chance of getting a head and a 50% chance of getting a tail. If we flip a coin 1000 times, the observed ratio of heads to tails will be very close to the expected 50 : 50 ratio. If, however, we flip a coin only 10 times, there is a good chance that we will obtain not exactly 5 heads and 5 tails, but rather maybe 7 heads and 3 tails or 8 tails and 2 heads. This kind of deviation from an expected ratio due to limited sample size is referred to as **sampling error.**

Sampling error arises when gametes unite to produce progeny. Many organisms produce a large number of gametes but, when population size is small, a limited number of gametes unite to produce the individuals of the next generation. Chance influences which alleles are present in this limited sample and, in this way, sampling error may lead to **genetic drift,** or changes in allelic frequency. Because the deviations from the expected ratios are random, the direction of change is unpredictable. We can nevertheless predict the magnitude of the changes.

The magnitude of genetic drift   The effect of genetic drift can be viewed in two ways. First, we can see how it influences the change in allelic frequencies of a single population with the passage of time. Second, we can see how it affects differences that accumulate among series of populations. Imagine that we have 10 small populations, all beginning with the exact same allelic frequencies of $p = 0.5$ and $q = 0.5$. When genetic drift occurs in a population, allelic frequencies within the population will change but, because drift is random, the way in which allelic frequencies change in each population will not be the same. In some populations, p may increase as a result of chance. In other populations, p may decrease as a result of chance. In time, the allelic frequencies in the 10 populations will become different: the populations will genetically diverge. As time passes, the change in allelic frequency within a population and the genetic divergence among populations are due to the same force—the random change in allelic frequencies. The magnitude of genetic drift can be assessed either by examining the change in allelic frequency within a single population or by examining the magnitude of genetic differences that accumulate among populations.

The amount of genetic drift can be estimated from the variance in allelic frequency. Variance, s^2, is a statistical measure that describes the degree of variability in a trait (see pp. 307–308 in Chapter 11). Suppose that we observe a large number of separate populations, each with N individuals and allelic frequencies of p and q. After one generation of random mating, genetic drift expressed in terms of the variance in allelic frequency among the populations (s_p^2) will be

$$s_p^2 = \frac{pq}{2N} \qquad (12.16)$$

The amount of change resulting from genetic drift (the variance in allelic frequency) is determined by two parameters: the allelic frequencies (p and q) and the population size (N).

Genetic drift will be maximal when p and q are equal (each 0.5). For example, assume that a population consists of 50 individuals. When the allelic frequencies are equal ($p = q = 0.5$), the variance in allelic frequency (s_p^2) will be (0.5 × 0.5)/(2 × 50) = 0.0025. In contrast, when $p = 0.9$ and $q = 0.1$, the variance in allelic frequency will be only 0.0009. Genetic drift will also be higher when the population size is small. If $p = q = 0.5$, but the population size is only 10 instead of 50, then the variance in allelic frequency becomes (0.5 × 0.5)/(2 × 10) = 0.0125, which is five times as great as when population size is 50.

This divergence of populations through genetic drift is strikingly illustrated in the results of an experiment carried out by Peter Buri on fruit flies (**Figure 12.11**). Buri examined the frequencies of two alleles (bw^{75} and bw) that affect eye color in the flies. He set up 107 replicate populations, each consisting of 8 males and 8 females. He began each population with a frequency of bw^{75} equal to 0.5. He allowed the fruit flies within each replicate to mate randomly and, each generation, he randomly selected 8 male and 8 female flies to be the parents of the next generation. He followed the changes in the frequencies of the two alleles over 19 generations. In one replicate population, the average frequency of bw^{75} (p) over the 19 generations was 0.5312. We can use Equation 25.16 to calculate the expected variance in allelic frequency due to genetic drift. The frequency of the bw allele (q) will be $1 - p = 1 - 0.53125 = 0.46875$. The population size ($N$) equals 16. The expected variance in allelic frequency will be

$$\frac{pq}{2N} = \frac{0.53125 \times 0.46875}{2 \times 16} = 0.0156$$

which was very close to the actual observed variance of 0.0151.

The effect of population size on genetic drift is illustrated by a study conducted by Luca Cavalli-Sforza and his colleagues. They studied variation in blood types among villagers in the Parma Valley of Italy, where the amount of migration between villages was limited. They found that variation in allelic frequency was greatest between small isolated villages in the upper valley but decreased between larger villages and towns farther down the valley. This result is exactly what we expect with genetic drift: there should be more genetic drift and thus more variation among villages when population size is small.

For ecological and demographic studies, population size is usually defined as the number of individuals in a group. The evolution of a gene pool depends, however, only on those individuals who contribute genes to the next generation. Population geneticists usually define population size as the equivalent number of breeding adults, the **effective population size** (N_e). Several factors determine the equivalent number of breeding adults, including the sex ratio, variation between individuals in reproductive success, fluctuations in population size, the age structure of the population, and whether mating is random.

Experiment

Question: What effect does genetic drift have on the genetic composition of populations?

Methods   Buri examined the frequencies of two alleles (bw^{75} and bw) that affect *Drosophila* eye color in 107 replicate small populations over 19 generations.

Results

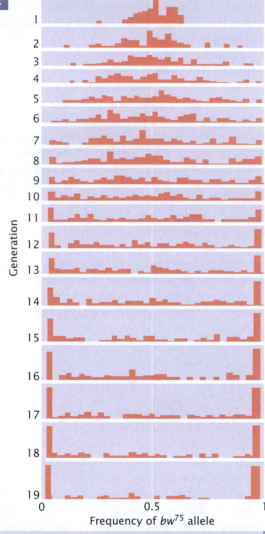

Frequency of bw^{75} allele

Conclusion: As a result of genetic drift, allelic frequencies in the different populations diverged and often became fixed for one allele or the other.

12.11 Populations diverge in allelic frequency and become fixed for one allele as a result of genetic drift. In Buri's study of two eye-color alleles (bw^{75} and bw) in *Drosophila*, each population consisted of 8 males and 8 females and began with the frequency of bw^{75} equal to 0.5.

CONCEPTS

Genetic drift is change in allelic frequency due to chance factors. The amount of change in allelic frequency due to genetic drift is inversely related to the effective population size (the equivalent number of breeding adults in a population).

✔ CONCEPT CHECK 8

Which of the following statements is an example of genetic drift?

a. Allele *g* for fat production increases in a small population because birds with more body fat have higher survivorship in a harsh winter.

b. Random mutation increases the frequency of allele *A* in one population but not in another.

c. Allele *R* reaches a frequency of 1.0 because individuals with genotype *rr* are sterile.

d. Allele *m* is lost when a virus kills all but a few individuals and just by chance none of the survivors possess allele *m*.

Causes of genetic drift   All genetic drift arises from sampling error, but there are several different ways in which sampling error can arise. First, a population may be reduced in size for a number of generations because of limitations in space, food, or some other critical resource. Genetic drift in a small population for multiple generations can significantly affect the composition of a population's gene pool.

A second way that sampling error can arise is through the **founder effect,** which is due to the establishment of a population by a small number of individuals; the population of bighorn sheep at the National Bison Range, discussed in the introduction to this chapter, underwent a founder effect. Although a population may increase and become quite large, the genes carried by all its members are derived from the few genes originally present in the founders (assuming no migration or mutation). Chance events affecting which genes were present in the founders will have an important influence on the makeup of the entire population.

A third way in which genetic drift arises is through a **genetic bottleneck,** which develops when a population undergoes a drastic reduction in population size. A genetic bottleneck developed in northern elephant seals (**Figure 12.12**). Before 1800, thousands of elephant seals were found along the California coast, but the population was devas-

12.12 Northern elephant seals underwent a severe genetic bottleneck between 1820 and 1880. Today, these seals have low levels of genetic variation. *[PhotoDisc.]*

tated by hunting between 1820 and 1880. By 1884, as few as 20 seals survived on a remote beach of Isla de Guadelupe west of Baja, California. Restrictions on hunting enacted by the United States and Mexico allowed the seals to recover, and there are now more than 30,000 seals in the population. All seals in the population today are genetically similar, because they have genes that were carried by the few survivors of the population bottleneck.

The effects of genetic drift Genetic drift has several important effects on the genetic composition of a population. First, it produces change in allelic frequencies within a population. Because drift is random, allelic frequency is just as likely to increase as it is to decrease and will wander with the passage of time (hence the name genetic drift). **Figure 12.13** illustrates a computer simulation of genetic drift in five populations over 30 generations, starting with $q = 0.5$ and maintaining a constant population size of 10 males and 10 females. These allelic frequencies change randomly from generation to generation.

A second effect of genetic drift is to reduce genetic variation within populations. Through random change, an allele may eventually reach a frequency of either 1 or 0, at which point all individuals in the population are homozygous for one allele. When an allele has reached a frequency of 1, we say that it has reached **fixation.** Other alleles are lost (reach a frequency of 0) and can be restored only by migration from another population or by mutation. Fixation, then, leads to a loss of genetic variation within a population. This loss can be seen in the northern elephant seals described earlier. Today, these seals have low levels of genetic variation; a study of 24 protein-encoding genes found no individual or population

differences in these genes. A subsequent study of sequence variation in mitochondrial DNA also revealed low levels of genetic variation. In contrast, the southern elephant seal had much higher levels of mitochondrial DNA variation. The southern elephant seals also were hunted, but their population size never dropped below 1000; therefore, unlike the northern elephant seals, they did not experience a genetic bottleneck.

Given enough time, all small populations will become fixed for one allele or the other. Which allele becomes fixed is random and is determined by the initial frequency of the allele. If the population begins with two alleles, each with a frequency of 0.5, both alleles have an equal probability of fixation. However, if one allele is initially common, it is more likely to become fixed.

A third effect of genetic drift is that different populations diverge genetically with time. In Figure 12.13, all five populations begin with the same allelic frequency ($q = 0.5$) but, because drift is random, the frequencies in different populations do not change in the same way, and so populations gradually acquire genetic differences. Eventually, all the populations reach fixation; some will become fixed for one allele, and others will become fixed for the alternative allele.

The three results of genetic drift (allelic frequency change, loss of variation within populations, and genetic divergence between populations) take place simultaneously, and all result from sampling error. The first two results take place *within* populations, whereas the third takes place *between* populations.

CONCEPTS

Genetic drift results from continuous small population size, the founder effect (establishment of a population by a few founders), and the bottleneck effect (population reduction). Genetic drift causes a change in allelic frequencies within a population, a loss of genetic variation through the fixation of alleles, and genetic divergence between populations.

Natural Selection

A final process that brings about changes in allelic frequencies is natural selection, the differential reproduction of genotypes (see p. 320 in Chapter 11). Natural selection takes place when individuals with adaptive traits produce a greater number of offspring than that produced by others in the population. If the adaptive traits have a genetic basis, they are inherited by the offspring and appear with greater frequency in the next generation. A trait that provides a reproductive advantage thereby increases with the passage of time, enabling populations to become better suited to their environments—to become better adapted. Natural selection is unique among evolutionary forces in that it promotes adaptation (**Figure 12.14**).

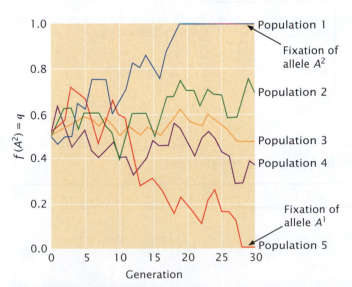

12.13 Genetic drift changes allelic frequencies within populations, leading to a reduction in genetic variation through fixation and genetic divergence among populations. Shown here is a computer simulation of changes in the frequency of allele A^2 (q) in five different populations due to random genetic drift. Each population consists of 10 males and 10 females and begins with $q = 0.5$.

12.14 Natural selection produces adaptations, such as those seen in the polar bears that inhabit the extreme Arctic environment. These bears blend into the snowy background, which helps them in hunting seals. The hairs of their fur stay erect even when wet, and thick layers of blubber provide insulation, which protects against subzero temperatures. Their digestive tracts are adapted to a seal-based carnivorous diet. [Digital Vision.]

Fitness and the selection coefficient   The effect of natural selection on the gene pool of a population depends on the fitness values of the genotypes in the population. **Fitness** is defined as the relative reproductive success of a genotype. Here, the term *relative* is critical: fitness is the reproductive success of one genotype compared with the reproductive successes of other genotypes in the population.

Fitness (W) ranges from 0 to 1. Suppose the average number of viable offspring produced by three genotypes is

Genotypes:	A^1A^1	A^1A^2	A^2A^2
Mean number of offspring produced:	10	5	2

To calculate fitness for each genotype, we take the mean number of offspring produced by a genotype and divide it by the mean number of offspring produced by the most prolific genotype:

$$A^1A^1 \qquad\qquad A^1A^2$$
$$\text{Fitness } (W): \quad W_{11} = \frac{10}{10} = 1.0 \quad W_{12} = \frac{5}{10} = 0.5$$

$$A^2A^2$$
$$W_{22} = \frac{2}{10} = 0.2 \qquad\qquad (12.17)$$

The fitness of genotype A^1A^1 is designated W_{11}, that of A^1A^2 is W_{12}, and that of A^2A^2 is W_{22}. A related variable is the **selection coefficient** (s), which is the relative intensity of selection against a genotype. We usually speak of selection for a particular genotype, but keep in mind that, when selection is *for* one genotype, selection is automatically *against* at least one other genotype. The selection coefficient is equal to $1 - W$; so the selection coefficients for the preceding three genotypes are

$$\qquad\qquad\qquad A^1A^1 \qquad A^1A^2 \qquad A^2A^2$$
$$\text{Selection coefficient } (1 - W): \; s_{11} = 0 \quad s_{12} = 0.5 \quad s_{22} = 0.8$$

CONCEPTS

Natural selection is the differential reproduction of genotypes. It is measured as fitness, which is the reproductive success of a genotype compared with other genotypes in a population.

✔ CONCEPT CHECK 9

The average numbers of offspring produced by three genotypes are: $GG = 6$; $Gg = 3$, $gg = 2$. What is the fitness of Gg?

a. 3
b. 0.5
c. 0.3
d. 0.27

Table 12.4 Method for determining changes in allelic frequency due to selection

	A^1A^1	A^1A^2	A^2A^2
Initial genotypic frequencies	p^2	$2pq$	q^2
Fitnesses	W_{11}	W_{12}	W_{22}
Proportionate contribution of genotypes to population	p^2W_{11}	$2pqW_{12}$	q^2W_{22}
Relative genotypic frequency after selection	$\dfrac{p^2W_{11}}{\overline{W}}$	$\dfrac{2pqW_{12}}{\overline{W}}$	$\dfrac{q^2W_{22}}{\overline{W}}$

Note: $\overline{W} = p^2W_{11} + 2pqW_{12} + q^2W_{22}$

Allelic frequencies after selection: $p' = f(A^1) = f(A^1A^1) + \frac{1}{2}f(A^1A^2)$; $q' = 1 - p$.

Table 12.5 Formulas for calculating change in allelic frequencies with different types of selection

Type of selection	Fitness Values			
	A^1A^1	A^1A^2	A^2A^2	Change in q
Selection against a recessive trait	1	1	$1-s$	$\dfrac{-spq^2}{1-sq^2}$
Selection against a dominant trait	1	$1-s$	$1-s$	$\dfrac{-spq^2}{1-s+sq^2}$
Selection against a trait with no dominance	1	$1-\frac{1}{2}s$	$1-s$	$\dfrac{-\frac{1}{2}spq}{1-sq}$
Selection against both homozygotes (overdominance)	$1-s_{11}$	1	$1-s_{22}$	$\dfrac{pq\,(s_{11}p - s_{22}q)}{1 - s_{11}p^2 - s_{22}q^2}$

The general selection model   Differential fitness among genotypes with the passage of time leads to changes in the frequencies of the genotypes, which, in turn, lead to changes in the frequencies of the alleles that make up the genotypes. We can predict the effect of natural selection on allelic frequencies by using a general selection model, which is outlined in **Table 12.4**. Use of this model requires knowledge of both the initial allelic frequencies and the fitness values of the genotypes. It assumes that mating is random and that the only force acting on a population is natural selection. The general selection model can be used to calculate the allelic frequencies after any type of selection. It is also possible to work out formulas for determining the change in allelic frequency when selection is against recessive, dominant, and codominant traits, as well as traits in which the heterozygote has highest fitness (**Table 12.5**).

Worked Problem

Let's apply the general selection model in Table 12.4 to a real example. Alcohol is a common substance in rotting fruit, where fruit-fly larvae grow and develop; larvae use the enzyme alcohol dehydrogenase (Adh) to detoxify the effects of this alcohol. In some fruit-fly populations, two alleles are present at the locus that encodes ADH: Adh^F, which encodes a form of the enzyme that migrates rapidly (fast) on an electrophoretic gel; and Adh^S, which encodes a

form of the enzyme that migrates slowly on an electrophoretic gel. Female fruit flies with different Adh genotypes produce the following numbers of offspring when alcohol is present:

Genotype	Mean number of offspring
Adh^F/Adh^F	120
Adh^F/Adh^S	60
Adh^S/Adh^S	30

a. Calculate the relative fitnesses of females having these genotypes.

b. If a population of fruit flies has an initial frequency of Adh^F equal to 0.2, what will the frequency be in the next generation when alcohol is present?

• Solution

a. First, we must calculate the fitnesses of the three genotypes. Fitness is the relative reproductive output of a genotype and is calculated by dividing the mean number of offspring produced by that genotype by the mean number of offspring produced by the most prolific genotype. The fitnesses of the three Adh genotypes therefore are:

Genotype	Mean number of offspring	Fitness
Adh^F/Adh^F	120	$W_{FF} = \dfrac{120}{120} = 1$
Adh^F/Adh^S	60	$W_{FS} = \dfrac{60}{120} = 0.5$
Adh^S/Adh^S	30	$W_{SS} = \dfrac{30}{120} = 0.25$

	Adh^F*Adh*^{F}	*Adh*^{F}*Adh*^{S}	*Adh*^{S}*Adh*^{S}
Initial genotypic frequencies:	$p^2 = (0.2)^2 = 0.04$	$2pq = 2(0.2)(0.8) = 0.32$	$q^2 = (0.8)^2 = 0.64$
Fitnesses:	$W_{FF} = 1$	$W_{FS} = 0.5$	$W_{SS} = 0.25$
Proportionate contribution of genotypes to population:	$p^2W_{FF} = 0.04(1) = 0.04$	$2pqW_{FS} = (0.32)(0.5) = 0.16$	$q^2W_{SS} = (0.64)(0.25) = 0.16$

b. To calculate the frequency of the *Adh*^{F} allele after selection, we can apply the table method. On the first line of the table above, we record the initial genotypic frequencies before selection has acted. If mating has been random (an assumption of the model), the genotypes will have the Hardy–Weinberg equilibrium frequencies of p^2, $2pq$, and q^2. On the second row of the table above, we put the fitness values of the corresponding genotypes. The proportion of the population represented by each genotype after selection is obtained by multiplying the initial genotypic frequency times its fitness (third row of Table 12.4). Now the genotypes are no longer in Hardy–Weinberg equilibrium.

The mean fitness ($\overline{W}$) of the population is the sum of the proportionate contributions of the three genotypes: $\overline{W} = p^2W_{11} + 2pqW_{12} + q^2W_{22} = 0.04 + 0.16 + 0.16 = 0.36$. The mean fitness $\overline{W}$ is the average fitness of all individuals in the population and allows the frequencies of the genotypes after selection to be obtained.

The frequency of a genotype after selection will be equal to its proportionate contribution divided by the mean fitness of the population ($p^2W_{11}/\overline{W}$ for genotype A^1A^1, $2pqW_{12}/\overline{W}$ for genotype A^1A^2, and $q^2W_{22}/\overline{W}$ for genotype A^2A^2) as shown in the fourth line of Table 12.4. We can now add these values to our table as shown below:

	Adh^F*Adh*^{F}	*Adh*^{F}*Adh*^{S}	*Adh*^{S}*Adh*^{S}
Initial genotypic frequencies:	$p^2 = (0.2)^2 = 0.04$	$2pq = 2(0.2)(0.8) = 0.32$	$q^2 = (0.8)^2 = 0.64$
Fitnesses:	$W_{FF} = 1$	$W_{FS} = 0.5$	$W_{SS} = 0.25$
Proportionate contribution of genotypes to population:	$p^2W_{FF} = 0.04(1) = 0.04$	$2pqW_{FS} = (0.32)(0.5) = 0.16$	$q^2W_{SS} = (0.64)(0.25) = 0.16$
Relative genotypic frequency after selection:	$\dfrac{p^2W_{FF}}{\overline{W}} = \dfrac{0.04}{0.36} = 0.11$	$\dfrac{2pqW_{FS}}{\overline{W}} = \dfrac{0.16}{0.36} = 0.44$	$\dfrac{q^2W_{SS}}{\overline{W}} = \dfrac{0.16}{0.36} = 0.44$

After the new genotypic frequencies have been calculated, the new allelic frequency of *Adh*^{F} (p') can be determined by using the now-familiar formula of Equation 12.4:

$$p' = f(Adh^F) = f(Adh^F/Adh^F) + \tfrac{1}{2}f(Adh^F/Adh^S)$$
$$= 0.11 + \tfrac{1}{2}(0.44) = 0.33$$

and that of q' can be obtained by subtraction:

$$q' = 1 - p'$$
$$q' = 1 - p' = 1 - 0.33 = 0.67$$

We predict that the frequency of *Adh*^{F} will increase from 0.2 to 0.33.

> **?** For more practice with the selection model, try Problem 33 at the end of the chapter.

The results of selection The results of selection depend on the relative fitnesses of the genotypes. If we have three genotypes (A^1A^1, A^1A^2, and A^2A^2) with fitnesses W_{11}, W_{12}, and W_{22}, we can identify six different types of natural selection (**Table 12.6**). In type 1 selection, a dominant allele A^1 confers a fitness advantage; in this case, the fitnesses of genotypes A^1A^1 and A^1A^2 are equal and higher than the fitness of A^2A^2 ($W_{11} = W_{12} > W_{22}$). Because both the heterozygote and the A^1A^1 homozygote have copies of the A^1 allele and produce more offspring than the A^2A^2 homozygote does, the frequency of the A^1 allele will increase with time, and the frequency of the A^2 allele will decrease. This form of selection, in which one allele or trait is favored over another, is termed **directional selection.**

Type 2 selection (see Table 12.6) is directional selection against a dominant allele A^1 ($W_{11} = W_{12} < W_{22}$). In this case, the A^2 allele increases and the A^1 allele decreases. Type 3 and type 4 selection also are directional selection but, in these cases, there is incomplete dominance and the heterozygote

Table 12.6 Types of natural selection

Type	Fitness Relation	Form of Selection	Result
1	$W_{11} = W_{12} > W_{22}$	Directional selection against recessive allele A^2	A^1 increases, A^2 decreases
2	$W_{11} = W_{12} < W_{22}$	Directional selection against dominant allele A^1	A^2 increases, A^1 decreases
3	$W_{11} > W_{12} > W_{22}$	Directional selection against incompletely dominant allele A^2	A^1 increases, A^2 decreases
4	$W_{11} < W_{12} < W_{22}$	Directional selection against incompletely dominant allele A^1	A^2 increases, A^1 decreases
5	$W_{11} < W_{12} > W_{22}$	Overdominance	Stable equilibrium, both alleles maintained
6	$W_{11} > W_{12} < W_{22}$	Underdominance	Unstable equilibrium

Note: W_{11}, W_{12}, and W_{22} represent the fitnesses of genotypes A^1A^1, A^1A^2, and A^2A^2, respectively.

has a fitness that is intermediate between the two homozygotes ($W_{11} > W_{12} > W_{22}$ for type 3; $W_{11} < W_{12} < W_{22}$ for type 4). When A^1A^1 has the highest fitness (type 3), the A^1 allele increases and the A^2 allele decreases with the passage of time. When A^2A^2 has the highest fitness (type 4), the A^2 allele increases and the A^1 allele decreases with time. Eventually, directional selection leads to fixation of the favored allele and elimination of the other allele, as long as no other evolutionary forces act on the population.

Two types of selection (types 5 and 6) are special situations that lead to equilibrium, where there is no further change in allelic frequency. Type 5 selection is referred to as **overdominance** or heterozygote advantage. Here, the heterozygote has higher fitness than the fitnesses of the two homozygotes ($W_{11} < W_{12} > W_{22}$). With overdominance, both alleles are favored in the heterozygote, and neither allele is eliminated from the population. Initially, the allelic frequencies may change because one homozygote has higher fitness than the other; the direction of change will depend on the relative fitness values of the two homozygotes. The allelic frequencies change with overdominant selection until a stable equilibrium is reached, at which point there is no further change. The allelic frequency at equilibrium ($\hat{q}$) depends on the relative fitnesses (usually expressed as selection coefficients) of the two homozygotes:

$$\hat{q} = f(A^2) = \frac{s_{11}}{s_{11} + s_{22}} \qquad (12.18)$$

where s_{11} represents the selection coefficient of the A^1A^1 homozygote and s_{22} represents the selection coefficient of the A^2A^2 homozygote.

The last type of selection (type 6) is **underdominance,** in which the heterozygote has lower fitness than both homozygotes ($W_{11} > W_{12} < W_{22}$). Underdominance leads to an unstable equilibrium; here, allelic frequencies will not change as long as they are at equilibrium but, if they are disturbed from the equilibrium point by some other evolutionary force, they will move away from equilibrium until one allele eventually becomes fixed.

CONCEPTS

Natural selection changes allelic frequencies; the direction and magnitude of change depend on the intensity of selection, the dominance relations of the alleles, and the allelic frequencies. Directional selection favors one allele over another and eventually leads to fixation of the favored allele. Overdominance leads to a stable equilibrium with maintenance of both alleles in the population. Underdominance produces an unstable equilibrium because the heterozygote has lower fitness than those of the two homozygotes.

✔ CONCEPT CHECK 10

How does overdominance differ from directional selection?

Change in the allelic frequency of a recessive allele due to natural selection   The rate at which selection changes allelic frequencies depends on the allelic frequency itself. If an allele (A^2) is lethal and recessive, $W_{11} = W_{12} = 1$, whereas $W_{22} = 0$. The frequency of the A^2 allele will decrease with time (because the A^2A^2 homozygote produces no offspring), and the rate of decrease will be proportional to the frequency of the recessive allele. When the frequency of the allele is high, the change in each generation is relatively large but, as the frequency of the allele drops, a higher proportion of the alleles are in the heterozygous genotypes, where they are immune to the action of natural selection (the heterozygotes have the same phenotype as the favored homozygote). Thus, selection against a rare recessive allele is very inefficient and its removal from the population is slow.

The relation between the frequency of a recessive allele and its rate of change under natural selection has an important implication. Some people believe that the medical treatment of

Table 12.7 Effects of different evolutionary forces on allelic frequencies within populations

Force	Short-Term Effect	Long-Term Effect
Mutation	Change in allelic frequency	Equilibrium reached between forward and reverse mutations
Migration	Change in allelic frequency	Equilibrium reached when allelic frequencies of source and recipient population are equal
Genetic drift	Change in allelic frequency	Fixation of one allele
Natural selection	Change in allelic frequency	Directional selection: fixation of one allele
		Overdominant selection: equilibrium reached

patients with rare recessive diseases will cause the disease gene to increase, eventually leading to degeneration of the human gene pool. This mistaken belief was the basis of eugenic laws that were passed in the early part of the twentieth century prohibiting the marriage of persons with certain genetic conditions and allowing the involuntary sterilization of others. However, most copies of rare recessive alleles are present in heterozygotes, and selection against the homozygotes will have little effect on the frequency of a recessive allele. Thus, whether the homozygotes for a recessive trait reproduce or not has little effect on the frequency of the disorder.

Mutation and natural selection Recurrent mutation and natural selection act as opposing forces on detrimental alleles; mutation increases their frequency and natural selection decreases their frequency. Eventually, these two forces reach an equilibrium, in which the number of alleles added by mutation is balanced by the number of alleles removed by selection.

The frequency of a recessive allele at equilibrium ($\hat{q}$) is equal to the square root of the mutation rate divided by the selection coefficient:

$$\hat{q} = \sqrt{\frac{\mu}{s}} \qquad (12.19)$$

For selection acting on a dominant allele, the frequency of the dominant allele at equilibrium can be shown to be

$$\hat{q} = \frac{\mu}{s} \qquad (12.20)$$

Achondroplasia is a common type of human dwarfism that results from a dominant gene. People with this condition are fertile, although they produce only about 74% as many children as are produced by people without achondroplasia. The fitness of people with achondroplasia therefore averages 0.74, and the selection coefficient (s) is $1 - W$, or 0.26. If we assume that the mutation rate for achondroplasia is about 3×10^{-5} (a typical mutation rate in humans), then we can predict that the equilibrium frequency for the achondroplasia allele will be

$$\hat{q} = \frac{0.00003}{0.26} = 0.0001153$$

This frequency is close to the actual frequency of the disease.

CONCEPTS

Mutation and natural selection act as opposing forces on detrimental alleles: mutation tends to increase their frequency and natural selection tends to decrease their frequency, eventually producing an equilibrium.

CONNECTING CONCEPTS

The General Effects of Forces That Change Allelic Frequencies

You now know that four processes bring about change in the allelic frequencies of a population: mutation, migration, genetic drift, and natural selection. Their short- and long-term effects on allelic frequencies are summarized in **Table 12.7.** In some cases, these changes continue until one allele is eliminated and the other becomes fixed in the population. Genetic drift and directional selection will eventually result in fixation, provided these forces are the only ones acting on a population. With the other evolutionary forces, allelic frequencies change until an equilibrium point is reached, and then there is no additional change in allelic frequency. Mutation, migration, and some forms of natural selection can lead to stable equilibria (see Table 12.7).

The different evolutionary forces affect both genetic variation within populations and genetic divergence between populations. Evolutionary forces that maintain or increase genetic variation within populations are listed in the upper-left quadrant of **Figure 12.15.**

	Within populations	Between populations
Increase genetic variation	Mutation Migration Some types of natural selection	Mutation Genetic drift Some types of natural selection
Decrease genetic variation	Genetic drift Some types of natural selection	Migration Some types of natural selection

12.15 Mutation, migration, genetic drift, and natural selection have different effects on genetic variation within populations and on genetic divergence between populations.

These forces include some types of natural selection, such as over-dominance, in which both alleles are favored. Mutation and migration also increase genetic variation within populations because they introduce new alleles to the population. Evolutionary forces that decrease genetic variation within populations are listed in the lower-left quadrant of Figure 12.15. These forces include genetic drift, which decreases variation through fixation of alleles, and some forms of natural selection such as directional selection.

The various evolutionary forces also affect the amount of genetic divergence between populations. Natural selection increases divergence between populations if different alleles are favored in the different populations, but it can also decrease divergence between populations by favoring the same allele in the different populations. Mutation almost always increases divergence between populations because different mutations arise in each population. Genetic drift also increases divergence between populations because changes in allelic frequencies due to drift are random and are likely to change in different directions in separate populations. Migration, on the other hand, decreases divergence between populations because it makes populations similar in their genetic composition.

Migration and genetic drift act in opposite directions: migration increases genetic variation within populations and decreases divergence between populations, whereas genetic drift decreases genetic variation within populations and increases divergence among populations. Mutation increases both variation within populations and divergence between populations. Natural selection can either increase or decrease variation within populations, and it can increase or decrease divergence between populations.

An important point to keep in mind is that real populations are simultaneously affected by many evolutionary forces. We have examined the effects of mutation, migration, genetic drift, and natural selection in isolation so that the influence of each process would be clear. However, in the real world, populations are commonly affected by several evolutionary forces at the same time, and evolution results from the complex interplay of numerous processes.

CONCEPTS SUMMARY

- Population genetics examines the genetic composition of groups of individuals and how this composition changes with time.
- A Mendelian population is a group of interbreeding, sexually reproducing individuals, whose set of genes constitutes the population's gene pool. Evolution takes place through changes in this gene pool.
- A population's genetic composition can be described by its genotypic and allelic frequencies.
- The Hardy–Weinberg law describes the effects of reproduction and Mendel's laws on the allelic and genotypic frequencies of a population. It assumes that a population is large, randomly mating, and free from the effects of mutation, migration, and natural selection. When these conditions are met, the allelic frequencies do not change and the genotypic frequencies stabilize after one generation in the Hardy–Weinberg equilibrium proportions p^2, $2pq$, and q^3, where p and q equal the frequencies of the alleles.
- Nonrandom mating affects the frequencies of genotypes but not those of alleles.
- Inbreeding, a type of positive assortative mating, increases the frequency of homozygotes while decreasing the frequency of heterozygotes. Inbreeding is frequently detrimental because it increases the appearance of lethal and deleterious recessive traits.
- Mutation, migration, genetic drift, and natural selection can change allelic frequencies.

- Recurrent mutation eventually leads to an equilibrium, with the allelic frequencies being determined by the relative rates of forward and reverse mutation. Change due to mutation in a single generation is usually very small because mutation rates are low.
- Migration, the movement of genes between populations, increases the amount of genetic variation within populations and decreases the number of differences between populations.
- Genetic drift is change in allelic frequencies due to chance factors. Genetic drift arises when a population consists of a small number of individuals, is established by a small number of founders, or undergoes a major reduction in size. Genetic drift changes allelic frequencies, reduces genetic variation within populations, and causes genetic divergence among populations.
- Natural selection is the differential reproduction of genotypes; it is measured by the relative reproductive successes (fitnesses) of genotypes. The effects of natural selection on allelic frequency can be determined by applying the general selection model. Directional selection leads to the fixation of one allele. The rate of change in allelic frequency due to selection depends on the intensity of selection, the dominance relations, and the initial frequencies of the alleles.
- Mutation and natural selection can produce an equilibrium, in which the number of new alleles introduced by mutation is balanced by the elimination of alleles through natural selection.

IMPORTANT TERMS

genetic rescue (p. 334)
Mendelian population (p. 334)
gene pool (p. 334)
genotypic frequency (p. 335)
allelic frequency (p. 335)

Hardy–Weinberg law (p. 337)
Hardy–Weinberg equilibrium (p. 337)
positive assortative mating (p. 341)
negative assortative mating (p. 341)
inbreeding (p. 341)

outcrossing (p. 341)
inbreeding coefficient (p. 341)
inbreeding depression (p. 342)
equilibrium (p. 345)
migration (gene flow) (p. 345)

sampling error (p. 347)
genetic drift (p. 347)
effective population size (p. 347)
founder effect (p. 348)

genetic bottleneck (p. 348)
fixation (p. 349)
fitness (p. 350)
selection coefficient (p. 350)

directional selection (p. 352)
overdominance (p. 353)
underdominance (p. 353)

ANSWERS TO CONCEPT CHECKS

1. There are fewer alleles than genotypes, and so the gene pool can be described by fewer parameters when allelic frequencies are used. Additionally, the genotypes are temporary assemblages of alleles that break down each generation; rather than the genotypes, the alleles are passed from generation to generation in sexually reproducing organisms.

2. a

3. c

4. 0.10

5. b

6. b

7. c

8. d

9. b

10. In overdominance, the heterozygote has the highest fitness. In directional selection, one allele or one trait is favored over another.

WORKED PROBLEM

1. A recessive allele for red hair (r) has a frequency of 0.2 in population I and a frequency of 0.01 in population II. A famine in population I causes a number of people in population I to migrate to population II, where they reproduce randomly with the members of population II. Geneticists estimate that, after migration, 15% of the people in population II consist of people who migrated from population I. What will be the frequency of red hair in population II after the migration?

• **Solution**

From Equation 12.14, the allelic frequency in a population after migration (q'_{II}) is

$$q'_{II} = q_I(m) + q_{II}(1 - m)$$

where q_I and q_{II} are the allelic frequencies in population I (migrants) and population II (residents), respectively, and m is the proportion of population II that consist of migrants. In this problem, the frequency of red hair is 0.2 in population I and 0.01 in population II. Because 15% of population II consists of migrants, $m = 0.15$. Substituting these values into Equation 12.14, we obtain

$$q'_{II} = 0.2(0.15) + (0.01)(1 - 0.15) = 0.03 + 0.0085 = 0.0385$$

which is the expected frequency of the allele for red hair in population II after migration. Red hair is a recessive trait; if mating is random for hair color, the frequency of red hair in population II after migration will be

$$f(rr) = q^2 = (0.0385)^2 = 0.0015$$

COMPREHENSION QUESTIONS

Section 12.1

1. What is a Mendelian population? How is the gene pool of a Mendelian population usually described?

Section 12.2

2. What are the predictions given by the Hardy–Weinberg law?

*3. What assumptions must be met for a population to be in Hardy–Weinberg equilibrium?

4. What is random mating?

*5. Give the Hardy–Weinberg expected genotypic frequencies for (a) an autosomal locus with three alleles, and (b) an X-linked locus with two alleles.

Section 12.3

6. Define inbreeding and briefly describe its effects on a population.

Section 12.4

7. What determines the allelic frequencies at mutational equilibrium?

*8. What factors affect the magnitude of change in allelic frequencies due to migration?

9. Define genetic drift and give three ways in which it can arise. What effect does genetic drift have on a population?

*10. What is effective population size? How does it affect the amount of genetic drift?

11. Define natural selection and fitness.

12. Briefly describe the differences between directional selection, overdominance, and underdominance. Describe the effect of each type of selection on the allelic frequencies of a population.

13. What factors affect the rate of change in allelic frequency due to natural selection?

Section 12.4

*14. Compare and contrast the effects of mutation, migration, genetic drift, and natural selection on genetic variation within populations and on genetic divergence between populations.

APPLICATION QUESTIONS AND PROBLEMS

Section 12.1

15. How would you respond to someone who said that models are useless in studying population genetics because they represent oversimplifications of the real world?

*16. Voles (Microtus ochrogaster) were trapped in old fields in southern Indiana and were genotyped for a transferrin locus. The following numbers of genotypes were recorded, where T^E and T^F represent different alleles.

$T^E T^E$	$T^E T^F$	$T^F T^F$
407	170	17

Calculate the genotypic and allelic frequencies of the transferrin locus for this population.

17. [DATA ANALYSIS] Jean Manning, Charles Kerfoot, and Edward Berger studied the frequencies at the phosphoglucose isomerase (GPI) locus in the cladoceran Bosmina longirostris. At one location, they collected 176 animals from Union Bay in Seattle, Washington, and determined their GPI genotypes by using electrophoresis (J. Manning, W. C. Kerfoot, and E. M. Berger. 1978. Evolution 32:365–374).

Genotype	Number
$S^1 S^1$	4
$S^1 S^2$	38
$S^2 S^2$	134

Determine the genotypic and allelic frequencies for this population.

18. Orange coat color of cats is due to an X-linked allele (X^O) that is codominant with the allele for black (X^+). Genotypes of the orange locus of cats in Minneapolis and St. Paul, Minnesota, were determined, and the following data were obtained:

$X^O X^O$ females	11
$X^O X^+$ females	70
$X^+ X^+$ females	94
$X^O Y$ males	36
$X^+ Y$ males	112

Calculate the frequencies of the X^O and X^+ alleles for this population.

Section 12.2

19. A total of 6129 North American Caucasians were blood typed for the MN locus, which is determined by two codominant alleles, L^M and L^N. The following data were obtained:

Blood type	Number
M	1787
MN	3039
N	1303

Carry out a chi-square test to determine whether this population is in Hardy–Weinberg equilibrium at the MN locus.

20. [DATA ANALYSIS] Most black bears (Ursus americanus) are black or brown in color. However, occasional white bears of this species appear in some populations along the coast of British Columbia. Kermit Ritland and his colleagues determined that white coat color in these bears results from a recessive mutation (G) caused by a single nucleotide replacement in which guanine substitutes for adenine at the melanocortin 1 receptor locus (mcr1), the same locus responsible for red hair in humans (K. Ritland, C. Newton, and H. D. Marshall. 2001. Current Biology 11:1468–1472). The wild-type allele at this locus (A) encodes black or brown color. Ritland and his colleagues collected samples from bears on three islands and determined their genotypes at the mcr1 locus.

Genotype	Number
AA	42
AG	24
GG	21

a. What are the frequencies of the A and G alleles in these bears?

b. Give the genotypic frequencies expected if the population is in Hardy–Weinberg equilibrium.

c. Use a chi-square test to compare the number of observed genotypes with the number expected under Hardy–Weinberg equilibrium. Is this population in Hardy–Weinberg equilibrium? Explain your reasoning.

21. Genotypes of leopard frogs from a population in central Kansas were determined for a locus (M) that encodes the enzyme malate dehydrogenase. The following numbers of genotypes were observed:

Genotype	Number
M^1M^1	20
M^1M^2	45
M^2M^2	42
M^1M^3	4
M^2M^3	8
M^3M^3	6
Total	125

   a. Calculate the genotypic and allelic frequencies for this population.

   b. What would the expected numbers of genotypes be if the population were in Hardy–Weinberg equilibrium?

22. Full color (D) in domestic cats is dominant over dilute color (d). Of 325 cats observed, 194 have full color and 131 have dilute color.

   a. If these cats are in Hardy–Weinberg equilibrium for the dilution locus, what is the frequency of the dilute allele?

   b. How many of the 194 cats with full color are likely to be heterozygous?

23. Tay–Sachs disease is an autosomal recessive disorder. Among Ashkenazi Jews, the frequency of Tay–Sachs disease is 1 in 3600. If the Ashkenazi population is mating randomly for the Tay–Sachs gene, what proportion of the population consists of heterozygous carriers of the Tay–Sachs allele?

24. In the plant *Lotus corniculatus*, cyanogenic glycoside protects the plant against insect pests and even grazing by cattle. This glycoside is due to a simple dominant allele. A population of *L. corniculatus* consists of 77 plants that possess cyanogenic glycoside and 56 that lack the compound. What is the frequency of the dominant allele responsible for the presence of cyanogenic glycoside in this population?

*25. Color blindness in humans is an X-linked recessive trait. Approximately 10% of the men in a particular population are color blind.

   a. If mating is random for the color-blind locus, what is the frequency of the color-blind allele in this population?

   b. What proportion of the women in this population are expected to be color blind?

   c. What proportion of the women in the population are expected to be heterozygous carriers of the color-blind allele?

*26. The human MN blood type is determined by two codominant alleles, L^M and L^N. The frequency of L^M in Eskimos on a small Arctic island is 0.80. If the inbreeding coefficient for this population is 0.05, what are the expected frequencies of the M, MN, and N blood types on the island?

Section 12.3

27. Demonstrate mathematically that full-sib mating ($F = \frac{1}{4}$) reduces the heterozygosity by $\frac{1}{4}$ with each generation.

Section 12.4

28. The forward mutation rate for piebald spotting in guinea pigs is 8×10^{-5}; the reverse mutation rate is 2×10^{-6}. If no other evolutionary forces are assumed to be present, what is the expected frequency of the allele for piebald spotting in a population that is in mutational equilibrium?

29. [DATA ANALYSIS] For a period of 3 years, Gunther Schlager and Margaret Dickie estimated the forward and reverse mutation rates for five loci in mice that encode various aspects of coat color by examining more than 5 million mice for spontaneous mutations (G. Schlager and M. M. Dickie. 1966. *Science* 151:205–206). The numbers of mutations detected at the dilute locus are as follows:

	Number of gametes examined	Number of mutations detected
Forward mutations	260,675	5
Reverse mutations	583,360	2

Calculate the forward and reverse mutation rates at this locus. If these mutations rates are representative of rates in natural populations of mice, what would the expected equilibrium frequency of dilute mutations be?

*30. In German cockroaches, curved wing (cv) is recessive to normal wing (cv^+). Bill, who is raising cockroaches in his dorm room, finds that the frequency of the gene for curved wings in his cockroach population is 0.6. In the apartment of his friend Joe, the frequency of the gene for curved wings is 0.2. One day Joe visits Bill in his dorm room, and several cockroaches jump out of Joe's hair and join the population in Bill's room. Bill estimates that, now, 10% of the cockroaches in his dorm room are individual roaches that jumped out of Joe's hair. What is the new frequency of curved wings among cockroaches in Bill's room?

31. A population of water snakes is found on an island in Lake Erie. Some of the snakes are banded and some are unbanded; banding is caused by an autosomal allele that is recessive to an allele for no bands. The frequency of banded snakes on the island is 0.4, whereas the frequency of banded snakes on the mainland is 0.81. One summer, a large number of snakes migrate from the mainland to the island. After this migration, 20% of the island population consists of snakes that came from the mainland.

   a. If both the mainland population and the island population are assumed to be in Hardy–Weinberg equilibrium for the alleles that affect banding, what is the frequency of the allele for bands on the island and on the mainland before migration?

b. After migration has taken place, what is the frequency of the banded allele on the island?

32. Pikas are small mammals that live at high elevation in the talus slopes of mountains. Most populations located on mountain tops in Colorado and Montana in North America are isolated from one another, because the pikas don't occupy the low-elevation habitats that separate the mountain tops and don't venture far from the talus slopes. Thus, there is little gene flow between populations. Furthermore, each population is small in size and was founded by a small number of pikas.

A group of population geneticists propose to study the amount of genetic variation in a series of pika populations and to compare the allelic frequencies in different populations. On the bases of the biology and the distribution of pikas, predict what the population geneticists will find concerning the within- and between-population genetic variation.

33. In a large, randomly mating population, the frequency of the allele (*s*) for sickle-cell hemoglobin is 0.028. The results of studies have shown that people with the following genotypes at the beta-chain locus produce the average numbers of offspring given:

Genotype	Average number of offspring produced
SS	5
Ss	6
ss	0

a. What will the frequency of the sickle-cell allele (*s*) be in the next generation?

b. What will the frequency of the sickle-cell allele be at equilibrium?

34. Two chromosomal inversions are commonly found in populations of *Drosophila pseudoobscura*: Standard (*ST*) and Arrowhead (*AR*). When treated with the insecticide DDT, the genotypes for these inversions exhibit overdominance, with the following fitnesses:

Genotype	Fitness
ST/ST	0.47
ST/AR	1
AR/AR	0.62

What will the frequencies of *ST* and *AR* be after equilibrium has been reached?

*35. In a large, randomly mating population, the frequency of an autosomal recessive lethal allele is 0.20. What will the frequency of this allele be in the next generation if the lethality takes place before reproduction?

36. The fruit fly, *Drosophila melanogaster*, normally feeds on rotting fruit, which may ferment and contain high levels of alcohol. Douglas Cavener and Michael Clegg studied allelic frequencies at the locus for alcohol dehydrogenase (*Adh*) in experimental populations of *D. melanogaster* (D. R. Cavener and M. T. Clegg. 1981. *Evolution* 35:1–10). The experimental populations were established from wild-caught flies and were raised in cages in the laboratory. Two control populations were raised on a standard cornmeal–molasses–agar diet. Two ethanol populations were raised on a cornmeal–molasses–agar diet to which was added 10% ethanol. The four populations were periodically sampled to determine the allelic frequencies of two alleles at the alcohol dehydrogenase locus, Adh^S and Adh^F. The frequencies of these alleles in the experimental populations are shown in the graph.

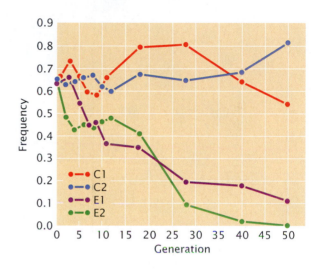

a. One the basis of these data, what conclusion might you draw about the evolutionary forces that are affecting the *Adh* alleles in these populations?

b. Cavener and Clegg measured the viability of the different *Adh* genotypes in the alcohol environment and obtained the following values:

Genotype	Relative viability
Adh^F/Adh^F	0.932
Adh^F/Adh^S	1.288
Adh^S/Adh^S	0.596

Using these relative viabilities, calculate relative fitnesses for the three genotypes. If a population has an initial frequency of $p = f(Adh^F) = 0.5$, what will the expected frequency of Adh^F be in the next generation on the basis of these fitness values?

37. A certain form of congenital glaucoma is caused by an autosomal recessive allele. Assume that the mutation rate is 10^{-5} and that persons having this condition produce, on the average, only about 80% of the offspring produced by persons who do not have glaucoma.

a. At equilibrium between mutation and selection, what will the frequency of the gene for congenital glaucoma be?

b. What will the frequency of the disease be in a randomly mating population that is at equilibrium?

CHALLENGE QUESTION

Section 12.4

38. The Barton Springs salamander is an endangered species found only in a single spring in the city of Austin, Texas. There is growing concern that a chemical spill on a nearby freeway could pollute the spring and wipe out the species. To provide a source of salamanders to repopulate the spring in the event of such a catastrophe, a proposal has been made to establish a captive breeding population of the salamander in a local zoo. You are asked to provide a plan for the establishment of this captive breeding population, with the goal of maintaining as much of the genetic variation of the species as possible in the captive population. What factors might cause loss of genetic variation in the establishment of the captive population? How could loss of such variation be prevented? With the assumption that only a limited number of salamanders can be maintained in captivity, what procedures should be instituted to ensure the long-term maintenance of as much of the variation as possible?

13 | Evolutionary Genetics

Some chimpanzees, like humans, have the ability to taste phenylthiocarbamide (PTC), whereas others cannot taste it. Recent research indicates that the PTC taste polymorphism evolved independently in humans and chimpanzees . [Jong & Petra Wegner/Animals Animals-Earth Scenes.]

TASTER GENES IN SPITTING APES

Almost every student of biology knows about the taster test. The teacher passes out small pieces of paper impregnated with a compound called phenylthiocarbamide (PTC), and the students, following the teacher's instructions, put the paper in their mouths. The reaction is always the same: a number of the students immediately spit the paper out, repelled by the bitter taste of the PTC. A few students, however, can't taste the PTC and continue to suck on the paper, wondering what all the spitting is about. Variation among individuals in a trait such as the ability to taste PTC is termed a polymorphism.

The ability to taste PTC is inherited as an autosomal dominant trait in humans. The frequencies of taster and non-taster alleles have been estimated in hundreds of human populations worldwide. Almost all populations have both tasters and nontasters; the frequency of the two alleles varies widely.

PTC is not found in nature, but the ability to taste it is strongly correlated with the ability to taste other naturally occurring bitter compounds, some of which are toxic. The ability to taste PTC has also been linked to dietary preferences and may be associated with susceptibility to certain diseases, such as thyroid deficiency. These observations suggest that natural selection has played a role in the evolution of the taster trait. Some understanding of the mechanism of the evolution of the taster trait was gained when the famous population geneticist R.A. Fisher and his colleagues took a trip to the zoo.

Fisher wondered whether other primates also might have the ability to taste PTC. To answer this question, he prepared some drinks with different levels of PTC and set off for the zoo with his friends, fellow biologists Edmund (E. B.) Ford and Julian Huxley. At the zoo, the PTC-laced drinks were offered to eight chimpanzees and one orangutan. Fisher and his friends were initially concerned that they might not be able to tell whether the apes could taste the PTC. That concern disappeared, however, when the first one sampled the drink and immediately spat on Fisher. Of the eight chimpanzees tested, six were tasters and two were nontasters.

The observation that chimpanzees and humans both have the PTC taste polymorphism led Fisher and his friends to assume that the polymorphism arose in a common

361

ancestor of humans and chimpanzees who passed it on to both species. However, they had no way to test their hypothesis. Sixty-five years later, geneticists armed with the latest techniques of modern molecular biology were able to determine the origin of the PTC taste polymorphism and test the hypothesis of Fisher and his friends.

Molecular studies reveal that our ability to taste PTC is controlled by alleles at the *TAS2R38* locus, a 1000-bp gene found on chromosome 7. This locus encodes receptors for bitter compounds and is expressed in the cells of our taste buds. One common allele encodes a receptor that allows the ability to taste PTC; an alternative allele encodes a receptor that does not respond to PTC.

Recent research has demonstrated that PTC taste sensitivity in chimpanzees also is controlled by alleles at the *TAS2R38* locus. However, much to the surprise of the investigators, the taster alleles in humans and chimpanzees are not the same at the molecular level. In the human taster and nontaster alleles, nucleotide differences at three positions affect which amino acids are present in the taste-receptor protein. In chimpanzees, however, none of these nucleotide differences are present. Instead, a mutation in the initiation codon produces the nontaster allele. This substitution eliminates the normal initiation codon, and the ribosome initiates translation at an alternative downstream initiation codon, resulting in the production of a shortened protein receptor that fails to respond to PTC.

What these findings mean is that Fisher and his friends were correct in their observation that humans and chimpanzees both have PTC taste polymorphism but were incorrect in their hypothesis about its origin: humans and chimpanzees independently evolved the PTC taste polymorphism.

This chapter is about the genetic basis of evolution. As illustrated by the story of the PTC taster polymorphism, evolutionary genetics has a long history but has been transformed in recent years by the application of powerful techniques of molecular genetics. In Chapter 12, we considered the evolutionary forces that bring about change in the allelic frequencies of a population: mutation, migration, genetic drift, and selection. In this chapter, we examine some specific ways that these forces shape the genetic makeup of populations and bring about long-term evolution. We begin by looking at how the process of evolution depends on genetic variation and how much genetic variation is found in natural populations. We then turn to the evolutionary changes that bring about the appearance of new species and how evolutionary histories (phylogenies) are constructed. We end the chapter by taking a look at patterns of evolutionary change at the molecular level.

13.1 Organisms Evolve Through Genetic Change Occurring Within Populations

Evolution is one of the foundational principles of all of biology. Theodosius Dobzhansky, an important early leader in the field of evolutionary genetics, once remarked "Nothing in biology makes sense except in the light of evolution." Indeed, evolution is an all-encompassing theory that helps to make sense of much of natural world, from the sequences of DNA found in our cells to the types of plants and animals that surround us. The evidence for evolution is overwhelming. Evolution has been directly observed numerous times; for example, hundreds of different insects evolved resistance to common pesticides following their introduction after World War II. Evolution is supported by the fossil record, comparative anatomy, embryology, the distribution of plants and animals (biogeography), and molecular genetics.

In spite of its vast importance to all fields of biology, evolution is often misunderstood and misinterpreted. In our society, the term *evolution* frequently refers to any type of change. However, biological **evolution** refers only to a specific type of change—genetic change taking place in a group of organisms. Two aspects of this definition should be emphasized. First, evolution includes genetic change only. Many nongenetic changes take place in living organisms, such as the development of a complex intelligent person from an original single-celled zygote. Although remarkable, this change isn't evolution, because it does not include changes in genes. The second aspect to emphasize is that evolution takes place in *groups* of organisms. An individual organism does not evolve; what evolves is the gene pool common to a group of organisms.

Evolution can be thought of as a two-step process. First, genetic variation arises. Genetic variation has its origin in the processes of mutation, which produces new alleles, and recombination, which shuffles alleles into new combinations. Both of these processes are random and produce genetic variation continually, regardless of evolution's need for it. The second step in the process of evolution is the increase and decrease in the frequencies of genetic variants. Various evolutionary forces discussed in Chapter 12 cause

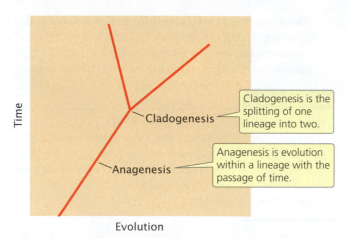

13.1 Anagenesis and cladogenesis are two different types of evolutionary change. Anagenesis is change within an evolutionary lineage; cladogenesis is the splitting of lineages (speciation).

some alleles in the gene pool to increase in frequency and other alleles to decrease in frequency. This shift in the composition of the gene pool common to a group of organisms constitutes evolutionary change.

We can differentiate between two types of evolution that take place within a group of organisms connected by reproduction. **Anagenesis** refers to evolution taking placing in a single group (a lineage) with the passage of time (**Figure 13.1**). Another type of evolution is **cladogenesis,** the splitting of one lineage into two. When a lineage splits, the two branches no longer have genes in common and evolve independently of one another. New species arise through cladogenesis.

CONCEPTS

Biological evolution is genetic change that takes place within a group of organisms. Anagenesis is evolution that takes place within a single lineage; cladogenesis is the splitting of one lineage into two.

✔ CONCEPT CHECK 1

Briefly describe how evolution takes place as a two-step process.

13.2 Many Natural Populations Contain High Levels of Genetic Variation

Because genetic variation must be present for evolution to take place, evolutionary biologists have long been interested in the amounts of genetic variation in natural populations and the forces that control the amount and nature of that variation. For many years, they could not examine genes directly and were limited to studying the phenotypes of organisms. Although it was impossible to quantify genetic variation directly, studies of phenotypes suggested that many

populations of organisms harbor considerable genetic variation. Populations of organisms in nature exhibit tremendous amounts of phenotypic variation: frogs vary in color pattern, birds differ in size, butterflies vary in spotting patterns, mice differ in coat color, humans vary in blood types, to mention just a few. Crosses revealed that a few of these traits were inherited as simple genetic traits (**Figure 13.2**) but, for most traits, the precise genetic basis was complex and unknown, preventing early evolutionary geneticists from quantifying the amount of genetic variation in natural populations.

As discussed in Chapter 11, the ability of a population to respond to selection (the response to selection) depends on the narrow-sense heritability, which is the measure of the additive genetic variation of a trait within a population. Many organisms respond to artificial selection carried out by humans, suggesting that the populations of these organisms contain much additive genetic variation. For example, humans have used artificial selection to produce numerous breeds of dogs that vary tremendously in size, shape, color, and behavior (see Figure 11.20). Early studies of chromosome variations in *Drosophila* and plants suggested that genetic variation in natural populations also is plentiful and widespread.

Normal homozygotes

Heterozygotes

Recessive bimacula phenotype

13.2 Early evolutionary geneticists were forced to rely on the phenotypic traits that had a simple genetic basis. Variation in the spotting patterns of the butterfly *Panaxia dominula* is an example.

Molecular Variation

The tremendous advances in molecular genetics in recent years have made it possible to investigate evolutionary change directly by analyzing protein and nucleic acid sequences. These molecular data offer a number of advantages for studying the process and pattern of evolution:

Molecular data are genetic. Evolution results from genetic change with time. Many anatomical, behavioral, and physiological traits have a genetic basis, but the relation between the underlying genes and the trait may be complex. Protein and nucleic acid sequence variation has a clear genetic basis that is often easy to interpret.

Molecular methods can be used with all organisms. Early studies of population genetics relied on simple genetic traits such as human blood types, banding patterns in snails, or spotting patterns in butterflies (see Figure 13.2), which are restricted to a small group of organisms. However, all living organisms have proteins and nucleic acids; so molecular data can be collected from any organism.

Molecular methods can be applied to a huge amount of genetic variation. An enormous amount of data can be accessed by molecular methods. The human genome, for example, contains more than 3 billion base pairs of DNA, which constitutes a large pool of information about our evolution.

All organisms can be compared with the use of some molecular data. Trying to assess the evolutionary history of distantly related organisms is often difficult because they have few characteristics in common. The evolutionary relationships between angiosperms were traditionally assessed by comparing floral anatomy, whereas the evolutionary relationships of bacteria were determined by their nutritional and staining properties. Because plants and bacteria have so few structural characteristics in common, evaluating how they are related to one another was difficult in the past. All organisms have certain molecular traits in common, such as ribosomal RNA sequences and some fundamental proteins. These molecules offer a valid basis for comparisons among all organisms.

Molecular data are quantifiable. Protein and nucleic acid sequence data are precise, accurate, and easy to quantify, which facilitates the objective assessment of evolutionary relationships.

Molecular data often provide information about the process of evolution. Molecular data can reveal important clues about the process of evolution. For example, the results of a study of DNA sequences have revealed that one type of insecticide resistance in mosquitoes probably arose from a single mutation that subsequently spread throughout the world.

The database of molecular information is large and growing. Today, this database of DNA and protein sequences can be used for making evolutionary comparisons and inferring mechanisms of evolution.

CONCEPTS

Molecular techniques and data offer a number of advantages for evolutionary studies. Molecular data are genetic in nature and can be investigated in all organisms; they provide potentially large data sets, allow all organisms to be compared by using the same characteristics, are easily quantifiable, and provide information about the process of evolution.

Protein Variation

The initial breakthrough in quantifying genetic variation in natural populations came with the application of electrophoresis to population studies. This technique separates macromolecules, such as proteins or nucleic acids, on the basis of their size and charge. In 1966, Richard Lewontin and John Hubby extracted proteins from fruit flies, separated the proteins by electrophoresis, and stained for specific enzymes. An examination of the pattern of bands on gels enabled them to assign genotypes to individual flies and to quantify the amount of genetic variation in natural populations. In the same year, Harry Harris quantified genetic variation in human populations by using the same technique. Protein variation has now been examined in hundreds of different species by using protein electrophoresis (**Figure 13.3**).

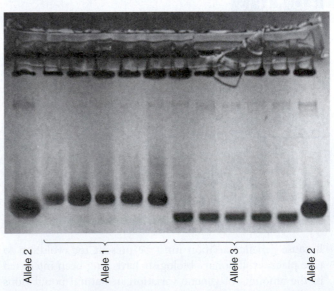

13.3 Molecular variation in proteins is revealed by electrophoresis. Tissue samples from *Drosophila pseudoobscura* were subjected to electrophoresis and stained for esterase. Esterases encoded by different alleles migrate different distances. Shown on the gel are homozygotes for three different alleles.

Measures of genetic variation   The amount of genetic variation in populations is commonly measured by two parameters. The **proportion of polymorphic loci** is the proportion of examined loci in which more than one allele is present in a population. If we examined 30 different loci and found two or more alleles present at 15 of these loci, the percentage of polymorphic loci would be $^{15}/_{30} = 0.5$. The **expected heterozygosity** is the proportion of individuals that are expected to be heterozygous at a locus under the Hardy–Weinberg conditions, which is $2pq$ when there are two alleles present in the population. The expected heterozygosity is often preferred to the observed heterozygosity because expected heterozygosity is independent of the breeding system of an organism. For example, if a species self-fertilizes, it may have little or no heterozygosity but still have considerable genetic variation, which the expected heterozygosity will represent. Expected heterozygosity is typically calculated for a number of loci and is then averaged over all the loci examined.

The proportion of polymorphic loci and the expected heterozygosity have been determined by protein electrophoresis for a number of species (**Table 13.1**). About one-third of all protein loci are polymorphic, and expected heterozygosity averages about 10%, although there is considerable diversity among species. These measures actually underestimate the true amount of genetic variation though, because protein electrophoresis does not detect some amino acid substitutions; nor does it detect genetic variation in DNA that does not alter the amino acids of a protein (synonymous codons and variation in noncoding regions of the DNA).

Explanations for protein variation   By the late 1970s, geneticists recognized that most populations possess large amounts of genetic variation, although the evolutionary significance of this fact was not at all clear. Two opposing hypotheses arose to account for the presence of the extensive molecular variation in proteins. The **neutral-mutation hypothesis** proposed that the molecular variation revealed by protein electrophoresis is adaptively neutral; that is, individuals with different molecular variants have equal fitness at realistic population sizes. This hypothesis does not propose that the proteins are functionless; rather, it suggests that most variants are functionally equivalent. Because these variants are functionally equivalent, natural selection does not differentiate between them, and their evolution is shaped largely by the random processes of genetic drift and mutation. The neutral-mutation hypothesis accepts that natural selection is an important force in evolution but views selection as a process that favors the "best" allele while eliminating others. It proposes that, when selection is important, there will be *little* genetic variation.

The **balance hypothesis** proposes, on the other hand, that the genetic variation in natural populations is maintained by selection that favors variation (balancing

Table 13.1   Proportion of polymorphic loci and heterozygosity for different organisms, as determined by protein electrophoresis

Group	Number of Species	Proportion of Polymorphic Loci		Heterozygosity	
		Mean	SD*	Mean	SD*
Plants	15	0.26	0.17	0.07	0.07
Invertebrates (excluding insects)	28	0.40	0.28	0.10	0.07
Insects (excluding *Drosophila*)	23	0.33	0.20	0.07	0.08
Drosophila	32	0.43	0.13	0.14	0.05
Fish	61	0.15	0.01	0.05	0.04
Amphibians	12	0.27	0.13	0.08	0.04
Reptiles	15	0.22	0.13	0.05	0.02
Birds	10	0.15	0.11	0.05	0.04
Mammals	46	0.15	0.10	0.04	0.02

*SD, standard deviation from the mean.

Source: After L. E. Mettler, T. G. Gregg, and H. E. Schaffer, *Population Genetics and Evolution*, 2d ed. (Englewood Cliffs, N.J.: Prentice Hall, 1988), Table 9.3. Original data from E. Nevo, Genetic variation in natural populations: Patterns and theory, *Theoretical Population Biology* 13:121–177, 1978.

selection). Overdominance, in which the heterozygote has higher fitness than that of either homozygote, is one type of balancing selection. Under this hypothesis, the molecular variants are not physiologically equivalent and do not have the same fitness. Instead, genetic variation within natural populations is shaped largely by selection, and, when selection is important, there will be *much* variation.

Many attempts to prove one hypothesis or the other failed, because precisely how much variation was actually present was not clear (remember that protein electrophoresis detects only some genetic variation) and because both hypotheses are capable of explaining many different patterns of genetic variation. The results of recent studies that provide direct information about DNA sequence variation demonstrate that much variation at the level of DNA has little obvious effect on the phenotype and is therefore likely to be neutral.

CONCEPTS

The application of electrophoresis to the study of protein variation in natural populations revealed that most organisms possess large amounts of genetic variation. The neutral-mutation hypothesis proposes that most molecular variation is neutral with regard to natural selection and is shaped largely by mutation and genetic drift. The balance hypothesis proposes that genetic variation is maintained by balancing selection.

✔ CONCEPT CHECK 2

Which statement is true of the neutral-mutation hypothesis?

a. All proteins are functionless.
b. Natural selection plays no role in evolution.
c. Most molecular variants are functionally equivalent.
d. All of the above.

DNA Sequence Variation

The development of techniques for isolating, cutting, and sequencing DNA in the past 25 years has provided powerful tools for detecting, quantifying, and investigating genetic variation. The application of these techniques has provided a detailed view of genetic variation at the molecular level.

Restriction-site variation   Among the first techniques for detecting and analyzing genetic variation in DNA sequences was the use of restriction enzymes. Each restriction enzyme recognizes and cuts a particular sequence of DNA nucleotides known as that enzyme's restriction site. Variation in the presence of a restriction site is called a restriction fragment length polymorphism (RFLP). Each restriction enzyme recognizes a limited number of nucleotide sites in a particular piece of DNA but, if a number of different restriction enzymes are used and the sites

recognized by the enzymes are assumed to be random sequences, RFLPs can be used to estimate the amount of variation in the DNA and the proportion of nucleotides that differ between organisms. RFLPs can also be used to analyze the genetic structure of populations and to assess evolutionary relationships among organisms. RFLPs were widely used in evolutionary studies before the development of rapid and inexpensive methods for directly sequencing DNA, and restriction analysis is still employed today in studies of molecular evolution. However, the use of restriction enzymes to analyze DNA sequence variation gives an incomplete picture of the underlying variation, because it detects variation only at restriction sites.

In an evolutionary application of RFLPs, Nicholas Georgiadis and his colleagues studied genetic relationships among African elephants (*Loxodonta africana;* **Figure 13.4**) from 10 protected areas of Africa. Mitochondrial DNA (mtDNA) was extracted from samples collected from 270 elephants and amplified with the polymerase chain reaction (PCR). The mtDNA was then cleaved with 10 different restriction enzymes, and RFLPs were detected with gel electrophoresis. The degree of genetic differences among elephants from different sites was measured from the sequence variation revealed by the RFLPs. The results of the study showed that the elephant populations are genetically differentiated across Africa, but there is no significant regional subdivision in their genetic structure. From the patterns of variation, the researchers concluded that the elephants have a complex population history, with subdivided populations that exhibit intermittent gene flow.

Microsatellite variation   Microsatellites are short DNA sequences that exist in multiple copies repeated in tandem. Variation in the number of copies of the repeats is common, with individual organisms often differing in the number of repeat copies. Microsatellites can be detected by using PCR. Pairs of primers are used that flank a region of

13.4 Restriction fragment length polymorphisms have been used to study population structure and gene flow among populations of the African elephant, *Loxodonta africana*. [Digital Vision.]

repeated copies of the sequence. The DNA fragments that are synthesized in the PCR reaction vary in length, depending on the number of tandem repeats present. DNA from an individual organism with more repeats will produce a longer amplified segment. After PCR has been completed, the amplified fragments are separated with the use of gel electrophoresis and stained, producing a series of bands on a gel. The banding patterns that result represent different alleles (variants in the DNA sequence) and can be used to quantify genetic variation, assess genetic relationships among individual organisms, and quantify population genetic differences. An advantage of using microsatellites is that the PCR reaction can be used on very small amounts of DNA and is rapid. The amplified fragments can be fluorescently labeled and detected by a laser, allowing the process to be automated.

David Coltman and his colleagues used microsatellite variation to study paternity in bighorn sheep (**Figure 13.5**; also see the introduction to Chapter 12) and showed that sport hunting of trophy rams has reduced the weight and horn size of the animals. Samples of blood, hair, and ear tissue were collected from bighorn sheep at Ram Mountain in Alberta, Canada—a population that has been monitored since 1971. DNA was extracted from the tissue samples and amplified with PCR, revealing variation at 20 microsatellite loci. On the basis of the microsatellite variation, paternity was assigned to 241 rams, and the family relationships of the sheep were worked out. Using these family relationships and the quantitative genetic techniques described in Chapter 11, the geneticists were able to show that ram weight and horn size had high heritability and exhibited a strong positive genetic correlation (see pp. 323–324 in Chapter 11). Trophy

hunters selectively shoot rams with large horns, often before they are able to reproduce. This selective pressure has produced a response to selection: the rams are evolving smaller horns. Between 1971 and 2002, horn size in the population decreased by about one-quarter. Because of the positive genetic correlation between horn size and body size, the body size of rams also is decreasing. Unfortunately, the killing of trophy rams with large horns has led to a decrease in the very traits that are prized by the hunters. This study illustrates the use of microsatellites in an evolutionary study that has practical application.

Variation detected by DNA sequencing   The development in the past 10 years of techniques for rapidly and inexpensively sequencing DNA have made this type of data an important tool in population and evolutionary studies. DNA sequence data often reveal processes that influence evolution and are invaluable for determining the evolutionary relationships of different organisms. The use of PCR for producing the DNA used in the sequencing reactions means that data can be obtained from a very small initial sample of DNA.

An example of the use of DNA sequence data to decipher evolutionary relationships is the unusual case of HIV infection in a dental practice in Florida. In July 1990, the U.S. Centers for Disease Control and Prevention (CDC) reported that a young woman in Florida (later identified as Kimberly Bergalis) had become HIV positive after undergoing an invasive dental procedure performed by a dentist who had AIDS. Bergalis had no known risk factors for HIV infection and no known contact with other HIV-positive persons. The CDC acknowledged that Bergalis might have acquired the infection from her dentist. Subsequently, the dentist wrote to all of his patients, suggesting that they be tested for HIV infection. By 1992, seven of the dentist's patients had tested positive for HIV, and this number eventually increased to ten.

Originally diagnosed with HIV infection in 1986, the dentist began to develop symptoms of AIDS in 1987 but continued to practice dentistry for another 2 years. All of his HIV-positive patients had received invasive dental procedures, such as root canals and tooth extractions, in the period when the dentist was infected. Among the seven patients originally studied by the CDC (patients A–G, **Table 13.2**), two had known risk factors for HIV infection (intravenous drug use, homosexual behavior, or sexual relations with HIV-infected persons), and a third had possible but unconfirmed risk factors.

To determine whether the dentist had infected his patients, the CDC conducted a study of the molecular evolution of HIV isolates from the dentist and from the patients. HIV undergoes rapid evolution, making it possible to trace the path of its transmission. This rapid evolution also allows HIV to develop drug resistance quickly, making the development of a treatment for AIDS difficult.

13.5 Microsatellite variation has been used to study the response of bighorn sheep to selective pressure on horn size due to trophy hunting. [Eyewire.]

Table 13.2   HIV-positive persons included in study of HIV isolates from a Florida dental practice

Person	Sex	Known Risk Factors	Average Differences in DNA Sequences (%)	
			From HIV from Dentist	From HIV from Controls
Dentist	M	Yes		11.0
Patient A	F	No	3.4	10.9
Patient B	F	No	4.4	11.2
Patient C	M	No	3.4	11.1
Patient E	F	No	3.4	10.8
Patient G	M	No	4.9	11.8
Patient D	M	Yes	13.6	13.1
Patient F	M	Yes	10.7	11.9

Source: After C. Y. Ou et al., *Science* 256:1165–1171, 1992, Table 1.

Blood specimens were collected from the dentist, the patients, and a group of 35 local controls (other HIV-infected people who lived within 90 miles of the dental practice but who had no known contact with the dentist). DNA was extracted from white blood cells, and a 680-bp fragment of the *envelope* gene of the virus was amplified by PCR. The fragments from the dentist, the patients, and the local controls were then sequenced and compared.

The divergence between the viral sequences taken from the dentist, the seven patients, and the controls is shown in Table 13.2. Viral DNA taken from patients with no confirmed risk factors (patients A, B, C, E, and G) differed from the dentist's viral DNA by 3.4% to 4.9%, whereas the viral DNA from the controls differed from the dentist's by an average of 11%. The viral sequences collected from five patients (A, B, C, E, and G) were more closely related to the viral sequences collected from the dentist than to viral sequences from the general population, strongly suggesting that these patients acquired their HIV infection from the dentist. The viral isolates from patients D and F (patients with confirmed risk factors), however, differed from that of the dentist by 10.7% and 13.6%, suggesting that these two patients did not acquire their infection from the dentist.

An analysis of the evolutionary relationships of the viral sequences (**Figure 13.6**) confirmed that the virus taken from the dentist had a close evolutionary relationship to viruses taken from patients A, B, C, E, and G. The viruses from patients D and F, with known risk factors, were no more similar to the virus from the dentist than to viruses from local controls, indicating that the dentist most likely infected five of his patients, whereas the other two patients probably acquired their infections elsewhere. Of three additional HIV-positive patients that have been identified since 1992, only one has viral sequences that are closely related to those from the dentist.

The study of HIV isolates from the dentist and his patients provides an excellent example of the relevance of evolutionary studies to real-world problems. How the dentist infected his patients during their visits to his office remains a mystery, but this case is clearly unusual. A study of almost 16,000 patients treated by HIV-positive health-care workers failed to find a single case of confirmed transmission of HIV from the health-care worker to the patient.

CONCEPTS

Variation in DNA nucleotide sequence can be analyzed by using restriction fragment length polymorphisms, microsatellites, and data from direct DNA sequencing.

✔ CONCEPT CHECK 3

What are some of the advantages of using microsatellites for evolutionary studies?

13.3 New Species Arise Through the Evolution of Reproductive Isolation

The term *species* literally means kind or appearance; **species** are different kinds or types of living organisms. In many cases species differences are easy to recognize: a horse is clearly a different species from a chicken. Sometimes, however, species differences are not so clear cut. Some species of *Plethodon* salamanders are so similar in appearance that they can be distinguished only by looking at their proteins or genes.

The concept of a species has two primary uses in biology. First, a species is a name given to a particular type of organism. For effective communication, biologists must use a standard set of names for the organisms that they study, and species names serve that purpose. When a geneticist talks about conducting crosses with *Drosophila melanogaster*, other biologists immediately understand which organism

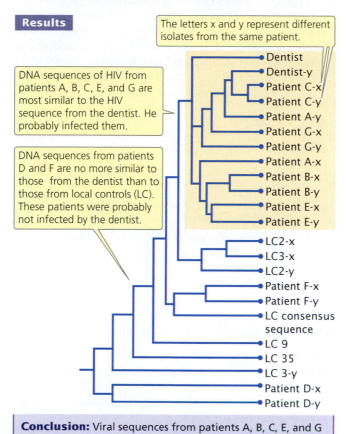

Experiment

Question: Did the dentist transmit HIV to his patients?

Methods Viral samples were collected from the dentist, patients, and local controls. Gene fragments were sequenced and compared.

Results

The letters x and y represent different isolates from the same patient.

- Dentist
- Dentist-y
- Patient C-x
- Patient C-y
- Patient A-y
- Patient G-x
- Patient G-y
- Patient A-x
- Patient B-x
- Patient B-y
- Patient E-x
- Patient E-y
- LC2-x
- LC3-x
- LC2-y
- Patient F-x
- Patient F-y
- LC consensus sequence
- LC 9
- LC 35
- LC 3-y
- Patient D-x
- Patient D-y

DNA sequences of HIV from patients A, B, C, E, and G are most similar to the HIV sequence from the dentist. He probably infected them.

DNA sequences from patients D and F are no more similar to those from the dentist than to those from local controls (LC). These patients were probably not infected by the dentist.

Conclusion: Viral sequences from patients A, B, C, E, and G cluster with those of the dentist, indicating a close evolutionary relationship. Sequences from patients D and F, along with those of local controls, are more distantly related.

13.6 Evolutionary tree showing the relationships of HIV isolates from a dentist, seven of his patients (A through G), and other HIV-positive persons from the same region (local controls, LC). The phylogeny is based on DNA sequences taken from the *envelope* gene of the virus. [*After C. Ou et al., Molecular epidemiology of HIV transmission in a dental practice, Science 256:1167, 1992.*]

was used. The second use of the term species is in an evolutionary context: a species is considered an evolutionarily independent group of organisms.

The Biological Species Concept

What kinds of differences are required to consider two organisms different species? A widely used definition of species is the **biological species concept,** first fully developed by evolutionary biologist Ernst Mayr in 1942. Mayr was primarily interested in the biological characteristics that are responsible for separating organisms into independently evolving units. He defined a species as a group of organisms whose members are capable of interbreeding with one another but are reproductively isolated from the members of other species. In other words, members of the same species have the biological potential to exchange genes, and members of different species cannot exchange genes. Because different species do not exchange genes, each species evolves independently.

Not all biologists adhere to the biological species concept, and there are several problems associated with it. In practice, most species are distinguished on the basis of phenotypic (usually anatomical) differences. Biologists often assume that phenotypic differences represent underlying genetic differences; if the phenotypes of two organisms are quite different, then they probably cannot and do not interbreed in nature.

Reproductive Isolating Mechanisms

The key to species differences under the biological species concept is reproductive isolation—biological characteristics that prevent genes from being exchanged between different species. Any biological factor or mechanism that prevents gene exchange is termed a **reproductive isolating mechanism.**

Prezygotic reproductive isolating mechanisms **Prezygotic reproductive isolating mechanisms** prevent gametes from two different species from fusing and forming a hybrid zygote. In **ecological isolation,** members of two species do not encounter one another and therefore do not reproduce with one another, because they have different ecological niches, living in different habitats and interacting with the environment in different ways. For example, some species of forest-dwelling birds feed and nest in the forest canopy, whereas other species confine their activities to the forest floor. Because they never come into contact, these birds are reproductively isolated from one another. Other species are separated by **behavioral isolation,** differences in behavior that prevent interbreeding. Many male frogs attract females of the same species by using a unique, species-specific call. Two closely related frogs may use the same pond but never interbreed because females are attracted only to the call of their own species.

Another type of prezygotic reproductive isolation is **temporal isolation,** in which reproduction takes place at different times of the year. Some species of plants do not exchange genes, because they flower at different times of the year. **Mechanical isolation** results from anatomical differences that prevent successful copulation. This type of isolation is seen in many insects, in which closely related species differ in their male and female genitalia, and so copulation is physically impossible. Finally, some species are separated by **gametic isolation,** in which mating between individuals of different species takes place, but the gametes do not form zygotes. Male gametes may not survive in the female reproductive tract or may not be attracted to female gametes. In other cases, male and female gametes meet but are too incompatible to fuse to form a zygote. Gametic isolation is seen in many plants, where pollen from one species cannot fertilize the ovules of another species.

Postzygotic reproductive isolating mechanisms Other species are separated by **postzygotic reproductive isolating mechanisms,** in which gametes of two species fuse and form a zygote, but there is no gene flow between the two species, either because the resulting hybrids are inviable or sterile or because reproduction breaks down in subsequent generations.

If prezygotic reproductive isolating mechanisms fail or have not yet evolved, mating between two organisms of different species may take place, with the formation of a hybrid zygote containing genes from two different species. In many cases, such species are still separated by **hybrid inviability,** in which incompatibility between genomes of the two species prevents the hybrid zygote from developing. Hybrid inviability is seen in some groups of frogs, in which mating between different species and fertilization take place, but the resulting embryos never complete development.

Other species are separated by **hybrid sterility,** in which hybrid embryos complete development but are sterile. Donkeys and horses frequently mate and produce a viable offspring—a mule—but most mules are sterile; thus, there is no gene flow between donkeys and horses (but see Problem 43 at the end of Chapter 9). Finally, some closely related species are capable of mating and producing viable and fertile F_1 progeny. However, genes do not flow between the two species because of **hybrid breakdown,** in which further crossing of the hybrids produces inviable or sterile offspring. The different types of reproductive isolating mechanisms are summarized in **Table 13.3**.

Table 13.3 Types of reproductive isolating mechanisms

Type	Characteristic
Prezygotic	Mechanisms Before a Zygote Has Formed
Ecological	Differences in habitat; individuals do not meet
Temporal	Reproduction takes place at different times
Mechanical	Anatomical differences prevent copulation
Behavioral	Differences in mating behavior prevent mating
Gametic	Gametes incompatible or not attracted to each other
Postzygotic	Mechanisms After a Zygote Has Formed
Hybrid inviability	Hybrid zygote does not survive to reproduction
Hybrid sterility	Hybrid is sterile
Hybrid breakdown	F_1 hybrids are viable and fertile, but F_2 are inviable or sterile

CONCEPTS

The biological species concept defines a species as a group of potentially interbreeding organisms that are reproductively isolated from the members of other species. Under this concept, species are separated by reproductive isolating mechanisms, which may intervene before a zygote is formed (prezygotic reproductive isolating mechanisms) or after a zygote is formed (postzygotic reproductive isolating mechanisms).

✔ CONCEPT CHECK 4

Which statement is an example of postzygotic reproductive isolation?

a. Sperm of species A dies in the oviduct of species B before fertilization can take place.

b. Hybrid zygotes between species A and B are spontaneously aborted early in development.

c. The mating seasons of species A and B do not overlap.

d. Males of species A are not attracted to the pheromones produced by the females of species B.

Modes of Speciation

Speciation is the process by which new species arise. In regard to the biological species concept, speciation comes about through the evolution of reproductive isolating mechanisms—mechanisms that prevent the exchange of genes between groups of organisms.

There are two principle ways in which new species arise. **Allopatric speciation** arises when a geographic barrier first splits a population into two groups and blocks the exchange of genes between them. The interruption of gene flow then leads to the evolution of genetic differences that result in reproductive isolation. **Sympatric speciation** arises in the absence of any external barrier to gene flow; reproductive isolating mechanisms evolve within a single population. We will take a more detailed look at both of these mechanisms next.

Allopatric speciation Allopatric speciation is initiated when a geographic barrier splits a population into two or more groups and prevents gene flow between the isolated groups (**Figure 13.7a**). Geographic barriers can take a number of forms. Uplifting of a mountain range may split a population of lowland plants into separate groups on each side of the mountains. Oceans serve as effective barriers for many types of terrestrial organisms, separating individuals on different islands from one another and from those on the mainland. Rivers often separate populations of fish located in separate drainages. The erosion of mountains may leave populations of alpine plants isolated on separate mountain peaks.

After two populations have been separated by a geographic barrier that prevents gene flow between them,

(a)

Population

An original population…

Geographical barrier

…is split into two populations by a geographic barrier to gene flow.

(b)

Genetic differentiation

The populations acquire genetic differences over time owing to selection, genetic drift, and mutations,…

…which lead to the evolution of reproductive isolating mechanisms (RIMs).

(c)

Secondary contact

If the populations come into contact again, RIMs prevent gene flow between them.

Selection for prezygotic RIM

If postzygotic RIMs have evolved, selection will strengthen prezygotic RIMs, leading to different species.

Species A   Species B

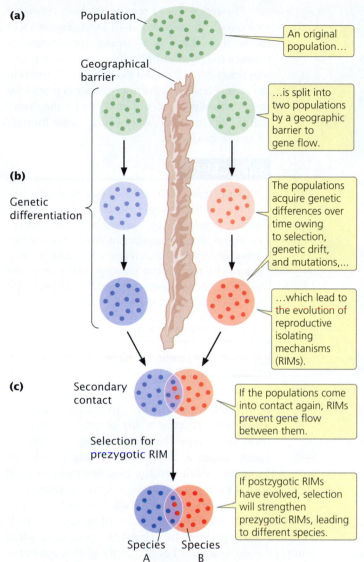

13.7 Allopatric speciation is initiated by a geographic barrier to gene flow between two populations.

they evolve independently (**Figure 13.7b**). The genetic isolation allows each population to accumulate genetic differences that are not found in the other population through natural selection, unique mutations, and genetic drift (if the populations are small). These genetic differences eventually lead to prezygotic and postzygotic isolation. It is important to note that prezygotic isolation and postzygotic isolation arise simply as a consequence of genetic divergence.

If the geographic barrier that once separated the two populations disappears or individuals are able to disperse over it, the populations come into secondary contact (**Figure 13.7c**). At this point, several outcomes are possible. If limited genetic differentiation has taken place during the separation of the populations, reproductive isolating mechanisms may not have evolved or may be incomplete. Genes will flow

between the two populations, eliminating any genetic differences that did arise, and the populations will remain a single species.

A second possible outcome is that genetic differentiation during separation leads to prezygotic reproductive isolating mechanisms; in this case, the two populations are different species. A third possible outcome is that, during their time apart, some genetic differentiation took place between the populations, leading to incompatibility in their genomes and postzygotic isolation. If postzygotic isolating mechanisms have evolved, any mating between individuals from the different populations will produce hybrid offspring that are inviable or sterile. Individuals that mate only with members of the same population will have higher fitness than that of individuals that mate with members of the other population; so natural selection will increase the frequency of any trait that prevents interbreeding between members of the different populations. With the passage of time, prezygotic reproductive isolating mechanisms will evolve. In short, if some postzygotic reproductive isolation exists, natural selection will favor the evolution of prezygotic reproductive isolating mechanisms to prevent wasted reproduction by individuals mating with members of the other population.

A number of variations in this general model of allopatric speciation are possible. Many new species probably arise when a small group of individuals becomes geographically isolated from the main population; for example, a few individuals of a mainland population might migrate to a geographically isolated island. In this situation, founder effect and genetic drift play a larger role in the evolution of genetic differences between the populations.

An excellent example of allopatric speciation can be found in Darwin's finches, a group of birds on the Galápagos Islands; these finches were discovered by Charles Darwin in his voyage aboard the *Beagle*. The Galápagos are an archipelago of islands located some 900 km off the coast of South America (**Figure 13.8**). Consisting of more than a dozen large islands and many smaller ones, the Galápagos formed from volcanoes that erupted over a geological hot spot that has moved eastward in the past 3 million years. Thus, the islands to the east (San Cristóbal and Española) are older than those to the west (Isabela and Fernandina). With the passage of time, the number of islands in the archipelago increased as new volcanoes arose.

Darwin's finches consist of 14 species that are found on various islands in the Galápagos archipelago (**Figure 13.9**). The birds vary in the shape and sizes of their beaks, which are adapted for eating different types of food items. Genetic studies have demonstrated that all the birds are closely related and evolved from a single ancestral species that

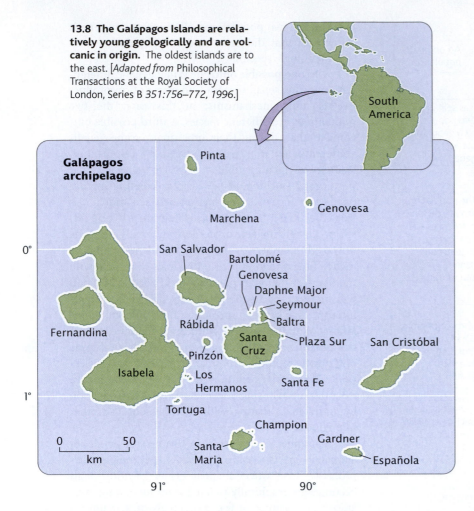

13.8 The Galápagos Islands are relatively young geologically and are volcanic in origin. The oldest islands are to the east. [*Adapted from* Philosophical Transactions at the Royal Society of London, Series B *351:756–772, 1996.*]

Galápagos archipelago

Pinta

Genovesa

Marchena

San Salvador
Bartolomé
Genovesa
Daphne Major
Seymour
Rábida
Baltra
Fernandina
Santa Cruz
Plaza Sur
San Cristóbal
Pinzón
Isabela
Los Hermanos
Santa Fe
Tortuga
Champion
Gardner
Santa Maria
Española

South America

0°

1°

0   50
km

91°   90°

migrated to the islands from the coast of South America some 2 million to 3 million years ago. The evolutionary relationships among the 14 species, based on studies of microsatellite data, are depicted in the evolutionary tree shown in Figure 13.9. Most of the species are separated by a behavioral isolating mechanism (song in particular), but some of the species can and occasionally do hybridize in nature.

The first finches to arrive in the Galápagos probably colonized one of the larger eastern islands. A breeding population became established and increased with time. At some point, a few birds dispersed to other islands, where they were effectively isolated from the original population, and established a new population. This population underwent genetic differentiation owing to genetic drift and adaptation to the local conditions of the island. It eventually became reproductively isolated from the original population. Individual birds from the new population then dispersed to other islands and gave rise to additional species. This process was repeated many times. Occasionally, newly evolved birds dispersed to an island where another species was already present, giving rise to secondary contact between the species. Today, many of the islands have more than one resident finch.

The age of the 14 species has been estimated with data from mitochondrial DNA. **Figure 13.10** shows that there is a strong correspondence between the number of bird species present at various times in the past and the number of islands in the archipelago. This correspondence is one of the most compelling pieces of evidence for the theory that the different species of finches arose through allopatric speciation.

CONCEPTS

Allopatric speciation is initiated by a geographic barrier to gene flow. A single population is split into two or more populations by a geographic barrier. With the passage of time, the populations evolve genetic differences, which bring about reproductive isolation. After postzygotic reproductive isolating mechanisms have evolved, selection favors the evolution of prezygotic reproductive isolating mechanisms.

✔ **CONCEPT CHECK 5**

What role does genetic drift play in allopatric speciation?

Sympatric speciation Sympatric speciation arises in the absence of any geographic barrier to gene flow; reproductive isolating mechanisms evolve within a single interbreeding population. Sympatric speciation has long been controversial within evolutionary biology. Ernst Mayr believed that sympatric speciation was impossible, and he demonstrated that many apparent cases of sympatric speciation could be explained by allopatric speciation. More recently, however, evidence has accumulated that sympatric speciation can and has arisen under special circumstances. The difficulty with sympatric speciation is that isolating mechanisms arise as a *consequence* of genetic differentiation, which takes place only if gene flow between groups is interrupted. But, without reproductive isolation (or some external barrier), how can gene flow be interrupted? How can genetic differentiation arise within a single group that is freely exchanging genes?

Most models of sympatric speciation assume that genetic differentiation is initiated by strong disruptive selection taking place within a single population. One homozygote (A^1A^1) is strongly favored on one resource (perhaps the plant species that is host to an insect) and the other homozygote (A^2A^2) is favored on a different resource (perhaps a different host plant). Heterozygotes (A^1A^2) have low fitness on both resources. In this situation, natural selection will favor genotypes at other loci that cause assortative mating (matings between like individuals, see Chapter 12), and so no matings take place between A^1A^1 and A^2A^2, which would produce A^1A^2 offspring with low fitness.

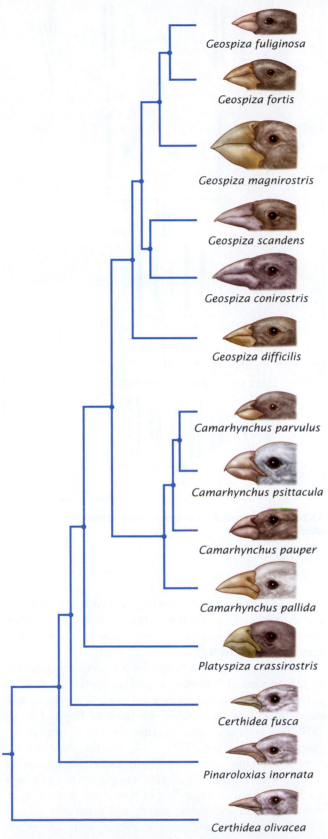

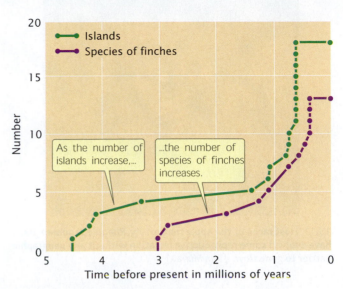

13.10 The number of species of Darwin's finches present at various times in the past corresponds with the number of islands in the Galápagos archipelago. [*Data from P. R. Grant, B. R. Grant, and J. C. Deutsch. Speciation and hybridization in island birds. Philosophical Transactions of the Royal Society of London Series B 351:765–772, 1996.*]

13.9 Darwin's finches consist of 14 species that evolved from a single ancestral species that migrated to the Galápagos Islands and underwent repeated allopatric speciation. [*After B. R. Grant and P. R. Grant. Bioscience 53:965–975, 2003.*]

Now imagine that alleles at a second locus affect mating behavior, such that C^1C^1 individuals prefer mating only with other C^1C^1 individuals, and C^2C^2 individuals prefer mating with other C^2C^2 individuals. If alleles at the A locus are non-randomly associated with alleles at the C locus so that only A^1A^1 C^1C^1 individuals and A^2A^2 C^2C^2 individuals exist, then gene flow will be restricted between individuals using the different resources, allowing the two groups to evolve further genetic differences that might lead to reproductive isolation and sympatric speciation.

The difficulty with this model is that recombination quickly breaks up the nonrandom associations between genotypes at the two loci, producing individuals such as A^1A^1 C^2C^2, which would prefer to mate with A^2A^2 C^2C^2 individuals. This mating would produce all A^1A^2 offspring, which do poorly on both resources. Thus, even limited recombination will prevent the evolution of the mating-preference genes.

Sympatric speciation is more probable if the genes that affect resource utilization also affect mating preferences. It is apparently the case in host races, populations of specialized insects that feed on different host plants. Guy Bush studied what appeared to be initial stages of speciation in host races of the apple maggot fly (*Rhagoletis pomonella*, **Figure 13.11**). The flies of this species feed on the fruits of a specific host tree. Mating takes place near the fruits, and the flies lay their eggs on the ripened fruits, where their larvae grow and develop. *R. pomnella* originally existed only on fruits of hawthorn trees, which are native to North

13.11 Host races of the apple maggot fly, *Rhagoletis pomenella*, have evolved some reproductive isolation without any geographic barrier to gene flow. [Tom Murray.]

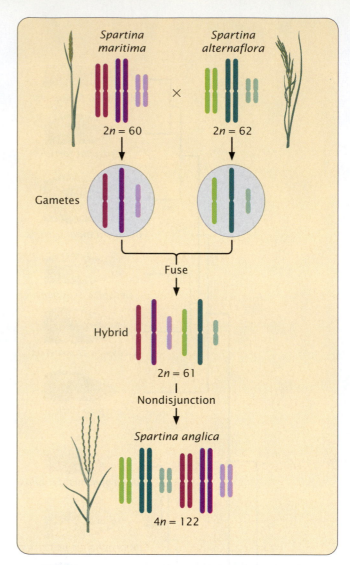

13.12 *Spartina anglica* arose sympatrically through allopolyploidy.

America; 150 years ago, *R. pomnella* was first observed on cultivated apples, which are related to hawthorns but a different species. Infestations of apples by this new apple host race of *R. pomnella* quickly spread, and, today, many apple trees throughout North America are infested with the flies.

The apple host race of *R. pomnella* probably originated when a few flies acquired a mutation that allowed them to feed on apples instead of the hawthorn fruits. Because mating takes place on and near the fruits, flies that utilize apples are more likely to mate with other flies utilizing apples, leading to genetic isolation between flies using hawthorns and those utilizing applies. Indeed, Bush found that some genetic differentiation has already taken place between the two host races. Flies lay their eggs on ripening fruit, and there has been strong selection for the flies to synchronize their reproduction with the period when their host species has ripening fruit. Apples ripen several weeks earlier than hawthorns. Correspondingly, the peak mating period of the apple host races is 3 weeks earlier than that of the hawthorn race. These differences in the timing of reproduction between apple and hawthorn races have further reduced gene flow—to about 2%—between the two host races and have led to significant genetic differentiation between them. All of it has evolved in the past 150 years. Although genetic differentiation has taken place between apple and hawthorn host races of *R. pomnella* and some degree of reproductive isolation has evolved between them, reproductive isolation is not yet complete and speciation has not fully taken place.

Speciation through polyploidy A special type of sympatric speciation takes place through polyploidy (see Chapter 9). Polyploid organisms have more than two genomes ($3N$, $4N$, $5N$, etc.). As discussed in Chapter 9, allopolyploidy often arises when two diploid species hybridize, producing $2N$ hybrid offspring. Nondisjunction in one of the hybrid offspring produces a $4N$ tetraploid. Because this

tetraploid contains exactly two copies of each chromosome, it is usually fertile and will be reproductively isolated from the two parental species by differences in chromosome number (see Figure 9.28).

Numerous angiosperm species are allopolyploids, and so this process is common in plants. Speciation through polyploidy was observed when it led to a new species of salt-marsh grass that arose along the coast of England about 1870. This polyploid contains genomes of the European salt grass *Spartina maritima* ($2N = 60$) and the American salt grass *S. alternaflora* ($2N = 62$, **Figure 13.12**). Seeds from the American salt grass were probably transported to England in the ballast of a ship. Regardless of how it got there, *S. alternaflora* grew in an English marsh and eventually crossed with *S. maritima*, producing a hybrid with $2N = 61$. Nondisjunction in the hybrid then led to chromosome doubling, producing a new species *S. anglica* with $4N = 122$ (see **Figure 13.12**). This new species subsequently spread along the coast of England.

Sympatric speciation arises within a single interbreeding population without any geographic barrier to gene flow. Sympatric speciation may arise under special circumstances, such as when resource use is linked to mating preference (in host races) or when species hybridization leads to allopolyploidy.

Genetic Differentiation Associated with Speciation

As we have seen, genetic differentiation leads to the evolution of reproductive isolating mechanisms, which restrict gene flow between populations and lead to speciation. How much genetic differentiation is required for reproductive isolation to take place? This question has received considerable study by evolutionary geneticists, but, unfortunately, there is no universal answer. Some newly formed species differ in many genes, whereas others appear to have undergone divergence in just a few genes.

One group of organisms that have been extensively studied for genetic differences associated with speciation is the genus *Drosophila*. The *Drosophila willistoni* group consists of at least 12 species found in Central and South America in various stages in the process of speciation. Using protein electrophoresis, Francisco Ayala and his colleagues genotyped flies from different geographic populations (populations with limited genetic differences), subspecies (populations with considerable genetic differences), sibling species (newly arisen species), and nonsibling species (older species). For each group, they computed a measure of genetic similarity, which ranges from 1 to 0 and represents the overall level of genetic differentiation (**Table 13.4**). They found that there was a general decrease in genetic similarity as flies evolve from geographic populations to subspecies to sibling species to nonsibling species. These data suggest that considerable genetic differentiation at many loci is required for speciation to arise. A study of *D. simulans* and *D. melanogaster,* two species that produce inviable hybrids when crossed, suggested that at least 200 genes contribute to the inviability of hybrids between the two species.

Table 13.4 Genetic similarity in groups of the *Drosophila willistoni* complex

Group	Mean Genetic Similarity
Geographic populations	0.970
Subspecies	0.795
Sibling species	0.517
Nonsibling species	0.352

However, other studies suggest that speciation may have arisen through changes in just a few genes. For example, *D. heteroneura* and *D. silvestris* are two species of Hawaiian fruit flies that exhibit behavioral reproductive isolation. The isolation is determined largely by differences in head shape; *D. heteroneura* has a hammer-shaped head with widely separated eyes that is recognized by females of the same species but rejected by *D. silvestris* females. Genetic studies indicate that only a few loci (about 10) determine the differences in head shape.

Some newly arising species have a considerable number of genetic differences; others have few genetic differences.

13.4 The Evolutionary History of a Group of Organisms Can Be Reconstructed by Studying Changes in Homologous Characteristics

The evolutionary relationships among a group of organisms are termed a **phylogeny.** Because most evolution takes place over long periods of time and is not amenable to direct observation, biologists must reconstruct phylogenies by inferring the evolutionary relationships among present-day organisms. The discovery of fossils of ancestral organisms can aid in the reconstruction of phylogenies, but the fossil record is often too poor to be of much help. Thus, biologists are often restricted to the analysis of characteristics in present-day organisms to determine their evolutionary relationships. In the past, phylogenetic relationships were reconstructed on the basis of phenotypic characteristics—often, anatomical traits. Today, molecular data, including protein and DNA sequences, are frequently used to construct phylogenetic trees.

Phylogenies are reconstructed by inferring changes that have taken place in homologous characteristics. Such characteristics evolved from the same character in a common ancestor. For example, the front leg of a mouse and the wing of a bat are homologous structures, because both evolved from the forelimb of an early mammal that was an ancestor to both mouse and bat. Although these two anatomical features look different and have different functions, close examination of their structure and development reveals that they are indeed homologous. And, because mouse and bat have these homologous features and others in common, we know that they are both mammals. Similarly, DNA sequences are homologous if two present-day sequences evolved from a single sequence found in an ancestor. For example, all eukaryotic organisms have a gene for cytochrome *c,* an enzyme that helps carry out oxidative respiration. This gene is assumed to have arisen in a single organism in the distant past and was then passed down to descendants of that early ancestor. Today, all copies of the

of documented cases in which genes are transferred from bacteria to eukaryotes. The extent of horizontal gene transfer among eukaryotic organisms is controversial, with few well-documented cases. Horizontal gene transfer can obscure phylogenetic relationships and make the reconstruction of phylogenetic trees difficult.

> **CONCEPTS**
>
> New genes may evolve through the duplication of exons, shuffling of exons, duplication of genes, and duplication of whole genomes. Genes can be passed among distantly related organisms through horizontal gene transfer.

CONCEPTS SUMMARY

- Evolution is genetic change taking place within a group of organisms. It is a two-step process: (1) genetic variation arises, and (2) genetic variants change in frequency.

- Anagenesis refers to change within a single lineage; cladogenesis is the splitting of one lineage into two.

- Molecular methods offer a number of advantages for the study of evolution.

- The use of protein electrophoresis to study genetic variation in natural populations showed that most natural populations have large amounts of genetic variation in their proteins. The neutral-mutation hypothesis proposes that molecular variation is selectively neutral and is shaped largely by mutation and genetic drift. The balance hypothesis proposes that molecular variation is maintained largely by balancing selection.

- Variation in DNA sequences can be assessed by analyzing restriction fragment length polymorphisms, microsatellites, and data from direct sequencing.

- A species can be defined as a group of organisms that are capable of interbreeding with one another and are reproductively isolated from the members of other species.

- Species are prevented from exchanging genes by reproductive isolating mechanisms, either before a zygotes has formed (prezygotic reproductive isolation) or after a zygote has formed (postzygotic reproductive isolation).

- Allopatric speciation arises when a geographic barrier prevents gene flow between two populations. With the passage

of time, the two populations acquire genetic differences that may lead to reproductive isolating mechanisms.

- Sympatric speciation arises when reproductive isolation exists in the absence of any geographic barrier. It may arise under special circumstances.

- Some species arise only after populations have undergone considerable genetic differences; others arise after changes have taken place in only a few genes.

- Evolutionary relationships (a phylogeny) can be represented by a phylogenetic tree, consisting of nodes that represent organisms and branches that represent their evolutionary connections.

- Two different approaches to constructing phylogenetic trees are the distance approach and the parsimony approach.

- Different parts of the genome show different amounts of genetic variation. In general, those parts that have the least effect on function evolve at the highest rates.

- The molecular-clock hypothesis proposes a constant rate of nucleotide substitution, providing a means of dating evolutionary events by looking at nucleotide differences between organisms.

- Genome evolution takes place through the duplication and shuffling of exons, the duplication of genes to form gene families, whole-genome duplication, and the horizontal transfer of genes between organisms.

IMPORTANT TERMS

evolution (p. 362)
anagenesis (p. 363)
cladogenesis (p. 363)
proportion of polymorphic loci (p. 365)
expected heterozygosity (p. 365)
neutral-mutation hypothesis (p. 365)
balance hypothesis (p. 365)
species (p. 368)
biological species concept (p. 369)
reproductive isolating mechanism (p. 369)
prezygotic reproductive isolating
　mechanism (p. 369)

ecological isolation (p. 369)
behavioral isolation (p. 369)
temporal isolation (p. 369)
mechanical isolation (p. 369)
gametic isolation (369)
postzygotic reproductive isolating
　mechanism (p. 370)
hybrid inviability (p. 370)
hybrid sterility (p. 370)
hybrid breakdown (p. 370)
speciation (p. 370)
allopatric speciation (p. 370)

sympatric speciation (p. 370)
phylogeny (p. 375)
phylogenetic tree (p. 376)
node (p. 376)
branch (p. 376)
rooted tree (p. 376)
gene tree (p. 376)
molecular clock (p. 379)
exon shuffling (p. 380)
multigene family (p. 381)

ANSWERS TO CONCEPT CHECKS

1. First genetic variation arises. Then various evolutionary forces cause changes in the frequency of genetic variants.

2. c

3. Microsatellites are often highly variable among individuals. They can be amplified with the use of PCR, and so they can be detected with a small amount of starting DNA. Finally, the detection and analysis of microsatellites can be automated.

4. b

5. Genetic drift can bring about changes in the allelic frequencies of populations and lead to genetic differences among populations. Genetic differentiation is the cause of postzygotic and prezygotic reproductive isolation between populations that leads to speciation.

6. c

7. b

COMPREHENSION QUESTIONS

Section 13.1

1. How is biological evolution defined?

*2. What are the two steps in the process of evolution?

3. How is anagenesis different from cladogenesis?

Section 13.2

*4. As a measure of genetic variation, why is the expected heterozygosity often preferred to the observed heterozygosity?

5. Why does protein variation, as revealed by electrophoresis, underestimate the amount of true genetic variation?

6. What are some of the advantages of using molecular data in evolutionary studies?

*7. What is the key difference between the neutral-mutation hypothesis and the balance hypothesis?

8. Discuss some of the methods that have been used to study variation in DNA.

Section 13.3

*9. What is the biological species concept?

10. What is the difference between prezygotic and postzygotic reproductive isolating mechanisms. List the different types of each.

11. What is the basic difference between allopatric and sympatric modes of speciation?

*12. Briefly outline the process of allopatric speciation.

13. What are some of the difficulties with sympatric speciation?

*14. Briefly explain how switching from hawthorn fruits to apples has led to genetic differentiation and partial reproductive isolation in *Rhagoletis pomonella*.

Section 13.4

15. Draw a simple phylogenetic tree and identify a node, a branch, and an outgroup.

*16. Briefly describe the difference between the distance approach and the parsimony approach to the reconstruction of phylogenetic trees.

Section 13.5

17. Outline the different rates of evolution that are typically seen in different parts of a protein-encoding gene. What might account for these differences?

*18. What is the molecular clock?

19. What is exon shuffling? How can it lead to the evolution of new genes?

20. What is a multigene family? What processes produce multigene families?

*21. Define horizontal gene transfer. What problems does it cause for evolutionary biologists?

APPLICATION QUESTIONS AND PROBLEMS

Section 13.1

22. The following illustrations represent two different patterns of evolution. Briefly discuss the differences in these two patterns, particularly in regard to the role of cladogenesis in evolutionary change.

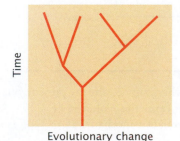
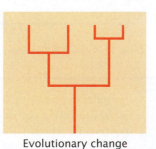

Evolutionary change   Evolutionary change

Section 13. 2

*23 DATA ANALYSIS Donald Levin used protein electrophoresis to study genetic variation in two species of wildflowers in Texas, *Phlox drummondi* and *P. cuspidate* (D. Levin. 1978. *Evolution* 32:245–263) *P. drummondi* reproduces only by cross fertilization, whereas *P. cuspidate* is partly self-fertilizing. The table at the top of page 730 gives allelic frequencies for several loci and populations studied by Levin.

  a. Calculate the percentage of polymorphic loci and expected heterozygosity for each species. For the expected heterozygosity, calculate the mean for all loci in a population, and then calculate the mean for all populations.

  b. What tentative conclusions can you draw about the effect of self-fertilization on genetic variation in these flowers?

For Problem 23

Species	Population	Allelic frequencies							
		Adh		*Got-2*		*Pgi-2*		*Pgm-2*	
		a	*b*	*a*	*b*	*a*	*b*	*a*	*b*
P. drummondii	1	0.10	0.90	0.02	0.98	0.04	0.96	1.0	0.0
P. drummondii	2	0.11	0.89	0.0	1.0	0.31	0.69	0.83	0.17
P. drummondii	3	0.26	0.74	1.0	0.0	0.14	0.86	1.0	0.0
P. cuspidate	1	0.0	1.0	0.0	1.0	0.0	1.0	0.0	1.0
P. cuspidate	2	0.0	1.0	0.0	1.0	0.0	1.0	0.0	1.0
P. cuspidate	3	0.91	0.09	0.0	1.0	0.0	1.0	0.0	1.0

Note: *a* and *b* represent different alleles at each locus.

Section 13.4

*24. How many rooted trees are theoretically possible for a group of 7 organisms? How many for 12 organisms?

25. [DATA ANALYSIS] Michael Bunce and his colleagues in England, Canada, and the United States extracted and sequenced mitochondrial DNA from fossils of Haast's eagle, a gigantic eagle that was driven to extinction 700 years ago when humans first arrived in New Zealand (M. Bunce et al. 2005. *Plos Biology* 3:44–46). Using mitochondrial DNA sequences from living eagles and those from Haast eagle fossils, they created the phylogenetic tree at the right.

On this phylogenetic tree, identify (a) all terminal nodes; (b) all internal nodes; (c) one example of a branch; and (d) the outgroup.

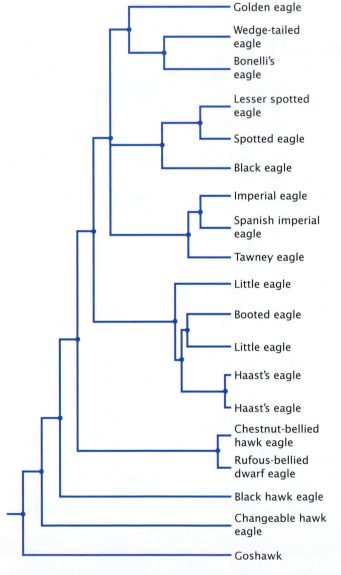

(After M. Bunce et al. *Plos Biology* 3:44–46, 2005.)

CHALLENGE QUESTION

Section 13.3

25. Explain why natural selection may cause prezygotic reproductive isolating mechanisms to evolve if postzygotic reproductive isolating mechanisms are already present, but natural selection can never cause the evolution of postzygotic reproductive isolating mechanisms.

Glossary

acceptor arm The arm in tRNA to which an amino acid attaches.

acentric chromatid Lacks a centromere; produced when crossing over takes place within a paracentric inversion. The acentric chromatid does not attach to a spindle fiber and does not segregate in meiosis or mitosis; so it is usually lost after one or more rounds of cell division.

acidic activation domain Commonly found in some transcriptional activator proteins, a domain that contains multiple amino acids with negative charges and stimulates the transcription of certain genes.

acrocentric chromosome Chromosome in which the centromere is near one end, producing a long arm at one end and a knob, or satellite, at the other end.

adaptive mutation Process by which a specific environment induces mutations that enable organisms to adapt to the environment.

addition rule States that the probability of any of two or more mutually exclusive events occurring is calculated by adding the probabilities of the individual events.

additive genetic variance Component of the genetic variance that can be attributed to the additive effect of different genotypes.

adjacent-1 segregation Type of segregation that takes place in a heterozygote for a translocation. If the original, nontranslocated chromosomes are N_1 and N_2 and the chromosomes containing the translocated segments are T_1 and T_2, then adjacent-1 segregation takes place when N_1 and T_2 move toward one pole and T_1 and N_2 move toward the opposite pole.

adjacent-2 segregation Type of segregation that takes place in a heterozygote for a translocation. If the original, nontranslocated chromosomes are N_1 and N_2 and the chromosomes containing the translocated segments are T_1 and T_2, then adjacent-2 segregation takes place when N_1 and T_1 move toward one pole and T_2 and N_2 move toward the opposite pole.

allele One of two or more alternate forms of a gene.

allelic frequency Proportion of a particular allele in a population.

allopatric speciation Arises when a geographic barrier first splits a population into two groups and blocks the exchange of genes between them. Compare **sympatric speciation.**

allopolyploidy Condition in which the sets of chromosomes of a polyploid individual possessing more than two haploid sets are derived from two or more species.

alternate segregation Type of segregation that takes place in a heterozygote for a translocation. If the original, nontranslocated chromosomes are N_1 and N_2 and the chromosomes containing the translocated segments are T_1 and T_2, then alternate segregation takes place when N_1 and N_2 move toward one pole and T_1 and T_2 move toward the opposite pole.

alternation of generations Complex life cycle in plants that alternates between the diploid sporophyte stage and the haploid gametophyte stage.

Ames test Test in which special strains of bacteria are used to evaluate the potential of chemicals to cause cancer.

amniocentesis Procedure used for prenatal genetic testing to obtain a sample of amniotic fluid from a pregnant woman. A long sterile needle is inserted through the abdominal wall into the amniotic sac to obtain the fluid.

amphidiploidy Type of allopolyploidy in which two different diploid genomes are combined, so that every chromosome has one and only one homologous partner and the genome is functionally diploid.

anagenesis Evolutionary change within a single lineage.

anaphase Stage of mitosis in which chromatids separate and move toward the spindle poles.

anaphase I Stage of meiosis I. In anaphase I, homologous chromosomes separate and move toward the spindle poles.

anaphase II Stage of meiosis II. In anaphase II, chromatids separate and move toward the spindle poles.

aneuploidy Change from the wild type in the number of chromosomes; most often an increase or decrease of one or two chromosomes.

anticipation Increasing severity or earlier age of onset of a genetic trait in succeeding generations. For example, symptoms of a genetic disease may become more severe as the trait is passed from generation to generation.

archaea One of the three primary divisions of life. Archaea consist of unicellular organisms with prokaryotic cells.

artificial selection Selection practiced by humans.

attachment site Special site on a bacterial chromosome where a prophage may insert itself.

autopolyploidy Condition in which all the sets of chromsomes of a polyploid individual possessing more than two haploid sets are derived from a single species.

autosome Chromosome that is the same in males and females; nonsex chromosome.

backcross Cross between an F_1 individual and one of the parental (P) genotypes.

bacterial colony Clump of genetically identical bacteria derived from a single bacterial cell that undergoes repeated rounds of division.

bacteriophage Virus that infects bacterial cells.

balance hypothesis Proposes that much of the molecular variation seen in natural populations is maintained by balancing selection that favors genetic variation.

Barr body Condensed, darkly staining structure that is found in most cells of female placental mammals and is an inactivated X chromosome.

base analog Chemical substance that has a structure similar to that of one of the four standard bases of DNA and may be incorporated into newly synthesized DNA molecules in replication.

base-excision repair DNA repair that first excises modified bases and then replaces the entire nucleotide.

base substitution Mutation in which a single pair of bases in DNA is altered.

behavioral isolation Reproductive isolation due to differences in behavior that prevent interbreeding.

biological species concept Defines a species as a group of organisms whose members are capable of interbreeding with one another but are reproductively isolated from the members of other species. Because different species do not exchange genes, each species evolves independently. Not all biologists adhere to this concept.

bivalent Refers to a synapsed pair of homologous chromosomes.

blending inheritance Early concept of heredity proposing that offspring possess a mixture of the traits from both parents.

branch Evolutionary connections between organisms in a phylogenetic tree.

branch migration Movement of a cross bridge along two DNA molecules.

broad-sense heritability Proportion of the phenotypic variance that can be attributed to genetic variance.

cell cycle Stages through which a cell passes from one cell division to the next.

cell line Genetically identical cells that divide indefinitely and can be cultured in the laboratory.

cell theory States that all life is composed of cells, that cells arise only from other cells, and that the cell is the fundamental unit of structure and function in living organisms.

centiMorgan Another name for map unit.

centriole Cytoplasmic organelle consisting of microtubules; present at each pole of the spindle apparatus in animal cells.

centromere Constricted region on a chromosome that stains less strongly than the rest of the chromosome; region where spindle microtubules attach to a chromosome.

centrosome Structure from which the spindle apparatus develops: contains the centriole.

checkpoint A key transition point at which progression to the next stage in the cell cycle is regulated.

chiasma (pl., **chiasmata**) Point of attachment between homologous chromosomes at which crossing over took place.

chorionic villus sampling (CVS) Procedure used for prenatal genetic testing in which a small piece of the chorion (the outer layer of the placenta) is removed from a pregnant woman. A catheter is inserted through the vagina and cervix into the uterus. Suction is then applied to remove the sample.

chromatin Material found in the eukaryotic nucleus; consists of DNA and proteins.

chromosome deletion Loss of a chromosome segment.

chromosome duplication Mutation that doubles a segment of a chromosome.

chromosome inversion Rearrangement in which a segment of a chromosome has been inverted 180 degrees.

chromosome mutation Difference from the wild type in the number or structure of one or more chromosomes; often affects many genes and has large phenotypic effects.

chromosome rearrangement Change from the wild type in the structure of one or more chromosomes.

chromosome theory of heredity States that genes are located on chromosomes.

cis configuration Arrangement in which two or more wild-type genes are on one chromosome and their mutant alleles are on the homologous chromosome; also called coupling configuration.

cladogenesis Evolution in which one lineage is split into two.

codominance Type of allelic interaction in which the heterozygote simultaneously expresses traits of both homozygotes.

coefficient of coincidence Ratio of observed double crossovers to expected double crossovers.

cohesin Molecule that holds the two sister chromatids of a chromosome together. The breakdown of cohesin at the centromeres enables the chromatids to separate in anaphase of mitosis and anaphase II of meiosis.

colony See **bacterial colony.**

competent cell Capable of taking up DNA from its environment (capable of being transformed).

complementation Two different mutations in the heterozygous condition are exhibited as the wild-type phenotype; indicates that the mutations are at different loci.

complementation test Test designed to determine whether two different mutations are at the same locus (are allelic) or at different loci (are nonallelic). Two individuals that are homozygous for two independently derived mutations are crossed, producing F_1 progeny that are heterozygous for the mutations. If the mutations are at the same locus, the F_1 will have a mutant phenotype. If the mutations are at different loci, the F_1 will have a wild-type phenotype.

complete linkage Linkage between genes that are located close together on the same chromosome with no crossing over between them.

complete medium Used to culture bacteria or some other microorganism; contains all the nutrients required for growth and synthesis, including those normally synthesized by the organism. Nutritional mutants can grow on complete medium.

concept of dominance Principle of heredity discovered by Mendel stating that, when two different alleles are present in a genotype, only one allele may be expressed in the phenotype. The dominant allele is the allele that is expressed, and the recessive allele is the allele that is not expressed.

concordance Percentage of twin pairs in which both twins have a particular trait.

concordant Refers to a pair of twins both of whom have the trait under consideration.

conditional mutation Expressed only under certain conditions.

conjugation Mechanism by which genetic material may be exchanged between bacterial cells. During conjugation, two bacteria lie close together and a cytoplasmic connection forms between them. A plasmid or sometimes a part of the bacterial chromosome passes through this connection from one cell to the other.

consanguinity Mating between related individuals.

continuous characteristic Displays a large number of possible phenotypes that are not easily distinguished, such as human height.

correlation Degree of association between two or more variables.

correlation coefficient Statistic that measures the degree of association between two or more variables. A correlation coefficient can range from −1 to +1. A positive value indicates a direct relation between the variables; a negative correlation indicates an inverse relation. The absolute value of the correlation coefficient provides information about the strength of association between the variables.

cotransduction Process in which two or more genes are transferred together from one bacterial cell to another. Only genes located close together on a bacterial chromosome will be cotransduced.

cotransformation Process in which two or more genes are transferred together during cell transformation.

coupling configuration *See* **cis configuration.**

cross bridge In a heteroduplex DNA molecule, the point at which each nucleotide strand passes from one DNA molecule to the other.

crossing over Exchange of genetic material between homologous but nonsister chromatids.

cytokinesis Process by which the cytoplasm of a cell divides.

cytoplasmic inheritance Inheritance of characteristics encoded by genes located in the cytoplasm. Because the cytoplasm is usually contributed entirely by only one parent, most cytoplasmically inherited characteristics are inherited from a single parent.

deamination Loss of an amino group (NH_2) from a base.

deletion Mutation in which one or more nucleotides are deleted from a DNA sequence.

deletion mapping Technique for determining the chromsomal location of a gene by studying the association of its phenotype or product with particular chromosome deletions.

depurination Break in the covalent bond connecting a purine base to the 1′-carbon atom of deoxyribose, resulting in the loss of the purine base. The resulting apurinic site cannot provide a template in replication, and a nucleotide with another base may be incorporated into the newly synthesized DNA strand opposite the apurinic site.

diakinesis Fifth substage of prophase I in meiosis. In diakinesis, chromosomes contract, the nuclear membrane breaks down, and the spindle forms.

dicentric bridge Structure produced when the two centromeres of a dicentric chromatid are pulled toward opposite poles, stretching the dicentric chromosome across the center of the nucleus. Eventually,the dicentric bridge breaks as the two centromeres are pulled apart.

dicentric chromatid Chromatid that has two centromeres; produced when crossing over takes place within a paracentric inversion. The two centromeres of the dicentric chromatid are frequently pulled toward opposite poles in mitosis or meiosis, breaking the chromosome.

dihybrid cross A cross between two individuals that differ in two characteristics—more specifically, a cross between individuals that are homozygous for different alleles at the two loci (*AA BB* × *aa bb*); also refers to a cross between two individuals that are both heterozygous at two loci (*Aa Bb* × *Aa Bb*).

dioecious organism Belongs to a species whose members have either male or female reproductive structures.

diploid Possessing two sets of chromosomes (two genomes).

diplotene Fourth substage of prophase I in meiosis. In diplotene, centromeres of homologous chromosomes move apart, but the homologs remain attached at chiasmata.

directional selection Selection in which one trait or allele is favored over another.

direct repair DNA repair in which modified bases are changed back into their original structures.

discontinuous characteristic Exhibits only a few, easily distinguished phenotypes. An example is seed shape in which seeds are either round or wrinkled.

discordant Refers to a pair of twins of whom one twin has the trait under consideration and the other does not.

displaced duplication Chromosome rearrangement in which the duplicated segment is some distance from the original segment, either on the same chromosome or on a different one.

dizygotic twins Nonidentical twins that arise when two different eggs are fertilized by two different sperm; also called fraternal twins.

dominance genetic variance Component of the genetic variance that can be attributed to dominance (interaction between genes at the same locus).

dominant Refers to an allele or a phenotype that is expressed in homozygotes (*AA*) and in heterozygotes (*Aa*); only the dominant allele is expressed in a heterozygote phenotype.

dosage compensation Equalization in males and females of the amount of protein produced by X-linked genes. In placental mammals, dosage compensation is accomplished by the random inactivation of one X chromosome in the cells of females.

double fertilization Fertilization in plants; includes the fusion of a sperm cell with an egg cell to form a zygote and the fusion of a second sperm cell with the polar nuclei to form an endosperm.

double-strand-break model Model of homologous recombination in which a DNA molecule undergoes double-strand breaks.

down mutation Decreases the rate of transcription.

Down syndrome (trisomy 21) Characterized by variable degrees of mental retardation, characteristic facial features, some retardation of growth and development, and an increased incidence of heart defects, leukemia, and other abnormalities; caused by the duplication of all or part of chromosome 21.

ecological isolation Reproductive isolation in which different species live in different habitats and interact with the environment in different ways. Thus, their members do not encounter one another and do not reproduce with one another.

Edward syndrome (trisomy 18) Characterized by severe retardation, low-set ears, a short neck, deformed feet, clenched fingers, heart problems, and other disabilities; results from the presence of three copies of chromosome 18.

effective population size Effective number of breeding adults in a population; influenced by the number of individuals contributing genes to the next generation, their sex ratio, variation between individuals in reproductive success, fluctuations in population size, the age structure of the population, and whether mating is random.

egg Female gamete.

environmental variance Component of the phenotypic variance that is due to environmental differences among individual members of a population.

epigenetics Phenonmena due to alterations to DNA that do not include changes in the base sequence; often affect the way in which the DNA sequences are expressed. Such alterations are often stable and heritable in the sense that they are passed from one cell to another.

episome Plasmid capable of integrating into a bacterial chromosome.

epistasis Type of gene interaction in which a gene at one locus masks or suppresses the effects of a gene at a different locus.

epistatic gene Masks or suppresses the effect of a gene at a different locus.

equilibrium Situation in which no further change takes place; in population genetics, refers to a population in which allelic frequencies do not change.

eubacteria One of the three primary divisions of life. Eubacteria consist of unicellular organisms with prokaryotic cells and include most of the common bacteria.

eukaryote Organism with a complex cell structure including a nuclear envelope and membrane-bounded organelles. One of the three primary divisions of life, eukaryotes include unicellular and multicellular forms.

evolution Genetic change taking place in a group of organisms.

exon shuffling Process, important in the evolution of eukaryotic genes, by which exons of different genes are exchanged and mixed into new combinations, creating new genes that are mosaics of other preexisting genes.

expanding trinucleotide repeat Mutation in which the number of copies of a trinucleotide (or some multiple of three nucleotides) increases in succeeding generations.

expected heterozygosity Proportion of individuals that are expected to be heterozygous at a locus when the Hardy–Weinberg assumptions are met.

expressivity Degree to which a trait is expressed.

familial Down syndrome Caused by a Robertsonian translocation in which the long arm of chromosome 21 is translocated to another chromosome; tends to run in families.

fertilization Fusion of gametes, or sex cells, to form a zygote.

fetal cell sorting Separation of fetal cells from maternal blood. Genetic testing on the fetal cells can provide information about genetic diseases and disorders in the fetus.

F factor Episome of *E. coli* that controls conjugation and gene exchange between *E. coli* cells. The F factor contains an origin of replication and genes that enable the bacterium to undergo conjugation.

F_1 (first filial) generation Offspring of the initial parents (P) in a genetic cross.

F_2 (second filial) generation Offspring of the F_1 generation in a genetic cross; the third generation of a genetic cross.

first polar body One of the products of meiosis I in oogenesis; contains half the chromosomes but little of the cytoplasm.

fitness Reproductive success of a genotype compared with that of other genotypes in a population.

fixation Point at which one allele reaches a frequency of 1. At this point, all members of the population are homozygous for the same allele.

forward mutation Alters a wild-type phenotype.

founder effect Sampling error that arises when a population is established by a small number of individuals; leads to genetic drift.

fragile site Constriction or gap that appears at a particular location on a chromosome when cells are cultured under special conditions. One fragile site on the human X chromosome is associated with mental retardation (fragile-X syndrome) and results from an expanding trinucleotide repeat.

frameshift mutation Alters the reading frame of a gene.

fraternal twins Nonidentical twins that arise when two different eggs are fertilized by two different sperm; also called dizygotic twins.

frequency distribution Graphical way of representing values. In genetics, usually the phenotypes found in a group of individuals are displayed as a frequency distribution. Typically, the phenotypes are plotted on the horizontal (x) axis and the numbers (or proportions) of individuals with each phenotype are plotted on the vertical (y) axis.

G_0 (gap 0) Nondividing stage of the cell cycle.

G_1 (gap 1) Stage in interphase of the cell cycle in which the cell grows and develops.

G_2 (gap 2) Stage of interphase in the cell cycle that follows DNA replication. In G_2, the cell prepares for division.

gain-of-function mutation Produces a new trait or causes a trait to appear in inappropriate tissues or at inappropriate times in development.

gametic isolation Reproductive isolation due to the incompatibility of gametes. Mating between members of different species takes place, but the gametes do not form zygotes. Seen in many plants, where pollen from one species cannot fertilize the ovules of another species.

gametophyte Haploid phase of the life cycle in plants.

gene Genetic factor that helps determine a trait; often defined at the molecular level as a DNA sequence that is transcribed into an RNA molecule.

gene flow Movement of genes from one population to another; also called migration.

gene interaction Interactions between genes at different loci that affect the same characteristic.

gene mutation Affects a single gene or locus.

gene pool Total of all genes in a population.

generalized transduction Transduction in which any gene may be transferred from one bacterial cell to another by a virus.

gene tree Phylogenetic tree representing the evolutionary relationships among a set of genes.

genetic bottleneck Sampling error that arises when a population undergoes a drastic reduction in population size; leads to genetic drift.

genetic correlation Phenotypic correlation due to the same genes affecting two or more characteristics.

genetic counseling Educational process that attempts to help patients and family members deal with all aspects of a genetic condition.

genetic drift Change in allelic frequency due to sampling error.

genetic–environmental interaction variance Component of the phenotypic variance that results from an interaction between genotype and environment. Genotypes are expressed differently in different environments.

genetic map Map of the relative distances between genetic loci, markers, or other chromosome regions determined by rates of recombination; measured in percent recombination or map units.

genetic marker Any gene or DNA sequence used to identify a location on a genetic or physical map.

genetic maternal effect Determines the phenotype of an offspring. With genetic maternal effect, an offspring inherits genes for the characteristics from both parents, but the offspring's phenotype is determined not by its own genotype but by the nuclear genotype of its mother.

genetic rescue Introduction of new genetic variation into an inbred population that often dramatically improves the health of the population in an effort to increase its chances of long-term survival.

genetic variance Component of the phenotypic variance that is due to genetic differences among individual members of a population.

genic balance system Sex-determining system in which sexual phenotype is controlled by a balance between genes on the X chromosome and genes on the autosomes.

genic interaction variance Component of the genetic variance that can be attributed to genic interaction (interaction between genes at different loci).

genic sex determination Sex determination in which the sexual phenotype is specified by genes at one or more loci, but there are no obvious differences in the chromosomes of males and females.

genome Complete set of genetic instructions for an organism.

genomic imprinting Differential expression of a gene that depends on the sex of the parent that transmitted the gene. If the gene is inherited from the father, its expression is different from that if it is inherited from the mother.

genotype The set of genes possessed by an individual organism.

genotypic frequency Proportion of a particular genotype.

germ-line mutation Mutation in a germ-line cell (one that gives rise to gametes).

germ-plasm theory States that cells in the reproductive organs carry a complete set of genetic information.

G_2/M (gap 2/mitotic) checkpoint Important point in the cell cycle near the end of G_2. After this checkpoint has been passed, the cell undergoes mitosis.

goodness-of-fit chi-square test Statistical test used to evaluate how well a set of observed values fit the expected values. The probability associated with a calculated chi-square value is the probability that the differences between the observed and the expected values may be due to chance.

G_1/S (gap 1/synthesis) checkpoint Important point in the cell cycle. After the G_1/S checkpoint has been passed, DNA replicates and the cell is committed to dividing.

gynandromorph Individual organism that is a mosaic for the sex chromosomes, possessing tissues with different sex-chromosome constitutions.

haploid Possessing a single set of chromosomes (one genome).

haploinsufficient gene Must be present in two copies for normal function. If one copy of the gene is missing, a mutant phenotype is produced.

Hardy–Weinberg equilibrium Frequencies of genotypes when the conditions of the Hardy–Weinberg law are met.

Hardy–Weinberg law Important principle of population genetics stating that, in a large, randomly mating population not affected by mutation, migration, or natural selection, allelic frequencies will not

change and genotypic frequencies stabilize after one generation in the proportions p^2 (the frequency of AA), $2pq$ (the frequency of Aa), and q^2 (the frequency of aa), where p equals the frequency of allele A and q equals the frequency of allele a.

hemizygosity Possession of a single allele at a locus. Males of organisms with XX-XY sex determination are hemizygous for X-linked loci, because their cells possess a single X chromosome.

heritability Proportion of phenotypic variation due to genetic differences. *See also* **broad-sense heritability** and **narrow-sense heritability.**

hermaphroditism Condition in which an individual organism possesses both male and female reproductive structures. True hermaphrodites produce both male and female gametes.

heterogametic sex The sex (male or female) that produces two types of gametes with respect to sex chromosomes. For example, in the XX-XY sex-determining system, the male produces both X-bearing and Y-bearing gametes.

heterokaryon Cell possessing two nuclei derived from different cells through cell fusion.

heterozygote screening Tests members of a population to identify heterozygous carriers of a disease-causing allele who are healthy but have the potential to produce children with the disease.

heterozygous Refers to an individual organism that possesses two different alleles at a locus.

histone Low-molecular-weight protein found in eukaryotes that complexes with DNA to form chromosomes.

homogametic sex The sex (male or female) that produces gametes that are all alike with regard to sex chromosomes. For example, in the XX-XY sex-determining system, the female produces only X-bearing gametes.

homologous pair of chromosomes Two chromosomes that are alike in structure and size and that carry genetic information for the same set of hereditary characteristics. One chromosome of a homologous pair is inherited from the male parent and the other is inherited from the female parent.

homozygous Refers to an individual organism that possesses two identical alleles at a locus.

horizontal gene transfer Transfer of genes from one organism to another by a mechanism other than reproduction.

hybrid breakdown Reproductive isolating mechanism in which closely related species are capable of mating and producing viable and fertile F_1 progeny, but genes do not flow between the two species, because further crossing of the hybrids produces inviable or sterile offspring.

hybrid inviability Reproductive isolating mechanism in which mating between two organisms of different species take place and hybrid offspring are produced but are not viable.

hybrid sterility Hybrid embryos complete development but are sterile; exemplified by mating between donkeys and horses to produce a mule, a viable but usually sterile offspring.

hypostatic gene Gene that is masked or suppressed by the action of a gene at a different locus.

identical twins Twins that arise when a single egg fertilized by a single sperm splits into two separate embryos; also called monozygotic twins.

inbreeding Mating between related individuals that takes place more frequently than expected on the basis of chance.

inbreeding coefficient Measure of inbreeding; the probability (ranging from 0 to 1) that two alleles are identical by descent.

inbreeding depression Decreased fitness arising from inbreeding; often due to the increased expression of lethal and deleterious recessive traits.

incomplete dominance Refers to the phenotype of a heterozygote that is intermediate between the phenotypes of the two homozygotes.

incomplete linkage Linkage between genes that exhibit some crossing over; intermediate in its effects between independent assortment and complete linkage.

incomplete penetrance Refers to a genotype that does not always express the expected phenotype. Some individuals possess the genotype for a trait but do not express the phenotype.

incorporated error Incorporation of a damaged nucleotide or mismatched base pair into a DNA molecule.

independent assortment Independent separation of chromosome pairs in anaphase I of meiosis; contributes to genetic variation.

induced mutation Results from environmental agents, such as chemicals or radiation.

induction Stimulation of the synthesis of an enzyme by an environmental factor, often the presence of a particular substrate.

in-frame deletion Deletion of some multiple of three nucleotides, which does not alter the reading frame of the gene.

in-frame insertion Insertion of some multiple of three nucleotides, which does not alter the reading frame of the gene.

inheritance of acquired characteristics Early notion of inheritance proposing that acquired traits are passed to descendants.

insertion Mutation in which nucleotides are added to a DNA sequence.

integrase Enzyme that inserts prophage, or proviral, DNA into a chromosome.

intercalating agent Chemical substance that is about the same size as a nucleotide and may become sandwiched between adjacent bases in DNA, distorting the three-dimensional structure of the helix and causing single-nucleotide insertions and deletions in replication.

interchromosomal recombination Recombination among genes on different chromosomes.

interference Degree to which one crossover interferes with additional crossovers.

intergenic suppressor mutation Occurs in a gene (locus) that is different from the gene containing the original mutation.

interkinesis Period between meiosis I and meiosis II.

interphase Period in the cell cycle between the cell divisions. In interphase, the cell grows, develops, and prepares for cell division.

intrachromosomal recombination Recombination among genes located on the same chromosome.

intragenic mapping Maps the locations of mutations within a single locus.

intragenic suppressor mutation Occurs in the same gene (locus) as the mutation that it suppresses.

inverted repeats Sequences on the same strand that are inverted and complementary.

karyotype Picture of an individual organism's complete set of metaphase chromosomes.

kinetochore Set of proteins that assemble on the centromere, providing the point of attachment for spindle microtubules.

Klinefelter syndrome Human condition in which cells contain one or more Y chromosomes along with multiple X chromosomes (most commonly XXY but may also be XXXY, XXXXY, or XXYY). Persons with Klinefelter syndrome are male in appearance but frequently possess small testes, some breast enlargement, and reduced facial and pubic hair; often taller than normal and sterile, most have normal intelligence.

leptotene First substage of prophase I in meiosis. In leptotene, chromosomes contract and become visible.

lethal allele Causes the death of an individual organism, often early in development, and so the organism does not appear in the progeny of a genetic cross. Recessive lethal alleles kill individual organisms that are homozygous for the allele; dominant lethals kill both heterozygotes and homozygotes.

lethal mutation Causes premature death.

LINE *See* **long interspersed element.**

linkage group Genes located together on the same chromosome.

linked genes Genes located on the same chromosome.

locus Position on a chromosome where a specific gene is located.

lod (logarithm of odds) score Logarithm of the ratio of the probability of obtaining a set of observations, assuming a specified degree of linkage, to the probability of obtaining the same set of observations with independent assortment; used to assess the likelihood of linkage between genes from pedigree data.

loss-of-function mutation Causes the complete or partial absence of normal function.

Lyon hypothesis Proposed by Mary Lyon in 1961, this hypothesis proposes that one X chromosome in each female cell becomes inactivated (a Barr body) and suggests that which X becomes inactivated is random and varies from cell to cell.

lysogenic cycle Life cycle of a bacteriophage in which phage genes first integrate into the bacterial chromosome and are not immediately transcribed and translated.

lytic cycle Life cycle of a bacteriophage in which phage genes are transcribed and translated, new phage particles are produced, and the host cell is lysed.

mapping function Relates recombination frequencies to actual physical distances between genes.

map unit (m.u.) Unit of measure for distances on a genetic map; 1 map unit equals 1% recombination.

maternal blood testing Testing for genetic conditions in a fetus by analyzing the blood of the mother. For example, the level of α-fetoprotein in maternal blood provides information about the probability that a fetus has a neural-tube defect.

mean Statistic that describes the center of a distribution of measurements; calculated by dividing the sum of all measurements by the number of measurements; also called the average.

mechanical isolation Reproductive isolation resulting from anatomical differences that prevent successful copulation.

Answers to Selected Questions and Problems

Chapter 1

1. In the Hopi culture, albinos were considered special and given special status. Because extensive exposure to sunlight could be damaging or deadly, Hopi male albinos did no agricultural work. Albinism was a considered a positive trait rather than a negative physical condition, allowing albinos to have more children and thus increasing the frequency of the allele. Finally, the small population size of the Hopi tribe may have helped increase the allele frequency of the albino gene owing to chance.

3. Genetics plays important roles in the diagnosis and treatment of hereditary diseases, in breeding plants and animals for improved production and disease resistance, and in producing pharmaceuticals and novel crops through genetic engineering.

5. Transmission genetics: The inheritance of genes from one generation to the next, gene mapping, characterization of the phenotypes produced by mutations.

   Molecular genetics: The structure, organization, and function of genes at the molecular level.

   Population genetics: Genes and changes in genes in populations.

8. Pangenesis theorizes that information for creating each part of an offspring's body originates in each part of the parent's body and is passed through the reproductive organs to the embryo at conception. Pangenesis suggests that changes in parts of the parent's body may be passed to the offspring's body. The germ-plasm theory, in contrast, states that the reproductive cells possess all of the information required to make the complete body; the rest of the body contributes no information to the next generation.

10. Preformationism is the idea that an offspring results from a miniature adult form that is already preformed in the sperm or the egg. All traits would thus be inherited from only one parent, either the father or the mother, depending on whether the homunculus (the preformed miniature adult) resided in the sperm or the egg.

13. Gregor Mendel

16. Genes are composed of DNA nucleotide sequences and are located at specific positions in chromosomes.

18. (**a**) Transmission genetics; (**b**) population genetics; (**c**) population genetics; (**d**) molecular genetics; (**e**) molecular genetics; (**f**) transmission genetics.

19. Genetics is old in the sense that humans have been aware of hereditary principles for thousands of years and have applied them since the beginning of agriculture and the domestication of plants and animals. It is very young in the sense that the fundamental principles were not uncovered until Mendel's time, and the structure of DNA and the principles of recombinant DNA were discovered within the past 60 years.

21. (**a**) Pangenesis postulates that specific particles (called gemmules) carried genetic information from all parts of the body to the reproductive organs, and then the genetic information was conveyed to the embryo where each unit directs the formation of its own specific part of the body. According to the germ-plasm theory, gamete-producing cells found within the reproductive organs contain a complete set of genetic information that is passed to the gametes. The theories are similar in that both propose that genetic information is contained in discrete units that are passed on to the offspring. They differ in where that genetic information resides. In pangenesis, it resides in different parts of the body and must travel to the reproductive organs. In the germ-plasm theory, all the genetic information is already in the reproductive cells.
    (**b**) Preformationism holds that the sperm or egg contains a miniature preformed adult called a homunculus. In development, the homunculus grows to produce an offspring. Only one parent contributes genetic traits to the offspring. Blending inheritance requires contributions of genetic material from both parents. The genetic contributions from the parents blend to produce the genetic material of the offspring. Once blended, the genetic material cannot be separated for future generations.
    (**c**) The theory of inheritance of acquired characteristics postulates that traits acquired in a person's lifetime alter the genetic material and can be transmitted to offspring. Our modern theory of heredity indicates that offspring inherit genes located on chromosomes passed from their parents. These chromosomes segregate in meiosis in the parent's germ cells and are passed into the gametes.

22. (**a**) Both cell types have lipid bilayer membranes, DNA genomes, and machinery for DNA replication, transcription, translation, energy metabolism, response to stimuli, growth and reproduction. Eukaryotic cells have a nucleus containing chromosomal DNA and possess internal membrane-bounded organelles.
    (**b**) A gene is a basic unit of hereditary information, usually encoding a functional RNA or polypeptide. Alleles are variant forms of a gene, arising through mutation.
    (**c**) The genotype is the set of genes or alleles inherited by an organism from its parent(s). The expression of the genes of a particular genotype, through interaction with environmental factors, produces the phenotype, the observable trait.
    (**d**) Both are nucleic acid polymers. RNA contains a ribose sugar, whereas DNA contains a deoxyribose sugar. RNA also contains uracil as one of the four bases, whereas DNA contains thymine. The other three bases are common to both DNA and RNA. Finally, DNA is usually double stranded, consisting of two complementary strands, whereas RNA is single stranded with regions of internal base pairing that form complex secondary structures.
    (**e**) Chromosomes are structures consisting of DNA and associated proteins. The DNA contains the genetic information.

26. All genomes must have the ability to store complex information and to vary. The blueprint for the entire organism must be contained within the genome of each reproductive cell. The information has to be in the form of a code that can be used as a set of instructions for

assembling the components of the cells. The genetic material of any organism must be stable, be replicated precisely, and be transmitted faithfully to the progeny but must be capable of mutating.

28. (a) Having the genetic test removes doubt about the potential for the disorder: you are either susceptible or not. Knowing about the potential of a genetic disorder enables you to make life-style changes that might lessen the effect of the disease or lessen the risk. The types and nature of future medical tests could be positively affected by the genetic testing, thus allowing for early warning and screening for the disease. The knowledge could also enable you to make informed decisions regarding offspring and the potential of passing the trait to your offspring. Additionally, by knowing what to expect, you could plan your life accordingly.

Reasons for not having the test typically concern the potential for testing positive for the susceptibility to the genetic disease. If the susceptibility is detected, there is potential for discrimination. For example, your employer (or possibly future employer) might consider you a long-term liability, thus affecting employment options. Insurance companies may not want to insure you for the condition or its symptoms, and social stigmatization regarding the disease could be a factor. Knowledge of the potential future condition could lead to psychological difficulties in coping with the anxiety of waiting for the disease to manifest.

(b) There is no "correct" answer, but some of the reasons for wanting to be tested are: the test would remove doubt about the susceptibility, particularly if family members have had the genetic disease; either a positive or a negative result would allow for informed planning of life style, medical testing, and family choices in the future.

Chapter 2

1.

Prokaryotic cell	Eukaryotic cell
No nucleus	Nucleus present
No paired chromosomes	Paired chromosomes common
Typically a single circular chromosome containing a single origin of replication	Typically multiple linear chromosomes containing centromeres, telomeres, and multiple origins of replication
Single chromosome is replicated, with each copy moving to opposite sides of the cell	Chromosomes require mitosis or meiosis to segregate to the proper location
Histone proteins are complexed to DNA	No histone proteins complexed to DNA

3. The events are (i) a cell's genetic information must be copied; (ii) the copies of the genetic information must be separated from one another; and (iii) the cell must divide.

6.

Metacentric   Submetacentric   Acrocentric   Telocentric

8. Prophase: The chromosomes condense and become visible, the centrosomes move apart, and microtubule fibers form from the centrosomes.

Prometaphase: The nucleoli disappear and the nuclear envelope begins to disintegrate, allowing the cytoplasm and nucleoplasm to join. The sister chromatids of each chromosome are attached to microtubles from the opposite centrosomes.

Metaphase: The spindle microtubules are clearly visible, and the chromosomes arrange themselves on the metaphase plate of the cell.

Anaphase: The sister chromatids separate at the centromeres after the breakdown of cohesin protein, and the newly formed daughter chromosomes move to the opposite poles of the cell.

Telophase: The nuclear envelope re-forms around each set of daughter chromosomes. Nucleoli reappear. Spindle microtubules disintegrate.

10. In the mitotic cell cycle, the genetic material is precisely copied so that the two resulting cells contain the same genetic information. In other words, the daughter cells have genomes identical with each other and with that of the mother cell.

14. Meiosis includes two cell divisions, thus producing four new cells (in many species). The chromosome number of a haploid cell produced by meiosis I is half the chromosome number of the original diploid cell. Finally, the cells produced by meiosis are genetically different from the original cell and genetically different from each other.

16.

Mitosis	Meiosis
A single cell division produces two genetically identical progency cells.	Two cell divisions usually result in four progeny cells that are not genetically identical.
Chromosome number of progeny cells and original cell remain the same.	Daughter cells are haploid and have half the chromosomal complement of the original diploid cell as a result of the separation of homologous pairs in anaphase I.
Daughter cells and original cell are genetically identical. No crossing over or separation of homologous chromosomes takes place.	Crossing over in prophase I and separation of homologous pairs in anaphase I produce daughter cells that are genetically different from each other and from the original cell.
Homologous chromosomes do not synapse.	Homologous chromosomes synapse prophase I.
In metaphase, individual chromosomes line up on the metaphase plate.	In metaphase I, homologous pairs of chromosomes line up on the metaphase plate. Individual chromosomes line up in metaphase II.
In anaphase, sister chromatids separate.	In anaphase I, homologous chromosomes separate. Sister chromatids separate in anaphase II.

The most important difference is that mitosis produces cells genetically identical with each other and with the original cell, resulting in the orderly passage of information from one cell to its progeny. In contrast, by producing progeny that do not contain pairs of homologous chromosomes, meiosis results in the reduction of chromosome number from the original cell. Meiosis also allows for genetic variation through crossing over and the random assortment of homologous chromosomes.

22. (a) 12 chromosomes and 24 DNA molecules; (b) 12 chromosomes and 24 DNA molecules; (c) 12 chromosomes and 24 DNA molecules; (d) 12 chromosomes and 24 DNA molecules; (e) 12 chromosomes and 12 DNA molecules; (f) 6 chromosomes and 12 DNA molecules; (g) 12 chromosomes and 12 DNA molecules; (h) 6 chromosomes and 6 DNA molecules.

24. The diploid number of chromosomes is six. The left-hand cell is in meiosis I; the middle cell is in anaphase of mitosis; the right-hand cell is in anaphase II of meiosis.

26.

Event	Mitosis	Meiosis I	Meiosis II
Does crossing over occur?	No	Yes	No
What separates in anaphase?	Sister chromatids	Homologous pairs pairs of chromosomes	Sister chromatids
What lines up on the metaphase plate?	Individual chromosomes	Homologous pairs of chromosomes	Individual chromosomes
Does cell division usually takes place?	Yes	Yes	Yes
Do homologs pair?	No	Yes	No
Is genetic variation produced?	No	Yes	No

27. (a) Cohesin is needed to hold the sister chromatids together until anaphase of mitosis. If cohesin fails to form early in mitosis, the sister chromatids could separate before anaphase. As a result, chromosomes would not segregate properly to daughter cells. (b) Shugoshin protects cohesin proteins from degradation at the centromere during meiosis I. Cohesin at the arms of the homologous chromosomes is not protected by shugoshin and is broken in anaphase I, enabling the two homologs to separate. If shugoshin is absent during meiosis, then the cohesin at the centromere may be broken, allowing for the separation of sister chromatids along with the homologs in anaphase I and leading to the improper segregation of chromosomes to daughter cells. (c) If shugoshin does not break down, the cohesin proteins at the centromere will remain protected from degradation. The intact cohesin will prevent the sister chromatids from separating in anaphase II of meiosis; as a result, sister chromatids will not separate, leading to daughter cells that have too many chromosomes or too few.

30. The progeny of an organism whose cells contain more homologous pairs of chromosomes should be expected to exhibit more variation. The number of different combinations of chromosomes that are possible in the gametes is $2n$, where n is equal to the number of homologous pairs of chromosomes. For the fruit fly with four pairs of chromosomes, the number of possible combinations is $2^4 = 16$. For *Musca domestica* with six pairs of chromosomes, the number of possible combinations is $2^6 = 64$.

31. (a) Metaphase I

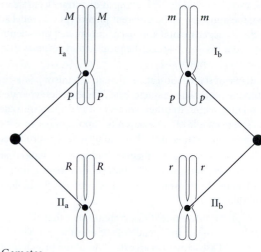

(b) Gametes

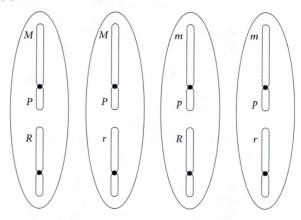

34. (a) No. The first polar body and the secondary oocyte are the result of meiosis I. In meiosis I homologous chromosomes segregate; thus, both the first polar body and the secondary oocyte will contain only one homolog of each original chromosome pair, and the alleles of some genes in these homologs will differ. Additionally, crossing over in prophase I will have generated new and different arrangements of genetic material for each homolog of the pair.
(b) No. The second polar body and the ovum will contain the same members of the homologous pairs of chromosomes that were separated in meiosis I and produced by the separation of sister chromatids in anaphase II. However, the sister chromatids of each pair are no longer identical, because they underwent crossing over in prophase I and thus contain genetic information that is not identical.

37. Most male animals produce sperm by meiosis. Because meiosis takes place only in diploid cells, haploid male bees do not undergo meiosis. Male bees can produce sperm but only through mitosis. Haploid cells that divide mitotically produce haploid cells.

Chapter 3

1. Mendel was successful for several reasons. He chose to work with a plant, *Pisum sativum*, that was easy to cultivate, grew relatively rapidly, and produced many offspring whose phenotypes were easy to determine, which allowed Mendel to detect mathematical ratios of progeny phenotypes. The seven characteristics that he chose to

study exhibited only a few distinct phenotypes and did not show a range of variation. Finally, by looking at each trait separately and counting the numbers of the different phenotypes, Mendel adopted a reductionist experimental approach and applied the scientific method. From his observations, he proposed hypotheses that he was then able to test empirically.

3. The principle of segregation states that an organism possesses two alleles for any particular characteristic. These alleles separate in the formation of gametes. In other words, one allele goes into each gamete. The principle of segregation is important because it explains how the genotypic ratios in the haploid gametes are produced.

8. Walter Sutton's chromosome theory of inheritance states that genes are located on chromosomes. The independent segregation of pairs of homologous chromosomes in meiosis provides the biological basis for Mendel's two principles of heredity.

9. The principle of independent assortment states that alleles at different loci segregate independently of one another. The principle of independent assortment is an extension of the principle of segregation: the principle of segregation states that the two alleles at a locus separate; according to the principle of independent assortment, when these two alleles separate, their separation is independent of the separation of alleles at other loci.

13. (a) The parents are RR (orange fruit) and rr (cream fruit). All the F_1 are Rr (orange). The F_2 are $1\ RR : 2\ Rr : 1\ rr$ and have an orange-to-cream phenotypic ratio of $3 : 1$.
(b) Half of the progeny are homozygous for orange fruit (RR) and half are heterozygous for orange fruit (Rr).
(c) Half of the progeny are heterozygous for orange fruit (Rr) and half are homozygous for cream fruit (rr).

15. (a) Although the white female guinea pig gave birth to the offspring, her eggs were produced by the ovary from the black female guinea pig. The transplanted ovary produced only eggs containing the allele for black coat color. Like most mammals, guinea pig females produce primary oocytes early in development, and thus the transplanted ovary already contained primary oocytes produced by the black female guinea pig.
(b) The white male guinea pig contributed a w allele, and the white female guinea pig contributed the W allele from the transplanted ovary. The offspring are thus Ww.
(c) The transplant experiment supports the germ-plasm theory. According to the germ-plasm theory, only the genetic information in the germ-line tissue in the reproductive organs is passed to the offspring. The production of black guinea pig offspring suggests that the allele for black coat color was passed to the offspring from the transplanted ovary in agreement with the germ-plasm theory. According to the pangenesis theory, the genetic information passed to the offspring originates at various parts of the body and travels to the reproductive organs for transfer to the gametes. If pangenesis were correct, then the guinea pig offspring should have been white. The white-coat alleles would have traveled to the transplanted ovary and then into the white female's gametes. The absence of any white offspring indicates that the pangenesis hypothesis is invalid.

16. (a) Female parent is $i^B i^B$; male parent is $I^A i^B$.
(b) Both parents are $i^B i^B$.
(c) Male parent is $i^B i^B$; female parent is $I^A I^A$ or, possibly, $I^A i^B$, but a heterozygous female in this mating is unlikely to have produced eight blood-type-A kittens owing to chance alone.
(d) Both parents are $I^A i^B$.
(e) Either both parents are $I^A I^A$ or one parent is $I^A I^A$ and the other

parent is $I^A i^B$. The blood type of the offspring does not allow a determination of the precise genotype of either parent.
(f) Female parent is $i^B i^B$; male parent is $I^A i^B$.

19. (a) Sally (Aa), Sally's mother (Aa), Sally's father (aa), and Sally's brother (aa); (b) $\frac{1}{2}$; (c) $\frac{1}{2}$.

21. Use h for the hairless allele and H for the dominant allele for the presence of hair. Because H is dominant to h, a rat terrier with hair could be either homozygous (HH) or heterozygous (Hh). To determine which genotype is present in the rat terrier with hair, cross this dog with a hairless rat terrier (hh). If the terrier with hair is homozygous (HH), then no hairless offspring will be produced by the testcross. However, if the terrier is heterozygous (Hh), then $\frac{1}{2}$ of the offspring will be hairless.

24. (a) $\frac{1}{18}$; (b) $\frac{1}{36}$; (c) $\frac{11}{36}$; (d) $\frac{1}{6}$; (e) $\frac{1}{4}$; (f) $\frac{3}{4}$.

25. (a) $\frac{1}{128}$; (b) $\frac{1}{64}$; (c) $\frac{7}{128}$; (d) $\frac{35}{128}$; (e) $\frac{35}{128}$.

27. Parents:

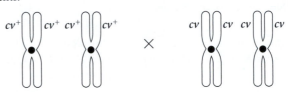

F_1 generation:

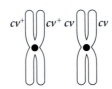

F_2 generation:

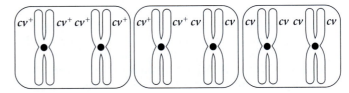

28. (a) In the F_1 black guinea pigs (Bb), only one chromosome possesses the black allele, and so the number of copies present at each stage are: G_1, one black allele; G_2, two black alleles; metaphase of mitosis, two black alleles; metaphase I of meiosis, two black alleles; after cytokinesis of meiosis, one black allele but only in half of the cells produced by meiosis. (The remaining half will not contain the black allele.)
(b) In the F_1 brown guinea pigs (bb), both homologs possess the brown allele, and so the number of copies present at each stage are: G_1, two brown alleles; G_2, four brown alleles; metaphase of mitosis, four brown alleles; metaphase I of meiosis, four brown alleles; metaphase II, two brown alleles; after cytokinesis of meiosis, one brown allele.

30. (a) $\frac{9}{16}$ black and curled, $\frac{3}{16}$ black and normal, $\frac{3}{16}$ gray and curled, and $\frac{1}{16}$ gray and normal.
(b) $\frac{1}{4}$ black and curled, $\frac{1}{4}$ black and normal, $\frac{1}{4}$ gray and curled, $\frac{1}{4}$ gray and normal

31. (a) $\frac{1}{2}\ (Aa) \times \frac{1}{2}\ (Bb) \times \frac{1}{2}\ (Cc) \times \frac{1}{2}\ (Dd) \times \frac{1}{2}\ (Ee) = \frac{1}{32}$
(b) $\frac{1}{2}\ (Aa) \times \frac{1}{2}\ (bb) \times \frac{1}{2}\ (Cc) \times \frac{1}{2}\ (dd) \times \frac{1}{4}\ (ee) = \frac{1}{64}$
(c) $\frac{1}{4}\ (aa) \times \frac{1}{2}\ (bb) \times \frac{1}{4}\ (cc) \times \frac{1}{2}\ (dd) \times \frac{1}{4}\ (ee) = \frac{1}{256}$
(d) No offspring having this genotype. The $Aa\ Bb\ Cc\ dd\ Ee$ parent

cannot contribute a *D* allele, the *Aa bb Cc Dd Ee* parent cannot contribute a *B* allele. Therefore, their offspring cannot be homozygous for the *BB* and *DD* gene loci.

34. (**a**) Gametes from *Aa Bb* individual:

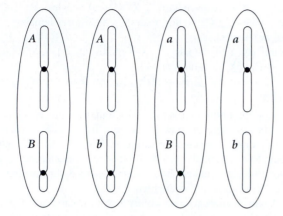

Gametes from *aa bb* individual:

(**b**) Progeny at G₁:

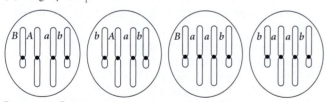

Progeny at G₂:

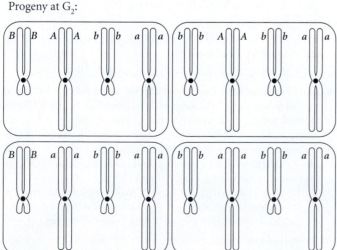

Progeny at prophase of mitosis:

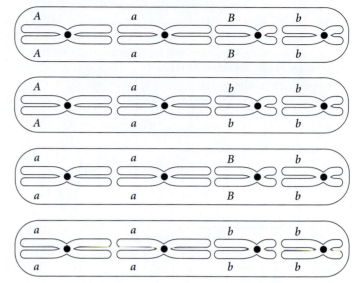

The order of chromosomes on the metaphase plate can vary.

35. (**a**) The burnsi × burnsi cross produced both burnsi and pipiens offspring, suggesting that the parents were heterozygous with each possessing a *burnsi* allele and a *pipiens* allele. The cross also suggests that the *burnsi* allele is dominant over the *pipiens* allele. The progeny of the burnsi × pipiens crosses suggest that each of the crosses was between a homozygous recessive frog (pipiens) and a heterozygous dominant frog (burnsi). The results of both crosses are consistent with the brunsi phenotype being recessive to the pipiens phenotype.
(**b**) Let *P* represent the *burnsi* allele and *p* represent the *pipiens* allele.

$$\text{burnsi } (Pp) \times \text{burnsi } (Pp)$$
$$\text{burnsi } (Pp) \times \text{pipiens } (pp)$$
$$\text{burnsi } (Pp) \times \text{pipiens } (pp)$$

c. For the burnsi × burnsi ($Pp \times Pp$) cross, we would expect a phenotypic ratio of 3 : 1 in the offspring. A chi-square test to evaluate the fit of the observed numbers of progeny with an expected 3 : 1 ratio gives a chi-square value of 2.706 with 1 degree of freedom. The probability associated with this chi-square value is between 0.1 and 0.05, indicating that the differences between what we expected and what we observed could have been generated by chance.

For the first burnsi × pipiens ($Pp \times pp$) cross, we would expect a phenotypic ratio of 1 : 1. A chi-square test comparing observed and expected values yields $\chi^2 = 1.78$, df = 1, $P > 0.05$. For the second burnsi × pipiens ($Pp \times pp$) cross, we would expect a phenotypic ratio of 1 : 1. A chi-square test of the fit of the observed numbers with those expected with a 1 : 1 ratio yields $\chi^2 = 0.46$, df = 1, $P > 0.05$. Thus, all three crosses are consistent with the predication that burnsi is dominant over pipiens.

38. (**a**) For the cross between a heterozygous F₁ plant (*Cc Ff*) and a homozygous recessive plant (*cc ff*), we would expect a phenotypic ratio of 1 : 1 : 1 : 1 for the different phenotypic classes. A chi-square test comparing the fit of the observed data with the expected 1 : 1 : 1 : 1 ratio yields a chi-square value of 35 with df = 3 and $P < 0.005$.
(**b**) From the chi-square value, it is unlikely that chance produced the differences between the observed and expected ratios, indicating that the progeny do not occur in a 1 : 1 : 1 : 1 ratio.
(**c**) The number of plants with the *cc ff* genotype is much less than expected. The *cc ff* genotype is possibly sublethal. In other words,

California poppies with the homozygous recessive genotypes are possibly less viable than the other possible genotypes.

40. The first geneticist has identified an allele for obesity that he believes to be recessive. Let's call his allele for obesity o_1 and the normal allele O_1. On the basis of the crosses that the geneticist performed, the allele for obesity appears to be recessive.

Cross 1 with possible genotype:

$$\text{Obese } (o_1 o_1) \times \text{Normal } (O_1 O_1)$$
$$\downarrow$$

F_1 All normal $(O_1 o_1)$

Cross 2 with possible genotypes:

F_1 Normal $(O_1 o_1) \times$ Normal $(O_1 o_1)$
$$\downarrow$$

F_2 8 normal $(O_1 O_1 \text{ and } O_1 o_1)$
              2 obese $(o_1 o_1)$

Cross 3 with possible genotypes:

$$\text{Obese } (o_1 o_1) \times \text{Obese } (o_1 o_1)$$
$$\downarrow$$

F_1 All obese $(o_1 o_1)$

Let's call the second geneticist's allele for obesity o_2 and the normal allele O_2. The cross between obese mice from the two laboratories produced only normal mice. The alleles for obesity from both laboratories are recessive. However, they are located at different gene loci. Essentially, the obese mice from the different laboratories have separate obesity genes that are independent of one another.

The likely genotypes of the obese mice are as follows:

$$\text{Obese mouse 1 } (o_1 o_1 \, O_2 O_2) \times \text{Obese mouse 2 } (O_1 O_1 \, o_2 o_2)$$
$$\downarrow$$

F_1 All normal $(O_1 o_1 \, O_2 o_2)$

Chapter 4

1. Females produce larger gametes than those produced by males.

5. The pseudoautosomal region is a region of similarity between the X and Y chromosomes that is responsible for pairing the X and Y chromosomes in meiotic prophase I. Genes in this region are present in two copies in males and females and are thus inherited like autosomal genes, whereas other Y-linked genes are passed on only from father to son.

6. Diploid insects are female, whereas haploid insects are male. Fertilized eggs develop into females, and eggs that are not fertilized develop as males.

10. Males show the phenotypes of all X-linked traits, regardless of whether the X-linked allele is normally recessive or dominant. Males inherit X-linked traits from their mothers, pass X-linked traits to all of their daughters, and, through their daughters, to their daughters' descendants, but not to their sons or their sons' descendants.

15. Y-linked traits appear only in males and are always transmitted from fathers to sons, thus following a strict paternal lineage. Autosomal male-limited traits also appear only in males, but they can be transmitted to sons through their mothers.

16. (a) Female; (b) male; (c) male, sterile; (d) female; (e) male; (f) female; (g) metafemale; (h) male; (i) intersex; (j) female; (k) metamale, sterile; (l) metamale; (m) intersex.

21. Her father's mother and her father's father, but not her mother's mother or her mother's father.

23. (a) Yes; (b) yes; (c) no; (d) no.

24. (a) F_1: $\frac{1}{2}$ X^+Y (gray males), $\frac{1}{2}$ X^+X^y (gray females); F_2: $\frac{1}{4}$ X^+Y (gray males), $\frac{1}{4}$ X^yY (yellow males), $\frac{1}{4}$ X^+X^y (gray females), $\frac{1}{4}$ X^+X^+ (gray females).
(b) F_1: $\frac{1}{2}$ X^yY (yellow males), $\frac{1}{2}$ X^+X^y (gray females); F_2: $\frac{1}{4}$ X^+Y (gray males), $\frac{1}{4}$ X^yY (yellow males), $\frac{1}{4}$ X^+X^y (gray females), $\frac{1}{4}$ X^yX^y (yellow females).
(c) F_2: $\frac{1}{4}$ X^+Y (gray males), $\frac{1}{4}$ X^yY (yellow males), $\frac{1}{4}$ X^+X^+ (gray females), $\frac{1}{4}$ X^+X^y (gray females).
(d) $\frac{1}{8}$ gray males, $\frac{3}{8}$ yellow males, $\frac{5}{16}$ gray females, and $\frac{3}{16}$ yellow females.

25. Because color blindness is a recessive trait, the color-blind daughter must be homozygous recessive. Because the color blindness is X-linked, John has grounds for suspicion. Normally, their daughter would have inherited John's X chromosome. Because John is not color blind, he could not have transmitted an X chromosome with a color-blind allele to his daughter.

A remote alternative possibility is that the daughter is XO, having inherited a recessive color-blind allele from her mother and no sex chromosome from her father. In that case, the daughter would have Turner syndrome. A new X-linked color-blind mutation also is possible, albeit even less likely.

If Cathy had a color-blind son, then John would have no grounds for suspicion. The son would have inherited John's Y chromosome and the color-blind X chromosome from Cathy.

27. Because Bob must have inherited the Y chromosome from his father, and his father has normal color vision, a nondisjunction event in the paternal lineage cannot account for Bob's genotype. Bob's mother must be heterozygous X^+X^c because she has normal color vision, and she must have inherited a color-blind X chromosome from her color-blind father. For Bob to inherit two color-blind X chromosomes from his mother, the egg must have arisen from a nondisjunction in meiosis II. In meiosis I, the homologous X chromosomes separate, and so one cell has the X^+ chromosome and the other has X^c. The failure of sister chromatids to separate in meiosis II would then result in an egg with two copies of X^c.

31. (a) Male parent is X^+Y; female parent is X^+X^m.
(b) Male parent is X^mY; female parent is X^+X^+.
(c) Male parent is X^mY; female parent is X^+X^m.
(d) Male parent is X^+Y; female parent is X^mX^m.
(e) Male parent is X^+Y; female parent is X^+X^+.

32. F_1: $\frac{1}{2}$ Z^bZ^+ (normal males); $\frac{1}{2}$ Z^bW (bald females).
F_2: $\frac{1}{4}$ Z^+Z^b (normal males), $\frac{1}{4}$ Z^+W (normal females), $\frac{1}{4}$ Z^bZ^b (bald males), $\frac{1}{4}$ Z^bW (bald females).

33. (a) 1; (b) 0; (c) 0; (d) 1; (e) 1; (f) 2; (g) 0; (h) 2; (i) 3.

38. (a) F_1: all males are have miniature wings and red eyes ($X^mY\,s^+s$), and all females have long wings and red eyes ($X^+X^m\,s^+s$). F_2: $\frac{3}{16}$ male, normal, red; $\frac{1}{16}$ male, normal, sepia; $\frac{3}{16}$ male, miniature, red; $\frac{1}{16}$ male, miniature, sepia; $\frac{3}{16}$ female, normal, red; $\frac{1}{16}$ female, normal, sepia; $\frac{3}{16}$ female, miniature, red; $\frac{1}{16}$ female, miniature, sepia.
(b) F_1: all females have long wings and red eyes ($X^{m+}X^m\,s^+s$), and all males have long wings and red eyes ($X^{m+}Y\,s^+s$).
F_2: $\frac{3}{16}$ males, long wings, red eyes; $\frac{1}{16}$ males, long wings, sepia eyes; $\frac{3}{16}$ males, mini wings, red eyes; $\frac{1}{16}$ males, mini wings, sepia eyes; $\frac{3}{8}$ females, long wings, red eyes; $\frac{1}{8}$ long wings, sepia eyes.

40. The trivial explanation for these observations is that this form of color blindness is an autosomal recessive trait. In that case, the father

would be a heterozygote, and we would expect equal proportions of color-blind and normal children of either sex.

    If, on the other hand, we assume that this form of color blindness is an X-linked trait, then the mother is $X^c X^c$ and the father must be X^+Y. Normally, all the sons would be color blind, and all the daughters should have normal vision. The most likely way to have a daughter who is color blind would be for her not to have inherited an X^+ from her father. The observation that the color-blind daughter is short in stature and has failed to undergo puberty is consistent with Turner syndrome (XO). The color-blind daughter would then be X^cO.

Chapter 5

1. In incomplete dominance, the phenotype of the heterozygote is intermediate between the phenotypes of the two homozygotes. In codominance, both alleles are expressed and both phenotypes are manifested simultaneously.

2. In incomplete penetrance, the expected phenotype of a particular genotype is not expressed. Environmental factors, as well as the effects of other genes, may alter the phenotypic expression of a particular genotype.

5. A complementation test is used to determine whether two different recessive mutations are at the same locus (are allelic) or at different loci. The two mutations are introduced into the same individual organism by crossing homozygotes for each of the mutants. If the progeny show a mutant phenotype, then the mutations are allelic (at the same locus). If the progeny show a wild-type (dominant) phenotype, then the mutations are at different loci and are said to complement each other because each of the mutant parents can supply a functional copy (or dominant allele) of the gene mutated in the other parent.

6. Cytoplasmically inherited traits are encoded by genes in the cytoplasm. Because the cytoplasm is usually inherited from a single (most often the female) parent, reciprocal crosses do not show the same results. Cytoplasmically inherited traits often show great variability because different egg cells (female gametes) may have differing proportions of cytoplasmic alleles owing to random sorting of mitochondria (or plastids in plants).

11. Continuous characteristics, also called quantitative characteristics, exhibit many phenotypes with a continuous distribution. They result from the interaction of multiple genes (polygenic traits), the influence of environmental factors on the phenotype, or both.

13. (a) The results of the crosses indicate that cremello and chestnut are pure-breeding traits (homozygous). Palomino is a hybrid trait (heterozygous) that produces a 2 : 1 : 1 ratio when palominos are crossed with each other. The simplest hypothesis consistent with these results is incomplete dominance, with palomino as the phenotype of the heterozygotes resulting from chestnuts crossed with cremellos.
(b) Let C^B = chestnut, C^W = cremello. The parents and offspring of these crosses have the following genotypes: chestnut = $C^B C^B$; cremello = $C^W C^W$; palomino = $C^B C^W$.

14. (a) $\frac{1}{2} L^M L^M$ (type M), $\frac{1}{2} L^M L^N$ (type MN); (b) all $L^N L^N$ (type N); (c) $\frac{1}{2} L^M L^N$ (type MN), $\frac{1}{4} L^M L^M$ (type M), $\frac{1}{4} L^N L^N$ (type N); (d) $\frac{1}{2} L^M L^N$ (type MN), $\frac{1}{2} L^N L^N$ (type N); (e) all $L^M L^N$ (type MN).

17. (a) The 2 : 1 ratio in the progeny of two spotted hamsters suggests lethality, and the 1 : 1 ratio in the progeny of a spotted hamster and a hamster without spots indicates that spotted is a heterozygous phenotype. If S and s represent the locus for white spotting, spotted hamsters are Ss and solid-colored hamsters are ss. One-quarter of the zygotes expected from a mating of two spotted hamsters are SS, embryonic lethal, and missing from the progeny, resulting in the 2 : 1 ratio of spotted to solid progeny.
(b) Because spotting is a heterozygous phenotype, it should not be possible to obtain Chinese hamsters that breed true for spotting.

24. The child's genotype has an allele for blood-type B and an allele for blood-type N that could not have come from the mother and must have come from the father. Therefore, the child's father must have an allele for type B and an allele for type N. George, Claude, and Henry are eliminated as possible fathers because they lack an allele for either type B or type N.

25. (a) All walnut ($Rr\,Pp$); (b) $\frac{1}{4}$ walnut ($Rr\,Pp$), $\frac{1}{4}$ rose ($Rr\,pp$), $\frac{1}{4}$ pea ($rr\,Pp$), $\frac{1}{4}$ single ($rr\,pp$); (c) $\frac{9}{16}$ walnut ($R\_\,P\_$), $\frac{3}{16}$ rose ($R\_\,pp$), $\frac{3}{16}$ pea ($rr\,P\_$), $\frac{1}{16}$ single ($rr\,pp$); (d) $\frac{3}{4}$ rose ($R\_\,pp$), $\frac{1}{4}$ single ($rr\,pp$); (e) $\frac{1}{4}$ walnut ($Rr\,Pp$), $\frac{1}{4}$ rose ($Rr\,pp$), $\frac{1}{4}$ pea ($rr\,Pp$), $\frac{1}{4}$ single ($rr\,pp$); (f) $\frac{1}{2}$ rose ($Rr\,pp$), $\frac{1}{2}$ single ($rr\,pp$).

29. (a) Labrador retrievers vary in two loci, B and E. Black dogs have dominant alleles at both loci ($B\_\,E\_$), brown dogs have $bb\,E\_$, and yellow dogs have $B\_\,ee$ or $bb\,ee$. Because all the puppies were black, they must all have inherited a dominant B allele from the yellow parent and a dominant E allele from the brown parent. The brown female parent must have been $bb\,EE$, and the yellow male must have been $BB\,ee$. The black puppies were all $Bb\,Ee$.
(b) Mating two yellow Labradors will produce all yellow puppies. Mating two brown Labradors will produce either all brown puppies, if at least one of the parents is homozygous EE, or $\frac{3}{4}$ brown and $\frac{1}{4}$ yellow if both parents are heterozygous Ee.

31. Let A and B represent the two loci. The F_1 heterozygotes are $Aa\,Bb$. The F_2 are: $A\_\,B\_$ disc-shaped, $A\_\,bb$ spherical, $aa\,B\_$ spherical, $aa\,bb$ long.

Chapter 6

1. The three factors are: (i) controlled mating experiments are impossible; (ii) humans have a long generation time, and so tracking the inheritance of traits for more than one generation takes a long time; and (iii) the number of progeny per mating is limited, and so phenotypic ratios are uncertain.

3. Autosomal recessive trait: affected males and females arise with equal frequency from unaffected parents; often appears to skip generations; unaffected child of an affected parent is a carrier.

    Autosomal dominant trait: affected males and females arise with equal frequency from a single affected parent; does not usually skip generations.

    X-linked recessive trait: affects males predominantly and is passed from an affected male through his unaffected daughter to his grandson; not passed from father to son.

    X-linked dominant trait: affects males and females; is passed from an affected male to all his daughters but not to his sons; is passed from an affected woman (usually heterozygous for a rare dominant trait) equally to half her daughters and half her sons.

    Y-linked trait: affects males exclusively; is passed from father to all sons.

6. Monozygotic and dizygotic. Monozygotic twins arise when a single fertilized egg splits into two embryos in early embryonic development divisions. They are genetically identical. Dizygotic twins arise from two different eggs fertilized time by two different sperm. They have, on the average, 50% of their genes in common.

9. Genetic counseling provides assistance to clients by interpreting the results of genetic testing and diagnosis; providing information about relevant disease symptoms, treatment, and progression; assessing and calculating the various genetic risks that the person or couple faces; and helping clients and family members cope with the stress of decision-making and facing up to the drastic changes in their lives that may be precipitated by a genetic condition.

17. (a)

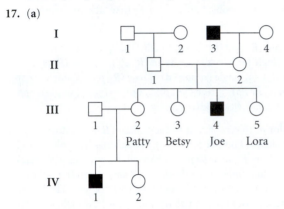

(b) X-linked recessive; (c) zero; (d) $\frac{1}{4}$; (e) $\frac{1}{4}$.

21. (a) Autosomal dominant. The trait must be autosomal because affected males pass the trait to both sons and daughters. It is dominant because it does not skip generations, all affected individuals have affected parents, and it is extremely unlikely that multiple unrelated individuals mating into the pedigree are carriers of a rare trait.
(b) X-linked dominant. Superficially, this pedigree appears similar to the pedigree in part *a* in that both males and females are affected, and it appears to be a dominant trait. However, closer inspection reveals that, whereas affected females can pass the trait to either sons or daughters, affected males pass the trait only to all daughters.
(c) Y-linked. The trait affects only males and is passed from father to son. All sons of an affected male are affected.
(d) X-linked recessive or sex-limited autosomal dominant. Because only males show the trait, the trait could be X-linked recessive, Y-linked, or sex-limited. We can eliminate Y-linkage because affected males do not pass the trait to their sons. X-linked recessive inheritance is consistent with the pattern of unaffected female carriers producing both affected and unaffected sons and affected males producing unaffected female carriers, but no affected sons. Sex-limited autosomal dominant inheritance is also consistent with unaffected heterozygous females producing affected heterozygous sons, unaffected homozygous recessive sons, and unaffected heterozygous or homozygous recessive daughters. The two remaining possibilities of X-linked recessive versus sex-limited autosomal dominant could be distinguished if we had enough data to determine whether affected males have both affected and unaffected sons, as expected from autosomal dominant inheritance, or whether affected males have only unaffected sons, as expected from X-linked recessive inheritance. Unfortunately, this pedigree shows only two sons from affected males. In both cases, the sons are unaffected, consistent with X-linked recessive inheritance, but two male progeny are not enough to conclude that affected males cannot produce affected sons.

(e) Autosomal recessive. Unaffected parents produced affected progeny, and so the trait is recessive. The affected daughter must have inherited recessive alleles from both unaffected parents, and so the trait must be autosomal. If it were X-linked, her father would show the trait.

23. (a) X-linked recessive; (b) $\frac{1}{4}$; (c) $\frac{1}{2}$.

27. Migraine headaches: genetic and environmental. Markedly greater concordance in monozygotic twins, who are 100% genetically identical, than in dizygotic twins, who are 50% genetically identical, is indicative of a genetic influence. However, only 60% concordance for monozygotic twins indicates that environmental factors also play a role.

Eye color: genetic. The concordance is greater in monozygotic twins than in dizygotic twins. Moreover, the monozygotic twins have 100% concordance for this trait, indicating that environment has no detectable influence.

Measles: no detectable genetic influence. There is no difference in concordance between monozygotic and dizygotic twins. Some environmental influence can be detected because monozygotic twins show less than 100% concordance.

Clubfoot: genetic and environmental. The reasoning for migraine headaches applies here. A strong environmental influence is indicated by the high discordance in monozygotic twins.

High blood pressure: genetic and environmental. Reasoning is similar to that for clubfoot.

Handedness: no genetic influence. The concordance is the same in monozygotic and dizygotic twins. Environmental influence is indicated by the less than 100% concordance in monozygotic twins.

Tuberculosis: no genetic influence. Concordance is the same in monozygotic and dizygotic twins. The importance of environmental influence is indicated by the very low concordance in monozygotic twins.

Chapter 7

1. Recombination means that meiosis generates gametes with allelic combinations that differ from the original gametes inherited by an organism. Recombination may be caused by the independent assortment of loci on different chromosomes or by a physical crossing over between two loci on the same chromosome.

2. (a) Nonrecombinant gametes; (b) 50% recombinant and 50% nonrecombinant; (c) more than 50% of the gametes nonrecombinant, and less than 50% recombinant.

5. For genes in coupling configuration, two wild-type alleles are on the same chromosome and the two mutant alleles are on the homologous chromosome. For genes in repulsion, the wild-type allele of one gene and the mutant allele of the other gene are on the same chromosome, and vice versa on the homologous chromosome. The two arrangements have opposite effects on the results of a cross. For genes in coupling configuration, most of the progeny will be either wild type for both genes or mutant for both genes, with relatively few that are wild type for one gene and mutant for the other. For genes in repulsion, most of the progeny will be mutant for only one gene and wild type for the other, with relatively few recombinants that are wild type for both or mutant for both.

8. The farther apart two loci are, the more likely there will be double crossovers between them. The calculated recombination frequency

will underestimate the true crossover frequency because the double-crossover progeny are not counted as recombinants.

10. Positive interference indicates that a crossover inhibits or interferes with the occurrence of a second crossover nearby. Negative interference suggests that a crossover event can stimulate additional crossover events in the same region of the chromosome.

13. (a) ½ banded, yellow; ½ unbanded, brown.
(b) ¼ banded, yellow; ¼ banded, brown; ¼ unbanded, yellow; ¼ unbanded, brown.
(c) 40% banded, yellow; 40% unbanded, brown; 10% banded, brown; 10% unbanded, yellow.

15. The genes are linked and have not assorted independently.

17. (a) Both plants are *Dd Pp*.
(b) Yes. Map distance = 3.8 m.u.
(c) The two plants have different coupling configurations. In plant A, the dominant alleles *D* and *P* are coupled; one chromosome is __D____P__ and the other is __d____p__ . In plant B, they are in repulsion; its chromosomes are __D____p__ and __d____P__ .

22.

	Genotype	Body color	Eyes	Bristles	Proportion
(a)	*e⁺ ro⁺ f⁺*	normal	normal	normal	20%
	e⁺ ro⁺ f	normal	normal	forked	20%
	e ro f⁺	ebony	rough	normal	20%
	e ro f	ebony	rough	forked	20%
	e⁺ ro f⁺	normal	rough	normal	5%
	e⁺ ro f	normal	rough	forked	5%
	e ro⁺ f⁺	ebony	normal	normal	5%
	e ro⁺ f	ebony	normal	forked	5%
(b)	*e⁺ ro⁺ f⁺*	normal	normal	normal	5%
	e⁺ ro⁺ f	normal	normal	forked	5%
	e ro f⁺	ebony	rough	normal	5%
	e ro f	ebony	rough	forked	5%
	e⁺ ro f⁺	normal	rough	normal	20%
	e⁺ ro f	normal	rough	forked	20%
	e ro⁺ f⁺	ebody	normal	normal	20%
	e ro⁺ f	ebony	normal	forked	20%

24.

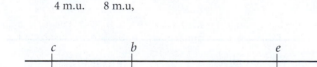

a   *g*   *d*

4 m.u.   8 m.u,

c   *b*   *e*

10 m.u.   18 m.u.

Gene *f* is unlinked to either of these groups, on a third linkage group.

27. (a) *V* is the middle gene.
(b) The *Wx–V* distance = 7 m.u. and the *Sh–V* distance = 30 m.u. The *Wx–Sh* distance is the sum of these two distances, or 37 m.u.
(c) Coefficient of coincidence = 0.80; interference = 0.20

30.

Genotype	Body	Eyes	Wings	Proportion
b⁺ pr⁺ vg⁺	normal	normal	normal	407%
b pr vg	black	purple	vestigial	407%
b⁺ pr⁺ vg	normal	normal	vestigial	63%
b pr vg⁺	black	purple	normal	63%
b⁺ pr vg	normal	purple	vestigial	28%
b pr⁺ vg⁺	black	normal	normal	28%
b⁺ pr vg⁺	normal	purple	normal	2%
b pr⁺ vg	black	normal	vestigial	2%

32.

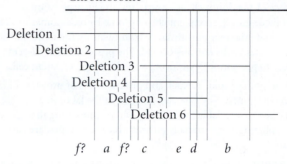

The location of *f* is ambiguous; it could be in either location shown above.

34. Enzyme 1 is on chromosome 9; enzyme 2 is on chromosome 4; enzyme 3 is on the X chromosome.

Chapter 8

3.

Types of matings	Outcomes
F⁺ × F⁻	Two F⁺ cells
Hfr × F⁻	One F⁺ cell and one F⁻ cell
F′ × F⁻	Two F′ cells

The F factor contains a number of genes that take part in the conjugation process, including genes necessary for the synthesis of the sex pilus. The F factor has an origin of replication that enables the factor to be replicated in the conjugation process and genes for opening the plasmid and initiating the chromosome transfer.

4. To map genes by conjugation, an Hfr strain is mixed with an F⁻ strain. The two strains must have different genotypes and must remain in physical contact for the transfer to take place. The conjugation process is interrupted at regular intervals. The chromosomal transfer from the Hfr strain always begins with a part of the integrated F factor and proceeds in a linear fashion. Transfer of the entire chromosome takes approximately 100 minutes. The time required for individual genes to be transferred is relative to their positions on the chromosome and the direction of transfer initiated by the F factor. Gene distances are typically mapped in minutes. The genes that are transferred by conjugation to the recipient must be incorporated into the recipient's chromosome by recombination to be expressed.

In transformation, the relative frequency at which pairs of genes are transferred, or cotransformed, indicates the distance between the two genes. Closer gene pairs are cotransformed more frequently. As in interrupted conjugation, the donor DNA must be incorporated into the recipient cell's chromosome by recombination. Physical contact of the donor and recipient cells is not needed. The recipient cell takes up the DNA directly from the environment, and so the

DNA from the donor strain must be isolated and broken up before transformation can take place.

The transfer of DNA by transduction requires a viral vector. DNA from the donor cell is packaged into a viral protein coat. The viral particle containing the bacterial donor DNA then infects a recipient bacterial cell. The donor bacterial DNA is incorporated into the recipient cell's chromosome by recombination. Only genes that are close together on the bacterial chromosome can be cotransduced. Therefore, the rate of cotransduction, like the rate of cotransformation, is an indication of the physical distances between genes on the chromosome.

These three processes all involve the uptake by the recipient cell of a piece of the donor chromosome and the incorporation of some of that piece into the recipient chromosome by recombination. They also calculate the mapping distance by measuring the frequency with which recipient cells are tansformed. The processes use different methods to get donor DNA incorporated into the recipient cell.

6. Reproduction is rapid, asexual, and produces lots of progeny. Their genomes are small. They are easy to grow in the laboratory. Techniques are available for isolating and manipulating their genes. Mutant phenotypes, especially auxotrophic phenotypes, are easy to measure.

10. In generalized transduction, randomly selected bacterial genes are transferred from one bacterial cell to another by a virus. In specialized transduction, only genes from a particular locus on the bacterial chromosome are transferred to another bacterium. The process of specialized transduction requires lysogenic phages that integrate into specific locations on the host cell's chromosome. When the phage DNA excises from the host chromosome and the excision process is imprecise, the phage DNA will carry a small part of the bacterial DNA. The hybrid DNA must be injected by the phage into another bacterial cell during another round of infection.

The transfer of DNA by generalized transduction requires that the host DNA be broken down into smaller pieces and that a piece of the host DNA is packaged into a phage coat instead of phage DNA. The defective phage cannot produce new phage particles in a subsequent infection, but it can inject the bacterial DNA into another bacterium or recipient. Through a double-crossover event, the donor DNA can become incorporated into the bacterial recipient's chromosome.

11. To conduct the complementation test, Benzer infected cells of *E. coli* K with large numbers of the two mutant phage types. For successful infection, each mutant phage must supply the gene product or protein missing in the other mutant phage. Complementation will result only if the mutations are at separate loci. If the two mutations are at the same locus, there will be no complementation of gene products, and no plaques will be produced on the *E. coli* K lawns.

12. A retrovirus is able to integrate its RNA genome into its host cell's DNA genome through the action of the enzyme reverse transcriptase, which can synthesize complementary DNA from either an RNA or a DNA template. Reverse transcriptase uses the retroviral single-stranded RNA as a template to synthesize double-stranded DNA. The newly synthesized DNA molecule can then integrate into the host chromosome to form a provirus.

15. By using low doses of antibiotics for 5 years, Farmer Smith selected for bacteria that are resistant to the antibiotics. The doses used killed sensitive bacteria but not moderately sensitive or slightly resistant bacteria. As time passed, only resistant bacteria remained in his pigs

because any sensitive bacteria were eliminated by the low doses of antibiotics.

In the future, Farmer Smith should continue to use the vitamins but should use the antibiotics only when a sick pig requires them. In this manner, he will not be selecting for antibiotic-resistant bacteria, and the chances of successful treatment of his sick pigs will be greater.

18. In each of the Hfr strains, the F factor has been inserted into a different location in the chromosome. The orientation of the F factor in the strains varies as well.

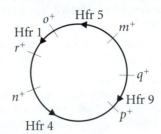

19. Distances between genes are in minutes.

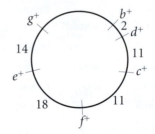

28. Number of plaques produced by $c^+ m^+ = 460$; by $c^- m^- = 460$; by $c^+ m^- = 40$ (recombinant); by $c^+ m^- = 40$ (recombinant).

30. (a) The recombination frequency between r_2 and h is $^{13}/_{180} = 0.072$, or 7.2%. The RF between r_{13} and h is $^3/_{216} = 0.014$, or 1.4%.

(b)

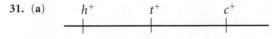

1.4 m.u.   5.8 m.u.

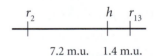

7.2 m.u.   1.4 m.u.

31. (a)

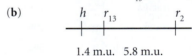

(b)

7.1 m.u.   24.1 m.u.

(c) Coefficient of coincidence $= (6 + 5)/(0.071 \times 0.241 \times 942) = 0.68$. Interference $= 1 - 0.68 = 0.32$.

Chapter 9

1. Chromosome rearrangements:

Deletion: Loss of a part of a chromosome.

Duplication: Addition of an extra copy of a part of a chromosome.

Inversion: A part of a chromosome is reversed in orientation.

Translocation: A part of one chromosome becomes incorporated into a different (nonhomologous) chromosome.

Aneupoloidy: Loss or gain of one or more chromosomes, causing the chromosome number to deviate from $2n$ or the normal euploid complement.

Polyploidy: Gain of entire sets of chromosomes, causing the chromosome number to change from $2n$ to $3n$ (triploid), $4n$ (tetraploid), and so on.

2. The expression of some genes is balanced with the expression of other genes; the ratios of their gene products, usually proteins, must be maintained within a narrow range for proper cell function. Extra copies of one of these genes cause that gene to be expressed at proportionately higher levels, thereby upsetting the balance of gene products.

5. A paracentric inversion does not include the centromere; a pericentric inversion includes the centromere.

9. Translocations can produce phenotypic effects if the translocation breakpoint disrupts a gene or if a gene near the breakpoint is altered in its expression because of relocation to a different chromosomal environment (a position effect).

13. Primary Down syndrome is caused by the spontaneous, random nondisjunction of chromosome 21, leading to trisomy 21. Familial Down syndrome most frequently arises as a result of a Robertsonian translocation of chromosome 21 with another chromosome, usually chromosome 14.

14. Uniparental disomy is the inheritance of both copies of a chromosome from the same parent. It may arise from a trisomy condition in which the early embryo loses one of the three chromosomes, and the two remaining chromosomes are from the same parent.

16. In autopolyploidy, all sets of chromosomes are derived from a single species. In allopolyploidy, the sets of chromosomes are derived from two or more different species. Autopolyploidy may arise through nondisjunction in an early $2n$ embryo or through meiotic nondisjunction that produces a gamete with extra sets of chromosomes. Allopolyploidy is usually preceded by hybridization between two different species, followed by chromosome doubling.

18. (a) Duplications; (b) polyploidy; (c) deletions; (d) inversions; (e) translocations.

19. (a) Tandem duplication of AB; (b) displaced duplication of AB; (c) paracentric inversion of DEF; (d) deletion of B; (e) deletion of FG; (f) paracentric inversion of CDE; (g) pericentric inversion of ABC; (h) duplication and inversion of DEF; (i) duplication of CDEF, inversion of EF.

22. (a) $\frac{1}{3}$ Notch white-eyed females, $\frac{1}{3}$ wild-type females, and $\frac{1}{3}$ wild-type males; (b) $\frac{1}{3}$ Notch red-eyed females, $\frac{1}{3}$ wild-type females, and $\frac{1}{3}$ white-eyed males; (c) $\frac{1}{3}$ Notch white-eyed females, $\frac{1}{3}$ white-eyed females, and $\frac{1}{3}$ white-eyed males.

24. (a)

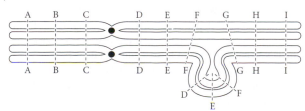

(b)

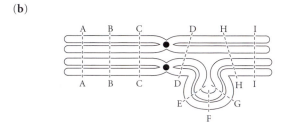

(c)

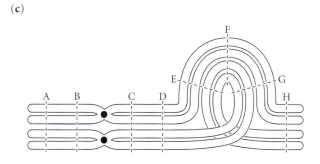

(d)

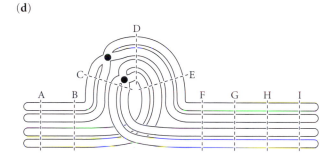

27. (a)

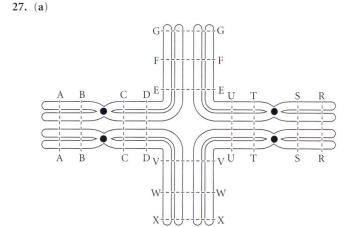

(b) Alternate:

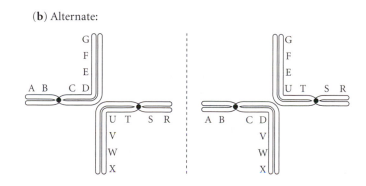